Handbook of Optical Fibers

Gang-Ding Peng
Editor

Handbook of Optical Fibers

Volume 3

With 1544 Figures and 89 Tables

Editor
Gang-Ding Peng
Photonics and Optical Communications
School of Electrical Engineering and Telecommunications
University of New South Wales
Sydney, NSW, Australia

ISBN 978-981-10-7085-3 ISBN 978-981-10-7087-7 (eBook)
ISBN 978-981-10-7086-0 (print and electronic bundle)
https://doi.org/10.1007/978-981-10-7087-7

This Springer imprint is published by the registered company Springer Nature Singapore Pte Ltd.
The registered company address is: 152 Beach Road, #21-01/04 Gateway East, Singapore 189721, Singapore

Preface

This research- and application-oriented book covers main topical areas of optical fibers. The selection of the chapters is weighted on technological and application-specific topics, very much a reflection of where research is heading to and what researchers are looking for. Chapters are arranged in a user-friendly format essentially self-contained and with extensive cross-references. They are organized in the following sections:

Optical Fiber Communication
Solitons and Nonlinear Waves in Optical Fibers
Optical Fiber Fabrication
Active Optical Fibers
Special Optical Fibers
Optical Fiber Measurement
Optical Fiber Devices
Optical Fiber Device Measurement
Distributed Optical Fiber Sensing
Optical Fiber Sensors for Industrial Applications
Polymer Optical Fiber Sensing
Photonic Crystal Fiber Sensing
Optical Fiber Microfluidic Sensors

Several sections and chapters of the book show how diverse optical fiber technologies are becoming now. We envisage that many new optical fibers under development will find important future applications in telecommunications, sensing, and so on. Though we have been trying to cover most relevant and important topics in optical fibers, some topics may have not been presented.

All the authors are either pioneers or leading researchers in their respective areas. Their chapters have reflected well the excellent research work, technology deployment, and commercial application of their own and others. Hence this handbook, as a new entry to the Springer Nature's Major Reference Works (MRWs), will be useful for researchers, academics, engineers, and students to access expertly summarized specific topics on optical fibers for research, education, and learning purposes.

There could be technical and grammatical errors in this book. Please feel free to send your correction, advice, and feedback to us. One key feature of the Springer Nature's MRWs is to ensure continuing update and improvements.

I would take this opportunity to express by deepest gratitude to all my colleagues, either as section editors or authors, for their hard work and great contribution to this book. I also would like to thank the Springer Nature editors and staff, especially Dr. Stephen Siu Wai Yeung and Dr. Juby George, for their kind and professional support throughout this book project.

July 2019

Gang-Ding Peng

Contents

Volume 1

Volume 2

About the Editor

Gang-Ding Peng
Photonics and Optical Communications
School of Electrical Engineering
and Telecommunications
University of New South Wales
Sydney, NSW, Australia

Gang-Ding Peng received his B.Sc. degree in physics from Fudan University, Shanghai, China, in 1982, and the M.Sc. degree in applied physics and Ph.D. in electronic engineering from Shanghai Jiao Tong University, Shanghai, China, in 1984 and 1987, respectively. From 1987 through 1988 he was a lecturer at Jiao Tong University. He was a postdoctoral research fellow in the Optical Sciences Centre of the Australian National University, Canberra, from 1988 to 1991. He has been working at the University of NSW in Sydney, Australia, since 1991; was a Queen Elizabeth II Fellow from 1992 to 1996; and is currently a professor in the same university. He is a fellow and life member of both Optical Society of America (OSA) and The International Society for Optics and Photonics (SPIE). His research interests include silica and polymer optical fibers, optical fiber and waveguide devices, optical fiber sensors, and nonlinear optics.

He has worked in research and teaching in photonics and fiber optics for more than 30 years and maintained a high research profile internationally.

Section Editors

Part I: Optical Fiber Communication

Ming-Jun Li Corning Incorporated, Corning, NY, USA

Chao LU Department of Electronic and Information Engineering, The Hong Kong Polytechnic University, Hong Kong SAR, China

Part II: Solitons and Nonlinear Waves in Optical Fibers

Boris A. Malomed Faculty of Engineering, Department of Physical Electronics, School of Electrical Engineering, Tel Aviv University, Tel Aviv, Israel

ITMO University, St. Petersburg, Russia

Part III: Optical Fiber Fabrication

Hairul Azhar Bin Abdul Rashid Faculty of Engineering, Multimedia University, Cyberjaya, Malaysia

Part IV: Active Optical Fibers

Kyunghwan Oh Department of Physics, Institute of Physics and Applied Physics, Yonsei University, Seoul, Republic of Korea

Part V: Special Optical Fibers

Perry Shum Nanyang Technological University, Singapore, Singapore

Zhilin Xu Center for Gravitational Experiments, School of Physics, Huazhong University of Science and Technology, Wuhan, China

Part VI: Optical Fiber Measurement

Jianzhong Zhang Key Lab of In-fiber Integrated Optics, Ministry of Education, Harbin Engineering University, Harbin, China

Part VII: Optical Fiber Devices

John Canning *i*nterdisciplinary Photonics Laboratories (*i*PL), Global Big Data Technologies Centre (GBDTC), Tech Lab, School of Electrical and Data Engineering, University of Technology Sydney, Sydney, NSW, Australia

Tuan Guo Institute of Photonics Technology, Jinan University, Guangzhou, China

Part VIII: Optical Fiber Device Measurement

Yanhua Luo Photonics and Optical Communications, School of Electrical Engineering and Telecommunications, University of New South Wales, Sydney, NSW, Australia

Key Laboratory of Optoelectronic Devices and Systems of Ministry of Education and Guangdong Province, Shenzhen University, Shenzhen, China

Part IX: Distributed Optical Fiber Sensing

Yosuke Mizuno Institute of Innovative Research, Tokyo Institute of Technology, Yokohama, Japan

Part X: Optical Fiber Sensors for Industrial Applications

Tong Sun OBE School of Mathematics, Computer Science and Engineering, City, University of London, London, UK

Part XI: Polymer Optical Fiber Sensing

Ginu Rajan School of Electrical, Computer and Telecommunications Engineering, University of Wollongong, Wollongong, Australia

School of Electrical Engineering and Telecommunications, UNSW, Sydney, Australia

Part XII: Photonic Crystal Fiber Sensing

D. N. Wang College of Optical and Electrical Technology, China Jiliang University, Hangzhou, China

Part XIII: Optical Fiber Microfluidic Sensors

Yuan Gong Key Laboratory of Optical Fiber Sensing and Communications (Ministry of Education of China), University of Electronic Science and Technology of China, Chengdu, Sichuan, China

Contributors

John S. Abbott Corning Incorporated, Corning, NY, USA

Hairul Azhar Bin Abdul Rashid Faculty of Engineering, Multimedia University, Cyberjaya, Malaysia

Kazi S. Abedin OFS Laboratories, Somerset, NJ, USA

Muhammad Rosdi Abu Hassan Centre for Optical Fibre Technology (COFT), School of Electrical, Electronic Engineering, Nanyang Technological University, Singapore, Singapore, Singapore

Claudia Aichele Department of Fiber Optics, Leibniz Institute of Photonic Technology (Leibniz IPHT), Jena, Germany

Shaif-Ul Alam Optoelectronics Research Centre (ORC), University of Southampton, Southampton, UK

Eliathamby Ambikairajah School of Electrical Engineering and Telecommunications, UNSW, Sydney, Australia

Ghafour Amouzad Mahdiraji School of Engineering, Taylor's University, Subang Jaya, Selangor, Malaysia

Flexilicate Sdn. Bhd., University of Malaya, Kuala Lumpur, Malaysia

S. A. Babin Institute of Automation and Electrometry SB RAS, Novosibirsk, Russia

Novosibirsk State University, Novosibirsk, Russia

John Ballato Center for Optical Materials Science and Engineering Technologies (COMSET) and the Department of Materials Science and Engineering, Clemson University, Clemson, SC, USA

A. Barthélémy XLIM, UMR CNRS 7252, Université de Limoges, Limoges, France

Kishore Bhowmik HFC Assurance, Operate and Maintain Network, NBN, Melbourne, VIC, Australia

Scott R. Bickham Corning Incorporated, Corning, NY, USA

Lúcia Bilro Instituto de Telecomunicações, Campus Universitário de Santiago, Aveiro, Portugal

L. Brun Faiveley Brecknell Willis, Somerset, UK

John Canning *i*nterdisciplinary Photonics Laboratories (*i*PL), Global Big Data Technologies Centre (GBDTC), Tech Lab, School of Electrical and Data Engineering, University of Technology Sydney, Sydney, NSW, Australia

J. Carlton City, University of London, London, UK

Christophe Caucheteur Electromagnetism and Telecommunication Department, University of Mons, Mons, Belgium

Quan Chai Key Laboratory of In-Fiber Integrated Optics, Ministry Education of China, Harbin Engineering University, Harbin, China

I. S. Chekhovskoy Novosibirsk State University, Novosibirsk, Russia

Institute of Computational Technologies SB RAS, Novosibirsk, Russia

Jianping Chen State Key Laboratory of Advanced Optical Communication Systems and Networks, Department of Electronic Engineering, Shanghai Jiao Tong University, Shanghai, China

Xin Chen Corning Incorporated, Corning, NY, USA

Y. Chen City, University of London, London, UK

Baokai Cheng Department of Electrical and Computer Engineering, Center for Optical Materials Science and Engineering Technologies (COMSET), Clemson University, Clemson, SC, USA

Yushi Chu Key Laboratory of In-Fiber Integrated Optics, Ministry Education of China, Harbin Engineering University, Harbin, China

Photonics and Optical Communications, School of Electrical Engineering and Telecommunications, UNSW, Sydney, NSW, Australia

*i*nterdisciplinary Photonics Laboratories (*i*PL), Global Big Data Technologies Centre (GBDTC), Tech Lab, School of Electrical and Data Engineering, University of Technology Sydney, Sydney, NSW, Australia

J. Doug Coleman Corning Incorporated, Corning, NY, USA

Matteo Conforti CNRS, UMR 8523 – PhLAM – Physique des Lasers Atomes et Molécules, University of Lille, Lille, France

Kevin Cook *i*nterdisciplinary Photonics Laboratories (*i*PL), Global Big Data Technologies Centre (GBDTC), Tech Lab, School of Electrical and Data Engineering, University of Technology Sydney, Sydney, NSW, Australia

V. Couderc XLIM, UMR CNRS 7252, Université de Limoges, Limoges, France

Katrina D. Dambul Faculty of Engineering, Multimedia University, Cyberjaya, Selangor, Malaysia

S. Das Fiber Optics and Photonics Division, CSIR-Central Glass and Ceramic Research Institute, Kolkata, India

A. Dhar Fiber Optics and Photonics Division, CSIR-Central Glass and Ceramic Research Institute, Kolkata, India

Mingjie Ding Photonics and Optical Communications, School of Electrical Engineering and Telecommunications, University of New South Wales, Sydney, NSW, Australia

Fabrizio Di Pasquale Institute of Communication, Information and Perception Technologies (TECIP), Scuola Superiore Sant'Anna, Pisa, Italy

Yongkang Dong National Key Laboratory of Science and Technology on Tunable Laser, Harbin Institute of Technology, Harbin, China

John D. Downie Corning Incorporated, Corning, NY, USA

Peter Dragic Department of Electrical and Computer Engineering, University of Illinois at Urbana-Champaign, Urbana, IL, USA

D. Dutta Fiber Optics and Photonics Division, CSIR-Central Glass and Ceramic Research Institute, Kolkata, India

M. Fabian City, University of London, London, UK

Ghazal Fallah Tafti Photonics and Optical Communications, School of Electrical Engineering and Telecommunications, UNSW, Sydney, NSW, Australia

Desheng Fan Photonics and Optical Communications, School of Electrical Engineering and Telecommunications, University of New South Wales, Sydney, NSW, Australia

Xinyu Fan State Key Laboratory of Advanced Optical Communication Systems and Networks, Department of Electronic Engineering, Shanghai Jiao Tong University, Shanghai, China

Andrea Fasano DTU Mekanik, Department of Mechanical Engineering, Technical University of Denmark, Lyngby, Denmark

M. P. Fedoruk Novosibirsk State University, Novosibirsk, Russia

Institute of Computational Technologies SB RAS, Novosibirsk, Russia

Dimitrios J. Frantzeskakis Department of Physics, National and Kapodistrian University of Athens, Athens, Greece

Alexander Fuerbach MQ Photonics Research Centre, Department of Physics and Astronomy, Macquarie University, North Ryde, NSW, Australia

C. Gerada The University of Nottingham, Nottingham, UK

Anahita Ghaznavi Photonics and Optical Communications, School of Electrical Engineering and Telecommunications, UNSW, Sydney, NSW, Australia

Chao-Yang Gong Key Laboratory of Optical Fiber Sensing and Communications (Ministry of Education of China), University of Electronic Science and Technology of China, Chengdu, Sichuan, China

Yuan Gong Key Laboratory of Optical Fiber Sensing and Communications (Ministry of Education of China), University of Electronic Science and Technology of China, Chengdu, Sichuan, China

K. T. V. Grattan City, University of London, London, UK

Jian Guo Shandong Key Laboratory of Optical Fiber Sensing Technologies, Qilu Industry University (Laser Institute of Shandong Academy of Sciences), Jinan, China

Tuan Guo Institute of Photonics Technology, Jinan University, Guangzhou, China

Tetsuya Hayashi Optical Communications Laboratory, Sumitomo Electric Industries, Ltd., Yokohama, Kanagawa, Japan

Jun He Key Laboratory of Optoelectronic Devices and Systems of Ministry of Education and Guangdong Province, College of Physics and Optoelectronic Engineering, Shenzhen University, Shenzhen, China

Guangdong and Hong Kong Joint Research Centre for Optical Fibre Sensors, Shenzhen University, Shenzhen, China

Hoi Lut Ho Department of Electrical Engineering, The Hong Kong Polytechnic University, Hong Kong, China

Theodoros P. Horikis Department of Mathematics, University of Ioannina, Ioannina, Greece

Sheng-Lung Huang Graduate Institute of Photonics and Optoelectronics, and Department of Electrical Engineering, National Taiwan University, Taipei, Taiwan

Darren D. Hudson MQ Photonics Research Centre, Department of Physics and Astronomy, Macquarie University, North Ryde, NSW, Australia

Georges Humbert XLIM Research Institute, UMR 7252 CNRS, University of Limoges, Limoges, France

Ezra Ip NEC Laboratories America, Princeton, NJ, USA

Stuart D. Jackson Department of Engineering, MQ Photonics Research Centre, School of Engineering, Macquarie University, North Ryde, NSW, Australia

S. Javdani City, University of London, London, UK

Taofei Jiang National Key Laboratory of Science and Technology on Tunable Laser, Harbin Institute of Technology, Harbin, China

Wei Jin Department of Electrical Engineering, The Hong Kong Polytechnic University, Hong Kong, China

Yongmin Jung Optoelectronics Research Centre (ORC), University of Southampton, Southampton, UK

S. I. Kablukov Institute of Automation and Electrometry SB RAS, Novosibirsk, Russia

Gerd Keiser Boston University, Boston, MA, USA

A. V. Kir'yanov Centro de Investigaciones en Optica, Guanajuato, Mexico

K. Krupa Department of Information Engineering, University of Brescia, Brescia, Italy

A. Kudlinski CNRS, UMR 8523 – PhLAM – Physique des Lasers Atomes et Molécules, University of Lille, Lille, France

Elizabeth Lee Precision Measurements Group, Singapore Institute of Manufacturing Technology, Singapore, Singapore

Ming-Jun Li Corning Incorporated, Corning, NY, USA

Changrui Liao College of Optoelectronic Engineering, Shenzhen University, Shenzhen, China

Sascha Liehr Division 8.6 "Fibre Optic Sensors", Bundesanstalt für Materialforschung und –prüfung (BAM), Berlin, Germany

Chupao Lin College of Optoelectronic Engineering, Shenzhen University, Shenzhen, China

Horng Sheng Lin Universiti Tunku Abdul Rahman, Sungai Long Campus, Kajang, Malaysia

Florian Lindner Department of Fiber Optics, Leibniz Institute of Photonic Technology (Leibniz IPHT), Jena, Germany

Deming Liu School of Optical and Electronic Information, Next Generation Internet Access National Engineering Laboratory (NGIAS), Huazhong University of Science and Technology, Wuhan, Hubei, P. R. China

Tongyu Liu Laser Institute, Qilu University of Technology-Shandong Academy of Science, Jinan, Shandong, China

Xin Long State Key Laboratory of Advanced Optical Communication Systems and Networks, Department of Electronic Engineering, Shanghai Jiao Tong University, Shanghai, China

Jiaqi Luo Precision Measurements Group, Singapore Institute of Manufacturing Technology, Singapore, Singapore

Yanhua Luo Photonics and Optical Communications, School of Electrical Engineering and Telecommunications, University of New South Wales, Sydney, NSW, Australia

Key Laboratory of Optoelectronic Devices and Systems of Ministry of Education and Guangdong Province, Shenzhen University, Shenzhen, China

Faisal Rafiq Mahamd Adikan Flexilicate Sdn. Bhd., University of Malaya, Kuala Lumpur, Malaysia

Integrated Lightwave Research Group, Department of Electrical Engineering, Faculty of Engineering, University of Malaya, Kuala Lumpur, Malaysia

Sergejs Makovejs Corning Incorporated, Ewloe, UK

Boris A. Malomed Faculty of Engineering, Department of Physical Electronics, School of Electrical Engineering, Tel Aviv University, Tel Aviv, Israel

ITMO University, St. Petersburg, Russia

Christos Markos DTU Fotonik, Department of Photonics Engineering, Technical University of Denmark, Lyngby, Denmark

G. Millot ICB, UMR CNRS 6303, Université de Bourgogne, Dijon, France

Fedor Mitschke Institut für Physik, Universität Rostock, Rostock, Germany

S. Z. Muhamad Yassin Photonics Laboratory, Telekom Research and Development, Cyberjaya, Malaysia

A. Mussot CNRS, UMR 8523 – PhLAM – Physique des Lasers Atomes et Molécules, University of Lille, Lille, France

Hossein Najafi Institute for Applied Laser, Photonics and Surface Technologies (ALPS), Bern University of Applied Sciences, Burgdorf, Switzerland

Rogério Nogueira Instituto de Telecomunicações, Campus Universitário de Santiago, Aveiro, Portugal

Ricardo Oliveira Instituto de Telecomunicações, Campus Universitário de Santiago, Aveiro, Portugal

Nasr Y. M. Omar Faculty of Engineering, Multimedia University, Cyberjaya, Malaysia

M. Pal Fiber Optics and Photonics Division, CSIR-Central Glass and Ceramic Research Institute, Kolkata, India

M. C. Paul Fiber Optics and Photonics Division, CSIR-Central Glass and Ceramic Research Institute, Kolkata, India

Gang-Ding Peng Photonics and Optical Communications, School of Electrical Engineering and Telecommunications, University of New South Wales, Sydney, NSW, Australia

Jiankun Peng National Engineering Laboratory for Fiber Optic Sensing Technology (NEL-FOST), Wuhan University of Technology, Wuhan, China

Sönke Pilz Institute for Applied Laser, Photonics and Surface Technologies (ALPS), Bern University of Applied Sciences, Burgdorf, Switzerland

Soo Yong Poh Integrated Lightwave Research Group, Department of Electrical Engineering, Faculty of Engineering, University of Malaya, Kuala Lumpur, Malaysia

Haifeng Qi Shandong Key Laboratory of Optical Fiber Sensing Technologies, Qilu Industry University (Laser Institute of Shandong Academy of Sciences), Jinan, China

Ginu Rajan School of Electrical, Computer and Telecommunications Engineering, University of Wollongong, Wollongong, Australia

School of Electrical Engineering and Telecommunications, UNSW, Sydney, Australia

Yun-Jiang Rao Key Laboratory of Optical Fiber Sensing and Communications (Ministry of Education of China), University of Electronic Science and Technology of China, Chengdu, Sichuan, China

P. H. Reddy Academy of Scientific and Innovative Research (AcSIR), IR-CGCRI Campus, Kolkata, India

A. A. Reduyk Novosibirsk State University, Novosibirsk, Russia

David J. Richardson Optoelectronics Research Centre (ORC), University of Southampton, Southampton, UK

Valerio Romano Institute for Applied Laser, Photonics and Surface Technologies (ALPS), Bern University of Applied Sciences, Burgdorf, Switzerland

Institute of Applied Physics (IAP), University of Bern, Bern, Switzerland

A. M. Rubenchik Lawrence Livermore National Laboratory, Livermore, CA, USA

Kay Schuster Department of Fiber Optics, Leibniz Institute of Photonic Technology (Leibniz IPHT), Jena, Germany

Filipa Sequeira Instituto de Telecomunicações, Campus Universitário de Santiago, Aveiro, Portugal

O. V. Shtyrina Novosibirsk State University, Novosibirsk, Russia

Institute of Computational Technologies SB RAS, Novosibirsk, Russia

O. S. Sidelnikov Novosibirsk State University, Novosibirsk, Russia

D. V. Skryabin Department of Nanophotonics and Metamaterials, ITMO University, St Petersburg, Russia

Department of Physics, University of Bath, Bath, UK

Yang Song Department of Electrical and Computer Engineering, Center for Optical Materials Science and Engineering Technologies (COMSET), Clemson University, Clemson, SC, USA

Zhiqiang Song Shandong Key Laboratory of Optical Fiber Sensing Technologies, Qilu Industry University (Laser Institute of Shandong Academy of Sciences), Jinan, China

Marcelo A. Soto Institute of Electrical Engineering, EPFL Swiss Federal Institute of Technology, Lausanne, Switzerland

Dan Sporea National Institute for Laser, Plasma and Radiation Physics, Center for Advanced Laser Technologies, Măgurele, Romania

Biao Sun Precision Measurements Group, Singapore Institute of Manufacturing Technology, Singapore, Singapore

Qizhen Sun School of Optical and Electronic Information, Next Generation Internet Access National Engineering Laboratory (NGIAS), Huazhong University of Science and Technology, Wuhan, Hubei, P. R. China

Tong Sun OBE School of Mathematics, Computer Science and Engineering, City, University of London, London, UK

Ming Tang Wuhan National Lab for Optoelectronics (WNLO) and National Engineering Laboratory for Next Generation Internet Access System (NGIA), School of Optical and Electronic Information, Huazhong University of Science and Technology (HUST), Wuhan, China

Lei Teng National Key Laboratory of Science and Technology on Tunable Laser, Harbin Institute of Technology, Harbin, China

A. Tonello XLIM, UMR CNRS 7252, Université de Limoges, Limoges, France

Stefano Trillo Department of Engineering, University of Ferrara, Ferrara, Italy

S. K. Turitsyn Novosibirsk State University, Novosibirsk, Russia

Aston Institute of Photonic Technologies, Aston University, Birmingham, UK

Sonja Unger Department of Fiber Optics, Leibniz Institute of Photonic Technology (Leibniz IPHT), Jena, Germany

M. Vidakovic City, University of London, London, UK

S. Wabnitz Novosibirsk State University, Novosibirsk, Russia

Department of Information Engineering, University of Brescia, Brescia, Italy

National Institute of Optics INO-CNR, Brescia, Italy

Chao Wang School of Electrical Engineering, Wuhan University, Wuhan, Hubei, China

D. N. Wang College of Optical and Electrical Technology, China Jiliang University, Hangzhou, China

Min Wang National Engineering Laboratory for Fiber Optic Sensing Technology (NEL-FOST), Wuhan University of Technology, Wuhan, China

School of Electronic and Electrical Engineering, Wuhan Textile University, Wuhan, China

Weijia Wang National Engineering Laboratory for Fiber Optic Sensing Technology (NEL-FOST), Wuhan University of Technology, Wuhan, China

Weitao Wang Shandong Key Laboratory of Optical Fiber Sensing Technologies, Qilu Industry University (Laser Institute of Shandong Academy of Sciences), Jinan, China

Wenyu Wang Photonics and Optical Communications, School of Electrical Engineering and Telecommunications, University of New South Wales, Sydney, NSW, Australia

Yiping Wang Key Laboratory of Optoelectronic Devices and Systems of Ministry of Education and Guangdong Province, College of Physics and Optoelectronic Engineering, Shenzhen University, Shenzhen, China

Guangdong and Hong Kong Joint Research Centre for Optical Fibre Sensors, Shenzhen University, Shenzhen, China

Lei Wei School of Electrical and Electronic Engineering, Nanyang Technological University, Singapore, Singapore

Jianxiang Wen Key Laboratory of Specialty Fiber Optics and Optical Access Networks, Shanghai University, Shanghai, China

Aleksander Wosniok 8.6 Fibre Optic Sensors, Federal Institute for Materials Research and Testing (BAM), Berlin, Germany

Getinet Woyessa DTU Fotonik, Department of Photonics Engineering, Technical University of Denmark, Lyngby, Denmark

Gui Xiao Photonics and Optical Communications, School of Electrical Engineering and Telecommunications, UNSW, Sydney, NSW, Australia

Hai Xiao Department of Electrical and Computer Engineering, Center for Optical Materials Science and Engineering Technologies (COMSET), Clemson University, Clemson, SC, USA

Limin Xiao Advanced Fiber Devices and Systems Group, Key Laboratory of Micro and Nano Photonic Structures (MoE), Department of Optical Science and Engineering Fudan University, Shanghai, China

Key Laboratory for Information Science of Electromagnetic Waves (MoE), Fudan University, Shanghai, China

Shanghai Engineering Research Center of Ultra-Precision Optical Manufacturing, Fudan University, Shanghai, China

Fei Xu National Laboratory of Solid State Microstructures and College of Engineering and Applied Sciences, Nanjing University, Nanjing, Jinagsu, P. R. China

Binbin Yan State Key Laboratory of Information Photonics and Optical Communications, Beijing University of Posts and Telecommunications, Beijing, China

Zhijun Yan School of Optical and Electronic Information, Next Generation Internet Access National Engineering Laboratory (NGIAS), Huazhong University of Science and Technology, Wuhan, Hubei, P. R. China

Fan Yang Department of Electrical Engineering, The Hong Kong Polytechnic University, Hong Kong, China

Jun Yang Key Lab of In-Fiber Integrated Optics, Ministry Education of China, Harbin Engineering University, Harbin, China

College of Science, Harbin Engineering University, Harbin, China

Minghong Yang National Engineering Laboratory for Fiber Optic Sensing Technology (NEL-FOST), Wuhan University of Technology, Wuhan, China

Xia Yu Precision Measurements Group, Singapore Institute of Manufacturing Technology, Singapore, Singapore

Zhangjun Yu Key Lab of In-Fiber Integrated Optics, Ministry Education of China, Harbin Engineering University, Harbin, China

College of Science, Harbin Engineering University, Harbin, China

Lei Yuan Department of Electrical and Computer Engineering, Center for Optical Materials Science and Engineering Technologies (COMSET), Clemson University, Clemson, SC, USA

Libo Yuan Key Lab of In-Fiber Integrated Optics, Ministry Education of China, Harbin Engineering University, Harbin, China

College of Science, Harbin Engineering University, Harbin, China

Zulfadzli Yusoff Multimedia University, Persiaran Multimedia, Cyberjaya, Malaysia

Amirhassan Zareanborji Photonics and Optical Communications, School of Electrical Engineering and Telecommunications, UNSW, Sydney, NSW, Australia

Chen-Lin Zhang Key Laboratory of Optical Fiber Sensing and Communications (Ministry of Education of China), University of Electronic Science and Technology of China, Chengdu, Sichuan, China

Hongying Zhang Institute of Photonics and Optical Fiber Technology, Harbin University of Science and Technology, Harbin, China

Jianzhong Zhang Key Lab of In-fiber Integrated Optics, Ministry of Education, Harbin Engineering University, Harbin, China

Lei Zhang College of Optical Science and Engineering, Zhejiang University, Hangzhou, China

Lin Zhang Aston Institute of Photonic Technologies, Aston University, Birmingham, UK

Chun-Liu Zhao College of Optical and Electrical Technology, China Jiliang University, Hangzhou, China

Qiancheng Zhao Photonics and Optical Communications, School of Electrical Engineering and Telecommunications, UNSW, Sydney, NSW, Australia

Dengwang Zhou National Key Laboratory of Science and Technology on Tunable Laser, Harbin Institute of Technology, Harbin, China

Feng Zhu College of Optoelectronic Engineering, Shenzhen University, Shenzhen, China

E. A. Zlobina Institute of Automation and Electrometry SB RAS, Novosibirsk, Russia

Weiwen Zou State Key Laboratory of Advanced Optical Communication Systems and Networks, Department of Electronic Engineering, Shanghai Jiao Tong University, Shanghai, China

Part IX
Distributed Optical Fiber Sensing

Distributed Rayleigh Sensing

41

Xinyu Fan

Contents

Abstract

Rayleigh backscattering (RBS) in optical fiber is a fundamental phenomenon caused by random fluctuations in the index profile along the fiber length. Optical reflectometry is the best tool to obtain RBS signals with a distributed way along

---

X. Fan (✉)
State Key Laboratory of Advanced Optical Communication Systems and Networks, Department of Electronic Engineering, Shanghai Jiao Tong University, Shanghai, China
e-mail: fan.xinyu@sjtu.edu.cn

G.-D. Peng (ed.), *Handbook of Optical Fibers*,
https://doi.org/10.1007/978-981-10-7087-7_5

the fiber and is widely used as a nondestructive measurement at one end of the fiber. With the help of optical reflectometry, RBS signals are used for distributed fiber-optic sensing with temperature/strain/vibration information along the fiber. Understanding the mechanisms of RBS provides a powerful technique for static and vibration sensing used for applications such as structural health monitoring and damage assessment analysis. The technology based on RBS signals to extract the environmental perturbation is mature, but researches on obtaining a high spatial resolution together with a long measurement range are still very active, promoting the technology to be used in more industrial applications with strict requirements on these parameters such as monitoring the optical fiber inside the aircraft wings with a spatial resolution of better than 1 mm over several 100 m.

In this book chapter, after a first description of RBS mechanism in optical fibers, the working principle of static and vibration measurement based on RBS signals is provided. Then, different kinds of optical reflectometry based on time-domain, frequency-domain, and coherence-domain techniques are introduced in details. In the final section, advanced methods to improve both the spatial resolution and the measurement range are presented which pave the way for new horizons in high-end applications.

Keywords

Rayleigh backscattering (RBS) · Static measurement · Vibration measurement · Optical reflectometry · Optical time-domain reflectometry (OTDR) · Optical frequency-domain reflectometry (OFDR) · Optical coherence-domain reflectometry (OCDR) · Optical low coherence reflectometry (OLCR) · Pulse compression · Synthesis of optical coherence function (SOCF) · Spatial resolution · Measurement range · Sensitivity

Introduction

Optical fiber sensing technology has become one of the most important intelligent monitoring methods used in many applications because of its advantages of cost-efficiency, lightweight, corrosion resistance, immunity to electromagnetic interference (EMI), etc. Distributed fiber-optic sensing technology has been applied for monitoring infrastructure such as bridges, aircraft wings, and oil pipelines because of the abovementioned advantages. After more than 40 years of development, the distributed fiber-optic sensing technology has made great progress in many applications with much improved measurement range, spatial resolution, and sensitivity.

As the core technique of distributed fiber-optic sensing technology, reflectometry technique, which is capable of nondestructive measurement of the optical fiber, is usually realized by obtaining Rayleigh scattering, Brillouin scattering, and Raman scattering along the fiber. Especially for Rayleigh backscattering (RBS) signals, which include reflectivity, refractive index, polarization distribution information of the optical fiber link, may be used to determine various abnormal "events," for example, the positions of fiber splicing, bending, fracture, and corrosion defects.

Furthermore, the RBS signals generated by using a coherent source provide temperature, strain, and vibration information along the fiber.

To extract this information along fiber, there are mainly three kinds of optical reflectometry techniques based on time-domain, coherence-domain, and frequency-domain methods. The advantage of time-domain technique is its long measurement range (several tens of kilometers to even 100 km), but its spatial resolution is limited to meter level. The coherence-domain technique has an opposite performance comparing to the time-domain one. Its spatial resolution can be very high but with a limited measurement range, for example, micrometer-level resolution over meter-level range. The frequency-domain technique has a moderate performance such as cm-level resolution over km-level range, or mm-level resolution over hundred-meter-level range, or sub-mm-level resolution over ten-meter-level range.

For some high-end applications such as the diagnosis for fiber-to-the-home (FTTH) networks, the requirements on reflectometry performance is to have mm or sub-mm spatial resolution together with a measurement range of several kilometers or even several tens of kilometers. This is very challenging and researchers have made many attempts to push the reflectometry performance.

This book chapter first presents the physical mechanisms of RBS in optical fibers and describes the basic working principle of RBS-based distributed static sensing and vibration sensing. All three kinds of optical reflectometry techniques are then described in details. Some state-of-the-art reflectometry techniques with better performance for high-end applications are also described in this chapter.

Rayleigh Scattering in Optical Fibers

Physical Mechanisms

Rayleigh scattering is an elastic scattering phenomenon that the lightwave is scattered by the particles with a radius that is much smaller than the wavelength of the lightwave. The intensity of the Rayleigh scattering is proportional to the fourth power of the lightwave wavelength. In an optical fiber, the refractive index may have fluctuations due to the limitation of the fiber manufacturing technique. These refractive index fluctuations work as the small scatters, and therefore the Rayleigh scattering can be observed in optical fiber (Marcuse 1974). Thanks to the specific structure of the optical fiber, the Rayleigh scattering with a backward direction can be captured by the fiber and transmitted backwards, and it is called Rayleigh backscattering (RBS) (Brinkmeyer 1980), as schematically shown in Fig. 1.

RBS is one of the main factors that contribute to the fiber loss, which is unwelcome in most of transmission applications. However, by utilizing the RBS-related information, different kinds of optical reflectometry are designed and applied to different applications. By injecting a series of optical probe pulses, an optical time-domain reflectometry (OTDR) can obtain the information of a hundred-kilometer-long fiber about loss and damage, with a spatial resolution of several meters. If we modulate the probe lightwave with a more complex form such as

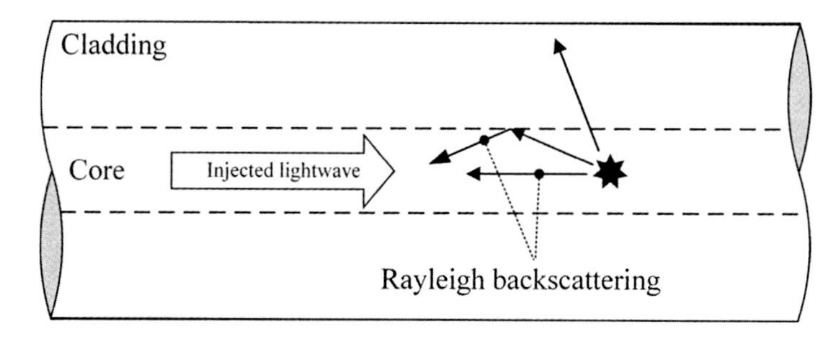

Fig. 1 Schematic of Rayleigh backscattering

linear frequency modulation, we obtain an optical frequency-domain reflectometry (OFDR), which has a spatial resolution as high as hundreds of micrometers and a limited measurement range.

Here, we take the RBS in a single mode fiber (SMF) as an example to explain. The optical source has an output electrical field as

$$E_0(t) = A_0 e^{-j[\omega_0 t + \varphi(t)]} \tag{1}$$

where A_0 is the amplitude of the output lightwave, ω_0 is the optical angular frequency, and $\varphi(t)$ is the phase term. Here, we ignore the output power fluctuation of the optical source.

Firstly, we consider the situation that the optical source has a completely incoherent output. Under this situation, the lightwave will not interfere, and therefore the phase term in Eq. 1 is disordered and may be omitted for convenience. After the incoherent lightwave is injected into the fiber, the RBS lightwave from the ith scatter can be expressed as

$$E_{\text{scatter}}(t) = A_0 R_i e^{-2\alpha z} e^{-j\omega_0(t-2nz/c)} \tag{2}$$

where R_i and z are the reflectivity and the position of the scatter respectively, α is the attenuation coefficient of the fiber, n is the effective refractive index of the fiber, and c is the speed of light. Since the lightwave is incoherent, the RBS lightwave from different scatters can overlap without interfering. Taking a general assumption that the spatial distribution of the scatters is uniform and dense, the RBS lightwave from a position z should be the summation of the RBS lightwave from the scatters at the position z, which can be expressed as

$$E_S(t, z) = A_0 \left(\sum_i R_i \right) e^{-2\alpha z} e^{-j\omega_0(t-2nz/c)} = A_0 R(z) e^{-2\alpha z} e^{-j\omega_0(t-2nz/c)} \tag{3}$$

where $R(z)$ is the RBS coefficient of the fiber at the position z. This coefficient depends on many factors, such as fiber structure, fiber homogeneity, and wavelength of the input lightwave. Usually in an SMF, the RBS coefficient is $-60 \sim -70$ dB/m. If we inject a continuous lightwave into the fiber, the RBS lightwave can be described as

$$E_{RBS}(t) = A_0 \int R(z) e^{-2\alpha z} e^{-j\omega_0 (t - 2nz/c)} dz \tag{4}$$

A corresponding experimental result is shown in Fig. 2a, where a super luminescent diode is employed as an incoherent optical source and a 10-km-long SMF is used for the measurement.

With a coherent laser source, the situation becomes more complex. Under this situation, the phase term in Eq. 1 cannot be ignored, and the RBS lightwave from the ith scatter can be expressed as

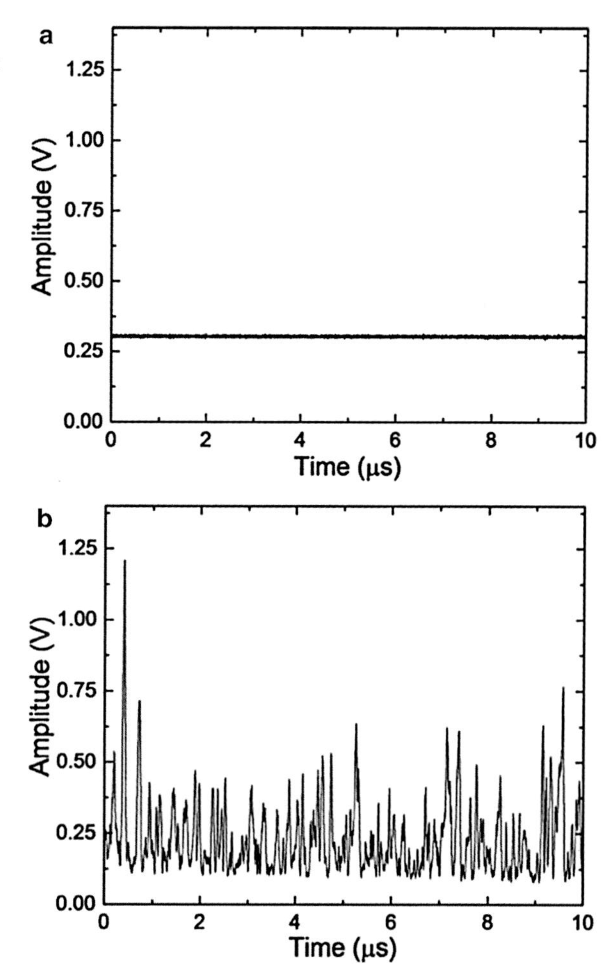

Fig. 2 The intensity of RBS lightwave with (**a**) incoherent optical source and (**b**) coherent optical source

$$E_{\text{scatter}}(t) = A_0 R_i e^{j\theta_i} e^{-2\alpha z} e^{-j[\omega_0(t-2nz/c)+\varphi(t-2nz/c)]} \tag{5}$$

where θ_i is the phase shift of the ith scatter. The RBS lightwave from different scatters interferes, and the RBS lightwave from a position z can be expressed as

$$\begin{aligned} E_S(t,z) &= A_0 \left(\sum_i R_i e^{j\theta_i} \right) e^{-2\alpha z} e^{-j[\omega_0(t-2nz/c)+\varphi(t-2nz/c)]} \\ &= A_0 R(z) e^{j\theta(z)} e^{-2\alpha z} e^{-j[\omega_0(t-2nz/c)+\varphi(t-2nz/c)]} \end{aligned} \tag{6}$$

where $\theta(z)$ is the phase shift of the fiber at position z. Here, we still adopt the same assumption that the spatial distribution of scatters is uniform and dense. Comparing with the result in Eq. 3, here the RBS coefficient $R(z)$ and phase shift $\theta(z)$ are random, due to that the reflectivity and phase shift of scatters are randomly distributed. Furthermore, the RBS lightwave from different positions may interfere when the distance is shorter than the coherent length of the laser source, which leads to an unpredictable RBS lightwave. An experimental result is shown in Fig. 2b, where a fiber laser is employed as a coherent optical source with a 100-km-long coherent length. The intensity of RBS lightwave shows a stochastic characteristic, as discussed above.

Static Measurement

The RBS signals generated by using a coherent optical source provide very useful information for sensing the static parameters along the fiber.

If the temperature changes or a strain is applied to the fiber, the phase also changes as the refractive index or the length is changed. However, Rayleigh pattern would be unchanged if the relationship $n_1 v_1 = n_2 v_2$ holds between two measurements, where $n_{1,2}$ is the refractive index, and $v_{1,2}$ is the optical frequency of the optical source. When the refractive index of the fiber is changed, one can change the optical frequency to keep the Rayleigh pattern unchanged and the corresponding sensing of parameter changes could be achieved by surveying the change of the optical frequency. The principle is shown in Fig. 3. By this means, Rayleigh pattern allows for very sensitive static measurements of refractive index variations, which can be used for very high resolution measurements for temperature, strain, and birefringence. Koyamada et al. (2009) achieve a distributed temperature sensing with a resolution of 0.01 K and a long sensing distance up to 8 km, which is highly competitive among distributed sensing schemes in 2015; Zhou et al. achieve a strain sensing with a resolution of 10 nε in 2015; Soto et al. achieve birefringence measurements of both polarization–maintaining fiber (PMF) with high birefringence and ordinary SMF for telecommunication in 2015. In traditional method, one obtains optical frequency-related Rayleigh pattern by

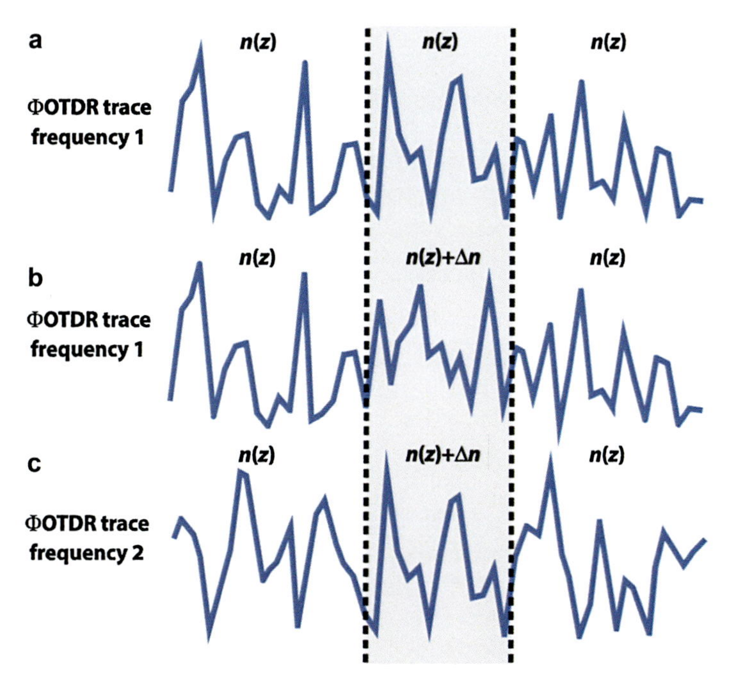

Fig. 3 Rayleigh pattern obtained in different conditions (different optical frequency or different refractive index)

precisely sweeping the frequency of the pulses step by step and employing a correlation algorithm to calculate the parameters.

Vibration Measurement

The RBS signals obtained along the fiber by using a highly coherent optical source have a jagged form as shown in Fig. 2b. With a frequency-stabled optical source and an undisturbed fiber, a stable RBS trace can be obtained, since the phase of backscattered lightwave from each scattering center depends on the wavelength of optical source and the distance of scattering centers. However, any external disturbance leads to the strain of fiber, which changes the scattering centers' distance considerably and finally results in a variation of RBS intensity. Therefore, by detecting and comparing the successive RBS intensity traces, external vibration can be obtained.

The intensity variation of RBS lightwave does not linearly depend on the external vibration, which is undesirable in vibration sensing since it may cause the higher-order harmonics, and the sensitivity of optical intensity to vibration is rather low at a long distance. On the other hand, the optical phase of RBS lightwave was introduced into vibration sensing, since the optical phase variation has a directly proportional relationship to the vibration amplitude. There were some researches to obtain the phase signal of RBS lightwave through the interference pattern among RBS lightwave from different pulses or different positions.

For both static measurement and vibration measurement, reflectometry is a fundamental technique to provide the RBS signals along the fiber. In this chapter, reflectometry techniques based on time domain, frequency domain, and coherence domain will be described in details. To obtain a better spatial resolution together with a longer measurement range, advanced methods are also described in this chapter.

Time-Domain Optical Reflectometry

Optical Time-Domain Reflectometry (OTDR)

The conceptual schematic of OTDR is shown in Fig. 4. In an OTDR, an optical probe pulse is injected into the fiber through a coupling device, and the RBS lightwave is coupled out and detected by a detector. By analyzing the output of the detector, the distributed attenuation characteristics of the fiber can be obtained.

With different optical sources and detectors, the OTDR has many configurations to meet different demands. Here, we show some traditional types:

- A typical OTDR (Barnoski and Jensen 1976):
  - Based on: a pulsed optical source and a photo-detector (PD)
  - Application: fiber attenuation measurement; optical sensors multiplexing
- A polarization OTDR (Rogers 1981):
  - Based on: a typical OTDR and a polarization analyzer
  - Application: polarization-related characteristics measurement such as polarization mode dispersion or birefringence
- A photon-counting OTDR (Wegmuller et al. 2004):
  - Based on: an ultrashort optical pulse generator and an avalanche photodiode

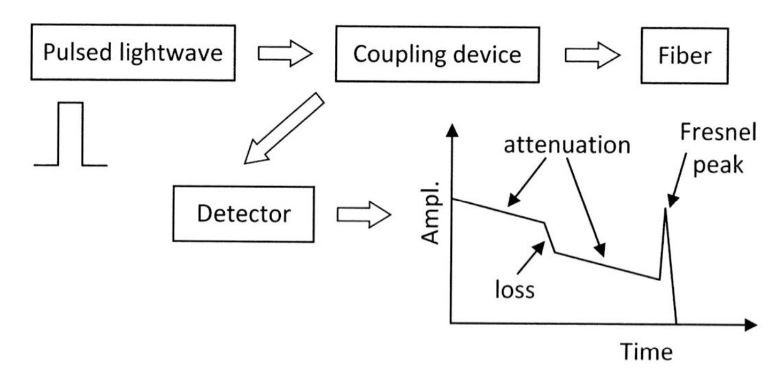

Fig. 4 Conceptual schematic of OTDR

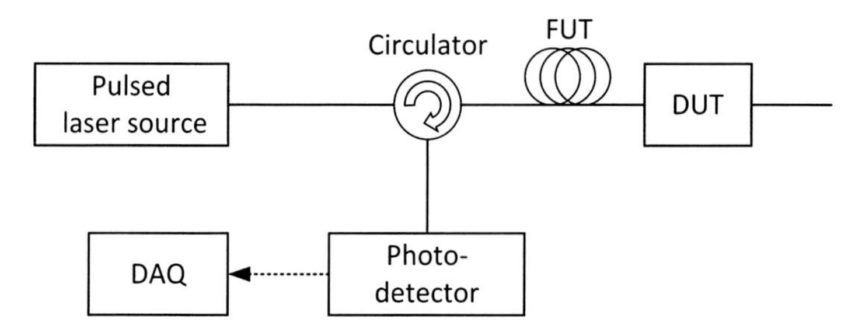

Fig. 5 Schematic of an all-fiber OTDR

- Application: realizing a long measurement distance (several tens of kilometer) with a high spatial resolution (centimeter level)

- A phase-sensitive OTDR (Shatalin et al. 1998):
  - Based on: a coherent optical source
  - Application: distributed strain/temperature/vibration/acoustic sensing

Since all the configurations have a same basic principle, we start from a very typical OTDR with an all-fiber configuration. The schematic of the system is shown in Fig. 5. The OTDR uses an incoherent optical source with a pulsed output, which can be expressed as

$$E_{\text{pluse}}(t) = E_0 A(t) e^{-j\omega_0 t} \tag{7}$$

where E_0 is the amplitude of the pulse and A(t) is the waveform of the probe pulse. As discussed in the former section, the RBS lightwave from the position z can be described as

$$E_R(t, z) = E_0 \sqrt{R(z)} A(t - 2z/v) e^{-j\omega_0 t} \tag{8}$$

where R is the RBS coefficient and $v = c/n$ is the speed of light propagating in the fiber. Therefore, the output of the PD can be expressed as

$$I(t) \propto E_0^2 \int_{v(t-\Delta t/2)/2}^{v(t+\Delta t/2)/2} R(z) dz \tag{9}$$

Here, we assume that the pulse has a rectangular shape and a duration time of Δt. That is, the spatial resolution of OTDR equals to the width of the optical pulse, which is $v\Delta t$. Figure 6 shows a typical trace measured by a commercial OTDR with a spatial resolution of 10 m, and the FUT is a 25-km-long SMF.

The performance of OTDR is evaluated by three aspects: dynamic range, spatial resolution, and the length of dead zone. The OTDR dynamic range is the difference of RBS lightwave power at the initial part and the noise power, and it indicates the

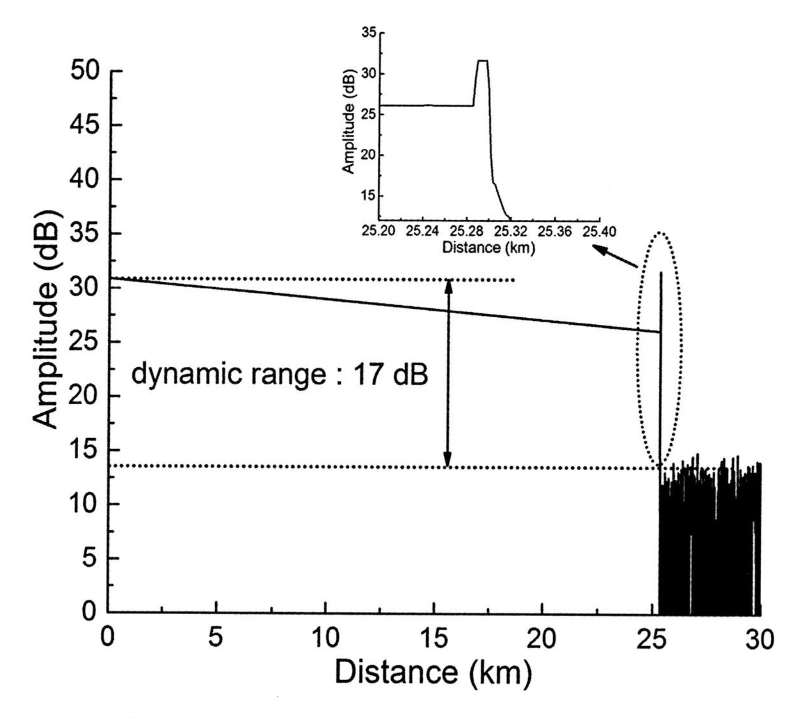

Fig. 6 A typical OTDR trace of a 25-km single mode fiber

SNR performance and the system measurement range. By using averaging, optical amplifier or coding technique, the dynamic range can be much improved. The spatial resolution shows the ability of distinguishing two adjacent reflective events, which only depends on the pulse duration time of the probe lightwave. The dead zone is caused by the limited PD performance. With a high-reflection event, the PD may reach a saturation status and need more time to cool down, without the capability of responding to another event at this time, and caused the dead zone after a strong reflection event.

Phase-Sensitive OTDR (ϕ-OTDR)

OTDR has a good performance in measuring the static characteristics such as reflectivity and attenuation but is difficult to measure the dynamic parameters such as the vibration. The polarization-sensitive OTDR (P-OTDR) is proposed for distributed dynamic sensing, but suffering from the transitivity of the state of polarization (SOP) which makes POTDR difficult for multi-events sensing. Recently, phase-sensitive OTDR (Φ-OTDR) is proposed for distributed dynamic sensing and attracted much attention, since it is sensitive, simple, and stable. In a Φ-OTDR system, a highly coherent optical source is used as the optical source, and its electrical field can be expressed as

$$E_0(t) = E_0 A(t) e^{-j[\omega t + \theta_0]} \tag{10}$$

where $A(t)$ is the waveform of probe pulse, and θ_0 is the phase of the optical probe pulse. Here, the probe pulse is supposed to have a rectangular shape and a duration time of Δt. The lightwave scattered from individual scattering centers interfere coherently, and the RBS lightwave can be described as

$$E_R(t) = R(t) e^{j\theta_R(t)} E_0 e^{j[\omega t + \theta_0]}$$

$$\text{where } R(t) e^{j\theta_R(t)} = \int_{v(t-\Delta t/2)/2}^{v(t+\Delta t/2)/2} r(z) e^{j\theta(z)} e^{j[\omega z/v]} dz \tag{11}$$

Here, $r(z)$ and $\theta(z)$ are the reflectivity and phase shift of the fiber at position z, respectively. Due to the interference, the intensity-distance trace of the RBS lightwave has a jagged form. With an external disturbance exerted on the fiber, the relative distance of the scatters experiences a change, leading to the intensity and phase variation of the RBS lightwave. The external disturbance can be monitored by extracting this intensity/phase change of RBS lightwave.

The principle of the intensity-extracted Φ-OTDR is shown in Fig. 7. Assuming that the Φ-OTDR system is stable, the intensity-distance trace should keep constant. However, with external disturbances exerted on the fiber, the trace may have changes at the positions of disturbances. By monitoring the changes, the external disturbance can be measured, and therefore Φ-OTDR can be used for distributed vibration sensing. Figure 8 shows an example of distributed vibration sensing with time-domain result and frequency-domain result (Martins et al. 2013). In this experiment, vibrations with different frequencies are exerted on the fiber using a piezoelectric

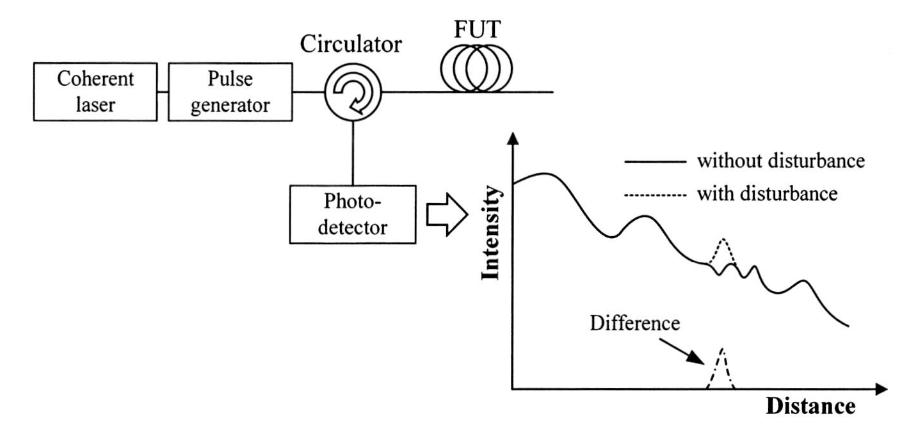

Fig. 7 Principle of the intensity-extracted Φ-OTDR

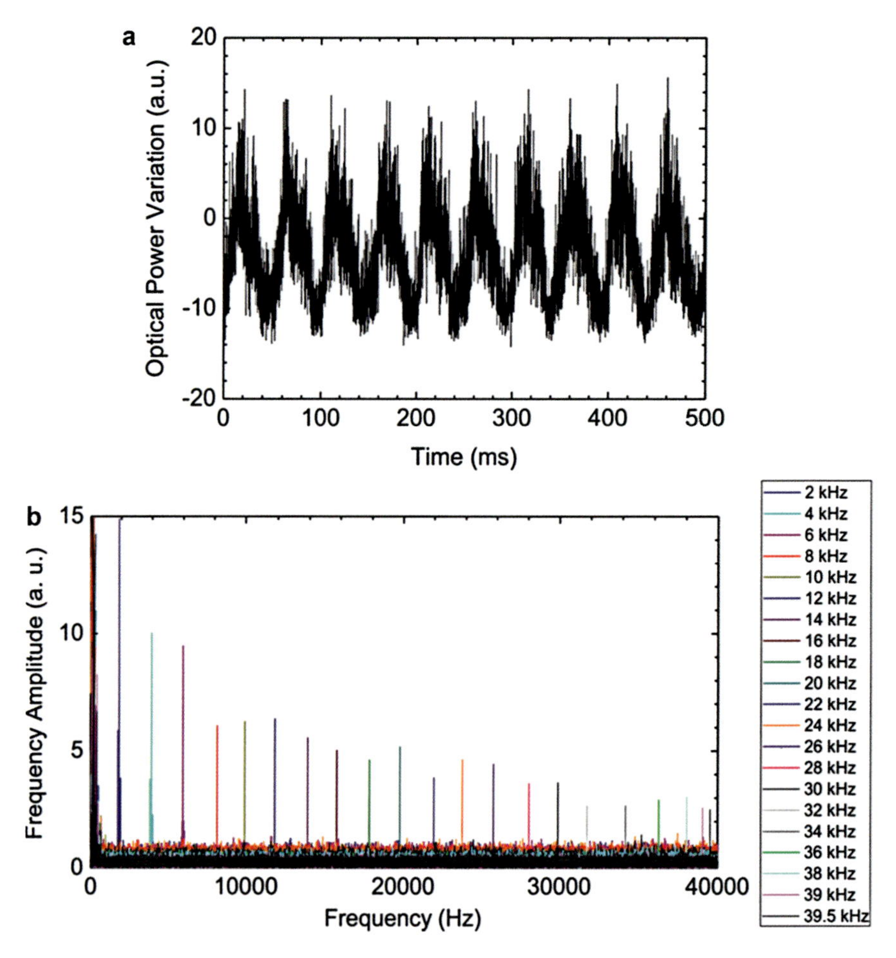

Fig. 8 (**a**) The measurement result of a 20-Hz vibration and (**b**) the FFT spectra of vibrations with different frequencies (Martins et al. 2013)

transducer, and the results show that the intensity-extracted Φ-OTDR has the capability of measuring vibrations.

However, the intensity-extracted method has a main disadvantage that the amplitude of vibration is not proportional to the variation of the optical intensity. In order to obtain the accurate time-domain information of vibration, the phase signal of RBS lightwave is extracted by measuring the phase difference between two continuous sections with a proper interval.

For Φ-OTDR, the frequency response to vibration is limited by the measurement distance, since the interval of pulse series must be larger than the round-trip time of the FUT. Specifically speaking, with a 10-km measurement distance, the frequency response is less than 5 kHz; while with a 40-km measurement distance the frequency response is limited to 1.25 kHz. Actually, by modulating the injected probe pulse,

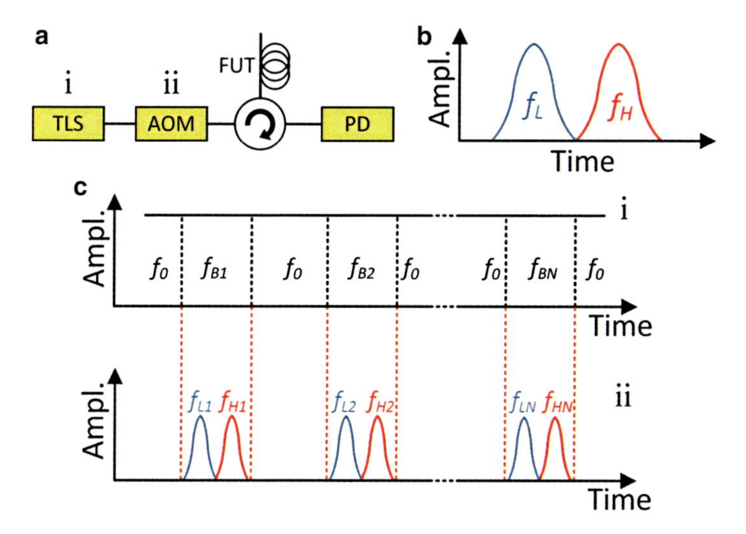

Fig. 9 (**a**) Conceptual schematic of FDM-Φ-OTDR; (**b**) conceptual schematic of the pulse pair; (**c**) the modulation on TLS and AOM (Yang et al. 2017)

a frequency division multiplexing (FDM) method can be used for overcoming the tradeoff relationship between measurement distance and frequency response. The conceptual schematic (Yang et al. 2017) is shown in Fig. 9. A tunable laser source (TLS) is employed as the optical source and an acousto-optic modulator (AOM) is used to generate pulse pairs with frequency spacing, as shown in Fig. 9b. The RBS lightwave of a pulse pair can be expressed as

$$E(t) = E_L(t-\tau)\cos\left[\omega_L(t-\tau) + \varphi_L(t-\tau)\right] + E_H(t)\cos\left[\omega_H(t) + \varphi_H(t)\right] \tag{12}$$

where ω_L and ω_H are the frequencies of the two pulses, τ is the pulse width. Supposed that only one pulse pair is injected into FUT and the bandwidth of PD is larger than $|\omega_L\text{-}\omega_H|$, the signal received by PD can be described as

$$\begin{aligned} I(t) = I_L(t-\tau) + I_H(t) + 2\sqrt{I_L(t-\tau)\, I_H(t)}\cos\left[(\omega_H - \omega_L)t\right. \\ \left. + \varphi_H(t) - \varphi_L(t-\tau)\right] \end{aligned} \tag{13}$$

With a vibration exerted on the fiber, the first pulse is influenced by the vibration, while the second pulse is not affected, which introduces a phase change in the beat frequency term in Eq. (13). After the second pulse is also influenced by the vibration, the phase change disappeared. Therefore, by extracting the phase signal of beat frequency in Eq. (13), the vibration can be measured and located.

With different frequency spacing, different pulse pairs generate different beat frequency signals, but the RBS lightwave from different pulse pairs may interfere, and generate a lot of undesired frequency components. In order to distinguish different pulse pairs in frequency domain, the TLS is modulated and a frequency offset is introduced to each pulse pair, as shown in Fig. 9c. Each pulse pair has a specific frequency spacing which is less than the PD bandwidth, while the frequency offset difference between different pulse pairs is set much larger than the PD bandwidth in order to avoid the undesired interference. Therefore, the beat frequency from different pulse pairs can be distinguished by using a digital bandpass filter. By synthesizing the extracted phase signal from different beat frequencies, the vibration frequency response can be enhanced. An experiment is accomplished with a 10-km measurement distance and four times multiplexing, corresponding to a 20-kHz frequency response under this range. The measurement result of a 6-kHz vibration is shown in Fig. 10, and the measured vibration signal shows a high SNR and a linear response.

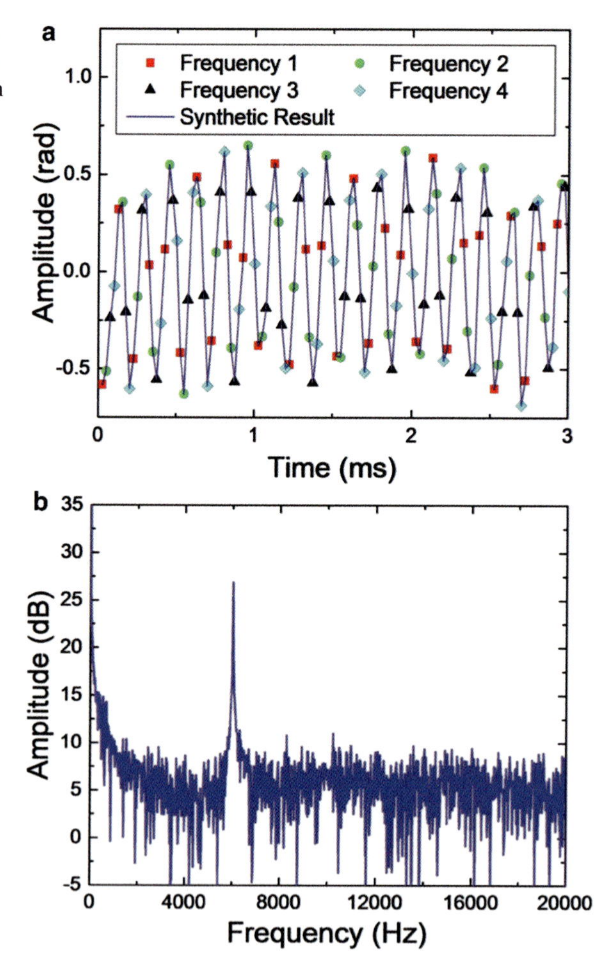

Fig. 10 The 6-kHz vibration measurement result with four pulse pairs in (**a**) time domain and (**b**) frequency domain (Yang et al. 2017)

Coherent OTDR (COTDR)

OTDR has shown the capability in distributed sensing, including static sensing and dynamic sensing. However, in the practical communication link with a distance of thousands of kilometers, Erbium-doped fiber amplifier (EDFA) is widely used to amplify the signals, which leads to a large amplified spontaneous emission (ASE) noise. The direct detection scheme used in OTDR cannot distinguish the ASE noise and the RBS lightwave, which limits the measurement SNR. Furthermore, when multiple EDFAs are cascaded in the system, the ASE noise may have a very high level, further preventing OTDR from being applied in the long-range link monitoring.

In order to measure the long-range communication link such as the submarine optical fiber cable, coherent OTDR (COTDR) is proposed with coherent detection schemes (King et al. 1987). The conceptual schematic of coherent detection is shown in Fig. 11. For a coherent detection system, a local reference lightwave is used for mixing with the signal lightwave to be detected. The mixed lightwave is then injected into the balanced photodetector (BPD), and the beat signal can be obtained from the BPD output. In order to escape from the DC noise, the frequency of local reference lightwave is slightly different from the frequency of signal lightwave to have a heterodyne detection. Therefore, the local reference lightwave with an optical frequency of f_{LO} and the signal lightwave with an optical frequency of f_S can be described as

$$\begin{aligned} E_{LO} &= E_{LO} \exp\left(j 2\pi f_{LO} t\right) \\ E_S &= E_S \exp\left(j 2\pi f_S t\right) \end{aligned} \tag{14}$$

where E_{LO} and E_S are the amplitude of the local reference lightwave and signal lightwave, respectively. The electric field of the mixed lightwave can be described as

$$E = E_{LO} + E_S = E_{LO} \exp\left(j 2\pi f_{LO} t\right) + E_S \exp\left(j 2\pi f_S t\right) \tag{15}$$

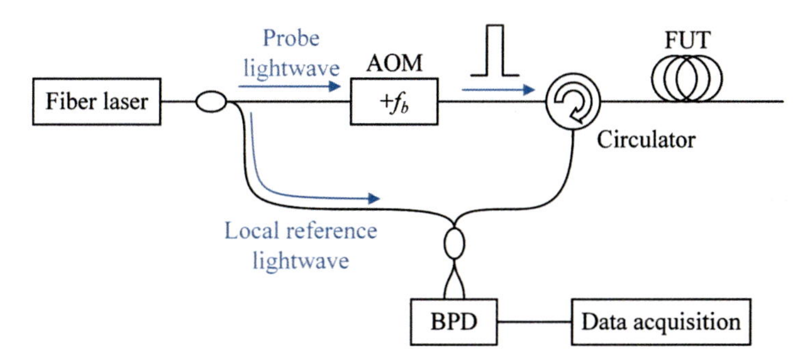

Fig. 11 Conceptual schematic of COTDR

Therefore, the BPD output can be expressed as

$$I = kEE^* = k\left[E_{LO}^2 + E_S^2 + 2E_{LO}E_S\cos 2\pi\left(f_{LO} - f_S\right)t\right] \tag{16}$$

where k is the BPD responsivity. The AC component of the output can be expressed as

$$I_{AC} = 2kE_{LO}E_S\cos 2\pi\left(f_{LO} - f_S\right)t \tag{17}$$

Equation 17 shows that the amplitude of the output is directly proportional to the amplitudes of both signal and local reference lightwave. Therefore, the SNR of the system can be expressed as

$$\frac{S}{N} = \frac{2k^2E_S^2E_{LO}^2}{2ei_dB + 2ekE_{LO}^2B + 2ekE_N^2B} \tag{18}$$

where e is electron charge, i_d is the dark current of the detector, B is the bandwidth of the detector, and E_N is the effective amplitude of the other noise in the system such as thermal noise. The three terms in the denominator stand for the dark current noise, shot noise, and the other noise. Under the most common situation to introduce a large local power E_{LO}^2, the shot noise is much larger than the other noise. Thus, the SNR can be simplified to

$$\frac{S}{N} = \frac{2kE_S^2}{eB} \tag{19}$$

A schematic of COTDR system is shown in Fig. 11. The highly coherent lightwave from the laser is divided into two beams by using an optical coupler: the probe lightwave for launching into the FUT and the local reference lightwave for coherent detection. An AOM is then used to modulate the probe lightwave into a pulse with a frequency shift of f_b. The RBS lightwave returned from FUT is mixed with the local lightwave in another 3 dB coupler, and the beat signal is detected by using a BPD. Usually, the laser source employed in COTDR should have a linewidth less than 10 kHz and a good frequency stability, which can help reduce the bandwidth of RBS signal and enhance SNR.

The performance of COTDR is usually evaluated by several aspects: dynamic range, spatial resolution, measurement time, and reflectivity measurement accuracy. The dynamic range of COTDR is much larger than that of OTDR since it may realize the shot-noise limited detection. Since COTDR is usually used for measuring long-haul submarine lines, which leads to a large pulse round-trip time, the measurement time may reach hours or days for a large average times over 2^{18}. Furthermore, due to the application of narrow linewidth laser and coherent detection, the coherent Rayleigh noise and polarization mismatch result in the reflectivity trace in a jagged form, which may influence the measurement of loss and reflection events. By

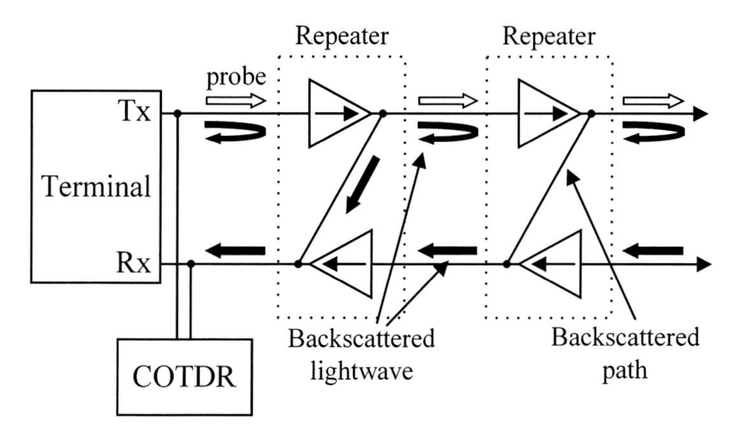

Fig. 12 Measurement block diagram of COTDR for long-haul submarine optical amplifier transmission lines (Iida et al. 2012)

using frequency hopping and wavelength shift averaging technique, the reflectivity measurement accuracy can be improved (Shimizu et al. 1992).

As discussed above, a COTDR is often used for fault location and characterization in long-haul submarine lines with optical amplifiers. It should be noted that the optical isolator used to eliminate multireflection effects in EDFA blocks the returning propagation of a backscattering light. Therefore, a bypass designed repeaters is used for COTDR, as shown in Fig. 12. In most cases, the interval of repeaters is very large, which require a large dynamic range of COTDR to realize the complete measurement. Recently, a frequency division multiplexing (FDM) COTDR technique is proposed for enhancing the dynamic range without increasing measurement time (Iida et al. 2012). The principle of FDM-COTDR is shown in Fig. 13. Instead of a single frequency probe in COTDR, FDM-COTDR uses multi-frequency probes. Each frequency component is serially arranged in the time domain, thus avoiding both peak power increase and nonlinear interaction as a result of four-wave mixing. The RBS lightwave from the probe of different frequency components can be distinguished by using a digital bandpass filter. By employing FDM method, the system can perform several independent measurements within a single round-trip time, therefore enhancing the dynamic range without increasing measurement time. The system setup of FDM-COTDR is shown in Fig. 14. Comparing with COTDR, FDM-COTDR additionally uses a phase modulator and an arbitrary waveform generator to generate a multifrequency probe. By using 40 frequency components, the dynamic range is enhanced by 8 dB (Iida et al. 2012), as shown in Fig. 15.

Since COTDR can also be regarded as a Φ-OTDR with coherent detection, it has attracted much attention in distributed dynamic sensing, especially distributed vibration/acoustic sensing (Lu et al. 2010). In COTDR, the phase signal can be simply extracted from the beat frequency by using I/Q demodulation, which is more convenient and reliable. Furthermore, COTDR has a larger dynamic range, leading

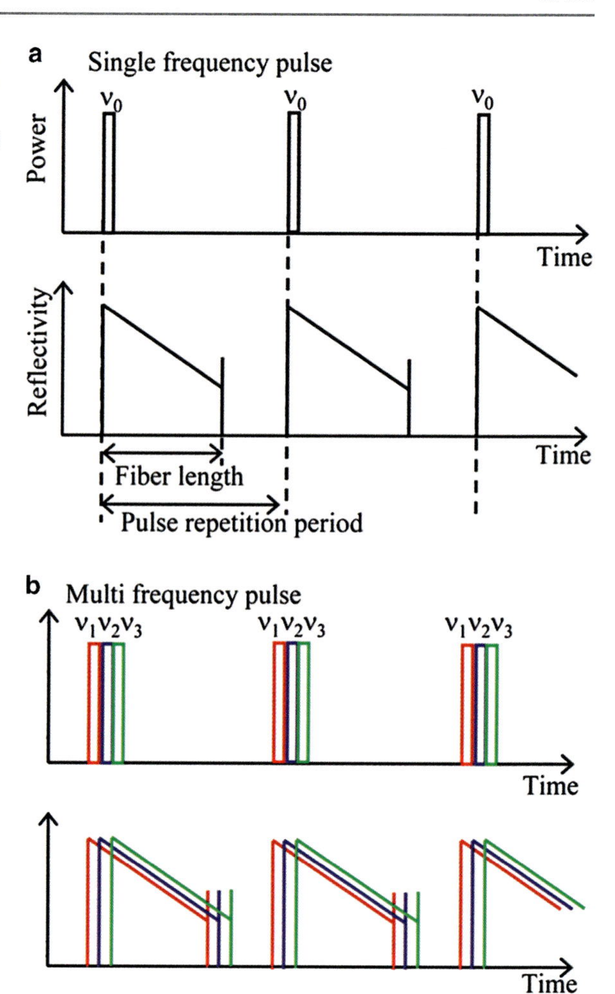

Fig. 13 Schematic diagrams of measurement principle. (**a**) Conventional C-OTDR. (**b**) FDM-OTDR (Iida et al. 2012)

to a longer measurement distance. Finally, as introduced above, compared to the conventional Φ-OTDR, COTDR is much more suitable for applying a FDM method. Figure 16 shows a system setup of FDM-COTDR for distributed vibration sensing by using FDM technique and phase extraction technique (Wang et al. 2015a). The system realized a bandwidth of 175 kHz over a 10-km measurement distance with a SNR of 16 dB. Figure 17 shows a measurement result of a vibration signal with the frequency swept from 7 kHz to 167 kHz. The performance of COTDR in distributed vibration sensing is much better than that of Φ-OTDR with direct detection in terms of SNR and frequency response.

However, there are still two problems of COTDR. First, the laser source frequency drift may result in an unwelcome low-frequency noise, which can be suppressed by using a frequency-stabilized laser source. Second, the fading noise, which consists of polarization fading noise and coherent Rayleigh noise, may

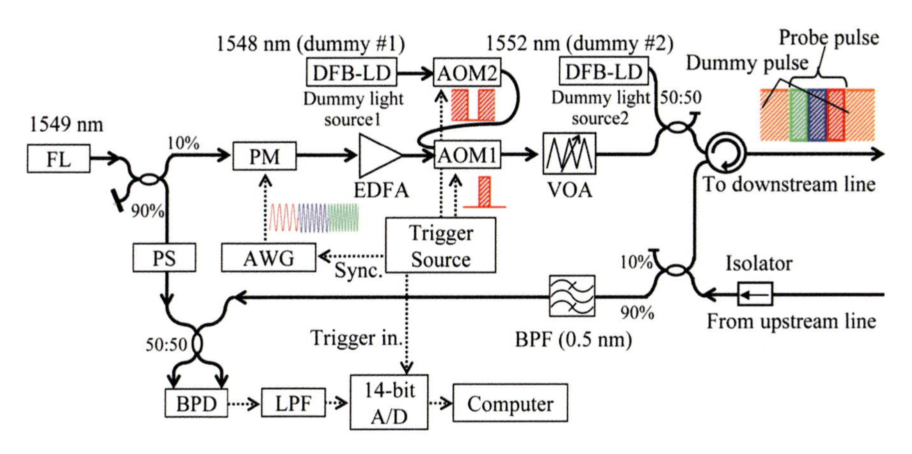

Fig. 14 Experimental setup of FDM-COTDR (Iida et al. 2012)

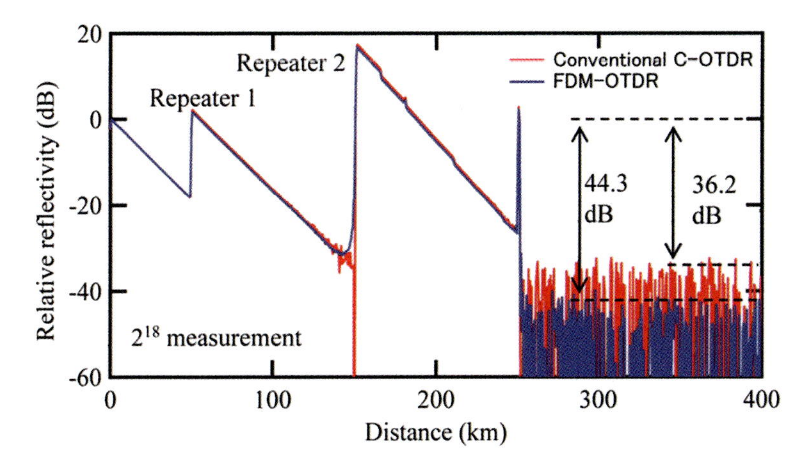

Fig. 15 COTDR traces of 250 km long-haul submarine optical amplifier transmission line for both conventional C-OTDR (red) and FDM-OTDR (blue line) with 2^{18} measurements (Iida et al. 2012)

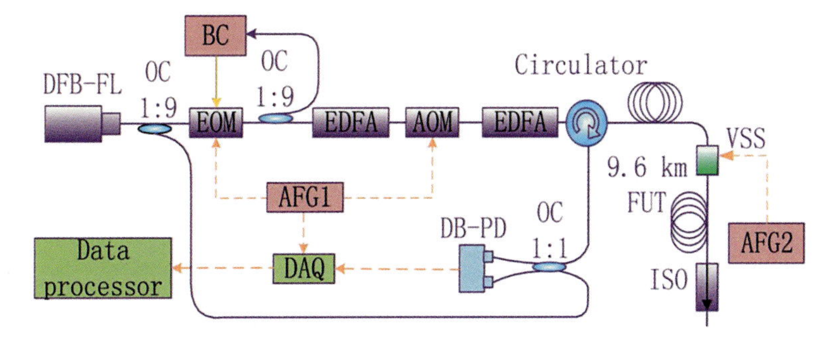

Fig. 16 The system setup of a FDM-COTDR for distributed vibration sensing (Wang et al. 2015a)

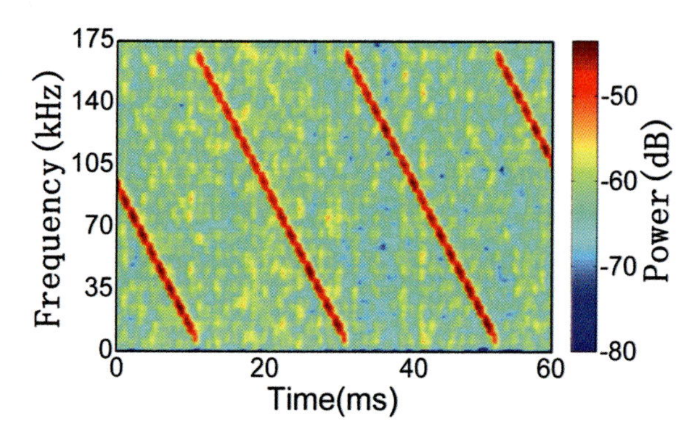

Fig. 17 STFT spectrum of the rebuilt chirp signal (Wang et al. 2015a)

deteriorate the COTDR performance. The fading noise results in some points on the fiber with a very limited SNR and thus causes a large phase extraction error. These points are called "dead zone" since the external disturbances cannot be measured at these points. Recently, some researches are trying to eliminate the dead zone, such as using polarization diversity receiver, phase noise analysis, or to use an inner-pulse frequency-division method (Yang et al. 2016; Chen et al. 2017).

Frequency-Domain Optical Reflectometry

OFDR was firstly proposed by W. Eickhoff and R. Ulrich in 1981. It is able to provide a spatial resolution below 100 μm (Soller et al. 2005), which is nearly comparable with optical low coherence reflectometry (OLCR), and does not require mechanical scanning stage. There are two conventional frequency-domain reflectometric techniques including incoherent optical frequency-domain reflectometry (I-OFDR) (Spillman et al. 1988; Liehr et al. 2009) and coherent optical frequency-domain reflectometry (C-OFDR) (Eickhoff and Ulrich 1981). In this section, we will introduce in detail the basic principle and some applications of I-OFDR and C-OFDR.

Incoherent Optical Frequency-Domain Reflectometry (I-OFDR)

Similar to OTDR, which obtains the trace by measuring the impulse response in the time domain, I-OFDR conducts it in the frequency domain. It is achieved by measuring the frequency-domain transfer function and then performing inverse Fourier transform (IFT) to it. The basic measurement setup of I-OFDR is shown in Fig. 18 (Liehr et al. 2009). The key component is the electrical vector network analyzer (VNA), which generates a frequency-swept sinusoidal signal and measures the frequency response of the system. The radio frequency (RF) signals generated by the VNA are split to reference signals and modulation signals. The light source is directly intensity-modulated by the electrical signal and then launched into the

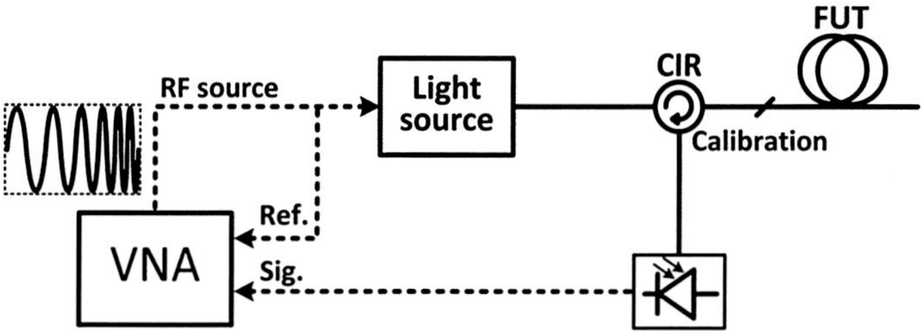

Fig. 18 Schematic of the I-OFDR setup (Liehr et al. 2009)

FUT. The backscattered and reflected light from this FUT is detected by a PD and the generated electrical signal is fed to the VNA to calculate the complex transfer function of the setup. A calibration process is necessary since the measurement provides the transfer function of the whole signal path including the electrical components. After dividing the measurement trace and the calibration data, the transfer function of the FUT can be obtained. Then the frequency-domain transfer function can be converted to time domain (the time can be easily scaled to distance) after the IFT manipulation.

The spatial resolution of I-OFDR is inversely proportional to the frequency modulation range ($f_{\max} - f_{\min}$) and can be given by

$$\delta z = \frac{c}{2n_g\left(f_{\max} - f_{\min}\right)} \tag{20}$$

where n_g is the group refractive index. The measurement range, which is closely related to the frequency modulation step Δf, can be written as

$$z_m = \frac{c}{2n_g \Delta f} \tag{21}$$

According to Eq. 20, it is obvious that a wide frequency modulation range may help to obtain a high spatial resolution. Normally, a frequency modulation span of several hundred MHz, which means a submeter spatial resolution in I-OFDR, can be easily achieved. If a wideband VNA is utilized in the system, a centimeter-level spatial resolution is promising to be obtained.

I-OFDR is superior to OTDR in spatial resolution, so it can be used in some fields where high spatial resolution is required. Moreover, unlike OTDR, which launches a single pulse into the FUT, I-OFDR uses an intensity-modulated continuous wave (CW) lightwave, so a high SNR can be obtained with a relatively low optical power. In terms of measurement range, I-OFDR may work beyond the laser coherence length because of its incoherent characteristics. Most recently, a measurement range of 151 km has been achieved with the assistance of Kerr phase-interrogator, while the spatial resolution maintains 11.2 cm (Baker et al. 2014). I-OFDR has been widely used in various applications. Besides the diagnosis for fiber links, it can also be used for distributed sensing, such as temperature or strain measurement by extracting the useful information from the RBS signals (Liehr et al. 2009).

Although I-OFDR provides high performance, it is not cost-efficient because the expensive VNA is used in the system. Besides, in order to obtain a high spatial resolution, the frequency modulation range is expected to be very broad, which means a wideband receiver is required. To overcome this problem, C–OFDR, which requires a low receiving bandwidth, is also widely studied. In the next subsection, we will introduce the C-OFDR technique in detail.

Coherent Optical Frequency-Domain Reflectometry (C-OFDR)

Conventional C-OFDR

The basic configuration of C-OFDR is illustrated in Fig. 19. The C-OFDR system is mainly composed of a linearly chirped light source and a two-beam interferometer. The frequency-swept lightwave is split into two beams, which are used as the reference light and the probe light. A polarization controller is utilized to adjust the SOP of two beams. The reference light and the backscattered light are mixed at the 50/50 optical coupler and then detected by a BPD. The electronic signals are collected by an analog-to-digital converter (ADC) and then digitally processed using a computer.

For quantitative analysis, by assuming that the slope of the frequency chirp is perfectly linear in time, and the frequency sweeping span is broad enough to obtain the desired spatial resolution, the optical field can be described as

$$E(t) = E_0 e^{j\left(\omega_0 t + \pi\gamma t^2 + \theta(t)\right)} \tag{22}$$

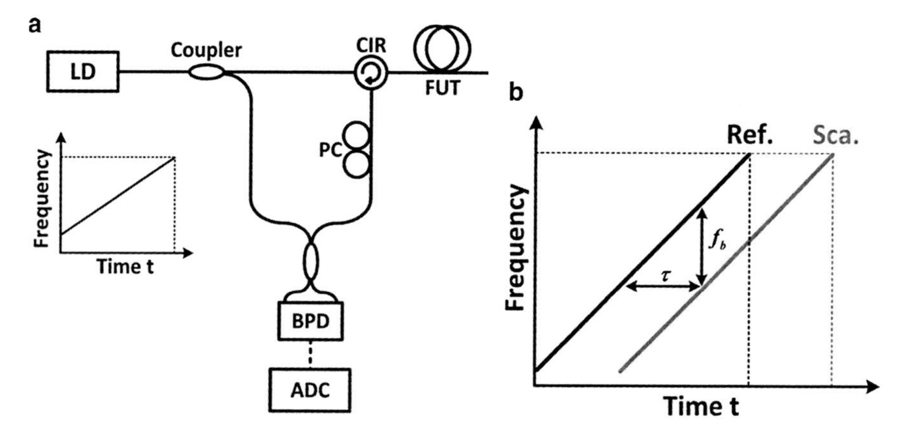

Fig. 19 (**a**) Basic configuration of C-OFDR system; (**b**) graphical illustration of the generation of beat frequency

where E_0, ω_0, γ, and $\theta(t)$ are the amplitude, the initial optical frequency, the frequency sweep rate of the light source, and the randomly phase fluctuation caused by laser phase noise, respectively. For simplicity, we consider the case of only one single reflection event from the FUT. The reference light $E(t)$ and the backscattered light $\sqrt{R}E\,(t - \tau_{FUT})$, which is delayed by τ_{FUT}, are combined in an optical coupler and coherently detected by a BPD, where R is the reflectivity of the FUT. The photocurrent can be represented by Venkatesh and Sorin (1993):

$$
\begin{aligned}
I(t) &= \left| E(t) + \sqrt{R}E\,(t - \tau_{FUT}) \right|^2 \\
&= E_0^2 \left\{ 1 + R + 2\sqrt{R}\cos\left[\omega_b t + \omega_0 \tau_{FUT} - \tfrac{1}{2}\omega_b \tau_{FUT} + \theta(t) - \theta\,(t - \tau_{FUT}) \right] \right\}
\end{aligned}
\tag{23}
$$

where the beat frequency ω_b can be given by $\omega_b = 2\pi\gamma\tau_{FUT}$.

After some algebraic manipulations, the one-side spectral density of the electrical signal, which is a more useful expression related to experimental measurements, can be given by Venkatesh and Sorin (1993):

$$
\begin{aligned}
S_I^1(f) = (1+R)^2\delta(f) + 2\mathrm{Re}^{-\left(2\tau_{FUT} / \tau_c \right)}\delta\,(f - f_b) + \frac{2R\tau_c}{1 + \pi^2\tau_c^2 (f - f_b)^2} \\
\left\{ 1 - e^{-\left(2\tau_{FUT} / \tau_c \right)} \left[\cos 2\pi\,(f - f_b)\,\tau_{FUT} + \sin\frac{2\pi\,(f - f_b)\,\tau_{FUT}}{\pi\tau_c\,(f - f_b)} \right] \right\}
\end{aligned}
\tag{24}
$$

where $f_b = \omega_b/2\pi = \gamma\tau_{FUT} = 2nL\gamma/c$. Here, τ_c is the laser source coherence time, n is the effective refractive index of fiber core, L is the length of FUT, and c is the light velocity in vacuum. The first term of the spectral density function is a delta function at DC, which is mainly dependent on the value of reflectivity R. The second term is a delta function at the beat frequency, which we called "coherent term." This term provides the position information and the backscattered intensity, but its amplitude has an exponential term $e^{-(2\tau_{FUT}/\tau_c)}$, which means it decayed rapidly when the measurement distance approaches half of the laser coherence length. Therefore, the measurement range of C-OFDR is normally limited by the coherence characteristics of the light source utilized in the system. The third term represents the distribution of phase noise around the beat frequency, which we called "phase noise term." It becomes dominant and the spectrum of the beat frequency deteriorates to a Lorentzian shape when FUT length is longer than half of the laser coherence length.

C-OFDR provides a high spatial resolution, which is inversely proportional to the frequency sweeping span of the light source. Similar to Eq. 20, the spatial resolution of C-OFDR can be given by

$$
\Delta L = c / 2n\Delta F \tag{25}
$$

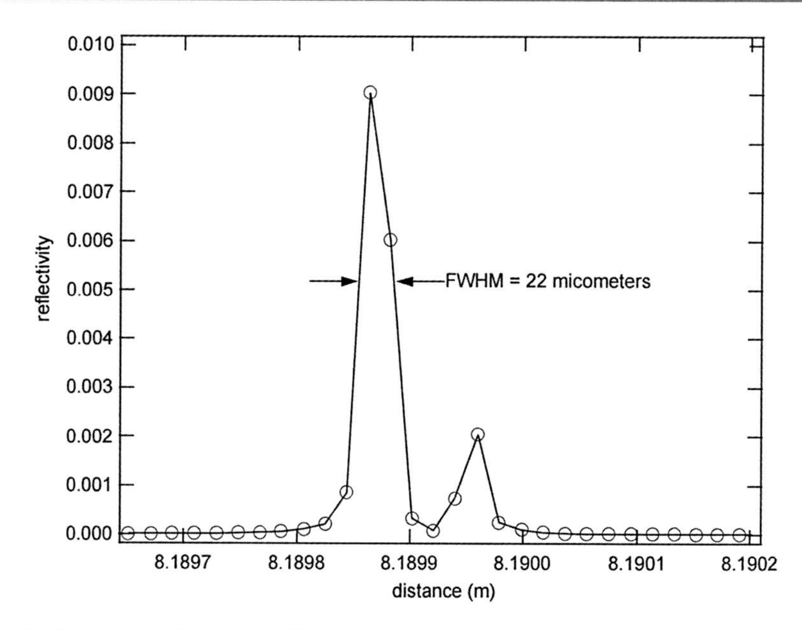

Fig. 20 Measured reflectivity details of polished fiber (PC) terminations, showing a 100 μm separation between the two fiber ends and 22 μm full width half maximum resolution of the individual peaks (Soller et al. 2005)

where ΔF is the frequency sweeping span of the light source. It is clear that a broader span helps to obtain a higher spatial resolution. Currently, the realization of optical frequency sweeping can be achieved by inner-cavity tuning of a tunable laser, or external modulation of a CW lightwave. For the inner-cavity tuning technique, an ultrawide sweeping span can be obtained, which means a high spatial resolution is promising to be achieved. As shown in Fig. 20, a spatial resolution of 22 μm has been realized (Soller et al. 2005), corresponding to a frequency sweeping span of 50 THz (about 40 nm in wavelength).

Although micrometer-level spatial resolution has been obtained, the measurement range of C-OFDR is limited (normally a few tens of meters) since the inner-cavity tuning brings a broad spectral linewidth and a bad sweeping linearity. On the other hand, one can use external modulation with frequency-swept RF signal to sweep the optical frequency ultra-linearly, and several centimeters spatial resolution over 5 km has been obtained by using this method (Koshikiya et al. 2008). Moreover, some techniques have been proposed to further extend the measurement range of C-OFDR. Here, we mainly introduce two typical methods called phase-noise-compensated (PNC-OFDR) (Fan et al. 2009) and time-gated digital (TGD-OFDR) (Liu et al. 2015).

Phase-Noise Compensated OFDR (PNC-OFDR)

Figure 21 shows the schematic of the PNC-OFDR technique. Besides the main interferometer, an auxiliary interferometer, which is used for phase-noise compen-

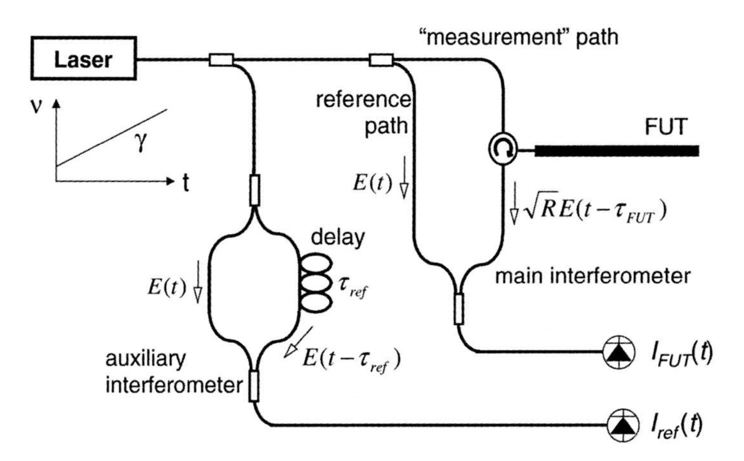

Fig. 21 Schematic of PNC-OFDR technique (Fan et al. 2009)

sation, has been incorporated into the system. The main interferometer provides a beat signal, produced by the optical interference between two signals originating from the same linearly chirped source. For an optical source with an ideally linear frequency sweeping slope, the optical field can be expressed as Eq. 22. For the beat signal, ignoring all the DC terms, it can be written as

$$I_{FUT}(t) = 2\sqrt{R}E_0 \cos\left[2\pi\gamma\tau_{FUT}t + \theta(t) - \theta\left(t - \tau_{FUT}\right) + C_{FUT}\right] \tag{26}$$

where C_{FUT} is a constant. Similarly, for the auxiliary interferometer, the beat signal can be written as

$$\begin{aligned} I_{REF}(t) &\propto \cos\left[2\pi\gamma\tau_{REF}t + \theta(t) - \theta\left(t - \tau_{REF}\right) + C_{REF}\right] \\ &= \cos\left[X_1(t)\right] \end{aligned} \tag{27}$$

where C_{REF} is a constant and $X_1(t)$ is the phase term of reference signal I_{REF}. Considering the scattered signal I_{FUT} sample based on timing that corresponds to certain increments in the phase term of the reference signal, for example, π increments, the sampled signal becomes

$$I_{FUT}\left(t_M\right) = 2\sqrt{R}E_0 \cos\left[\left(\pi\tau_{FUT}/\tau_{REF}\right)M + \Phi\left(t_M\right) + C\right] \tag{28}$$

where t_M is the sampling time decided from the reference signal, M is the sampling index, C is a constant, and the phase term $\Phi(t_M)$ is the randomly sampled collection of $\Phi(t)$, which can be given by

$$\Phi(t) = \left[\theta(t) - \theta\left(t - \tau_{FUT}\right)\right] - \frac{\tau_{FUT}}{\tau_{REF}}\left[\theta(t) - \theta\left(t - \tau_{REF}\right)\right] \tag{29}$$

It is obvious that the phase term $\Phi(t)$ is cancelled out when τ_{REF} is equal to τ_{FUT}, meaning the phase noise–induced adverse impact can be totally mitigated.

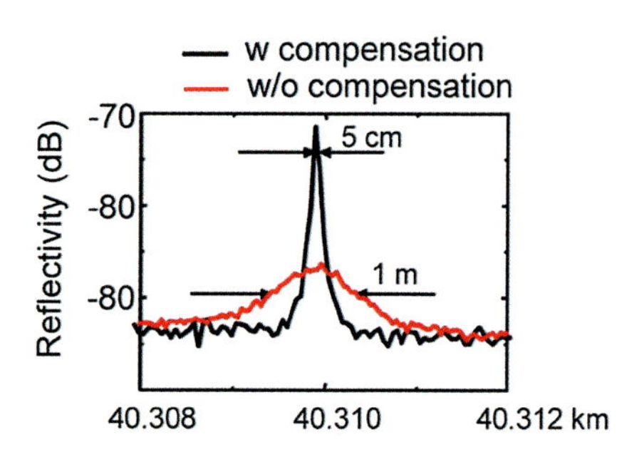

Fig. 22 Measured reflection profiles at 40 km with and without phase noise compensation (Ito et al. 2012)

By using this method, the phase noise can be compensated when the measurement distance is shorter than the laser coherence length. For reflection events beyond laser coherence length, concatenately generated phase (CGP) technique can be used to generate the required reference phase signal (Fan et al. 2009). Figure 22 shows the measured reflection profiles at 40 km. After phase noise compensation, the spatial resolution improves from 1 m to 5 cm. PNC-OFDR technique firstly extends the measurement range to tens of kilometers while keeping a centimeter-level spatial resolution.

Time-Gated Digital OFDR (TGD-OFDR)

Besides phase-noise compensation, improving the frequency sweeping rate is another method to suppress the phase noise. However, a higher laser chirp rate together with a long measurement distance generates a very high beat frequency, which in most cases goes beyond the sampling rate of available data acquisition devices. The TGD-OFDR technique can realize high frequency sweeping rate independent of the length of FUT. The schematic of the TGD-OFDR system is shown in Fig. 23a. The lightwave from a laser source is split into two beams by a coupler. The probe beam (upper) is frequency-swept and time-gated by an AOM, while the reference beam (lower) remains a frequency-stable continuous lightwave. The scattered lightwave interferes with the reference lightwave, and then produces beat signals, corresponding to different locations in FUT. As shown in Fig. 23b, although beat signals (gray solid line) generated by reflected lightwave at different locations along FUT have the same chirp rate and range, they appear within different time window. To retrieve the reflections along the FUT from the detected beat signals, a digital implementation of frequency-to-distance mapping is used. An equivalent reference (black solid line) is generated digitally, which has the same chirp rate with the chirped probe lightwave. The frequency difference between the detected beat signal and the equivalent reference is proportional to the corresponding time delay. Then Fourier transform is conducted to calculate the spectrum of the beat signal, completing the distance-to-frequency mapping as that in typical OFDR systems.

By using TGD-OFDR technique, a spatial resolution of 1.6 m over entire 110-km range has been experimentally demonstrated (Liu et al. 2015). The spatial

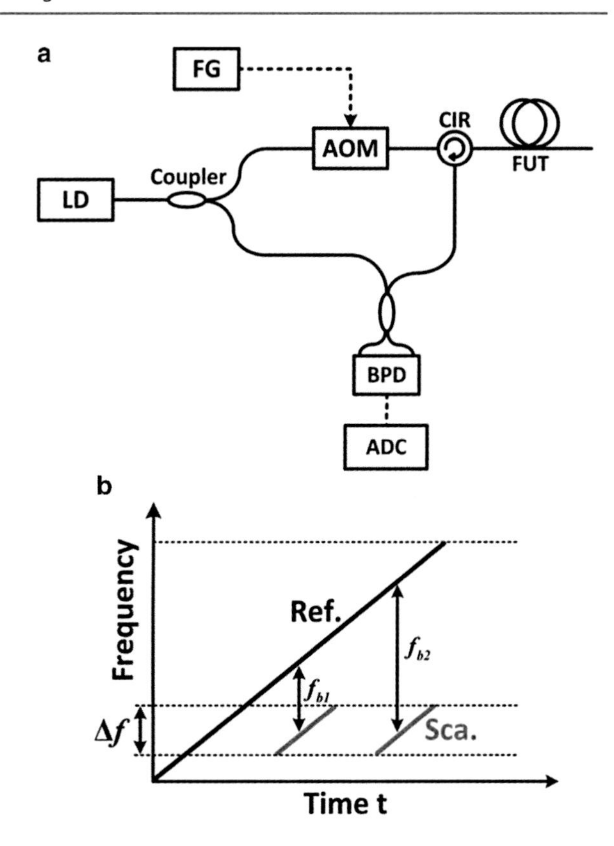

Fig. 23 (**a**) Schematic diagram of TGD-OFDR system; (**b**) graphical illustration of the generation of beat frequency (Liu et al. 2015)

resolution of TGD-OFDR is limited by the frequency chirp span of the AOM, which is normally narrower than 100 MHz. Other modulators such as single-sideband modulator (SSBM) can be utilized to expand the frequency sweeping span, and centimeter-level spatial resolution is promising to be obtained. However, the sweeping span is still limited by the performances of synthesized RF frequency sweeper, and normally narrower than 10 GHz. High-order modulation technique has been proposed to magnify the optical frequency chirp span (Xu et al. 2014), and injection-locking technique is utilized to remove the overlap between adjacent high-order sidebands (Wang et al. 2017a). Finally, a spatial resolution of 4 mm over entire 10 km has been obtained with the proposed frequency swept light source. Moreover, other nonlinear effects such as four-wave-mixing process can also be used to further expand the frequency sweeping span to 100 GHz (the frequency residual error after linear fitting is 160 kHz) (Wang et al. 2017a). A millimeter-level spatial resolution over kilometer-level measurement range can be obtained with this light source because of its narrow linewidth and high linearity.

Based on the abovementioned C-OFDR technique, many measurement systems have been developed. In the state-of-the-art C-OFDR systems, micrometer-level spatial resolution over tens of meters, millimeter-level spatial resolution

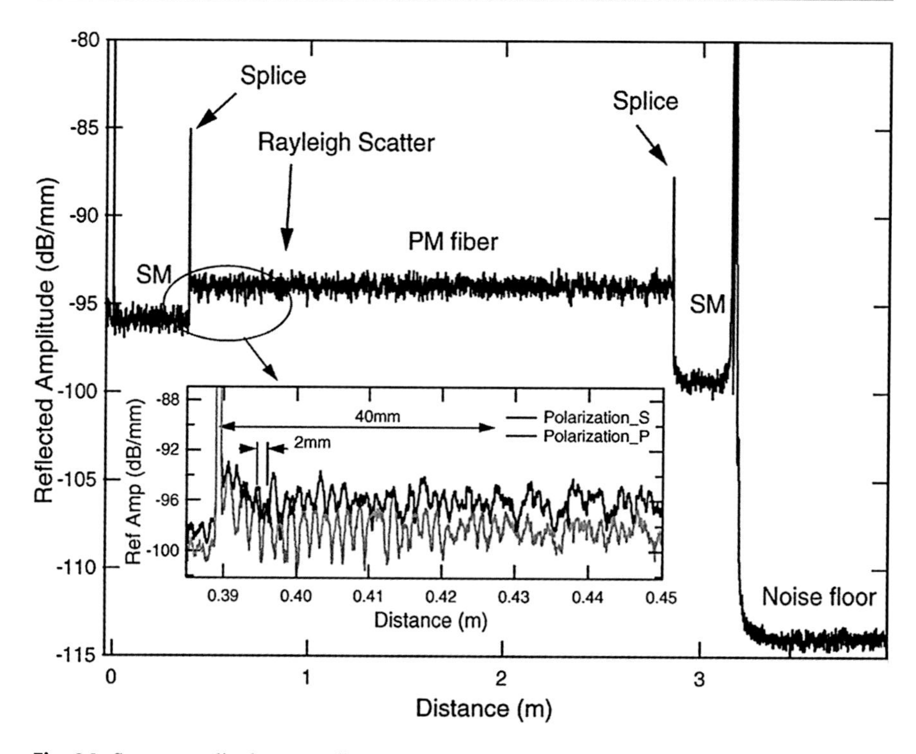

Fig. 24 Scatter amplitude versus distance measurement of the PM fiber (Froggatt et al. 2006)

over 10 km, and centimeter-level spatial resolution over tens of kilometers have been achieved. Benefiting from the high spatial resolution, it can be used to measure the beat length of PMF (Froggatt et al. 2006). Figure 24 shows the measured RBS signals. A periodically intensity fluctuation, which is caused by the existing birefringence in PMF, can be clearly observed. Other parameters of the optical fiber, like polarization-mode-dispersion (PMD) and bending induced birefringence, can also be measured with C-OFDR since it provides high spatial resolution.

C-OFDR can be used to measure temperature, strain, vibration (dynamic strain), or other disturbances applied to FUT via the useful information extracted from the RBS signals. As shown in Fig. 25a, the temperature information can be extracted from the RBS lightwave obtained with C-OFDR (Ohno et al. 2016). The measurement range can be extended to 100 km after some simple algebraic manipulations, and the temperature sensitivity is higher than 0.01°. Besides static measurement, C-OFDR is also capable of measuring vibration benefiting from the high measurement speed. Figure 25b shows the measured vibration (Zhou et al. 2012), and the strain sensitivity and the response rate are relatively high and promising to be further improved.

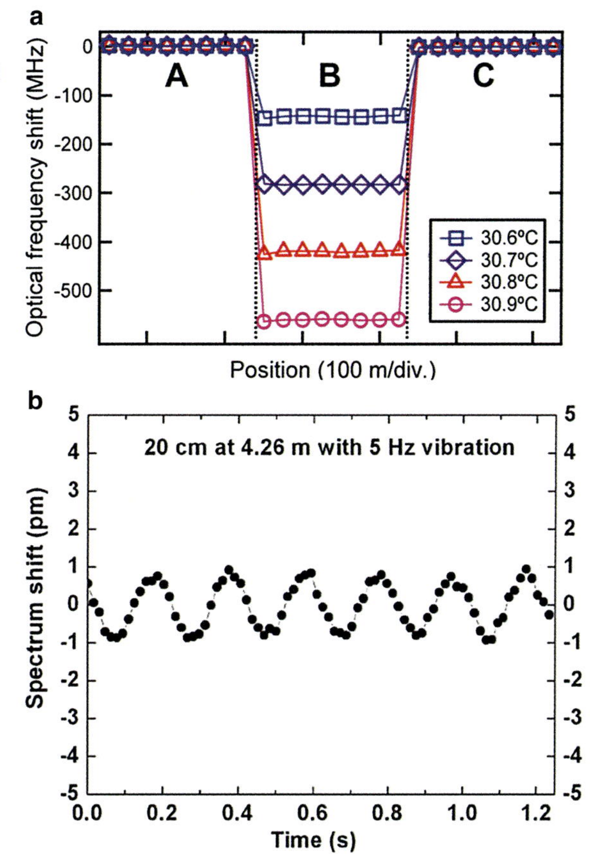

Fig. 25 (**a**) Measured temperature (Ohno et al. 2016) and (**b**) vibration signal (Zhou et al. 2012) with C-OFDR

Coherence-Domain Optical Reflectometry (OCDR)

Optical Low Coherence Reflectometry (OLCR)

Principle of OLCR

OLCR often appears in the literature as white-light interferometry, whose implementation scheme is usually composed of a Michelson interferometer, and its basic principle was first demonstrated by Albert A. Michelson more than 100 years ago. OLCR with a spatial resolution better than 1 mm benefits manufactures of microstructure optical components. Different from the other reflectometry techniques, OLCR uses broadband light sources which have short coherence lengths. In the OLCR implementation, low-coherence light source becomes uncorrelated when the delay between the two arms is beyond the coherence length. In that case, the light beams from a low-coherence light source are able to interfere with each other only for ultrashort delay. The Michelson interferometer used in the OLCR demonstrations (Sorin and Baney 1992) is shown in Fig. 26. The light beam from

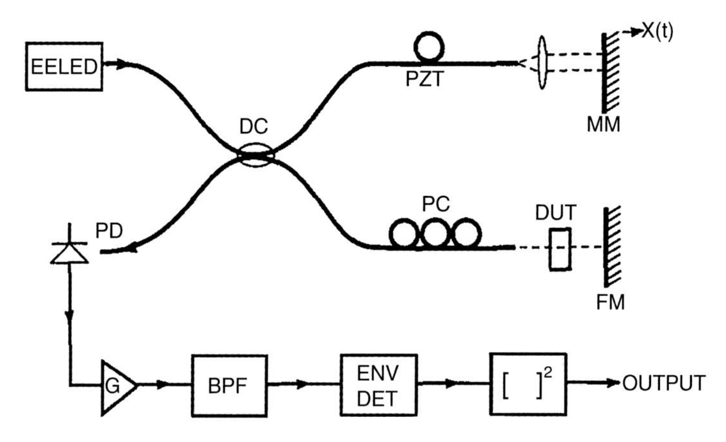

Fig. 26 Experimental setup for OLCR measurements using Michelson interferometer (Sorin and Baney 1992)

a wideband light source is incident on an optical coupler, then the optical coupler splits the optical beam into the detection arm and the reference arm. The distance of the mirror in the reference arm can be adjusted with an electrical motor. The reflected signals from the two arms are then detected by the PD. The photocurrent for a single reflection can be expressed as

$$I_d(t) = \Re\left[P_{ref} + P_{dut} + 2\sqrt{P_{ref}P_{dut}}\,|V(t)|\cos\Delta\phi(t)\right] \tag{30}$$

where P_{ref} and P_{dut} are the detected powers from reference arm and detection arm, respectively, $\Delta\phi$(t) is the relative phase shift between the reference arm and the detection arm, and $V(t)$ is the temporal coherence function. The last term gives the amplitude of the interference fringes. In the case of coherent light sources, the fringes remain constant, while the laser sources used in OLCR are incoherent sources or partly coherent sources, such as edge-emitting light-emitting diode (EELED), or ASE of EDFA, or superluminescent diode (SLD); hence, these light sources provide interference signals within a finite time delay. The fringes of OLCR depend on the optical time delay set by the position of the reference mirror and the position of device under test (DUT) which carries the information about the structure of the sample. $V(t)$ is the Fourier transform of the power spectral density $S(f)$ of the light source. According to the Fourier transform theory, a broadband light source should have a narrow coherence length; this means that when the reference mirror is far away from the zero-delay position, the signals from the reference arm and detection arm remain uncorrelated. The optical phases from the reference arm and the detection arm are uncorrelated and random, while the temporal coherence function is equal to zero. If the mirror position is set so that the time delay is smaller than the coherence length of the laser sources, the interference fringe is shown in Fig. 27.

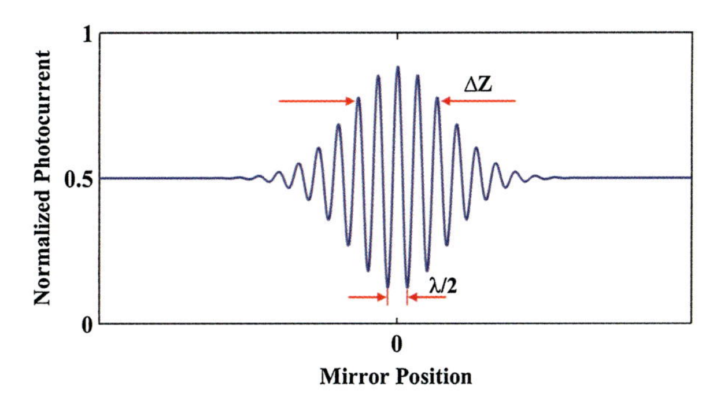

Fig. 27 Illustration of the fringes from an OLCR measurement when the low-coherence source has a Gaussian wavelength shape

In the conventional implementation of an OLCR system, a piezoelectric phase modulator or a motor stage is used to generate a Doppler frequency shift for avoiding the *1/f* noise. For the case when using a motor stage to drive the reference mirror with a constant velocity of υ, the interference signals may introduce a Doppler frequency of f_d, which can be expressed by

$$f_d = \frac{2\upsilon}{\lambda} \tag{31}$$

where λ is the central wavelength of the laser source. Then in the receiving processing part, a bandpass filter with the central frequency of f_d is added. By using this heterodyne detection method, a sensitivity just several dB above the shot noise may be obtained.

Spatial Resolution and Measurement Range

The spatial resolution is determined by the width of the fringe envelope, which can be expressed as

$$\Delta z \cong \frac{\lambda^2}{2n\Delta\lambda} \tag{32}$$

where n is the refractive index of the DUT and $\Delta\lambda$ is the wavelength range of the laser source. For example, in an OLCR system a typical broadband source with 100 nm wavelength range centered at 1550 nm has a spatial resolution of about 7.5 μm. The measurement resolution is one of the key factors of OLCR. Especially for some high-end detection applications, it is of great significance to distinguish the adjacent abnormal reflectors. In order to be observed, the spacing of adjacent reflections must be larger than the spatial resolution. The EELED used in OLCR has a bandwidth larger than 100 nm, and the spatial resolution can vary from about 5 to 100 μm. As for the ASE of EDFAs, the spectrum of EDFA can be extended from C to L band, with a wavelength range of about 100 nm and an average power of more than 100 mW, which makes ASE of EDFA quite suitable for OLCR considering the spatial resolution and the sensitivity. It is reported by Clivaz et al. in 1992 that by using a fluorescence light source generated using a Ti: Al_2O_3 crystal, a spatial resolution of 1.9 μm has been achieved. Another possible optical source is the femtosecond pulsed laser. As the femtosecond laser technique develops, its wavelength range can cover 1–2 μm range. Meanwhile, the femtosecond pulsed lasers can achieve very high reflection sensitivities because of their high average powers and very low relative intensity noise (RIN). However, the relatively high cost and the fact that high peak power may damage the detection sample are two main drawbacks when femtosecond pulsed laser is used as the optical source for OLCR.

The measurement range of OLCR is determined by the scanning range of the reference arm. The movable distance is usually provided by the fiber delay line or the reflection mirror. The usual movable distance is several tens of centimeters to 1 m, and this is adequate for testing many optical components. However, for some applications with much longer distances, researchers tried to extend the measurement range. One of the straightest ways is to use several pairs of delay fibers to add additional fiber length at different measurements, which is very time-consuming. Another elegant method is to add a recirculating optical delay line into the reference arm. In this scheme, a fiber resonator is placed in the reference arm, which generates delayed replicas of the low coherence source. By using this method, a measurement range of >150 m is finally demonstrated. As mentioned above, the proposal of using femtosecond laser as the optical source may be a good choice for longer measurement range if combining the noise-like phase modulation method. With the noise-like phase modulation by an arbitrary waveform generator, the signals from the detection arm and the reference arm which have been synchronized are added and yield a correlation gating at an arbitrary delay. By carefully adjusting the gating delay, a measurement range beyond 1 km can be reached.

Nevertheless, as the spatial resolution and measurement range increases, the effects of chromatic dispersion should be taken into consideration. For example, when the spatial resolution is about 10 μm, the measurement range cannot exceed several meters using the phase modulation method mentioned above. However, the method of using additional delay fiber may cancel the chromatic dispersion effects.

Reflection Sensitivity and Phase Noise

The measurement reflection sensitivity of an OLCR system is decided by optical source and the receiver. The source power finally determines the amount of the power reflected in the detection and reference arm and the bandwidth of the receiver determines the amount of the noise. The SNR can be expressed as (Sorin and Baney 1992):

$$SNR = \frac{2\Re^2 P_{ref} P_{dut}}{4kT\Delta f/R_{eff} + 2q\Re P_{ref}\Delta f + (RIN)\Re^2 P_{ref}^2 \Delta f} \tag{33}$$

where P_{ref} and P_{dut} are detected powers from the reference and detection arms, R_{eff} and Δf are the effective noise resistance and measurement bandwidth of the

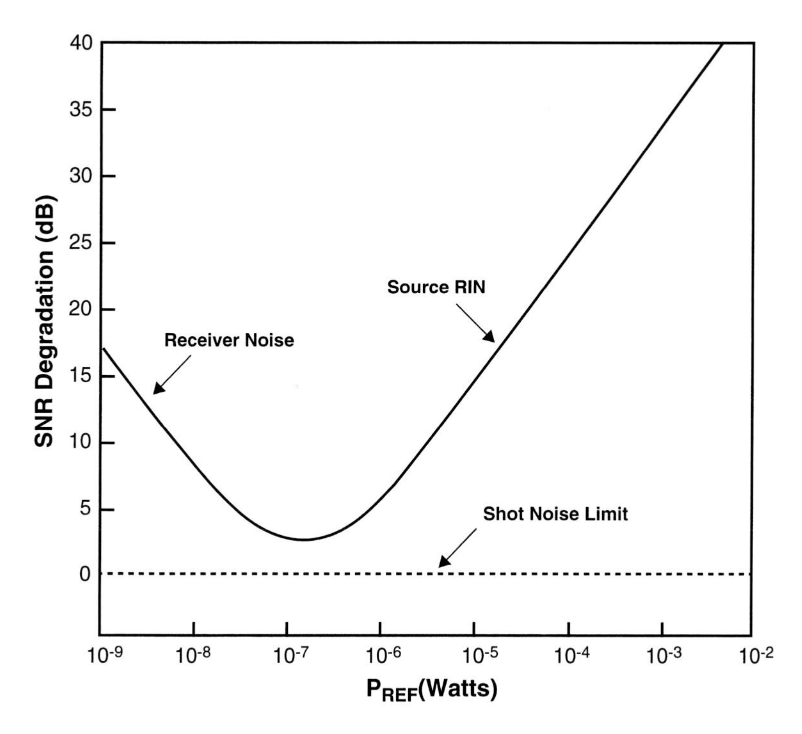

Fig. 28 The SNR of OLCR as limited by intensity noise and thermal noise (Sorin and Baney 1992)

receiver, and (*RIN*) is the intensity noise of the optical source. For wideband laser sources, when the input power is quite small, the dominant noise is mainly determined by the thermal noise, and the (*RIN*) intensity noise term becomes dominant when the input power exceeds microwatts, which is shown in Fig. 28.

When the high power laser is used, it is easy to achieve a shot-noise-limited sensitivity. Meanwhile, experiments show that only when the strong reflection is relatively weak, the shot-noise-limited sensitivity can be reached, otherwise the sensitivity of the system does not necessarily increase with the increasing power of input laser which is shown in Fig. 29a. The proposal of using index matching oil can greatly suppress the end reflection, and finally a sensitivity of −140 dB is reached (Takada et al. 1991) and shown in Fig. 29b.

Optical Coherence-Domain Reflectometry Based on Synthesis of Coherence Function (OCDR-SOCF)

OLCR is a special case of OCDR, which has a micrometer-order spatial resolution, while its distributed measurement speed, as well as the measurement range, is limited by the mechanical scanning as described in the previous section. On the other hand, an OCDR based on the synthesis of optical coherence function (OCDR-SOCF) could provide a spatial resolution better than centimeter-order and a

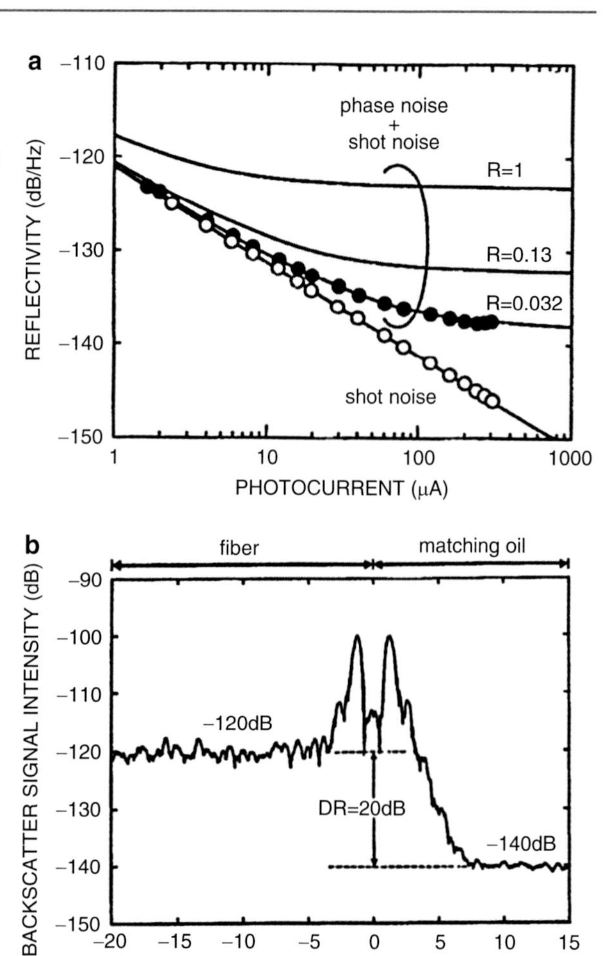

Fig. 29 (**a**) Minimum detectable reflectivity versus total mean photocurrent for reference light. (**b**) –140 dB sensitivity with 14 μm spatial resolution achieved (Takada et al. 1991)

distributed measurement time shorter than 10-s without mechanical scanning. Here we describe OCDR-SOCF in this section.

In the OCDR-SOCF system, the optical frequency of LD is directly controlled by the injection current with an appropriate waveform, and the coherence function, which corresponds to the time-averaged spectrum of the laser source, is then synthesized. The fundamental setup is shown in Fig. 30 (Hotate and He 2006). The OCDR system mainly consists of a frequency-modulated LD and an interferometer. An AOM is used as the frequency shifter for heterodyne detection, which may remove the DC noise. The reference light and the reflected light from the DUT are mixed at the 50/50 optical coupler and then detected by a BPD. The electronic signals are collected by an ADC and then digitally processed using a computer.

In order to obtain a high spatial resolution, the full width half maximum (FWHM) of the coherence peak is expected to be narrow enough. At the early stage of

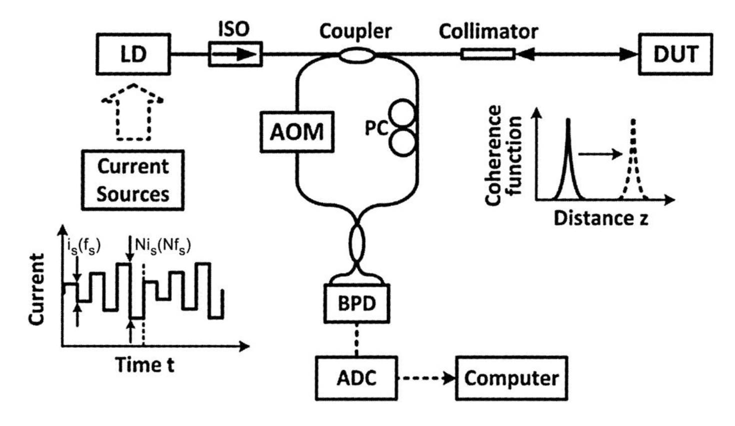

Fig. 30 Basic configuration of the SOCF based OCDR system (Hotate and He 2006)

OCDR-SOCF technology, the oscillation frequency of LD is modulated by utilizing a waveform shown in the down-left corner of Fig. 30. The synthesized coherence function can be given by

$$\gamma(f_s, z) = \frac{\cos\left[\frac{(N+1)\pi f_s z}{c}\right]\sin\left(\frac{N\pi f_s z}{c}\right)}{N\sin\left(\frac{\pi f_s z}{c}\right)} \tag{34}$$

where N is the number of modulation step, f_s is the frequency interval between two adjacent steps, c is the light velocity in vacuum, and z is the position in DUT. Assuming a relatively large N, the coherence function becomes a delta-like shape. The relationship between the position z_s and the frequency spacing f_s is given as

$$z_s = \frac{i_d c}{f_s} \tag{35}$$

where i_d is a natural number corresponding to the number of a periodic delta-like function used for the measurement. The coherence peak can be swept by changing the frequency spacing f_s in the waveform. The spatial resolution of OCDR is defined as the FWHM of the delta-like coherence peak. It can be approximately given by

$$\delta_z = \frac{c}{2f_m} \tag{36}$$

where f_m is the total frequency shift of the LD. Considering the case that the optical source is a tunable DBR laser whose tunable frequency range is 380 GHz, δ_z becomes 390 μm in air. The OCDR achieved by stepwise modulation is promising to realize millimeter-level spatial resolution, but there are some deterioration factors, such as parasitic intensity modulation and frequency nonlinearity, which greatly degrade the performance. Although nonlinearity compensation can be employed, the process is rather complicated and not cost-efficient.

To overcome these difficulties, the sinusoidal modulation method is studied and widely used for OCDR-SOCF. As shown in Fig. 31, LD is directly modulated by injecting a sinusoidal current to synthesize periodical coherence peaks, which can be swept by changing the modulation frequency to realize distributed measurement. The measurement range, which corresponds to the spacing between correlation peaks, is given by

$$D = \frac{c}{2nf_2} \tag{37}$$

where n is the refractive index of the fiber, and f_2 is the modulation frequency of LD. The spatial resolution is described as

$$\Delta d \cong \frac{0.76c}{\pi n f_1} \tag{38}$$

where f_1 is the modulation amplitude of the LD. It is clear that a larger modulation amplitude helps to obtain a higher spatial resolution. As shown in Fig. 31b, the low sidelobe suppression ratio (SLSR) is about 5 dB and may cause inaccuracy in measurements. Some apodization techniques have been proposed, and a dynamic range of near 40-dB has been realized with half-wave intensity modulation.

Besides stepwise modulation method and sinusoidal modulation method, frequency comb generation and coherence synchronization technique can also be used to synthesize coherence peaks. OCDR based on optical frequency comb (OCDR-OFC) is capable of providing high dynamic range (>45 dB) (He et al. 2010), but the optical frequency comb (OFC) is relatively difficult to generate and control. The coherence synchronization scheme can realize millimeter-level spatial resolution over 100-m measurement range (Okamoto et al. 2016), but it is not cost-efficient since several optical modulators and signal generators are incorporated in the system. Actually, some simplified OCDR systems have been developed, without AOM and reference path (Shizuka et al. 2016), which greatly reduced the complexity of OCDR system and made it more practicable.

OCDR with high spatial resolution, random access, and short measurement time has great potential in diagnosing optical devices, optical fiber subscriber networks, and fiber-optic assembly modules. By utilizing a broadband tunable superstructure grating distributed Bragg reflector laser diode (SSG-DBR-LD) as the light source, and controlling injection modulation current precisely, a 24-μm spatial resolution has been realized (Hotate et al. 2004), as shown in Fig. 32. The high-spatial-resolution OCDR system can combine with optical-coherence tomography (OCT) to realize cross-sectional imaging in scattering objects. The synthesized optical-coherence tomography (SOCT) can obtain a longitudinal reflectivity distribution with neither mechanical scanning nor numerical computation.

In terms of measurement range, unlike OFDR, which can only work within the coherence length of LD, OCDR is capable of providing high performance beyond

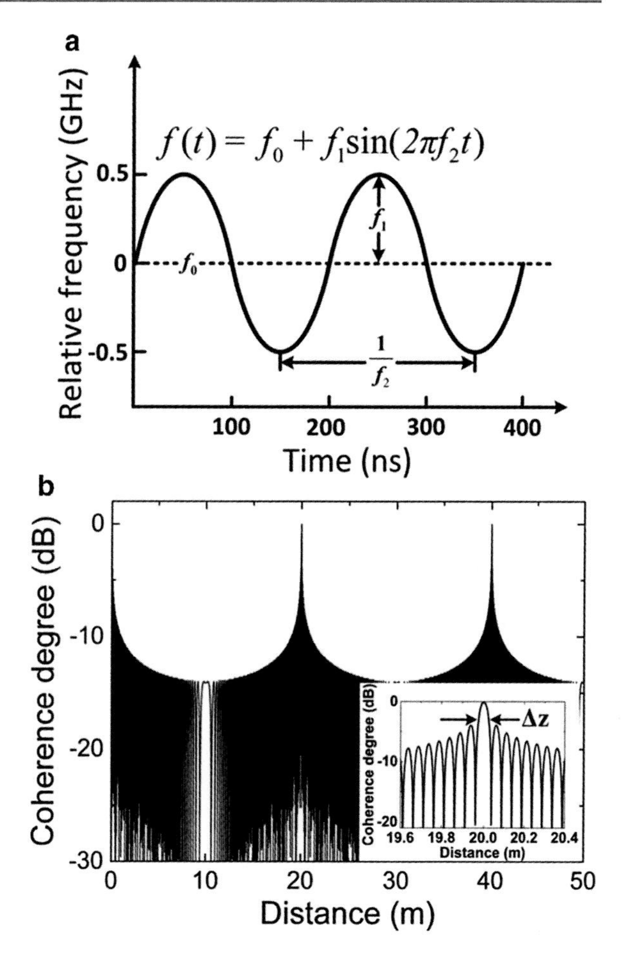

Fig. 31 OCDR achieved by sinusoidal frequency modulation. The inset at the bottom right corner of (**b**) is the extended figure around the coherence peak (Hotate and He 2006)

the coherence length (Kashiwagi and Hotate 2009). A spatial resolution of 19 cm at the distance of 5 km has been realized with a commercial DFB-LD, making it suitable for diagnosing middle-haul and long-haul system. For the diagnosis of fiber-optic assembly modules, 0.1-dB reflectance resolution is required in order to distinguish an out-of-spec fusion-splice. The major hurdle to this target comes from the interference of Rayleigh scatterings and multiple reflections in the fiber, which is also known as the coherent speckle. A method called wavelength domain averaging technique (Shimizu et al. 1992) can suppress the coherence speckle effect for high reflectance resolution.

OCDR provides high spatial resolution, random access, high measurement speed, and low complexity. However, it is still relatively difficult to achieve submeter spatial resolution over tens of kilometers, although temporal gating scheme and other techniques have been proposed. In order to fulfill requirements of both high

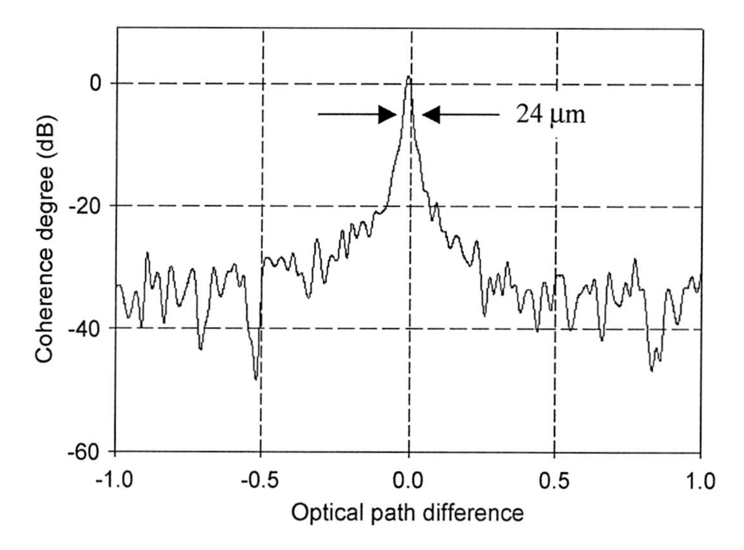

Fig. 32 Demonstration of 24-μm spatial resolution (Hotate et al. 2004)

spatial resolution and ultralong measurement range, some advanced methods have been developed. In the next section, we will introduce these methods.

Advanced Methods to Improve Both the Spatial Resolution and the Measurement Range

Optical reflectometry described in the former sections may either realize an ultrahigh spatial resolution but with a limited measurement range such as OLCR, or realize a long measurement range while having limitations on spatial resolution such as OTDR. However, for some high-end applications such as FTTH access network diagnosis, the requirements on both spatial resolution and measurement range are very strict. For example, bend-insensitive fiber is widely installed in FTTH networks, but the bending in optical fiber links is very difficult to locate by using reflectometry to detect loss for this kind of fiber, which shortens the service lifetime of the optical fiber if the bending is ignored and just as it is. In order to locate the bending position, a reflectometry capable of measuring the polarization change with a high spatial resolution is required. Usually, a millimeter spatial resolution is necessary to locate the bending, and a measurement range of several kilometers is required for these FTTH-related applications. Moreover, the vision of telecommunication carriers is to establish a remote online optical fiber line detection and monitoring system at the telecom buildings. In order to realize the management, optical identification (OID) devices are usually necessary to be embedded into the fiber connectors or optical network devices. The number of OID coding is set to 8 bits (256 codes), 16 bits (65,536 codes), or even more.

Spatial multiplexing and wavelength multiplexing technology are two representative technologies of OID manufacturing, but the wavelength multiplexing method is still at the stage of scientific research because of its difficulty in realizing. Due to the limited space in the connector whose length is less than two centimeters, spatial multiplexing technology restricts each bit length, and therefore optical reflectometry equipped at the buildings need to have millimeter or sub-mm spatial resolution with a measurement range of several kilometers or even several tens of kilometers. This is very challenging, and researchers have made many attempts to push the reflectometry performance.

OFDR technology has great advantages on ultrahigh spatial resolution, which depends on the wavelength tuning range of light source. If the light source has a wide-enough tuning range, it will achieve a very high spatial resolution. By using a light source with a tunable range of 40 nm, assisted by an auxiliary interferometer for the sweep nonlinearity compensation, OFDR realized a spatial resolution of 22 μm and a measurement range of 35 m. However, limited by the poor coherence length of the tunable light source, it is very difficult to achieve a long measurement range while keeping the spatial resolution. How to extend OFDR measurement range while maintaining its high spatial resolution is always a hot research topic for experts and scholars in related fields.

OTDR technology has great advantages on long measurement range, but the improvement of its spatial resolution is very difficult. In order to improve its spatial resolution, researchers used mode-locked laser as the signal source, but restricted by the detector bandwidth, only the signal with a pulse width greater than 10 ps can be detected, and therefore achieved mm-level spatial resolution. Because the responsivity of the wideband PD is low, the noise floor is high, and the average power of the short pulse is very low, the RBS signals of the high-resolution OTDR is still under the noise of 45 dB, which greatly limits the applications of OTDR. By adopting Geiger mode single-photon detector to OTDR technology, which is also called ν-OTDR, a spatial resolution of 1.5 cm is realized (Bethea et al. 1988). With the continuous development of single-photon detector, a spatial resolution of 5 cm over 50 km measurement range is achieved by the abovementioned method (Legre et al. 2007). By converting 1.5 μm optical signal to 875 nm using wavelength-conversion method, and using the advantages of single-photon detector at this wavelength, a spatial resolution of 10 cm over a measurement range of 217 km is realized (Shentu et al. 2013). After the superconducting nanowire single-photon detector is used for the detection, the spatial resolution may reach 4.0 mm, which is limited by the timing jitter (42 ps) of the detection system (Zhao et al. 2013). This timing jitter is a fundamental limit for OTDR to obtain better spatial resolution. Moreover, since the bandwidth of the ultrashort pulse signal is very large, the pulse will be broadened by the influence of optical fiber dispersion when propagating with a long distance.

In order to get rid of the strict requirement of OFDR technology for laser coherence and overcome the shortage of spatial resolution in OTDR technology, researchers tried to adopt new schemes in recent years. Chaotic laser is proposed as a signal source for optical reflectometry, and a spatial resolution of 8.2 cm over 100 km measurement range is obtained (Wang et al. 2015b). By adopting a monolithic integrated chaotic laser with a bandwidth of 40 GHz used as the optical source, the performance is improved to 2.6 mm/47 km (Zhang et al. 2017). As mentioned in the former section, a scheme of TGD-OFDR was proposed and realized a spatial resolution of 1.6 m over 110 km measurement range (Liu et al. 2015). Actually, the chaotic laser scheme and TGD-OFDR scheme can be regarded as the pulse compression technique, which has no high requirements for laser coherence, and therefore, it is very promising to be used for long-distance measurement. The problem of this technique is that the spatial resolution of mm-level should be realized with the help of a receiving system having a bandwidth of about 100 GHz. Therefore, the spatial resolution is ultimately limited by the receiving system.

As a conclusion, there are different limitations for OFDR, OTDR, and pulse compression technique. For OFDR, it has advantages on spatial resolution and sensitivity, and has no high requirements on receiving bandwidth, but it is very vulnerable to the phase noise of both optical source and the fiber for long-distance measurement. For OTDR, to achieve a high spatial resolution, there is a high demand for bandwidth and sensitivity of the detection system to detect the ultrashort pulse, and the dispersion effect should be carefully dealt with for long-distance measurement. The pulse compression technique has better spatial resolution over OTDR and is immune to phase noise of optical source compared with OFDR but also has very strict requirements on the bandwidth of the receiving system. Therefore, there are three advanced methods to deal with these problems. The first one is to develop an optical source with ultrahigh linear frequency sweeping rate used for OFDR (Wang et al. 2017a). The second one is to use linear optical sampling (LOS) technique to measure ultrashort pulse used for OTDR and adopt dispersion compensation scheme (Wang et al. 2017b). The third one is to use LOS technique for characterizing the ultrawide linearly chirped signals for pulse compression (Wang et al. 2017c).

Wideband Ultra-linearly Swept Optical Source for OFDR

In an OFDR system, the spatial resolution is given by Eq. 25 where ΔF is the frequency sweeping span. According to this equation, a narrower spatial resolution requires a larger ΔF. By utilizing the high-order sideband modulation, an enlarged span can be obtained. However, it is still limited because of the overlap between two adjacent high-order sidebands as shown in Fig. 33. In order to further broaden the frequency sweeping span, injection-locking technique is introduced to extract the frequency-swept lightwave from the overlapped high-order sidebands.

The electric field of the optical comb (master laser) which is generated by high-power RF modulation can be decomposed into its Fourier components

$$E(t) \propto \left[\sum_{n=-\infty}^{\infty} J_n(\beta)\, e^{i2\pi\left(v_0 + n f_m\right)t} \right] e^{i\theta(t)} \tag{39}$$

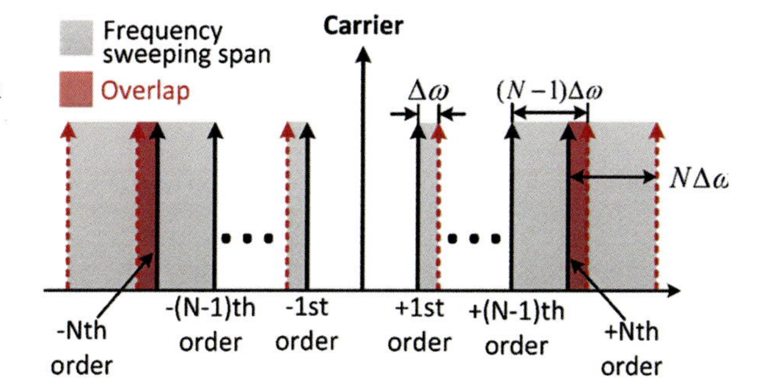

Fig. 33 Schematic illustration of the spectrum for the frequency sweep with high-order sidebands of external modulation (Wang et al. 2017a)

where J_n are n-th order Bessel functions of the first kind, β is the frequency modulation index, v_0 is the frequency of unmodulated optical carrier, f_m is the frequency of the RF signal, and $\theta(t)$ is the random phase fluctuations of the optical carrier. When the slave laser is injection-locked to the K-th sideband, its emission fields can be given by

$$E_K(t) \propto e^{i\left[2\pi\left(v_0+Kf_m\right)t+\phi(t)\right]} \tag{40}$$

Equation 40 means that the desired sideband can be obtained by utilizing injection-locking technique. Different from the conventional optical bandpass filter to extract the necessary wavelength, the center wavelength of the injection-locked slave laser can be changed very quickly by controlling its working current, which means that a synchronous current control to the slave laser may maintain a stable optical injection locking to follow the frequency sweep.

The injection-locking technique has solved the problem of overlap between two adjacent sidebands and broadened the frequency sweeping span. However, the frequency sweeping span is still limited by the electronic bottleneck. An effective solution is to use all-optical processing technique, for example, using optical nonlinear effect to generate new frequency components and further broaden the sweeping span. Let us consider the degenerated FWM process, the generated first-order idler has an electric field E_i which can be described by

$$E_i = C{A_p}^2 A_s \exp\left[j\left(2\omega_p - \omega_s\right)t + 2\left(\phi_p - \phi_s\right)\right] \tag{41}$$

where ω_p, ϕ_p, and A_p are respectively the frequency, phase, and amplitude of the pump; ω_s, ϕ_s, and A_s are respectively the frequency, phase, and amplitude of the seed, and C is a constant related to FWM efficiency. After the FWM process, the pump, the seed, and the first-order idler (idler-1) will have a frequency relationship given by $\omega_{i\text{-}1} = 2\omega_p - \omega_s$. Considering a pump with a frequency sweeping span of Δω, the obtained idler-1 will have a frequency of $2(\omega_p + \Delta\omega) - \omega_s$. Therefore, the frequency sweeping span of the idler-1 increases to double the pump sweeping span, as shown in Fig. 34a. With a cascaded configuration to obtain multistage FWM, the frequency sweeping span can be further broadened. As shown in Fig. 34b, the

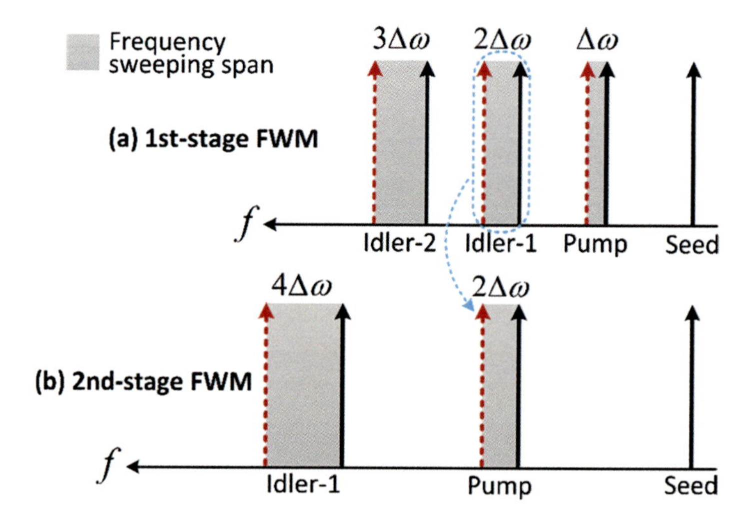

Fig. 34 Schematic illustration of the principle. (**a**) Broadening the frequency sweeping using first stage FWM and (**b**) second stage FWM (Wang et al. 2017a)

idler-1 generated by the first stage FWM is filtered as the pump for the second stage FWM. Similarly, the idler-1 of the second stage FWM will have a further doubled frequency sweeping span.

Note that the high-order sideband modulation and FWM have a capability of broadening the frequency sweeping span, but it will also magnify the laser phase noise and bring more phase noise if the RF sweeps nonlinearly. In order to mitigate the effect of the phase noise and extend the measurement range, a PNC technique may be introduced in the system.

By using the principles described above, in an experiment (Wang et al. 2017a), a wideband ultra-linearly swept optical source is realized with a sweeping span of 100 GHz. This source is used as the optical source for OFDR systems, and a 2-km SMF is used as the FUT, a spatial resolution of 1.1 mm at the far end of the FUT was realized as shown in Fig. 35. The measurement range of the proposed system is now limited by the memory of the ADC, which is very promising to be further expanded. It is believed that this system is suitable for applications where both high spatial resolution and relatively long measurement range are required.

LOS Technique-Assisted OTDR (LOS-OTDR)

To realize an ultrahigh spatial resolution by using OTDR, it is necessary to find a technique with a large detection bandwidth. Linear optical sampling (LOS) technique (Dorrer et al. 2003), known for the capability of observing the complex amplitude response of ultrashort optical pulses using slow electronics with a low bandwidth, and its shot-noise limited sensitivity, has been used in many fields for

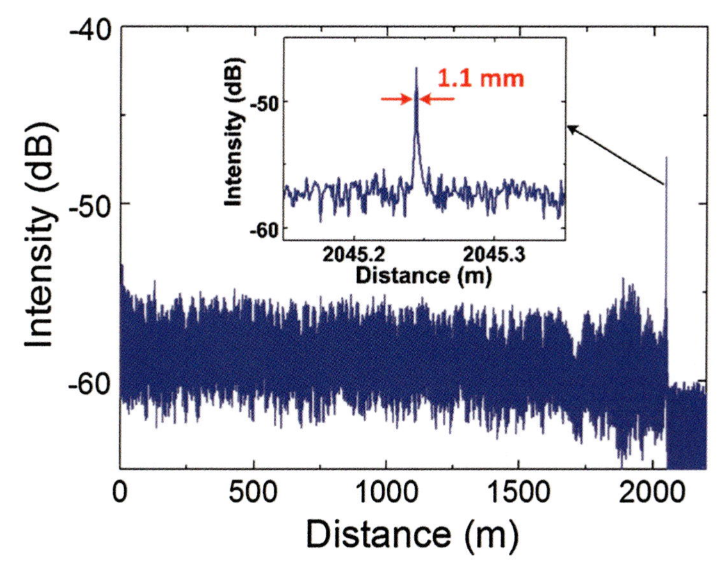

Fig. 35 Measured reflection trace. The inset shows the details of reflection peak at the end of the FUT (Wang et al. 2017a)

monitoring the waveform in high-speed transmission systems. The signal under test (SUT) interferes with the sampling signal launched from the mode-locked laser (MLL) and is then detected by the ordinary PDs. Because the timing jitter of the MLL is several femtoseconds, the sampling rate can be up to 100 TS/s, and this makes it possible to measure ultrashort pulse launched from pulsed lasers and its reflected lightwave from optical fibers. In addition, MLLs usually have a linewidth better than 1 kHz, which makes it a promising optical source for long-range sensing applications. Therefore, the adoption of LOS technique in reflectometry system has the potential to realize an ultrahigh spatial resolution with a long measurement range (Wang et al. 2017b).

The setup for LOS-OTDR is similar to a conventional OTDR but with a different receiving technique. For this system, an MLL is acting as a pulse generator. The ultrashort pulses are launched through the circulator into the FUT and backscattered to the third port of the circulator. To characterize the backscattered signal, another MLL can be used as the sampling laser.

By using this principle, the measurement range is limited to the distance corresponding to the sampling window. To solve this problem, a method can be used to adjust the sampling window and calculate the real position using a simple linear relationship since the reflection position changes in the new window. With this method, a proof-of-concept experiment is demonstrated with a measurement of a 900-m SMF. Another solution to locate the reflection events without this adjustment is to make a long sampling window by decreasing the repetition rate of the sampling laser using pulse-selecting technique. For example, if the repetition rate is decreased

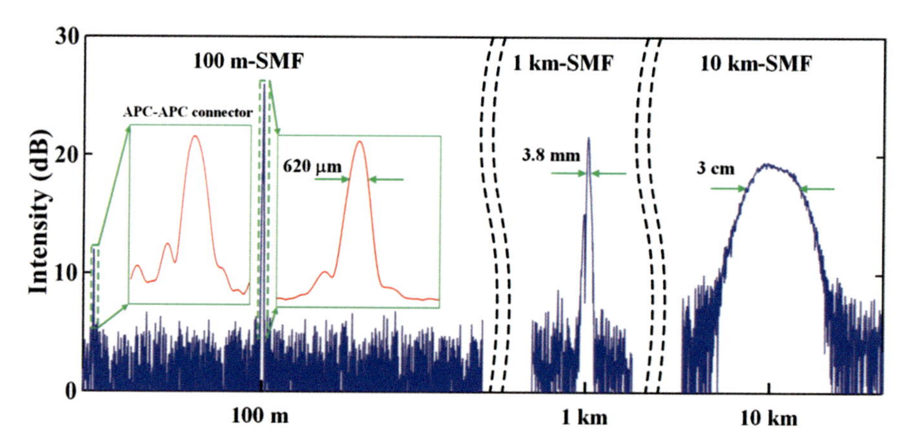

Fig. 36 Spatial resolution achieved at different distances (Wang et al. 2017b)

to 10 kHz, the corresponding detection distance would be increased to 10 km. As shown in Fig. 36, by using this method, experiments show that the FWHM of the reflection peak at 100 m is 620 μm, and broadened to 3.8 mm at 1 km, and finally becomes 3 cm after passing through a 10-km SMF.

It should also be mentioned that the RBS signal is not measured since it is −100 dB for a spatial resolution of 1 mm, which is below the noise floor. The pulse broadening is mainly caused by CD effect as it propagates in the FUT. As the spatial resolution increases, it is necessary to deal with the influence of dispersion which depends on the wavelength range of the probe pulses and the measurement distance.

Since LOS technique has the capability to obtain both intensity and phase information of the signal, digital compensation in the frequency domain is a good method to mitigate the degrading effects caused by CD effect. By adopting a digital CD compensation algorithm at long distances, a spatial resolution of 340 μm is achieved at 10 km. This technique shows a new application prospect for OTDR system in ultrahigh resolution distributed applications such as OID for PONs or precisely locating the abnormal reflections in aircrafts.

Pulse Compression Technique with the Aid of LOS Technique

Pulse compression techniques based on matched filtering method using linearly chirped pulses (LCPs) or phase coded pulses are widely used in modern radar systems and distributed fiber sensing systems for high spatial resolution and large dynamic range. In some other applications such as biomedical imaging, LCPs are used as optical sources for extracting depth of the organization sample, which show a great prospect in tumor detection. In particular, for long distance applications such as coherent radar systems and distributed fiber sensing systems, pulse compression techniques allow to transmit long pulses minimizing the transmitted peak power while maintaining a high spatial resolution.

Conventionally, LCPs are generated by using direct digital synthesizers (DDSs) or microwave generators. However, state-of-the-art DDSs have a maximum frequency limited to few GHz, which results in a small time-bandwidth product (TBWP). To deal with this problem, a couple of photoelectric techniques have been proposed to generate chirped or phase-modulated pulses taking advantage of the high speed and broad bandwidth of optical signals. Phase modulation using conventional modulators based on photoelectric process is a popular method, whereas the necessity of using modulators greatly limits the bandwidth of the generated signals and makes the system complicated, and the TBWP is usually less than 100 due to the limited modulation index. Another promising method is to use the quadratic phase modulation effects of chromatic dispersion for generating LCPs. In this system, ultrashort optical pulses launching from an MLL are sent into a dispersion compensating fiber (DCF) or a linearly chirped fiber Bragg grating (LCFBG) and are stretched to be several nanoseconds. Microwave signals may be generated based on heterodyning a LCP with a continuous wave (CW) laser or two linear pulses with different chirp rates. However, mainly limited by the PD bandwidth, photonic-assisted LCPs are usually restricted to ~60 GHz. For applications such as frequency modulated continuous wave (FMCW) light detection and ranging (LIDAR) system and OFDR, they directly deduce target range from the beat frequency generated from the pulse reflected by the target and the reference pulse. In this case, the detection bandwidth may be compressed greatly. However, the nonlinearity of the frequency sweep deforms the range resolution or the spatial resolution. Since the pulse compression technique does not require a very strict linear relationship of the frequency sweep, it enables a high spatial resolution for optical sensing applications if one realizes a large frequency sweep and characterizes it with a receiving technique having a high detection-bandwidth.

LOS technique, known for its capability of observing the complex amplitude response of large bandwidth optical signals using slow electronics with low bandwidth and shot-noise limited sensitivity, has been used in many fields for monitoring the waveform in high-speed transmission systems. To achieve a large TBWP, LCFBG with a large dispersion coefficient is a good option. Therefore, to use LCFBG for generating LCP signal, and then to detect it in combination with LOS technique, making it possible to acheive LCP with a large bandwidth and TBWP.

The schematic diagram of this method is shown in Fig. 37. An ultrashort optical pulse train with tens of nanometers wavelength range from a femtosecond laser is rectangular shaped through an optical bandpass filter. An LCFBG is utilized as a dispersive element to implement a wavelength-to-time mapping effect.

By taking advantage of the capability to characterize the ultrawide frequency-range signals using the LOS technique, it is easy to generate and detect sub-THz-range linearly chirped signal, and achieve a sub-millimeter spatial resolution with a low-bandwidth photodetector and an ADC, breaking the limitation of the electronic bottleneck.

Figure 37 illustrates an experimental configuration to generate and detect the LCPs. The lower femtosecond pulse laser (FSPL) which launches pulses with

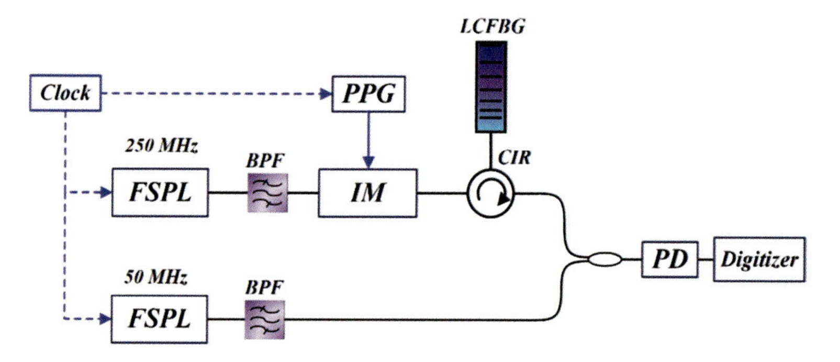

Fig. 37 Experimental configuration to generate and detect LCPs (Wang et al. 2017b)

50-MHz repetition rate acts as the sampling laser. The upper FSPL which has a wavelength range centered at 1550 nm and has a repetition rate of 250 MHz is used as the seed pulse source. By using an external clock to synchronize both FSPLs and a pulse pattern generator (PPG), an intensity modulator (IM) serves as a pulse picker. As a result, the repetition rate of the upper FSPL is reduced to be 50 MHz with -3 dBm average power. The sampling rate is set to be 150 TS/s by setting the sampling pulse slightly different from an integer multiple of the LCPs. The dispersive element is an LCFBG operating from 1528 nm to 1565 nm incorporated into an optical circulator to operate in reflection mode. The signal is detected by a BPD and captured by a 10-bit real-time oscilloscope with a sampling rate of 1 GS/s.

A first experiment shows a sweeping range of 450 GHz when a 10-m long LCFBG with 1981 ps/nm dispersion coefficient is chosen as the dispersion element, and the pulse is then stretched to a temporal duration of 10 ns. The generated LCP is characterized by using LOS technique, and then the data are used to calculate the autocorrelation of the pulse. The main lobe of the autocorrelation has an FWHM of 2.4 ps, corresponding to a pulse compression ratio of 4167.

To extend the measurement range, an intensity modulator may be adopted to act as a temporal gate. By increasing the length of the pulse pattern, the time apertures may be increased freely. However, the dynamic range is restricted by the extinction ratio of the intensity modulator, and the average power has been degraded due to the use of temporal gating technique. For long time apertures measurement, there is a scheme to use noise-like phase modulation to yield a correlation gate at arbitrary delay with an average power kept to be high. Taking advantage of the delta function property of its autocorrelation function, pseudo-random binary sequence (PRBS) is used in the experiment to generate similar correlation gate since it is simple.

Based on LCFBG or the chromatic dispersion effect, a sub-THz-range linearly chirped signal has been generated and detected with the help of LOS technique with low bandwidth electronics, breaking the limitation of the electronic bottleneck. Only restricted by the wavelength range of the sampling laser, the maximum frequency range achieved by Wang et al. in 2017c is 700 GHz, corresponding to 120 μm spatial resolution in optical fiber and 180 μm spatial resolution in free space. In addition, long time aperture chirped pulse train is achieved with PRBS phase modulation, which makes it possible to enable long measurement range as well as ultrahigh spatial resolution.

As a stabilized femtosecond laser with a coherence length of hundreds of kilometers has the potential to cover a frequency range of 100 THz, as well as with the dispersion element, this technique has the opportunity to achieve several micrometers spatial resolution and ultralong measurement range with the help of pulse compression technique. It features the generation and detection of large bandwidth LCPs and paves the way for new horizons in ultrahigh spatial resolution and long measurement range LIDAR, distributed fiber sensing, and biomedical imaging.

Conclusion

This book chapter presented the physical mechanisms of Rayleigh backscattering (RBS) in optical fibers and described the basic working principle of RBS-based distributed static sensing and vibration sensing. Reflectometry techniques, which are fundamental methods to provide the RBS signals along the fiber, have been described in details. These techniques include time-domain, frequency-domain, and coherence-domain methods. For some high-end applications, additional challenge is to obtain a better spatial resolution together with a longer measurement range. Some state-of-the-art reflectometry techniques have also been described in this chapter.

References

C. Baker, Y. Lu, J. Song, et al., Incoherent optical frequency domain reflectometry based on a Kerr phase-interrogator. Opt. Express **22**(13), 15370–15375 (2014)

M.K. Barnoski, S.M. Jensen, Fiber waveguides: A novel technique for investigating attenuation characteristics. Appl. Optics **15**, 2112–2115 (1976)

C.G. Bethea, S. Cova, G. Ripamonti, B.F. Levine, High-resolution and high-sensitivity optical-time-domain reflectometer. Opt. Lett. **13**, 233–235 (1988)

E. Brinkmeyer, Analysis of the backscattering method for single-mode optical fibers. J. Opt. Soc. Am. **70**, 1010–1012 (1980)

D. Chen, Q. Liu, Z. He, Phase-detection distributed fiber-optic vibration sensor without fading-noise based on time-gated digital OFDR. Opt. Express **25**(7), 8315–8325 (2017)

X. Clivaz, F. Marquis-Weible, R.P. Salathe, Optical low coherence reflectometry with 1.9 mu m spatial resolution. Electron. Lett. **28**(16), 1553–1555 (1992)

C. Dorrer, D.C. Kilper, H.R. Stuart, G. Raybon, M.G. Raymer, Linear optical sampling. IEEE Photon. Technol. Lett. **15**, 1746–1748 (2003)

W. Eickhoff, R. Ulrich, Optical frequency domain reflectometry in single-mode fiber. Appl. Phys. Lett. **39**(9), 693–695 (1981)

X. Fan, Y. Koshikiya, F. Ito, Phase noise compensated optical frequency domain reflectometry. IEEE J. Quantum Electron. **45**, 594–602 (2009)

M.E. Froggatt, D.K. Gifford, S. Kreger, et al., Characterization of polarization-maintaining fiber using high-sensitivity optical-frequency-domain reflectometry. J. Lightwave Technol. **24**(11), 4149–4154 (2006)

Z. He, H. Takahashi, K Hotate, Optical coherence-domain reflectometry by use of optical frequency comb[C]//Conference on Lasers and Electro-Optics. Optical Society of America, 2010. CFH4

K. Hotate, Z. He, Synthesis of optical-coherence function and its applications in distributed and multiplexed optical sensing. J. Lightwave Technol. **24**(7), 2541–2557 (2006)

K. Hotate, K. Makino, M. Ishikawa, et al., High-spatial-resolution fiber optic distributed force sensing with synthesis of optical coherence function//Optical Technologies for Industrial, Environmental, and Biological Sensing. Proc. SPIE 5272, Industrial and Highway Sensors Technology (8 March 2004), 5272:157–163 (2004)

H. Iida, Y. Koshikiya, F. Ito, K. Tanaka, High-sensitivity coherent optical time domain reflectometry employing frequency-division multiplexing. J. Lightwave Technol. **30**, 1121–1126 (2012)

F. Ito, X. Fan, Y. Koshikiya, Long range coherent OFDR with light source phase noise compensation. J. Lightwave Technol. **30**, 1015–1024 (2012)

M. Kashiwagi, K. Hotate, Long range and high resolution reflectometry by synthesis of optical coherence function at region beyond the coherence length[J]. IEICE Electron. Exp. **6**(8), 497–503 (2009)

J. King, D. Smith, K. Richards, P. Timson, R. Epworth, S. Wright, Development of a coherent OTDR instrument. J. Lightwave Technol. **5**, 616–624 (1987)

Y. Koshikiya, X. Fan, F. Ito, Long range and cm-level spatial resolution measurement using coherent optical frequency domain reflectometry with SSB-SC modulator and narrow linewidth fiber laser. J. Lightwave Technol. **26**(18), 3287–3294 (2008)

Y. Koyamada, M. Imahama, K. Kubota, et al., Fiber-optic distributed strain and temperature sensing with very high measurand resolution over long range using coherent OTDR. J. Lightwave Technol. **27**(9), 1142–1146 (2009)

M. Legré, R. Thew, H. Zbinden, N. Gisin, High resolution optical time domain reflectometer based on 1.55μm up-conversion photon-counting module. Opt. Express **15**, 8237–8242 (2007)

S. Liehr, N. Nöther, K. Krebber, Incoherent optical frequency domain reflectometry and distributed strain detection in polymer optical fibers. Meas. Sci. Technol. **21**(1), 017001 (2009)

Q. Liu, X. Fan, Z. He, Time-gated digital optical frequency domain reflectometry with 1.6-m spatial resolution over entire 110-km range. Opt. Express **23**(20), 25988–25995 (2015)

Y. Lu, T. Zhu, L. Chen, X. Bao, Distributed vibration sensor based on coherent detection of phase-OTDR. J. Lightwave Technol. **28**, 3243–3249 (2010)

D. Marcuse, Rayleigh scattering and the impulse response of optical fibers. Bell Syst. Tech. J. **53**, 705–715 (1974)

H.F. Martins, S. Martin-Lopez, P. Corredera, M.L. Filograno, O. Frazão, M. González-Herráez, Coherent noise reduction in high visibility phase-sensitive optical time domain reflectometer for distributed sensing of ultrasonic waves. J. Lightwave Technol. **31**, 3631–3637 (2013)

S. Ohno, D. Iida, K. Toge, et al., Long-range measurement of Rayleigh scatter signature beyond laser coherence length based on coherent optical frequency domain reflectometry [J]. Opt. Express **24**(17), 19651–19660 (2016)

T. Okamoto, D. Iida, K. Toge, et al., Optical correlation domain reflectometry based on coherence synchronization: Theoretical analysis and proof-of-concept [J]. J. Lightwave Technol. **34**(18), 4259–4265 (2016)

A.J. Rogers, Polarization-optical time domain reflectometry: A technique for the measurement of field distributions. Appl. Optics **20**, 1060–1074 (1981)

S.V. Shatalin, V.N. Treschikov, A.J. Rogers, Interferometric optical time-domain reflectometry for distributed optical-fiber sensing. Appl. Optics **37**, 5600–5604 (1998)

G.-L. Shentu, Q.-C. Sun, X. Jiang, X.-D. Wang, J.S. Pelc, M.M. Fejer, Q. Zhang, J.-W. Pan, 217 km long distance photon-counting optical time-domain reflectometry based on ultra-low noise up-conversion single photon detector. Opt. Express **21**, 24674–24679 (2013)

K. Shimizu, T. Horiguchi, Y. Koyamada, Characteristics and reduction of coherent fading noise in Rayleigh backscattering measurement for optical fibers and components. J. Lightwave Technol. **10**, 982–987 (1992)

M. Shizuka, S. Shimada, N. Hayashi, Y. Mizuno, K. Nakamura, Optical correlation-domain reflectometry without optical frequency shifter. Appl. Phys. Express **9**(3), 032702 (2016)
B. Soller, D. Gifford, M. Wolfe, et al., High resolution optical frequency domain reflectometry for characterization of components and assemblies [J]. Opt. Express **13**(2), 666–674 (2005)
W.V. Sorin, D.M. Baney, A simple intensity noise reduction technique for optical low-coherence reflectometry. IEEE Photon. Technol. Lett. **4**(12), 1404–1406 (1992)
M.A. Soto, X. Lu, H.F. Martins, et al., Distributed phase birefringence measurements based on polarization correlation in phase-sensitive optical time-domain reflectometers. Opt. Express **23**(19), 24923–24936 (2015)
W.B. Spillman, P.L. Fuhr, B.L. Anderson, Performance of integrated source/detector combinations for smart skins incoherent optical frequency domain reflectometry distributed fibre optic sensors, in Fiber Optic Smart Structures and Skins, ed. by E. Udd, Proc. SPIE986, 106–118 (1988)
K. Takada, A. Himeno, K. Yukimatsu, Phase-noise and shot-noise limited operations of low coherence optical time domain reflectometry. Appl. Phys. Lett. **59**(20), 2483–2485 (1991)
S. Venkatesh, W.V. Sorin, Phase noise considerations in coherent optical FMCW reflectometry[J]. J. Lightwave Technol. **11**(10), 1694–1700 (1993)
Z. Wang, Z. Pan, Z. Fang, Q. Ye, B. Lu, H. Cai, et al., Ultra-broadband phase-sensitive optical time-domain reflectometry with a temporally sequenced multi-frequency source. Opt. Lett. **40**, 5192–5195 (2015a)
Z. Wang, M. Fan, L. Zhang, H. Wu, D. Churkin, Y. Li, X. Qian, Y. Rao, Long-range and high-precision correlation optical time-domain reflectometry utilizing an all-fiber chaotic source. Opt. Express **23**, 15514–15520 (2015b)
B. Wang, X. Fan, S. Wang, J. Du, Z. He, Long-range millimeter-resolution OFDR based on 100 GHz linear frequency-sweep of optical source by injection-locking technique and cascaded FWM process. Opt. Express **25**, 3514–3524 (2017a)
S. Wang, X. Fan, Z. He, Ultra-high resolution optical reflectometry based on linear optical sampling technique with digital dispersion compensation. IEEE Photonics J **9**, 6804710 (2017b)
S. Wang, X. Fan, B. Wang, G. Yang, Z. He, Sub-THz-range linearly chirped signals characterized using linear optical sampling technique to enable sub-millimeter resolution for optical sensing applications. Opt. Express **25**, 10224–10233 (2017c)
M. Wegmuller, F. Scholder, N. Gisin, Photon-counting OTDR for local birefringence and fault analysis in the metro environment. J. Lightwave Technol. **22**, 390–400 (2004)
D. Xu, J. Du, X. Fan, Z He, 10-times broadened fast optical frequency sweeping for high spatial resolution OFDR[C]//Optical Fiber Communication Conference. Optical Society of America, 2014. W3D. 2
G. Yang, X. Fan, S. Wang, B. Wang, Q. Liu, Z. He, Long-range distributed vibration sensing based on phase extraction from phase-sensitive OTDR. IEEE Photonics Journal **8**, 1–12 (2016)
G. Yang, X. Fan, Q. Liu, Z. He, Increasing the frequency response of direct-detection phase-sensitive OTDR by using frequency division multiplexing, in *Proceedings of Optical Fiber Sensors Conference (OFS26)*, (Jeju, 2017)
L. Zhang, B. Pan, G. Chen, D. Lu, L. Zhao, Long-range and high-resolution correlation optical time-domain reflectometry using a monolithic integrated broadband chaotic laser. Appl. Optics **56**, 1253–1256 (2017)
Q. Zhao, J. Hu, X. Zhang, L. Zhang, T. Jia, L. Kang, J. Chen, P. Wu, Photon-counting optical time-domain reflectometry with superconducting nanowire single-photon detectors, in Superconductive Electronics Conference (ISEC), 2013 IEEE 14th International, 2013, pp. 1–3
D.P. Zhou, Z. Qin, W. Li, et al., Distributed vibration sensing with time-resolved optical frequency-domain reflectometry. Opt. Express **20**(12), 13138–13145 (2012)
L. Zhou, F. Wang, X. Wang, et al., Distributed strain and vibration sensing system based on phase-sensitive OTDR. IEEE Photon. Technol. Lett. **27**(17), 1884–1887 (2015)

Distributed Raman Sensing

42

Marcelo A. Soto and Fabrizio Di Pasquale

Contents

M. A. Soto (✉)
Institute of Electrical Engineering, EPFL Swiss Federal Institute of Technology, Lausanne, Switzerland
e-mail: marcelo.soto@epfl.ch

F. Di Pasquale
Institute of Communication, Information and Perception Technologies (TECIP), Scuola Superiore Sant'Anna, Pisa, Italy
e-mail: f.dipasquale@santannapisa.it

G.-D. Peng (ed.), *Handbook of Optical Fibers*,
https://doi.org/10.1007/978-981-10-7087-7_6

Abstract

The Raman scattering effect constitutes one of the basic physical mechanisms exploited in optical fiber distributed temperature sensing. In particular Raman distributed temperature sensors (RDTS) have been developed for more than three decades, becoming today a mature technology that is widely applied to several strategic industrial fields. Making use of the thermally-activated spontaneous Raman scattering (SpRS) process, continuous measurements of a temperature profile over a sensing range of tens of kilometers can be obtained with high accuracy and meter-scale spatial resolution. Knowing the distributed temperature profile over large infrastructures provides a powerful technique for applications ranging from oil and gas to fire detection, and from energy production to transportation applications and environmental monitoring. Although this technology can be considered to be quite mature, research on Raman distributed temperature sensing is still active, with the main goal being extending the sensing distance while keeping high spatial resolution and a low cost of the system, and providing reliable and robust RDTS units able to operate in harsh environments.

In this book chapter, after a first description of the physical mechanisms behind Raman scattering, the working principle of RDTS system is provided along with a description of the most-common system configurations. Then, advanced techniques to improve the RDTS performance (e.g., pulse coding and image processing) are presented. In the final section, some examples of RDTS industrial applications are addressed, presenting several field trials which demonstrate the effectiveness of RDTS as practical monitoring solutions in a wide range of industrial fields.

Introduction

Optical fiber technologies are nowadays widely exploited worldwide not only for telecommunication (Agrawal 2010) but also for sensing applications (Yin et al. 2008; Culshaw and Kersey 2008). The low loss and large bandwidth capabilities of single-mode optical fibers (SMF) allow for terabit capacity transmission over ultra-long distances without optoelectronic regeneration. The same features make the optical fiber a very interesting element for distributed sensing, allowing for measurement ranges of tens of kilometers. In distributed sensing (Dakin and Kersey 1993), the optical fiber itself acts as a sensing element, thus providing the possibility to perform spatially resolved measurements of physical parameters like temperature, strain, and vibration over tens of kilometers distance, with meter-scale spatial resolution and high dynamic bandwidth. Fiber-optic sensor systems based on either SMF or multimode fibers (MMFs) have been developed in recent years, evolving from laboratory research to real industrial applications, thus gaining more and more attention in several strategic industrial fields.

This book chapter reports established information in the particular field of Raman-based distributed optical fiber sensing. This also includes the development of advanced sensing techniques to address current market requirements in terms of performance improvement.

Raman-based distributed temperature sensors (RDTS) (Dakin and Pratt 1985; Hartog and Leach 1985; Kikuchi et al. 1988) have been intensively studied in the last decades, due to their great advantages over conventional electrical sensors. RDTS have found successful applications in a wide range of industrial fields, including fire detection, power cable monitoring, leakage detection in the oil and gas industry, and environmental monitoring. Due to the intrinsic properties of dielectric optic fibers, RDTS can be easily installed along large infrastructures, being characterized by immunity to electromagnetic interference and by the possibility to operate in harsh environments, where the use of standard electronic sensors would not be possible. Typical examples of such hazardous environments can be found in power plants and distribution stations; along oil and gas pipelines; within tanks carrying explosive gases, liquids, or solids; and in general in all environments characterized by the presence of high magnetic and electric fields, nuclear radiations, and potentially explosive atmospheres.

Although optical frequency-domain reflectometry (OFDR) (Eickhoff and Ulrich 1981; Farahani and Gogolla 1999) can be used to interrogate the Raman scattering in the fiber with very high spatial resolution, most RDTS systems are based on optical time-domain reflectometry (OTDR) (Barnoski et al. 1977) measurements. In Raman-OTDR schemes, an optical pulse is sent into the optical fiber, and the spontaneous Raman backscattered light is detected at the transceiver. The temperature-dependent spontaneous Raman anti-Stokes backscattered signal is normalized to either the spontaneous Raman Stokes (Dakin and Pratt 1985) or Rayleigh (Hartog and Leach 1985) backscattered component to compensate for spurious losses along the sensing fiber and laser power fluctuations, thus providing a reliable measurement system.

Several measurement configurations can be adopted in Raman distributed sensing, including (i) single-ended schemes (Kikuci et al. 1988), in which the sensing fiber is interrogated from one fiber end only; (ii) double-ended or looped configurations (Fernandez et al. 2005), where the anti-Stokes and Stokes traces are alternately acquired in forward and backward directions and then properly averaged; (iii) multiwavelength schemes in which two pulsed lasers at different selected wavelengths are simultaneously used (Suh and Lee 2008); and (iv) schemes making use of only the Raman anti-Stokes component (Hwang et al. 2010; Soto et al. 2012). Each of these configurations has its own advantages and disadvantages, which will be thoroughly described in this chapter. Single-ended schemes allow, in principle, long sensing distances; however, they are intrinsically affected by wavelength-dependent losses (WDL) and their variation over time. In particular WDL-related issues are mainly relevant in harsh environments and can distort and bias the measured temperature traces. For instance, the presence of ionizing radiation can induce a wavelength-dependent radiation-induced attenuation (RIA) (Toccafondo et al. 2015a), which can lead to significant WDL (Kimura et al. 2001). Fiber losses can also be strongly affected in hot and humid environments, such as in geothermal, oil and gas well applications, where the differential wavelength-dependent attenuation of the fiber typically varies over time as a consequence of high temperature and hydrogen concentration (Jaaskelainen 2010). In all such cases, a loop configuration or a double-wavelength scheme is usually employed, providing inherent correction of WDL and local losses and leading therefore to highly reliable and robust systems for industrial applications.

The performance of RDTS systems is ultimately determined by the signal-to-noise ratio (SNR) of the traces. This imposes a well-known trade-off between the sensor parameters, e.g., sensing range, spatial resolution, and temperature resolution. In order to improve the SNR and performance of RDTS systems, advanced techniques have been proposed in the literature (e.g., optical pulse coding Nazarathy et al. 1989; Jones 1993; Park et al. 2006 and image processing Soto et al. 2016a). These methods provide attractive solutions, enabling sensing distances up to many tens of kilometers while maintaining meter-scale spatial resolutions. This way, industrial applications in which only distributed temperature measurements are required can strongly benefit from the availability of medium- to long-range RDTS systems. Many of those implementations can also strongly compete with Brillouin distributed fiber sensing (Horiguchi and Tateda 1989), in which measurements are intrinsically affected by the cross-sensitivity between strain and temperature.

This book chapter first provides a brief description of the Raman sensing physical mechanisms and introduces a theoretical model governing RDTS systems. Then different sensor configurations are introduced, including schemes based on single-end, loop, dual-wavelength, and using the Raman anti-Stokes component only. Then advanced techniques based on pulse coding and image processing to enhance the RDTS system performance are described. The book chapter concludes addressing the main industrial applications of RDTS systems. Several Raman DTS field trials will be reported, demonstrating the effectiveness of RDTS for fire detection, for leakage detection, and for high-energy physics experiments.

Raman Scattering in Optical Fibers

Physical Mechanisms

The physical mechanism that is exploited in RDTS systems to infer the temperature of a sensing fiber and of the surrounding environment is the so-called *Raman effect*. This corresponds to an inelastic optical scattering phenomenon discovered in liquids by C.V. Raman in 1922.

Light scattering phenomena (Boyd 2003) can be classified into elastic and inelastic, depending on the energy relation between incoming and scattered photons. In elastic scattering the scattered photons maintain the same energy as the incoming photons, while in inelastic scattering, the scattered photons are shifted in frequency compared to the incoming photon, as they receive or give energy from or to the inhomogeneous medium in which the light is propagating. By definition, the components of the scattered light which are downshifted in frequency are called

Stokes components, and their energy is lower than that of the incident photons. On the other hand, the components of the scattered light which are upshifted in frequency are called *anti-Stokes* components, and their energy is larger than that of the incident photons as they receive energy from the medium.

Light scattering in media can also be classified as *spontaneous* and *stimulated* scattering processes. In spontaneous light scattering, the properties of the medium are unmodified by the presence of the incident lightwave; this occurs only when the intensity of incident light is lower than a given threshold. On the other hand, when the intensity of the incident light increases above the threshold, the properties of the material are modified, and the scattering process turns to be nonlinear. In such a case, the scattering is generated by the material fluctuations induced by the intense propagating light, and the process is known as stimulated scattering.

Raman scattering can be classified as inelastic as it involves the generation of new light spectral components in the presence of a light propagating through a gas, liquid, or solid. In case of silica fibers, Raman scattering can be seen as an interaction between light (photons) and molecular vibrations (phonons) of the silica glass. More specifically, if some energy is transferred from the incoming photons to the glass phonons, then the scattered photons will possess less energy than the incoming ones. This results in the generation of a new light component with a lower frequency (longer wavelength), which is known as Raman Stokes component. On the contrary, if some energy is transferred from the glass phonons to the incoming photons, then the emerging photons will possess more energy. This results in the generation of a new light component at a higher frequency (smaller wavelength), which is called *Raman anti-Stokes* component. As schematically shown in Fig. 1, the generated Raman components (Stokes and anti-Stokes) are spectrally separated symmetrically around the incoming light frequency. The difference between the pump frequency and the frequency corresponding to the maximum Stokes and anti-Stokes intensities is called *Raman frequency shift* and depends on the specific material. For silica glass, this Raman frequency shift is approximately 13.2 THz.

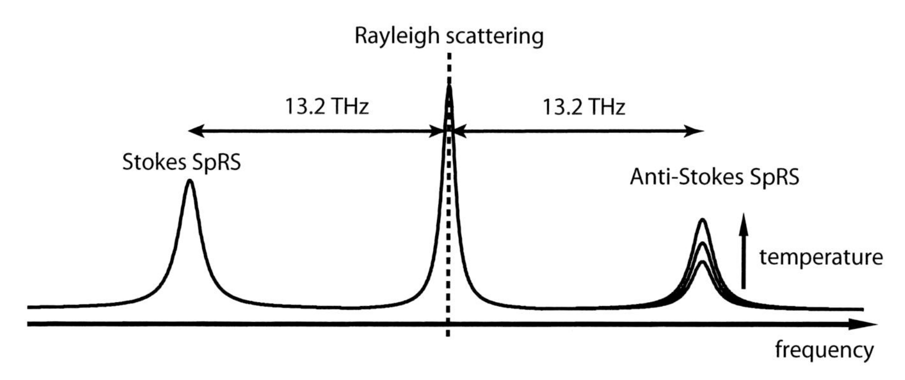

Fig. 1 Raman scattering in glass involves the generation of a strongly temperature-dependent anti-Stokes light component and a slightly temperature-dependent Stokes component

Since molecular vibrations are strongly connected to temperature, spontaneous Raman scattering (SpRS) is inherently temperature-dependent and can then be effectively employed in silica glass fibers to provide temperature sensing (Dakin and Pratt 1985; Hartog and Leach 1985).

The intensity of the spontaneous anti-Stokes light exhibits in fact a strong dependence on the temperature, while the intensity of the spontaneous Stokes light is found to be only slightly temperature dependent. In order to provide robust temperature measurements that are not significantly dependent on either fiber loss variations or pump power fluctuations, the power ratio of the Raman anti-Stokes over the Raman Stokes light is typically used (also the ratio of Raman anti-Stokes over Rayleigh scattering is sometimes employed).

Note that while stimulated Raman scattering can be effectively exploited for optical amplification (Islam 2004), distributed temperature sensing, on the contrary, requires the use of pump power levels below the stimulated Raman scattering threshold in order to avoid any modification of the material properties by the light propagation which would inevitably impair the temperature measurement.

Spontaneous Raman Scattering

Raman scattering is generated by the interaction of light with resonant modes of the molecules in the medium. There are two different kinds of interaction, one with vibrational modes (originating vibrational Raman scattering) and another with rotational modes (originating rotational Raman scattering). From these two, the interaction with vibrational modes is stronger and dominates the Raman scattering process. Furthermore, vibrational Raman scattering generates a frequency shift one order of magnitude larger than that due to rotational Raman scattering.

Raman scattering can be effectively described in terms of quantum energy levels as schematically shown in Fig. 2a, b. A molecule initially in the so-called ground state can be excited by an incident pump photon at frequency ω_p to a higher energy vibrational state, through a virtual intermediate state. In such a process the incident photon at frequency ω_p is virtually absorbed, while a second photon is emitted at

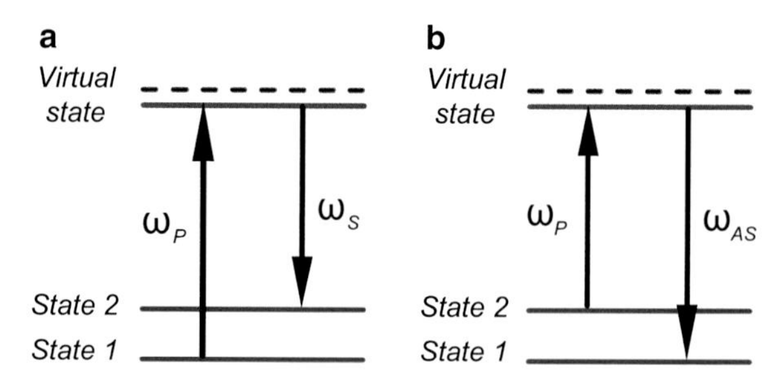

Fig. 2 Energy level representation of Raman (**a**) Stokes and (**b**) anti-Stokes scattering processes

lower frequency $\omega_S = \omega_p - \Omega_R$, where Ω_R is the frequency associated to the energy of the involved vibrational mode (see Fig. 2a). As the emitted photon has lower energy than the pump photon, the process corresponds to Raman Stokes scattering. It can also happen that, if the molecule already has some vibrational energy, the pump photon at frequency ω_p can absorb a quantum of energy from the medium, leading to the emission of a photon of higher energy at frequency $\omega_{AS} = \omega_p + \Omega_R$. This process corresponds to Raman anti-Stokes scattering as described in Fig. 2b.

Note that in the quantum mechanical model, vibrational modes are represented by optical phonons. Due to the rather high energy of the involved vibrational modes, the Raman frequency shift represents a significant fraction of the pump optical frequency, three orders of magnitude higher than the Brillouin frequency shift associated to the much lower energy associated to acoustic phonon interaction (Rogers 1999).

The Raman scattering spectrum of different materials typically exhibits several peaks, associated to different vibrational modes, which can be interpreted as a fingerprint of the specific material, explaining the basic concept widely exploited in Raman spectroscopy (Tolles et al. 1977). For each molecular vibration, two Raman components can be observed, related to absorption of energy from the incident photon (Stokes process) and to the absorption of energy from the medium (anti-Stokes process), respectively. The Raman anti-Stokes process can only occur though thermal excitation of the molecules, and it is then expected to vanish at absolute zero temperature (0 K). This is because at zero absolute, the medium has no energy, and therefore there are no phonons (vibrational energy) to interact and transfer energy to the optical wave. On the contrary, energy transitions from the ground state to a higher energy level are allowed at zero absolute, enabling the existence of Raman Stokes process. This feature results in an asymmetry between the intensities of the two Raman scattering components (Islam 2004).

The transition rates of the Stokes (W_S) and anti-Stokes (W_{AS}) processes can be mathematically described as:

$$W_S \propto N_0 \left(1 + N_\Omega\right) \tag{1}$$

$$W_{AS} \propto N_0 N_\Omega \tag{2}$$

where N_0 is the incident photon number, which is proportional to the pump intensity, and N_Ω is the Bose-Einstein thermal population factor, given by (Islam 2004):

$$N_\Omega = \left[\exp\left(\frac{h\Delta\upsilon}{k_B T}\right) - 1\right]^{-1} \tag{3}$$

where $\Delta\upsilon = {}^{\Omega_R}\!/_{2\pi}$ is the frequency separation between Raman anti-Stokes and pump signal, h is the Planck constant, k_B is the Boltzmann constant, and T is the absolute temperature in Kelvin.

The ratio between the anti-Stokes and Stokes intensities is then proportional to the factor $\exp\left(\frac{h\Delta\upsilon}{k_B T}\right)$, pointing out that at high temperature values, the anti-Stokes and Stokes light intensities approach the same value, while the ratio tends to zero at $T = 0$ K.

Stimulated Raman Scattering

Spontaneous Raman scattering has been described in the previous section assuming that the pump light intensity is low enough not to perturb the optical properties of the medium. However, when a high intensity electromagnetic field is applied to the medium, the consequent inharmonic motion of the bound electrons affects the material optical properties inducing the so-called *stimulated Raman scattering* (SRS) effect. SRS is a highly efficient coherent nonlinear process, originated by the third-order susceptibility of the optical fiber (Agrawal 2010; Boyd 2003).

In SRS, molecular vibrations modulate the refractive index of the medium at its natural vibration frequency Ω_R, generating two sidebands around the pump frequency ω_p. When another optical field propagates in the medium at the Stokes frequency ω_S, it beats with the pump wave, modulating the total vibrational intensity. The process excites molecular vibrations at frequency $\Omega_R = \omega_p - \omega_S$, which in turn coherently amplify the Stokes wave. Such beating and molecular oscillations enhance each other, leading to a highly efficient nonlinear effect which amplifies the Stokes signal in both forward and backward directions with respect to the pump.

Note that SRS can be initiated by either a probe wave propagating along the optical fiber in forward or backward directions or by the spontaneous Raman scattering light. In the first case, the beating between the two propagating fields and the consequent excited vibrational modes provide an efficient mechanism which can be effectively exploited for optical amplification (Islam 2004). This mechanism has been widely exploited for distributed optical amplification in long-haul fiber optic transmission systems (Namiki and Emori 2001).

By assuming steady-state conditions, the pump and stimulated Raman Stokes intensities can be described by the following propagation equations:

$$\frac{d}{dz} I_p(z) = -\frac{\upsilon_p}{\upsilon_s} g_R(\upsilon)\, I_p(z) I_s(z) - \alpha_p I_p(z) \tag{4}$$

$$\pm \frac{d}{dz} I_s(z) = \pm \left[g_R(\upsilon)\, I_p(z) I_s(z) - \alpha_s I_s(z) \right] \tag{5}$$

where $g_R(\upsilon)$ represents the Raman gain coefficient, υ_p and υ_s are the pump and Stokes frequencies, and α_p and α_s are the fiber attenuation coefficients at the pump and Stokes frequencies. The symbol $\pm$ represents forward and backward SRS light components, respectively. Note that the Raman gain coefficient depends on the fiber core composition and extends over a broad range of frequencies with an amorphous spectral shape. For silica fibers $g_R(\upsilon)$ results from the contribution of many frequency vibrations giving rise to a large continuum spectrum characterized by a broad peak at a frequency separation of around 13 THz from the pump frequency.

Equations 4 and 5 can be solved analytically under the assumption of undepleted pump approximation; for the forward propagating stimulated Stokes component, the analytical solution leads to:

$$I_s(z) = I_s(0) \exp\left[g_R(\upsilon) I_p(0) L_{\text{eff}} - \alpha_s z\right] \quad (6)$$

where $I_s(0)$ is the input Stokes intensity at the near fiber end ($z = 0$), and L_{eff} is the fiber effective length, defined as:

$$L_{\text{eff}} = \frac{1}{\alpha_p}\left[1 - \exp\left(-\alpha_p z\right)\right] \quad (7)$$

A similar expression can also be obtained for the backward propagating stimulated Stokes component:

$$I_s(z) = I_s(L) \exp\left[g_R(\upsilon) I_p(0) \exp(-\alpha_s z) L_{\text{eff}} - \alpha_s (L - z)\right] \quad (8)$$

where $I_s(L)$ is the input Stokes intensity at the far fiber end ($z = L$), L being the fiber length.

Note that when no Raman Stokes component is injected into the fiber, SRS can be generated and built up from the thermally-activated spontaneous Raman scattering. In any case SRS can be considered as a nonlinear process that takes place when the Raman pump exceeds a given threshold power, defined as the input pump power at which the Stokes power equals the pump power at the fiber output. Following this definition, and assuming that the polarization of the pump and Stokes light is the same along the entire optical fiber, the Raman threshold for forward SRS can be expressed as:

$$P_{Th}^{SRS} = 16\frac{A_{\text{eff}}}{g_R L_{\text{eff}}} \quad (9)$$

where A_{eff} is the fiber effective core area. A similar equation can also be derived for the backward SRS process:

$$P_{Th}^{SRS} = 20\frac{A_{\text{eff}}}{g_R L_{\text{eff}}} \quad (10)$$

Note that Eqs. 9 and 10 have been derived under steady-state conditions and assuming that the interacting optical waves are equally and fully polarized. In the case of using fully depolarized signals, the threshold increases by a factor of 2 with respect to these expressions. Furthermore, in the case of using a pulsed pump signal, different SRS threshold powers can be defined. For the backward SRS threshold power, the effective interaction length becomes equal to half the spatial pulse width, thus leading to a significant increase in the threshold level. In this case the most restrictive interaction turns out to be the forward Raman scattering, due to the continuous interaction between the pump and co-propagating Raman components along the optical fiber. However, in this case, the fiber dispersion and the consequent walk-off distance ultimately define the SRS threshold power (Wait et al. 1997). In particular if the walk-off distance is shorter than the effective length defined by the physical fiber length, the SRS threshold turns out to increase accordingly. Thus in such a case, the maximum interaction length L_{eff} is determined by the walk-off distance, given by:

$$L_W = \frac{W_0}{D\Delta\lambda} \tag{11}$$

where W_0 is the pulse duration, D the fiber dispersion, and $\Delta\lambda$ the separation between pump and Stokes wavelengths.

Raman-Based Distributed Temperature Sensors

Working Principle

Distributed optical fiber sensors based on Raman scattering make use of the temperature dependence of the spontaneous Raman Scattering process and in particular of the Raman anti-Stokes component (Dakin and Pratt 1985; Hartog and Leach 1985). This means that the peak power of the pump signal used for interrogation should be below the SRS threshold previously defined. The use of pump power levels above the SRS threshold modifies the material properties, making distributed temperature sensing impossible in such a condition.

In particular, spontaneous Raman scattering is a temperature-dependent process caused by thermally driven molecular vibrations, in which two spectral components shifted from the incoming light are generated. While the intensity of the Raman upshifted frequency component (i.e., the anti-Stokes light) exhibits a strong dependence on the temperature, the downshifted frequency component (i.e., the Stokes light) is only slightly temperature dependent.

The temperature sensitivity of the spontaneous Raman anti-Stokes scattering is actually a well-known parameter, being characterized by a quasi-linear relative sensitivity of 0.8%/K around room temperature. This value, however, slightly increases at low temperatures, for instance, to about 1.2%/K around −52 °C (Hartog and Leach 1985). Although the absolute power of the spontaneous Raman anti-Stokes component is more than 30 dB weaker than the Rayleigh backscattered light, its temperature dependence is sufficiently large to provide a reliable mechanism to perform distributed temperature sensing for real applications using an optical fiber.

In order to obtain the spatial information in RDTS systems, optical reflectometry techniques are typically exploited. While Raman temperature sensing with very high spatial resolution (e.g., in the centimeter scale) can be obtained with an approach based on optical frequency-domain reflectometry (OFDR) (Eickhoff and Ulrich 1981; Farahani and Gogolla 1999), most of commercial systems make use of OTDR interrogation methods (Barnoski et al. 1977) to obtain a temperature profile along many kilometers of fiber. In this chapter the description focuses on the Raman-OTDR technique, since this is the most common method for interrogation.

Raman-OTDR consists in launching short and high-power optical pulses into a sensing fiber and detecting the backscattered spontaneous Raman anti-Stokes signal as a function of time, with high temporal resolution. As in any OTDR system, the spatial information is obtained from the pulse round-trip time between the fiber input ($z = 0$) and a given position z along the fiber. Thus, knowing the group velocity v_g in the fiber, the temporal scale t of the time-domain Raman measurements can be straightforwardly converted into distance using the relation $z = v_g t/2$, where the factor 1/2 originates from the round-trip propagation. The spatial resolution of the system Δz is determined by the pulse duration T, so that $\Delta z = v_g T/2$.

In a well-designed RDTS system, the time-domain trace of the spontaneous anti-Stokes Raman scattering light shows an exponentially decaying behavior, resulting from the intrinsic fiber attenuation. Due to the round-trip propagation, the decay constant of the measured Raman backscattered intensity is equal to two times the fiber attenuation coefficient. This feature results in time-domain traces with very low amplitude at the end of the sensing fiber, thus imposing a well-known trade-off between sensing range, spatial resolution, and temperature resolution. Long sensing ranges indeed result in Raman anti-Stokes traces with low signal-to-noise ratio (SNR), especially at the end of the fiber. Since the temperature resolution is strictly determined by the SNR of the measured traces, long sensing ranges typically lead to poor temperature resolutions. A possibility to increase the SNR, and hence to improve the temperature resolution, is to use pulses with higher peak power; however, this power is limited by the onset of nonlinear effects in the fiber. Once the maximum allowed peak power is being used, the only simple alternative to increase the trace SNR is to use longer pulses. However, although longer pulses increase the intensity of the Raman anti-Stokes backscattered signal, they lead to worse spatial resolution. A common strategy used to mitigate the impact of noise on the traces, and increase SNR, is to acquire and average a large number of time-domain traces. Considering that the noise of the system is dominated by additive Gaussian noise, averaging N_{avg} consecutive traces leads to a noise reduction in a factor $\sqrt{N_{avg}}$, at the expense of a longer measurement time. For instance, averaging 10,000 traces leads to a 100-fold SNR improvement (= 20 dB). In order to obtain much larger SNR enhancements, a significantly larger number of traces has to be averaged. This might lead to very long measurement times, which could reach unacceptable levels for some applications. Special techniques to partially overcome this trade-off have been proposed in the last decade. Advanced methods for performance improvement in RDTS systems describes some of the most common interrogation methods employed to enhance the performance of RDTS sensors.

Since the working principle of RDTS systems consists in converting local temperature changes into local amplitude changes in the anti-Stokes Raman traces (Dakin and Pratt 1985; Hartog and Leach 1985), the sensor is, in principle, highly affected by local losses (such as splices, connectors, and bending losses) and eventual laser power fluctuations. These local losses and fluctuations of the pulse power launched into the sensing fiber can result in significant errors in the temperature measurements. Thus, while any local amplitude variation in the time-domain anti-Stokes trace can be easily misinterpreted as a local temperature change, laser power fluctuations can bias the entire measured distributed temperature profile. To overcome this problem and make RDTS more robust against trace amplitude variations (local or distributed), the Raman anti-Stokes trace is typically normalized by a temperature-independent OTDR signal, such as the Raman Stokes or Rayleigh intensity (Dakin and Pratt 1985; Hartog and Leach 1985). Strictly speaking, the Raman Stokes trace is still slightly dependent on local temperature variations; however, using this Stokes trace for normalization can still provide a temperature-sensitive ratio that mitigates the detrimental impact of local losses or laser power variations.

A generic scheme for RDTS system is shown in Fig. 3. To perform OTDR measurements, short, high-power optical pulses are launched into the sensing fiber. These optical pulses can be obtained from either a high-power pulsed laser or an externally modulated semiconductor laser diode operating in continuous wave combined with optical amplification. Pulses are then launched into the sensing fiber through an optical coupler or circulator. The scattering components reaching the fiber input in backward propagating direction are then sent into an optical receiver block. This unit is typically composed of wavelength-selective optical filters, separating the anti-Stokes component from the Stokes and Rayleigh backscattered lights. Either a single-receiver or a double-receiver scheme can be employed. If a single-receiver scheme is used, then an optical switch is required to alternatively measure the temperature-dependent anti-Stokes signal and the reference signal (being either the Stokes or the Rayleigh component), in a configuration similar to time-domain multiplexing detection, as shown in Fig. 3. However, in order

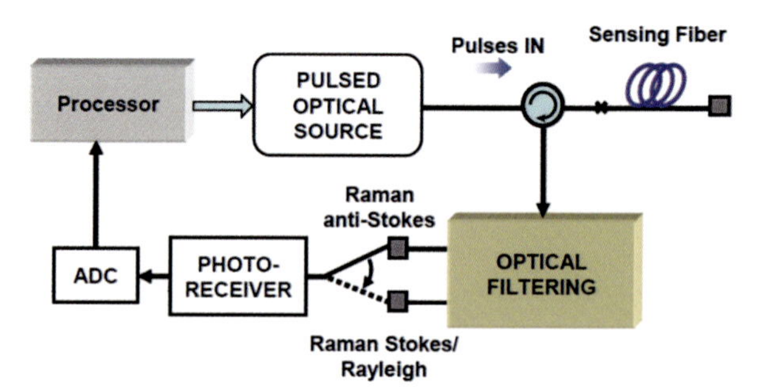

Fig. 3 Generic block diagram of a typical RDTS based on a single-receiver scheme

to increase the robustness of the system against the impact of potential laser power fluctuations, a double-receiver scheme is typically preferred. This allows the simultaneous measurement of both Raman anti-Stokes and reference signal, enabling an instantaneous suppression of any temperature-independent fluctuation of the backscattered intensity. Due to the low intensity of the spontaneous Raman scattering, the entire receiver stage is usually composed of an avalanche photodiode (APD), followed by a high-gain transimpedance amplifier (TIA), an optional low-noise voltage amplifier, and then a digitalization and processing unit.

The low optical power characterizing the spontaneous Raman backscattering components generated by short optical pulses limits considerably the accuracy of RDTS systems, especially in single-mode fibers. Considering that both the numerical aperture and the Raman capture factor of multimode fibers (MMFs) are usually larger than those of single-mode fibers, MMFs are typically preferred for high-performance RDTS systems. The larger effective area of MMFs actually makes the thresholds of nonlinear effects much higher than the ones in SMFs, thus allowing for the use of pulses with much larger peak power (approximately one order of magnitude larger) and larger Raman backscattered power being coupled back to the photo-receiver stage. Although the use of MMFs increases the SNR of the measured time-domain Raman traces, modal dispersion becomes an important parameter limiting the performance of the sensor, especially at long ranges. Consequently, in a RDTS using, for example, a pump pulse of 10 ns, a spatial resolution of about 1 m can be obtained along the first kilometers of sensing fiber; however, the accumulated effect of modal dispersion (on both pulse and backscattering) through a long sensing fiber impairs the spatial resolution, especially at distances beyond ~10 km. To limit the impact of modal dispersion, graded-index MMFs are required, while the large modal dispersion of step-index MMFs makes this kind of fibers completely inappropriate for long-range sensing with meter-scale spatial resolution.

Single-End Configuration

In principle, standard RDTS systems only need access to one fiber end, as previously described; however, for some applications, the access to both fiber ends can be required to increase reliability of the measurements, as will be discussed later in this section.

In single-end RDTS configurations, pump pulses are launched into the sensing fiber from one fiber end ($z = 0$), and the spontaneous Raman backscattered light is detected from the same fiber end, as schematically shown in Fig. 4. When the Raman-OTDR Stokes trace is used to normalize the anti-Stokes trace, the following temperature-dependent ratio is obtained:

$$R(z) = \frac{P_{\mathrm{AS}}(z)}{P_S(z)} = C_R \exp\left(\frac{-h\Delta\nu}{k_B T(z)}\right) \exp\left[-\int_0^z [\alpha_{\mathrm{AS}}(\xi) - \alpha_{\mathrm{S}}(\xi)]\, d\xi\right] \quad (12)$$

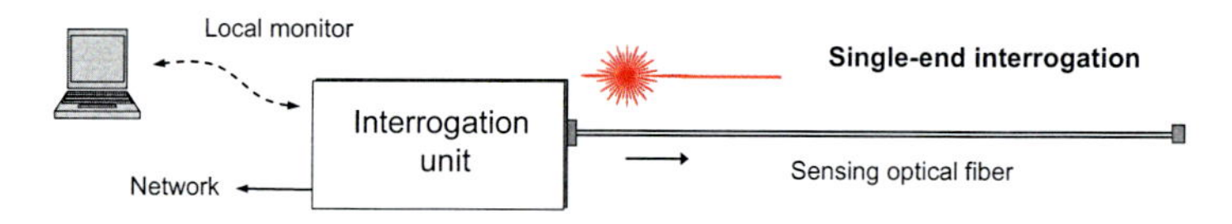

Fig. 4 Basic single-end Raman-OTDR interrogation scheme

where $\Delta\upsilon$, h, and k_B are the frequency separation between Raman anti-Stokes and pump signal, the Planck constant, and the Boltzmann constant, respectively; $T(z)$ is the local fiber temperature, α_{AS} and α_S are the attenuation coefficient at the anti-Stokes and Stokes wavelengths, and C_R is a constant that takes into account the differences in the Raman capture factor and the response of the receiver at the different wavelengths. Note that the ratio $R(z)$ depends on the differential wavelength-dependent loss (WDL) of the fiber. If this factor is properly characterized as a function of distance, Eq. 12 can be renormalized so that $R(z)$ turns out to be simply proportional to the factor $\exp\left(\frac{-h\Delta\upsilon}{k_B T(z)}\right)$. Thus, after calibrating the value of C_R, the distributed temperature profile $T(z)$ can be obtained as:

$$T(z) = \frac{-h\Delta\nu}{k_B}\left[\ln(R(z)) - \ln(C_R)\right]^{-1}. \tag{13}$$

Equation 13 provides a good estimation of the distributed temperature profile, assuming that the factor C_R and the WDL are properly characterized and do not change over time. Unfortunately, WDL are not always constant over time and might change depending on external environmental conditions. Consequently, the ratio anti-Stokes/Stokes (or anti-Stokes/Rayleigh) is expected to change during the sensor lifetime due to, for instance, the fiber aging process or to external harsh environmental conditions. This might lead to significant errors in the distributed temperature estimation. Such errors actually manifest as a gradual deviation of the temperature trace with respect to the real value; an effect that increases with the distance and turns to be a critical factor at long sensing distances. Thus, as a consequence of time-varying WDL, the single-end scheme shown in Fig. 4 and described by Eq. 13 turns out to be inappropriate for many applications since it requires a continuous calibration, which is not always possible in many real cases.

In order to make a RDTS system less vulnerable to time-varying WDL (generated either locally at specific fiber locations or distributed within long fiber sections), three different alternative configurations are typically implemented: (i) double-end (or loop), (ii) dual-wavelength, and (iii) anti-Stokes-only configurations. These schemes are described hereafter.

Double-End Configuration

In order to correct the abovementioned WDL issues, measurements can be performed in a double-end configuration (also called loop configuration), as schematically shown in Fig. 5. Compared to the scheme described in Fig. 3 for

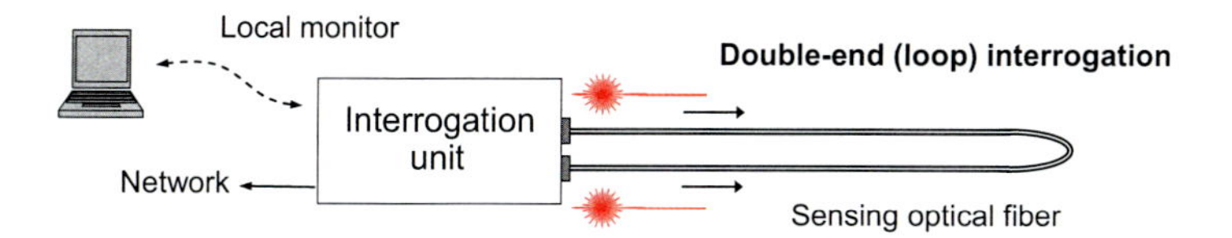

Fig. 5 Basic double-end (or loop) Raman-OTDR interrogation scheme

a single-end system, a double-end RDTS system includes an optical switch in front of the sensing fiber input, so that pump pulses can be alternatively sent into the sensing fiber from both ends. This way, the Raman anti-Stokes trace and the reference trace (i.e., either the Raman Stokes or the Rayleigh component) are measured in both forward and backward directions (Fernandez et al. 2005). Like in the single-end configuration, the ratio between the anti-Stokes to reference powers is first independently obtained in both forward ($R^{\text{For}}(z)$) and backward ($R^{\text{Back}}(z)$) directions. In the case of using the Stokes trace as a reference, this leads to:

$$R^{\text{For}}(z) = \frac{P_{\text{AS}}^{\text{For}}(z)}{P_S^{\text{For}}(z)} = C_R^{\text{For}} \exp\left(\frac{-h\Delta\nu}{k_B T(z)}\right) \exp\left[-\int_0^z \left[\alpha_{\text{AS}}(\xi) - \alpha_{\text{S}}(\xi)\right] d\xi\right] \tag{14}$$

$$R^{\text{Back}}(z) = \frac{P_{\text{AS}}^{\text{Back}}(z)}{P_S^{\text{Back}}(z)} = C_R^{\text{Back}} \exp\left(\frac{-h\Delta\nu}{k_B T(z)}\right) \exp\left[-\int_z^L \left[\alpha_{\text{AS}}(\xi) - \alpha_{\text{S}}(\xi)\right] d\xi\right] \tag{15}$$

where C_R^{For} and C_R^{Back} are constants that take into account the differences in the Raman capture factor and the response of the receiver at the different wavelengths for forward and backward propagation directions, respectively. To eliminate the longitudinal impact of the WDL, the geometric mean of these two ratios is then calculated. Thus, the ratio in loop configuration $R^{\text{Loop}}(z)$ is obtained as:

$$\begin{aligned} R^{\text{Loop}}(z) &= \sqrt{R^{\text{For}}(z) R^{\text{Back}}(z)} \\ &= \left[C_R^{\text{For}} C_R^{\text{Back}}\right]^{1/2} \exp\left(\frac{-h\Delta\nu}{k_B T(z)}\right) \exp\left[-\left(\alpha_{\text{AS}} - \alpha_{\text{S}}\right) \frac{L}{2}\right] \end{aligned} \tag{16}$$

The temperature profile can then be estimated inverting Eq. 16, so that:

$$T(z) = \frac{-h\Delta\nu}{k_B}\left[\ln\left(\frac{R(z)}{C_R^{\text{Loop}}}\right) + \left(\alpha_{\text{AS}} - \alpha_{\text{S}}\right)\frac{L}{2}\right]^{-1} \tag{17}$$

where $C_R^{\text{Loop}} = \left(C_R^{\text{For}} C_R^{\text{Back}}\right)^{1/2}$ is a constant to be calibrated.

Note that the same expression is obtained if the procedure consists in first calculating the geometrical mean of the forward and backward time-domain traces and then normalizing the averaged anti-Stokes trace by the averaged reference trace.

Equations 16 and 17 point out that, although the ratio in loop configuration $R^{\mathrm{Loop}}(z)$ depends on the WDL, this dependence is constant for any fiber position. Indeed the geometric mean of forward and backward traces cancels out all loss factors dependent on the fiber position z. This feature leads to self-calibrated temperature measurements, thus making the RDTS system more robust against loss variations occurring during the sensor lifetime. Thus, even though the RDTS system requires access to both fiber ends, this configuration has been demonstrated to provide a simple and effective solution to compensate differential attenuation issues (Fernandez et al. 2005).

Note, however, that although the use of traces in forward and backward directions compensates the effects of WDL on the temperature profile (i.e., the possible distortions in the trace), differential losses can still offset the entire measured distributed temperature profile. To correct such effect, a calibration fiber spool is typically used in RDTS systems. The temperature of this calibration spool (usually placed inside the interrogation unit) can be precisely measured by an electronic sensor, thus providing a simple and effective method to auto-compensate possible errors in the temperature profile obtained by the sensor.

Dual-Wavelength Configuration

As previously mentioned, the double-end RDTS configuration can be used to correct for the differential attenuation between Raman anti-Stokes and Stokes (or Rayleigh) wavelengths, which can induce large errors in the temperature evaluation. RDTS based on double-end configurations, although being robust and reliable, are however limited in terms of sensing range compared to single-end systems. An alternative technique to mitigate the impact of varying differential attenuation consists in the use of a single-end scheme combined with a multi-laser configuration (Suh and Lee 2008).

This scheme has the same advantages of the double-end configuration to address issues related to WDL; however, it also enables longer sensing ranges due to the single-end access. In this case, two laser sources at different wavelengths are alternatively used. The wavelength of the first excitation source is chosen so that it coincides with the wavelength of the Stokes component of the second excitation source (Suh and Lee 2008). This way, the anti-Stokes component originated by the first excitation source coincides with the wavelength of the second source. Two measurement modes can be defined in such a way that the Raman anti-Stokes trace generated by the primary source can be corrected by the Raman Stokes trace induced by a secondary source. Since the pump pulse originated by the first source and its respective anti-Stokes component experience the same round-trip attenuation of the

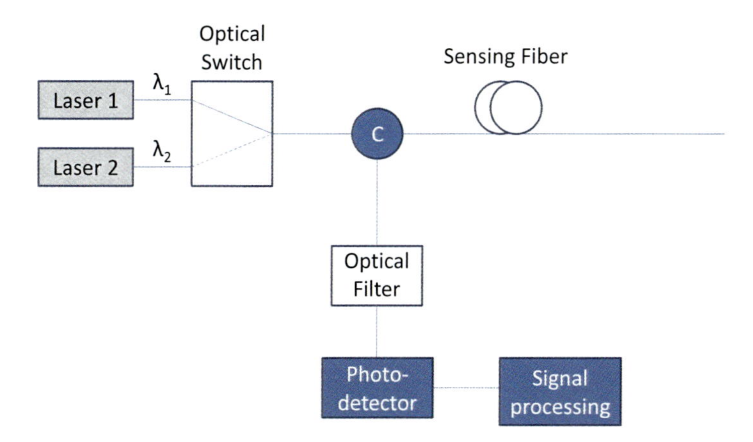

Fig. 6 Basic dual-laser single-end Raman-OTDR interrogation scheme

pump pulse of the second source and its Stokes component, the effects of distributed differential attenuation are fully eliminated.

Figure 6 shows a schematic of a dual-laser RDTS system in which a measurement source at wavelength λ_1 and a calibration source at wavelength λ_2 can be alternatively coupled into the sensing fiber using an optical switch. The backscattered light components generated by the measurement and calibration sources can be filtered, photo-detected, and then processed to extract the temperature profile along the sensing fiber.

As the Raman Stokes attenuation profile of the calibration light source, obtained in calibration mode, is only slightly sensitive to temperature effects, it can be used to correct the anti-Stokes profile generated by the measurement source during the measurement mode. The temperature profile can then be estimated using the same expression used in single-end configuration, as described in Eq. 13.

Anti-Stokes-Only Configuration

As previously mentioned, most-developed RDTS systems utilize the ratio between the spontaneous Raman anti-Stokes intensity and either the Raman Stokes or Rayleigh backscattering intensity in a single-end fiber configuration. Although they represent the most-used solution in long-range applications due to their extended sensing distance, sensors based on single-end are intrinsically affected by variations of the differential wavelength-dependent losses of the fiber. In addition to the use of a double-end scheme or dual-wavelength system to eliminate or mitigate the impact of WDL, another alternative is to use only the Raman anti-Stokes component.

This can be implemented in either single-end (Hwang et al. 2010) or double-end (Soto et al. 2012) scheme. In the two cases, traces are measured in forward and backward directions, and the geometric mean between traces is calculated, similarly to the conventional double-end scheme. In the single-end case, pulses are launched into a unique fiber end, while a reflective mirror at the far fiber end allows for the pump pulse and backscattering to propagate into the sensing fiber in the two opposite directions (Hwang et al. 2010). On the other hand, in the double-end scheme, pump pulses are alternatively sent into the sensing fiber from both fiber ends, so that anti-Stokes traces are straightforwardly obtained in both forward and backward directions (Soto et al. 2012). Calculating the geometric mean of these traces, all loss factors depending on the fiber position are canceled out, leading to self-calibrated temperature measurements that do not depend on the optical loss variations during the sensor lifetime.

The anti-Stokes Raman intensity traces measured in both forward $\left(I_{\mathrm{AS}}^{\mathrm{For}}\right)$ and backward $\left(I_{\mathrm{AS}}^{\mathrm{Back}}\right)$ directions can be written as:

$$I_{\mathrm{AS}}^{\mathrm{For}}(z) = C_{\mathrm{AS}}^{\mathrm{For}}\left\{\exp\left(\frac{-h\Delta\nu}{k_B T(z)}\right) - 1\right\}^{-1} \exp\left[-\int_0^z \left[\alpha_{\mathrm{AS}}(\xi) + \alpha_{\mathrm{P}}(\xi)\right] d\xi\right] \tag{18}$$

$$I_{\mathrm{AS}}^{\mathrm{Back}}(z) = C_{\mathrm{AS}}^{\mathrm{Back}}\left\{\exp\left(\frac{-h\Delta\nu}{k_B T(z)}\right) - 1\right\}^{-1} \exp\left[-\int_z^L \left[\alpha_{\mathrm{AS}}(\xi) + \alpha_{\mathrm{P}}(\xi)\right] d\xi\right] \tag{19}$$

where α_{P} is the fiber loss coefficient at the pulse wavelength. So that the geometric mean results in:

$$\begin{aligned} I_{\mathrm{AS}}^{\mathrm{Loop}}(z) &= \sqrt{I_{\mathrm{AS}}^{\mathrm{For}}(z) I_{\mathrm{AS}}^{\mathrm{Back}}(z)} \\ &= C_{\mathrm{AS}}^{\mathrm{Loop}}\left\{\exp\left(\frac{-h\Delta\nu}{k_B T(z)}\right) - 1\right\}^{-1} \exp\left[-\frac{1}{2}\int_0^L \left[\alpha_{\mathrm{AS}}(\xi) + \alpha_{\mathrm{P}}(\xi)\right] d\xi\right] \end{aligned} \tag{20}$$

where $C_{\mathrm{AS}}^{\mathrm{Loop}} = \left(C_{\mathrm{AS}}^{\mathrm{For}} C_{\mathrm{AS}}^{\mathrm{Back}}\right)^{1/2}$ is a constant to be calibrated.

In practice, it is difficult to estimate the absolute fiber temperature using Eq. 20, basically because of possible inaccuracies when determining the coefficient $C_{\mathrm{AS}}^{\mathrm{Loop}}$, value that can differ from fiber to fiber. Therefore, the sensor has to be calibrated, by normalizing the anti-Stokes trace in loop (obtained by geometric mean) with another mean anti-Stokes trace at a known reference temperature $T_{\mathrm{ref}}(z)$. Thus the distributed temperature profile can be estimated as:

$$T(z) = \left\{\frac{k_B}{h\Delta\nu} \ln\left[\frac{I_{\mathrm{AS,ref}}^{\mathrm{Loop}}(z, T_{\mathrm{ref}})}{I_{\mathrm{AS}}^{\mathrm{Loop}}(z, T)}\left[\exp\left(\frac{h\Delta\nu}{k_B T_{\mathrm{ref}}(z)}\right) - 1\right] + 1\right]\right\}^{-1} \tag{21}$$

This equation points out that the use of only anti-Stokes traces in loop configuration allows for the compensation of both local and distributed losses along the sensing fiber, in a similar way as the standard loop configuration does. This means that in the calculation, there is no z-dependent factor tilting or distorting the temperature profile as a function of distance. However, laser power fluctuations and changes in the fiber attenuation (with respect to the initial calibration) induce a constant z-independent offset in the final temperature profile. In order to compensate for this offset, a fiber spool placed inside the RDTS can be used. This way, the temperature of this internal fiber spool (measured with an additional electronic sensor) is then used to correct the offset induced in the entire temperature profile.

Advanced Methods for Performance Improvement in RDTS Systems

The fundamental limitation of conventional Raman-OTDR measurements is the existing trade-off between spatial resolution and SNR (limiting the measurand resolution and the maximum measurement distance). Actually, any increase in the SNR leads to an enhancement of the dynamic range of the intensity traces, which can be used (i) to extend the measurement distance, (ii) to improve the spatial resolution, (iii) to enhance the measurand resolution, and/or (iv) to reduce the measurement time. As previously mentioned, the SNR of RDTS measurements can be enhanced by averaging a large number of Raman-OTDR traces. Thus, an SNR improvement of a factor $\sqrt{N_{avg}}$ can be achieved by averaging N_{avg} traces. This procedure, however, leads to an increment of the measurement time. In many cases, a very long number of averages would be required to reach a given target dynamic range, thus leading to unacceptable acquisition times. A first approach in the system design to reduce the measurement time and achieve high dynamic ranges in conventional Raman-OTDR is to improve the noise characteristics of the receiver and/or maximize the energy of the pulses launched into the sensing fiber. Note that the receiver improvement can be achieved using, for example, low-noise avalanche photodiodes (APD) instead of standard PIN photodiodes. However, this feature is not sufficient to fully overcome all fundamental limitations in the trace SNR. On the other hand, the backscattered Raman signal power (and hence, the SNR of the measurements) can be increased by launching into the fiber pulses with higher energy, i.e., with longer pulse duration and/or higher peak power. The pulse peak power can be increased to some extent until the onset of nonlinear effects, which impose a limit to the maximum usable peak power. Once the peak pulse power is optimized, the only way to enhance the dynamic range of the Raman-OTDR measurements would be increasing the pulse width. Unfortunately, longer pulses lead to an unavoidable degradation of the spatial resolution.

In this section two main approaches to improve the dynamic range of RDTS systems, while keeping high spatial resolution, are described. The first approach corresponds to radar-based pulse coding techniques, which have been readapted to RDTS systems. This technique allows one to increase the total power launched into the fiber without activating nonlinear effects and thus enables a significant SNR enhancement while keeping sharp spatial resolution. The second approach is based on image denoising techniques to reduce noise from RDTS measurements and thus to increase the performance of the sensor.

Optical Pulse Coding

Optical pulse coding is based on spread-spectrum techniques, such as those ones used in radar systems, and consists in spreading the interrogating signal (classically a single-pulse) in time domain (Soto et al. 2007; Bolognini et al. 2007). In this way, the average power (and energy) of the interrogating signal as well as the respective Raman backscattered signal is increased. This is achieved by launching into the sensing fiber pulse sequences (so-called *codewords*) with some particular properties, which allow one to obtain the impulse response of the fiber using a particular decoding process. Measurements with higher dynamic range and improved SNR can then be achieved after decoding. The spatial resolution of the decoded trace is determined by the duration of each pulse contained in the sequences.

One of the differences between RDTS and radar systems is that radars allow the use of bipolar coding techniques (i.e., sequences containing -1's and 1's levels), while only unipolar codes (i.e., containing 0's and 1's) are suitable for OTDR-based systems, such as RDTS. This issue limits the amount of pulse coding schemes that can be effectively used in RDTS systems. Among all the existing types of codes, two of them have been demonstrated to offer real benefits in RDTS: complementary-correlation Golay codes (Soto et al. 2007) and Simplex codes (Bolognini et al. 2007).

Complementary-Correlation Golay Codes

Golay codes make use of the full advantages of correlations by interrogating the sensing fiber with a pulse sequence pair that has complementary autocorrelation properties. The basic principle consists in launching into the sensing fiber a defined pulse sequence (code) and correlating the measured Raman-OTDR traces with the respective pulse sequence used for interrogation. The process allows for measurements of the fiber response with an enhanced SNR, thus overcoming the well-known trade-off between SNR and spatial resolution. To understand the working principle of complementary-correlation Golay (CC-Golay) codes, let us express the measured Raman-OTDR electrical signal $s(t)$ as the convolution ($\otimes$) of the interrogating code sequence $p(t)$, the receiver impulse response $r(t)$, and the backscattering impulse response of the fiber $h(t)$, so that:

$$s(t) = p(t) \otimes r(t) \otimes h(t). \tag{22}$$

If the measured coded Raman trace $s(t)$ is correlated (*) with the respective launched coding signal $p(t)$, the following expression is obtained (Nazarathy et al. 1989):

$$s(t) * p(t) = [p(t) \otimes r(t) \otimes h(t)] * p(t) = [p(t) * p(t)] \otimes [r(t) \otimes h(t)] . \tag{23}$$

This equation points out that the backscattering fiber response *h(t)* can be accurately recovered depending on the autocorrelation properties of the code sequence (assuming a fast-enough receiver response). Therefore, it is the length of the autocorrelation function (ideally a delta function) which determines the spatial resolution. This is essentially defined by the bandwidth of the RDTS system.

Some attempts in OTDR applications have used sequences based on pseudorandom or Barker codes; however, they have not been demonstrated to be fully suitable solutions, since their autocorrelation function exhibits sidelobes that introduce distortions and overlap the backscattering information coming from different fiber locations. Golay codes use, however, a pulse sequence pair with complementary sidelobes. In particular, if an L-bit complementary code pair is used, represented by the bipolar sequences A_k and B_k, they satisfy the following expression (Nazarathy et al. 1989):

$$(A_k * A_k) + (B_k * B_k) = 2L\delta_k . \tag{24}$$

The main characteristic of Golay codes is that even though the individual autocorrelations of A_k and B_k have sidelobes, as respectively shown in Fig. 7a, b, when they are added together, the sidelobes cancel out, and the main correlation peak reinforces to a level equal to $2L$ (see Fig. 7c).

Golay codes are, in principle, defined as bipolar codes (Nazarathy et al. 1989), containing codewords with -1 and 1 elements, which are not suitable for conventional direct-detection OTDR-based systems as RDTS (due to the impossibility to generate optical pulses with negative intensity). In order to make Golay codes suitable for RDTS systems, the bipolar sequences must be transmitted on a bias equal to half of the peak power, converting the two complementary codewords A_k and B_k into four unipolar sequences. Note that this is exactly the same approach used in classical coded OTDR systems (Nazarathy et al. 1989). Hence, in order to perform Raman-OTDR measurements based on Golay codes, the following procedure must be adopted:

1. Four probe sequences must be launched into the fiber and their respective backscattering signal must be detected. The four codewords are the following:
   (a) The unipolar version of the first Golay pair sequence A_k, defined as:

$$u_k^A = 0.5\,(1 + A_k)\,, \tag{25}$$

   (b) The 1's complement of the previous sequence (i.e., swapping 1's and -1's). This is given by:

$$u_k^{\overline{A}} = 0.5\,(1 - A_k)\,, \tag{26}$$

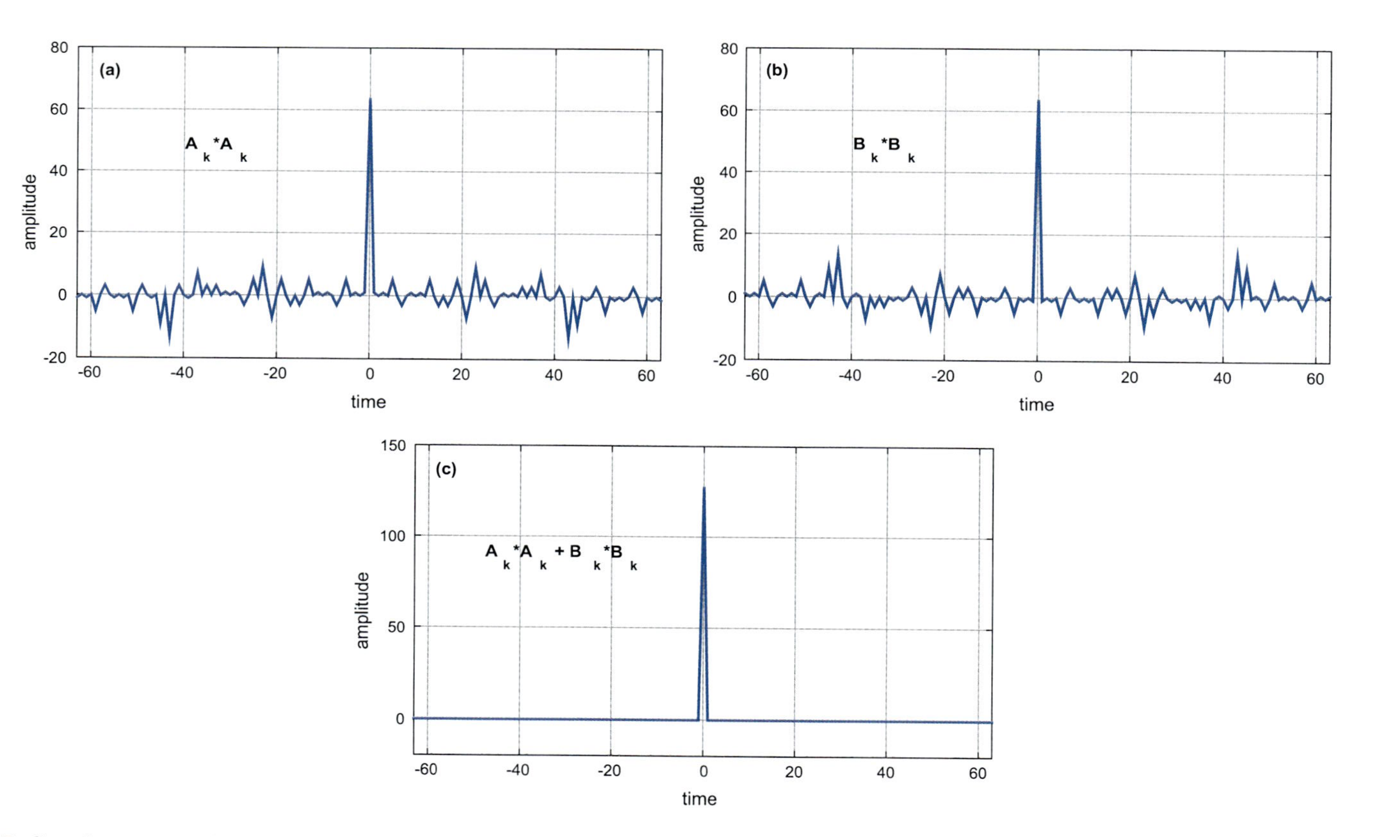

Fig. 7 Complementary code autocorrelations. (**a**) and (**b**) are the individual autocorrelation of each sequence of 64-bit Golay codes. (**c**) Sum of the autocorrelation functions (replotted following concept presented in Nazarathy et al. 1989)

(c) The unipolar version of the second Golay pair sequence B_k, defined as:

$$u_k^B = 0.5\,(1 + B_k)\,, \tag{27}$$

(d) The 1's complement of the previous sequence, given by:

$$u_k^{\overline{B}} = 0.5\,(1 - B_k)\,. \tag{28}$$

2. The backscattered signals of the first two measurements (given by Eqs. 25 and 26) are subtracted from each other:

$$x_k^A = h_k \otimes u_k^A - h_k \otimes u_k^{\overline{A}}$$

and then correlated with the bipolar version of the first Golay sequence A_k:

$$y_k^A = x_k^A * A_k$$

where h_k is the discrete-time backscattering impulse response of the fiber (in this case the receiver response has been neglected, considering it as a delta function).

3. Similarly, the backscattered signals resulting from the last two measurements (given by Eqs. 27 and 28) are subtracted from each other:

$$x_k^B = h_k \otimes u_k^B - h_k \otimes u_k^{\overline{B}}$$

and then correlated with the bipolar version of the second Golay sequence B_k:

$$y_k^B = x_k^B * B_k$$

4. The results of both correlations (y_k^A and y_k^B) are summed up to obtain the estimated signal of the backscattering fiber response. This procedure leads to:

$$\begin{aligned} y_k &= y_k^A + y_k^B = h_k \otimes A_k * A_k + h_k \otimes B_k * B_k \\ &= h_k \otimes [A_k * A_k + B_k * B_k] = 2\mathrm{Lh}_k. \end{aligned} \tag{29}$$

where Eq. 24 has been used.

Note that the signal amplitude of the decoded trace is then $y_k = 2\mathrm{Lh}_k$; however, the above-described decoding process results in a noise buildup factor equal to $2\sqrt{L}$ with respect to the noise σ_{trace} of the single acquisition trace. This means that the noise in the decoded trace can be written as:

$$\sigma_{\mathrm{decoded\_trace}} = 2\sqrt{L}\sigma_{\mathrm{trace}}. \tag{30}$$

Therefore, the SNR of the decoded Raman-OTDR trace is:

$$\mathrm{SNR}_{\mathrm{Golay}} = \sqrt{L}\frac{h_k}{\sigma_{\mathrm{trace}}} \tag{31}$$

In order to estimate the benefits of using Golay codes in RDTS, the achieved SNR (given by Eq. 31) has to be compared to the SNR obtained by using conventional single-pulse RDTS with the same pulse peak power and same measurement time. This means that while Golay needs the acquisition of four coded traces obtained with each four unipolar sequences, the conventional RDTS can acquire and average four single-pulse traces in the same acquisition time. Under this condition the SNR of the conventional RDTS system can be written as:

$$\mathrm{SNR}_{\mathrm{Pulsed}} = \frac{2h_k}{\sigma_{\mathrm{trace}}}. \tag{32}$$

where the factor 2 comes from the SNR improvement obtained by the four averaged single-pulse traces.

The SNR improvement provided by optical pulse coding techniques is called *coding gain*, which in the case of Golay codes is given by the ratio between Eqs. 31 and 32, resulting in the following expression (Nazarathy et al. 1989):

$$\mathrm{Gain}_{\mathrm{Golay}} = \frac{\mathrm{SNR}_{\mathrm{Golay}}}{\mathrm{SNR}_{\mathrm{Pulsed}}} = \frac{\sqrt{L}}{2}. \tag{33}$$

Note that if N_{avg} time traces are averaged in a single-pulse RDTS, the additional SNR enhancement provided by Golay coding and described in Eq. 33 takes into account that the total acquisition time is the same, i.e., that each unipolar Golay-coded Raman-OTDR trace is averaged $N_{\mathrm{avg}}/4$ times.

Several iterative methods for the construction of complementary code pairs have been proposed by M.J.E. Golay in 1961. One of those procedures, known as appending, is based on the procedure described by the following operation:

$$\begin{bmatrix} A \\ B \end{bmatrix} \rightarrow \begin{bmatrix} A \mid B \\ A \mid \overline{B} \end{bmatrix}, \tag{34}$$

where $\overline{B}$ represents the complement of B and | denotes the concatenation of sequences. Therefore, starting from the one-element Golay pair, sequences of length equal to any power of two can be easily derived:

$$\begin{bmatrix} 1 \\ 1 \end{bmatrix} \rightarrow \begin{bmatrix} 1 & 1 \\ 1 & -1 \end{bmatrix} \rightarrow \begin{bmatrix} 1 & 1 & 1 & -1 \\ 1 & 1 & -1 & 1 \end{bmatrix} \rightarrow \begin{bmatrix} 1 & 1 & 1 & -1 & 1 & 1 & -1 & 1 \\ 1 & 1 & 1 & -1 & -1 & -1 & 1 & -1 \end{bmatrix}, \tag{35}$$

This procedure may be continued on to generate sequences with any length equal to a power of two. The construction of complementary sequences in which the length is not a power of two is also possible, as described in Golay (1961);

however, it is not always possible to find a suitable complementary pair of series for any arbitrary length. Nevertheless, independently of their length, a pair of complementary sequences always allows one to obtain four unipolar sequences that can be utilized in RDTS applications. Some attempts have also proposed the use of only three codewords for OTDR applications (Healey 1989; Nazarathy et al. 1990), reducing then the measurement time; however, it is very difficult to find suitable pairs in such cases, which may not even exist for particular lengths (Healey 1989). In Nazarathy et al. (1990), a procedure for the generation of such three sequences is proposed; nevertheless, the resulting codewords are not very efficient since they are affected by a penalty of 3 dB with respect to the coding gain defined by Eq. 33 (under similar measurement conditions, such as code length and measurement time).

Simplex Codes

Simplex codes correspond to a coding technique in which pulse sequences are derived from the Hadamard matrix (Jones 1993; Harwit and Sloane 1979), which is a bipolar matrix with particular orthogonal properties. Such a matrix has been widely used in spectroscopy, providing an efficient method to improve the SNR of the measurements. However, due to its bipolar elements (i.e., −1's and 1's), it is impossible to use the Hadamard matrix directly in OTDR-based applications using direct detection, such as RDTS. Thus, in order to obtain suitable pulse sequences, Simplex codes require Hadamard transform (Soto et al. 2007; Harwit and Sloane 1979), which allows for the calculation of an S-matrix (Simplex matrix) containing unipolar elements and orthogonal rows. As a matter of fact, each row of the S-matrix defines a Simplex codeword (Jones 1993).

S-matrices can be constructed using methods involving quadratic residues, maximal-length shift-register sequences, and twin primes (Jones 1993). It has been demonstrated that the optimum S-matrix, which minimizes the standard deviation of the decoded signal, can be derived from the Hadamard matrix by deleting the first row and the first column and then replacing 1's by 0's and −1's by 1's. Thus, Simplex coding can be easily implemented in Raman-OTDR applications by turning the laser on and off, according to the sequences of 1's and 0's defined by the S-matrix.

The basic working principle of Simplex coding can be better understood with the following example based on an S-matrix of order 3 (see Fig. 8). Let us define $\psi_1(t)$ as the Raman-OTDR trace resulting from a single interrogating pulse $P_1(t)$ and $\psi_2(t)$ and $\psi_3(t)$ as the traces resulting from the time-delayed probe pulses $P_2(t) = P_1(t-\tau)$ and $P_3(t) = P_1(t-2\tau)$, as depicted in Fig. 8a. With τ being the duration of the pulses, the following relationships are satisfied (Lee et al. 2004):

$$P_2(t) = P_1(t-\tau) \rightarrow \psi_2(t) = \psi_1(t-\tau), \tag{36}$$

$$P_3(t) = P_1(t-2\tau) \rightarrow \psi_3(t) = \psi_1(t-2\tau), \tag{37}$$

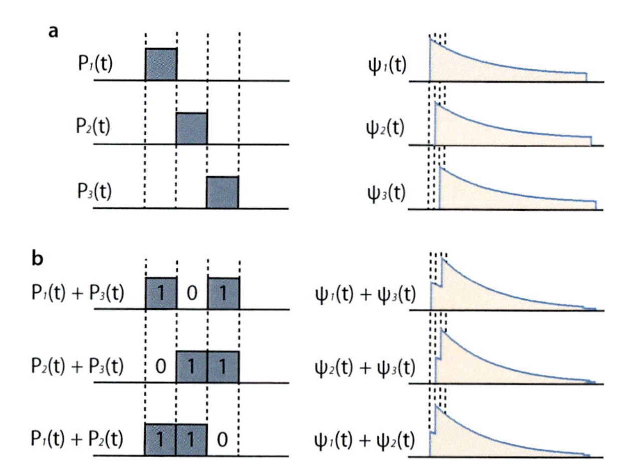

Fig. 8 Example of Simplex coding technique for OTDR-based applications using an S-matrix of order 3. (**a**) Probe pulse and respective fiber response for different time delays. (**b**) Linear combination of pulses, conforming 3-bit Simplex codes (replotted following concept presented in Lee et al. 2004)

Under linear conditions, the Simplex-coded Raman-OTDR traces $\eta_1(t)$, $\eta_2(t)$, and $\eta_3(t)$ can be measured by launching into the fiber three different Simplex sequences (codewords), according to Fig. 8b:

$$P_1(t) + P_3(t) \rightarrow \eta_1(t) = \psi_1(t) + \psi_3(t) + e_1(t), \tag{38}$$

$$P_2(t) + P_3(t) \rightarrow \eta_2(t) = \psi_2(t) + \psi_3(t) + e_2(t), \tag{39}$$

$$P_1(t) + P_2(t) \rightarrow \eta_3(t) = \psi_1(t) + \psi_2(t) + e_3(t), \tag{40}$$

which in matrix notation can be written as:

$$\begin{pmatrix} \eta_1(t) \\ \eta_2(t) \\ \eta_3(t) \end{pmatrix} = \mathbf{S_3} \begin{pmatrix} \psi_1(t) \\ \psi_2(t) \\ \psi_3(t) \end{pmatrix} + \begin{pmatrix} e_1(t) \\ e_2(t) \\ e_3(t) \end{pmatrix}, \tag{41}$$

where $e_1(t)$, $e_2(t)$, and $e_3(t)$ represent the noise amplitude of each measurement and $\mathbf{S_3}$ is the S-matrix of order 3, given by:

$$\mathbf{S_3} = \begin{pmatrix} 1\ 0\ 1 \\ 0\ 1\ 1 \\ 1\ 1\ 0 \end{pmatrix}. \tag{42}$$

In order to recover the single-pulse Raman-OTDR trace $\psi_1(t)$, Hadamard transform has to be applied to the measured coded traces, according to Lee et al. (2004):

$$\begin{pmatrix} \widehat{\psi}_1(t) \\ \widehat{\psi}_2(t) \\ \widehat{\psi}_3(t) \end{pmatrix} = \mathbf{S_3}^{-1} \begin{pmatrix} \eta_1(t) \\ \eta_2(t) \\ \eta_3(t) \end{pmatrix} = \begin{pmatrix} \psi_1(t) \\ \psi_2(t) \\ \psi_3(t) \end{pmatrix} + \mathbf{S_3}^{-1} \begin{pmatrix} e_1(t) \\ e_2(t) \\ e_3(t) \end{pmatrix}, \tag{43}$$

where $\widehat{\psi}_{\mathrm{i}}(t)$ (with $i = 1, 2, 3$) corresponds to an estimation of the single-pulse trace $\psi_{\mathrm{i}}(t)$. Thus, three different decoded traces can be obtained, which can be inversely time-shifted with multiples of τ and then averaged to give rise to a single decoded Raman-OTDR trace with improved SNR. Noise analysis demonstrates that in this case (i.e., Simplex codes of order 3), an SNR enhancement of $2/\sqrt{3}$ can be achieved in comparison to single-pulse Raman-OTDR measurements (with the same measurement time) (Lee et al. 2004).

The previous example can be extended to the generalized case of code length equal to L, where coded-Raman-OTDR traces measured at the fiber input ($z = 0$) can be represented by the following expression:

$$\begin{pmatrix} \eta_1(t) \\ \vdots \\ \eta_i(t) \\ \vdots \\ \eta_L(t) \end{pmatrix} = \mathbf{S_L} \begin{pmatrix} \psi_1(t) \\ \vdots \\ \psi_i(t) \\ \vdots \\ \psi_L(t) \end{pmatrix} + \begin{pmatrix} e_1(t) \\ \vdots \\ e_i(t) \\ \vdots \\ e_L(t) \end{pmatrix}, \tag{44}$$

where $\eta_i(t)$ is the i-th coded Raman trace (with $i \in [1, L]$), $\psi_i(t)$ is the single-pulse Raman trace delayed by an i-th multiple of the bit duration with respect to $\psi_1(t)$, and $e_i(t)$ is the amplitude of the uncorrelated zero-mean noise added to the measurement of the i-th coded trace.

The decoding process can be carried out by linear processing (Hadamard transform) of the traces as follows:

$$\begin{pmatrix} \widehat{\psi}_1(t) \\ \vdots \\ \widehat{\psi}_i(t) \\ \vdots \\ \widehat{\psi}_L(t) \end{pmatrix} = \mathbf{S_L}^{-1} \begin{pmatrix} \eta_1(t) \\ \vdots \\ \eta_i(t) \\ \vdots \\ \eta_L(t) \end{pmatrix} = \begin{pmatrix} \psi_1(t) \\ \vdots \\ \psi_i(t) \\ \vdots \\ \psi_L(t) \end{pmatrix} + \mathbf{S_L}^{-1} \begin{pmatrix} e_1(t) \\ \vdots \\ e_i(t) \\ \vdots \\ e_L(t) \end{pmatrix}, \tag{45}$$

where $\mathbf{S_L}$ is the S-matrix of order L and $\widehat{\psi}_i(t)$ is the estimated single-pulse Raman-OTDR trace $\psi_i(t)$. If every row is inversely time-shifted with multiples of the pulse duration τ, and introducing the matrix $\mathbf{T_L}$ (corresponding to the normalized matrix of $\mathbf{S_L}^{-1}$), the following expression is obtained (Lee et al. 2004):

$$\begin{pmatrix} \widehat{\psi}_1(t) \\ \vdots \\ \widehat{\psi}_i\left(t+(i-1)\,\tau\right) \\ \vdots \\ \widehat{\psi}_L\left(t+(L-1)\,\tau\right) \end{pmatrix} = \begin{pmatrix} \psi_1(t) \\ \vdots \\ \psi_i\left(t+(i-1)\,\tau\right) \\ \vdots \\ \psi_L\left(t+(L-1)\,\tau\right) \end{pmatrix} + \frac{2}{L+1}\mathbf{T_L} \begin{pmatrix} e_1(t) \\ \vdots \\ e_i\left(t+(i-1)\,\tau\right) \\ \vdots \\ e_L\left(t+(L-1)\,\tau\right) \end{pmatrix}, \quad (46)$$

where:

$$\mathbf{T_L} = \frac{L+1}{2}\mathbf{S_L}^{-1}, \quad (47)$$

It is worth pointing out that an interesting property of the matrix $\mathbf{T_L}$ is that the sum of elements of each row is always -1, so that:

$$\sum_{i=1}^{L} T_{i,k} = -1, \quad (48)$$

Considering that $\psi_i(t+(i-1)\tau)=\psi_1(t)$ for all $i\in[1,L]$, each row of the left-side term of Eq. 46 actually corresponds to different estimations of $\psi_1(t)$, so that:

$$\begin{aligned} \widehat{\psi}_1(t) &= \psi_1(t) + \tfrac{2}{L+1}\sum_{k=1}^{L} T_{1,k}e_k(t), \\ &\vdots \\ \widehat{\psi}_1\left(t+(i-1)\,\tau\right) &= \psi_1(t) + \tfrac{2}{L+1}\sum_{k=1}^{L} T_{i,k}e_k\left(t+(i-1)\,\tau\right), \\ &\vdots \\ \widehat{\psi}_1\left(t+(L-1)\,\tau\right) &= \psi_1(t) + \tfrac{2}{L+1}\sum_{k=1}^{L} T_{L,k}e_1\left(t+(L-1)\,\tau\right) \end{aligned} \quad (49)$$

Thus, all these traces can then be averaged according to:

$$\frac{1}{L}\sum_{i=1}^{L}\widehat{\psi}_1\left(t+(i-1)\,\tau\right) = \psi_1(t) + \frac{2}{L\,(L+1)}\sum_{i=1}^{L}\sum_{k=1}^{L} T_{i,k}e_k\left(t+(i-1)\,\tau\right), \quad (50)$$

Hence, the noise variance of the final decoded trace can be estimated as (Lee et al. 2004):

$$\sigma_{\text{decoded}}^2 = E\left\{\left(\frac{1}{L}\sum_{i=1}^{L}\widehat{\psi}_1\left(t+(i-1)\tau\right)-\psi_1(t)\right)^2\right\}$$
$$= \frac{4}{L^2(L+1)^2}E\left\{\left(\sum_{i=1}^{L}\sum_{k=1}^{L}T_{i,k}e_k\left(t+(i-1)\tau\right)\right)^2\right\} \tag{51}$$
$$= \frac{4}{(L+1)^2}\left\{\sigma_{\text{trace}}^2 - \frac{1}{L^2}\sum_{i=1}^{L}(L-i)\,R_N\left(i\tau\right)\right\},$$

where $E\{x^2\}$ represents the variance of x. In this case the noise in the receiver, affecting the coded traces, has been considered to be an independent identically distributed (i.i.d.) and a wide-sense stationary (w.s.s.) random process with zero mean and variance σ_{trace}^2, leading to:

$$\begin{aligned} &E\left\{e_i\left(t+\zeta\right)\right\} = 0,\\ &E\left\{e_i^2\left(t+\zeta\right)\right\} = \sigma_{\text{trace}}^2,\\ &E\left\{e_i(t)e_j\left(t+\zeta\right)\right\} = 0, \qquad \forall i \neq j,\\ &E\left\{e_i(t)e_i\left(t+\zeta\right)\right\} = R_i\left(\zeta\right) = R_N\left(\zeta\right), \qquad \forall i = 1,2,\ldots,L, \end{aligned} \tag{52}$$

Note that in the case of an ideal receiver (with infinite bandwidth), $R_N(\zeta) = 0$ for any $\zeta \neq 0$, so that Eq. 51 can be simplified to (Lee et al. 2004):

$$\sigma_{\text{decoded}}^2 = E\left\{\left(\frac{1}{L}\sum_{i=1}^{L}\widehat{\psi}_1\left(t+(i-1)\tau\right)-\psi_1(t)\right)^2\right\} = \frac{4\sigma_{\text{trace}}^2}{(L+1)^2}, \tag{53}$$

To perform a fair comparison of the SNR enhancement provided by Simplex codes, it should be considered that, during the time required to perform the L coded Raman-OTDR measurements (one for each Simplex codeword), the standard RDTS system can average L single-pulse OTDR traces. This averaging reduces the noise in a factor of $\sqrt{L}$, so that the SNR enhancement provided by Simplex coding (i.e., its coding gain) can be expressed as (Jones 1993; Lee et al. 2004):

$$\text{Gain}_{\text{Simplex}} = \frac{\text{SNR}_{\text{Simplex}}}{\text{SNR}_{\text{Pulsed}}} = \frac{\sqrt{\sigma_{\text{trace}}^2/L}}{\sqrt{4\sigma_{\text{trace}}^2/(L+1)^2}} = \frac{L+1}{2\sqrt{L}}. \tag{54}$$

Figure 9 compares the coding gain of Simplex coding (reported in Eq. 54) with the one provided by Golay codes (given by Eq. 33). The figure points out that Simplex codes offer a slightly higher SNR enhancement with respect to Golay codes (when considering the same measurement time); however, when long codewords (>32 bit) are used, both techniques exhibit practically the same coding gain.

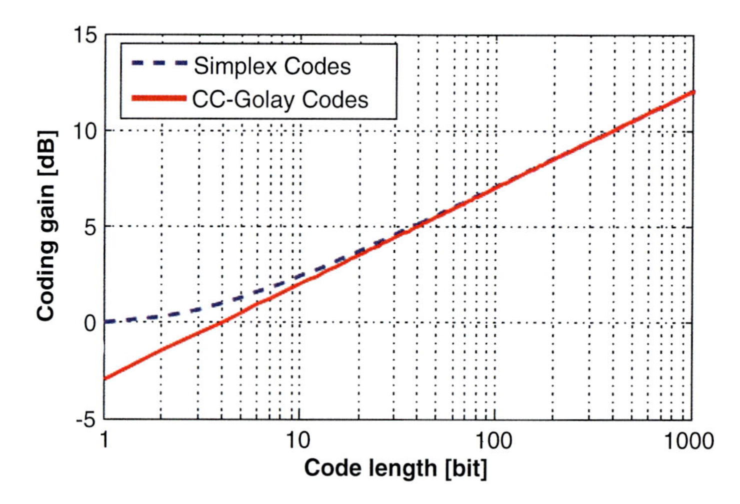

Fig. 9 Comparison of the coding gain provided by both Simplex codes and complementary-correlation Golay codes

As an example of the pulse coding technique applied to RDTS, a high-performance scheme based on Simplex codes is here described (see Fig. 10). The implementation in this case is based on 255-bit Simplex codes, direct detection, and the additional use of lumped Raman amplification to further extend the sensing range.

It must be mentioned that the use of a typical erbium-doped fiber amplifier (EDFA) is not suitable for boosting the optical power of a coded pulse sequence. This is because the amplification of a burst of pulses results in EDFA gain saturation, which leads to an unequal amplification of the pulses in the sequence (Bolognini et al. 2007). Such distorted pulse sequences do not satisfy the power uniformity required for linear pulse coding techniques, since they can induce severe distortions in the decoded time-domain traces. In this case, a co-pumped lamped Raman amplifier (LRA) is used due to its fast response time with minimum gain saturation issues (Bolognini et al. 2007).

A Fabry-Perot laser diode centered at 1550 nm and 80 mW average power has been used in this system. The continuous-wave light from the laser has been intensity modulated by a computer-controlled OTDR board containing a digital signal processor able to generate pulse patterns of Simplex coding with 100 ns single-bit pulse width. The input pulses are coupled into the sensing fiber using an optical circulator, and the backscattered Raman anti-Stokes and Rayleigh components are filtered and coupled into the receiver. The receiver stage is composed of a high-sensitivity InGaAs avalanche photodiode (APD) followed by a 3 MHz high-gain transimpedance amplifier (TIA). This reduced electrical bandwidth limits the spatial resolution to approximately 17 m. In addition, lumped Raman amplification is used to produce undistorted high-power coded pulses and allow for additional sensing

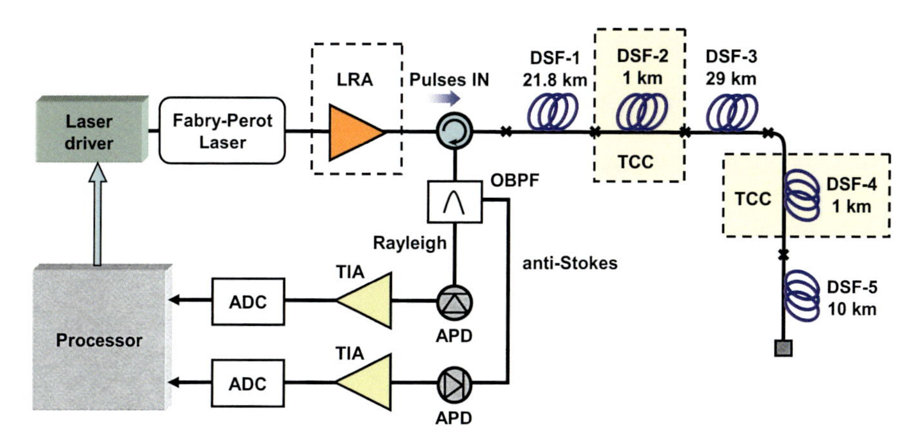

Fig. 10 Experimental setup implementing RDTS with Simplex-coded OTDR (Replotted from Bolognini et al. 2007)

distance enhancement. Five spools of single-mode dispersion-shifted fibers have been used in this case to compose a total sensing range of 62.8 km, as shown in Fig. 10.

The temperature resolutions of the conventional Raman-OTDR, the 255-bit Simplex-coded Raman-OTDR, and 255-bit Simplex-coded Raman-OTDR combined with lumped Raman amplification are compared in Fig. 11. For instance, considering a target temperature resolution of 5 K, results point out that the maximum sensing range of the conventional Raman-OTDR is limited to ~15 km, while 255-bit Simplex-coded Raman-OTDR allows for about 30 km measurement range (i.e., 15 km increase). On the other hand, the use of discrete Raman amplification in combination to the coded Raman-OTDR system allows for a measurement range of ~40 km (i.e., ~10 km further increase). These experimental results are consistent with the theoretical coding gain expected from Eq. 54, predicting a 9 dB coding gain and 15 km one-way sensing distance enhancement. Also the net gain provided by the lumped Raman amplifier (6 dB) accounts for the 10 km sensing range enhancement.

Cyclic Simplex Codes

The previously described coding techniques based on Simplex or Golay codes exploit a set of different sequences (i.e., codewords) of short (about 10 ns for a meter-scale spatial resolution) non-return-to-zero optical pulses to increase the energy launched into the sensing fiber. However, the required high repetition rate of the laser pulses (~100 MHz for a 10 ns pulse duration) is not achievable in all kinds of RDTS implementations. For instance, high-performance RDTS systems operating over multimode fibers can make use of very high-power optical pulses (reaching typically a few tens of W), and in many cases high-power pulsed lasers are used. Most of those lasers, such as rare-earth-doped fiber or passive

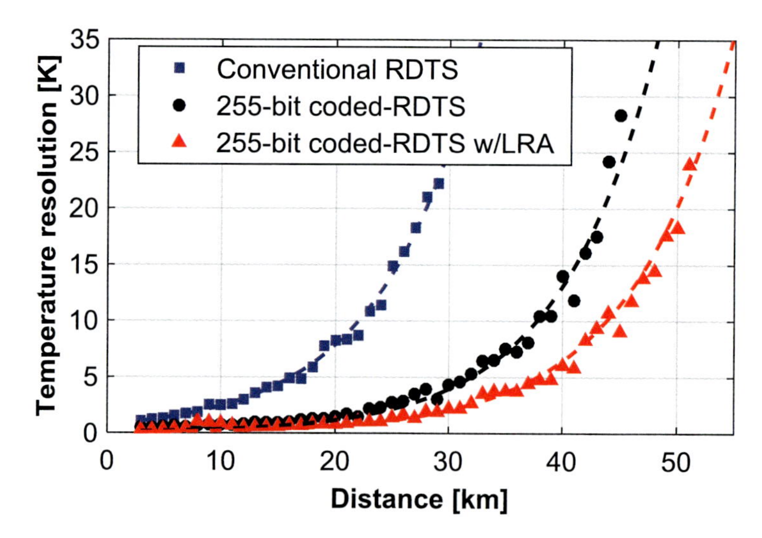

Fig. 11 Temperature resolution versus distance (exponential fit shown in *dashed lines*) using conventional Raman-OTDR (*squares*), coded Raman-OTDR (*circles*), and coded Raman-OTDR combined with lumped Raman amplification (*triangles*) (Replotted using data from Bolognini et al. 2007)

Q-switched ones, can effectively operate at a maximum repetition rate limited by a few hundred kHz, making the use of the classical pulse coding implementation schemes impossible. Although coding can be implemented using semiconductor lasers and an external intensity modulation, the maximum peak power is however highly limited in this case. Indeed, optical amplification cannot be easily used to boost the power of coded pulse sequences due to intensity distortions induced by the transient behavior of the optical amplifier, especially in Erbium-doped fiber amplifiers.

In order to make pulse coding compatible with high-power pulsed laser technologies, cyclic coding schemes can be used for interrogation (Baronti et al. 2010; Soto et al. 2011a, b). In this case, the optical fiber is interrogated with a multi-pulse pattern having a slow repetition rate and a repetition period of the entire sequence equal to the fiber round-trip time. This way the code sequence turns out to be longitudinally distributed along the entire sensing fiber.

The description of the code is as follows (Baronti et al. 2010). First, let us consider an L-bit binary pattern $P = \{p_0, p_1, \ldots, p_{L-1}\}$, with $p_i = 1, 0$ for $i \in [0, L-1]$. Suppose that the acquisition of the backscattered trace is divided into L consecutive intervals and that an optical pulse is launched into the sensing fiber at the beginning of each interval if the relevant pattern bit is equal to 1. Suppose also that this triggering scheme is periodically repeated. According to the pattern P, backscattered traces resulting from delayed pulses overlap along the fiber, as shown, for example, in Fig. 12 for $L = 7$ and $P = \{0, 1, 1, 1, 0, 1, 0\}$.

Given the system sampling period T_S, the length of the backscattered trace can be expressed as the number of sampled points N_S. Then, let H be the number of samples

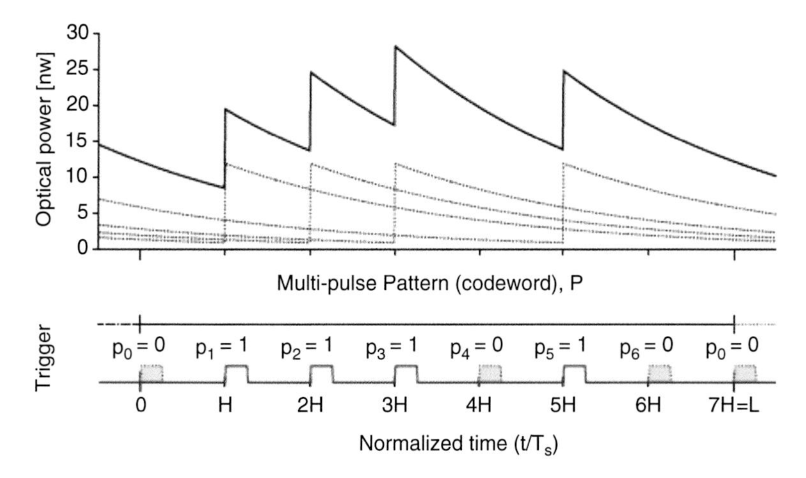

Fig. 12 Trace overlap in the case of distributed optical coding technique with $M = 7$ and $P = \{0, 1, 1, 1, 0, 1, 0\}$

within each of the L intervals, so that $N_S = \mathrm{HL}$. The array of the acquired samples can be denoted as $y[i + \mathrm{jH}]$, with the index $i \in [0, H - 1]$. In addition, the array of the single-pulse response samples to be recovered can be denoted as $x[i + \mathrm{jH}]$. In this way, j scans intervals, whereas i scans the samples within the j-th interval. It can be observed from Fig. 12 that each sampled data point $y[i + \mathrm{jH}]$ has the contribution of the backscattered traces originated by the pulses launched in the j-th interval and in the $L - 1$ previous ones. Hence, the following relationship holds for y and x samples:

$$y\,[i + \mathrm{jH}] = \sum_{k=0}^{L-1} p_{|j-k|_L}\, x\,[i + \mathrm{kH}]\,. \tag{55}$$

In particular, for a given i, Eq. 55 turns into a linear system of L equations (one for each j value) with a cyclic coefficient matrix (Song and Golomb 1994). In fact, its first row ($j = 0$) can be denoted as $\{p_0, p_1, p_2 \ldots, p_{L-2}, p_{L-1}\}$, while any other row is the right-shifted copy of the previous one.

From a noise point of view, y and x samples can be reasonably considered as uncorrelated random variables. The noise affecting the recovered x samples is determined by the linear system itself, i.e., by the inverse of the system coefficient matrix. Hence, P should be chosen so that a good noise reduction can be achieved. For this, as the coefficient matrix is cyclic by construction, it is sufficient to use a matrix derived from the cyclic code theory and choose the pattern P accordingly.

Considering cyclic Simplex codes, a cyclic binary coefficient matrix of order $L = 4n - 1$ (with $n = 1, 2, \ldots$) can be built using the method reported in (Song and Golomb 1994). Being σ_y and σ_x the standard deviation of any y and x sample, respectively, the achieved SNR enhancement, i.e., the coding gain $\mathrm{Gain}_{\mathrm{Simplex}}$, is given by the following relation:

$$\text{Gain}_{\text{Simplex}} = \frac{\sigma_y}{\sigma_x} = \frac{L+1}{2\sqrt{L}}. \tag{56}$$

This is similar to the gain reported in Eq. 54 for standard Simplex codes.

The properties of cyclic Simplex coding, based on quasi-periodic bit sequences, are such that real-time decoding can be achieved in less than the round-trip fiber transit time. This feature can be a key factor to obtain fast temperature measurements, avoiding the use of long averaging and reducing the processing overhead time.

An example of RDTS system exploiting cyclic pulse coding in single-mode fiber has been implemented to demonstrate long-distance temperature sensing with meter-scale spatial resolution (Soto et al. 2011a). Figure 13 shows the scheme of the experimental setup. The light source is a high-power rare-earth-doped fiber laser generating pulses of 10 ns at 1550 nm, with a maximum repetition rate of 250 kHz. The cyclic Simplex codeword is generated using an acousto-optic modulator acting as a chopper to cancel out pulses when a bit "0" is required and letting pulses through when a bit "1" is expected. The backscattered light from 26 km single-mode fiber is properly filtered to separate the Stokes and anti-Stokes components, which are then detected by two highly sensitive APDs. The repetition rate of the laser has been set to 230 kHz, allowing 71 bits to be allocated along the whole 26-km-long sensing fiber.

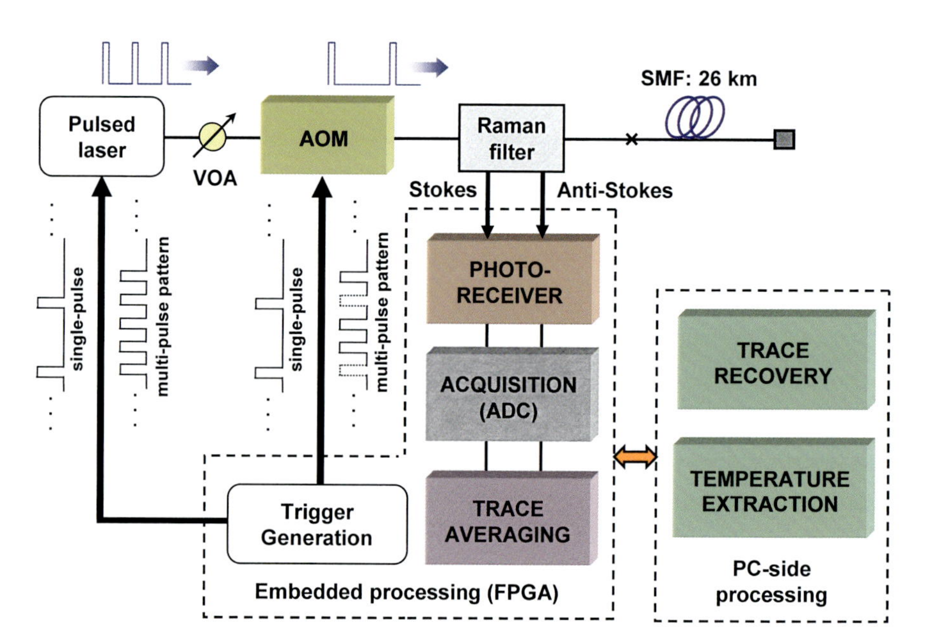

Fig. 13 Experimental setup for RDTS system using cyclic pulse coding on single-mode fiber (Soto et al. 2011a)

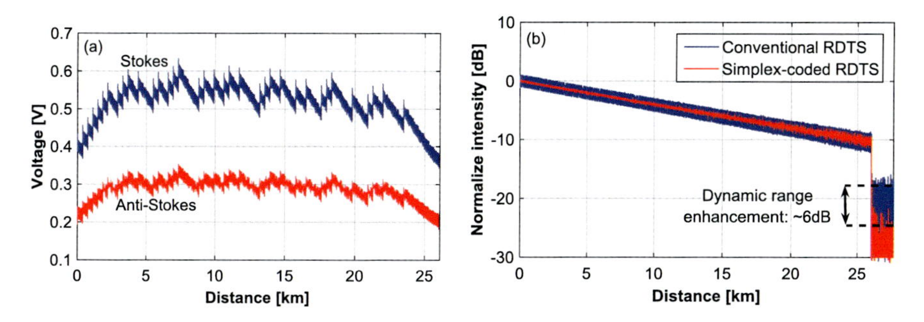

Fig. 14 (**a**) 71-bit cyclic-Simplex-coded Stokes and anti-Stokes traces. (**b**) Anti-Stokes traces obtained with conventional RDTS and cyclic-coded RTDS (*decoded trace*) (Soto et al. 2011a)

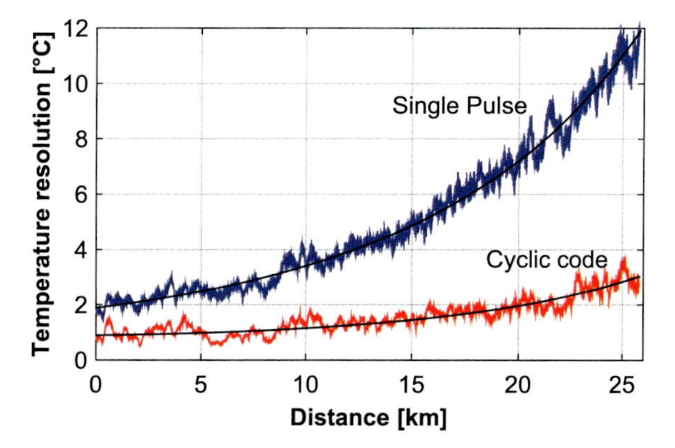

Fig. 15 Temperature resolution versus distance, for single-pulse RDTS and cyclic-Simplex-coded RDTS (Soto et al. 2011a)

Figure 14a shows the coded Raman Stokes and anti-Stokes traces, clearly illustrating the distributed nature of the cyclic coding method. Figure 14b compares the anti-Stokes traces obtained with conventional single-pulse RDTS and after decoding the cyclic-coded RTDS trace. The SNR enhancement is obtained by calculating the standard deviation of both normalized traces, leading to an experimental coding gain of 6 dB, which is in good agreement with the expected theoretical value of 6.3 dB.

Finally Fig. 15 compares the temperature resolution of the single-pulse RDTS and the cyclic-coded RDTS, both obtained with a measurement time of 30 s. The lower SNR of the standard RDTS has a clear impact on the final temperature resolution of the sensor (~12 °C at 26 km distance). The use of cyclic coding improves this resolution down to 3 °C at the same distance. A spatial resolution of ~1.0 m for the two measurements has been verified by heating up to 55 °C a few meters of fiber at the end of the 26 km spool and observing the 10–90% response distance to the temperature step.

Results demonstrate that cyclic pulse coding enables the use of single-mode fibers and high-power pulsed lasers for long-range temperature sensing. The use of cyclic coding in RDTS systems has two main advantages: (i) avoids the use of multimode fibers, and therefore overcomes the spatial resolution degradation induced by modal dispersion, and (ii) improves the poor SNR of standard RDTS operating over single-mode fibers. This way, long-distance sensing with meter or submeter spatial resolutions can be effectively achieved over tens of km of single-mode fiber, obtaining a temperature resolution of only a few degrees within a measurement time of a few minutes.

Image Processing Techniques

Another approach to reduce the impact of noise present in Raman-OTDR traces and to improve the performance of RDTS systems is the use of digital signal processing methods. In principle, a linear low-pass digital filter can be applied to the data points to reduce noise when the electrical bandwidth of the measurements is larger than the required bandwidth defined by the useful signal. In this case, the noise reduction is however simply defined by the ratio between signal bandwidth and electrical bandwidth, thus being highly limited in a properly designed system. This approach is constrained exactly in the same way as it is the design of the electrical bandwidth of the RDTS system, thus being affected by the well-known trade-off between spatial resolution (bandwidth) and measurement noise. In order to overcome this trade-off, smarter signal processing techniques have to be used. An example is the use of wavelet denoising, which has been employed to remove noise from Raman-OTDR traces (Saxena et al. 2015). An important SNR improvement has been demonstrated using this technique. Recently, a two-dimensional (2D) approach based on image processing has been proposed to eliminate noise and improve the SNR of RDTS systems, resulting in a much enhanced sensor performance when compared to the use of traditional one-dimensional (1D) approaches (Soto et al. 2016a, b).

This 2D denoising method consists in creating a two-dimensional data structure (a matrix M) containing the data points acquired by a RDTS system and then processing this matrix as an image corrupted by noise (Soto et al. 2016a, b). In contrast to other distributed fiber sensors (e.g., Brillouin distributed sensors), where the measured data is by essence two-dimensional (Soto et al. 2015, 2016a), RDTS systems measure one-dimensional (1D) data. Measured traces are stored in two unidimensional arrays, one array containing the amplitude of the anti-Stokes signal and another array containing the amplitude of either the Raman Stokes or Rayleigh signal. Therefore, in order to apply image processing methods, the 2D matrix used for processing has to be generated from the 1D Raman-OTDR traces. In particular two 2D data structures, matrices M_{aS} and M_S (one for the anti-Stokes and another for the Stokes – or Rayleigh – component), are generated in the distance-time (z,T_i) domain by stacking consecutive 1D Raman-OTDR traces obtained from sequential measurements, T_i being the moment of the acquisition of the i-th trace (Soto et al. 2016a, b). This way, each row of the 2D matrices represents an independent measurement of the Raman anti-Stokes or Stokes trace as a function of distance.

It should be noted that the principle of Raman distributed sensing is to measure quasi-static temperature changes, in which the measurand (i.e., the temperature) slowly changes when compared to the acquisition time. Therefore consecutive Raman-OTDR traces are typically expected to be highly correlated, containing a high level of redundancy in the time and distance domains. Image processing techniques can smartly exploit this redundancy to eliminate large amount of noise while keeping high levels of details (i.e., securing high spatial and temporal resolutions).

The concept behind this approach considers that the intensity of the Raman backscattered light measured at position z and acquisition time T_i can be directly mapped into the intensity of a monochromatic image. This way, data stored in the matrices M_{aS} and M_S, and containing the consecutive Raman-OTDR traces, can be represented by a function $f(x, y)$, where x and y represent the spatial coordinates on the obtained monochromatic image. Then, processing the 2D data in the matrices M_{aS} and M_S by image denoising techniques can remove the white-additive noise present in the measured Raman-OTDR traces.

Among several image processing techniques, two categories of denoising methods and approaches can be identified that can be used with RDTS data (Soto et al. 2015, 2016a): (i) *pixel-wise methods*, in which the processing is applied directly on the pixel values of the image (i.e., directly on the RDTS data), and (ii) *transform-based techniques*, in which denoising is performed in a transform domain. Independently of the approach, all methods share the same feature: large amount of noise can be removed from the data while preserving the detail information.

Pixel-wise algorithms are usually based on the use of sliding 2D windows (sometimes also called neighborhoods), corresponding to the set of all pixels at (x, y) that surround a certain pixel (x', y') to be processed. Although a large window leads to high noise removal, this could also eliminate image details. One of the most basic methods of noise removal is known as Gaussian filtering, which is based on replacing the value of $f(x', y')$ at the center of a window by a weighted average of pixel values $f(x, y)$ inside the window, where the weights are given by a two-dimensional Gaussian function centered at (x', y'). Gaussian filters are 2D linear filters, and therefore, any increase in the width of the Gaussian function could lead to the unwanted removal of image details. Their performance and limitations are practically the same as one-dimensional digital low-pass filters. A more sophisticated version of weighted averages is known as nonlocal means (NLM) algorithm (Buades et al. 2005), which has demonstrated to be very efficient for noise reduction in RDTS systems and also in other distributed fiber sensors. Compared to 2D Gaussian filters, the nonlocal means algorithm averages data inside a window centered at (x', y'), using a weighting factor for each pixel at (x, y) depending on the Euclidean distance between defined small neighborhoods around (x', y') and (x, y). An exponential decaying factor, depending on the noise standard deviation, is also used. The NLM method can be considered as an improvement with respect to Gaussian filters, especially regarding the preservation of edges, texture, and fine structures.

The second approach for image denoising, compatible for processing RDTS data, is based on transform-based methods, which use the frequency domain to separate the components of an image associated with high-frequency noise from the components containing relevant information. An example of this category is the use of 2D discrete cosine transform (DCT) (Soto et al. 2015; Buades et al. 2005), which converts the values of each sliding window into the frequency domain, then discards the components that are smaller than a certain threshold level, and finally converts the result back to the spatial domain. Another powerful algorithm for image denoising is the wavelet denoising, which makes use of 2D discrete wavelet transform (DWT) (Buades et al. 2005; Mallat 1989). This method decomposes an image into subversions containing different levels of detail and applies to each of them a certain threshold method to eliminate noise. Several parameters, such as the wavelet basis function, the threshold level, and the number of decomposition levels, must be adjusted in a 2D DWT; all these parameters have a direct impact on the efficiency of the noise removal.

In order to verify the effectiveness of image denoising on RDTS systems, the experimental setup described in Fig. 16 has been implemented. In this case, the light generated by a distributed-feedback laser operating at 1552 nm is intensity modulated by an electro-optic modulator (EOM) to generate optical pulses of ~10 ns. An erbium-doped fiber amplifier (EDFA) is then used to boost the peak pulse power up to about 4 W (at the sensing fiber input). Pulses are launched into a 9-km-long 50/125 μm graded-index multimode fiber through a wavelength-selective filter (WSF), which is also used to separate the spontaneous Stokes and anti-Stokes Raman components backscattered from the sensing fiber into two branches. The two Raman components are simultaneously measured using two parallel 50 MHz avalanche photodetectors, followed by a data acquisition system (DAQ) connected to a computer. As a result of modal dispersion, a spatial resolution of 2 m has been verified at 9 km distance.

In this case a sampling interval equivalent to 0.5 m/pt is used in the acquisition, leading to traces with 18,000 longitudinal points covering the entire 9-km-long sensing range. Each pair of anti-Stokes and Stokes traces is acquired every 35 s. Two

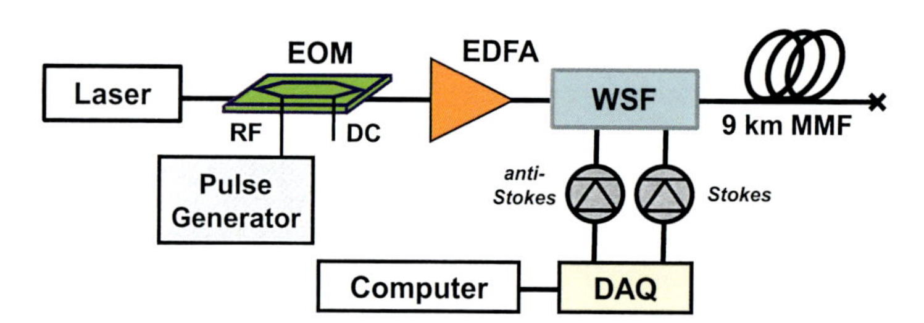

Fig. 16 Conventional Raman distributed fiber sensor, using a 9-km-long sensing fiber and 2 m spatial resolution (Soto et al. 2016b)

2D data structures conformed by 21 consecutive traces are used for image denoising. This way, two 2D matrices of 21 × 18,000 points are generated and updated for each new pair of Raman traces that is measured.

Among the different possible image denoising techniques suitable for RDTS systems, in this case the concept has been demonstrated using the method of *nonlocal means*. This image denoising algorithm removes noise from the RDTS data by exploiting the high level of redundancy existing in the measurements. Contrarily to traditional linear filtering techniques, this algorithm applies a nonlocal weighted average of pixels (data points) inside the image, where the weighting factors are defined based on the level of similarity between different sections of the image. This similarity is evaluated by calculating the Euclidean distance between small 2D windows, called *similarity windows*. The maximum size of the similarity window is determined by the size of the details to be preserved. This means that the similarity window has to be smaller than the number of data points defined by the spatial and temporal resolutions of the sensor. Therefore, a similarity window of 3 × 3 is used to secure a spatial resolution of 2 m (corresponding to four longitudinal points). A window size larger than the dimension defined by the target resolutions leads to loss of spatial and/or temporal resolving capabilities in the sensor. The algorithm actually searches for repetitive 2D patterns over the entire RDTS data, by comparing all possible combinations of similarity windows. This typically leads to large processing times. In order to reduce this time, the search of similarities is limited only to a partial section of the data, defining searching windows of 21 × 21 points centered into the data points being processed. The third parameter of the NLM algorithm is the smoothing parameter h, which is defined to be ten times the noise standard deviation.

Figure 17a shows the evaluation of the SNR as a function of distance for the raw data (blue curve) and denoised data (red curve), when using 2^{12} averaged traces. Results demonstrate that the SNR of 26.2 dB obtained at ~9 km distance with the raw data can be remarkably improved up to 39.8 dB after NLM denoising, representing an SNR enhancement of 13.6 dB. As a consequence, the temperature resolution of 0.5 K, obtained from the raw data at the fiber end, can be improved down to 22 mK by NLM, as shown in Fig. 17b. The processing time of each matrix is about 1 s using a conventional computer with a 3.5 GHz processor and 8 GB RAM.

To verify that the image denoising does not distort the temporal and spatial evolution of the measurements and no relevant information is eliminated, the longitudinal evolution of the measured temperature has been analyzed within a 10 m hotspot, while its time evolution has been analyzed over 50 min. Figure 18a shows that the applied 2D image denoising has no detrimental impact on the hotspot measurement, securing a spatial resolution of 2 m and a correct temperature measurement at the end of the sensing range. On the other hand, Fig. 18b shows the time evolution of the hotspot, demonstrating that no temporal distortion or delay is induced by the image processing method.

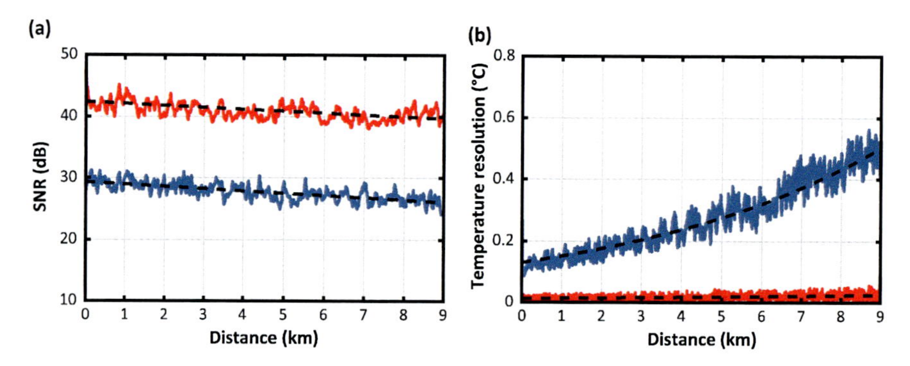

Fig. 17 Impact of NLM image denoising on the performance of a RDTS system. (**a**) Evaluation of the SNR versus distance for raw data (*blue curve*) and for denoised data (*red curve*). (**b**) Temperature resolution versus distance for raw data (*blue curve*) and for denoised data (*red curve*) (Soto et al. 2016a)

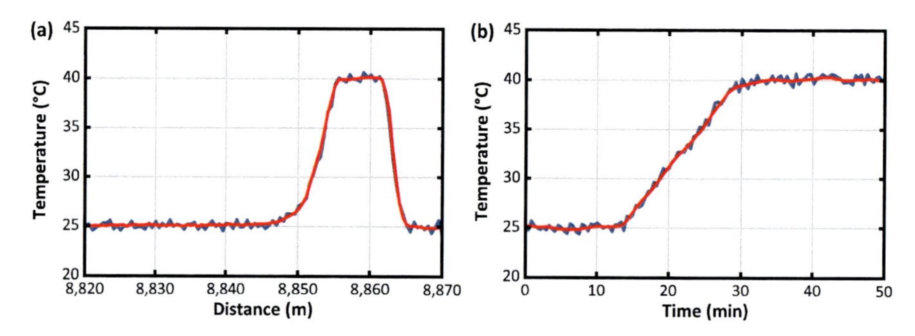

Fig. 18 Spatial and temporal responses of the RDTS system for the raw data (*blue curve*) and for the NLM-denoised data (*red curve*). (**a**) 10-m-long hotspot. (**b**) Time evolution of the hotspot within 50 min (each measurement is obtained every 35 s) (Soto et al. 2016a)

Industrial Applications of Raman Distributed Temperature Sensors

Raman distributed temperature sensors are today widely employed in many industrial fields. These include pipelines and power cables monitoring, fire detection in tunnels, wells and reservoirs monitoring, geological and hydraulic engineering, and, in general, applications where a continuous temperature profile along an extended physical system or a large infrastructure can be effectively related to its health conditions or give information about the external environment. In addition, since the sensing element is simply given by a dielectric silica optical fiber, which is interrogated through light propagation, RDTS systems provide significant competitive advantages compared to conventional sensors, especially under harsh environmental conditions. In particular, they can be safely and conveniently installed in all environments that are not compatible with standard sensors exploiting conductors, like copper wires, or electrical measurements. Examples of such environments are given by temperature sensing in proximity of high voltages, such as in power plants or in distribution stations; along pipelines, plants, or tanks carrying explosive gases, liquids, or solids; in the presence of magnetic fields; and, in general, in hazardous environments, or whenever issues associated to electromagnetic interference and nuclear radiation are encountered.

Among the several industrial applications of RDTS systems, a few examples have been selected in this section among the most attractive in terms of market opportunities, namely, leakage detection (such as in oil, gas, steam, and hot water pipelines) and fire detection (such as in tunnels and industrial plants). In the following, examples of three different RDTS field trials are described. These have been carried out in independent collaborations: (i) with SAIPEM, a large engineering and construction company operating in the oil and gas market, (ii) with the Italian railway network operator (RFI), and (iii) with the European Organization for Nuclear Research (CERN). The presented experimental results clearly demonstrate the effectiveness of RDTS as a practical and reliable monitoring solution, even under extremely harsh conditions such as the mixed-field radiation environment at CERN.

Leakage Detection in Pipelines

In this section a RDTS field validation for oil leakage detection in soil is reported (Signorini et al. 2014). The capability of the distributed Raman sensor in detecting and locating, with high accuracy and spatial resolution, drop leakages in soil is demonstrated through a hot water leakage simulation in a field trial at the SAIPEM facilities in Italy (ENI group).

Although several industrial applications already benefit from distributed fiber optic sensing techniques, there are still technological improvements and in-field issues to be solved in order to make the technology effective and reliable in a wide range of applications. Considering in particular the oil and gas market sector, in which a strong growth of RDTS in field applications has been observed, specific requirements should be met in order to provide an effective solution competitive with respect to other technologies. In particular the use of RDTS systems for leakage detection should guarantee the capability to detect and locate drop leakages in soil over extremely long distances with high spatial resolution.

This field trial is based on a Raman DTS system operating at 1064 nm on graded-index multimode fibers over a maximum sensing distance of 5 km and with meter-scale spatial resolution. Figure 19 reports a schematic structure of the field trial performed in July 2013 at the SAIPEM facilities in Fano, Italy, including details of the wells in which oil leakage is simulated, the RDTS reading unit within the control room, and the taps used to control the hot water leakage simulation in four

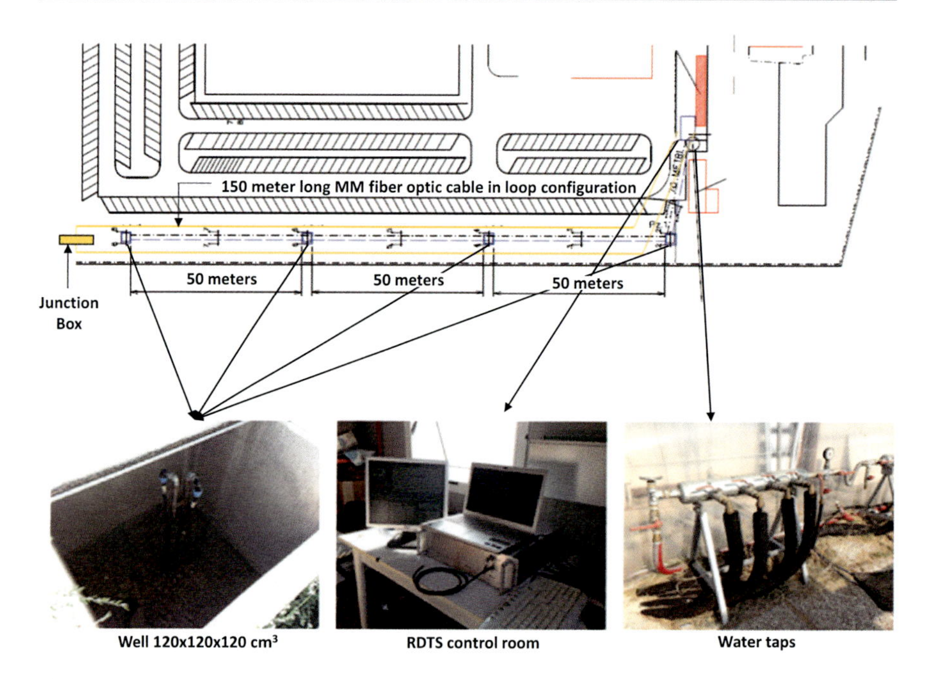

Fig. 19 Schematic diagram of the field test setup (Signorini et al. 2014)

different locations. Note that a cable including multimode graded-index fibers is installed underground within a plastic pipe.

The used RDTS system described in Fig. 19 only requires in-field calibration, in either single-end or loop configuration. This provides reliable results even when using previously installed cables and avoids unpractical laboratory pre-calibration procedures. The capability of the distributed Raman sensor in detecting and locating, with high accuracy and spatial resolution, drop leakages in soil is demonstrated by simulating oil spilling in the ground through water leakages in each of the four wells. Leakages are equally spaced along the pipeline, being separated 50 m from each other and individually controlled by water taps shown in Fig. 19.

Figure 20 shows the temperature profile at different times after a leakage of hot water at 40 °C is induced in the first well through the water tap control. The total length of the cable is 300 m, and only ~170 m are laid underground (the cable includes two graded-index multimode fibers spliced together to allow loop configuration measurements).

Note that the field trial has been carried out on a hot summer day (Friday, July 26, 2013) as evident from Fig. 20, clearly showing 110 m of fiber cable left on the bobbin after the fourth well reaches the high outside temperature of ~30 °C. The portion of cable in the air route just outside the room control is heated up to ~38 °C due to direct sunlight exposure. Figure 20 also clearly points out the effectiveness of the RTDS system in detecting the temperature increase induced by

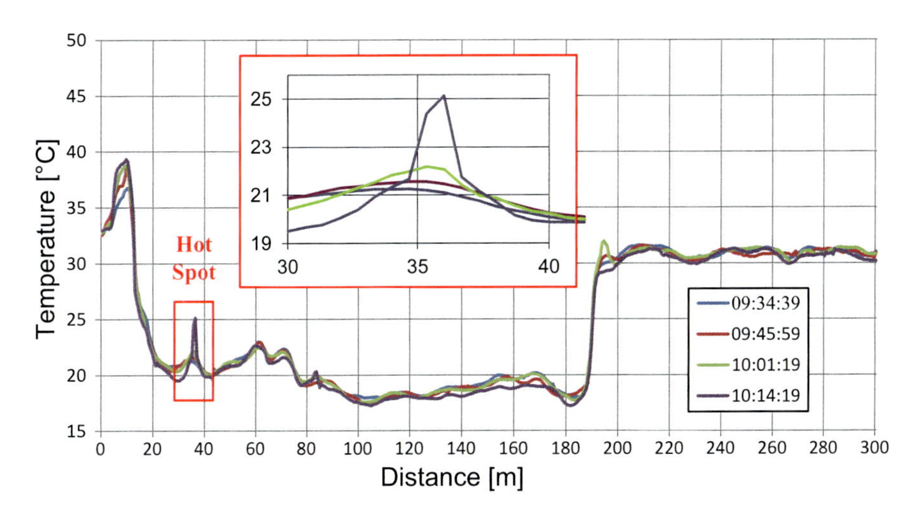

Fig. 20 Hotspot corresponding to a leakage detection within well n.1 positioned at ~35 m from the DTS control room (Signorini et al. 2014)

the hot water leakage within the well. Note that the measured RDTS temperature profile, experimentally verified during the field trial using a thermocouple, was slowly increasing in time during monitoring and did not reach the hot water temperature of ~40 °C. This is due to the fact that the multicore fiber cable used in the loop configuration has not been buried directly in soil in the SAIPEM facilities but installed inside a plastic pipe positioned underground and passing through the four wells.

This is also clearly illustrated in Fig. 21, where the temperature within the first well is reported versus time after the start of the leakage. A good agreement is observed when comparing the temperature measured with the RDTS system and a thermocouple placed inside the same well. These results clearly demonstrate the potential of the distributed sensing technology to effectively detect and locate leakages along a pipeline through a hot water leakage simulation.

Note that recent developments in Raman DTS technology are going toward the capability to perform temperature measurement over tens of kilometers. This feature can address the challenging market requirements of undersea power cables and pipeline monitoring, in which the use of standard single-mode fibers is required to avoid the spatial resolution degradation induced by intermodal dispersion in multimode fibers. Moreover, due to a trade-off between backscattering coefficient and fiber attenuation, ultra-long sensing distances can only be achieved by operating in the third optical telecommunication window at ~1550 nm. The lower power threshold imposed by stimulated Raman scattering in standard single-mode fibers also requires the use of lower peak power levels, and consequently advanced techniques, as described in Advanced methods for performance improvement in RDTS systems, are needed to reach this challenging goal.

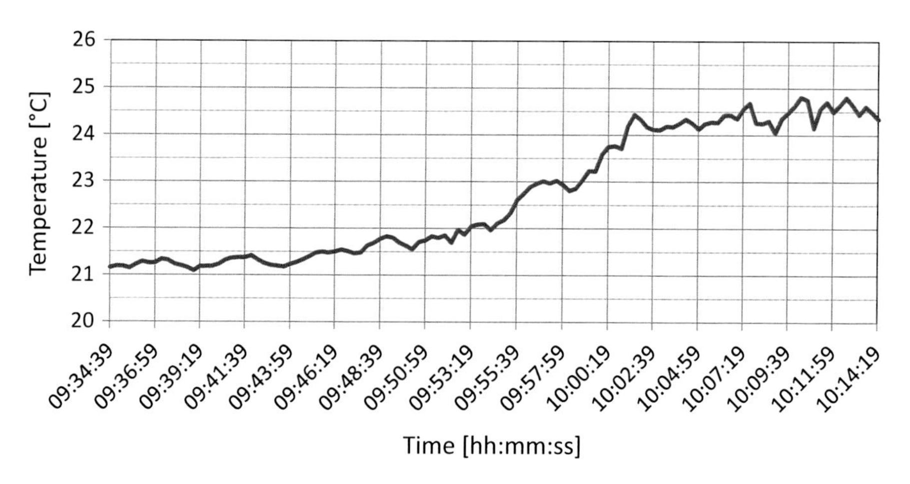

Fig. 21 Temperature versus time during the hot water leakage within well n.1 positioned at ~35 m from the RDTS control room (Signorini et al. 2014)

Experimental results based on cyclic Simplex coding techniques have confirmed the possibility to achieve temperature measurement with meter-scale spatial resolution over more than 50 km distance along a standard single-mode fiber, with a few minutes measurement time and a few degrees temperature resolution (Soto et al. 2011b). This technology can effectively compete with Brillouin distributed fiber sensors in terms of cost and performance for leakage detection and power cable monitoring.

Fire Detection in Railway Tunnels

Safety systems in modern transportation infrastructures require highly reliable and extremely fast linear heat detection systems. Among the several applications requiring such fire detection systems with a continuous measurement capability over long distances, one of the most relevant is fire alarm in railway and highway tunnels. RDTS has been proved to be an ideal cost-effective solution for fire detection in long tunnels, especially in hazardous environments, due to its robustness, reliability, and immunity to electromagnetic interferences. RDTS systems operating in loop configurations and with an internal reference fiber spool are particularly suitable to satisfy all such requirements. The used fiber cable must be specifically designed to be compliant with low-smoke-zero-halogen (LSZH) specifications. The fiber optic fire detection system described in this section has been tested in a field trial with the Italian railway network (RFI) (Signorini et al. 2016). The fire detection system is based on multimode graded-index fibers and has been integrated with a proprietary RFI telecommunication network, in order to provide automatic activation of supervisor and protective systems, based on specific temperature information on the different zones under monitoring.

The system provides the capability to define several zones within and outside the tunnel and to identify suitable alarm criteria with specific threshold values in each of them, allowing one to quickly detect fire without false alarms. In particular key information on the affected area is provided by the maximum temperature values as well as by the spatial and temporal temperature gradients defined in each zone. The selected field trial site was a tunnel on the west coast of Italy, where a specially designed cable including eight multimode graded-index optical fibers (50/125) has been installed. This goes from the RDTS control room to the full tunnel length, passing through an underground route and an air route (exposed to sunlight and with notable temperature excursions within a day-night cycle). The fibers have been properly fusion-spliced at both fiber ends in order to create a multiple round-trip path simulating a very long tunnel. The RDTS system operates in loop configuration at 1550 nm, with a maximum sensing distance up to 25 km (12.5 km in loop). The system performance has been first evaluated in lab conditions, confirming its capability to detect temperature variations over a sensing distance up to 25 km, with a response time down to a few seconds, temperature resolution of less than 1 °C, and 2 m spatial resolution.

Figure 22 reports a typical temperature resolution of the RDTS system working in loop configuration, with different measurement times. The proposed RDTS solution for fire detection is operating in the field using loop configuration and with an internal reference fiber spool for temperature measurement stabilization. As previously theoretically described, the double-end configuration cancels out all loss factors dependent on the fiber position z, leading to a self-calibrated temperature measurement system which is robust to loss variations occurring during the sensor lifetime. The temperature of the additional internal fiber spool is measured by a thermocouple, providing a reference to compensate for variation in the temperature sensitivity and consequent deviation from the actual temperature.

The fiber cable is suitable for fire detection in harsh environments. In particular it has been specially designed to be moisture-, corrosion-, and rodent-resistant and compliant with LSZH specifications. The system is integrated with the RFI proprietary communication network, a key requirement from RFI in order to provide automatic activation of protective systems, based on specific temperature information on the different zones under monitoring. Figure 23 describes the specific field trial cable installation scheme. The cable goes from the control room, where the RDTS reading unit is located, to the tunnel (1.85 km length), passing through an underground route and an air route for a total path of approximately 2 km. The fiber in the air route is directly exposed to sunlight, thus being affected by notable temperature excursions within a day-night cycle.

The eight multimode graded-index fibers (50/125 μm) within the cable have been properly fusion-spliced at both fiber ends, in order to create a multiple round-trip path simulating a 16-km-long tunnel. Figure 23 also shows some pictures of the cable, cable termination, and splice protection at the tunnel exit. Figure 24a reports the distributed temperature profile measured along the multimode fiber

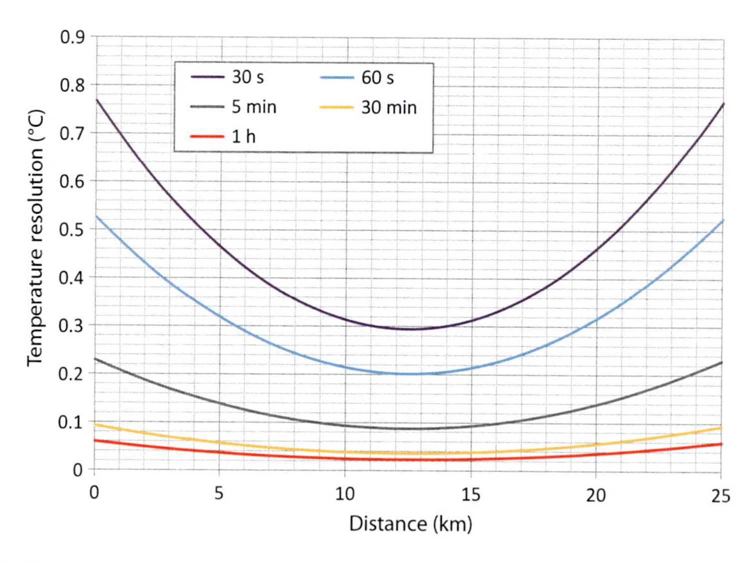

Fig. 22 Temperature resolution versus distance in loop configuration for different measurement times

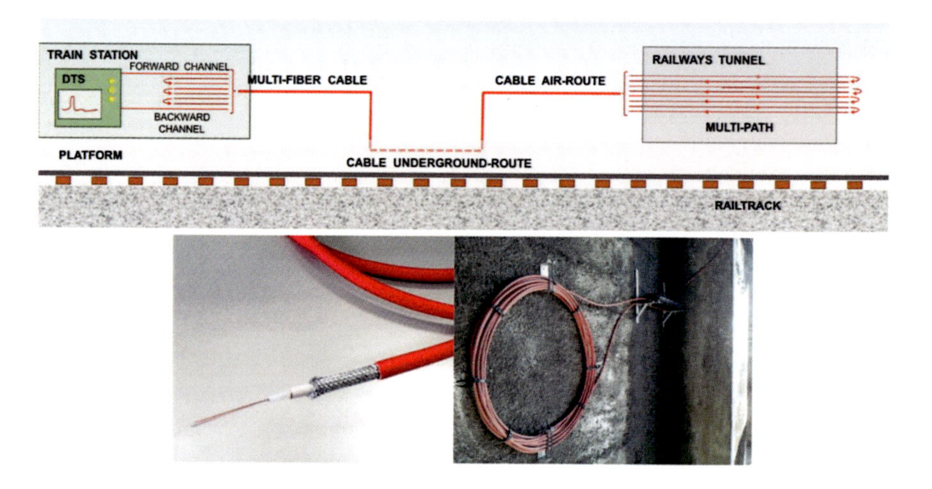

Fig. 23 RFI field trial scheme and details of the cable and installation (Signorini et al. 2016)

cable, showing good periodicity along the multi-round-trip measurement path. The consistency of the measurement with the specific fiber configuration periodicity confirms then the effectiveness and range capabilities of the used RDTS system. Figure 24b reports in detail a single round-trip temperature profile along the tunnel and within the cable route connecting the railway station, where the RDTS is located, to the tunnel. The measurement data confirm the reliability of the RDTS measurement system, showing good symmetry in the forward and backward

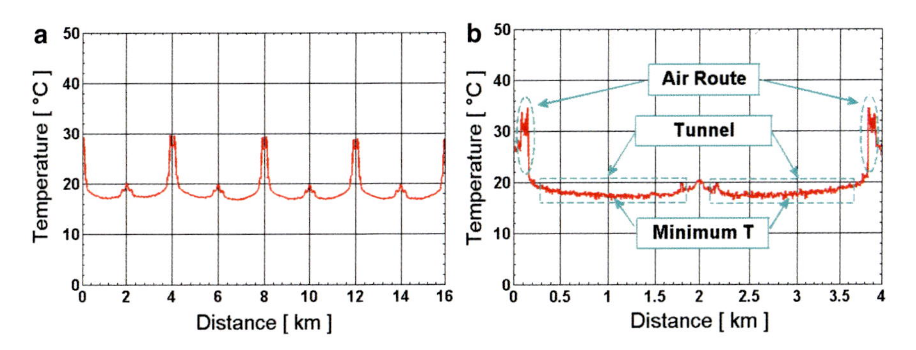

Fig. 24 Temperature profiles (**a**) along the multi-fiber cable and (**b**) for one single round-trip (Signorini et al. 2016)

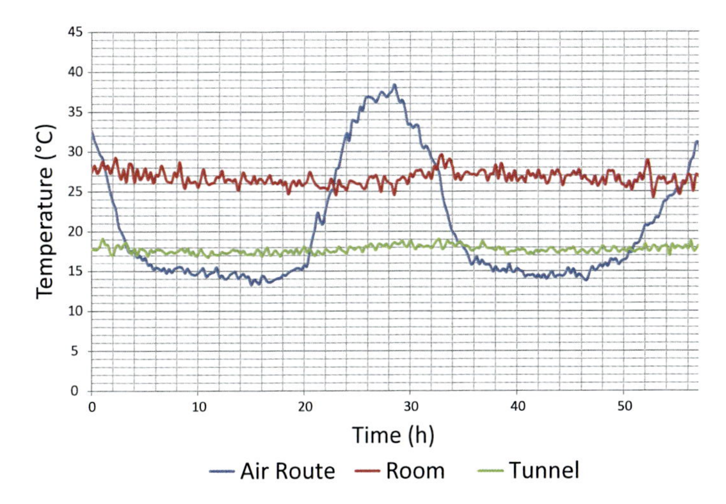

Fig. 25 Temperature monitoring in three locations along the optical fiber (57 h span) (Signorini et al. 2016)

directions. Results also point out the presence of high temperature values in the air route, due to direct sun exposition, and almost constant temperature along the tunnel, being only slightly affected by the external environment near the tunnel entrances.

The temperature variation versus time is reported in Fig. 25, for a time interval of 57 h, in three different locations along the fiber cable, i.e., inside the control room (red line), in the air route section (blue line), and in the middle of the tunnel (green line). Results in Fig. 25 show an interesting behavior: the temperature inside the tunnel is almost constant during day and night at standard operating conditions, while significant temperature variations can be observed along the air route.

Although it was not possible to carry out real fire simulations since the tunnel is under regular train circulation, the field trial has however enabled the measurement of local temperature increase due to the presence of trains occasionally stopping inside the tunnel and induced by the heat generation from electric train engines.

This field trial for fire detection in railway tunnels using RDTS systems clearly points out the potential of the distributed Raman sensing technology for railway infrastructure safety. The RDTS technology is also under investigation for railway track temperature monitoring in order to prevent buckling. The presented results confirm high spatial resolutions, long sensing distance, fast response time, and a clear advantage of distributed sensing with respect to conventional electrical sensors.

Temperature Measurement in High-Radiation Environments

In order to validate the use of RDTS systems in extremely harsh conditions, this last section describes a field trial performed at the CERN high-energy accelerator mixed (CHARM)-field facility, recently developed for testing devices within accelerator representative highly radiative environments (Toccafondo et al. 2015b). This field test has been carried out to evaluate the reliability of a RDTS system in loop configuration over radiation-tolerant multimode optical fibers, when operating on a harsh environment. In this case the fiber is exposed to a mixed-field radiation, including protons, neutrons, photons, and other particles, which can potentially alter the fiber material properties.

The use of RDTS systems in the CERN accelerators complex requires a preliminary validation in a representative radiation environment. The CHARM facility recently constructed at CERN and currently in full operation is schematically shown in Fig. 26. This provides a variety of particle energy spectra which can be representative of several radiation environments, allowing large acceleration factors up to around 10^9. CHARM can be used to test electronics, optics, and photonics equipment to be installed inside particle accelerators and also to test devices and systems to be used for space-, atmospheric-, and ground-level applications. A 24 GeV/c beam is focused on a copper target generating a mixed radiation field composed of protons, neutrons, pions, photons, muons, and other particles. Different desired shielding and target configuration can be remotely controlled according to the needs in terms of particle spectra and intensity.

In order to validate the RDTS performance in such environment, a radiation-tolerant multimode graded-index (50/125 μm) optical fiber has been installed in the facility in such a way that it can be interrogated from both sides from the same rack in the technical room, implementing then a RDTS in loop configuration.

The red lines in Fig. 26 represent the 115-m-long path of the optical fiber inside the facility. In order to expose the fiber to different radiation environments and respective intensities, shielding blocks are inserted along the path which runs at two different heights from the ground, 95 cm and 280 cm (back and forth). The fiber can then reach a great variety of total accumulated doses depending on its position and on its height, with dose rates ranging from a few μGy/s up to a few tens of mGy/s.

In the field trial, a RDTS system operating at 1064 nm on graded-index multimode fibers, over a maximum sensing distance of 5 km and 1 m spatial resolution, has been used in loop configuration. A temperature measurement campaign of

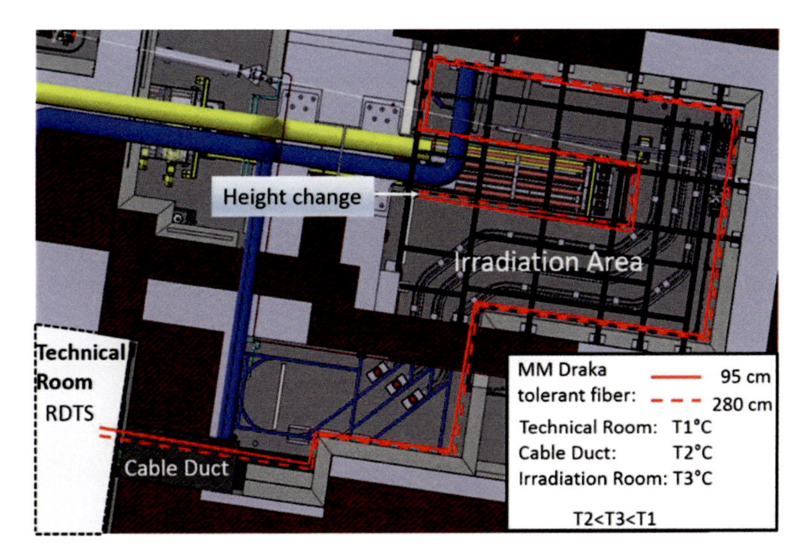

Fig. 26 2D view of the CHARM facility including the path of the radiation-tolerant multimode fiber (Toccafondo et al. 2015b)

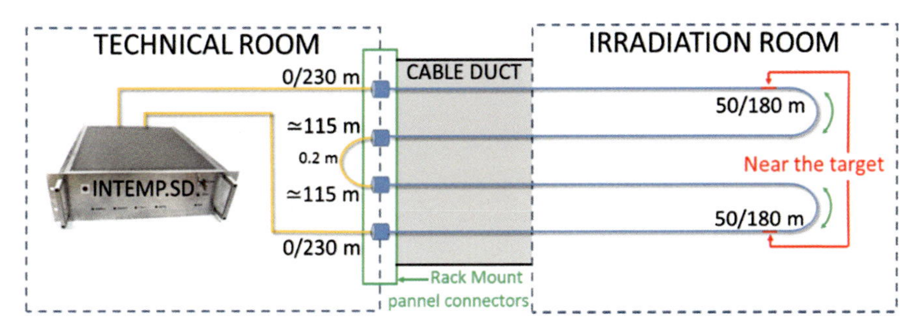

Fig. 27 RDTS control unit and fiber optic layout (Toccafondo et al. 2015b)

24 h has been performed. The radiation-tolerant multimode optical fiber has been exposed to total absorbed doses up to 1.25 kGy inside the CHARM mixed-field radiation environment. A schematic diagram of the experiment is reported in Fig. 27, showing the distances along the radiation-tolerant fiber path from the rack to which the RDTS is connected. The radiation-tolerant multimode optical fiber used for distributed temperature measurements has first been characterized in terms of RIA versus wavelength at different total absorbed doses, up to 11 kGy. By means of a Gammamat TK1000 ^{60}Co source, the fiber is γ-ray irradiated at a dose rate of 0.23 Gy/s and at constant temperature of 24 °C.

The wavelength-dependent RIA at four different total doses, corresponding to four different times during irradiation, is shown in Fig. 28. It is clearly evident that the Raman Stokes and anti-Stokes light components are affected differently depending on the total dose the optical fiber is exposed to.

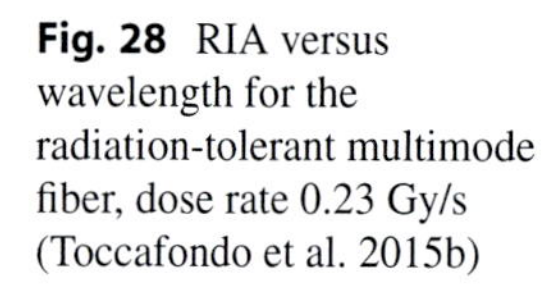

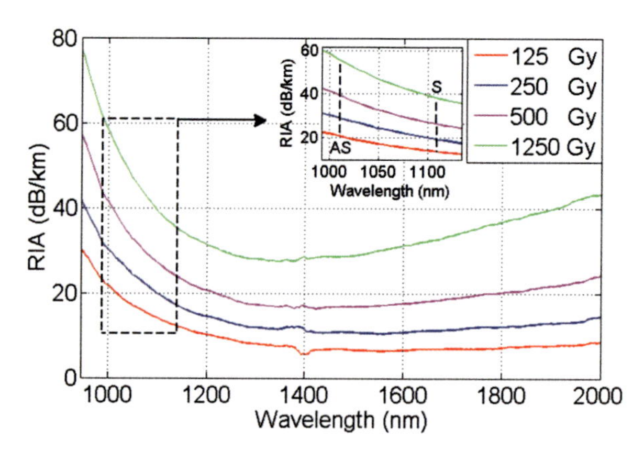

Fig. 28 RIA versus wavelength for the radiation-tolerant multimode fiber, dose rate 0.23 Gy/s (Toccafondo et al. 2015b)

The typical environment of hadron accelerator installations has been generated by focusing a 24 GeV/c beam on a copper target inside the CHARM facility, generating then a mixed radiation field composed of protons, neutrons, photons, muons, and other particles. The total accumulated doses along the fiber path after 24 h of irradiation ranged from tens of Gy up to the highest value of 1.25 kGy at positions 50 m and 180 m, where the fiber passes near the target, as shown in Fig. 27. Note that the dose values at those locations have been estimated by measurements performed nearby by conventional passive radiation dosimeters, and they represent challenging conditions for the RDTS measurement.

Figure 29 shows the temperature profiles measured by the RDTS at the beginning and the end of the irradiation within the 24 h measurement campaign. Note that, although high spatial gradients in the dose rate are present near the target, no significant temperature variations are observed in the figure. The RDTS spatial sampling interval is 0.5 m, allowing to fully resolve the higher exposed zone which is about 2 m wide. Also note that the symmetry of the temperature profiles in Fig. 29 confirms the double-end fiber deployment geometry within the facility. The insets in Fig. 29 are zooming the fiber sections corresponding to the higher exposed test positions, where the radiation levels ranged from 600 to 1250 Gy. No significant temperature variations are observed during the whole measurement test as expected since the irradiation area is kept at constant temperature by a ventilation system. The only temperature variations which can be observed are those along the path in the cable duct, which was exposed to external daily temperature changes.

The successful temperature measurements previously shown in Fig. 29 clearly demonstrate that the RDTS operating in loop configuration on radiation-tolerant multimode fibers can effectively correct for wavelength-dependent losses induced by a mixed-field radiation environment. This provides reliable and robust temperature profile with high spatial resolution.

Note that the effectiveness of Raman-based temperature measurements in a mixed-field radiation environment is important for safety purposes in large accelerators and nuclear power plants, as well as for the realization of distributed

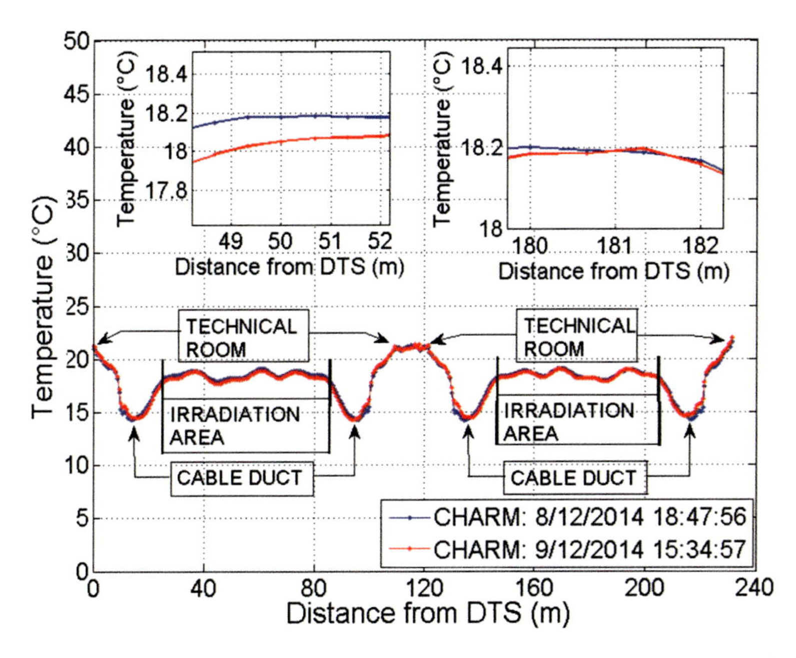

Fig. 29 Temperature profiles along the radiation-tolerant fiber at the beginning and the end of the irradiation process (Toccafondo et al. 2015b)

optical fiber radiation sensing systems currently being studied at CERN. A reliable temperature profile is indeed crucial in these systems for calibration and correction of the temperature dependence of the RIA on selected radiation sensitive fibers.

Conclusion

This book chapter has first presented descriptions of the physical mechanism behind Raman scattering in optical fibers and of the basic working principle of Raman distributed temperature sensors, along with the most common system configurations. Advanced techniques, based on pulse coding and image processing, for enhancing the performance of RDTS systems have also been described to address the current research trends in this well-established technology. A final section has reviewed specific industrial applications, in which RDTS systems can show significant comparative advantages with respect to conventional electronic sensors.

Raman distributed temperature sensing constitutes today a mature and very successful photonic technology with a broad range of industrial applications including oil and gas, transportation, energy, and environmental monitoring. Although widely used in several industrial fields, the application of RDTS technology still requires attention and further improvement concerning critical issues, like the proper fiber optic cabling design, the long-term reliability and stability of the interrogation unit, its power consumption, and robustness against harsh environmental conditions. Additional technical challenges include the ability to reach longer sensing distances on single-mode fiber with short measurement times and the capability to achieve submeter-scale spatial resolution. The possible full solid-state integration of the sensor reading unit and its potential low-cost mass-scale production using silicon photonics foundries is another important aspect which could lead to significant future progress in this sensing technology, in terms of robustness, reliability, cost reduction, and application range.

References

G.P. Agrawal, *Fiber-Optic Communication Systems*, 4th edn. (Wiley, New York, 2010)

M.K. Barnoski, M.D. Rourke, S.M. Jensen, R.T. Melville, Optical time domain reflectometer. Appl. Opt. **16**(9), 2375–2379 (1977)

F. Baronti, A. Lazzeri, R. Roncella, R. Saletti, A. Signorini, M.A. Soto, G. Bolognini, F. Di Pasquale, SNR enhancement of Raman-based long-range distributed temperature sensors using cyclic simplex codes. Electron. Lett. **46**(17), 1221–1223 (2010)

G. Bolognini, J. Park, M.A. Soto, N. Park, F. Di Pasquale, Analysis of distributed temperature sensing based on Raman scattering using OTDR coding and discrete Raman amplification. Meas. Sci. Technol. **18**(10), 3211–3218 (2007.) Special Issue: Optical Fibre Sensors

R.W. Boyd, *Nonlinear Optics*, 2nd edn. (Academic, San Diego, 2003)

A. Buades, B. Coll, J.M. Morel, A review of image denoising methods, with a new one. Multiscale Model. Simul. **4**(2), 490–530 (2005)

B. Culshaw, A. Kersey, Fiber-optic sensing: a historical perspective. J. Lightwave Technol. **26**(9), 1064–1078 (2008)

J.P. Dakin, A.D. Kersey, Distributed optic fiber sensors. Proc. SPIE **1797**, 76 (1993)

J.P. Dakin, D.J. Pratt, Distributed optical fibre Raman temperature sensor using a semiconductor light source and detector. Electron. Lett. **21**(13), 569–570 (1985)

W. Eickhoff, R. Ulrich, Optical frequency domain reflectometry in single-mode fiber. Appl. Phys. Lett. **39**, 693 (1981)

M.A. Farahani, T. Gogolla, Spontaneous Raman scattering in optical fibers with modulated probe light for distributed temperature Raman remote sensing. J. Lightwave Technol. **17**(8), 1379–1391 (1999)

A.F. Fernandez, P. Rodeghiero, B. Brichard, F. Berghmans, A.H. Hartog, P. Hughes, K. Williams, A.P. Leach, Radiation-tolerant Raman distributed temperature monitoring system for large nuclear infrastructures. IEEE Trans. Nucl. Sci. **52**(6), 2689–2691 (2005)

M.J.E. Golay, Complementary series. IRE Trans. Inf. Theory **7**(2), 82–87 (1961)

A.H. Hartog, A.P. Leach, Distributed temperature sensing in solid-core fibres. Electron. Lett. **21**(23), 1061–1062 (1985)

M. Harwit, N.J.A. Sloane, *Hadamard Transform Optics* (Academic, New York, 1979)

P. Healey, Complementary code sets for OTDR. Electron. Lett. **25**(11), 692–693 (1989)

T. Horiguchi, M. Tateda, BOTDA – nondestructive measurement of single-mode optical fiber attenuation characteristics using Brillouin interaction: theory. J. Lightwave Technol. **7**(8), 1170–1176 (1989)

D. Hwang, D.-J. Yoon, I.-B. Kwon, D.-C. Seo, Y. Chung, Novel auto-correction method in a fiber-optic distributed-temperature sensor using reflected antistokes Raman scattering. Opt. Express **18**(10), 9747–9754 (2010)

N.M. Islam, *Raman Amplifiers for Telecommunications 1: Physical Principles* (Springer, New York, 2004)

M. Jaaskelainen, Temperature monitoring of geothermal energy wells. Proc. SPIE **7653**, 765303 (2010)

M.D. Jones, Using simplex codes to improve OTDR sensitivity. IEEE Photon. Technol. Lett. **5**(7), 822–824 (1993)

K. Kikuchi, T. Naito, T. Okoshi, Measurement of Raman scattering in single-mode optical fiber by time-domain reflectometry. IEEE J. Quantum Electron. **24**(10), 1973–1975 (1988)

K. Kikuci, T. Naito, T. Okoshi, Measurement of Raman scattering in single-mode optical fiber by time-domain-reflectometry. IEEE J. Quantum Electron. **24**(10), 1973–1975 (1988)

A. Kimura, E. Takada, K. Fujita, M. Nakazawa, H. Takahashi, S. Ichige, Application of a Raman distributed temperature sensor to the experimental fast reactor JOYO with correction techniques. Meas. Sci. Technol. **12**(7), 966–973 (2001)

D. Lee, H. Yoon, N.Y. Kim, H. Lee, N. Park, Analysis and experimental demonstration of simplex coding technique for SNR enhancement of OTDR, in *Proceedings of IEEE LTIMC 2004* (2004)

S.G. Mallat, A theory for multiresolution signal decomposition: the wavelet representation. Pattern Anal. Mach. Intell. IEEE Trans. on **11**(7), 674–693 (1989)

S. Namiki, Y. Emori, Ultrabroad-band Raman amplifiers pumped and gain-equalized by wavelength-division-multiplexed high-power laser diodes. IEEE J. Sel. Top. Quantum Electron. **7**(1), 3–16 (2001)

M. Nazarathy, S.A. Newton, R.P. Giffard, D.S. Moberly, F. Sischka, W.R. Trutna, S. Foster, Real-time long-range complementary correlation optical time-domain reflectometer. J. Lightwave Technol. **7**(1), 24–38 (1989)

M. Nazarathy, S.A. Newton, W.R. Trutna, Complementary correlation OTDR with three code-words. Electron. Lett. **26**(1), 70–71 (1990)

J. Park, G. Bolognini, D. Lee, P. Kim, P. Cho, F. Di Pasquale, N. Park, Raman-based distributed temperature sensor with simplex coding and link optimisation. IEEE Photon. Technol. Lett. **18**, 1879–1881 (2006)

A. Rogers, Distributed optical fiber sensing. Meas. Sci. Technol. **1**(8), 75–99 (1999)

M.K. Saxena et al., Raman optical fiber distributed temperature sensor using wavelet transform based simplified signal processing of Raman backscattered signals. Opt. Laser Technol. **65**, 14–24 (2015)

A. Signorini, T. Nannipieri, L. Gabella, F. Di Pasquale, G. Latini, D. Ripari, Raman distributed temperature sensor for oil leakage detection in soil: a field trial and future trends, 23rd International Conference on Optical Fiber Sensors 2014 (Santander, Spain, 2014)

A. Signorini, T. Nannipieri, F. Di Pasquale, E. Fedeli, E. Marzilli, Fire detection in long railway tunnels using high performance Raman based optical fiber sensors, 11th WCRR (World Congress on Railway Research) (Milan, 29th May, 2nd June 2016)

H.Y. Song, S.W. Golomb, Some new constructions for simplex codes. IEEE Trans. Inf. Theory **40**(2), 504–507 (1994)

M.A. Soto, P.K. Sahu, S. Faralli, G. Bolognini, F. Di Pasquale, B. Nebendahl, C. Rueck, Distributed temperature sensor system based on Raman scattering using correlation-codes, IEE Electron. Lett. **43**(16), 862–864 (2007)

M.A. Soto, T. Nannipieri, A. Signorini, A. Lazzeri, F. Baronti, R. Roncella, G. Bolognini, F. Di Pasquale, Raman-based distributed temperature sensor with 1 m spatial resolution over 26 km SMF using low-repetition-rate cyclic pulse coding. Opt. Lett. **36**(13), 2557–2559 (2011a)

M.A. Soto, T. Nannipieri, A. Signorini, A. Lazzeri, F. Baronti, R. Roncella, G. Bolognini, F. Di Pasquale, Advanced cyclic coding technique for long-range Raman DTS systems with meter-scale spatial resolution over standard SMF, in *IEEE Sensors Conference 2011* (Limerick, Ireland, 2011b), paper 1767

M.A. Soto, A. Signorini, T. Nannipieri, S. Faralli, G. Bolognini, F. Di Pasquale, Impact of loss variations on double-ended distributed temperature sensors based on Raman anti-stokes signal only. J. Lightwave Technol. **30**(8), 1215–1222 (2012)

M.A. Soto, J.A. Ramírez, L. Thévenaz, Intensifying Brillouin distributed fibre sensors using image processing, in *Proceedings of SPIE 9634, 24th International Conference on Optical Fibre Sensors* (2015), 96342D

M.A. Soto, J.A. Ramírez, L. Thévenaz, Intensifying the response of distributed optical fibre sensors using 2D and 3D image restoration. Nat. Commun. **7**, 10870 (2016a)

M.A. Soto, J.A. Ramírez, L. Thévenaz, Reaching millikelvin resolution in Raman distributed temperature sensing using image processing, in *Proceedings of SPIE 9916, 6th European Workshop on Optical Fibre Sensors* (2016b), 99162A

K. Suh, C. Lee, Auto-correction method for differential attenuation in a fiber-optic distributed-temperature sensor. Opt. Lett. **33**(16), 1845–1847 (2008)

I. Toccafondo, T. Nannipieri, A. Signorini, E. Guillermain, J. Kuhnhenn, M. Brugger, F. Di Pasquale, Raman distributed temperature measurement at CERN high energy AcceleRator mixed field facility (CHARM). IEEE Phot. Technol. Lett. **27**(20), 2182–2185 (2015a)

I. Toccafondo, T. Nannipieri, A. Signorini, E. Guillermain, J. Kuhnhenn, M. Brugger, F. Di Pasquale, Raman distributed temperature measurement at CERN high energy AcceleRator mixed field facility (CHARM). IEEE Photon. Technol. Lett. **27**(20), 2182–2185 (2015b)

I. Toccafondo, Y.E. Marin, E. Guillermain, J. Kuhnhenn, J. Mekki, M. Brugger, F. Di Pasquale, Distributed optical fiber radiation sensing in a mixed-field radiation environment at CERN, to be published in J. Lightwave Technol. **35**(16), 3303–3310 (2017)

W.M. Tolles, J.W. Nibler, J.R. McDonald, A.B. Harvey, A review of the theory and application of coherent anti-stokes Raman spectroscopy (CARS). Appl. Spectrosc. **31**(4), 253–271 (1977)

P.C. Wait, K.D. Souza, T.P. Newson, A theoretical comparison of spontaneous Raman and Brillouin based fibre optic distributed temperature sensors. Opt. Commun. **144**, 17–23 (1997)

S. Yin, P.B. Ruffin, F.T.S. Yu, *Fiber Optic Sensors*, 2nd edn., ed. by CRC Press (Taylor and Francis Group, Boca Raton, FL, 2008)

Distributed Brillouin Sensing: Time-Domain Techniques

43

Marcelo A. Soto

Contents

M. A. Soto (✉)
Institute of Electrical Engineering, EPFL Swiss Federal Institute of Technology, Lausanne, Switzerland
e-mail: marcelo.soto@epfl.ch

G.-D. Peng (ed.), *Handbook of Optical Fibers*,
https://doi.org/10.1007/978-981-10-7087-7_7

Abstract

Distributed optical fiber sensors based on spontaneous and stimulated Brillouin scattering have been a subject of intense research and industrial developments for almost 30 years. Combining interrogation methods based on optical time-domain reflectometry and the dependence of Brillouin scattering on environmental variables, such as temperature and strain, high-performance distributed sensing techniques have been developed over the last decades for a wide range of industrial applications. This chapter presents a comprehensive description of the fundamentals of time-domain techniques exploited for distributed Brillouin optical fiber sensing. This includes the basic principles and limitations of different classical configurations. Theoretical descriptions of sophisticated techniques to overcome the fundamental limitations of classical Brillouin time-domain schemes are also presented. In this way, the most-common advanced approaches to reach high spatial resolution, dynamic, and long-range distributed Brillouin sensing are thoroughly described from theoretical and practical points of view. The material presented in this chapter is intended to serve as a guideline to design and implement state-of-the-art distributed Brillouin optical fiber sensors exploiting time-domain interrogation approaches.

Introduction

The use of Brillouin scattering to perform distributed measurements along an optical fiber was initially proposed in the late 1980s as an alternative method to the standard optical time-domain reflectometry (OTDR) for the characterization of the fiber attenuation versus distance. Nevertheless, the dependence of Brillouin scattering on strain and temperature has made this inelastic scattering process much more attractive for distributed sensing applications. A large majority of schemes exploits time-domain interrogation to read the features of the Brillouin scattering generated at each position over a sensing fiber with high spatial resolution. Launching a short optical pulse into the fiber and reading the backscattered light, a full map of the Brillouin scattering generated locally along the optical fiber can be obtained. This way, changes in Brillouin scattering features are employed to retrieve a distributed profile of the temperature and strain over the sensing fiber.

Since its first implementation in the late 1980s, the capabilities of distributed Brillouin sensors have been steadily improved over decades. Nowadays, the technology has reached some level of maturity, resulting in commercially available

Brillouin sensing instruments exhibiting high performance and providing a reliable monitoring tool for a wide range of applications. The critical factors limiting the performance of classical Brillouin sensing configurations have been clearly identified over the last years. This has driven researches to develop novel sensing schemes to overcome some of the fundamental limitations and push the performance of Brillouin distributed sensing beyond the state of the art.

There exist practically two approaches to implement distributed Brillouin fiber sensors: (i) using spontaneous Brillouin scattering (SpBS) and (ii) using stimulated Brillouin scattering (SBS). Theoretical descriptions of SpBS and SBS are presented in section "Brillouin Scattering in Optical Fibers" of this chapter, in which the main parameters of Brillouin scattering are described. Time-domain distributed sensing schemes using SpBS are based on an interrogation method called *Brillouin optical time-domain reflectometry* (BOTDR), which is addressed in section "Brillouin Optical Time-Domain Reflectometry (BOTDR)." On the other hand, time-domain sensors exploiting SBS make use of an interrogation method named *Brillouin optical time-domain analysis (BOTDA)*, which is described in section "Brillouin Optical Time-Domain Analysis (BOTDA)."

In addition, section "Limitations and Design Considerations in Brillouin Time–Domain Sensing" addresses the main factors limiting the performance of Brillouin time-domain sensing techniques, while sections "Time-Domain Methods for High Spatial Resolution Brillouin Sensing," "Dynamic Brillouin Distributed Sensing," and " Long-Range Brillouin Distributed Sensing" describe theoretical and practical aspects of different approaches to overcome limiting factors and allow for high spatial resolution, dynamic, and long-range Brillouin sensing.

Brillouin Scattering in Optical Fibers

Light scattering in optical fibers occurs as a consequence of fluctuations in the optical properties of the fiber material (i.e., of the silica glass in standard optical fibers) (Boyd 2003). In the case of spontaneous scattering, these fluctuations in the material optical properties are independent of the incident light beam and are typically initiated by mechanical or thermal excitation. This condition only occurs when the optical power of the incident light is low. When the power of the incident beam increases beyond a certain critical level, the optical properties of the material turn out to be gradually modified by the light beam, and therefore the scattering process becomes nonlinear. In such a case, scattering is generated by material fluctuations induced by the incident light; this process is known as stimulated scattering (Boyd 2003; Agrawal 2007).

This section is devoted to the theoretical description of spontaneous and stimulated Brillouin scattering. The temperature and strain dependences that are commonly exploited for distributed sensing are also discussed.

Spontaneous Brillouin Scattering (SpBS)

The propagation of light through an optical fiber, and the origin of spontaneous scattering processes, can be well-described by the following general wave equation (Boyd 2003; Agrawal 2007):

$$\nabla^2 \mathbf{E} - \frac{1}{c^2}\frac{\partial^2 \mathbf{E}}{\partial t^2} = \mu_0 \frac{\partial^2 \mathbf{P}}{\partial t^2} \tag{1}$$

where $\mathbf{P}$ is the polarization field, $\mathbf{E}$ is the wave electric field, c is the speed of light in vacuum, and μ_0 is the magnetic permittivity in the vacuum. In a linear propagation regime, the polarization $\mathbf{P}$ is proportional to the wave electric field $\mathbf{E}$ according to $\mathbf{P} = \epsilon_0 \mathbf{E} + \Delta\epsilon_{ik}\mathbf{E}$, where ϵ_0 is the mean value of the dielectric constant of the medium and $\Delta\epsilon_{ik}$ corresponds to variations in the dielectric tensor due to inhomogeneities and anisotropies (Boyd 2003). The right-hand side term in Eq. 1 and the variations of the dielectric tensor $\Delta\epsilon_{ik}$ are the responsible for the origin of spontaneous scattering processes in an optical fiber. The fluctuations in the dielectric tensor $\Delta\epsilon_{ik}$ can be decomposed into a scalar component $\Delta\epsilon$ and a tensor component $\Delta\epsilon_{ik}^{(t)}$ (Boyd 2003). The scalar term gives rise to Rayleigh and Brillouin scattering, while the tensor term is responsible for Rayleigh-wing and Raman scattering. To describe the origin of spontaneous Brillouin scattering (SpBS), let us only focus on the scalar term. In particular, variations in thermodynamic quantities, such as density or temperature, contribute to the scalar dielectric constant fluctuations $\Delta\epsilon$, determining the generation of both Brillouin and Rayleigh scatterings. Dielectric constant fluctuations depends much more strongly on density than on temperature variations, so that they can be simply described as $\Delta\epsilon \approx (\partial\epsilon/\partial\rho)_T \Delta\rho$, where $\Delta\rho$ represents the density fluctuations in the material, and the subindex T indicates that the derivative is calculated at constant temperature. Density fluctuations $\Delta\rho$ are in turn determined by variations in the entropy Δs and pressure $\Delta\rho$ inside the optical fiber. Considering entropy s and pressure p statistically independent variables, the mean-square value of the scalar dielectric constant fluctuations can be written as (Boyd 2003)

$$\left\langle (\Delta\epsilon)^2 \right\rangle \approx \left(\frac{\partial\epsilon}{\partial\rho}\right)_T^2 \left\langle (\Delta\rho)^2 \right\rangle = \left(\frac{\partial\epsilon}{\partial s}\right)_p^2 \left\langle (\Delta s)^2 \right\rangle + \left(\frac{\partial\epsilon}{\partial p}\right)_s^2 \left\langle (\Delta p)^2 \right\rangle. \tag{2}$$

According to the Landau-Placzek theory, the first right-hand side term in Eq. 2 represents non-propagative entropy fluctuations (i.e., static and frozen in the medium) that give rise to Rayleigh scattering, while the second term corresponds to moving pressure fluctuations (called *acoustic wave*) originating Brillouin scattering.

From the thermodynamic fluctuation theory (Boyd 2003; Cummins and Gammon 1966), the mean-square value of density, entropy, and pressure fluctuations can be, respectively, written as $\langle(\Delta\rho)^2\rangle/\rho^2 = k_B T \beta_T / V$, $\langle(\Delta s)^2\rangle = k_B C_P \rho V$, and $\langle(\Delta p)^2\rangle = k_B T / V \beta_S$, where k_B is the Boltzmann constant, β_T and β_S are the isothermal and adiabatic compressibilities, T is the temperature, and C_P is the specific heat per unit mass of the medium (Cummins and Gammon 1966). Thus, considering that the intensity of the light backscattered in a volume V is proportional to the mean-square value of the dielectric constant fluctuations, the total backscattered light intensity I_s can be written as (Cummins and Gammon 1966)

$$I_s = I_0 \frac{\pi^2}{r^2 \lambda_0^4} k_B T \beta_T V \left(\rho \frac{\partial \epsilon}{\partial \rho} \right)_T^2 = I_0 \frac{\pi^2 V}{r^2 \lambda_0^4} \left\{ \left(\frac{\partial \epsilon}{\partial T} \right)_P^2 \frac{k_B T^2}{C_P \rho} + \left(\rho \frac{\partial \epsilon}{\partial \rho} \right)_s^2 k_B T \beta_S \right\}, \quad (3)$$

where I_o is the intensity of the incident light of wavelength λ_0, and r is the distance from the scattering to the observation point. The first right-hand side term corresponds to the intensity of Rayleigh backscattering, while the second term represents the total intensity of spontaneous Brillouin scattering. The ratio between Rayleigh and Brillouin scattering intensities is called Landau-Placzek ratio (LPR) (Boyd 2003; Cummins and Gammon 1966) and can be written as

$$\mathrm{LPR} = \frac{I_{\mathrm{Rayleigh}}}{I_{\mathrm{SpBS}}} = \frac{\beta_T - \beta_S}{\beta_S}. \quad (4)$$

As mentioned before, one of the fundamental differences between Rayleigh and spontaneous Brillouin scattering is related to the motion of the density fluctuations originating these two processes. Rayleigh scattering is generated by density variations that are static (frozen) in the medium, while SpBS is originated by a moving pressure wave (sound wave) of frequency Ω and wavevector $\mathbf{q}$. The spatiotemporal evolution of the pressure wave originating spontaneous Brillouin scattering can be described as (Boyd 2003)

$$\frac{\partial^2 \Delta p}{\partial t^2} - \Gamma \nabla^2 \frac{\partial \Delta p}{\partial t} - \upsilon_a \nabla^2 \Delta p = 0, \quad (5)$$

where Γ is the acoustic wave damping factor and υ_a is the sound velocity in the medium, which can be expressed in terms of thermodynamic variables as

$$\upsilon_a = \sqrt{\frac{K}{\rho}} = \sqrt{\frac{1}{C_S \rho}}, \quad (6)$$

where K is the bulk modulus. A solution of Eq. 5 is a pressure wave of wavelength $\Lambda = 2\pi \upsilon_a / \Omega$, described as

$$\Delta p(\mathbf{r}, t) = \Delta p_0 e^{i(\mathbf{q} \cdot \mathbf{r} - \Omega t)} + c.c., \quad (7)$$

where Δp_0 is the maximum amplitude of the pressure wave and *c.c.* denotes the complex conjugated of the previous term (Boyd 2003). This pressure wave induces fluctuations in the scalar dielectric constant $\Delta\epsilon$, leading to a polarization field in Eq. 1 that can be expressed as

$$\mathbf{P}(\mathbf{r},t) = \epsilon_0 \left(\frac{\partial \epsilon}{\partial \rho}\right)\left(\frac{\partial \rho}{\partial p}\right)_s \Delta p\,(\mathbf{r},t)\, E_0\,(z,t) = \epsilon_0 \gamma_e C_s \Delta p\,(\mathbf{r},t)\,\mathbf{E_0}\,(\mathbf{r},t), \quad (8)$$

where γ_e is the electrostrictive constant.

Considering an incident optical field of frequency ω_0 and wavevector $\mathbf{k}$, described as

$$\mathbf{E_0}\,(\mathbf{r},t) = E_0 e^{i(\mathbf{k}\cdot\mathbf{r}-\omega_0 t)} + c.c., \quad (9)$$

and replacing Eqs. 8 and 9 into Eq. 1, it is found that the scattered field obeys the following wave equation:

$$\nabla^2\mathbf{E} - \frac{n^2}{c^2}\frac{\partial^2\mathbf{E}}{\partial t^2} = -\frac{\gamma_e C_s}{c^2}\Big\{(\omega_0-\Omega)^2 E_0 \Delta p^* e^{i(\mathbf{k}-\mathbf{q})\cdot\mathbf{r}-i(\omega_0-\Omega)t} + (\omega_0+\Omega)^2 E_0 \Delta p e^{i(\mathbf{k}+\mathbf{q})\cdot\mathbf{r}-i(\omega_0+\Omega)t} + c.c.\Big\}. \quad (10)$$

Equation 10 points out that, contrarily to Rayleigh scattering, Brillouin scattered light is affected by a Doppler effect induced by the acoustic wave motion, resulting in a spectral shift of the Brillouin scattered light. This feature makes Brillouin scattering to be an inelastic process, resulting in two spectral components, as shown in Fig. 1 and as described by the right-hand side terms in Eq. 10. Depending on the direction of the acoustic wave propagation with respect to the propagation of the incident light beam, the scattered light can be spectrally shifted toward either higher or lower frequencies. By definition, the component of the scattered light that is downshifted in frequency is called *Stokes*, and the one upshifted is called *anti-Stokes* (Boyd 2003; Agrawal 2007).

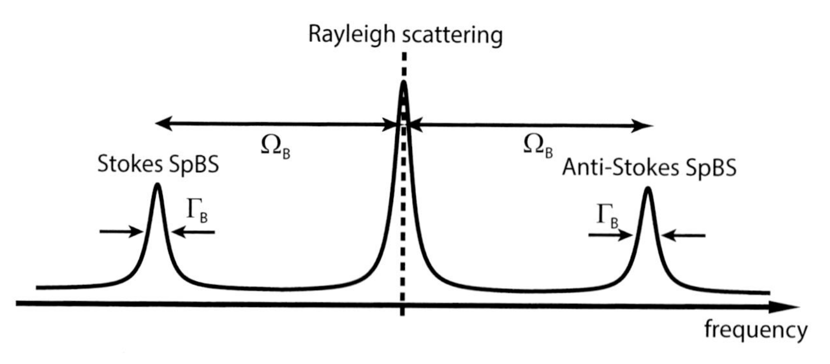

Fig. 1 Spectrum of backscattered light, showing a central Rayleigh component and two spontaneous Brillouin scattering lines, corresponding to Stokes and anti-Stokes processes

The first component of the right-hand side term in Eq. 10, describes a scattered wave with wavevector $\mathbf{k}' = \mathbf{k} - \mathbf{q}$ and frequency $\omega' = \omega_0 - \Omega$, corresponding to the Brillouin Stokes component. The second term, however, describes a scattered wave with wavevector $\mathbf{k}' = \mathbf{k} + \mathbf{q}$ and frequency $\omega' = \omega_0 + \Omega$, representing the Brillouin anti-Stokes component. In both cases the frequencies and the wave vectors of the two involved optical waves and respective acoustic wave are coupled through the following dispersion relations (Boyd 2003):

$$\omega_0 = |\mathbf{k}| \frac{c}{n}, \quad \omega' = |\mathbf{k}'| \frac{c}{n}, \quad \Omega = |\mathbf{q}| \upsilon_a. \tag{11}$$

These expressions define very strict phase-matching conditions for Brillouin scattering to occur, as depicted in Fig. 2. Conservation of both momentum and energy must be simultaneously satisfied, and therefore only very specific frequencies and wave vectors give origin to efficient coupling. This occurs basically in two conditions. The first one is when the acoustic wave moves in the same direction of the incident field, thus downshifting the frequency of the scattered wave due to Doppler effect (in the Stokes process). And the second one is when the acoustic wave moves in opposite direction with respect to the incident optical wave, upshifting the frequency of the scattered light (in the anti-Stokes process).

Considering that the frequency of the acoustic wave Ω is much smaller than the optical frequencies ω_0 and ω', it is possible to assume that $|\mathbf{k}| \approx |\mathbf{k}'|$, for both Stokes and anti-Stokes components. Then, the wavevector of the acoustic waves can be both written as $|\mathbf{q}| = 2|\mathbf{k}| \sin(\theta/2)$, where θ is the angle between the scattered and incident waves. Thus, based on the dispersion relations in Eq. 11, the acoustic frequency Ω, and therefore the frequency shift of the scattered light $|\omega_0 - \omega'|$, can be expressed as

$$\Omega = 2\,|\mathbf{k}|\, \upsilon_a \sin\left(\frac{\theta}{2}\right) = \frac{2n\omega_0\upsilon_a}{c} \sin\left(\frac{\theta}{2}\right). \tag{12}$$

Equation 12 points out that the acoustic frequency nullifies for forward scattering ($\theta = 0$) and is maximum for backscattering ($\theta = \pi$). The maximum frequency is called *Brillouin frequency shift* ν_B (BFS) and is given by

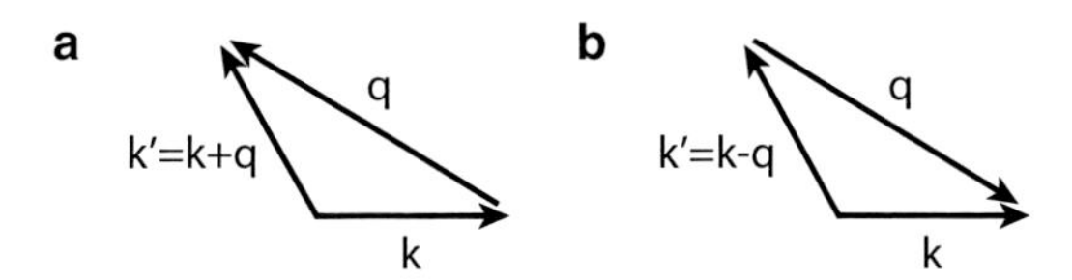

Fig. 2 Relation between wave vectors of incident light **k**, scattered light **k**′ and acoustic wave **q**. (**a**) Brillouin anti-Stokes process, (**b**) Brillouin Stokes process

$$\nu_B = \frac{\Omega_B}{2\pi} = \frac{2n\nu_0 v_a}{c} = \frac{2n v_a}{\lambda_0}, \tag{13}$$

where $\lambda_0 = c/\nu_0$ is the wavelength of the incident light at an optical frequency $\nu_0 = \omega_0/2\pi$ (Boyd 2003). Note that ν_B can be positive or negative, as shown in Fig. 1, depending on the sign of the wavevector associated with the interacting acoustic wave, thus representing Brillouin anti-Stokes or Stokes processes.

Equation 12 points out that the spectrum of Brillouin scattering is, in part, determined by the angle θ between the incident and scattered waves; however it must be pointed out that the spectral width is also given by the decay time of the acoustic wave. In optical fibers, the capture angle θ is very small, and therefore the Brillouin spectral width is basically determined by the damping factor Γ of the acoustic wave. Since the acoustic wave decays exponentially over time, the SpBS spectrum has a Lorentzian-shaped spectrum (Boyd 2003; Agrawal 2007), with a full-width at half-maximum (FWHM) $\Gamma_B = 2\pi\Delta\nu_B$ expressed as

$$\Gamma_B = q\Gamma = \frac{1}{\tau_p} = \frac{2}{\tau_A}, \tag{14}$$

where τ_p is the decay time associated with the acoustic energy (so-called *phonon lifetime*), and $\tau_A = 2\tau_p$ is the decay time of the acoustic wave amplitude. In standard silica optical fibers, $\tau_p \approx 6$ ns, corresponding to an acoustic amplitude decay time $\tau_A \approx 12$ ns, which leads to a SpBS spectral width $\Gamma_B \approx 2\pi \cdot 27$ Mrad/s at 1550 nm.

Stimulated Brillouin Scattering (SBS)

When the optical properties of the medium are modified by a propagating light, the induced scattering is said to be nonlinear. In the case of Brillouin scattering, the stimulation process is driven by the phenomenon of *electrostriction* (Boyd 2003; Agrawal 2007), a process by which a material gets compressed in the presence of an electromagnetic field.

Stimulated Brillouin scattering is initiated by the interference (beating) between two counter-propagating light beams of different frequencies, as illustrated in Fig. 3. The beating field generates a pressure wave that modulates the refractive index of the fiber through electrostriction. When the beating frequency matches the Brillouin frequency of the medium (acoustic mode), both the pressure wave created by electrostriction and the spontaneously generated acoustic wave will vibrate in-phase and at the same frequency, being in full resonance. This resonance effect only takes place when the acoustic and optical waves fulfill the same phase-matching conditions as described for SpBS in Eq. 11. Under this resonance condition, the acoustic wave is reinforced by electrostriction. However, it must be pointed out that due to phase-matching conditions, only the acoustic wave associated with Brillouin Stokes process is reinforced in stimulated Brillouin scattering, while the pressure

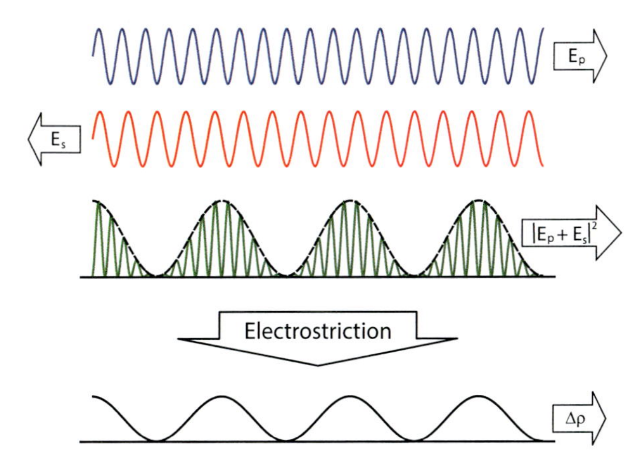

Fig. 3 Stimulation of a Brillouin-induced acoustic wave by means of electrostriction

wave created by the anti-Stokes process quickly quenches the associated SpBS acoustic wave.

Due to the motion of the pressure wave generated by electrostriction, the refractive index modulation (so-called *Brillouin grating*) also moves along the fiber at the sound velocity. This means that the incoming light is reflected by this Brillouin grating, being downshifted in frequency due to Doppler effect, with a frequency shift matching the acoustic wave frequency. The phase-matching also makes the reflected signal to be coupled in phase into the counter-propagating signal, thus enhancing the Brillouin backscattering intensity. Stronger backscattered light leads to stronger interference and stronger electrostriction, which in turn leads to a further enhancement of the acoustic wave. The process self-sustains in resonance condition, resulting in amplification of the Brillouin Stokes signal.

There are two mechanism by which electrostriction can be induced in an optical fiber. The first one is by the interference generated by the incoming light and the spontaneously generated Brillouin scattering. In this case waves are inherently in the right phase-matching condition to make the process resonate. As a result, the weak SpBS Stokes component turns out to be amplified by efficient power transfer from the incoming light beam (commonly called *pump wave*); for this reason this process is typically referred as *amplified spontaneous Brillouin scattering*. The second mechanism is to deliberately inject into the optical fiber a weak optical signal (commonly called *signal wave*) in opposite direction with respect to the pump, at an optical frequency matching the SpBS frequency. Since in this case the frequency difference between the optical waves matches the Brillouin frequency of the fiber (i.e., the acoustic wave frequency), electrostriction is efficiently generated.

A formal mathematical description of SBS can be presented based on the wave Eq. 1. However, contrarily to the spontaneous linear regime, in this case the polarization field $\mathbf{P}$ is dominated by a nonlinear term determined by changes in the material susceptibility $\Delta\chi$ in the presence of the optical field, so that it can be expressed as $\mathbf{P} = \mathbf{P}_{\mathrm{NL}} = \Delta\chi\mathbf{E} = \epsilon_0 C_T \gamma_e^2 |\mathbf{E}|^2 \mathbf{E}$. Furthermore, the acoustic wave evolution in this case can be described by a similar expression as in Eq. 5 but adding a driving force given by electrostriction, so that

$$\frac{\partial^2 \Delta p}{\partial t^2} - \Gamma \nabla^2 \frac{\partial \Delta p}{\partial t} - \upsilon_a \nabla^2 \Delta p = \nabla \cdot \mathbf{f}, \tag{15}$$

where $\mathbf{f} = -\frac{1}{2}\epsilon_0 \gamma_e \nabla \left\langle \mathbf{E}^2 \right\rangle$ (Boyd 2003; Agrawal 2007). Note that the time average of the squared electric field $\langle \mathbf{E}^2 \rangle$ results from the slow response time of the molecular displacement in the material, which cannot respond at optical frequencies but only to the envelope of the interference (beating) between pump and signal.

To provide a formal mathematical description of the coupled wave equations for the three waves (pump, signal, and acoustic waves) interacting by stimulated Brillouin scattering, Eqs. 1 and 15 have to be combined, including the nonlinear polarization $\mathbf{P}_{\mathrm{NL}}$. This yields the following set of coupled equations (Boyd 2003; Agrawal 2007):

$$\frac{\partial E_p}{\partial z} + \frac{1}{\upsilon_g}\frac{\partial E_p}{\partial t} = i\frac{1}{2} g_2 E_s \rho - \frac{\alpha}{2} E_p, \tag{16}$$

$$\frac{\partial E_s}{\partial z} - \frac{1}{\upsilon_g}\frac{\partial E_s}{\partial t} = -i\frac{1}{2} g_2 E_p \rho^* + \frac{\alpha}{2} E_s, \tag{17}$$

$$\frac{\partial \rho}{\partial t} + \Gamma_A \rho = i g_1 E_p E_s^*, \tag{18}$$

where E_p, E_s and ρ are the slow-varying (envelope) amplitudes of the pump, signal, and acoustic wave, v_g is the group velocity, α is the fiber attenuation (assumed to be the same for pump and signal), g_1 and g_2 are the electrostrictive and elasto-optic coupling constants, and Γ_A is the frequency detuning factor defined as

$$\Gamma_A = i\left(\Omega_B^2 - \Omega^2 - i\Omega\Gamma_B\right)/\left(2\Omega\right), \tag{19}$$

where Ω_B is the (angular) Brillouin frequency defined in Eq. 13, Ω is the pump-signal frequency offset, and Γ_B is the Brillouin FWHM defined in Eq. 14. Note that in Eq. 18 the propagation of the acoustic wave has been neglected during interaction, thus allowing us to drop the spatial derivative in Eq. 15.

The system of coupled wave Eqs. 16–18 fully describes stimulated Brillouin scattering in optical fibers, in time and space domains, on the assumption of plane-wave interaction and same polarization for pump and signal. This system of equations does not have an analytical solution, and thus, numerical methods must be typically employed to find a proper solution. In the case of weak interaction (so-called *small-gain approximation*) and undepleted pump conditions, the set of equations can be solved using a perturbation method approach along with Laplace transform, so that an analytical solution can be obtained under very specific conditions (Beugnot et al. 2011). It is important to point out that solving the full set of equations is only meaningful when the interaction process takes place along very short times (shorter than the acoustic wave amplitude decay time $\tau_A \approx 12$ ns in silica optical fibers). In the case of longer interaction times, the temporal derivatives in Eqs. 16–18 can be neglected, thus leading to a steady-state acoustic wave amplitude given as (Boyd 2003)

$$\rho = i\frac{g_1}{\Gamma_A}E_p E_s^*, \tag{20}$$

while the equations for the optical waves can be expressed as

$$\frac{\partial E_p}{\partial z} = -\frac{g_1 g_2}{2\Gamma_A}|E_s|^2 E_p - \frac{\alpha}{2}E_p, \tag{21}$$

$$\frac{\partial E_s}{\partial z} = -\frac{g_1 g_2}{2\Gamma_A^*}\left|E_p\right|^2 E_s + \frac{\alpha}{2}E_s. \tag{22}$$

Note that Γ_A in Eqs. 20–22 is a complex number, indicating that stimulated Brillouin scattering induces amplitude and phase changes in both pump and signal waves. Indeed the term g_1g_2/Γ_A can be expressed as a sum of real and imaginary parts as

$$\frac{g_1 g_2}{\Gamma_A} = g_{B0}\frac{1}{1+(2\Delta\Omega/\Gamma_B)^2} - \mathrm{i}g_{B0}\frac{2\Delta\Omega/\Gamma_B}{1+(2\Delta\Omega/\Gamma_B)^2} \tag{23}$$

where $g_{B0} = 2g_1g_2/\Gamma_B$ is the Brillouin gain coefficient and $\Delta\Omega = \Omega_B - \Omega$ is the detuning between the Brillouin frequency Ω_B and the pump-signal frequency difference Ω. The real part of Eq. 23 gives rise to amplification of the signal (and depletion of the pump), with a linear Brillouin gain $g_B(\Delta\Omega) = \mathrm{Re}\{g_1g_2/\Gamma_A\}$, while the imaginary part $\mathrm{Im}\{g_1g_2/\Gamma_A\}$ represents the phase shift induced by SBS. The spectral response of the real and imaginary parts are commonly called linear *Brillouin gain spectrum* (BGS) and *Brillouin phase spectrum* (BPS), respectively. Figure 4 shows the BGS and BPS shapes around the resonance frequency. Note that at zero detuning $\Delta\Omega = 0$, only the real part exists, leading to a maximum gain defined by the linear Brillouin gain coefficient g_{B0}. On the other hand, the SBS-induced phase shift nullifies at $\Delta\Omega = 0$ and maximizes at $\Delta\Omega = \pm\Gamma_B/2$.

When the power transferred between pump and signal is small (i.e., in a small-gain approximation), the first component of the right-hand side term in Eq. 21 can be neglected, resulting in a pump intensity evolution equal to $I_p(z) = I_{pi}e^{-\alpha z}$, where I_{pi} is the input intensity of the pump launched into the fiber at position $z = 0$. Hence, the solution of Eq. 22 for the signal wave can be written in terms of optical intensities as

$$I_s(z) = I_{si}e^{g_B(\Delta\Omega)I_{pi}\exp(-\alpha z)L_{\mathrm{eff}}-\alpha(L-z)}, \tag{24}$$

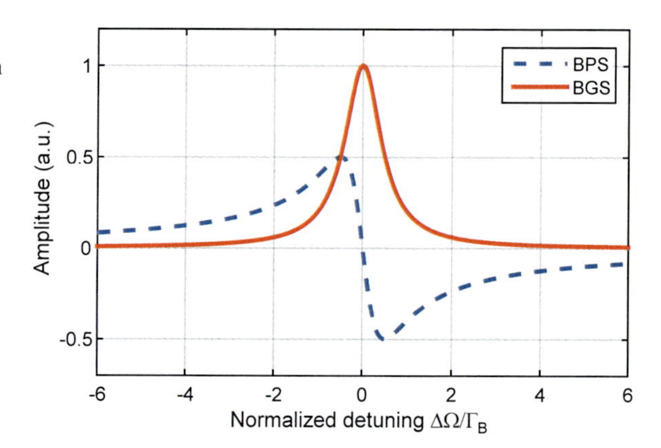

Fig. 4 Brillouin gain spectrum (BGS) and Brillouin phase spectrum (BPS) generated by a CW pump

where I_{si} is the input intensity of the signal launched into the fiber at position $z = L$, L is the fiber length, and $L_{\text{eff}} = (1 - e^{-\alpha(L-z)})/\alpha$ is the effective length in the fiber section from z to L.

Note that I_{si} in Eq. 24 represents the input power of a signal injected into the optical fiber from the far fiber end ($z = L$); however it can also represent the spontaneous Brillouin scattering light generated by the pump. Indeed, when a strong CW pump power is launched into the fiber, the SpBS light generated along the entire fiber results to be amplified by the pump itself. This leads to *amplified spontaneous Brillouin scattering*, whose onset occurs at a critical power, referred as the *threshold* power for SBS, which is described as (Agrawal 2007)

$$P_{\text{cri}}^{\text{SBS}} = \frac{21 A_{\text{eff}}}{g_B(\Omega_B) L_{\text{eff}}}. \tag{25}$$

For a long single-mode optical fiber ($g_B \approx 2 \cdot 10^{-11}$m/W, $A_{\text{eff}} = 80$ μm^2, $L_{\text{eff}} \approx \alpha^{-1} \approx 22$ km), this threshold value is $P_{\text{cri}}^{\text{SBS}} \approx 5$ mW. As any nonlinear effect, this critical power increases in short fibers with an inverse proportion to the effective length. It must be however pointed out that Eq. 25 is simply a definition, indicating the input pump power that leads to a Stokes signal with the same optical power at $z = 0$ than the pump at the far end ($z = L$) (Agrawal 2007). Nevertheless, efficient electrostriction and SBS process already occur at lower pump powers.

Temperature and Strain Dependence

As described, the origin of Brillouin scattering is highly linked to thermodynamic quantities, such as temperature and density of the propagating medium. This makes both spontaneous and stimulated processes highly dependent on environmental quantities such as temperature and strain.

In particular the intensity of the spontaneous Brillouin scattering process depends on temperature and strain (Cummins and Gammon 1966), as it can be inferred from the second term in Eq. 3. The temperature dependence is indeed explicit in such an expression, while the strain dependence is implicit through the dependence of the Brillouin intensity on the material density. In the case of silica optical fibers, it has been demonstrated that changes in the fiber temperature ΔT and strain $\Delta\varepsilon$ induce changes in the SpBS intensity ΔI_{SpBS} according to the following relation:

$$\frac{\Delta I_{\mathrm{SpBS}}}{I_{\mathrm{SpBS}}} = C_{T,I_{\mathrm{SpBS}}}\Delta T + C_{\varepsilon,I_{\mathrm{SpBS}}}\Delta\varepsilon \tag{26}$$

where $C_{T,I_{\mathrm{SpBS}}} \approx 0.36\%/°\mathrm{C}$ and $C_{\varepsilon,I_{\mathrm{SpBS}}} \approx 8 \cdot 10^{-4}\%/\mu\varepsilon$ are the temperature and strain sensitivities of the SpBS intensity, respectively (Maughen et al. 2001). Although these values can change from fiber to fiber, the strain dependence of the SpBS intensity is indeed always very low to provide reliable sensing (e.g., a change of ~450 $\mu\varepsilon$ produces the same effect than a temperature variation of 1 °C); and therefore, the SpBS intensity is normally only exploited for temperature sensing, offering very low performance when used for strain sensing.

On the other hand, the Brillouin frequency shift described in Eq. 13 also depends on temperature and strain. Although the refractive index of the fiber depends on those physical quantities, the BFS response is mostly dominated by the temperature and strain dependence of the acoustic velocity in the fiber. It must be pointed out that the mentioned strain dependence is related to longitudinal strain, and not to strain induced by lateral pressure, which actually has a negligible effect on the BFS. A linear dependence of the BFS changes on temperature and longitudinal strain changes has been verified in a wide range, so that (Maughen et al. 2001)

$$\frac{\Delta\Omega_B}{2\pi} = C_{T,\mathrm{BFS}}\Delta T + C_{\varepsilon,\mathrm{BFS}}\Delta\varepsilon. \tag{27}$$

where $C_{T,\mathrm{BFS}} \approx 1.07$ MHz/°C and $C_{\varepsilon,\mathrm{BFS}} \approx 0.046$ MHz/$\mu\varepsilon$ are the temperature and strain sensitivities of the Brillouin frequency in standard optical fibers. Although similar sensitivities can be found for different optical fibers, quite different values can be found in some cases due to thermally-induced strain in the fiber coating.

The following sections of this chapter focus on time-domain techniques to measure the features of Brillouin scattering along an optical fiber. As it will be discussed, the intensity and frequency shift associated with spontaneous Brillouin scattering are commonly measured using *optical time-domain reflectometry* (OTDR), in a method called *Brillouin OTDR* (BOTDR). On the other hand, stimulated Brillouin scattering is exploited to perform very accurate measurements of the Brillouin frequency profile along the optical fiber, using a method called *Brillouin optical time-domain analysis* (BOTDA). These two methods allow for reliable distributed sensing of temperature and strain using optical fibers.

Brillouin Optical Time-Domain Reflectometry (BOTDR)

Basic Concepts

The temperature and strain dependence of SpBS can be exploited as an efficient method for distributed sensing employing optical fibers. In principle, the intensity and frequency shift of the SpBS light are both temperature and strain dependent, as described in section "Temperature and Strain Dependence." This makes possible different configurations for distributed sensing. In the time-domain approach, the features of SpBS are interrogated using *optical time-domain reflectometry* (OTDR) (Aoyama et al. 1981), in which a short optical pulse is launched into a sensing fiber and the SpBS light is measured in detection (see Fig. 5). This is called *Brillouin optical time-domain reflectometry* (BOTDR) (Shimizu et al. 1993).

In this approach, a short optical pulse (typically of a few tens of ns) is launched into an optical fiber from a given fiber end (position defined as $z = 0$), as shown in Fig. 5. While the optical pulse propagates through the fiber, spontaneous Brillouin scattering is locally generated at each fiber position with a spatial confinement determined by the pulse duration T. This pulse duration defines a very short Brillouin interaction length, which in turn determines the spatial resolution of the sensor as $\Delta z = v_g T/2$, where v_g is the group velocity of the fiber and the factor 2 accounts for the round-trip propagation. The generated SpBS light propagates back to the fiber input ($z = 0$) where a photodetector is placed. The SpBS intensity is measured by a photo-receiver as a function of time, originating a trace commonly named *BOTDR trace*. Knowing the group velocity of the fiber, the time-domain axis of the measurement can be straightforwardly converted into distance over the sensing fiber as $z = v_g t/2$. As in any OTDR-based system, only a single pulse must propagate through the sensing fiber to enable the easy localization of the SpBS that is generated at each fiber position. Pulses can be however launched into the fiber with a time interval longer than the light round trip time $T_{\text{roundtrip}} = 2L/v_g$ determined by the fiber length L. This way, consecutive BOTDR traces can be acquired and

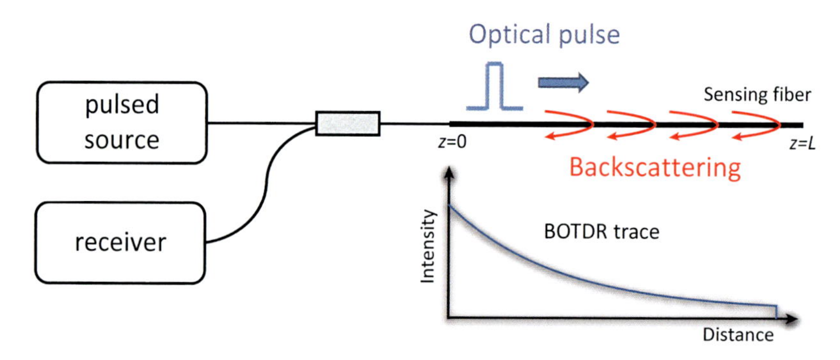

Fig. 5 Principle of Brillouin optical time-domain reflectometry

averaged to reduce noise on the measured traces. By averaging N_{avg} traces, the trace SNR is enhanced by a factor $\sqrt{N_{avg}}$, at the expense of an increment in the data acquisition time (by a factor N_{avg}).

This BOTDR interrogation method can be used to measure either the total intensity or the BFS associated to SpBS as a function of distance. In this section these two approaches for BOTDR sensing are described. The first one uses a simple direct-detection scheme to measure the total SpBS intensity versus distance (Wait and Newson 1996a; Souza and Newson 2001). Contrarily, the second approach employs an optical coherent detection in combination with electrical demodulation to measure BOTDR traces corresponding to different SpBS spectral components (Maughen et al. 2001; Shimizu et al. 1993). By scanning the SpBS spectrum, a full map of the SpBS spectral shape versus distance is then obtained, from which the BFS profile is retrieved by an analysis of the Brillouin peak frequency at each fiber location.

BOTDR Based on Landau-Placzek Ratio

Principle

The simplest implementation of a SpBS-based distributed fiber sensor consists in using optical time-domain reflectometry, as described in Fig. 5, in combination with the direct detection of the Brillouin backscattered light intensity. The principle simply requires sending a short optical pulse through the sensing fiber and detecting in time domain the total SpBS intensity generated in the fiber. Based on this BOTDR approach, a distributed profile of the temperature and strain changes along the fiber can be obtained by quantifying the relative changes in the SpBS intensity at each fiber location (Wait and Newson 1996a; Souza and Newson 2001).

To compensate for any undesired fluctuation of the detected SpBS power, for instance, originated from laser power fluctuations and local losses (such as splices and connectors), the intensity of the Rayleigh backscattered light is used for normalization. This means that, in addition to the BOTDR traces, also Rayleigh OTDR traces must be acquired. As described in Eq. 4, the ratio of the Rayleigh scattering intensity to the spontaneous Brillouin backscattered intensity is known as Landau-Placzek ratio (LPR) (Cummins and Gammon 1966) and provides, in principle, both temperature and strain information along the sensing fiber. For a mono-component glass silica fiber, the LPR can be expressed as (Wait and Newson 1996a)

$$\mathrm{LPR}(z) = \frac{I_{\mathrm{Rayleigh}}(z)}{I_{\mathrm{SpBS}}(z)} = \frac{T_f}{T(z)} \left\{ \beta_T \rho v_a^2(z) - 1 \right\}, \tag{28}$$

where $I_{\mathrm{Rayleigh}}(z)$ and $I_{\mathrm{SpBS}}(z)$ are the local intensities of Rayleigh and spontaneous Brillouin scattering (both Stokes and anti-Stokes components), respectively, β_T is the isothermal compressibility, ρ the silica density in the fiber, T_f the fictive temperature (temperature at which the thermodynamic density fluctuations become fixed in the fiber), and υ_a is the local acoustic velocity, which is temperature and strain dependent.

For a multicomponent glass, Eq. 28 must be modified to account for local fluctuations in the composition, increasing significantly the complexity of the analysis. However, in such a case, it has been demonstrated that the reciprocal temperature dependence of the LPR is still preserved. Although the acoustic wave velocity υ_a depends on both temperature and strain, this dependence is very small compared to the reciprocal temperature term explicitly shown in Eq. 28, which is actually the term ruling the temperature dependence of the LPR (Wait and Newson 1996a). Compared to the temperature dependence, LPR is very poorly sensitive to strain variations, and hence LPR-based BOTDR systems are basically used only for distributed temperature sensing.

Note that, thanks to the small frequency separation between Brillouin and Rayleigh signals (~10.8 GHz for silica fibers at 1550 nm), these two backscattering components are practically affected by the same fiber attenuation coefficient. This means that the exponentially decaying behavior of the OTDR traces over distance is fully canceled out when calculating the LPR.

Taking into account that the precise knowledge of several fiber parameters is required to calculate the absolute temperature employing Eq. 28, the use of such an expression is not practical for real applications. A more practical approach is to compare the LPR profile measured at the unknown temperature $T(z)$ with a reference LPR curve obtained at a reference temperature $T_R(z)$. Thus, to reliably determine the actual temperature profile $T(z)$ along the sensing fiber, the following equation can be used (Souza and Newson 2001):

$$T(z) = \frac{1}{K_T}\left(1 - \frac{\mathrm{LPR}\,(T(z))}{\mathrm{LPR}\,(T_R(z))}\right) + T_R(z), \tag{29}$$

where K_T is the temperature sensitivity of the sensor. Note that in real-field scenarios it not always possible to perform a calibration at a completely uniform temperature profile, and hence the z-dependence of $T_R(z)$ (and of the reference LPR) provides a robust calibration for all locations over the entire sensing length.

Note that to make use of the LPR, only linear spontaneous Brillouin scattering must be activated in the fiber. Stimulated Brillouin scattering actually induces gain and loss of the Stokes and anti-Stokes components, thus disrupting the linear behavior required by the LPR, and biasing the estimated temperature trace.

Generic LPR-Based BOTDR Scheme

Figure 6 shows a generic LPR-based BOTDR scheme. A continuous-wave laser can be used as optical source, followed by a pulse shaper device (such an electro-optic intensity modulator) that generates short optical pulses with high extinction ratio. The pulse duration must be typically of the order of a few tens of nanoseconds to secure a spatial resolution of a few meters (e.g., a pulse of 10 ns results in a spatial resolution of 1 m). Optical amplification, such as the use of an erbium-doped fiber amplifier (EDFA), is typically employed to boost the peak power of the pulse.

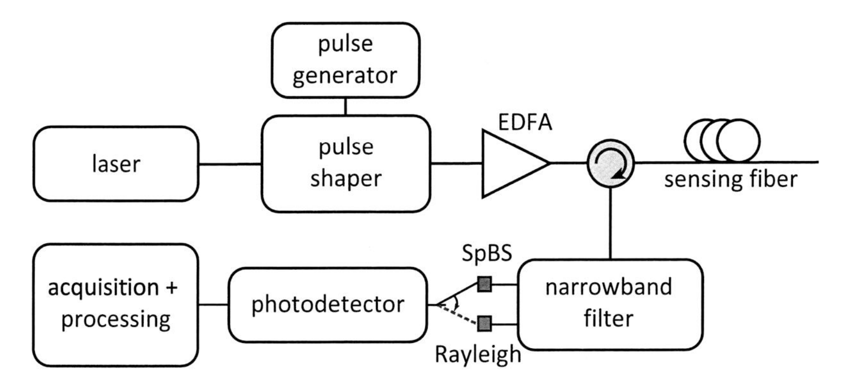

Fig. 6 Block diagram of a generic LPR-based BOTDR sensor scheme

While a high pulse power is typically required to increase the SpBS intensity, and therefore to enhance the SNR of the measured BOTDR traces, the maximum peak pump power that can be launched into the fiber is limited by the onset of nonlinear effects (see section "Constraints in the Pump Pulse Due to Nonlinear Effects").

The Rayleigh and SpBS components originated through the sensing fiber are then extracted by using an optical circulator and an optical filter. Considering that spontaneous Brillouin scattering is characterized by similar Stokes and anti-Stokes intensities, the measurement of only one of the Brillouin components is sufficient to calculate the LPR. Rayleigh and Brillouin components must be separated by a narrowband optical filter before photo-detection, using, for instance, a narrowband fiber Bragg grating (FBG).

One of the main difficulties arising in this intensity-based BOTDR approach comes from the different requirements for the optical source when measuring both Rayleigh and Brillouin backscattered light. Actually, to properly separate the weak Brillouin component from the strong Rayleigh backscattered light at the receiver side, the linewidth of the optical source must be significantly narrower than the frequency separation between both components (i.e., <10.8 GHz in silica fibers). Note that this requirement is essential to achieve a reliable optical filtering to separate Rayleigh and Brillouin components. Unfortunately, the use of a narrowband source generates a random interference pattern in the Rayleigh scattering trace as a result of the coherent interference of the Rayleigh components backscattered from the different scattering points in the fiber. In the context of OTDR systems, this phenomenon is known as *coherent Rayleigh noise* (CRN) (Souza 2006) and introduces coherent amplitude speckles and fading in the backscattered Rayleigh trace. Note that this coherent noise is not present in the Brillouin traces thanks to the random phase fluctuations associated with SpBS (Souza 2006). However, because of imperfections in the optical filtering at the receiver side, Brillouin intensity measurements can still be slightly contaminated by CRN, due to residual Rayleigh component leaking into the photodetector. This effect must be eliminated using narrow optical filters with high out-of-band rejection.

The coherent fluctuations of the Rayleigh intensity trace exhibit a random pattern along the sensing fiber; however, they are static over time (unless environmental conditions change the phase and polarization of the Rayleigh backscattering light). Therefore, conventional signal averaging techniques cannot be effectively employed to reduce the random trace fluctuations induced by CRN.

A possible solution to reduce CRN from the Rayleigh OTDR traces is to use a broadband optical source. In this way, the LPR-based BOTDR system can be implemented using two independent optical sources (Wait and Newson 1996b): one narrowband laser for Brillouin intensity measurements (narrow enough to allow optical separation of the spontaneous Brillouin scattering from the Rayleigh signal) and a second broader source to measure broadband (incoherent) Rayleigh OTDR traces with reduced coherent Rayleigh speckles.

Alternative schemes based on a single narrowband laser can also be employed to reduce CRN. Considering that the coherent interference and, hence, the fiber positions of the coherent speckles depend on the laser optical frequency, a smoother Rayleigh OTDR trace can be obtained by averaging coherent Rayleigh traces obtained with pulses at different optical frequencies. This technique is known as *wavelength-averaging method* and has proven to be efficient to eliminate coherent patterns on Rayleigh OTDR traces (Souza 2006). Another equivalent solution to broaden the linewidth of a narrowband laser is dithering its driving current. This approach allows for the use of a single narrow laser, enabling reliable OTDR trace measurements of the incoherent Rayleigh intensity (when applying current dithering) and of the SpBS intensity (when no current dithering is applied).

BOTDR Based on Brillouin Frequency Shift

Principle

Another approach for BOTDR sensing is based on the detection of BFS variations induced by strain and temperature changes. In this case, a high-power optical pulse is launched into the sensing fiber; however, compared to the LPR approach, here the spectrum of SpBS is measured, rather than its intensity. Instead of using direct detection, an optical coherent detection is required to precisely scan the Brillouin spectrum and to measure the BOTDR traces at different SpBS spectral components (Maughen et al. 2001; Shimizu et al. 1993). For this, the weak spontaneous Brillouin scattering originated in the sensing fiber is first mixed at the receiver with a strong CW optical local oscillator (OLO), as shown in Fig. 7 (note that in this figure the OLO is spectrally located at the same pump optical frequency). Assuming parallel states of polarization of the SpBS and OLO, this heterodyning process results in an electrical (current) beating signal at the photodetector output described as

$$i_{\mathrm{beat}}(t) = 2\Re\sqrt{P_{\mathrm{SpBS}}(t)P_{\mathrm{OLO}}}\cos\left(2\pi f_{\mathrm{beat}} + \varphi_{\mathrm{beat}}\right), \tag{30}$$

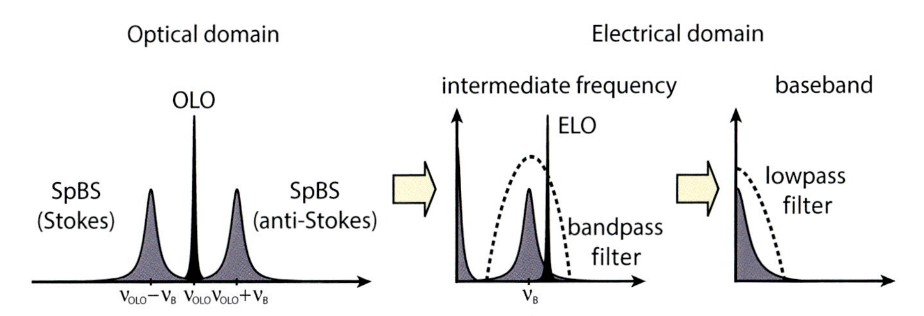

Fig. 7 Principle of optical heterodyne detection followed by electrical demodulation to scan and reconstruct the SpBS spectral response in a BFS-based BOTDR sensor

where $\Re$ is the responsivity of the photodetector, $P_{\mathrm{SpBS}}(t)$ is the time-dependent SpBS power generated along the sensing fiber, P_{OLO} is the (constant) power of the OLO, $f_{\mathrm{beat}} = |\nu_{\mathrm{SpBS}} - \nu_{\mathrm{OLO}}|$ is the beating frequency (commonly called *intermediate frequency*) corresponding to the difference between the optical frequencies of the SpBS and OLO, and $\varphi_{\mathrm{beat}} = |\varphi_{\mathrm{SpBS}} - \varphi_{\mathrm{OLO}}|$ is the phase difference between SpBS and OLO.

Note that this heterodyne detection process amplifies the SpBS signal and shifts the information from the optical domain to the electrical domain, as shown in Fig. 7, providing at the same time-efficient electric filtering of the Brillouin signal. The spectrum of the beating signal is centered on an intermediate frequency given by the frequency difference between the SpBS spectral peak and OLO. This beating signal is selected by an electrical band-pass filter, and then demodulated in the electrical domain, for instance, by mixing it with an electrical local oscillator (ELO), which translates the beating signal to baseband. With the help of a low-pass filter, the envelope of the beating signal is then obtained for different SpBS spectral components.

To measure time-domain traces at difference frequencies within the SpBS spectrum, the frequency of the electrical local oscillator is scanned within a predefined spectral range. The spatial resolution of the system is defined by the pulse width; however, it must be ensured that the bandwidths of the band-pass filter (used to select the beating signal at intermediate frequency) and the low-pass filter (to select the portion of the scanned spectrum of the baseband signal) are large enough to avoid degradation of the spatial resolution. This coherent detection allows for the reconstruction of the SpBS spectrum versus distance over the fiber. Note that the time traces at different SpBS spectral components can also be obtained using other schemes, as, for instance, employing I/Q detection making use of a 90° optical hybrid and balanced detection.

After reconstructing the Brillouin spectral shape at each position, the temperature and strain information is estimated along the fiber by retrieving the distributed BFS profile. This BFS profile can be obtained by fitting a polynomial or Lorentzian curve to the measured local SpBS spectrum at each fiber position (details on the impact of this fitting can be found in section "Error on the Estimated Brillouin Frequency Shift"). Knowing the temperature and strain sensitivities of the fiber, the BFS profile can then be converted into strain or temperature along the fiber by simply inverting Eq. 27. Note however that discriminating temperature from strain is not possible with a single BFS profile. This indeed represents one of the main limitations of Brillouin sensing techniques.

To provide long-term stability to the system, the OLO is typically obtained from the same optical source employed to generate the pulses launched into the sensing fiber. Although there exist some attempts of using homodyne detection of the SpBS light for BOTDR sensing, the use of heterodyne detection offers much better performance because it avoids the presence of low-frequency noise. Considering that a unique optical source is used in this case to generate the optical pulses and the OLO, this scheme is usually called *self-heterodyne BOTDR* (Shimizu et al. 1993). This is today the most common and reliable method for BOTDR sensing.

Different alternatives exist for the implementation of this heterodyne detection system. The most straightforward approach is to directly mix the SpBS coming from the fiber with the OLO (both being spectrally separated by the BFS of the fiber), as depicted in Fig. 7. This scheme however requires a photo-receiver and electrical demodulation system able to operate at high frequency (frequencies larger that the BFS, i.e., $>\sim$10.8 GHz in silica fibers) (Maughen et al. 2001). A possible alternative is to shift the optical frequency of either the optical pulse or OLO using a frequency translation device, such an electro-optic modulator or a high-speed acousto-optic modulator. This spectral shift reduces the optical frequency difference between OLO and one of the SpBS components. This way more sensitive low-frequency photo-receivers can be used, while the operating frequency of the electrical demodulation system can also be significantly reduced.

Note that optical coherent detection is highly sensitive to the relative polarization between the OLO and SpBS reaching the detector. The amplitude of the beating signal is maximum only when OLO and SpBS have the same polarization and nullifies with orthogonal polarizations. Due to the low birefringence of standard optical fibers, the polarization of SpBS typically varies randomly along the optical fiber, leading to fluctuations (fading) in the measured BOTDR trace amplitude. These fading reduce the accuracy of Brillouin spectral measurements and therefore must be avoided by using a polarization diversity scheme. This can be implemented, for example, using a polarization scrambler or polarization switch in the OLO branch, in combination to the averaging of BOTDR traces measured with different OLO polarizations.

Compared to the use of direct-detection LPR-based systems, the use of coherent self-heterodyne detection of the spontaneous Brillouin spectrum has the following advantages (Shimizu et al. 1993):

1. Considering that the Brillouin frequency of a standard single-mode fiber is about 10.8 GHz (at 1550 nm), heterodyne detection combined with electrical filtering allows for an excellent separation of the Brillouin and Rayleigh components.

This filtering eliminates any potential contamination of the Brillouin signal with the coherent Rayleigh speckles resulting from the use of a narrowband laser. This is indeed an important feature since SpBS is very weak compared to Rayleigh scattering, making this spectral separation very difficult to achieve in the optical domain. The narrow electrical filtering also enables a higher precision in the SpBS spectral reconstruction compared to the use of optical filters.

2. The environmental information in a BFS-based BOTDR sensor is contained in the frequency domain rather than in the SpBS intensity, as in a LPR-based sensor. This makes the system inherently more robust against external perturbations affecting the backscattered spontaneous Brillouin power.
3. Mixing the weak SpBS with a strong OLO allows for a high receiver sensitivity, making the detection ideally shot-noise-limited. This represents a significant improvement compared to the sensitivity of direct-detection schemes, where the load resistance thermal noise typically determines the sensitivity.

Generic Self-Heterodyne BOTDR Scheme

Figure 8 shows a generic schematic of a BOTDR sensor in which the SpBS spectrum is measured by a self-heterodyne detection combined with an electrical coherent demodulator. In this scheme a single optical source, corresponding to a narrowband semiconductor laser (typically of ~1 MHz linewidth or narrower), is used. The laser light is divided into two beams by an optical splitter. One of these beams is used as an optical local oscillator, which must be depolarized

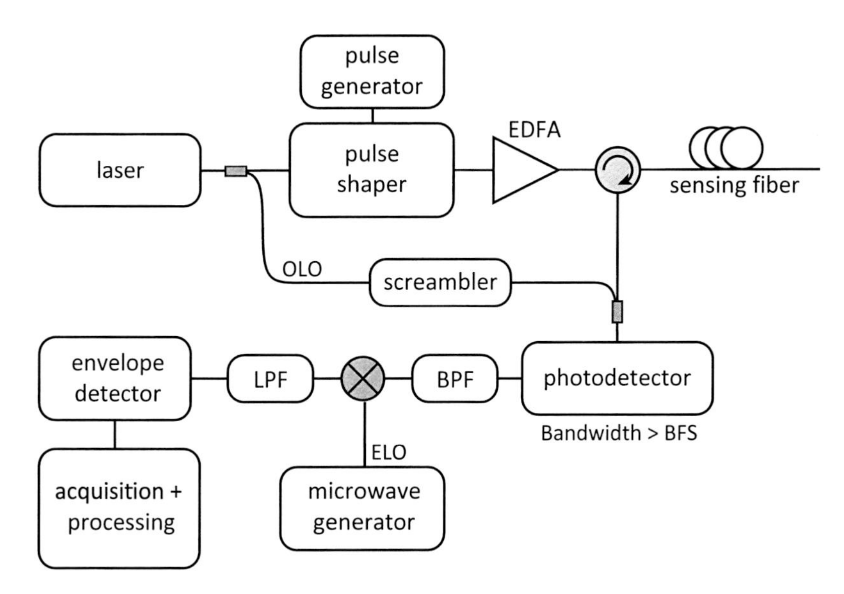

Fig. 8 Block diagram of a generic BFS-based BOTDR sensor scheme, using optical heterodyne detection and an electrical demodulator based on coherent electrical detection

(e.g., by using a polarization scrambler) before being sent into the receiver stage. The other beam is used to generate the optical pulse by using an intensity-modulating device that provides high extinction ratio, such as an electro-optic modulator or a semiconductor optical amplifier. Pulses are typically amplified by an EDFA and then launched into the sensing fiber through a three-port optical circulator, which also allows for coupling the backscattered signal into the receiver. Limitations to the peak power and extinction ratio of the pulses are described in sections "Constraints in the Pump Pulse Due to Nonlinear Effects" and "Requirements for the Pulse Extinction Ratio," respectively.

In the receiver side, the backscattering light is mixed with the depolarized CW optical local oscillator, whose power must be ideally adjusted to be just below the saturation point of the detector. The electrical beating signal is selected by a band-pass filter (BPF) and then demodulated by an electrical coherent detection scheme. For this, an electrical local oscillator (ELO) at a frequency close to the BFS of the fiber is used to scan the Brillouin spectrum in a frequency range of typically a few hundred MHz (this range is basically determined by the maximum temperature and strain variations to be measured). The electrical local oscillator is used to convert the SpBS spectrum at intermediate frequency into baseband, so that then by employing a low-pass filter (LPF) and an envelope detector, time traces are measured at different SpBS spectral components.

Brillouin Optical Time-Domain Analysis (BOTDA)

Principle

Another approach to perform distributed sensing is to retrieve the Brillouin frequency profile along the sensing fiber exploiting stimulated Brillouin scattering. Compared to the SpBS approach used in BOTDR, in this case, in addition to a short optical pump pulse, a narrowband continuous-wave light (commonly called *probe wave*) is also launched into the fiber but in opposite direction with respect to the pulse propagation (Horiguchi et al. 1995). By definition, the fiber end by which the pulse is injected is referred as the *near fiber end* ($z = 0$), while the end by which the CW probe is launched is referred as the *far fiber end* ($z = L$, L being the sensing fiber length). When the frequency difference between pump and probe matches the resonance condition for Brillouin scattering, efficient power transfer between pump and probe takes place as a result of the SBS interaction.

Unlike BOTDR systems, where the SpBS response is spectrally scanned using heterodyne detection and electrical filtering, in this case, the SBS spectral response is probed and analyzed at specific optical frequencies making use of this narrowband CW probe. This method is referred as *Brillouin optical time-domain analysis* (BOTDA) (Horiguchi et al. 1995) and allows for a very precise interrogation of the Brillouin spectral shape, which is sampled with a spectral accuracy mainly given by the linewidth of the CW probe (being usually $\sim$1 MHz or narrower).

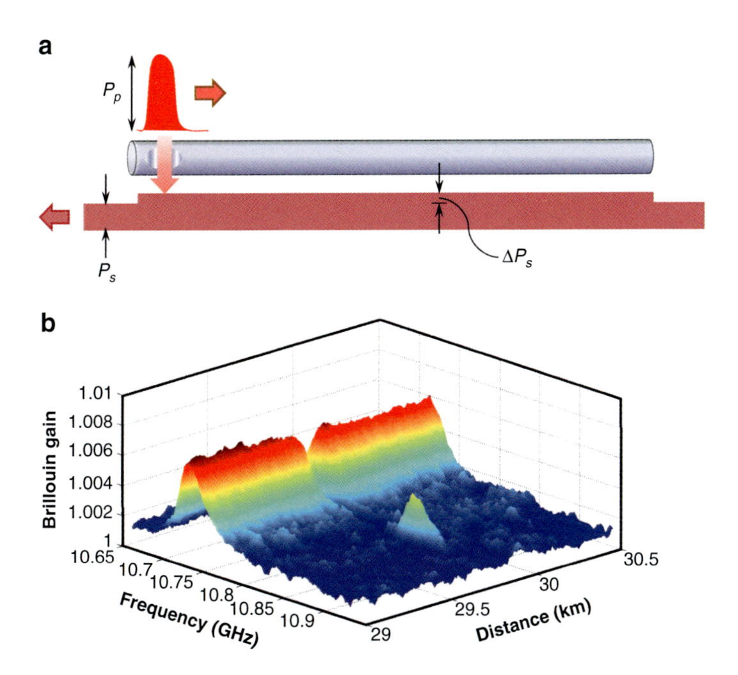

Fig. 9 Principle of BOTDA sensing. (**a**) Pump-probe SBS interaction giving rise to a local power transfer ΔP_s (Soto and Thévenaz 2013a). (**b**) Reconstructed BGS shape versus distance over the sensing fiber

To reconstruct the Brillouin gain spectrum, defined by the real part of Eq. 23, as a function of distance, the pump and probe frequency difference must be swept within a range of a few hundred MHz around the BFS of the fiber (i.e., around ~10.8 GHz, at 1550 nm in standard single-mode fibers). As shown in Fig. 9a, CW probe power variations resulting from the Brillouin interaction with the pulse are measured as a function of time, for different pump-probe frequency offsets. Note that, SBS makes the contrast of the detected signal power ΔP_s to be typically much higher than the one of SpBS-based BOTDR systems, leading normally to time traces with higher SNR. Following an OTDR approach, the time-domain axis of the measurement is converted into distance over the fiber using the group velocity v_g, while the spatial resolution Δz of the sensor is determined by the duration T of the pump pulse, so that $\Delta z = v_g T/2$. As a result of the time-domain acquisition of these CW probe intensity variations at different scanned pump-probe frequency offsets, a full map of the Brillouin gain spectrum can be obtained at each location of the sensing fiber, as shown in Fig. 9b.

Note that depending on the pump and probe optical frequencies, power can be transferred from the pump to the probe or vice versa. Thus, there exist two possible configurations for a BOTDA sensor:

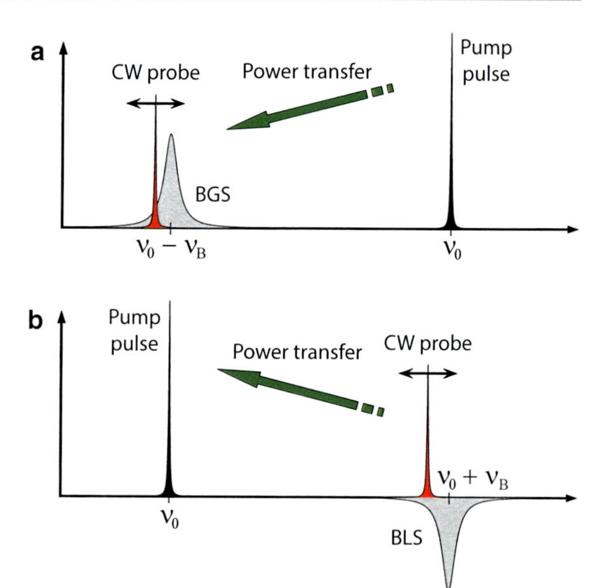

Fig. 10 Pump-probe power transfer in a (**a**) Brillouin gain configuration and (**b**) Brillouin loss configuration

1. Brillouin gain configuration: In this case, the pump optical frequency is higher than the frequency of the CW probe involved in the SBS interaction (Fig. 10a). Thus, power is transferred from the pump to the CW probe, resulting in amplification of the probe and depletion of the pump.
2. Brillouin loss configuration: In this scheme the optical frequency of the pump pulse is lower than the frequency of the CW probe (Fig. 10b). Power is transferred from the CW probe to the pump, thus attenuating the CW probe. The local loss induced into the probe gives information of the local Brillouin spectral shape, which in this case is called *Brillouin loss spectrum* (BLS).

For the sake of clarity, and without loss of generality, the descriptions in this chapter are mostly dedicated to the Brillouin gain configuration, bearing in mind that the analysis hereafter presented is equivalently valid for the Brillouin loss configuration, simply assuming a negative gain on the probe.

Conventional BOTDA Scheme

In principle, a BOTDA system requires two narrowband optical beams (Horiguchi et al. 1995). One of them is used to generate the pump pulse, while the other is used as a CW probe signal. Two independent laser sources could be used, but one of the main practical difficulties in this case is associated with the long-term relative frequency stability between the two light emissions. On the one hand, a laser linewidth much narrower than the natural BGS width (~27 MHz) is required to avoid detrimental broadening of the measured BGS shape; while, on the other hand, the narrowband requirement together with long-term stability are critical to secure a precise pump-probe frequency scanning. Since the BGS must be scanned with high precision (usually with steps of $\sim$1 MHz), any drift in the lasers frequency difference can easily impair the quality of the BGS reconstruction.

To enhance the long-term stability of the pump-probe frequency offset, a scheme using a single laser and a frequency shifter is normally employed (Soto and Thévenaz 2013a; Niklès et al. 1997). Figure 11a illustrates a simple, generic BOTDA sensor scheme using a single laser. The light from this single narrowband laser is split into two arms to produce both the pulsed pump and the CW probe. In the figure, the upper branch shows the generation of the pump pulse, in which the CW laser light is intensity modulated by a fast pulse-shaping device, such as an electro-optic modulator, a semiconductor optical amplifier, or any other modulating device generating optical pulses with high extinction ratio. The pulse peak power is typically boosted by an optical amplifier (e.g., by an EDFA) to launch high-peak power pulses into the fiber. This peak power is constrained by the onset of nonlinear effects. As explained in section "Modulation Instability (MI)," modulation instability (MI) is the main limiting effect, constraining the pulse peak power to a maximum of $\sim$100 mW, in the case of long sensing fibers ($L > \alpha^{-1} \approx 22$ km). For shorter fibers, this power limit increases with an inverse proportion to the fiber effective length.

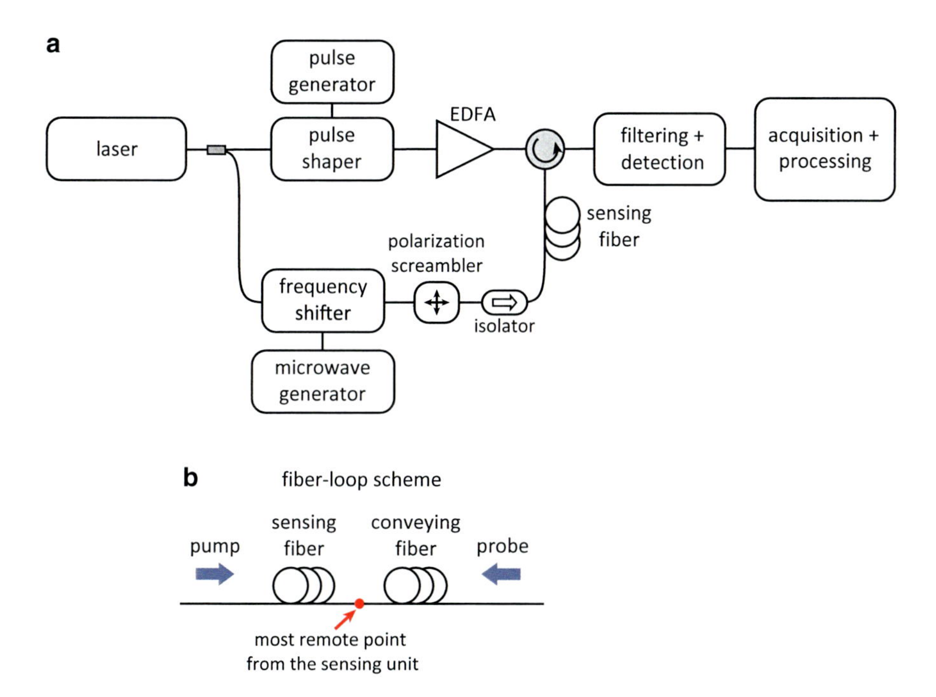

Fig. 11 (**a**) Generic schematic of a conventional BOTDA sensor. (**b**) Fiber-loop configuration

On the other hand, the lower branch in Fig. 11a shows the generation of the probe signal using an optical frequency-shifting device, in which the pump-probe frequency offset is precisely controlled to scan the BGS of the sensing fiber. One of the most common implementations uses an electro-optic modulator driven by a microwave signal to generate two spectral sidebands around the CW laser optical frequency, while also enabling efficient carrier suppression (Niklès et al. 1997). This configuration is called *dual-sideband probe scheme* and is highly preferred in BOTDA sensing since it enhances the robustness of the system against pump depletion issues, as it will described in section "Pump Depletion and Systematic Errors in a Dual-Sideband Probe Scheme." Note that in this dual-sideband scheme, the carrier of the intensity-modulated probe has the same optical frequency than the pump pulse, and therefore to avoid interference between these two counter-propagating waves, carrier suppression in the intensity modulation of the probe is essential. This can be simply obtained using a proper DC bias voltage in the electro-optic modulator (Niklès et al. 1997).

Note that SBS highly depends on the relative polarization between the pump and probe. Thus, since the polarization state of these two signals varies randomly during propagation over a standard single-mode fiber, measured traces are typically affected by polarization fading, which lead to unreliable measurements. To eliminate the harmful impact of these polarization fading, a polarization diversity scheme (usually composed of a polarization scrambler or polarization switch) is required in either the pump or probe arm.

In the receiver stage, a narrowband optical filter, such as a fiber Bragg grating (FBG), is used before photo-detection to select one of the probe sidebands launched into the fiber and to filter out all unwanted spectral components: i.e., to eliminate the Rayleigh backscattered light generated by the pump, the second unwanted probe sideband, and optical noise resulting from optical amplification. The selected probe sideband can be either the lower or higher frequency band of a dual-sideband scheme, thus defining a Brillouin gain or loss configuration. This selected sideband, containing the distributed SBS information along the fiber, is then detected by a photo-receiver followed by a data acquisition and processing unit.

It must be mentioned that this description only takes into account the basic building elements that are strictly essential for the basic operation of a BOTDA sensor. There are certainly countlessly diversified possible implementations, including in some cases complex hardware sophistications to improve the quality of the measurements. Some of those schemes will be discussed later in this chapter.

Note that a BOTDA sensor requires access to both fiber ends, representing a significant penalty compared to BOTDR systems. Although a fiber of length L can provide a sensing range equal to L, the real remoteness of the sensor is limited to $L/2$, which is the real maximum distance between the sensing unit and the farthest sensing point. To increase the real remoteness of the sensor, a fiber scheme commonly called *fiber-loop configuration* can be used (see Fig. 11b) (Dong et al. 2012; Soto et al. 2014a). Compared to the classical fiber scheme, where the entire fiber light is used for sensing, in the fiber-loop scheme, only half of optical fiber length is used for sensing (this is the actual *sensing fiber*), while the other half (commonly named as *carry-over fiber* or *conveying fiber*) is only used to send the probe signal up to the most remote sensing location, corresponding to the farthest end of the sensing fiber. Since a sensing length L requires a total fiber length equal to $2L$ in this case, the performance of the system is massively affected (Soto and Thévenaz 2013a; Soto et al. 2014a). This is because the probe cumulates twice fiber losses in this case, and, hence, very low SNR must be expected in implementations targeting long and ultra-long sensing ranges.

The description of the sensor hereafter presented is, in principle, referred to the classical BOTDA scheme in which the entire optical fiber defines the sensing section (Fig. 11a). However, the entire analysis is equally applicable to the fiber-loop configuration, simply considering that the probe power at the end of the conveying fiber corresponds to the input probe power in the sensing fiber.

Mathematical Description of the BOTDA Working Principle

Sensor Response

To obtain a mathematical description of the time traces measured by a BOTDA sensor, a perturbation method can be used to solve the system of coupled Eqs. 21 and 22 describing SBS in CW regime. For a fixed pump-probe frequency offset $\Delta\nu$, and assuming parallel states of polarization for pump and probe, the amount of power transferred from the pump $P_p(z)$ to the probe $P_s(z)$ in a Brillouin gain scheme is given by

$$\Delta P_s(\Delta\nu, z) = P_s(z)\left[\exp\left(\frac{g_B(\Delta\nu, z)}{A_{\text{eff}}} P_P(z)\Delta z\right) - 1\right], \tag{31}$$

where $g_B(\Delta\nu, z)$ represents the local BGS at the set pump-probe frequency offset $\Delta\nu = \Delta\Omega/2\pi$ as defined by the real part of Eq. 23, A_{eff} is the fiber effective area, and Δz is the interaction length, which defines the spatial resolution of the sensor (Soto and Thévenaz 2013a). Note that depending on the relative pump and probe optical frequencies, $g_B(\Delta\nu, z)$ can be defined as positive or negative to represent either a Brillouin gain or loss configuration.

The quantity described by Eq. 31 corresponds to the power variation that the CW probe experiences at a location z and depends on the used pump-probe frequency offset $\Delta\nu$. This term corresponds to the *local sensor response*, in which the gain value $g_B(\Delta\nu, z)$ is assumed to be constant over the spatial resolution. Since Eq. 31 is derived from a perturbation method using the SBS coupled Eqs. 21–22 in CW regime, this response is, in principle, only valid for pulses longer than the acoustic response time $\tau_A \approx 12$ ns. However, considering that the most advanced BOTDA configurations for high spatial resolution make use of pre-excited acoustic waves (see section "Time-Domain Methods for High Spatial Resolution Brillouin Sensing"), Eq. 31 can also be safely used for submeter spatial resolution systems, provided that the pre-excitation is long enough to let the acoustic wave reach its steady-state value.

To avoid longitudinal and spectral distortions in the measurements, the net SBS gain and loss experienced by the probe and pump over the entire sensing fiber must be maintained below a certain level. This is particularly critical for the pump pulse, which is continuously depleted (or amplified) by the SBS interaction with the CW probe over the entire sensing range. This pump depletion (in Brillouin gain configuration) or excess amplification (in Brillouin loss configuration) results in nonlocal effects and systematic errors that must be avoided (see section "Constraints in the Probe Signal of BOTDA Sensors"). This means that most BOTDA sensors (especially those operating over km ranges) must work in a small-gain regime, i.e., fulfilling the condition $\Delta P_s(z)/P_s(z) \ll 0.1$. Under this condition, the local sensor response defined in Eq. 31 can be linearized and simplified as (Soto and Thévenaz 2013a)

$$\Delta P_s\left(\Delta\nu, z\right) = \frac{g_B\left(\Delta\nu, z\right)}{A_{\text{eff}}} P_p(z) P_s(z) \Delta z. \tag{32}$$

In most well-designed medium- to long-range BOTDA sensors, this relative power transfer is actually in the order of a few percent along the first kilometers of fiber (for spatial resolutions of a few meters). Under this small-gain regime, pump depletion can be neglected, while the absolute value of the sensor response turns out to be similar in gain and loss configurations. Under a negligible depletion condition, the pump power only decays exponentially over distance due to the linear fiber loss. Similarly, the small-gain approximation makes the probe power P_s negligibly modified by the Brillouin interaction and therefore also predominantly ruled by the linear loss. Losses experienced by pump and probe are identical in normal conditions owing to their very close spectral vicinity; however, for the sake of generality, some of the following equations are expressed using two distinct loss coefficients. This distinction is mostly irrelevant from a practical viewpoint, but it is helpful for the physical interpretation. Under these realistic and practical assumptions, the local sensor response $\Delta P_s(z)$ can be expressed as a function of the input pump and probe powers (P_{pi} and P_{si}) as (Soto and Thévenaz 2013a)

$$\begin{aligned}\Delta P_s\left(\Delta\nu, z\right) &= \frac{g_B\left(\Delta\nu, z\right)}{A_{\text{eff}}} P_{pi} \exp\left(-\alpha_p z\right)\; P_{\text{si}}\; \exp\left[-\alpha_s\left(L - z\right)\right] \Delta z \\ &= \frac{g_B\left(\Delta\nu, z\right)}{A_{\text{eff}}} P_{pi}\; P_{si}\; \exp\left(-\alpha L\right) \Delta z, \quad \text{where } \alpha = \alpha_p = \alpha_s.\end{aligned} \tag{33}$$

Note that Eq. 33 indicates that the local sensor response, i.e., the power that is locally transferred between pump and probe, is z-independent and constant over the entire sensing fiber when the linear Brillouin gain $g_B(\Delta\nu, z)$ is uniform. This behavior results from the counter-propagative configuration required for the interacting optical waves in stimulated Brillouin scattering, as shown in Fig. 12. Indeed, the local power transfer is scaled by the product of local pump and probe

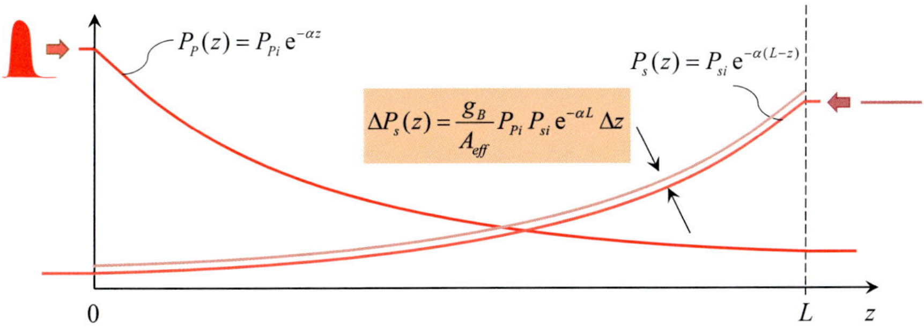

$P_p(z) = P_{Pi}\, e^{-\alpha z}$

$P_s(z) = P_{si}\, e^{-\alpha(L-z)}$

$\Delta P_s(z) = \frac{g_B}{A_{eff}} P_{Pi} P_{si}\, e^{-\alpha L} \Delta z$

0 L z

Fig. 12 Pump and probe in a BOTDA sensor decay exponentially over distance due to the linear fiber loss. Due to this longitudinal evolution, the product of the local pump and probe powers at each position z over the fiber is constant and independent of position z (Soto and Thévenaz 2013a)

powers, which turns out to be constant at any position z. In other words, while the pump power is large at position $z = 0$, the probe power is low since it has been fully attenuated after propagation through the entire fiber length L. The power situation of the two waves is exactly reversed in the same proportion at the far fiber end $z = L$, and for uniform attenuation, it can be generalized that these powers are in exact inverse proportion at any position z, as described in Eq. 33.

It should be considered that the power locally transferred at each fiber position z, and described in Eq. 33, is measured at the near fiber end ($z = 0$). Therefore, the probe signal locally amplified or depleted at position z must propagate from z to 0 before detection. This propagation results in an attenuation of the probe over this distance, leading to a *sensor response* $\Delta P_s^0(\Delta\nu, z)$ measured at the near fiber end described as (Soto and Thévenaz 2013a)

$$\Delta P_s^0(\Delta\nu, z) = \Delta P_s(z) \exp(-\alpha_s z) = \frac{g_B(\Delta\nu, z)}{A_{\text{eff}}} P_{pi} \exp\left(-\alpha_p z\right) P_{si} \exp(-\alpha_s L)\, \Delta z. \tag{34}$$

This expression actually corresponds to the *measured sensor response* and is proportional to the local gain $g_B(\Delta\nu, z)$, the spatial resolution Δz, the local pump power $P_{pi} \exp(-\alpha_p z)$, and the CW probe power $P_{si} \exp(-\alpha_s L)$ reaching the near fiber end ($z = 0$) in absence of SBS interaction. To calculate the local Brillouin gain/loss associated with this response as a function of the fiber position z, the measured sensor response $\Delta P_s^0(\Delta\nu, z)$ must be divided by the CW probe power measured in the absence of SBS interaction, yielding

$$G_s(\Delta\nu, z) = \frac{\Delta P_s^0(\Delta\nu, z)}{P_{si} \exp(-\alpha_s L)} = \frac{g_B(\Delta\nu, z)}{A_{\text{eff}}} P_{pi} \exp\left(-\alpha_p z\right) \Delta z. \tag{35}$$

Considering that pump and probe experience the same linear attenuation, Eq. 34 indicates that the more distant the position z, the smaller the measured sensor response. The worst case corresponds to the trace amplitude at the far fiber end ($z = L$). Consequently, under standard BOTDA conditions, the sensor accuracy must be always evaluated at the end of the sensing range, where the measured sensor response is (Soto and Thévenaz 2013a)

$$\Delta P_s^0 (z = L) = \frac{g_B}{A_{\text{eff}}} P_{pi} \; P_{si} \; \exp(-2\alpha L) \, \Delta z. \tag{36}$$

Equation 36 points out that the sensor response corresponding to the end of the sensing fiber and measured at $z = 0$ depends on the round-trip fiber attenuation and therefore is scaled by the factor exp($-2\alpha L$). This factor defines how the sensing distance impacts on the signal-to-noise ratio of BOTDA traces, in particular at the far end where the lowest absolute signal amplitude is measured. This behavior indicates that extending the sensing distance of a long-range BOTDA sensor by a distance ΔL leads to an SNR reduction in a factor exp($-2\alpha \Delta L$), provided that given (optimal) pump and probe powers are launched into the fiber. This factor is slightly different in short fibers since input pump and probe powers can be modified depending on the effective fiber length, thus also impacting on the overall measurement SNR.

Brillouin Gain Spectral Shape

Note that the natural Brillouin spectrum described by the real part of Eq. 23 corresponds to a Lorentzian-shaped profile that is obtained from CW pump-probe interaction. However, the spectral shape measured by a Brillouin time-domain sensor (not only BOTDA but also BOTDR) does not necessarily correspond to this natural BGS shape, but is highly dependent on the temporal and spectral features of the pump pulse. In particular, the measured BGS in a time-domain Brillouin sensor is determined by the convolution of the natural Lorentzian-shaped BGS and the normalized power spectral density of the pump pulse (Fellay et al. 1997; Alem et al. 2017).

Brillouin time-domain sensors require the use of an ideal rectangular-shaped pump pulse (with steep rising and falling times) to avoid impairments originated by self-phase modulation, as it will be clarified in section "Self-Phase Modulation (SPM)." Using this pulse temporal shape, the Brillouin spectral profile measured by the sensor can be fully described by a closed-form mathematical expression. The normalized power spectral density of a rectangular pulse of duration T is given by a sinc-square function as $S_p(\Omega) = T \; \text{sinc}^2(T\Omega/2\pi)$, which convolved with the natural BGS shape in Eq. 23 results in the following measured BGS shape (Alem et al. 2017):

$$g_T(\Delta\Omega) = g_B(\Delta\Omega)\left(1 - \frac{2\left(\Gamma_B^2 - 4\Delta\Omega^2\right)\left(1 - e^{-\Gamma_B T/2}\cos(\Delta\Omega T)\right) + 8\Gamma_B \Delta\Omega e^{-\Gamma_B T/2}\sin(\Delta\Omega T)}{\Gamma_B T\left(\Gamma_B^2 + 4\Delta\Omega^2\right)}\right), \tag{37}$$

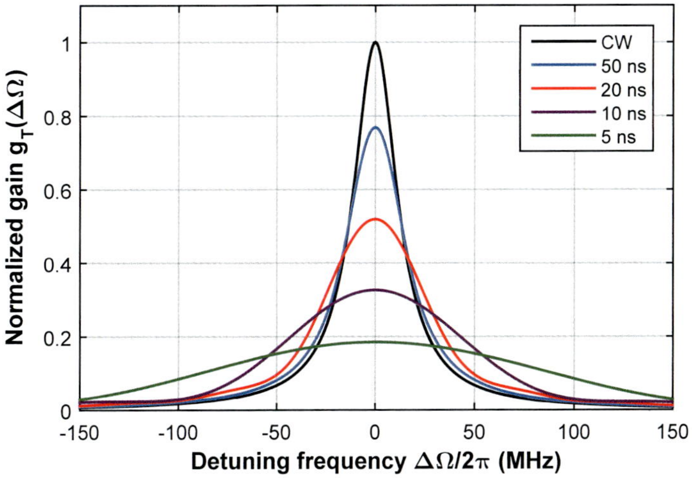

Fig. 13 Brillouin gain spectrum for different pump pulse durations, normalized to coefficient g_{B0}

where $\int_{-\infty}^{\infty} S_p(\Omega)\, d\Omega = 1$, $\Delta\Omega = 2\pi\Delta\nu$ is the frequency detuning around the Brillouin frequency, $g_B(\Delta\Omega)$ is the natural Lorentzian-shaped BGS, and Γ_B is its FWHM ($\Gamma_B \sim 2\pi \cdot 27$ Mrad/s in standard fibers at 1550 nm).

Figure 13 shows the BGS described in Eq. 37 for different pump pulse durations. The figure points out that the spatial resolution of the sensor (typically in the range from 1 to 10 m, corresponding to pulse widths in the range from 10 to 100 ns) has two significant effects on the measured BGS spectral shape:

1. Maximum Brillouin gain. When the pump pulse duration is of the order of the SBS acoustic wave response time (~12 ns), the maximum SBS gain at the Brillouin frequency (i.e., at $\Delta\Omega = 0$) is reduced due to the incomplete activation of the acoustic field. In particular, the maximum Brillouin gain that can be achieved with a rectangular pump pulse of duration T is (Alem et al. 2017)

$$g_{\max} = g_{B0}\left(1 - \frac{2\left(1 - e^{-\Gamma_B T/2}\right)}{\Gamma_B T}\right), \tag{38}$$

where g_{B0} is the Brillouin gain coefficient ($\sim 3 \cdot 10^{-11}$ m/W in standard silica fibers).

Figure 14a illustrates the dependence of the gain factor $g_{\max}/g_{B0}$ on the pump pulse width, as given by Eq. 38. The figure shows that the Brillouin gain reduces significantly for pump pulses shorter than 50 ns, becoming extremely low for pulses shorter than the acoustic response time (i.e., $T < \tau_A \approx 12$ ns). For very short pulses ($T \ll \tau_A$), the maximum gain actually reduces in direct proportion

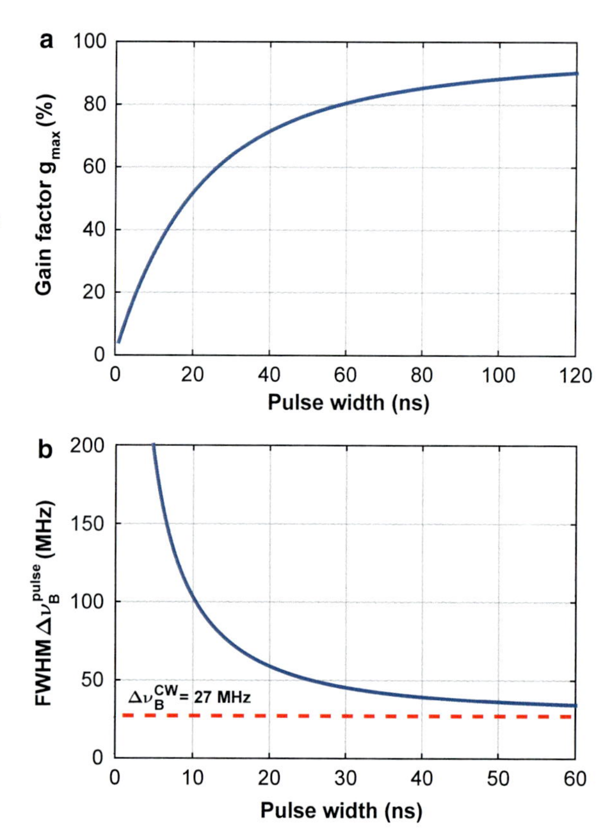

Fig. 14 Impact of the pump pulse width (spatial resolution) on the maximum gain and width of the BGS obtained by a time-domain Brillouin sensor (assuming a natural FWHM of 27 MHz). (**a**) Gain reduction factor with respect to the CW maximum gain g_{B0} and (**b**) FWHM vs pulse width (Alem et al. 2017)

to the pulse width (i.e., $g_{max} \propto T$), resulting in a big penalty for the performance of the sensor.

2. Spectral full-width at half-maximum: Note in Fig. 13 that the partial activation of the acoustic wave with spatial resolutions below ∼5 m (i.e., pulse widths <50 ns) does not only reduce the maximum Brillouin gain but also broadens the spectrum of the measured BGS shape. Based on Eq. 37, the FWHM of the BGS $\Delta\nu_B^{\text{pulse}}$ measured with a pulsed pump can be expressed as (Alem et al. 2017)

$$\Delta\nu_B^{\text{pulse}} = \Delta\nu_B^{\text{CW}} \sqrt{\frac{2 + \sqrt{4 + \psi^2}}{\psi}}, \tag{39}$$

where $\psi = \Gamma_B T/2 + e^{-\Gamma_B T/2} - 1$ and $\Delta\nu_B^{\text{CW}} = \Gamma_B/2\pi$ is the FWHM of the BGS measured in continuous wave. Figure 14b shows this broadening effect on the measured BGS when using relatively short optical pulses. The figure indicates that, for instance, a spatial resolution of 2 m (i.e., pump duration of $T = 20$ ns) leads to a $\Delta\nu_B^{\text{pulse}} = \sim 60$ MHz, while this width broadens up to ∼100 MHz for a spatial resolution of 1 m. When using very short pulses ($T \ll \tau_A$), the exponential function $e^{-\Gamma_B T/2}$ can be approximated by the first three terms of its Taylor expansion, so that $\psi = \Gamma_B^2 T^2/8$, which leads to a BGS FWHM being proportional to the inverse of the pulse duration, as $\Delta\nu_B^{\text{pulse}} = 2\sqrt{2}/\pi T \approx 0.9/T$ (Alem et al. 2017).

The reduction of the maximum gain and broadening of the measured BGS when using pump pulses shorter than the acoustic wave response time ($T < 12$ ns) increases dramatically the frequency uncertainty on the BFS estimation, as indicated later in Eq. 40. This imposes serious limitations to the best spatial resolution that can be attained with a Brillouin time-domain distributed sensor (both BOTDA and BOTDR). It is usual to consider that the best spatial resolution is ultimately limited down to 1 m (corresponding to a pulse width of 10 ns) (Fellay et al. 1997), while the use of pump pulses shorter than 10 ns normally results in inaccurate and unreliable measurements. To achieve spatial resolutions below 1 m, dedicated techniques are required. While many of them focus on frequency or correlation domain methods, time-domain schemes for high spatial resolution typically exploit approaches to pre-activate the Brillouin acoustic wave and avoid the low gain and broadening effects depicted in Fig. 14 (see section "Time-Domain Methods for High Spatial Resolution Brillouin Sensing" for a description on high spatial resolution techniques).

Error on the Estimated Brillouin Frequency Shift

In order to retrieve the temperature and strain profile along a sensing fiber, the local BFS (i.e., peak frequency of the local BGS) must be first estimated at each fiber location. A distributed profile of temperature or strain is then estimated by converting the BFS profile into the measurand using the respective sensitivity coefficients (inverting Eq. 27). The uncertainty in the temperature and strain estimations is therefore ultimately determined by the accuracy on the determination on the BFS profile along the fiber.

A conventional approach to retrieve the peak frequency of the local Brillouin spectrum (in both BOTDR and BOTDA) at each fiber position z consists in fitting the measured spectrum, so that the local BFS is obtained from the fitting parameters, as illustrated in Fig. 15. The process typically considers only data points above a given fraction of the peak value. These points are then used, for instance, in a quadratic least-square fitting process, from which the local BFS is estimated. When only the data points within the Brillouin FWHM $\Delta\nu_B^{\text{pulse}}$ are used in the fitting and Gaussian additive noise dominates the measurement, the uncertainty σ_ν on the BFS estimation can be written as (Soto and Thévenaz 2013a)

$$\sigma_\nu(z) = \frac{1}{\text{SNR}(z)}\sqrt{\frac{3}{4}\delta \cdot \Delta\nu_B^{\text{pulse}}}, \tag{40}$$

where SNR(z) is the local SNR of the BOTDA trace at position z and at the BGS peak frequency, $\Delta\nu_B^{\text{pulse}}$ is the FWHM of the measured BGS, as defined in Eq. 39, and δ is the scanning frequency step.

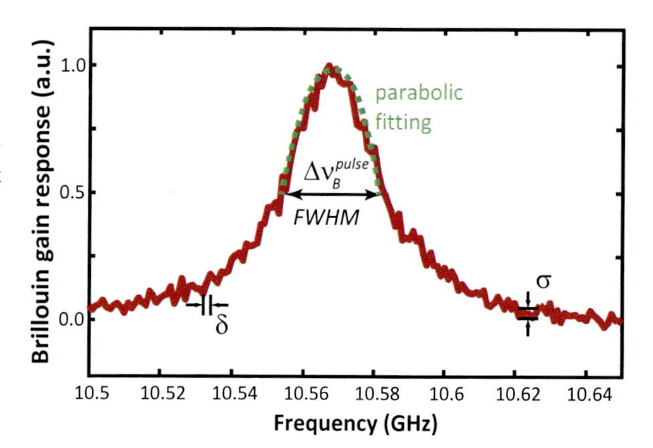

Fig. 15 Typical local BGS shape measured by a Brillouin sensor (after normalization). Noise on the signal (σ) induces uncertainty in the BFS estimation, subject to statistical errors that depend on the Brillouin FWHM $\left(\Delta\nu_B^{\text{pulse}}\right)$ and the frequency step (δ) used to scan the gain spectral response (Soto and Thévenaz 2013a)

Note that the validity of Eq. 40 is subjected to the condition $\delta \ll \Delta\nu_B^{\text{pulse}}$, which is equivalent to require a large number of spectral samples. Another important condition is that a prior rough estimate of the local BGS peak frequency has been performed, so that all spectral points used in the fitting are evenly distributed around the expectation BFS value (Soto and Thévenaz 2013a). These conditions are typically satisfied in classical measurements.

Note that Eq. 40 defines important relations between the uncertainty in the BFS estimation and measurement conditions. On the one hand, this expression points out an implicit dependence of the BFS uncertainty σ_ν on the spatial resolution. This is due to the dependence of the maximum gain (affecting the SNR) and of the BGS width on the pulse duration, as described in Eqs. 38 and 39. On the other hand, Eq. 40 also indicates the existence of a strict inverse proportionality between the BFS uncertainty and SNR. This dependence constitutes the most relevant relation defining the performance of a Brillouin distributed sensor (not only in BOTDA but also in other schemes, such as BOTDR). Hence the natural SNR reduction caused by the fiber attenuation leads to an unavoidable increase in the BFS uncertainty over distance (Soto and Thévenaz 2013a).

It becomes evident from Eq. 40 that there is a strict trade-off between all the parameters of the sensor:

- Spatial resolution: Due to the pulse width dependence of both SNR, linked through the maximum gain as described in Eq. 38, and $\Delta\nu_B^{\text{pulse}}$ as described in Eq. 39
- Measurement distance: Through the impact of the fiber loss on the SNR
- Frequency uncertainty: Defining the temperature and strain resolutions
- Measurement time: Implicitly dependent on the number of scanned frequencies and averaged traces (which in turn determines the SNR)

For a constant SNR condition, improving any of these parameters leads to an inevitable degradation of one or more of the other parameters. For instance, extending the sensing range by 7.5 km leads to a 3 dB SNR reduction at the far fiber end, which results in a twofold increase in the measurement uncertainty. To overcome this SNR penalty, a longer pump pulse could be used, thus worsening the spatial resolution; or the number of averaged traces could be increased by a factor 4, at the expense of a fourfold increase in the measurement time.

Figure of Merit of BOTDA Sensors

To evaluate the performance of a BOTDA sensing scheme, all critical parameters (e.g., spatial resolution, sensing range, measurement time, and measurand resolution) must be considered. Taking into account that all these parameters are interrelated through the sensor response and the SNR on the traces, any combination of them subject to the same SNR condition must lead to the same performance. An objective metric to evaluate the performance of BOTDA sensors can be expressed by the following figure of merit (FoM) (Soto and Thévenaz 2013a):

$$\mathrm{FoM} = \frac{(\alpha L_{\mathrm{eff}})^2 \exp\left[(2+f_l)\,\alpha L\right]}{\Delta z\,\sqrt{N_{\mathrm{Tr}} N_{\mathrm{avg}}}} \frac{\sqrt{\delta\,\Delta\nu_B^{\mathrm{pulse}}}}{\sigma_\nu}, \tag{41}$$

where consistent units have to be used and Δz must be in meters. Note that in Eq. 41, L is the sensing fiber length, α is the fiber attenuation, $L_{\mathrm{eff}} = (1 - \exp(-\alpha L))/\alpha$ is the fiber effective length, N_{avg} is the number of averages used for each time trace, N_{Tr} is the number of traces required for each pump-probe frequency offset (e.g., $N_{\mathrm{Tr}} = 1$ for conventional BOTDA, $N_{\mathrm{Tr}} = 2$ for a differential pulse method, N_{Tr} can also be much larger in the case of using pulse coding techniques – see section "Optical Pulse Coding" for more details), and f_l is a parameter that takes into account the fiber configuration: $f_l = 0$ for the standard fiber configuration (i.e., when the total fiber length is equal to the sensing length L) and $f_l = 1$ for the fiber-loop configuration, in which only half of the total fiber length is employed for sensing (i.e., when the total fiber length is equal to $2L$). Note that the factor $(\alpha L_{\mathrm{eff}})^2$ accounts for the potential power increase that can be allowed in short-range BOTDA schemes and resulting from the increase in the threshold power $\propto 1/(\alpha L_{\mathrm{eff}})$ for all nonlinear effects at short distances.

This figure of merit highlights all the relevant parameters that have to be considered to evaluate the performance of a BOTDA sensor. As a simple reference, this FoM is equal to 1 for a typical sensor of distance range L = 30 km, spatial resolution $\Delta z = 2$ m, and BFS error $\sigma_\nu = 1.3$ MHz, obtained with a standard BOTDA scheme (i.e., $f_l = 0$ and $N_{\mathrm{Tr}} = 1$), $N_{\mathrm{avg}} = 2000$ temporal trace averaging and with a scanning frequency step $\delta = 1$ MHz, in a standard single-mode fiber with $\Delta\nu_B^{\mathrm{pulse}} = 60$ MHz (for 2 m spatial resolution) and an attenuation coefficient $\alpha = 0.2$ dB/km.

Equation 41 represents an objective metric to evaluate the performance of a BOTDA sensor, and therefore any novel sensing scheme bringing real progress to the field, in terms of enhancement of the capabilities of the sensor, must lead to a larger FoM.

Limitations and Design Considerations in Brillouin Time-Domain Sensing

To maximize the response of a Brillouin distributed sensor and avoid distortions in the measured traces, a careful optimization of the input signals is required in both BOTDR and BOTDA systems. On the one hand, the ideal pump pulse must fulfill several necessary conditions: (i) the pulse temporal shape has to be ideally rectangular, with sharp rising and falling edges, (ii) it must have very high extinction ratio, (iii) the peak pump power must remain below the onset of nonlinear effects, and (iv) the pulse width cannot be arbitrarily reduced to reach very sharp spatial resolutions. On the other hand, in the case of BOTDA, the probe power launched in opposite direction to the pump must also fulfill very specific requirements. While a high-power probe is highly preferred to increase the SNR of the traces, the maximum power allowed in the fiber is limited to avoid distortion and systematic errors resulting from *nonlocal effects.*

This section presents a thorough description of the factors conditioning the optimal design of the optical signals used in Brillouin time-domain sensors. First, the limitations imposed by nonlinear effects and extinction ratio to the pump pulse in BOTDR and BOTDA schemes are described. Then, specific constraints to the CW probe signal in a BOTDA sensor are discussed. The material presented in this section can be used as a guideline to design and optimize a robust Brillouin time-domain distributed sensor.

Constraints in the Pump Pulse Due to Nonlinear Effects

Modulation Instability (MI)

Modulation instability occurs in optical fibers as a result of the interplay between nonlinear Kerr effect and the group velocity dispersion in anomalous regime (i.e., when the second-order dispersion coefficient β_2 is negative) (Agrawal 2007; Alem et al. 2015). In the process tiny perturbations of a continuous-wave optical field, including those originated by noise, create small refractive index variations that induce a nonlinear phase shift in the CW beam through Kerr effect. The presence of anomalous dispersion reinforces the intensity perturbation profile, leading to a self-sustained parametric amplification of the perturbations during light propagation. This parametric amplification can reach a bandwidth of several tens or even hundreds of GHz, thus providing a very fast response ($\sim$10 ps) when compared to the pulse duration used in Brillouin sensing (typically >10 ns). For such pulse durations, MI behaves in steady-state regime, simplifying the analysis here presented.

In a time-domain Brillouin sensor, MI manifests as a gradual power transfer from the pump pulse to two symmetric spectral sidebands around the pump optical frequency, as the pulse propagates down the fiber (Alem et al. 2015; Alahbabi et al. 2004; Foaleng and Thévenaz 2011). The process leads to a MI gain spectrum that, for a lossless fiber, is defined as (Alem et al. 2015)

$$G_{\mathrm{MI}}(\Omega) = 1 + \frac{\sinh^2\left(2\gamma P_0 L \sqrt{(\Omega/\Omega_c)^2 \left(1 - (\Omega/\Omega_c)^2\right)}\right)}{2(\Omega/\Omega_c)^2 \left(1 - (\Omega/\Omega_c)^2\right)}, \tag{42}$$

where γ is the nonlinear coefficient of the fiber and P_0 is the pump power, Ω is the frequency detuning with respect of the pump, and $\Omega_c = \sqrt{4\gamma P_0/|\beta_2|}$ is the cutoff frequency. Figure 16a shows the MI gain spectrum described in Eq. 42 for three different pump powers, in the case of a standard single-mode fiber (SMF) operating at 1550 nm and quantum-level background noise. The figure point out that the MI gain spectrum has two symmetric sidelobes around the pump frequency, with a maximum gain equal to $1 + 2\sinh^2(\gamma P_0 L)$, which takes place at frequencies $\Omega_{\max} = \pm\Omega_c/\sqrt{2} = \pm\sqrt{2\gamma P_0/|\beta_2|}$ (Alem et al. 2015; Alahbabi et al. 2004).

If, in addition to the optical pump generating MI, another wave is launched into the fiber with an optical frequency falling inside the MI gain bandwidth, this signal will be amplified during propagation. In Brillouin optical time-domain sensors (both BOTDR and BOTDA), a high-power optical pulse is used to interrogate the Brillouin response along the fiber. Typically the pulse power is boosted by an optical amplifier, such as an EDFA. Optical amplification generates amplified spontaneous emission (ASE) noise, which co-propagates with the pump pulse, thus seeding efficient MI inside the fiber. This process results in parametric amplification of the noise, which gradually depletes the pump as it propagates down the fiber.

The most harmful impact of MI on a Brillouin sensor is the eventual severe depletion that the pump can experience, thus leading to a significant reduction of the sensor response after a given critical distance (Alem et al. 2015; Alahbabi et al. 2004; Foaleng and Thévenaz 2011), as shown in Fig. 16b for different input pump powers. This phenomenon imposes significant limitations to the maximum power of both the pump pulse and optical noise that can be injected into the fiber. Thus, the maximum pump power is determined by the critical power of MI, expressed as (Alem et al. 2015)

$$P_{\mathrm{crit}} = \frac{\sigma_{\mathrm{crit}}}{2\gamma L_{\mathrm{eff}}}, \text{ with } \sigma_{\mathrm{crit}} = \ln\left(\frac{R_D\sqrt{2\pi\,|\beta_2|}}{S_n\gamma\sqrt{L_{\mathrm{eff}}}}\right)\sigma_{\mathrm{crit}}, \tag{43}$$

where σ_{crit} represents the critical MI gain integrated over the entire fiber. This critical gain depends on the spectral density S_n of the noise co-propagating with the pulse, the interaction length (defined by the nonlinear effective length L_{eff}), and the maximum level of depletion R_D tolerated by the sensor, defined as

$$R_D(z) \triangleq \frac{P_0\exp(-\alpha z) - P(z)}{P_0\exp(-\alpha z)} = 1 - \frac{P(z)}{P_0\exp(-\alpha z)}, \tag{44}$$

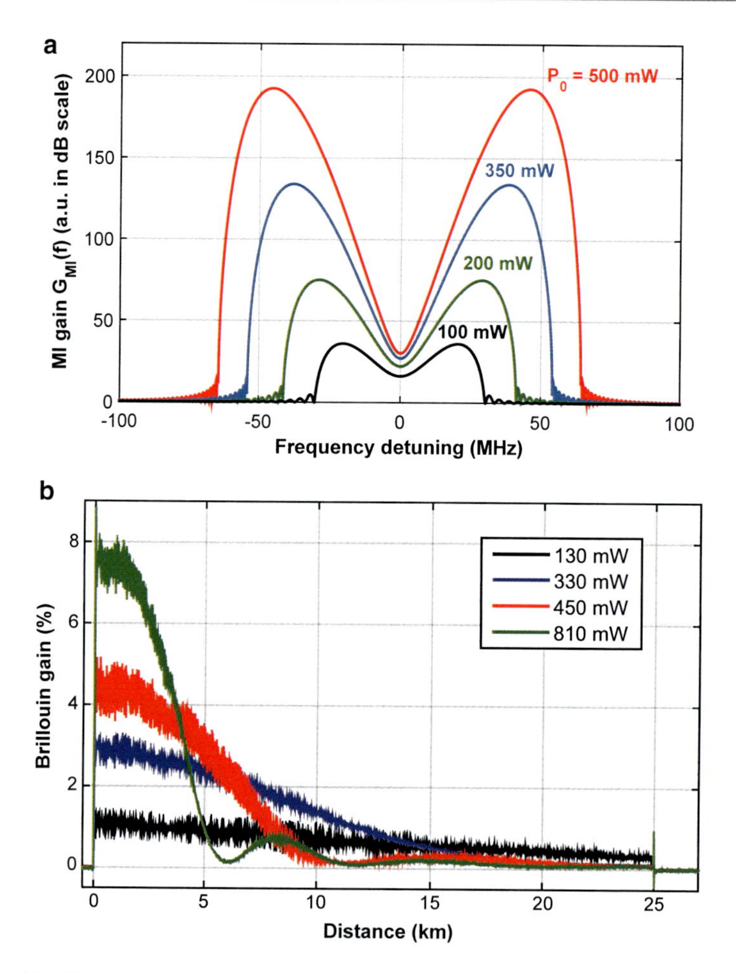

Fig. 16 (**a**) MI gain spectrum of a 25 km lossless fiber for different input powers, using typical SMF values $\beta_2 = -22$ ps^2/km and $\gamma = 1.8$ W^{-1}/km. (**b**) Impact of the pump depletion induced by MI on BOTDA traces, over a 25 km fiber for different pump powers (Replot using data from Alem et al. 2015)

where $P(z)$ is the actual pump power at position z (i.e., depleted by MI) and $P_0 \exp(-\alpha z)$ is the local pump power expected without MI.

Note that the transcendental expression for σ_{crit} in Eq. 43 does not have a closed-form solution, and hence σ_{crit} must be found numerically before being used to calculate the critical power. Using a long standard single-mode fiber with parameters $\alpha = 0.2$ dB/km, $\beta_2 = -22$ ps^2/km and $\gamma = 1.8$ W^{-1}/km, $L_{\text{eff}} \approx \alpha^{-1} \approx 22$ km, and an EDFA launching a background noise into the fiber of $S_n = -121$ dBm/Hz, a critical gain $\sigma_{\text{crit}} = 7.15$ is found for a maximum depletion of 10% at 25 km distance. According to Eq. 43, this critical gain defines a maximum critical power $P_{\text{crit}} = \sim 135$ mW. This value can, however, vary depending on the noise features of the optical amplifier used to boost the pump power, the total fiber length, and the tolerated level of depletion. Based on standard values of the noise power spectral density in EDFAs, it is commonly considered that the power limit imposed by MI to the pump pulse power is ~100 mW (in long optical fibers $L > \alpha^{-1} \approx 22$ km). This power level is the lowest value when compared to the onset of other nonlinear effects and corresponds to the most stringent limit imposed to the pump power of both BOTDR and BOTDA sensors (Alem et al. 2015; Alahbabi et al. 2004; Foaleng and Thévenaz 2011).

Amplified Forward Spontaneous Raman Scattering

The impact of modulation instability in a Brillouin sensor can be fully eliminated by using pump and probe signals operating in the normal dispersion regime ($\beta_2 > 0$) of the fiber. In this case, however, the maximum usable pump power is still limited by another nonlinear effect: *amplified spontaneous forward Raman scattering* (Fellay et al. 1999).

Raman scattering originates from the interaction of the incoming light with vibrational modes of the molecules in the fiber (Boyd 2003; Agrawal 2007). During this light-matter interaction, the incoming light transfers energy to the medium (in a Stokes process) or receives energy from the vibrational modes (in an anti-Stokes process). In the Raman scattering process, Stokes and anti-Stokes components are simultaneously generated through the fiber in both forward and backward directions. From these two directions, the most detrimental effect on a Brillouin sensor is caused by the forward Raman interaction, because in this case Raman components co-propagate with the pump pulse interacting along the entire sensing fiber length. If a high pump power is used, the thermally-activated Raman scattering (i.e., the spontaneous Raman scattering) seeds the process and reinforces the light-matter interaction, resulting in a self-seeded nonlinear process. This phenomenon is called *amplified spontaneous forward Raman scattering* or simply *forward stimulated Raman scattering*.

The most deleterious impact of stimulated Raman scattering (SRS) on the sensor performance is the pump depletion that can occur during propagation over long sensing distances (Foaleng and Thévenaz 2011; Fellay et al. 1999), as shown in Fig. 17a. As in the case of modulation instability, the occurrence of pump depletion decreases the sensor response at long distances, reducing SNR and making impossible to reliably detect local SBS interactions at long distances. To avoid those detrimental effects, the maximum pump power that can be launched into the fiber is limited by the onset of amplified spontaneous forward Raman scattering to the following critical power (Agrawal 2007; Foaleng and Thévenaz 2011):

$$P_{\text{crit}} = \frac{16 A_{\text{eff}}}{g_R L_{\text{eff}}}, \tag{45}$$

where g_R is the Raman scattering coefficient ($g_R \approx 10^{-13}$ m/W for standard single-mode fibers), A_{eff} is the effective area, and L_{eff} is the effective fiber length. Note that due to the use of a pump pulse of duration T and to the large spectral separation $\Delta\lambda$

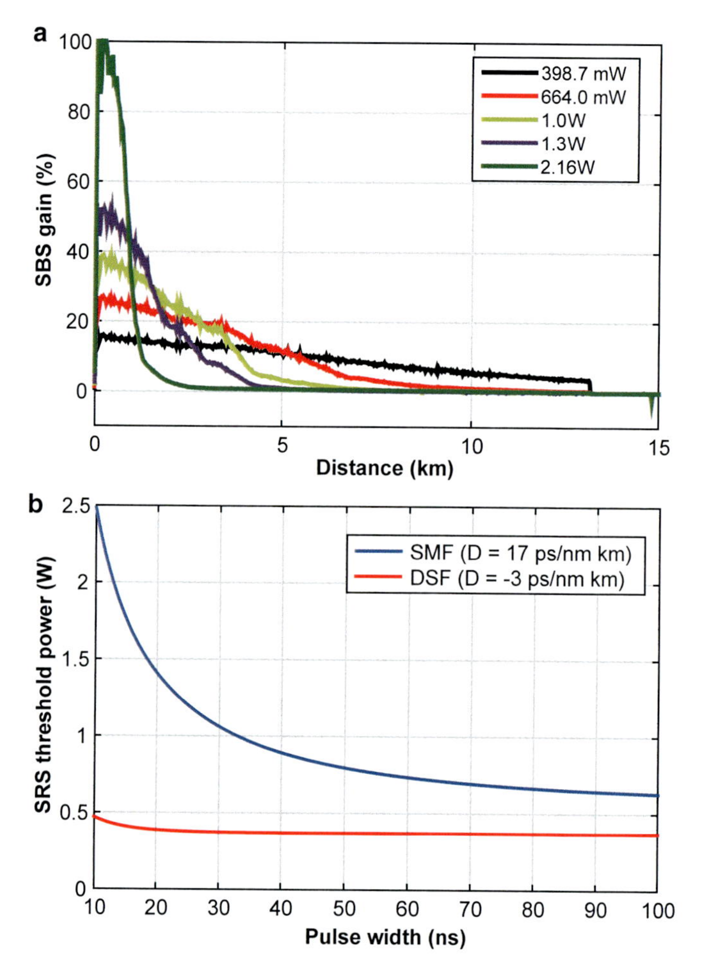

Fig. 17 (**a**) Impact of Raman scattering on BOTDA traces over a 13-km-long DSF with normal dispersion at 1550 nm (Replot using data from Foaleng and Thévenaz 2011). (**b**) SRS threshold versus pulse width, for a standard SMF ($A_{\mathrm{eff}} = 80\ \mu\mathrm{m}^2$, $D = 17$ ps/nm km) and a DSF with normal dispersion ($A_{\mathrm{eff}} = 50\ \mu\mathrm{m}^2$, $D = -3$ ps/nm km), considering $g_R \approx 10^{-13}$ m/W and $L > \alpha^{-1} \approx 22\,\mathrm{km}$

between Raman components and pump wavelength ($\Delta\lambda \approx 100$ nm at 1550 nm), the effective interaction length of forward Raman scattering is actually determined by the walk-off distance $L_W = T/(D \cdot \Delta\lambda)$, rather than on the physical length of the fiber. Here D is the dispersion coefficient at the pump wavelength. This walk-off distance makes the maximum power limit in Eq. 45 depend on the pump pulse width T, i.e., on the spatial resolution of the sensor (Wait et al. 1997).

Figure 17b shows the critical power of amplified spontaneous forward Raman scattering (also called *SRS threshold*) as a function of the pulse width for a standard single-mode fiber (SMF) and a dispersion-shifted fiber (DSF), assuming a long fiber length ($L > \alpha^{-1} \approx 22$ km). It can be observed that in a SMF, the critical power for SRS highly depends on the pulse duration; this is because of the large walk-off of the fiber at 1550 nm. In the case of a very long pulse, this threshold level tends to ~580 mW (CW pump condition), while increases up to ~2.5 W for 10 ns pulses. As mentioned before, a Brillouin sensor operating over a SMF reaches the limit imposed by MI before than this forward SRS critical power. However, when using a sensing fiber in the normal dispersion region, commonly being a DSF with zero-dispersion wavelength above 1550 nm, the Raman threshold shown in Fig. 17b becomes the lowest limit constraining the maximum pump power of a Brillouin sensor (Foaleng and Thévenaz 2011). This power limit is ~370 mW, being almost independent of the pulse width due to the small walk-off of the fiber at 1550 nm. Only a small increase in this critical power can be seen with a 10 ns pulse, imposing a power limit of ~470 mW in this case. Note that slightly different values for this limit can be found in the literature due to the large variety of DSFs with different dispersion coefficients at 1550 nm.

Self-Phase Modulation (SPM)

When a high-power pump pulse propagates through the sensing fiber in a Brillouin time-domain sensor, the fiber refractive index is modified by the optical intensity as $n = n_1 + n_2 I(t)$, where n_1 and n_2 are the linear and nonlinear index coefficients, respectively. The second term indicates that the temporal profile of the pulse intensity $I(t)$ induces a time-varying nonlinear phase shift $\phi_{NL}(t)$ across the optical pulse itself. This phenomenon is known as *self-phase modulation* (Boyd 2003; Agrawal 2007). The time dependence of the SPM-induced nonlinear phase shift implies that the instantaneous frequency of the pulse changes, originating a frequency chirp described as

$$\Delta\omega(t) = \frac{d\phi_{NL}(z,t)}{dt} = -n_2 \frac{\omega}{c} z \frac{dI(t)}{dt}, \tag{46}$$

where ω is the central optical frequency of the pulse, and c is the speed of light in vacuum (Foaleng et al. 2011). Note that the SPM-induced chip increases linearly over distance and is enhanced for signals showing continuous temporal transitions. It must be noted that SPM only affects the phase of the optical pulse, inducing spectral broadening, while its temporal profile remains unchanged. This means that in presence of SPM, the Brillouin spectral width measured by the sensor could be highly affected depending on the amount of nonlinear phase shift accumulated over the sensing distance, but the spatial resolution remains unaffected.

The ideal pulse shape is that one showing a constant intensity (i.e., a flat top), like a rectangular pulse, so that $\Delta\omega(t) = 0$ when assuming ideally short rising and falling edges (Foaleng et al. 2011). Any other pulse shape leads to chirp variations that highly depend on the exact pulse temporal profile. An interesting comparison can be made with Gaussian pulses, which have a spectral width that is one half the spectral width of rectangular pulses (for the same temporal width). In principle, this feature might lead to narrower BGS measurements. However, this is only valid for the first kilometers of fiber, since the smooth temporal shape $E(t) = A\,\exp(-t^2/\tau^2)$ leads to an SPM-induced chirp that increases with distance according to $\Delta\omega(t) = 4\gamma z P t^2/\tau^2 \exp(-2t^2/\tau^2)$, where τ is the $1/e$ temporal width, P is the pulse peak power, and γ is the nonlinear fiber coefficient. This SPM-induced chirp can lead to a significant spectral broadening when the pulse propagates over long optical fibers (Foaleng et al. 2011).

To exemplify the impact of SPM on BGS measurements, Fig. 18a shows how the effective Brillouin spectral width measured by a Brillouin sensor increases linearly over distance and as a function of the peak pulse power, when using 30 ns FWHM Gaussian pulses. Figure 18b compares this behavior with the one resulting when

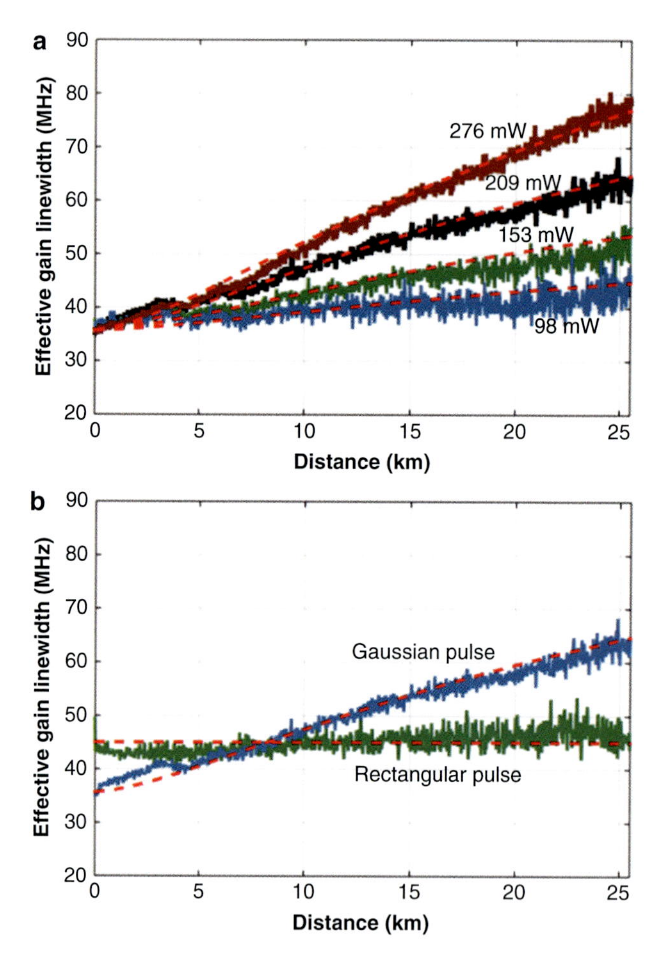

Fig. 18 Brillouin gain spectral broadening due to SPM over a 25.5 km standard fiber. (**a**) Using Gaussian-shaped pulses of 30 ns FWHM with different peak powers. (**b**) Comparison of the effective BGS linewidth resulting from using Gaussian and rectangular pulses of the same temporal width (30 ns FWHM) (Replot using data from Foaleng et al. 2011)

using a rectangular pulse of the same duration. Note that the Gaussian pulse leads to an initial narrower spectrum and then broadens with distance, while the BGS measured with a rectangular pulse having short (~500 ps) rising and falling edges is not affected by SPM (Foaleng et al. 2011).

For the sensor design point of view, it might be interesting to have a mathematical expression indicating the maximum input pump power allowed in the fiber to avoid SPM. Nevertheless, due to the countless possibilities of pulse temporal and spectral shapes, it becomes practically impossible to establish a universal expression for this. It is however important to note that, from the practical viewpoint, in order to mitigate SPM, it is sufficient to use pulse shapes as rectangular as possible (and with minimum chirp), bearing in mind that small deviations from the ideal shape might be still allowed, since in most of the cases other nonlinear effects, such as modulation instability, could impose more stringent restrictions to the pulse peak power.

Requirements for the Pulse Extinction Ratio

The sensor response defined in Eq. 34 assumes that the pump pulse has infinite extinction ratio. However, in real BOTDA implementations, the pump pulse is shaped by a non-ideal intensity-modulating device (usually an electro-optic modulator), which actually leaks a small amount of power in the *off* state of the pulse, as illustrated in Fig. 19a. This small CW leakage component propagates down the sensing fiber and interacts with the CW probe, generating a continuous SBS response that accumulates along the fiber according to (Soto and Thévenaz 2014)

$$\Delta P_s^{\text{leak}} = \frac{g_B}{A_{\text{eff}}} k^{-1} P_{pi} P_{si} \exp(-\alpha L) L_{\text{eff}}, \tag{47}$$

where P_{pi} is the input peak power of the pump pulse, P_{si} is the input CW probe power, k is the pulse extinction ratio, and L_{eff} is the fiber effective length. As a consequence, this CW SBS interaction adds to the measured BOTDA traces a continuous level that is proportional to the mean Brillouin gain of the fiber at the set pump-probe frequency offset, providing no relevant z-dependent information to the time-domain traces. Consequently, the measured spectrum turns out to be given by the superposition of two gain peaks (Soto and Thévenaz 2014): one corresponding to the mean BGS integrated along the entire sensing fiber (resulting from the CW leak component) and another containing the useful local Brillouin gain information (created by the pump pulse). At long sensing distances, the pulse contribution $\Delta P_s^0(z)$ given by Eq. 34 might become similar or even lower than the continuous contribution of the ΔP_s^{leak} component, which may originate significant errors on the BFS estimation if not compensated. A first-order approximation indicates that the pulse extinction ratio has to be larger than the number of resolved points along the fiber (i.e., $k \geq L/\Delta z$, where L is the sensing length and Δz is the spatial resolution). However, this approximation neglects the fiber attenuation and the reduction of the sensor response over distance. To avoid measurements errors, the sensor response

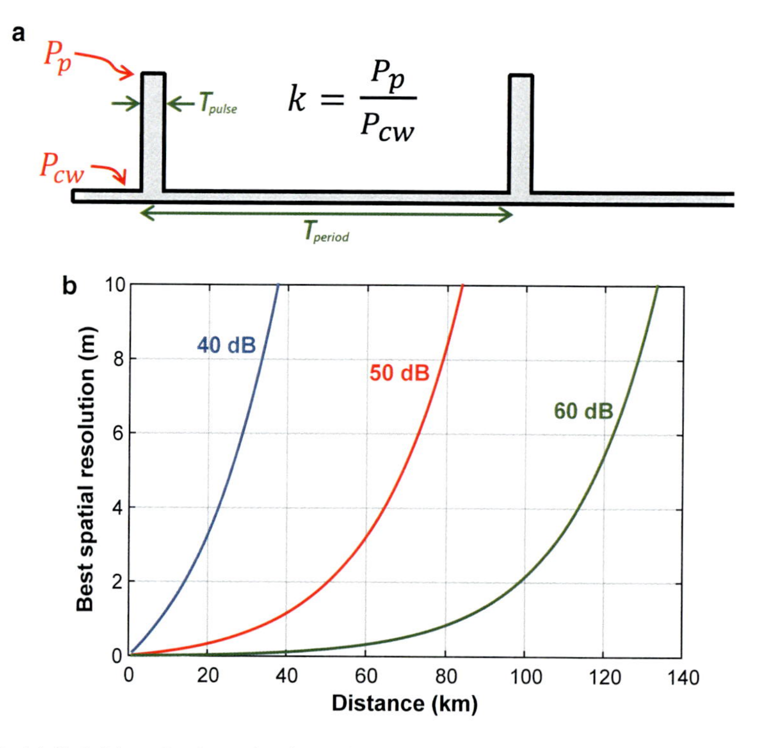

Fig. 19 (**a**) Definition of pulse extinction ratio as a function of the pulse peak power and CW leakage power. (**b**) Best attainable spatial resolution versus sensing range, for three different pulse extinction ratios (note that no dedicated signal processing is considered in this case)

near the far fiber end $\Delta P_s^0\,(z = L)$ must be larger than the response to the leak ΔP_s^{leak}. Thus, to fulfill the condition $\Delta P_s^0\,(z = L) \geq \Delta P_s^{\text{leak}}$, the requirements for the pulse extinction ratio can be written as (Soto and Thévenaz 2014)

$$k \geq \frac{L_{\text{eff}}}{\Delta z\ \exp\left(-\alpha L\right)}. \tag{48}$$

This expression indicates that to reach reliable sensing over, for instance, 50 km with 1 m spatial resolution, an extinction ratio $k \geq 53$ dB is required. This value goes well beyond the extinction ratio attainable with electro-optic modulators (typically limited to ~40 dB). To increase the pulse extinction ratio, several methods can be used: for instance, using semiconductor optical amplifiers, cascaded electro-optic modulators, or radio-frequency (RF) shaping methods. This way, extinction ratios of more than 60 dB can be obtained.

Note that Eq. 48 can be used to estimate the best attainable spatial resolution as a function of the sensing range and pulse extinction ratio (Soto and Thévenaz 2014). Figure 19b shows the best attainable spatial resolution with different extinction ratios. The figure points out that the use of an electro-optic modulator with

$k = 40\,\mathrm{dB}$ provides only limited sensing distances (e.g., a range of 25 km for the best resolution of 5 m), while much longer distances could be achieved (maintaining the spatial resolution) by increasing the pulse extinction ratio. It is however worth mentioning that signal processing and a proper normalization of the traces could be helpful to reduce the unwanted and detrimental impact of the DC leak component.

Constraints in the Probe Signal of BOTDA Sensors

In addition to the distortions arising from pump-induced nonlinear effects, BOTDA sensors are also affected by unwanted nonlinear interactions initiated by the CW probe. They can also lead to pump depletion and, in some cases, to huge distortions of the pump pulses, thus impairing significantly the measurement quality.

In principle, the maximum probe power that can be launched into the fiber is ultimately limited by the onset of *amplified spontaneous Brillouin scattering*. The onset of this phenomena occurs at a critical probe power ~5 mW for long standard fibers ($L > \alpha^{-1} \approx 22$ km), as defined by Eq. 25. This power level represents the ultimate limit for the probe power, not only at the input of the sensing fiber, but at the input of the conveying fiber in a loop configuration. Using a larger probe power leads to a noisy CW probe, which degrades the overall performance of the sensor (Geinitz et al. 1999; Thévenaz et al. 2013).

It must however be pointed out that the use of probe powers much lower than this critical level can already induce severe distortions in the measured BOTDA traces. This is a result of the cumulated power transfer that the pump pulse experiences while interacting with the CW probe over the entire sensing range. This is intrinsic to the measurement principle behind the BOTDA technique and might result in a non-negligible net depletion/amplification of the pump pulse. This cumulative effect is the origin of so-called *nonlocal effects*, which in turn lead to BGS distortions and *systematic errors* on the BFS determination (Geinitz et al. 1999; Thévenaz et al. 2013).

In the following, details on the origin and impact of those nonlocal effects and systematic errors are described for single-probe and dual-probe BOTDA configurations.

Pump Depletion and Systematic Errors in a Single-Probe Scheme

As described in section "Sensor Response," the CW probe in a BOTDA sensor is locally amplified by the pump pulse with a gain factor $G_s(z) = g_B(z)/A_{\text{eff}} P_p(z)\Delta z$. The interaction is confined only to half the pulse length, defining the spatial section Δz where the local SBS gain takes place. Due to the counter-propagative nature of the interaction, the total gain locally experienced by the probe turns out to be very small (e.g., <1% for 1 m spatial resolution, when using a pump power of ~100 mW). However, the net impact of the SBS interaction on the pump power is very different, since the power transfer occurring at each fiber position cumulates on the pump over the entire sensing length as a result of the interaction with the CW probe.

This integrated gain/loss effect on the pump pulse can be quantified by solving the SBS coupled Eqs. 21 and 22 using a perturbation approach. This calculation leads to a pump power at position z, defined as (Geinitz et al. 1999)

$$P_P(z, \Delta\nu) = P_{pi} \exp\left(-\alpha_p z\right) \exp\left\{ \int_0^z \frac{-g_B(\xi, \Delta\nu)}{A_{\text{eff}}} P_{si} \exp\left[-\alpha_s (L - \xi)\right] d\xi \right\}, \tag{49}$$

where $\Delta\nu$ is the pump-probe frequency offset. Note that the negative sign in the integral factor represents the depletion experienced by the pump in a Brillouin gain configuration (positive g_B); however, the equation is also valid to describe the pump excess amplification in a Brillouin loss configuration (negative g_B).

Equation 49 points out that the local pulse power at a given fiber position z strongly depends on the input probe power and the cumulated SBS interaction taking place between the near fiber end ($z = 0$) and the observation point at position z. This means that when using strong probe powers, the pump power will carry a non-negligible amount of information related to the BGS profile at different fiber sections, while it will also become highly dependent on the pump-probe frequency offset. As a consequence, the local pump-probe SBS interaction at a position z will not only depend on the local BGS as normally expected, but also on nonlocal BGS information. This phenomenon is called *nonlocal effects* in BOTDA and leads to severe distortions in the measured BGS shape. These distortions originate errors on the BFS estimation (Geinitz et al. 1999; Thévenaz et al. 2013); however, these are not stochastic but systematic errors, which manifest biasing the BFS profile. Hence errors cannot be reduced by trace averaging and cannot be evaluated by the standard deviation calculated over consecutive traces.

The worst-case scenario for the occurrence of pump depletion is when the BFS profile is uniform over the entire sensing fiber. This is because in this case the integral in Eq. 49 can be extended over the entire fiber length, resulting in strong pump *depletion* (or *excess amplification* in a loss configuration). Although real fibers do not have a perfectly uniform BFS profile, variations of a few MHz are common and can still be considered uniform enough compared to the BGS width, thus also contributing efficiently to the cumulative effect on the pump. Under this uniform BFS scenario, the output pump power at $z = L$ can be expressed as

$$P_P(L, \Delta\nu) = P_{pi} \exp\left(-\alpha_p L\right) \exp\left\{ \frac{-g_B(\Delta\nu)}{A_{\text{eff}}} P_{si} L_{\text{eff}} \right\}, \tag{50}$$

This expression indicates that the pump power can be highly depleted (or amplified) if a high-power probe is sent down a long optical fiber. It must be noted that the pump depletion/amplification also highly depends on the pump-probe frequency offset $\Delta\nu$, and therefore higher levels of depletion/amplification are expected when scanning frequencies near the average BFS of the fiber. In extreme cases, a strong depletion can lead to a *spectral hole burning* (centered at the mean

BFS), thus making it impossible to reliably estimate the BFS at long distances. Although spectral hole burning can be avoided by reducing the probe power or using a shorter fiber, errors on the BFS estimation can still occur due to the dependence of the pump power on the pump-probe frequency difference. If, under this condition, the BFS of the fiber changes along a very short fiber section, near the far fiber end, the error ν_e on the BFS estimation can be described as (Thévenaz et al. 2013)

$$\nu_e \cong \frac{-\xi d\,\Delta\nu_B}{\left[1+4\xi^2\right]^2 - 2d\left[1+2\xi^2\right]}, \tag{51}$$

where $\xi = \delta\nu/\Delta\nu_B$ is the BFS variation $\delta\nu$ induced near the far fiber end normalized by the FWHM $\Delta\nu_B$ of the undistorted BGS, and $d = (P_{p0} - P_p)/P_{p0}$ is the level of depletion, where P_{p0} and P_p are the pump powers without and with SBS interaction. Figure 20a shows the normalized error originated from nonlocal effects as a function of the normalized BFS shift ξ for different levels of depletion. The figure indicates that when the BFS of the short fiber section is larger than the BFS of the previous long segment (i.e., $\delta\nu > 0$), the estimated BFS of the short section is upshifted with respect to the real value. On the other hand, when $\delta\nu < 0$, the BFS measured over the short fiber section is downshifted compared to the real BFS. Under a weak depletion scenario ($d < 0.2$), the error ν_e vanishes for $\xi = 0$ (i.e., under no BFS variation) and for $\xi \gg 1$ (i.e., when the BGS of the two fiber sections do not overlap), while ν_e is maximum when $\xi = \delta\nu/\Delta\nu_B \approx 0.26$. Figure 20b shows the maximum error $\nu_e(\xi \approx 0.26)$ induced by nonlocal effects as a function of the depletion level. From Eq. 51, the maximum tolerated depletion $d_{\max}$ to secure a frequency error below ν_e can be defined as (Thévenaz et al. 2013)

$$d_{\max} = \frac{\left(1+4\xi^2\right)\left[16\xi e^2 - \left(1+4\xi^2\right)e\right]}{12\xi e^2 - 2\left(1+2\xi^2\right)e + \xi}, \tag{52}$$

where $e = \nu_e/\Delta\nu_B$ is the frequency error normalized to the BGS FWHM. Equation 52 indicates that to secure systematic errors below 1 MHz, a maximum depletion of 19.4% is tolerated in the system (assuming $\Delta\nu_B = 27$ MHz). However, if this error has to remain below 0.1 MHz, then depletion must be below 2.3%.

Combining the definition of depletion d and Eq. 52, the maximum probe power $P_{\mathrm{si}}^{\max}$ that can be launched into the sensing fiber to secure a given level of depletion d can be expressed as (Thévenaz et al. 2013)

$$P_{\mathrm{si}}^{\max} = -\ln\left(1-d\right)\frac{A_{\mathrm{eff}}}{g_B L_{\mathrm{eff}}}. \tag{53}$$

This equation indicates that the maximum probe power allowed at the sensing fiber input depends only on the fiber parameters and the tolerated level of depletion, being completely independent of the pump power. Using standard fiber conditions (i.e., $g_B = 2 \cdot 10^{-11}$m/W, $A_{\mathrm{eff}} = 80\ \mu\mathrm{m}^2$, and $L_{\mathrm{eff}} \approx \alpha^{-1} \approx 22$ km), Eq. 53 indicates that the probe power must not exceed 40 μW (≈ –14 dBm) to ensure a pump

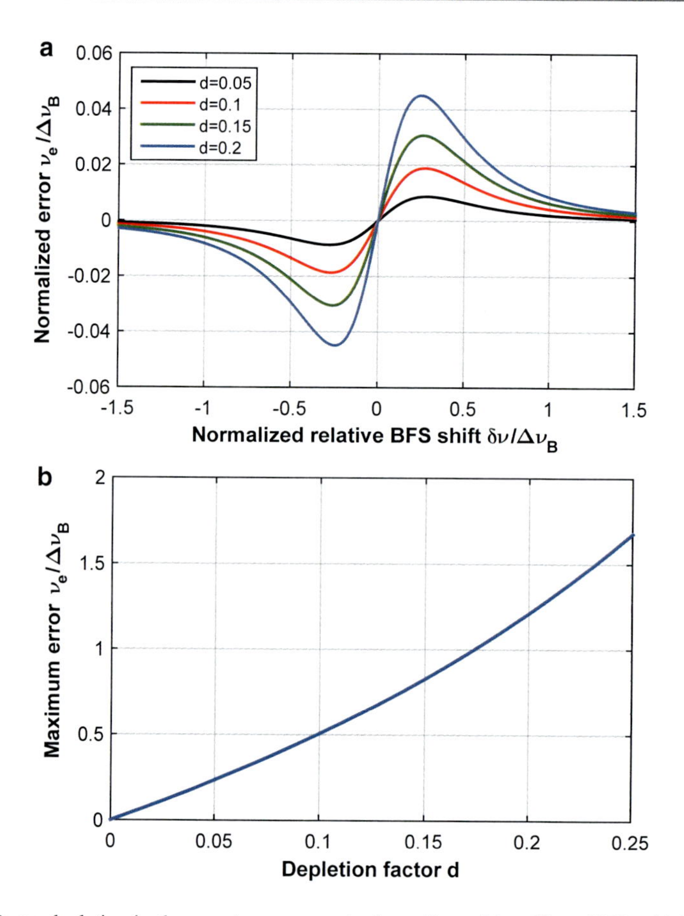

Fig. 20 Pump depletion in the worst-case scenario: long fiber with uniform BFS, with BFS shift $\delta\nu$ in a final very short fiber section. (**a**) BFS error ν_e vs relative BFS shift $\delta\nu$ (both normalized to the BGS FWHM $\Delta\nu_B$), for different depletion levels and (**b**) Maximum normalized error (at $\delta\nu/\Delta\nu_B \approx 0.26$) vs depletion

depletion below 20%. This limit represents a significantly reduced level compared to the ultimate power limit defined by the onset of amplified spontaneous Brillouin scattering (~5 mW), pointing out that nonlocal effects and systematic errors impose the most fundamental limitation to the maximum probe power that can be used in a BOTDA sensor. Note that this is the probe power limit at the input of the sensing fiber, and, hence, higher probe powers can still propagate over the conveying fiber of a fiber/loop configuration.

Pump Depletion and Systematic Errors in a Dual-Sideband Probe Scheme

Nonlocal effects and systematic errors are critical in a single-probe BOTDA scheme; however, they can be highly mitigated using a dual-sideband scheme. As illustrated

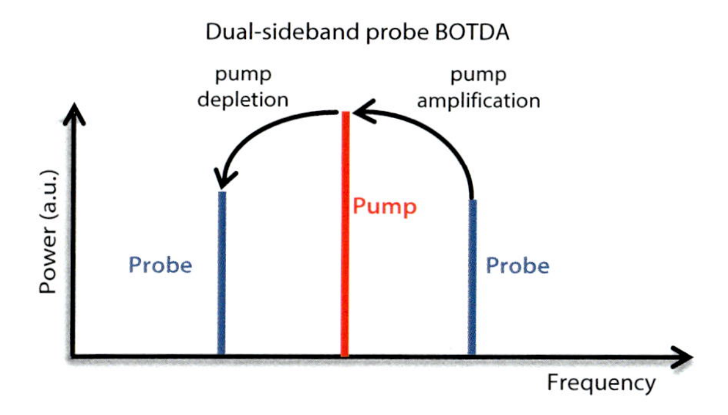

Fig. 21 Power transfer in a dual-sideband BOTDA scheme

in Fig. 21, in this case the lower-frequency probe sideband depletes the pump power, while the upper-frequency probe sideband amplifies the pump, thus compensating the cumulative depletion/amplification on the pump (Thévenaz et al. 2013). In a first-order approximation, depletion can be perfectly compensated if the two probe sidebands have exactly the same optical power.

Any power difference between sidebands would lead to a net gain or loss on the pump power, thus potentially resulting in systematic errors. Under this assumption, Eq. 53 can be readapted for a dual-sideband probe configuration, making $P_{\mathrm{si}}^{\max}$ equal to the power difference between bands (Thévenaz et al. 2013), i.e., $\Delta P_{\mathrm{si}}^{\max} = \left|P_{\mathrm{si}}^{L} - P_{\mathrm{si}}^{U}\right|$, where P_{si}^{L} and P_{si}^{U} are the input power of the lower and upper sidebands, respectively. However, this approach is only valid if the SBS interaction negligibly modifies the probe sidebands. A more precise expression can be established considering that probe sidebands have different powers as a result of the gain and loss induced by the pump. This way, the maximum total probe power $\left(P_{\mathrm{si}}^{L} + P_{\mathrm{si}}^{U}\right)^{\max}$ that can be launched into the sensing fiber can be described as a function of the tolerable depletion d as (Thévenaz et al. 2013)

$$\left(P_{\mathrm{si}}^{L} + P_{\mathrm{si}}^{U}\right)^{\max} = \frac{d}{\left(g_B/A_{\mathrm{eff}}\right)^2 P_{pi}\,(1-d)\exp\left(-\alpha L\right)\Delta z L}. \tag{54}$$

Contrarily to the single-probe case, Eq. 54 indicates that the maximum allowed probe power does not only depend on the fiber parameters and level of tolerable depletion but also on the input pump power and the spatial resolution of the sensor. This is because the SBS-induced power difference between sidebands depends on the pump power and spatial resolution, both of which determine the amount of net depletion/amplification experienced by the pump. This is typically called *second-order pump depletion.*

Using standard fiber parameters (i.e., $g_B = 2 \cdot 10^{-11}$ m/W, $A_{\mathrm{eff}} = 80$ μm^2, and $L_{\mathrm{eff}} \approx \alpha^{-1} \approx 22$ km), a spatial resolution $\Delta z = 1$ m and an input pump power $P_{pi} = 100$ mW, Eq. 54 indicates that the maximum probe power is limited to ~4.9 mW per sideband in this case (Thévenaz et al. 2013). This value however turns out to decrease with an inverse proportion to the spatial resolution Δz. Nevertheless, the power limit is in this case approximately 100-fold times higher than the power limit imposed by the first-order depletion in a single-sideband probe scheme.

Temporal and Spectral Distortion of the Pump Pulse in a Classical Dual-Sideband Scheme

The analysis presented in section "Pump Depletion and Systematic Errors in a Dual-Sideband Probe Scheme" is valid under the assumption that the gain and loss experienced by the pump and generated by each probe sideband compensate perfectly each other for all scanned pump-probe frequency offsets. This condition is, however, not satisfied when using the most common dual-sideband probe scheme based on an electro-optic modulator (Niklès et al. 1997; Dominguez-Lopez et al. 2016). In this scheme, an intensity modulator creates two spectrally symmetric sidebands around the pump optical frequency. As shown in Fig. 22 (top), when the pump-probe frequency offset matches the local BFS of the fiber, the Brillouin gain and loss spectra generated, respectively, by the upper and lower probe sidebands

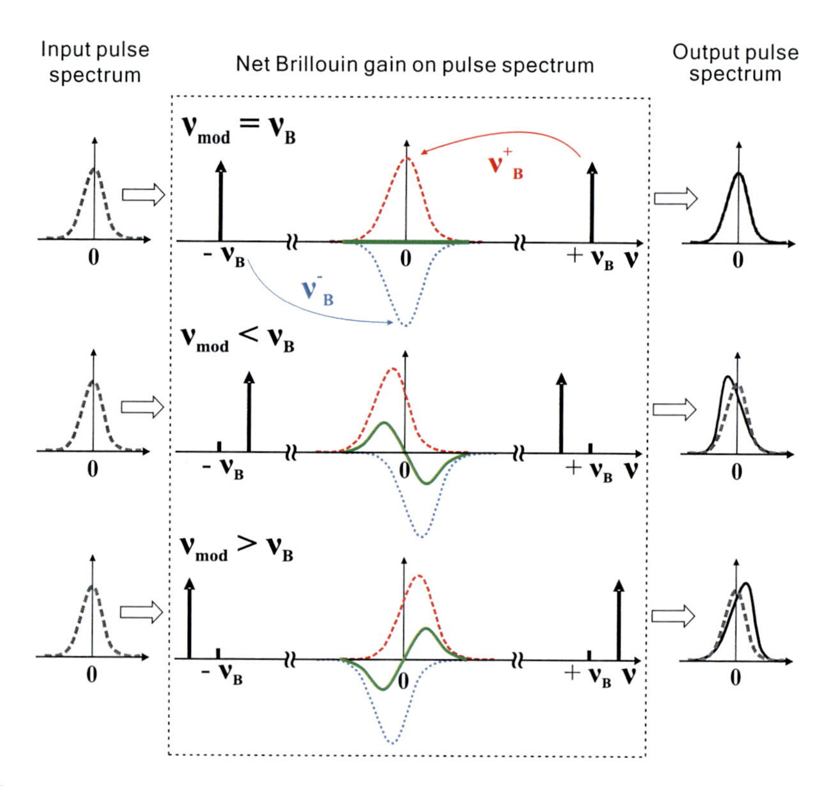

Fig. 22 Net Brillouin gain on the pump pulse while scanning the pump-probe frequency offset in a classical dual-sideband probe BOTDA. *Gray dotted lines*, input pulse spectrum; *red dotted lines*, BGS generated by the upper probe sideband; *blue dotted lines*, BLS generated by the lower probe sideband; *green solid lines*, net BGS experienced by the pulse; *black solid line*, output pulse spectrum after SBS interaction (Dominguez-Lopez et al. 2016)

perfectly compensate each other, resulting in a zero net gain on the pump. However, during the frequency scanning, the two probe waves are swept toward opposite spectral directions, creating gain and loss spectra that do not fully overlap (central and lower plots in Fig. 22). This effect results in an uneven spectral compensation of the gain-loss probe spectra in case of detuning, originating a non-zero net gain on the pump. Hence, the pulse spectrum becomes asymmetrically distorted and downshifted or upshifted in frequency as the pulse propagates through the fiber, depending on whether the scanned pump-probe frequency offset is higher or lower than the dominant BFS of the fiber.

The asymmetric spectral distortion also results in deformations of the temporal pulse shape (Dominguez-Lopez et al. 2016). These deformations manifest as large temporal broadening and pulse amplification. This translates into significant changes in the pulse energy, which, combined with the strong dependence of the distortions on the scanned pump-probe frequency offset, leads to BGS or BLS measurements with high levels of deformation. At low probe power levels, i.e., below 0.5 mW/sideband for a long fiber ($L > \alpha^{-1} \approx 22$ km), the phenomenon manifests as a moderate narrowing of the BGS and broadening of the BLS, with minor impact on the peak amplitude. However, when the input probe power is increased beyond 1 mW/sideband, the net asymmetric gain/loss on the pulse increases with a nonlinear dependence on the probe power. These distortions result in measured BGS and BLS having two strong peaks around the central dominant BFS of the fiber (Dominguez-Lopez et al. 2016).

To minimize the spectral and temporal distortions of the pulses, the SBS interaction has to be reduced by decreasing the input probe power. In the case of long sensing fibers ($L > \alpha^{-1} \approx 22$ km), the maximum probe power allowed at the input of the sensing fiber is about −5 dBm/sideband to secure minimum distortion during scanning. This constitutes the maximum probe power that can be launched into the sensing fiber in a classical dual-sideband BOTDA scheme. Advanced BOTDA configuration can however be employed to reduce pulse distortions by decreasing the accumulated net gain on the pump pulse.

Time-Domain Methods for High Spatial Resolution Brillouin Sensing

When the pump pulse used in a Brillouin time-domain sensor is shorter than the acoustic wave response time (i.e., $< \tau_A \approx 12$ ns), the incomplete activation of the acoustic field results in a large broadening of the Brillouin spectrum and a reduction of its peak amplitude (see section "Brillouin Gain Spectral Shape"). This transit behavior of the acoustic wave imposes an ultimate limit to the best attainable spatial resolution of classical Brillouin time-domain methods such as BOTDR and BOTDA. This limit is 1 m, corresponding to a pulse width of 10 ns (Fellay et al. 1997).

Most of the existing techniques to reach submeter spatial resolutions, especially resolutions of a few cm or below, are based on frequency and correlation domains.

However, there are still a few methods exploiting time-domain techniques, in which submeter resolution is achieved by pre-exciting the acoustic field (Beugnot et al. 2011). This section summarizes some basic concepts related to the acoustic wave transient behavior and describes the most common time-domain techniques for submeter resolution.

Acoustic Wave Transient Behavior: Basic Concepts

In order to better understand the limitations of the spatial resolution in a Brillouin time-domain sensor, it turns out helpful to analyze the transient behavior of the acoustic wave within the first 10 ns. The evolution of the acoustic field can be obtained by solving the set of coupled Eqs. 16–18 describing SBS in the time and space domains. Figure 23 shows the acoustic wave amplitude as a function of time and frequency detuning. The acoustic field is first activated by a pulse of 10 ns. The figure points out that the acoustic field has a very broad spectrum during the first nanoseconds of activation. Then, this acoustic field is reinforced at the central resonance frequency, resulting in a spectral narrowing, while the peak amplitude grows exponentially over time. The Brillouin spectrum measured by a time-domain sensor is given by the integral of this acoustic wave over the pulse duration. This means that if the pump pulse is very short (e.g., <5 ns), the Brillouin effect will be dominated by a weak acoustic wave with broad spectrum (Beugnot et al. 2011). In Fig. 23 the acoustic wave activation ceases after 10 ns, leading then to an exponential decay of the field amplitude. The use of longer pulses ($\gg \tau_A$) makes however the acoustic spectrum to evolve toward the natural Lorentzian shape of 27 MHz FWHM characterizing Brillouin scattering in CW regime.

In order to realize Brillouin sensing with high spatial resolution using time-domain interrogation techniques, approaches based on the pre-activation of the

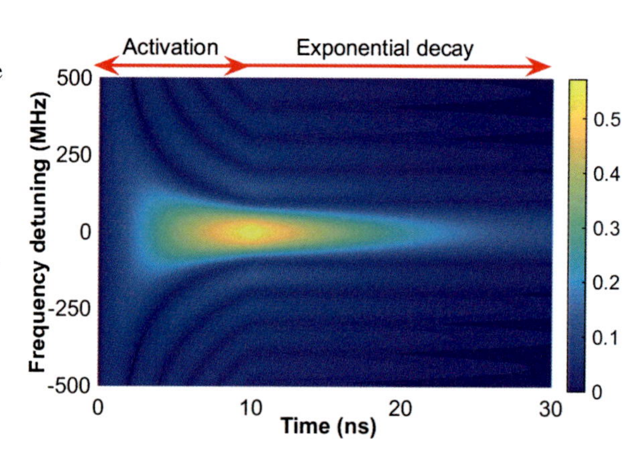

Fig. 23 Transient behavior of the Brillouin acoustic wave over time as a function of the frequency detuning. The acoustic field is activated during 10 ns, time in which the acoustic wave builds up increasing its amplitude and narrowing its spectrum. After 10 ns the acoustic field decays exponentially

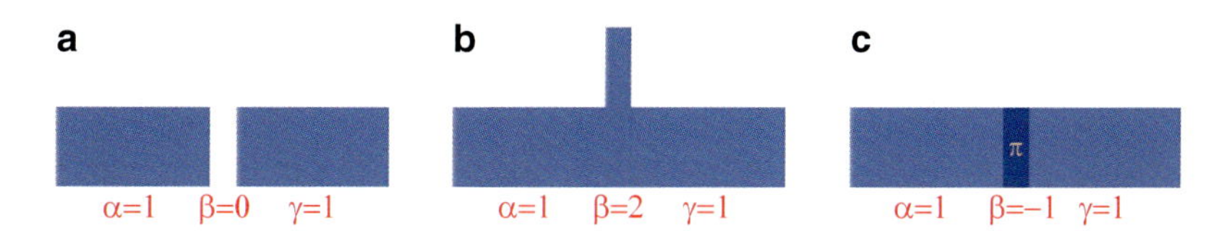

Fig. 24 Pulses commonly used to pre-activate a Brillouin acoustic wave: (**a**) *dark*, (**b**) *bright*, and (**c**) π-phase pulses

acoustic wave are used. The concept can be explained assuming the use of a CW pump beam (or equivalently assuming a very long pump pulse), so that the acoustic wave has enough time to build up until reaching a steady-state condition (i.e., activating a strong and narrowband acoustic field). High spatial resolution is achieved by perturbing the acoustic field with a very short optical pulse ($T < 10$ ns) in the middle of the CW beam (Beugnot et al. 2011). This perturbation can be induced using three different kinds of short pulses: (i) a *dark pulse* (Fig. 24a), characterized by the absence of light within the CW light beam; (ii) a *bright pulse* (Fig. 24b), corresponding to an intensity pulse on top of the CW beam; and (iii) a *phase pulse* (Fig. 24c) in which a π-phase shift is induced in the CW beam during a very short time. These pulses can be represented mathematically as a three-section pulse as (Beugnot et al. 2011)

$$\begin{aligned} E_p(z,t) = E_p^0 \Big[& \alpha u\left(t - \frac{z}{v_g}\right) + (\beta - \alpha)\, u\left(t - t_0 - \frac{z}{v_g}\right) \\ & + (\gamma - \beta)\, u\left(t - t_0 - \delta\tau_p - \frac{z}{v_g}\right)\Big], \end{aligned} \tag{55}$$

where α, β and γ are complex numbers representing the amplitude and phase of the three sections of the pulse, t_0 is the time when the second pulse section enters the fiber, and $\delta\tau_p$ is the duration of the second pulse section (corresponding to the pulse section of interest in this analysis). For these three pulses of interest, $\alpha = \gamma = 1$, while $\beta = 0$ for a dark pulse, $\beta = 2$ for a bright pulse, and $\beta = -1$ for a π-phase pulse.

The effects of these three pulses on the probe wave can be obtained by solving the SBS coupled Eqs. 16–18. In the transit regime, this set of equations must be solved numerically; however, assuming an undepleted pump and a small-gain regime, the system of equations can also be solved analytically by applying a perturbation method along with Laplace transform (Beugnot et al. 2011). In this perturbation method, the probe power can be expressed as $P_s(z,t) = \left|E_s^0 + e_s(z,t)\right|^2 = \left|E_s^0\right|^2 + 2\,\mathrm{Re}\left\{E_s^0 e_s(z,t)\right\}$, corresponding to the sum of a constant level and a small time-varying term. An analytical expression for $e_s(z,t)$ can be obtained in the Laplace domain, so that the probe intensity variations can be evaluated as (Beugnot et al. 2011)

$$
\begin{aligned}
e_s(z,t) = \frac{g_B(z_0)\, I_p^0 E_s^0}{2\Gamma_A^*} \Bigg\{ & \alpha^2 \left[u\left(t - \frac{2z_0 - z}{v_g}\right) - u\left(t - t_0 - \frac{2z_0 - z}{v_g}\right)\right] \\
& + \beta \left[\beta^* - (\beta^* - \alpha)\exp\left[-\Gamma_A^*\left(t - t_0 - \frac{2z_0 - z}{v_g}\right)\right]\right]\left[u\left(t - t_0 - \frac{2z_0 - z}{v_g}\right)\right. \\
& \left. - u\left(t - t_0 - \delta\tau_p - \frac{2z_0 - z}{v_g}\right)\right] \\
& + \gamma\left[\gamma^* - \left[\beta^*\left(1 - \exp\left(\Gamma_A^* \delta\tau_p\right)\right) - \left(\alpha - \gamma^* \exp\left(\Gamma_A^* \delta\tau_p\right)\right)\right]\right. \\
& \left. \exp\left[-\Gamma_A^*\left(t - t_0 - \frac{2z_0 - z}{v_g}\right)\right]\right] u\left(t - t_0 - \delta\tau_p - \frac{2z_0 - z}{v_g}\right)\Bigg\}
\end{aligned}
\tag{56}
$$

where $g_B(z_0)$ is the Brillouin gain coefficient at a given position z_0, and $I_p^0 = \left|E_p^0\right|^2$. Figure 25 shows the intensity variations of the probe wave caused by a conventional short intensity pulse ($\alpha = \gamma = 0, \beta = 1$) and a bright, dark, and π-phase pulses of duration $\delta\tau_p = 5$ ns. The figure points out that bright and dark pulses (Fig. 25b, c) induce larger probe intensity variations compared to the conventional short intensity pulse (Fig. 25a); while the phase pulse produces an even larger contrast, as shown in Fig. 25d. This is because the π-phase pulse turns out to be reflected out of phase in the Brillouin grating generated by the acoustic

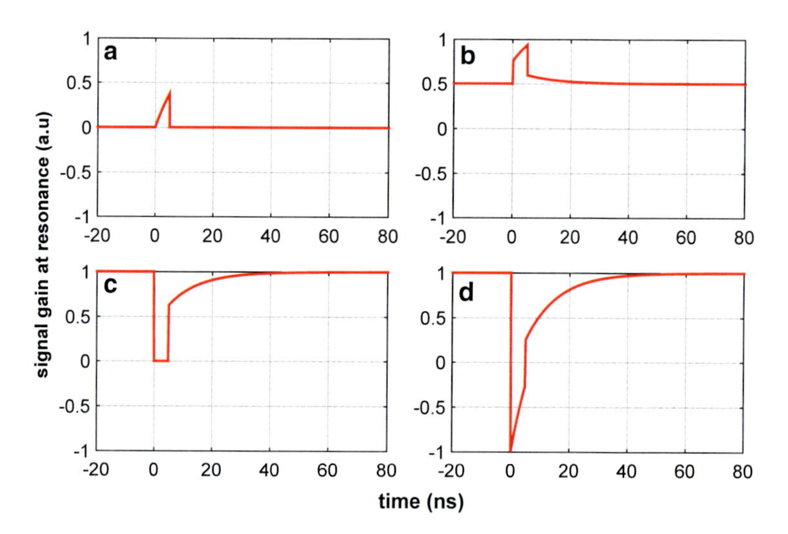

Fig. 25 Probe variations as a function of time at the resonance peak frequency, using a pre-activated acoustic wave and short pulses of $T = 5$ ns. Effects of (**a**) a conventional intensity pulse ($\alpha = 0, \beta = 1, \gamma = 0$), (**b**) a bright pulse ($\alpha = 0.5, \beta = 1, \gamma = 0.5$), (**c**) a dark pulse ($\alpha = 1, \beta = 0, \gamma = 1$), and (**d**) a π-phase pulse ($\alpha = 1, \beta = -1, \gamma = 1$) (Replot based on the concept presented in Beugnot et al. 2011)

field, which has no time to readapt its phase to the new condition imposed by the abrupt phase variation (Beugnot et al. 2011).

Early developments of high spatial resolution systems were realized based on dark pulses; nevertheless, it was later found out that the strong CW pump level induces a large amplification of the probe, which leads to strong pump depletion over a few km of sensing range. Another practical problem of this CW approach is that the use of a high-power CW pump also generates a large amount of spontaneous Brillouin scattering, which accumulates along the sensing fiber inducing high-amplitude noise affecting the measurements.

Differential Pulse BOTDA Techniques

The early techniques described in section "Acoustic Wave Transient Behavior: Basic Concepts," based on the acoustic wave pre-activation, have been further developed using long optical pulses instead of a CW beam. This way, pump depletion and noise issues are highly mitigated. In this case, the principle is based on the use of two long intensity pulses of duration T, being one of them abruptly disturbed by a short intensity or phase change of duration $\delta\tau_p$, just before the end of the pulse, as illustrated in Fig. 26. This approach is called *differential pulse method*, because traces obtained by these two pulses are subtracted one from each other, thus resulting in measurements with a spatial resolution defined by the differential pulse $\delta\tau_p$.

There exist two methods exploiting this differential pulse approach, depending on the type of short pulse used:

1. *Differential pulse-width pair BOTDA* (DPP-BOTDA), in which the short pulse corresponds to a dark pulse of duration $\delta\tau_p$. Equivalently, it can be considered that DPP-BOTDA uses two intensity pulses of different widths, one of duration $T - \delta\tau_p$ and a second one of duration T (Li et al. 2008).
2. *Brillouin echoes distributed sensing* (BEDS), in which the short pulse is a π-phase-shifted pulse. In this case the two intensity pulses have the same width T, while one of them is π-phase shifted just before the end of the pulse (Foaleng et al. 2010).

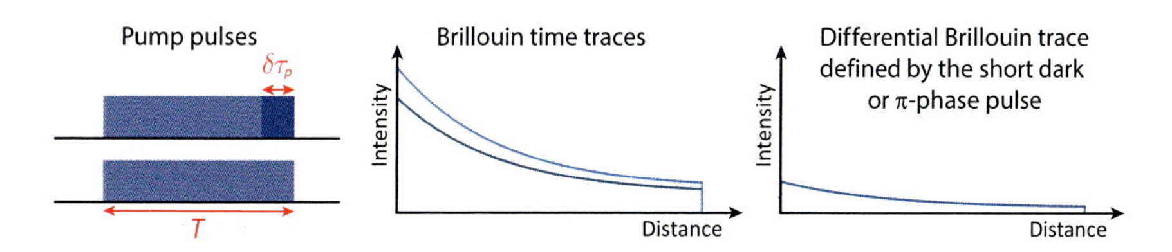

Fig. 26 Principle of differential pulse methods. Two pulses of width T are used, one of them having a short perturbation of duration $\delta\tau_p$ (a dark or π-phase pulse). They generate two slightly different Brillouin time traces, which after subtraction lead to a time trace with high spatial resolution $\Delta z = v_g\delta\tau_p/2$. The process is repeated at each scanned pump-probe frequency offset, resulting in a distributed measurement of the BGS with high spatial resolution

Let us first analyze the effect of the first pulse, which is the one composed by a long section of duration $T-\delta\tau_p$ and a second very short section of width $\delta\tau_p$ (where $\delta\tau_p < 10$ ns). This second section corresponds to the dark or π-phase pulse. If the first section is long enough, so that $T-\delta\tau_p \gg \tau_A$, the Brillouin acoustic wave will have enough time to be fully activated, generating an acoustic field of high-amplitude and narrow spectrum. This long acoustic excitation gives rise to BOTDA measurements showing a narrow BGS, whose FWHM is basically determined by the pulse width $T-\delta\tau_p$ as described in Eq. 39. During the second (short) pulse section, the acoustic field is disturbed either by a dark pulse (in DPP-BOTDA) (Li et al. 2008) or a π-phase pulse (in BEDS) (Foaleng et al. 2010), in the same way as described in Fig. 25. This short disturbance of the acoustic field allows the system to interrogate the Brillouin spectrum with a very short spatial resolution. However, in order to reach this high spatial resolution, the Brillouin gain on the probe produced by the first pulse section must be eliminated. This is performed by subtracting the BOTDA trace obtained by the second pulse (with constant amplitude and phase) from the BOTDA trace obtained by the first pulse, as shown in Fig. 26.

Since a well-designed Brillouin sensor behaves as a linear system, the common Brillouin gain produced by the intensity pulses is eliminated by this subtraction, thus only remaining the effect of the dark or π-phase pulse. The main advantage of this differential pulse approach is that spatial resolutions of a few cm can be obtained while maintaining a narrow BGS shape. This way, the detrimental impact of the BGS broadening occurring when using short pulses is fully eliminated, thanks to the pre-activation of the acoustic wave.

Note that DPP-BOTDA and BEDS systems require a layout very similar to a classical BOTDA scheme, as shown in Fig. 11a. However, in the case of BEDS systems, an additional phase modulator must be added in cascade (before or after) to the intensity modulator to generate the pump. Also a proper synchronization between intensity and phase pulses is required. For a proper operation of the technique, one of the main requirements is that pulses must have very sharp rising and falling edges (i.e., with slopes $\ll\delta\tau$) to allow high-frequency components to interrogate the Brillouin response with cm resolution. In addition, the electrical bandwidth of the detection and acquisition systems must be sufficiently large to resolve fast probe intensity transitions cause by longitudinal changes of the measurand over centimeter scales.

It must be pointed out that pulse distortions must be avoided in order to keep the sharp edges of the pump pulses all over the sensing fiber. It then becomes critical to mitigate nonlocal effects as much as possible. Distortions resulting from the frequency scanning procedure in a dual-sideband BOTDA scheme (Dominguez-Lopez et al. 2016), as described in section "Temporal and Spectral Distortion of the Pump Pulse in a Classical Dual-Sideband Scheme," can lead to significant pulse temporal broadening, with a consequent enlargement of the rising and falling times of the pulses. The presence of nonlocal effects makes practically impossible to obtain very sharp spatial resolutions (e.g., 1–2 cm) in a DPP-BOTDA scheme over long optical fibers (e.g., tens of km). To avoid such distortions, the input probe power must be maintained below −5 dBm/sidebands, which usually leads to differential traces with very low SNR. Alternatively, advanced techniques can be required to mitigate those nonlocal effects and allow higher probe powers at the sensing fiber input with the aim of increasing the SNR of the measurements.

Dynamic Brillouin Distributed Sensing

General Concepts

Classical Brillouin time-domain interrogation techniques require a broad scanning of the Brillouin spectrum, which together to trace averaging leads to a measurement time of usually a few minutes. This long acquisition time constrains classical time-domain techniques to only quasi-static measurements. To exploit Brillouin time-domain methods for dynamic distributed sensing, special attention must be paid to all factors limiting the acquisition speed. The measurement time of classical BOTDA sensors is mostly determined by four factors (Peled et al. 2012):

1. Time-of-flight (or round-trip time): This is the time required for the light to propagate down the entire sensing fiber and come back to the receiver. It determines the minimum repetition interval of the pump pulses launched into the fiber. Considering a fiber of length L and group velocity v_g, the round-trip time is defined as $T_{\text{roundtrip}} = 2L/v_g$.
2. Number of averages (N_{avg}): In order to improve the SNR, a large number of traces is typically averaged, mainly for long-range sensing. Trace averaging needs N_{avg} pump pulses to be consecutively launched into the sensing fiber (with a time interval $>T_{\text{roundtrip}}$) for each scanned pump-probe frequency offset. Note that, in addition, to eliminate the polarization dependence of the Brillouin gain, a polarization scrambler combined with averaging is commonly employed. The number of averages is normally of the order of 100 or even 1000 and can be ultimately reduced down to $N_{\text{avg}} = 2$ in a classical BOTDA scheme using a polarization switch for polarization fading compensation.
3. Number of scanned frequencies (N_{freq}): To properly reconstruct the BGS shape, a minimum number of pump-probe frequency offsets must be scanned around the average BFS, covering at least the BGS width. The use of small frequency steps leads to a better measurand accuracy and lower uncertainty (see Eq. 40) (Soto and Thévenaz 2013a); however, this increases the number of scanned traces and leads to large measurement time. On the other hand, a large scanning range is typically required to extend the temperature or strain range to be measured. Typically $N_{\text{freq}} = 100$ to 400 frequencies are scanned.
4. Switching time for changing the pump-probe frequency offset (T_{switch}): The frequency scanning process requires changing the optical frequency of either the pump or probe signal. This is typically performed in milliseconds or even longer time scales.

In long-range sensing, the time-of-flight and the number of averages are the dominant terms determining the overall measurement time (Peled et al. 2012). Actually, while the time-of-flight is inherently determined by the fiber length, a large number of averages are normally required with long fibers in order to reach a target SNR. In short-range sensing, the number of averaged traces can be substantially reduced since, in addition to the lower impact of the fiber attenuation, large pump and probe powers can be used without activating nonlinear effects. In this case the dominant factor determining the measurement speed is the time required to switch between consecutive scanned frequencies (Peled et al. 2012). Furthermore, it must be noticed that the need for scanning the BGS shape imposes a common limitation to the acquisition time of any kind of classical BOTDA measurement.

Excluding any additional overhead time due to hardware limitations or signal processing, and considering the four abovementioned factors, the measurement time of a BOTDA sensor can be written as

$$T_{\text{meas}} = N_{\text{avg}} N_{\text{freq}} T_{\text{roundtrip}} + N_{\text{freq}} T_{\text{switch}} \tag{57}$$

where it has been assumed that N_{avg} time-domain traces are consecutively acquired and averaged before changing the pump-probe frequency offset, as it is in most of real implementations.

Note that the sampling frequency $f_s = 1/T_{\text{meas}}$ of the measurand is highly dependent on the pulse repetition rate. Hence, the capabilities of the sensor to detect dynamic changes of a given physical variable (usually dynamic strain, rather than temperature) highly depend on the fiber length. For this reason dynamic (strain) distributed sensing normally targets short ranges, thus allowing for high dynamic sensing capabilities.

Several approaches exist to minimize the impact of one, or many, of the factors in Eq. 57. In the following, the most common techniques for dynamic Brillouin time-domain sensing are described.

Fast BOTDA (F-BOTDA)

Conventional BOTDA implementations make use of an electronic synthesizer to generate a microwave signal (e.g., at ~10.8 GHz) and to scan the pump-probe frequency in a range of a few hundred MHz. The sweeping time T_{switch} of these synthesizers is typically of the order of 1 ms (or higher) per frequency, imposing an overhead time (given by to the second term in Eq. 57) of a few hundred milliseconds just for frequency sweeping (Peled et al. 2012). This makes dynamic sensing practically impossible when using a classical BOTDA scheme.

A method called *Fast BOTDA* (F-BOTDA) (Peled et al. 2012) focuses on reducing this switching time, while keeping the first term in Eq. 57 limited to short fibers and small number of averages. The method makes use of a pump pulse at a fixed optical frequency, and a probe signal whose optical frequency is swept with

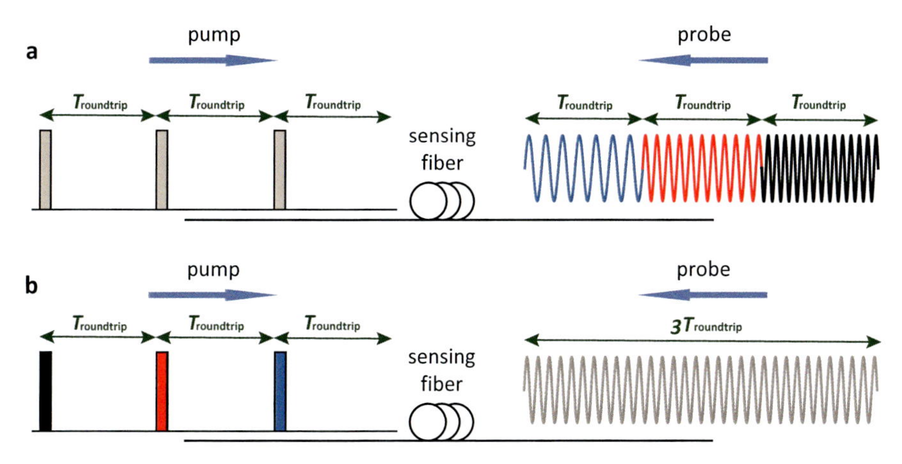

Fig. 27 Principle of the Fast BOTDA technique, using three scanned frequencies for illustration purposes (normally a few hundreds of frequencies are scanned). (**a**) A single pump frequency interacting with probe waves at different frequencies. (**b**) A fixed-frequency probe interacts with pump pulses of different optical frequencies (Replot based on the concept presented in Peled et al. 2012)

a time interval $T_{roundtrip} = 2L/v_g$ (if no averaging is used) by using a fast arbitrary waveform generator (AWG) that replaces the electronic synthesizer of the classical BOTDA scheme (see Fig. 27a). The sequence of probe frequencies needs to be directly written in the memory of the fast AWG, which must be able to change frequencies in nanosecond scale. A second alternative is to use a single-frequency probe interacting with pump pulses with different optical frequencies, as illustrated in Fig. 27b. Both schemes are conceptually equivalent, having only differences in the implementation. Here below the first approach is described.

In the case shown in Fig. 27a, pulses are sent over the sensing fiber with a well-controlled synchronization, so that each pump pulse meets a unique probe frequency during propagation. The time-domain sensor response resulting from consecutive pump-probe interactions at different frequencies is acquired in a *single measurement* in order to optimize the acquisition time. This results in a very long time trace composed of all BOTDA traces originated from different pump-probe frequency offsets, which are appended one after another. Averaging can be performed by repeating the set of probe frequencies launched into the fiber, or alternatively by sweeping the probe frequency every $N_{avg}T_{roundtrip}$, so that N_{avg} traces are measured before switching to the next probe frequency.

Note that for a proper operation of the method, the AWG has to fulfill two main requirements:

1. The AWG must have a large memory to contain all points describing the time-domain waveforms associated with the entire set of scanned frequencies.

2. It must allow a tunability of a few hundred MHz around a central frequency of about 10.8 GHz, with the ability to change frequencies in a few ns. In general, it is difficult and expensive to obtain this broad frequency range operation and fast sweeping time at such a high central frequency. As an alternative, an AWG operating at a lower central frequency (e.g., around 1 GHz) can be employed combined with frequency up-conversion to increase the output frequency.

Note that F-BOTDA can reach effective sampling rates beyond 1 kHz over short sensing ranges. For example, a theoretical sampling rate of 10 kHz can be achieved by scanning 100 frequencies over a 100-m-long fiber, with no averaging and assuming no impact of polarization. If a polarization switch is used to compensate polarization fading, the sampling rate is reduced down to 5 kHz, which, by the Nyquist sampling criterion, makes possible measurements of dynamic strains up to 2.5 kHz.

Slope-Assisted BOTDA (SA-BOTDA)

Classical (Gain) Amplitude-Based SA-BOTDA

Another approach to increase the measurement speed is to reduce the time associated with the first term in Eq. 57, while the second one is fully eliminated. This can be achieved by suppressing completely the frequency scan and interrogating the Brillouin interaction at a single pump-probe frequency offset matching the half-gain point of the BGS (Bernini et al. 2009; Peled et al. 2011). In this way, fast BFS changes induced by dynamic strain are converted into amplitude variations, as shown in Fig. 28. A simple linear gain-to-strain conversion factor can be used, as long as the maximum strain deviation remains within the linear slope region of the BGS. This method is called *slope-assisted BOTDA* (SA-BOTDA) (Peled et al. 2011).

Note that the interaction at the half-gain point of the BGS normally cannot be secured for the entire sensing fiber because of the natural BFS nonuniformity.

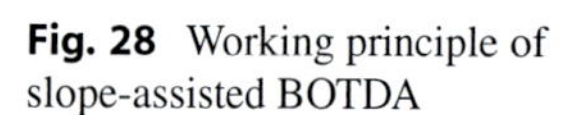

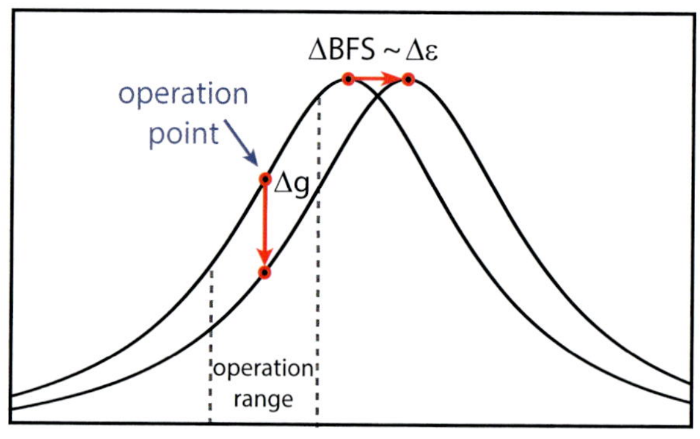

Fig. 28 Working principle of slope-assisted BOTDA

This nonuniformity makes the classical BOTDA scheme, using a CW single-frequency probe wave, unsuitable for this slope-assisted approach.

A first implementation of this principle in Bernini et al. (2009) made use of pulsed pump and probe signals to precisely access the half-gain point of the BGS at a given desired interrogation position, which could be determined by the fiber location where both pump and probe pulses meet inside the fiber. Nevertheless, this approach is limited to the interrogation of only a single fiber position and requires a longitudinal scan of the interrogation point by changing the relative delay between pump and probe pulses. This scanning is however very time consuming and reduces significantly the dynamic capabilities of the system.

To achieve truly dynamic and distributed sensing using this SA-BOTDA technique, a CW probe wave of variable optical frequency must be used. In particular, the temporal evolution of the probe optical frequency must be tailored based on the longitudinal variations of the BFS profile plus (or menus) half the Brillouin spectral width (Peled et al. 2011). In this way, and provided a proper synchronization between the pump pulse and the frequency-modulated probe, the pump pulse propagating through a fiber meets the CW probe at a local frequency offset matching exactly the half-gain point of the BGS at all fiber locations.

For the implementation, a first classical BOTDA scanning is required to acquire a reference (typically nonuniform) BFS profile $\nu_B^{\text{ref}}(z)$. Then, making use of an arbitrary waveform generator, the probe frequency ν_{probe} is tailored as $\nu_{\text{probe}} = \nu_{\text{pump}} \pm \nu_B^{\text{ref}}(z) \pm \Delta\nu_B^{\text{pulse}}/2$, where ν_{pump} is the pump optical frequency and $\Delta\nu_B^{\text{pulse}}$ is the FWHM of the BGS, which is essentially determined by the spatial resolution, as described in Eq. 39. Note that the sign $\pm$ in front of the reference BFS profile $\nu_B^{\text{ref}}(z)$ represents whether a Brillouin gain ($-$) or loss ($+$) configuration is used, while the sign $\pm$ in front of $\Delta\nu_B^{\text{pulse}}/2$ depends on whether the operation point is set at the low-frequency ($-$) or high-frequency ($+$) slope of the spectrum.

The SA-BOTDA scheme requires the same hardware as a classical BOTDA, with the exception of the microwave generator which in this case must be replaced by a fast AWG. This AWG must have the ability to change its output frequency in nanosecond scales to avoid impairments on the spatial resolution.

To secure a well-behaved sensing operation, the maximum strain amplitude must induce a BGS shift smaller than the spectral region defined by the linear slope of the BGS (see Fig. 28), so that a simple linear gain-to-strain conversion factor could be used (Bernini et al. 2009; Peled et al. 2011). Note that this conversion factor depends not only on the conventional strain sensitivity $C_{\varepsilon,\text{BFS}}$ of the BFS but also on the BGS slope at the half-gain point ($\pm\Delta\nu_B/2$). This way, strain variations $\Delta\varepsilon$ can be quantified from gain variations $\Delta g(\pm\Delta\nu_B/2)$ as

$$\Delta\varepsilon = \frac{-1}{C_{\epsilon,\text{BFS}} \cdot dg\left(\Delta\nu\right)/d\Delta\nu|_{\Delta\nu=\pm\Delta\nu_B/2}} \Delta g\left(\pm\Delta\nu_{\text{B}}/2\right) \tag{58}$$

Note that the spatial resolution defines both the BGS slope and the linear spectral region of the BGS. For instance, a maximum dynamic range of 600 $\mu\varepsilon$ can be achieved using long pump pulses (defining a FWHM of 27 MHz and a linear slope of <30 MHz). This dynamic range can be expanded when using shorter spatial resolutions, with the penalty of a reduced sensitivity, i.e., of a smaller gain-to-strain conversion factor.

Since the method eliminates the frequency scan, a significant improvement in the measurement speed can be achieved with SA-BOTDA sensors. Considering a time-of-flight $T_{\text{roundtrip}}$ and N_{avg} averaged traces, the measurement time turns out to be equal to $T_{\text{meas}} = N_{\text{avg}} T_{\text{roundtrip}}$, representing a time reduction in a factor N_{freq} compared to a classical BOTDA measurement (neglecting the elimination of the second term in Eq. 57). Hence, the sampling frequency of a SA-BOTDA system can be written as $f_s = v_g/(2N_{\text{ave}}L)$. Considering no averaging, this sampling corresponds to a maximum measurable strain frequency equal to 1 MHz over a 50-m-long fiber or 50 kHz over a 1 km fiber. Note that those values are usually halved when using a polarization switch for fading compensation.

SA-BOTDA Based on Phase-Shift Response

One of the main penalties of the classical SA-BOTDA method lies on the amplitude dependence of the measurements. This dependence implies that any amplitude change in the measured time-domain traces, resulting from laser power fluctuations or bending/local losses, can be wrongly interpreted as a change in the strain amplitude. This can lead to an offset in the measured strain values. A completely different approach to eliminate the power dependence of SA-BOTDA measurements, and also to extend the dynamic range of the measured strain, is based on measurements of the Brillouin phase-shift in addition to the Brillouin gain (Urricelqui et al. 2012).

As shown in Fig. 29, the method requires a pulsed pump interacting with a phase-modulated probe. The modulating frequency of the probe must be larger than the FWHM of the BGS, so that only one of the probe sidebands is affected by the SBS interaction with the pump pulse. This way, and considering only the first-order

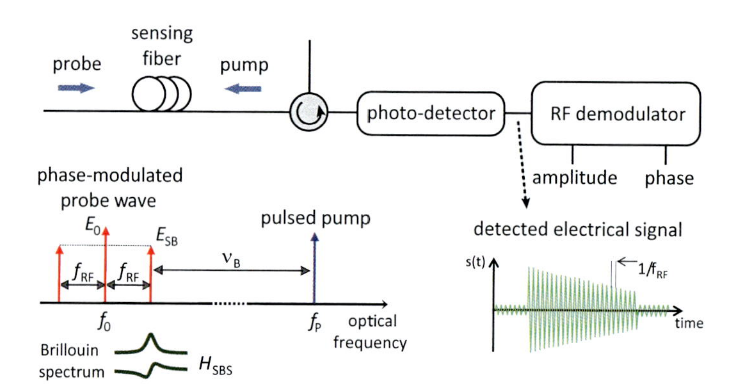

Fig. 29 Scheme of a SA-BOTDA sensor based on phase-modulated probe and RF phase-shift response (Replot based on the concept presented in Urricelqui et al. 2012)

modulation sidebands, the optical field at the input of the photodetector can be expressed as (Urricelqui et al. 2012)

$$E(t) = -E_{SB}e^{-i2\pi(f_0 - f_{RF})t} + E_0 e^{-i2\pi f_0 t} + E_{SB} H_{SBS}(f_0 + f_{RF}, z)\, e^{-i2\pi(f_0 + f_{RF})t}, \tag{59}$$

where E_0 and E_{SB} are the field amplitudes of the carrier and sidebands of the phase-modulated probe, f_0 is the optical frequency of the probe carrier, f_{RF} is the modulation frequency, and $H_{SBS} = e^{G_{SBS} + i\varphi_{SBS}} \approx (1 + G_{SBS})\, e^{i\varphi_{SBS}}$ is the complex Brillouin gain/loss spectrum, where G_{SBS} and φ_{SBS} are the Brillouin amplitude and phase responses given by the real and imaginary parts of Eq. 23, respectively.

The resulting radio-frequency (RF) signal at the output of the photodetector can be written as (Urricelqui et al. 2012)

$$\begin{aligned} i_{RF}(t) &= \Re E_0 E_{SB}\left[(1 + G_{SBS})\cos(2\pi f_{RF} t + \varphi_{SBS}) - \cos(2\pi f_{RF} t)\right] \\ &\approx \frac{4\Re E_0 E_{SB} g_{B0} \Delta\nu_B}{\sqrt{\Delta\nu_B^2 + (2\Delta\nu)^2}} \cos\left(2\pi f_{RF} t - \arctan\left(\frac{2\Delta\nu}{\Delta\nu_B}\right)\right). \end{aligned} \tag{60}$$

where $\Re$ is the responsivity of the photodetector.

Note that the phase-shift spectral response of the resulting RF signal is $\varphi_{RF} = \arctan(2\Delta\nu/\Delta\nu_B)$. For a small pump-probe detuning $\Delta\nu$, this response shows a linear behavior (i.e., $\varphi_{RF} \propto \Delta\nu$), thus enabling a linear conversion between induced phase-shift and strain variations. Thus, instead of using the slope of the BGS as in the classical SA-BOTDA, in this case the slope of the RF phase-shift spectral response $d\varphi_{RF}/d\nu$ is employed in the conversion. It must be pointed out that the linear spectral region of the RF phase response is broaden than the BGS slope, thus allowing for measurements of larger strain amplitudes. Since the RF phase shift is independent of the pump and probe powers, the method is robust against variations of the detected power. This feature also makes the method to be partially robust against pump depletion issues, resulting in an interesting approach not only for dynamic but also for long-range Brillouin sensing.

Sweep-Free BOTDA (SF-BOTDA)

Another approach to eliminate the time associated with the frequency scan is to measure simultaneously multiple Brillouin interactions at different pump-probe frequency offsets. This method, called *sweep-free BOTDA* (SF-BOTDA) (Voskoboinik et al. 2011), interrogates different spectral positions of the BGS in a *single measurement*, allowing for the reconstruction of the entire BGS shape.

Single-Frequency Pump Interacting with Multitone Probe

A first implementation of this approach in Chaube et al. (2008) used a single-frequency pulse interacting with a counter-propagating optical frequency comb, as shown in Fig. 30a. Each tone of the comb simultaneously probes different

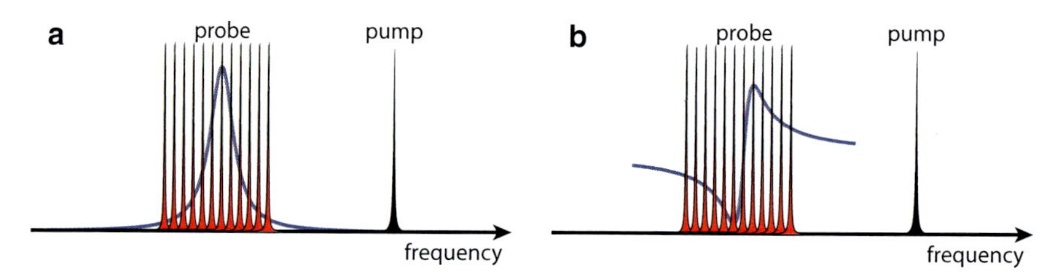

Fig. 30 Single-frequency pump interacting with a multitone probe to measure in a single shot (**a**) the BGS and (**b**) the BPS

spectral positions of the Brillouin spectrum generated by the pump pulse. Note that either the amplitude (i.e., gain or loss) (Chaube et al. 2008) or the phase response (Jin et al. 2017) of the Brillouin interaction can be measured with this approach, as also illustrated in Fig. 30b. By a simple analysis of the Brillouin gain (or loss) or phase shift experienced by each tone of the probe, the Brillouin gain/loss or phase spectrum can be fully reconstructed, with a frequency sampling step equal to the frequency spacing of the probe comb. To determine the level of amplification/attenuation or phase shift of each probe tone (normally separated by a few MHz), a coherent detection scheme is required. This way, each tone of the probe optical comb is converted to microwave frequencies and demodulated in the electrical domain. Using a fast analogue-to-digital conversion, the Brillouin response (amplitude and/or phase) is extracted by digital signal processing, typically involving digital filtering and fast Fourier transform. Note that when detecting the Brillouin phase spectrum, special techniques for frequency offset recovery and phase noise compensation are required to minimize the impact of phase noise on the BPS reconstruction (Jin et al. 2017).

In these two approaches, a narrow spectral separation between the comb tones would be preferred for a proper reconstruction of the BGS/BLS or BPS profile. However the temporal response of the SBS interaction, determined by the spatial resolution of the system, imposes important constraints to the frequency spacing between tones. In particular, the SBS interaction results in spectral broadening of each of the interacting spectral lines (Chaube et al. 2008). This spectral broadening is proportional to the spectral width of the pulses, and therefore, to avoid cross talk between adjacent tones, a frequency spacing larger than two times the pulse spectral width is required. This imposes a fundamental limitation to all these schemes in which a single pump tone is used to interact with a multitone probe, resulting in a strict trade-off between spatial resolution and sampling frequency step (associated with the BFS uncertainty through Eq. 40).

Multitone Pump Interacting with Multitone Probe

To overcome the trade-off between spatial resolution and frequency sampling step described in section "Single-Frequency Pump Interacting with Multitone Probe," a multitone pump can be used to interact with a multitone probe signal (Voskoboinik et al. 2011). In this case different pump-probe frequency offsets

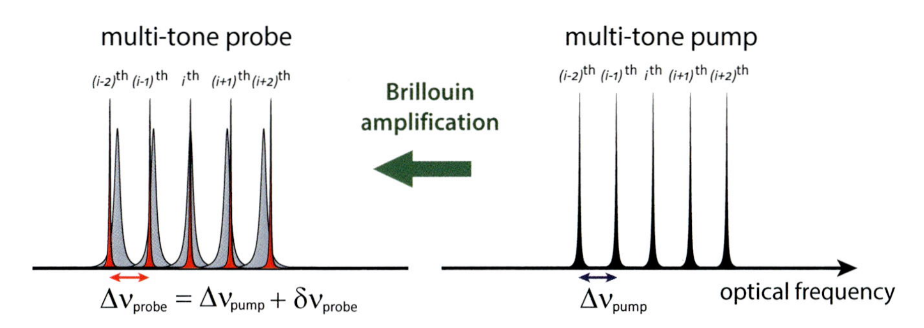

Fig. 31 Sweep-free BOTDA using a multifrequency pump interacting with a multifrequency probe (Replot based on the concept presented in Voskoboinik et al. 2011)

are simultaneously interrogated by using multiple independent pump-probe pairs, as shown in Fig. 31. This technique requires a pump composed of N spectral components, each one generating an independent Brillouin gain (or loss) spectrum. To avoid overlapping of the independent BGS replicas, pump tones must be separated by a frequency difference $\Delta\nu_{\text{pump}}$ much larger than the Brillouin gain spectral width. This way, the independent (spectrally parallel) BGSs can be simultaneously probed with multiple probe tones, which are launched in counter-propagating direction with slightly different spacing $\Delta\nu_{\text{probe}}$ compared to the pump components. Defining $\Delta\nu_{\text{probe}} = \Delta\nu_{\text{pump}} \pm \delta\nu_{\text{probe}}$, with $\Delta\nu_{\text{pump}}$ and $\Delta\nu_{\text{probe}}$ being much larger than the BGS width, each probe tone can interact with a unique pump tone (Voskoboinik et al. 2011). As a result, each pump-probe pair interrogates a different spectral region of the BGS, allowing for the reconstruction of the BGS with a sampling frequency step equal to $\delta\nu_{\text{probe}}$. The spatial resolution only restricts the minimum spectral spacing between pump tones $\Delta\nu_{\text{pump}}$, while the frequency step $\delta\nu_{\text{probe}}$ remains independent of the pulse spectral width.

Note that a large number of tones result in a better BGS spectral sampling, thus increasing the precision of the measurand. However, the spectral arrangement of pump and probe tones, and therefore also the number of spectral lines, are conditioned basically by two main factors:

1. In a Brillouin gain (loss) configuration, the highest (lowest) optical frequency of the probe tones has to be lower (higher) than the lowest (highest) optical frequency of the pump tones. This condition avoids detrimental overlapping between pump and probes components and therefore unwanted pump-probe cross-interactions. This restriction also implicitly defines a maximum available bandwidth for the spectral allocation of the pump and probe lines, being constrained to a maximum spectral range equal to the BFS of the fiber (i.e., $\sim$10.8 GHz).
2. The spectral separation between lines in the two frequency combs must be much larger than the BGS full-width at half-maximum in order to avoid

unwanted cross-interaction between different pump and probe components. This way, the use of a frequency spacing of at least two or three times the BGS spectral width can normally secure the interaction of a single pump tone with a single-probe tone through a unique and independent Brillouin acoustic wave.

In principle, the pump-wavelength dependence of the BFS can lead to slightly shifted BGS replicas, thus resulting in an effective spectral sampling step different from $\delta\nu_{\text{probe}}$. However, since the maximum spectral range for the pump and probe is constrained to the fiber BFS, the spectral shift of the BGS replicas turns out to be only a few kilohertz. This spectral shift is negligible compared to the typical BFS uncertainty of the measurements, and therefore it is possible to safely assume that all BGS replicas are exactly identical.

One of the main limitations of the method is imposed by intertone nonlinear interactions, such as four-wave mixing (FWM) or modulation instability, which can occur in the sensing fiber. A solution to reduce the FWM efficiency and avoid cross talk between FWM components and pump-probe tones is to use unequal frequency spacing, for both pump and probe optical combs. In this case, however the frequency of each comb tone must be precisely allocated to secure that adjacent pump-probe pairs of lines allow for a uniform spectral sampling of the BGS/BLS. Although the use of unequal frequency spacing is helpful to reduce the cross talk induced by FWM, this approach does not eliminate the mutual seeding of modulation instability generated by the multiple high-power pump pulses. Indeed, considering that the maximum spectral range covered by the pump and probe tones is limited by the BFS of the fiber (i.e., to $\sim$10.8 GHz), it is typically expected that all spectral components of the comb fall inside the MI gain bandwidth (see section "Modulation Instability (MI)"), representing still a source of distortion for the measured Brillouin time traces.

To avoid seeding nonlinear intertone interactions, a wavelength-dependent delay has to be applied to the pump, so that pulses at different optical frequencies do not overlap with each other in time domain (Voskoboinik et al. 2011). This sequential launching of the pump lines avoids intertone interactions, however, leads to time-shifted replicas of the BOTDA trace. To avoid ruining the spatial resolution, traces must be realigned by applying the reverse wavelength-dependent delay originally applied to the pump pulses. This alignment process can be performed optically or digitally by post-processing the temporal traces obtained by coherent detection.

The Brillouin gain/loss information carried by each tone of the probe comb is retrieved by using coherent detection, followed by a fast real-time acquisition system and the use of signal processing. Using fast Fourier transform and digital filtering, each of the probe tones can be separated digitally and then analyzed. Applying inverse Fourier transform to each independent frequency component, followed by temporal synchronization, the local BGS can be then retrieved as a function of distance (Voskoboinik et al. 2011).

Long-Range Brillouin Distributed Sensing

One of the most interesting features of Brillouin sensors is their possibility to perform distributed measurements over very long sensing distances (i.e., tens of kilometers), with high spatial resolution (i.e., meter or submeter scales). This gives unique opportunities to this technology to monitor large structures where other distributed fiber sensing techniques cannot operate. However, as indicated in Eq. 34, the response of a BOTDA sensor decays exponentially with distance due to the linear fiber loss experienced by the pump power during propagation. This effect, together with the losses experienced by the probe, can significantly reduce the SNR of the measurements at long sensing distances. This SNR reduction as a function of distance indicates that the fiber attenuation is indeed the factor that imposes the most stringent limitation to the performance of a BOTDA sensor, especially for long ranges. A trivial strategy to boost the SNR of the measurements, and thus allows for an enhancement of the figure of merit of the sensor, is to increase the pump and probe powers launched into the sensing fiber. However, as described in "Limitations and Design Considerations in Brillouin Time-Domain Sensing", the onset of nonlinear effects limits the maximum pump and probe powers allowed into the sensing fiber, imposing a serious trade-off between the parameters of the sensor. The only alternative to extend the sensing range of a conventional BOTDA scheme is then to use long spatial resolutions; however, this alternative generally leads to similar figure of merits, providing no real improvement to the sensor performance.

In order to overcome the SNR limitations existing in standard BOTDA configurations, and thus to obtain a real enhancement of the figure of merit of the system, dedicated advanced techniques must be used. This section reviews some of the most common techniques used in BOTDA sensing to reach very long sensing distances. Although techniques are here described in the context of BOTDA sensing, many of them are equally applicable to boost the performance of BOTDR sensors.

Optical Amplification

Distributed Raman Amplification (DRA)

One of the methods that has proved to be very effective in enhancing the sensing capabilities of BOTDA systems is distributed Raman amplification. This technique makes use of stimulated Raman scattering to provide distributed optical amplification to the BOTDA pump and probe along the sensing fiber. As a result, the pump-probe Brillouin interaction is enhanced, thus leading to a significant SNR improvement of the measured BOTDA traces. The technique, so-called *Raman-assisted BOTDA* (RA-BOTDA) (Rodríguez-Barrios et al. 2010; Martin-López et al. 2010), allows for different possible implementations, including, for instance, schemes using forward, backward, or bidirectional Raman pumping, with

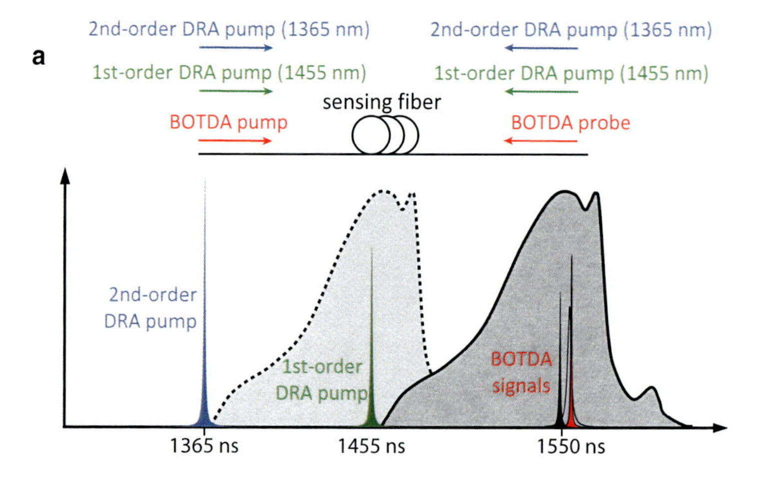

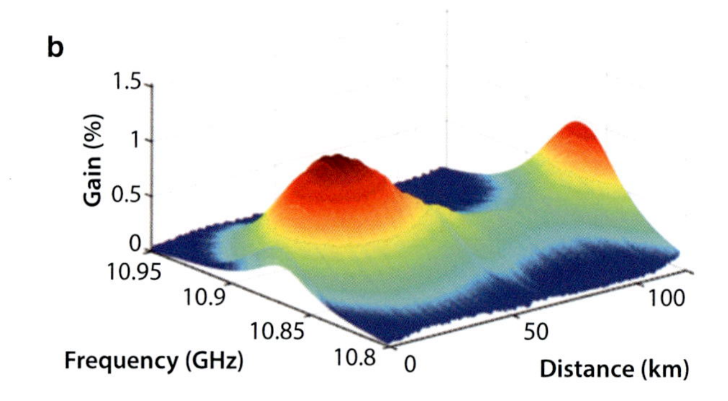

Fig. 32 Principle of a Raman-assisted BOTDA sensor. (**a**) Wavelength allocation in a multi-pumping scheme. (**b**) BGS measured by a BOTDA system employing a first-order bidirectional Raman pumping scheme over 120-km-long fiber

first-, second-, or even higher-order schemes, as shown in Fig. 32a. Certainly, the single-end access of BOTDR makes, in principle, only possible the use of forward-pumping schemes (Cho et al. 2003).

In a first-order Raman-assisted BOTDA scheme, one or more high-power CW Raman pumps are launched into the sensing fiber by one or both fiber ends, at a wavelength around 1450 nm (Rodríguez-Barrios et al. 2010; Soto et al. 2011). Pumping at that wavelength produces a broadband optical amplification in the 1550 nm band, amplifying both pump and probe waves interacting in the BOTDA sensor. Figure 32b shows a typical BGS measurement versus distance, when first-order bidirectional Raman amplification is used. Note that in this case the maximum Brillouin interaction does not occur at the fiber input, as in a classical BOTDA scheme, but occurs a few kilometers inside the fiber. This is because the Brillouin pump is amplified by the co-propagating Raman pump during the first kilometers (normally along the fiber effective length, corresponding to ~20 km). After ~20 km distance, the BOTDA pump is attenuated by the fiber, being then amplified again by the backward propagating Raman pump over the last 20 km. Note also that the weaker Brillouin interaction does not occur at the end of the fiber, but a few kilometers (~20 km) before the end, which represents the fiber location with the lowest SNR and where the worst measurand resolution is found over the entire fiber length.

In order to extend the sensing distance, second-order Raman pumping schemes can be used (Martin-López et al. 2010). As shown in Fig. 32a, in this case additional Raman pumps operating at ~1365 nm can be injected into the sensing fiber (also following a forward, backward, or bidirectional pumping scheme). These second-order pumps provide distributed Raman gain to the first-order pumps at 1455 nm, which in turn amplify the BOTDA signals.

Distributed Raman amplification helps in compensating the detrimental impact of attenuation along the sensing fiber. Although the longitudinal uniformity of the distributed optical amplification is an important factor, the optimization of all the input powers in a RA-BOTDA must take into account different aspects to avoid performance degradation in the system (Soto et al. 2011). On the one hand, potential trace distortions resulting from detrimental nonlinear effects and depletion of the BOTDA pump power must be avoided. On the other hand, sources of noise originated from the distributed Raman amplification process must be kept under control; among them the most important issues are related to the double Rayleigh scattering, amplified spontaneous emission, and relative intensity noise (RIN) transfer from Raman pumps to BOTDA signals.

In particular, the optical power of the waves launched at the near fiber end ($z = 0$), i.e., the Brillouin pump and the forward-propagating Raman pump(s), must be optimized considering two main design rules (Soto et al. 2011):

1. The maximum power that the Brillouin pump can reach inside the sensing fiber must be kept below the onset of nonlinear effects. This means that the input power of the Brillouin pump pulse must be reduced compared to the classical BOTDA scheme, so that the limit power of ~100 mW imposed by modulation instability is reached a few kilometers away from the fiber input.
2. The proper design of the input powers should also push the maximum Brillouin pump level to a deep location inside the sensing fiber, so that the Brillouin pump power at the far fiber end is maximized.

Note that the power optimization for the BOTDA pump and forward-propagating Raman pump(s) is mostly determined by the longitudinal evolution of the Brillouin pump power through the fiber, having as a main objective the maximization of the pulse power at the far fiber end. Noise-related issues are less restrictive in this case. Even though an efficient RIN transfer from the Raman pump to the Brillouin pump is expected (owing to the co-propagative nature of this Raman interaction), this is not a real constraint for the power optimization because the noise acquired by the BOTDA pump is not efficiently transferred by SBS to the probe signal. Indeed, the counter-propagative nature of SBS averages the effect that the noise in the Brillouin pump has on the probe, thus limiting RIN transfer in this case. This implies that the RIN transferred from the forward-propagating Raman pump can be neglected in the design of a Raman-assisted BOTDA sensor.

The situation is however different for the design of the backward-propagating signals, i.e., the BOTDA probe and the backward Raman pump(s). In this case not only distortions of the BOTDA traces must be avoided, but also the SNR of the traces must be taken into account. The design must consider two aspects (Soto et al. 2011):

1. The maximum power levels that the probe signal reaches inside the fiber has to be low enough to avoid depletion of the Brillouin pump power (being typically weak at the end of long sensing fibers).
2. Optical noise must be reduced to maximize the optical signal-to-noise ratio of the probe reaching the receiver. For this, noise resulting from double Rayleigh scattering, amplified spontaneous emission, and RIN transfer from the backward-propagating Raman pump to the probe must be minimized.

Note that the pump-to-signal RIN transfer is one of the major sources of impairments in co-propagating Raman amplification. This is of crucial importance when optimizing the backward-propagating signals in a RA-BOTDA sensor, so that an effective design must consider the technology employed for Raman pumping. On the one hand, fiber Raman lasers can provide high Raman gain; however, they typically exhibit high RIN levels, making them less attractive to assist BOTDA sensors. On the other hand, semiconductor lasers are characterized by low RIN levels, and although they typically have lower optical powers, they are more suitable as a backward-propagating Raman pump. In addition, note that due to the polarization dependence of the Raman gain, Raman pumps must be always depolarized before being injected into the fiber.

To further mitigate the impact of pump-to-probe RIN transfer in a Raman-assisted BOTDA sensor, specifically dedicated methods can also be employed. Besides the use of low-RIN Raman pumps (Soto et al. 2011), there have been mainly three complementary methods proposed in the literature to further reduce the impact of RIN:

1. Use of a *vector-BOTDA system* (Angulo-Vinuesa et al. 2014), where the BOTDA information contained in the probe is shifted to a high-frequency region (>500 MHz) using phase modulators and coherent detection. Thanks to the reduced cutoff frequency of the RIN transfer function with respect to the spectral shift set in the vector-BOTDA system, a significant noise reduction can be achieved.
2. Use of *balanced detection* (Dominguez-Lopez et al. 2014), in which the two probe waves of a dual-sideband BOTDA are separately launched into each of the inputs of a single balanced detector. During Raman amplification, the RIN of the backward Raman pump is simultaneously transferred to the two probe bands,

originating a common noise. This common noise is here canceled out by balanced detection. The system also profits from the push-pull effect of the Brillouin gain and loss traces, thus further increasing the SNR.
3. Use of a *digital notch filter* (Angulo-Vinuesa et al. 2012) in combination with Fourier transform to eliminate frequency components associated with the RIN transferred from the backward-propagating pump to the probe.

Distributed Brillouin Amplification (DBA)

The purpose of using distributed Brillouin amplification in BOTDA sensing is similar to the use of distributed Raman amplification (Urricelqui et al. 2015), although a different physical mechanism is exploited for amplification. There are essentially two main differences between DBA and DRA: (i) Compared to stimulated Raman scattering, which occurs in forward and backward directions in an optical fiber, the SBS process exploited in DBA only takes place in backward direction; therefore, only a backward pumping scheme is allowed in this case. (ii) Stimulated Brillouin scattering offers a narrow amplification band, and hence special considerations must be taken into account to secure reliable amplification of the BOTDA signals.

To implement a DBA-assisted BOTDA sensor, an additional CW pump signal has to be injected into the sensing fiber to provide Brillouin amplification, in principle, to either the BOTDA pump or probe. By launching a CW DBA pump from the far end of the fiber, as shown in Fig. 33, efficient amplification of the BOTDA pump can be achieved, compensating for the impact of the fiber attenuation (Urricelqui et al. 2015). To obtain a reliable and stable Brillouin amplification of the pump pulses, and

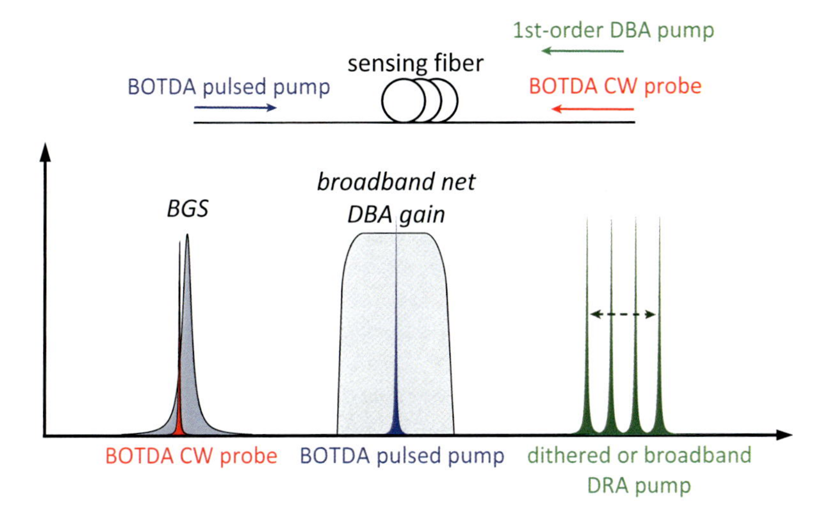

Fig. 33 Principle of a DBA-assisted BOTDA sensor. The figure shows the spectral allocation of BOTDA signals and DBA pump. The DBA pump must be broadband or dithered in the frequency domain in order to provide a broad net Brillouin gain around the pump pulse spectrum

to eliminate the polarization dependence of SBS, a polarization diversity scheme is required, for instance, making use of a polarization scrambler.

Due to the higher Brillouin coefficient in silica fibers compared to the Raman scattering coefficient, a CW pump power of only a few mW is required in this case (instead of the Watt levels required in DRA). This represents a significant advantage of Brillouin amplification with respect to Raman amplification. Another advantage of this scheme is that the counter-propagative nature of the SBS process actually makes issues related to pump-to-signal RIN transfer practically negligible in a DBA-assisted BOTDA scheme.

The narrow DBA band (typically much narrower than the natural FWHM of ~27 MHz, for a high-gain regime) imposes some significant differences in the implementation compared to RA-BOTDA systems. There exist essentially three main limitations imposed by the narrow Brillouin amplification band:

1. SBS, being a temperature- and strain-dependent process, the Brillouin amplification band provided by the CW pump to the BOTDA pump can shift according to the temperature and strain profile along the fiber. This means that the net gain on the BOTDA pump pulse turns out to be dependent on the environmental conditions. This situation must be avoided to enhance the reliability of the system.
2. The narrow Brillouin gain spectral width created by a CW DBA pump is typically narrower than the spectrum of the BOTDA pump pulse (usually of several tens of MHz). This feature leads to an inefficient amplification of the pulses, causing spectral and temporal distortions of the pump.
3. A highly nonuniform BFS profile combined with narrowband DBA causes inefficient distributed optical amplification over the sensing fiber if a CW single-frequency DBA pump is used.

To increase the robustness of the system against these limitations, techniques to broaden the spectral the DBA band can be used. A possible approach is to modulate the DBA pump optical frequency (preferably with a triangular shape), applying a peak frequency deviation of several tens of MHz or even a few hundred MHz. As a result of this triangular frequency modulation, a broadened and flat Brillouin gain spectrum can be obtained. Another alternative is to use a frequency comb as a DBA pump to provide broadband gain to the BOTDA pump. To secure a uniform distributed amplification of the BOTDA pump, the central frequency of the DBA pump comb must be tune at the average BFS of the sensing fiber.

Like in RA-BOTDA schemes, higher-order Brillouin pumping schemes can also be employed by launching, for instance, a second-order DBA pump into the fiber from the near end, i.e., by the same end by which the BOTDA pump is injected. This second-order DBA pump amplifies the counter-propagating first-order DBA pump, which in turn amplifies the BOTDA pump. In this case, however, attention must be paid to avoid nonlinear cross-interactions between the BOTDA pump and the co-propagating second-order DBA pump.

Although DBA can provide very large net gain and compensate the impact of fiber attenuation, the temporal shape of the BOTDA pump pulse could turn out to be highly distorted along the DBA process (Urricelqui et al. 2015). This distortion could be associated with the acoustic wave activation in SBS and to the phase distortion introduced by DBA. In particular, the implementation of a broadband, flattop Brillouin gain spectrum offers an ideal amplitude response to the amplification process; however, the nonlinear phase response of the amplifier can highly alter the pulse temporal shape. Distortions can be maintained below a given tolerated level by using moderate net Brillouin gains; however, for long sensing ranges and/or high DBA pump powers, large net Brillouin gain are commonly expected. This large DBA gain results in a slow light effect, inducing in some cases large temporal deformations of the pump pulse. Although some effects on the spatial resolution could be expected due to the delay induced by slow light, this effect should normally be small compared to the impairments introduced by pulse distortions. In long-range sensing, pulse distortions are critical since the combination of a high-peak pump power and a non-rectangular pulse shape may induce SPM during pulse propagation (see section "Self-Phase Modulation (SPM)"). SPM broadens the pulse spectrum and the BGS measured by the BOTDA sensor, and therefore the DBA-induced pulse distortions must be eliminated. This can be partially achieved using differential pulse measurements based on DPP-BOTDA (see section "Differential-Pulse BOTDA Techniques") (Urricelqui et al. 2015).

In-Line Discrete Amplification

As an alternative to distributed amplification, in-line optical amplifiers such as EDFAs can be placed through the sensing fiber to amplify both pump and probe signals, thus compensating for the SNR reduction caused by the fiber attenuation (Dong et al. 2012). A convenient implementation of in-line amplification stages (also called *optical repeaters*) is depicted in Fig. 34a. This scheme uses two optical circulators and two EDFAs to independently amplify the counter-propagating pump and probe signals. Many of such blocks (repeaters) can be regularly placed along the sensing fiber, thus defining fiber spans of several tens of kilometers each. However, one of the main limitations of this technique is the impact of the noise that accumulates over multiple fiber spans. In particular, the low-to-moderate MI sidebands generated along the first fiber span enter the first repeater, being amplified and then launched into the second fiber span. This means that strong MI bands propagate along the second fiber span, thus seeding efficient MI and reducing the maximum pump power allowed in this second span. This accumulative effect on the MI bands over several spans makes cascaded in-line amplification less efficient, especially when using a large number of repeaters. To limit the amount of MI, the number of optical amplification stages is typically constrained to 1 or 2.

A potential alternative to minimize the impact of MI is depicted in Fig. 34b, in which another fiber, parallel to the sensing fiber, is used to carry a copy of the pump pulse but at a very low-power level (Gyger et al. 2014). The propagation of this low-power replica of the pump pulse reduces the amount of MI generated over

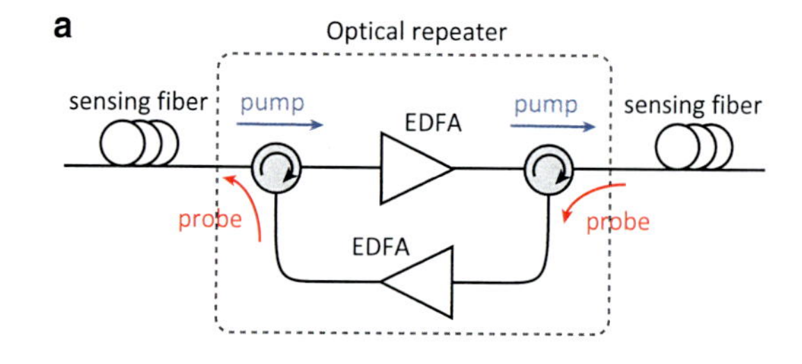

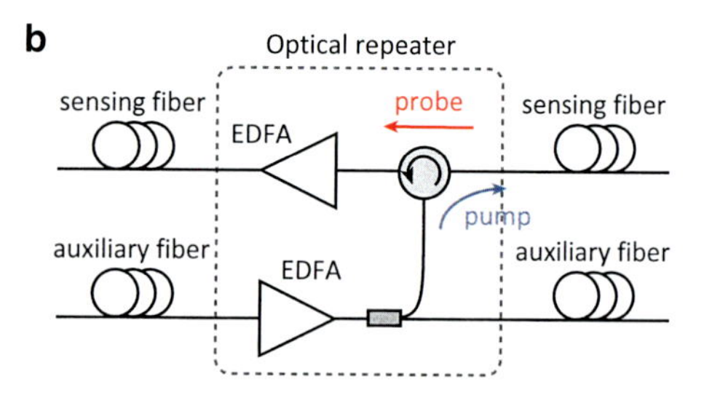

Fig. 34 Possible schemes for sensing range extension using in-line discrete optical amplification. (**a**) Simple configuration, however this scheme amplifies MI bands generated in a previous fiber span and launch them into the following span. (**b**) Scheme that avoids amplification of MI bands on the pump over consecutive spans

this auxiliary fiber. At the end of each fiber span, the residual pump power and generated MI bands propagating over the sensing fiber are dropped, and the signal is replaced by a clean pump pulse taken from the auxiliary fiber. In the process, the low-power pulse in this auxiliary fiber is amplified to a high-power level before being injected into the following span of the sensing fiber. This approach prevents MI bands from growing over the entire sensing range thus avoiding pump depletion and trace distortions. This way, since the pump pulse at the input of each fiber span is clean from MI bands, the maximum power allowed at the input of each stage is always the same (typically ~100 mW when using long fiber spans) (Gyger et al. 2014).

Optical Pulse Coding

The SNR of the time-domain traces measured by a Brillouin sensor highly depends on the sensor response. This response depends on the input pump and probe powers and the Brillouin interaction length. While the input powers are limited by nonlinear effects, the interaction length is determined by the spatial resolution in a classical scheme. Optical pulse coding techniques allows for an extension of the interaction length in both BOTDR and BOTDA sensors (Soto et al. 2008, 2010a), while permitting the system to keep a high spatial resolution.

Pulse coding uses sequences of short optical pulses which are launched into the sensing fiber in bursts at a repetition rate given by the fiber length, as shown in Fig. 35. It must be noted that all pulses in the sequences have the same width, which defines the spatial resolution of the measurements. The use of a large number

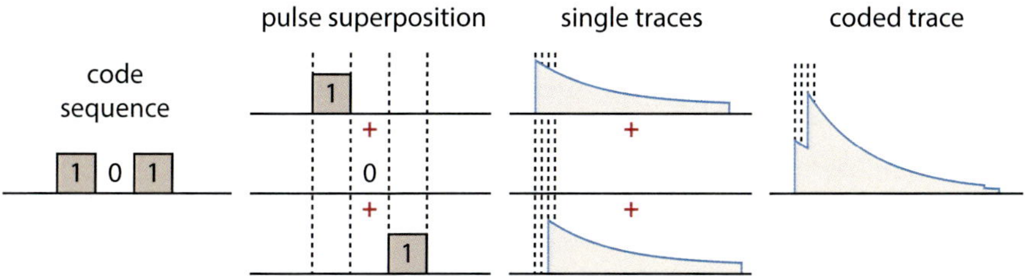

Fig. 35 Principle of optical pulse coding. Code sequences are composed of delayed identical pulses. *Coded traces* are represented as the linear superposition of delayed single-pulse traces (as long as the small-gain approximation is still valid)

of pulses increases the total optical power propagating through the sensing fiber without increasing the peak pump power. This leads to a higher SpBS intensity in a BOTDR sensor (Soto et al. 2008), and higher Brillouin gain in a BOTDA system (Soto et al. 2010a), while keeping the peak pump power below the limit imposed by nonlinear effects.

Under linear operation conditions, the measured coded time-domain traces contain the superposition of traces resulting from individual pulses delayed by multiples of the pulse width. This means that no reliable information can be extracted directly from coded traces. To recover the single-pulse response of the fiber, a dedicated decoding procedure is required. The decoding process generates a Brillouin time trace having an enhanced SNR, and with the correct spatial resolution (i.e., given by a single pulse width instead of by the sequence width). Note that in a classical approach, coding and decoding are carried out for each independent pump-probe frequency offset. This way, the decoded traces with enhanced SNR offer the possibility to improve any of the sensor parameters, i.e., to extend the sensing range, improve the spatial resolution, reduce the measurement time, and/or reduce the measurand uncertainty over the entire sensing fiber.

Unipolar Codes

A traditional approach for optical pulse coding in fiber sensing is based on the intensity modulation of the pump beam, enabling a natural incorporation of unipolar code sequences (containing 0's and 1's) into Brillouin distributed sensors (Soto et al. 2008, 2010a). This kind of codes is called *unipolar codes* or *binary codes*, in which elements 0's and 1's represent the *off* and *on* state of the light, respectively (see Fig. 35). Among the existing codes, two of them have proven to bring major benefits to classical Brillouin sensors; they are (i) *Simplex codes* (Jones 1993) and (ii) *complementary-correlation Golay codes* (Nazarathy et al. 1989).

Simplex code sequences are formed as a linear combination of pulses. The coded Brillouin trace measured by each code sequence corresponds to the linear superposition of single-pulse Brillouin traces delayed according to the pulse sequence patterns (see Fig. 35). Note that this concept of linear superposition is valid only when the small-gain approximation is satisfied, otherwise a linearization process is required before decoding. To allow for a proper coding-decoding process, sequences are defined by the rows of a so-called *S-matrix*, which is a unipolar version of a modified Hadamard matrix (Jones 1993). Rows in this matrix are orthogonal and can be easily generated using quadratic residues methods, shift register sequences, and twin primes. Note that the S-matrix is a $L \times L$ square matrix defining L orthogonal sequences of L bits each. This leads to L independent coded traces, which can be described as (Jones 1993)

$$\boldsymbol{\eta}(t) = \mathbf{S} \cdot \boldsymbol{\psi}(t) + \mathbf{e}(t) \tag{61}$$

where $\boldsymbol{\eta}(t) = [\eta_1(t), \ldots, \eta_L(t)]^T$ is a vector containing the L measured coded traces $\eta_i(t)$, $\mathbf{S}$ is the S-matrix, $\boldsymbol{\psi}(t) = [\psi_1(t), \ldots, \psi_L(t)]^T$ is a vector of replicas of the single-pulse response $\psi_1(t)$ with different delays, and $\mathbf{e}(t) = [e_1(t), \ldots, e_L(t)]^T$ is a matrix of uncorrelated Gaussian zero-mean noise. The decoding process is represented as a matrix inversion system, so that decoded traces are obtained as (Jones 1993)

$$\widehat{\boldsymbol{\psi}}(t) = \mathbf{S}^{-1} \cdot \boldsymbol{\eta}(t). \tag{62}$$

Note that each row of $\widehat{\boldsymbol{\psi}}(t)$ represents a decoded (estimated) time-shifted version of the single-pulse fiber response $\psi_1(t)$. After a proper time synchronization, decoded traces can be averaged to further reduce noise. This way a decoded trace, equivalent to the single-pulse response, is retrieved with enhanced SNR. The SNR enhancement obtained by pulse coding is called *coding gain*, which for Simplex codes is given as (Jones 1993)

$$C_{\text{gain}} = \frac{L+1}{2\sqrt{L}} \rightarrow \approx \frac{\sqrt{L}}{2} \text{ for long code lengths, i.e. for } L > 64. \tag{63}$$

Note that, since the S-matrix is a square matrix, L different sequences of L bits each must be used. This, however, does not represent a penalty to the measurement time of the system, since the coding gain in Eq. 63 already compares the SNR obtained with Simplex coding (using L different sequences) with respect to the single-pulse case using L averaged traces (i.e., taking into account similar acquisition time conditions).

On the other hand, complementary-correlation Golay codes use pulse sequences with particular correlation properties (Nazarathy et al. 1989). Pulse coding is in this case based on two bipolar sequences A_k and B_k (i.e., containing elements -1's and $+1$'s), whose autocorrelation functions have the same correlation peak, but complementary-correlation sidelobes. Therefore, sidelobes cancel out each other when the autocorrelation functions are summed up together, so that $\{(A_k * A_k) + (B_k * B_k)\} = 2L\delta_k$, where $*$ represents the correlation function, L is the number of bits, and δ_k is the Kronecker delta function (Nazarathy et al. 1989) The basic use of Golay codes in Brillouin time-domain sensors requires first to convert

A_k and B_k into four unipolar sequences, as $u_k^A = 0.5\,(1 + A_k)$, $u_k^{\overline{A}} = 0.5\,(1 - A_k)$, $u_k^B = 0.5\,(1 + B_k)$, and $u_k^{\overline{B}} = 0.5\,(1 - B_k)$. Using these four sequences, then four coded traces are obtained; each one can be denoted as the temporal convolution $\otimes$ between the pulse sequence and the fiber impulse response h_k. Thus, finally a decoded trace y_k can be obtained by the following decoding calculation:

$$y_k = \left(\underbrace{h_k \otimes u_k^A}_{\substack{\text{trace measured} \\ \text{with } u_k^A}} - \underbrace{h_k \otimes u_k^{\overline{A}}}_{\substack{\text{trace measured} \\ \text{with } u_k^{\overline{A}}}} \right) * A_k + \left(\underbrace{h_k \otimes u_k^B}_{\substack{\text{trace measured} \\ \text{with } u_k^B}} - \underbrace{h_k \otimes u_k^{\overline{B}}}_{\substack{\text{trace measured} \\ \text{with } u_k^{\overline{B}}}} \right) * B_k$$
$$y_k = h_k \otimes A_k * A_k + h_k \otimes B_k * B_k = h_k \otimes \{(A_k * A_k) + (B_k * B_k)\} = 2Lh_k \tag{64}$$

Equation 64 points out that the coding-decoding process is equivalent to interrogate the fiber impulse response h_k with the sum of the autocorrelations of the two sequences. This leads to a coding gain equal to

$$C_{\text{gain}} = \frac{\sqrt{L}}{2}, \tag{65}$$

One of the fundamental requirements to apply linear pulse coding techniques is to maintain a linear system. This means that the response of the fiber must be the same for each pulse in the sequence, so that a traditional linear decoding process could be used. This restriction implies that the gain coefficient $g_B(\Delta\nu, z)$ in the sensor response (see Eq. 34) must be the same for all pump pulses. This condition normally cannot be satisfied when using pulses of a few tens of ns (corresponding to a few meters spatial resolution), due to the large cumulated gain induced by the pulse sequences. In case the small-gain approximation cannot be satisfied (i.e., when using a large number of bits, high peak power and long pulses) a linearization of the coded traces is required before decoding. Furthermore, the inertial behavior of the acoustic wave generated by a given pulse typically affects the acoustic field activation produced by other pulses in the code sequence (Soto et al. 2010b). This effect is illustrated in Fig. 36a, which shows a 3 bit sequence {1,0,1} of 10 ns non-return-to-zero (NRZ) pulses. It is possible to observe that for the first 10 ns pulse, the acoustic wave amplitude rises from the zero level up to an intermediate level (not fully activated). During the second bit (being a "0"), the activation of the acoustic wave ceases, and its amplitude decays exponentially. However the 10 ns time slot is not long enough for the acoustic wave to completely vanish, and therefore, the activation of the acoustic field during the third bit (bit "1") rises from an upper level compared to the first one. Consequently, the Brillouin gain induced by the third bit turns out to be larger than the one obtained from the first one. This behavior of the acoustic wave becomes more critical with long pulse sequences, resulting in a

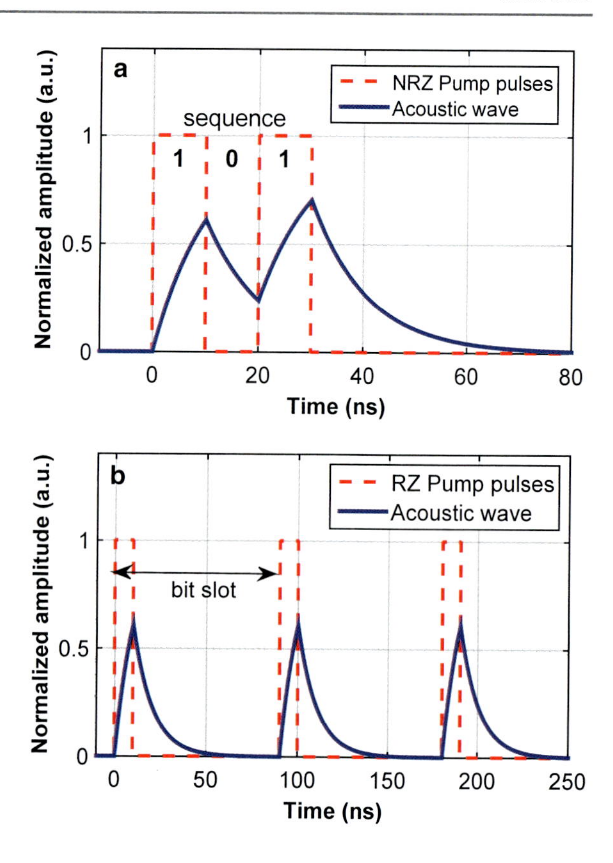

Fig. 36 Pulse coding using (**a**) NRZ and (**b**) RZ pulses. Bit patterning effect is eliminated with RZ format (Soto et al. 2010b)

bit patterning effect that destroys the linearity of the code (Soto et al. 2010b). To solve this issue, bits must be separated by a time interval larger than three to four times the acoustic wave response time $\tau_A \approx 12$ ns. This time interval (>50 ns) gives enough time to the acoustic wave to be fully damped before the next bit initiates a new acoustic wave activation. This is exemplified in Fig. 36b with a bit sequence {1,1,1}, when using a bit slot of 90 ns. This pulse format is called *return-to-zero* (RZ) format, and its use is essential to keep the linearity of the code in a Brillouin time-domain sensor (Soto et al. 2010b).

Note that so far the description of pulse coding has assumed that bit sequences are sent down the fiber in burst. However, there is also the possibility to spread the code over the entire sensing fiber using *cyclic codes*, in which the repetition period equals the fiber round-trip time. This approach is based on the use of a single-pulse pattern of length $L = 4n - 1$ (being n an integer number) and having cyclic properties. Examples of this kind of codes are *cyclic Simplex codes* (Iribas et al. 2017), which have the same coding gain of unipolar Simplex codes, as defined in Eq. 63. Compared to the use of bursts, this format offers important advantages in terms of pump depletion, thus avoiding potential distortions occurring with long unipolar codes and high probe powers.

Time-Frequency (or Colored) Codes

One of the limitations of classical unipolar codes is the reduction of the threshold power for SBS, limiting the peak pump power to levels even below 100 mW for specific spatial resolutions and code lengths.

An alternative coding scheme to partially alleviate some of the limitations of (single-frequency) unipolar codes is to use *time-frequency codes*, also called *colored codes* (Le Floch et al. 2012; Sauser et al. 2013). In this case, in addition to the intensity coding scheme, pulses with different optical frequencies are used. Compared to single-frequency codes, where coding and decoding processes are independently applied to each scanned frequency, colored codes intrinsically combine intensity-based coding and the frequency scanning required in a Brillouin sensor. This feature imposes one of the main restrictions to this type of codes: the code length L implicitly defines the number of scanned frequencies, which must also be equal to L. This implies that large codes are commonly required to have a sufficiently large spectral scanning range.

Note that, since each pulse of the code sequence has a different optical frequency, they activate SBS with different detuning frequencies, resulting in an efficient increment of the SBS power threshold. This allows colored codes to reach the maximum peak power imposed by nonlinear effects such as modulation instability. This kind of codes also mitigates pump depletion issues and nonlocal effects, which might result very significant in single-frequency unipolar codes. This mitigation arises from the fact that the CW probe in this case interacts with pump pulses at different frequencies, reducing the overall pump-probe power transfer when compared to the interrogation of the Brillouin peak frequency using single-frequency codes.

There exist different methods to generate colored codes. A possible alternative is to use *block-circulant matrix with circulant blocks* (BMCB) (Le Floch et al. 2012), following a pattern in which the frequency of each bit slot is defined by the code itself. Under this scheme, a code length L defines L^2 pulse sequences, as illustrated in Fig. 37a for the case of $L = 3$ bits. The coding gain of this

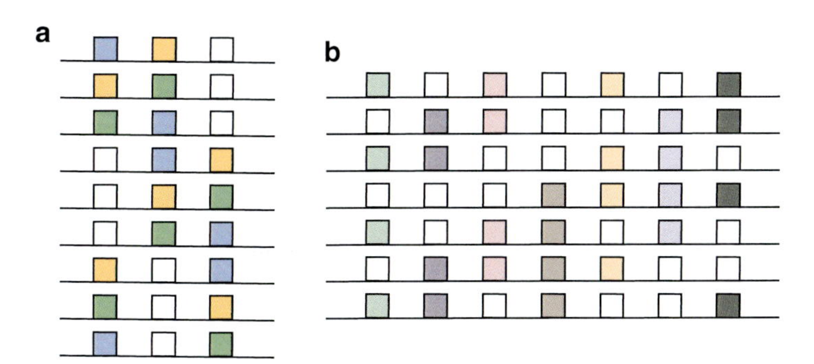

Fig. 37 (**a**) 3-bit BMCB-based time-frequency code. (**b**) 7-bit colored Simplex codes. *Colors* represent different optical frequencies for the pulses, while the *white color* is for the absence of the intensity pulse

BMCB-based time-frequency code is $C_{\text{gain}} = \sqrt{L/2}$, representing a 1.5 dB higher SNR enhancement when compared to unipolar codes. Alternatively, colored codes can also be constructed following a Simplex coding structure, where simply 1's and 0's indicate the presence or absence of an optical pulse, while each time slot is associated with a predefined optical frequency (Sauser et al. 2013), as shown in Fig. 37b for $L = 7$ bits. Colored Simplex codes have the same gain as the classical approach, i.e., $C_{\text{gain}} = (L + 1)/2\sqrt{L}$.

Bipolar Codes

Compared to BOTDR systems, BOTDA sensors offer a unique possibility to extend the types of coding schemes to be used. In particular, combining Brillouin gain and loss processes, the effect of positive (+1's) and negative (−1's) code elements can be achieved. This allows the implementation of *bipolar codes* (Soto et al. 2013b), where optical pulses are still generated by intensity modulation, but the positive and negative sign associated with each code element is encoded in the optical frequency of the pulses. The implementation requires a central CW probe signal interacting with a pump signal composed of two optical frequencies, spectrally located around the probe optical frequency, as depicted in Fig. 38a. As also shown in Fig. 38b,

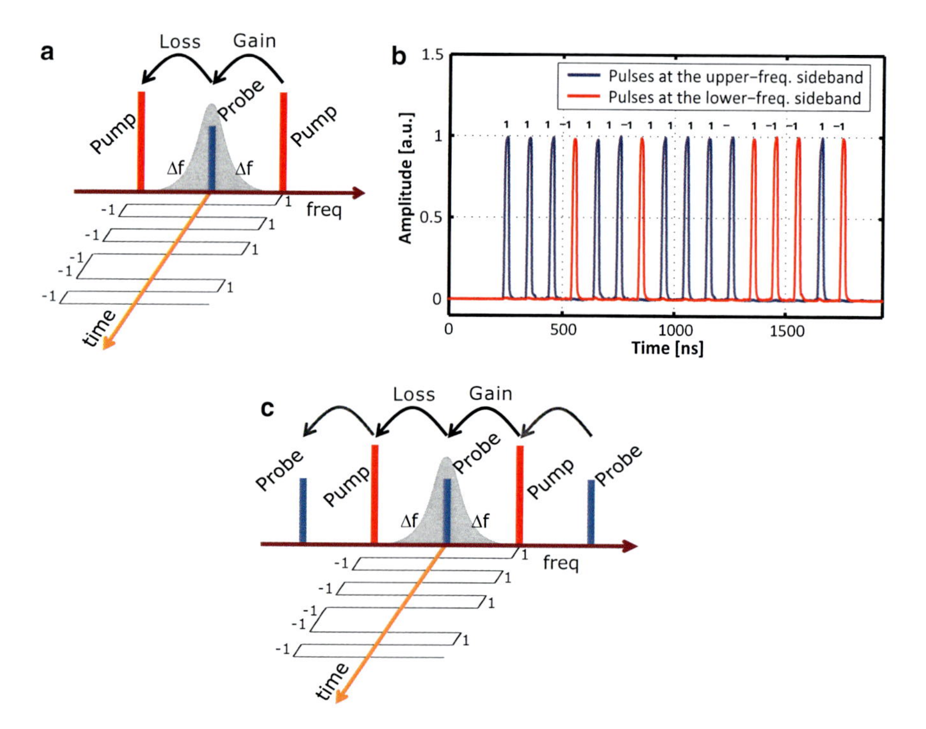

Fig. 38 Bipolar codes in BOTDA. (**a**) Spectral (**b**) temporal allocation of +1's and −1's (Soto et al. 2013b). (**c**) Scheme using three probe bands to increase robustness against pump depletion/amplification (Yang et al. 2016)

bipolar codes are implemented so that "+1" elements are allocated at the upper pump sideband, inducing Brillouin gain on the probe, while "−1" elements are placed at the lower pump sideband, inducing Brillouin loss on the probe. Note that, while the single-probe scheme shown in Fig. 38a is conceptually sufficient to describe the implementation of bipolar coding in BOTDA sensors, this scheme is highly affected by depletion of the upper pump sideband and amplification of the lower pump sideband, thus potentially impairing the sensor performance. To increase the robustness against those pump depletion/amplification issues, a scheme using three probe bands is preferred (see Fig. 38c) (Yang et al. 2016), so that the extra probe bands compensate the depletion/amplification that the pump tones experience during propagation.

Bipolar codes that have demonstrated significant SNR improvement in BOTDA sensors are *bipolar Golay codes* (Nazarathy et al. 1989). Compared to unipolar Golay codes, in this case only two pulse sequences are used. These correspond to the original bipolar sequences A_k and B_k defining Golay codes. These sequences result in two coded traces, denoted as $h_k \otimes A_k$ and $h_k \otimes B_k$, respectively. The decoding process requires to correlate each coded traces with the respective bipolar sequence launched into the fiber and then sum up the results. Thus, the decoded trace is given as

$$\begin{aligned} y_k &= \left(\underbrace{h_k \otimes A_k}_{\substack{\text{trace measured} \\ \text{with } A_k}} \right) * A_k + \left(\underbrace{h_k \otimes B_k}_{\substack{\text{trace measured} \\ \text{with } B_k}} \right) * B_k \\ &= h_k \otimes \{(A_k * A_k) + (B_k * B_k)\} = 2Lh_k \end{aligned} \tag{66}$$

The coding gain of bipolar Golay codes is $C_{\text{gain}} = \sqrt{L}$ (Soto et al. 2013b). This represents an additional improvement of 3 dB to the SNR of BOTDA traces when compared to unipolar Golay codes (see Eq. 65) of the same length L.

Multiplexing Schemes

Time-Division Multiplexing

As described in section "Constraints in the Probe Signal of BOTDA Sensors," to avoid the harmful impact of nonlocal effects and systematic errors, the probe power launched into the fiber must be kept below a certain threshold level. This level is about −5 dBm per sideband in a dual-sideband scheme (see section "Temporal and Spectral Distortion of the Pump Pulse in a Classical Dual-Sideband Scheme") (Dominguez-Lopez et al. 2016). To boost the probe power with the aim of increasing the measurement SNR and avoid nonlocal effects, a possible solution is to reduce the fiber section where pump and probe interact. For this, a long probe pulse can be used instead of the classical CW probe. The probe pulse width defines the interaction length where the pump is depleted (or amplified), while the relative delay between pump and probe pulses defines the interaction region inside the fiber. The Brillouin distribution over the entire fiber is retrieved by measuring different fiber sections by controlling the delay between pump and probe pulses. This technique is known as *time-division multiplexing-based BOTDA* (Dong et al. 2011).

For the implementation, the sensing fiber is divided into several sections of length L_i. For the i-th section, an effective interaction length equal to $L^i_{\mathrm{eff}} = [1 - \exp(\alpha L_i)]/\alpha$ is defined. In each of these sections the net gain/loss experienced by the pump pulse $G_i = g_B P_{si} L^i_{\mathrm{eff}}/A_{\mathrm{eff}}$ (where P_{si} is the probe power entering the i-th section) must be low enough so that pump depletion/excess amplification is avoided. While each of the fiber sections could in principle be defined of the same length, a proper design can be made to maintain the same net gain/loss in each fiber segment. As a consequence, fiber sections with different length could be defined depending on the local probe power. This way, the probe power attenuation along the sensing fiber allows the use of long sections in the first half of the sensing range, where the probe is weak after attenuation, while shorter sections can be defined at far distances, where the probe power is strong.

The method allows in principle the use of high probe powers, increasing the measurement SNR without the detrimental impact of nonlocal effects. The only penalty is the longer measurement time required for the acquisition of consecutive time traces over different fiber sections. It must be however pointed out that the use of larger probe powers also allows for the reduction of the number of averages, and therefore big penalties in the overall measurement time are not always expected.

Frequency-Division Multiplexing

The purpose of *frequency-division multiplexing* (Dong et al. 2012) is similar to the one previously described for time-division multiplexing, i.e., to reduce the Brillouin interaction length to avoid pump depletion issues and to allow for an increase in the probe power launched into the fiber with the aim of enhancing the SNR of the time traces.

In this case a conventional BOTDA scheme is employed, making use of a pulsed pump and a CW probe signal. However, the sensing fiber is selected such that the entire sensing range is divided into sections having very different Brillouin frequencies. The BFS difference between sections must be ideally larger than the BGS full-width at half-maximum. This way, pump and probe interact only along a reduced fiber length, instead of interacting along the entire sensing fiber. The length of each fiber segment can follow the same rule used for time-domain multiplexing, i.e., short fiber sections can be used at long distances owing to the strong local probe power, while longer fiber sections can be placed at shorter distances where the probe power is lower.

The only penalty of this technique is the enlarged pump-probe frequency scanning range that is required to cover the Brillouin gain spectrum of all the fiber sections. This typically leads to longer measurement times; however, the acquisition time can be optimized up to some extent by scanning different spectral ranges.

Time and Frequency Pump-Probe Multiplexing

There is also another technique based on time and frequency multiplexing that uses very different approach compared to the previous ones (Soto et al. 2014b). The aim in this case is to boost the total power of the pump and probe without activating nonlinear effects. The method makes use of a multifrequency pump pulse and a matching multifrequency CW probe, such that each tone of the pump interacts with a unique tone of the probe, as shown in Fig. 39a. For this, the pump and probe spectral lines must have rigorously the same frequency spacing, allowing for spectrally parallel SBS interactions between single pump-probe pairs. This frequency-division multiplexing approach can be considered as a parallelization of the Brillouin interaction in the fiber.

The maximum power of each tone (for both pump and probe) allowed into the fiber is exactly the same as in a conventional BOTDA scheme. Therefore the use of N spectral tones for the pump and probe turns out to be equivalent to N parallel Brillouin interactions, leading to an increase in the sensor response by a factor N.

It should be however noted that if high-power pulses at different optical frequencies are simultaneously launched into the fiber, harmful nonlinear processes (such as FWM and MI) will be seeded during propagation. These nonlinear cross-interactions lead to distortions of the measured BOTDA traces, impairing

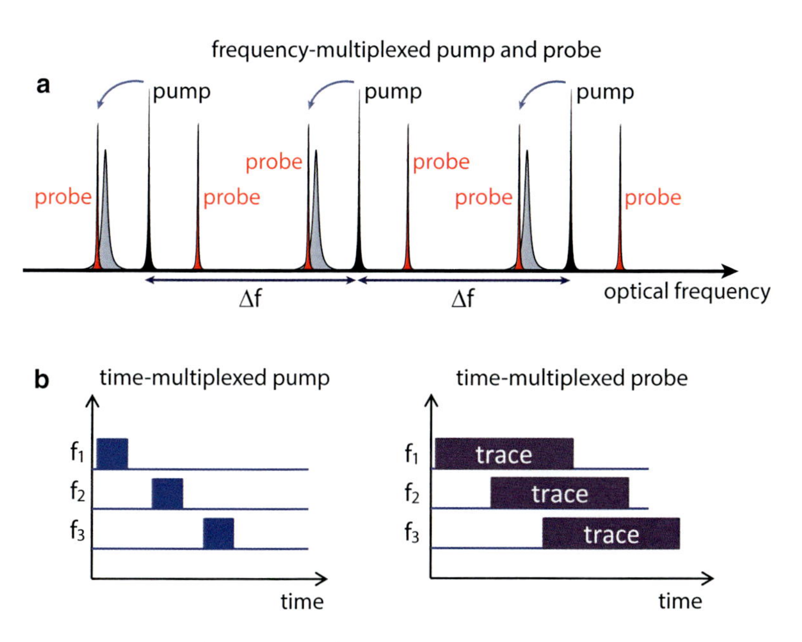

Fig. 39 (**a**) Pump-probe frequency-multiplexing scheme. (**b**) Time-multiplexing of pump and probe

significantly the sensor response. To maximize the pulse power and avoid nonlinear cross-interactions among the pump tones, a wavelength-dependent delay must be applied to the N pump pulses, as shown in Fig. 39b. This delay must be similar or longer than the single-pulse duration to secure that pulses do not overlap in time and location anywhere along the fiber (Soto et al. 2014b). However, this gives rise to N similar time-shifted replicas of the BOTDA trace, as illustrated in Fig. 39b. To avoid temporal jamming due to the temporal shift of the pump tones, the N probe signals containing the distributed Brillouin response must be properly rearranged in time before photo-detection by applying a reverse wavelength-dependent delay to the one originally set to the corresponding pump pulse replicas. This temporal realignment of the BOTDA traces (carried by each probe component) is essential to avoid impairing the spatial resolution of the sensor. Although this second temporal shift is preferably performed optically before photo-detection (to allow conventional direct detection), it can also be carried out in the electrical domain or using signal processing after high-frequency photo-detection.

The use of this time-frequency multiplexing of pump and probe increases the sensor response and SNR of the measurements. Nevertheless, the scalability of the method is essentially limited by two potential effects (Soto et al. 2014b): (i) four-wave mixing between the probe components and (ii) chromatic dispersion. While the impact of FWM can be considerably reduced using unequally spaced spectral components, special techniques can also be employed to mitigate the impact of chromatic dispersion. Indeed, pump and probe tones can be spectrally allocated following different possible configurations. This may allow for the optimization of the frequency spacing, thus maximizing the number of tones while minimizing the effect of chromatic dispersion. It must be pointed out that the use of a large number of tones, and/or a large spectral separation between lines, can also induce a spectral broadening of the measured BGS, as a result of the wavelength dependence of the BFS. However, a proper allocation of the spectral tones can also minimize this spectral broadening.

Digital Signal Processing

Most of the previous techniques for long-range sensing require the modification of the conventional BOTDA scheme. However, there is also the possibility of using signal processing techniques to boost the SNR of the traces measured by classical BOTDR and BOTDA schemes. Hereafter some approaches are presented.

One-Dimensional (1D) Processing

The simplest signal processing method to enhance the measurement SNR is the use of *digital low-pass filters* to reduce noise in the Brillouin sensor data. This approach only bring benefits if the electrical bandwidth of the acquisition system is larger than the optimal one required to achieve a given target spatial resolution. In that case, the measurement bandwidth can be reduced digitally according to the desired resolution. However, in a well-designed system (i.e., when the electrical bandwidth

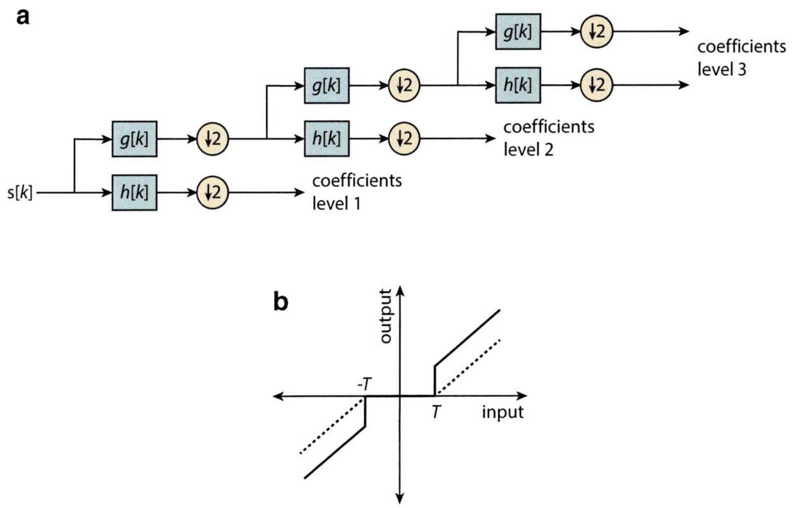

Fig. 40 (**a**) Filter bank representation of the discrete wavelet transform, for a three-level decomposition; $h[k]$ and $g[k]$ are high-pass and low-pass filters, and $\downarrow 2$ represents a down-sampling operator by factor 2. (**b**) Nonlinear thresholding function; continuous line, hard thresholding; dashed line, soft thresholding

matches the spatial resolution requirement), this method only provides a marginal or even null SNR improvement. In such a case, noise reduction using a digital (linear) filter is only effective allowing a degradation of the spatial resolution.

To partially overcome this trade-off between spatial resolution and SNR enhancement, nonlinear signal processing methods can be employed. A technique that has shown to be efficient in noise reduction without affecting the spatial resolution of the sensor is the use of a *wavelet denoising* algorithm (Mallat 1999; Farahani et al. 2012). This unidimensional processing method can be used to eliminate noise from the Brillouin time-domain traces (independently for each scanned frequency), or from the local BGS measured at each fiber location.

Wavelet denoising requires three steps. In the first step, discrete wavelet transform (DWT) is used to decompose the signal into multiple frequency bands and obtain coefficients having different levels of precision. Using filter banks (low-pass and high-pass filters) and down-sampling, the signal is decomposed as illustrated in Fig. 40a. For each of these bands, i.e., for each decomposition level, the DWT results in wavelet coefficients $w_{i,j}$ described as (Mallat 1999)

$$w_{j,k} = \int_{-\infty}^{\infty} f(t)\psi_{j,k}(t)\mathrm{d}t, \text{ with } \psi_{j,k}(t) = \frac{1}{\sqrt{2^j}}\psi\left(\frac{t-k2^j}{2^j}\right) \tag{67}$$

where the set of $\psi_{j,k}(t)$ conforms an orthonormal basis formed by dilatation (scaling) and translation (shift) of a mother wavelet $\psi(t)$ and j and k are the scaling and translation factors. In the case of the Brillouin sensing data, the DWT results in wavelet coefficients of large amplitude, which are associated with the signal itself (i.e., to the time-domain trace, or local spectrum), and small amplitude coefficients that are associated with noise.

As a second step, a nonlinear denoising technique called *wavelet thresholding* (*shrinkage*) (Mallat 1999) is then applied to eliminate wavelet coefficients associated with noise, as illustrated in Fig. 40b. To distinguish noise from signal, a threshold level T is defined. Wavelet coefficients having an amplitude below this threshold are associated with noise and are eliminated (set to zero) from the wavelet representation. On the other hand, the amplitude of those coefficients above the threshold (associated with the useful signal) remain unmodified (this is called *hard thresholding*) or are slightly reduced by an amount equal to T (this is called *soft thresholding*). One of the main differences between hard and soft thresholding is that soft thresholding usually leads to smoother signals, while hard thresholding better preserve sharp transitions of the signals. Note that shrinkage is applied independently to the different levels of the signal decomposition, and in principle the threshold value can be different for each level, allowing the elimination of noise at any spectral position, while keeping all frequency components of the original signal. There are several methods to define the threshold level; however this is typically determined as a function of the noise standard deviation σ. A common definition is the universal threshold, described as $T = \sigma\sqrt{2\log(N)}$, where N is the number of samples. Note that this is not always an optimal threshold, and sometimes better denoising can be achieve using lower threshold levels (Mallat 1999).

As a last step, once wavelet thresholding has been applied, inverse discrete wavelet transform is used to reconstruct a denoised version of the signal using only the remaining high-amplitude wavelet coefficients.

Two-Dimensional (2D) Processing: Image (Denoising) Processing

The classical measurement process in a Brillouin distributed sensor results in a map of the local BGS as a function of distance, as shown in Fig. 41a. Data points are commonly stored in a two-dimensional matrix $\mathbf{g}[z, \Delta f]$, in which each data point contains the measured Brillouin response at a given position z and frequency offset Δf, as depicted in Fig. 41b. This two-dimensional data structure makes possible the use of 2D signal processing techniques, such as *image denoising* algorithms, to eliminate noise from the matrix $\mathbf{g}[z, \Delta f]$ by processing each position-frequency pair $[z, \Delta f]$ as a noisy pixel of an image (Soto et al. 2016, 2017). Following this 2D approach, linear and nonlinear image denoising can be efficiently used. Compared to 1D methods, these 2D algorithms are commonly more efficient in terms of noise removal since they can smartly exploit all data patterns existing in the two dimensions of the measured data $\mathbf{g}[z, \Delta f]$.

In the case of linear image filters, they commonly use 2D neighborhood (local) operators to reduce noise from noisy pixels in an image. The process of denoising the BOTDA data $\mathbf{g}[z, \Delta f]$ can be represented by a two-dimensional convolution resulting in a filtered matrix data $\mathbf{g}_f[z, \Delta f] = \mathbf{h}[z, \Delta f] \otimes \mathbf{g}[z, \Delta f]$, where

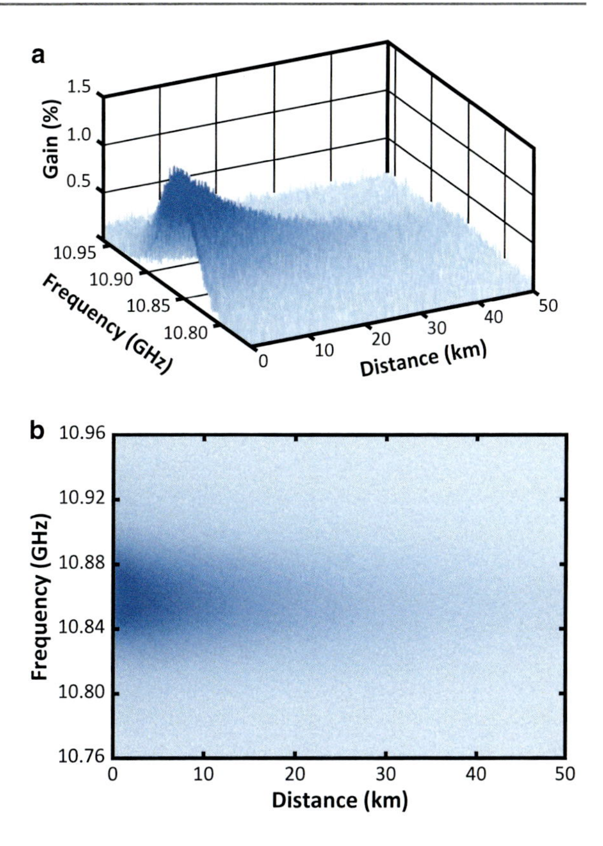

Fig. 41 Image denoising for BOTDA data enhancement. (**a**) 3D map of the BGS measured vs distance. (**b**) *Top view* of noisy data. *Darker blue* tones represent higher SBS gain (Soto et al. 2016)

$\mathbf{h}[z, \Delta f]$ is the bidimensional impulse response of the filter (spatial kernel in the $[z, \Delta f]$ space). Due to the linear and space-invariant properties of this kind of filters, 2D linear image denoising can also be expressed in the Fourier domain as $\mathbf{G_f}[u, v] = \mathbf{H}[u, v]\mathbf{G}[u, v]$, where $\mathbf{H}[u, v]$ and $\mathbf{G}[u, v]$ are the 2D Fourier transform of $\mathbf{h}[z, \Delta f]$ and $\mathbf{g}[z, \Delta f]$, respectively. Hence, linear image denoising applied to the Brillouin sensor data can be implemented following two equivalent approaches (Soto et al. 2017): (i) in the *spatial domain* by processing the data points directly over the $[z, \Delta f]$ space and (ii) in the *frequency domain* by using direct and inverse 2D Fourier transforms.

Linear image filters are simply 2D low-pass filters, and therefore their performance is highly determined by the trade-off between noise removal capability and spatial over-smoothing. Indeed, a linear filter that removes large amount of noise typically blurs and over-smooths the BOTDA data, leading to a loss of high-frequency details commonly associated with the high spatial resolving capability of the sensor. Examples of this kind of filters are the *2D mean filter* (also known as *moving averaging filter*) and the *2D Gaussian filter*.

To partially overcome the trade-off between spatial smoothing and noise removal, nonlinear image filters can be used. Even though their design and implementation are normally more complex than 2D linear filters, they can potentially offer considerably higher SNR enhancement while keeping all high-frequency details of the data. Since they involve the use of a nonlinear function for denoising, which in many cases require space-variant functions, the filtering process cannot be equivalently represented in both space and frequency domain using Fourier analysis. Based on this feature, there exist two independent categories of nonlinear image denoising methods (Soto et al. 2017): (i) *pixel-wise techniques*, in which a nonlinear denoising operator is applied directly on the $[z, \Delta f]$ space of the Brillouin data, and (ii) *transformation-based methods*, in which the nonlinear filtering function is applied in a domain representing the frequency content of the data, followed by a process to convert back the denoised data to the original $[z, \Delta f]$ space. Examples of the first category are the *median filter*, *bilateral filter*, and *nonlocal means* (NLM); while transform-based techniques (second category) can be based, for example, on *2D discrete cosine transform* or *2D discrete wavelet transform*. Although the amount of noise removed with these nonlinear image denoising algorithms highly depends on the data itself (e.g., on the existing patters and redundancies, as well as on the noise level), SNR enhancements of ~14 dB have been demonstrated with NLM and wavelet denoising over a 50-km-long classical BOTDA sensor scheme (Soto et al. 2016).

Three-Dimensional (3D) Processing: Video (Denoising) Processing

The working principle of Brillouin distributed fiber sensing requires that the entire Brillouin gain information (i.e., including the full frequency scan and averaging) must be acquired in a time shorter than the temporal evolution of the measurand. This feature inherently implies that consecutive BOTDA measurements contain some level of correlated information. A natural extension of the 2D denoising approach described in section "Two-Dimensional (2D) Processing: Image (Denoising) Processing" is to consider also the temporal axis in the processing. This way, 3D processing techniques can be used. Powerful 3D denoising techniques based on *video denoising algorithms* can therefore be exploited for BOTDA data denoising (Soto et al. 2016). In this case, each BOTDA measurement $\mathbf{g}[z, \Delta f, t_k]$ is processed as a frame of a video sequence, where t_k is the time of the k-th acquisition. Compared to image denoising, this 3D processing approach provides larger noise removal possibilities since video denoising can effectively exploit patterns of information found not only in the $[z, \Delta f]$ space but also in the temporal dimension.

Some possible methods for video denoising are the *3D nonlocal means* and *3D wavelet denoising*. A key feature of video denoising is that the processing takes into account the nonstationary features of the data over time and hence can easily deal with the motion of pixels among different frames. This feature represents a key advantage to minimize delay and over-smoothing of the temporal evolution of the measurand. Using a 3D NLM algorithm, processing ten consecutive frames $\mathbf{g}[z, \Delta f, t_k]$, an SNR enhancement of 20.7 dB has been demonstrated over a 50-km-long BOTDA sensor (2 m spatial resolution, 1.4 dB original SNR and 42 s measurement time) (Soto et al. 2016). It must be noted that, like image denoising algorithm, the SNR improvement achieved by video denoising highly depends on the patterns found in the BOTDA measurements, and therefore very different levels of SNR improvement can be expected in real-case scenarios depending on specific sensing conditions.

References

G.P. Agrawal, *Nonlinear Fiber Optics*, 4th edn. (Academic, San Diego, 2007)

M.N. Alahbabi, Y.T. Cho, T.P. Newson, P.C. Wait, A.H. Hartog, Influence of modulation instability on distributed optical fiber sensors based on spontaneous Brillouin scattering. J. Opt. Soc. Am. B **21**(6), 1156–1160 (2004)

M. Alem, M.A. Soto, L. Thévenaz, Analytical model and experimental verification of the critical power for modulation instability in optical fibers. Opt. Express **23**(23), 29514–29532 (2015)

M. Alem, M. A. Soto, M. Tur, L. Thévenaz, Analytical expression and experimental validation of the Brillouin gain spectral broadening at any sensing spatial resolution, in *Proc. SPIE 10323, 25th International Conference on Optical Fiber Sensors*, 103239J (2017)

X. Angulo-Vinuesa, S. Martin-Lopez, P. Corredera, M. Gonzalez-Herraez, Raman-assisted Brillouin optical time-domain analysis with sub-meter resolution over 100 km. Opt. Express **20**, 12147–12154 (2012)

X. Angulo-Vinuesa, D. Bacquet, S. Martin-Lopez, P. Corredera, P. Szriftgiser, M. Gonzalez-Herraez, Relative intensity noise transfer reduction in Raman-assisted BOTDA systems. IEEE Photon. Technol. Lett. **26**(3), 271–274 (2014)

K.-I. Aoyama, K. Nakagawa, T. Itoh, Optical time domain reflectometry in a single-mode fiber. IEEE J. Quantum Electron. **QE-17**(6), 862–868 (1981)

R. Bernini, A. Minardo, L. Zeni, Dynamic strain measurement in optical fibers by stimulated Brillouin scattering. Opt. Lett. **34**, 2613–2615 (2009)

J.-C. Beugnot, M. Tur, S.F. Mafang, L. Thévenaz, Distributed Brillouin sensing with sub-meter spatial resolution: modeling and processing. Opt. Express **19**(8), 7381–7397 (2011)

R.W. Boyd, *Nonlinear Optical*, 2nd edn. (Academic, San Diego, 2003)

P. Chaube, B.G. Colpitts, D. Jagannathan, A.W. Brown, Distributed fiber-optic sensor for dynamic strain measurement. IEEE Sensors J. **8**(7), 1067–1072 (2008)

Y.T. Cho, M. Alahbabi, M.J. Gunning, T.P. Newson, 50-km single-ended spontaneous-Brillouin-based distributed-temperature sensor exploiting pulsed Raman amplification. Opt. Lett. **28**, 1651–1653 (2003)

H.Z. Cummins, R.W. Gammon, Rayleigh and Brillouin scattering in liquids: the Landau—Placzek ratio. J. Chem. Phys. **44**(7), 2785–2796 (1966)

A. Dominguez-Lopez, A. Lopez-Gil, S. Martín-López, M. Gonzalez-Herraez, Strong cancellation of RIN transfer in a Raman-assisted BOTDA using balanced detection. IEEE Phot. Technol. Lett. **26**(18), 1817–1820 (2014)

A. Dominguez-Lopez, Z. Yang, M.A. Soto, X. Angulo-Vinuesa, S. Martin-Lopez, L. Thévenaz, M. Gonzalez-Herraez, Novel scanning method for distortion-free BOTDA measurements. Opt. Express **24**(10), 10188 (2016)

Y. Dong, L. Chen, X. Bao, Time-division multiplexing-based BOTDA over 100km sensing length. Opt. Lett. **36**, 277–279 (2011)

Y. Dong, L. Chen, X. Bao, Extending the sensing range of Brillouin optical time-domain analysis combining frequency-division multiplexing and in-line EDFAs. J. Lightwave Tech. **30**(8), 1161–1167 (2012)

M. Farahani, M. Wylie, E. Castillo-Guerra, B. Colpitts, Reduction in the number of averages required in BOTDA sensors using wavelet denoising techniques. J. Lightwave Technol. **30**, 1134–1142 (2012)

A. Fellay, L. Thévenaz, M. Facchini, M. Niklès, P. Robert, Distributed sensing using stimulated Brillouin scattering: towards ultimate resolution, in *OSA Technical Digest Series vol. 16, 12th International Conference on Optical Fiber Sensors*. (1997), p. 324–327

A. Fellay, L. Thévenaz, M. Facchini, P. Robert, Limitation of Brillouin time-domain analysis by Raman scattering, in *Proceeding of the 5th Optical Fibre Measurement Conference*. (1999), p. 110–113

S. M. Foaleng, L. Thévenaz, Impact of Raman scattering and modulation instability on the performances of Brillouin sensors, in *Proc. SPIE 7753, 21st International Conference on Optical Fiber Sensors*, 77539V (2011)

S.M. Foaleng, M. Tur, J.-C. Beugnot, L. Thévenaz, High spatial and spectral resolution long-range sensing using Brillouin echoes. J. Lightwave Technol. **28**(20), 2993–3003 (2010)

S.M. Foaleng, F. Rodríguez-Barrios, S. Martin-Lopez, M. González-Herráez, L. Thévenaz, Detrimental effect of self-phase modulation on the performance of Brillouin distributed fiber sensors. Opt. Lett. **36**, 97–99 (2011)

E. Geinitz, S. Jetschke, U. Röpke, S. Schröter, R. Willsch, H. Bartelt, The influence of pulse amplification on distributed fibre-optic Brillouin sensing and a method to compensate for systematic errors. Meas. Sci. Technol. **10**(2), 112–116 (1999)

F. Gyger, E. Rochat, S. Chin, M. Niklès, L. Thévenaz, Extending the sensing range of Brillouin optical time-domain analysis up to 325 km combining four optical repeaters, in *Proc. SPIE 9157, 23rd International Conference on Optical Fibre Sensors*, 91576Q (2014)

T. Horiguchi, K. Shimizu, T. Kurashima, M. Tateda, Y. Koyamada, Development of a distributed sensing technique using Brillouin scattering. J. Lightwave Technol. **13**(7), 1296–1302 (1995)

H. Iribas, A. Loayssa, F. Sauser, M. Llera, S. Le Floch, Cyclic coding for Brillouin optical time-domain analyzers using probe dithering. Opt. Express **25**, 8787–8800 (2017)

C. Jin, L. Wang, Y. Chen, N. Guo, W. Chung, H. Au, Z. Li, H.-Y. Tam, C. Lu, Single-measurement digital optical frequency comb based phase-detection Brillouin optical time domain analyzer. Opt. Express **25**, 9213–9224 (2017)

M.D. Jones, Using simplex codes to improve OTDR sensitivity. IEEE Phot. Technol. Lett. **5**(7), 822–824 (1993)

S. Le Floch, F. Sauser, M. A. Soto, L. Thévenaz, Time/frequency coding for Brillouin distributed sensors, in *Proc. SPIE 8421, OFS2012 22nd International Conference on Optical Fiber Sensors*, 84211J (2012)

W. Li, X. Bao, Y. Li, L. Chen, Differential pulse-width pair BOTDA for high spatial resolution sensing. Opt. Express **16**(26), 21616–21625 (2008)

S. G. Mallat, *A Wavelet Tour of Signal Processing*. (Academic, 1999)

S. Martin-López, M. Alcon-Camas, F. Rodríguez-Barrios, P. Corredera, J.D. Ania-Castanón, L. Thévenaz, M. González-Herráez, Brillouin optical time-domain analysis assisted by second-order Raman amplification. Opt. Express **18**(18), 18769–18778 (2010)

S.M. Maughen, H.H. Kee, T.P. Newson, Simultaneous distributed fibre temperature and strain sensor using microwave coherent detection of spontaneous Brillouin backscatter. Meas. Sci. Technol. **12**(7), 834–842 (2001)

M. Nazarathy, S.A. Newton, R.P. Giffard, D.S. Moberly, F. Sischka, W.R. Trutna, S. Foster, Real-time long range complementary correlation optical time domain reflectometer. J. Lightwave Technol. **7**(1), 24–38 (1989)

M. Niklès, L. Thévenaz, P.A. Robert, Brillouin gain spectrum characterization in single-mode optical fibers. J. Lightwave Technol. **15**(10), 1842–1851 (1997)

Y. Peled, A. Motil, L. Yaron, M. Tur, Slope-assisted fast distributed sensing in optical fibers with arbitrary Brillouin profile. Opt. Express **19**, 19845–19854 (2011)

Y. Peled, A. Motil, M. Tur, Fast Brillouin optical time domain analysis for dynamic sensing. Opt. Express **20**, 8584–8591 (2012)

F. Rodríguez-Barrios, S. Martín-López, A. Carrasco-Sanz, P. Corredera, J.D. Ania-Castanón, L. Thévenaz, M. González-Herráez, Distributed Brillouin fiber sensor assisted by first-order Raman amplification. J. Lightwave Technol. **28**(15), 2162–2172 (2010)

S. Le Floch, F. Sauser, M. Llera, M. A. Soto, L. Thévenaz, Colour simplex coding for brillouin distributed sensors, in *Proc. SPIE 8794, Fifth European Workshop on Optical Fibre Sensors*, 879437 (2013)

K. Shimizu, T. Horiguchi, Y. Koyamada, T. Kurashima, Coherent self-heterodyne detection of spontaneously Brillouin-scattered light waves in a single-mode fiber. Opt. Lett. **18**(3), 185–187 (1993)

M.A. Soto, L. Thévenaz, Modeling and evaluating the performance of Brillouin distributed optical fiber sensors. Opt. Express **21**(25), 31347–31366 (2013a)

M.A. Soto, S. Le Floch, L. Thévenaz, Bipolar optical pulse coding for performance enhancement in BOTDA sensors. Opt. Express **21**(14), 16390–16397 (2013b)

M. A. Soto, L. Thévenaz, Towards 1'000'000 resolved points in a distributed optical fibre sensor, in *Proc. SPIE 9157, 23rd International Conference on Optical Fibre Sensors*, 9157C3 (2014)

M.A. Soto, P.K. Sahu, G. Bolognini, F. Di Pasquale, Brillouin-based distributed temperature sensor employing pulse coding. IEEE Sensors J. **8**(3), 225–226 (2008)

M.A. Soto, G. Bolognini, F. Di Pasquale, L. Thévenaz, Simplex-coded BOTDA fiber sensor with 1 m spatial resolution over a 50 km range. Opt. Lett. **35**(2), 259–261 (2010a)

M.A. Soto, G. Bolognini, F. Di Pasquale, Analysis of pulse modulation format in coded BOTDA sensors. Opt. Express **18**(14), 14878–14892 (2010b)

M.A. Soto, G. Bolognini, F. Di Pasquale, Optimization of long-range BOTDA sensors with high resolution using first-order bi-directional Raman amplification. Opt. Express **19**(5), 4444–4457 (2011)

M.A. Soto, X. Angulo-Vinuesa, S. Martin-Lopez, S.-H. Chin, J.D. Ania-Castañon, P. Corredera, E. Rochat, M. Gonzalez-Herraez, L. Thevenaz, Extending the real remoteness of long-range Brillouin optical time-domain fiber analyzers. J. Lightwave Technol. **32**(1), 152–162 (2014a)

M.A. Soto, A.L. Ricchiuti, L. Zhang, D. Barrera, S. Sales, L. Thévenaz, Time and frequency pump-probe multiplexing to enhance the signal response of Brillouin optical time-domain analyzers. Opt. Express **22**, 28584–28595 (2014b)

M.A. Soto, J.A. Ramírez, L. Thévenaz, Intensifying the response of distributed optical fibre sensors using 2D and 3D image restoration. Nat. Commun. **7**, 10870 (2016)

M. A. Soto, J. A. Ramírez, L. Thévenaz, Image and video denoising for distributed optical fibre sensors, in *Proc. SPIE 10323, 25th International Conference on Optical Fiber Sensors*, 103230K (2017)

K.D. Souza, Significance of coherent Rayleigh noise in fibre-optic distributed temperature sensing based on spontaneous Brillouin scattering. Meas. Sci. Technol. **17**(5), 1065–1069 (2006)

K.D. Souza, T.P. Newson, Improvement of signal-to-noise capabilities of a distributed temperature sensor using optical preamplification. Meas. Sci. Technol. **12**(7), 952–957 (2001)

L. Thévenaz, S.F. Mafang, J. Lin, Effect of pulse depletion in a Brillouin optical time-domain analysis system. Opt. Express **21**(12), 14017–14035 (2013)

J. Urricelqui, A. Zornoza, M. Sagues, A. Loayssa, Dynamic BOTDA measurements based on Brillouin phase-shift and RF demodulation. Opt. Express **20**(24), 26942–26949 (2012)

J. Urricelqui, M. Sagues, A. Loayssa, Brillouin optical time-domain analysis sensor assisted by Brillouin distributed amplification of pump pulses. Opt. Express **23**, 30448–30458 (2015)

A. Voskoboinik, O.F. Yilmaz, A.W. Willner, M. Tur, Sweep-free distributed Brillouin time-domain analyzer (SF-BOTDA). Opt. Express **19**, B842–B847 (2011)

P.C. Wait, T.P. Newson, Landau Placzek ratio applied to distributed fibre sensing. Opt. Commun. **122**, 141–146 (1996a)

P.C. Wait, T.P. Newson, Reduction of coherent noise in the Landau Placzek ratio method for distributed fibre optic temperature sensing. Opt. Commun. **131**, 285–289 (1996b)

P.C. Wait, K.D. Souza, T.P. Newson, A theoretical comparison of spontaneous Raman and Brillouin based fibre optic distributed temperature sensors. Opt. Commun. **144**, 17–23 (1997)

Z. Yang, M.A. Soto, L. Thévenaz, Increasing robustness of bipolar pulse coding in Brillouin distributed fiber sensors. Opt. Express **24**, 586–597 (2016)

Distributed Brillouin Sensing: Frequency-Domain Techniques

44

Aleksander Wosniok

Contents

Abstract

The substantial progresses in fiber-optic communications in combination with the increasing economic and political interest in structural health monitoring have led to a commercial establishment of distributed Brillouin sensing. The sensor systems, mainly based on the time-domain techniques, have been successfully implemented in the areas such as pipeline leak detection, geohazard effects, and ground movement detection.

A. Wosniok (✉)
8.6 Fibre Optic Sensors, Federal Institute for Materials Research and Testing (BAM), Berlin, Germany
e-mail: aleksander.wosniok@bam.de

G.-D. Peng (ed.), *Handbook of Optical Fibers*,
https://doi.org/10.1007/978-981-10-7087-7_8

This chapter introduces a further advancement in the area of Brillouin sensing in the frequency domain. The so-called Brillouin optical frequency-domain analysis (BOFDA) offers crucial perspectives in terms of dynamic range and cost efficiency.

The main principle of the frequency-domain approach takes advantage of the reversibility between the time and frequency domain given by a Fourier transform in the analysis of linear systems. The hereby presented overview gives a summary of the benefits and challenges of frequency-domain measurements closely tied to the narrowband recording of the complex transfer function. This function relates the counterpropagating pump and probe laser light along the sensor fiber providing the pulse response of the measurement system by applying the inverse Fourier transform (IFT). The strain or temperature distribution can be then determined from the retrieved Brillouin frequency shift (BFS) profile along the fiber.

Introduction

The sensing methods picked out as a central theme of the chapter constitute a direct alternative to well-known time-domain techniques. Contrary to the pulse-based time-domain approaches, in the frequency-domain measurements, only continuous wave light (cw light) of a narrow-linewidth laser is coupled into the sensor fiber. Thereby, all spatially resolved information of the Brillouin gain is included in a sinusoidal modulation of either pump or probe laser light intensity recorded as a complex transfer function of the system. By analyzing the Brillouin backscattered light at different modulation frequencies and at different spectral shifts between pump and probe laser light, the strain and temperature distribution along the whole sensor fiber length can be determined applying the inverse Fourier transform (IFT) to the recorded complex transfer function. The use of the IFT presumes linearity and time invariance of the measured Brillouin gain in the fiber under test. This condition is ensured by using low-intensity laser light coupled into the sensor fiber.

The measuring procedure via the frequency-domain transfer function can be seen as a decomposition of the laser pulses into their spectral components. These components are single-frequency harmonic signals corresponding with the frequency values of the performed sinusoidal modulation of laser light intensity.

Due to the time-frequency duality given by the Fourier transform, the two concepts of measurement in the time as well as in the frequency domain remain comparable in dynamic range, spatial resolution, and measurement accuracy. However, on the way toward practical realization, frequency-domain techniques offer an essential low-cost potential, as is explained in greater detail below.

Frequency-Domain Measurements

The distributed fiber-optic sensing methods presented in this section use the time-frequency duality of the Fourier transform for linear and time-invariant systems. In contrast to the pulse-based time-domain sensor systems, the Brillouin interaction in the frequency domain takes place exclusively between counterpropagating cw light waves of narrow-linewidth lasers coupled into the sensor fiber. At the same time, a progressive sinusoidal amplitude modulation of either the cw pump or the counterpropagating cw probe light provides the spatially resolved information on the local strains and temperature gradients along the sensor fiber. The progressive sinusoidal amplitude modulation is to be understood as a Fourier pulse decomposition replacing therefore the pulse modulation in time-domain measurements.

This section aims to give a detailed description of the theoretical backgrounds on the so-called Brillouin optical frequency-domain analysis (BOFDA) as an alternative to the standard time-domain BOTDA systems. The underlying system theory is only applicable to linear systems. The nonlinearities caused on the one hand by the measuring principle and on the other hand by the stimulated Brillouin effect itself are to be suppressed. These nonlinearities are discussed below, and it is shown how the sensor system can be best linearized.

Completing recent advances in the frequency-domain research, a separate subsection is devoted to a reflectometry technique called Brillouin optical frequency-domain reflectometry (BOFDR). The Brillouin optical frequency-domain reflectometry features lower measurement accuracies than BOFDA but allows measurements with one-end access by detecting the spontaneous Brillouin scattering from a sinusoidally modulated pump light.

Theoretical Principles

The Brillouin optical frequency-domain analysis BOFDA is based on the measurement of a complex transfer function H that relates the amplitudes and phases of counterpropagating pump and probe light waves along a sensor fiber (Garus et al. 1996, 1997). As shown in Fig. 1 and in Fig. 2, the technique can be implemented by an amplitude modulation of either the probe (loss method) or the pump light (gain method). In both basic configurations, the counterpropagating continuous wave (pump or probe wave) adopts the amplitude modulation at the same frequency f_m as the excitation (probe and pump wave, respectively) bearing the system information $H(f_m)$. For the distributed measurement, the modulation frequency is progressively swept by a reference signal of the vector network analyzer VNA with M equidistant frequency steps Δf_m, which transferred into the time domain using an IFT yields the discrete pulse response $h(t_n f_D)$ for a frequency difference f_D between the pump and probe waves (Gogolla 2000):

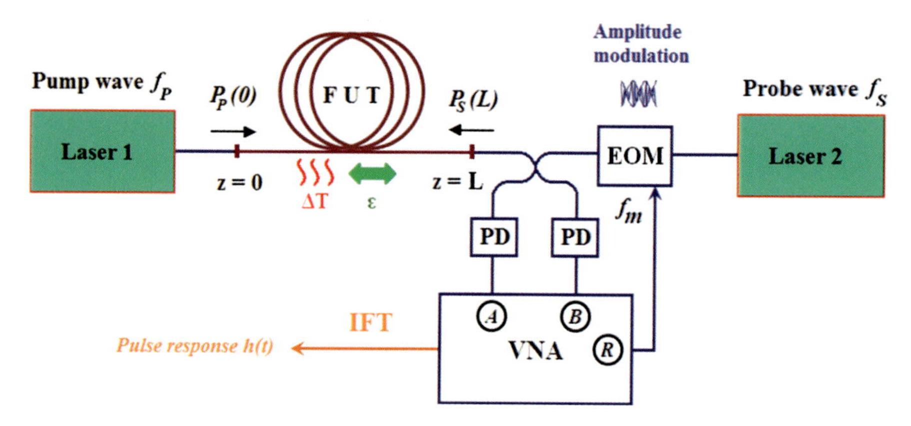

Fig. 1 BOFDA basic configuration – loss method. *VNA* vector network analyzer, *EOM* electro-optic modulator, *PD* photo diode, *FUT* fiber under test

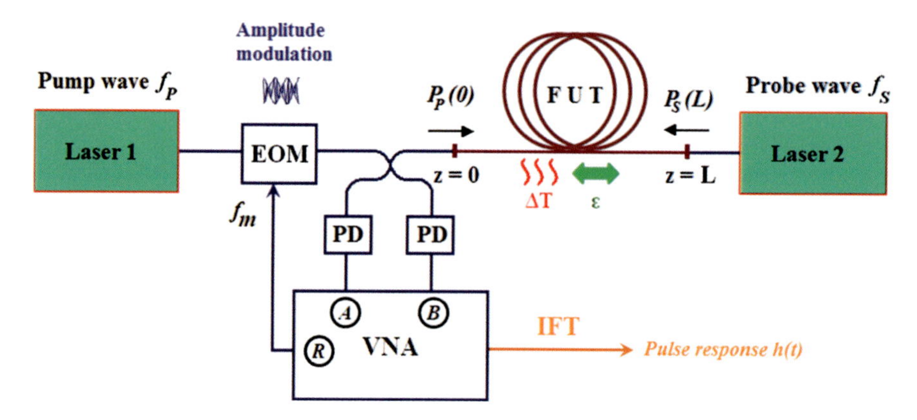

Fig. 2 BOFDA basic configuration – gain method. *VNA* vector network analyzer, *EOM* electro-optic modulator, *PD* photo diode, *FUT* fiber under test

$$h\left(t_{n}, f_{D}\right) = \frac{1}{M} \sum_{m=0}^{M-1} H\left(f_{m}, f_{D}\right) \exp\left(2\pi i f_{m} t_{n}\right) \tag{1}$$

where: $t_n = \frac{n}{M\,\Delta f_m}$ und $n = 0, 1, 2, \ldots, M-1$.

By converting the time of detection t_n into the location z_n,where the Brillouin interaction occurred, the substitution $t_n = \frac{2 z_n n_{\text{gr}}}{c_0}$ in $h(t_n, f_D)$ should be performed. Therefore, the spatial pulse response $h(z_n, f_D)$ can be easily determined at each location z_n in the fiber using the proportionality factor $\frac{c_0}{n_{\text{gr}}}$ with the vacuum speed of light c_0 and the group refractive index of the fiber core n_{gr}.

The pump and probe waves interact in the fiber by means of stimulated Brillouin scattering (SBS). The Brillouin gain is maximal, if the frequency difference $f_D = f_P - f_S$ between the pump frequency f_P and the probe frequency f_S is equal to the characteristic Brillouin frequency f_B of the fiber. Since the frequency f_B depends on the temperature and strain in the fiber, the fully distributed Brillouin gain spectra (BGSs) giving information about temperature gradients and local strain along the fiber under test (FUT) can be recorded as an IFT of the complex transfer function H by varying the frequency difference f_D in the range of f_B.

The loss method (see Fig. 1) corresponds to a configuration with the probe signal (Stokes light) being modulated in amplitude by an electro-optic modulator EOM. The complex baseband transfer function $H_L(f_m, f_D)$ is expressed by:

$$H_L(f_m, f_D) = \frac{\widehat{P}_P(f_m, f_D)}{\widehat{P}_S(f_m)} \exp\left[\mathrm{i}\Phi_P(f_m, f_D) - \mathrm{i}\Phi_S(f_m)\right] \tag{2}$$

with the alternating part $\widehat{P}_P$ of the transmitted modulated pump power and the probe excitation $\widehat{P_S}$ considering the signal phases Φ_P for the pump and Φ_S for the probe wave.

The gain method (see Fig. 2) relies on the modulation of pump signal giving the following transfer function $H_G(f_m, f_D)$:

$$H_G(f_m, f_D) = \frac{\widehat{P}_S(f_m, f_D)}{\widehat{P}_P(f_m)} \exp\left[\mathrm{i}\Phi_S(f_m, f_D) - \mathrm{i}\Phi_P(f_m)\right] \tag{3}$$

with the alternating part $\widehat{P}_S$of the transmitted modulated probe power and the pump excitation$\widehat{P}_P$.

From a metrological point of view, there are any significant differences between the basic configurations presented in Fig. 1 and in Fig. 2. Since the gain of a weak probe signal is easier detectable than the loss of the relatively high pump signal, the following considerations relate only to the gain method. However, the principle validity of the described theory holds also for the loss method.

In the gain method, for the pump power P_P coupled into the fiber at $z = 0$, the following relation applies:

$$P_P(t, f_m)\big|_{z=0} = \overline{P}_P + \widehat{P}_P(f_m)\cos\left[2\pi f_m t + \phi_P(f_m)\right] \tag{4}$$

Due to an AC coupling in the photo diode amplifier, both the DC component $\overline{P}_P$ of the pump power and the DC component $\overline{P}_S(f_m, f_D)$ of the adopted amplitude modulation of the cw probe signal according to:

$$P_S(t, f_m, f_D)|_{z=0} = \overline{P}_S(f_m, f_D) + \widehat{P}_S(f_m, f_D)\cos\left[2\pi f_m t + \phi_S(f_m, f_D)\right] \tag{5}$$

are not measured, so that Eq. 3 is valid for the recorded complex transfer function H_G.

The frequency dependence of the amplitude $\widehat{P}_P(f_m)$and the phase $\phi_P(f_m)$ traces back to the frequency-specific behavior of hardware components in the measurement setups presented in the section "Measurement Setups."

The amplitude $\widehat{P}_S(f_m, f_D)$ is strongly dependent on the modulation frequency f_m, basically caused by interference effects due to superposition of modulation signals and by the limited full width at half maximum (FWHM) of the Brillouin gain of about 35 MHz, which ultimately leads to a low-pass behavior. The function $\phi_S(f_m, f_D)$ results from the phase delay time of the modulation components of the probe (Stokes) power along the sensor length. The mentioned physical effects thus indicate the dependence of the two quantities $\widehat{P}_S$ and ϕ_S on the temperature and strain conditions of the sensor fiber, which is detected by the complex transfer function H_G according to Eq. 3.

In order that the IFT of Eq. 3 corresponds to the pulse response given by Eq. 1, the prerequisite of the system linearity must be fulfilled as good as possible. A method to achieve the required quasilinearity is based on the following considerations. There is an exponential (nonlinear) relation between the pump power P_P coulped into the fiber with the length L at $z = 0$ and the probe power P_S coupled into the opposite fiber end at $z = L$ given by simplified proportionality:

$$P_S(0) \sim P_S(L)\exp\left(\mathrm{GP}_P(0)\right) \tag{6}$$

with G representing a Brillouin gain parameter. The accurate composition of G can be neglected here for the sake of simplicity. Putting expression (5) for modulated probe signal into Eq. 6, the following relation can be taken as a second-order Taylor polynomial approximation:

$$\begin{aligned} P_S(0) \sim P_S(L)\Big[& 1 + G\overline{P}_P m_P \cos(2\pi f_m t) + \frac{1}{4}\left(G\overline{P}_P m_P\right)^2 \\ & + \frac{1}{4}\left(G\overline{P}_P m_P\right)^2 \cos(2\pi 2 f_m t)\Big] \end{aligned} \tag{7}$$

where the last term is a harmonic wave to be suppressed in the adopted modulation of the probe signal.

Particularly, this can be achieved both by a low pump power $P_P(0)$ and by its low degree of modulation $m_P = \frac{\widehat{P}_P}{\overline{P}_P}$ (less than 1). The value m_P can be primarily determined selecting an appropriate amplification of the reference output signal of the network analyzer VNA (setting of $\widehat{P}_P$).

Measurement Range and Spatial Resolution

The feasible BOFDA measurement range is specified on the one hand mathematically by the properties of the discrete inverse Fourier transform (DIFT) and on the other hand physically by the optical attenuation along the sensor fiber.

According to Bracewell (1999), a DIFT generally features a periodic continuation of the pulse response *h*(t). Therefore, *h*(t) represents a periodic function whose period P_F has its seeds in the discretization of the scanned transfer function $H(f_m)$ and can thus be expressed by the following relation:

$$P_F = \frac{1}{\Delta f_m} = \frac{M}{\left(f_m^{\max} - f_m^{\min}\right)} \tag{8}$$

where $f_m^{\max}$ and $f_m^{\min}$ are the maximum and minimum modulation frequencies, respectively.

If the length-dependent scan time of the sensor fiber is greater than P_F,the adjacent pulse responses overlap, which leads to a distortion of the measurement signal. To avoid this, the maximum sensor length $z_{\max}$ must not exceed the following value:

$$z_{\max} = \frac{c_0}{2n_{\mathrm{gr}}\Delta f_m} = \frac{Mc_0}{2n_{\mathrm{gr}}\left(f_m^{\max} - f_m^{\min}\right)} \tag{9}$$

In practice, $z_{\max}$ is additionally limited by the occurrence of undesired optical losses both due to critical sensor bending (microbending and macrobending effects in optical fibers) and due to bad splice and plug connections.

The frequency-limited measurement of H can be treated as a multiplication of a transfer function $H^{'}$ recorded over an infinite frequency range with a normalized rectangular function rect(f_m) going from $f_m^{\min}$to $f_m^{\max}$.

Accordingly, this means the following convolution in the time domain by applying IFT $\mathcal{F}^{-1}$:

$$h(t) = h'(t)^* \mathcal{F}^{-1}\left[\mathrm{rect}\left(f_m\right)\right] \tag{10}$$

which results under the assumptions $f_m^{\max} \gg f_m^{\min}$ and $f_m^{\min} \to 0$ in:

$$h(t) = f_m^{\max} \cdot h'(t) * \mathrm{sinc}\left(2\pi f_m^{\max} t\right) \tag{11}$$

Consequently, the width ∂t of the main peak of the sinc function $\mathrm{sinc}(t) = \frac{\sin(t)}{t}$ sets the maximum spatial resolution ∂z of BOFDA:

$$\partial z = \frac{c_0}{2n}\partial t = \frac{c_0}{2n f_m^{max}} \tag{12}$$

In order to achieve the spatial resolution of 1 m, a stepwise amplitude modulation of the optical pump wave with the modulation frequencies f_m up to 100 MHz is therefore required.

However, the crucial limitation of the spatial resolution ∂z is due to the low-pass behavior of the modulated optical signals mentioned above. This manifests itself as a degradation of signal-to-noise ratio (SNR) of the recorded BGSs. In other words, the Brillouin gain spectra become narrower and noisier as the values f_m increase. Figure 3 shows that the amplitude modulation of the pump wave with the modulation frequencies f_m up to 100 MHz (spatial resolution of 1 m) leads to a good approximation of Lorentz-shaped profiles of the BGSs. The spatial resolution of 1 m proves to be a reasonable limitation to meet measurement accuracy requirements at acceptable measurement times whose specifics are explained closer separately in the next subsection.

The spatial resolution ∂z describes the ability of a BOFDA system to distinguish small-sized events along a sensor fiber with distinct temperature-strain state (different BFSs). The occurrence of two events within the lengths below ∂z leads to a distortion of the Brillouin spectra depending on the BFS difference between the separate events related to the FWHM of the Brillouin gain. Figure 4 illustrates two different cases of two separate events occurring below the spatial resolution limit, what simultaneously impairs the measurement accuracy in determining distributed BFS values.

Measurement Time

The measurement time in BOFDA method is conditioned by a narrowband filtering of the measurement signals using VNA and by propagation time of the light signals in the sensor fiber. This means, for example, 0.3 ms for the filter bandwidth of 3 kHz and 0.1 ms for a typical sensor length of 10 km. Since according to Eqs. 9 and 12 10^4 frequency modulation steps are needed to achieve the spatial resolution of 1 m, the required measurement time is approximately 4 s for each setting $f_D = f_P - f_S$. The frequency range $f_D^{max} - f_D^{min}$ to be swept in order to record spatially resolved BGSs carrying information about strain and temperature distribution along the fiber is dictated mainly by the expected strain level. As shown in Fig. 3, this means a range of about 300 MHz for strain values of about 0.5% (5000 $\mu\varepsilon$). If it is taken into account that in case of a precise measurement a frequency step Δf_D of 1 MHz should be set, the total measurement time would be around 20 min. Therefore, in consideration of relatively long total measurement times in BOFDA, a compromise between a sensor length, measurement accuracy, and spatial resolution, which are over Δf_D, Δf_m,and f_m^{max} the basis for the total measurement time, is to be found in most applications.

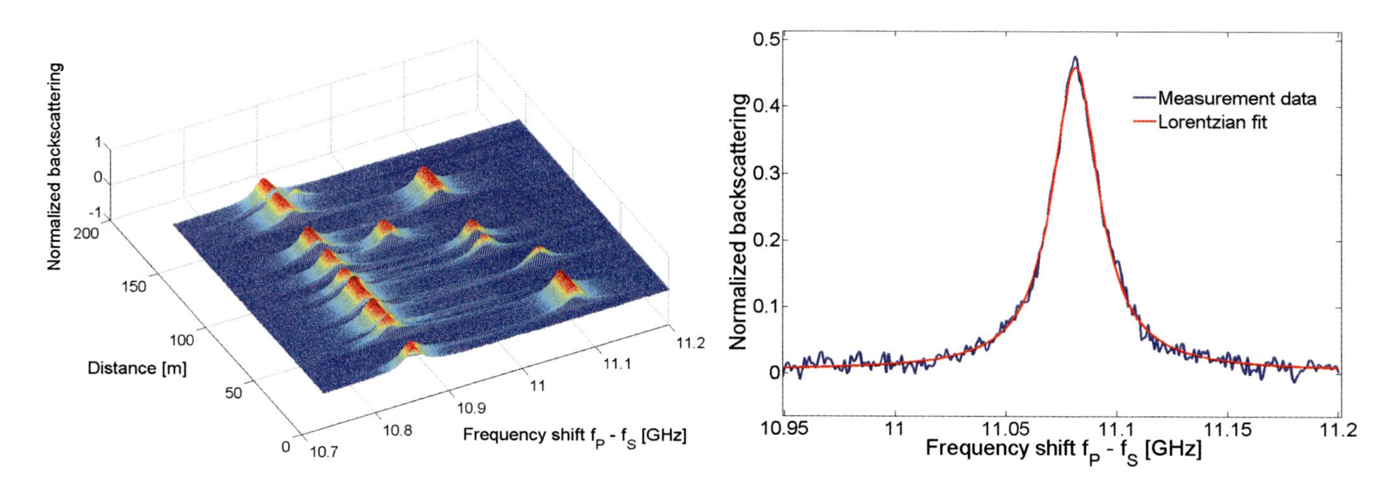

Fig. 3 Spatially resolved Brillouin gain spectra of an inhomogeneously strained fiber (160 m) measured at $f_m^{\max} = 100$ MHz. *Left*: 3D measurement diagram (spatial resolution: 1 m). *Right*: Brillouin gain spectrum at $z = 20$ m

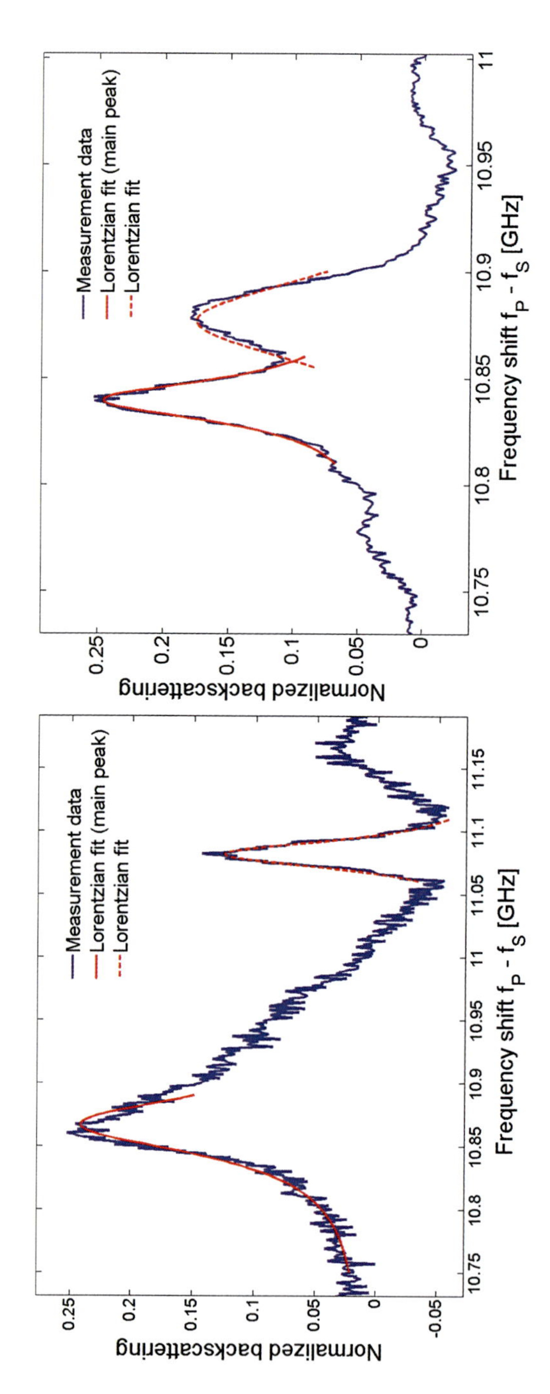

Fig. 4 Brillouin gain spectra for two different events (different BFSs) occurring below the spatial resolution of 1 m. *Left*: BFS difference greater than the FWHM of the Brillouin gain. *Right*: BFS difference smaller than the FWHM of the Brillouin gain

As explained above, the data acquisition time using VNA is decisively limited by the setting time of its analog filters, which is inversely proportional to the bandwidth of the filters. However, a measurement of the transfer function via VNA is not absolutely necessary. A VNA is even clearly overdesigned for recording the single-scattering parameter S_{21} corresponding to the transfer function $H(f_m)$ in the BOFDA setups presented in section "Measurement Setups." In Nöther (2010), a concept of a digital BOFDA system is presented in detail. Since here the complete digital processing can be performed off-line, the digital BOFDA features a significant advantage of data acquisition time reduction (Wosniok et al. 2009). A comprehensive introduction to the digital concept of BOFDA is provided in the section "Digital Processing Technique."

Measurement Accuracy

All local strains ε and temperature changes ΔT can be determined from the distributed BGSs recorded along a whole sensor fiber as already shown in Fig. 3 left. Each individual spectrum of a standard single-mode fiber (SMF), used commonly as a distributed sensor, features a Lorentz profile (see Fig. 3 right) giving an explicit value f_B of BFS defined as a frequency at a maximum of BGS. The value f_B is in turn sensitive to both longitudinal strain and temperature as follows:

$$f_B(T, \varepsilon) = C_T \cdot (T - T_R) + C_\varepsilon \cdot \varepsilon + f_{B0}\left(T_R = 20^\circ\mathrm{C}, 0\right) \tag{13}$$

in which C_T and C_ε describe the Brillouin thermal and strain coefficients, respectively.

The strain and temperature accuracy is thus directly related to the issue of uncertainty in BFS determination. The measurement of f_B can be negatively affected on the one hand by Brillouin gain weakening along the fiber and on the other hand by distortion of Brillouin gain spectra themselves. The Brillouin gain weakening can be caused both by extrinsic factors such as microbending and macrobending effects, bad splice, or plug connections and by measurement-related factors such as polarization setting or choice of frequency step Δf_D. Apart from physical reasons of too high $f_m^{\max}$ value described above, the distortion of Brillouin gain spectra finds its origin in discrete inverse Fourier transform artifacts.

Due to the metrological relevance of SBS polarization effects, the polarization adjustment is separately presented in the section "Polarization Effects." In general, the Brillouin gain depends strongly on polarization conditions of pump and probe wave represented by a polarization factor γ. The polarization factor γ can have values between 0 and 1. $\gamma = 0$ corresponds to the situation where the polarization vectors of the two counterpropagating waves (pump and probe) are perpendicular to each other, so that no interaction takes place (complete failure of SBS). $\gamma = 1$ represents a maximal Brillouin amplification achieved by parallel orientation of polarization vectors of pump and probe wave. Therefore, a misalignment of polarization vectors $\gamma \approx 0$ results in weak-gained Brillouin spectra. In order to avoid this polarization effect, which can occur on several points along the sensor fiber due to the ever-present birefringence, the critical polarization states can be successfully controlled by polarization controllers and scramblers as described in "Digital Processing Technique."

The distortion of Brillouin gain spectra referring to the discrete inverse Fourier transform artifacts can be ascribed to two effects: overlapping and leakage effect (Gogolla 2000). The former effect defines the measurement range characterized in "Measurement Range and Spatial Resolution." The overlapping means that a mismatch set value Δf_m and the real sensor length lead to an overlap of high-order periodic continuation of the pulse response to the first-order pulse response $h(t)$ which contains the measurement data to be determined. In this way, an additional defective signal component results from this overlap distorting the recorded BGSs ultimately. This negative effect can easily be avoided by choosing a suitable frequency step Δf_m which arises from Eq. 9 for a given sensor length $z_{\max}$.

The frequency-limited detection of the transfer function H also manifests itself by the leakage effect inducing overshoots in the pulse response $h(t)$. The ideal case of detecting the measurement signals in the unlimited modulation frequency range corresponds to a Dirac pulse $\delta(t)$ in the time domain. According to Eq. 11, the real case of the frequency-limited measurement leads in turn to a convolution of the ideal pulse response $h'(t)$ with a sinc function in the time domain. As shown in Fig. 5 with IFT[rect(f_m)], the sinc function features a relatively narrow width of the main lobe and high side lobes. The occurrence of the mentioned overshoots is attributed to the high side lobes of the sinc function. In order to suppress this negative effect, the measured transfer function $H(f_m)$ is to be multiplied by a suitable window function, which additionally results in a corresponding convolution in the time domain. However, the simultaneous requirements for a narrow width of the main lobe (influence on the spatial resolution) and the lowest possible side lobes (influence on the measurement accuracy) of the window function can practically not be fulfilled. A large number of window functions (Riemann, Bohman, Hanning-Poisson, Hamming, Kaiser-Bessel, etc.) are available in the literature (Harris 1978) which offers specific advantages for different tasks in the signal analysis. In the method of BOFDA, the Kaiser-Bessel functions, shown in Fig. 5, with the selectable window parameter β in the range of 3 to 4 proves to be the most suitable.

As a result of the above-described optimization steps, BOFDA setups can provide measurement accuracies in the range of 0.5 MHz using SMFs as a distributed sensor, which corresponds to 0.5 °C for temperature measurement and 10 $\mu\varepsilon$ for monitoring of mechanical deformation, respectively. Thus, BOTDA and BOFDA are comparable in terms of achievable measurement accuracies.

Compared to the Lorentzian BGS of the SMFs, the BGSs of the multimode silica fibers (MMFs) proved to be detectably wider and little distorted (Minardo et al. 2014). This behavior slightly impairs the BFS determination and thus the measurement accuracy in MMFs. With respect to issues of accuracy of the BFS determination, unwanted spectrum distortion can be traced back to the SBS caused

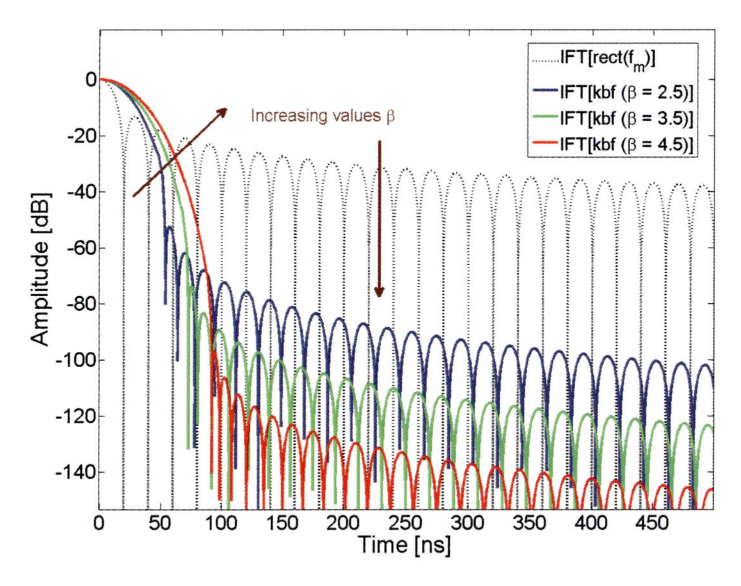

Fig. 5 Inverse Fourier transform of a normalized rectangular function and the normalized Kaiser-Bessel function for different β parameters. The upper cutoff frequency of all calculated window functions is 25 MHz

by higher-order acoustic modes supported by corresponding high-order optical modes in a MMF.

In some real applications, an access to both sensor fiber ends required for BOFDA operation can be considerably hindered or even impossible. A way out of such a problem in practice is given by a one-end accessibility using Brillouin optical frequency-domain reflectometry technique (BOFDR) presented below. By detecting the spontaneous Brillouin scattering, BOFDR measurements remain in fact free from signal distortion related to acoustic wave modulation at high frequencies f_m (Minardo et al. 2016) characteristic for BOFDA low-pass behavior. However, the measurement accuracy of BOFDR, in the range of a few MHz at best, is fundamentally limited by the nature of the spontaneous Brillouin scattering providing substantially lower SNRs (Minardo et al. 2016) and wider BGSs (Yaniay et al. 2002) compared to BOFDA based on SBS. With a focus on application-relevant high requirements regarding measurement precision, individual optimization steps of BOFDA approaches are discussed in detail in section "Measurement Setups."

One-End-Access Measurements

Brillouin sensing systems are usually based on SBS generated by coupling frequency-shifted laser light beams into both ends of a sensor fiber. Compared to such Brillouin analyzers (BOFDA systems), a reflectometry approach proposed in Minardo et al. (2016) allows for a Brillouin configuration, in which a light beam is injected only into one fiber end. As described in the section above, the advantage of one-end accessibility in BOFDR is realized at the expense of the measurement accuracy derived from the character of the spontaneous Brillouin scattering used in BOFDR. In particular, the measurement accuracy related to estimation error of BFS decreases with the square root of the FWHM of the measured BGS (Soto and Thévenaz 2013). With the value of about 90 MHz (Yaniay et al. 2002), BGS in case of the spontaneous Brillouin scattering is considerably wider than its 35-MHz-wide SBS counterpart. Moreover, the FWHM of the measured BGS increases with the bandwidth of the electrical band-pass filter BPF used in the simplified BOFDR setup shown in Fig. 6 according to Minardo et al. (2016). Since the spatial resolution of BOFDR given by Eq. 12, in analogy to BOFDA, is also linked to the BPF bandwidth, a trade-off must be made between the measurement accuracy and the spatial resolution in choosing BPF bandwidth. The choice of BPF bandwidth in the range of 100 MHz corresponds to a spatial resolution of 1 m, and it allows for reasonable SNR values.

Similar to BOFDA, the operation of BOFDR presented in Fig. 6 is based on amplitude modulation of the pump wave by sweeping the modulation frequency f_m. Due to the spontaneous Brillouin scattering in the sensor fiber, the Brillouin backscattered signal comprises both Stokes and anti-Stokes component. The letter can be filtered out by the use of a narrowband fiber Bragg grating (FBG) after propagating back to the fiber input, while the Stokes component is converted into an electrical signal by a photodetector PD at the beat frequency f_B. After conversion to the electrical domain, the electrical signal of PD at the value f_B of BFS is frequency-downshifted to the intermediate frequency (IF) band, by mixing it with the tunable

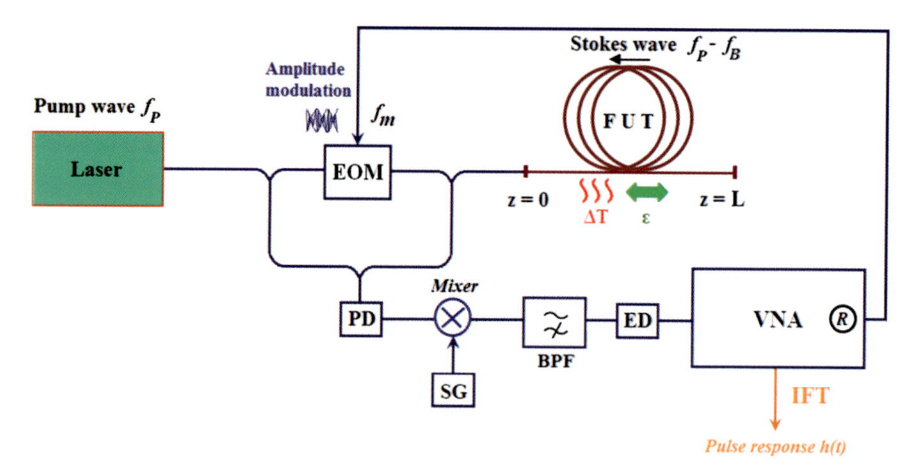

Fig. 6 BOFDR basic configuration. *VNA* vector network analyzer, *EOM* electro-optic modulator, *PD* photo diode, *SG* signal generator, *BPF* band-pass filter, *ED* envelope detector, *FUT* fiber under test

signal provided by a microwave signal generator SG. After the following filtering and amplification process by the use of a BPF, the baseband signal is recovered by passing the IF signal through an envelope detector ED connected to the VNA. In order to record spatially resolved BGSs carrying information about strain and temperature distribution along the fiber, the frequency of SG is swept in the range of f_B according to expected temperature gradients and local strain values along the sensor fiber.

All frequency-domain approaches feature a progressive sinusoidal amplitude modulation of the cw laser light coupled into a fiber. This also leads to a joint conclusion about the measurement range mathematically expressed by Eq. 9. In fact, the BOFDR technique proves to be indeed inherently more vulnerable to additional optical losses due to bending effects along the sensor fiber than its BOFDA counterpart described in further detail below.

Advantages of Frequency-Domain Technique

The frequency-domain analysis methods for distributed strain and temperature measurement along even several tens of kilometers offer some benefits compared to the standard BOTDA concept. An important asset of BOFDA can be directly derived from the measurement principle based on a narrow-bandwidth operation. The BOFDA measurement bandwidth is specified by a narrowband filter of VNA, typically in the kHz range around the modulation frequency f_m, whereas the BOTDA measurement bandwidth of at least 100 MHz is necessary to resolve pulses of several nanoseconds. Since SNR is inversely proportional to the measurement bandwidth, BOFDA enables a significant improvement of SNR up to 50 dB. However, this statement should not be overestimated by considering that, relating to pump depletion properties, significantly higher laser powers directly contributing to SNR improvement can be used in BOTDA (Gogolla 2000). Moreover, for noise suppression in BOTDA, a few thousand individual measurements are averaged (Bao et al. 1995).

Another advantage of BOFDA is that this technique overcomes the need for use of expensive high-speed electronics for the purpose of generation and detection of short laser pulses, which in turn reduces system costs.

From the theoretical point of view, the narrow-bandwidth operation in case of BOFDA should result in relatively long measurement time. The associated problem of the long setting times of analog filters in a VNA's internal circuitry discussed in the section "Measurement Time" can be solved by adopting of a digital setup approach as described in detail in the section "Digital Processing Technique." Such digitization combines two benefits: on the one hand, it reduces the measurement time, and on the other hand, it offers a decisive cost advantage by replacing an expensive VNA with a digital circuit. In addition, the existing low-cost potential of BOFDA is also intensified by the use of low-power lasers which are operated below the pump threshold (prerequisite of the system linearity).

Measurement Setups

The measuring procedures of BOFDA follow an accurate algorithmic pattern of generation of a progressive sinusoidal amplitude modulation of the pump wave (gain method) by sweeping the modulation frequency f_m for a selected constant laser frequency difference f_D of counterpropagating pump and probe signals. The values of the transfer function $H_G(f_m,f_D)$ recorded by the sweep provide a complete signal trace $h(z,f_D)$ over the whole fiber length using DIFT. Repeating this sweeping process of f_m for each value of f_D set in the range of characteristic Brillouin frequency shift f_B, three-dimensional images of BGSs can be determined as already shown in Fig. 3 left.

This section focuses on the advancements of BOFDA setups by presenting two hardware solutions for the realization of the f_D scanning to be performed for the purpose of recording distributed BGSs along a sensor fiber. The starting point for implementation of the measurement algorithm is given here by a two-laser setup. As a result of adoption of the idea of modulating the laser's amplitude with an RF signal with a frequency in the range of f_B corresponding the sideband technique proposed in Nikles et al. (1997), the relaxation and setting time of a control loop to drive two separate lasers can be reduced using only one laser source in the BOFDA setup.

This section also gives information on crucial optimization steps for increasing measurement accuracy (polarization adjustment) and improving the measurement time with a simultaneous system cost reduction using digital processing.

Two-Laser Configuration

The first investigation on the optical frequency-domain technique referred to a distributed reflectometry method for measurement of linear backscattering (Ghafoori-Shiraz and Okoshi 1986). This approach was adopted in the 1990s for Brillouin sensing (Garus et al. 1997; Gogolla 2000; Krebber 2001). Within this research, numerical models of SBS for BOFDA and laboratory measurements along 11-km-long optical fiber with a spatial resolution of 1 m were presented. The BOFDA setup described here in this section is based on advancements achieved by investigations within the program "Risk Management of Extreme Flood Events (RIMAX)" of the German Federal Ministry of Education and Research (2005–2009) offering new perspectives in terms of dynamic range and measurement accuracy.

Figure 7 shows a complete BOFDA laboratory setup based on two separate Nd:YAG lasers used as frequency-shifted pump and probe (Stokes) signal sources.

In the measurement setup above, two separate functional blocks can be identified.

The first functional group serves to control and adjust the frequency differences $f_D = f_P - f_S$, at which the transfer functions are recorded by a VNA. This control loop is composed of a photodetector PD (upper left) with a bandwidth of 25 GHz, an electrical spectrum analyzer SA for the detection of the current beat frequency

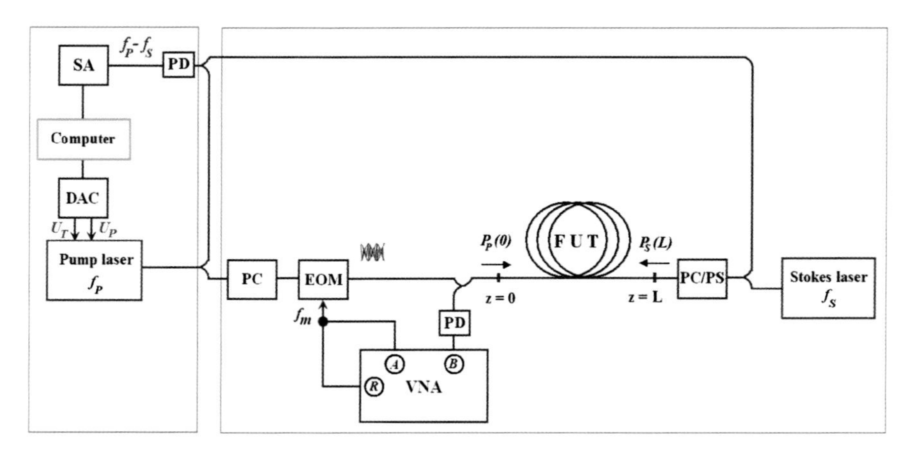

Fig. 7 Basic two-laser configuration of BOFDA. *VNA* vector network analyzer, *EOM* electro-optic modulator, *PD* photo diode, *DAC* digital-to-analog converter, *SA* electrical spectrum analyzer, *PC/PS* polarization controller/polarization scrambler, *FUT* fiber under test

f_D, and a digital-to-analog converter DAC, directly developed for voltage control of the pump laser frequency f_P.

Both Nd:YAG lasers, emitting at the wavelength of 1.319 μm and with a bandwidth of 5 kHz, are tunable by means of electrical signals. Such frequency tuning is achieved by altering the emission wavelength of the crystal resonator in two ways simultaneously. First, rough tuning in a range of 30 GHz is carried out by a peltier element acting as a thermoelectric cooler (TEC) over a voltage U_T attached to the laser crystal and characterized by a coefficient of 3.8 GHz/V. Second, fast frequency tuning over a voltage U_P for fine adjustments in a range of 30 MHz is realized by means of a piezoelectric element mounted on the laser crystal and characterized by a coefficient of 1.0 MHz/V.

The spectrum analyzer communicates with a control computer over a GPIB interface. As presented in Nöther et al. (2008a, b) and described in detail in Nöther (2010), the two required voltage values U_T und U_P are computed by a special algorithmic calculation and sent to a digital-to-analog converter over a serial port RS-232. If the current laser frequency difference f_D controlled by the spectrum analyzer meets the accuracy requirements of Δf_D (here 150 kHz) for f_D, then the measurement of the transfer function is being continued by the VNA for the set frequency difference f_D.

As shown in Fig. 8 (Wosniok 2013), the initial operating point $f_D(U_{TP}, U_{TS})$ should be favorably set centrally between two adjacent laser mode hops around the natural BFS of 12.8 GHz at the wavelength of 1.319 μm. It is defined by a rough tuning via TEC of the pump laser (control voltage U_{TP}) or of the probe (Stokes) laser (control voltage U_{TS}). Figure 8 depicts the setting of the initial operating point by setting U_{TP}, when the rough tuning of the Stokes laser is switched off ($U_{TS} = 0$).

The vector network analyzer VNA is the core element of the second block in the BOFDA setup presented in Fig. 7. The VNA is used for the purpose of recording

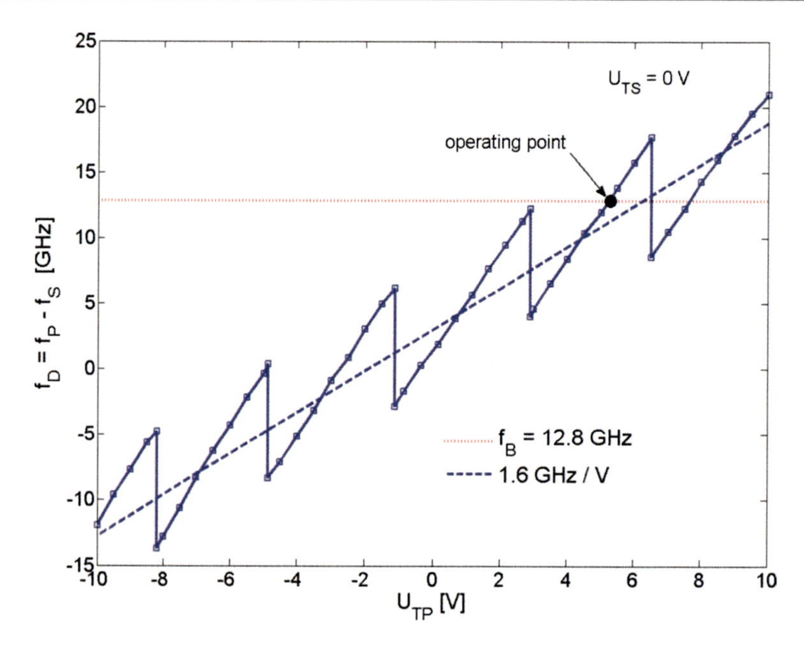

Fig. 8 Setting the laser operating point between mode hops. U_{TP} voltage on TEC of the pump laser, U_{TS} voltage on TEC of the probe (Stokes) laser

individual measurement signals $H(f_m,f_D)$ and providing RF input of the EOM with an electrical modulation signal at the adjustable frequency f_m at the same time. The electrical modulation signal generated by the internal VNA oscillator serves also directly as the reference signal containing information about amplitude and phase of the system input (pump signal). The electrical information about system output (probe signal) needed for calculating the scattering parameter S_{21} – transfer function $H(f_m,f_D)$ – is received using a photodetector with a bandwidth of 1 GHz, whose AC-coupled output is connected to the VNA as shown in Fig. 7.

The polarization controller in the pump path serves to allow for the best possible state of linear polarization required to maximize the extinction ratio of the EOM. The polarization controller or scrambler in the probe path improves the average Brillouin gain along the sensor fiber.

The generation of the beat frequency $f_P - f_S$ by means of two laser sources according to Fig. 7 is hampered by the still existing drift of the laser frequency, which impairs the measurement times by the necessary adjustment of the pump laser frequency f_P in order to control the frequency difference f_D (adverse relaxation and setting time of a control loop). In general, the reasons for the frequency fluctuations are the digital-to-analog converter noise, temperature changes in laser resonator, or vibrations. The environmental parameters such as air temperature, humidity, and pressure can also increase the drifting in a slow way. The drifting problem can be circumvented by applying the sideband technique described below, which downsizes the BOFDA setup from two laser sources to one.

One-Laser Configuration

In the implementation of the sideband technique (Nikles et al. 1997) for BOFDA, the cost-effectiveness of the measurement system can be additionally enhanced by the use of a convenient laser diode widely adopted in optical communication applications. Figure 9 shows a BOFDA laboratory setup built up on one single semiconductor distributed feedback laser diode (DFB laser diode) operating at the wavelength of 1.55 μm and with a bandwidth of 300 kHz.

Since the value f_B of the BFS is inverse proportional to the wavelength of the pump wave, the natural BFS changes here from 12.8 GHz achieved by Nd:YAG laser source at the wavelength of 1.319 μm to 10.9 GHz. This inverse proportionality also applies to the thermal $C_T = \frac{\delta f_B}{\delta T}$ and strain $C_\varepsilon = \frac{\delta f_B}{\delta \varepsilon}$ coefficients appearing in Eq. 13 as summarized in Table 1 for a standard single-mode fiber SMF-28e.

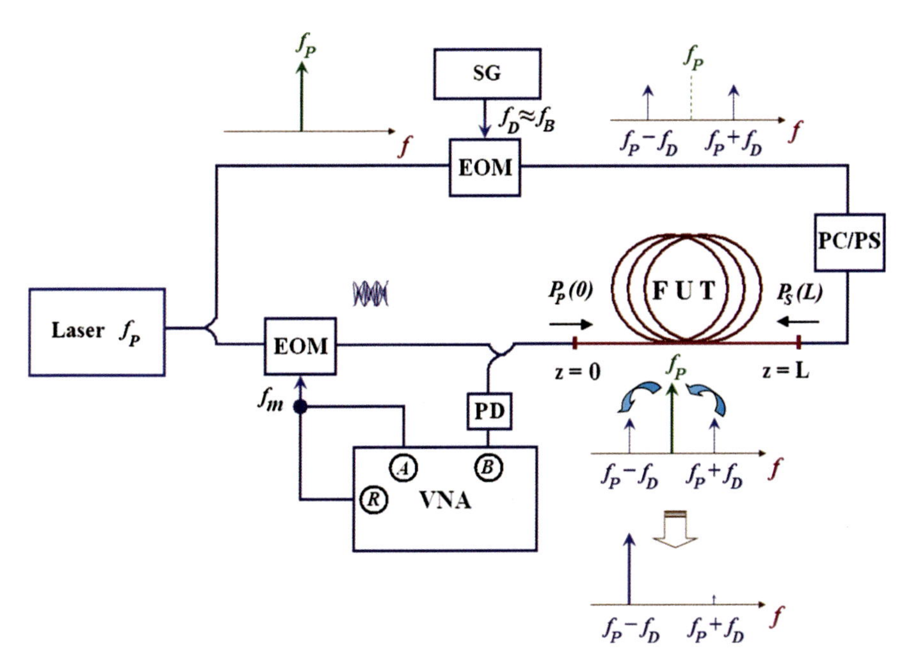

Fig. 9 Basic one-laser configuration of BOFDA using the sideband technique. *VNA* vector network analyzer, *EOM* electro-optic modulator, *PD* photo diode, *SG* signal generator, *PC/PS* polarization controller/polarization scrambler

Table 1 Thermal and strain coefficients of the SMF-28e in the second ($\lambda_P = 1.319$ μm) and third ($\lambda_P = 1.55$ μm) near-infrared spectral windows

Wavelength λ_P[μm]	C_T [MHz/°C]	C_ε[MHz/10^3 με]
1.319	1.3	56
1.55	1.1	48

Compared to Fig. 7, the one-laser configuration presented in Fig. 9 requires an additional EOM to modulate the optical power in amplitude using a signal generator SG at a frequency $f_D = f_P - f_S$ in the range of f_B ($\approx$ 10.9 GHz). Such amplitude modulation gives rise to new sidebands in the optical spectrum. The lower sideband at the frequency $f_P - f_D = f_S$ acts as the frequency-downshifted probe signal playing the same role as the Stokes laser in the two-laser configuration. The upper sideband at the frequency $f_P + f_D = f_S + 2f_D$ amplifies the counterpropagating pump signal at the frequency f_P in the sensor fiber. Since the intensity of the upper sideband is very low compared to the pump intensity, the intensity increase at f_P due to this interaction is minor and can be neglected. On the contrary, the upper sideband, here in the role of a pump signal to the modulated counterpropagating pump wave at the frequency f_P, is significantly depleted due to the high intensity differences of the two interacting signals.

By setting a DC voltage of the additional EOM to a minimum optical transmission, so that no light is transmitted in absence of the amplitude modulation, the carrier at the frequency f_P can be suppressed. As a result, only the interaction between the modulated pump wave and the counterpropagating lower sideband as the probe signal is relevant from the viewpoint of the measurement principle.

The two EOMs (called also Mach-Zehnder modulators) are operated with different DC bias voltages to adjust the operating points of the modulators in accordance with their separate requirements. As mentioned above, a DC voltage of the EOM should be set to a minimum optical transmission to suppress the carrier, whereas the modulation of the pump wave is carried out in the linear range, i.e., in the middle between the maximal and minimal transmission.

Since the signal quality of the spatially resolved BGSs recorded using the described one-laser configuration can be degraded by nonlinear effects that arise from undesired interaction between the carrier and the sidebands, the DC voltage is to be adjust to a small drift of the modulator transfer function during the measurement process. Such a drift can result in a change of the optical signal, especially in case of the additional EOM in the probe path. Here, the modulated optical signal can be seriously affected by a rebuilding of the suppressed carrier if the DC voltage is not corrected.

Polarization Effects

The dependence of the Brillouin gain on the states of polarization (SOPs) of the two counterpropagating pump and probe waves can generally be summarized into a polarization factor γ described in the section "Measurement Accuracy." If $\gamma(z)$ assumes values close to zero along the fiber, a significant deterioration of the accuracy in the determination of BFS values f_B is to be expected due to weak measuring signals. As already mentioned in the first section, even "dead zones" can occur at each individual measurement point with $\gamma = 0$, where no SBS is observed. To eliminate these unacceptable sensor states, $\gamma(z) \approx 0$ must be avoided at any location z along the fiber.

In conventional optical fibers used in telecommunication, the polarization state of a light wave propagating through the fiber changes continuously due to the inherent birefringence of the fiber (Nöther et al. 2008b). The stochastic nature of the physical effect of the birefringence makes the control of the polarization states along the measured section so difficult. Already during fiber manufacturing process, slight and random deviations from an ideal circular cross-section of the fiber core occur, which results in minor differences in the refractive index ($n_x > n_y$) along the two main axes x (slow axis) and y (fast axis) causing the birefringence B:

$$B = n_x - n_y \tag{14}$$

Local birefringence is additionally affected by mechanical stress caused by different dopant concentration in the fiber core and cladding as well as by torsion applied to the fiber during the manufacturing process (elasto-optic changes in effective refractive indices). In spite of the random character of the linear birefringence formation, some individual key factors influencing the birefringence are quantitatively described in Rashleigh (1983). According to Rashleigh (1983), also external effects, such as compressive or tensile forces, contribute to the birefringence. For sensor technology purposes, a very important finding is the fact that a macrobending of optical fibers also results in variation of the material birefringence properties as:

$$B \sim \left(\frac{r}{R}\right)^2 \tag{15}$$

where r and R are the fiber and bending radius, respectively.

Conditioned by the way of sensor installation, the random fluctuations of the state of polarization as well as the optical losses can be restricted. This in turn can increase the measuring accuracy.

Due to the birefringence B given by Eq. (14), a light wave with the wavelength λ injected into the sensor fiber will adopt all possible states of polarization within the beat length L_B defined as:

$$L_B = \frac{\lambda}{B} \tag{16}$$

In low-birefringence fibers, like standard telecommunication single-mode silica optical fibers typically used as distributed sensors, the beat length is below 100 m. According to Foschini and Poole (1991), the development of light's SOPs in this type of fiber can be described as a three-dimensional Brownian motion. Since the BOFDA measurement ranges of several kilometers largely exceed the beat length L_B, the SOPs are evenly distributed over the Poincaré sphere with good approximation. This leads to the SBS mixing efficiencies of $\widetilde{\gamma} = \frac{2}{3}$ for coinciding polarizations of counterpropagating pump and probe (Stokes) waves and $\widetilde{\gamma} = \frac{1}{3}$ for orthogonal polarizations. Therefore, even in the case of a simple rotation of one of

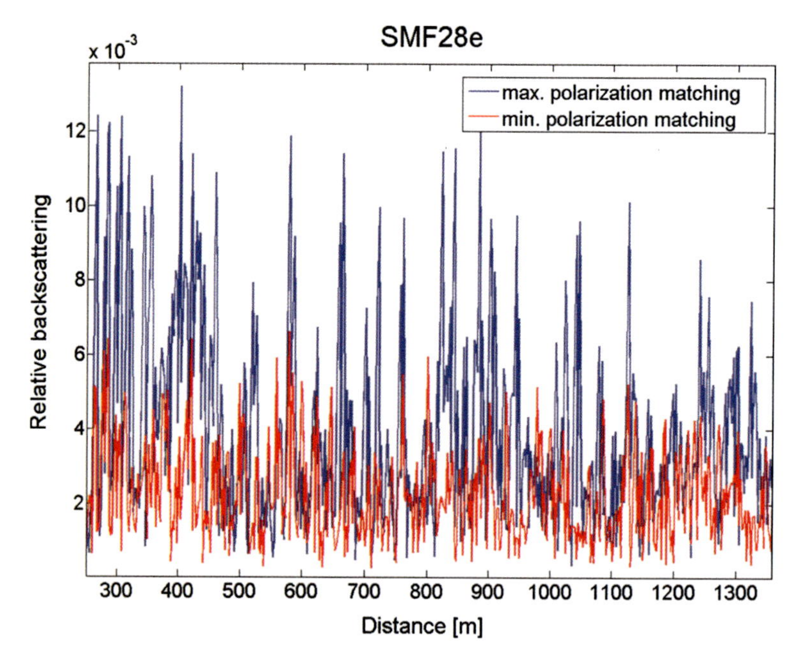

Fig. 10 Polarization matching of the Brillouin gain distribution along a SMF-28e for linearly polarized pump and probe light

the two polarization vectors, an average Brillouin gain of the measurement signals backscattered in the SBS process can be enhanced by factor 2. Such a polarization adjustment of the Brillouin gain along a standard SMF-28e fiber is shown in Fig. 10. The relative SOPs of the interacting light waves have been matched here by turning the polarization direction of the linearly polarized Stokes wave (probe signal) by means of a polarization controller (PC/PS in the probe path used as a PC) in the BOFDA setup according to Fig. 7.

Figure 10 illustrates significant fluctuations in the backscattered distributed Brillouin signals occur for measurements with predefined SOPs of the pump and the probe light coupled into the two opposite fiber ends. In order to obtain a homogeneous Brillouin gain over the entire fiber length, two approaches to overcome the polarization-related signal fluctuations are presented in this section.

The first approach is based on an averaging procedure of the BOFDA measurement at a fixed number of SOPs that are distributed over the Poincaré sphere. The violet curve in Fig. 11 represents the result of such an averaging of two signal traces recorded at orthogonal SOPs and at the frequency difference $f_P - f_S = f_B$. This averaged signal trace shows that an averaging over more different SOPs is necessary to smooth the Brillouin trace over the whole length of the sensor fiber which automatically increases the measurement time.

The second approach which can be used to cope with the polarization-related fluctuations is to scramble one of the two SOPs of the interacting light waves. In

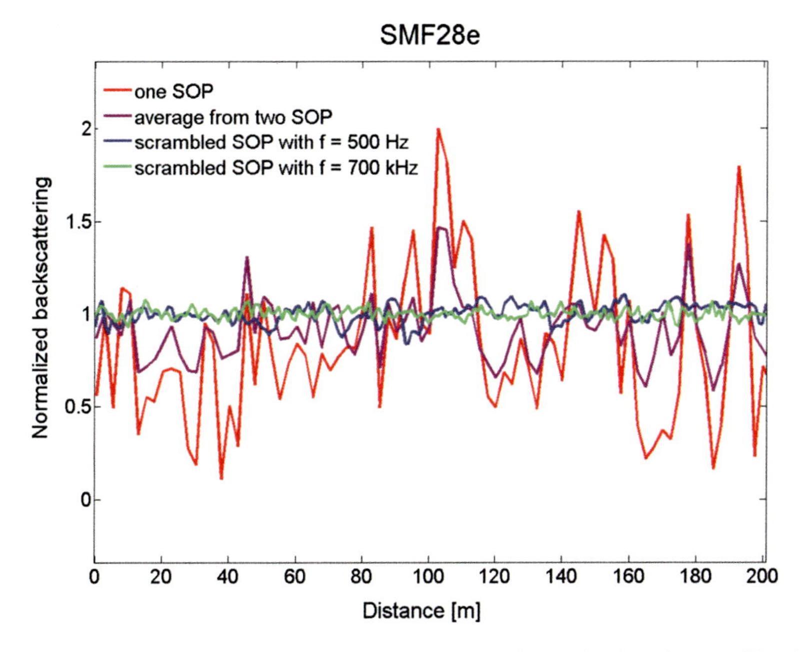

Fig. 11 Polarization-related fluctuations of the Brillouin gain distribution along a 200-m-long SMF-28e measured by the BOFDA system. All measurements were performed at a constant pump and Stokes optical power, VNA's bandwidth of 3 kHz, and at constant temperature conditions of the unstained FUT

this method, the polarization direction of either the pump or the probe (Stokes) wave is rapidly rotated by means of a polarization scrambler PS before coupling into the fiber, whereas the counterpropagating light wave is injected into the fiber at the opposite fiber end as a linearly polarized wave. Figure 11 points out that the polarization-related fluctuations of the spatially resolved measurement signals can be effectively suppressed by the choice of higher rotational frequencies f_R of the polarization rotation. The enhancement of the rotational frequency f_R from 500 Hz to 700 kHz presented in Fig. 11 results in a visible increase of the polarization independency of the measured BOFDA signals, which is reflected in the improvement of the corresponding standard deviation from 5.2% to 2.9%. In general, in order to obtain a low polarization disturbance, the following condition must be fulfilled (Gogolla 2000):

$$f_R > 10\ B_F \tag{17}$$

Narrowing of the filter bandwidth B_F within the VNA to eliminate polarization fluctuations leads directly to an increase of the measurement time (the settling time of the VNA's analog filters is inversely proportional to their bandwidth – see section "Measurement Time"). When seen from this aspect, the increase of f_R seems to be the most efficient for further elimination of the polarization fluctuations. At the same

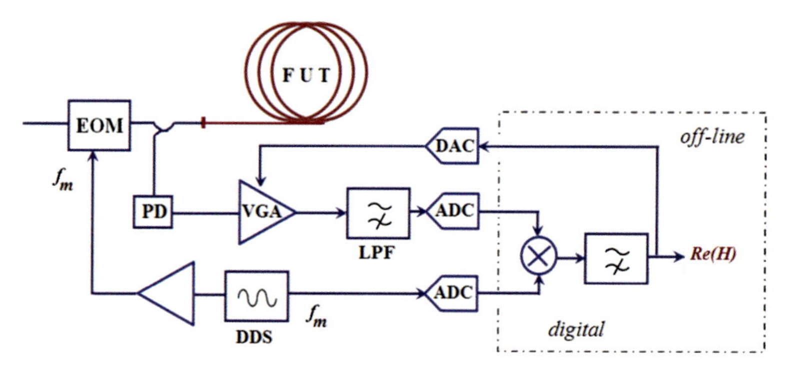

Fig. 12 Digital optical signal processing for BOFDA. *DDS* digital data synthesis, *ADC* analog-to-digital converter, *DAC* digital-to-analog converter, *LPG* low-pass filter, *VGA* variable gain amplifier

time, compared to the common scrambling modules, the electro-optic polarization scramblers offer substantially higher rotational frequencies f_R up to about 20 GHz.

Digital Processing Technique

The central idea of the digital concept is to replace the overdesigned and expensive VNA with a digital setup recording the complex transfer function. The general concept of a cost-effective digital signal processing is given in Fig. 12 (Wosniok et al. 2009). Similarly to the analog solutions discussed above, the amplitude modulation of the pump signal is carried out by an electro-optic modulator; however, the modulation frequency f_m is generated by a digital synthetizer using a direct data synthesis (DDS) circuit (Nöther 2010). This modulation signal is used both to drive the electro-optic modulator and to be directly converted back into the digital domain giving a reference signal for calculation of the transfer function H. The signal to be measured with the information about the Brillouin gain along the sensor fiber is detected by a photo diode PD, limited in its bandwidth by an anti-aliasing low-pass filter LPF, and finally digitalized by an analog-to-digital converter ADC. In this way, the digital data can be transferred to a computer where a definite off-line fast Fourier transform (FFT) is to be performed.

Since DDC and ADCs are standard electronic devices, the cost of the measurement components are significantly lower for the digital implementation. The off-line signal processing enables fast measurements at a high dynamic range.

Conclusion

This chapter provides a compact overview of fundamental principles of the Brillouin frequency-domain techniques for structural health monitoring applications. Starting from the theoretical backgrounds on measurement in the frequency domain, the

BOFDA approaches using the one- and the two-laser configurations were presented in detail. With the focus on real-life temperature and strain measurement applications, the advantages and limitations of distributed Brillouin sensing based on both the highly accurate frequency-domain analysis and the practically advantageous one-end-access reflectometry were discussed.

References

X. Bao, J. Dhliwayo, N. Heron, D.J. Webb, D.A. Jackson, J. Lightwave Technol. **13**(7), 1340–1348 (1995)

R. Bracewell, *The Fourier Transform and Its Applications* (McGraw-Hill, New York, 1999)

G.J. Foschini, C.D. Poole, J. Lightwave Technol. **9**(11), 1439–1456 (1991)

D. Garus, K. Krebber, F. Schliep, T. Gogolla, Opt. Lett. **21**(17), 1402–1404 (1996)

D. Garus, T. Gogolla, K. Krebber, F. Schliep, J. Lightwave Technol. **15**(4), 654–662 (1997)

H. Ghafoori-Shiraz, T. Okoshi, J. Lightwave Technol. **4**(3), 316–322 (1986)

T. Gogolla, *Theoretische Untersuchung der Brillouin-Wechselwirkung in Lichtleitfasern zur kontinuierlich verteilten Temperatur-und Dehnungsmessung auf Basis der Frequenzbereichsanalyse* (Shaker, Aachen, 2000)

F.J. Harris, Proc. IEEE **66**, 1 (1978)

K. Krebber, *Ortsauflösende Lichtleitfaser-Sensorik für die Temperatur und Dehnung unter Nutzung der stimulierten Brillouin-Streuung basierend auf der Frequenzbereichsanalyse* (RUB, Bochum, 2001)

A. Minardo, R. Bernini, L. Zeni, Opt. Express **22**, 14 (2014)

A. Minardo, R. Bernini, R. Ruiz-Lombera, J. Mirapeix, J.M. Lopez-Higuera, L. Zeni, Opt. Express **24**(26), 29994–30001 (2016)

M. Nikles, L. Thevenaz, P. Robert, J. Lightwave Technol. **15**(10), 1842–1851 (1997)

N. Nöther, *Distributed Fiber Sensors in River Embankments: Advancing and Implementing the Brillouin Optical Frequency Domain Analysis* (BAM, Berlin, 2010)

N. Nöther, A. Wosniok, K. Krebber, Proc. SPIE **6933**, 69330T-1–69330T-9 (2008a)

N. Nöther, A. Wosniok, K. Krebber, Proc. SPIE **7003**, 700303.1–700303.9 (2008b)

S. Rashleigh, J. Lightwave Technol. **1**(2), 312–331 (1983)

M.A. Soto, L. Thévenaz, Opt. Express **21**(25), 31347–31366 (2013)

A. Wosniok, *Untersuchungen zur Unterscheidung der Einflussgrößen Temperatur und Dehnung bei Anwendung der verteilten Brillouin-Sensorik in der Bauwerksüberwachung* (TUB, Berlin, 2013)

A. Wosniok, N. Nöther, K. Krebber, Procedia Chem. **1**(1), 397–400 (2009)

A. Yaniay, J.-M. Delavaux, J. Toulouse, J. Lightwave Technol. **20**(8), 1425–1432 (2002)

Distributed Brillouin Sensing: Correlation-Domain Techniques

45

Weiwen Zou, Xin Long, and Jianping Chen

Contents

Abstract

The Brillouin based distributed sensors have great potentials in many fields such as smart materials and structers. The correlation domain technique developed in the last two decades is based on the measurement for the correlated-peaks along the optical fiber. It have attracted much attention as its unparalleled advantages in ultra high resolution and the ability to achieve dynamic measurement, which are difficult for the traditional time domain techniques. Both spontaneous Brillouin scattering and stimulated Brillouin scattering can be utilized in correlation domain techniques. Efforts have been paid to improve the performances including the effective sensing points enlargement and noise suppression.

W. Zou (✉) · X. Long · J. Chen
State Key Laboratory of Advanced Optical Communication Systems and Networks, Department of Electronic Engineering, Shanghai Jiao Tong University, Shanghai, China
e-mail: wzou@sjtu.edu.cn

G.-D. Peng (ed.), *Handbook of Optical Fibers*,
https://doi.org/10.1007/978-981-10-7087-7_9

Moreover, the correlation domain techniques can be well combined with other techniques such as time domain techniques or Brillouin dynamic grating techniques.

Preface

Distributed Brillouin optical fiber sensors have attracted much attention and are thought potential due to their unparalleled abilities to measure the environment (such as temperature and strain) along an optical fiber. The capacity to detect continuously along the optical fiber is superior to the pointed or multiplexed fiber-optic sensors based on fiber Bragg grating and/or inline Fabry-Perot resonator. The Brillouin scattering utilized for optical fiber sensing is a nonlinear optical process and has many advantages such as high accuracy due to the narrowband of the Brillouin gain spectrum (BGS), simultaneously sensitive to both temperature and strain, and immunity to the electro-magnetic interference. Among the Brillouin-based distributed optical sensing techniques, the correlation-domain technique utilizes continuous lightwave as pump and measures the Brillouin gain spectrum information from the correlation peak. Compared to the conventional pulse-based time-domain technique, it can achieve centimeter or even millimeter resolution and fast dynamic measurement, which is difficult for time-domain technique due to the pulsed-pump-based sensing mechanism.

This chapter introduces the distributed Brillouin sensing correlation-domain techniques in optical fibers. The basic principle, working mechanism of two types of correlation-domain techniques, and methods proposed to improve the sensing ability are demonstrated, respectively.

Distributed Brillouin-Based Sensors

Brillouin Scattering in Optical Fibers

Brillouin scattering is a "photon-phonon" interaction as annihilation of a pump photon creates a Stokes photon and a phonon simultaneously (Agrawal 2012). The created phonon is the vibrational modes of atoms, also called a propagation density wave or an acoustic phonon/wave due to the electrostriction effect. In a silica-based optical fiber, Brillouin Stokes wave propagates dominantly backward (Ippen and Stolen 1972) although very partially forward (Shelby et al. 1985). The frequency (~9–11 GHz) of Stokes photon at ~1550-nm wavelength is dominantly downshifted due to Doppler shift associated with the forward movement of created acoustic phonons (Fig. 1).

Brillouin scattering is basically categorized into two types: spontaneous Brillouin scattering (SpBS) and stimulated Brillouin scattering (SBS) in optical fibers. In principle, the SpBS is caused by a noise fluctuation and influences the pump wave (E_p). The SBS occurs when the pump power for SpBS is beyond the so-called Brillouin threshold value (P_{th}) or when two coherent waves with a

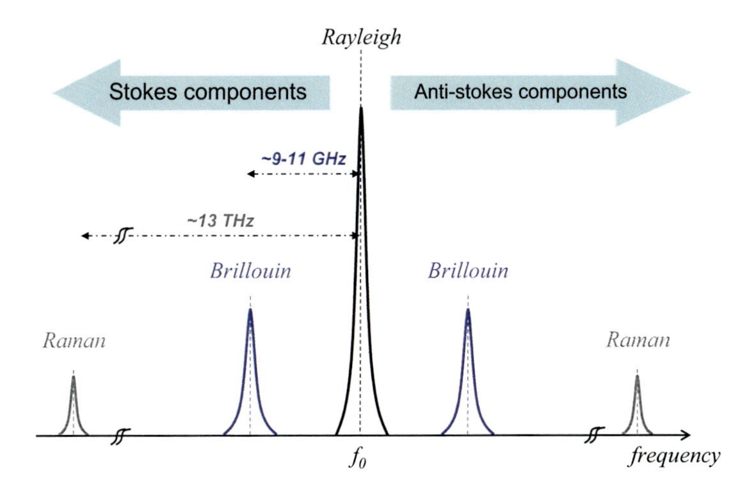

Fig. 1 Schematic spectrum of scattered light resulting from three scattering processes in optical fibers (After Zou et al. (2015). From Mizuno et al. (2008, 2010), Song et al. (2008, 2009b), and Elooz et al. (2014))

frequency difference equivalent to the phonon's frequency are counter-propagated. Brillouin scattering dynamics in optical fibers are generally governed by the following coupling equations (Gaeta and Boyd 1991; Jenkins et al. 2007):

$$\begin{cases} \left(\frac{1}{v_g}\frac{\partial}{\partial t} + \frac{\partial}{\partial z}\right) E_p = -\frac{\alpha}{2} E_p + i\kappa_1 \rho E_s \\ \left(\frac{1}{v_g}\frac{\partial}{\partial t} - \frac{\partial}{\partial z}\right) E_s = -\frac{\alpha}{2} E_s + i\kappa_1 \rho^* E_s \\ \left(\frac{\partial}{\partial t} + \frac{\Gamma_B}{2} + 2\pi i \nu_B\right) \rho = i\kappa_2 E_p {E_s}^* + N \end{cases}, \tag{1}$$

where E_p and E_s stand for the normalized slowly varying fields of pump and Stokes (or probe) lightwaves, respectively; ρ denotes the acoustic (or phonon) field in terms of the material density distribution; N represents the random fluctuation or white noise in position and time (Gaeta and Boyd 1991); v_g is the group light velocity in the fiber; α is the fiber's propagation loss; Γ_B is the damping rate of the acoustic wave, which equals to the reciprocal of the phonon's lifetime ($1/\Gamma_B = \tau_\rho = \sim 10$ ns) and is related to the acoustic linewidth ($\Delta \nu_B = \Gamma_B/\pi$) (Pine 1969); and κ_1 and κ_2 are the coupling coefficients among E_p, E_s, and ρ. If SpBS is considered, it is reasonable to assume E_s is sufficiently small so that the second term of N dominates in the right side of Eq. 1. In contrast, the first term of $i\kappa_2 E_p {E_s}^*$ dominates for SBS.

Concerning the SBS power transfer between pump lightwave P_p and probe lightwave P_s under the assistance of the acoustic wave and the so-called acousto-optic effect, Eq. 1 can be rewritten as

$$\begin{cases} \left(\frac{1}{v_g}\frac{\partial}{\partial t} + \frac{\partial}{\partial z} + \alpha\right) P_p = -g(\nu) P_p P_s \\ \left(\frac{1}{v_g}\frac{\partial}{\partial t} - \frac{\partial}{\partial z} + \alpha\right) P_s = +g(\nu) P_p P \end{cases} \tag{2}$$

where the sign difference between the right hands means that the pump power is reduced or depleted but the probe (Stokes) power is increased or amplified. And $g(v)$ is called BGS, the key phraseology to represent Brillouin scattering in optical fibers. It denotes the spectral details of the light amplification from strong pump wave to weak counter-propagating probe/Stokes wave in SBS or those of the noise-initialized scattered phonons in SpBS.

There are three basic parameters of Brillouin frequency shift (BFS, ν_B), Brillouin gain peak value (g_{B0}), and Brillouin linewidth ($\Delta\nu_B$) in the main-peak BGS. ν_B is defined as

$$\nu_B = \frac{2n_{eff} V_a}{\lambda} \tag{3}$$

where λ is the light wavelength, n_{eff} is the effective refractive index of the fiber, and V_a is the effective acoustic velocity of the fundamental acoustic mode.

Brillouin gain peak value g_{B0} is determined by

$$g_{B0} = \frac{4\pi {n_{eff}}^8 {p_{12}}^2}{\lambda^3 \rho_0 c \nu_B \Delta\nu_{B0}} \tag{4}$$

where ρ_0 is the density of silica glass (~2202 kg/m^3) and p_{12} the photoelastic constant (~0.271). In most silica-based fibers, the peak gain value of g_{B0} lies in the range of $1.5{\sim}3 \times 10^{-11}$ m/W (Nikles et al. 1997).

$\Delta\nu_{B0}$ in silica optical fibers with a typical value of 30–40 MHz is characteristic of SpBS. However, in the SBS process, it was theoretically proved that $\Delta\nu_B$ strongly depends on the pump power, which is expressed as follows (Gaeta and Boyd 1991; Yeniay et al. 2002):

$$\Delta\nu_B = \Delta\nu_{B0}\sqrt{\frac{\ln 2}{G_s}} \tag{5}$$

where G_s is the single pass gain experienced by the weak probe wave from the strong pump wave.

Principle of Brillouin-Based Distributed Sensing

Sensing of Measurands

As a nondestructive attenuation measurement technique for optical fibers, the SBS distributed measurement could measure attenuation distribution along the fiber having no break from an interrogated optical power as a function of time, but it has much higher signal-to-noise ratio (more than ~10 dB) than optical time-domain reflectometry (OTDR) due to SBS high gain. For this measurement SBS process was performed by injecting an optical pulse source and a continuous-wave (CW) light into two ends of fiber under test (FUT). When the frequency difference of the pulse pump and CW probe is tuned offset around ν_B of the FUT, the CW probe power experiences Brillouin gain from the pulse light through SBS process. Furthermore it was found that this nondestructive attenuation measurement can be extended into a frequency-resolved technique because ν_B of optical fibers has linear dependence on measurands of strain and temperature as follows (Horiguchi et al. 1989; Kurashima et al. 1990):

$$\nu_B - \nu_{B0} = A \cdot \delta\varepsilon + B \cdot \delta T \tag{6}$$

where ν_{B0} is measured at room temperature (25 °C) and in the "loose state" as a reference point, $\delta\varepsilon$ is the applied strain, and δT is the temperature change. The "loose state" means that the FUT is laid freely in order to avoid any artificial disturbances. A *(or* C_ν^{ε}*)* is the strain coefficient in a unit of MHz/$\mu\varepsilon$, and B *(or* C_T^{ε}*)* is the temperature coefficient in a unit of MHz/ °C. Fig. 2 illustrates the characterized strain or temperature dependence in a standard single-mode fiber (SMF), where the BGS always moves toward higher ν_B and its gain reduces or increases when $\delta\varepsilon$ or δT is increased, respectively. At 1550 nm, $A = 0.04$–0.05 MHz/$\mu\varepsilon$ and $B = 1.0$–1.2 MHz/°C, which depends on the fiber's structure and jackets. Note that Eq. 6 is the basic sensing mechanism of Brillouin-based distributed sensors.

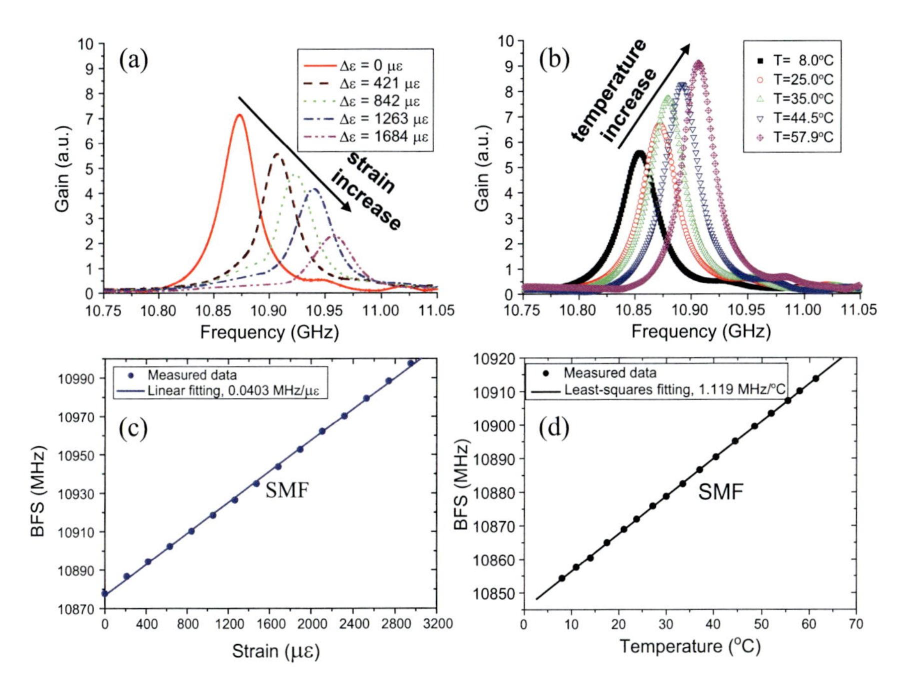

Fig. 2 (**a**) Strain and (**b**) temperature dependences of BGS in SMF; (**c**) strain and (**d**) temperature dependences of Brillouin frequency shift ν_B in SMF (From Song et al. (2008), Zou et al. (2009b), and Mizuno et al. (2010))

Sensing of Location

Besides the sensing of measurands, the mapping of spontaneous or stimulated Brillouin scattering process (not just nondestructive attenuation measurement (Horiguchi and Tateda 1989)) is another key issue to realize distributed optical fiber sensing (Horiguchi et al. 1990; Kurashima et al. 1993; Horiguchi et al. 1993). Two different mapping ways, as schematically illustrated in Fig. 3, were proposed. One is to repeat the localized BGS in scanned positions along the FUT; the other is to repeat the Brillouin interaction under different frequency offset. There are three different mapping or position-interrogation techniques, including time domain (Horiguchi et al. 1990; Kurashima et al. 1993; Horiguchi et al. 1993; Niklès et al. 1996; Shimizu et al. 1994; Bao et al. 1993), frequency domain (Garcus et al. 1997; Garus et al. 1996), and correlation domain (Hotate and Hasegawa 2000; Hotate and Tanaka 2002; Mizuno et al. 2008). Regarding the injection ways of optical fields, there are two opposite groups, i.e., analysis versus reflectometry. The analysis is two-end injection based on SBS; while the reflectometry is one-end injection based on SpBS. Note that there is an additional method between analysis and reflectometry, called one-end analysis (Niklès et al. 1996; Song and Hotate 2008; Zou et al. 2010).

The basic principle of time-domain sensing technique is the "time-of-flight" phenomenon in FUT. For two-end or one-end analysis, named Brillouin optical time-domain analysis (BOTDA) (Horiguchi et al. 1990; Horiguchi et al. 1993; Bao et al. 1993), one of pump and probe waves is pulsed in time and the other is continuous wave (CW). Subsequently, they are successively interacted along the FUT during the time of flight of the pulsed wave. In contrast, for one-end reflectometry, called Brillouin optical time-domain reflectometry (BOTDR) (Kurashima et al. 1993; Shimizu et al. 1994), the pump wave is pulsed in time, and the SpBS Stokes wave is reflected along the FUT during the pump's time of flight. The spatial resolution (ΔZ_{TD}) of time-domain distributed sensing is physically determined by the pulse width (τ) (Horiguchi and Tateda 1989):

$$\Delta Z_{TD} = \frac{\tau c}{2n} \tag{7}$$

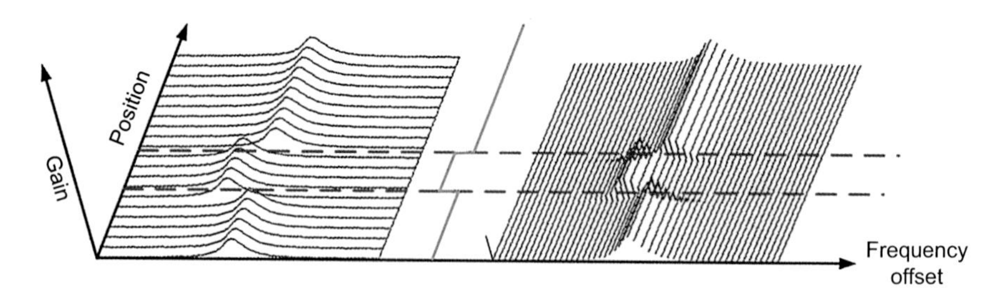

Fig. 3 Schematic of sensing of location or mapping of BGS (After Zou et al. (2015). From Song et al. (2009b))

where n is the group velocity of the pulse. The BGS mapping is realized by repeating the above measurement when the spectrum of the reflected Stokes in BOTDR is processed or the optical frequency offset between the pump and probe in BOTDA is tuned around the BFS ν_B.

There are two kinds of correlation-domain sensing techniques, nominated Brillouin optical correlation-domain analysis (BOCDA) (Hotate and Hasegawa 2000; Hotate and Tanaka 2002) and Brillouin optical correlation-domain reflectometry (BOCDR) (Mizuno et al. 2008; Mizuno et al. 2010a). In experiment, the BOCDR and BOCDA can be executed by substituting a distributed feedback laser diode (DFB-LD) driven by a function generator (such as in a sinusoidal function) for the light source. The optical frequency offset between pump and probe or between scattered Stokes and optical oscillator changes with time as well as position, deviating from the preset constant frequency offset around the BFS ν_B. The spatial resolution of BOCDA and BOCDR are both determined by (Hotate and Hasegawa 2000)

$$\Delta Z_{CD} = \frac{c}{2nf_m} \cdot \frac{\Delta \nu_B}{\pi \Delta f} \tag{8}$$

where f_m is the modulation frequency of the sinusoidal function, Δf the modulation depth, and $\Delta \nu_B$ the Brillouin linewidth defined in Eq. 5.The maximum measurement length (or sensing range, L_{CD}) is decided by the distance between two neighboring correlation peaks (Hotate and Hasegawa 2000):

$$L_{CD} = \frac{c}{2nf_m} \tag{9}$$

Because of the difference of the physical pictures between time domain and correlation domain, their sensing performance is different. For example, the spatial resolution of BOTDA/BOTDR was typically limited to be ~1 m by the lifetime of acoustic phonons (10 ns) and the nature of intrinsic Brillouin linewidth. However, BOCDA/BOCDR is of CW nature free from this limitation, and their spatial resolution can be ~cm-order (Hotate and Tanaka 2002; Mizuno et al. 2009a) or even ~mm-order (Song et al. 2006a).

Correlation-Domain Technique

BOCDA

In an original BOCDA technique, the frequencies of the pump and the counter-propagating probe lightwaves are both sinusoidally modulated. As shown in Fig. 4, at specific positions along the optical fiber, these two lightwaves undergo synchronous frequency modulation, and the beat frequency maintained a constant value (Hotate and Ong 2002). Furthermore, if this time-invariant frequency is

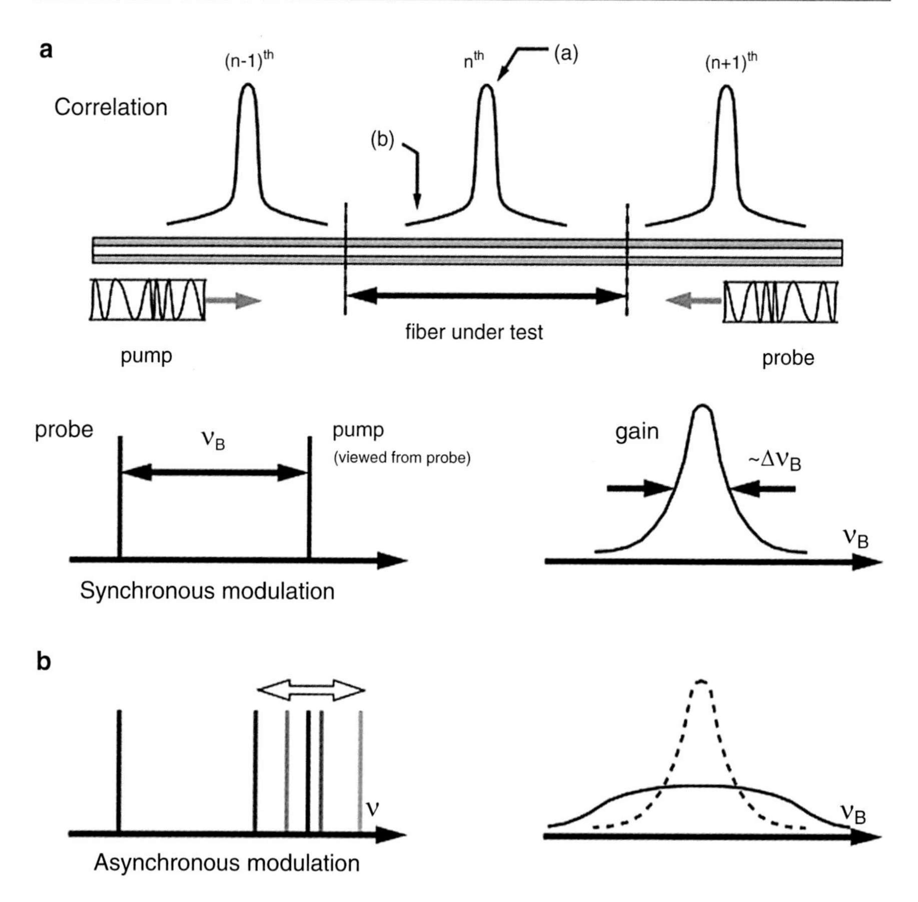

Fig. 4 Illustration of the correlation of the sinusoidally modulated pump and probe lightwaves. The Brillouin gain spectra are different between (**a**) correlation-peak positions and (**b**) non-correlation positions (After Hotate and Ong (2002). From Mizuno et al. (2008), Zou et al. (2009b), and Song et al. (2009b))

close to the local BFS (∼11 GHz for a standard single-mode fiber), the acoustic wave will be strongly and continuously stimulated. For these positions, they are called "correlation-peak points" or "correlation-peak positions." The Brillouin gain spectrum for correlation-peak position possesses narrow bandwidth of Lorentz shape (see Fig. 4a). As for the non-correlation positions, the pump and probe lightwaves experience asynchronous frequency modulations. From the view of the pump lightwave, the frequency of the probe lightwave swings broadly around. The beat frequency needed to maintain the acoustic wave is not satisfied, resulting a flat weak Brillouin gain spectrum (see Fig. 4b).

By sweeping the frequency difference between two lightwaves detecting the probe lightwave, the measured gain spectrum is the integration of the BGS along the optical fiber, which mainly comes from the correlation-peak positions and the non-correlation positions which contribute few to the final result. If only one peak exists along the optical fiber, the BFS of the measured BGS is a function of the measurands from the correlation-peak position.

Considering a sinusoidal frequency modulation is modulated on both pump and probe lightwaves, the modulation frequency is f_m, the frequency deviation is Δf, and the mean frequency difference between two lightwaves is Δv. As the two lightwaves are counter-propagating along an optical fiber, the temporal delay τ between them depends on the longitudinal position as

$$\tau = \frac{2\Delta x}{v_g} \tag{10}$$

where Δx represents the distance from correlation peak (it could be any one). Then the beat frequency between these two lightwaves can be written as

$$\begin{aligned} f_{beat}(t) &= \Delta v + \Delta f \left\{ \sin\left(2\pi f_m t\right) - \sin\left[2\pi f_m \left(t - \tau\right)\right] \right\} \\ &= \Delta v + 2\Delta f \cos\left(\pi f_m \tau\right) \sin\left[\pi f_m \left(2t - \tau\right)\right] \\ &= \Delta v + 2\Delta f \cos\left(2\pi f_m \frac{\Delta x}{v_g}\right) \sin\left[2\pi f_m \left(t - \frac{\Delta x}{v_g}\right)\right] \end{aligned} \tag{11}$$

The beat frequency $f_{beat}(t)$ is still a frequency modulated signal with modulation frequency f_m and frequency deviation $2\Delta f \cos(2\pi f_m \Delta x/v_g)$. To ensure the correlation peak, the beat frequency between pump and probe lightwaves should be restricted in an acceptable bandwidth to generate a strong and stable acoustic wave, which can be considered as the FWHM of the intrinsic BGS Δv_B. If condition $\Delta x << v_g/f_m$ is satisfied, this bandwidth restriction can be expressed as

$$2\pi f_m \Delta f \frac{\Delta x}{v_g} \leq \Delta v_B \tag{12}$$

It can be considered that the final measured gain spectrum is mainly contributed by the positions that satisfy Eq. 12. Thus the spatial resolution of the correlation-domain technique is given by

$$\Delta z = \frac{v_g \Delta v_B}{2\pi f_m \Delta f} \tag{13}$$

which is the same as Eq. 8.

Thus the spatial resolution depends on both the modulation frequency f_m and the frequency deviation Δf, which can be very small by carefully selecting these two values. For instance, if $f_m = 50$ MHz and $\Delta f = 1$ GHz and the Δv_B is usually around 30 MHz, the calculated sensing resolution according to Eq. 13 is around 2 cm, which is ten times better than the basic pump-pulsed-based time-domain techniques.

On the other hand, the 0th correlation peak is always located at the center of the optical fiber, i.e., the zero path difference point for two lightwaves. While the other non-0th correlation-peaks' position are periodically distributed along the fiber and are dependent on the parameter of the frequency modulation. According to Eq. 11, the distance d_m between the adjacent correlation peaks will be given by

$$d_m = \frac{v_g}{2f_m} \tag{14}$$

By scanning this modulation frequency, the positions of the non-0th correlation peaks can be changed continuously along the optical fiber, and the distributed information can be obtained. It should be noted that, as mentioned above, the correlation-domain sensing needs to ensure that only one correlation peak contributes to the final gain spectrum's measurement. This means that the sensing range for this technique cannot exceed d_m; otherwise two or more correlation peaks occur along the optical fiber and the measurement will be confused. According to the definition of Eq. 14, the sensing range of a basic BOCDA technique is dependent on the modulation frequency, usually several to tens of meters in applications, which is very short compared with the time-domain techniques.

Figure 5 gives an illustration explaining the main configuration of the basic BOCDA technique (Yamauchi and Hotate 2004). Sinusoidal frequency modulation is implemented by direct modulation onto the laser, and the generated lightwave is divided into two arms by a coupler. The loop configuration ensures the counter-propagation between two arms, and the frequency shifter is used to tune the frequency different. Thus the beat frequency between pump and probe lightwaves is close to the BFS of the FUT. Tunable delay is utilized to shift the correlation peak to be measured. Isolator and circulator limit the optical fiber that the stimulated Brillouin scattering occurs, i.e., the FUT, to ensure one and only one correlation peak exists. The amplified probe lightwave is finally received by the photodetector, and the gain spectrum is obtained by simply scanning the beat frequency of the frequency shifter.

Unlike the time-domain techniques, the measurements of correlation-domain system are continuously processed due to the CW of both two lightwaves and independent on the optical fiber length. For the sensing at one particular position, the measurement period is mainly limited by sweeping time of the frequency shifter in

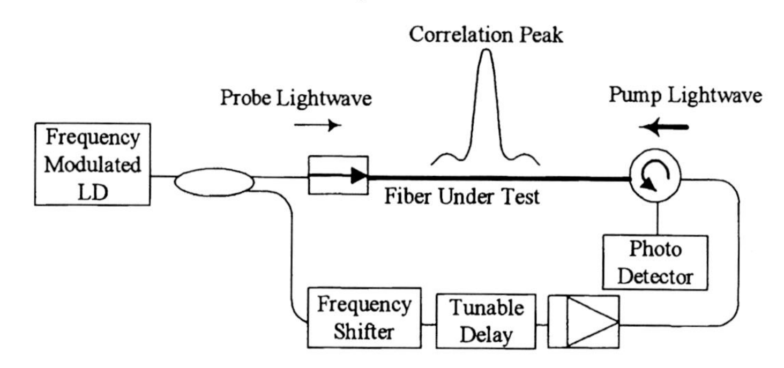

Fig. 5 The basic configuration of the BOCDA technique (After Yamauchi and Hotate (2004). From Mizuno et al. (2008) and Zou et al. (2009a))

order to obtain the BGS, which can be done in milliseconds or even faster. Moreover, thanks to the CW design, the acoustic wave is well maintained and measured Brillouin gain is strong enough. Thus the averaging work is not necessary for a correlation-domain technique. In this way, the Brillouin-based correlation-domain sensors have the ability to implement the dynamic measurement.

BOCDR

Like other Brillouin-based distributed sensing techniques, such as BOTDA and BOTDR, the correlation-domain techniques also include two types depending on whether the spontaneous or stimulated Brillouin scattering is measured. The following introduced BOCDR utilizes the spontaneous Brillouin scattering by the pump lightwave to implement the distributed sensing. The most advantage for BOCDR compared with BOCDA technique is the ability to work with one-end access to the fiber.

Figure 6 is a schematic of a basic BOCDR system (Mizuno et al. 2008). The lightwave generated by the light source is divided into pump and reference lightwaves by an optical coupler. The pump lightwave is launched into the FUT from one end, and the reflected Stokes lightwave by the SpBS is directed into a heterodyne receiver composed of two balanced photodiodes (PDs), while the reference lightwave is used as an optical local oscillator to be processed with the Stokes lightwave. The electrical beat signal of the two lightwaves is monitored by an electrical spectrum analyzer (ESA). Since there is a frequency difference of about 11 GHz between the Stokes light and the reference light, this configuration is called self-heterodyne scheme. Recently, an ultrahigh-speed configuration of BOCDR with a strain sampling rate of up to 100 kHz has been developed (Mizuno et al. 2016).

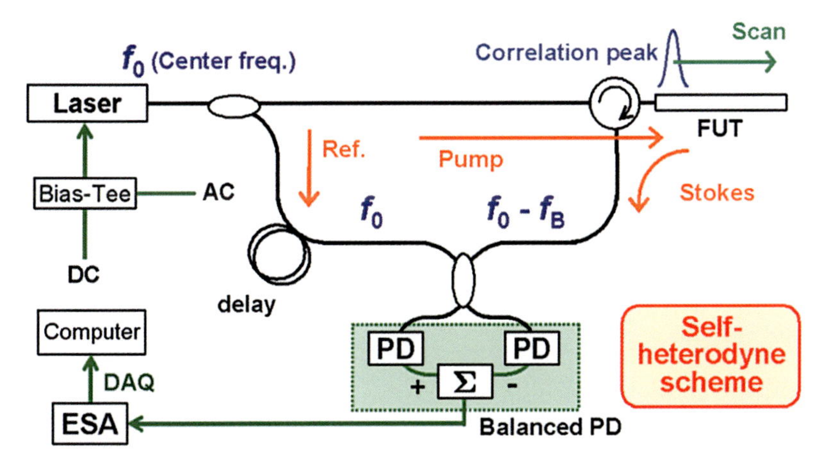

Fig. 6 Conceptual schematic of BOCDR technique (After Mizuno et al. (2008). From Elooz et al. (2014) and Song et al. (2009b))

Unlike the BOCDA system, the measured BGS of the correlation-peak position is directly obtained without frequency sweeping. In order to resolve the position in the FUT, the optical frequency of the light beam from the laser is directly modulated in a sinusoidal wave by modulating the injection current to the laser. From the viewpoint of time averaging, the correlation (or coherence) function is synthesized into a series of periodical peaks, whose period is inversely proportional to the frequency of the sinusoidal modulation f_m, which is similar to the case in BOCDA scheme. By controlling f_m to leave only one correlation peak within the range of the FUT, the Brillouin scattering generated at the position correspondent to the peak has high correlation with the reference lightwave and then gives high heterodyne output. The peak frequency observed in the ESA gives the BFS caused at the position. By sweeping f_m, the correlation peak is scanned along the FUT to obtain the distribution of BGS or BFS.

Figure 7 shows the measured distributed BGS along the FUT by a fabricated basic BOCDR system as an example (Mizuno et al. 2008). In the system, a 100-m-long SMF is utilized as the FUT. The modulation frequency f_m is set as 457.4–458.4 kHz and the frequency deviation Δf is 5.4 GHz. A 0.2-% strain is applied onto a 50-cm-long section along the FUT, while the system's spatial resolution should be 40 cm according to Eq. 13, which is also suitable for BOCDR system. The overall sampling rate of the BGS measurement for a single position was 50 Hz, which is much higher than that of the time-domain techniques (typical measurement time several minutes).As it is shown in the distribution of the BGSs, the BFSs are shifted to around 11 GHz away from the original 10.8 GHz within a small section. The corresponding BFS along the FUT is plotted in Fig. 8. A 50-cm section is found with BFS shifted, which agrees with the experimental setup. The measured strain can be determined by the frequency shift of the BFS and the strain coefficient of the

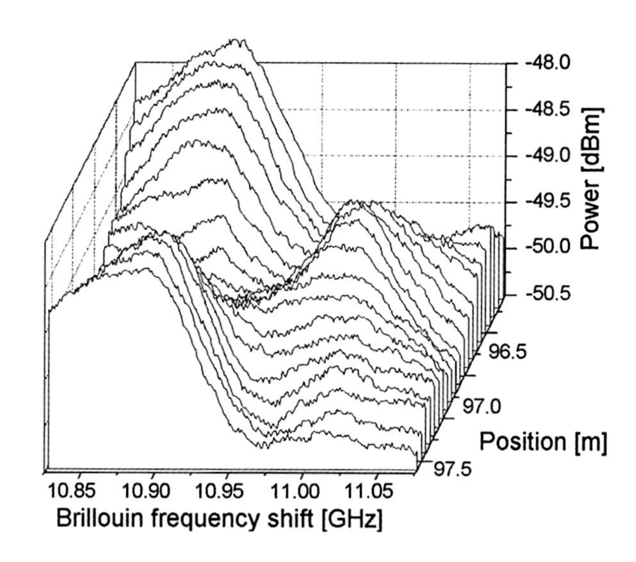

Fig. 7 Distribution of the BGS along the FUT (After Mizuno et al. (2008, 2010))

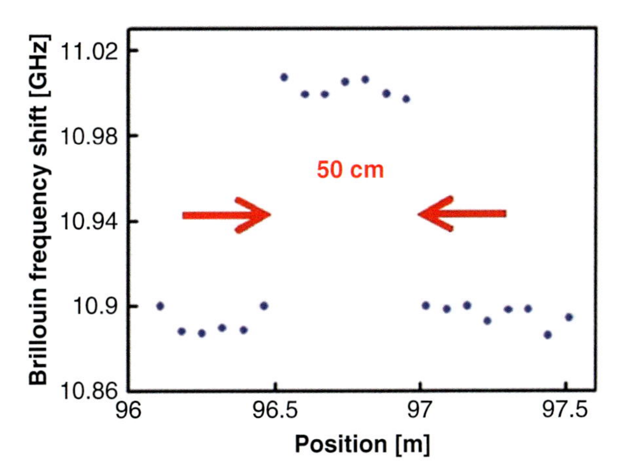

Fig. 8 Distribution of the BFS (peak of the BGS) along the FUT (After Mizuno et al. (2008))

SMF. The change of the BFS was about 100 MHz, which is in good agreement with the applied strain of 0.2%.

Advances in Distributed Brillouin Correlation-Domain Sensing

Effective Sensing Points Enlargement

Despite the high spatial resolution, the basic BOCDA/BOCDR systems can only measure the optical fiber no more than hundreds of meters. The measurement range of the Brillouin-based distributed correlation-domain technique is limited by the "one-peak" condition and given as defined as Eq. 14. According to that, if lower modulation frequency f_m is applied, longer detection range we get. It seems easy to achieve a long sensing range as applying low f_m isn't difficult. However, noted that f_m also affects the spatial resolution as Eq. 13, lower f_m results in a worse sensing resolution. It is meaningless to give up the original incomparable resolution just in order to enlarge the measurement range. Thus this modulation parameter introduces a trade-off problem on the range-resolution issue. Through dividing the measurement range by the spatial resolution, factor f_m is eliminated as

$$N = \frac{d_m}{\Delta z} = \frac{\pi \Delta f}{\Delta v_B} \tag{15}$$

where N is called the effective sensing points.

Meanwhile, the original proposed correlation-domain scheme measures the local BGS along the optical fiber, and all information from the correlation-peak positions will be integrated. A better way to enlarge the measurement range is to allow many correlation peaks along the fiber under test but only detect the gain spectrum from single one of them. Many efforts have been made utilizing this basic principle to enlarge the sensing range, such as double modulations (Mizuno et al. 2010b; Zou et al. 2011), double lock-in amplifiers (Song and Hotate 2006), temporal gating (Kannou et al. 2003; Yamashita et al. 2012; Mizuno et al. 2009b), sensing fiber with different types (Jeong et al. 2011, 2012), and combination with the time-domain technique (Elooz et al. 2014; London et al. 2016; Denisov et al. 2016). In the following several schemes will be introduced.

Double Modulations

The basic idea of using double-modulation scheme is to generate two-pattern correlation-peaks distribution (Mizuno et al. 2010b). In this way, the spatial resolution and measurement rang are decided by different modulation parameters. Figure 9 shows the experimental setup of a BOCDR system based on the double-modulation scheme. Two frequency modulations are applied on the pump and reference lightwaves, respectively. The modulation frequencies are f_0 ($+f_\varepsilon$) and mf_0, where f_0 is a fundamental frequency, m is an integer, and f_ε ($\sim$0.5 kHz) is needed to avoid beating between the two frequencies, which causes large fluctuations of BGS. The amplitude of the frequency modulation at f_0, denoted as Δf_1, is set to be several hundreds of MHz (difficult to measure accurately due to the frequency characteristics of the laser circuit). The amplitude at mf_0, denoted as Δf_m, is about 5.4 GHz (a little lower than a half of BFS in silica fibers). Figure 10 shows the correlation-peaks distributions caused by the above two frequency modulations. For f_0 ($+f_\varepsilon$) case, the correlation peaks have a larger distance from each other, while in mf_0 case, the correlation peaks are of narrow linewidth. As the final measured result

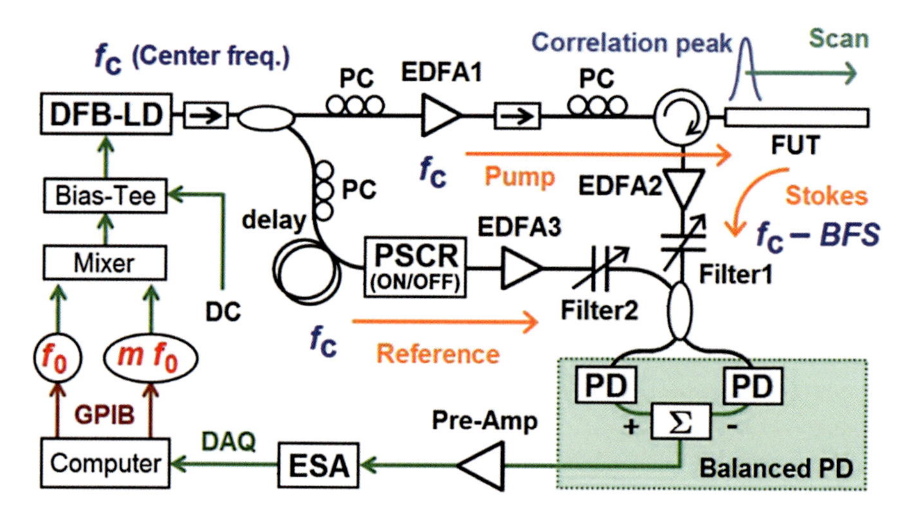

Fig. 9 Experimental setup of the BOCDR technique with double-modulation scheme. DAQ, data acquisition; DC, direct current; EDFA, erbium-doped fiber amplifier; ESA, electrical spectrum analyzer; FUT, fiber under test; GPIB, general-purpose interface bus; PC, polarization controller; PD, photodetector; PSCR, polarization scrambler (After Mizuno et al. (2010b))

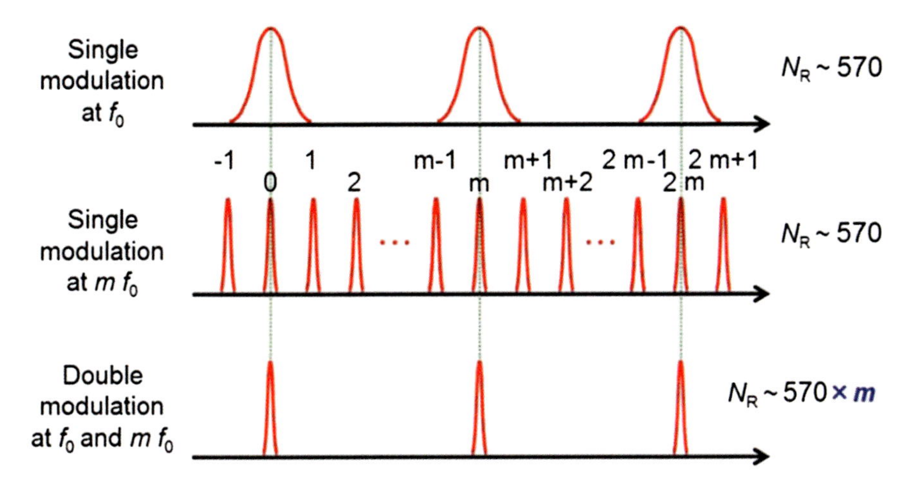

Fig. 10 Operating principle of double-modulation scheme (After Mizuno et al. (2010b))

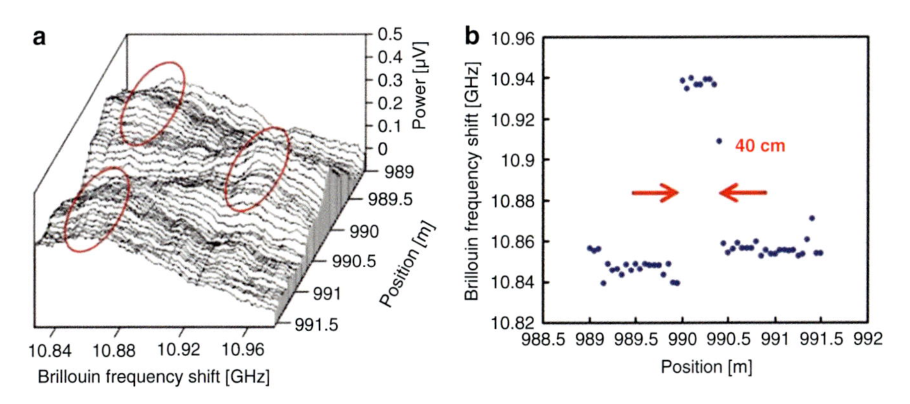

Fig. 11 Measured distributions of (**a**) BGS and (**b**) BFS with the double-modulation scheme when the noise-floor compensation technique was employed ($N = 5690$) (After Mizuno et al. (2010b))

comes from the correlation-peaks distribution combined by these two cases, high spatial resolution and long detection range can be simultaneously achieved. The N is increased by m times according to the theoretical analysis.

Figure 11 gives the experimental results obtained by the double-modulation scheme with f_0 and $10f_0$ ($\Delta f_{10} = 5.4$ GHz) using a noise-floor compensation technique (Mizuno et al. 2009c). A strain of 0.15% was applied to a 40-cm section (990.0–990.4 m). With f_0 swept from 68.362 kHz to 68.474 kHz, the spatial resolution and the measurement range were 26.5 cm and 1.51 km, respectively, corresponding to N of 5690. The measured results are shown in Figs. 11a, b. The strain-applied 40-cm section was successfully detected. The measurement accuracy was about +/− 15 MHz, corresponding to about +/− 260 $\mu\varepsilon$.

Double Lock-in Amplifiers

According to Eq. 15, a direct way to increase the effective points N is to apply the frequency modulation with larger Δf (Song and Hotate 2006). However, it has been practically limited to around 5 GHz, less than the half of the ν_B of the fiber, over which the spectrum of the pump and the probe waves starts to overlap. If this condition is not satisfied, the backward reflection of the pump lightwave acts as a terrible noise for probe lightwave due to the indistinguishable spectrum. Besides, in case of using a typical EOM to generate a stable frequency difference between the pump and the probe waves, the system additionally suffers the reduction of signal if $\Delta f > \nu_B$, where the overlap of the spectrum starts between the two sidebands from the EOM. Since they give opposite effects (gain and loss) on the BGS measurement, this condition results in a strong decrease of probe signal by the counterbalancing of Brillouin gain and loss.

In order to avoid the unwanted overlaps, the reflection of the pump lightwave must be filtered out from the probe lightwave. A usual way is to utilize an optical filter before the detection (see Fig. 12a). However, this method cannot distinguish the reflection component whose frequency is within the probe's spectrum. Figure 12b demonstrates a schematic of the double lock-in detection, while the previous single lock-in configuration is shown in Fig. 12a. In addition to the lock-in detection with the pump wave, another lock-in detection is applied to the probe wave at different lock-in frequency, which removes the backward reflection of pump waves from the probe signal without using any optical filter. Therefore, the double lock-in detection works effectively even in the situation when no optical filtering is available due to a large modulation ($\Delta f > \nu_B/2$) of the laser source.

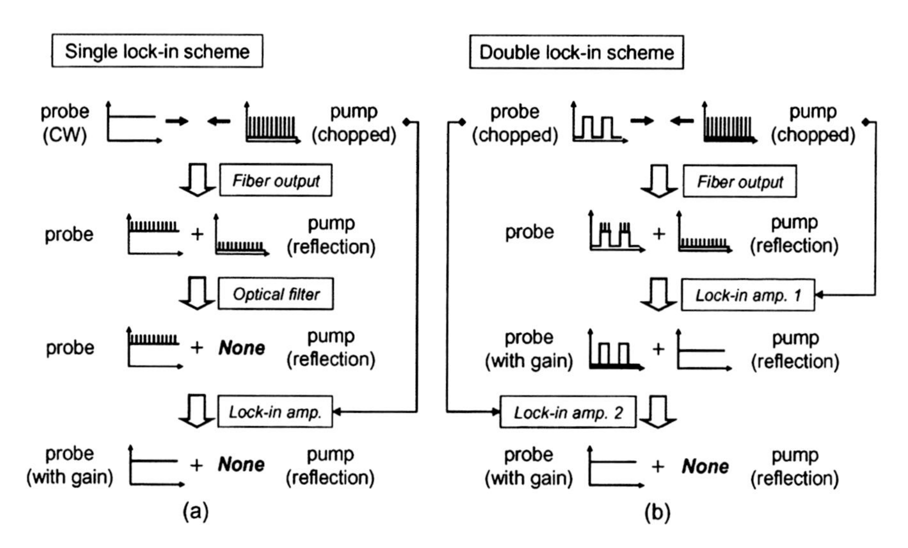

Fig. 12 Measurement schemes for the BOCDA system using (**a**) single lock-in amplifier and (**b**) double lock-in amplifiers. Note that the optical filter in (**a**) is not available if the laser modulation amplitude is larger than $\nu_B/2$ (After Song and Hotate (2006))

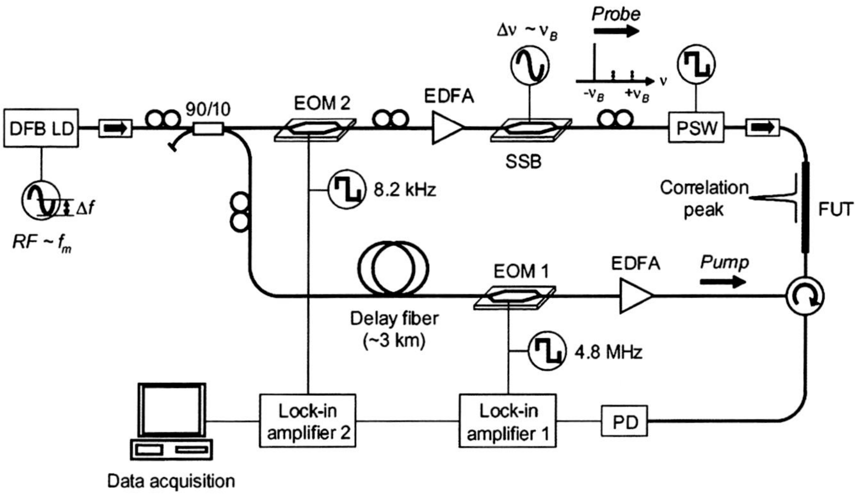

Fig. 13 Experimental setup for the BOCDA system using double lock-in amplifiers and an SSB modulator. PSW, polarization switch (After Song and Hotate (2006))

The experimental setup is depicted in Fig. 13. Based on the original BOCDA scheme with single lock-in amplifier, the probe branch is also chopped using an additional EOM, and the second lock-in amplifier is used to filter out the reflection of the pump lightwave. The modulation frequency f_m of the laser was 310–320 kHz depending on the position of the correlation peak in the fiber, which corresponds to the measurement range of more than 320 m according to Eq. 14. The amplitude of the frequency modulation was 15.5 GHz which was larger than ν_B of conventional fibers ($\sim$10.8 GHz), and the spatial resolution of the measurement was calculated to be about 20 cm from Eq. 13. The N of sensing points is about 1500 according to Eq. 15.

Experimental results are shown in Fig. 14 with double lock-in amplifiers comparing with the results from single lock-in amplifier scheme. BGSs measured for two positions within SMF (upper) and DSF (lower) are demonstrated in Fig. 14a. It is clear that for the single lock-in detection case, the measured BGS is seriously affected by the noise with a large background dc component, which can be attributed to the backward reflection of the pump lightwave. Especially in the SMF, the wanted BGS is buried in the noise and the BFS cannot be determined, while the double lock-in detection scheme can effectively suppress the reflection noise and the BFS is more easily valued. Figure 14b shows the plot of the Brillouin peak frequencies around one of the DSF sections for the single lock-in (dashed line) and double lock-in (solid line) configurations. A large fluctuation of peak frequencies is observed in the case of single lock-in detection due to the noises. These results confirm the effectiveness of the double lock-in detection in a long range measurement.

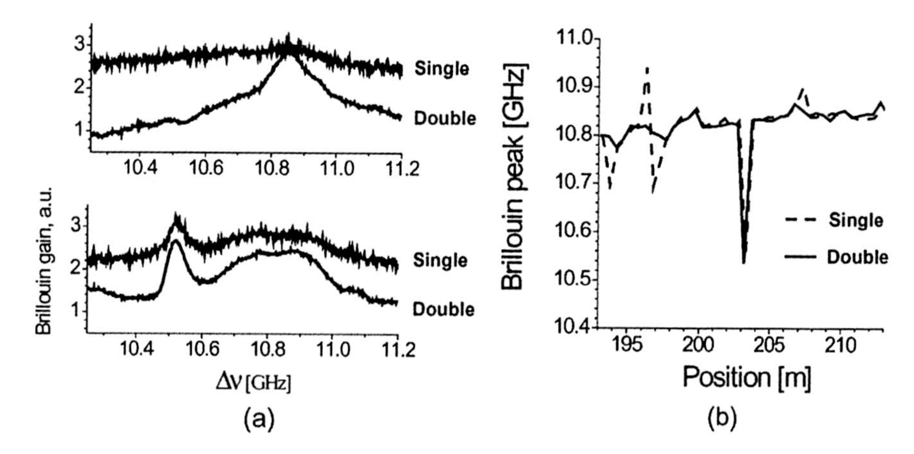

Fig. 14 Comparison of the data in the case of the single and the double lock-in detections. (**a**) The BGS graphs corresponding to the SMF (upper) and the DSF (lower) sections of the FUT. (**b**) Measured Brillouin peak frequencies around the DSF section of the FUT in the single (dashed line) and the double (solid line) lock-in systems. (After Song and Hotate (2006))

Combination with Time-Domain Technique

The effective points N defined in Eq. 15 is obtained from the definition of Eqs. 13 and 14, which give the original spatial resolution and measurement range, respectively (Elooz et al. 2014). When f_m is large, the distance between the correlation peaks is small and the detection range is limited. Otherwise the measured BGS will be the integration of several BGSs of correlation peaks distributed along the FUT. So if there is a way to distinguish the BGS from each correlation peak, the measurement range of the system is no longer limited by Eq. 14, and thus a large N can be achieved. Luckily, the time-domain techniques happen to have the ability to translate the distribution information along the FUT, and thus the combination of these two techniques is raised.

Denote the complex envelopes of the pump and probe lightwaves as $A_p(z,t)$ and $A_s(z,t)$, respectively, where z denotes position along a fiber of length L and t represents time. The pump wave enters the fiber at $z = 0$ and propagates in the positive z direction, whereas the signal wave propagates from $z = L$ in the negative z direction. In the proposed scheme, the signal envelope at its point of entry into the fiber is modulated by a phase sequence c_n with a symbol duration T that is much shorter than the acoustic lifetime:

$$A_s\,(z = L, t) = a_s \sum_n c_n rect\left[\frac{t - nT}{T}\right] \equiv A_{s0}(t) \tag{16}$$

$$A_p\,(z = 0, t) = a_p rect\left(\frac{t}{\theta}\right) \sum_n c_n rect\left[\frac{t - nT}{T}\right] \equiv A_{p0}(t) \tag{17}$$

where a_p and a_s are the constant magnitudes of the pump and probe lightwaves and θ is the duration of the pump amplitude pulse.

The magnitude of the acoustic field at a given location and time is given by

$$Q(z,t) = jg_1 \int_0^t \exp\left[-\Gamma_A (t-a)\right] A_p \left(a - \frac{z}{v_g}\right) A_s{}^* \left[a - \frac{z}{v_g} - \Delta(z)\right] da \tag{18}$$

where g_1 is a parameter which depends on the electrostrictive coefficient, the speed of sound, and the density of the fiber, vg is the group velocity of light in the fiber, and the position-dependent temporal offset Δz is defined as $\Delta z = (2z\text{–}L)/v_g$.

The distribution and variation of the acoustic wave along the FUT are important for the understanding and analysis of a Brillouin-based system. Equation 18 can be solved with simple waveforms given for pump and probe lightwaves, or it can be just numerically integrated, subject to the boundary conditions of Eq. 16 and Eq. 17. Under the conditions (including the modulation of the pump and probe lightwaves) described above, the magnitude of the acoustic wave over a 6-m-long fiber is illustrated in Fig. 15. The BFSs along the FUT are the same, and the frequency difference between the pump and probe is chosen to match this value. A perfect Golomb code ($N = 127$, $T = 200$ ps) is used in the phase modulation of both pump and probe, and a 26 ns-long amplitude pulse was superimposed on the

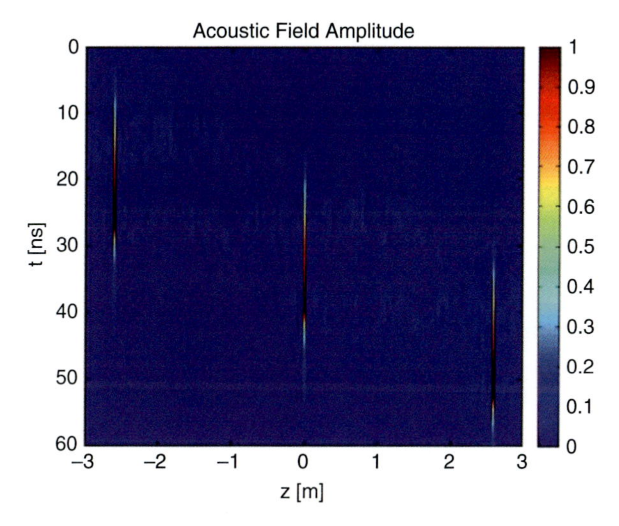

Fig. 15 Simulated magnitude of the acoustic wave density fluctuations (in normalized units), as a function of position and time along a 6-m-long fiber section. Both pump and signal waves are co-modulated by a perfect Golomb phase code that is 127 bits long, with symbol duration of 200 ps. The pump wave was further modulated by a single amplitude pulse of 26 ns duration. The acoustic field, and hence the SBS interaction between pump and signal, is confined to discrete and periodic narrow correlation peaks. The peaks are built up sequentially one after another with no temporal overlap (After Elooz et al. (2014))

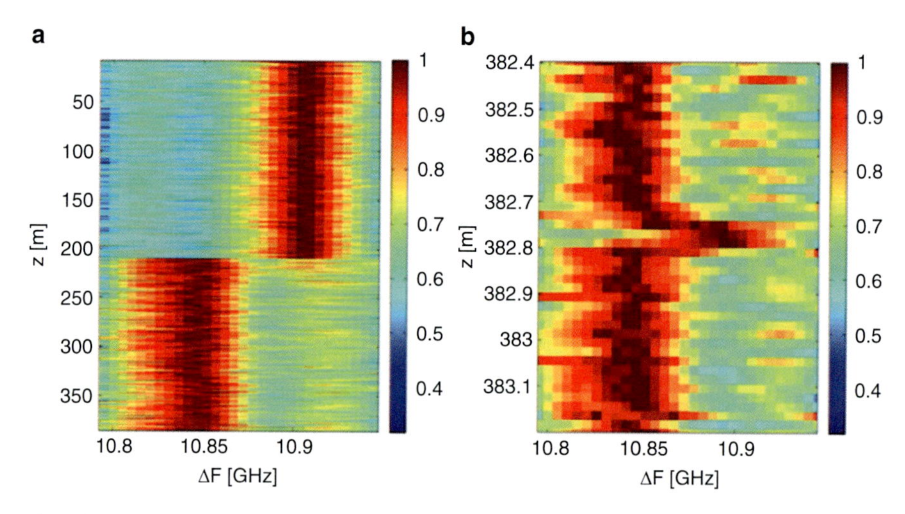

Fig. 16 Measured Brillouin gain map as a function of frequency offset between pump and signal and position along a 400-m-long fiber under test. The fiber consisted of two sections, each approximately 200 m long, with Brillouin shifts at room temperature of approximately 10.84 GHz and 10.90 GHz, respectively. A 5-cm-long hot spot was located toward the output end of the pump wave. The map was reconstructed using only 127 scans per frequency offset, according to the combined BOTDA/BOCDA method. The complete map is shown on panel (**a**), and a zoom-in on the hot spot region is shown on panel (**b**) (After Elooz et al. (2014))

phase-modulated pump wave. As it is shown, three correlation peaks are generated along the fiber with same distance, which agrees with the basic theory of a BOCDA scheme. However, these correlation peaks do not last for a long period; instead they only exist for no more than 20 ns. It should be noted that the "moving" direction of the probe lightwave is from the right-top to the left-button in a z-t distribution map. The Brillouin gain comes from the gathering of the interaction between the pump and acoustic wave all along the probe lightwave's path. In this way, the detected probe gain can distinguish the information from different correlation peaks due to their short existing duration, resulting in the allowance for the multiple correlation peaks in a correlation-domain technique.

The experimental measured BGS distribution along the FUT is given in Fig. 16. Measurement is implemented over a 400-m-long fiber with a 5-cm-long hot spot section. The spatial resolution is 2 cm due to the theory, and the hot spot section is clearly detected (see Fig. 16b), which means that the effective point N reaches 20,000.

Noise Suppression

As it is discussed in section "Principle of Brillouin-Based Distributed Sensing," the Brillouin-based distributed sensors include two matters: sensing of the location and sensing of the measurands. The effective point N associated with the measurement range and spatial resolution is the key performance on the location sensing, answering the question that how long and how precise we can measure the distributed information along the FUT. The other problem is that how accurate and how large we can determine the environment parameters (i.e., BFS) which change at specific position along the fiber. In usually, the BFS is determined by the frequency with biggest Brillouin gain in a BGS. However, it is difficult to achieve a high accuracy, and even sometimes the BFS cannot be determined from the BGS due to lots of background noises such as the reflection of the pump lightwave. A simple way to suppress the background noise in Brillouin-based distributed sensing applications is the use of Brillouin gain and loss effects (Zou et al. 2012). In this modified scheme, a dual-parallel Mach–Zehnder modulator is sufficient to generate both the Brillouin gain and loss effect and the SNR is improved by 3 dB. Besides, efforts have been paid to focus on the noise suppression in the Brillouin-based correlation-domain sensors. In the following, we will introduce the suppression of these noises to enhance the sensing performance based on the intensity modulation.

As it is illustrated in Fig. 17a, the measured signal in a basic BOCDA system is the integration of the local BGS distributed along the optical fiber. As the acoustic wave is only maintained within the correlation peaks, the detected gain spectrum

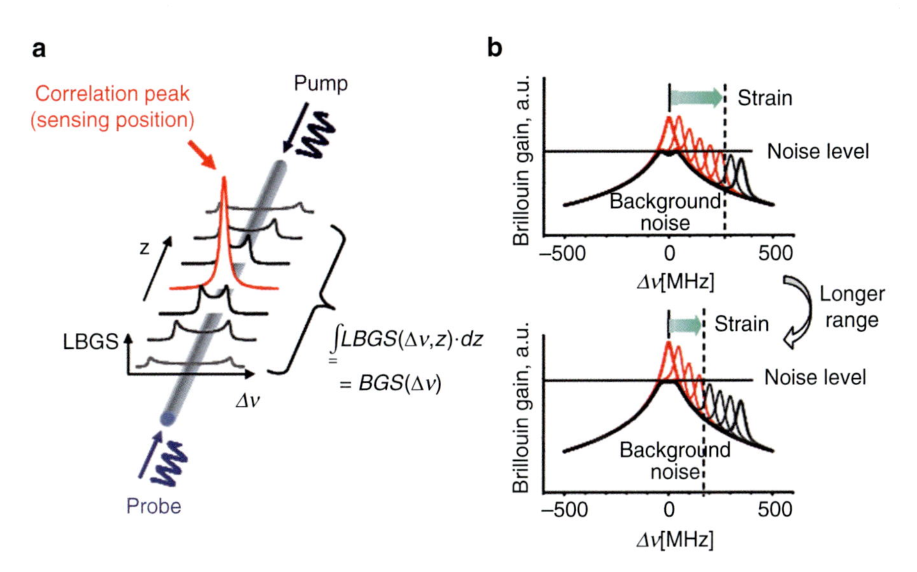

Fig. 17 (**a**) Schematic of a Brillouin optical correlation-domain analysis (BOCDA) system. Measured Brillouin gain spectrum (BGS) is the sum of local BGS's (LBGS); $\Delta\nu$, frequency offset between pump and probe waves. (**b**) Variation of the BGS in response to the applied strain to the sensing (correlation peak) position. Note that the maximum measurable strain (dashed line) is limited by the peak of the background noise and that the measurable strain limit is decreased in longer measurement range (lower) than the shorter case (upper). $\Delta\nu$ is the relative frequency offset with the initial value set to zero (After Song et al. (2006b))

is considered mainly contributed by the BGS at these positions. However, when the measurement range is enlarged, the BGSs of the non-correlation-peak positions tend to stack higher background noise for the final measured signal, which restricts the maximum measurable strain and the measurement range of the BOCDA system by leading to the failure of sensing over certain limits (see Fig. 17b).

For the purpose of suppression and the modification of the background noise of the BGS, intensity modulation (IM) can be used to modify the optical spectra of pump and probe lightwaves. Three different intensity modulation waveforms are used to generate different optical power spectra as shown in Fig. 18a. Each spectral shape can be characterized by different power distribution between the side and the center of the initial spectrum (No IM) of the sinusoidal frequency modulation. The calculated waveform was applied to the IM using an arbitrary function generator which was synchronized to the frequency modulation as depicted in Fig. 18b. The other optical spectra (IM 2, IM 3) were produced by manipulating the offset and the amplitude of the modulation waveform of the transmittance used for the IM 1.

Figure 19 shows the measured BGS's in different modulation schemes. As is clearly seen in the case of no intensity modulation (No IM), the real signal from the DSF section is lower than the noise peak, so cannot be detected properly in this condition. When the IMs are applied, strong suppression of the noise peak is observed on the DSF sections compared to the signal amplitude in all cases (IM 1~3). At the same time, a large dip is observed at the center of the BGS in the cases of IM 1 and IM 2, which might be the effect of over-suppression. This feature gives a problem in the peak detection of normal position by "absorbing" the signal as shown in the BGSs of their SMF sections.

Experiments have been carried out to verify the scheme. Distributed measurements with and without the intensity modulation by 10-cm step on the FUT using

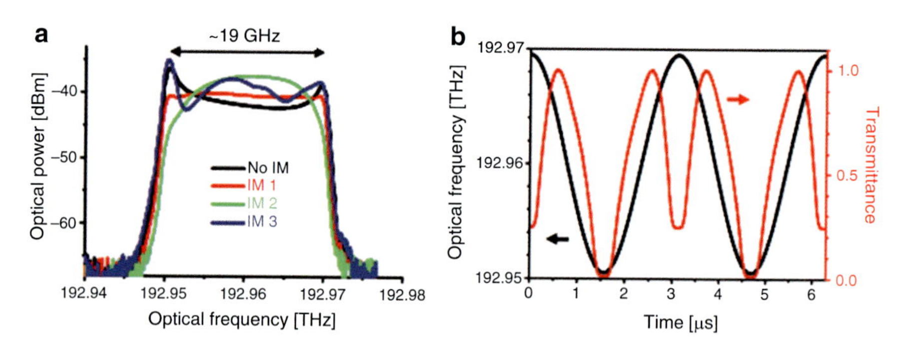

Fig. 18 (**a**) Power spectra measured by an optical spectrum analyzer with intensity modulation Schemes (IM 1~3) applied in addition to the initial frequency modulation (No IM). (**b**) Time waveforms showing the synchronization between the frequency modulation of the LD (black) and the transmittance of the intensity modulator (red) applied to generate a flat-top spectrum (IM 1) shown in (**a**). Note that the other waveforms (IM 2~3) are synchronized in the same way (After Song et al. (2006b))

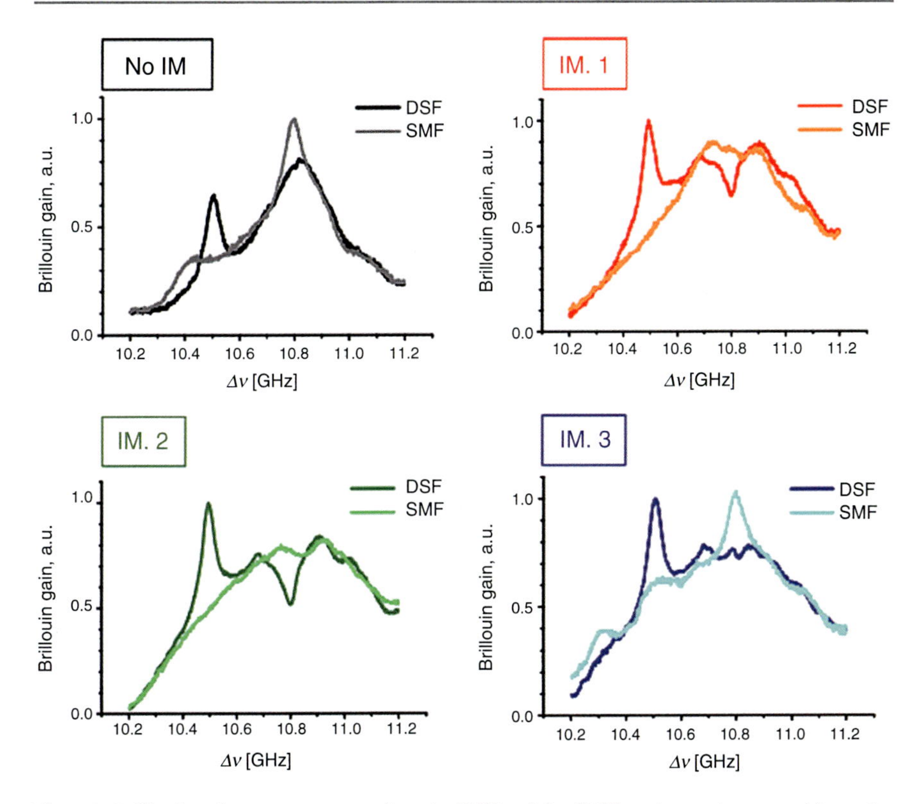

Fig. 19 Brillouin gain spectra measured on the DSF and the SMF sections using several intensity modulation schemes shown in Fig. 18 (After Song et al. (2006b))

the same experimental parameters are performed. The distributed BFS along the FUT is given in Fig. 20. As it can be seen, the BFS change within the small section of the DSF is detected by measurement using IM 3.

Strain and Temperature Discrimination

As explained in section "Sensing of Measurands," the sensing of the measurands are all dependent on the measurement of the localized BFS. This common mechanism results in a heavy trouble for any Brillouin-based sensors in discriminating the response to strain from that to temperature by using a single piece of fiber (Song et al. 2008). In current practices, two fibers are used to discriminate the strain and the temperature: the first one is embedded or bonded at the target material/structure to feel the total effects of strain and temperature, while the second fiber is placed beside the first one and kept in loose condition so that it feels the effect of temperature only; then the strain and the temperature can be calculated by mathematics. In the following, a new method based on the Brillouin dynamic gratings (BDG) for

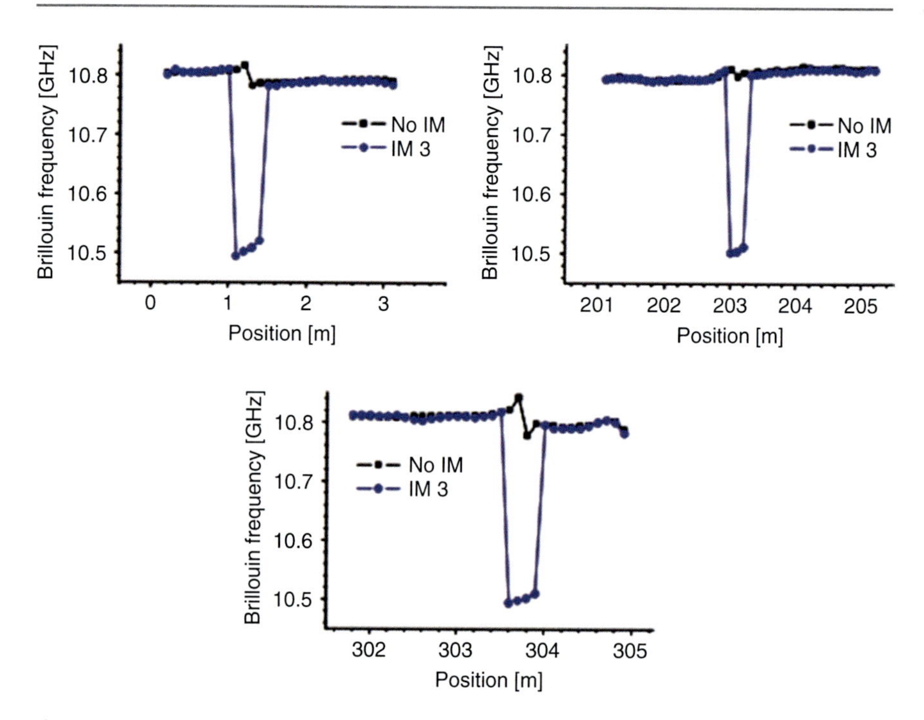

Fig. 20 Result of distributed measurement on the fiber under test near the DSF sections. The DSF sections are properly detected only in the optimum intensity modulation (IM 3). The measurement inaccuracy of the ν_B at each position was about +/−3 MHz (After Song et al. (2006b))

complete discrimination of strain and temperature by use of only one piece of Panda-type polarization-maintaining fiber (PMF) is demonstrated.

Principle of the BDG

As given in Eq. 3, the BGS is determined by the effective refractive index in an optical fiber. While in a PMF (or any medium with birefringence), optical waves with two principal polarizations (i.e., x and y polarization) experience different ν_B owing to their different refractive indexes. Considering that the acoustic wave generated by SBS is a longitudinal one that is free of the transversal polarization, an interesting condition can be reached that the x- and the y-polarized optical waves in a PMF show the same ν_B at different wavelengths. When the dispersion of the acoustic wave is ignored, the condition is expressed by following equations:

$$n_x v_x = n_y v_y \tag{19}$$

where $n_{x,y}$ and $v_{x,y}$ are the refractive indexes and the optical frequencies in x and y polarizations, respectively.

Since the SBS-induced acoustic waves can be viewed as moving gratings for the reflection of the pump wave without polarization dependence, it is expected that acoustic waves generated by SBS between the x-polarized pump and Stokes waves at the optical frequency v_x will show strong reflectance to the y-polarized pump wave at the frequency of v_y. Considering that the intensity and the wavelength of the acoustic waves are easily tuned by controlling the x-polarized "writing" beams, one may expect the SBS in a PMF to play a role of a tunable dynamic grating.

For the writing of the dynamic grating, a 1550-nm laser diode was used as a light source, and the output power was divided by a 50/50 coupler. A SSBM and a microwave synthesizer were used to generate the Stokes wave (pump2) of the writing beams, and the output was amplified and polarized by an EDFA and an x polarizer. The Brillouin pump wave (pump1) of the writing beams was prepared by amplifying the original wave with the same polarization as that of pump2. Pump1 and the pump2 were launched into a PMF in opposite direction to each other through polarization beam combiners (PBC1, PBC2). For a reading beam (probe), a tunable laser with an operating wavelength near 1550 nm was used as a light source after being polarized in the y-axis. The output was launched into the PMF in the direction of the pump1 through a polarization-maintaining circulator and PBC1. The transmitted power of the probe was measured using a power meter, and the backreflected spectrum was monitored using an OSA through a y polarizer (see Fig. 21).

For the detection of the dynamic grating, the frequency of the probe was tuned at the higher frequency region while monitoring the spectrum with the OSA, and the result is shown in Fig. 22. When Δv (the frequency difference between pump1 and the probe) was ~72.6 GHz, a large reflection of the probe was observed (black curve) as a result of the dynamic grating at the frequency detuned from the probe by the same amount as that between pump1 and pump2. When one of the pumps (pump1) was turned off, the dynamic grating disappeared as depicted by the gray curve although the probe was still propagated as confirmed by the Rayleigh scattering seen at the probe frequency. In both cases, the x-polarized pumps were

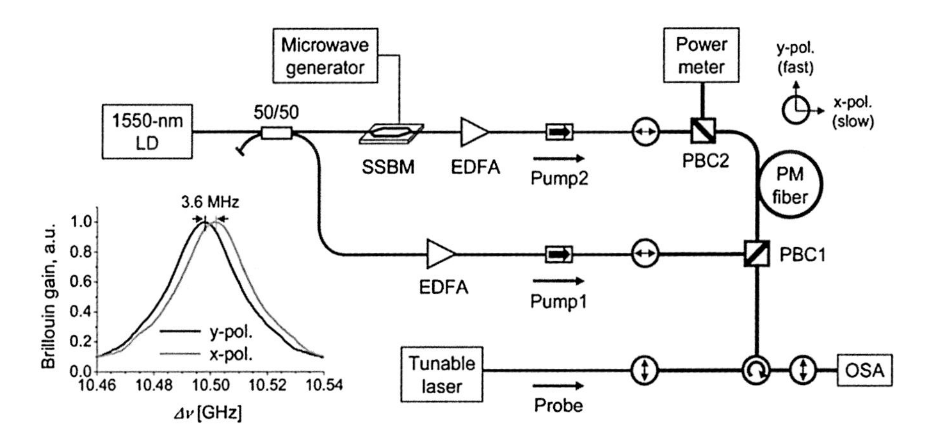

Fig. 21 Experimental setup (After Song et al. (2008))

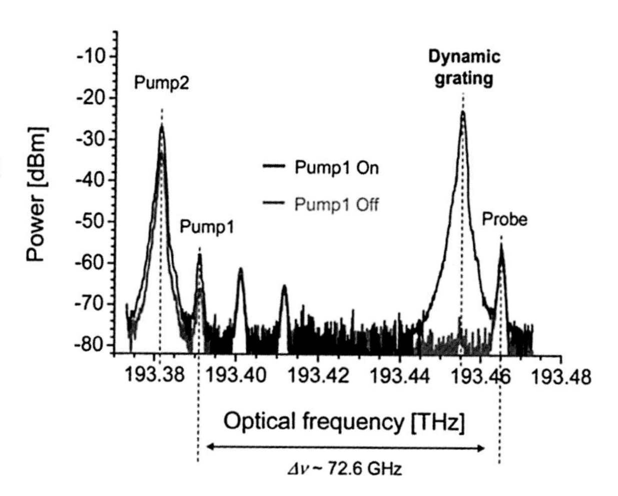

Fig. 22 Optical spectra monitored by an OSA in the generation of dynamic rating. The gray curve corresponds to the case that one (pump1) of the writing beams is turned off, and the black curve with both writing beams turned on (After Song et al. (2008)))

observed in spite of the use of the y polarizer in front of the OSA, which originated from the finite extinction ratio (~20 dB) of the polarizing components. The small peaks near pump1 correspond to the first- and second-order anti-Stokes waves that were suppressed in the SSBM used for the generation of pump2.

The principle of the completely discrimination of the strain and temperature is that the coefficients of the BFS changed by the strain and temperature are strongly independent. Figure 23 shows the experimental results as the evidence. It demonstrates that for the strain variation, the BFS and the frequency deviation $\Delta\nu$ behave similarly, while for the temperature change, they react in opposite direction. All the dependences show excellent linearity; thus by linear fitting, we get the strain coefficient and the temperature coefficient of $C\nu\varepsilon = +0.03938$ MHz/με and $C\nu T = +1.0580$ MHz/°C for the BFS ν_B and $Cf\varepsilon = +0.8995$ MHz/με and $CfT = -55.8134$ MHz/°C for the frequency deviation f_{yx}, respectively.

Combination with BDG Technique

In the following, we show that the dynamic grating can be localized in an arbitrary position along the PMF by using a correlation-based CW technique (Zou et al. 2009b). A distributed measurement of the dynamic grating spectrum (DGS) is demonstrated with 1.2-m spatial resolution and 110-m measurement range. In experiment, temperature-induced changes in both the BFS of BGS and the frequency deviation $\Delta\nu$ of DGS are measured in heated segments cascaded along a 110-m PMF.

Figure 24 shows the experimental setup for distributed generation and detection of the DGS in a ~110-m-long PMF. The output from a 1549 nm distributed feedback laser diode (DFB-LD1) is equally divided into probe and pump beams. The probe beam is prepared by downshifting its frequency using a SSBM and a microwave synthesizer. It is amplified via an EDFA and launched into the PMF after passing through an x polarizer and a polarization-maintaining isolator (PM-ISO). The pump beam is chopped by an EOM and amplified by a high-power EDFA, which is

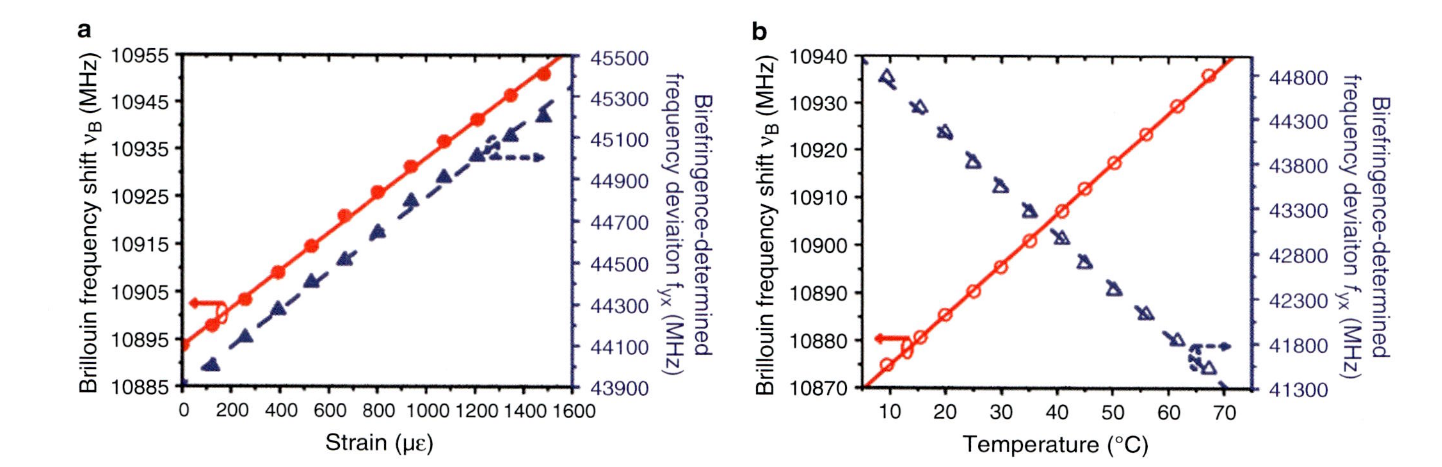

Fig. 23 Measured strain and temperature coefficients. (**a**) Strain dependence. (**b**) Temperature dependence. Circles denote the experimental results for Brillouin frequency shift (νB) in left vertical axes, and triangles correspond to the birefringence-determined frequency deviation ($\Delta\nu$) in right vertical axes, respectively (After Zou et al. (2009a))

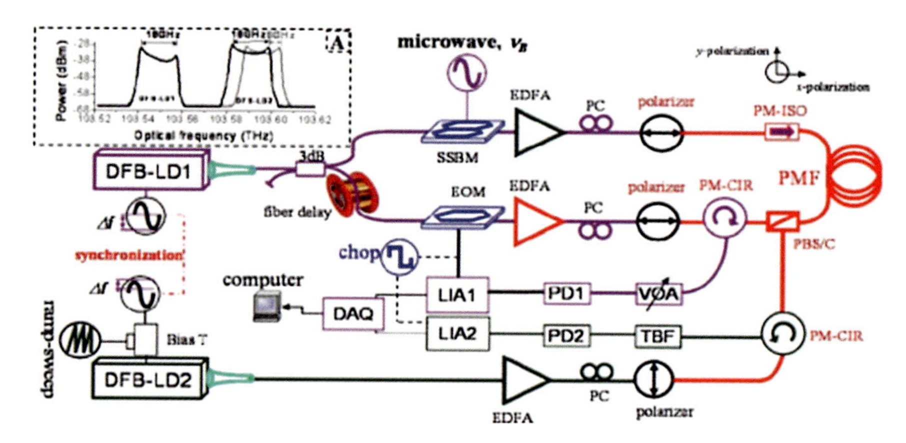

Fig. 24 (Color online) Experimental configuration for distributed generating and measuring the BGS and DGS in a PMF. The abbreviations are explained in the text body. Inset A depicts the optical power spectra of two modulated DFB-LDs both with $2\Delta f = 18$ GHz, where the gray curve has 6 GHz offset from the black curve corresponding to a reduced dc injection current to DFB-LD2 (After Zou et al. (2009b))

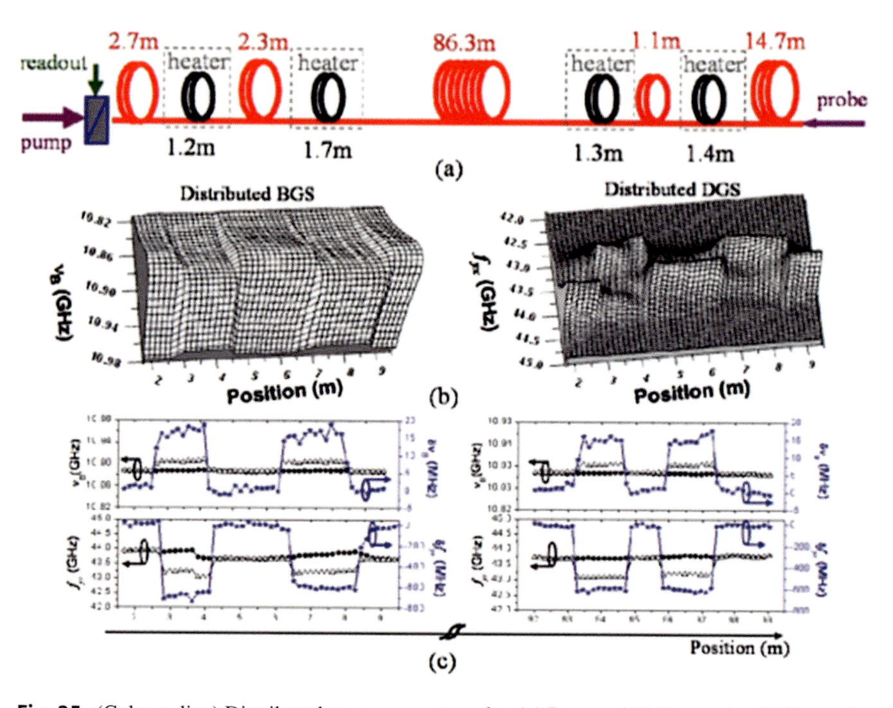

Fig. 25 (Color online) Distributed measurement results. (**a**) Prepared PMF sample. (**b**) Examples of 3D distribution of measured BGS and DGS from 2 m to 9 m when the heater is turned on. (**c**) Summarization of detected BFS B (upper) or frequency deviation fyx (lower) and their temperature-induced changes near the heated segments. Solid-circle line (open-triangle line) corresponds to the heater being turned off (on); solid-star line indicates the temperature-induced changes between the heater's on and off states (After Zou et al. (2009b))

launched into the other end of the PMF after passing through an x polarizer, a PM circulator (PM-CIR), and a PM beam splitter–combiner (PBS–C). Used as the readout beam, the output from the second laser diode (DFB-LD2) is launched into the PMF after passing through a y polarizer and the PBS–C to readout the SBS-generated DGS. The frequency of the readout beam is deviated by $\Delta\nu$ from the pump beam by tuning the DFB-LD2's DC injection current.

The distributed measurement ability is verified as demonstrated in Fig. 25. Four heated segments cascaded along the PMF sample are prepared (see Fig. 25a). For the distributed measurement, the modulation frequency f_m is scanned from 860 kHz to 930 kHz with a step of ~100 Hz corresponding to 16 cm; the microwave frequency to SSBM and the y-polarized carrier frequency are ramp swept for the characterization of BGS and DGS, respectively. As an example, the detected distribution of BGS and DGS from ~2 m to ~9 m is illustrated in Fig. 25b. The initial DGS is not uniform along the fiber owing to the irregularity of the local birefringence introduced during the fiber fabrication. Figure 25c summarizes the measured ν_B and $\Delta\nu$ along the fiber sample when the heater (a hot plate) with ΔT = ~15 °C is turned off (solid-circle line) or turned on (open-triangle line); their differences between the heater's on–off states are also shown (solid-star line). The opposite responses of the ν_B and $\Delta\nu$ to ΔT can be clearly observed.

Conclusion

We have introduced the principle of the Brillouin-based correlation-domain distributed sensing techniques. The basic ideas for BOCDA/BOCDR are to establish the correlation peaks with strong and stable acoustic wave and then to measure the gain spectrum from these correlation-peak positions. High spatial resolution and the ability to achieve dynamic measurement are the main advantages for correlation-domain techniques compared with the traditional time-domain techniques. Works to enlarge the measurement range and to suppress the background noise have been demonstrated.

References

G.P. Agrawal, *Nonlinear Fiber Optics*, 5th edn. (Academic, New York, 2012)

X. Bao, D.J. Webb, D.A. Jackson, 32-km distributed temperature sensor based on Brillouin loss in an optical fiber. Opt. Lett. **18**, 1561–1563 (1993)

A. Denisov, M.A. Soto, L. Thévenaz, Going beyond 1000000 resolved points in a Brillouin distributed fiber sensor: theoretical analysis and experimental demonstration. Light: Sci. Appl. **5**, e16074 (2016)

D. Elooz, Y. Antman, N. Levanon, A. Zadok, High-resolution long-reach distributed Brillouin sensing based on combined time-domain and correlation-domain analysis. Opt. Express **22**, 6453–6463 (2014)

A.L. Gaeta, R.W. Boyd, Stochastic dynamics of stimulated Brillouin scattering in an optical fiber. Phys. Rev. A **44**, 3205–3209 (1991)

D. Garcus, T. Gogolla, K. Krebber, F. Schliep, Brillouin optical-fiber frequency-domain analysis for distributed temperature and strain measurements. J. Lightwave Technol. **15**, 654–662 (1997)

D. Garus, K. Krebber, F. Schliep, T. Gogolla, Distributed sensing technique based on Brillouin optical-fiber frequency-domain analysis. Opt. Lett. **21**, 1402–1404 (1996)

T. Horiguchi, M. Tateda, BOTDA-nondestructive measurement of single-mode optical fiber attenuation characteristics using Brillouin interaction: Theory. J. Lightwave Technol. **7**, 1170–1176 (1989)

T. Horiguchi, T. Kurashima, M. Tateda, Tensile strain dependence of Brillouin frequency shift in silica optical fibers. IEEE Photon. Technol. Lett. **1**, 107–108 (1989)

T. Horiguchi, T. Kurashima, M. Tateda, A technique to measure distributed strain in optical fibers. IEEE Photon. Technol. Lett. **2**, 352–354 (1990)

T. Horiguchi, T. Kurashima, Y. Koyamada, Measurement of temperature and strain distribution by Brillouin frequency shift in silica optical fibers, in *Fibers' 92*, (International Society for Optics and Photonics, 1993), pp. 2–13

K. Hotate, T. Hasegawa, Measurement of Brillouin gain Spectrum distribution along an optical fiber using a correlation-based technique–proposal, experiment and simulation. IEICE Trans. Electron. **83**, 405–412 (2000)

K. Hotate, S.S. Ong, Distributed fiber Brillouin strain sensing by correlation-based continuous-wave technique~ cm-order spatial resolution and dynamic strain measurement. Proc. SPIE **4920**, 299–310 (2002)

K. Hotate, M. Tanaka, Distributed fiber Brillouin strain sensing with 1-cm spatial resolution by correlation-based continuous-wave technique. IEEE Photon. Technol. Lett. **14**, 179–181 (2002)

E.P. Ippen, R.H. Stolen, Stimulated Brillouin scattering in optical fibers. Appl. Phys. Lett. **21**, 539–541 (1972)

R.B. Jenkins, R.M. Sova, R.I. Joseph, Steady-state noise analysis of spontaneous and stimulated Brillouin scattering in optical fibers. J. Lightwave Technol. **25**, 763–770 (2007)

J.H. Jeong, K. Lee, J.M. Jeong, S.B. Lee, Measurement range expansion in Brillouin optical correlation-domain analysis system, *in Proceedings of the International Conference on Lasers and Electro-Optics*, vol. 1 (2011)

J.H. Jeong, K. Lee, J.-M. Jeong, S.B. Lee, Measurement range enlargement in Brillouin optical correlation domain analysis using multiple correlation peaks. J. Opt. Soc. Korea **16**, 210 (2012)

M. Kannou, S. Adachi, K. Hotate, Temporal gating scheme for enlargement of measurement range of Brillouin optical correlation domain analysis for optical fiber distributed strain measurement, *in Proceedings of the 16th International Conference on Optical Fiber Sensors*, vol. 454 (2003)

T. Kurashima, T. Horiguchi, M. Tateda, Thermal effects of Brillouin gain spectra in single-mode fibers. IEEE Photon. Technol. Lett. **2**, 718–720 (1990)

T. Kurashima, T. Horiguchi, H. Izumita, S.-i. Furukawa, Y. Koyamada, Brillouin optical-fiber time domain reflectometry. IEICE Trans. Commun. **76**, 382–390 (1993)

Y. London, Y. Antman, E. Preter, N. Levanon, A. Zadok, Brillouin optical correlation domain analysis addressing 440 000 resolution points. J. Lightwave Technol. **34**, 4421–4429 (2016)

Y. Mizuno, W. Zou, Z. He, K. Hotate, Proposal of Brillouin optical correlation-domain reflectometry (BOCDR). Opt. Express **16**, 12148–12153 (2008)

Y. Mizuno, Z. He, K. Hotate, One-end-access high-speed distributed strain measurement with 13-mm spatial resolution based on Brillouin optical correlation-domain reflectometry. IEEE Photon. Technol. Lett. **21**, 474–476 (2009a)

Y. Mizuno, Z. He, K. Hotate, Measurement range enlargement in Brillouin optical correlation-domain reflectometry based on temporal gating scheme. Opt. Express **17**, 9040–9046 (2009b)

Y. Mizuno, Z. He, K. Hotate, Stable entire-length measurement of fiber strain distribution by Brillouin optical correlation-domain reflectometry with polarization scrambling and noise-floor compensation. Appl. Phys. Express **2**, 062403 (2009c)

Y. Mizuno, W.W. Zou, Z.Y. He, K. Hotate, Operation of Brillouin optical correlation domain reflectometry: Theoretical analysis and experimental validation. J. Lightwave Technol. **28**, 3300–3306 (2010a)

Y. Mizuno, Z. He, K. Hotate, Measurement range enlargement in Brillouin optical correlation-domain reflectometry based on double-modulation scheme. Opt. Express **18**, 5926–5933 (2010b)

Y. Mizuno, N. Hayashi, H. Fukuda, K.Y. Song, K. Nakamura, Ultrahigh-speed distributed Brillouin reflectometry. Light Sci. Appl. **5**(12), e16184 (2016)

M. Niklès, L. Thévenaz, P.A. Robert, Simple distributed fiber sensor based on Brillouin gain spectrum analysis. Opt. Lett. **21**, 758–760 (1996)

M. Nikles, L. Thevenaz, P.A. Robert, Brillouin gain spectrum characterization in singlemode optical fibers. J. Lightwave Technol. **15**, 1842–1851 (1997)

A.S. Pine, Brillouin scattering study of acoustic attenuation in fused quartz. Phys. Rev. **185**, 1187–1193 (1969)

R.M. Shelby, M.D. Levenson, P.W. Bayer, Resolved forward Brillouin scattering in optical fibers. Phys. Rev. Lett. **54**, 939 (1985)

K. Shimizu, T. Horiguchi, Y. Koyamada, T. Kurashima, Coherent self-heterodyne Brillouin OTDR for measurement of Brillouin frequency shift distribution in optical fibers. J. Lightwave Technol. **12**, 730–736 (1994)

K.Y. Song, K. Hotate, Enlargement of measurement range in a Brillouin optical correlation domain analysis system using double lock-in amplifiers and a single-sideband modulator, *in Proceedings of the International Conference on OSA/OFC* (2006)

K.-Y. Song, K. Hotate, Brillouin optical correlation domain analysis in linear configuration. IEEE Photon. Technol. Lett. **20**, 2150–2152 (2008)

K.Y. Song, Z.Y. He, K. Hotate, Distributed strain measurement with millimeter-order spatial resolution based on Brillouin optical correlation domain analysis. Opt. Lett. **31**, 2526–2528 (2006a)

K.Y. Song, Z. He, K. Hotate, Optimization of Brillouin optical correlation domain analysis system based on intensity modulation scheme. Opt. Express **14**, 4256–4263 (2006b)

K.Y. Song, W. Zou, Z. He, K. Hotate, All-optical dynamic grating generation based on Brillouin scattering in polarization-maintaining fiber. Opt. Lett. **33**(9), 926–928 (2008)

R.K. Yamashita, W. Zou, Z. He, K. Hotate, Measurement range elongation based on temporal gating in Brillouin optical correlation domain distributed simultaneous sensing of strain and temperature. IEEE Photon. Technol. Lett. **24**, 1006–1008 (2012)

T. Yamauchi, K. Hotate, Performance evaluation of Brillouin optical correlation domain analysis for fiber optic distributed strain sensing by numerical simulation. Optics East. International Society for Optics and Photonics. **5589**, 164–174 (2004)

A. Yeniay, J.-M. Delavaux, J. Toulouse, Spontaneous and stimulated Brillouin scattering gain spectra in optical fibers. J. Lightwave Technol. **20**, 1425 (2002)

W. Zou, Z. He, K. Hotate, Complete discrimination of strain and temperature using Brillouin frequency shift and birefringence in a polarization-maintaining fiber. Opt. Express **17**(3), 1248–1255 (2009a)

W. Zou, Z. He, K.Y. Song, K. Hotate, Correlation-based distributed measurement of a dynamic grating spectrum generated in stimulated Brillouin scattering in a polarization-maintaining optical fiber. Opt. Lett. **34**(7), 1126–1128 (2009b)

W. Zou, Z. He, K. Hotate, Single-end-access correlation-domain distributed fiber optic sensor based on stimulated Brillouin scattering. J. Lightwave Technol. **28**, 2736–2742 (2010)

W. Zou, Z. He, K. Hotate, Enlargement of measurement range by double frequency modulations in one-laser Brillouin correlation domain distributed discrimination system, *in Proceedings of the International Conference on Lasers and Electro-Optics*, vol. 1 (2011)

W. Zou, C. Jin, J. Chen, Distributed strain sensing based on combination of Brillouin gain and loss effects in Brillouin optical correlation domain analysis. Appl. Phys. Express **5**(8), 082503 (2012)

W. Zou, X. Long, J. Chen, Brillouin scattering in optical fibers and its application to distributed sensors, in *Advances in Optical Fiber Technology: Fundamental Optical Phenomena and Applications*, (InTech, Croatia, 2015)

Part X
Optical Fiber Sensors for Industrial Applications

Optical Fiber Sensors for Remote Condition Monitoring of Industrial Structures

46

Tong Sun OBE, M. Fabian, Y. Chen, M. Vidakovic, S. Javdani, K. T. V. Grattan, J. Carlton, C. Gerada, and L. Brun

Contents

T. Sun OBE (✉)
School of Mathematics, Computer Science and Engineering, City, University of London, London, UK
e-mail: t.sun@city.ac.uk

M. Fabian · Y. Chen · M. Vidakovic · S. Javdani · K. T. V. Grattan · J. Carlton
City, University of London, London, UK

C. Gerada
The University of Nottingham, Nottingham, UK

L. Brun
Faiveley Brecknell Willis, Somerset, UK

G.-D. Peng (ed.), *Handbook of Optical Fibers*,
https://doi.org/10.1007/978-981-10-7087-7_19

Abstract

Optical fibers have been explored widely for their sensing capability to meet increasing industrial needs, building on their success in telecommunications. This chapter provides a review of research activities at City, University of London in response to industrial challenges through the development of a range of optical fiber Bragg grating (FBG)-based sensors for transportation structural monitoring. It includes the instrumentation of marine propellers using arrays of FBGs mapped onto the surface of propeller blades to allow for capturing vibrational modes, with reference to simulation data. The research funded by the EU Cleansky programme enables the development of self-sensing electric motor drives to support 'More Electric Aircraft' concept. The partnership with Faiveley Brecknell Willis in the UK enables the integration of FBG sensors into the railway current-collecting pantographs for real-time condition monitoring when they are operating under 25 kV conditions.

Keywords

Fiber Bragg Gratings · Marine propellers · Optical fiber sensors · Self-sensing electrical motors · Smart railway current-collecting pantographs · Structural condition monitoring

Fiber Bragg Grating (FBG)-Based Sensing Technology

An intensive review of the use of fiber optics for structural condition monitoring has been undertaken (Grattan and Meggitt 1998) showing a number of techniques, amongst which the most widely used are Fiber Bragg grating (FBG)-based techniques (Kerrouche et al. 2009). FBGs produce wavelength encoded signals which are not susceptible to instrumental drift and environmental interference and have proven to be more robust and reliable, suitable for operation in harsh working conditions.

Optical fiber Bragg gratings (FBGs) are used as a basis for simultaneous temperature and strain measurement. A FBG is a structure with the refractive index of the fiber core being periodically modulated and reflects the light at a wavelength termed the Bragg wavelength (λ_B) that satisfies the Bragg condition, given in Eq. 1

$$\lambda_B = 2n_{\text{eff}}\Lambda \tag{1}$$

where n_{eff} is the effective refractive index of the fiber core and Λ is the grating period, where *both* are affected by strain and/or temperature variations, a feature that is reflected in the sensor design.

The underpinning sensing mechanism of a FBG is that its Bragg wavelength (λ_B) shift is determined by the change in surrounding temperature and/or strain applied as described in Eq. 2 (Pal et al. 2005).

$$\frac{\Delta \lambda B}{\lambda B} = (1 - Pe)\,\varepsilon + [(1 - Pe)\,\alpha + \zeta]\,\Delta T \tag{2}$$

where P_e is the photoelastic constant of the fiber, ε is the strain induced on the fiber, α is the fiber thermal expansion coefficient and ζ is the fiber thermal-optic coefficient. The first term of Eq. 2 represents the longitudinal strain effect on the FBG and the second term represents the thermal effect, which comprises a convolution of thermal expansion of the material and the thermal-optic effect. Equation 2 also indicates clearly the cross-sensitivity of a FBG to strain and to temperature, therefore when strain measurement is required using a FBG, its temperature effect is required to be compensated through optimizing the sensor design and sensor packaging.

One of the key features that FBG-based sensors have demonstrated is their multiplexing capabilities. Fig. 1 shows a typical FBG-based sensor layout based on wavelength-division-multiplexing (WDM) (Othonos and Kalli 1999) with each grating (sensor point) being encoded with a specific wavelength. This characteristic is of particular importance for monitoring large-scale and/or critical structures that require densely distributed sensors, for simultaneous multi-point multi-parameter measurement yet with limited number of fibers ('wires'). Compared to conventional strain gauge-based techniques, the FBG-based quasi-distributed sensing approach has shown significant advantages in terms of the ease of handling/installation and integration of a large number of sensing points, i.e., FBG strain/temperature sensors, coupled to a single source and interrogated by a single detector. In addition, there is no need to post-process the FBG raw data obtained due to their high signal-to-noise ratio compared to those from strain gauges. The WDM scheme, illustrated in Fig. 1, can be used very effectively to address a number of gratings, yielding not just the strain/vibration/force and temperature values of multiple sensors but also, through prior calibration, their physical locations on the target structures.

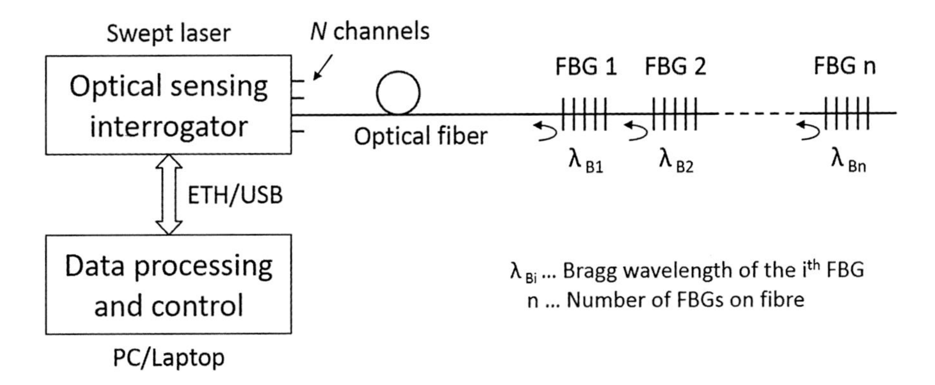

Fig. 1 Quasi-distributed FBG sensor system using wavelength-division-multiplexing technique

FBG-Based Sensors for Monitoring Full-Scale Marine Propellers

Background

For marine propellers, their blades rotate slowly compared for example to blades in fans, jets and in turbines and hence, the centrifugal forces on the blades and therefore the stiffness increase that would result can be ignored, as it is minimal. In a series of experiments and subsequent calculations carried out by Conn (1939), it was concluded that it is mainly the flexural frequencies that are affected by the centrifugal forces. The flexural frequencies of a rotating blade can be calculated from the following relationship:

$$f^2 = f_0^2 + k\Omega^2 \tag{3}$$

where f is natural frequency of the rotating blade (Hz), f_0 is the natural frequency of a non-rotating blade, Ω is the angular velocity (in revolutions per second (rps), this normally having a value of around 5–10 in the case of marine propellers) and k is a constant which has the values 0.35 and 1.35 for vibration parallel and perpendicular to the blade breadths. It can be seen from Eq. 3 that the effect of rotation on the frequencies of the blades is negligible. This was confirmed in work carried out by Castellini and Santolini (1998), where the natural frequencies of a small-scale, *model* propeller underwater were measured using a non-contact tracking laser vibrometer. They concluded that under rotating conditions, the bending modes were observed to be the most important vibration modes, since excitation due to hydrodynamic effects, gas bubbles or cavitation induces hardly any torsion effects on the blade structure.

The principal effect of immersing a propeller in water is the reduction in the frequencies at the particular mode at which the vibration occurs. However, this reduction is not constant and appears to be greater for lower modes of vibration, when compared to higher modes. In order to investigate this effect, Carlton (2012) has defined the frequency reduction ratio as

$$\Lambda = \frac{\text{frequency of mode in water}}{\text{frequency of mode in air}} \tag{4}$$

Considering a blade as a system with a single degree of freedom, the relationship between the motion of such a system under an undamped situation (while the stiffness remains unchanged) can be presented as a simple mass ratio equation, as shown in Eq. 5:

$$\Lambda = \left(\frac{M_b}{M_b + M_w}\right)^{1/2} \tag{5}$$

where M_b is the equivalent mass of the blade and M_w is the added mass of water.

This research explores the potential of using a FBG-based sensor network for direct measurement of natural frequencies and mode shapes of vibrations of a full-scale marine propeller of modern design in water and in air. Compared to the conventional methods using accelerometers or strain gauges, this FBG-based approach shows significant advantages in terms of being minimally invasive on the blade itself and its performance, yet allowing for a very large number of sensing points to be investigated simultaneously. There is no electrical hazard with the use of these sensors in air, or more particularly in highly conductive sea water, unlike the case with electronic strain gauges. Compared to the point-to-point laser-based method, this approach is advantageous as it is insensitive to the refractive effect arising from directing a laser beam with high precision to a specific part of the surface of blade – indeed there is no need for the blade to be visible to the operator.

Experimental Setup

The propeller selected for investigation in this experiment is a left-handed propeller designed for a twin-screw ship. The fixed pitch propeller blades had a diameter of 1900 mm with a variable pitch distribution of the blade. Table 1 presents the principal characteristics of propeller blade geometry and the material properties of the propeller.

Figure 2a shows a photograph of the propeller immersed in a water tank of base size 4 × 4 meters (and wall height of 2 meters) and the 5 blades of the propeller are instrumented with a total of 335 sensing points with each blade being mapped with 67 sensors and their locations are pre-determined by simulation results as shown in Fig. 2b.

All the FBGs used in this work are fabricated at City with an excimer laser-based fabrication system using phase mask technique. In each channel seven FBGs with different wavelengths, set to be between 1525 and 1565 nm, were designed and fabricated to ensure that there was no spectral overlap from one sensor to the next, even when each sensor responds over its maximum range of vibration-induced strain to avoid any ambiguity in the measurement. These FBG-based channels then formed the network of sensors on each blade as shown in Fig. 2b – a pattern that was repeated for each of the five blades of the propeller.

The sensor location however, as illustrated in Fig. 2b, is designed to be as far as possible normal to the chordal lines, to enable the results obtained from any

Table 1 Characteristics of propeller under test

Diameter	1900 mm
Mean pitch	1631 mm
Expanded area ratio	0.765
Modulus of elasticity, E	121 GPa
Poisson ratio, ν	0.33
Density, ρ	7650 kgm^3

a

b

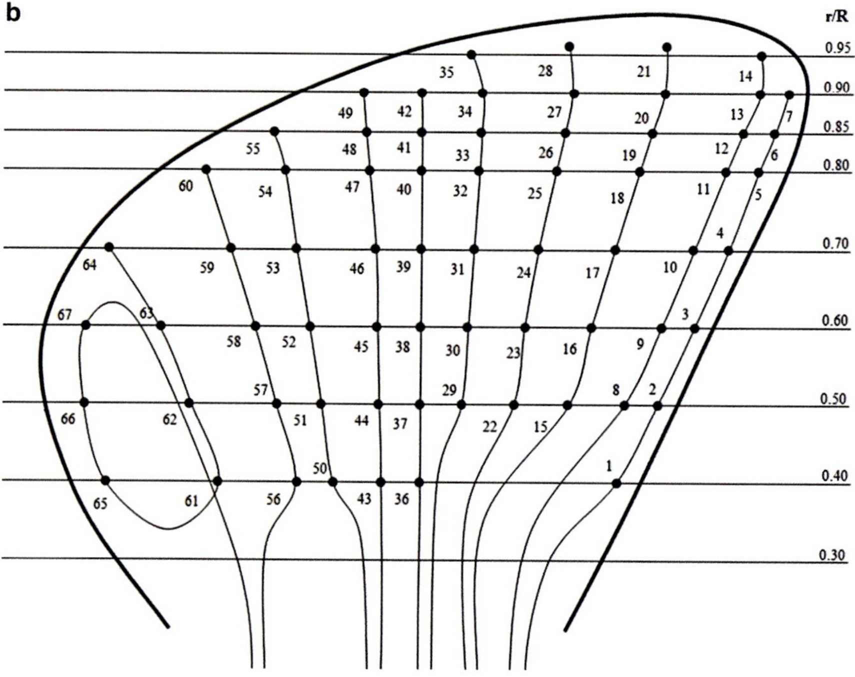

Fig. 2 (**a**) Left-handed propeller, instrumented with 335 sensors, in water tank; (**b**) an expanded view of a typical blade showing the sensor location points (numbered) used for the optical sensor vibrational analysis

individual sensor to be as closely comparable to those obtained from the other sensors. The modes of vibration of the blades, in air, were excited by striking the blades at various known locations on the propeller with a hammer with a relatively hard rubber tip, in order to excite the vibration of the blade across the full vibrational frequency range. Since the sensing interrogator unit was able to read data from 4 fibers simultaneously, first the optical fiber array next to the trailing edge of the blades was used as a reference for normalizing the captured amplitudes of the strain data. Therefore, when capturing data in this way, a series of three tests was necessary to obtain all the data for all the sensing locations for the blade shown in Fig. 2a. This was not difficult to do quickly and reproducibly. Then the amplitudes determined from the data from each set of tests were normalized to those obtained from the 1st sensor of the first fiber array, to allow the amplitude then to be comparable.

Mode orders were tracked using a strain mode shape based (SMSB) method. Hence the strain mode shapes were generated using the normalized data obtained from each sensor location and the results were mapped onto an expanded surface of the blade and cross-compared with data obtained from across the different blades. In order to compare the experimental results with those from a simulation carried out, these results were then compared to those obtained from an extensive Finite Element (FE) analysis (Javdani et al. 2016), carried out using Abaqus software.

Experimental Results and Discussions

Both the theoretical simulation and the experimental vibration tests in water were undertaken to determine the required water level to be used above the propeller blade, so that the natural frequencies are not affected by the depth of immersion. A deep tank was available and a series of preliminary tests was carried out using different water levels, namely 1350 , 1450 and 1550 mm, measured for convenience from the bottom of the water tank. The propeller in these tests was placed on a solid base of height 500 mm from the bottom of the tank, leaving the propeller blades to be immersed in 425 mm, 475 mm and 525 mm of water respectively from the free water surface. Further, tests were performed on the steady blades excited via the use of an impact hammer, in the conventional way. The results of these preliminary tests for a representative blade, Blade 3, are shown in Fig. 3. It can be seen from the results that increasing the water level from 1350 mm to 1550 mm does not affect the natural frequencies measured. Therefore it was considered unnecessary to increase the water level further and further tests in the tank were conducted using a 1350 mm water level.

A direct comparison was made of the results of the simulation and the experimental measurements, for each blade and for air and water, looking at the natural frequencies of vibration of the blade. Figure 4 shows a direct comparison of one of these representative sets of experimental and simulation results – this being obtained for blade 3 and acting as an illustration. The results show an excellent agreement for the first 12 natural frequencies, illustrating fully the capability of the specially designed optical FBG-based sensor network in capturing the vibration behaviour of a complex structure, such as a marine propeller. This has been done over a wide frequency range, at multiple positions, and both in air and water. Such a great level of details cannot be easily achieved with conventional sensors and under water.

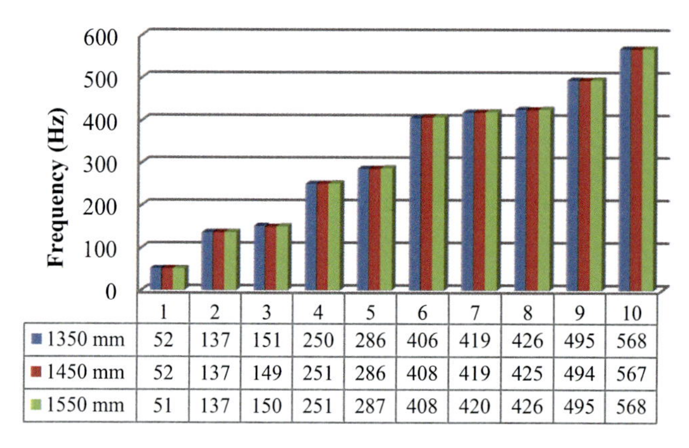

	1	2	3	4	5	6	7	8	9	10
1350 mm	52	137	151	250	286	406	419	426	495	568
1450 mm	52	137	149	251	286	408	419	425	494	567
1550 mm	51	137	150	251	287	408	420	426	495	568

Fig. 3 Natural frequencies measured at different water levels (shown below the x-axis). Natural frequencies (Hz) up to the 10th mode (the *top* numbers on the x-axis) are presented in the table, for three different water levels, 1350, 1450 and 1550 mm from the tank *bottom*

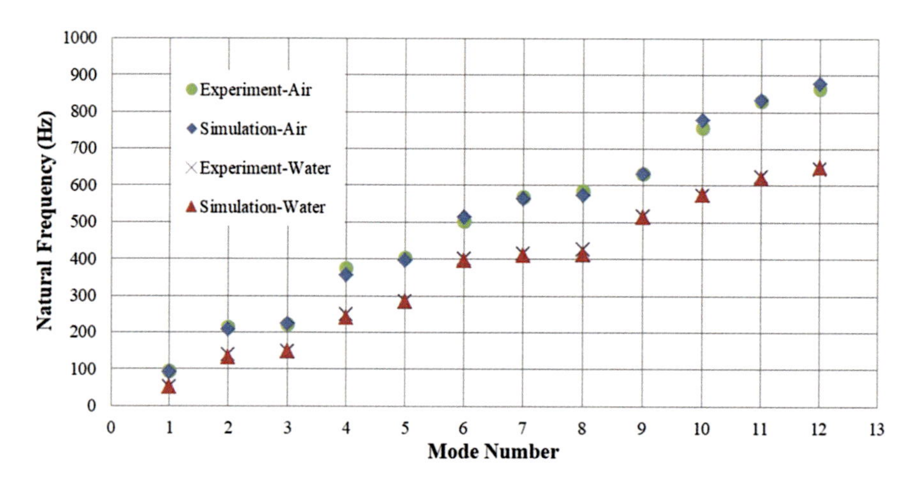

Fig. 4 Natural frequencies obtained from a series of tests, both experimental and simulation, both in air and in water, for blade 3

Conclusion

As detailed above, a FBG-based sensor network provides a novel solution to acquire accurate information from arrays of sensing points for better analysis and thus better understanding of the displacement mode shapes of a full-sized actual marine propeller. This allows monitoring in real-time of the associated Bragg wavelength shifts from arrays (networks) of FBG sensors. The data obtained have a shown a very good agreement with those obtained from simulation using finite element techniques, within experimental error. It was also confirmed that the location of the excitation on the blades, as expected, directly affected the amplitudes of the various frequencies that were detected in the experimental work.

Self-Sensing Electric Motor

Introduction

The 'more electric' concept in many areas of engineering has increased significantly the demand on reliability, power density and manufacturing efficiency of rotating electrical machines. To address this ever growing demand for new and reliable designs, electrical machines are increasingly required to be monitored in real-time with the data obtained being used for both model validation and prototype diagnostics. The latter helps to identify potential modes of failure and thus ensure the drive's reliability, as requested by machine owners or end users. If a conventional approach were to be adopted to achieve such multipoint, multi-parameter measurements, it would involve a drastic increase of component count thus reducing the overall reliability of the system in question. Further to this, due to the relatively large size of insulated conventional sensors, the resulting system will potentially occupy a spatial envelope larger than the drive itself. This work thus aims to address the above challenges, by replacing such conventional sensors with an integrated optical fiber-based, quasi-distributed, sensing system in electrical machines for real-time monitoring. Such a novel approach takes full advantage of the fiber sensors' reduced spatial envelope and immunity to electromagnetic interference.

One of the first efforts made in the direction of introducing an optical fiber sensor for motor and drive applications was to exploit Rayleigh backscattering in conjunction with a fiber having its outer cladding modified at intervals for a quasi-distributed temperature measurement system using an optical time domain reflectometry (OTDR) (Boiarski and Kurmer 1997). Since then different optical sensing techniques were applied to monitor end-winding vibratory behaviour (Kung et al. 2011), stator housing vibration (Corres et al. 2006), thermal effects (De Morais Sousa et al. 2012) and torque (Swart et al. 2006), for instance. In a previous report, the authors introduced a stator wave and rotor speed tracking system based on fiber Bragg grating (FBG) sensors (Fabian et al. 2015).

This research aims to explore the 'all-in-one' sensing concept underpinned by the FBG technique for simultaneous measurement of all the key parameters required, and operated by a single sensing interrogation unit thus eliminating the need for individual sensor systems for each parameter and therefore significantly reducing the complexity of electrical machine condition monitoring.

Principle of Operation

In order to evaluate the all-in-one sensing concept a PMAC machine was instrumented with a total of 48 FBGs at specific locations within the motor. The principle exploited to measure vibrations, the rotor speed and its position, the stator wave frequency and the spinning direction is based on the spatial modulation of the air-gap flux in the stator core of an induction machine. The resulting stator teeth displacement can be measured in the form of strain using FBGs as previously reported by the authors (Fabian et al. 2015). The method employed to measure torque is based on a differential wavelength approach where two FBGs are attached to the rotor shaft at an angle of ±45° with respect to the spinning axis (Swart et al. 2006). In this configuration, the difference between the two FBG reflection peak wavelengths is a measure for torque whereas their mid-point is an indicator for the temperature at that location. The dynamic Bragg wavelength shifts of all 48 FBGs were captured simultaneously using a Micron Optics SM130 sensing interrogator unit, at a sampling rate of 2000 Hz. The DC components of the transient signals were used for thermal analysis (and torque) whereas the AC components were used to determine stator vibrations and phase shifts necessary extract the dynamic parameters.

Instrumentation of Self-Sensing Permanent Magnet (PM) Motor

A self-sensing PMAC electric motor, instrumented with 48 FBG sensing points with 36 installed in the stator and 12 on the rotor, is shown in Fig. 5. The connection between the sensing fiber integrated into the rotor and the FBG interrogator is achieved using a fiber-optic rotary joint via a shaft adapter created using a 3-D printer.

The instrumentation of the stator is as follows. Two fibers of 12 FBGs each were routed along the stator windings for thermal profiling, two FBGs in each stator slot as shown in Fig. 6a, b. The fibers were looped around several times at either end of the stator core. A third fiber of 12 FBGs was circumferentially mounted on the stator core with each FBG placed in between adjacent stator teeth (Fig. 6c) to measure vibrations, the rotor speed and its position, the stator wave frequency and the spinning direction.

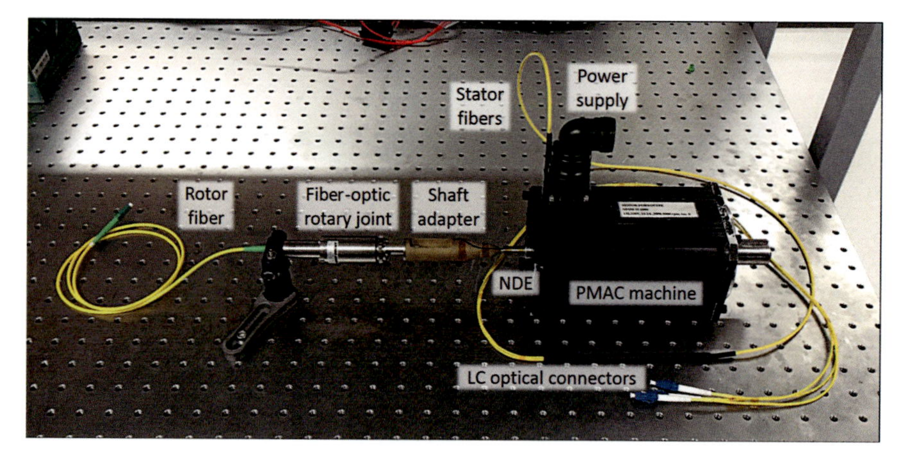

Fig. 5 PM machine instrumented with 48 FBGs on 4 fibers measuring rotor speed, torque, vibration, the rotor magnet temperatures and stator end-winding temperatures

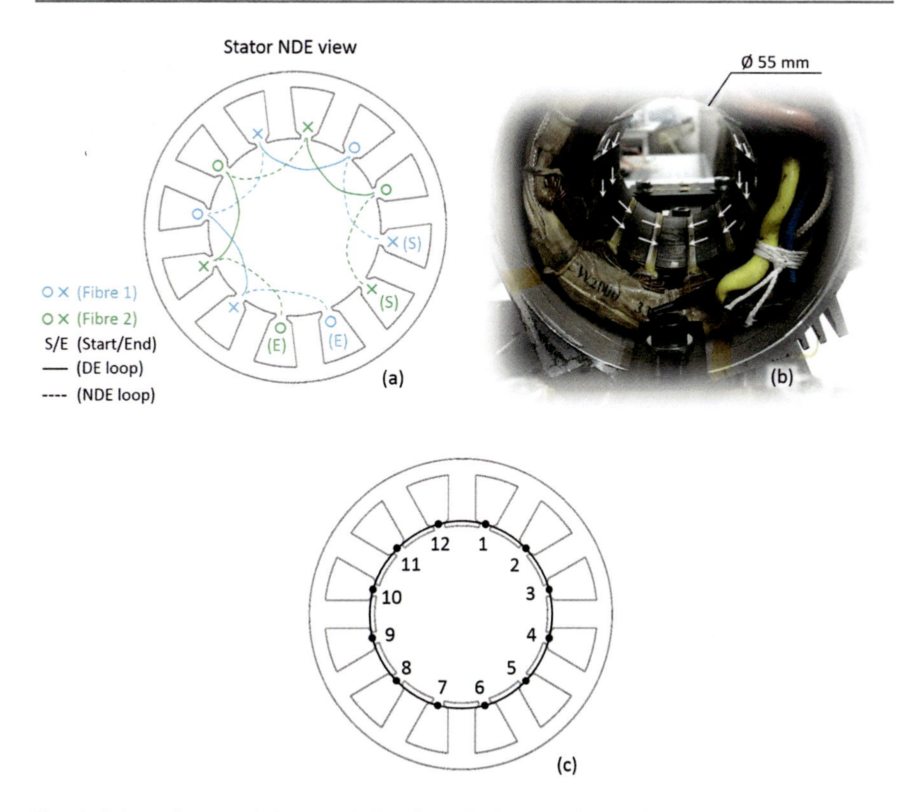

Fig. 6 Schematic (**a**) and photograph (**b**) of the distribution of 24 FBGs along the stator windings for their thermal profiling. (**c**) Schematic of the 12 circumferentially mounted FBGs used to measure stator vibrations that result in a number of parameters possible to be extracted

The rotor instrumentation is as follows. The remaining 12 FBGs were distributed across the rotor surface, one on each of the ten magnets, again for thermal profiling, and the other two on the rotor shaft for simultaneous torque and temperature monitoring. This is illustrated in Fig. 7. The rotor fiber was interrogated by means of a fiber-optic rotary joint which allows for the continuous monitoring of the rotor condition while spinning.

After the completion of the instrumentation and PMAC motor assembly discussed above, the FBGs used for temperature monitoring were calibrated by putting the motor in a climate chamber, as shown in Fig. 8, and running a pre-programmed temperature cycle (20–70 °C in steps of 10 °C).

The collected data from each FBG at each temperature step were averaged and then individually fitted using a linear least squares algorithm as shown in Fig. 9.

Figure 9a shows the temperature-dependent Bragg wavelength shifts of the 34 FBGs (24 in the stator and 10 the rotor), used for rotor and stator temperature profiling, at the temperature intervals mentioned above. It can be observed from Fig. 9a that the FBGs attached to the stator end-windings exhibit approximately

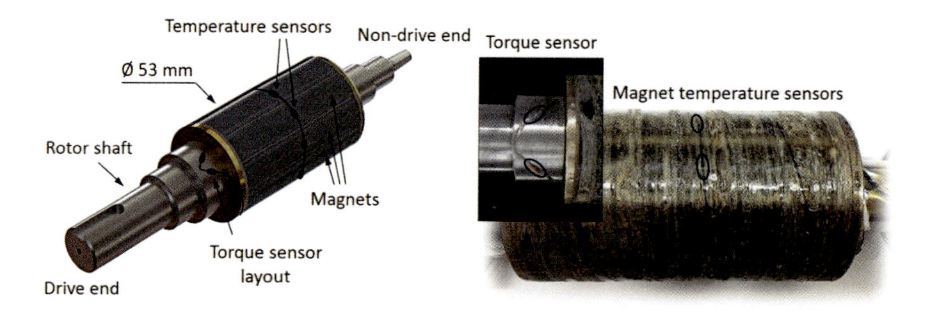

Fig. 7 Schematic and photograph of the instrumented PMAC rotor. One fiber of 12 FBGs is attached to the rotor, one to each of the 10 rotor magnets and 2 FBGs on the rotor shaft for torque measurement

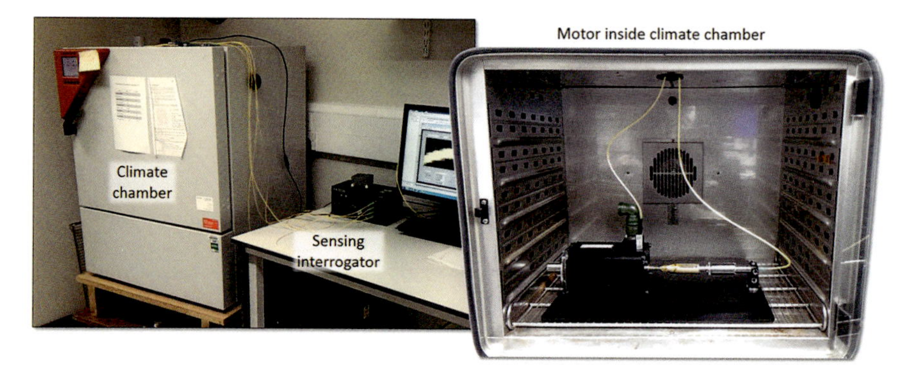

Fig. 8 Temperature calibration of the integrated FBG sensors in a climate chamber

three times the wavelength shift as the ones attached to the rotor magnets. This is due to the different thermal expansion coefficients of the end-winding material and the magnet. Figure 9b shows the fitted curves of two exemplary FBGs, with one installed on the end-winding of the stator and other on the rotor, to highlight their different temperature sensitivities (30 pm/°C for the stator FBG vs. 10 pm/°C for the rotor FBG). It is also clear from the fitted curves that the temperature induced Bragg wavelength shifts are highly linear.

With this information known, the absolute temperature T can be derived from the monitored Bragg wavelength λ_B of an FBG using Eq. 6 which is modified from Eq. 2, where c_T is the temperature coefficient or sensitivity of the FBG in *nm/°C,* obtained through calibration, and λ_{T0} is the±45 T0 °C.

$$T = \frac{1}{c_T}(\lambda_B - \lambda_{T0}) \quad (6)$$

The test bed was set up at the Institute for Aerospace Technology on the University of Nottingham's Jubilee Campus. The load motor was purchased from

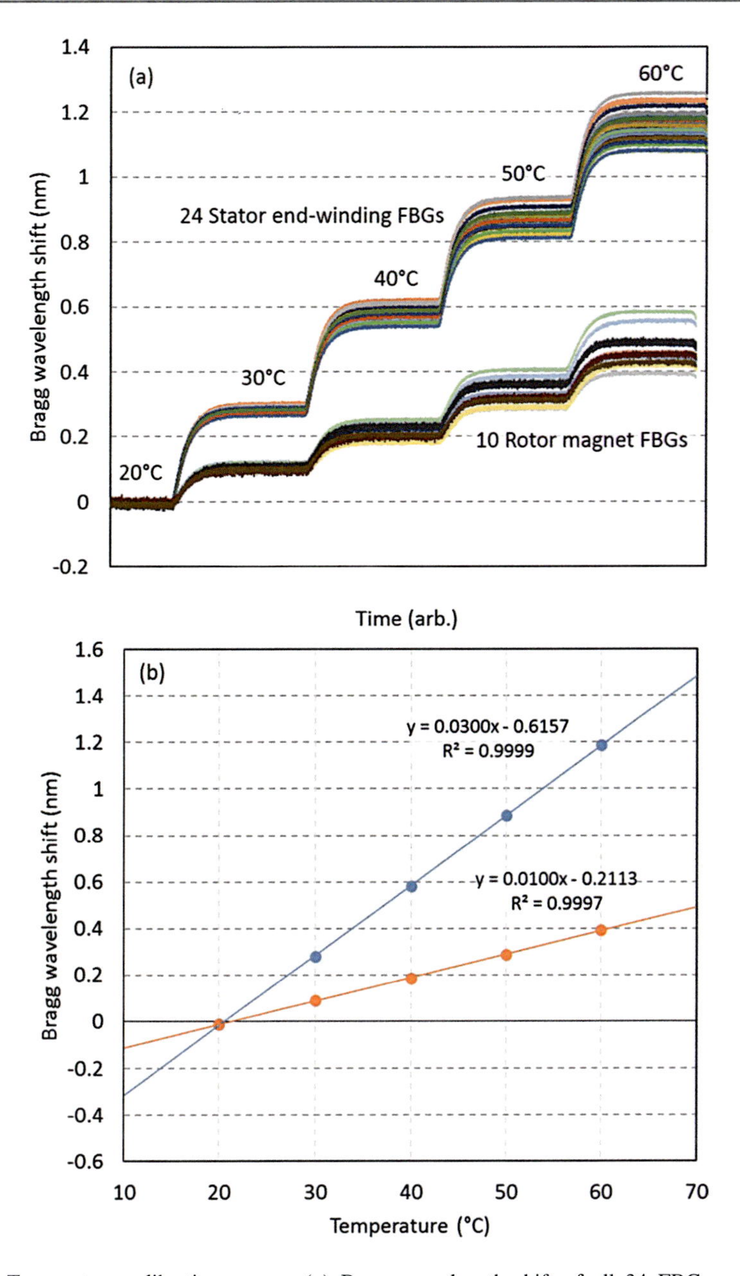

Fig. 9 Temperature calibration curves. (**a**) Bragg wavelength shift of all 34 FBGs used for temperature profiling; (**b**) Linear fitting of the temperature-dependent Bragg wavelength shifts of two exemplary FBGs

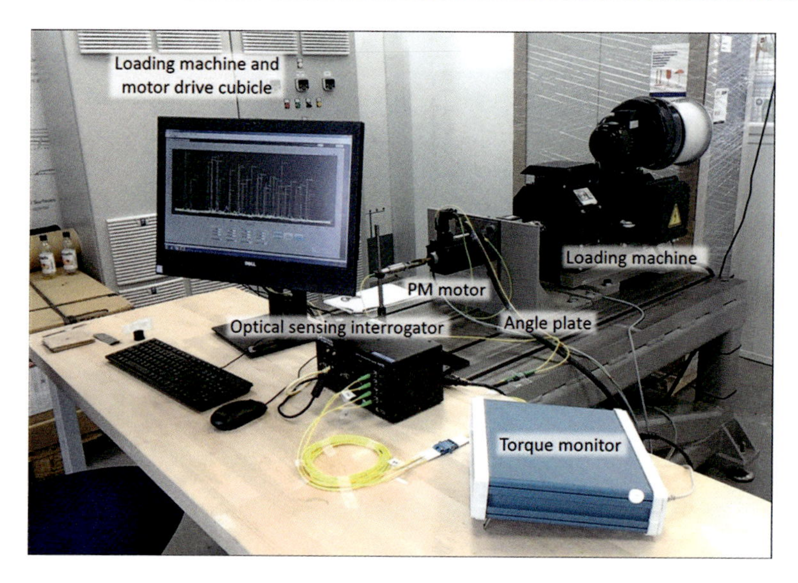

Fig. 10 Photograph of the assembled test bench and corresponding instrumentation hardware at the Institute for Aerospace Technology, Nottingham University

City Rewinds and Drives, and its corresponding 55 kW drive cubicle from Emerson Industrial Automation which also includes the drive for the PM machine. The torque transducer and couplings were sourced from Magtrol and smaller bits and pieces (encoder, cabling, bolts, etc.) from standard suppliers in the field. The test bench, as shown in Fig. 10, was thus assembled and commissioned by the technical team at the University of Nottingham who also made the shaft guard. As shown in the figure, the test bench includes the PMAC motor drive under test and load drive, integrated with both optical and conventional sensors.

Results and Discussions

Figure 11a shows the frequency response of one of the circumferentially mounted FBGs with the machine spinning at 16.7 Hz excitation. The first spectral feature at 16.7 Hz represents the rotor speed and the second (main) feature at 167 Hz corresponds to the stator wave frequency. Either of the two can be used to extract the rotor speed and convert it to rotations per minute (rpm). Other vibrational information, as evident from Fig. 11a, gives machine developers and engineers an important insight into the vibratory behaviour of a machine's design. Since vibrations are also an early indicator for impending machine failure, the constant monitoring of vibrations is of high importance in increased reliability environments.

Figure 11b shows the rotor speed obtained from the FBG data against a reference sensor with the rotor speed being varied between 1000 and 1600 rpm. It is clear from

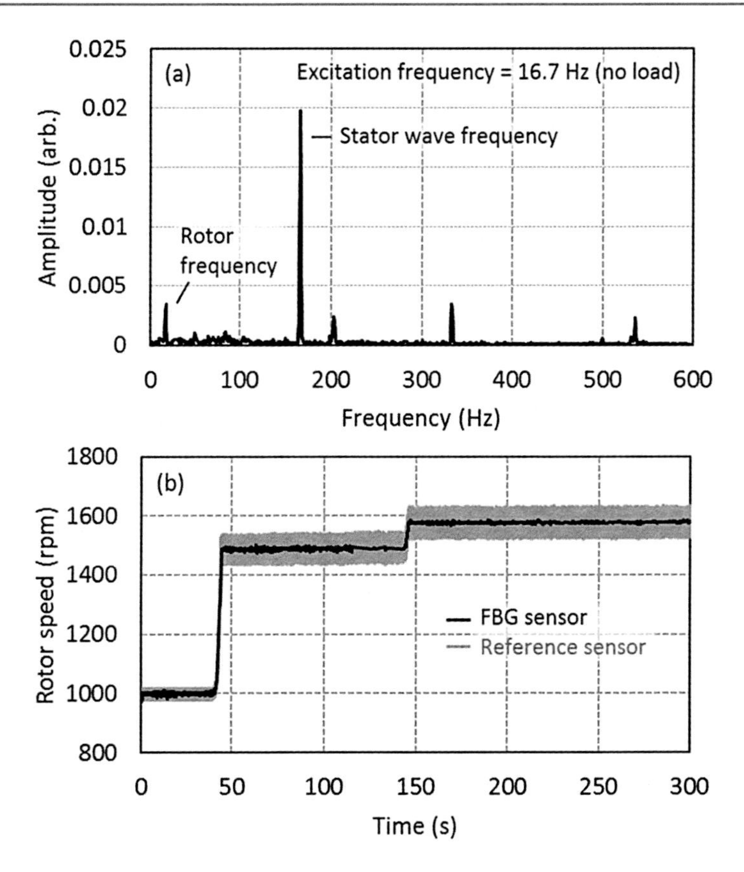

Fig. 11 (**a**) Frequency response of one of the circumferentially mounted FBGs in the case of a spinning rotor under no load. (**b**) Rotor speed versus time obtained from the FBG sensor and from a reference sensor

Fig. 10b that the FBG approach very closely matched the reference sensor signal at a much improved signal-to-noise ratio.

Figure 12a shows the dynamic responses of four of the circumferentially mounted FBGs highlighting the phase shift between them. This phase shift can be used to track the rotor position with regard to a reference point, i.e., acting like a conventional encoder. The phase shift between any two of those FBGs can also be used to determine the spinning direction of rotor, a positive phase shift indicating rotation in one direction and a negative phase shift rotation in the other direction.

Figure 12b shows the differential mode wavelength shift (the distance between the two FBG reflection peaks) of the torque sensor layout at varied levels of torque up to 2 Nm. Again, a very close correlation between the FBG approach and the reference sensor has been achieved with a linear torque – wavelength shift relationship (21.5 pm/Nm). In practise it is challenging to realise an angle of

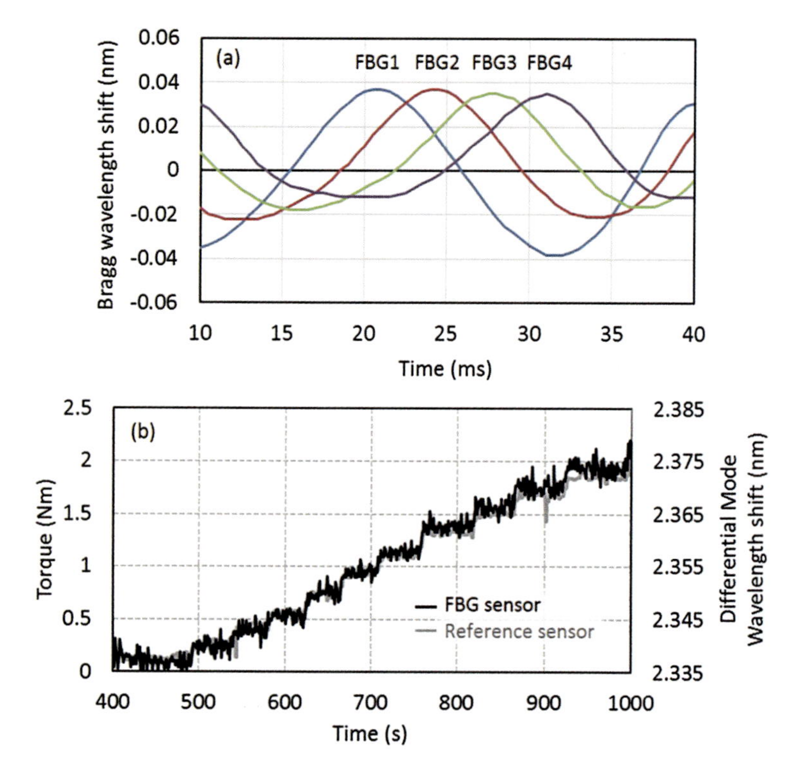

Fig. 12 (**a**) Exemplary phase shifts of the four of the circumferentially mounted FBGs. (**b**) Differential mode wavelength shift of the FBG torque sensor compared with the data of a reference torque transducer

exactly 90° between the two FBGs meaning that the differential mode wavelength will experience some sort of temperature dependence. However, this is easily compensated for using the torque-independent mid-point wavelength as an indicator for the temperature.

Figure 13 shows a typical screenshot of the motor monitoring application GUI at runtime. The top left graph shows the stator end-winding temperatures versus time and the rotor temperatures are shown in the graph below it. The temperature data are also visualised on the right in the form of colour coded 3-D models of both the stator and the rotor. The two graphs in the centre show the torque data and rotor speed. The latter is extracted from the spectral response of one of the circumferentially mapped FBGs. The vibration signature of the machine is shown in the bottom graph. In the shown case, the first spectral feature corresponding to the mechanical rotor frequency (rotor speed) and the second feature to the stator wave frequency. The FFT window length can be changed with the resulting frequency resolution being shown above the graph. The buttons in the top right to pause the data acquisition, to log the data and to terminate the application and are self-explanatory. In addition

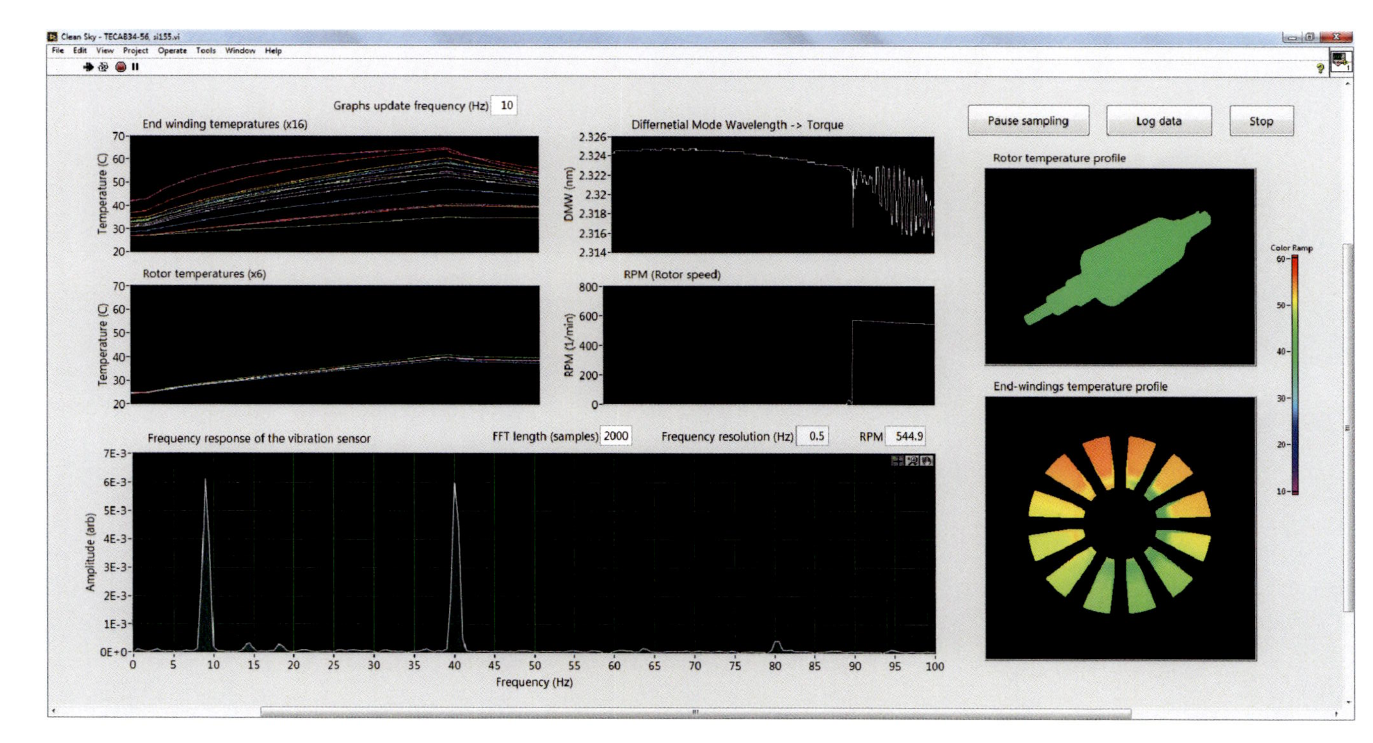

Fig. 13 Screenshot of the motor monitoring application at runtime showing key motor parameters

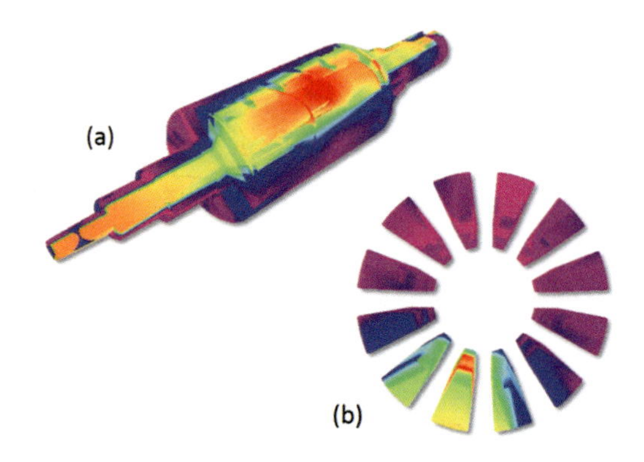

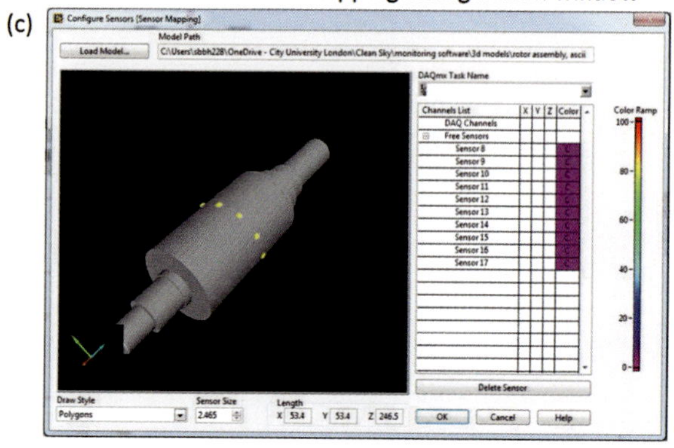

Fig. 14 Sensor data mapping in software through 3D models of rotor and stator. (**a**, **b**) Localised heating of fibers of the same sensor layout as the actual ones attached to the rotor and stator end-windings to evaluate the mapping algorithm. (**c**) Sensor mapping configuration window in LabVIEW™

to the data logging function, the contents of each graph can be exported to Excel at runtime with only two mouse clicks. This GUI design has been made flexible and can easily be edited by the end user to accommodate possible changes in the testing environment required by industry.

Figure 14 shows temperature data mapped in a 3-D format for both the rotor and stator as included in Fig. 13. In the mapping configuration window of Fig. 14, the sensor locations were marked on the model surface and a 1-D array containing the temperature data is wired to the VI. The algorithm developed has been validated by using fibers of identical FBG layout as the ones used in the motor that were locally heated. Figure 14a, b show the resulting screenshot of the 3-D real-time visualisation. The data obtained are also managed and prepared for the next-stage integration into the motor control software.

Conclusion

It was shown that when placing a network of FBGs at certain locations within an electrical machine, comprehensive condition monitoring can be performed at a high level of accuracy. Multiple parameters can be extracted from the FBG data by using appropriate data processing and compensation algorithms. The proposed all-in-one sensor system has the potential to replace conventional systems that require a separate sensor/system for each parameter to be monitored. It reduces the component count and spatial envelope of a test environment as all sensing elements are confined within the machine with minimum external wiring/coupling as opposed to conventional sensors. Future work will focus on the implementation of an active feedback control system using the sensor data to control a machine's speed under different load conditions, for instance.

Smart Pantograph

Introduction

The pantograph is a critical, roof-mounted part of a modern electric train, tram or electric bus to collect power through an overhead catenary wire and successful current collection requires a reliable pantograph-catenary contact with a *steady* force under *all conditions*, as the train travels along the line. The pantograph operates in a particularly harsh environment, being exposed to all weathers as its carbon strip rubs along the OLE at speeds up to 125 mph and at 25,000 volts conditions: monitoring its condition in real-time has posed a real technical challenge to the rail industry and optical fiber sensing provides an effective solution.

This research exploits fully the key advantages of FBGs and their suitability for pantograph condition monitoring, in terms of their immunity to electromagnetic interference, ease of multiplexing, small size and lightweight. The major drawback, however, of using a FBG-based technique is its cross-sensitivity to strain and to temperature, therefore a significant amount of effort has been made to compensate the temperature effect when a FBG is used for strain measurement. Camolli et al. (2008) reported the use of two single FBG sensors on separate fibers, where one of the FBGs is used for temperature compensation. This approach is based on the assumption that the temperature distribution is uniform along the pantograph, however this is not necessarily the case in a real time situation. The other FBG-based sensor system (Wagner et al. 2014) deploys the use of aluminum boxes confining 3 FBG sensors within a small footprint, with one strain-free FBG for temperature compensation. Each pan-head is instrumented with two boxes which increases the mass and consequently affects aerodynamic force when the train moves at high speeds. Embedding the FBG sensors between carbon and aluminum has been reported by Schroder et al. (2013). All the reported FBG techniques require either an additional fiber or an additional FBG for temperature compensation. Considering the high temperature sensitivity, which is one order of magnitude higher than that of its strain sensitivity, it is challenging to remove the temperature effect in a satisfactory way and this forms the core of this research.

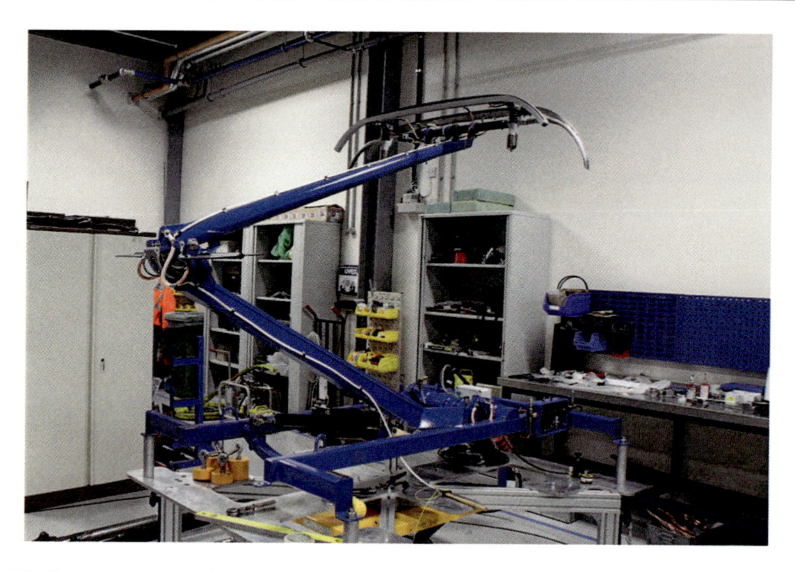

Fig. 15 Smart pantograph integrated with an FBG array

Temperature Compensated Contact Force Measurement

Figure 15 shows a pantograph integrated with arrays of FBGs, which have been designed and configured to allow both real-time measurement of the contact force and contact location and facilitate a closed-loop control to avoid unexpected failure of pantograph during operation. This is a joint development between City University of London and Faiveley Brecknell Willis in the UK.

This research exploits a novel temperature compensation method by using a package-based sensor design where three FBG-based sensor packages are integrated into three different locations of a pantograph, both for real-time measurement of the contact force and contact location and for temperature compensation, as illustrated in Fig. 15. Given the small footprint of each package, it is observed that the FBGs in the same package experience the same scale of temperature variations. This effect has been exploited in this research for effective temperature compensation.

To verify the above sensor design idea, Fig. 16 shows an experimental setup created for the evaluation of the developed temperature compensation method for the contact force measurement under high current conditions. As shown in the figure, a current supply was connected to both sides of the pantograph, allowing for a step change of current from 0 to 1500 A and then from 1500 A to 0 A to be applied. The higher current applied induces temperature change in the range from 25 °C to 55 °C over a period of 9 m, as recorded by three thermocouples, which were co-located with 3 FBG packages as shown in Fig. 16. To speed up the cooling process, a fan is used. During the whole measurement process, there is no contact force applied to the pantograph.

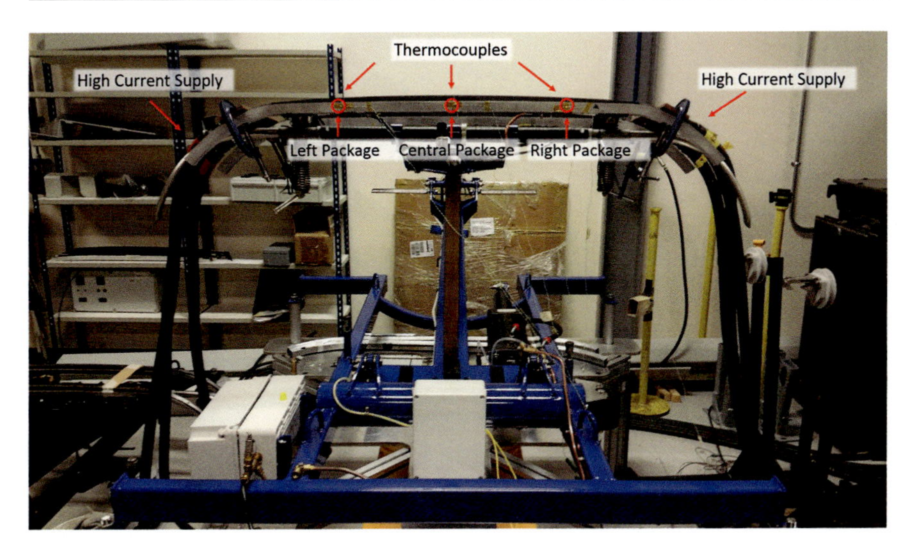

Fig. 16 Pantograph driven by a DC current, changing from 0 to 1500A and then from 1500 to 0A, when the contact force is zero

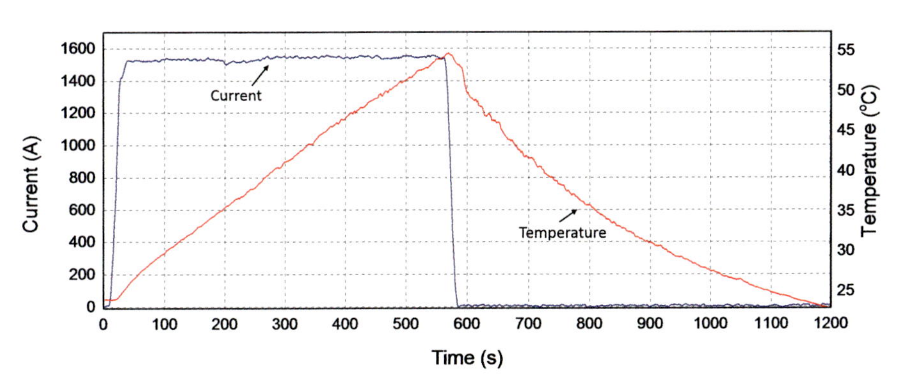

Fig. 17 Temperature change recorded by the central thermocouple and current value applied to the pantograph during the experimental tests

Experimental Results and Discussions

Figure 17 shows the current applied and the temperature change recorded by the thermocouple which is co-located with the central FBG package. The current supply is switched off when 55 °C is reached and a fan is used to accelerate the cooling process.

Figure 18a shows the wavelength shifts of three FBGs, i.e., FBG_4, FBG_5 and FBG_6, confined in the central package and located at the central area of the pantograph when a step change of current is applied to the pantograph from 0 to 1500 A and then from 1500 A to 0 A. The wavelength shift of each FBG within the same package experiences the change in applied strain (from the contact force)

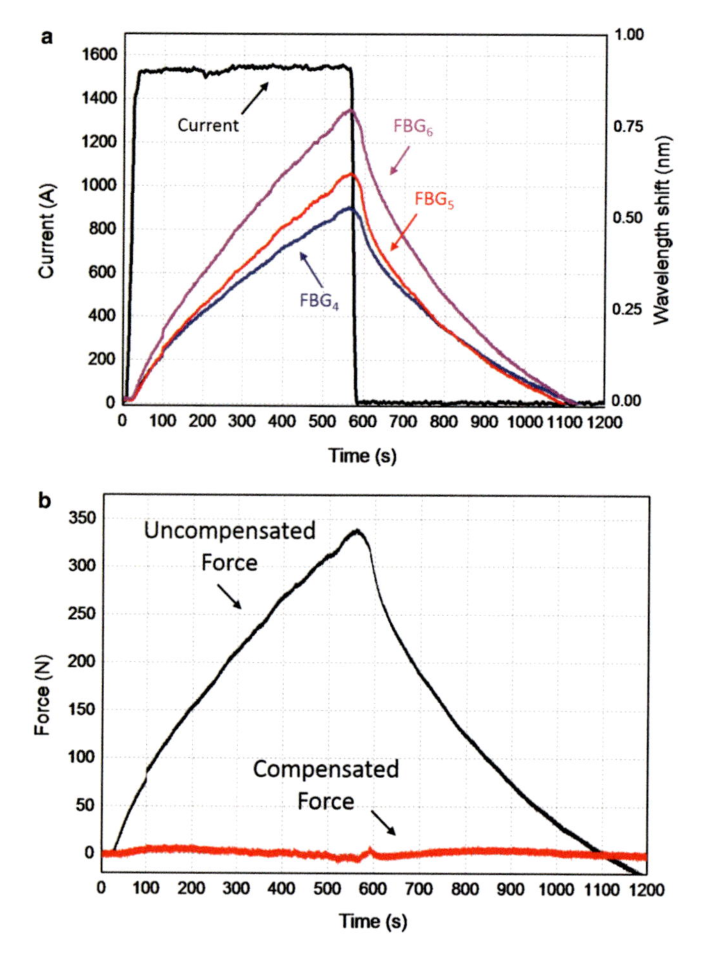

Fig. 18 (**a**) Wavelength shifts of FBGs 4, 5, 6 in the Central package recorded during the experimental tests; (**b**) The contact force measured by FBG packages before and after temperature compensation

and in temperature. The latter however can be subtracted based on the FBGs within the same package under the temperature variation and this underpins the algorithm developed for temperature compensation at City University of London. The black curve in Fig. 18b shows the contact force calculated from the data recorded by the central FBG package without considering temperature compensation. The red curve in Fig. 18b, however, shows clearly the contact force to be zero after the implementation of the temperature compensation algorithm developed and this agrees well with the test condition as the pantograph was electrified but without being in contact with the overhead line equipment (OLE). It is noticeable that temperature compensated FBG-sensor system removes the effect from the temperature changes and provides the information of the force being unaffected by the applied current.

Further to the above temperature-compensated contact force measurement using each FBG package, the contact location of the OLE against the pantograph can thus be obtained by calculating the ratio of the contact forces measured simultaneously by the three FBG packages integrated into the pantograph at three different, yet known locations.

Conclusion

This research explores a novel sensor design through integration of FBG sensor packages into a railway current-collecting pantograph for its remote condition monitoring. This is designed to remove temperature effect in the strain/force measurement and under the circumstances that the temperature effect is more dominant. The positive outcomes obtained from the field tests, by driving the pantograph using high currents, have further confirmed the effectiveness of this method used for temperature compensation. The research is still on-going and more vehicle tests will be undertaken in the near future to evaluate extensively the smart pantograph developed in industrial settings.

Summary

A range of FBG-based optical fiber sensor systems have been developed and evaluated, showing promise for wide industrial applications. These sensors have been designed to provide real-time measurement of a suite of key parameters that would help engineers to diagnose structural conditions thus to improve structural integrity and reliability through improved maintenance.

References

A.A. Boiarski, J.P. Kurmer, in *Electric Power Research Institute, TR-101950-V2, 2487-02, Final Report* (1997)

J.S. Carlton, Chapter 21, in *Propeller Blade Vibration in Marine Propeller and Propulsion*, 3rd edn., (Butterworth-Heinemann, Oxford, 2012), pp. 421–429

P. Castellini, C. Santolini, Measurement **24**(1), 43–54 (1998)

L. Comolli, G. Bucca, M. Bocciolone, A. Collina, in *Proceeding of SPIE,* vol. 7726 (2008)

J.F.C. Conn, in *Proceedings of the Trans. IESS*, 225–255 (1939)

J. Corres, J. Bravo, F.J. Arregui, I.R. Matias, IEEE Sensors J. **6**(3), 605–612 (2006)

K. De Morais Sousa, A.A. Hafner, H.J. Kalinowski, J.C.C. Da Silva, IEEE Sensors J. **12**(10), 3054–3061 (2012)

M. Fabian, J. Borg Bartolo, M. Ams, C. Gerada, T. Sun, K.T.V. Grattan, in *Proceedings of the SPIE* 9634 (2015), 963417 4 pp

S. Javdani, M. Fabian, J.S. Carlton, T. Sun, K.T.V. Grattan, IEEE Sensors J. **16**(4), 946–953 (2016)

A. Kerrouche, W.J.O. Boyle, T. Sun, K.T.V. Grattan, Sensors Actuators A Phys. **151**(2), 107–112 (2009)

P. Kung, L.Wang, M.I. Comanici, in *Proceedings of the IEEE Electric Insulation Conference* (2011), pp. 10–14

K.T.V. Grattan, B.T. Meggitt (ed.), *Optical Fiber Sensor Technology*, vol. 3 (Kluwer, London, 1998). ISBN:978-1-4419-4736-9

A. Othonos, K. Kalli, *Fiber Bragg Gratings: Fundamentals and Applications in Telecommunications and Sensing* (Artech House, Boston, 1999)

S. Pal, Y. Shen, J. Mandal, T. Sun, K.T.V. Grattan, IEEE Sensors J. **5**(6), 1462–1468 (2005)

K. Schroder, W. Ecke, M. Kautz, S. Willett, M. Jenzer, T. Bosselmann, Opt. Lasers Eng. **51**, 172–179 (2013)

P.L. Swart, A.A. Chtcherbakov, A.J. Van Wyk, Meas. Sci. Technol. **17**(5), 1057–1064 (2006)

R. Wagner, D. Maicz, W. Viel, F. Saliger, C. Saliger, R. Horak, T. Noack, in *7th European Workshop on Structural Health Monitoring* (2014)

Optical Fiber Sensor Network and Industrial Applications

47

Qizhen Sun, Zhijun Yan, Deming Liu, and Lin Zhang

Contents

Abstract

For many of sensing applications, multiplexed sensor networks which can map the sensing signal of a large structure or surveying at complex conditions are required, greatly promoting the development of the fiber optic sensor network with large capacity. In this chapter, three typical fiber optic sensor networks

Q. Sun (✉) · Z. Yan · D. Liu
School of Optical and Electronic Information, Next Generation Internet Access National Engineering Laboratory (NGIAS), Huazhong University of Science and Technology, Wuhan, Hubei, P. R. China
e-mail: qzsun@mail.hust.edu.cn

L. Zhang
Aston Institute of Photonic Technologies, Aston University, Birmingham, UK

G.-D. Peng (ed.), *Handbook of Optical Fibers*,
https://doi.org/10.1007/978-981-10-7087-7_20

and their applications will be introduced. Firstly, the ultra-weak fiber Bragg grating (UWFBG) sensor networks with ultra-large capacity for quasi-distributed and continuous distributed sensing in a single fiber link are investigated, which is realized by the multiplexing of UWFBGs or UWFBG based Fabry-Parot interferometers (FPI). Secondly, special fiber grating sensor networks with advanced functions and competitive performances, including the tilted fiber grating (TFG) sensors or distributed Bragg grating fiber laser (DBRFL) sensors multiplexed in a single fiber, are investigated. Thirdly, fiber optic sensors passive optical networks (SPON) with good adaptability, high extendibility, and great flexibility are comprehensively studied, which includes the star topology SPON and the tree topology SPON for colored sensors and colorless sensors accessing. For each type of sensor network, the sensor structures, networking mechanisms, system architectures, demodulation methods, and typical sensing performances are systematically discussed. Moreover, the developed systems or equipment and field tests for a wide range of commercial and industrial applications, especially for resource exploration, geophysics, infrastructure, medical diagnosis, food quality, and security control, are presented.

Keywords

Fiber optic sensor network · Fiber grating sensor network · Fiber optic distributed sensing · Senor passive optical network

Introduction

Due to the distinct advantages of light weight, small size, high sensitivity, immunity to electromagnetic interference, and ease to network, there is a high demand for smart optical fiber sensor technologies due to increasingly application needs in a wide range of sectors, including civil engineering, aerospace, maritime, energy, and defense industries, as well as in medical, environmental, and food sectors. Recent market analysis by ElectroniCast has reported that the global market value for fiber optic sensors was projected to $3.38 billion in 2016 and will increase to more than $5.98 billion in 2026 (ElectroniCast consultants 2017). For many of sensing applications, multiplexed sensor networks which can map the sensing signal of a large structure (e.g., oil-gas well, pipeline, bridge, border, aircraft wing, etc.) or geophysical surveying at complex conditions are required, for which single or pairs of sensors are not sufficient. Therefore, the fiber optic sensor network with large capacity is becoming an inevitable tendency for the sensing industry.

The fiber optic sensor network mainly includes point fiber sensor array and distributed fiber sensor system, of which the sensor units can be multiplexed by specific schemes, including time division multiplexing (TDM), wavelength division multiplexing (WDM), frequency division multiplexing (FDM), space division multiplexing (SDM), or their combinations. Apart from seeing many successful commercial deployments of fiber sensors, novel and function-enhanced fiber sensors have been developed by utilizing specially modified structures and speciality fibers.

In this chapter, three typical fiber optic sensor networks will be introduced, including the ultra-weak fiber Bragg grating (UWFBG) sensor network (section "Ultra-Weak Fiber Bragg Grating (UWFBG) Sensor Network and Applications"), special fiber grating sensor network (section "Special Fiber Grating Sensor Network and Applications"), and fiber optic sensors passive optical network (SPON) (section "Fiber Optic Sensors Passive Optical Network (SPON) and Applications"). The sensor structures, networking mechanisms, system architectures, demodulation methods, and the sensing performances will be discussed. Moreover, a wide range of industrial applications, especially for resource exploration, geophysics, infrastructure, medical diagnose, food quality, and security control will be presented.

Ultra-Weak Fiber Bragg Grating (UWFBG) Sensor Network and Applications

Over the last decade, one of the most versatile and broadly researched and developed optical fiber sensor platforms is the in-fiber gratings, owing that the modulation pattern of the refractive index (RI) in fiber grating is sensitive to external parameters such as temperature, strain, and surrounding RI, resulting in the nominal wavelength shift. Until now, fiber Bragg gratings written on standard fiber have been widely used for measuring temperature, strain and force, pressure, vibration, liquid level, displacement, twist and torsion, bending and loading, current and magnetic field, chemicals and biochemical, etc., which are showing great potential and broad market prospects in industrial fields.

The wavelength encoded nature of the information facilitates WDM for sensor networking, achieved by assigning individual sensors to a different slice of the available source spectrum (Fallon 2000), as illustrated in Fig. 1. This outstanding advantage makes fiber gratings become ideal candidates for many applications. Except for WDM, FBG can also be multiplexed by TDM or the combination of them to build a sensor network along one fiber for large area measurement.

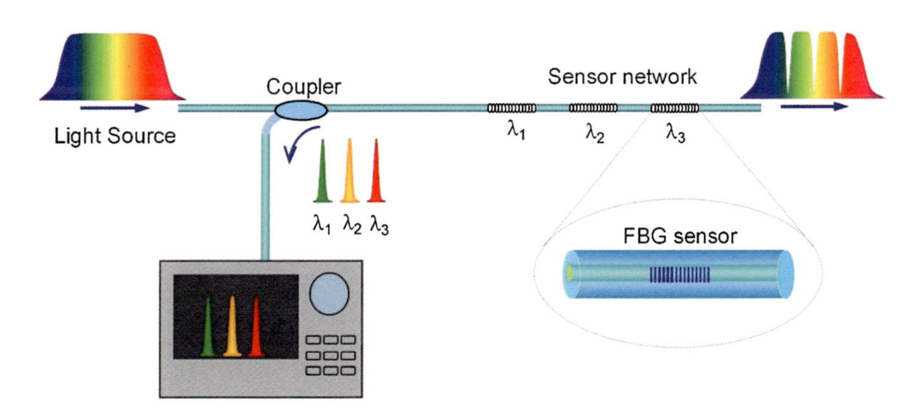

Fig. 1 Schematic of the fiber grating sensor network

However, the insertion loss induced by number of the FBGs is high, and therefore, the multiplexing capacity is limited to be about several hundreds, typically less than 100. Recently, UWFBG -based sensor network was proposed and investigated. The reflectivity of FBG is smaller than 1%, even decreased to about 10^{-5}, so the insertion loss into the fiber can be ignored. Consequently, there is enormous potential in the multiplexing capacity along single fiber. In this section, four types of UWFBG-based sensor network and the typical applications will be discussed.

TDM-Based Quasi-Distributed Sensor Network

Based on the narrow bandwidth and weak reflectivity of UWFBG, identical gratings can be multiplexed in time domain to improve the multiplexing capacity and sensing distance greatly. Figure 2a presents the working principle of the identical UWFBG multiplexed sensor network for quasi-distributed sensing. When pulse signal incidents into the fiber, a pulse sequence will be reflected and different pulse corresponds to different UWFBGs. Meanwhile, the central wavelength of the reflected pulse presents the value of the sensing parameter of it. It means that the UWFBG sensors can be interrogated in both time domain and wavelength domain, achieving multi-point synchronous precision measurement and positioning. The propagation of the optical wave in the fiber is similar with the backscattering of fiber, such as Rayleigh scattering, Brillion scattering and Raman scattering which can implement long distance and distributed sensing. Hence the sensing system is named as the microstructure-OTDR, i.e., M-OTDR, because the UWFBGs can be considered as longitudinal distributed microstructures. While the reflectivity of the UWFBG is several orders of magnitude higher than backscattering light as shown in Fig. 2b, it is a perfect candidate for improving the signal to noise ratio (SNR) of the backscattering light in fiber as the sensing point, resulting in the higher measurement precision and greatly shorter response time. Meanwhile, by networking the identical UWFBGs through TDM, the multiplexing capacity in a single fiber can be greatly improved to 1000 due to the relatively lower insertion loss (Zhang et al. 2012a). Along with the decrease of the UWFBG reflectivity, the effect of the cross-talk induced by multi-reflection between the gratings will be weakened gradually (Hu et al. 2014). When the UWFBG reflectivity is about −40 dB with the central wavelength of 1550.9 nm, the multiplexed number of gratings could reach up to1642, showing a low transmission loss (Wang et al. 2016).

Owing to the advantages of large multiplexing capacity, high measurement accuracy, and long-sensing distance, this sensing network has a great potential for health monitoring of bridges, dams, tunnels, and other distributed sensing applications. For example, the network made up of 6108 UWFBGs with two wavelength bands in a 10 km fiber was developed and the distributed temperature measurement was conducted by using a temperature test chamber. The experimental results were shown in Fig. 3, exhibiting the red shift of peak wavelength with the increase of temperature, measurement accuracy of 0.5 °C, and good linear response with the coefficient around 10.68 pm/°C at any gratings (Yang et al. 2016).

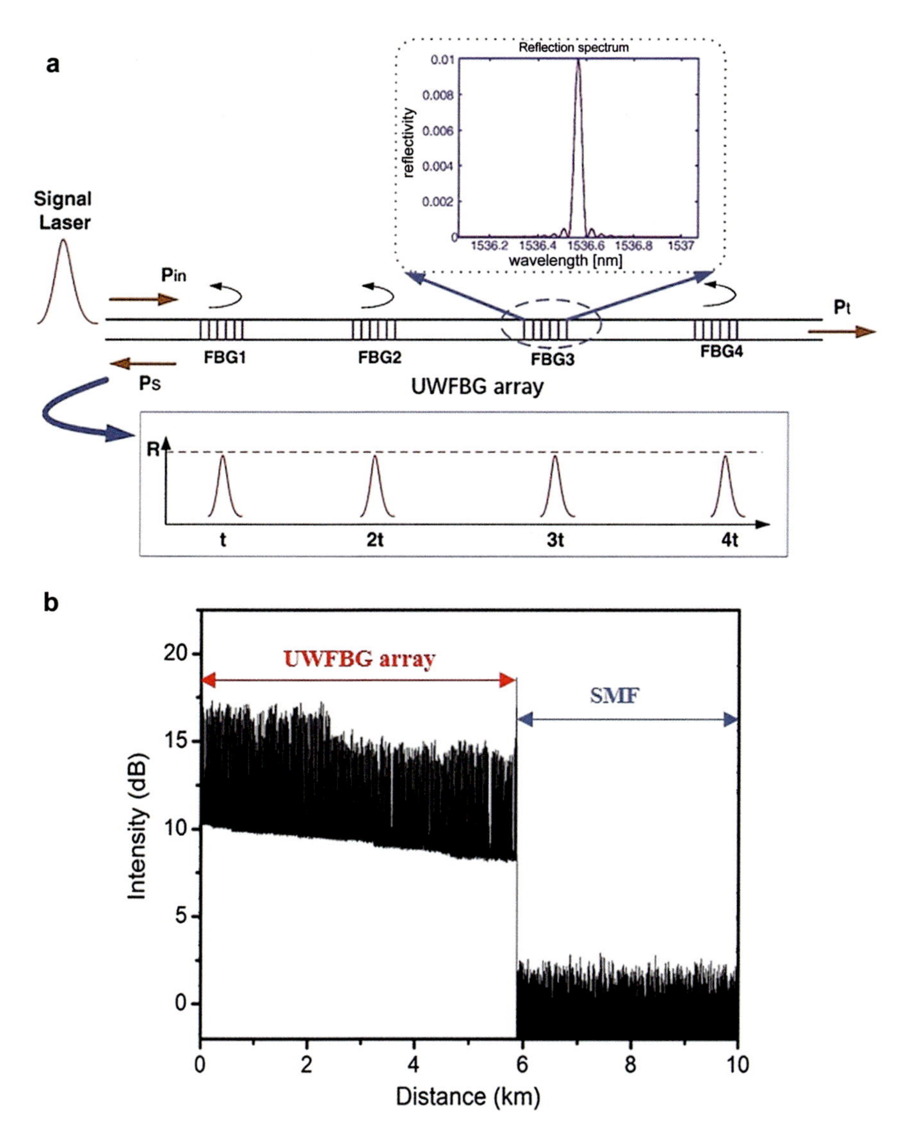

Fig. 2 Schematic of the identical UWFBG multiplexed sensor network: (**a**) working principle; (**b**) OTDR trace of the UWFBG array. (Copyright 2019 Springer)

TDM-Based Continuous-Distributed Sensor Network

Although high SNR can be realized with UWFBG, fiber between two neighboring UWFBGs becomes dead zone of the sensing system. To detect the event occurred in this section, the UWFBGs based fiber as the sensing link and the coherent OTDR as the demodulation scheme were combined to realize wideband

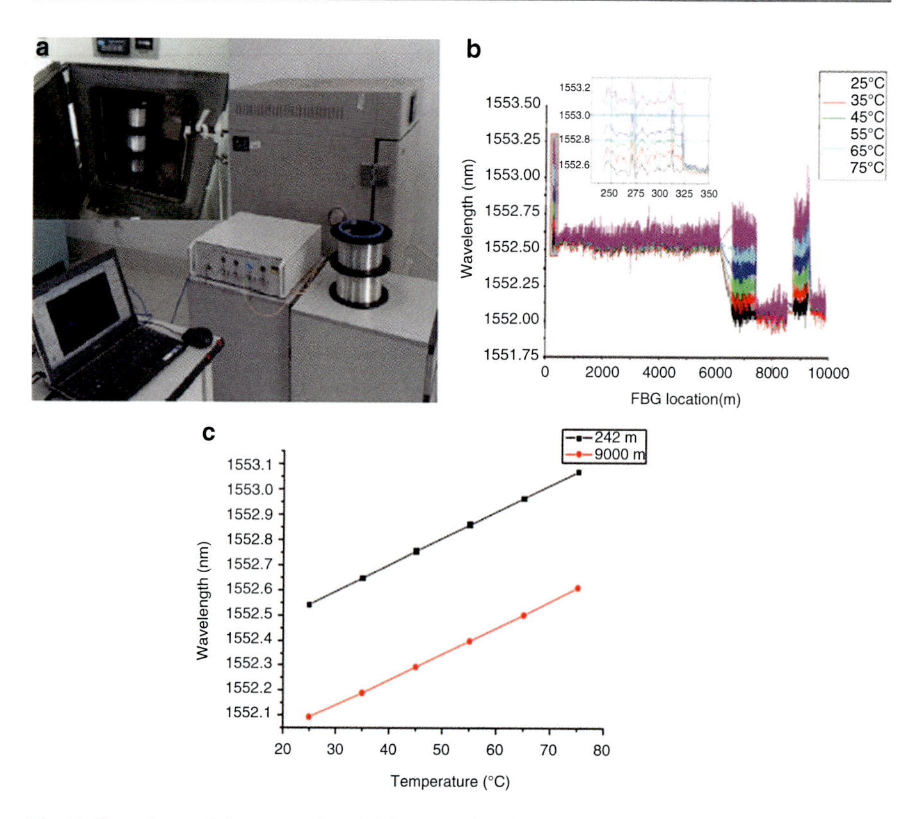

Fig. 3 Experimental demonstration: (**a**) images of the interrogation system of an UWFBG array, (**b**) temperature sensing performance of the UWFBG array, and (**c**) the temperature response of two UWFBGs located 242 and 9000 m. (Copyright 2019 Springer)

and high sensitive fully-distributed sensing. As illustrated in Fig. 4, light from the laser source is modulated into pulses and frequency shifted by Δf after passing through the acoustical optical modulator (AOM). After amplified by EDFA and filtering process, reflected pulses from the UWFBG are combined with the local oscillator and received by the balanced photo detector (BPD). The reflected signal contains the intensity and the phase information of every UWFBG. Phase change on the sensing fiber induced by external parameters is obtained by calculating the phase difference between every two adjacent UWFBGs through differential cross-multiplying (DCM) algorithm. And the event location is identified by searching the peaks of the backscattered light; therefore, the spatial resolution is determined by spatial interval of UWFBGs (Ai et al. 2017).

Based on the above setup, distributed vibration, acoustic wave, strain, and temperature, detection were explored. Specifically, as the phase change induced by the temperature and the vibration event would occupy different frequency band, after the low pass filter (LPF) and the high pass filter (HPF), the vibration and temperature

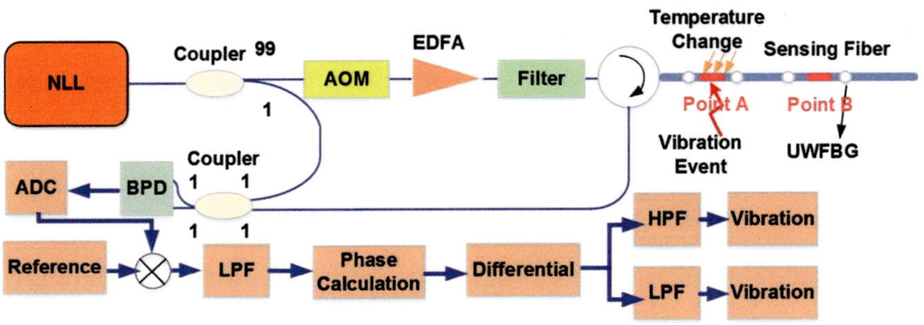

Fig. 4 Configuration of the distributed sensing system based on UWFBGs sensing link. (Copyright 2019 OSA)

change can be measured simultaneously (Ai et al. 2018). The experimental results are displayed in Fig. 5, which demonstrate that the sensor system can accurately map the temperature distribution and trace the multiple vibrations along the sensing fiber, as well as simultaneously detect the temperature change and vibration signal occurred at the same position of 752 m.

In addition, the sensor network provides excellent sensing performance for distributed acoustic sensing (DAS) with wideband covering from static to ultrasonic range. When no strain was applied on the sensing fiber, phase noise spectral densities at low frequency and high frequency were recorded and presented in Fig. 6a, b. It is clear that the phase noise above 1 Hz is as low as 7×10^{-4} rad, which means that the system can respond to the acoustical signal at ultra-low frequency region. To evaluate the sensitivity to acoustical waves, static strain test was conducted. As shown in Fig. 6c, sensitivity of 4.393 rad/$\mu\varepsilon$ as well as good linear relationship with $R^2 > 0.9998$ were achieved. From the phase noise spectral density above, the strain resolution over 1 Hz can be deduced to be lower than 0.16 nε. Meanwhile, the demodulated acoustic distribution along the sensing fiber was investigated. The experimental results in Fig. 7a, b prove that the DAS system owned wide response band from 0.1 Hz to 45 kHz, with successful recovery of the acoustic wave.

It should be emphasized that not only the UWFBG with certain wavelength selection but also the ultra-weak chirped FBG(UWCFBG) with wideband selection or the local abrupt change point of RI without wavelength selection can be served as the backscattering enhanced microstructure, which are inscribed in the fiber through UV or femtosecond lasers exposure.

The M-OTDR based DAS system have been widely applied in industrial fields such as borehole survey (Yamate et al. 2017; Mateeva et al. 2014), seismic recording (Ni et al. 2005; Jousset et al. 2018), and rail crack detection (Fan et al. 2019). Figure 8 presents our field test conducted in an oilfield (Fig. 8a). A 1 km long UWFBG array fiber cable was deployed into a cased borehole with a weight bar to pull the fiber cable down to the borehole (Fig. 8b). An explosive source was used to generate seismic energy on the surface with different

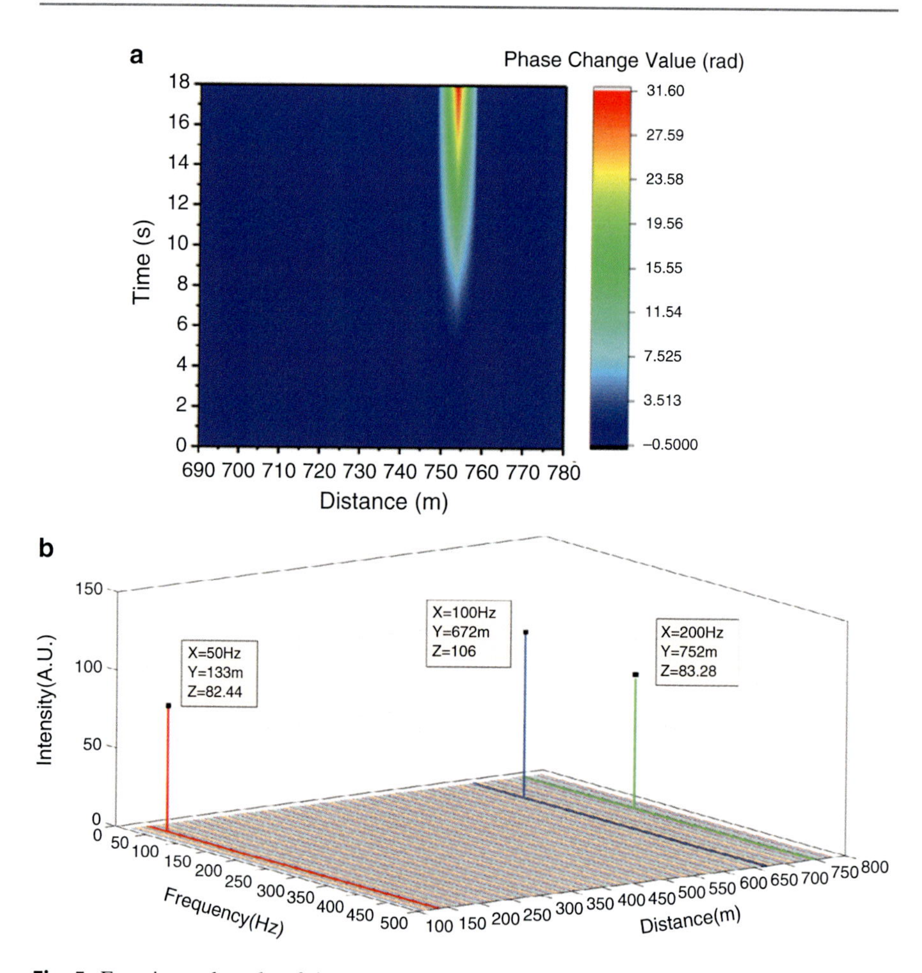

Fig. 5 Experimental results of simultaneous temperature and vibration measurement for UWFBG array based full-distributed sensor network: (**a**) phase distribution along the sensing fiber when the temperature at Point A changes; (**b**) vibration frequency distribution along the sensing fiber when multiple vibrations were applied on different positions

offset distance to the wellhead. From Fig. 8c, it can be seen that the fiber DAS system based on UWFBG sensor network acquired borehole seismic data with good quality, where both of the upgoing wave and downgoing wave can be clearly detected.

Figure 9 illustrates the field test of fiber DAS system to record seismic data. From Fig. 9a–c, it can be seen that the distance between the seismic signal and the seismometer or DAS system is about 8.7 km, and 500 m SNR enhanced sensing fiber cable was buried underground with the depth of about 20 cm to record the seismic wave transmission. Figure 9d clearly shows the excited wave by heavy hammer near the fiber, and the comparison of the recorded seismic data

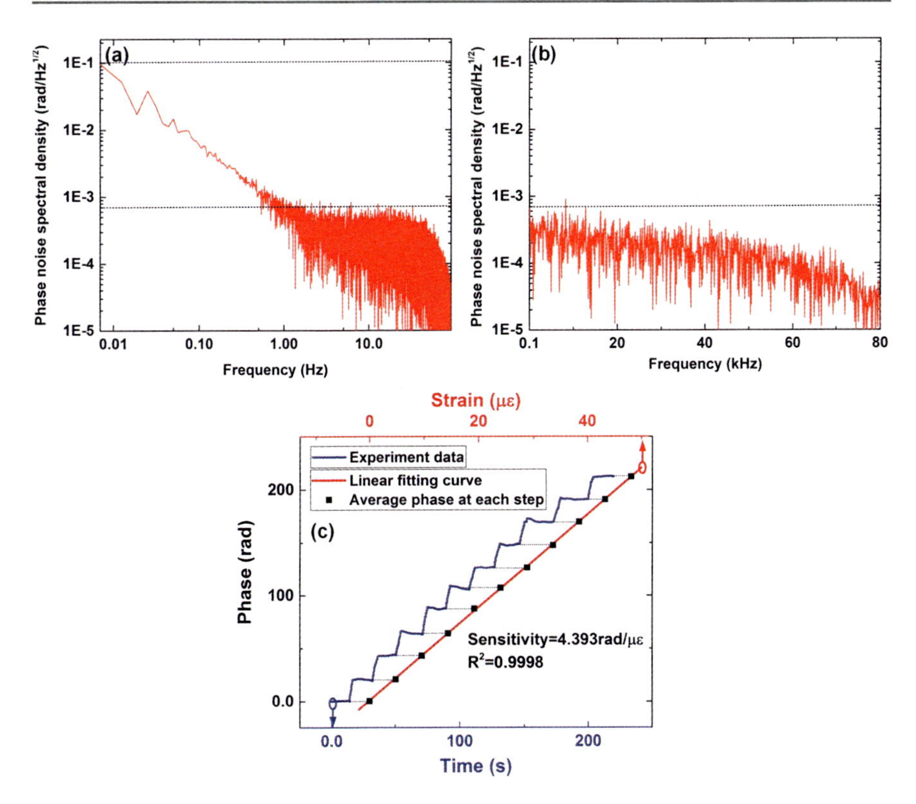

Fig. 6 Noise floor and sensitivity of UWFBG array fiber based DAS: (**a**) phase noise spectral density of low frequency less than 10 Hz; (**b**) phase noise spectral density from 0.1 to 80 kHz; (**c**) relationship between phase change and applied strain

from electrical seismometer and fiber DAS in Fig. 9f demonstrates that the fiber DAS can detect the seismic data with high sensitivity, high accuracy as well as wider response frequency band. In addition, as presented in Fig. 9e, the fiber DAS system successfully recorded the vehicle movements in this area all the time, which can used for analyzing the traffic condition and the noise pollution in the city.

Figure 10 shows our application test for rail crack detection, where the UWFBG array fiber cable was laid on rail waist (i.e., in the middle of the rail track, which is also called rail web) (Fig. 10a). When the train went through a crack on the rail, a strong acoustic source was excited at this position owing to the interaction between the wheels and uneven rail, and then propagated both in forward and backward directions. As depicted in Fig. 10b, c, by analyzing the temporal and spatial distribution of sound waves recorded by the DAS system, the intersection point of the forward and backward wave propagation traces can be found, which corresponds to the accurate location of the crack. The method has great prospect in enhancing railway safety.

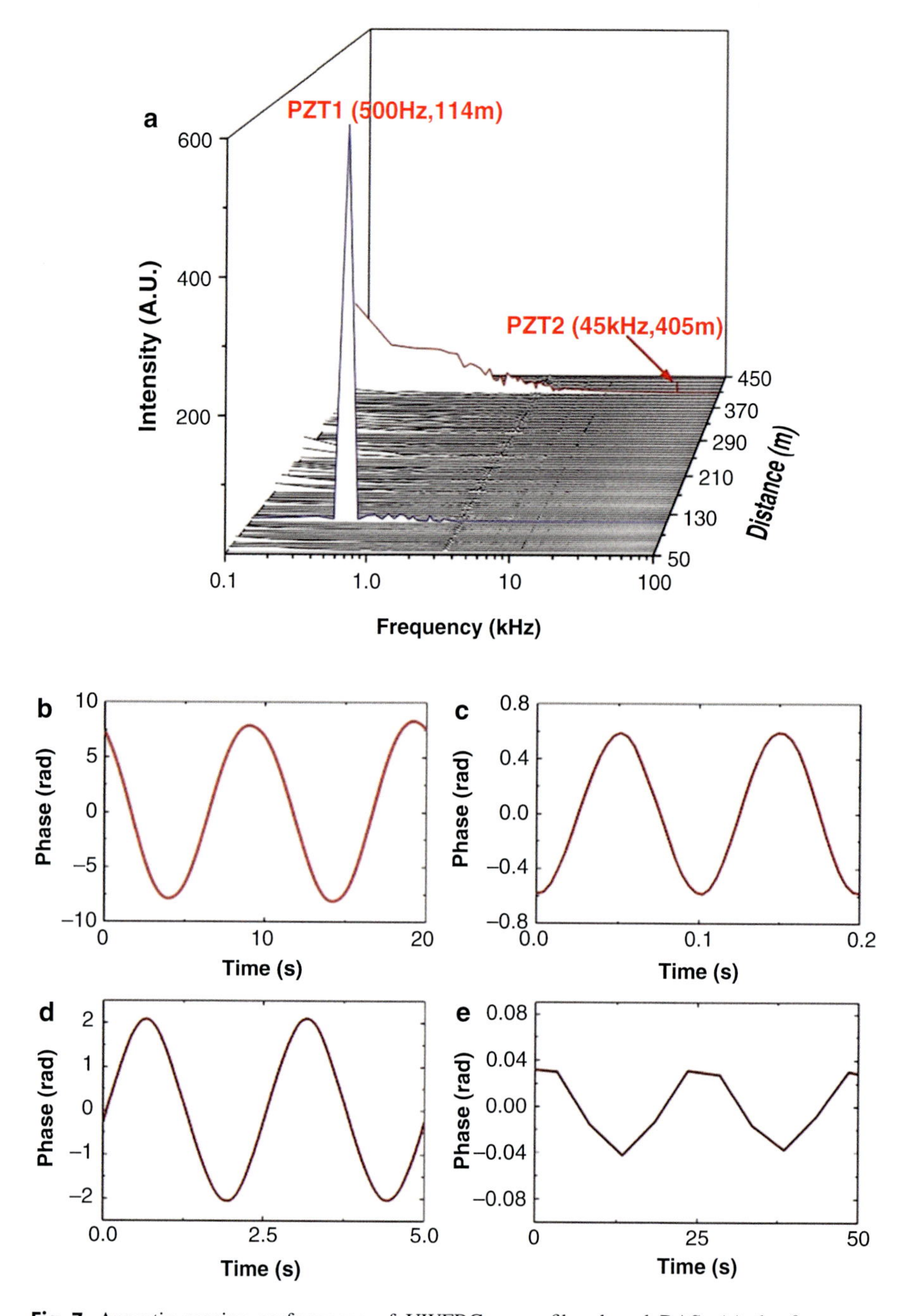

Fig. 7 Acoustic sensing performance of UWFBG array fiber based DAS: (**a**) the frequency spectrum along the fiber when 500 Hz and 45 kHz acoustic signals were applied through PZTs; demodulated waveform of acoustic signal with frequency of (**b**) 0.1 Hz; (**c**) 10 Hz; (**d**) 400 Hz; (**e**) 40 kHz

Fig. 8 Field test of borehole survey: (**a**) photograph of the well site; (**b**) photograph of the borehole and fiber cable; (**c**) recorded borehole seismic data excited by explosive source

WDM/FDM-Based Quasi-Distributed Sensor Network with High Spatial Resolution

The locating principle in TDM scheme described above is still based on the time delay tracking, so the spatial resolution is relatively larger than 1 m, which is not enough for special applications. In order to further increase the multiplexing capacity and improve the spatial resolution of the UWFBG sensor network, a fiber microstructure as shown in Fig. 11a was proposed and designed, which can be considered as Fabry-Pérot interferometer (FPI) composed of two closely spaced UWFBGs. Owing to the weak reflectivity of the gratings, one microstructure can be considered as a low-finesse FPI, of which the reflectivity R_S can be simplified as a

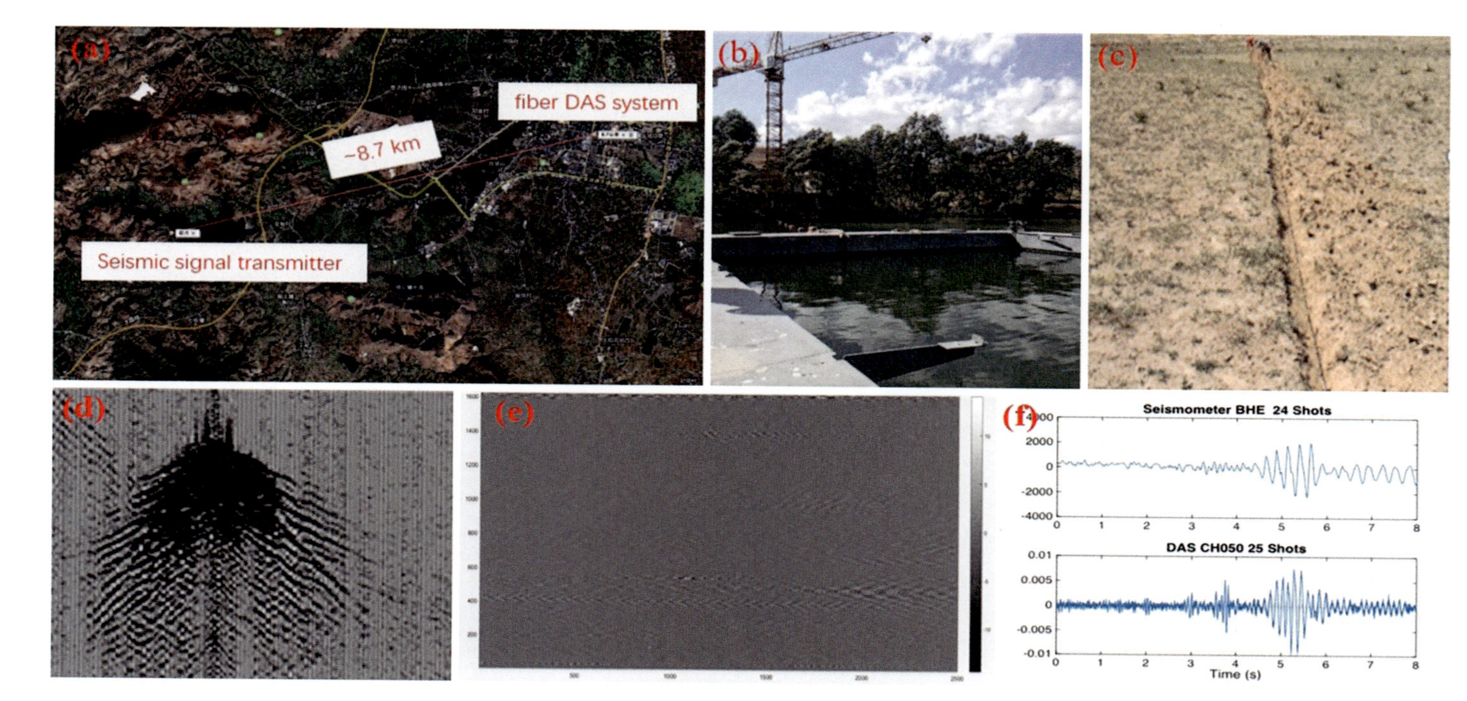

Fig. 9 Field test of seismic recording: (**a**) google map of the test site, where the distance between the seismic signal transmitter and the fiber DAS system is about 8.7 km; (**b**) photograph of the seismic signal transmitter; (**c**) photograph of the shallow buried sensing fiber cable; (**d**) recorded seismic data excited by heavy hammer signal near the fiber cable; (**e**) recorded seismic data excited by vehicle movements; (**f**) comparison of the recorded seismic data from electrical seismometer and fiber DAS, which was excited by the distant seismic signal transmitter

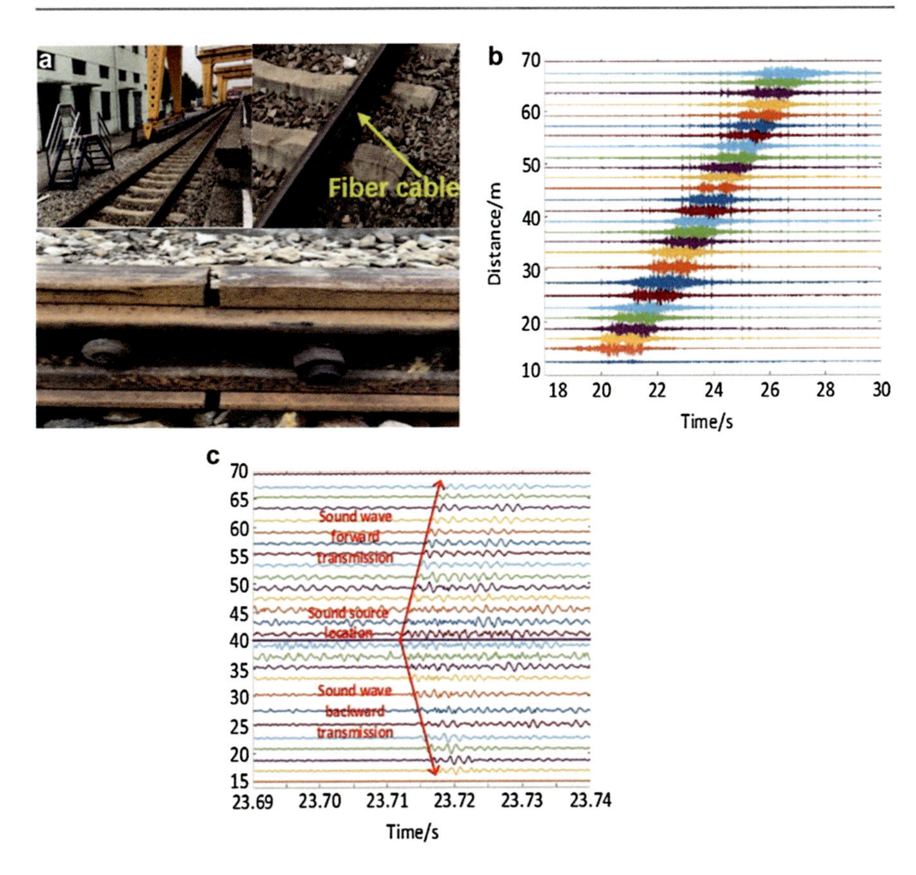

Fig. 10 Field test of railway monitoring: (**a**) photographs of the railway attached with fiber cable and crack on the track; (**b**) acoustic wave distribution along the railway measured by fiber DAS system; (**c**) Zoom of certain section in (**b**) after filtering. (Copyright 2019 OSA)

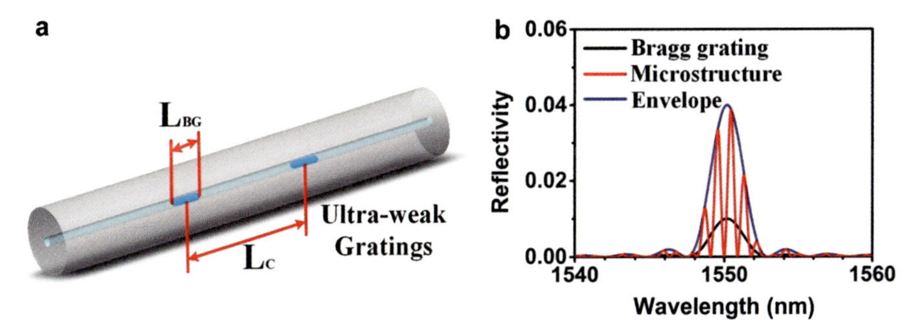

Fig. 11 (**a**) The configuration of the microstructure; (**b**) the reflection spectrum. (Copyright 2019 OSA)

two-beam interferometer (Miridonov et al. 1998):

$$R_S = 2R_G\left[1 + \cos\left(4\pi n_{\mathrm{eff}} L_C/\lambda\right)\right] \tag{1}$$

where n_{eff} is the effective refractive index, L_C is the cavity length of the microstructure, R_G is the reflectivity of the UWFBG, and λ is the operating wavelength. The free spectral range (FSR) of the resonant peak, determined by L_C, is associated with the frequency v_c of the microstructures in the Fourier frequency domain:

$$v_c = \frac{2n_{\mathrm{eff}} L_C}{\lambda_B{}^2} \tag{2}$$

The backward scattering spectrum R_S of the microstructure is an UWFBG envelop modulated by F-P comb filtering as depicted in Fig. 11b, which is related to the Bragg wavelength λ_B and the frequency v_c. Therefore, the microstructures can achieve wavelength and frequency encoding simultaneously (Zhang et al. 2019).

The multiplexing capacity of the microstructures is analyzed for high spatial resolution distributed sensing, which can be calculated through the number of WDM channels and FDM channels as follows:

$$N = N_{\mathrm{WDM}} \cdot N_{\mathrm{FDM}} = \frac{\delta\lambda_s}{\delta\lambda_B} \cdot \frac{L_m}{\Delta L_C} = \frac{n_{\mathrm{eff}} \delta\lambda_s L_m}{2\lambda_B{}^2} \tag{3}$$

Where $\delta\lambda_B$ is the operating bandwidth of the microstructures, L_m and ΔL_C are respectively the maximum cavity length of the microstructure and the cavity length difference between the microstructures in adjacent FDM channels, and $\delta\lambda_s$ is the bandwidth of the interrogation system. Theoretically, over 1000 microstructures with spatial interval less than 1 cm can be high-densely multiplexed along the single fiber (Li et al. 2012a).

In order to achieve high resolution and fast response, a demodulation system is developed by combining the fiber Fabry-Perot tunable filter (FFP-TF) tuning with a parallel signal processing algorithm. As shown in Fig. 12, the parallel processing flow of demodulation scheme contains four parts. Firstly, the modulated optical signal from the sensor array is scanned using a FFP-TF controlled synchronously by the amplified electronic signal. Secondly, the analog signal received by photodetector is collected and converted to digital signal. Thirdly, the digital signal is separated into n groups with the central wavelengths of $\lambda_1, \lambda_2 \ldots \lambda_n$, respectively, and then the data of n groups are synchronously processed in the FPGA with hardware Fast Fourier Transform (FFT). Through the integration of $\tilde{R}(v) = \int_{-\infty}^{\infty} R(v) \exp(-2\pi i \lambda v)\, d\lambda$, where λ is the wavelength, and $R(v)$ is FFT spectra with the peak frequency v of $F_1, F_2 \ldots F_m$, corresponding to different cavity lengths, the component of each frequency channels is filtered. Finally, the Inverse Fast Fourier Transform (IFFT), i.e., $R(\lambda) = (1/2\pi) \int_{-\infty}^{\infty} \tilde{R}(v) \exp(2\pi i \lambda v)\, dv$ (Zhang et al. 2012b), is performed for the m units simultaneously. As a result, all the sensor parameters

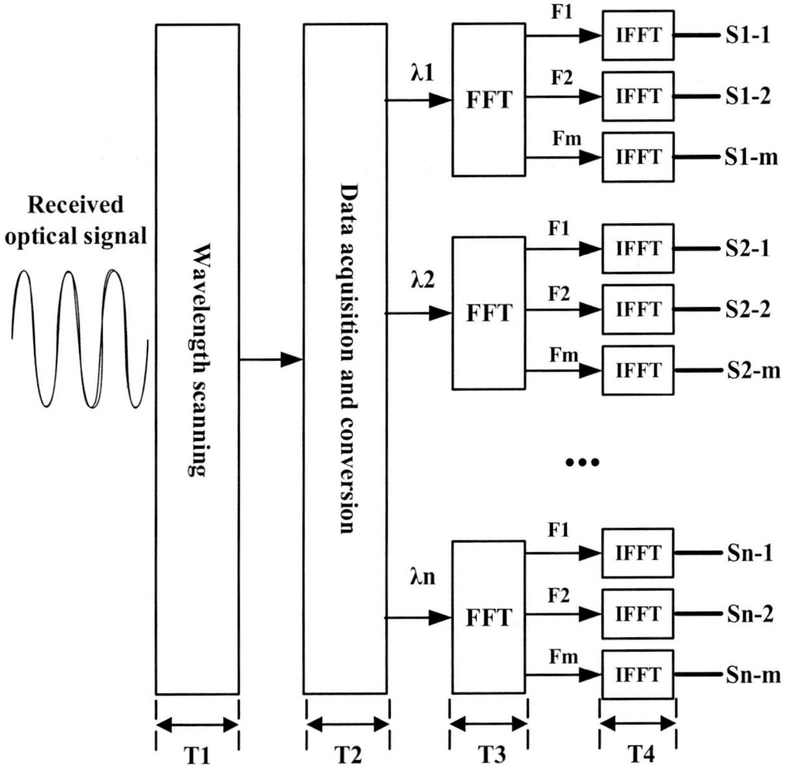

Fig. 12 Schematic of the parallel processing flow for the TDM/WDM microstructure sensor network. (Copyright 2019 IEEE)

can be recovered with high speed. According to this mechanism, a demodulation system with high wavelength resolution of 3 pm and high speed of 500 Hz in the C band was realized, demonstrating the possibility of quickly retrieving the sensing information for the TDM/WDM microstructure sensor network (Cheng et al. 2018).

Beneficial from the high spatial resolution and large capacity of the TDM/WDM microstructure sensor network , a high resolution manometry (HRM) was developed for measuring the pressure and motility of the gastrointestinal tract (Samo et al. 2016), by inserting the packaged sensing fiber into the gastrointestinal tract. Figure 13 illustrates the photographs of the HRM device with gastrointestinal tract pressure monitoring, packaged TDM/WDM microstructure sensing fiber with spatial resolution less than 1 cm, and the schematic of the HRM testing.

Because the bare fiber is only sensitive to the axial strain but almost insensitive to the lateral strain, pressure transducer is necessary for the sensing fiber to enhance the pressure sensitivity. As displayed in Fig. 14a, biocompatible silicon rubber

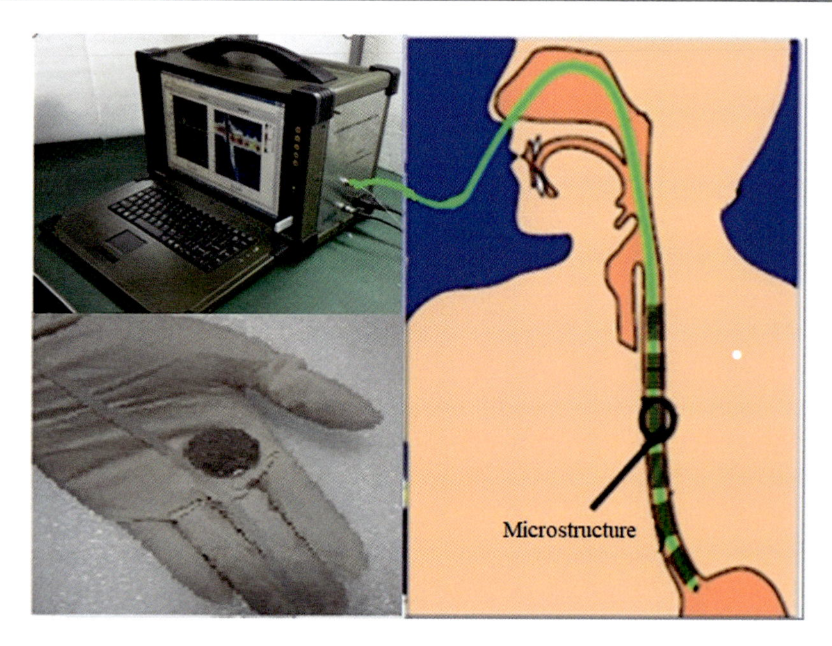

Fig. 13 HRM system based on TDM/WDM microstructure sensor network

was utilized to package the sensing fiber as a pressure transducer and protective coating. Regarding the linear deformation model of the silicon rubber under small pressure, the effective elastic modulus E_{eff} of the composite structure can be calculated as:

$$E_{\text{eff}} = \frac{A_f E_f + A_p E_p}{A_f + A_p} \tag{4}$$

where A_f, A_p are the cross-sectional area of the fiber and the polymer, and E_f, E_p are the elastic modulus of the fiber and the polymer, respectively. When the fiber was encapsulated in the center of the polymer with the diameter of 3 mm, the pressure sensitivity of the sensor was elevated to 2.22 nm/Mpa, which is much more sensitive than the bare FBGs sensor of 3 pm/Mpa. Further, the dynamic pressure response was investigated by tracking the pressure waves along the fiber. Figure 14b illustrates the experimental setup, where a 2 cm/s pressure wave was simulated by rolling a 100 g cylindrical metal stick over the packaged fiber, equivalent to the gastrointestinal tract pressure in vivo. The map of the real-time response is depicted in Fig. 14c, achieving the measurement of velocity, orientation, and value of the pressure wave, which provides information for the diagnosing clinician to find out the motility of the gastrointestinal tract clearly (Zhang et al. 2019 to be published).

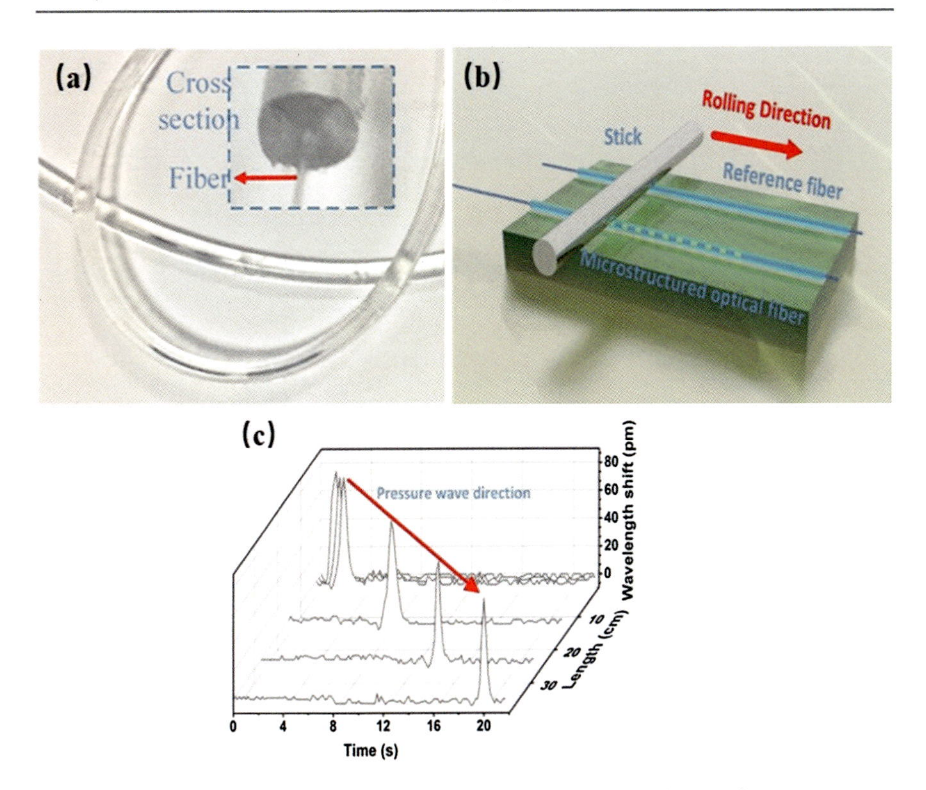

Fig. 14 (**a**) The photo of the packaged sensing fiber; (**b**) experimental set-up for pressure wave measurement; (**c**) the map of the real-time pressure response

Quasi-Distributed Sensor Network Based on 3D Encoded Microstructures

Based on the low-finesse FPI composed of two closely spaced UWFBGs, WDM/FDM/TDM quasi-distributed sensor network is further investigated to realize huge multiplexing capacity in a single fiber. On account of the FPI structure and weak reflectivity, the microstructure can be 3D encoded with different wavelength, frequency, and time slot, named as Wi, fj, and Tk, which are realized by choosing different central wavelengths λ_B of UWFBGs, different spatial distances between the UWFBGs pairs (defined as the cavity length L_C), and delay fiber with certain length.

The configuration of the WDM/FDM/TDM quasi-distributed sensor network is described in Fig. 15, including the microstructured optical fiber and the central office for demodulation. When a probe light pulse is launched into the sensing fiber, microstructures with same time code are first located and distinguished through the time delay of received pulses roughly. The demodulation module analyzing the spectrum to obtain the wavelength and frequency information can be used to locate every single microstructure. Owing to this 3D encoding mechanisms,

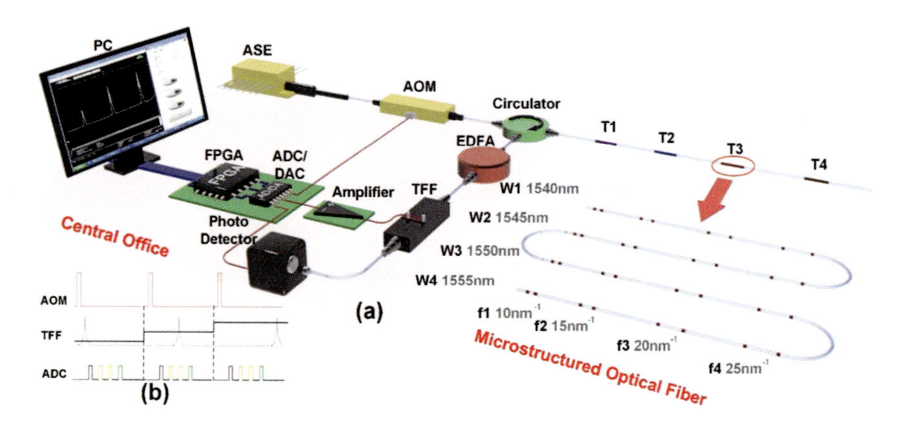

Fig. 15 Configuration of the WDM/FDM/TDM quasi-distributed sensor network: (**a**) experimental setup; (**b**) driving signals of AOM, TFF, and ADC in M-OTDR. (Copyright 2019 Springer)

microstructures spaced closely to each other can be further separated through wavelength and frequency encoding within one time domain, overcoming the restriction of spatial resolution in TDM scheme. The multiplexing capacity could reach to 18,000 when the reflectivity of the UWFBG is as lower as −40 dB (Sun et al. 2017). It can be seen that the quasi-distributed sensor network based on 3D encoded microstructures is able to provide enough sensing points and very flexible configuration for different kind of sensing applications.

As shown in Fig. 15a, the demodulation scheme is based on a tunable FP filter (TFF) and 3D decode through spectrum analysis. The probe light from Amplified spontaneous Emission (ASE) source is modulated into pulses by the acoustic optical modulator (AOM). The modulated light pulses are directed into the sensing fiber. The back-scattered pulses carried with sensing parameters are amplified by an erbium-doped optical fiber amplifier (EDFA) and filtered by the TFF. Avalanche photo detector (APD) transfers the optical signals into electrical ones. The DAC&ADC module (composed of an NI 5781 adapter module and an NI 7962 FPGA module) synchronously controls the modulate time of the AOM and the TFF through DAC, as well as the sequential logic of electrical data captured by ADC. The driving voltage of AOM is a series of voltage pulses with the width of 200 ns, while that of the TFF is modulated in a sawtooth wave (Wang et al. 2018a). It should be noted that scanning nonlinearity and temperature sensitivity of the TFF will seriously affect the demodulation accuracy and stability.

To resolve this issue, an improved demodulation scheme with self-calibration to actively compensate the error induced by TFF was proposed, as illustrated in Fig. 16a. A wavelength calibration unit with multiple reference FBGs is utilized with 3-order polynomial fitting to auto-calibrate the real-time relationship of TFF, and thus to eliminate the demodulation error. Note that the reference FBGs, which can be replaced with any optical filter device, are placed in an incubator chamber to keep the wavelengths constant. Pre-scanning and Polynomial fitting of the TFF function

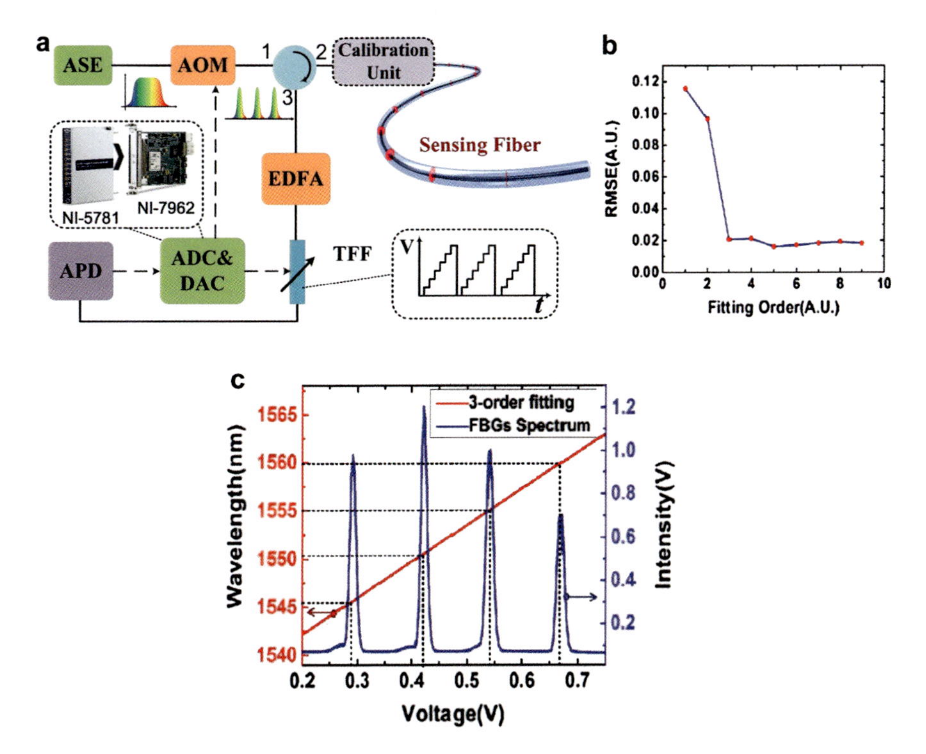

Fig. 16 Schematics of the demodulation platform with self-calibration: (**a**) system configuration; (**b**) the RMSE at different Polynomial fitting orders from 1 to 9; (**c**) spectrum of reference FBG s and the fitting curve. (Copyright 2019 IEEE)

is carried out by launching the tunable laser. Fittings with different fitting orders are established, and the root-mean-square errors (RMSE) are depicted in Fig. 16b, which indicates that linear fitting will induce greater demodulation error and 3-order polynomial fitting is enough for the demodulation. The detected spectrum of FBGs at a random moment and its real-time fitting curve are depicted in Fig. 16c, compensating the demodulation deviation (Wang et al. 2018a). The experimental results demonstrated that the demodulated wavelength deviation was only 6 pm when the temperature of TFF changes for 9.3°C, the wavelength demodulation resolution was 1 pm, and the long-term demodulation precision was 3 pm, which could provide reliable measurements in large engineering projects.

Special Fiber Grating Sensor Network and Applications

In comparison with ultraviolet (UV) lasers, femtosecond lasers may be a more powerful and versatile inscription source and have opened up a new territory for grating and microstructured fiber devices. Apart from seeing many successful commercial deployments of fiber grating sensors, novel and function-enhanced grating-based sensor structures have been developed for unique and high function sensors, by utilizing specially modified or tailored grating structures and special fibers. The exploitation of special grating structures and fibers and the combination with micro-nano features have broadened the sensor technology field for novel and high function optical fiber sensors. In the following, several special fiber grating sensors for network and their applications will be studied.

Tilted Fiber Grating (TFG) Sensor Network and Applications for near Infrared Detection (NID)

In some applications, such as food safety, petrochemical process control, and pharmaceutical production quality control and agriculture area, the indirect detection method was not suitable any longer. So far, the main detection method used in those areas was analytical chemistry technique, which included near infrared spectroscopy, gas/fluid chromatography, nuclear magnetic resonance etc. Among them, near infrared (NIR) technology owns many advantages such as rapid process, nondestructive, noninvasive, chemical free, universal application, and suitable for process analytical technology and quality control (Woodcock et al. 2008; Jamrógiewicz 2012; Cleve et al. 2000).

In NIR detection field, silica optical fiber has been employed as light energy transmitting optical fiber for application in harsh environment, owing to its low transmission loss at the NIR bandwidth. However, it has always been the focus of scientific research and exploration how to use the fiber to realize the distributed multi-point NID. There are several techniques to lead light transmitted inside the fiber core to out of fiber, including side-polished fiber, taped fiber, and fiber grating. Using side-polished fiber and taped fiber, light could interact with analyte by evanescence wave, which has very limited detection depth. The radiation of fiber grating offers a power controllable and effective method to achieve the analyte detection. Specifically, the 45° TFG is the most effective radiation fiber grating, by which the light transmitting inside the fiber core could be partially coupled out. Figure 17a presents the working principle of the 45°-TFG based on Brewster law, in which the light of transverse electric (TE) polarization is coupled out of fiber core and into radiation modes, and the light of transverse magnetic (TM) polarization still transmits inside the fiber core. Hence, the 45°-TFG can be treated as an ideal in-fiber power taping device. According to the previous analysis (Yan et al. 2011, 2013), the taping ratio of TE polarization depends on UV-induced index modulation and the length of grating, which are easily adjusted by controlling the exposing time and grating length. Figure 17b shows the simulated results of taping ratio with different index modulations and grating lengths at the wavelength of 1500 nm.

The most organic molecules have their fundamental characteristic absorption band at infrared area; their combine band would be located at the NIR area. In the mathematic, the light propagation in absorbing materials can be described using a complex valued refractive index. The real part of the refractive index indicates the phase velocity, while the imaginary part indicates the amount of absorption

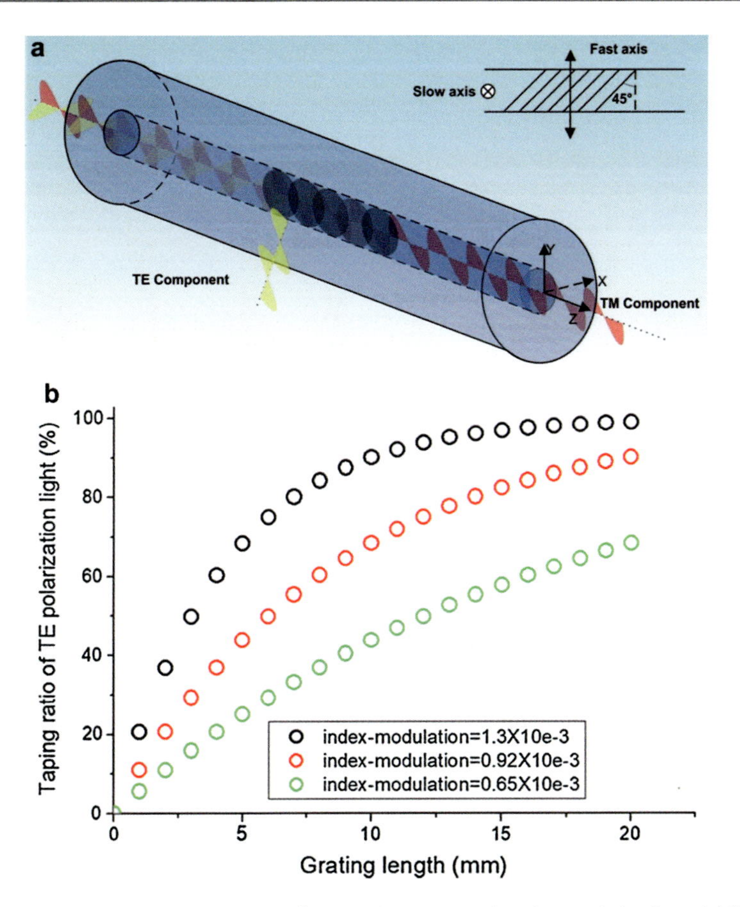

Fig. 17 (**a**) The structure diagram of 45°-TFG; (**b**) taping ratio of TE polarization of 45°-TFG with different grating lengths

loss when the electromagnetic wave propagates through the material. In terms of the absorption mechanism, the 45°-TFGs-based NIR detecting system consisting of 45°-TFGs, low-cost photodetectors, and single-wavelength laser diodes would be potentially used for distributed NIR detecting system. As shown in Fig. 18, the developed prototype 45°-TFG NIR detection system is an optical fiber transmission system, which can be used for multi-point detection. Each detection unit contains one major probe grating and one reference grating. During the measuring process, the light coupled from the reference 45°-TFG is directly reflected to the detector without any interaction with sample. While the light coupled from the probing 45°-TFG is launched to the surface of sample, the reflected/scattered light signal will give information on the probe content due to absorption.

A range of flour samples with different moisture levels were subjected to the measurement under this system. The different moisture level flour samples were prepared by leaving the fully dried (0% moisture level) flour sample in the air for

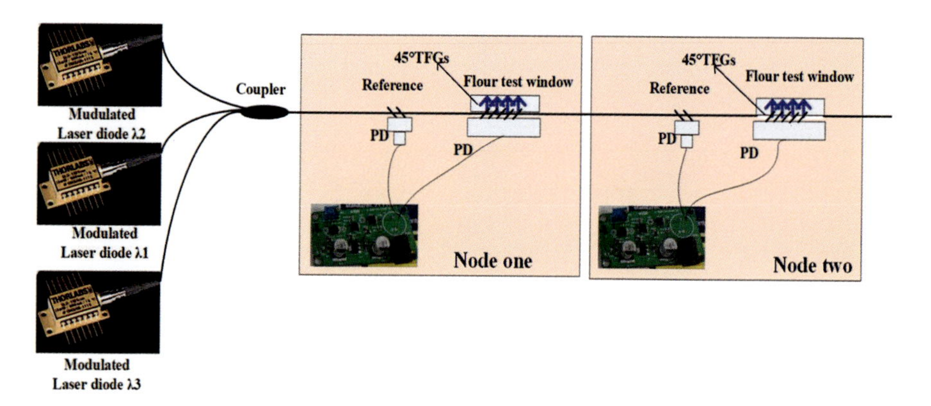

Fig. 18 Configuration of 45°-TFG-based distributed NIR detecting system

a period, then by weighting the mass increment of sample, the moisture level of flour sample was calculated. Fully dried flour sample was prepared by keeping the flour sample at 80 °C oven for 24 h. To make sure the testing flour sample has uniform density and maintains at certain moisture level, in the experiment, 1g weight dried flour sample was filled into small container and left in air to absorb water and then sealed into a small container. Figure 19a shows NIR absorption spectra of flour sample with 10% moisture and 11.4% protein content measured from 1260 to 1580 nm, indicating a similar shape with the one shown in the reported research (Manley 2014). The flour samples were also tested with different moisture levels from 0% to 15.6% at 1450 nm (see in Fig. 19b). The result shows Log(1/R) ($R = \frac{I_{\text{reflectance}}}{a \times I_{\text{reference}}}$, where $I_{\text{reflectance}}$ is the signal intensity reflected from the surface of sample; $I_{\text{reference}}$ is the signal intensity of reference light; a is the correction factor.) is linearly proportion to the moisture level with high sensitivity around 0.014/% and low mean square error around 0.977.

Figure 20 displays the first 45°-TFG-based NIR detection prototype system, which has been installed in the UK Warburtons baking production line for flour moisture test. The benefits of this 45°-TFGs-based NIR detecting system are low loss, compact structure, collimation free, and suitable for long distance, distributed operation. In future, the NIR system would be optimized to compensate the measuring error caused by physical property of sample, such as particle size, and achieve multicomponent measurement.

Er^{3+} Doped Fiber Grating Sensor Network and Applications

Compared to the passive Bragg gratings, fiber grating lasers, which are realized by inscribing gratings in Er^{3+} doped fiber (EDF), not only possess a higher SNR but also offer a large-scale multiplexing capability. Actually, the fiber grating laser sensors mainly adopt the distributed feedback fiber laser (DFBFL) and the

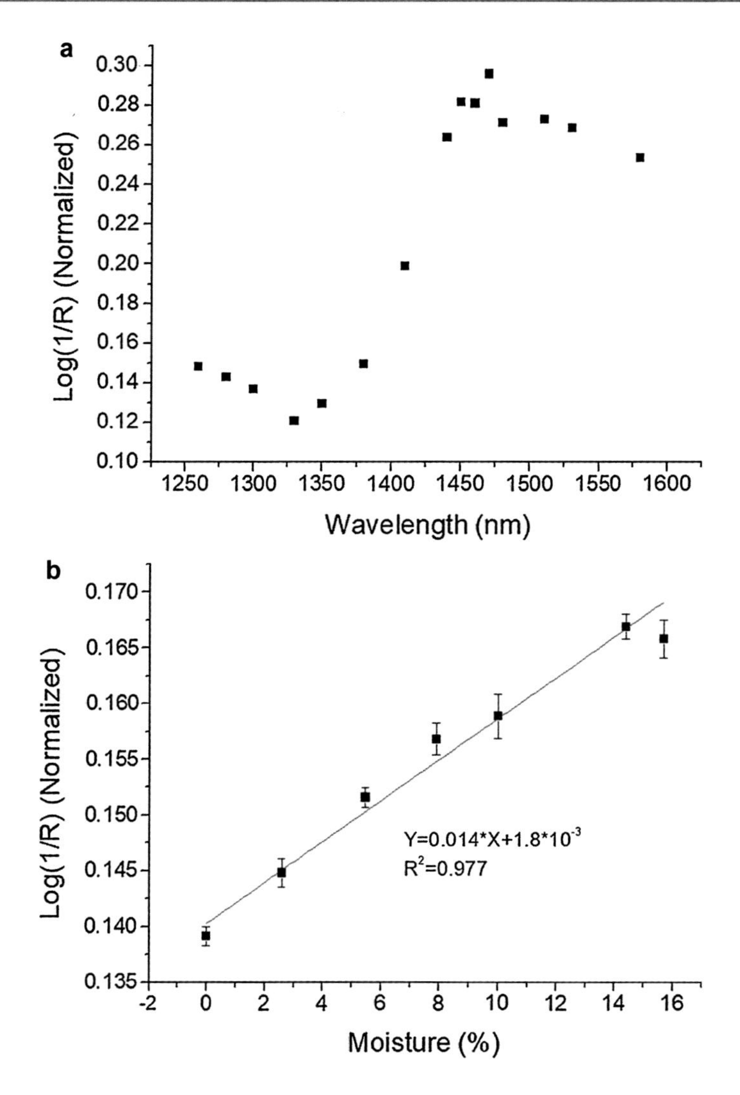

Fig. 19 (**a**) The NIR absorption spectra of flour sample with 10% moisture level and 11.4 protein content; (**b**) the measurement for flour sample with different moisture level at 1450 nm

distributed Bragg reflector fiber laser (DBRFL) as the sensing elements, which can be categorized into wavelength encode sensor and polarizer encode sensor with respect to their detection mechanism. Specifically, the DBRFL operates in single longitude mode with two orthogonal polarization modes and converts the measurand into change in the polarization modes beat frequency. Thanks to the advantages of easy interrogation, high measurement resolution, absolute encoding, and so on, the DBRFL sensors have attracted considerable interest in recent years.

Fig. 20 Flour moisture detection through a compact 45°-TFG-based distributed NIR detecting system

As illustrated in Fig. 21a, the DBRFL sensor is composed of two wavelengths matched Bragg gratings pair written in a short length of EDF directly. Due to the fiber birefringence, the laser always operates in two orthogonal polarization states, and when the laser output is monitored by a photodetector (PD), the two polarization modes will generate a beat signal in the radio frequency domain (Wo et al. 2012). The beat frequency is given as (Guo et al. 2011):

$$\Delta\nu = Bc/n_{\mathrm{eff}}\lambda_0 \tag{5}$$

where c is the light speed in vacuum, λ_0 is the laser wavelength, n_{eff} and B are the average refractive index and birefringence of the optical fiber, respectively. By carefully optimizing the cavity parameters such as the absorption of EDF, reflectivity, Bragg wavelength and bandwidth of the gratings, single-frequency lasing can be easily achieved when the cavity length reduces to several centimeters (see Fig. 21b). Meanwhile, very low noise floor of the beat signal can be achieved, where the noise power spectral density exhibits a C/f profile with the factor C about 3.923×10^4 (Hz^2) (as shown in Fig. 21c). When an external measurand is applied on the laser cavity, the fiber birefringence will change linearly due to the elasto-optic effect, as well as the beat frequency of the two polarization modes.

Similarly to the passive FBG sensors, the DBRFL sensors can be multiplexed in a single fiber through WDM and FDM . A sensor array consisting of as many as 16 DBRFLs have been presented, which were wavelength multiplexed by inscribing fiber gratings with different pitches, and frequency multiplexed in RF domain by controlling the intracavity birefringence. The lasing wavelengths range from 1528.8

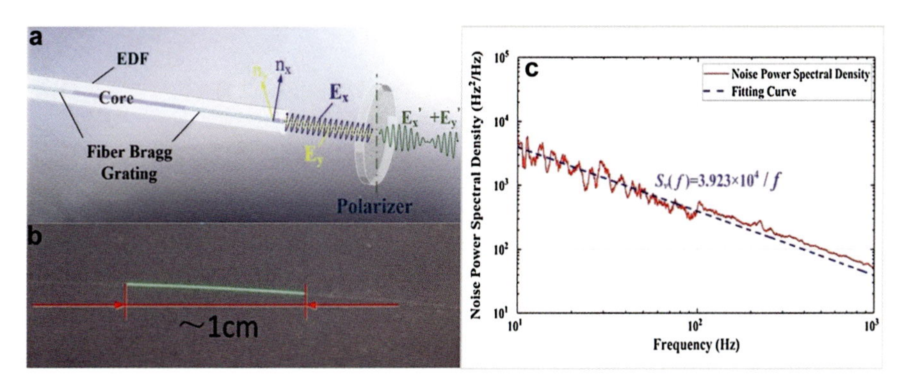

Fig. 21 Schematic diagram of DBRFL: (**a**) structure and principle of DBRFL; (**b**) photo of the short cavity DBRFL; (**c**) frequency noise power spectral density of the DBRFL

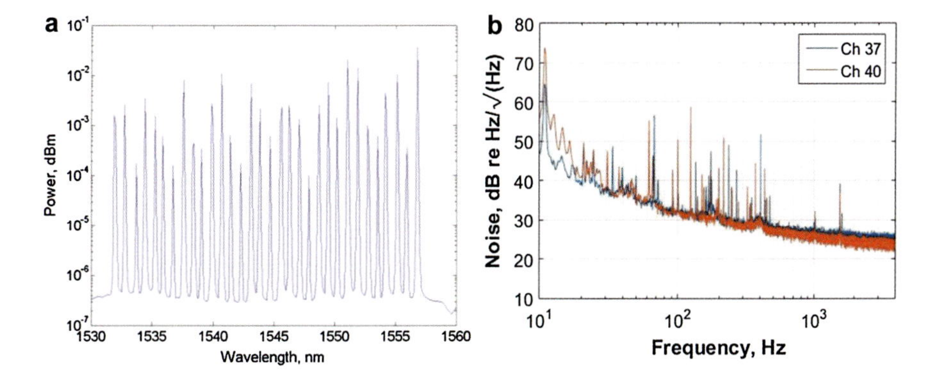

Fig. 22 The optical spectrum (**a**) and frequency noise spectra (**b**) of the multiplexed DBRFL sensor network. (Copyright 2019 OSA)

to 1566.4 nm with the spacing of 2.4 nm and SNR higher than 35 dB, and the output beat frequencies ranges from 140 MHz to 1.7 GHz, controlled by the CO_2 laser side irradiation to the laser cavity (Jin et al. 2014). Besides, through specially design the DBRFLs to eliminate out of band reflections (minimize laser-laser interactions), decrease the laser pump threshold (~1 mW at 1480 nm) and energy absorption per device (<0.4 dB), 32 DBRFL sensors were successfully wavelength multiplexed on a single fiber with output wavelengths matched to channels 26–57 of the C-band 100 GHz ITU grid (Harrison and Foster 2017). The optical spectrum of the multiplexed fiber laser array is shown in Fig. 22a, and Fig. 22b presents the frequency noise spectra of different DBRFLs. Certainly, the DBRFL is a credible candidate for large capacity sensing network.

The DBRFL sensor network has been successfully demonstrated for measuring different parameters, such as lateral force, vibration, acceleration, displacement, bending, ultrasound, current, etc. One typical application is developed for pulse and respiration monitoring for body health (Sun et al. 2014). As depicted in Fig. 23a,

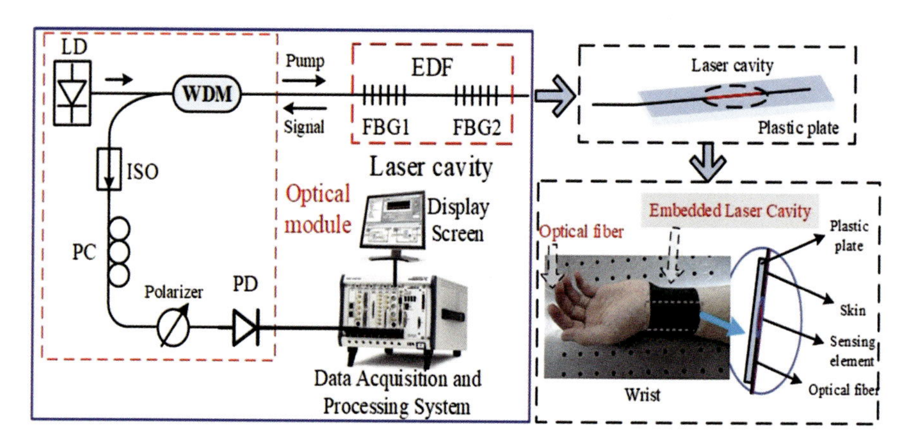

Fig. 23 The schematic diagrams of: (**a**) DBRFL based sensor system; (**b**) the packaged sensing device for pulse monitoring; (**c**) the assembling and fixing method for the wrist pulse monitoring. (Copyright 2019 SPIE)

the laser cavity with the physical length of only 10 mm is selected as the sensing element. A 980 nm laser diode is launched to illuminate the laser cavity through a 980/1550 nm WDM. The backward lasing output is injected into a data acquisition and real-time processing system (NI PXIe-1082) through a PD via an isolator (ISO), a polarization controller (PC), and fiber polarizer. By adjusting the PC, the beating signal intensity of the two orthogonal polarization modes could be maximized. The beat frequency signal received by the data process system is then acquired and displayed using the application program based on LabVIEW platform. To fabricate the transducer for arterial pulse monitoring, the laser cavity is first attached onto an elastic and flexible plastic plate with epoxy adhesive, and then tightly fixed onto the arterial pulse position (for example wrist artery) of a person with the help of textile belt, as displayed in Fig. 23b, c. The detail in Fig. 23c presents how the plastic plate is attached onto the wrist pulse. The fiber is kept straight during the packaging process to eliminate any bending effects.

As illustrated in Fig. 24a, the beat signal changes with time periodically and the maximum frequency change is about 10 MHz. In this arterial pulse waveform, the sharp rising edge indicates a contraction behavior of the artery, while the slow falling edge represents the artery relaxation behavior. Also, the incisura of the dicrotic notch to the end of diastole of the artery pulse wave is clearly depicted in Fig. 24b, which is accordant with the practical situation. Actually, the dicrotic notch is a very important and meaningful characteristic parameter for the cardiovascular disease diagnosis. From the pulse waveform, pulse rate could also be obtained. Each time when the beat signal reaches the systolic peak, the pulse beat count is incremented once. The investigation results in Fig. 24c indicate that the average pulse rate of the sample is around 75 times per minute.

Further, by using a 10×27 mm^2 solid plastic plate to support another laser cavity, the DBRFL sensor network can be simultaneously employed for monitoring

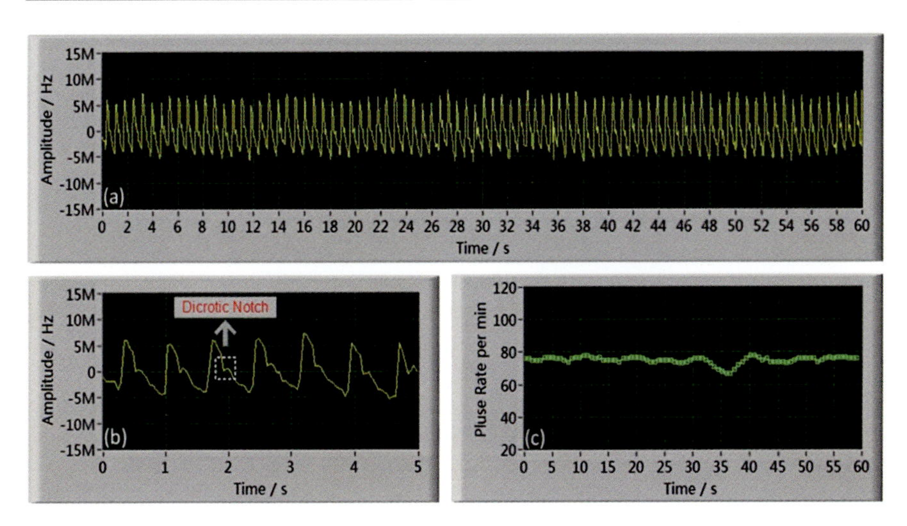

Fig. 24 Recorded wrist arterial pulse for tested person: (**a**) waveform within 1 min; (**b**) enlarged 5-s fragment from (**a**); (**c**) calculated average pulse rate during 1-min recording. (Copyright 2019 SPIE)

respiration activity (Wo et al. 2014), as shown in Fig. 25a. The textile belt is fastened tight on the abdomen position of the tested person, and then the respiration movement including both the contraction and relaxation behavior will lead the elongation of the flexible textile, which change the transverse pressure applied onto the laser cavity. In this case, the respiration strength, as well as the breath rate could be obtained simply by monitoring the frequency change of the beat signal.

Figure 25b, d give the recorded the respiration waveforms for the two persons within 1 min when the DBRFL sensor is subjected to the elongation of the abdominal circumference during respiration movements. It can be seen that the beat signal changes with time periodically and the maximum frequency change is about 8 MHz for subject A and 4 MHz for subject B, and the inhalation and exhalation behavior of the respiration can be clearly observed. In addition, the respiration rate can be obtained through Fourier transformation of the respiration waveform. Figure 25c, e indicate that the breath rates of the two persons are 12.60 and 19.32 times per minutes, respectively. The excellent performance of the DBRFL sensor network paves a new way for monitoring multiple healthy parameters.

Fiber Optic Sensors Passive Optical Network (SPON) and Applications

Most of us are familiar with the topology of FTTX, which consists of the optical line terminal (OLT), the passive optical network, and the active optical network units (ONU). The downstream is broadcasted from OLT to every ONUs, while the upstream is directionally transmitted from certain ONU to OLT. According to

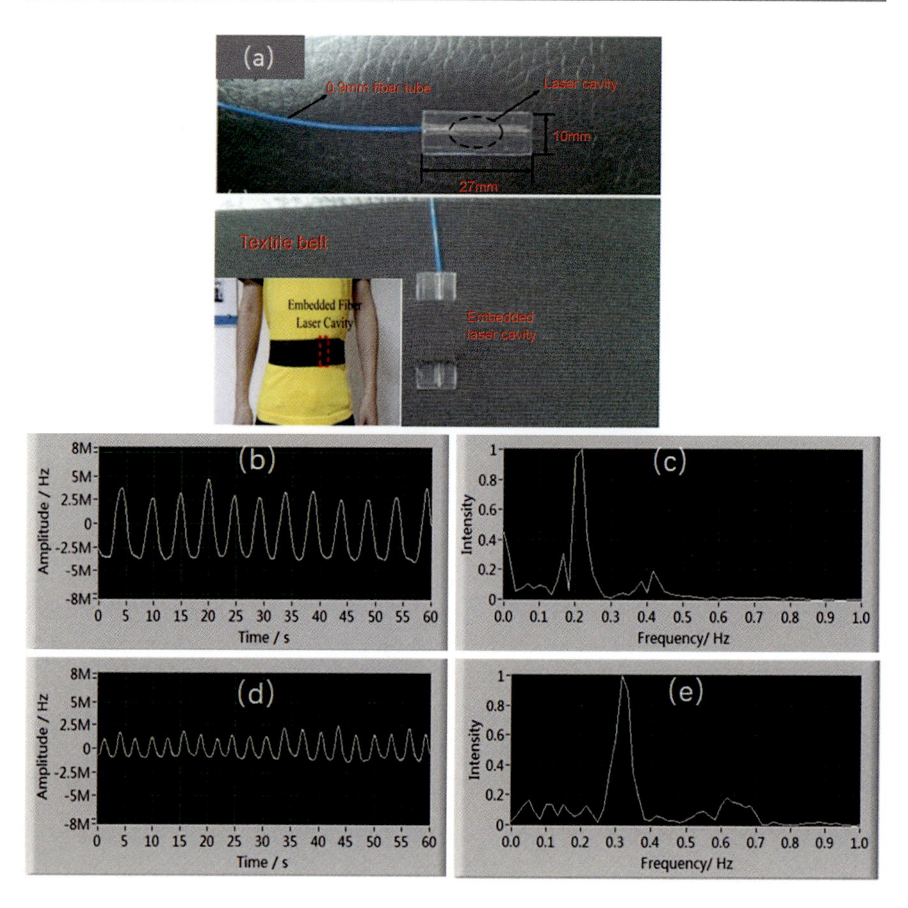

Fig. 25 (**a**) Fixture of the laser cavity for respiration monitoring; recorded respiration waveform for subject A (**b**); and subject B (**d**); frequency spectrum of respiration waveform for subject A (**c**); and subject B (**e**). (Copyright 2019 SPIE)

this configuration, the similar architecture for fiber sensor network, named as fiber optic sensors passive optical network (SPON), was proposed (Liu et al. 2016). As represented in Fig. 26, the SPON is also composed of three main parts, OLT, ODN, and optic sensor units (OSU). However, it is different with the communication PON. Firstly, the transmission data from OLT is always analog, and is broadcasted to every OSU from ODN. Secondly, the OSN is always passive due to the characteristics of optical sensor. Hence, in order to make the SPON work effectively, the OSU should be of self-feedback, that is, it can reflect the modulated signal which carries the sensing information to the OLT through the ODN, and then the modulated signal can be demodulated at OLT to achieve the sensing parameters. In this network architecture, the OSU can be multi-functions, multi-structures, and multi-points, as long as the OLT can provide appropriate light source, as well as signal demodulation processing. And the OSUs can be multiplexed through WDM, TDM, FDM, and

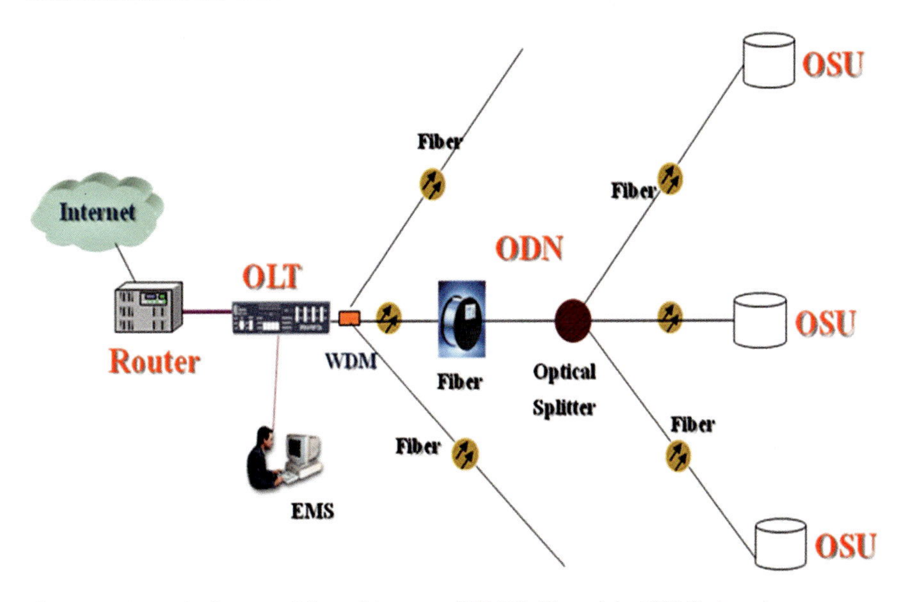

Fig. 26 Schematic diagram of the architecture of SPON. (Copyright 2019 Springer)

their combinations. Therefore, the SPON is of large capacity, good adaptability, high extendibility, and great flexibility as the communication PON. Next, several types of SPON architectures including the star topology and tree topology networks, and their practical applications will be presented.

TDM-Based Fiber Optic Acoustic SPON

Acoustic sensors have been widely used in scientific research and industrial applications, such as the study of vector sound field, acoustic imaging, underwater detecting, nondestructive testing (NDT) for large structures, partial discharge (PD) detection in power transformers, and photoacoustic gas detection. In addition to high sensitivity, the real-time multipoint measurement is also a common demand in the above applications. Acoustic sensor array can map the distribution of the whole sound field, thus to better locate and identify the characteristic events. Optical fiber acoustic sensors have unique advantages of multiplexing, high sensitivity and excellent immunity to EMI. In this part, a diaphragm-based optical fiber acoustic sensor array with the help of TDM and coherent phase detection will be reported. A prototype sensor array was developed and successfully applied in several fields.

The system configuration of the fiber acoustic SPON is illustrated schematically in Fig. 27a. Multiple sensor tips are connected by a 1*N coupler with single mode fiber (SMF) delay line (DL) in different length, and multiplexed through TDM, constructing a star topology network. Figure 27b depicts the structure of one

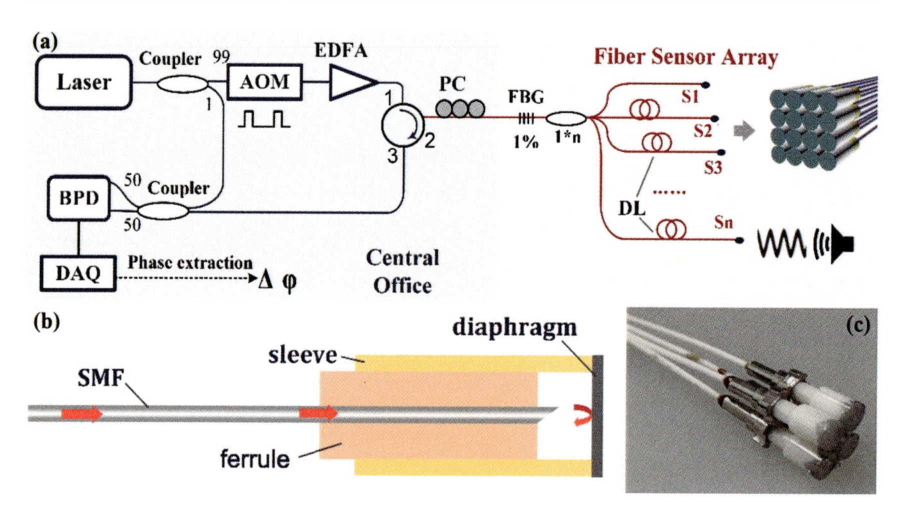

Fig. 27 (**a**) Schematic of multi-point acoustic sensing system using coherent detection. (**b**) Schematic of the sensor tip; (**c**) photograph of 2 × 2 sensors array. (Copyright 2019 OSA)

acoustic OSU in detail. A ceramic sleeve with the side opening is tightly assembled to a ceramic ferrule, in which an optical fiber pigtail is fixed central. The facet of fiber pigtail is polished with an angle of 8° to reduce the Fresnel reflection from the fiber end. A 10-layer graphene diaphragm is smoothly glued on the top of the sleeve with UV-curable adhesive. The extremely thin graphene diaphragm is covered with PMMA, with the total thickness of about 5 nm. The sensor tip is about 2.5 mm in diameter, while the distance between graphene diaphragm and fiber facet is several hundred micrometers (see the photograph of sensor tips in Fig. 27c). When the acoustic wave is applied on the sensor tip, the diaphragm vibrates accordingly, resulting in the length change between the diaphragm and fiber facet. In this way, the optical phase modulation of the reflected light from the diaphragm is achieved (Wang et al. 2018b).

As the reflection signal from graphene diaphragm is weak, coherent phase detection technique for improving the signal-to-noise ratio (SNR) is employed for the sensor array, which up-converts the acoustic signal into the high sideband frequency to reduce the noise. Hence, the laser with narrow linewidth outputs continuous light of 1550 nm, which is split into the probe light and the local-oscillator light by 99:1. The probe light is modulated into pulses with the width of 10 ns through an acoustical optical modulator (AOM), as well as being frequency shifted by 200 MHz. After amplified by an EDFA, the probe light pulses launch into the sensing fiber through a circulator. The injected pulses are partly reflected by a FBG before entering into the sensor array. Note that the weak FBG here serves as the referenced reflection point with fixed optical phase. The reflected pulses from sensor array are carried with acoustic signals and interfere with the local-oscillator light. The generated beat frequency signals are received by a balanced photo detector (BPD) and collected by a data acquisition card (DAQ module from National

Instrument with sample rate of 2 GS/s). Due to the different time delay of reflected pulses, beat frequency signals from different sensor tips can be distinguished in time domain (Wang et al. 2018b).

For *TDM*-PON, the capacity is mainly determined by the power budget, and the theoretical capacity *N* can be deduced as follow:

$$N \leq \sqrt{\frac{r\,(1-R)}{5\Delta L \alpha_R S_{\mathrm{n}}}} \tag{6}$$

Where *R* and *r* are the reflectivity of the weak *FBG* and graphene diaphragm, respectively; ΔL is the length of injected light pulse; α_R and S_n represent the Rayleigh scattering coefficient and the ratio of backscattered Rayleigh scattering, which is about 0.032×10^{-3}/m and 10^{-3}, respectively in SMF. Assuming that $R = 1\%$ and $r = 2\%$, the maximum network capacity will reach to 248 and can be further increased by using a sensing diaphragm with higher reflectivity (Wang et al. 2018b).

The sensing performances were experimentally investigated and presented in Fig. 28. The result in Fig. 28a exhibits relative high sensitivity of larger than −136 dB re 1 rad/μPa within the full measured frequency range from 300 Hz to 15 kHz. And a resonance peak appears at 1.2 kHz with the sensitivity up to −119 dB re 1 rad/μPa. As seen from Fig. 28b, the SNR is about 37.68 dB, corresponding to the MDP of only 75 μPa/Hz$^{1/2}$. The fluctuation of the output signal is about 0.66 dB (see Fig. 28c), proving high measurement stability under the large variation of temperature. In addition, the curves of different frequencies in Fig. 28d exhibit excellent and flat acoustic response with wide directivity, where the normalized acoustic sensitivity is optimal at 0° and only decreases to 0.72(−2.85 dB) within the angle range of ±90°. Although the frequency response of the sensor is not flat and broadband, it can be optimized by specially designing the diaphragm. The excellent MDP, high temperature stability, wide directivity, and large multiplexing capacity make the acoustic SPON favorable for acoustic detections in industrial applications (Wang et al. 2018b).

The fiber acoustic SPON can realize multipoint acoustic detection, providing the typical application for sound source localization. Three sensor tips can form the smallest sensor array system for two-dimensional (2D) sound source localization. As shown in the Fig. 29a, b, three sensors are placed at the three vertices of a square. A point acoustic source is set in the square and emits the sound wave at one specific moment, then the sound waves propagate through medium and detected by the sensor array simultaneously. Then the waveforms measured by three sensor tips have similar morphological characteristics but different in timing. While the time difference of arrival (TDOA) can help to calculate the distance difference Δx_{ij} of every two sensors to the sound source, hyperbolic positioning algorithm is adopted to locate the sound (Fu et al. 2014). The intersection of the two solid hyperbola lines corresponding to Δx_{23} and Δx_{13} is the location of the sound source. Through this way, the coordinate is calculated to be A′, which is very close to the actual

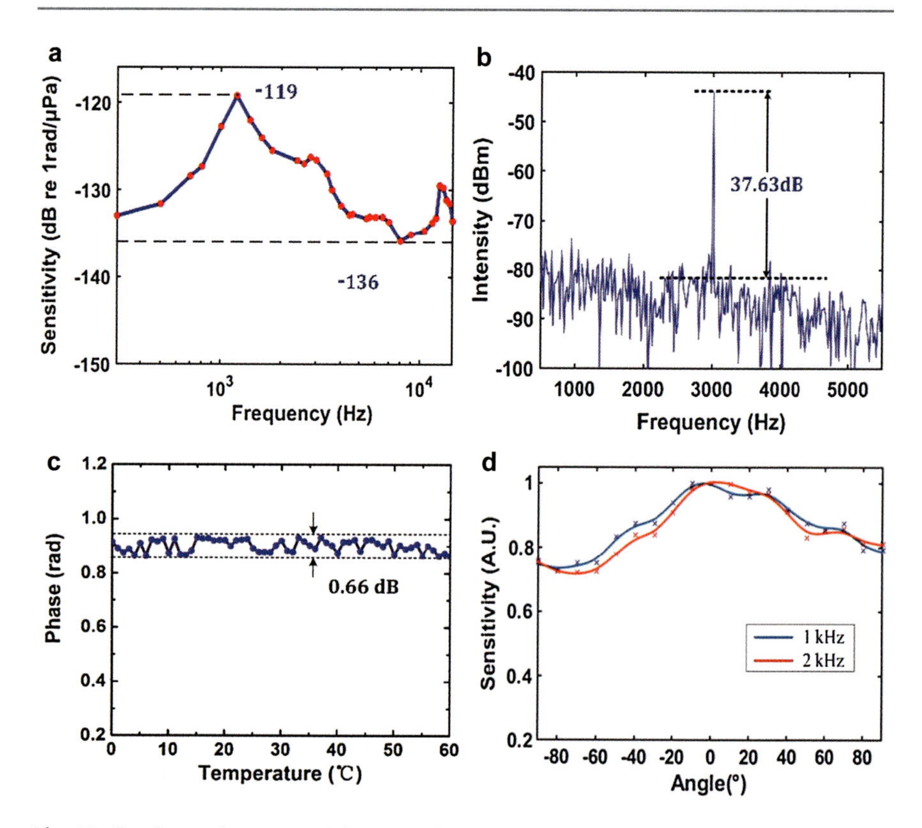

Fig. 28 Sensing performances of the sensor tip: (**a**) Frequency response from 300 Hz to 15 kHz; (**b**) Power spectrum of the measured signals when an acoustic pressure level of 25.7 mPa at 3 kHz was applied. (**c**) Temperature stability from 0 to 60 °C; (**d**) Normalized sensitivity with different acoustic incident angle. (Copyright 2019 OSA)

position A. The multiple tested results in Fig. 29c demonstrates that the maximum positioning deviation is 3.55 cm, which can be further improved by increasing the sampling rate, and the number of sensor tips.

Fiber optic hydrophone is the key device in the applications of underwater detection (Vidakovic et al. 2016; Kim et al. 2017). In the light of the above sensing principle and networking mechanism, a polymer diaphragm-based fiber optic hydrophone was developed, with a metal sleeve to resist the hydrostatic pressure and a 20 μm thickness PDMS diaphragm as the sensing part (see Fig. 30a). This sensor tip can meet the demands of high SNR, low frequency, and large multiplexing capacity to suppress the ocean noise. To calibrate the hydrophone in low frequency, a calibrator based on a vibrating column of liquid was built. As shown in Fig. 30b, d, the calibrator consists of a vibration exciter at the bottom and a half meter height tank, which uses an open column of liquid so that the wavelength of acoustic wave can be larger than the length of the column for low frequency detection. As seen from Fig. 30c, the undertest hydrophone and the

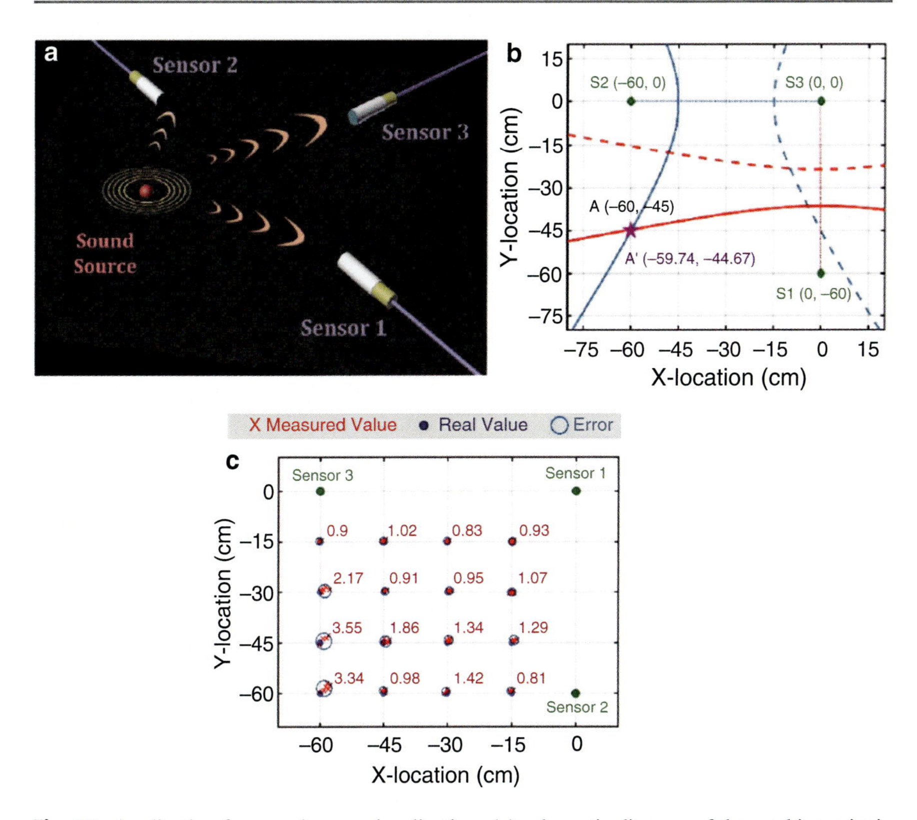

Fig. 29 Application for sound source localization: (**a**) schematic diagram of the working principle; (**b**) result of sound source localization with hyperbolic positioning technique; (**c**) positioning error assessment of sound source localization. (Copyright 2019 OSA)

reference hydrophone are placed at a same depth in the tank to achieve the reliable calibration.

The sensing performance of the polymer diaphragm-based fiber optic hydrophone was evaluated. From Fig. 31a, b, it can be seen that the fiber hydrophone could recover the acoustic signal with high fidelity and high SNR of 60 dB. As shown in Fig. 31c, the sensitivity of the air acoustic signal is as high as −120 dB re 1 rad/μPa with a flat frequency response from 50 to 800 Hz, while the sensitivity of the underwater acoustic signal is about −135 dB from 50 to 500 Hz with the non-uniformity less than 2 dB, which corresponds to the MDP of 300 μPa. Notably, the degradation of the sensitivity and resonant frequency underwater are mainly ascribed to the flow resistance of the water. Also, the response of the hydrophone in water at 100 Hz is plotted in Fig. 31d, where the underwater acoustic response is linear along with the pressure below 180 Pa, corresponding to a large dynamic range up to 118 dB.

Moreover, the fiber acoustic SPON can be used for pulse wave monitoring and diagnosis, which is an important tool in Traditional Chinese Medicine (TCM)

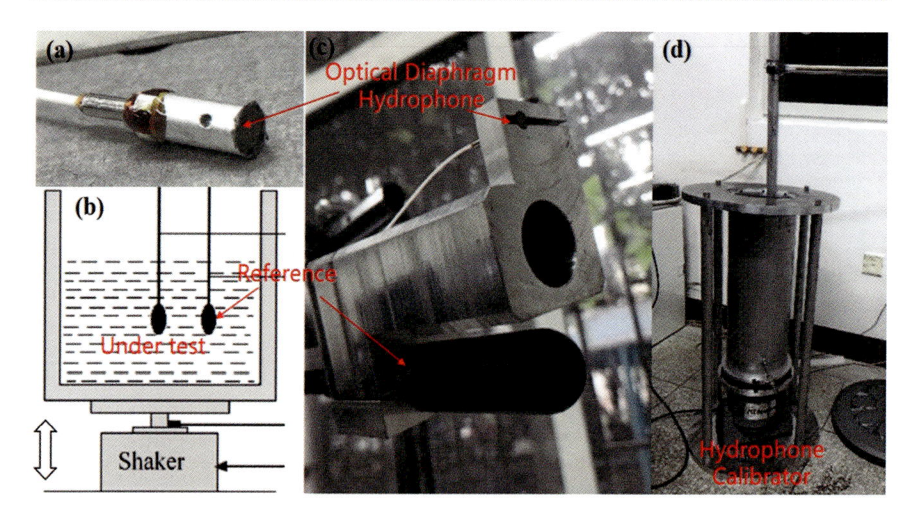

Fig. 30 (**a**) Photograph of the polymer diaphragm based hydrophone; (**b**) schematic of the calibration platform for low frequency underwater acoustic detection; (**c**) photograph of the diaphragm based hydrophone and the referenced hydrophone installation; (**d**) photograph of the calibration equipment

(McVeigh et al. 1999). As illustrated in Fig. 32a, b, the sensor tip is made of an aluminum diaphragm with 12 μm in thickness and a fiber pigtail with about 1 cm in diameter, to provide high-precision pulse wave monitoring. Wrist radial artery is chosen as measurement site and the fiber sensor tip is embedded into a sports-wristband to form a wearable device. The radial artery is usually chosen as the pulse monitoring location because the support of the radius facilitates the fixation of contact pulse sensor. Heart pumps blood and generates pressure signals in the artery which is detected by noninvasive fiber sensors as described in Fig. 32c, reflecting in the regular expansion of blood vessel walls.

A typical pulse waveform is plotted in Fig. 33a. Generally, the pulse cycle is divided into systolic phase and diastolic phase, the junction of which is called the dicrotic notch. In systolic phase, the blood is pumping from heart, resulting in a sharp rise in blood pressure which corresponds to the main wave in pulse waveform. And the predicrotic wave is formed by the multiple reflections of arterial wall. In diastolic phase, the aortic valve closes. The elasticity of the vessels causes a small amount of blood pumping back to the heart, creating a brief rise in blood pressure, appearing as dicrotic wave. From the pulse waveform, the widely used assessment indicators including augmentation index (AIx), average blood pressure(Pm), systolic blood *pressure* (SBP), diastolic blood pressure (DBP), blood viscosity (V), and the morphological description index of pulse waveform(K) can be calculated according to the formula in Fig. 33a (Avolio et al. 2009).

The pulse waveforms of four subjects aged between 22 and 24 years old were measured by the developed system, of which the details were restored with high-fidelity as shown in Fig. 33b. Then the pulse morphological parameters can be quantitatively analyzed in time domain.

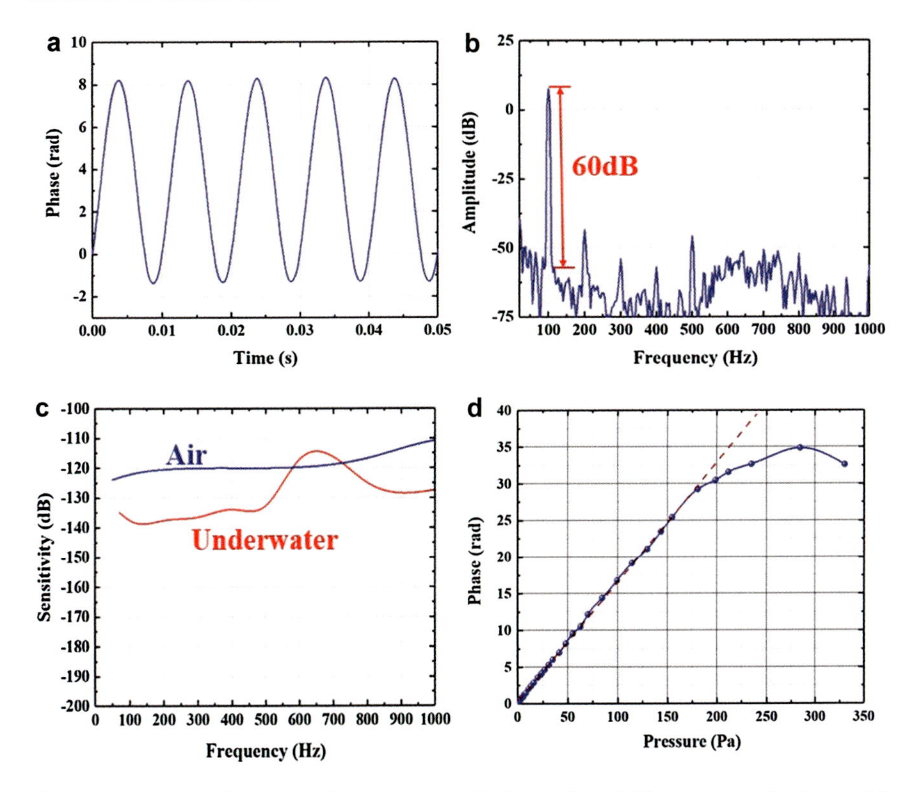

Fig. 31 Sensing performance of the polymer diaphragm-based fiber optic hydrophone: (**a**) acoustic wave recover at the frequency of 100 Hz; (**b**) frequency spectrum of the recovered signal in (**a**); (**c**) frequency responses from 50 Hz to 1 kHz when the fiber optic hydrophone was placed in air and underwater; (**d**) underwater acoustic response of the fiber optic at 100 Hz

The pulse parameters were calculated and listed in Table 1, according to the definitions in Fig. 33a. It is believed that the lower values of the parameters K, V, and AIx mean better peripheral vascular resistance, blood viscosity, and arterial compliance, respectively. Hence, the cardiovascular health conditions of subject A and B should be slightly better than C and D, though all of them stayed healthy concerning cardiovascular conditions. The pulse analysis results were consistent with the physical examination reports, preliminarily verifying the accuracy of pulse diagnosis by optical fiber sensing system and showing great potential in early cardiovascular diseases indicating.

WDM/TDM-Based Fiber Optic SPON

Except for the TDM-based SPON, hybrid WDM/TDM SPON is another mature architecture, which has higher deployment flexibility and potential larger

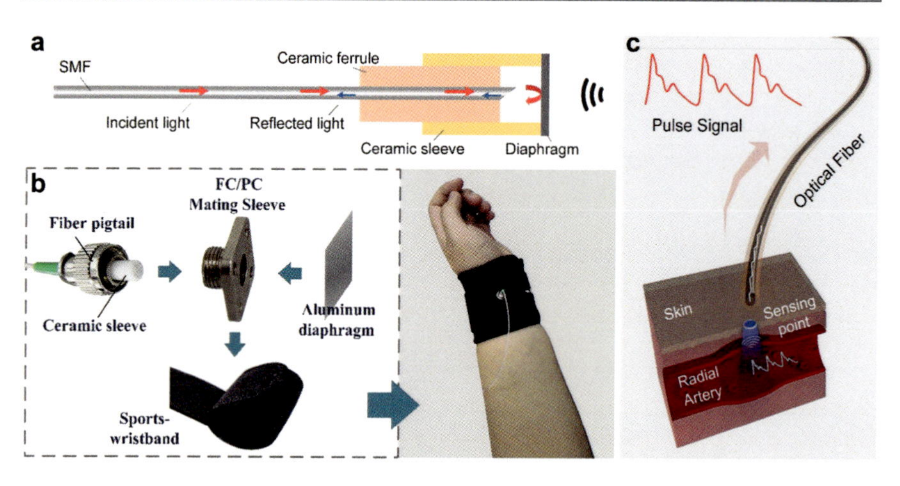

Fig. 32 (**a**) Schematic diagram of fiber optic pulse sensor; (**b**) structural design of wearable pulse sensor; (**c**) the transmission of pulse wave signals

multiplexing capacity. Here two hybrid multiplexing SPONs respectively for colored sensors and colorless sensors accessing, and their applications for micro-vibration monitoring, will be mainly discussed.

The first WDM/TDM SPON is designed for colored fiber sensors multiplexing, especially the fiber grating sensors with wavelength selectivity. Take the DBRFL sensors as the example, of which the structure, principle, as well as the networking in a single fiber have been introduced above. Actually, they can also be multiplexed in SPON to further increase the capacity. Figure 34 exhibits the schematic of the proof-of-concept sensing network, which is constructed by high-speed multi-channel optical switch (OS) for TDM and different lasing wavelengths of DBRFLs in every fiber link for WDM. Generally, the multiplexing capacity is limited by the pump budget and the switching speed of OS. In the central office, the TDM is realized by an 8 switching channels optical switch with only 100 ns switching time. The WDM is achieved by a dense wavelength division multiplexer (DWDM) with 16 wavelengths matched to channels 17–32 of C-band 200 GHz ITU grid. With the help of low pump threshold (<7.74 dBm) and low energy absorption (<0.8 dB) of the DBRFL, the 980 nm laser diode with high laser power (>28 dBm) supports serial pumping over 16 sensors. The interrogation system consists of 16 parallel demodulation channels for receiving the signals from 16 wavelength channels, and each channel can interrogate 8 channels by time-polling. Consequently, the DBRFL SPON can provide a multiplexing capacity of 128 (Liu et al. 2018).

For dynamic sensing, an interrogation system with high resolution and high response speed was developed to measure the beat frequency generated from the DBRFL sensor network. As exhibited in Fig. 35, the interrogation system consists of Radio-Frequency (RF) receiving part, digital processing part, frequency discriminator part, frequency source part, and Input/Output (I/O) part.

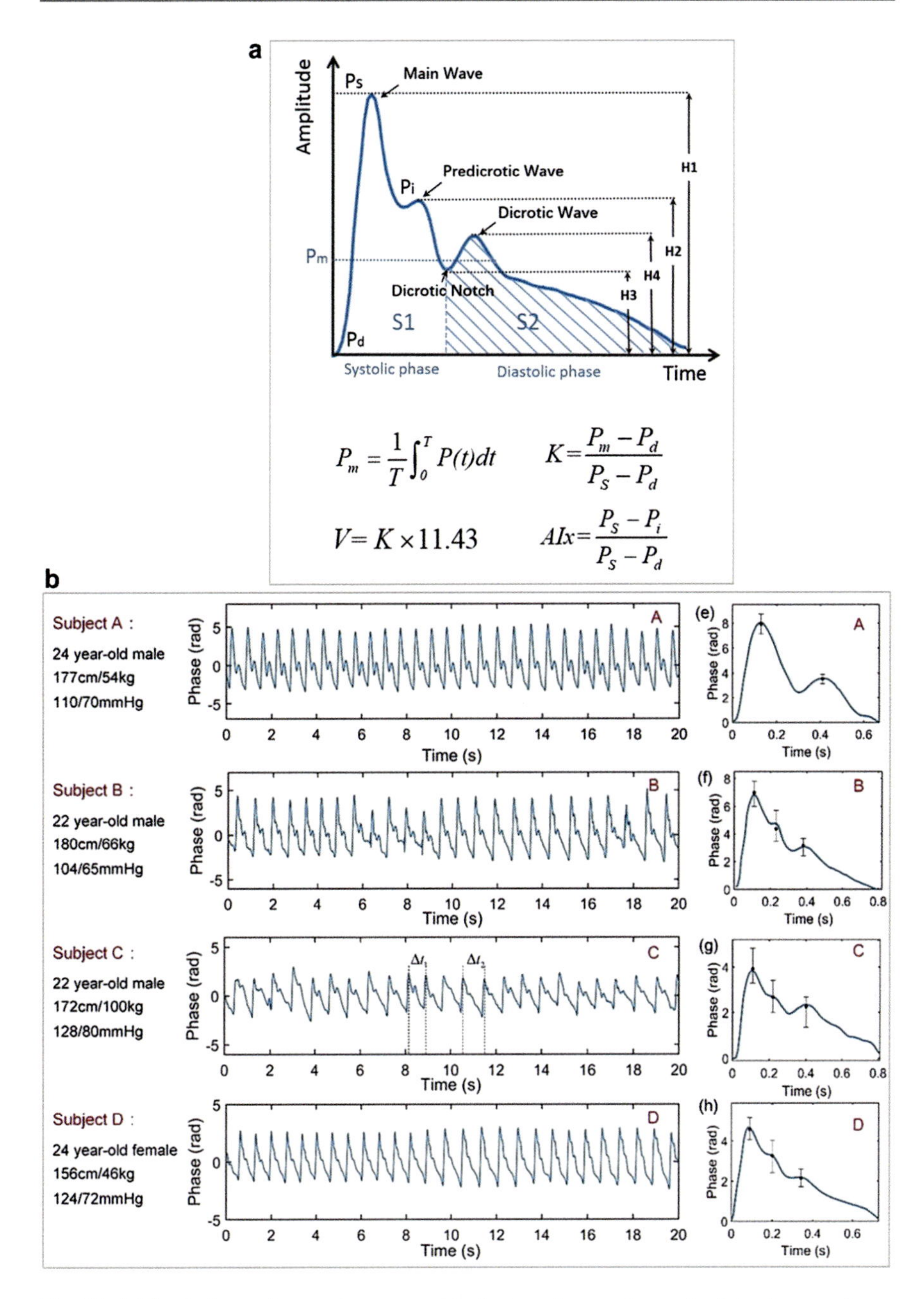

Fig. 33 (**a**) Typical waveform and assessment indicators of one pulse cycle; (**b**) pulse waveforms in 20 s (left) detected for four subjects and the statistical results in one pulse cycle (right)

Table 1 Analytical results of clinical pulse diagnosis

Indicator	Subject			
	A	B	C	D
T (s)	0.69	0.79	0.83	0.73
MBP (mmHg)	85.4	80.1	102.7	92.6
HR (bpm)	87	76	72	82
K value	0.384	0.387	0.474	0.397
V value	4.39	4.42	5.42	4.54
AIx	NA	−0.303	−0.298	−0.225

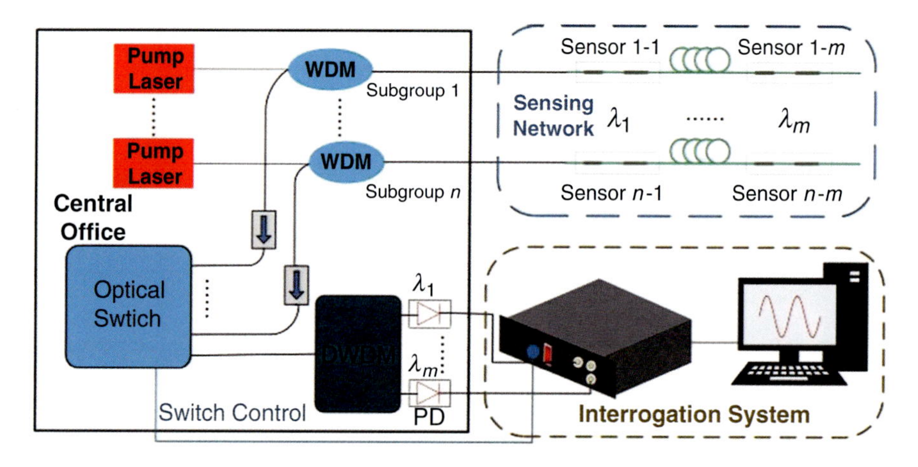

Fig. 34 Schematic diagram of the acceleration sensing network

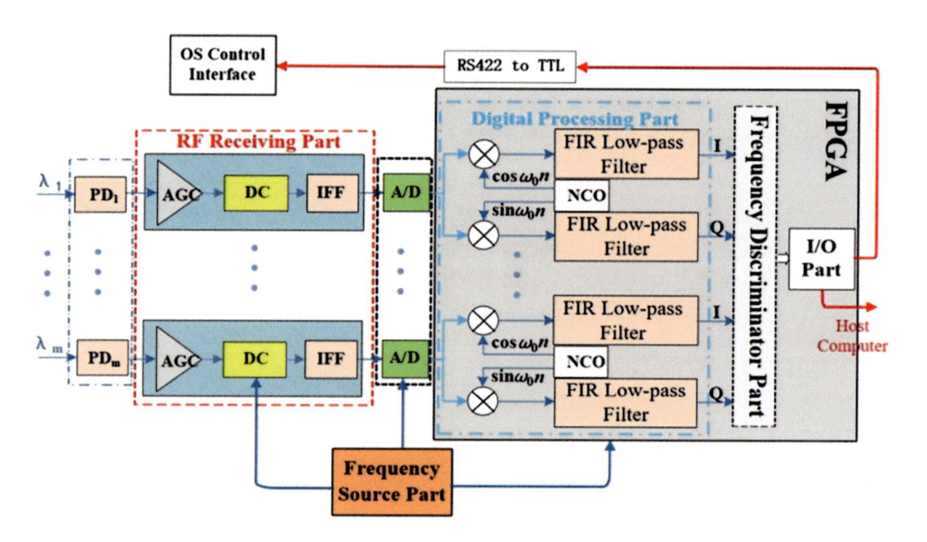

Fig. 35 Schematic of the interrogation system (*AGC* automatic gain control, *DC* down converter, *IFF* intermediate frequency filter, *NCO* numerically controlled oscillator)

The Radio-Frequency receiving part working as super-heterodyne receiver converts the original ultra-high frequency (UHF) signal to intermediate frequency (IF) signal by mixing the original signal and the local oscillator signal. Subsequently, the digital processing part mixes the digital IF signal with two orthogonal signals controlled by NCO to generate the orthogonal in-phase (I) and quadrature (Q) signals (Song 2016). Next, the frequency discriminator part extracts the frequency information from the I/Q signals through the arctangent operation and differentiator. Both the digital processing part and frequency discriminator part are realized by a high-performance FPGA (Xilinx Virtex-7) to ensure the processing speed. Beneficial from parallel demodulation of WDM channels and high switching speed of OS, the response bandwidth and frequency resolution of the interrogation system reached to DC-1 kHz and 4.7 Hz/$\sqrt{\text{Hz}}$ above 10 Hz, respectively.

Based on the DBRFL SPON, a fiber optic accelerometers network was developed, by adopting the cantilever-mass transducer (Guan et al. 2010). The field test was carried out, where the tested sensors were mounted on the vibration exciter (MB Dynamic PM25) along with a commercial piezoelectric accelerometer (PCB, Inc., Model 393B31) for calibration (see Fig. 36a). The resolution of this PZT accelerometer is about 4 ng/$\sqrt{\text{Hz}}$ at 100 Hz, which is the highest commercially available resolution to the best of our knowledge. The vibration exciter provides a sinusoidal vibration with adjustable frequency and amplitude. Firstly, three sensor elements were chosen to test the independence of the sensor in the network. As shown in Fig. 36b, when S12 was applied with a sinusoidal vibration at 90 Hz, while S11 and S21 were isolated the vibration, the crosstalk between S12 and S11 or S21 was lower than −70 dB. Secondly, the sensitivity and linearity of the DBRFL acceleration sensor were investigated. Taking the accelerations at 90 Hz as the example, the linear fitting result in Fig. 36c shows that the sensitivity of the sensor was 1.1604 GHz/g and the R-Square was over 0.9999. Finally, the measurement precision of the DBRFL accelerometer was tested. Theoretically, the precision is defined as the closeness of agreement between independent test results, which can be expressed by the effect of random errors on a measurement and quantified as the standard deviation of repeated measurements on the same sample using the same method (Menditto et al. 2007). Usually, Bland-Altman plot method is used to evaluate the precision (Cecconi et al. 2009). The distribution of multiple test results, illustrated in Fig. 36d, demonstrates the precision as high as 1.305 μg, which is regarded by the value of ±1.96 times of the standard deviation (95% confidence intervals) around the bias.

Another WDM/TDM SPON is served for colorless fiber optic sensors multiplexing. Figure 37a presents a typical tree topology network to access arbitrary fiber optic sensors, which is intrinsically the same as the optical fiber communication PON (Li et al. 2012b). Here, TDM is realized by employing fiber delay lines of different length in different link, and WDM is achieved through ODN (usually DWDM) connected to every OSU. The OLT of the hybrid SPON (HSPON) generates high-power, wideband pulse light, which is downward transmitted to each OSU through TDM/WDM distribution network and then receives the upward

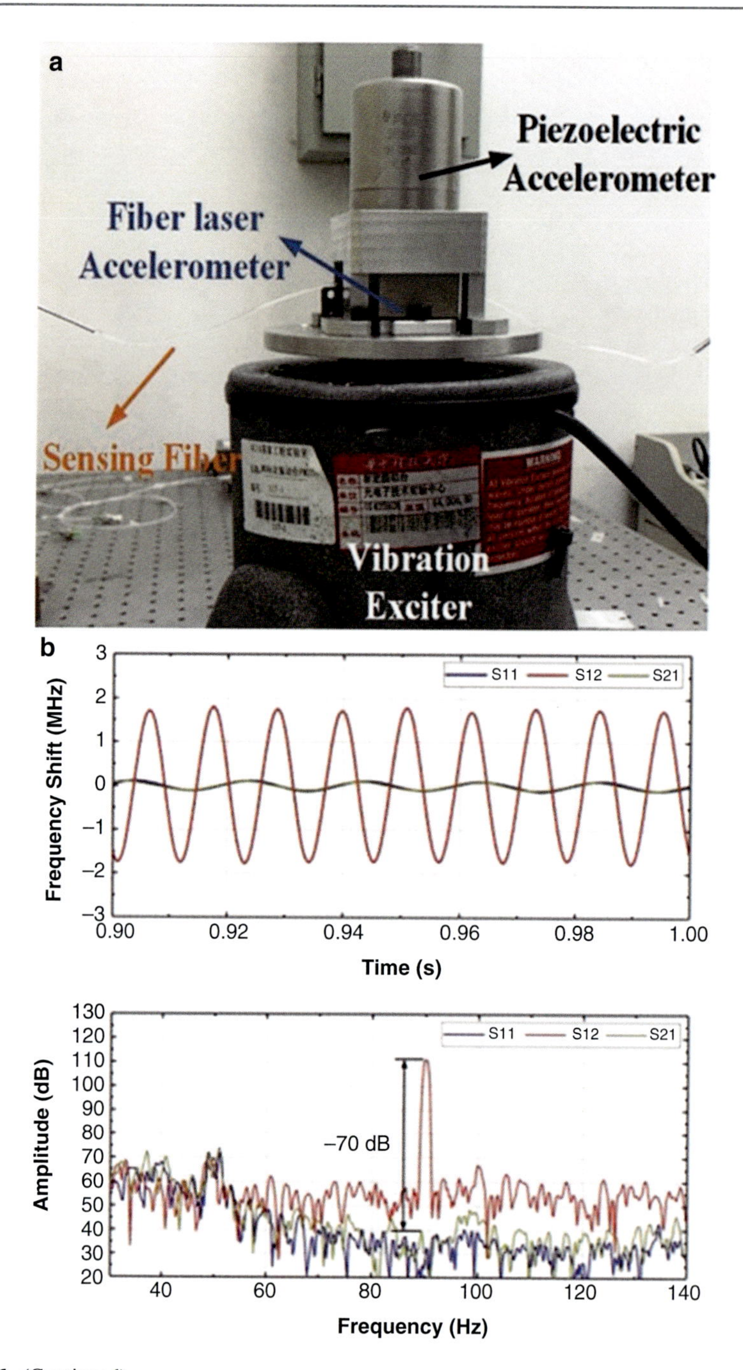

Fig. 36 (Continued)

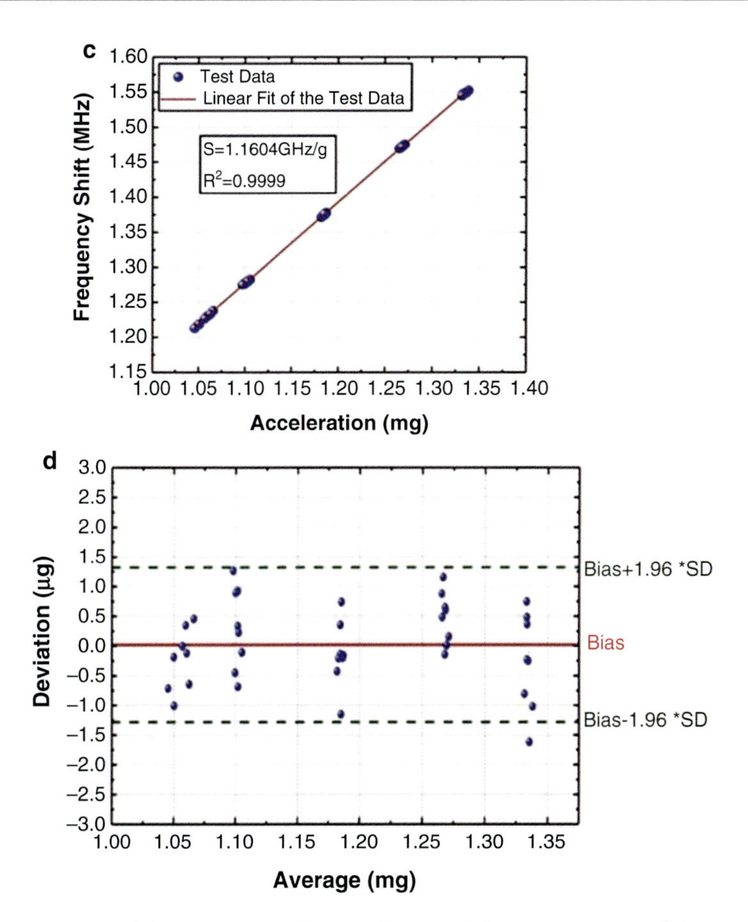

Fig. 36 (**a**) Photo of field test setup; (**b**) waveforms and frequency spectra of three sensors; (**c**) relationship between acceleration and beat frequency shift of DBRFLaccelerometer; (**d**) bland-Altman plot for DBRFL accelerometer and PZT accelerometer as reference

sensing optical signal from OSUs. The maximum capacity of the SPON is mainly determined by the spectrum bandwidth and power of light source, transmission loss of fiber link, and the channel interval of DWDM. According to the performance of the commercial devices, the capacity can reach over $8 \times 32 = 256$. In this configuration, the OSU can be interferometer-based vibration sensor, optical spectrum absorption-based gas sensor, Faraday Effect-based circulator sensor, etc., as well as flexibly deployed. Therefore, this WDM/TDM SPON can be widely utilized in many application fields such as multi-zone monitoring, multi-parameter detection, and multi-function monitoring.

A WDM/TDM SPON for perimeter intrusion detection was designed and developed. As displayed in Fig. 37b, the OSU is an unbalanced Mach-Zehnder Interferometer (MZI) followed by a Sagnac structure, which can detect the intrusion

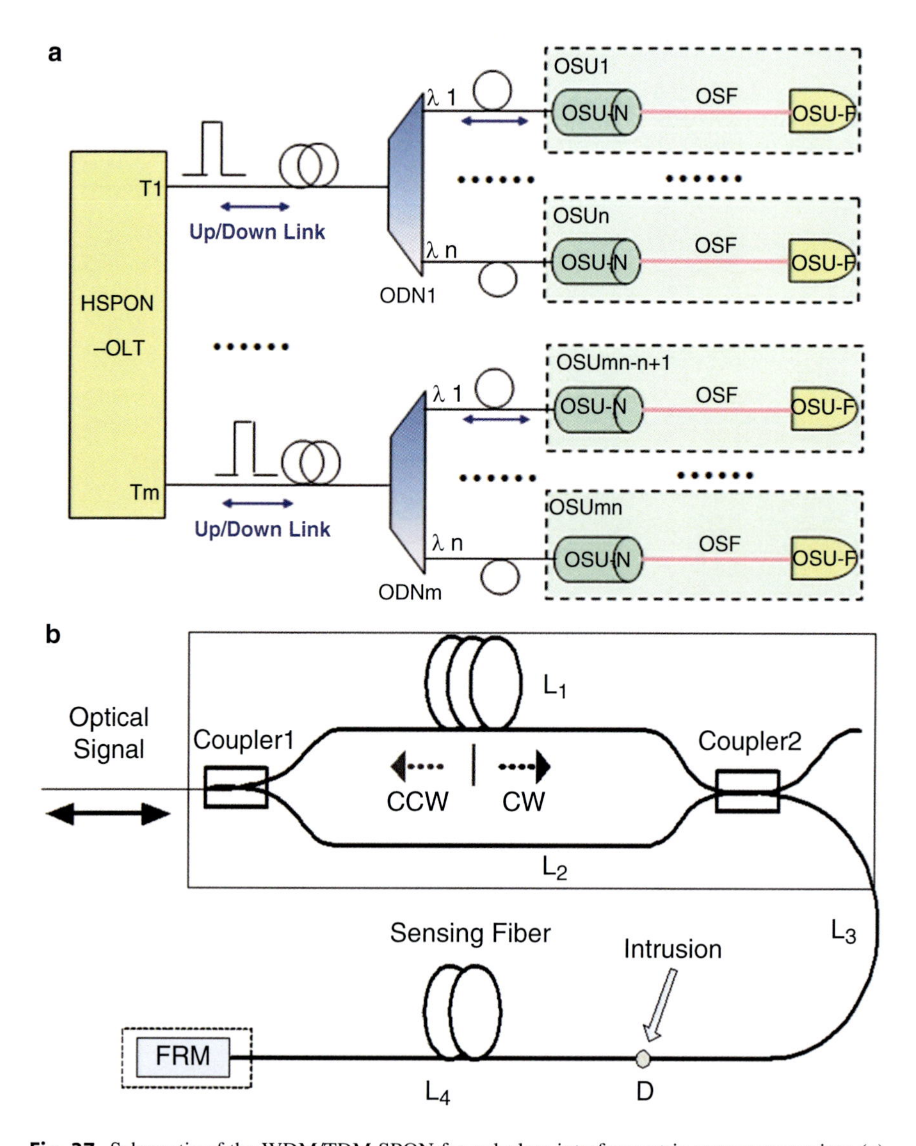

Fig. 37 Schematic of the WDM/TDM SPON for colorless interferometric sensors accessing: (**a**) architecture of the SPON; (**b**) structure of the interferometric sensor for vibration monitoring. (Copyright 2019 OSA)

in long distance without blind area. Here, the unbalanced MZI amplifies the response of intrusion in the far area and the Faraday rotation mirror (FRM) at the end of the sensing fiber is to overcome the polarization induced fading, as well as reflect the sensing signal to the OLT.

In terms of the principle discussed above, a perimeter intrusion monitoring system was explored, as illustrated in Fig. 38. The capacity of the system is 256, and the maximum sensing distance is 128 km. The fiber cable can be deployed on the ground or hung on the fence to sense the micro-vibration induced by the perimeter intrusion. The response time is less than 2 s. The missing alarm is nearly 0, due to its high sensitivity of the sensing structure, while the false alarm is only 4% benefiting from the intelligent data processing method. Different events such as walking, scrambling, and fiber cable damage can be detected and identified.

Besides, the fiber optic intrusion monitoring system can be used for cultural relics protection. Since cultural relics are usually located at remote rural areas without electricity supply, the conventional security monitoring techniques are difficult to implement it, while the fiber optic sensor presents great superiority for that. Figure 39a presents the system configuration. The host is the OLT, and the ODN is composed of WDM and TDM modules. The optical fiber micro vibration sensor is OSU, which is designed for underground condition by winding the optical fiber on the cone structure, to collect all the vibration signals near the sensor, and then transmit it to the cylinder roll protected by the intelligent materials. The field test demonstrated that the OSU can monitor the dig within the area of 25 m radius and the blasting within the area of 100 m radius. This system has been widely applied in many ancient tombs in China including Bailsman Tombs, Xianling Tombs, and so on, by deeply burying the OSUs underground (see Fig. 39b).

Fig. 38 Photographs of the HSPON-OLT (left) and the application software of the perimeter intrusion monitoring system (right). Inset: photo of the field test with fiber cables deployed on the ground and fence. Copyright 2019 OSA

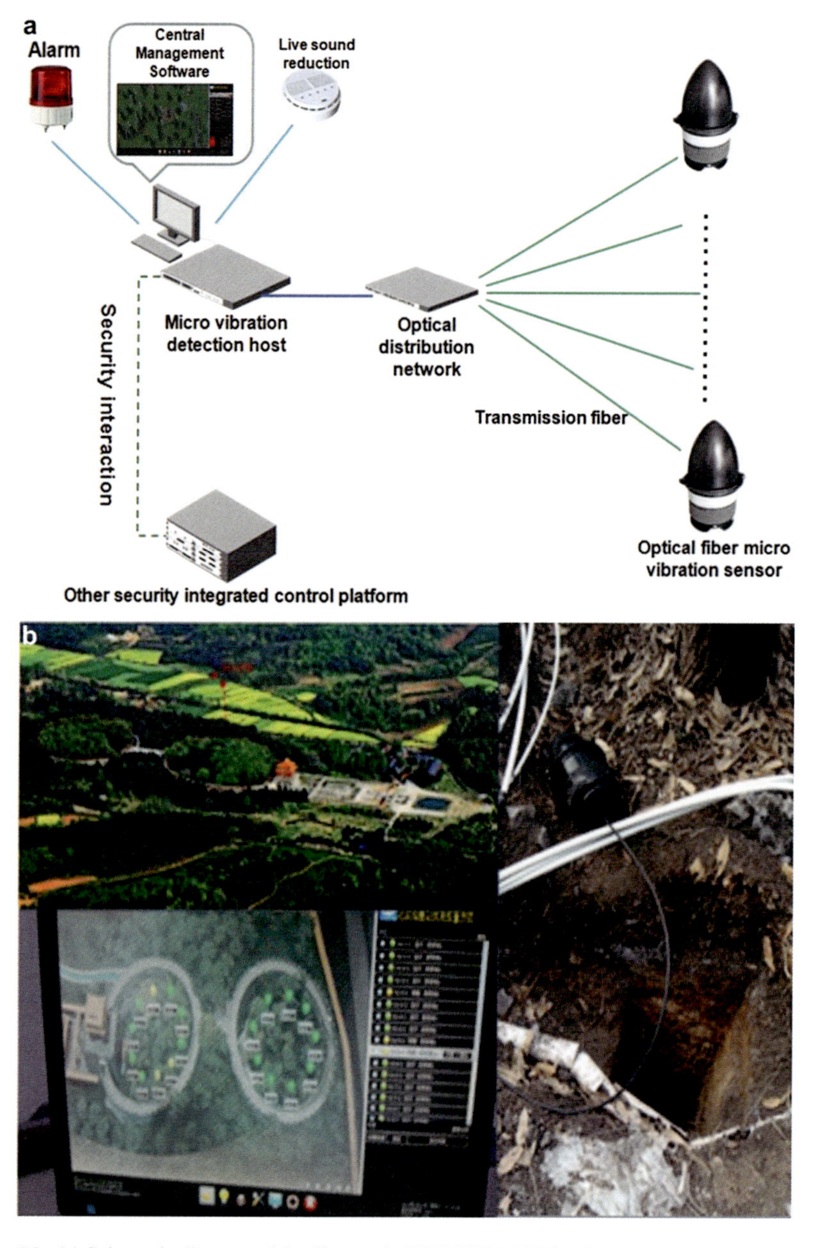

Fig. 39 (**a**) Schematic diagram of the fiber optic WDM/TDM SPON for cultural relics protection; (**b**) photograph of the field test in Xianling Tombs in China

Conclusion

In this chapter, the key technologies and industrial applications of three types of fiber optic sensor network are introduced, including the UWFBG sensor network, special fiber grating sensor network, and the fiber optic SPON. Through TDM, WDM, FDM, SDM, and their combinations, a large number of fiber optic sensors can be accessed and merged to the network. For the sensor network integrated in a single optical fiber, both quasi-distributed sensor and continuous-distributed sensor are realized by employing UWFBGs, TFGs, and DBRFLs. Moreover, the fiber optic SPON with star or tree topology provides the support of any colored or colorless fiber optic sensors accessing. The sensor structures, networking mechanisms, system architectures, demodulation methods, and typical sensing performances are systematically discussed. Besides, the developed systems or equipment for measuring the temperature, strain, acoustic, pressure, acceleration, micro-vibration, near infrared absorption, pulse wave, etc., as well as the field tests for a wide range of industrial applications such as resource exploration, geophysics, infrastructure, medical diagnose, food quality, and security control are presented. Beneficial from the large capacity, good adaptivity, and high flexibility, the fiber optic sensor network is becoming increasingly important in industrial applications, especially the sensing for large structures and surveying at complex conditions.

References

F. Ai, Q. Sun, W. Zhang, T. Liu, Z. Yan, D. Liu, in *Optical Fiber Communication Conference, OSA Technical Digest*, San Diego, California (2017)

F. Ai, H. Li, T. He, Z. Yan, D. Liu, Q. Sun, in *Optical Fiber Communication Conference, OSA Technical Digest*, San Diego, California (2018)

A.P. Avolio, M. Butlin, A. Walsh, Physiol. Meas. **31**, 1 (2009)

M. Cecconi, A. Rhodes, J. Poloniecki, A. Rhodes, Crit. Care **13**, 1 (2009)

J.W. Cheng, Q.Z. Sun, F. Ai, Y.Y. Luo, W. Zhang, X.L. Li, D.M. Liu, IEEE Photonics J. **10**, 3 (2018)

E. Cleve, E. Bach, E. Schollmeyer, Anal. Chim. Acta **420**, 143–247 (2000)

ElectroniCast Consultants (2017). https://www.electronicastconsultants.com. Accessed May 2017

R.W. Fallon, in *Doctoral Dissertation,* ed. by R.W. Fallon (Aston University, 2000), p. 18

C.Z. Fan, F. Ai, Y.J. Liu, Z.J. Xu, G. Wu, W. Zhang, C. Liu, Z.J. Yan, D.M. Liu, Q.Z. Sun, in *Optical Fiber Communication Conference, OSA Technical Digest* San Diego, California (2019)

T. Fu, Y. Liu, K.T. Lau, J. Leng, Compos. Part B-Eng. **66**, 420–429 (2014)

B.O. Guan, X.S. Sun, Y.N. Tan, in *European Workshop on Optical Fibre Sensors*, Porto, Portugal 76530Z (2010)

T. Guo, A.C. Wong, W.S. Liu, B.O. Guan, C. Lu, H.Y. Tam, Opt. Express **19**, 3 (2011)

J. Harrison, S. Foster, in *Optical Sensors 2017*, New Orleans, Louisiana (2017)

C.Y. Hu, H.Q. Wen, W. Bai, J. Lightwave Technol. **32**, 7 (2014)

M. Jamrógiewicz, J. Pharm. Biomed. Anal. **66**, 7 (2012)

L. Jin, Y.Z. Liang, M.P. Li, L.H. Cheng, J. Li, B.O. Guan, J. Lightwave Technol. **32**, 22 (2014)

P. Jousset, T. Reinsch, T. Ryberg, H. Blanck, A. Clarke, Nat. Commun. **9**, 2509 (2018)

Y. Kim, Y. Noh, K. Kim, IEEE Commun. Lett. **21**, 11 (2017)

X.L. Li, Q.Z. Sun, D. Liu, R.B. Liang, J.J. Zhang, J.H. Wo, P.P. Shum, D.M. Liu, Opt. Express **20**, 11 (2012a)

X.L. Li, Q.Z. Sun, J.H. Wo, M.L. Zhang, D.M. Liu, J. Lightwave Technol. **30**, 8 (2012b)
D.M. Liu, Q.Z. Sun, P. Lu, L. Xia, C.T. Sima, Photon. Sens. **6**, 1 (2016)
T. Liu, D.D. Lv, Y.Y. Luo, W.G. Lu, Z.J. Yan, K. Wang, C.Q. Li, D.M. Liu, Q.Z. Sun, in *Conference on Lasers and Electro-Optics*, San Jose, California (2018)
M. Manley, Chem. Soc. Rev. **43**, 24 (2014)
A. Mateeva, J. Lopez, H. Potters, J. Mestayer, B. Cox, D. Kiyashchenko, Geophys. Prospect. **62**, 4 (2014)
G.E. McVeigh, C.W. Bratteli, D.J. Morgan, C.M. Alinder, S.P. Glasser, S.M. Finkelstein, J.N. Cohn, Hypertension **33**, 6 (1999)
A. Menditto, M. Patriarca, B. Magnusson, Accred. Qual. Assur. **12**, 1 (2007)
S.V. Miridonov, M.G. Shlyagin, D. Tentori, Fiber Optic. Laser Sens. Appl. **3541**, 33–40 (1998)
S. Ni, H. Kanamori, D. Helmberger, Nature **434**, 582 (2005)
S. Samo, D.A. Carlson, D.L. Gregory, S.H. Gawel, J.E. Pandolfino, P.J. Kahrilas, Clinical Gastroenterology and Hepatology **15**, 3 (2016)
B.S. Song, in *System-Level Techniques for Analog Performance Enhancement*, ed. by B. S. S. By, (Springer, Cham, 2016), pp. 195–225
Q.Z. Sun, J.H. Wo, H. Wang, D.M. Liu, in *Proceedings of SPIE, Conference on Optical Fiber Sensors* San Diego, California (2014)
Q.Z. Sun, F. Ai, D.M. Liu, J.W. Cheng, H.B. Luo, K. Peng, Y.Y. Luo, Z.J. Yan, P.P. Shum, Sci. Rep. **7**, 41137:1–8 (2017)
M. Vidakovic, I. Armakolas, T. Sun, J. Carlton, K.T.V. Grattan, J. Lightwave Technol. **34**, 18 (2016)
Z. Wang, H.Q. Wen, C.Y. Hu, W. Bai, Y.T. Dai, Chin. Opt. Lett. **14**, 1 (2016)
J.Y. Wang, F. Ai, J.W. Cheng, W. Zhang, Z.J. Yan, D.M. Liu, Q.Z. Sun, IEEE Photonic. Tech. Lett. **30**, 18 (2018a)
J.Y. Wang, F. Ai, Q.Z. Sun, T. Liu, H. Li, Z.J. Yan, D.M. Liu, Opt. Express **26**, 19 (2018b)
J.H. Wo, M. Jiang, M.X. Malnou, Q.Z. Sun, J.J. Zhang, P.P. Shum, D.M. Liu, Opt. Express **20**, 3 (2012)
J.H. Wo, H. Wang, Q.Z. Sun, P.P. Shum, D.M. Liu, J. Biomed. Opt. **19**, 1 (2014)
T. Woodcock, C. O'Donnell, G. Downey, J. Near Infrared Spectrosc. **16**, 1 (2008)
T. Yamate, G. Fujisawa, T. Ikegami, J. Lightwave Technol. **35**, 16 (2017)
Z. Yan, C. Mou, K. Zhou, X. Chen, L. Zhang, J. Lightwave Technol. **29**, 18 (2011)
Z. Yan, A. Adebayo, K. Zhou, L. Zhang, H. Fu, D. Robinson, Opt. Photonics J. **3**, 5 (2013)
M.H. Yang, W. Bai, H.Y. Guo, H.Q. Wen, H.H. Yu, D.S. Jiang, Photonic Sens **6**, 1 (2016)
M.L. Zhang, Q.Z. Sun, Z. Wang, X.L. Li, H.R. Liu, D.M. Liu, Opt. Commun. **285**, 13–14 (2012a)
Y. Zhang, X. Xie, H. Xu, Opto-Electron. Eng. **39**, 8 (2012b)
W. Zhang, X.L. Ni, J.Y. Wang, F. Ai, Y.Y. Luo, Z.J. Yan, D.M. Liu, Q.Z. Sun, J. Lightwave Technol. (2019). Accepted to be published

Fiber Optic Sensors for Coal Mine Hazard Detection

48

Tongyu Liu

Contents

Abstract

A number of health and safety hazards present in underground coal mines, which include methane gas explosion, coal combustion, rock roof collapse, and flooding. Methane gas and coal combustion have been recognized by the coal mine industry as two major hazards, which resulted in most of the heavy casualties and economic losses. Conventional catalytic methane gas sensors suffer from poor accuracy and cumbersome maintenance, which is the bottleneck of methane hazard prevention. Coal mine combustion monitoring has been relying on gas

T. Liu (✉)
Laser Institute, Qilu University of Technology-Shandong Academy of Science, Jinan, Shandong, China
e-mail: tongyu.liu@vip.iss-ms.com

G.-D. Peng (ed.), *Handbook of Optical Fibers*,
https://doi.org/10.1007/978-981-10-7087-7_24

tubing bundles system, which suffers from long-time delay and poor reliability. Semiconductor laser diode methane gas sensors have been developed which have the advantages of low-power consumption, 0–100% full detection range, high accuracy, and no need of recalibration. Fiber optic Raman distributed sensors have been deployed in coal mine goaf and successfully detected combustion hazard in early phase. The FOS-based mine hazard detection system offers unique advantages of intrinsic safety, multi-location and multi-parameter monitoring. The application of FOS on monitoring of methane, coal combustion, microseismic, equipment condition, and rescue information systems is discussed in this chapter, showing future trend of research in this area.

Keywords
Fiber optic sensor · Coal mine · Methane · Seismic · Equipment condition monitoring · Rescue · Hazard detection

Introduction

Coal is the primary energy source of many countries in the world. Annual coal production has been around 8 billion tons in recent years as shown in Fig. 1 (http://www.bp.com/en/global/corporate/energy-economics/statistical-review-of-world-energy.html). In China, coal provides around 65% of the total energy, and annual coal production in 2016 was around 3.7 billion tons. Over 95% of Chinese coal mines are underground mines. The depth of underground coal mines increases by an average of around 10 m per year, with the deepest mine reaching 1500 m below the ground level. The increase in mining depth significantly increases safety hazards such as large strata pressure and rock or methane burst. Over 55% of the Chinese coal mines are gassy, hence methane is big safety hazard. The casualty figures between 2001 and 2016 are shown in Fig. 2.

Coal mine hazards consists of the following major types, (i) methane gas outburst and explosion; (ii) roof fall and coal and rock outburst; (iii) coal spontaneous combustion; (iv) water flooding; and (v) hazards caused by electrical machinery failures. Around 80% of the large casualty accidents, where more than 10 people died in a single incident have been caused by methane explosion. Therefore, major effort has been made on methane monitoring and control. Thanks to the coal mine methane gas monitoring systems, which have been compulsively implemented by the government work safety administrative authority since 2001, the coal mine work safety standard has been consistently improved and symbolized by the reduction of casualty figures. In 2002, China produced 1.4 billion tons of coal with casualty figure peaked at 6995. The casualty in 2016 dropped to 538 while the production increased to 3.5 billion tons, corresponding to an improvement of 33 times on the casualty rate per million ton of coal production from 4.9 in 2002 to 0.15 in 2016.

In order to achieve the ultimate zero casualty goal on coal mine work safety, it is necessary to have all the safety hazards monitored on-line so that accidents can be dealt with and prevented in advance. First of all, the ambient methane

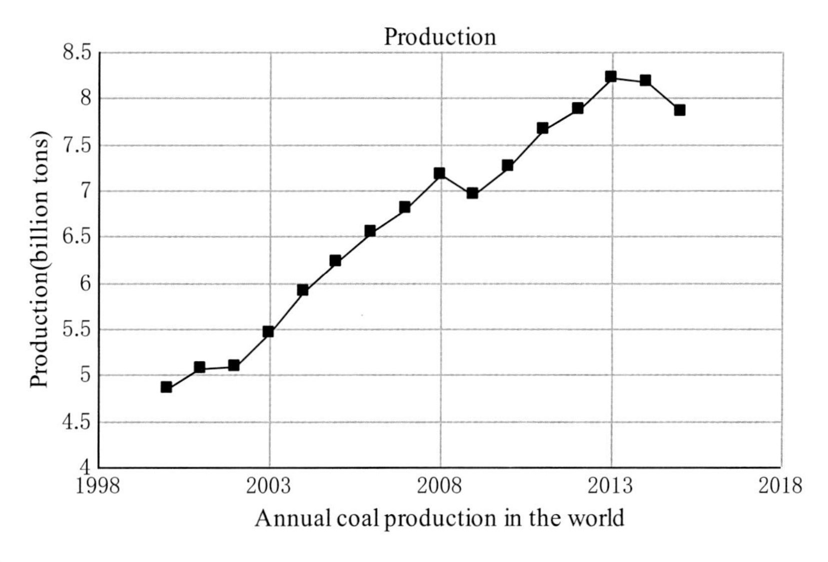

Fig. 1 Global annual coal production (billion tons) since year 2000

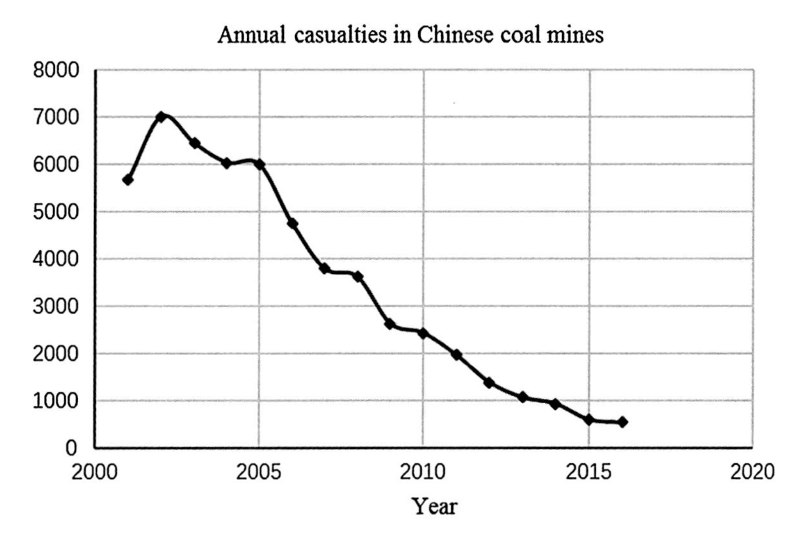

Fig. 2 Annual casualties caused by coal mine accidents in China

gas distribution and dynamics in all key locations of the coal mines need to be continuously monitored with high accuracy and reliability. Fiber optic methane sensors offer the potential of 0–100% full-range monitoring without recalibration. Secondly, sensors are required for monitoring coal mine goaf combustion, tunnel and rock stability, condition of hidden and storage water reservoirs, dynamic hazards such as micro-seismic and equipment conditions are also required, so that hazards of structural, equipment, and hidden-fire or flooding can be avoided. It is also necessary to install all passive fiber optic multisensor and communication network for emergency and rescue needs, when power supply may get cut and normal communication and electronic sensors may fail.

Coal Mine Methane Senor and Gas Monitoring System

Fiber Optic Methane Gas Sensors

Methane (CH_4) is by far the most important gas in coal mine safety monitoring, explosion occurs upon ignition when methane level in air reaches 5–16%. Conventional catalytic-type methane gas sensors contain a sensing resistor element with platinum heater, which will burn in combustible gases and causes resistance changes with gas concentration. A Wheatstone bridge circuit is used to monitor the difference between the active sensor resistance and the inert compensator resistor within the sensor. The measurement range of the catalyst methane gas sensors is small typically within 0–4%, and the sensitivity suffers from fading with time, which requires frequent recalibration and maintenance. The catalyst methane sensor is sensitive to inflammable gases hence lacks of selectivity. It also does not operate correctly at low oxygen level and prone to be poisoned. Infrared methane sensors using broadband IR LED source and photodiode detectors coated with IR filters to detect the methane absorption signal also suffer from interference of humidity and temperature drift. Semiconductor laser diode or LED emitting at near-infrared range have good selectivity and sensitivity and can be used with fiber optic sensor head (Iseki et al. 2000). Fiber optic methane sensor developed for landfill site monitoring by Culshaw et al. (1998) has created lots of interest for coal mine applications (Ni et al. 2008).

Figure 3 shows the methane gas absorption spectrum in the NIR range. It can be seen that a number of wavelengths can be used in the 1650 nm range. According to the Beer-Lambert beer's law, laser beam transmitting through the absorbing gas with concentration, c, and path length of L, the transmitting light power $I(\lambda)$ can be described by

$$I(\lambda) = I_0(\lambda)\exp(-\alpha(\lambda)CL) \tag{1}$$

where $I_o(\lambda)$ is the input light power, $\alpha(\lambda)$ is the gas absorption coefficient, which is affected by temperature and pressure as gas is compressible. Hence the gas concentration can be deduced by

$$C = \ln\left(I_0(\lambda)/I(\lambda)\right)/(\alpha(\lambda)L) \tag{2}$$

where $\alpha(\lambda) = S(T)P\psi(\lambda)$, P is ambient pressure, $\psi(\lambda)$ is the spectral line shape, and $S(T)$ is the temperature dependence of the absorption coefficient, which can be described by

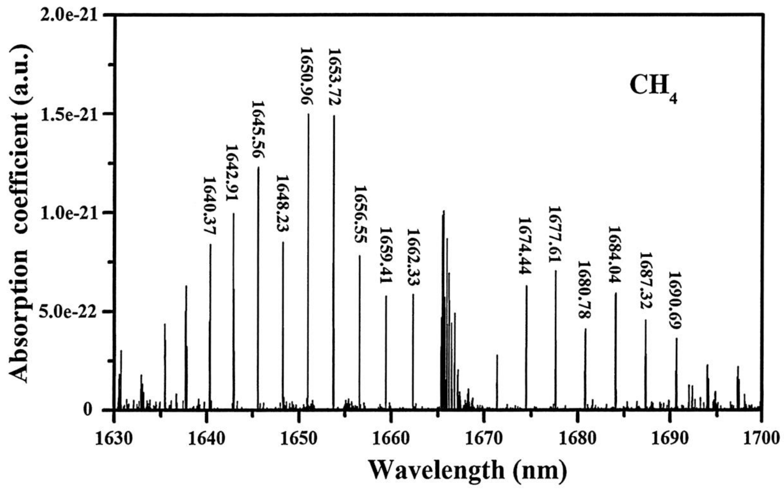

Fig. 3 NIR absorption spectrum of methane gas

$$S(T) = S\left(T_0\right) \frac{Q\left(T_0\right)}{Q(T)} \exp\left[-\frac{hcE_i^{n}}{k}\left(\frac{1}{T} - \frac{1}{T_0}\right)\right] \times \left[\frac{1 - \exp\left(-hc\nu_0 / kT\right)}{1 - \exp\left(-hc\nu_0 / kT_0\right)}\right] \quad (3)$$

S(T) can be characterized empirically for the operational temperature range using linear or polynomial fitting.

A typical laser methane sensor detection system is shown in Fig. 4. Light from the laser diode source emitting at around 1653 nm is split into two beams, one is directly to a photodiode amplified and converted to electrical voltage signal V_1 as a reference, the other goes through the sensor gas cell then delivered to the signal detection PD2 and amplified as signal voltage V_2. The laser diode wavelength can be modulated by sweeping the injection current using saw-tooth wave and gas absorption spectra are obtained. Both DFB and VECSEL laser diodes have been used as light source. DFB offers larger power, while VCSEL offers large tenability.

The ratio of V_1/V_2 across the sweeping wavelength range as the normalized spectra (As shown in Fig. 5) can be used to perform direct absorption analysis to determine the gas concentration.

While direct normalization method is simple and robust, the detection sensitivity can be enhanced by performing sinusoidal modulation to the laser driving current and detect the 2nd harmonic absorption spectra using phase locked detection. The laser driver modulation current and the absorption harmonic signal is illustrated in Fig. 6 with typical modulation frequency at 10 kHz and the absorption signal at 20 kHz (Iseki et al. 2000).

Multi-channel sensors can be achieved by simply splitting the laser output to multi fibers both for direct absorption and the 2nd harmonic scheme. Figure 7

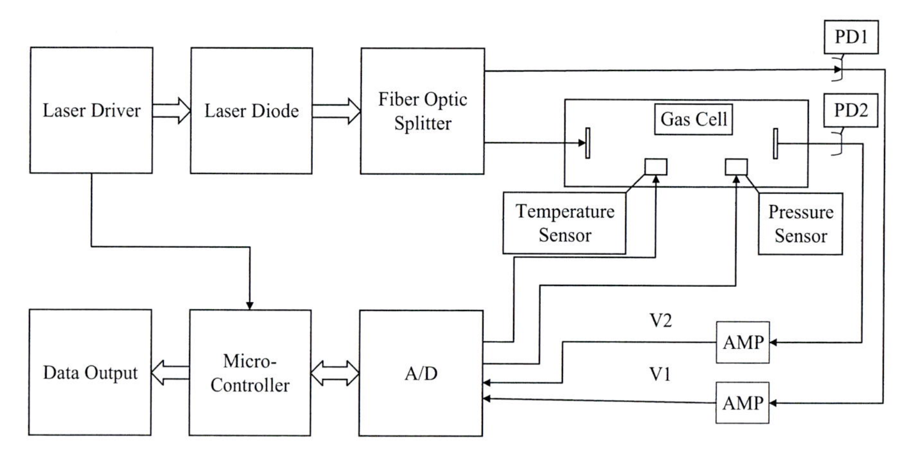

Fig. 4 Schematic diagram of the spectroscopic fiber optic gas sensor

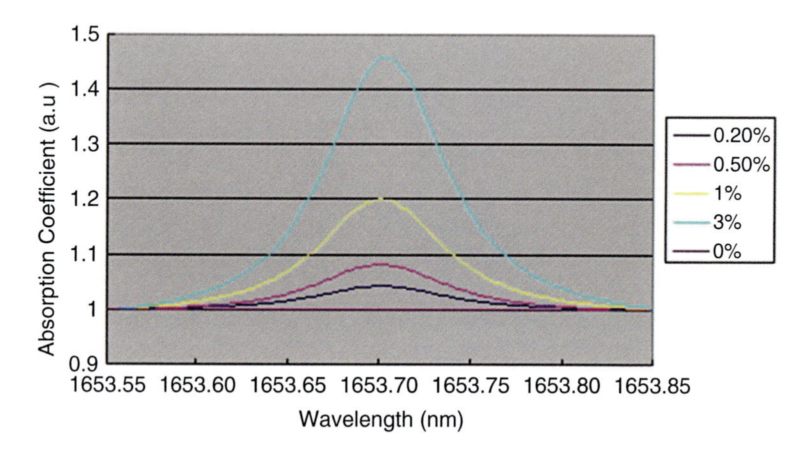

Fig. 5 Normalized typical NIR methane absorption spectra at different gas concentration

is a scheme for multi-channel detection where a 45-channel sensor array was demonstrated to monitor multi-point methane gas in a waste treatment field (Stewart et al. 2003).

The absorption strength of the gas spectra is affected by the absolute density of gas molecules in unit volume, which is affected by ambient temperature and pressure. Hence correction needs to be performed. Figure 8 was typical methane gas concentration measurement data from direct absorption analysis with a sensor gas cell length of 6.0 mm. It can be seen that measurement data decreases with temperature due to the reduced molecular density.

Typical technical requirement of a fiber optic methane gas senor is accuracy of +/−0.05% for the concentration range 0–4% and +/−5% of actual value for the range 4–100%.

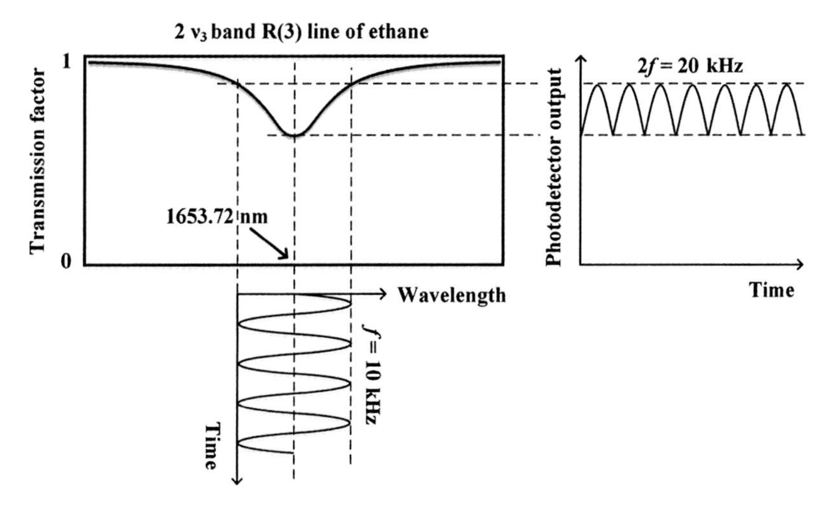

Fig. 6 The sinusoidal modulation of the laser driver current and the gas absorption harmonic signal

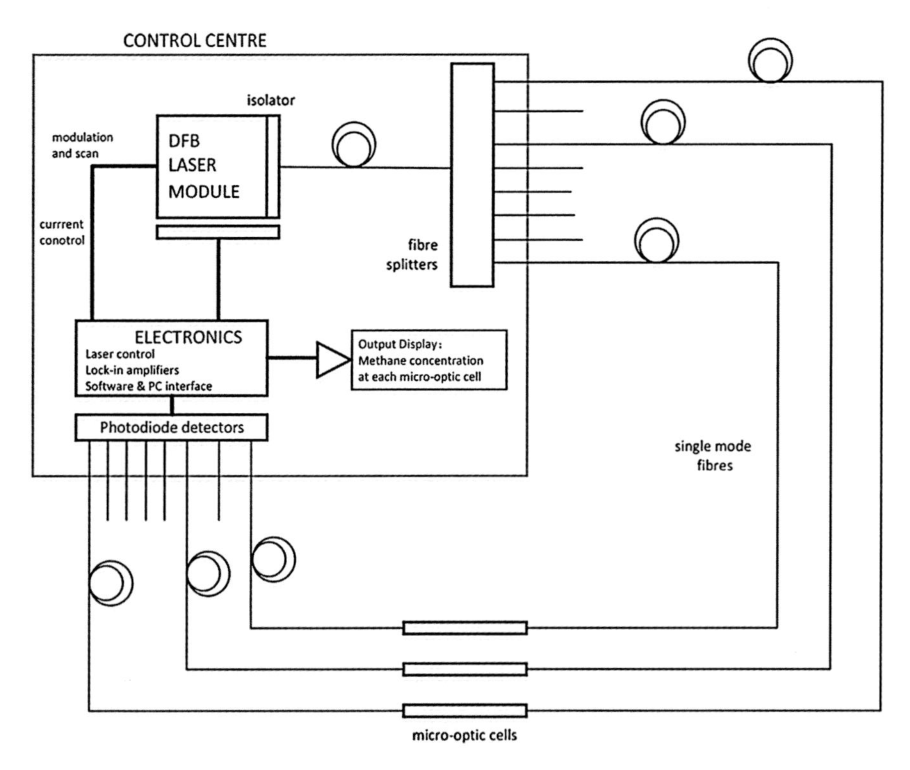

Fig. 7 Multi-channel fiber optic methane gas sensors using spatial multiplexing scheme

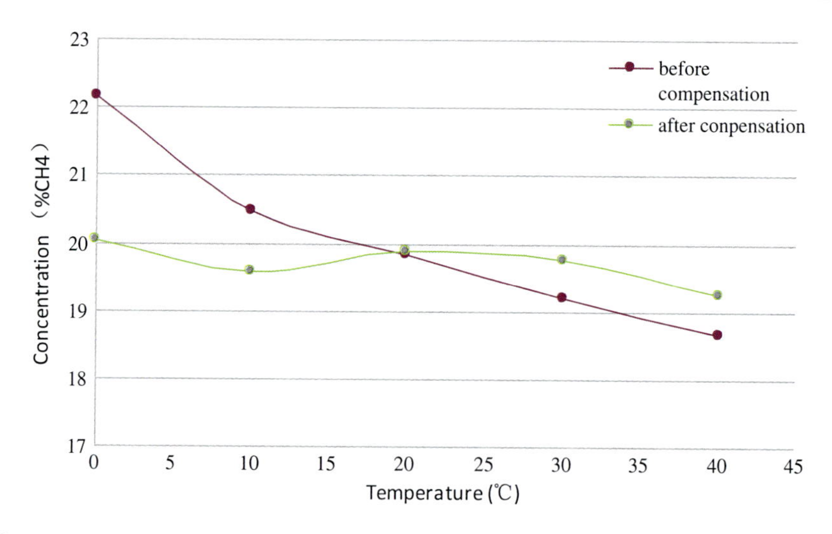

Fig. 8 Test concentration before and after temperature compensation

Coal Mine Methane Gas Online Monitoring System

Methane gas as a safety hazard needs to be managed in a comprehensive manner. For gassy mines, methane drainage is needed which involves drilling holes in the coal seam and pump the methane gas out first before the gas in the coal seam drops down to a safe level for coal production. Hence methane concentration and flow rate need to be monitored.

According to coal mine safety regulation, methane sensors are placed at critical locations such as top corner of the underground coal seam work-face where methane is prone to accumulate, near the coal cutting machine where methane is released with the production of the new coal cut off from the coal seam, and the in the ventilation exit tunnel at the work face. Currently Black-white methane sensors are dominant in use by the coal mine industry due to low cost. Laser and Infrared methane sensors have been gradually accepted by the industry. However, the fiber optic sensors are in early days for field application, mainly due to lack of industrial standard. Figure 9 is the block diagram of a coal mine methane monitoring system, which consists of sensors, sensor stations, data or signal transmitters or industrial network, data server, and display. Methane and other gas sensors are connected to the sensor station which provides DC power supply and signal/data communication interface. Sensor data via Sensor station and fiber optic transmission link for point to point or via the network switching connection in the industrial network feed to the data server at ground control data center. The coal seam methane gas alarming threshold typically set at 1.0% for audio alarming, while the sensor station will initiate power off switching signal to the work face production equipment when the ambient methane gas reaches at 1.5% or 2.0%.

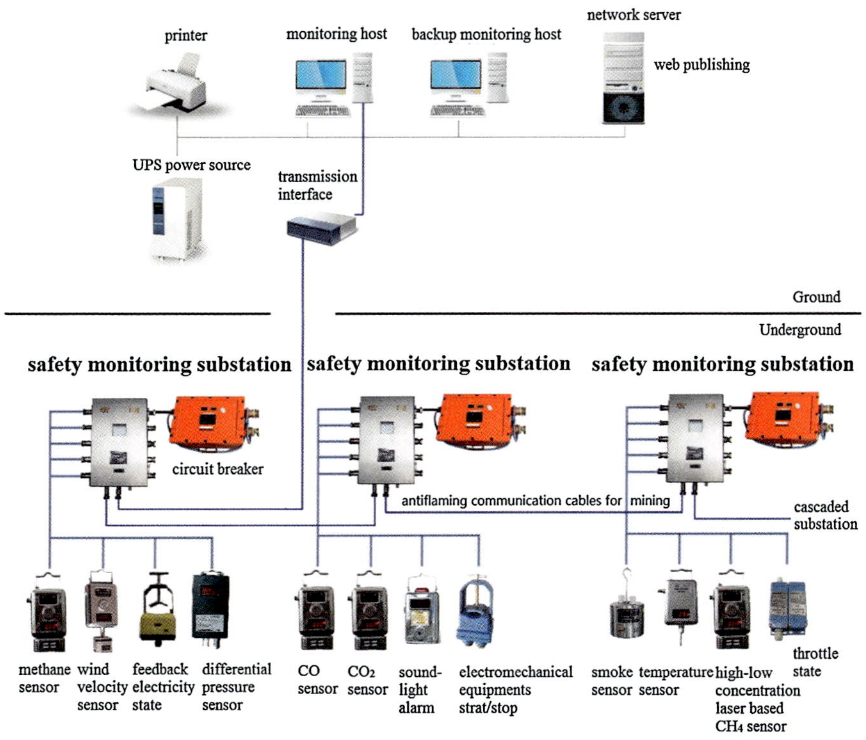

Fig. 9 Methane monitoring system block diagram

In case where field display is not necessary, fiber optic methane sensors are more advantageous than conventional electronic sensors. Typical example is in the methane drainage system, where fiber optic methane gas sensors can be used to monitor pipeline methane concentration and offer 0–100% full range capability. Regular re-calibration of the black-white methane sensor (typically once every 2 weeks) is very troublesome. The use of laser, infrared, or fiber optic methane sensors has extended the calibration or maintenance cycle to once every 6 months, which brings significant reduction on manpower for sensor maintenance and human error associated to the sensor calibration routine. Figure 10 shows the field record of methane gas monitoring data, where short spikes were sensor data reading response to 2.0% standard methane gas. The standard gas was used purely for the purpose of checking the accuracy of the laser methane sensors, recalibration, however, was not needed. Coal mine field application shows that the laser methane sensor continuously used for 5 years without recalibration.

Wireless signal transmission in underground coal mines can be more troublesome and unreliable than in open space on the ground, due to the absorption of EM energy by coal and rock in the tunnel wall, reflection from metal hydraulic rock support and machine tools, as well as the curvature in the tunnel. Nevertheless,

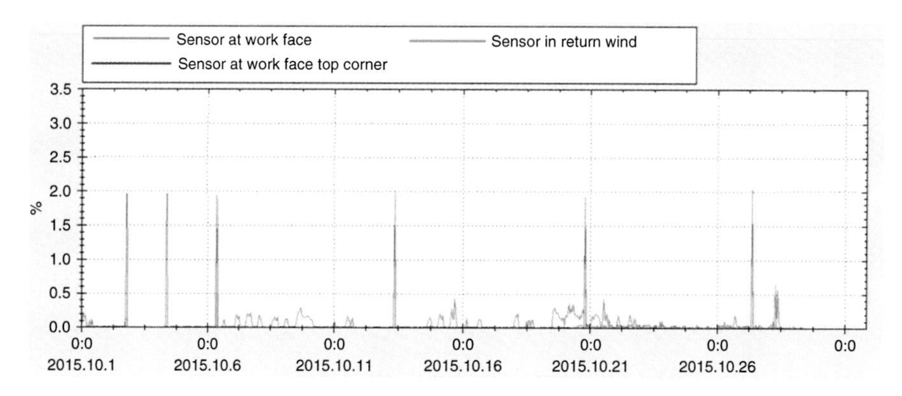

Fig. 10 Recording of field methane monitoring data where standard 2.0% methane gas were routinely used to test the sensors (X-axis: Dates; Y-axis: methane concentration (%))

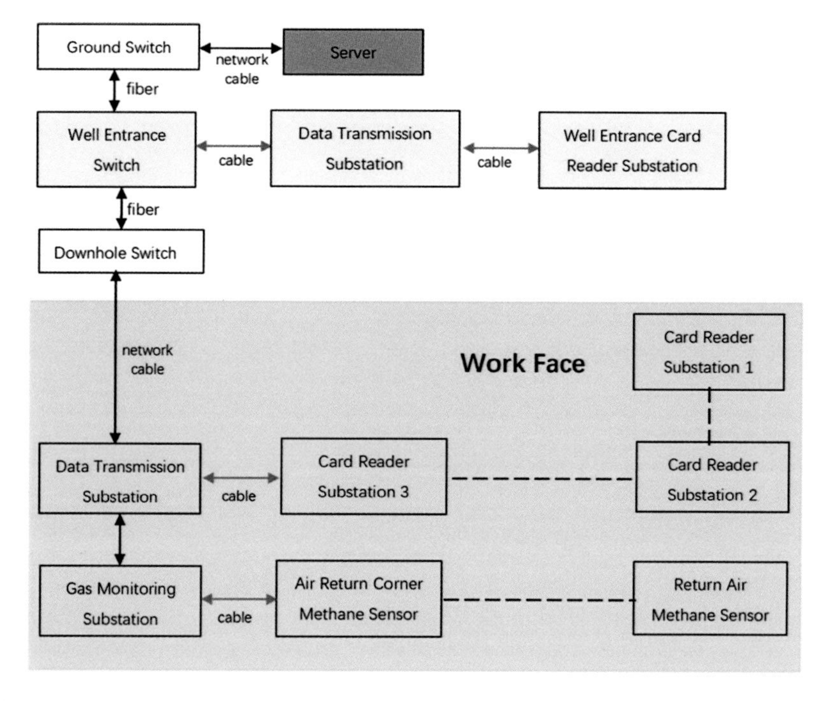

Fig. 11 Block diagram of coal mine underground wireless methane monitoring system

with wireless relay station installed in appropriate location, as illustrated in Fig. 11, robust wireless communication can be achieved which may transmit methane gas monitoring data from nearby mobile objects such as people or vehicles (Salankar and Suresh 2014). Laser absorption-based methane sensors are also used as mobile sensor in the wireless methane monitoring system.

Coal Mine Combustion Monitoring

Fire is one of the major hazards in coal mine production, which may be classified into two major categories: coal combustion and electrical fault. Coal combustion is due to coal residues in the goaf being oxidized and the heat accumulated which in turn accelerates the combustion and causes fire. The oxidization process is characterized by the presence and the increase of CO at the initial phase of oxidization, then the presence of C_2H_4 when the oxidization becomes severe and temperature rises, C_2H_2 is the final warning indicator where its presence means fire is imminent. Therefore, for early detection of combustion, CO is the key characteristic gas to monitor. O_2, CH_4, C_2H_2, C_2H_4, and CO_2 also need to be monitored. The required detection sensitivity for CO is 1 ppm, for both C_2H_4 and C_2H_2 is 0.01 ppm.

Laser Spectroscopy-Based Multi-Gas Sensors

Carbon monoxide has strong absorption in the IR region. As shown in Fig. 12, in the 4.6 μm and 4.75 μm region, the absorption lines of CO have coefficients of 4.5×10^{-19}. In order to use telecommunication laser device and standard single mode optical fiber, the harmonic absorption spectra in the NIR region are of interest. Figure 13 shows the absorption lines in the 1568 and 1583 nm wavelength regions. It can be seen that the NIR absorption is about four magnitudes weaker than the IR absorption.

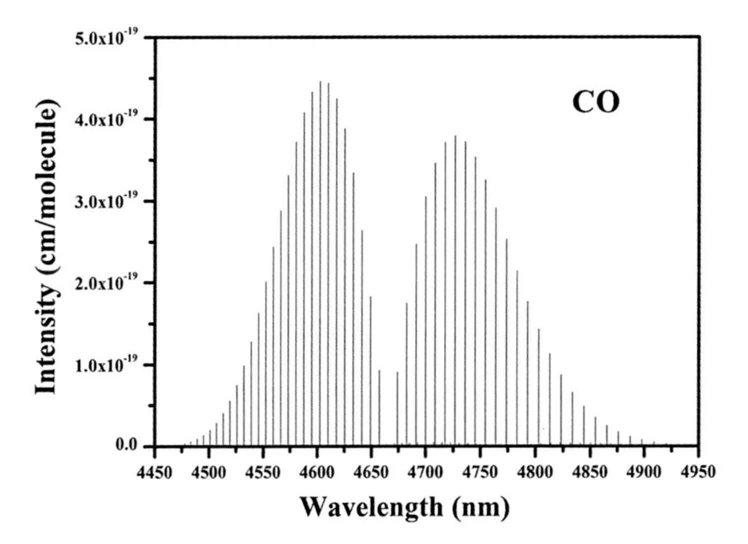

Fig. 12 The IR spectrum of CO gas

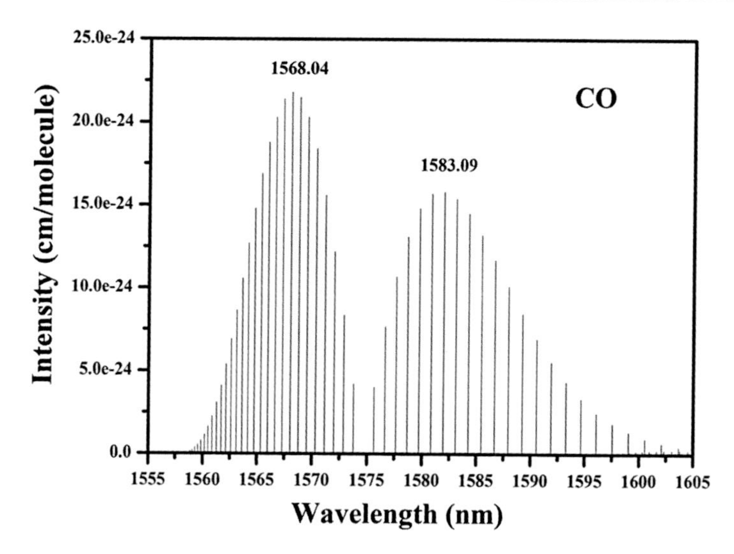

Fig. 13 The CO harmonic spectrum at 1.568 and 1.583 um region

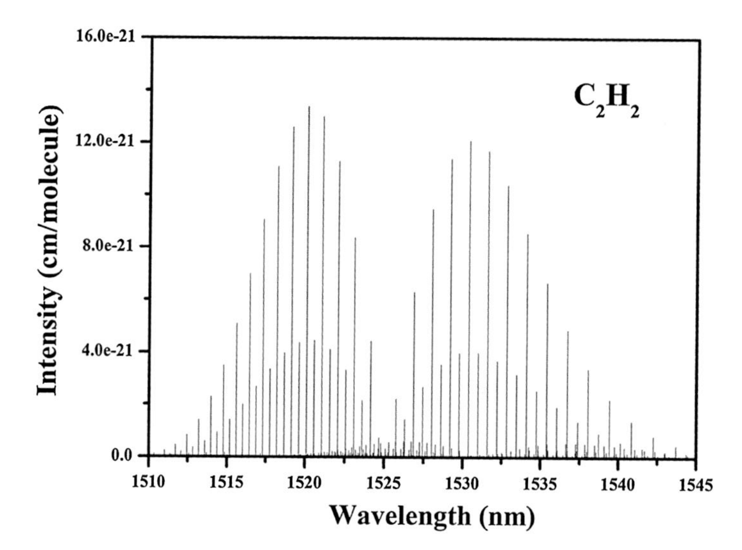

Fig. 14 NIR spectrum of C_2H_2

The NIR spectra of C_2H_2 are shown in Fig. 14, where it can be seen that the absorption band is in the region 1520–1535 nm. The C_2H_4, C_2H_6, CO_2, and O_2 have absorption bands around 1626 nm, 1654 nm, 1538 nm, 760 nm, respectively.

In order to achieve sub-ppm sensitivity, multiple reflection long path length gas cell is required. The Harriet design which employs two lenses has the advantage of high tolerance of mechanical vibration. The laser multi-gas sensor block diagram is shown in Fig. 15, where light from semiconductor laser diodes emitting

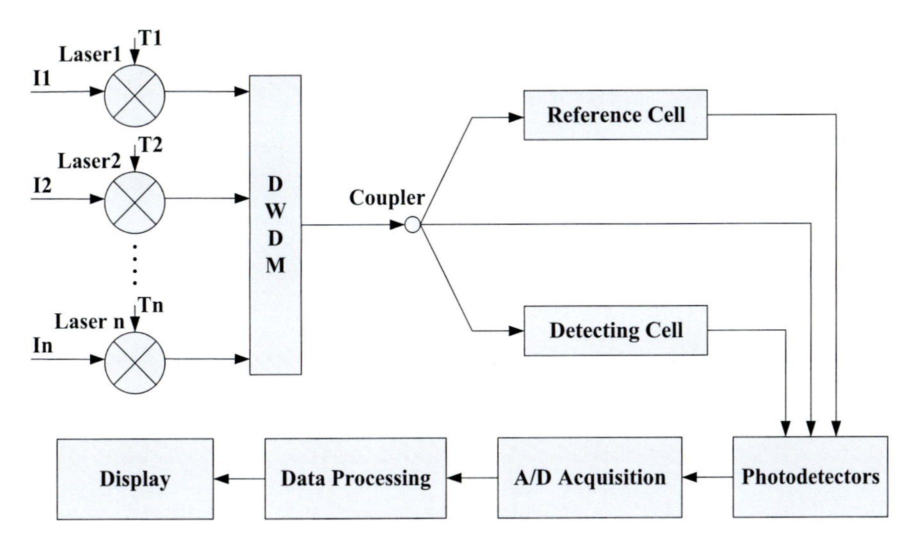

Fig. 15 Block diagram of laser spectroscopy-based multi-gas sensor system

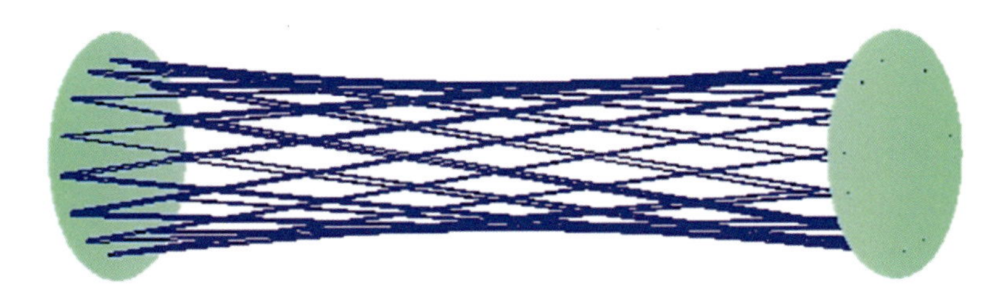

Fig. 16 Illustration of multi-path Harriet cell

at wavelengths centered, respectively, at the CO, C_2H_2, C_2H_4, and CO_2 NIR absorption wavelengths are combined via a DWDM wavelength combiner, then split into three fiber arms. One is directly linked to a photodiode and used as power reference; one is connected to the photodiode through a reference cell, which contains mixture of gases to be measured, and used as wavelength reference; the third one is connected to the sensor gas cell which is made of Harriet cell with optical path length of 20 m (see Fig. 16). Signal through the sensor gas cell is normalized by the power reference to eliminate the effect due to light source power fluctuation.

Typical normalized absorption signal of CO, C_2H_2, C_2H_4 are shown in Figs. 17, 18, and 19, respectively.

For single gas in N_2, the laser spectroscopy-based gas detection sensitivity and accuracy for CO, C_2H_2, and C_2H_4 can be 1 ppm, 0.1 ppm, and 0.1 ppm, respectively. However, when high concentration of CH_4, and or CO_2 present, it is important to compensate for the crosstalk to CO, C_2H_4.

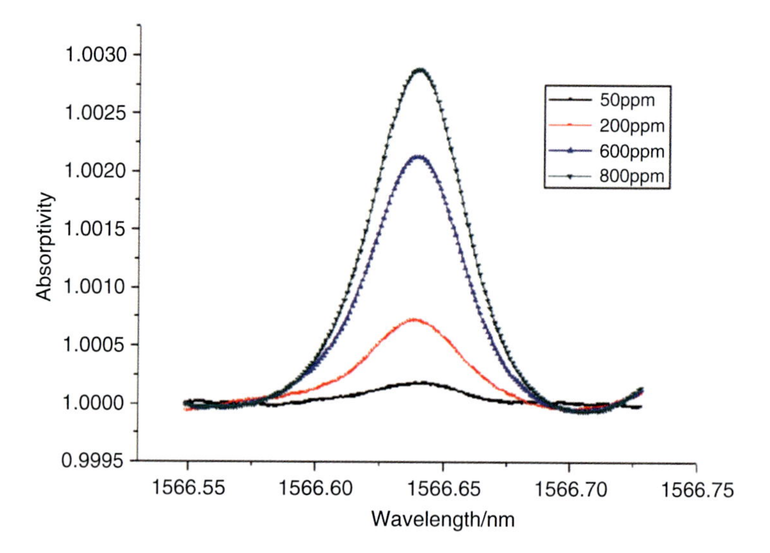

Fig. 17 Normalized absorption spectra of CO

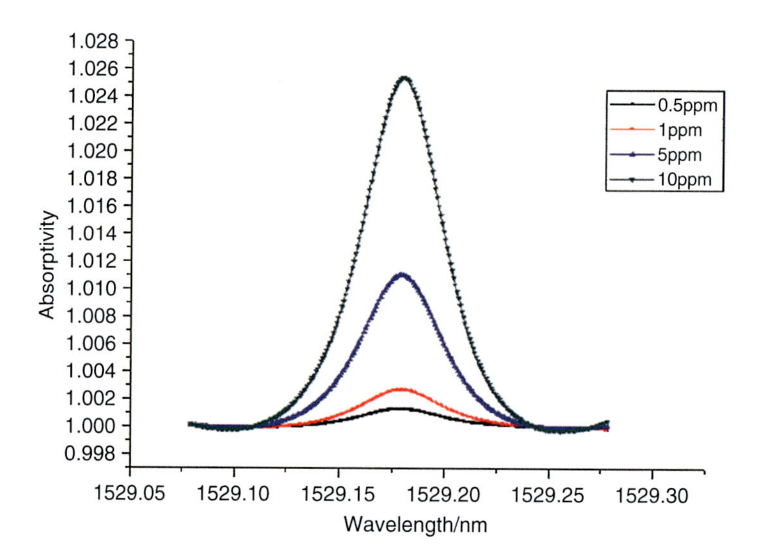

Fig. 18 Normalized absorption spectra of C_2H_2

Fiber Optic Distributed Temperature Sensor and Laser Multi-Gas Sensor-Based Coal Mine Combustion Monitoring System

Fiber optic Raman scatter-based distributed temperature sensor (DTS) (Dakin 1989) is increasingly used for coal mine goaf combustion monitoring. The length of the coal mine long wall workface is normally less than 6 km, hence multimode fiber

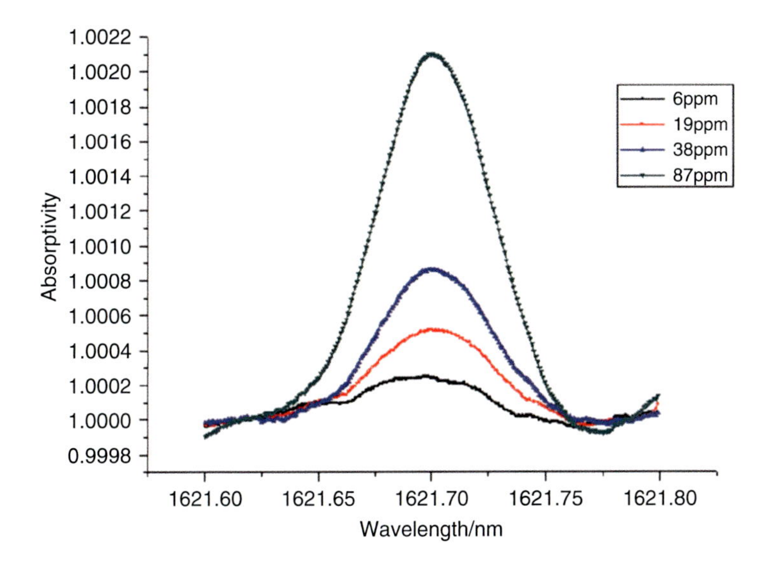

Fig. 19 Normalized absorption spectra of C_2H_4

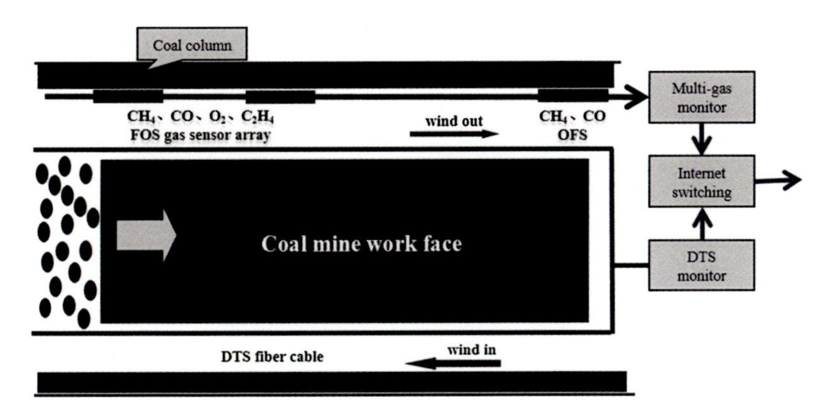

Fig. 20 Illustration of fiber optic sensor-based coal mine combustion monitoring system

cables can be used. The fiber optic temperature cables are placed along the wall of the ventilation tunnels in the coal seam production workface, which is left behind in the goaf as production in progress. More than 70% of coal mine goaf combustion occurs in the ventilation tunnel where air or oxygen leakage is more likely to occur due to faulty air seal and remaining coal residue in the ventilation tunnels. The DTS instrument is located nearby the underground power distribution chamber and DTS data is transmitted to the ground server via industrial network. The layout of the DTS goaf temperature monitoring fiber cable is illustrated in Fig. 20.

Conventional coal mine combustion detection system is based on gas analysis using laboratory Chromatography where the coal mine goaf gas samples are drawn

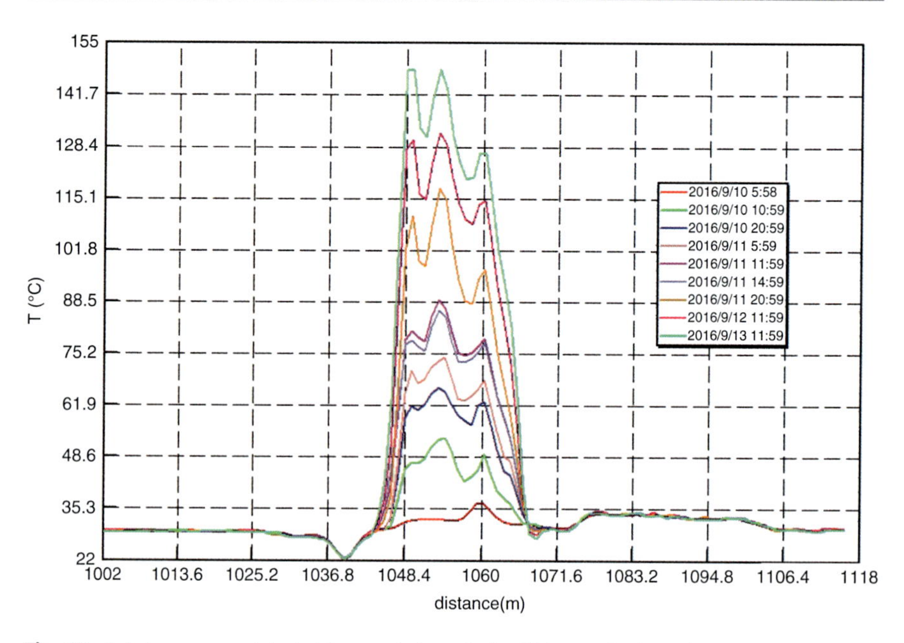

Fig. 21 A hot zone was detected in a sealed goaf of a China coal mine where temperature rose over 110 °C in 4 days

using a tubing buddle and pump system. The distance between goaf and the underground analysis laboratory can be 10 km or longer, hence the tubing bundle system is very troublesome to maintain and prone to error due to pipe line leakage or blockage. The fiber optic DTS provides convenient means to directly monitor temperature distribution and change with time. Data analysis determines the average temperature and its changes, the maximum temperature and its location (the hot zone) and changes with time. In case the DTS fiber cable is not directly in coal oxidization zone, the actual temperature may be deduced from the multi-gas data. Hence the combination of both DTS and multi-gas data offers early warning info for both the oxidization status and the hot zone location.

Figures 21 and 22 show temperature distribution inside a sealed coal mine goaf, where temperature rises in the hot zone was monitored before and after combustion treatment. The DTS monitoring data provides critical information which prevented coal combustion hazard. The combination of laser multi-gas sensor-based local tube bundle system and DTS will further enhance the capability of fire prevention.

Fiber Optic Sensor-Based Coal Mine Seismic Monitoring

With the increase of underground mining depth, so called coal mine dynamic hazards are also increasing, which is in the form of coal and/or rock burst and methane gas outburst for gassy mines. Monitoring data for rock stress, displacement,

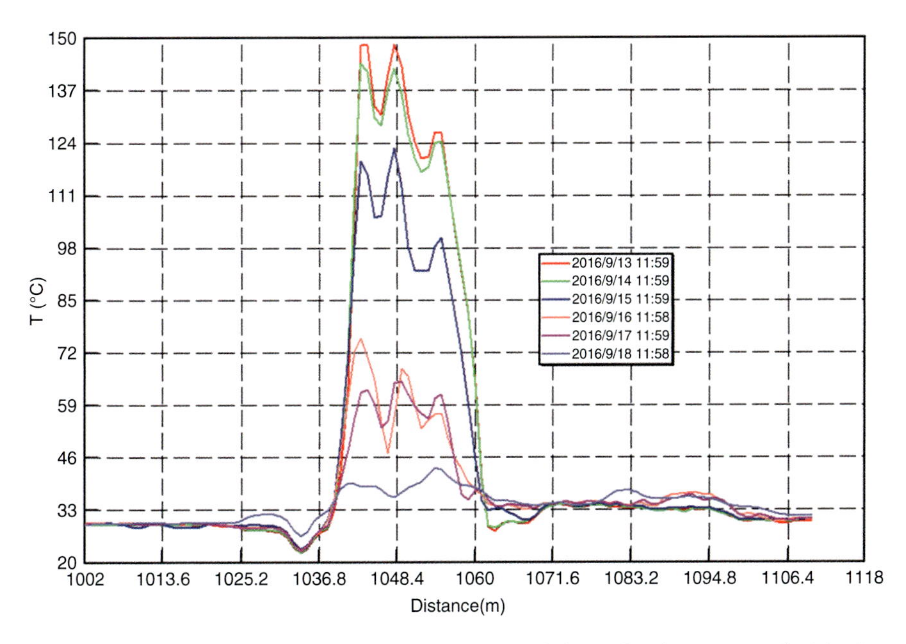

Fig. 22 The hot zone temperature dropped back to normal via combustion treatment by injecting N_2 in the goaf across a time period of 6 days

and micro-seismic event detection are crucial information for mining dynamic hazard prevention (Jiang et al. 2006). Coal mine micro-seismic events result from rock fracture which gives off elastic energy propagating in the coal seams or rocks. Acceleration sensors are placed in the boreholes in the coal seam or inside rocks to detect the micro-seismic compression or shear waves.

Fiber Optic Micro-Seismic Sensor and Interrogation System

Fiber Bragg Grating Accelerometer

A number of fiber optic accelerometers have been reported. Pedestedt and Jackson reported a highly sensitive interferometric accelerometer which consists of a compliant cylinder type fiber optic accelerometer (Pechstedt and Jackson 1995). For coal mine engineering use, low-cost and robustness may be more critical then sensitivity. The fiber Bragg grating and cantilever type accelerometers exhibits good frequency response and cross-sensitivity (Wu et al. 2009), hence were developed for coal mine applications (Wang et al. 2016). The schematic of the accelerometer is shown in Fig. 23. The FBG element experiences strain variation in response to vertical acceleration due to the movement of the mass and the cantilever. The selection of the mass at the end of the cantilever, the geometry, and material property of the cantilever determines the resonant frequency and the sensitivity.

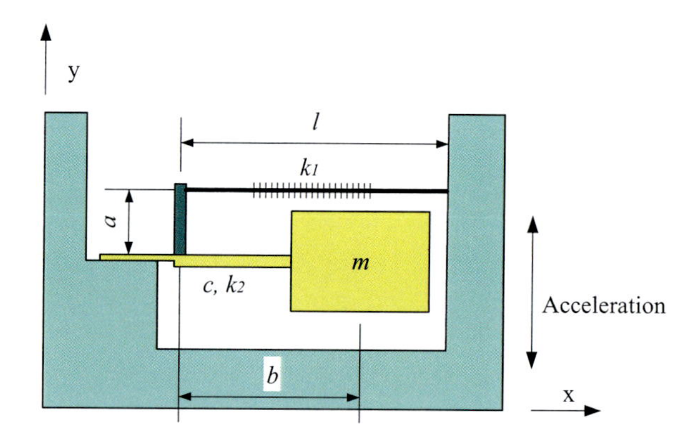

Fig. 23 Schematic of the cantilever-based FBG accelerometer design

With reference to Fig. 23, assuming the elastic coefficient of the cantilever is k_2, the inert mass is m, the elastic coefficient of the fiber Bragg grating is k_1, the seismic wave will cause acceleration change on the inert mass, which induces the strain variation on the leaf spring and the FBG sensor subsequently. The total elastic coefficient of the geophone can be expressed as,

$$k = k_2 + k_1\left(\frac{a}{b}\right)^2 \tag{4}$$

The natural angular frequency can be defined as:

$$\omega_0 = \sqrt{\frac{k}{m}} = \sqrt{\frac{k_2 + k_1\left(\frac{a}{b}\right)^2}{m}} \tag{5}$$

Assuming an external acceleration $a_g = A_g e^{i\omega t}$ and a relative damping coefficient δ, the strain experienced by the FBG can be expressed as:

$$\varepsilon = \frac{aa_g}{bl}\frac{1}{\sqrt{\left(\omega_0{}^2 - \omega^2\right)^2 + 4\delta^2\omega^2}} \tag{6}$$

When the damping ration $\zeta = \frac{\delta}{\omega_0} = 0.7$, the strain acceleration sensitivity K can be noted as:

$$K = \frac{a}{bl\omega_0^2} \tag{7}$$

The FBG wavelength shift is proportional to the strain experienced by the FBG. Considering the strain sensitivity for FBGs with peak wavelengths in the C band regime is about 1.2 $pm/\mu\varepsilon$ in general, so the accelerometer sensitivity (wavelength shift of FBG per unit acceleration) is given by:

$$S = \frac{\Delta\lambda}{a_g} = \frac{1.2 \cdot a}{bl\omega_0^2} \tag{8}$$

It can be seen from formula (5) and (8) that the natural angular frequency and the accelerometer sensitivity are determined by five parameters a, b, k_1, k_2, and m. By optimizing the parameters a, b, and m and choosing suitable material k_2, the geophone is designed to fit for micro-seismic measurement in coal mines. The frequency bandwidth is designed to be 0–240 Hz and the natural frequency is around 280 Hz. The accelerometer sensitivity is 220 pm/g.

FBG Edge Detection Dynamic Monitoring Scheme

The FBG interrogation is illustrated by Fig. 24a and b. The DFB laser (DFB LD) is driven by a constant current source (CCS) to control the light power. The emitted light is injected into the optical circulator and arrives at the FBG accelerometer. The reflected light from FBG accelerometer goes back to the optical circulator, and at last reflected light modulated by acceleration is converted into a voltage signal by a PIN trans-impedance amplifier (TIA). The resulting voltage is acquired by A/D card. The temperature of the DFB LD is automatically controlled by a system-on-chip MCU through a temperature controller (TC) and then the wavelength of the DFB LD can be adjusted to the −3 dB position from the reflection peak (Wang et al. 2010) (see Fig. 24a). When the FBG wavelength changes, the feedback control signal from the LPF will vary in proportion to the wavelength shift hence can be used as the acceleration signal.

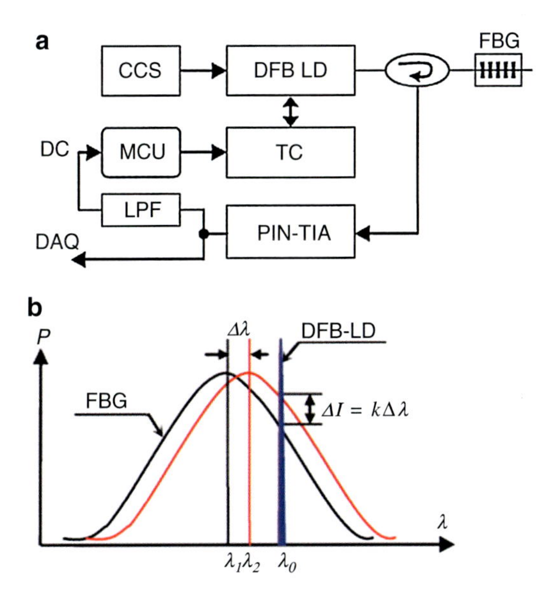

Fig. 24 Interrogation principle block diagram (**a**) and work principle diagram of FBG geophone measurement system (**b**)

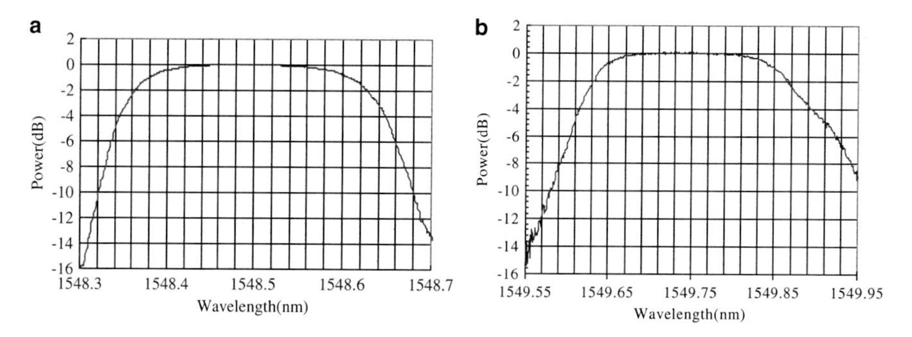

Fig. 25 Reflection spectra of FBG1 (**a**) and FBG2 (**b**)

As shown in Fig. 24b, the shape of the reflection spectrum of the FBG affects the detection sensitivity. Figure 25a and b show the reflectivity spectrum of FBG1 and FBG2, where the −3 dB power point is regarded as the working point. In Fig. 25a, when the wavelength changes 40 pm up and down nearby the operating point, the power changed 8.75 dB. In Fig. 25b, the corresponding power changed 5.2 dB. Experiments showed that FBG1 sensor has much higher detection sensitivity than FBG2.

Coal Mine Micro-Seismic Events Detection Based on Fiber Optic Accelerometers

FBG accelerometer sensors are installed such that they can cover the area to be monitored. The sensors system continuously record micro-seismic signals, which are above certain threshold level and are used to locate the source of the events. The micro-seismic signal recording includes several points before and after the trigger or threshold point. Both P-waves and S-waves are used for determining source properties.

The FBG accelerometer sensors were installed horizontally in coal mine. All the mounting holes located in rock roadway and are designed with 5° depression angle, 65 mm diameter, and a 5 m depth. The FBG accelerometer sensors are installed to the bottom of the hole with the help of installation rod and expansive cement was pumped into the installation hole to enable the sensors and coal rock integration. Figure 26a shows the photograph of the coal mine acceleration sensor and Fig. 26b shows the schematic of the acceleration sensor horizontally settled in coal mine.

The hypocenter location algorithm is based on the uniform velocity model. The travel-time equation can be expressed as,

$$\sqrt{(x_i - x)^2 + (y_i - y)^2 + (z_i - z)^2} - v\,(t_i - t) = 0 \tag{9}$$

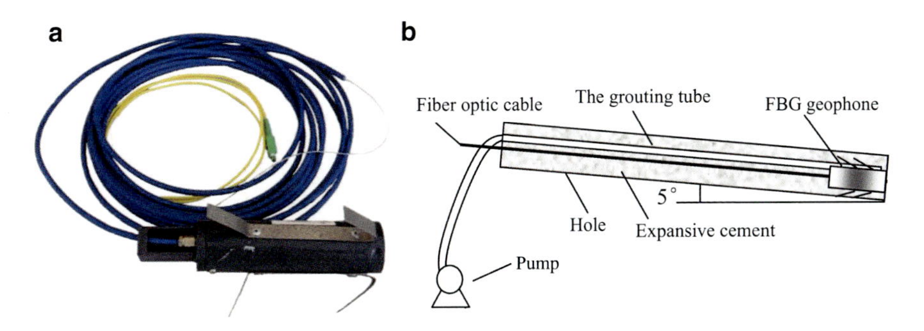

Fig. 26 Photograph of the coal mine acceleration sensor (**a**) and schematic of the acceleration sensor horizontally installed (**b**)

where, (x_i, y_i, z_i) is the coordinate of sensor i, v is the P-wave velocity, t_i is the arrival time at each geophone, (x, y, z) is the source coordinate, and t is the time hypocenter occurred.

The precision for determining the first seismic wave arrival time is very important for accurately locating the source. In the FBG accelerometer monitoring system, a time-window energy ratio algorithm was developed to pick up the first arrivals automatically.

When signal magnification and other parameters of the same event in each channel are relatively stable, the corresponding recorded vibration duration is also more stable. With higher vibration intensity, the vibration duration is longer, and vice versa. Using vibration duration to represent micro-seismic energy, the calculation process does not rely on amplitude, frequency, or any other kinetic parameters. The vibration duration is almost independent of the source location result. Based on this relationship, we can establish the relation between micro-seismic energy and vibration duration.

$$\lg E = a + b \lg T \tag{10}$$

and

$$\lg T = (\lg T_1 + \lg T_2 + \cdots + \lg T_N) / N \tag{11}$$

Where E is micro-seismic event energy, T is event signal duration, N is the number of effective channel of one event, $T_1 \ldots T_N$ is signal duration of each channel, the unit is second. To determine the unknown coefficients a and b, we use the event data with energy higher than 4500 J recorded by Poland SOS system in Xinglongzhuang coal mine from April 1, 2013 to April 6, 2013 to calibrate energy and magnitude of the corresponding fiber optics micro-seismic events. The fitting figure is shown in Formula (12) below.

$$\lg E = 5.0825 + 3.6555 \lg T \tag{12}$$

The precision of the first arrival picking up is very important to precision location of the source. Picking of the first arrivals of seismic events is carried out in the time-domain. In the FBG geophone monitoring system, an improved time-window energy ratio algorithm was used to pick first arrivals automatically.

The definition of the improved sliding time-window energy ratio A_I can be calculated as below (Wang et al. 2016):

$$A_1 = \frac{\left[\sum_{t=T_0}^{T_2} x^2(t)\right]^{\frac{1}{2}} + 3 \times \left[\sum_{t=0}^{T} x^2(t)\right]^{\frac{1}{2}} / n}{\left[\sum_{t=T_1}^{T_0} x^2(t)\right]^{\frac{1}{2}} + 3 \times \left[\sum_{t=0}^{T} x^2(t)\right]^{\frac{1}{2}} / n} \tag{13}$$

Where A_I is energy eigenvalue, T_I is the start point of the time-window, T_0 is the middle point and T_2 is the end point, $x(t)$ is the acceleration amplitude recorded by the monitoring system, T is duration of each channel, and n is the sample points quantity of each channel. The recorded signals before the first arrival point are considered as noise signal and those after are considered as effective signal; the time-window energy ratio is very large between them. When the time-window width of the periodic wave equals to two times the period, the time-window energy ratio represents the condition of the two adjacent periods. The time-window width can be expressed with the interval of two adjacent peaks (valleys) around the maximum amplitude. Signal dominant frequency f_0 corresponds to the maximum amplitude point, f_s is the sample frequency, the time-window width TWL can be calculated as:

$$\mathrm{TWL} = \frac{2 \times f_s}{f_0} \tag{14}$$

It is worth noting that the calibrated algorithm A_1 must be combined with Eq. 14, adopting TWL as an automatically picking time window width. It may need manual correction to prevent it from the influence of the waveform distortion after automatic picking. Micro-seismic detection software has been developed to describe both the location and energy level of the micro-seismic event on a mining map as shown in Fig. 27.

Figure 28 shows the distribution plan of the micro-seismic events observed in a China coal mine over the period of 2 months during May 14–July 23, 2013. It can be seen that the micro-seismic events were mainly distributed in the No. 10303 working face during production. The distribution of the mine earthquake has obvious regional characteristics and is closely related to the mining activities in the region. Accuracy of +/−10 m can be achieved for micro-seismic source location.

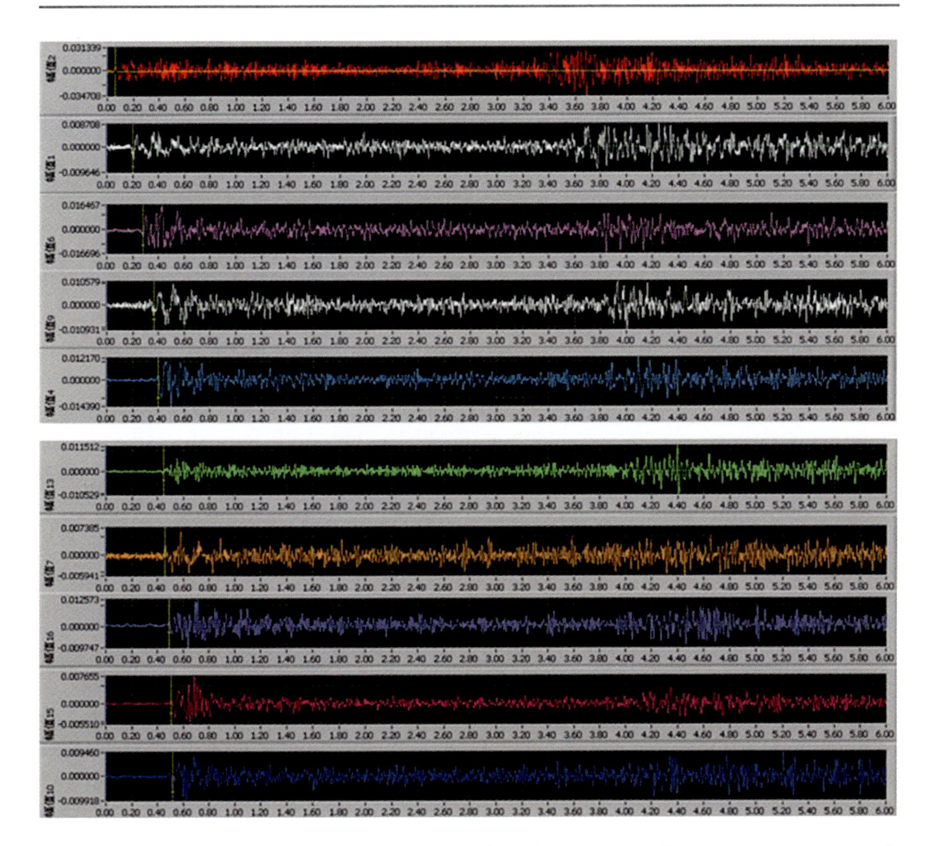

Fig. 27 Micro-seismic waveforms detected by the fiber optic accelerometer sensor network installed in a coal mine

Fiber Optic Sensors-Based Coal Mine Equipment Condition Monitoring System

With the increased level of automation, large power equipment are widely used such as coal cutter, road header in tunnel development,air compressor, water pump, conveyer belt, hoist system, etc. Temperature and vibration are two key parameters which are characteristic for equipment condition monitoring. FBG element is widely used for temperature and vibration sensing. While industrial practice still relies on routine survey, online monitoring is increasingly popular. The vibration sensors are also accelerometer type and have similar designs as described in section "Fiber Bragg Grating Accelerometer." The FBG interrogator is based on telecom product of tunable laser diodes, which can interrogate FBG array sensors at 1 kHz for a reasonably low cost.

Figure 29 shows a typical fiber Bragg grating vibration sensor for equipment condition monitoring. The field deployment of vibration sensors has been implemented

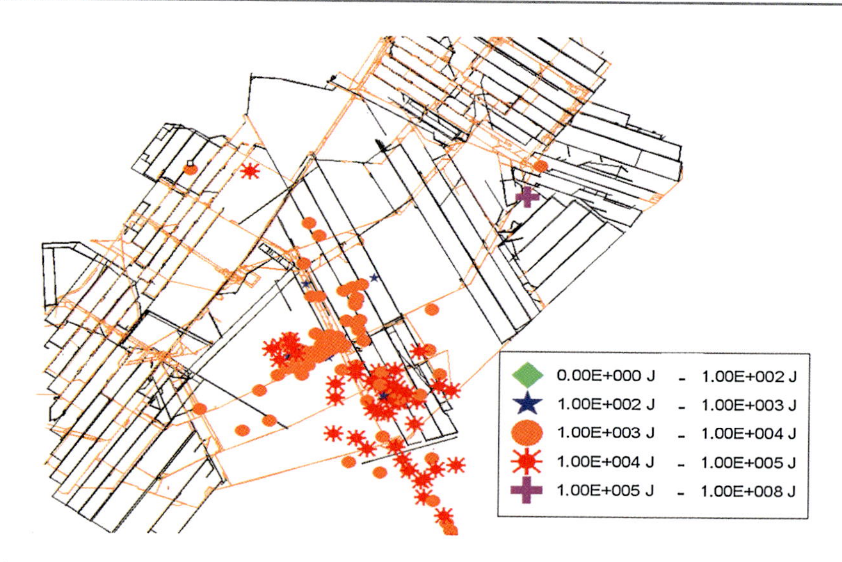

Fig. 28 Micro-seismic events distribution

Fig. 29 (**a**) Photograph of a fiber optic vibration sensor, (**b**) FO sensor for shaft monitoring

to monitor mechanical vibration of air compressors, water pumps, conveyor belt mandrels, hoist systems, etc. Typical data for a faulty gearbox shows second harmonic vibration signal that is present in the signal spectrum as shown in Fig. 30. Based on the predefined vibration intensity threshold and frequency characteristics, early diagnosis of equipment faults can be achieved.

All Fiber Optic Coal Mine Emergency Rescue Information System

One of the reasons that cause severe human casualties in coal mine accidents such as flooding or roof collapse is that the trapped persons in the accident lost contact with rescue people on the ground, and the rescue operation lacks information about the hazardous area due to the power cut-off which often occurs in an incident.

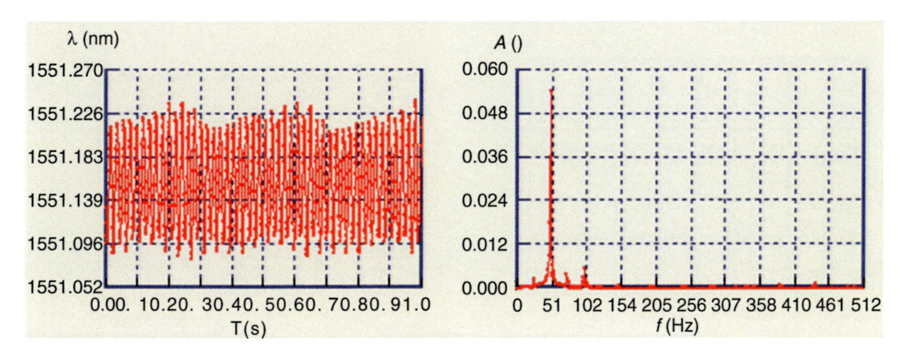

Fig. 30 Typical characteristic of faulty gearbox where second harmonic components present in the vibration signal, the software automatically record the alarm

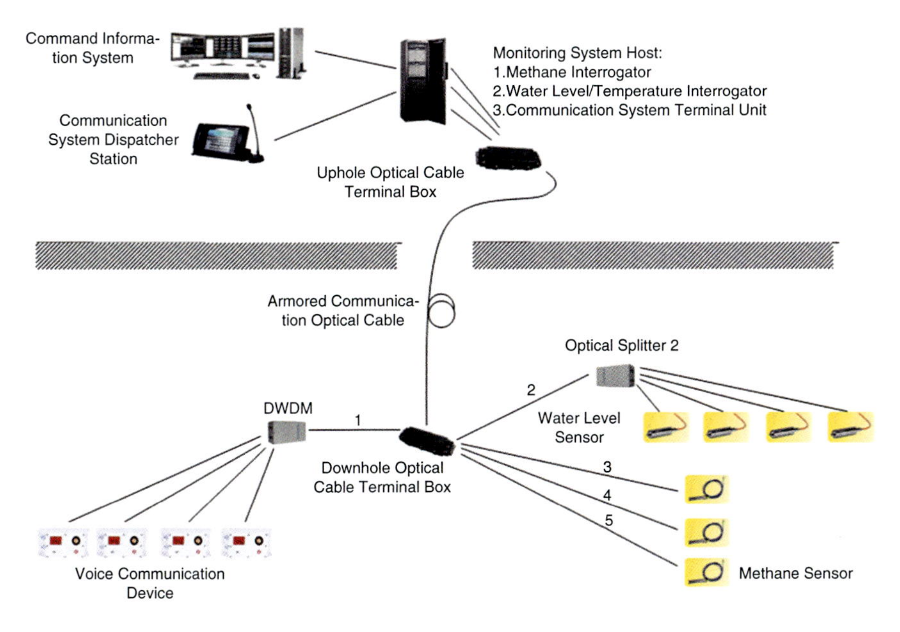

Fig. 31 Fiber optic coal mine rescue information system

Fiber optic multi-parameter sensors have been developed as novel coal mine rescue information system which consists of only the passive fiber optic sensing devices in the underground mine and the interrogators are installed in the control room on the ground. The system consists of fiber optic methane sensors, CO sensors, acoustic sensors, water level sensors, as well as temperature and micro-seismic sensors. The block diagram of the coal mine rescue information system is shown in Fig. 31. In case of emergency situation, even when the power supply to the underground mine is cut off, the system still continuously monitors the methane gas, temperature, water level, and rock stability, and provides key information for making timely decisions for rescue operation.

Summary and Future Trend

Coal mine presents one of the most hazardous scenarios in work places, where there are inflammable and explosive and toxic gases, combustion flooding, roof collapse, and many large equipment. The area is large, the environment is complex, and numbers of objects to be monitored is vast. Added to the complexity is that sometimes power supply is not readily available and conventional electronic sensors are not convenient to use. Fiber optic physical and gas sensors have been developed which offer unique advantages including intrinsic safety, long transmission distance, large area coverage, and multi-parameter sensing. In this chapter, we briefly described some of the key fiber optic sensors and typical applications for methane, combustion, rock-burst, equipment condition monitoring, and emergency information system. The system has been deployed in many Chinese coal mines and successfully prevented some coal combustion hazards, equipment fault prevention, etc.

With the advance of FOS technology on distributed acoustic sensors, wind flow sensors, gas sensors, humidity, and water quality sensors as well as wearable optical sensors, the demand is increasing to ensure the full coverage of coal mine environmental monitoring, and equipment and personnel monitoring is realized. This can be achieved through the development of Internet of Things for enhanced safety and intelligent big data analysis so that all hazards can be monitored and prevented. As the coal mines become safer and safer, the emphasis has been shifted more towards healthy environment and work conditions, equipped with portable laser-based coal dust sensors, as well as wind and humidity sensors.

Acknowledgment The work presented here was supported by funding from Chinese Ministry of Science and Technologies, Department of Science & technology, Shandong Province, national grants for international collaboration centre on OFS and IOT for safety, as well as funding support by Shandong Academy of Science, China. The information presented here have been supported by the colleagues at Laser Institute, Shandong Academy of Science and Micro-Sensor Photonics Ltd., as well as Yankuang Coal Mine Group and many industrial partners. Special thanks to Yubin Wei, Guangdong Song, Yanfang Li, Guangxian Jin, Jie Hu, Binxin Hu, Jinyu Wang, Jiqiang Wang, Tingting Zhang, Weisong Zhao, Lin Zhao, Chengxiang Song, Chang Wang, Jiasheng Ni, Zhaowei Wang, Guofeng Dong, Junpeng Ma, Xiangjun Meng, and Zengyu Zhao for their contribution and support in this work.

References

B. Culshaw, G. Steward, F. Dong, Fibre optic techniques for remote spectroscopic methane detection – Form concept to system realization. Sensors Actuators B **51**, 25–37 (1998)

J.P. Dakin, Chapter 15 – Distributed optical fiber sensor systems, in *Optical Fiber Sensors: Systems and Applications*, vol. 2, ed. by B. Culshaw, J. Dakin (Artech House, Boston, 1989), pp. 575–587

T. Iseki, H. Tai, K. Kimura, A portable remote methane sensor using a tunable diode laser. Meas. Sci. Technol. **11**(6), 594–602 (2000)

F. Jiang, S. Yang, Y. Cheng, et al., A study on microseismic monitoring of rock burst in coal mine [J]. Chin. J. Geophys. **49**(5), 1511–1516 (2006). (in Chinese)

J. Ni, J. Chang, T. Liu, Y. Li, Y. Zhao, Q. Wang, Fiber methane gas sensor and its application in methane outburst prediction in coal mine. J. Electron. Sci. Technol. China **6**(4), 373–376 (2008)

R.D. Pechstedt, D.A. Jackson, Design of a compliant-cylinder-type fiber-optic accelerometer: theory and experiment. Appl. Opt. **34**(16), 3009–3017 (1995)

P.A. Salankar, S.S. Suresh, Zigbee based underground mines parameter monitoring system for rescue and protection. IOSR J. VLSI Signal Process. (IOSR-JVSP) **4**(4), Ver. I (Jul–Aug. 2014), 32–36. e-ISSN: 2319 – 4200, p-ISSN No: 2319 – 4197 (2014)

"Statistical Review of World Energy", http://www.bp.com/en/global/corporate/energy-economics/statistical-review-of-world-energy.html

G. Stewart, B. Culshaw, W. Johnstone, G. Whitenett, K. Atherton, A. McLean, Optical fibre sensors and networks for environmental monitoring. Manag. Environ. Qual.: Int. J. **14**(2), 181–190 (2003)

J.-Y. Wang, T.-Y. Liu, C. Wang, X.-H. Liu, D.-H. Huo, J. Chang, A micro-seismic fiber Bragg Grating (FBG) sensor system based on distributed feedback laser. Meas. Sci. Technol. **21**, 094012 (6pp) (2010)

J. Wang, B. Hu, G. Song, L. Jiang, T. Liu, Design and application of Fiber Bragg Grating (FBG) geophone for higher sensitivity and wider frequency range. Measurement **79**, 228–235 (2016)

J. Wu, V. Masek, M. Cada, The possible use of fiber Bragg grating based accelerometers for seismic measurements, in *Proceedings Canadian Conference on Electrical and Computer Engineering*, May (3–6), 2009, pp. 860–863 (2009)

Optical Fiber Sensors in Ionizing Radiation Environments

49

Dan Sporea

Contents

Abstract

The present chapter addresses the use of optical fiber sensors (OFSs) in ionizing radiation environments. In this context, OFS research reflects (i) either the reliability of such sensors under radiation exposure and their capability to operate within prescribed limits (sensitivity, dynamic range, S/N, reproducibility, linearity, etc.) in order to act as transducers for specific quantities (temperature, humidity, strain, etc.) (ii) or, if they are subject to some degradation, the benefit of

D. Sporea (✉)
National Institute for Laser, Plasma and Radiation Physics, Center for Advanced Laser Technologies, Măgurele, Romania
e-mail: dan.sporea@inflpr.ro

G.-D. Peng (ed.), *Handbook of Optical Fibers*,
https://doi.org/10.1007/978-981-10-7087-7_25

the irradiation-induced changes for developing radiation dosimeters or radiation monitors.

OFSs are classified in extrinsic and intrinsic sensors, according to the role played by the optical fiber in the system. In the first case, the optical fiber acts as a light guide of the radiation-generated optical signal, while in the second situation, the fiber material constitutes the detecting medium, where light is produced under radiation exposure. Depending on the type of intrinsic sensor considered, some devices are structured inside the fiber and form the sensor. Intrinsic OFSs are further divided into discrete, quasi-distributed, and distributed sensing configurations. Discussion on OFSs in this chapter covers different fiber structures and materials, as they are reported in literature. Basic operating principles of these sensors are introduced to the reader excepting the cases when such concepts are detailed in other chapters. For reader's convenience, an extended list of references was included in order to set the scene for a better understanding of the benefits and limits of employing OFSs in such circumstances. In most of the presented cases, applications of optical fiber sensors in radiation dosimetry and/or radiation monitoring are mentioned.

The chapter targeted audience is formed by university students, technical personnel, and experts in specific fields (medicine, nuclear and space industries, operators of ionizing radiation sources) interested in the use of optical fibers for remote monitoring and control and in radiation reach environments, looking for the benefit associated with such sensors: immunity to electromagnetic fields, lack of fire risks, small size, low mass, and capability to handle multiparameter, multiplexed, or distributed measurements.

Keywords
Distributed radiation monitoring · Extrinsic and intrinsic sensors · Ionizing radiation · Optical fiber gratings · Optical fiber sensor · Radiation effects · Radiation dosimetry

Introduction

The performance of optical fibers and optical fiber sensors under ionizing radiation can be judged from two perspectives: (i) they are deteriorated under irradiation, in which case when they are considered as radiation dosimetry, and (ii) they are surviving radiation exposure, so that they can be used for data transmission, remote sensing, and control in radiation environments (Sporea et al. 2012a). In this chapter both aspects are discussed, in order to provide to the reader the basic background on these issues.

A classification of the optical fiber-based ionizing radiation detectors is given in Fig. 1. These sensors are described as belonging to two major categories: (i) intrinsic sensors, for which the optical fiber constitutes the sensing element, and (ii) extrinsic sensors, in which case the optical fiber is used as a coupling path for

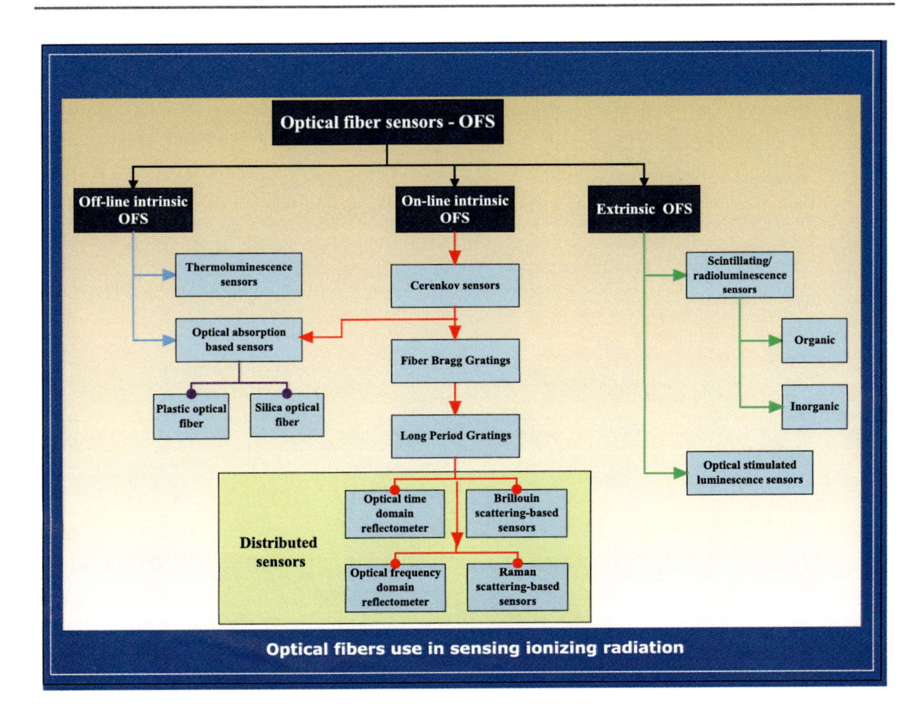

Fig. 1 The classification of optical fiber sensors used in radiation environments

the radiation-produced optical signal. In addition, the first category can be further divided into (i) off-line intrinsic sensors for which the radiation effect(s) is (are) measured after the irradiation process is ended and (ii) online intrinsic sensors in which the radiation effect(s) is (are) evaluated in real time, during the irradiation.

In setups where the optical fiber forms only a link between the sensing element and the optical detector, the detection of the ionizing radiation can be performed either directly through radioluminescence or scintillation or by using the optical signal generated as a luminescence signal upon the excitation with light of the irradiated material. *Radioluminescence* is defined as the "luminescence excited by impact of radioactive particles" (https://www.merriam-ebster.com/dictionary/radioluminescence), *scintillation* being "a flash of light produced in a phosphor by an ionizing event" (https://www.merriam-webster.com/dictionary/scintillation). The current definition of *luminescence* is "the low-temperature emission of light, as by a chemical or physiological process" (https://www.merriam-webster.com/dictionary/luminescence). The material employed to convert the ionizing radiation energy into an optical signal can be organic or inorganic, crystalline or amorphous.

Exposed to ionizing radiation, optical fibers exhibit radiation-induced attenuation (RIA) or radiation-induced luminescence (RIL) (Sporea et al. 2012a). Both these effects can be used to detect ionizing radiation. Based on RIL, optical fibers can be incorporated into radiation sensors of the type previously mentioned. Silica, plastic, or sapphire optical fibers were embedded into real-time radiation fields monitoring systems. Attenuation of the optical radiation transmitted along an optical fiber, exposed to ionizing radiation, can be measured either online during irradiation or off-line, postirradiation. For such applications, both plastic and silica-based optical fibers were investigated.

Thermoluminescence (luminescence "developed in a previously excited substance upon gentle heating" – https://www.merriam-webster.com/dictionary/thermoluminescence) present in some types of optical fibers was also employed for radiation dosimetry.

Cerenkov radiation emitted in optical fibers during their irradiation was another studied phenomenon in relation to ionizing radiation detection. Traveling in a dielectric, transparent media, a charged particle, having a velocity greater than the optical radiation phase velocity in that medium, interacts with medium molecules and generates an optical radiation with continuum spectrum covering the UV to visible spectral range.

A special class of intrinsic optical fiber is represented by distributed optical fibers sensing systems such as optical time-domain reflectometers (OTDR), optical frequency domain reflectometers (OFDR), and Raman and Brillouin scattering-based instruments. Because such equipments are designed to monitor changes along the length of an optical fiber system, research was carried out to investigate their possible use on distributed radiation dosimetry.

In the context of device performances in radiation fields, it is on interest to discuss the behavior of fiber gratings (fiber Bragg gratings and long-period gratings) when they are exposed to radiation.

Subsequent sections of this chapter will introduce the reader to the operating principles, characteristics, and challenges related to the main types of optical fibers sensors which can be used under irradiation conditions.

Extrinsic Optical Fiber Sensors

Sensors Based on Radioluminescence or Scintillation

The basic structure of an extrinsic optical fiber sensors based on scintillation, generally referred to as the "butt-coupled" (Santos et al. 2014), is depicted in Fig. 2 and consists of a scintillating material fixed with an optical epoxy at the end of a polymethyl methacrylate (PMMA) optical fiber.

Different materials were used to produce the detection tip:

– Saint-Gobain Crystals and Detectors organic scintillating fibers, having light peak emission in different spectral bands: BCF-10 (blue, $\lambda = 432$ nm), BCF-12 (blue, $\lambda = 435$ nm), BCF-20 (green, $\lambda = 492$ nm), BCF-60 (green, $\lambda = 530$ nm), decay time between 2.7 and 7 ns, and emission efficiency of about 7000–8000 photons/MeV at proton excitation (Beierholm et al. 2008; Jang et al. 2011; Klein et al. 2012; Therriault-Proulx et al. 2011).

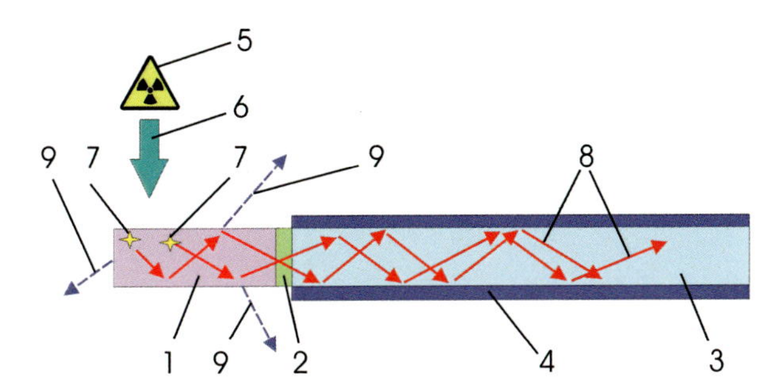

Fig. 2 Extrinsic optical fiber sensor type I: 1, scintillating/radioluminescent material; 2, optical epoxy; 3, core of the connecting plastic optical fiber; 4, cladding of the connecting plastic optical fiber; 5, ionizing radiation source; 6, ionizing radiation beam; 7, scintillating/radioluminescent event; 8, guided optical signal; 9, loosed optical signal

– Crystals of aluminum oxide (Al_2O_3:C) (Andersen et al. 2011; Buranurak and Andersen 2016). Ishikawa et al. used the BC-490 Plastic Scintillator Casting Resin from Saint-Gobain Ltd. to manufacture a hemisphere at the end of a plastic optical fiber (Ishikawa et al. 2015).

In another design the scintillating probe tip is formed by a piece of Ce^{+}-doped optical fiber (Mones et al. 2008). For all cases, the plastic optical fiber couples the emitted optical radiation to a detection system.

The sensors described above can be used only for dosimetry in a single point of the radiation field. In order to acquire data in multiple points by using a single guiding fiber, a system was designed by "cascading," along the light-collecting optical fiber, two or more detecting units formed by small pieces of scintillating materials having different emission wavelengths. By knowing the relative position of these scintillating units, the dose can be estimated in several spatially distinct points by monitoring with a spectrometer changes of the scintillating signal at the corresponding wavelengths (Duguay-Drouin 2016; Therriault-Proulx et al. 2012, 2013).

Mccarthy et al. investigated a scintillation-based sensor developed by embedding the end of a plastic optical fiber (PMMA) into a mixture of scintillating material and an optical epoxy. The fiber cladding was removed at its end where the sensing tip is mounted (Fig. 3). Several commercially available phosphors were tested, having different particle sizes and peak emission wavelength: GL47/N-C1 (ZnS:Ag/4 μm, $\lambda = 450$ nm), UKL59/N-R1 (Gd_2O_2S:Pr/8 μm, $\lambda = 513$ nm), UKL63/F-R1 (Gd_2O_2S:Eu/4 μm, $\lambda = 626$ nm), and UKL65/F-R1 (Gd_2O_2S:Tb/3.5 μm and 25 μm, $\lambda = 544$ nm).

Concerning the optical epoxy used, various options were investigated: Farnell Electronics EER1448RP250G ready mixed epoxy and resin, "Viafix" from Struers, and "EpoFix" manufactured also by Struers (McCarthy et al. 2011). To cover the plastic optical fiber end with the mixture of phosphorous scintillating material and epoxy solution, two techniques were employed: one based on a plastic cylinder-type

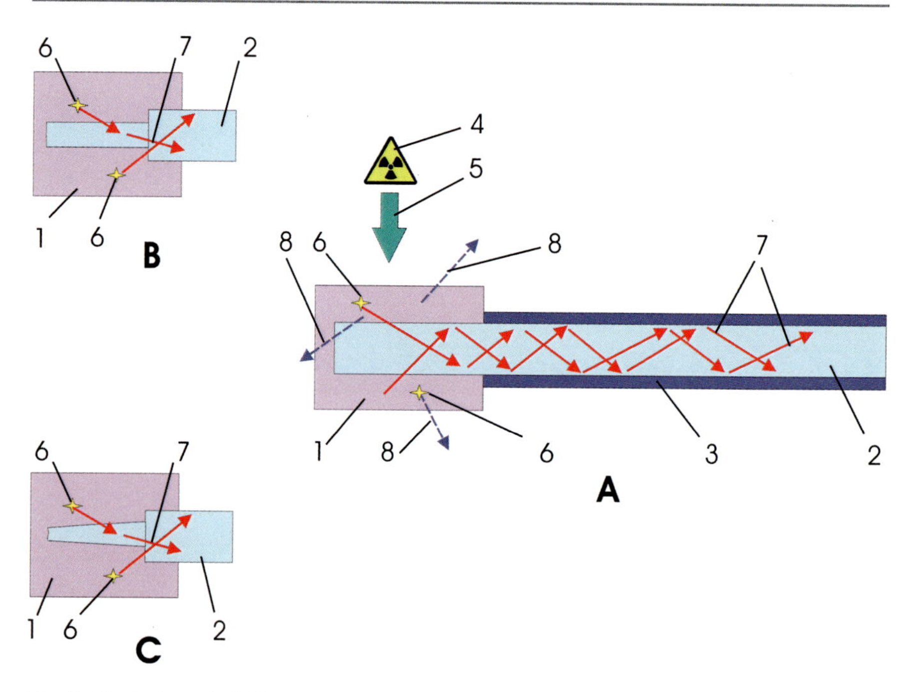

Fig. 3 Extrinsic optical fiber sensor type II: 1, scintillating/radioluminescent material; 2, core of the connecting plastic optical fiber; 3, cladding of the connecting plastic optical fiber; 4, ionizing radiation source; 5, ionizing radiation beam; 6, scintillating/radioluminescent event; 7, guided optical signal; 8, loosed optical signal. (**a**) Full fiber diameter implementation; (**b**) etched fiber-end implementation; (**c**) tapered fiber-end implementation

mold of 3 mm internal diameter and the other based on a heat shrink-type mold of 2 mm internal diameter. The coupling efficiency of the scintillating signal to the guiding PMMA fiber was improved either by covering the sensing tip with an Al foil or by painting it with TiO_2 paste (Sporea et al. 2014a).

The sensor head design optimization and the quality check of the implementation (concentricity of the fiber end and the scintillating cap, the presence of voids, the uniformity of the phosphorus grains distribution inside the cap) were tested by X-ray radiography, X-ray microtomography, and X-ray micro-fluorescence (Sporea et al. 2014b), some results being illustrated in Fig. 4.

Additional tests were carried out in order to evaluate (i) coupling efficiency of the optical signal to the PMMA fiber, (ii) the reproducibility of sensor's manufacturing process, (iii) the variation of the sensor responsivity along its length, and (iv) the sensor response to various operation conditions of the X-ray source. For these purposes, several sensors, produced by the two proposed techniques, using different phosphors and epoxies, without any coating on the sensing tip and with tips covered by two reflecting materials were investigated by X-ray fluorescence and radioluminescence and for several values of the X-ray source driving currents (Sporea et al. 2014a).

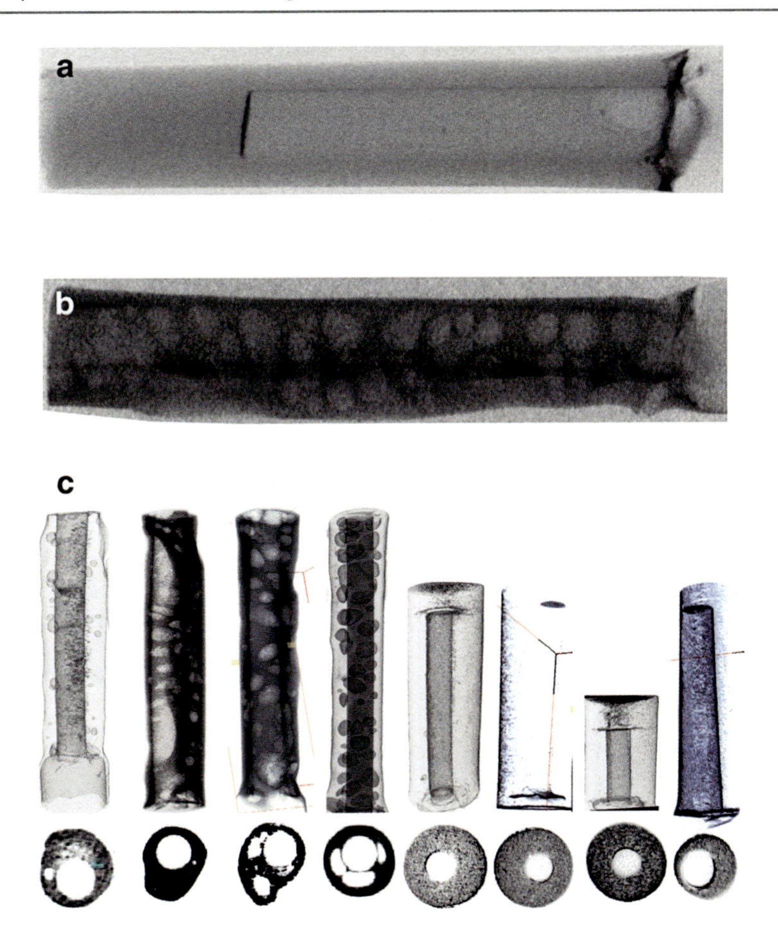

Fig. 4 Quality assurance of extrinsic optical fiber scintillating sensors: (**a**) X-ray radiography of the sensing tip; (**b**) X-ray fluorescence image of the sensing tip; (**c**) images of different sensors obtained by X-ray microtomography (Adapted after Sporea et al. 2014b)

The experimental setup is illustrated in Fig. 5 and consists in a miniature AMPTEK Inc. X-ray tube system, operating with an Ag target which constitutes the X-ray micro-source, a Silicon Drift Detector to monitor the X-ray fluorescence signal, and a system for the detection of the optical signal generated into the phosphorus under X-ray excitation. For spectral measurements (e.g., the radioluminescence signal) an Ocean Optics QE 65000 minispectrometer was used, while the signal integrated over the entire spectrum was measured with a Hamamatsu Multi-Pixel Photon Counter (MPPC), type C10507-11-100U. The investigations were done either by scanning the scintillating tip with a micro-X-ray beam along its length and measuring both the X-ray fluorescence signal and sensor's optical output or by scanning with X-ray beam the tip end and monitoring the optical signal for various operating conditions of the X-ray tube. The tests were performed with

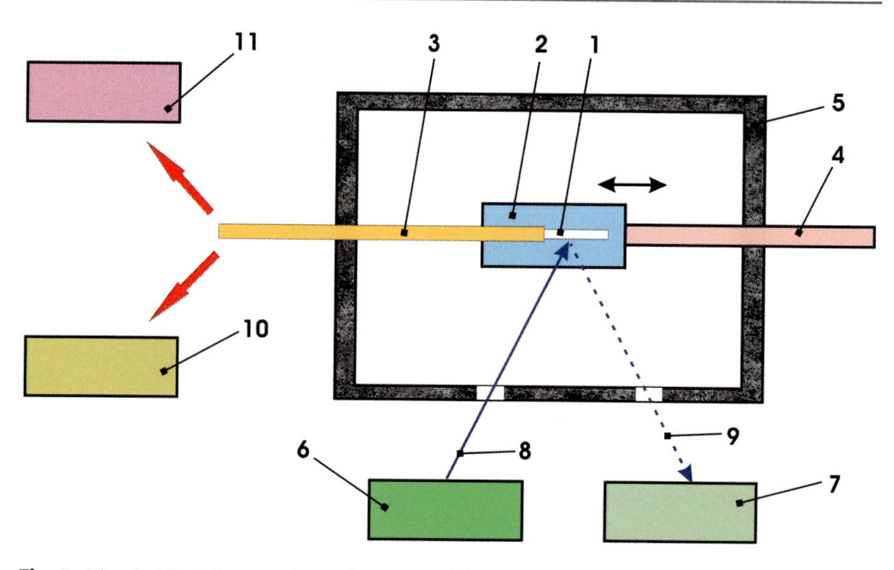

Fig. 5 The sketch of the experimental setup used for the characterization of sensor's response: 1, the tip of the extrinsic sensor; 2, translation stage; 3, PMMA connecting optical fiber; 4, translation stage guiding screw; 5, black box; 6, AMPTEK Inc. X-ray tube; 7, Silicon Drift Detector; 8, X-ray interrogation beam; 9, X-ray fluorescence signal; 10, optical fiber minispectrometer; 11, Hamamatsu Multi-Pixel Photon Counter (top view)

the sensor to be investigated placed into a black box in order to minimize the noise induced by the ambient light. This last type of evaluation was done for sensor's tips uncoated or coated with different optical reflecting materials. In order to localize the X-ray microbeam position along the sensor's head when scanned, this was wrapped at different lengths by copper wire, the detection of these wires by X-rays generating markers in the X-ray fluorescence signal.

The response of four sensors to the change in the operating condition (current and voltage) of the X-ray source is reported in Fig. 6. The effect of reflecting coating of the sensor tip on the light coupling efficiency to the guiding optical fiber is illustrated in Fig. 7, where the output optical signal is monitored as one of the sensors was scanned along its longitudinal axis by the X-ray microbeam. By scanning the sensor tip surface, a 3D map of its responsivity was derived in the case of a sensor without any coating, the same sensor wrapped in Al foil and the same sensor painted with TiO_2 (Fig. 8). These tests offer useful information for the sensor design and its manufacturing optimization (Sporea et al. 2014a). The reproducibility of the manufacturing process for six sensors developed through the abovementioned technique is shown in Fig. 9, for two subsequent measurements done under the same irradiation conditions, a fixed voltage and variable current steps of the X-ray source current.

A higher sensitivity of the sensor for low gamma radiation doses was obtained by modifying the polymer fiber end by embedding it into the scintillating material (gadolinium oxysulfide doped with europium or terbium – Gd_2O_2S:Eu, Gd_2O_2S:Tb), encapsulated into an externally black-painted micro-tube. The

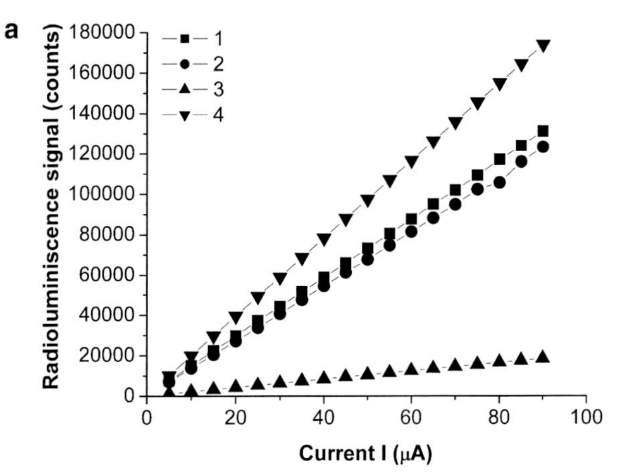

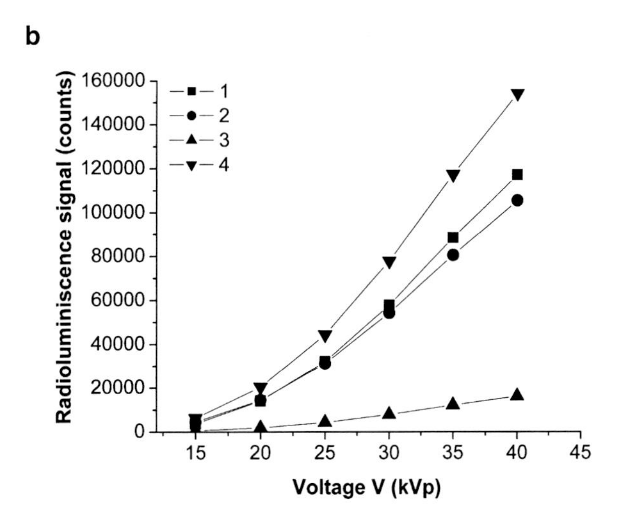

Fig. 6 The responsivity of four extrinsic sensors: 1, 55; 2, 56; 3, 57; and 4, 59, at $\lambda = 542$ nm, as a function of the X-ray source: (**a**) driving current, for the driving voltage $V = 40$ kVp; (**b**) driving voltage, for the driving current $I = 80$ μA (Sporea et al. 2014a)

polymer fiber diameter was reduced either by chemical etching (B in Fig. 3) or by thermomechanical tapering (C in Fig. 3) the fiber end in order to improve the collection scintillating signal by the guiding optical fiber. In this way, spectrometers with a lower sensitivity, and hence a reduced cost, can be used (de Andrés et al. 2017). This option was operated at dose rates as low as 0.5 Gy/h.

A solution to couple more scintillation signal into the plastic optical fiber in order to deliver it to the detector site was to make a small hole into the PMMA fiber core at its distal end and to fill this hole with the scintillating material mixed to optically transparent glue (Fig. 10). The sensor was applied to X-ray dosimetry, so the scintillating material chosen for this application was gadolinium oxysulfide doped with terbium (Gd_2O_2S:Tb) (O'Keeffe et al. 2016).

Because inorganic scintillating materials do not present water-equivalent characteristics, some problems arise in their clinical applications. This difficulty was overcame by using a dosimeter based on two different types of scintillating materials

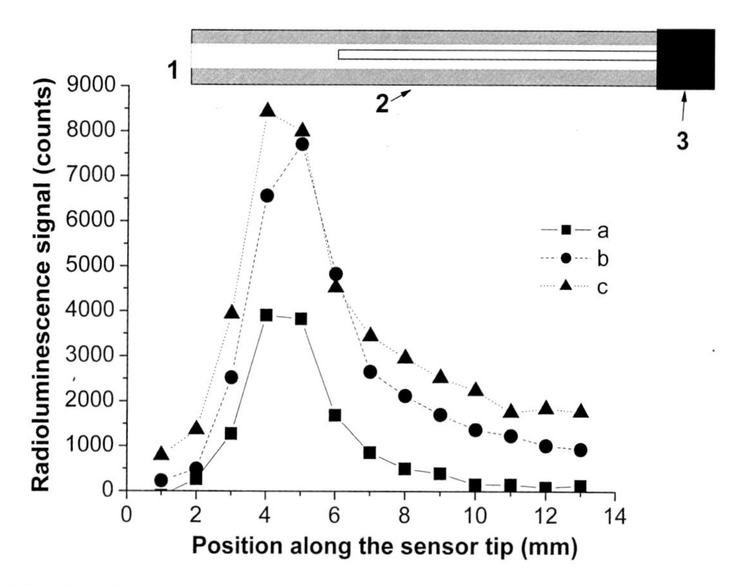

Fig. 7 The dependence of the detected optical signal on the position of the scanning X-ray microbeam along the sensor's tip for: a, uncoated sensor tip; b, sensor's tip wrapped in Al foil; c, sensor's tip painted with TiO_2 paste. The sensor's elements are as follows: 1, sensor tip; 2, reflecting materials; 3, PMMA optical fiber core (Sporea et al. 2014a)

(Gd_2O_2S:Tb and CsI:Tl) which fill the holes made in the core at the ends of two plastic optical fibers. Considering the distinct dose-depth characteristics of the two scintillating materials, calibration curves can be derived and correction factors are calculated (Qin et al. 2016).

A higher detection efficiency can be obtained if the devices described in Fig. 3 are modified by coating the exterior of the scintillating tip with titanium dioxide (TiO_2) (reflective wall type sensor) to achieve a higher confinement of the scintillating signal to the guiding plastic optical fiber and by encapsulating this tip into a black tape (Fig. 11) to shield the scintillating sensor head from the surrounding background light (Lee et al. 2010; Santos et al. 2014).

In one implementation, an inorganic scintillating material, such as a cerium-doped lutetium yttrium orthosilicate (LYSO) crystal, was mounted at the guiding optical fiber end. The crystal coated with Teflon-based reflector tape type BC-642 from Saint-Gobain Ceramic & Plastics was wrapped into an Al foil and black shielding tape. In this way, the collection of the scintillating signal by the PMMA optical fiber is improved, and the ambient light is prevented to be coupled to the light guide toward the detector (Yoo et al. 2013a).

In order to perform neutron dosimetry with high spatial resolution in a mixed gamma-neutron radiation field in a nuclear reactor, two optical fiber extrinsic sensors were used simultaneously. One similar to that shown in Fig. 11 made possible gamma-ray detection, while a sensor of the type presented in Fig. 12 was employed for thermal neutron dosimetry. The difference between the two

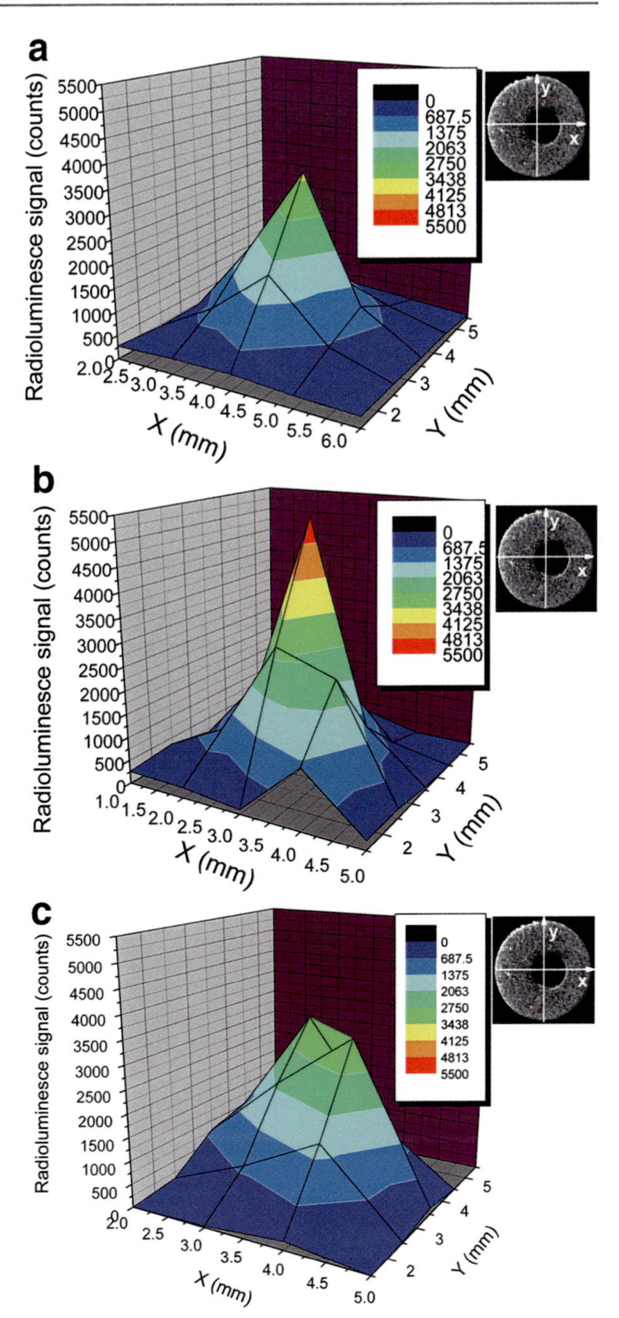

Fig. 8 The effect of the external reflector on the efficiency of the generated optical signal coupling to the optical fiber core: (**a**) sensor without a reflector; (**b**) the same sensor with an Al foil reflector; (**c**) the same sensor painted with TiO_2 (Sporea et al. 2014a)

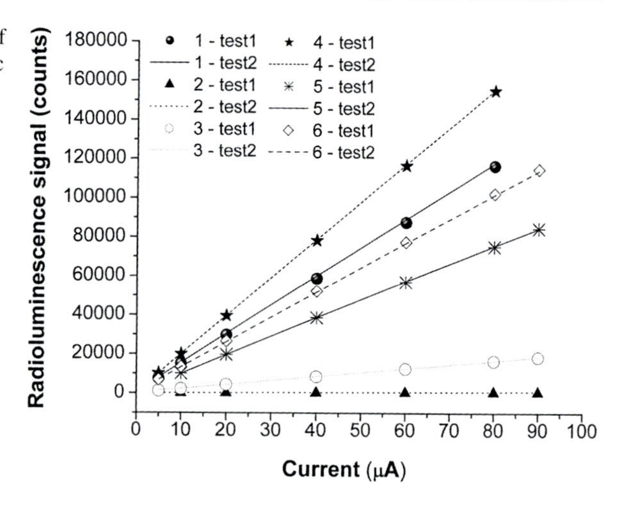

Fig. 9 The reproducibility of six scintillating-type extrinsic sensors as resulted after two tests, at fix voltage of $V = 40$ kVp and variable current of the AMPTEK X-ray source

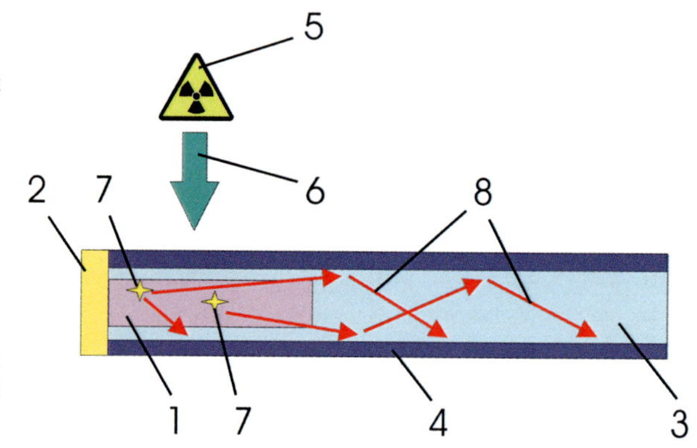

Fig. 10 Extrinsic optical fiber sensor type III: 1, scintillating/radioluminescent material; 2, optically reflecting cap; 3, core of the connecting plastic optical fiber; 4, cladding of the connecting plastic optical fiber; 5, ionizing radiation source; 6, ionizing radiation beam; 7, scintillating/radioluminescent event; 8, guided optical signal

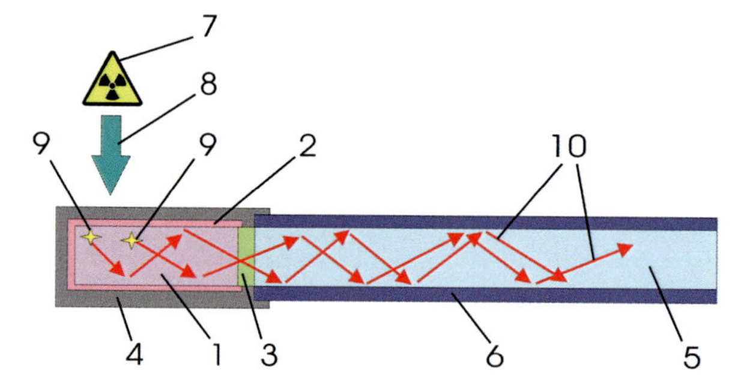

Fig. 11 Extrinsic optical fiber sensor type IV: 1, scintillating/radioluminescent material; 2, optically reflecting coating; 3, optical epoxy; 4, black cap; 5, core of the connecting plastic optical fiber; 6, cladding of the connecting plastic optical fiber; 7, ionizing radiation source; 8, ionizing radiation beam; 9, scintillating/radioluminescent event; 10, guided optical signal

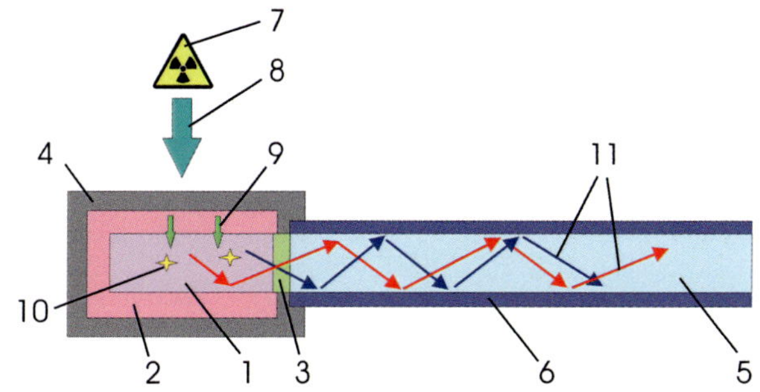

Fig. 12 Extrinsic optical fiber sensor type V: 1, scintillating/radioluminescent material; 2,^{6}LiF converter; 3, optical epoxy; 4, black cap; 5, core of the connecting plastic optical fiber; 6, cladding of the connecting plastic optical fiber; 7, ionizing radiation source; 8, mixed gamma-ray and thermal neutrons; 9, alpha particle generation; 10, scintillating/radioluminescent event; 11, guided optical signal

sensors resides in the fact that ^{6}LiF converter layer covers the scintillating fiber and emits alpha particles when expounded to thermal neutrons. Optical signal detection is done with a photomultiplier operating in the UV-visible spectral range (300–620 nm). By using this setup, a map of gamma-ray and thermal neutron distribution inside the nuclear fuel rode was performed (Jang et al. 2010).

The RL signal generated into a sapphire optical fiber irradiated by alpha particles is illustrated in Fig. 13. The amplitude changes of the RL signal follow the variation of the beam current; hence, a device can be developed for real-time radiation monitoring (Sporea and Sporea 2007). The sapphire optical fiber is irradiated by a beam of alpha particles in a vacuum interaction chamber. The RL signal is picked up by a MM optical fiber and is coupled to a mini optical fiber spectrometer through a vacuum feed-through. Different intensity counts displayed by the spectrometer (Fig. 13) are associated with different values of the beam current.

Investigations on RL emission in several IR optical materials (CaF_2, BaF_2, Al_2O_3, ZnSe) were also carried out under alpha particle irradiation with the same setup and coupling the optical signal with a MM optical fiber (Sporea et al. 2017). Two examples of the RL spectra are given in Fig. 14.

The RL signal generated was monitored in a erbium oxide sample irradiated by 1 MeV protons (flux 2–3 $\times$ 10^{17} p/(m^{2}s), at 20 °C and 10^{-7} torr. In a separate investigation, Er_2O_3 fixed at the end of a radiation-resistant optical fiber was inserted into the core of reactor, and mixed gamma-neutron irradiation (fast neutron flux 10^{19}–10^{21} n/(m^{2}s); thermal neutron flux (10^{19}–10^{22} n/(m^{2}s), temperature range 100–300 °C; gamma-ray dose rate 0.5–5 kGy/s) was performed (Shikama et al. 2006). In both cases, the optical signal was detected by an optical spectrum analyzer. An emission peak at $\lambda = 1560$ nm was observed under proton irradiation, while for neutron irradiation a broader emission spectrum was noticed (1600–2500 nm). It

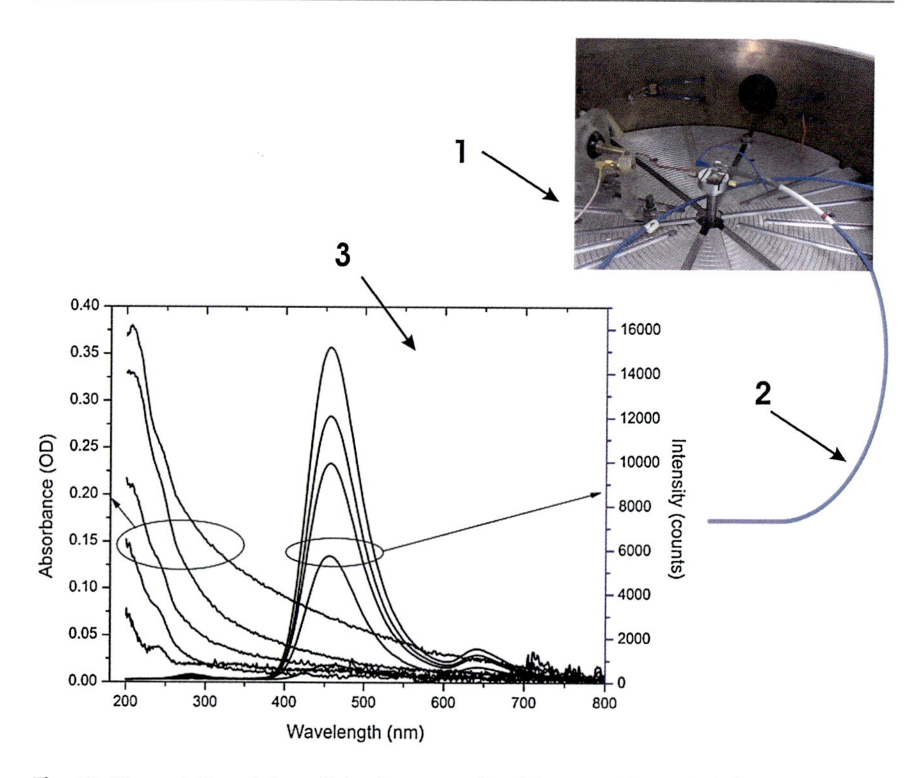

Fig. 13 The variation of the radioluminescence signal in a sapphire optical fiber during alpha particle irradiation: 1, interaction chamber; 2, coupling optical fiber; 3, optical fiber spectrometer (Adapted from Sporea et al. 2014a)

was suggested that these RL emission can be used for radiation dosimetry as they appear in the IR where lower irradiation-induced attenuation appears.

Cerium-, europium-, and ytterbium-doped silica optical fibers, tested under different irradiation conditions (X-ray system, CT scanner, protons) showed good linearity, stability, and reproducibility of the response to radiation and proved to be quite immune to humidity, temperature, and pressure (Chiodini et al. 2014).

More recently, silica radiation-resistant optical fibers doped with $Cu+$ or Ce^{3+} ions were investigated as RL emitters under 35–63 MeV proton beam exposure, at room temperature for possible use as radiation detector in proton therapy (Girard et al. 2017).

When exposed to ionizing radiation, the scintillating material emits light in the blue-green spectral region, signal which is coupled to the guiding optical fiber toward an optical detector. The detection of the optical signal guided by the plastic fiber is performed by various means such as photomultiplier tubes (PMT) (Beierholm et al. 2008; Jang et al. 2010, 2011; Yoo et al. 2013), red-blue-green photodiodes (Therriault-Proulx et al. 2011), luminescence readers (e.g., ME03 type) (Andersen et al. 2011), optical fiber minispectrometer (de

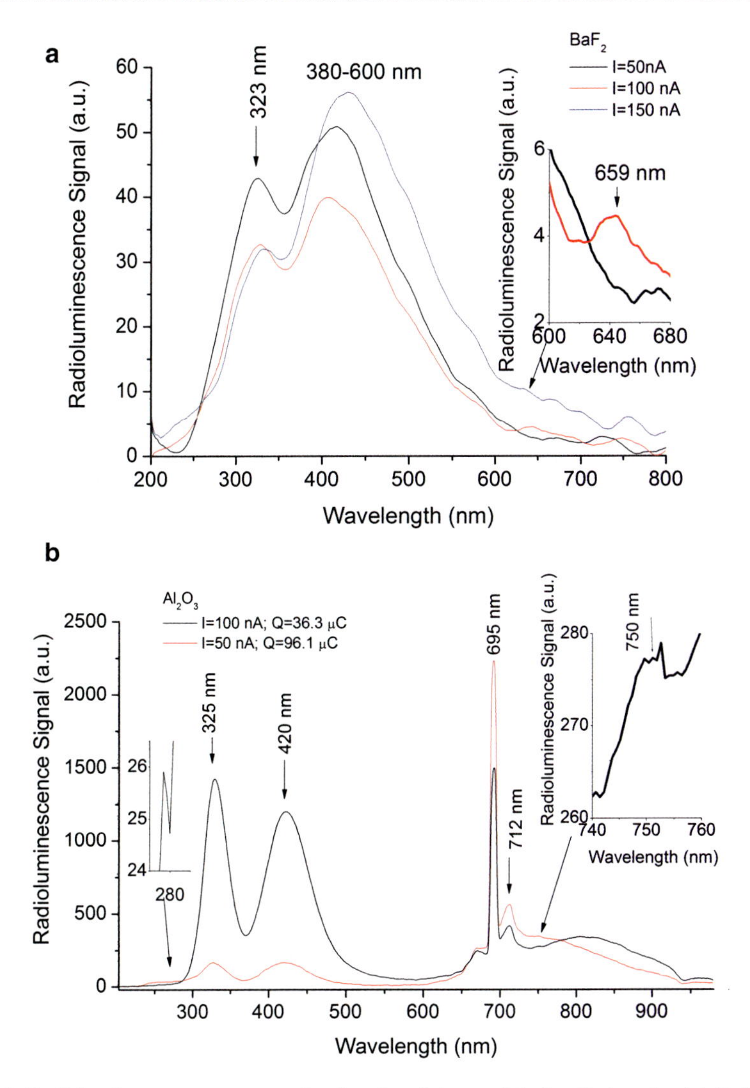

Fig. 14 Radioluminescence spectra emitted under alpha particle irradiations by (**a**) BaF_2; (**b**) Al_2O_3 (Sporea et al. 2017)

Andrés et al. 2017; McCarthy et al. 2011; Sporea et al. 2014a, b), CCD cameras (Klein et al. 2012), multi-pixel photon counter (MPPC) (O'Keeffe et al. 2016; Sporea et al. 2014a), temperature-stabilized PIN Si photodiodes (Rêgo et al. 2011), photodiode-amplifier system (Lee et al. 2010), and spectrograph connected to a CCD camera to run hyperspectral measurements (Therriault-Proulx et al. 2013).

By monitoring with a spectrometer the emission peak at $\lambda = 650$ nm generated in a commercial optical fiber irradiated by protons, the equivalence between the change of this signal and the absorbed dose was established, proving new opportunities for optical fiber-based dosimetry (Darafsheh et al. 2017).

When the optical fiber, either used as detecting element or to guide the optical signal toward the detector, is exposed to charge particles, the "stem" effect is present, as the radioluminescence generation is simultaneously accompanied by Cerenkov emission, for particle energies above the material-dependent Cerenkov threshold. This effect induces a degradation of the detection S/N because in some cases, the Cerenkov spectrum overlaps on the scintillation signal (Beaulieu and Beddar 2016; Beaulieu et al. 2013). For the "butt-coupled" sensor, the S/N evaluation, as referred to the Cerenkov radiation contribution, has to consider the optical and geometrical parameters of the sensor's tip and of the guiding optical fibers, as well as the characteristics of the optical fiber in respect to the generation of Cerenkov signal (Law et al. 2004).

Several methods to reject the Cerenkov component were proposed:

- Monochromatic or polychromatic "stem" effect removal techniques by subtracting the Cerenkov part of the optical signal using two of the three components of a red-blue-green photodiode (Therriault-Proulx et al. 2011)
- Spatial separation of the "stem" and the scintillating signals with a setup based on a CCD camera and dichroic mirror which makes possible the chromatic removal of the unwanted signal (Klein et al. 2012), a gated-counting detection scheme (Andersen et al. 2011; Beierholm et al. 2008)
- Dichroic mirror directing the two obtained signals (one dominated by the scintillation emission and one by the Cerenkov radiation) toward two separate detecting channels using PM-based photon counting heads (Ishikawa et al. 2015)
- Dummy plastic optical fiber, which does not include the scintillating material, exposed to the same radiation field as the sensor, so that the Cerenkov-induced signal is extracted from the overall signal but considering the signal in the dummy fiber (Lee et al. 2010; Liu et al. 2011)
- CCD color camera used for spectral discrimination between the scintillating signal superposed with the Cerenkov radiation on one side and the Cerenkov radiation on the other side (Liu et al. 2011)
- Detecting system composed of four optical fibers, one of them carrying the scintillation tip working in tandem with a similar one without the scintillation cap, and a set of two plastic optical fibers of different lengths to evaluate only the dose of Cerenkov radiation were employed in a subtraction setup to remove the Cerenkov contribution, to improve the S/N at detection and to evaluate simultaneously, in real time, both signals (Cerenkov and scintillation) (Yoo et al. 2013b)
- Hollow core optical fiber to transmit the scintillating signal from the sensor to the detecting system (Liu et al. 2011)

The types of the extrinsic sensors described above are targeting applications related to:

- Proton therapy (Andersen et al. 2011; Beaulieu and Beddar 2016; Jang et al. 2011)
- Dosimetry of thermal neutron in mixed gamma-neutron fields (Jang et al. 2010)
- Radiation protection dosimetry or dosimetry associated with radiation sources such as ^{60}Co, ^{125}I, ^{193}Ir, or linear accelerators in the 6–15 Mev energy range (de Andrés et al. 2017)
- Dosimetry in relation to intensity-modulated radiation therapy (IMRT) (Wootton et al. 2014) or volumetric-modulated arc therapy (VMAT) (Klein et al. 2012)
- Spectroscopy of gamma sources (Yoo et al. 2013a) and X-ray dosimetry (McCarthy et al. 2011; O'Keeffe et al. 2016; Sporea et al. 2014a, b)
- Radiology (Beaulieu and Beddar 2016; Rêgo et al. 2011)
- Patient dosimetry during brachytherapy (Beaulieu and Beddar 2016; Lee et al. 2010; Rêgo et al. 2011; Therriault-Proulx et al. 2013)
- Various ion dosimetries (He, C, Ne) (Broggio et al. 2007)
- Dosimetry associated with total body irradiation (Buranurak and Andersen 2016)

The main parameters of interest for such sensors are (i) the linearity of the optical output signal and the incident radiation energy loss, (ii) the spatial resolution, (iii) the dependency of the detected signal on the incidence angle of the excitation beam, (iv) the dose rate dependence of the optical signal, (v) the sensor response time, (vi) the temperature dependence of the output signal, (vii) the dynamic range, (viii) the reproducibility, (ix) the saturation level, and (x) the memory effect (Andersen et al. 2011; Beddar; Beierholm et al. 2008; Jang et al. 2011; Liu et al. 2011; McCarthy et al. 2011; Sporea et al. 2014b; Therriault-Proulx et al. 2013).

For organic BCF-12 and BCF-60 scintillating-based sensors, the temperature coefficients of the peak detected signal were found to be – 0.15%/K and – 0.55%/K, respectively (Buranurak et al. 2013). These values have to be considered in designing and calibrating such sensors for dosimetric measurements considering the fact that the temperature dependence functions on the considered wavelength range (Therriault-Proulx et al. 2015). The temperature coefficient was found to be independent on the incident energy change (Lee et al. 2015).

Sensors Based on Optically Stimulated Luminescence

Insulating materials or semiconductors can exhibit metastable states in the forbidden band because of fabrication-induced defects or defects due to dopants. When such materials are exposed to ionizing radiation, excited carriers (electron and holes) are trapped on these metastable states. The recombination of these carriers can be spontaneous, when a radioluminescence (RL) signal is emitted, or occurs with the emission of an optical signal, only upon an external stimulation (energy transfer) by heating the material (thermoluminescence (TL)) or by excitation with optical radiation (optically stimulated luminescence (OSL)) (Akselrod et al. 2007; Pradhan et al. 2008). Ideally, the RL signal can be used for dose rate measurements, because

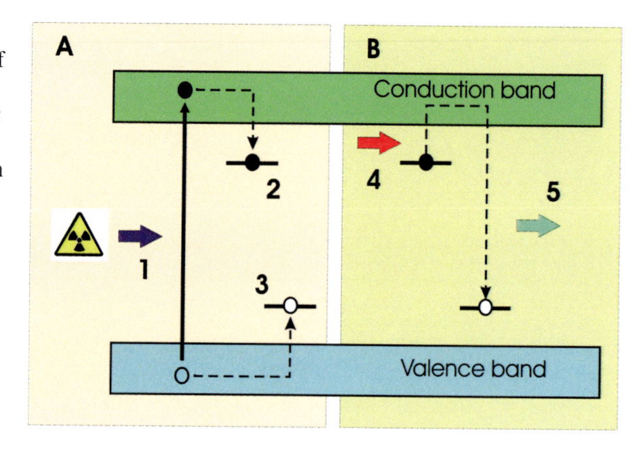

Fig. 15 The principle of OSL: panel (**a**) the process of excitation produced by ionizing radiation and carrier trapping; panel (**b**) the process of optical stimulation and luminescence emission: 1, ionizing radiation; 2, trapped electron; 3, trapped hole; 4, optical stimulus radiation; 5, luminescence signal (Adapted after S. Beddar)

it appears as long as the material is exposed to the ionizing radiation, while the OSL signal can serve for total dose evaluation, and it could be associated with an integrated effect of the irradiation, for the case that concurrent phenomena (e.g., fading) are not present until the interrogation is performed with the optical stimulation signal. OSL reading might occur at any time during or postirradiation (Aznar et al. 2004). The simplest representation of OSL is given in Fig. 15. Panel **a** depicts the excitation process by ionizing radiation accompanied by electrons and holes trapping on metastable stats in the forbidden band (energy levels located between the valence and the conduction bands). In panel **b** is the sketch of the optical simulation (4) followed by luminescence emission (5) after. Between the two events, electrons and holes can remain trapped on the forbidden band energy levels.

Depending on the method used to optically stimulate the irradiated sample, several methods were proposed (Akselrod et al. 2007; Bulur and Yeltik 2010; McKeever 2001; Mishra et al. 2015; Pradhan et al. 2008; Rawat et al. 2014):

- The continuous wave optical stimulation luminescence (CW-OSL), when the optical radiation is operating at constant intensity and the emitted optical signal is monitored continuously
- The linearly modulated optically stimulated luminescence (LM-OSL), when the "reading" optical signal intensity is increased in a linear way in time and the emitted optical signal is recorded during the stimulation
- The delayed optically stimulated luminescence (DOSL), case when the recording of the emitted optical signal is performed after the stimulation light is switched off

Other techniques are referred to as pulsed optical stimulation luminescence (POSL) and thermally assisted optical stimulation luminescence (TA-OSL). Pulsed operation provides a higher S/N at detection by applying a temporal discrimination between the excitation signal and the optical signal generated upon stimulation.

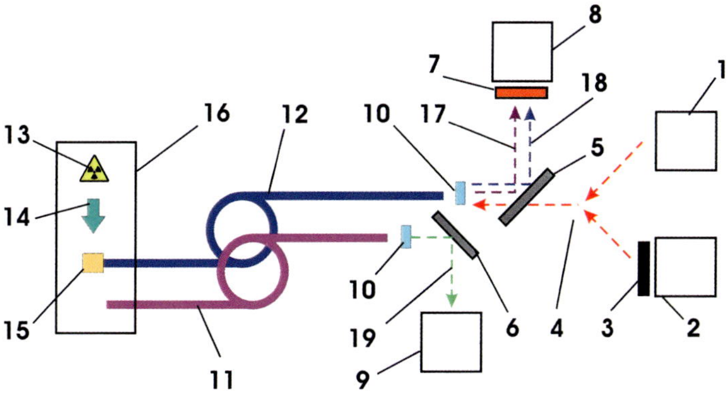

Fig. 16 The principle of OSL with stem effect removal: 1, CW operated or pulsed laser/LED; 2, CW laser; 3, optical shutter; 4, stimulating optical radiation; 5, dichroic mirror; 6, mirror; 7, optical filter; 8 and 9, optical detectors; 10, collimating optics; 11, dummy optical fiber; 12, active optical fiber; 13, ionizing radiation source; 14, ionizing radiation; 15, OSL probe; 16, irradiation zone; 17, OSL signal; 18, RL and Cerenkov signals in the active optical fiber; 19, RL and Cerenkov signals in the dummy optical fiber

TA-OSL employs probe heating during the stimulation process, which leads to an improved sensor sensitivity.

The principle of OSL setup is illustrated in Fig. 16 where several solutions are presented. The stimulation source can be a broadband light source with appropriate narrowband filters, a laser or a LED operating in CW or pulsed mode, or a CW laser in a shutter mounted at its exit aperture. A dichroic mirror and collimating optics are used to couple the stimulation optical radiation to the guiding optical fiber. At the distal end of this optical fiber is mounted the OSL probe which is subjected to ionizing radiation. As the probe is exposed to ionizing radiation, a radioluminescent signal is generated, and it propagates backward along the guiding optical fiber. In the meantime radioluminescence signal and Cerenkov radiation are produced because of the coupling optical fiber irradiation, and the stem effect is superposed on the useful RL coming from the probe. The dichroic mirror redirects the returned optical signal toward a detector. The removal of the stem effect can be done either by using an additional dummy optical fiber without the OSL probe at its end or by appropriate optical filtering of the detected signal. An alternative solution is temporal discrimination of the RL and the Cerenkov radiation. The OSL reading is performed after the irradiation is stopped, so no stem effect is present during this detection. Pulsed interrogation of the OSL probe during the radiation exposure provides an improved S/N through synchronized detection (Akselrod et al. 2007; Aznar et al. 2004; Gaza et al. 2005; McKeever et al. 2004; Mishra et al. 2015; Teichmann et al. 2016). In some implementations the laser and the detector are coupled to the guiding optical fibers through two separate optical links and an optical coupler (Akselrod et al. 2007; Benoit et al. 2008).

Various materials for OSL applications were studied (Bhatt 2011; Bulur and Yeltik 2010; Dotzler et al. 2007; Liu 2008; Pradhan et al. 2008): Al_2O_3:C; A-Al_2O_3; BeO; MgO; MgO:Tb; Mg_2SiO_4:Tb; MgS:Ce,Sm; $LiAlO_2$:Tb; $Li_2Al_4O_4$:Tb; $LiMgPO_4$:Tb,B; $KMgF_3$:Ce; KBr:In; KBr:Eu; KCl:Eu; $NaMgF_3$:Eu^{2+}; NaCl:Cu; $YAlO_3$:Mn; $RbCdF_3$:Mn; $RbMgF_3$:Mn; RbBr:Ti; $Y_3Al_5O_{12}$:C; SrS:Eu,Sm; RbBr:Ti; and CaS:Ce,Sm. Tests were carried out also on SiO_2:Cu^+ optical fiber, by observing the luminescent emission in the 450–600 nm spectral band, for a stimulating optical signal applied in the 700–900 nm range (Ranchoux et al. 2002).

The light source to provide the stimulation of trapped carriers can be a laser or LED (operation in CW mode or pulsed, e.g., 1 ms pulse duration at a frequency of 100 Hz), a broadband source operating in conjunction with narrowband filters, or a monochromator (McKeever 2001). In some approaches the pulsed operation is achieved by using a shutter to interrupt periodically the stimulating light beam. For pulsed regime the detection is synchronized with the excitation source pulses (Mishra et al. 2015).

In the case of OSL dosimeters, the parameters to be investigated in order to assess its performance as compared to other types of dosimeters are associated with sensor's response in relation to incident energy, dose rate, field size, angular incidence of the incident radiation, linearity, dose, temperature, fading, and reusability (Akselrod et al. 2007; Aznar et al. 2004; Yukihara and McKeever 2008; Yukihara et al. 2014; Zhydachevskii et al. 2016).

The minimum detectable dose (MDD), reflecting sensor's sensitivity, is dependent on the scintillating material used and on the measuring conditions: excitation wavelength, spectral sensitivity of the detector, power of the optical stimulus, and the coupling efficiency of the optical paths – stimulation and detection ones and selection of the integration time on the detector site. At an irradiance of about 72 mW/cm^2, for integration time of 1 s at the excitation wavelength $\lambda = 470$ nm, a sensitivity of $\approx$7 μGy was achieved (Rawat et al. 2014).

For low irradiation doses (between 25 and 2.5 mGy) and dose rate of $\approx$12.5 μGy/s, from a Sr^{90}/Y^{90} beta source a good linearity of the sensor response was obtained in a BeO sample, pre-heated at 160 °C (Bulur and Yeltik 2010). The sensitivity can be further increased by replacing the diffusing ceramics with transparent ones.

Benoit et al. reported a reproducibility of ±1.6% with a standard deviation of 0.8%, and a good linearity from 10 to 600 mGy, for one type of coupling optical fiber and dose from 0.1 to 6 Gy, for two other types of fiber, under ^{60}Co gamma and 6MV X-ray irradiations of a OSL fiber sensor based on SrS:Ce,Sm (Benoit et al. 2008).

By combining RL measurements during the irradiation with a Dual Phase Reference Summing (DPRS) technique, Teichmann et al. (2016) reached a good linear response for a BeO-based OSL irradiated by a Sr^{90}/Y^{90} source, under the dose rates between 0.017 and 3.6 Gy/h, for total doses from 1 mGy to 5 Gy. At a dose of 1 Gy and a source to probe distance of 40 cm, the stem effect is below 5% in a combined RL/OSL detection scheme. It can be further reduced by some of the methods listed in the section on "Sensors based on radioluminescence or scintillation." Alternative solutions are either to use an air-fill hollow core optical fiber for coupling the signal to the detection unit or to compare the values of the dose derived from RL measurements to those provided by the OSL in order to introduce in calculus the required corrections (Andersen et al. 2009).

A prototype of a RL/OSL instrument for medical radiation dosimetry using a Al_2O_3:C crystal showed reproducibility of 0.5% (1 SD) for RL measurements at 18 MV proton irradiation, 0.1% (1 SD) for OSL measurements under the same irradiation conditions, 0.3% (1 SD) for RL, and, respectively, 0.5% (1 SD) when OSL tests were done under 6 MV proton exposure. The RL signal had a linear variation with dose rate up to 600 MU/min. The variation of the signal with the angular incidence was found to be less than 2% (1SD) (Aznar et al. 2004).

The applications covered by OSL include (Andersen 2011; Aznar et al. 2004; McKeever 2001, McKeever et al. 2004; Yukihara et al. 2006, 2014; Yukihara and McKeever 2008):

- Retrospective dosimetry, when materials exposed to radiation are measured post factum
- Personnel dosimetry
- Environmental dosimetry
- Dosimetry for radiotherapy
- Space dosimetry
- Detection of radiological contaminants
- Mammography
- Proton and electron beam therapy
- Computed tomography
- Heavy-charged particle dosimetry
- Neutron dosimetry

Intrinsic Optical Fiber Sensors

Sensors Based on Thermoluminescence

According to Fig. 15 under ionizing radiation, carriers can be trapped on energy levels in the forbidden band. If energy is transferred to these carriers through heating and when they recombine, optical radiation emission occurs, leading to the phenomenon known as thermoluminescence. Studies were carried out to investigate the use of Ge-doped (Benabdesselam et al. 2013; Rahman et al. 2012) and Ge-B (Bradley et al. 2014) SiO_2 optical fibers as radiation detectors. The use of optical fibers as thermoluminescence sensors is classified as being off-line measuring technique, because the interrogation of the irradiated probe is performed postirradiation with dedicated instruments. In some cases, the postirradiation luminescence of the sample constitutes the phenomenon used to build the radiation sensor. As an example, in Fig. 17 are presented the optical signals emitted by two commercially available optical fibers irradiated by ^{60}Co gamma at dose rate 158 Gy/h.

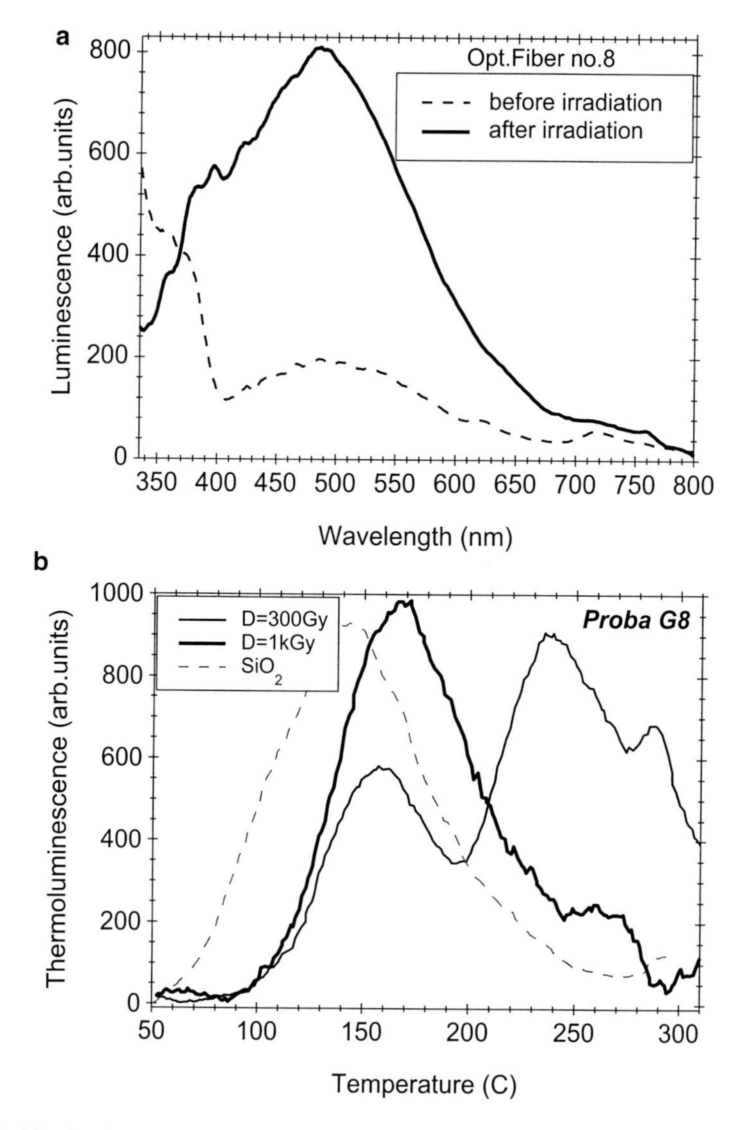

Fig. 17 The luminescence (**a**) and the thermoluminescence (**b**) signals after ^{60}Co gamma irradiation of a multimode step-index 400 μm silica core, visible/near-IR optical fiber, with acrylate coating, irradiated at 300 Gy (**a**); a 600 μm core diameter, deep UV-enhanced, polyimide jacket optical fiber irradiated up to 1 kGy (**b**) (Adapted after Sporea et al. 2012a)

For 90 kVp X-rays, a linear response with dose was noticed for doses between 1 and 35 Gy, while for proton and electron irradiations, the linearity extends up to 50 Gy. The TL signal was read with Harshaw Model-3500 TLD reader, under nitrogen atmosphere to avoid oxidation and to reduce triboluminescence contribution to the detected optical output.

Saeed and colleagues investigated the possible use of single clad neodymium-doped silica fiber for dosimetric use as thermoluminescent detectors and found the sensitivity of this material to be a little bit higher than 10% of classical TLD-100's sensitivity, for the case of 6 MeV proton irradiation (Saeed et al. 2013).

Tests performed on flat pure SiO_2 optical fibers irradiated by 6 MeV, 15 MeV, and 21 MeV electron beams demonstrated good reproducibility, independence of incident energy and dose rate, a good linearity, and low fading rates, for doses applied in medical therapy (Alawiah et al. 2013). Investigations were done on other types of undoped and doped silica optical fibers such as capillary and photonic crystal, irradiated by 6 and 20 MeV electron beams for doses varying from 0.5 to 8 Gy. The process of making the fiber in a flat form increases several times the sensor sensitivity as compared to capillary or undoped fiber ones (Bradley et al. 2015).

Sensors Based on Radiation-Induced Attenuation in Optical Fibers

In this section the application of the radiation-induced optical fibers optical attenuation as a method for radiation dosimetry is revised. The degradation of different types of optical fibers under irradiation with various ionizing radiation is not addressed in this chapter, because extended review publications are available on this subject (Berghmans et al. 2008; Di Francesca 2015; Girard and Marcandella 2010; Girard et al. 2013a; Sporea et al. 2012a).

Investigations on radiation effects induced optical attenuation, as a measure of integral dose, were performed in pure silica core, H_2-loaded (Sporea et al. 2012b), or doped optical fibers, by monitoring this parameter at specific wavelengths. Tests were done under gamma and X-rays, electron beam (Sporea et al. 2016a), or synchrotron irradiation (Sporea et al. 2014c). In some situations, the attenuation measurements were carried out off-line, while some other researches focused on online measurements. The recovery of the radiation-induced optical attenuation was monitored at room temperature or after sample heating (Sporea et al. 2014c). The effect of temperature on the dynamics of color centers during the irradiation was also studied (Sporea et al. 2016a).

An example of the setup for online/off-line optical attenuation evaluation is given in Fig. 18. A laptop controls, through a program written in LabVIEW, data acquisition and processing. Optical fiber samples are coupled one by one to the measuring setup composed of a UV-visible light source, an optical fiber attenuator, an optical fiber multiplexer, and a minispectrometer. The absorption spectra corresponding to subsequent doses are monitored, and the optical attenuation is computed for five wavelengths selected by the operator ($\lambda = 215$ nm; $\lambda = 229$ nm; $\lambda = 248$ nm; $\lambda = 265$ nm; $\lambda = 330$ nm). The dependence of the optical attenuation on the exposure dose can be used in radiation dosimetry.

Research to use pure core and doped (Er, P, Ge) optical fibers in gamma-ray radiation dosimetry, in the dose range 0.1–100 kGy, was carried out by Borgermans et al. (2001). The investigations highlighted the influence played by temperature and

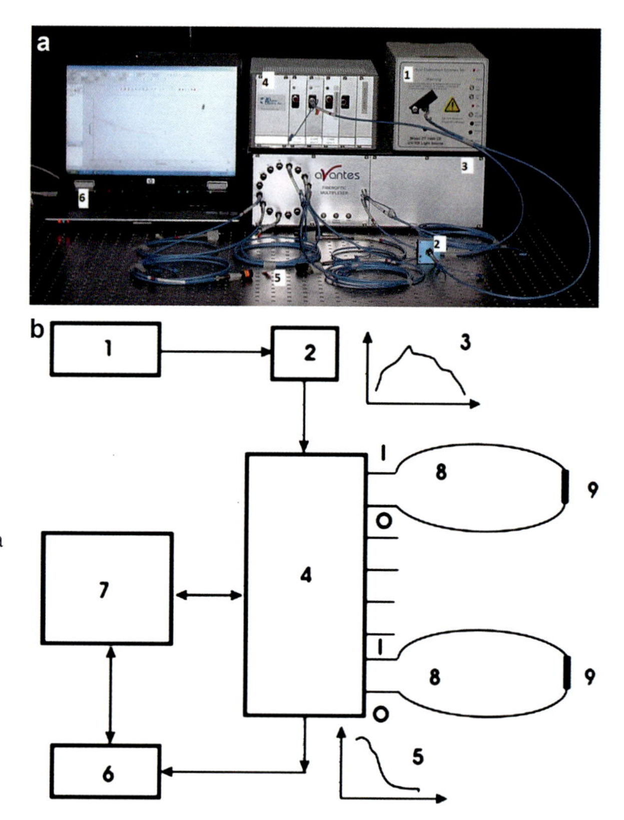

Fig. 18 The photo (**a**): 1, light source; 2, optical fiber attenuator; 3, optical fiber multiplexer; 4, optical fiber minispectrometer; 5, optical fiber sample; 6, laptop; the sketch (**b**) of the setup for irradiation-induced optical attenuation monitoring: 1, light source; 2, optical fiber attenuator; 3, spectrum of the light source; 4, optical fiber multiplexer; 5, optical fiber sample absorption spectrum; 6, optical fiber minispectrometer; 7, laptop; 8, connecting optical fibers; 9, irradiated optical fiber samples; I, multiplexer input; O, multiplexer output (Sporea et al. 2014c)

photobleaching on the radiation-induced attenuation dependence on dose. A more recent study suggested low-dose gamma radiation dosimetry with $GeO_2 + P_2O_5$ co-doped step-index multimode optical fibers, by monitoring with a photodiode the optical attenuation at $\lambda = 505$ nm and 560 nm (Ghosh et al. 2011). Two P-doped optical fibers were found to have a radiation sensitivity of 0.970 dB/m/Gy and 0.6631 dB/m/Gy, respectively, while the Ge-doped fiber with a radiation sensitivity of 0.16 dB/m/Gy (Ghosh et al. 2011).

Optical fiber-based dosimetry employing P-doped optical fibers was proposed for the evaluation of optical attenuation change in the UV and visible range (Girard et al. 2011). At $\lambda = 550$ nm, a sensitivity of 10 dB/m was found for a dose of 10 Gy, when the sample was exposed to 10 keV X-ray, for a total dose of 3 kGy.

Low-dose dosimetry was investigated with P-doped optical fibers irradiated by gamma-ray dose rates ≈6.4–6.6 mGy, up to a dose of 2 Gy, by measuring the radiation-induced optical attenuation at two wavelengths (413 nm and 470 nm), symmetrically located in respect to $\lambda = 440$ nm. Radiation-induced absorption bands, at wavelengths lower and higher than that $\lambda = 440$ nm, are associated with two different color centers. By the use of a two-wavelength approach, it is possible to establish a correlation between the dose and the difference between the amplitudes at the two wavelengths, as this difference does not depend on the dose rate and decreases slightly after the irradiation stops (Tomashuk et al. 2014).

Point and distributed radiation dosimetry was suggested based on small pieces of P-doped spliced between radiation pure-silica-core and F-doped-resistant optical fiber. Special consideration is required when implementing such sensors as their radiation sensitivity is dependent on dose and temperature during the irradiation process. Special "favorable" wavelengths for irradiation-induced attenuation have to be considered where temperature contribution is minimal (Girard et al. 2013b).

Apart from H_2-loading extensively discussed in literature (Girard et al. 2013a), the role played by O_2-loading of undoped-silica-core and F-doped-silica-core fibers on the radiation hardness of gamma-ray- and X-ray-irradiated optical fibers is still disputed (Di Francesca et al. 2014; Tomashuk et al. 2013).

More than 10 years ago, a new type of radiation dosimetry based on the radiation-induced attenuation in plastic optical fibers (POF) was proposed by O'Keeffe et al. by measuring the spectral attenuation before and after ^{60}Co gamma irradiation under dose rates of 0.5 Gy/h, 4.5 Gy/h, 40 Gy/h, 340 Gy/h, and 1.42 kGy/h and up to the total doses of 12 Gy, 108 Gy, 960 Gy, 8.16 kGy, and 34.08 kGy (O'Keeffe et al. 2004). Measurements focused on several wavelengths (550 nm, 580 nm, and 640 nm) where the optical attenuation indicated a linear dependence on the dose, for doses lower than 34 kGy. After several days from the irradiation end, the samples were tested one more time and a fading effect was noticed. More recently, online low-dose gamma-ray irradiations were reported by illuminating the tested POF with a LED and by monitoring the attenuation during the irradiation with a photodiode (Kovačević et al. 2013). In an alternative online investigation, two optical fiber spectrometers, covering the spectral range 400–1700 nm, were employed, as a Cytop® fiber was gamma irradiated (dose rate 5.7 kGy/h, the maximum dose for online measurements was 2.5 kGy, and 100 kGy for off-line tests). In addition, back-reflection measurements were done prior to and after irradiation with an OTDR (Stajanca et al. 2016). Minimum radiation-induced losses occurred at about 1 μm. As the spectral attenuation increases in the visible range, the tested Cytop® fiber can be recommended for radiation dosimetry.

Sensors Based on Cerenkov Radiation

In scintillation-based dosimetry, Cerenkov radiation can be present as a disturbing noise signal, contributing to S/N degradation at detection. In previous subsections, methods to reduce this contribution were mentioned. More recently, some studies suggested the possible development of a Cerenkov radiation-based dosimetry. In the case of X-ray exposure, this can be of interest in narrow beam stereotactic radiation therapy and surgery validation studies; for ^{60}Co irradiation, it can be applied to entrance and exit surface imaging dosimetry, while for electron treatment, it is recommended for surface dosimetric use (Glaser et al. 2014).

A Cerenkov radiation sensor for beta particle detection and discrimination of particle energy, between 174.57 and 264.06 keV, was designed by mounting at the PMMA guiding fiber end, instead of the scintillating material tip (Fig. 11), a PMMA holder covered at one end by a clear 11-μm-thick cling film. On the opposite face of the holder, the guiding PMMA fiber is introduced after the jacket was stripped. This holder is filled with 0.5–1.5 mm layer of distilled water or different transparent liquids having various refractive indexes (1.35, 1.40, 1.45, and 1.50 at $\lambda = 589.3$ nm). The optical fiber end is in contact with the liquid, and the Cerenkov radiation is detected by a PMT. Energy discrimination of the incident particles is obtained by changing the liquid inside the "cell" (Yoo et al. 2013c).

Zhang et al. reported the results of Cerenkov emission dosimetry imaging, performed under 6–18 MeV electron beam exposure and recording with a CMOS camera (Zhang et al. 2013).

A Cerenkov radiation sensor was developed in a plastic optical fiber (POF) having the core refractive index of 1.492 and the cladding refractive index of 1.402, for a NA = 051, for the dosimetry of proton irradiation at 173 MeV and 180 MeV, and dose rate 1–6 Gy/min (Jang et al. 2012). A linear response with the dose rate was observed.

Thermal neutron detection was performed with a Gd-foil (GD-143220, Nilaco Co.) acting as neutron converter and covering a rutile crystal (refractive index of 2.87 at $\lambda = 430$ nm), crystal used as medium for Cerenkov radiation detection (Jang et al. 2013). The assembly is mounted at the end of a PMMA optical fiber which guides the signal toward the detector, as in the case of scintillating sensors (Fig. 2).

Considering that the amount of Cerenkov radiation is higher in the UV range, the scintillating tip coupled to the end of the guiding PMMA fiber (Fig. 2) was replaced by a short piece of wavelength-shifting fiber (WSF) manufactured by Saint-Gobain Ceramic & Plastics. In this way, the Cerenkov radiation absorption in the UV spectral range is followed by an emission in visible; hence, an improved S/N is obtained at the detection (Jang et al. 2014).

The monitoring of Cerenkov emission in some high-purity silica core, F-doped cladding optical fibers of core diameter between 100 and 600 μm and polymer or Al coating, was performed during their exposure to radiation inside a nuclear reactor, in order to investigate the possible use in the evaluation of a nuclear reactor power. During the test run, both Cerenkov radiation and optical attenuation of the fibers were measured online with an optical spectrum analyzer. With the irradiation dose increase, the attenuation of the optical fibers not treated with H_2 increases too, so that Cerenkov emission, which depends on λ^{-3}, is difficult to detect below 800 nm. For H_2-loaded optical fiber, the Cerenkov radiation can be detected down to 450 nm, for moderate doses. In the case of higher doses, the attenuation band extends over the visible spectral range, the detection of the Cerenkov signal being possible only at wavelengths higher than 850 nm. These studies demonstrated the possibility to estimate the reactor power as function of the Cerenkov emission, because a linear relationship can be derived between the two quantities (Brichard et al. 2007).

Chaikovska and colleagues developed a beam loss monitor to be used to X-ray machine, based on the time-of-flight measurement of Cerenkov radiation (Chaikovska et al. 2014). The Cerenkov radiation-generated signal is detected at the two ends of the installed Hard Plastic Clad Silica (HPCS), and, in this way, the localization of the beam loss is performed. Another implementation (Naka et al. 2001) of the time-of-fly method was tested for the case of gamma-rays, fast neutrons, beta-rays, or charged particles, exhibiting a spatial resolution of 30 cm (^{90}Sr-^{90}Y beta-rays), 37 cm (^{137}Cs gamma-rays), and 13 cm (D-T neutrons) and detection efficiency of 0.11%, 1.6×10^{-5}%, and 1.2×10^{-4}%.

Fiber Bragg Gratings

Fiber Bragg gratings (FBG) were extensively studied under irradiation conditions for over 20 years, and good review papers are available addressing different grating designs (Gusarov and Hoeffgen 2013):

- Type I, Type IA, Type II, and Type IIA
- Chemical composition gratings
- Draw tower gratings (DTG)

produced in:

- Standard telecommunication Ge-doped fibers
- B/Ge co-doped non-hydrogenated fibers
- F-, N-, Sb-, Ti-, Pb-, and SnO_2-doped or H_2-loaded optical fibers

written by CW or pulsed (ns) UV laser radiation or by using UV/IR fs lasers (Grobnic et al. 2009; Gusarov et al. 2002, 2010; Henschel et al. 2007; Morana et al. 2015a).

Tests were performed under various irradiation conditions: gamma-ray, protons, mixed gamma-neutron, and X-ray (Gusarov et al. 2002; Gusarov and Hoeffgen 2013; Morana et al. 2015a).

FBGs written by UV fs laser in a B-co-doped Ge-silicate optical fiber produced by Fibercore Ltd., a GF1 optical fiber from Nufern, and a standard SMF28 optical fiber, which was sensitized by H_2-loaded, were tested under gamma irradiation as follows: the dose rate of 1.02 Gy/h, total doses 20 and 40 kGy; the dose rate of 0.84 Gy/h, total dose of 84 kGy, at controlled 35 °C. The sensors' spectral attenuation and wavelength peak were monitored with an optical fiber interrogator, in transmission and reflection. Long-term exposure indicated saturation of the wavelength peak shift (max. 15 pm for 40 kGy dose, independently of the fiber type or FBGs characteristics) and of the optical attenuation of FBGs. A recovery of the wavelength peak shift was noticed during the 200 h following the irradiation stop (Gusarov et al. 2010).

The effect of high-dose (3 MGy/SiO_2) X-ray irradiation of FBGs was investigated for gratings fabricated in Morana et al. (2015b):

(a) Standard SMF28 H_2-loaded fiber by phase mask using (i) a CW UV laser ($\lambda = 244$ nm) and (ii) a pulsed (ns) KrF UV laser ($\lambda = 248$ nm)

(b) Standard SFM28e by phase mark writing with a fs Ti:sapphire laser ($\lambda = 800$ nm)
(c) Standard SMF28 employing the point-by-point technique, with a fs laser operating at $\lambda = 800$ nm

The irradiation was performed with 10 keV X-rays, dose rate of 50 Gy (SiO_2)/s, at room temperature and at 100 °C, and up to total doses of 1 MGy and 3 MGy. FBGs monitoring was run with a tunable laser source and a high-resolution (1 pm) optical tester. FBGs engraved with fs laser radiation shown a better radiation resistivity as compared to FBGs produced by UV lasers. For the temperature span from 20 to 100 °C, these gratings can be used as temperature sensors with an error lower than 2 °C due to the irradiation. In addition, H_2-loading induced a higher radiation sensitivity of the tested sensors.

Results on the modification of the radiation sensitivity of Ge-doped-silica-core FBGs and of N-doped FBGs were reported by Butov et al. following an investigation on the H_2 contribution to this parameter, under gamma-ray irradiation up to 5 MGy (Butov et al. 2015). It was concluded that H_2-related vulnerability in N-doped fiber is related to hydrogen interaction with defects produced during the UV grating writing, while for Ge-doped ones, H_2 interacts mainly to defects generated upon irradiation.

Radiation-tolerant FBGs able to operate at high temperature (up to 230 °C) with a temperature error produced by irradiation in the range of 1.5 °C (corresponding to a wavelength peak shift of about 10 pm) were fabricated by fs laser exposure and thermal annealing at 750 °C. They were tested under X-ray irradiation to a total dose of 3 MGy, for a possible use for temperature monitoring in nuclear industry (Morana et al. 2014). A comparison of the radiation-induced temperature error (Δt) associated with the wavelength peak shift under X-ray irradiation to a dose of 1 MGy indicated that most radiation sensitive are FBGs inscribed by UV radiation in Ge-doped optical fibers ($\Delta t = +14$ °C), the most radiation resistant being the grating engraved by fs IR laser in pure silica core fibers ($\Delta t = -0.7$ °C) (Morana et al. 2015a).

Apart from the radiation tolerance of FBGs for possible use in temperature measurements, tests were run to investigate the variation of their strain sensitivity coefficient when they are exposed to ionizing radiation. Research on FBG written in Ge-doped, H_2-loaded optical fiber exposed to gamma radiation (dose rate 25 Gy/min, dose 105 kGy) indicated relative small alteration of the strain sensitivity coefficient (1.48–2.71%) (Jing et al. 2013).

Fernandez et al. reported the investigation on the chemical composition of FBG behavior, gratings obtained by periodically modification of the fluorine doping in the fiber core, subjected to mixed gamma and neutron radiation fields in a research reactor operating at 57 MW power, gamma dose rate of 5.8 MGy/h, maximum thermal flux of 6.3×10^{13} n/(cm^{2}s), the epithermal flux 3.3×10^{12} n/(cm^{2}s), and the fast flux (1 MeV equivalent) 7.41×10^{12} n/(cm^{2}s) (Fernandez et al. 2006). These gratings were selected for their capability to keep their stability at quite high temperatures (in excess of 1000 °C). For a gamma total dose of 500 MGy, a neutron fluence of 2.5×10^{19} n/cm^2, and at a mean temperature of 150 °C, the wavelength peak shift was around 5.8 nm, and a small broadening of the FBGs' full width at half maximum (FWHM) was noticed. Higher radiation exposure time and an increase of the operating temperature produced a pronounced degradation of the gratings.

Research on long-term survivability (8 years) of FBGs was done under a low flux in a nuclear reactor (total gamma dose 10 MGy, total neutron fluence 1.47×10^{17} n/cm^2, and temperature cycling between 10 and 80 °C), demonstrating small changes in optical fibers which were not subjected to H_2-loading and a moderate shift for the case of H_2-loaded fibers (Gusarov 2010).

Faustov et al. studied the radiation sensitivity to ^{60}C0 gamma radiation (total dose of 116 kGy) of several Type IA gratings prepared in different types of optical fibers: (i) Fibercore PS-1250/1500, (ii) B/Ge co-doped fiber; (iii) Fibercore SM-1500, (iv) high-Ge-doped fiber, and (v) B/Ge co-doped (Faustov et al. 2012). The Type IA grating fabricated in Fibercore PS-1250/1500 proved to be very radiation sensitive, exhibiting a wavelength peak shift of 190 pm, fact which recommend it for radiation dosimetry (Faustov et al. 2012).

By using FBGs engraved into Corning HI 1060 FLEX Specialty Fiber (Bragg wavelength $\lambda = 818.820$ nm, 1284.585 nm and 1515.795 nm), FBGs produced in Corning HI 780 Specialty Fiber ($\lambda = 657.191$ nm and 819.077 nm), and one FBG written in Siecor SMF 1528 fiber ($\lambda = 1542.105$ nm) their possible use for ionizing radiation measurement was evaluated as they were irradiated by gamma-rays to 100 kGy (Krebber et al. 2006).

FBGs produced in radiation-hardened optical fibers arranged into a 2D array were used for online beam monitoring of charged particle (e.g., electron beam), in a setup operating on the same principle as the laser beam analyzers (Fig. 19). When exposed to electron flux, the temperature of individual gratings increases proportionally to (i) the local energy each sensor receives and (ii) the integrated in time, and a wavelength peak shift of each FBG occurs. When the temperature reaches a specified limit, defined by the maximum permissible wavelength peak shift, a mechanical shutter interrupts the electron beam, and a fan, blowing air on the grating array, decreases the FBG temperature, producing a recovery of the wavelength shift. After the FBG matrix temperature arrives at the lowest programmed limit, the shutter opens, and the acquisition process continues with a new measuring cycle. In Fig. 19 the FBG array is located under the shutter. An example of the acquired 2D beam energy distribution is given in Fig. 20.

Experiments involving plastic fiber Bragg grating (PFBG) exposure to fast neutrons up to a dose of 720 Gy indicated a shift of the Bragg wavelength to 7 pm associated with the radiation-induced degradation of the polymer. The gratings were produced by phase mask technique in PMMA polymer with a 325 nm optical radiation from a CW-operated HeCd laser. The grating central wavelength was $\lambda = 850$ nm. The irradiation conditions were as follows: neutron emission rate of 2.2×10^6 ns^{-1} Ci^{-1} and peak thermal neutron flux of 10^6 n/s. The authors concluded that within some irradiation dose limits, PFBGs can be used as radiation detectors (Hamdalla and Nafee 2015).

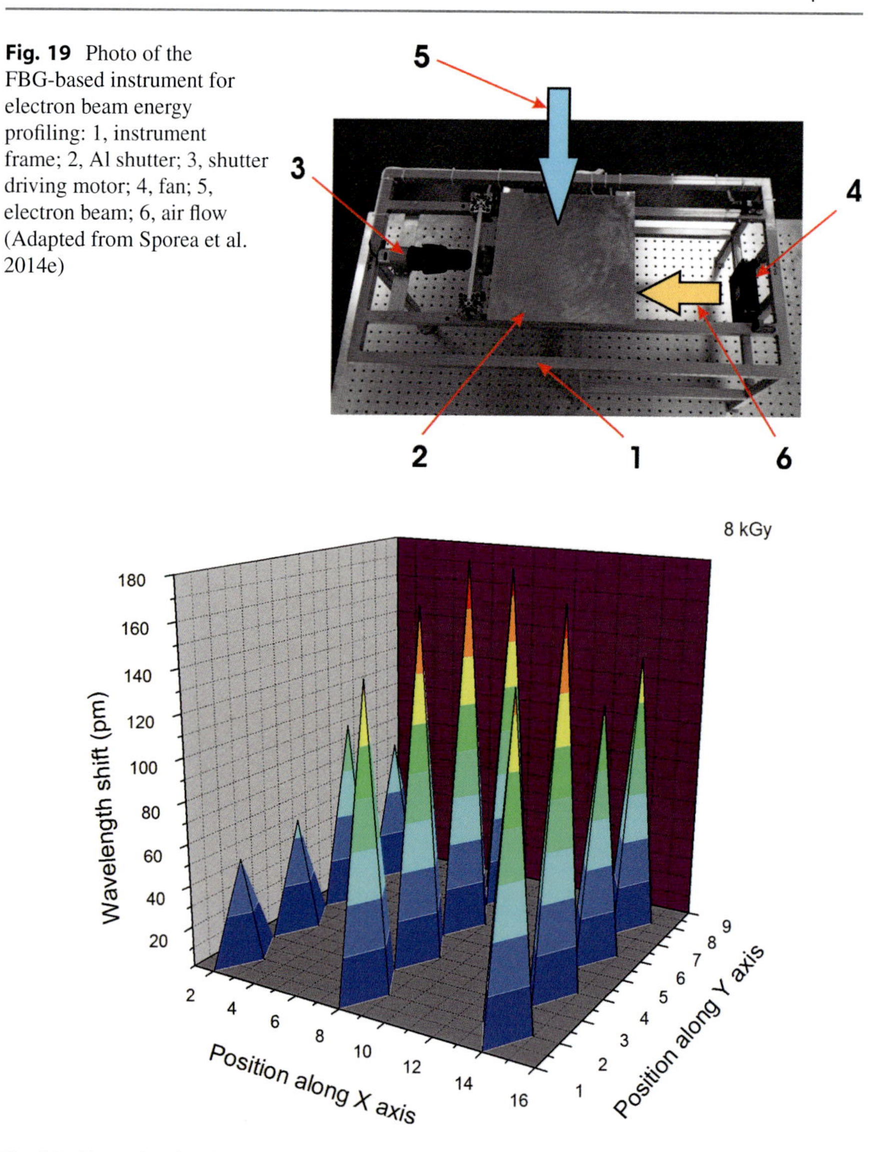

Fig. 19 Photo of the FBG-based instrument for electron beam energy profiling: 1, instrument frame; 2, Al shutter; 3, shutter driving motor; 4, fan; 5, electron beam; 6, air flow (Adapted from Sporea et al. 2014e)

Fig. 20 Example of real-time-acquired electron beam energy profile (Sporea et al. 2014e)

Long-Period Gratings

As indicated in the previous subsection, FBGs exhibit a high radiation resistance, a fact which recommend them for use in radiation environments for temperature or humidity monitoring. Few data are available on the behavior of long-period gratings (LPGs) under irradiation. Several studies were carried out on their performances under gamma-ray exposure.

Early work by Vasiliev and his group refers to gamma irradiation (dose rates of 5.4 and 6.6 Gy/s) of thermo-induced LPGs (period $\Lambda = 250$ μm, length $L = 25$ mm) written in an N-doped optical fiber by CO laser heating and LPGs engraved by UV exposure in Ge-doped fiber ($\Delta n = 0.018$, $\lambda_c = 1.04$ pm, $\Lambda = 150$ μm, $L = 25$ mm) (Vasiliev et al. 1998). The total dose was 1.47 MGy, reached by four subsequent irradiation steps (9.3, 17.8, 71.3, and 1370 kGy). No significant results were obtained as it concerns gamma-ray effects on LPGs, as far as no complete measurements were performed after the irradiation, and some recovery phenomenon was present.

A more detailed investigation was reported by Rego et al. working with LPGs developed by electric arc discharge (EAD) in pure silica core, F-doped cladding optical fibers from Oxford Electronics and Acreo (Rego et al. 2005). For a resonant wavelength $\lambda = 1550$ nm, the grating period was found to be $\approx$730 μm. Before the irradiation, the grating temperature coefficient was measured. For a temperature span from 20 to 100 °C, these coefficients were 28.9 ± 0.2 pm/°C (Oxford) and 49.9 ± 0.3 pm/°C (Acreo), respectively. Online measurements of LPG wavelength deep shift and attenuation were performed with broadband LEDs and a spectrum analyzer, as samples were irradiated at the dose rate of 1 kGy/h (H_2O), at the constant temperature 37.4 ± 0.1 °C, and up to a total dose of about 560 kGy. The spectra of the Oxford-based LPGs showed minor modifications during the irradiation. Postirradiation temperature coefficient of Oxford LPGs is close to the initial value, meaning that gamma irradiation does not affect this parameter. No significant change was observed for the strain sensitivity of LPGs written in Oxford fiber after they were exposed to gamma-ray.

Almost no wavelength shift or temperature coefficient variation was found, also the case of LPGs manufactured by periodic heating of communication-type SMF28 optical fiber with a CO_2 laser and irradiated by gamma-ray to a total dose of 5 kGy (Chaubey et al. 2007).

Chiral long-period gratings (CLPGs), made by Chiral Photonics Inc. using commercially optical fibers from ALCATEL, Corning, FiberLogix, FORC, Fujikura, and Nufern, were tested online at a ^{60}Co gamma source by monitoring with an interrogator RIA and the wavelength deep shift. Two sets of irradiation were run one at a dose rate of 0.87 Gy/s, total dose of 100 kGy, and one at 0.1 Gy/s, total dose of 20 kGy, in order to check the dose rate dependence of the irradiation-induced changes in LPGs (Henschel et al. 2010). For some of the CLPGs, the deep wavelength shift after 100 kGy irradiation was quite large (e.g., 10 nm) recommending them for radiation dosimetry.

Significant changes of the wavelength deep location after gamma-ray irradiation were noticed in turn around point (TAP) long-period gratings (LPGs) engraved in B/Ge co-doped photosensitive fiber from Fibercore, through exposure to CO_2 laser pulses. This type of grating proved to be very radiation sensitive. For a total dose of 6 kGy, the wavelength shift was 35 nm, while for a dose of 65 kGy, this parameter modified up to 80 nm (Kher et al. 2012).

More recently, extensive research was carried out on this subject, because the interest on using LPGs in radiation environments increased (i.e., nuclear waste

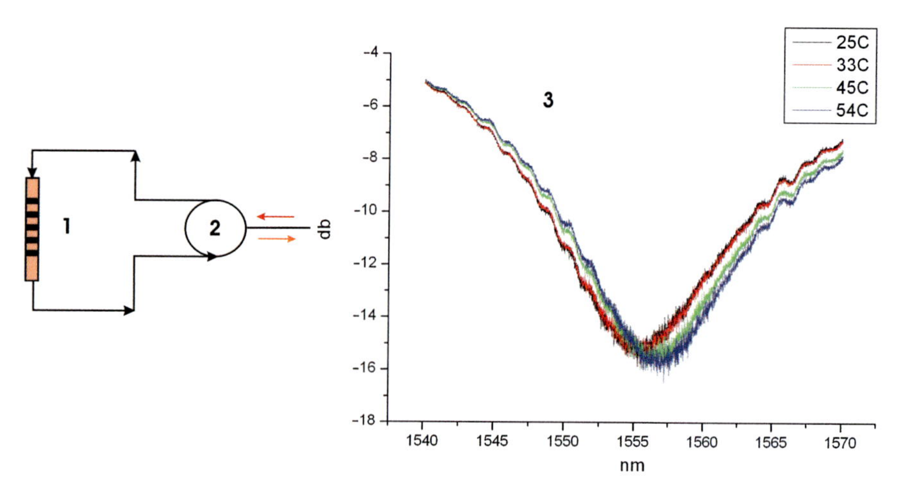

Fig. 21 The principle of LPG reading with an OFDR: 1, long-period grating; 2, fiber optic circulator; 3, optical frequency domain reflectometer

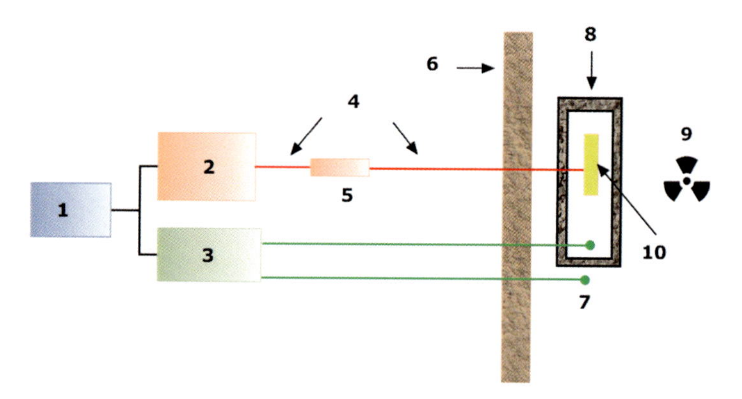

Fig. 22 The detection of LPG's wavelength deep shift with an interrogator: 1, laptop; 2, interrogator; 3, temperature data logger; 4, connecting optical fibers; 5, optical fiber circulator; 6, radiation shield; 7, thermocouple; 8, thermally insulating box; 9, gamma source; 10, LPG (Esposito et al. 2017)

repositories, high energy physics, and nuclear power plants). LPGs developed in various optical fibers were monitored online for the first time under gamma irradiation by using an optical frequency domain reflectrometer (OFDR) instead of an optical spectrum analyzer or interrogator (Sporea et al. 2014d, 2015, 2016b; Esposito et al. 2017). In the meantime, another novelty was introduced during these investigations by multiplexing LPGs and FNGs with OFDR reading. The LPG wavelength monitoring using this approach is illustrated in Fig. 21. An alternative scheme is based on an interrogator operating in conjunction with an optical fiber circulator (Fig. 22). LPGs were tested online under gamma radiation. The newly proposed method makes possible an improvement of the detection S/N.

A set of LPGs was written, by melting-drawing method based on CO_2 laser assisted by a micro-flame, in a fluorine F-doped 8.5 μm core diameter SM optical fiber, having a fluorine concentration of about 0.2 wt.% in the core and 1.8 wt.% in the cladding. These gratings have a length of 22 mm long, respectively, 25 mm, with a grating pitch of 730 μm, respectively, 720 μm. All LPGs were packed into a ceramic case to avoid any strain-induced variation of their signal during handling and irradiation. Before and after the irradiation, the temperature coefficient of each grating was measured in the laboratory, by heating the grating from room temperature to 65 °C. In several experiments, these LPGs were exposed to gamma-ray at dose rates of 4 Gy/h, 200 Gy/h, (Sporea et al. 2014d), 240 Gy/h (Sporea et al. 2015, 2016b), and 340 Gy/h (Sporea et al. 2015). Under various circumstances the values of total dose were 21.5 kGy, 33.9 kGy (Sporea et al. 2015, 2016b), 26 kGy (Sporea et al. 2016b), and 45 kGy (Sporea et al. 2014d). During the irradiation the sensors were fixed into a thermally insulating box, and their temperature was monitored to enable the correction of the detected signal against temperature variation. Very low-dose rates (e.g., 4 Gy/h) did not induce any change in the LPG parameters.

For one of the LPGs, the temperature sensitivity was modified after gamma irradiation from 27.7 to 29.3 pm/°C (Sporea et al. 2014d), while in a second case, these values were 50 pm/°C and 48 pm/°C, respectively (Sporea et al. 2015). The maximum resonant wavelength deep shift was 700 pm at a cumulated dose of 45 kGy (Sporea et al. 2014d). This shift recommends such a LPG for dosimetric use, as far as no strain or temperature stress act on it. A recovery of 600 pm was observed for one of the LPGs, during the first 120 h postirradiation, at room temperature (Sporea et al. 2014d).

Additional research focused on gamma irradiation of LPGs fabricated by EAD in standard Corning SMF28 Ge-doped fibers, manufactured by OZ Optics Ltd. and Thorlabs Inc., and in the radiation-resistant R1310-HTA fiber from Nufern Inc. The characteristics of the LPGs are (i) OZ-LPG, period $\Lambda = 628$ μm, three attenuation bands centered at 1437.0 nm (LP_{02}), 1475.3 nm (LP_{03}), and 1562.8 nm (LP_{04}), with depth of 6.3 dB, 22.3 dB, and 28.3 dB; (ii) Nufern-LPG, $\Lambda = 677$ μm, three attenuation bands, centered at 1418.3 nm (LP_{02}), 1461.8 nm (LP_{03}), and 1560.2 nm (LP_{04}), with depth of 2.4 dB, 13.7 dB, and 25.6 dB; and (iii) Thorlabs-LPG, $\Lambda = 646$ μm, three attenuation bands, centered at 1423.8 nm (LP_{02}), 1465.4 nm (LP_{03}), and 1554.0 nm (LP_{04}), with depth of 3.8 dB, 14.5 dB, and 26.1 dB (Esposito et al. 2017).

All measurements were performed online using the technique presented in Fig. 21, under a dose rate of 180 Gy/h, up to 35 kGy. Before and postirradiation the sensors' sensitivity to temperature and to external refractive index changes were evaluated in the laboratory. The investigation includes for the first time an experimental part and a simulation part targeting the assessment of radiation effects on this type of sensor. The model employed fits to the experimental data (Fig. 23).

By the end of the irradiation (35 kGy dose), the OZ-LPG and Thorlabs-LPG exhibited a wavelength deep shift of 3.7 nm and 3.9 nm, respectively. The LPG fabricated in the Nufern fiber proved to be more radiation sensitive, with a

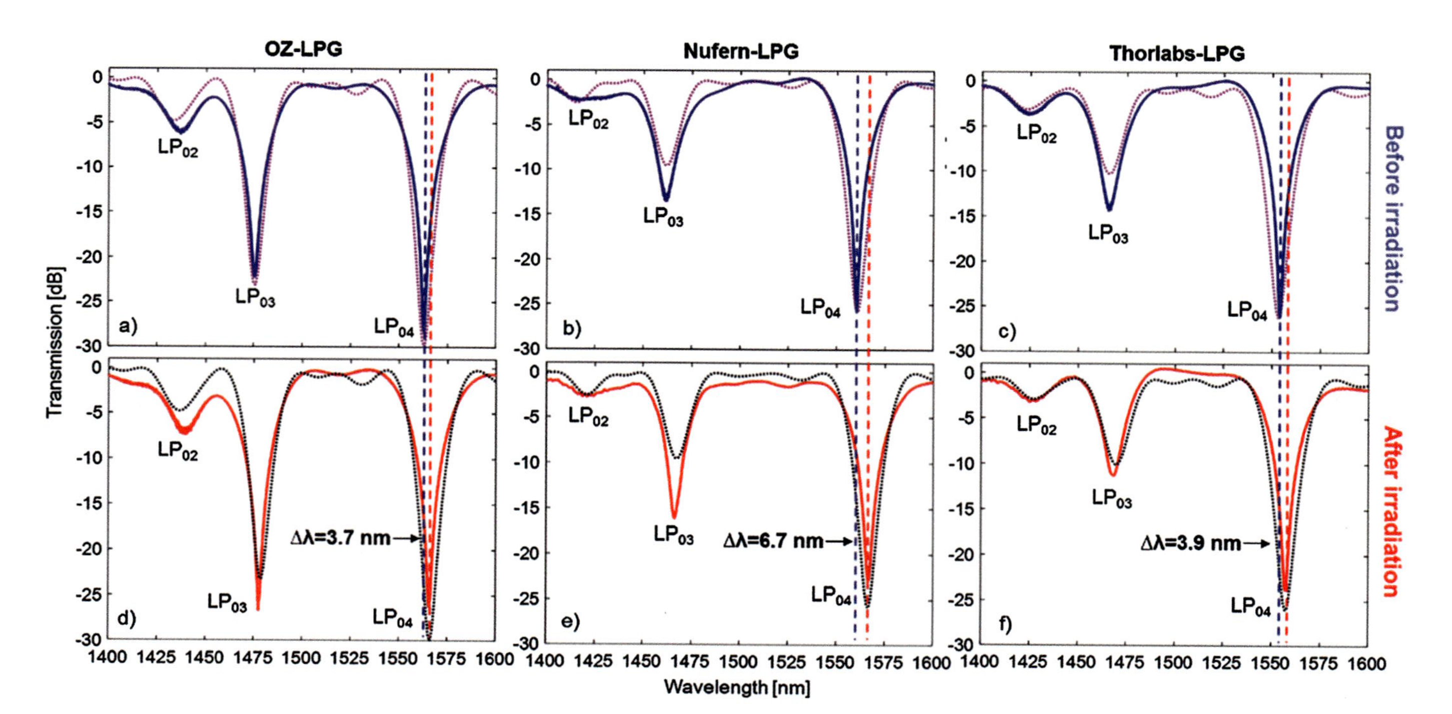

Fig. 23 Full spectra of gratings: comparison between experimental (solid lines) and numerical results (dotted lines), before (blue and magenta lines) and after gamma irradiation (red and black lines) of 35 kGy (Esposito et al. 2017)

wavelength shift of 6.7 nm. An increase of the temperature sensitivity was present after gamma irradiation for all LPGs: 3.3 pm/°C (OZ-LPG), 0.3 pm/°C (Nufern-LPG), and 1.4 pm/°C (Thorlabs-LPG).

More recently, LPGs were tested under mixed neutron-gamma radiation in a research nuclear reactor (Stăncălie et al. 2017). Systems based on optical time-domain reflectometry, optical frequency domain reflectometry, and Raman and Brillouin effects in optical fibers, under different configurations, were used to measure in a distributed manner: (i) some parameters (i.e., temperature, strain) in radiation environments or (ii) the irradiation dose. The following subsections refer briefly to some implementation examples.

Sensors Based on Optical Time-Domain Reflectometry

Following the outcomes of the early work performed by Friebele and colleagues (1984) on the use of optical time-domain reflectometry (OTDR) in radiation dosimetry, an intercomparison-based OTDR measurement was organized for the assessment of ^{60}Co radiation effects on the optical attenuation in SM and MM commercial communication fibers (West et al. 1994). The effects of (i) OTDR characteristics, (ii) launching conditions, (iii) photobleaching, and (iv) long-term measurements were investigated, in order to design appropriate recommendations for protocol development.

In another approach, the OTDR technique was combined with scintillating-based dosimetry for the design of a distributed radiation sensor (Takada et al. 1999). Scintillating tips (Gd_20_2S doped with rare earth ions, Pr^{3+}, Yb^{3+}, and Nd^{3+}) are attached to the ends of several optical fibers of different lengths. These fibers are coupled at the other end to a CCD camera, through an optical high-pass (>800 nm) filter, for the detection of IR scintillation emission. The OTDR instrument provides information on the localization of the optical attenuation as induced by radiation along the optical fibers guiding the scintillation signals.

Henschel and his team proposed the integration of the (i) OTDR monitoring of radiation-induced attenuation in optical fibers, along with (ii) measurements based on LEDs/power meter and (iii) those using a Mach-Zehnder interferometer for complex dosimetry at TESLA accelerator (Henschel et al. 2000).

A commercially available OTDR that was included into the diagnostic system of the Dortmund Electron Accelerator (DELTA) is a 1.5 GeV synchrotron light source, to evaluate beam losses produced by aperture limitations and those appearing during the injection (Schmidt et al. 2002).

Preliminary tests were carried out jointly by teams from CERN (Switzerland), Scuola Superiore Sant'Anna (Italy), and Fraunhofer-Institut für Naturwissenschaftlich-Technische Trendanalysen (Germany) to evaluate the feasibility of a distributed radiation sensor dedicated to the monitoring of the Large Hadron Collider (LHC). SM and MM P-doped optical fibers were subjected to gamma irradiation. RIA was measured at discrete wavelengths $\lambda = 830$ nm, 1312 nm, and 1570 nm (maximum dose 60 kGy). OTDR investigations were done for fiber lengths of 40 cm, 72 cm, 128 cm, and 240 cm, at two wavelengths, $\lambda = 850$ nm and 1300 nm, for the dose rate of 22 mGy/s, to reach the total dose of 300 Gy. The special resolution in this case was 30 cm (Toccafondo 2015).

Research was carried out, by Tunable Wavelength Coherent Optical Time-Domain Reflectometry (TW-COTDR), on radiation tolerance of communication grade, F-doped and N-doped optical fibers to gamma-ray exposure, performed at room temperature, irradiation done at 370 Gy(SiO_2)/h, total irradiation dose (TID) of 56 kGy, and at 25 kGy(SiO_2)/h, TID of 10 MGy (Planes et al. 2017). Tests were run to assess RIA contribution to fiber characteristic degradation, for possible application in strain and temperature monitoring in radiation environments. The F-doped fiber proved to be the most tolerant up to 56 kGy. Temperature and strain coefficients remained unchanged upon the radiation exposure.

Optical attenuation measurements ($\lambda = 700$–1700 nm) were complemented by OTDR investigations in near-IR ($\lambda = 1310$ nm and 1550 nm) for the evaluation of gamma-ray effects on perfluorinated polymer optical fiber (PF-POF) (Stajanca et al. 2016). The spectral window where this POF can be operated reliable under irradiation conditions corresponds to 1.1 μm. For other spectral ranges, the use of such plastic fibers for radiation dosimetry was suggested, these optical fibers having a sensitivity of 260 dBm^{-1}/kGy, over the 400–500 nm spectral band.

Sensors Based on Optical Frequency Domain Reflectometry

When the application of optical frequency domain reflectometer (OFDR) under ionizing radiation is discussed, two issues are of interest: (i) the assessment of OFDR reliability in measuring temperature distribution in such circumstances and (ii) the way the degradation of sensing optical fibers under irradiation can be utilized to build distributed dosimeter systems.

Evaluation of the feasibility of OFDR-based distributed temperature and strain measurements under gamma irradiation (dose rates from 10 to 30 kGy/h; total dose 10 MGy, ambient temperature up to 50 °C) employing radiation-resistant and radiation-sensitive optical fibers (standard communication, Ge-, GeP-, F-doped) was carried out by Rizzolo et al. (2015a). The results showed that the temperature and strain coefficients for such investigations are unaffected by gamma irradiation, the major limiting factor being RIA, which has acceptable values for the F-doped fiber. Researches in the field highlighted the role played by sensing optical fiber packaging during the exposure to ionizing radiation. A higher accuracy (0.5 °C) of distributed temperature reading can be obtained by thermal pre-treating of the fiber up to the highest value of the expected operating temperature (Rizzolo et al. 2015b).

Measurements done by optical frequency domain reflectometry on microstructured optical fibers (MOFs) irradiated by X-rays at dose rate of (3.2 ± 0.1) Gy(SiO_2)/s, up to the total irradiation dose of 1 MGy(SiO_2), indicated a high tolerance to radiation, providing new solutions for distributed temperature monitoring (Rizzolo et al. 2016). The OFDR tests were run in conjunction with RIA evaluation which made possible the selection of the appropriate MOF for the interrogation wavelength used. With the appropriate selection of the sensing MOF_2, the temperature measurement error achieved was as low as ~0.3 °C.

An experiment performed on P-doped radiation-sensitive optical fiber demonstrated that the limitation of the length over which distributed dosimetry can be performed depends on the OFDR Integrated Insertion Loss Dynamic Range (ILDR), as it is related to RIA values associated with a satisfactory noise level (Faustov et al. 2016).

Rayleigh-based OBR (Optical Backscatter Reflectometer™) measurements were performed to evaluate RIA in ten types (manufactured by Draka, Nufern, Corning, Fujikura) of SM optical fibers, after their irradiation at 1 kGy, 5 kGy, 20 kGy, 50 kGy, and 100 kGy gamma doses (Wosniok et al. 2016). The results can be used to evaluate postmortem the changes in the fiber core refractive index.

Tests on F-doped optical fibers used as sensing medium for distributed OFDR temperature monitoring in harsh environments (mixed gamma-ray, dose >2 MGy, and neutrons fluence about 10^{17} n/cm^2) indicated that the major difference as compared to simple gamma-ray irradiation is associated with defect formation. In spite of this fact, OFDR measurements can be implemented in such conditions as the temperature resolution achieved over 130 m fiber was 0.1 °C (Rizzolo et al. 2017).

Sensors Based on Raman Scattering

Researches covering the operation of Raman-based instruments in harsh environments refer to their possible use to monitor some parameters of interest (i.e., temperature).

The presence of H_2 (released at about 500 mmol/h), high temperature (maximum 90 °C), and gamma-ray (dose rates 1–10 Gy/h, maximum dose 10 MGy) expected for deep geological nuclear waste repository proved to affect strongly the performances of single-end Raman distributed temperature fiber-based sensors (RDTS), the limitations being related to the type of optical fibers employed (composition and coating). Several MM optical fibers (standard, step index, radiation-hardened, and radiation-resistant to H_2 diffusion) were tested at gamma doses 3, 6, and 10 MGy and temperature of 45–50 °C (Cangialosi et al. 2015). The conclusions of these investigations indicated that Ge- and P-doped optical fibers are not appropriate for Raman-based temperature sensing under irradiation conditions because RIA becomes too high and limits the length available to reliable measurements. H_2 diffusion can be limited by replacing the plastic coating with carbon coating.

A double-ended configuration adopted for RDTS indicated a higher radiation tolerance (gamma dose up to 300 kGy), in the case of special optical fibers available on the market (graded-index MM hermetic PYROCOAT® fibers, Corning acrylate coated graded-index MM) (Fernandez et al. 2005).

Under the assumption that the irradiation along the temperature sensing fiber is homogenous, a new data handling technique was proposed for RIA correction, which leads to an irradiation-induced temperature error equivalent to 2 °C, for a total dose of 10 MGy (Cangialosi et al. 2014).

In the situation that the irradiation is uniform along the fiber length, a simple correction can be done by using two reference thermocouples. If this is not the case, a loop setup has to be implemented to compensate for dose nonuniformity (Takada et al. 1998).

The reliability of a RDTS based on radiation-resistant MM optical fibers and working in a loop configuration was demonstrated to achieve a temperature resolution better than 1 °C; for a spatial resolution of 0.5 m, performances which might recommend the setup for operation at CHARM facility, where high radiation environments (photons, neutrons, protons) could degrade system parameters (Toccafondo et al. 2015).

Sensors Based on Brillouin Scattering

Brillouin scattering-based methods were used to evaluate degradation of gamma-irradiated optical fibers as candidates for distributed temperature and strain measurements (Alasia et al. 2006). RIA in standard telecom optical fibers irradiated to 9.9 MGy was measured with an OTDR, followed by tests carried out by Brillouin optical time-domain analysis (BOTDA). Up to the dose of 100 kGy, the frequency shift was about 5 MHz supporting the assumption of such methods used in radiation environments.

Additional investigations were performed on commercial telecommunication fibers, F-doped fibers, and highly Ge-doped fibers under gamma-ray exposure to a maximum dose of 10 MGY and dose rate of 28 kGy/h (Phéron et al. 2012). A BOTDA instrument having a resolution of ± 0.5 MHz and operating at $\lambda = 1550$ nm was used. RIA constitutes a major limitation in such measurements as it reduces the distance operating range of the equipment to several hundred meters. A lower induced attenuation at $\lambda = 1310$ nm as compared to that obtained after irradiation at $\lambda = 1550$ nm suggests the need to shift the measuring domain to lower wavelengths in order to have a higher sensitivity at the detection. The most promising fiber to be used in temperature and strain measurements through this method proved to be the F-doped fiber.

Wosniok et al. reported the results of a research on ten types of optical fibers, gamma irradiated at doses from 1 to 100 kGy and measured by Brillouin optical-fiber frequency-domain analysis (BOFDA), aiming to assess the use of this method for distributed radiation dosimetry based on optical fibers (Wosniok et al. 2016).

Other Optical Fiber Sensors

The variation of the thermo-optic coefficient of an optical fiber resonant cavity during the irradiation, and interrogated by high accuracy inteferrometry, was exploited to design a radiation dosimeter able to reach a resolution of 160 mGy, compatible with biomedical applications (Avino et al. 2013).

The modification with temperature of RIA produced in gamma-irradiated (10 kGy) Ge/P co-doped optical fibers was employed in the design of an all fiber temperature sensor, RIA being monitored at $\lambda = 850$ nm. The sensor's dynamic range covers the temperature interval from -40 to $+60$ °C, with a temperature sensitivity of 0.1146 dB/°C (Jing et al. 2015).

Studies were run to investigate radiation effects on optical fiber incorporated in fiber-optic gyroscope (FOG) to be used in spaceborn instrumentation. RIA and the degradation of fiber polarization-related performances are the major limitation concerning the employment of polarization-maintaining optical fibers in interferometric-type FOGs. Four polarization-maintaining optical fiber, having the core co-doped with Ge, P, and Ge, F, were investigated at $\lambda = 1310$ nm, as they were exposed to ^{60}Co gamma radiation to a total dose about 700 Gy. The main drawback for the mentioned application is related to RIA increase with dose (Wang et al. 2011). The radiation-induced mean wavelength shift in Ge-doped optical fiber was found to be lower than in the case of Ge, P co-doped fibers, under gamma irradiation of 700 Gy (Jing et al. 2012).

References

M.S. Akselrod, L. Better-Jensen, S.W.S. McKeever, Radiat. Meas. **41**, S78 (2007)
D. Alasia et al., Meas. Sci. Technol. **17**, 1091 (2006)
A. Alawiah et al., Proc. SPIE **8775**, 87750S (2013)
C.E. Andersen, in *AIP Conference Proceedings*, ed. by A. Rosenfeld, T. Kron, F. d'Errico, M. Moscovitch (American Institute of Physics, College Park, 2011), p. 1345, 100
C.E. Andersen et al., Med. Phys. **36**, 708 (2009)
C.E. Andersen et al., Radiat. Meas. **46**, 1090 (2011)
A.I. de Andrés, Ó. Esteban, M. Embid, Opt. Laser Technol. **93**, 201 (2017)
S. Avino et al., Appl. Phys. Lett. **103**, 184102 (2013)
M.C. Aznar et al., Phys. Med. Biol. **49**, 1655 (2004)
L. Beaulieu, S. Beddar, Phys. Med. Biol. **61**, R305 (2016)
L. Beaulieu et al., J. Phys. Conf. Ser. **444**, 012013 (2013)
S. Beddar., http://www.aapm.org/meetings/09ss/documents/32beddar-plasticdosimeters.pdf
A.R. Beierholm et al., Radiat. Meas. **43**, 898 (2008)
M. Benabdesselam, F. Mady, S. Girard, J. Non-Cryst. Solids **360**, 9 (2013)
D. Benoit et al., IEEE T. Nucl. Sci. **55**, 2154 (2008)
F. Berghmans et al., in *Optical Waveguide Sensing and Imaging*, ed. by W.J. Bock, I. Gannot, S. Tanev (Springer, Dordrecht, 2008), p. 127
B.C. Bhatt, Radiat. Prot. Environ. **34**, 6 (2011)
P. Borgermans et al., in *Proceedings of SPIE 4204, Fiber Optic Sensor Technology II,* ed. by B. Culshaw, J.A. Harrington, M.A. Marcus, M. Saad (SPIE, Bellingham, 2001), p. 151
D.A. Bradley et al., Radiat. Phys. Chem. **104**, 3 (2014)
D.A. Bradley et al., Appl. Radiat. Isot. **100**, 43 (2015)
B. Brichard et al., Meas. Sci. Technol. **18**, 3257 (2007)
D. Broggio et al., Nucl. Instr. Met. B **254**, 3 (2007)
E. Bulur, A. Yeltik, Radiat. Meas. **45**, 29 (2010)
S. Buranurak, C.E. Andersen, Radiat. Meas. **93**, 46 (2016)
S. Buranurak et al., Radiat. Meas. **56**, 307 (2013)
O. Butov et al., J. Appl. Phys. **118**, 074502 (2015)
C. Cangialosi et al., IEEE T. Nucl. Sci. **61**, 3315 (2014)

C. Cangialosi et al., J. Lightwave Technol. **33**, 2432 (2015)
I. Chaikovska, N. Delerue, A. Variola, in *Proc. IBIC 2014-3rd International Beam Instrumentation Conference,* ed. by D. Button, M. Montes-Loera, I. Martin, I. Costa, V.R.W. Schaaa, Paper TUPD23 (Stanford University, Stanford, 2014), p. 463
S. Chaubey et al., Sadhana **32**, 513 (2007)
N. Chiodini, A. Vedda, I. Veronese, Adv. Opt. **2014**, 9 pp (2017). https://doi.org/10.1155/2014/974584. Article ID 974584
A. Darafsheh et al., in *Proc. SPIE 10058, Optical Fibers and Sensors for Medical Diagnostics and Treatment Applications XVII*, ed. by I. Gannot (SPIE, Bellingham, 2017), p. 100580B
D. Di Francesca, *Roles of dopants, interstitial $O_2$and temperature in the effects of irradiation on silica-based optical fibers*, Doctoral dissertation, Saint Etienne (2015)
D. Di Francesca et al., IEEE T. Nucl. Sci. **61**, 3302 (2014)
C. Dotzler et al., Appl. Phys. Lett. **91**, 121910 (2007)
P. Duguay-Drouin, Caractérisation et optimisation d'un détecteur à scintillation à 2 points, PhD Thesis (2016), http://theses.ulaval.ca/archimede/fichiers/32457/32457.pdf
F. Esposito et al., Sci. Rep. **7**, 43389 (2017). https://doi.org/10.1038/srep43389
A. Faustov et al., IEEE T. Nucl. Sci. **59**, 1180 (2012)
A. Faustov et al., Results Phys. **6**, 86 (2016)
A.F. Fernandez et al., IEEE T. Nucl. Sci. **52**, 2689 (2005)
A.F. Fernandez et al., IEEE T. Nucl. Sci. **53**, 1607 (2006)
E.J. Friebele, C.G. Askins, M.E. Gingerich, Appl. Opt. **23**, 4202 (1984)
R.S. Gaza, W.S. McKeever, M.S. Akselrod, Med. Phys. **32**, 1094 (2005)
S. Ghosh et al., Appl. Opt. **50**, E80 (2011)
S. Girard, C. Marcandella, IEEE T. Nucl. Sci. **57**, 2049 (2010)
S. Girard et al., J. Non-Cryst. Solids **357**, 1871 (2011)
S. Girard et al., IEEE T. Nucl. Sci. **60**, 2015 (2013a)
S. Girard et al., IEEE T. Nucl. Sci. **60**, 4305 (2013b)
S. Girard et al., IEEE T. Nucl. Sci. **64**, 567 (2017)
A.K. Glaser et al., Phys. Med. Biol. **59**, 3789 (2014)
D. Grobnic et al., in *Proc. SPIE 7316. Defense, Security, and Sensing*, ed. by E. Udd, H.H. Du, A. Wang (SPIE, Bellingham, 2009), p. 73160C
A. Gusarov, IEEE T. Nucl. Sci. **57**, 2044 (2010)
A. Gusarov, S. Hoeffgen, IEEE T. Nucl. Sci. **60**, 2037 (2013)
A. Gusarov et al., Nucl. Instr. Met. B **187**, 79 (2002)
A. Gusarov, B. Brichard, D.N. Nikogosyan, IEEE T. Nucl. Sci. **57**, 2024 (2010)
T.A. Hamdalla, S.S. Nafee, Opt. Laser Technol. **74**, 167 (2015)
H. Henschel et al., Fiber optic radiation sensing systems for TESLA. TESLA Report 26 (2000)
H. Henschel et al., in *9th European Conference on Radiation and Its Effects on Components and Systems, 2007. RADECS 2007*, ed. by S. Girard, N. Richard (IEEE, New York, 2007), p. 1
H. Henschel et al., IEEE T. Nucl. Sci. **57**, 2915 (2010)
M. Ishikawa et al., J. Radiat. Res. **56**, 372 (2015)
K.W. Jang et al., J. Korean Phys. Soc. **56**, 1777 (2010)
K.W. Jang et al., Nucl. Instrum. Met. A **652**, 841 (2011)
K.W. Jang et al., Opt. Express **20**, 13907 (2012)
K.W. Jang et al., Opt. Express **21**, 14573 (2013)
K.W. Jang et al., Sensors **14**, 7013 (2014)
J. Jing et al., Chin. Phys. B **21**, 094220 (2012)
J. Jing, L. Song, S. Ning-Fang, Chin. Phys. B **23**, 014206 (2013)
J. Jing et al., *2nd International Workshop on Materials Engineering and Computer Sciences (IWMECS 2015)* (Atlantis Press, Paris, 2015), https://www.researchgate.net/profile/Jixun_Liu/publication/301462633_A_Novel_Temperature_Sensor_Based_on_Temperature-dependent_Attenuation_at_850nm_of_Irradiated_GeP_Co-doped_Fiber/links/5743139c08ae298602ee95-3d.pdf
S. Kher et al., IEEE Photon. Technol. Lett. **24**, 742 (2012)

D. Klein et al., Radiat. Meas. **47**, 921 (2012)
M. Kovačević et al., Opt. Laser Technol. **47**, 148 (2013)
K. Krebber, H. Henschel, U. Weinand, Meas. Sci. Technol. **17**, 1095 (2006)
S.H. Law et al., in *Biomedical Optics 2004* (SPIE, Bellingham, 2004), p. 105
B. Lee et al., IEEE T. Nucl. Sci. **57**, 1496 (2010)
B. Lee et al., Sensors **15**, 11012 (2015)
Y.-P. Liu, Chin. Phys. C **32**, 381 (2008)
P.Z.Y. Liu et al., Phys. Med. Biol. **56**, 5805 (2011)
D. McCarthy et al., IEEE Sensors **2011**, 121 (2011)
S.W.S. McKeever, Nucl. Instrum. Met. B **184**, 29 (2001)
S.W.S. McKeever et al., Radiat. Prot. Dosim. **109**, 269 (2004)
D.R. Mishra et al., Nucl. Instrum. Met. B **342**, 116 (2015)
E. Mones et al., Radiat. Meas. **43**, 888 (2008)
A. Morana et al., Opt. Lett. **39**, 5313 (2014)
A. Morana et al., J. Lightwave Technol. **33**, 2646 (2015a)
A. Morana et al., Opt. Express **23**, 8659 (2015b)
R. Naka et al., IEEE T. Nucl. Sci. **48**, 2348 (2001)
S. O'Keeffe et al., IEEE J. Sel. Top. Quantum Electr. **22**, 35 (2016)
S. O'Keeffe et al., *13th Intl. Plastic Optical Fibres Conference* (2004), http://citeseerx.ist.psu.edu/viewdoc/download?doi=10.1.1.59.3780&rep=rep1&type=pdf
X. Phéron et al., Opt. Express **20**, 26978 (2012)
I. Planes et al., Sensors **2017**(17), 396 (2017)
A.S. Pradhan, J.I. Lee, J.L. Kim, J. Med. Phys. **33**, 85 (2008)
Z. Qin et al., Biomed. Opt. Express **7**, 4919 (2016)
A.T. Rahman et al., Appl. Radiat. Isot. **70**, 1436 (2012)
G. Ranchoux et al., Radiat. Prot. Dosim. **100**, 255 (2002)
N.S. Rawat et al., Radiat. Meas. **71**, 212 (2014)
G. Rego et al., Appl. Opt. **44**, 6258 (2005)
F. Rêgo, L. Peralta, M.D.C. Abreu, arXiv preprint arXiv:1109.6545 (2011)
S. Rizzolo et al., IEEE T. Nucl. Sci. **62**, 2988 (2015a)
S. Rizzolo et al., Opt. Express **23**, 18997 (2015b)
S. Rizzolo et al., IEEE T. Nucl. Sci. **63**, 2038 (2016)
S. Rizzolo et al., IEEE T. Nucl. Sci. **64**, 61 (2017)
M.A. Saeed et al., Radiat. Phys. Chem. **91**, 98 (2013)
A.M.C. Santos, M. Mohammadi, V. S. Afshar, Opt. Express **22**, 4559 (2014)
G. Schmidt et al., in *Proceedings of EPAC*, ed. by J.-L. Laclare (CERN, Switzerland, 2002), p. 1969
T. Shikama et al., Meas. Sci. Technol. **17**, 1103 (2006)
D. Sporea, A. Sporea, Phys. Status Solidi C **4**, 1356 (2007)
D. Sporea et al., in *Selected Topics on Optical Fiber Technology*, ed. by M. Yasin, S.W. Harun, H. Arof (InTech, 2012a), p. 607
D. Sporea, A. Sporea, C. Oproiu, J. Nucl. Mater. **423**, 142 (2012b)
D. Sporea et al., Sensors **14**, 3445 (2014a)
D. Sporea et al., Sens. Actuat. A-Phys. **213**, 79 (2014b)
D. Sporea et al., Opt. Express **22**, 31473 (2014c)
D. Sporea et al., IEEE Photon. J. **6**, 1 (2014d)
D. Sporea et al., Sensors **14**, 15786 (2014e)
D. Sporea et al., Sensor Actuators A-Phys. **233**, 295 (2015)
D. Sporea, A. Sporea, C. Oproiu, in *Proc. International Conference on Advanced Materials for Science and Engineering (ICAMSE)*, ed. by T.-H. Meen, S.D. Prior, A. D. Kin-Tak Lam (IEEE, New York, 2016a), p. 691
D. Sporea et al., in *Proc. SPIE 9886, Micro-Structured and Specialty Optical Fibres IV*, ed. by K. Kalli, A. Mendez (SPIE, Bellingham, 2016b), p. 98861P
D. Sporea et al., Sci Rep **7**, 40209 (2017)
P. Stajanca et al., Opt. Mater. **58**, 226 (2016)

A. Stăncălie et al., Sci Rep. **7**, 15845 (2017)
E. Takada et al., J. Nucl. Sci. Technol. **35**, 547 (1998)
E. Takada et al., J. Nucl. Sci. Technol. **36**, 641 (1999)
T. Teichmann et al., Radiat. Meas. **90**, 201 (2016)
F. Therriault-Proulx et al., Med. Phys. **38**, 2176 (2011)
F. Therriault-Proulx et al., Phys. Med. Biol. **57**, 7147 (2012)
F. Therriault-Proulx, S. Beddar, L. Beaulieu, Med. Phys. **40**, 062101 (2013)
F. Therriault-Proulx, L. Wootton, S. Beddar, Phys. Med. Biol. **60**, 7927 (2015)
I. Toccafondo, *Distributed optical fiber radiation and temperature sensing at high energy accelerators and experiments*, Doctoral dissertation, CERN (2015)
I. Toccafondo et al., IEEE Photon. Tech. L. **27**, 2182 (2015)
A.L. Tomashuk et al., J. Lightwave Technol. **32**, 213 (2013)
A.L. Tomashuk et al., Opt. Express **22**, 16778 (2014)
S.A. Vasiliev et al., IEEE T. Nucl. Sci. **45**, 1580 (1998)
X. Wang et al., Chin. Opt. Lett. **9**, 060601 (2011)
R.H. West et al., J. Lightwave Technol. **12**, 614 (1994)
L. Wootton et al., Phys. Med. Biol. **59**, 647 (2014)
A. Wosniok et al., in *Proc. Sixth European Workshop on Optical Fibre Sensors (EWOFS'2016)*, ed. by E. Lewis (SPIE, Bellingham, 2016) p. 99162
W.J. Yoo et al., in *PIERS Proceedings*, August 12–15, 2013, Session 2PK, 862 (The Electromagnetics Academy, Cambridge, 2013a)
W.J. Yoo et al., Opt. Express **21**, 27770 (2013b)
W.J. Yoo et al., Appl. Radiat. Isot. **81**, 196 (2013c)
E.G. Yukihara, S.W.S. McKeever, Phys. Med. Biol. **53**, R351 (2008)
E.G. Yukihara et al., Radiat. Meas. **41**, 1126 (2006)
E.G. Yukihara, S.W.S. McKeever, M.S. Akselrod, Radiat. Meas. **71**(15) (2014)
R. Zhang et al., Phys. Med. Biol. **58**, 5477 (2013)
Y.A. Zhydachevskii et al., Radiat. Meas. **94**, 18 (2016)

Part XI
Polymer Optical Fiber Sensing

Polymer Optical Fiber Sensors and Devices

50

Ricardo Oliveira, Filipa Sequeira, Lúcia Bilro, and
Rogério Nogueira

Contents

Abstract

This chapter will present a review of polymer optical fiber-based sensors and techniques. The main characteristics of optical fibers will be briefly summarized, with special focus on POFs. Since POF end face termination plays an important role in the coupling efficiency, the recent technologies used to prepare the POF end face will be described. Finally, the most recent advances in sensing technologies employing POFs will be reviewed.

R. Oliveira (✉) · F. Sequeira · L. Bilro · R. Nogueira
Instituto de Telecomunicações, Campus Universitário de Santiago, Aveiro, Portugal
e-mail: oliveiraricas@av.it.pt; fsequeira@av.it.pt; lucia.bilro@av.it.pt; rnogueira@av.it.pt

G.-D. Peng (ed.), *Handbook of Optical Fibers*,
https://doi.org/10.1007/978-981-10-7087-7_1

Keywords

Optical fiber sensor · Polymer optical fiber (POF) · Microstructured polymer optical fiber (mPOF) · Polishing · Splicing · Intensity-based sensors · Wavelength-based sensors · Polarimetric-based sensors · Interferometric-based sensors · Multimode interference (MMI) · Polymer optical fiber Bragg grating (PFBG) · Long-period grating (LPG) · Tilted fiber Bragg grating (TFBG)

Introduction

Fiber optic sensors offer key advantages over other sensing technologies, which include immunity to electromagnetic interference, compact, lightweight, multiplexing capability, and higher sensitivity, among others. Polymer optical fibers (POFs) have been regarded as a viable alternative to silica fibers in a variety of sensing applications. The reason is related with the special physical and chemical properties of polymers when compared with silica, leading to have additional advantages. Examples of such are the much higher elongation, the low Young's modulus, and the higher thermo-optic coefficient. Additionally, polymers have biological compatibility, and they do not produce sharp edges when broken as it occurs with silica fibers. The polymer material to be used can also be tailored to fit the desired application, for example, if it needs to be humidity-sensitive or humidity-insensitive or if it needs to operate at high temperatures (i.e. near or above 100°C).

Polymer optical fibers may be found in two main groups, the ones based on large core, typically with diameters of hundreds of microns, which comprises multimode (MM) fibers, and the ones with small core, typically required for single-mode or few-mode operation, with few microns fibers diameters. While the use of large-core fibers benefits from applications requiring low-precision and low-cost components, the same does not apply for small-core POFs. In such fibers, the requirement of sources with a single transversal mode and the inherent precision needed for the fiber splicing may compromise the application. However, the inherent advantage of the use of special sensing techniques such as fiber Bragg gratings (FBGs) (that require SM behavior) has pushed these type of fibers for a more intense use in the recent years.

Single-Mode and Multimode Behavior in Polymer Optical Fibers

Nowadays, it is possible to find a variety of parameters to describe a specific optical fiber. However, some parameters may be more relevant than others depending on the application. The refractive index difference between core and cladding is one of those parameters that has a huge impact on the fiber optical properties. The most obvious is the maximum angle in which light can enter into an optical fiber. One advantage obtained when working with POFs is the high difference that can be obtained between the refractive index of the core and cladding. This happens because polymers are available in a wide range of refractive indices which vary from 1.32 for highly fluorinated acrylic-based materials to around 1.6 for some phenolic resins (Emslie 1988). For that reason, POFs have been reported with high numerical apertures, ranging from 0.2 up to 0.7 (Zubia and Arrue 2001). Indeed, this is a great advantage presented by POFs since the light coupling can be easily achieved. Nevertheless, as it is easy to produce POFs with large *NA*, it is also difficult to create fibers with SM profiles. The number of modes found in an optical fiber is related with the normalized frequency that is proportional to the numerical aperture and core radius and inversely proportional to the radiation wavelength. In order to occur SM operation, the normalized frequency needs to be lower than 2.405 (Ziemann et al. 2008). Considering a fiber with core radius equal to 4.2 μm and with a core and cladding refractive index of 1.4516 and 1.4473 at $\lambda = 1.3$ μm, respectively, one can estimate the wavelength at which the fiber will no longer support SM behavior. This is the cutoff wavelength (λ_c), and for the parameters defined before, the fiber will only support one mode when $\lambda_c > 1.23$ μm. In other words, for the occurrence of SM operation in a specific wavelength range, the core radius or refractive index contrast needs to be balanced. Regarding that, a larger core implies lower refractive index contrast and vice versa. One possible way to achieve low refractive index contrast is through doping; however, the dopant diffusion needs to be carefully controlled in order to have a stable refractive index. On the other hand, fibers with larger refractive index contrast may be developed through the use of two different materials. By doing that, the core radius needs to be smaller, leading to an increase of the scattering losses at the core-cladding interface (Large et al. 2008). This takes much more relevance when working at lower wavelengths since it implies smaller cores. One particular case may be found in POFs, since most of them have their transparent window localized at the visible region. For that reason, POFs with step-index (SI) profile have only been reported to be SM at the near-infrared region (Kuzyk et al. 1991).

The first SM-POF was demonstrated by Kuzyk and co-workers in the early 1990s (Kuzyk et al. 1991). The fiber was composed of 8 μm core and 125 μm cladding, operating at 1300 nm. The commercialization of the first SM-POFs was made by Paradigm Optics, Inc., being the fibers composed of a PMMA cladding and cores copolymerized with PMMA and polystyrene (PS) (<3%). The fibers were sold with trade names of SM-MORPOF02 and SM-MORPOF03 with λ_c around 750 nm and 1100 nm, respectively. Other SI-POFs based on other polymer materials have also been fabricated and reported in literature, to offer SM behavior, with low loss at the near-infrared region. One of those works may be found in Zhou et al. (2010), for a fiber composed of a perfluorinated graded index core and PMMA overcladding. The other work can be found in Woyessa et al. (2016), where the fiber is composed of TOPAS® grade 5013S-04 core and ZEONEX® 480R cladding. Alternatively, a fiber made of PMMA with a core composed of an array of tight and thin ZEONEX® 480R capillarity's was demonstrated to offer SM behavior in the visible region (Leon-Saval et al. 2012).

Microstructured Polymer Optical Fibers

The fibers presented before were composed of core and cladding, where the guiding mechanisms are normally based on total internal reflection. However, there exists another type of optical fiber, where the light guidance is controlled by air holes arranged in a special configuration in the cross section area of an optical fiber (see the example shown in Fig. 2). The use of these structures out of the central area allows the creation of a depressed index compared with the solid inner part of the fiber. Thus, the condition for total internal reflection or equivalently mode confinement is satisfied (Large et al. 2008). This approach was firstly explored to avoid chemical doping in fibers (Kaiser et al. 1973). After two decades, Philip Russell and his colleagues presented the microstructured silica fiber Knight et al. (1996). This type of fiber, conventionally known as "holey" fiber or photonic crystal fiber (PCF), has found many applications in different areas. In fact, the hole arrangement together with the hole dimension (d) can be designed to offer special characteristics, namely, the ability to remain single mode at all wavelengths for which fused silica is transparent having an endlessly single-mode operation. Additionally, the mode field can be concentrated in a small area, allowing to have high intensity inside the fiber that boosts highly nonlinear effects for supercontinuum generation (Dudley et al. 2006). Furthermore, if asymmetric structures are used, the fibers can offer high birefringence, maintaining the polarization state of the light traveling into the fiber (Ortigosa-Blanch et al. 2000). Another advantage of microstructured fibers is the possibility to adjust the dispersion behavior by controlling the air filing fraction (d/Λ), where Λ defines the distance between adjacent holes (Mortensen 2005). The potential of these fibers is still ongoing (Wong et al. 2012), and a prosperous future is thus expected.

Microstructured fibers have been made of silica, where the preform is essentially processed through the stack-and-draw-technology. Contrary to silica, polymers have low processing temperatures allowing the possibility to be extruded, casted, or easily drilled. In 2001, the first microstructured fiber in a POF (mPOF) was developed (van Eijkelenborg et al. 2001). In fact, almost any kind of hole pattern can be employed in a microstructured fiber, enabling the development of specific type of fibers for each application (Argyros et al. 2001). Until now, since the development of the first mPOF, many works have reported the use of different hole arrangements leading to the development of fibers with different characteristics, such as graded index, high birefringence, dual core, suspended core, and hollow core (van Eijkelenborg et al. 2003). Additionally, as it occurs with SI-POFs and GI-POFs, mPOFs can also be doped (Large et al. 2004). The capability of using different materials is also a reality, allowing the possibility to explore different characteristics presented by polymer materials, such as the ability to be humidity-sensitive (Zhang and Webb 2014) or humidity-insensitive (Woyessa et al. 2016) or the capability to resist to temperatures above 100 °C (Markos et al. 2013; Woyessa et al. 2016).

As described previously, SM operation in SI-POFs has only been reported at the infrared window. Contrary to SI-POFs, the hole dimensions and arrangement in

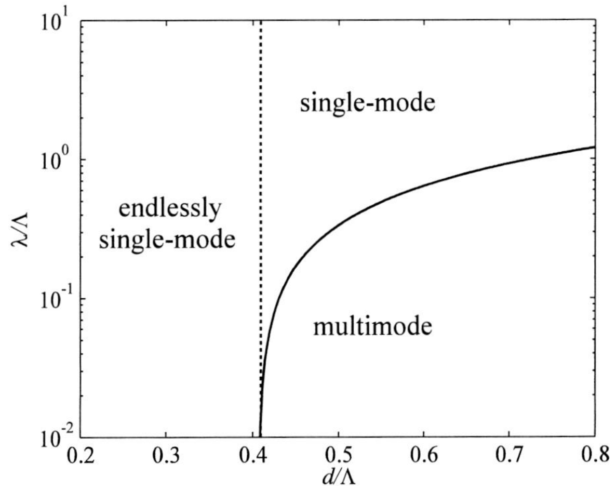

Fig. 1 Phase diagram for the single-mode-multimode operation, described in Eq. 1

mPOFs are easy to manipulate, and thus the capability to have SM operation in the visible region is achievable (van Eijkelenborg et al. 2001).

Regarding the cutoff in microstructured fibers, Mortensen (2002) has suggested a phase diagram for the single-mode-multimode operation regime (see Fig. 1), where the details of such boundary were later analyzed with his colleagues (Kuhlmey et al. 2002). The best fit equation describing such boundary transition is described as Kuhlmey et al. (2002):

$$\frac{\lambda^*}{\Lambda} \approx \alpha\left(\frac{d}{\Lambda} - \frac{d^*}{\Lambda}\right)^{\gamma} \tag{1}$$

where λ^* is the second-order cutoff, $\alpha = 2.80 \pm 0.12$ and $\gamma = 0.89 \pm 0.02$ are the fitting coefficients, and $d^*/\Lambda = 0.406$ is the air filling fraction (Kuhlmey et al. 2002; Mortensen et al. 2003).

For $d/\Lambda < d^*/\Lambda$, the microstructured fiber has the remarkable property of being so-called endlessly single mode (Birks et al. 1997). For $d/\Lambda > d^*/\Lambda$, the fiber supports a second-order mode at wavelengths $\lambda/\Lambda < \lambda^*/\Lambda$, and it is single mode for $\lambda/\Lambda > \lambda^*/\Lambda$ (Mortensen et al. 2003).

The normalized frequency in microstructured fibers (V_{PCF}) is due to the length scale of the problem, related to Λ, contrary to the core radius as it occurs for SI fibers. V_{PCF} is then defined as (Birks et al. 1997; Mortensen et al. 2003):

$$V_{\text{PCF}} = \frac{2\pi\Lambda}{\lambda}\sqrt{\left(n_{\text{co}}^2 - n_{\text{cl}}^2\right)} = \frac{2\pi\Lambda}{\lambda}\text{NA} \tag{2}$$

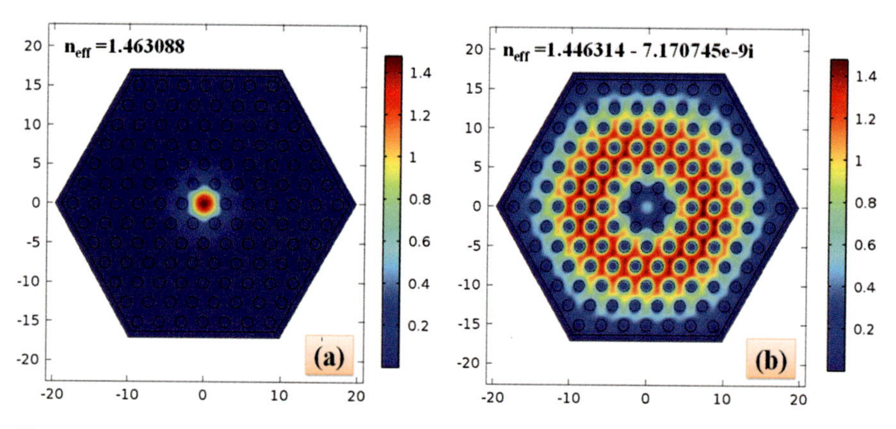

Fig. 2 Modes propagating in a PMMA-based SM-mPOF at 1550 nm region, with six hole layers, $d = 1.4\ \mu\text{m}$ and $\Lambda = 2.9\ \mu\text{m}$. (**a**) Fundamental mode (core region); (**b**) first mode propagating in the air-hole lattice region

where n_{co} and n_{cl} are the effective refractive index of the core mode and the first mode propagating into the microstructured region, respectively (see the example shown in Fig. 2).

Nevertheless, Mortensen et al. have shown that microstructured fibers only support SM behavior when $V_{PCF}\ (\lambda^*) < \pi$, contrary to the value found for the normalized frequency in SI fibers.

Despite the high characteristics offered by mPOFs, there are still difficulties when working with these fibers. Examples of that are the cleaving and splicing processes needed to have the fiber ready to work. For the cleaving, many problems have been reported such as crack formation and rough surface, which inevitably lead to high coupling losses. For the splicing methods, the conventional hot melting process reported for silica fibers cannot be employed due to the glass transition temperature of POFs. Since this topic is of high relevance, the most recent techniques used for the POF end face preparation will be described in this chapter.

Polymer Optical Fiber End Face Termination

Most of the fiber optical equipment such as sources and detectors are devoted for silica fibers. Therefore, when working with POFs, it is sometimes necessary to create a splice with a silica fiber. Splicing these two different fibers can be challenging especially when working with small diameter POFs. In fact, the splice needs to guarantee the efficiency of the alignment with the best mechanical stability and the lowest possible insertion loss.

It is well known that the POF end face quality is one of the main challenges when the splicing process is needed, especially in mPOFs. Thus, the best POF end face must be achieved, in order to avoid too much insertion loss and back reflections (i.e., return loss). In literature, several techniques have been reported, namely:

- Semiconductor dicing (SD) saw (Atakaramians et al. 2009)
- Focused-ion-beam milling (Atakaramians et al. 2009)
- Ultraviolet laser cleaving (Canning et al. 2002)
- Hot blade cleaving (Abdi et al. 2009; Law et al. 2006; Stefani et al. 2012a)
- Connectorization process (Abang et al. 2013; Abang and Webb 2012)
- Liquid nitrogen cleaving method (Ghirghi et al. 2014)

The techniques that use the SD saw, the FIB milling, the ultraviolet laser cleaving, and the liquid nitrogen cleaving method have shown to be viable processes in laboratory but hard to be implemented in operational conditions. Additionally, those works only reported the use of fibers with diameters >400 μm, which is high because several POFs have been described to have thinner diameters (>100 μm).

Hot Blade Cleaving Method

The most popular technique used to prepare the end face of a POF is the hot blade cleaving method. This technique was based on commercial available tools used to cleave large diameter POFs. Examples of those devices are the POF cutter block and the hot knife. The first is simply composed of a block drilled with several diameter holes where the fiber is inserted and later cleaved by pushing a blade against the POF transversal section. In the hot knife device, a heating system is fixed to a blade and used to soften the POF while it is being cleaved. These tools, however, are essentially suited to cleave with reasonable end face quality solid POFs with diameters higher than 1 mm. Although the concept of quality in large-core fibers is not the same when compared with small-core POFs (with cores that can be as small as 4 μm in some SM-POFs). For this type of fibers, it is necessary to use the sharpest possible blades and also to have control on different parameters, such as the temperature of the blade and plate and the velocity of the blade. Another key aspect to take into account is that POFs can exhibit different mechanical properties depending on the fiber structure, materials, drawing conditions, and also the thermal history. Thus, the effect of the blade on the POF is extremely dependent on these properties. Consequently, depending if the POF material is brittle or ductile, two different scenarios may result from the cleaving process at room temperature (Law et al. 2006). The main difference between them relates to the crack formation through the transversal axis of the fiber. Regarding a POF where the material exhibits brittle characteristics, one may find that the density of stresses at the contact between the edge of the blade and the POF material is much higher than the one found for a POF with ductile characteristics. Therefore, as the blade penetrates transversely into the POF, a crack formation will always occur for a POF with brittle material. However, for a ductile material, the density of stress at the edge of the blade is not sufficient to allow the crack formation, and thus the blade will be always in contact with the POF material (Law et al. 2006). One way to deal with the material properties of the fibers is through temperature. The phase transition from brittle to ductile can be achieved at a specific temperature, depending on the polymer material present on the POF and also on the drawing and thermal history of the fiber. Nevertheless, the velocity in which the blade is pushed down against the fiber needs also to be controlled, and for that, an optimization process regarding all the variables needs to be found. The demonstration of such aspects have already been reported in literature for POFs with different structures and materials (Abdi et al. 2009; Law et al. 2006; Stefani et al. 2012a), using homemade automated hot blade cleaving machines. The good results presented by this technique led several authors to develop this POF cleaving machine. An example of such device can be seen in Fig. 3.

It is worth to mention that this process needs to fine control these different parameters; otherwise, problems may occur such as core shift, surface roughness, and crack formation, among others. Examples of those can be seen in Fig. 4.

The proper cleaving parameters needed for a specific POF could only be accessed empirically. Therefore, it can be considered a time consumption process. However, it is still possible to achieve acceptable results. Examples of those can be seen on Fig. 5 for two mPOFs.

Saez-Rodriguez et al. (2015) demonstrated that a time-temperature equivalence principle can be used to cleave a POF at room temperature by applying a sawing motion instead of chopping as performed by the hot blade cleaving methods (Abdi et al. 2009; Law et al. 2006; Stefani et al. 2012a). The authors concluded that there are four key points that must be addressed to obtain a proper cleave, which are

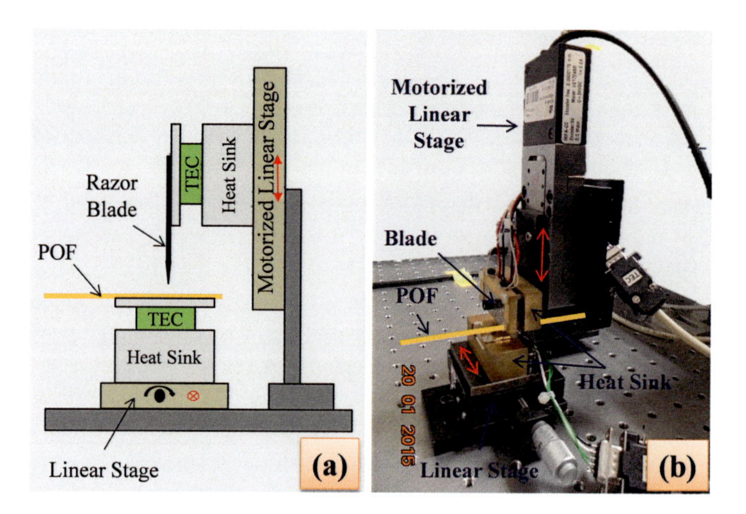

Fig. 3 Schematic (**a**) and photograph (**b**) of the homemade automated hot blade cleaving machine used to cleave POFs, based on two linear motors, two thermoelectric coolers (*TEC*), a brass plate, and a sharp blade

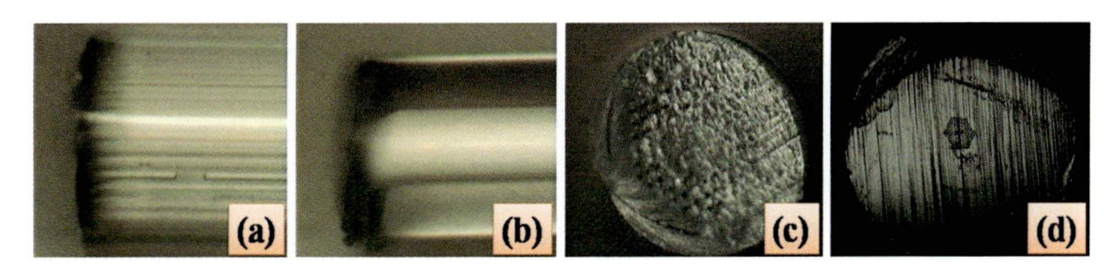

Fig. 4 (**a**, **b**) Transversal and (**c**, **d**) top end face microscope images of two microstructured POFs, MM-150 (**a, c**) and HiBi-mPOF (**b**, **d**), through an automated hot blade cleaving method and without proper parameters. The different end face defects are (**a**) burning, (**b**) core shift, (**c**) and (**d**) rough surface, with crack formation at the top of the mPOF shown in (**d**)

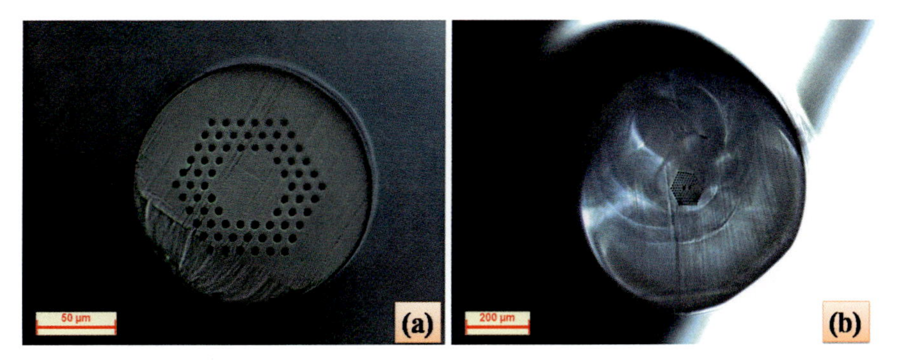

Fig. 5 POF end face images of a MM-mPOF (**a**) and an HiBi mPOF (**b**), prepared by the hot blade cleaving process using a blade velocity of 2 mm/s, blade temperature of 65 °C and 70 °C and plate temperature of 75 °C and 80 °C, respectively, for the MM-mPOF and the HiBi-mPOF (The fibers have been acquired from different suppliers)

sawing time, polymer stiffness, polymer toughness, and the sawing contribution. Despite the good results, the work was only focused on one POF sample.

Connectorization Process

Another popular POF end face termination method is the ferrule connector process (Abang et al. 2013; Abang and Webb 2012). This process is known to provide high-quality POF end face, and it is based on the already developed silica fiber connectorization technology. In this method, ferrules of physical contact connectors (FC-PC) with a bore ferrule diameter, matching the fiber diameter, are filled with a connector epoxy. The fiber is later inserted in the bore ferrule, and the epoxy is cured at room temperature. The fiber tip is then cleaved (diamond blade in case of silica fibers or sharp razor blade in case of POFs) and then polished. For that, different polishing films are used in descending order of grain size.

The POF fabrication process is still underdeveloped, and the control of the POF diameter is still challenging (Barton et al. 2004). In some types of fibers, these

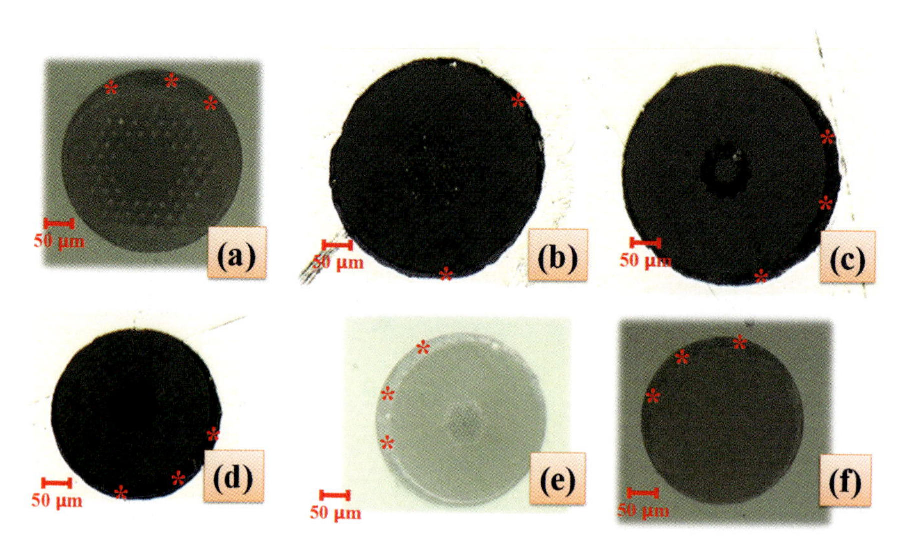

Fig. 6 Microscope images of the end face of the terminated POFs in FC/PC connectors. The symbol "*" marks the regions where there is a gap between the bore ferrule and the POFs. The mPOFs ((**a**) MM-150, (**b**) FM-250, (**c**) G3-250, (**d**) FM w/Rh6G, (**e**) SM-125) are from Kiriama Pty, Ltd., and the SI-POF ((f) SM-MORPOF03) is from Paradigm Optics, Inc.

diameters may fluctuate along the fiber length in about 10 μm or even more. Therefore, considering the connectorization process, whenever the POF does not match the diameter of the bore ferrule, it is necessary to perform an etching of the fiber in order to match the dimensions of the bore ferrule to be used (Abang et al. 2013). However, the etching process is difficult to control, and the core concentricity of the POF inside the bore ferrule is compromised in most of the times. This inevitably leads to poor coupling efficiency. Additionally, if a bore ferrule larger than the POF is used, the same could be observed (Lwin and Argyros 2009). Examples of the previously described problems may be seen for etched fibers in Fig. 6a, c, d and for fibers smaller than the bore ferrules in Fig. 6b, e, f. In all the cases, the core of the POFs is always out of the center.

Additionally to the concentricity of the POF inside the bore ferrule, one needs also to consider the concentricity of the core along the fiber length, as a result of the manufacturing process. This problem is not relevant for multimode POFs, but for SM-POFs, the same does not apply (Abang et al. 2013). Another disadvantage is that the polishing process is totally handmade which introduces manual errors to the process.

Fiber Polishing Process

The main disadvantages associated to the hot blade cleaving technique and the connectorization process, i.e., time consumption process for the first and core misalignment for the latter, can be minimized with a dual-step procedure based on the above two methods (Oliveira et al. 2015b). Initially POFs are cleaved by hand using a sharp razor blade previously heated to a temperature below the melting point of the fiber materials (around 70 °C or 80 °C). Thus, the creation of cracks in the transversal and longitudinal directions is avoided allowing the formation of a flat surface. Then, fibers are subjected to a polishing procedure using an automated fiber polishing machine. The cleaved POFs are then inserted in a bore ferrule and pulled down against the polishing machine film. The polishing process takes two 15 s cycles. The grain sizes of the polishing films are chosen to first roughly scratch the POF end face, removing imperfections from the hot blade process, and the second one to give a smooth termination. To avoid small dust particles at the end face of the fiber, specifically for the mPOFs, water spray is spread onto the second polishing film prior to the polishing step. An example of the results of this procedure may be seen in Fig. 7 (Oliveira et al. 2015b), for mPOFs (a–f), SI-POF(h), and GI-POF (i).

One advantage of this technique is that the core POFs are centered, which avoids constrains related with the core shift (see Fig. 4b). This reveals that the first polishing procedure removes a great amount of the POF terminal and inherently the defects associated with the initial hot blade cleaving process.

Sensing Techniques

Optical fiber sensors, in a simplistic way, are based on the modification of one or more properties of the light propagating through a fiber, or collected by the fiber, when the parameter to be measured changes. An interrogation scheme is then used to evaluate these changes in the optical signal. In this way, depending on the light property that is modified, optical fiber sensors can be divided into four main categories: intensity, phase, wavelength, and polarimetric-based sensors.

Intensity-Based Polymer Optical Fiber Sensors

Intensity-based sensors are one of the first optical fiber sensing detection schemes and are considered the most simple and cost-effective method. The key concept is based on the loss of light that is coupled into the optical fiber. From this point of view, the use of MM fibers with large-core diameters benefits the application since it enables an easy handling and installation and the use of low-cost connectors and components, which are less strict to geometrical tolerances.

Basically, the experimental setup for intensity-based detection schemes includes a light source (i.e., LED, OLED, halogen lamp, laser), the optical fiber, a light detector (photodetector or a spectrum analyzer, like a spectrometer, oscilloscope, or optical spectrum analyzer (OSA)), and a data acquisition device/software (see Fig. 8). LEDs and photodetectors are interesting solutions for low-cost and miniaturized intensity-based schemes. Yeh et al. reported a review aiming the evaluation of LED-based devices (range between UV and IR, 247–3800 nm) in chemical sensing

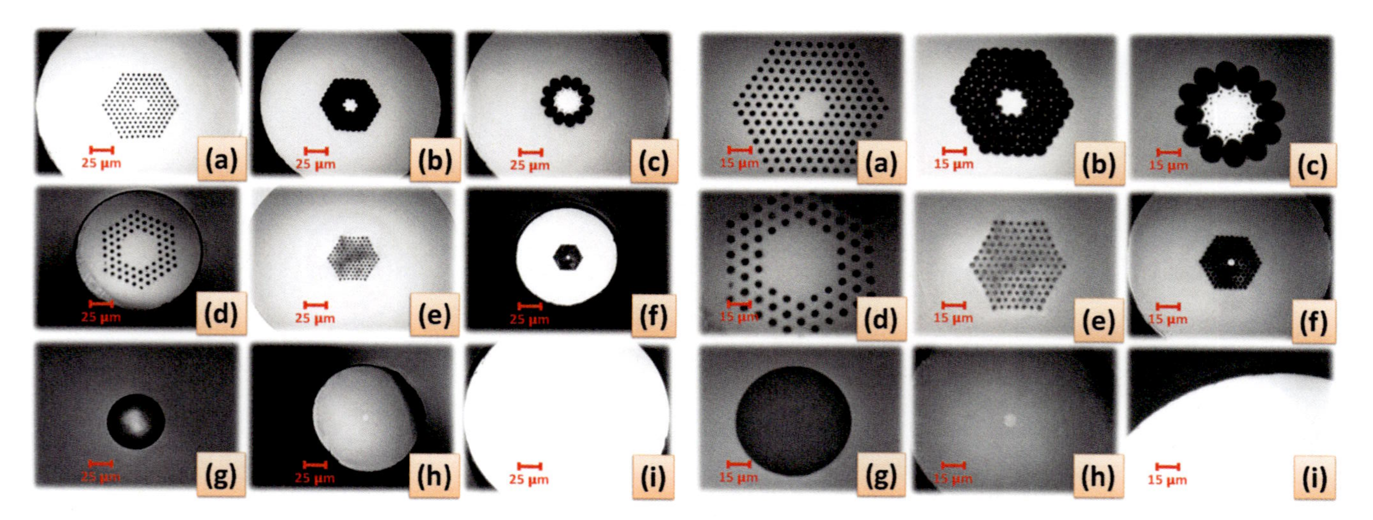

Fig. 7 Optical microscope images of several POFs with different structures and sizes, for two different magnification lens (M) (M = 10X for the left images and M = 20X for the right images). The letters are referred to (**a**) mPOF FM-250, (**b**) mPOF FM with Rh6G, (**c**) mPOF G3-250, (**d**) mPOF MM-150, (**e**) mPOF SM-340, (**f**) mPOF SM-125, (**g**) GI-POF (the GI-POF end face images only cover the fiber core), (**h**) SM-MORPOF03, and (**i**) MM-MORPOF01 (the lower MM-MORPOF01 image only shows a small portion of the fiber (core-cladding-air)) (Oliveira et al. 2015b)

Fig. 8 Schematic representation of an intensity-based detection scheme in optical fiber sensing

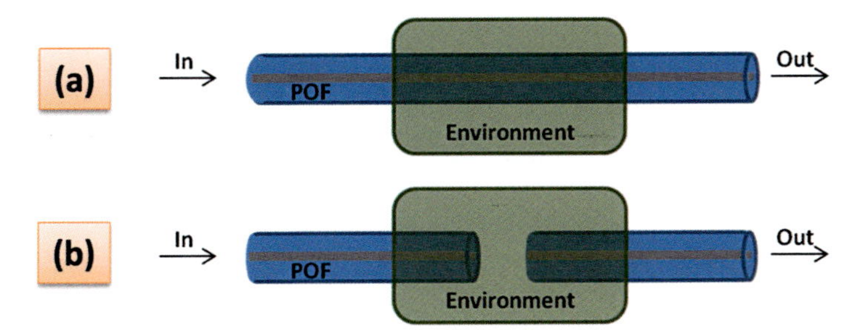

Fig. 9 Schematic representation of (**a**) intrinsic and (**b**) extrinsic POF sensors

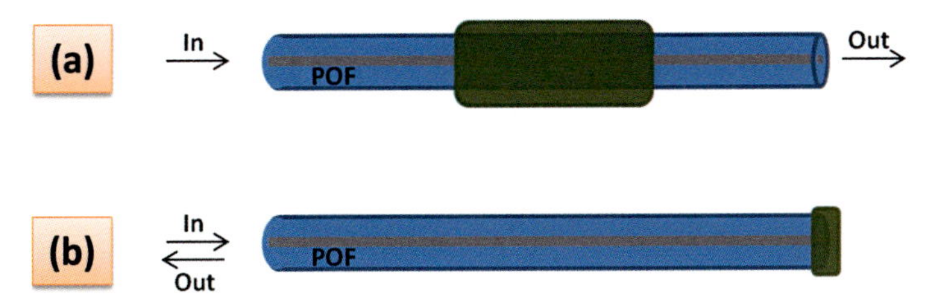

Fig. 10 Basic schematic representation of intrinsic (**a**) transmission and (**b**) reflection POF sensors

applications, showing their higher importance as light sources and that efforts are being made in order to increase the sensitivity of LED-based systems (Yeh et al. 2016).

Intensity-based optical fiber sensors (OFSs) can be classified as intrinsic and extrinsic, depending if the light propagates in the fiber and interacts with analyzed medium (see Fig. 9a) or if the fiber is used only as a waveguide and the light interacts directly with the sensing environment (see Fig. 9b), respectively. Figure 9 depicts the schematic representation of intrinsic and extrinsic OFSs, where "In" represents the input of light in the fiber and "Out" the output of the transmitted light.

In the case of extrinsic sensors, the light leaks out from the fiber in order to interact with the environment and reenters in the fiber (the same or in a different fiber). In the case of intrinsic sensors, the optical fiber is modified to work as a sensor, to detect changes in the environment. The light propagates in the fiber and interacts with sensitive/selective layers or external medium, after which can be transmitted to the other side (see Fig. 10a) or reflected back (see Fig. 10b). In transmission detection schemes, the light source and detector are placed on opposite sides of the fiber while in reflection detection schemes are placed on the same side.

In extrinsic configurations, as depicted in Fig. 9b, two longitudinally aligned fibers can be used as sensing elements due to the intensity modulation of the optical signal with the distance between the fibers or related with changes in the external medium (i.e., concentration, refractive index, intrinsic property of the analyte like absorption at a specific wavelength or fluorescence, etc.). When the distance between two fibers (L) is changed, the intensity of the signal that reaches the second fiber (I) will also change (being the ratio between the optical power (P) and the area (A), $P = I \times A$, valid if no light is absorbed or scattered by the medium), see Fig. 11. This basic principle can be used to detect the increase of the thickness of cracks, for example, or even small changes in the position by the misalignment of the fibers, which is largely used in the field of structural monitoring.

In the case where a medium exists between the fibers, there will be absorbance due to the interaction of light with the particles that constitute the external medium, and the concentration can be monitored through the detection of transmitted light variation, see Fig. 12. This absorbance (A) is related with the concentration (c) by the Beer-Lambert law:

$$-\log\left(\frac{P}{P_0}\right) = A = \varepsilon.L.c \tag{3}$$

where P_0 is the input power, P is the output power, ε is the absorptivity, L is the path length, and c is the analyte concentration. The Beer-Lambert law is not valid at high concentrations (>0.01 M) due to electrostatic interactions, if there are changes in refractive index at high analyte concentration, fluorescence or phosphorescence of the sample, and scattering of light due to particulates in the sample.

In the case that there is a refractive index variation in the external medium between the fibers, the irradiance will depend on the external refractive index (see Figure 13), since the area of the light cone will decrease with increasing refractive index leading to an increase of the detected transmitted light.

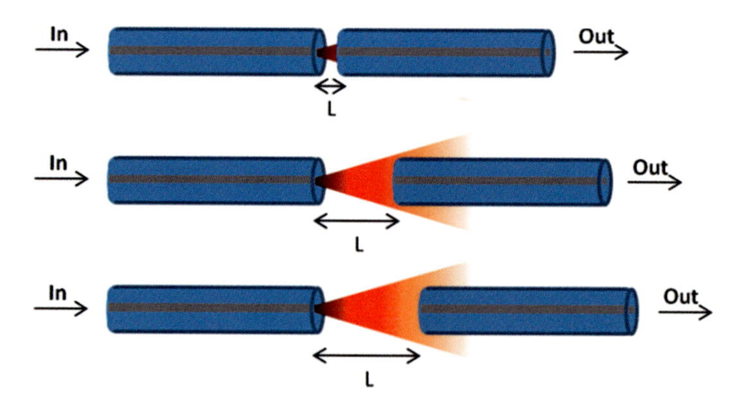

Fig. 11 Schematic representation of extrinsic transmission sensing configuration, with increasing distance (L) between the input and output fibers

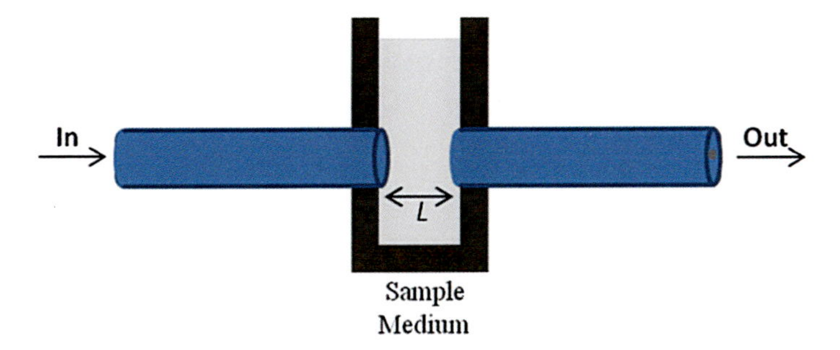

Fig. 12 Schematic representation of extrinsic transmission sensing configuration, with increasing concentration (c) in the external medium between the input and output fibers

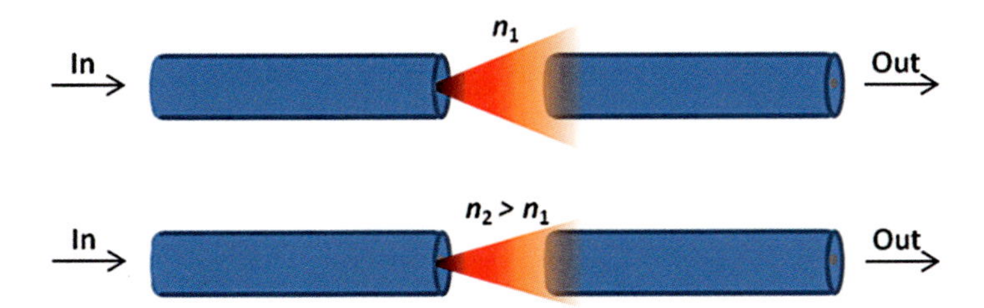

Fig. 13 Schematic representation of extrinsic transmission sensing configuration, with increasing refractive index (n) in the external medium between the input and output fibers

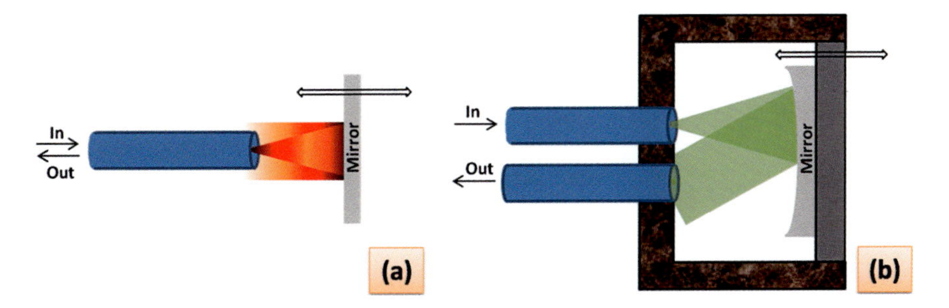

Fig. 14 Schematic representations of extrinsic reflection sensing configurations, with mirror displacement possibility

The above mentioned examples are also valid for reflection-based configurations. In this case, one single fiber can work as transmitter and receptor, or there can be two independent fibers (one working as the transmitter and the other as a receiver). In Fig. 14 schematic examples of two extrinsic reflection-based sensors are depicted with one (Fig. 14a) and two fibers (Fig. 14b). In these configurations, any physical parameter that can be responsible for the mirror displacement will cause a variation in the path length and, consequently, a variation in the irradiance reaching the output fiber.

These configurations are also valid for a constant distance between transmitter and target/receptor but with external medium variations, i.e., concentration or refractive index. Therefore, reflection-based configurations can be used to monitor absorption, fluorescence, light scattering, and refractive index.

Transmission intrinsic sensors can further be based on the interaction with the evanescent wave in order to monitor changes occurring at the surface of the fiber. Evanescent field consists in the penetration of the radiation into the fiber cladding and which energy decays exponentially with the distance from the core-cladding interface (penetration depth around hundreds of nanometers, $\sim \lambda$), see Fig. 15.

In a standard optical fiber, the interaction of the evanescent field with the external medium is negligible. In order to increase this interaction and monitor the changes occurring at the surface of the fiber, POFs can be easily manipulated by a wide range of simple physical and chemical techniques, depending on the desired sensing properties – tapering, etching, and polishing. POFs can be tapered by heating and stretching (see Fig. 16b), decreasing the thickness of the core and cladding in a specific region. Another simple procedure that can be performed is the removal of the fiber cladding, partially or totally (exposing the core). The cladding can be removed along the thickness of the fiber by applying mechanical polishing or chemical etching (see Fig. 16c). Side polishing allows removing the cladding in one side of the fiber, and a D-shape can be obtained (see Fig. 16d).

Depending on the aim of the application, one can chose to remove only the cladding (partial or totally) or also part of the core. In a fiber with an exposed core, the external medium acts as a substitute cladding, and the light that propagates in the fiber will interact with this external medium, and changes can be monitored.

In order to develop a POF chemical or biochemical sensor, besides the importance of sensitivity and resolution in the performance of the sensor, selectivity to the analyte of interest (target to be measured) is essential in order to guarantee its viability in real samples and for real applications. In the case that the analyte has an intrinsic optical property that can be used for sensing purposes, direct (label-free) measurement is performed (i.e., absorption at a specific wavelength, formation or consumption of chemical species by specific chemical reaction, etc.). If direct measurement is not possible, optically detectable species must be formed by interaction with adequate indicator reagent(s) or through the interaction with sensitive/selective layers.

The deposition of this layers in the POF's sensing region, often referred to as modified cladding, should allow the interaction of the analyte with the layer. This will cause a change in the optical properties which can be detected/measured by the variation of the intensity of the transmitted light. The characteristics of this layer,

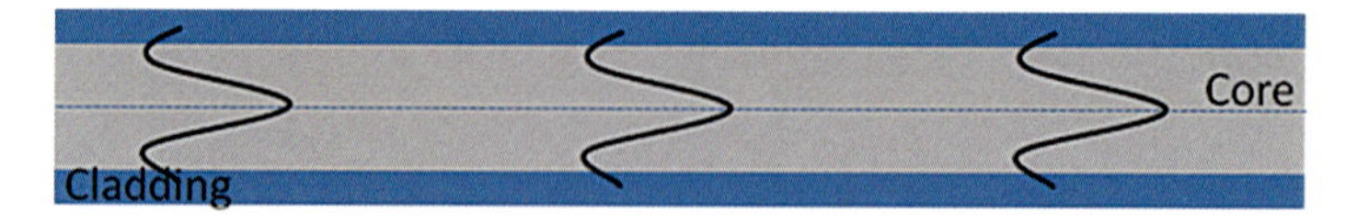

Fig. 15 Schematic representation of the evanescent wave in a standard optical fiber

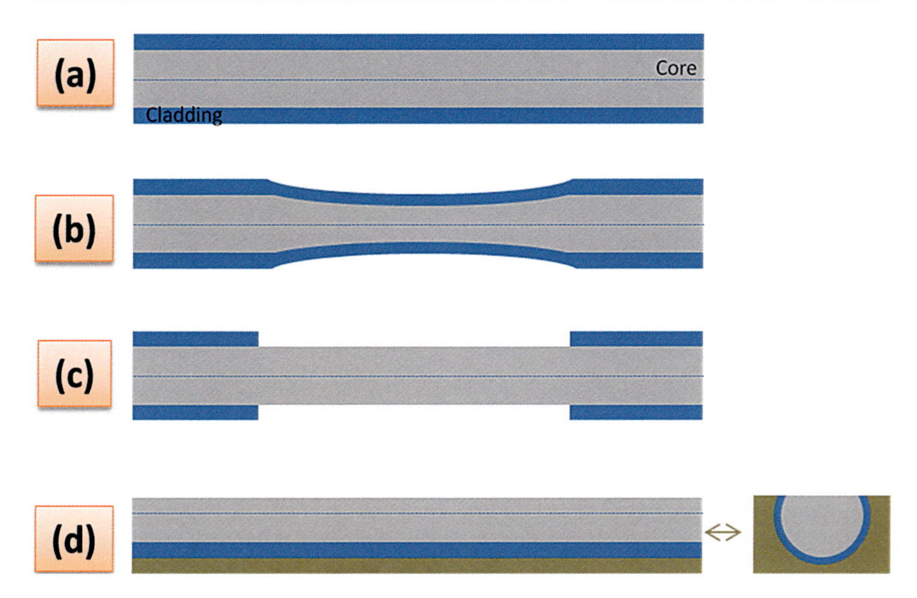

Fig. 16 Schematic representation of (**a**) standard, (**b**) tapered, (**c**) etched/polished, and (**d**) side-polished optical fiber

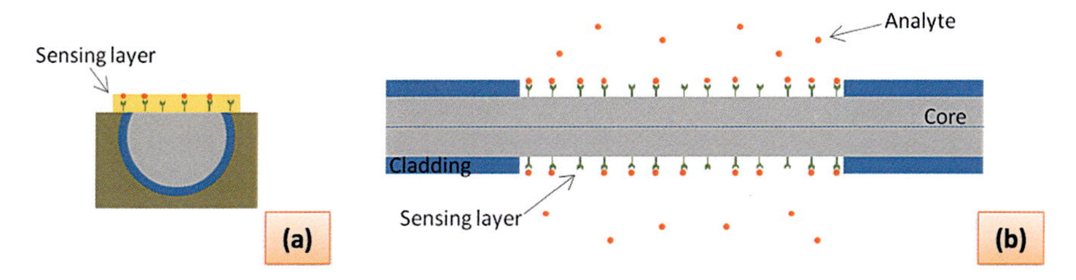

Fig. 17 Schematic representation of a selective sensing layer deposited on (**a**) D-shaped POF (side-polished) and (**b**) etched POF

namely, thickness, are highly important and must be optimized depending on the sensing properties that are aimed. Figure 17 depicts a schematic representation of a selective sensing layer deposited in a D-shaped POF (side-polished) and in an etched POF.

Sensitive layers with specific binding sites can be based on biological receptors (proteins, DNA, antibody-antigen interaction, cell receptors, etc.) or biomimetic detection (i.e., aptamers and molecularly imprinted polymers (MIPs)). MIPs have been used extensively as selective layers for optical fiber sensing as well as MIP nanoparticles. These synthetic polymers have highly selective recognition sites and the binding of the analyte with the MIP causes a variation in the polymer matrix that leads to the variation of the transmitted intensity allowing its optical detection. One interesting characteristic of MIPs is that they can be designed for the analyte of interest, although the process of optimization can be very time-consuming.

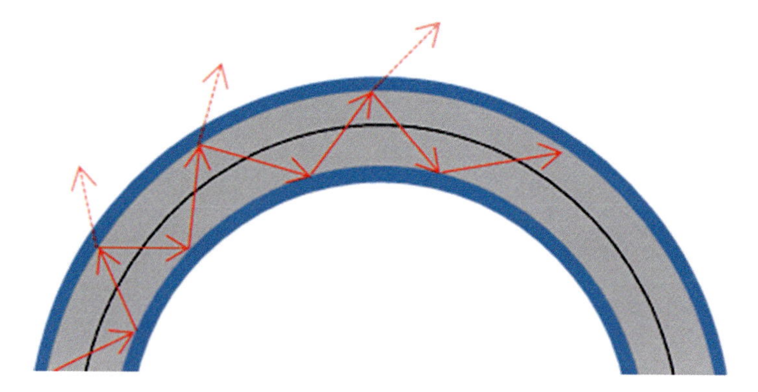

Fig. 18 Schematic representation of a POF with macrobending configuration

Another key factor used for sensing is the optical intensity variation due to macrobending or points of strain and pressure. This principle is also used in smart textiles, when POFs are introduced in textiles for sensing of several physical and chemical parameters. In terms of macrobending, it is easy to understand that when a fiber goes from a straight configuration to a curved one, there will be more radiation that leaks out from the fiber, see Fig. 18. There are several kinds of POF sensors that make use of macrobending in order to increase the sensitivity of the sensor to the variations in the external medium or sensitive deposited layers.

Fluctuation in the optical power signal can be a problem when dealing with intensity-based sensors. To overcome this issue, one simple procedure is to add a reference optical fiber into the experimental setup and use the normalized transmitted signal as output data (ratio between the transmitted signal from the sensor and the reference signal).

Applications

Nowadays, the development of POF sensors based on intensity configuration embraces several different areas, allowing the sensing of physical parameters, chemical and biochemical species and processes – strain, pressure, detection and monitoring of cracks, small displacements, air and liquid flow, human gait, cure monitoring, ionizing radiation, respiratory movements, temperature, pH, turbidity, liquid level, humidity, refractive index (RI), chemical and biochemical detection and monitoring (nitro aromatic explosives, dissolved oxygen, contaminants, fertilizers, alcohol, glucose, bacteria, red blood cells, etc.), among many others.

Not aiming to make an exhaustive description of POF intensity-based sensors, below are described some examples of practical applications reported recently.

Refractive Index Sensing

Several studies related to RI-POF sensors can be found in the literature. Basically the manufacturing process involves bending and/or the removal of the cladding and part of the core, after which this sensing region is placed in contact with solutions of different RI and the transmitted light in the output of the fiber is measured. The performance of the sensors is usually studied varying the following conditions: wavelength of the light source, waist diameter and total depth, length of the sensing region, macrobending, and number of tapered/polished regions. After some years studying the best sensor configuration for RI sensing, Liu and Feng presented an optimization scheme using a D-shape with 2 cm length and an excurvation structure, maintaining the conditions of bending radius and depth (5 cm and 500 μm, respectively); however, sensitivity and resolution of the sensor are not specified (Liu and Feng 2016). In the same year, Sequeira et al. reported the optimization of the length of a D-shaped POF for RI sensing, showing the length dependence of the sensing region for RI (1.33–1.39) using simple and low-cost methods. The best sensitivity (with resolution of 6.48×10^{-3} RIU) was obtained for 6 cm length and allows the application of this platform for chemical and biochemical evanescent field sensing (Sequeira et al. 2016).

Structural Health Monitoring

Optical fiber-based sensor systems are one of the most frequently used technologies for modern structural health monitoring in concrete structures for steel reinforcement corrosion, temperature, vibration, strain, and crack detection (Luo et al. 2016). Luo et al. employed tapered POFs (with 5 cm length and 480 μm diameter) as strain sensors to crack detection of a concrete beam, where the sensor was embedded and surface glued. The intensity of the transmitted light was modulated by the applied strain which allowed the crack detection (Luo et al. 2016).

Medicine

The first application of optical fibers in the medical field allowed the illumination of internal organs during endoscopic procedures (Taffoni et al. 2013). Nowadays there is a vast possibility of applications of optical fibers in the medical field and due to the characteristics inherent to POFs, they are also being widely used in vivo applications.

In the field of (Bio)Medical optics, some studies have been conducted in order to detect ionizing radiation. For example, Zhuang et al. reported an inorganic scintillating POF dosimeter for the measurement of ionizing radiation during radiotherapy applications. At the tip of the POFs (two parallel-paired fibers), each core was micro-machined to create a small hole, and two inorganic scintillating materials were embedded and sealed using epoxy adhesive. Both materials fluoresce immediately (sub ns range) when exposed to ionizing radiation (X-ray, electron beam, and high-energy photon). Therefore, the incident radiation dose can be calculated through the intensity of the visible fluorescent light that is measured by photodetectors (Zhuang et al. 2016).

The monitoring of the carotid pulse is a key factor in hypertension assessment and cardiovascular prevention. Leitão et al. reported the development of an extrinsic POF sensor based on reflection configuration for noninvasive monitoring of the carotid pulse with a sensitivity of 727 μV/μm. A circular rigid aluminum-coated reflector adhesive was taped in the subject skin, over the carotid surface.

The micro-movements of the carotid artery changed the distance of the reflector to the fiber, modulating the reflected light. In this case, the light leaked out of the fiber and reentered after being reflected at the patient skin by means of a reflector adhesive (Leitão et al. 2015).

Elias et al. reported in 2015 an early-stage investigation towards the development of oral cancer sensor using POF. In this work, a simple transmission POF evanescent wave sensor was developed for the detection of nitrite with different concentrations (3–100 μM) using Griess reaction; future work aims to improve the sensitivity and the experimental setup in order to detect nitrite in human saliva (Elias et al. 2015).

Environmental Sensing

Grassini et al. reported the development of two low-cost POF sensors for measuring the cumulative response of low concentration of pollutants in the atmosphere, through the deposition of a thin film of sensitive compounds in the exposed core of the fiber. The interaction between the sensitive layer and the gas changed the transmitted light that is measured at the photodiode. The first sensor was realized for monitoring indoor atmospheres for cultural heritage applications, where long time exposition of metallic artifacts to hydrogen sulfide (H_2S) leads to tarnishing and impairs their conservation and aspect. The other sensor was designed to monitor the concentration of hydrogen fluoride vapors (HF) inside the RPC (resistive plate chamber) muon detector of the Compact Muon Solenoid (CMS) experiment at CERN in Geneva, where HF is a reaction product of RPC use which should be monitored and removed. In order to measure the cumulative response of a gas concentration, the selected coatings must be affected by non-reversible chemical reactions and, at the same time, able to detect extremely low concentrations of specific pollutants (Grassini et al. 2015).

A tapered POF sensor for ammonia (0–50 ppm) at room temperature was developed in 2015 by Raj et al., through the coating of the POF core with silver nanoparticles/PVP/PVA. The gas sensing properties were studied using ammonia, methanol and ethanol gases and had shown a good selectivity towards ammonia which increases with silver concentration (Raj et al. 2015). In 2016, Khalaf et al. coated a side-polished POF with graphene/polyaniline nanocomposite also for ammonia sensing (0.25–1%) at room temperature; the sensing principle was based on the interaction between the evanescent wave and the sensing deposited layer which could be detected by the increase of absorbance with the increase of ammonia concentration in the range of (600–800) nm (Khalaf et al. 2016).

A disposable evanescent wave POF sensor for the detection of the mycotoxin citrinin and the herbicide 2,4-dichlorophenoxyacetic acid (2,4-D) was developed by Ton et al. in 2015, by coating a 4 cm polystyrene fiber with a MIP containing a fluorescent monomer (FIM). FIM was excited by the evanescent wave, and the fluorescence intensity was proportional to the analyte concentration. The sensor detected the analytes in the low nM range and exhibited specific and selective recognition (Ton et al. 2015). This sensor works in reflection due to a mirror placed in one end of the fiber, although the sensing layer was coated along the 4 cm fiber.

Cennamo et al. reported a newly work with the development of a MIP-POF sensor for the selective detection of dibenzyl disulfide (DBDS) in transformer oil (responsible for the corrosive properties of the oil); the sensor was based on two POFs optically coupled by means of a trench milled in between and filled with MIP. When the binding between the MIP and the DBDS occurs, the refractive index of the MIP increased, and therefore there was an increase in the relative output signal (ratio between the transmitted signal of the two fibers); this study was performed for three wavelengths (600 nm, 733 nm, and 752 nm), the saturation is reached at a concentration of about 0.5 ppm (2×10^{-6} M), and the lowest detection limit is about 0.013 ppm (5.3×10^{-8} M) at 752 nm (wavelength with the highest sensitivity at low concentration) (Cennamo et al. 2017). In the future it is aimed to replace the halogen lamp and spectrometers by an LED and photodetectors, for industrial applications and to decrease the cost of the detection system.

In resume, intensity-based fiber optic sensors offer numerous advantages in different areas; however, there are other fiber optic technologies that can be employed to enhance performance that intensity-based approaches cannot offer. Some of these technologies will be explored in the following subsections.

Interferometric Polymer Optical Fiber Sensors

Interferometry is a high-sensitive measurement technique that is capable to respond to small changes of an external parameter. Interferometric sensors have been implemented using silica optical fibers in a variety of configurations (Li et al. 2012).

When a lightwave of a given wavelength, λ, propagates inside an optical fiber of length L, the phase angle, Φ, at the end of the fiber is given by:

$$\Phi = \frac{2\pi L}{\lambda} = \frac{2\pi n_{\text{core}}}{\lambda_0} \tag{4}$$

where n_{core} defines the RI of the fiber core and λ_0 is the wavelength of the light in vacuum. If an external perturbation causes a change in the RI or in the length of the fiber, a phase change occurs and can be defined by:

$$\Delta\Phi = \frac{2\pi}{\lambda}\left(n_{\text{core}}\Delta L + L\Delta n_{\text{core}}\right) \tag{5}$$

A length variation (with constant RI) will induce a phase change of:

$$\Delta\Phi = \frac{2\pi}{\lambda} n_{\text{core}}\Delta L \tag{6}$$

When a lightwave is injected into two equal single-mode fibers, the power is split, but the phase remains the same. If the two optical fibers experience the same conditions, the lightwaves will recombine at the same phase angle and constructive interference will occur, giving the maximum intensity output. However, if the fibers are subjected for instance to different thermal or mechanical strains, they will recombine with a phase difference proportional to the different lengths the radiation traveled, and destructive interference will occur, causing the output intensity to decrease.

The special characteristics offered by POFs, namely, SM-POFs, lead Silva-López et al. in 2005 to the measurement of the optical phase sensitivity on a SM-POF at the visible region (Silva-López et al. 2005). The arrangement was based on a Mach-Zehnder interferometer where one arm was composed of a SM-POF and was subject to strain and temperature variations. The interference fringes were observed between the light that traveled down the fiber and light in the second free space arm of the interferometer. The reported sensitivities were $1.31 \pm 3 \times 10^5$ rad/m and -212 ± 26 rad/(m × K), respectively, for the strain and temperature, which were larger than those reported for silica fibers. Despite the good results, the strain characterization was performed only in a small strain range (0–0.04%), leaving unexplored the potential elongation range of POFs (between 30% and 60% (Jiang et al. 2002) depending on the drawing conditions and thermal history). For that reason, Kiesel et al. (2008) report in 2008 the operation a POF-based interferometer at 15.8% elongation, with similar sensitivities reported by Silva-López et al. (2005).

Other structures, such as Fabry-Pérot interferometers, have also been reported in POFs (Statkiewicz-Barabach et al. 2016; Webb et al. 2005). This kind of interferometer differs from the previously presented by not requiring a reference fiber; instead, the fiber presents two partially reflective mirrors. These partially reflective mirrors cause the light to resonant inside the cavity, to be transmitted at the second mirror, and to reach the detector. This detector will magnify the phase difference and double the sensitivity to phase differences when compared to other interferometer configurations. The first description of these structures in POFs was demonstrated by Webb et al. (2005), through the inscription of two 1-cm-long FBGs, separated by a 3 cm cavity. However, the reflection spectrum appeared with low finesse, which they assume to be related to the high attenuation of the polymer at the 1550 nm region. Probably by that reason, no characterization was reported at the time of publication. Recently, Statkiewicz-Barabach et al. (2016) reported the fabrication of a Fabry-Pérot structure at the 1300 nm with good fringe visibility. The creation of the fiber structures was done in a single step using the phase mask technique and placing an "obstacle" before the phase mask and at the middle of the irradiated area. By using a "blocking obstacle" with a dimension of 0.1 mm, they report a Fabry-Pérot structure with distance between fringes of about 1.7 nm, corresponding to a Fabry-Pérot cavity of 0.34 nm. The fiber structure was subsequently characterized to strain and temperature, obtaining sensitivities of 1.07 pm/$\mu\varepsilon$ and −25 pm/°C, respectively (Statkiewicz-Barabach et al. 2016).

Recently, one fiber optic sensing technique that is receiving attention by the scientific community is the fiber modal interferometry, commonly known as MMI (multimode interference). This device has been fabricated using a silica MM fiber sandwiched between two silica SM fibers producing a single-mode-multimode-single-mode (SMS) fiber structure (see Fig. 19). The inherent simple fabrication, the

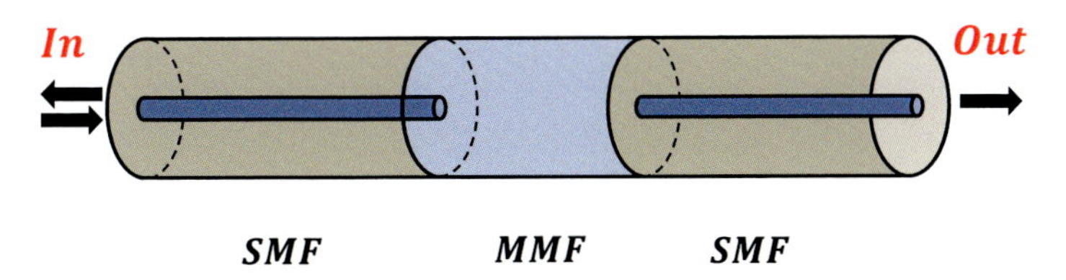

Fig. 19 Schematic of a multimode interference device, comprising a MM fiber between two SM fibers

low production cost, and the high sensitivity and compactness that can be obtained were used for the detection of a variety of parameters, such as strain (Oliveira et al. 2016e), temperature (Oliveira et al. 2016e), liquid level (Oliveira et al. 2016a), and refractive index (Oliveira et al. 2016e; Wu et al. 2011).

In this fiber device, the fundamental mode of the leading SM fiber will excite multiple modes on the MM fiber. As these modes have different propagation constants, they will interfere as they propagate along the MM fiber. This will allow the formation of multiple images of the input field, which are an exact replica of both phase and amplitude. The length in which these images are formed can be obtained from the restricted symmetric interference condition (Soldano and Pennings 1995), which is given by:

$$L = p\frac{3L_{\pi}}{4}, \quad p = 0, 1, 2, 3 \tag{7}$$

with L_{π} defining the beat length (Soldano and Pennings 1995), expressed as:

$$L_{\pi} \approx \frac{4n_{\text{MMF}}D_{\text{MMF}}^2}{3\lambda_0} \tag{8}$$

where λ_0 is the free space wavelength, with n_{MMF} and D_{MMF}, respectively, the effective refractive index and diameter of the fundamental mode of the MM fiber, considering air as external medium. If the MM fiber is cleaved where one of the images is being formed, a bandpass filter centered at λ_0 will be formed. The combination of Eqs. 7 and 8 provides the wavelength at which the filter will be centered:

$$\lambda_0 \approx p\frac{n_{\text{MMF}}D_{\text{MMF}}^2}{L} \tag{9}$$

The different opportunities of POFs when compared with silica fibers lead Huang et al. in 2012 to develop the first MMI structure incorporating a POF (Huang et al. 2012). The structure was composed of a PMMA multimode POF in the middle of two silica SM fibers and was intended for large strain measurement, revealing a sensitivity of -1.73 pm/$\mu\varepsilon$ in a 2% elongation. Later, Numata et al. used the same configuration for the characterization of strain and temperature using perfluorinated (Numata et al. 2014) and also partially chlorinated (Numata et al. 2015) MM-POFs, in their low-loss region. The reported sensitivities were as high as -112 pm/$\mu\varepsilon$ and 49.8 nm/°C, respectively, for strain and temperature, for the SMS structure based on perfluorinated fiber (Numata et al. 2014), and 4.47 pm//$\mu\varepsilon$ and 9.66 nm/°C, respectively, for the strain and temperature for the partially chlorinated POF (Numata et al. 2015). These values were much higher than the ones reported for silica fibers and the one presented by Huang et al. (2012) for the strain characterization, which according to the authors was probably related with the unique core polymer properties.

The fabrication of a MMI based on a MM-POF, composed of PMMA cladding and ZEONEX® 480R core material, was also developed (Oliveira et al. 2016c) (see Fig. 20). The sensor was designed to have a peak response in the 1550 nm region where most of devices exist. Similar to the previous works, this sensor was also characterized to strain and temperature (see Fig. 21a, b), reporting sensitivities of -2.77 pm/$\mu\varepsilon$ and 212.6 pm/°C, respectively, for the strain and temperature. Nevertheless, the water absorption capabilities offered by the PMMA cladding were used to characterize the structure to relative humidity variations (see Fig. 21c).

The results collected for the humidity characterization revealed a red shift with increasing relative humidity. The shift was due to the swelling of the PMMA material, which imposes stresses in different directions in the core MM-POF. Additionally, the uptake of water led inherently to a refractive index change of the PMMA. The combination of both effects around the polymer core affects directly the propagating modes, leading to a change in the peak or dips of the MMI spectra. The reported relative humidity sensitivity was 67 pm/% RH which was close to the

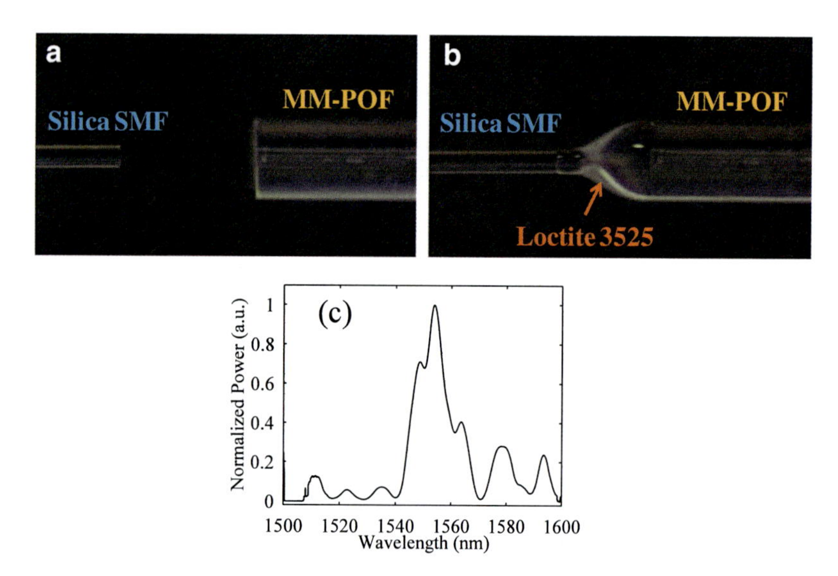

Fig. 20 (**a**, **b**) Splicing process between a silica-SMF and the MM-POF. (**c**) Normalized power at the output of the SMS structure (Oliveira et al. 2016c)

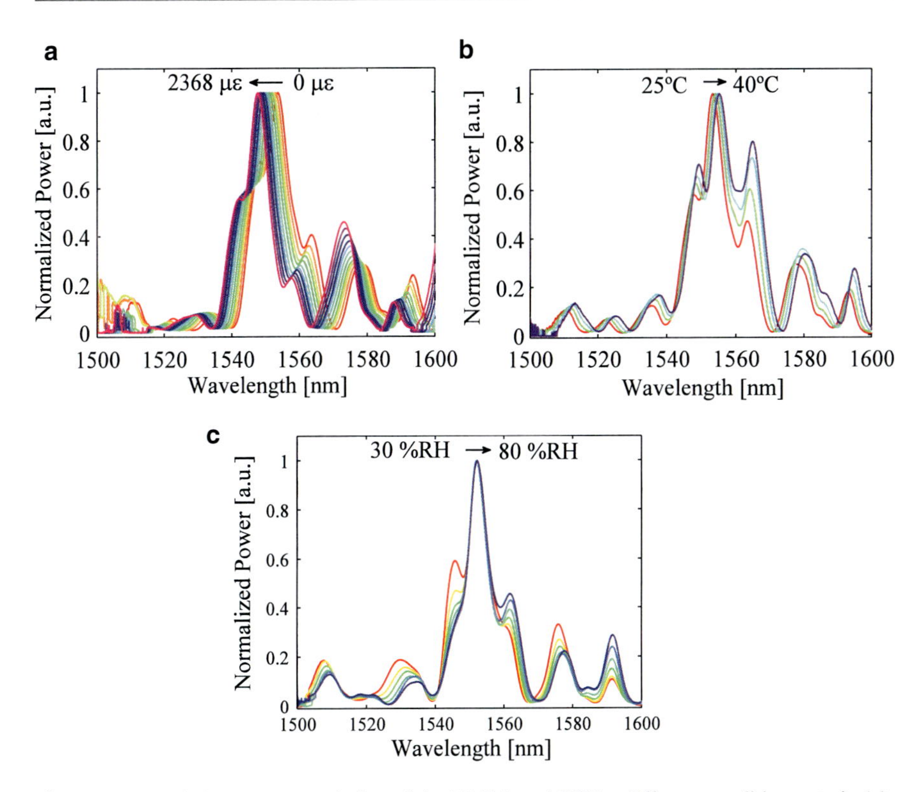

Fig. 21 Transmission spectra evolution of the MMI-based POF at different conditions: strain (**a**), temperature (**b**), and humidity (**c**) (Oliveira et al. 2016c)

ones found for FBGs written in PMMA-based POFs (Zhang and Webb 2014) and for a silica MMI structure coated with hygroscopic polymers (Gu et al. 2011).

MMI devices based on silica fibers have extensively been studied for the detection of refractive index (Oliveira et al. 2016e; Wu et al. 2011). For those sensors, it is necessary to expose the fiber core to the external environment. This has been simply done by using a MM fiber composed of a single material commonly known as no-core fiber or by simply etching the cladding of a MMF (Wu et al. 2011). PMMA-based polymer optical fibers based on FBG configurations have also been reported as concentration sensors (Ferreira et al. 2014; Zhang et al. 2012b). However, the Bragg wavelength shift is dependent on the swelling of the fiber and on the refractive index change of the polymer material due to the water uptake. Therefore, it is inevitably a time-dependent process (Zhang et al. 2012b), which can take several tens of minutes for the wavelength stabilization (Nogueira et al. 2015; Zhang et al. 2012). Contrary to PMMA, other polymers, like CYTOP®, cycloolefin polymers (COPs), and copolymers (COCs), provide low water absorption capabilities (Asahi Glass Co. Ltd. 2009; Woyessa et al. 2016; Yuan et al. 2011), and for that reason, they have been used in different

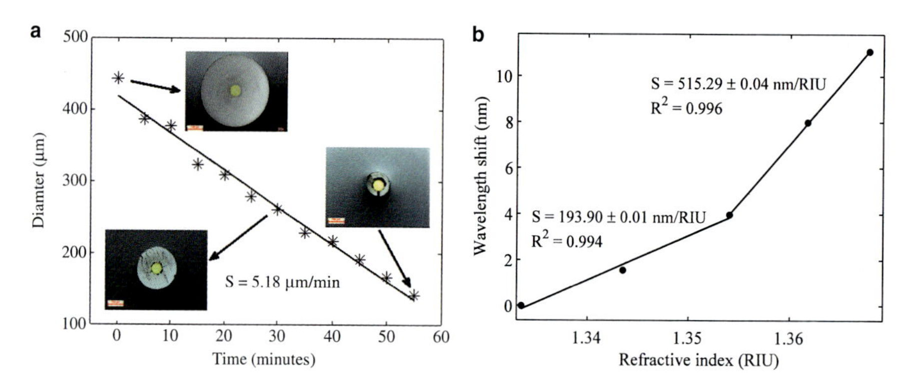

Fig. 22 (**a**) Etching process of the PMMA cladding; (**b**) wavelength shift of a MMI sensor based on a ZEONEX® 480R POF material (Oliveira et al. 2016d)

humidity-insensitive applications (Woyessa et al. 2016; Yuan et al. 2011). Etching the PMMA cladding layer allows to join the low water absorption capabilities offered by the COP material and the high sensitivity that can be achieved in an MMI configuration (see Fig. 22a).

The low water absorption offered by the COP polymer removes the problem associated with the time needed for the uptake of water and the inherent signal stabilization. The results concerning the wavelength shift performed for different refractive index solutions may be seen in Fig. 22b.

Wavelength-Based Polymer Optical Fiber Sensors

Wavelength-based sensors are a type of sensors that exhibit a change in the spectral domain when an external perturbation occurs. Among the different spectrally based fiber optic sensors are the fluorescence sensors, blackbody sensors, and Bragg grating sensors. Regarding the last, which are the most used ones, they are composed of a periodic index that can generate depending on the type of the structure, a reflection, transmission, or both responses. These responses are sensitive to a variety of parameters, and for that reason they have been employed in several applications. Inevitably, POFs offer a variety of opportunities over silica fibers, and for that reason this type of structures have become in the last years subject of great interest by the scientific community. For that reason, the most prominent developments regarding the implementation of FBGs will be described in the following three subsections, long-period gratings (LPGs), and tilted Bragg gratings (TFBGs) in POFs.

Fiber Bragg Gratings

Fiber Bragg gratings (FBG) consist in a periodic modulation of the refractive index along the length of an optical fiber and are produced through the exposure of the fiber to an intense optical interference pattern. The periodic structure in a FBGs acts like a selective mirror for the wavelength that satisfies the Bragg condition expressed as:

$$\lambda_{\text{Bragg}} = 2n_{\text{eff}}\Lambda \tag{10}$$

where Λ is the grating period and n_{eff} the effective refractive index of the waveguide mode. The first inscription of a permanent grating in an optical fiber was demonstrated by Hill et al. in 1978 (1978), by launching an intense Argon-ion laser into an optical fiber doped with germanium. Since then, the interest in these grating structures has increased, and the development of different inscription techniques has been reported. Nowadays, FBGs are commercially available and have become an important technology for optical communications and sensing applications.

In 1999, 8 years after the first demonstration of the SM behavior in POFs (Kuzyk et al. 1991), the group of Gang-Ding Peng and Pak L. Chu, motivated by the high elongation capabilities offered by polymers, reported the inscription of a polymer optical fiber Bragg grating (PFBG) in a SI fiber, either MM (Peng et al. 1999) or SM (Xiong et al. 1999), at the infrared region, using a 325 nm UV radiation. The high-tuning capabilities of such gratings were explored later, showing a wavelength shift of about 70 nm for an almost 5% increasing strain (Peng and Chu 2000). In 2005, 4 years after the production of the first mPOF (van Eijkelenborg et al. 2001), the group of David J. Webb reported the inscription of Bragg gratings at the infrared region in mPOFs, either SM or MM (Dobb et al. 2005), using the same methodology. PFBGs have been essentially written in PMMA fibers due to the high availability of this polymer. Nevertheless, PFBGs in other polymers such as TOPAS® (Johnson et al. 2011), ZEONEX® (Oliveira et al. 2016d) polycarbonate (Fasano et al. 2016), and also CYTOP® (Koerdt et al. 2016; Lacraz et al. 2015) have also been reported. Despite the high attenuation of POFs at the infrared region, most of the PFBGs have been reported at this spectral window. The reason is mainly due to the availability of sources and detectors at this region. Nevertheless, the inscription of PFBGs has already been demonstrated in different spectral regions, including the visible region (Bundalo et al. 2014; Marques et al. 2013). The writing methodology has fundamentally been done through the phase mask technique due to the easy implementation of the recording setup. Other inscription methods such as the point-by-point and line-by-line through a femtosecond laser have also been employed (Lacraz et al. 2015; Stefani et al. 2012b).

The description of photosensitivity in the most known polymer (PMMA) can be traced back to the early 1970s. Despite these works have been performed half a century ago, they are still subject of study in the current days (Sáez-Rodríguez et al. 2014). The possible mechanisms behind the photosensitivity in PFBGs, under the most preferred UV radiation (325 nm), are described as a competitive process between photodegradation and polymerization (Sáez-Rodríguez et al. 2014). Indeed, in addition to the polymer production process, the wavelength and power of the light source are key factors for the photosensitive mechanisms, leading to different chemical reactions inside the polymer chains.

The production of a PFBG under 325 nm UV radiation is a time consumption process that can take tens of minutes (Liu et al. 2002). The record inscription time is 7 min, being this value well below than most of the works reported in literature (Bundalo et al. 2014). Nevertheless, one disadvantage associated with the long exposure is the stability needed during the inscription process, which can be challenging due to the dimension of the structures being created.

Concerning the 248 nm UV radiation, very popular for the production of FBGs in silica fibers, Küper and Stuke described for the first time that at low fluences (below the damage threshold reported for PMMA (0.65 J/cm^2) (Srinivasan et al. 1986)), the polymer composition is modified due to photochemical-induced changes of the UV absorption (incubation effect) (Küper and Stuke 1989). Baker et al. (Baker and Dyer 1993), in a study based on the work done by Küper and Stuke, showed a small but significant change in the refractive index of a PMMA film using a KrF laser irradiation together with low repetition rates and also low fluences (40 mJ/cm^2 and 5 Hz), in order to minimize the ablative removal of PMMA. They show a refractive index change (RIC) with increasing number of pulses, presenting a value of 10^{-2} after 2000 pulses (Baker and Dyer 1993). Another interesting detailed work regarding the refractive index modification with 248 nm laser light at low fluences (fluence = 17 mJ/cm^2 and repetition rate = 5 Hz) was presented by Wochnowski et al. (2000). In his work it was concluded that the RIC was due to a Norrish type I photochemical reaction. This means that a complete separation of the side chain from the PMMA molecule took place, promoting a volume contraction by the van der Waals interactions and consequently an increase of the RIC (Wochnowski et al. 2000).

A PFBG can also be written with UV laser in an mPOF in few seconds (see Fig. 23) (Oliveira et al. 2015a). In the figure, a pulsed KrF UV laser operating at 248 nm was used with a fluence of 33 mJ/cm^2 together with low repetition rate of 1 Hz, to avoid rising of temperature and consequently damage of the polymer material.

This kind of setup can be used to produce Bragg gratings in several POFs, including SI-POFs, mPOFs (Nogueira et al. 2015), highly birefringent POFs (Oliveira et al. 2016b), and unclad POF (Oliveira et al. 2016d). The peak power evolution of some of those fibers can be seen in Fig. 24

PFBGs have been considered as an alternative to silica FBGs in different sensing applications. The reason is related with the advantages offered by POFs compared with silica. POFs can be elongated up to 40%, and therefore POFBGs with high-tuning capabilities have been reported (Peng and Chu 2000). An example of such behavior can be seen in Fig. 25a for a mPOF strained until 2% with a 35 nm wavelength shift.

POFs have a Young's modulus that is 30 times lower than that of silica, and thus, hydrostatic pressure sensors based on PFBGs have been reported to allow a sensitivity 52 times higher than the reported for silica FBGs (Bhowmik et al. 2015). As opposed to PFBGs, in silica FBGs, the pressure-induced grating period change (strain effect) dominates the change in refractive index, and as a result a negative sensitivity is observed. An example of such behavior can be seen in Fig. 25b for

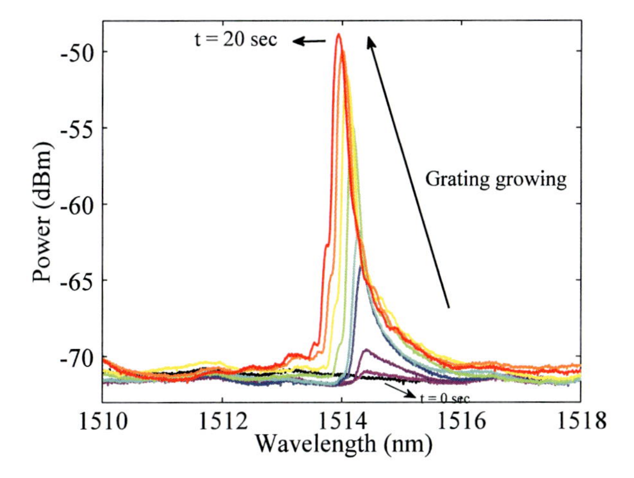

Fig. 23 Spectra evolution taken over time for a PFBG written in a few-mode mPOF-based PMMA, with a 248 nm KrF UV laser (Oliveira et al. 2015a)

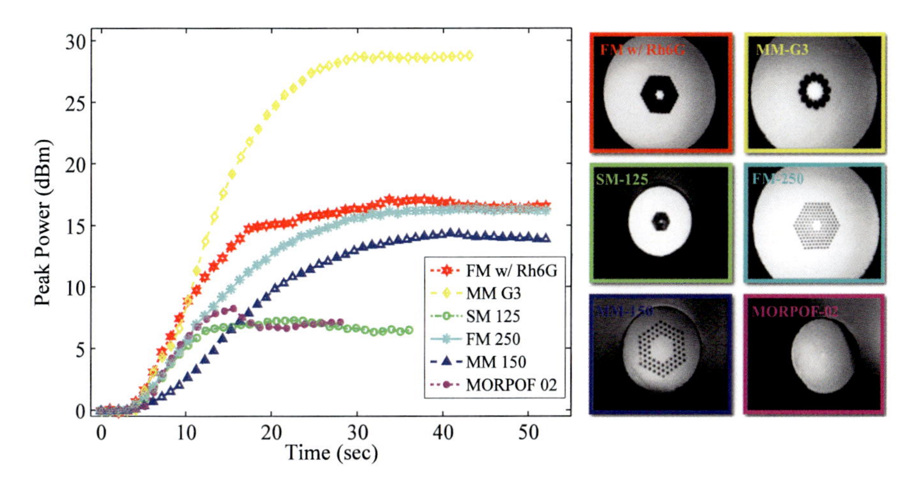

Fig. 24 Peak power evolution of different POFBGs (both SI and mPOF), during the inscription process (Nogueira et al. 2015)

a PFBG written in an mPOF. Most of the polymers have the capability to absorb the water content that surrounds the environment; in particular, PMMA can absorb 0.3%/day. This may be not interesting for the detection of other parameters, due to the inherent cross sensitivity. Despite this feature, PFBGs have been used to detect the relative humidity (Zhang et al. 2010; Zhang and Webb 2014); see an example of the wavelength shift with relative humidity for a PFBG in an mPOF in Fig. 25c. On the other hand, humidity sensitivity in silica FBGs can only be achieved through the deposition of an hygroscopic material onto the surface of the fiber (Gu et al. 2011). The water absorption capabilities of polymers has also been used for the detection

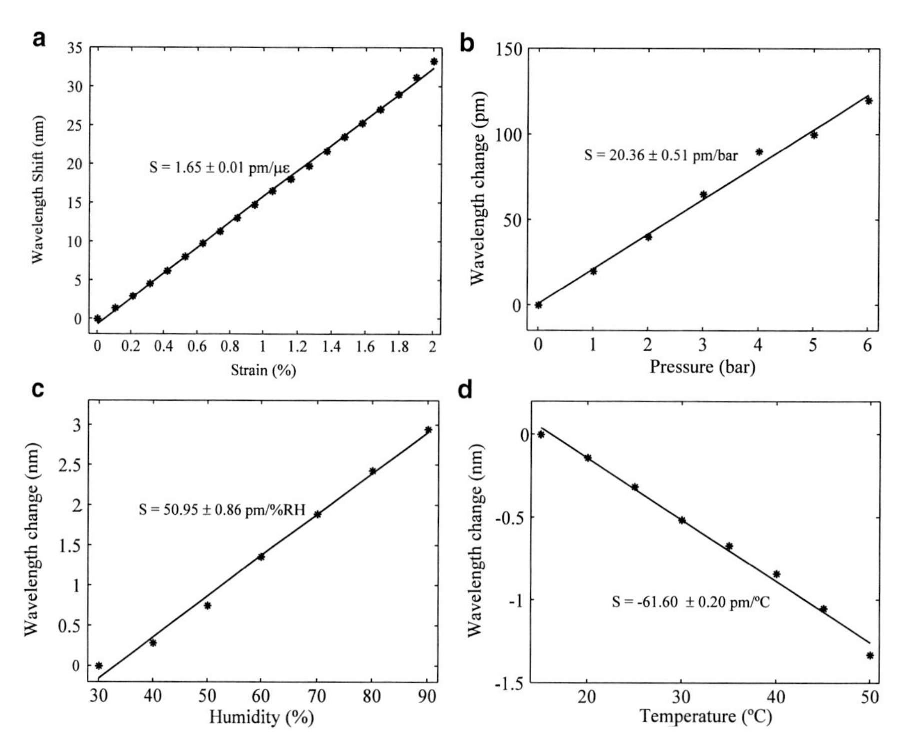

Fig. 25 (**a**) Wavelength shift versus (**a**) strain for a FM-250 mPOF (Oliveira et al. 2015a), (**b**) pressure for a SM-125 mPOF (Nogueira et al. 2015), (**c**) humidity for a MM-G3 (Nogueira et al. 2015), and (**d**) temperature for a FM w Rh6G (Nogueira et al. 2015)

of concentration in solutions (Ferreira et al. 2014; Zhang et al. 2012a) and also the detection of water in jet fuel (Zhang et al. 2009). Another interesting characteristic of PFBGs is the high sensitivity to temperature variations. The first work reporting PFBG temperature characterization was shown by Liu et al. (2001), presenting a tunability of presented a tunability of about 18 nm in a 50 °C range. In such work, the wavelength shift revealed a nonlinear behavior with higher sensitivity at high temperatures. In fact, when POFs are drawn (essentially at lower temperatures and higher tensions) residual stresses are retained in the polymer matrix. Those stresses can be easily removed by annealing the fiber at a temperature below the glass transition temperature of the polymer. By doing that, a linear behavior can be observed as demonstrated in Carroll et al. (2007) with a sensitivity of −52 pm/°C. It is worth to mention that the sensitivity of PFBGs to temperature is negative (see example in Fig. 25d for a PFBG written in a mPOF), contrary to silica fibers. This difference is due to the dominance of the negative thermo-optic coefficient over the thermal expansion coefficient (in silica fibers both effects are positive). Nonetheless, most of the works based on PFBGs have been reported using PMMA as the main material. However, other polymers such as TOPAS®, ZEONEX®, or polycarbonate have also been reported allowing the capability to resist at higher temperatures (Fasano et al. 2016; Markos et al. 2013; Oliveira et al. 2016d) and to be humidity-insensitive (Markos et al. 2013; Oliveira et al. 2016d).

Long-Period Gratings

Long-period gratings are fiber structures that are composed of a periodic modulation of the refractive index along the fiber axis. The period of these structures is of the order of hundreds of microns and allows the existence of coupling between the fundamental core mode and co-propagating cladding modes, following the next equation:

$$m\lambda = \Lambda_{LPG}\,(n_{co} - n_{cl}) \tag{11}$$

where m is the order of the coupling, λ is the wavelength, Λ_{LPG} is the grating period, and n_{co} and n_{cl} are the core and cladding effective mode indices, respectively.

The first LPG to be reported in POF was demonstrated in 2004, by mechanically pressing a rod with triangular groves over an mPOF (van Eijkelenborg et al. 2004). In 2005, Li et al. (2005) have reported the first stable LPG in a SI-POF. In their work, a mercury lamp through a 275 μm amplitude mask was used to inscribe a 3 dB grating depth LPG at the infrared region. One year later, LPGs have been imprinted in a stable way through heat and mechanical stress in mPOFs in the visible region (Hiscocks et al. 2006) (see the spectrum example shown in Fig. 26).

The special characteristics of POFs and the multiparameter capabilities offered by LPGs led several authors to use these structures to sense different parameters, such as strain (Steffen et al. 2009), temperature (Steffen et al. 2009), humidity (Steffen et al. 2009; Witt et al. 2011), and pressure (Statkiewicz-Barabach et al. 2013).

LPGs have also been subject of interest for respiratory movement monitoring (Krebber et al. 2008). Authors sewed the LPG onto an elastic textile fabric and performed 25 strain cycles between 0% and 1%. The feasibility of the results led the authors to consider the sensor suited for monitoring purposes.

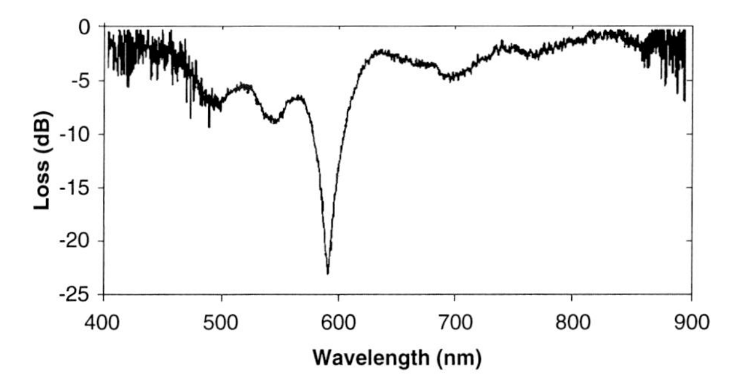

Fig. 26 Spectrum of an LPG imprinted through heat and mechanical stress in a SM-340 (structured sold by Kiriama Pty., Ltd.)

Tilted Fiber Bragg Gratings

Tilted fiber Bragg gratings are short-period gratings that possess all the advantages described for the well-developed FBG technology together with the ability to excite cladding modes resonantly. In a TFBG, the modulation of the refractive index is tilted with respect to the longitudinal axis of the fiber. Similar to LPGs, this tilt enables strong coupling between the forward core mode to the backward-propagating cladding modes. The reflected Bragg wavelength λ_{TFBG} and the cladding mode resonances λ^{i}_{clad} can be determined by the phase-matching condition, through the following equations:

$$\lambda_{TFBG} = \frac{2n_{eff,core}\Lambda}{\cos\,\theta};\lambda^{i}_{clad} = \frac{\left(n^{i}_{eff,core} + n^{i}_{eff,clad}\right)\Lambda}{\cos\,\theta} \tag{12}$$

where n_{eff}, $n^{i}_{eff,core}$, and $n^{i}_{eff,clad}$ are the effective indices of the core mode at λ_{TFBG}, the ith core mode, and the cladding mode at λ^{i}_{clad}, respectively. The grating period along the axis of the fiber, Λ_{TFBG}, is given by $\Lambda_{TFBG} = \Lambda/\cos\theta$, where θ is the tilt angle. Unlike the core mode, cladding modes are sensitive to external parameters such as bending and surrounding refractive index. For that reason, fiber optic sensors based on this technology have been subject of interest. TFBGs in silica fibers are a well-developed area (Guo et al. 2016); however, this technology has recently been demonstrated in POF (Hu et al. 2014). The reason for such delay is mainly related with the attenuation of POFs at the infrared region, where the SM behavior is easier to be achieved and also the need of two silica-POF couplings in order to see the TFBG in transmission. Despite these problems, Xuehao Hu et al. (2014) reported the inscription of TFBGs with different tilt angles in a slightly etched step-index SM-POF at the 1550 nm region. For that, they have used the phase mask technique, scanning a 325 nm HeCd laser trough a 2 cm phase mask length. One of those TFBGs, with tilt angle of 3°, was used for refractive index characterization showing a sensitivity of $\sim$13 nm/RIU in a refractive index range of 1.42–1.49 (similar to silica TFBGs) (Hu et al. 2014). Later, the same authors have reported the creation of a surface plasmon resonance (SPR)-based TFBG sensor by depositing a 50 nm gold layer on a SI-POF TFBG with ($\theta = 6°$) at the 1550 nm region (Hu et al. 2015). By using this technique, the authors were able to obtain a sensitivity of 553 $\pm$ 35 nm/RIU for a surrounding refractive index between 1.408 and 1.428. It is worth to remember that POFs, contrary to silica fibers, have better physical properties and improved biocompatibility. For that reason the development of this type of structures in POFs is very promising in medical or biological applications.

Polarimetric-Based Polymer Optical Fiber Sensors

Highly birefringent (HiBi) fibers are a special type of fibers where the geometry imposes two effective refractive index values, one for each of the orthogonal axes. Compared to conventional polarization maintaining (PM) fibers, birefringence arising from microstructural asymmetries has several interesting features. To create this kind of fiber structures, the symmetry of the hexagonal structure in the microstructured cladding needs to be broken. POFs are flexible to allow technologies such as drilling and casting methods allowing an easy implementation of this kind of structures.

Polarimetric optical fiber sensors based on HiBi fibers have been subject of great interest in the last years due to the potential high sensitivity (Chah et al. 2012; Statkiewicz et al. 2004). Several physical quantities can be measured with this kind of fiber optic device, namely, hydrostatic pressure, strain, vibration, temperature, acoustic wave, etc. The interferometric system commonly used to measure the sensitivity of the phase modal birefringence is shown in Fig. 27. It is composed of a broad band light source, polarizer, and analyzer placed at the ends of the tested fiber, aligned at 45° to its polarization axes. The two orthogonal polarized modes will be similarly excited in the fiber and will interfere at its output after passing through the analyzer. The spectral interference fringes are then analyzed with the use of an optical spectrum analyzer.

The polarimetric sensitivity of the fiber to an external parameter (X) represents the phase difference between polarization modes ($\varphi_x - \varphi_y$) induced by unit change of the parameter over unit length (L) of the fiber and can be expressed by Eq. 13 (Statkiewicz et al. 2004):

$$K_x = \frac{1}{L}\frac{d\left(\varphi_x - \varphi_y\right)}{dX} = \frac{2\pi}{\lambda}\left(\frac{d\Delta n}{dX} + \frac{\Delta n}{L}\frac{dL}{dX}\right) \tag{13}$$

Due to the flexibility in shaping fiber structures in polymers, mPOFs have been created in a variety of structures (van Eijkelenborg et al. 2003). The arrangement of those structures was also designed to provide mPOFs with high birefringence, and soon, polarimetric sensors based on this type of fibers have been reported. In 2010, Szczurowski et al. demonstrate for the first time the use of a polarimetric fiber optic sensor based on a dual core HiBi-mPOF with a birefringence in each core exceeding 10^{-4} at the visible region. They demonstrate the capability to measure hydrostatic pressure, strain, and temperature. The obtained polarimetric sensitivities were 72 rad/(m × MPa) and 3.1 rad/(mε × m), respectively, for the hydrostatic pressure and strain at 710 nm (Szczurowski et al. 2010). The sensitivity values obtained for the hydrostatic pressure were of opposite sign to the observed for silica

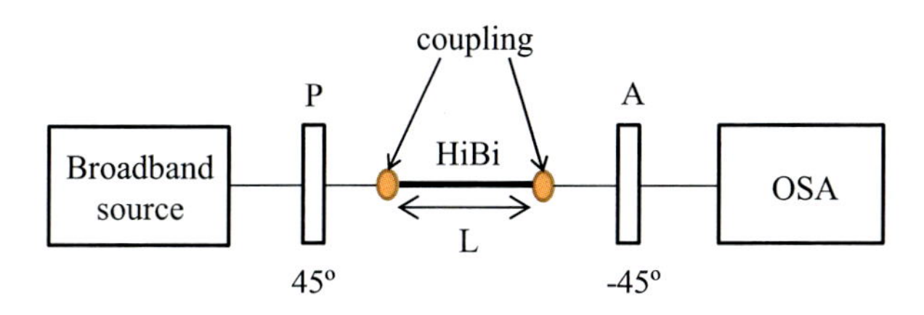

Fig. 27 Setup used to measure the polarimetric sensitivity

fibers, which they assume to be related to the negative stress-optic coefficient when compared with silica. Regarding the polarimetric sensitivity to strain, the value was lower than the one observed for bow-tie silica fibers and similar to the one obtained for elliptical core and microstructured silica fibers (Huang et al. 1990; Statkiewicz et al. 2004). Finally, the results obtained for the temperature tests have revealed a nonlinear response with high hysteresis, which they assume to be caused by partially release of stress built-in during the drawing process. Those results were also observed in later studies involving other mPOF structures (Martynkien et al. 2013).

Conclusions

Polymer optical fibers have been pointed as a valuable alternative to traditional silica fibers in different sensing areas. The extensive range of polymer materials together with the wide range of POF fabrication techniques opens new opportunities for the development of special applications. It is possible to create special POFs that can withstand to high elongation regimes, humidity-sensitive or humidity-insensitive, that can resist in harsh environments, among other features. While most of the intensity-based technologies employ large-core MM fibers, due to the easy integration of components and the low cost involved, other sensing technologies have been explored. Most of them require the operation of POFs in the SM regime which inevitably brings restrictions from the availability of sources and detectors at the visible region. Additionally, the inherent precision needed for such fibers prompts for a drawback regarding the splicing mechanisms used until now. The solution for this problem should be addressed in the near future. Also research will be focused in the development of new fibers structures and materials in order to overcome the still high attenuation observed in POFs. In conclusion, the market is still demanding for new solutions, and for that reason it is expected that POF-based sensors will continue to be a very active research topic in the next years.

Acknowledgements This work was supported by FCT-Fundação para a Ciência e Tecnologia through Portuguese national funds by project hiPOF (PTDC/EEI-TEL/7134/2014), project INITIATE and investigator grant IF/01664/2014.

References

A. Abang, D.J. Webb, Demountable connection for polymer optical fiber grating sensors. Opt. Eng. **51**(8), 80503 (2012). https://doi.org/10.1117/1.OE.51.8.080503

A. Abang, D. Saez-Rodriguez, K. Nielsen, O. Bang, D.J. Webb, Connectorisation of fibre Bragg grating sensors recorded in microstructured polymer optical fibre, in *Proceedings of the SPIE 8794, Fifth European Workshop on Optical Fibre Sensors*, (Krakow, 2013), p. 87943Q. https://doi.org/10.1117/12.2026796

O. Abdi, K.C. Wong, T. Hassan, K.J. Peters, M.J. Kowalsky, Cleaving of solid single mode polymer optical fiber for strain sensor applications. Opt. Commun. **282**(5), 856–861 (2009). https://doi.org/10.1016/j.optcom.2008.11.046

A. Argyros, I. Bassett, M. van Eijkelenborg, M. Large, J. Zagari, N.A. Nicorovici, . . . , C.M. de Sterke. Ring structures in microstructured polymer optical fibres. Opt. Express. **9**(13), 813–820 (2001). https://doi.org/10.1364/OE.9.000813

AGC Chemicals, Asahi Glass Co. Ltd., *Amorphous Fluoropolymer (CYTOP)* (Tokyo, 2009)

S. Atakaramians, K. Cook, H. Ebendorff-Heidepriem, V.S. Afshar, J. Canning, D. Abbott, T.M. Monro, Cleaving of extremely porous polymer fibers. IEEE Photon. J. **1**(6), 286–292 (2009). https://doi.org/10.1109/JPHOT.2009.2038796

A.K. Baker, P.E. Dyer, Refractive-index modification of polymethylmethacrylate (PMMA) thin films by KrF-laser irradiation. Appl. Phys. A Solids Surf. **57**(6), 543–544 (1993). https://doi.org/10.1007/BF00331756

G. Barton, M.A. van Eijkelenborg, G. Henry, M.C.J. Large, J. Zagari, Fabrication of microstructured polymer optical fibres. Opt. Fiber Technol. **10**(4), 325–335 (2004). https://doi.org/10.1016/j.yofte.2004.05.003

K. Bhowmik, G.-D. Peng, Y. Luo, E. Ambikairajah, V. Lovric, W.R. Walsh, G. Rajan, Experimental study and analysis of hydrostatic pressure sensitivity of polymer fibre Bragg gratings. J. Lightwave Technol. **33**(12), 2456–2462 (2015). https://doi.org/10.1109/JLT.2014.2386346

T.A. Birks, J.C. Knight, P.S. Russell, Endlessly single-mode photonic crystal fiber. Opt. Lett. **22**(13), 961–963 (1997). https://doi.org/10.1364/OL.22.000961

I.-L. Bundalo, K. Nielsen, C. Markos, O. Bang, Bragg grating writing in PMMA microstructured polymer optical fibers in less than 7 minutes. Opt. Express **22**(5), 5270–5276 (2014). https://doi.org/10.1364/OE.22.005270

J. Canning, E. Buckley, N. Groothoff, B. Luther-Davies, J. Zagari, UV laser cleaving of air–polymer structured fibre. Opt. Commun. **202**(1–3), 139–143 (2002). https://doi.org/10.1016/S0030-4018(01)01727-8

K.E. Carroll, C. Zhang, D.J. Webb, K. Kalli, A. Argyros, M.C. Large, Thermal response of Bragg gratings in PMMA microstructured optical fibers. Opt. Express **15**(14), 8844–8850 (2007). https://doi.org/10.1364/OE.15.008844

N. Cennamo, G. Testa, S. Marchetti, L. De Maria, R. Bernini, L. Zeni, M. Pesavento, Intensity-based plastic optical fiber sensor with molecularly imprinted polymer sensitive layer. Sensors Actuators B Chem. **241**, 534–540 (2017). https://doi.org/10.1016/j.snb.2016.10.104

K. Chah, N. Linze, C. Caucheteur, P. Mégret, P. Tihon, O. Verlinden, . . . , M. Wuilpart, Temperature-insensitive polarimetric vibration sensor based on HiBi microstructured optical fiber. Appl. Opt. **51**(25), 6130–6138 (2012). https://doi.org/10.1364/AO.51.006130

H. Dobb, D.J. Webb, K. Kalli, A. Argyros, M.C.J. Large, M.A. van Eijkelenborg, Continuous wave ultraviolet light-induced fiber Bragg gratings in few – and single-mode microstructured polymer optical fibers. Opt. Lett. **30**(24), 3296–3298 (2005). https://doi.org/10.1364/OL.30.003296

J.M. Dudley, G. Genty, S. Coen, Supercontinuum generation in photonic crystal fiber. Rev. Mod. Phys. **78**(4), 1135–1184 (2006). https://doi.org/10.1103/RevModPhys.78.1135

S.N. Elias, N. Arsad, S. Abubakar, Nitrite detection using plastic optical fiber (POF); an early stage investigation towards the development of oral cancer sensor using POF. Optik – Int. J. Light Electron Opt. **126**(21), 2908–2911 (2015). https://doi.org/10.1016/j.ijleo.2015.07.038

C. Emslie, Polymer optical fibres. J. Mater. Sci. **23**(7), 2281–2293 (1988). https://doi.org/10.1007/BF01111879

A. Fasano, G. Woyessa, P. Stajanca, C. Markos, K. Nielsen, H.K. Rasmussen, . . . , O. Bang, Fabrication and characterization of polycarbonate microstructured polymer optical fibers for high-temperature-resistant fiber Bragg grating strain sensors. Opt. Mater. Express **6**(2), 649–659 (2016). https://doi.org/10.1364/OME.6.000649

R. Ferreira, L. Bilro, C. Marques, R. Oliveira, R. Nogueira. Refractive index and viscosity: dual sensing with plastic fibre gratings, in *Proceedings of the SPIE 9157, 23rd International Conference on Optical Fibre Sensors*, (Santander, 2014), p. 915793. https://doi.org/10.1117/12.2058833

M.V.P. Ghirghi, V. Minkovich, A.G. Villegas, Polymer optical fiber termination with use of liquid nitrogen. IEEE Photon. Technol. Lett. **26**(5), 516–519 (2014). https://doi.org/10.1109/LPT.2013.2295885

S. Grassini, M. Ishtaiwi, M. Parvis, A. Vallan, Design and deployment of low-cost plastic optical fiber sensors for gas monitoring. Sensors **15**(1), 485–498 (2015). https://doi.org/10.3390/s150100485

B. Gu, M. Yin, A.P. Zhang, J. Qian, S. He, Optical fiber relative humidity sensor based on FBG incorporated thin-core fiber modal interferometer. Opt. Express **19**(5), 4140–4146 (2011). https://doi.org/10.1364/OE.19.004140

T. Guo, F. Liu, B.-O. Guan, J. Albert, Tilted fiber grating mechanical and biochemical sensors. Opt. Laser Technol. **78**, 19–33 (2016). https://doi.org/10.1016/j.optlastec.2015.10.007

K.O. Hill, Y. Fujii, D.C. Johnson, B.S. Kawasaki, Photosensitivity in optical fiber waveguides: application to reflection filter fabrication. Appl. Phys. Lett. **32**(10), 647–649 (1978). https://doi.org/10.1063/1.89881

M.P. Hiscocks, M.A. van Eijkelenborg, A. Argyros, M.C.J. Large, Stable imprinting of long-period gratings in microstructured polymer optical fibre. Opt. Express **14**(11), 4644–4649 (2006). https://doi.org/10.1364/OE.14.004644

X. Hu, C.-F.J. Pun, H.-Y. Tam, P. Mégret, C. Caucheteur, Tilted Bragg gratings in step-index polymer optical fiber. Opt. Lett. **39**(24), 6835–6838 (2014). https://doi.org/10.1364/OL.39.006835

X. Hu, P. Mégret, C. Caucheteur, Surface plasmon excitation at near-infrared wavelengths in polymer optical fibers. Opt. Lett. **40**(17), 3998–4001 (2015). https://doi.org/10.1364/OL.40.003998

S.Y. Huang, J.N. Blake, B.Y. Kim, Perturbation effects on mode propagation in highly elliptical core two-mode fibers. J. Lightwave Technol. **8**(1), 23–33 (1990). https://doi.org/10.1109/50.45925

J. Huang, X. Lan, H. Wang, L. Yuan, T. Wei, Z. Gao, H. Xiao, Polymer optical fiber for large strain measurement based on multimode interference. Opt. Lett. **37**(20), 4308–4310 (2012). https://doi.org/10.1364/OL.37.004308

C. Jiang, M.G. Kuzyk, J.-L. Ding, W.E. Johns, D.J. Welker, Fabrication and mechanical behavior of dye-doped polymer optical fiber. J. Appl. Phys. **92**(1), 4–12 (2002). https://doi.org/10.1063/1.1481774

I.P. Johnson, W. Yuan, A. Stefani, K. Nielsen, H.K. Rasmussen, L. Khan, ..., O. Bang, Optical fibre Bragg grating recorded in TOPAS cyclic olefin copolymer. Electron. Lett. **47**(4), 271–272 (2011). https://doi.org/10.1049/el.2010.7347

P. Kaiser, E.A.J. Marcatili, S.E. Miller, A new optical fiber. Bell Syst. Tech. J. **52**(2), 265–269 (1973). https://doi.org/10.1002/j.1538-7305.1973.tb01963.x

A.L. Khalaf, F.S. Mohamad, P.T. Arasu, A.A. Shabaneh, N.A. Rahman, H.N. Lim, ..., M.H. Yaacob, Modified plastic optical fiber coated graphene/polyaniline nanocomposite for ammonia sensing, in *2016 IEEE 6th International Conference on Photonics (ICP)*. (IEEE, 2016). https://doi.org/10.1109/ICP.2016.7510030

S. Kiesel, K. Peters, T. Hassan, M. Kowalsky, Large deformation in-fiber polymer optical fiber sensor. **20**(6), 2008–2010 (2008). https://doi.org/10.1109/LPT.2008.916929

J.C. Knight, T.A. Birks, P.S.J. Russell, D.M. Atkin, All-silica single-mode optical fiber with photonic crystal cladding. Opt. Lett. **21**(19), 1547–1549 (1996). https://doi.org/10.1364/OL.21.001547

M. Koerdt, S. Kibben, O. Bendig, S. Chandrashekhar, J. Hesselbach, C. Brauner, ..., L. Kroll, Fabrication and characterization of Bragg gratings in perfluorinated polymer optical fibers and their embedding in composites. Mechatronics **34**, 137–146 (2016). https://doi.org/10.1016/j.mechatronics.2015.10.005

K. Krebber, P. Lenke, S. Liehr, J. Witt, M. Schukar, Smart technical textiles with integrated POF sensors, *in Proceedings of SPIE 6933, Smart Sensor Phenomena, Technology, Networks, and Systems*, (California, 2008), p. 69330V. https://doi.org/10.1117/12.776758

B.T. Kuhlmey, R.C. McPhedran, C.M. de Sterke, Modal cutoff in microstructured optical fibers. Opt. Lett. **27**(19), 1684–1686 (2002). https://doi.org/10.1364/OL.27.001684

M.G. Kuzyk, U.C. Paek, C.W. Dirk, Guest-host polymer fibers for nonlinear optics. Appl. Phys. Lett. **59**(8), 902–904 (1991). https://doi.org/10.1063/1.105271

S. Küper, M. Stuke, UV-excimer-laser ablation of polymethylmethacrylate at 248 nm: Characterization of incubation sites with fourier transform IR- and UV-spectroscopy. Appl. Phys. A Solids Surf. **49**(2), 211–215 (1989). https://doi.org/10.1007/BF00616301

A. Lacraz, M. Polis, A. Theodosiou, C. Koutsides, K. Kalli, Femtosecond laser inscribed Bragg gratings in low loss CYTOP polymer optical fiber. IEEE Photon. Technol. Lett. **27**(7), 693–696 (2015). https://doi.org/10.1109/LPT.2014.2386692

M.C.J. Large, S. Ponrathnam, A. Argyros, N.S. Pujari, F. Cox, Solution doping of microstructured polymer optical fibres. Opt. Express **12**(9), 1966–1971 (2004). https://doi.org/10.1364/OPEX.12.001966

M.C.J. Large, L. Poladian, G.W. Barton, M.A. van Eijkelenborg, *Microstructured Polymer Optical Fibres. Microstructured Polymer Optical Fibres* (Springer US, Boston, 2008). https://doi.org/10.1007/978-0-387-68617-2

S.H. Law, J.D. Harvey, R.J. Kruhlak., M. Song, E. Wu, G.W. Barton, ..., M.C.J. Large, Cleaving of microstructured polymer optical fibres. Opt. Commun. **258**(2), 193–202 (2006). https://doi.org/10.1016/j.optcom.2005.08.011

C.S.J. Leitão, P.F. da Costa Antunes, J.A.M. Bastos, J. de Lemos Pinto, P.S. de Brito André, Plastic optical fiber sensor for noninvasive arterial pulse waveform monitoring. IEEE Sensors J. **15**(1), 14–18 (2015). https://doi.org/10.1109/JSEN.2014.2336594

S.G. Leon-Saval, R. Lwin, A. Argyros, Multicore composite single-mode polymer fiber. Opt. Express **20**(1), 141–148 (2012). https://doi.org/10.1364/OE.20.000141

Z. Li, H.Y. Tam, L. Xu, Q. Zhang, Fabrication of long-period gratings in poly(methyl methacrylate-co-methyl vinyl ketone-co-benzyl methacrylate)-core polymer optical fiber by use of a mercury lamp. Opt. Lett. **30**(10), 1117–1119 (2005). https://doi.org/10.1364/OL.30.001117

L. Li, L. Xia, Z. Xie, D. Liu, All-fiber Mach-Zehnder interferometers for sensing applications. Opt. Express **20**(10), 11109–11120 (2012). https://doi.org/10.1364/OE.20.011109

G. Liu, D. Feng, Evanescent wave analysis and experimental realization of refractive index sensor based on D-shaped plastic optical fiber. Optik – Int. J. Light Electron Opt. **127**(2), 690–693 (2016). https://doi.org/10.1016/j.ijleo.2015.10.129

H.Y. Liu, G.D. Peng, P.L. Chu, Thermal tuning of polymer optical fiber Bragg gratings. IEEE Photon. Technol. Lett. **13**(8), 824–826 (2001). https://doi.org/10.1109/68.935816

H. Liu, G. Peng, P. Chu, Polymer fiber Bragg gratings with 28-dB transmission rejection. IEEE Photon. Technol. Lett. **14**(7), 935–937 (2002). https://doi.org/10.1109/LPT.2002.1012390

D. Luo, Y. Yue, P. Li, J. Ma, L.L. Zhang, Z. Ibrahim, Z. Ismail, Concrete beam crack detection using tapered polymer optical fiber sensors. Measurement **88**, 96–103 (2016). https://doi.org/10.1016/j.measurement.2016.03.028

R. Lwin, A. Argyros, Connecting microstructured polymer optical fibres to the world, In *18th International Conference on Plastic Optical Fibers*. (Sydney, 2009)

C. Markos, A. Stefani, K. Nielsen, H.K. Rasmussen, W. Yuan, O. Bang, High-Tg TOPAS microstructured polymer optical fiber for fiber Bragg grating strain sensing at 110 degrees. Opt. Express **21**(4), 4758–4765 (2013). https://doi.org/10.1364/OE.21.004758

C.A.F. Marques, L.B. Bilro, N.J. Alberto, D.J. Webb, R.N. Nogueira, Narrow bandwidth Bragg gratings imprinted in polymer optical fibers for different spectral windows. Opt. Commun. **307**, 57–61 (2013). https://doi.org/10.1016/j.optcom.2013.05.059

T. Martynkien, P. Mergo, W. Urbanczyk, Sensitivity of birefringent microstructured polymer optical fiber to hydrostatic pressure. IEEE Photon. Technol. Lett. **25**(16), 1562–1565 (2013). https://doi.org/10.1109/LPT.2013.2271240

N.A. Mortensen, Effective area of photonic crystal fibers. Opt. Express **10**(7), 341–348 (2002). https://doi.org/10.1364/OE.10.000341

N.A. Mortensen, Semianalytical approach to short-wavelength dispersion and modal properties of photonic crystal fibers. Opt. Lett. **30**(12), 1455–1457 (2005). https://doi.org/10.1364/OL.30.001455

N.A. Mortensen, J.R. Folkenberg, M.D. Nielsen, K.P. Hansen, Modal cutoff and the V parameter in photonic crystal fibers. Opt. Lett. **28**(20), 1879–1881 (2003). https://doi.org/10.1364/OL.28.001879

R. Nogueira, R. Oliveira, L. Bilro, J. Heidarialamdarloo, New advances in polymer fiber Bragg gratings. Opt. Laser Technol. **78**(Part A), 104–109 (2015). https://doi.org/10.1016/j.optlastec.2015.08.010

G. Numata, N. Hayashi, M. Tabaru, Y. Mizuno, K. Nakamura, Ultra-sensitive strain and temperature sensing based on modal interference in perfluorinated polymer optical fibers. IEEE Photon. J. **6**(5), 6802306 (2014). https://doi.org/10.1109/JPHOT.2014.2352637

G. Numata, N. Hayashi, M. Tabaru, Y. Mizuno, K. Nakamura, Strain and temperature sensing based on multimode interference in partially chlorinated polymer optical fibers. IEICE Electron. Express **12**(2), 20141173 (2015). https://doi.org/10.1587/elex.12.20141173

R. Oliveira, L. Bilro, R. Nogueira, Bragg gratings in a few mode microstructured polymer optical fiber in less than 30 seconds. Opt. Express **23**(8), 10181–10187 (2015a). https://doi.org/10.1364/OE.23.010181

R. Oliveira, L. Bilro, R. Nogueira, Smooth end face termination of microstructured, graded-index, and step-index polymer optical fibers. Appl. Opt. **54**(18), 5629–5633 (2015b). https://doi.org/10.1364/AO.54.005629

R. Oliveira, S. Aristilde, J.H. Osorio, M.A.R. Franco, L. Bilro, R.N. Nogueira, C.M.B. Cordeiro, Intensity liquid level sensor based on multimode interference and fiber Bragg grating. Meas. Sci. Technol. **27**(12), 125104 (2016a). https://doi.org/10.1088/0957-0233/27/12/125104

R. Oliveira, L. Bilro, T.H.R. Marques, M. Napierala, T. Tenderenda, P. Mergo, . . . , R. Nogueira, Bragg Gratings Inscription in Highly Birefringent Microstructured POFs. IEEE Photon. Technol. Lett. **28**(6), 621–624 (2016b). https://doi.org/10.1109/LPT.2015.2503241

R. Oliveira, T.H.R. Marques, L. Bilro, C.M.B. Cordeiro, R.N. Nogueira, Strain, temperature and humidity sensing with multimode interference in POF. *The 25th International Conference on Plastic Optical Fibres*, OP35 (2016c)

R. Oliveira, T.H.R. Marques, L. Bilro, R. Nogueira, C.M.B. Cordeiro, Multiparameter POF Sensing based on Multimode Interference and Fiber Bragg Grating. J. Light. Technol. 1–8 (2016d). https://doi.org/10.1109/JLT.2016.2626793

R. Oliveira, J.H. Osorio, S. Aristilde, L. Bilro, R.N. Nogueira, C.M.B. Cordeiro, Simultaneous measurement of strain, temperature and refractive index based on multimode interference, fiber tapering and fiber Bragg gratings. Meas. Sci. Technol. **27**(7), 75107 (2016e). https://doi.org/10.1088/0957-0233/27/7/075107

A. Ortigosa-Blanch, J.C. Knight, W.J. Wadsworth, J. Arriaga, B.J. Mangan, T.A. Birks, P.S.J. Russell, Highly birefringent photonic crystal fibers. Opt. Lett. **25**(18), 1325–1327 (2000). https://doi.org/10.1364/OL.25.001325

G.D. Peng, P.L. Chu, Polymer optical fiber Photosensitivities and highly tunable fiber gratings. Fiber Integr. Opt. **19**(4), 277–293 (2000). https://doi.org/10.1080/014680300300001662

G.D. Peng, Z. Xiong, P.L. Chu, Photosensitivity and gratings in dye-doped polymer optical fibers. Opt. Fiber Technol. **5**(2), 242–251 (1999). https://doi.org/10.1006/ofte.1998.0298

D. Rithesh Raj, S. Prasanth, T.V. Vineeshkumar, C. Sudarsanakumar, Ammonia sensing properties of tapered plastic optical fiber coated with silver nanoparticles/PVP/PVA hybrid. Opt. Commun. **340**, 86–92 (2015). https://doi.org/10.1016/j.optcom.2014.11.092

D. Saez-Rodriguez, K. Nielsen, O. Bang, D. Webb, Simple room temperature method for polymer optical fibre cleaving. J. Lightwave Technol. **33**(23), 4712–4716 (2015). https://doi.org/10.1109/JLT.2015.2479365

D. Sáez-Rodríguez, K. Nielsen, O. Bang, D.J. Webb, Photosensitivity mechanism of undoped poly(methyl methacrylate) under UV radiation at 325 nm and its spatial resolution limit. Opt. Lett. **39**(12), 3421–3424 (2014). https://doi.org/10.1364/OL.39.003421

F. Sequeira, D. Duarte, L. Bilro, A. Rudnitskaya, M. Pesavento, L. Zeni, N. Cennamo, Refractive index sensing with D-shaped plastic optical fibers for chemical and biochemical applications. Sensors **16**(12), 2119 (2016). https://doi.org/10.3390/s16122119

M. Silva-López, A. Fender, W.N. MacPherson, J.S. Barton, J.D.C. Jones, D. Zhao, . . . , I. Bennion, Strain and temperature sensitivity of a single-mode polymer optical fiber. Opt. Lett. **30**(23), 3129–3131(2005). https://doi.org/10.1364/OL.30.003129

L.B. Soldano, E.C.M. Pennings, Optical multi-mode interference devices based on self-imaging: Principles and applications. J. Lightwave Technol. **13**(4), 615–627 (1995). https://doi.org/10.1109/50.372474

R. Srinivasan, B. Braren, R.W. Dreyfus, L. Hadel, D.E. Seeger, Mechanism of the ultraviolet laser ablation of polymethyl methacrylate at 193 and 248 nm: laser-induced fluorescence analysis, chemical analysis, and doping studies. J. Opt. Soc. Am. B **3**(5), 785–791 (1986). https://doi.org/10.1364/JOSAB.3.000785

G. Statkiewicz-Barabach, D. Kowal, M.K. Szczurowski, P. Mergo, W. Urbanczyk, Hydrostatic pressure and strain sensitivity of long period grating fabricated in polymer microstructured fiber. IEEE Photon. Technol. Lett. **25**(5), 496–499 (2013). https://doi.org/10.1109/LPT.2013.2244590

G. Statkiewicz-Barabach, P. Maniewski, P. Mergo, W. Urbanczyk, Fiber Bragg grating-based Fabry-Perot interferometer in polymer fiber, in *The 25th International Conference on Plastic Optical Fibres*. (UK 2016), p. OP22

G. Statkiewicz, T. Martynkien, W. Urbańczyk, Measurements of modal birefringence and polarimetric sensitivity of the birefringent holey fiber to hydrostatic pressure and strain. Opt. Commun. **241**(4), 339–348 (2004). https://doi.org/10.1016/j.optcom.2004.07.021

A. Stefani, K. Nielsen, H.K. Rasmussen, O. Bang, Cleaving of TOPAS and PMMA microstructured polymer optical fibers: core-shift and statistical quality optimization. Opt. Commun. **285**(7), 1825–1833 (2012a). https://doi.org/10.1016/j.optcom.2011.12.033

A. Stefani, M. Stecher, G.E. Town, O. Bang, Direct writing of fiber Bragg grating in microstructured polymer optical fiber. IEEE Photon. Technol. Lett. **24**(13), 1148–1150 (2012b). https://doi.org/10.1109/lpt.2012.2197194.

M. Steffen, M. Schukar, J. Witt, K. Krebber, M. Large, A. Argyros, Investigation of mPOF-LPGs for sensing applications, in *18th International Conference on Plastic Optical Fibers*. (Sydney, 2009), p. 25–26

M.K. Szczurowski, T. Martynkien, G. Statkiewicz-Barabach, W. Urbanczyk, D.J. Webb, Measurements of polarimetric sensitivity to hydrostatic pressure, strain and temperature in birefringent dual-core microstructured polymer fiber. Opt. Express **18**(12), 12076–12087 (2010). https://doi.org/10.1364/OE.18.012076

F. Taffoni, D. Formica, P. Saccomandi, G. Di Pino, E. Schena, Optical fiber-based MR-compatible sensors for medical applications: An overview. Sensors **13**(10), 14105–14120 (2013). https://doi.org/10.3390/s131014105

X.A. Ton, V. Acha, P. Bonomi, B. Tse Sum Bui, K. Haupt, A disposable evanescent wave fiber optic sensor coated with a molecularly imprinted polymer as a selective fluorescence probe. Biosens. Bioelectron. **64**, 359–366 (2015). https://doi.org/10.1016/j.bios.2014.09.017

M.A. van Eijkelenborg, A. Argyros, G. Barton, I.M. Bassett, M. Fellew, G. Henry, ..., J. Zagari Recent progress in microstructured polymer optical fibre fabrication and characterisation. Opt. Fiber Technol. **9**(4), 199–209 (2003). https://doi.org/10.1016/S1068-5200(03)00045-2

M.A. van Eijkelenborg, W. Padden, J.A. Besley, Mechanically induced long-period gratings in microstructured polymer fibre. Opt. Commun. **236**, 75–78 (2004). https://doi.org/10.1016/j.optcom.2004.03.004

M. van Eijkelenborg, M. Large, A. Argyros, J. Zagari, S. Manos, N. Issa, ..., N.A. Nicorovici, Microstructured polymer optical fibre. Opt. Express, **9**(7), 319–327 (2001). https://doi.org/10.1364/OE.9.000319

D. Webb, M. Aressy, A. Argyros, J.S. Barton, H. Dobb, ... M. Silva-López, Grating and interferometric devices in POF, in *14th International Conference on Polymer Optical Fibers*. (Hong Kong, 2005), p. 325–328

J. Witt, M. Breithaupt, J. Erdmann, K. Krebber, Humidity sensing based on microstructured POF long period gratings, in *Proceedings of the 20th International Conference on Plastic Optical Fibres*. (Bilbao, 2011), p. 409–414

C. Wochnowski, S. Metev, G. Sepold, UV-laser-assisted modification of the optical properties of polymethylmethacrylate. Appl. Surf. Sci. **154**, 706–711 (2000). https://doi.org/10.1016/S0169-4332(99)00435-3

G.K.L. Wong, M.S. Kang, H.W. Lee, F. Biancalana, C. Conti, T. Weiss, P.S.J. Russell, Excitation of orbital angular momentum resonances in helically twisted photonic crystal fiber. Science **337**(6093), 446–449 (2012). https://doi.org/10.1126/science.1223824

G. Woyessa, A. Fasano, A. Stefani, C. Markos, H.K. Rasmussen, O. Bang, Single mode step-index polymer optical fiber for humidity insensitive high temperature fiber Bragg grating sensors. Opt. Express **24**(2), 1253–1260 (2016). https://doi.org/10.1364/OE.24.001253

Q. Wu, Y. Semenova, P. Wang, G. Farrell, High sensitivity SMS fiber structure based refractometer – analysis and experiment. Opt. Express **19**(9), 7937–7944 (2011). https://doi.org/10.1364/OE.19.007937

Z. Xiong, G.D. Peng, B. Wu, P.L. Chu, Highly tunable Bragg gratings in single-mode polymer optical fibers. IEEE Photon. Technol. Lett. **11**(3), 352–354 (1999). https://doi.org/10.1109/68.748232

P. Yeh, N. Yeh, C.-H. Lee, T.-J. Ding. Applications of LEDs in optical sensors and chemical sensing device for detection of biochemicals, heavy metals, and environmental nutrients. Renew. Sust. Energ. Rev. **75**, 461–468 (2017). https://doi.org/10.1016/j.rser.2016.11.011

W. Yuan, L. Khan, D.J. Webb, K. Kalli, H.K. Rasmussen, A. Stefani, O. Bang, Humidity insensitive TOPAS polymer fiber Bragg grating sensor. Opt. Express **19**(20), 19731–19739 (2011). https://doi.org/10.1364/OE.19.019731

C. Zhang, X. Chen, D.J. Webb, G.-D. Peng. Water detection in jet fuel using a polymer optical fibre Bragg grating, in *Proceedings of the SPIE 7503, 20th International Conference on Optical Fibre Sensors*, (Edinburgh, 2009), p. 750380. https://doi.org/10.1117/12.848696

C. Zhang, W. Zhang, D.J. Webb, G.-D. Peng, Optical fibre temperature and humidity sensor. Electron. Lett. **46**(9), 643 (2010). https://doi.org/10.1049/el.2010.0879

W. Zhang, D.J. Webb, Humidity responsivity of poly(methyl methacrylate)-based optical fiber Bragg grating sensors. Opt. Lett. **39**(10), 3026–3029 (2014). https://doi.org/10.1364/OL.39.003026

W. Zhang, D.J. Webb, G.-D. Peng, Investigation into time response of polymer fiber Bragg grating based humidity sensors. J. Lightwave Technol. **30**(8), 1090–1096 (2012a). https://doi.org/10.1109/JLT.2011.2169941

W. Zhang, D.J. Webb, G.-D. Peng, Polymer optical fiber Bragg grating acting as an intrinsic biochemical concentration sensor. Opt. Lett. **37**(8), 1370–1372 (2012b). https://doi.org/10.1117/12.922279

G. Zhou, C.F.J. Pun, H.Y. Tam, A.C.L. Wong, C. Lu, P.K.A. Wai, Single-mode perfluorinated polymer optical fibers with refractive index of 1.34 for biomedical applications. IEEE Photon. Technol. Lett. **22**(2), 106–108 (2010). https://doi.org/10.1109/LPT.2009.2036377

Q. Zhuang, H. Yaosheng, M. Yu, L. Wei, L. Xianping, Z. Wenhui, . . . , L. Elfed, Water-equivalent fiber radiation dosimeter with two scintillating materials. Biomedical Optics Express **7**(12), 4919–4927 (2016). https://doi.org/10.1364/BOE.7.004919

O. Ziemann, J. Krauser, P.E. Zamzow, W. Daum, *Optical Short Range Transmission Systems. POF Handbook* (Springer, Berlin, 2008). https://doi.org/10.1007/978-3-540-76629-2

J. Zubia, J. Arrue, Plastic optical fibers: an introduction to their technological processes and applications. Opt. Fiber Technol. **7**(2), 101–140 (2001). https://doi.org/10.1006/ofte.2000.0355

Solid Core Single-Mode Polymer Fiber Gratings and Sensors

51

Kishore Bhowmik, Gang Ding Peng, Eliathamby Ambikairajah, and Ginu Rajan

Contents

K. Bhowmik (✉)
HFC Assurance, Operate and Maintain Network, NBN, Melbourne, VIC, Australia
e-mail: kishore_ete1@yahoo.com

G. D. Peng
Photonics and Optical Communications, School of Electrical Engineering and Telecommunications, University of New South Wales, Sydney, NSW, Australia

E. Ambikairajah
School of Electrical Engineering and Telecommunications, UNSW, Sydney, Australia

G. Rajan
School of Electrical, Computer and Telecommunications Engineering, University of Wollongong, Wollongong, Australia

School of Electrical Engineering and Telecommunications, UNSW, Sydney, Australia

G.-D. Peng (ed.), *Handbook of Optical Fibers*,
https://doi.org/10.1007/978-981-10-7087-7_4

Abstract

Research on single-mode polymer optical fibers Bragg gratings based on polymer optical fiber and their applications has considerably progressed in the recent years. This chapter provides an overview on recent research developments on solid core single-mode polymer fiber Bragg grating sensors, devices, and their applications.

Keywords

Polymer optical fiber (POF) · Fiber Bragg grating (FBG) · Polymer optical fiber Bragg grating (POFBG) · POFBG applications

Introduction

Over many years, sensing has become a key enabling technology for entertainment, health, transport, and many industrial applications. In different advanced applications in which sensitivity, size, and multiplexing are vital, sensors based on optical fibers can provide novel solutions. As a result, optical fiber sensor technology is developed as a powerful and rich technology that is currently being implemented in a wide range of applications (Culshaw and Kersey 2008; Bogue 2011; Lopez-Higuera et al. 2011). Optical fiber sensors can be described as a medium through which physical properties interact with light guided either through an optical fiber or by an optical fiber to an interaction region to produce a modulated optical signal with information related to the measurement parameters. One of the main advantages of fiber sensors is that they have the capability to measure a wide range of parameters based on the requirements of end users. Some other key advantages of fiber sensors include lightweight, electromagnetic interference (EMI) resistance, high sensitivity or bandwidth, multiplexing capability, and environmental robustness. So the technology in fiber sensors are progressively developing and becoming widespread.

Among different types of sensors based on optical fibers, fiber Bragg grating (FBG) is one of the most common technologies that has emerged as a promising sensing element owing to its potential for reliably measuring several physical parameters. FBG-based sensors appear to be useful for various applications such as strain, temperature, pressure, vibration, etc. (Kersey et al. 1997; Mora et al. 2000). Although, to date, most of the optical fiber sensors demonstrated were based on silica ones, however, silica-based FBGs still have some issues related to their materials, such as a high Young's modulus (70 GPa) and a failure strain range of only 5–10% (Kurkjian et al. 1989), which can considerably reduce the strain limit (Nellen et al. 2003), thereby making them disfavored by some applications.

Polymer optical fibers (POFs), also known as plastic optical fibers, have some unique advantages over conventional silica-based ones for sensing applications. The interest in POF has increased more recently mainly due to the attractive material properties compared to silica which includes larger elasticity capabilities, high numerical aperture, lower stiffness, and biocompatibility. Lightweight and low cost further add to the advantages of POF. Due to their low stiffness, POFs can be used in applications where high sensitivity is expected, and they also bend more easily than equivalent silica fibers with the same diameter. Moreover, as POFs are biocompatible in nature, a POF sensor can be used in biosensing applications as well. These unique properties are exploited and have been used to expand the envelope of sensing applications beyond those previously realized with silica optical fiber sensors.

FBG research and technology for silica fibers has reached a mature state, while it is still in a very early phase of research for the polymer fiber-based ones, with some of the fundamental properties and characteristics still unknown for polymer FBGs. Compared with silica FBGs, the research on polymer FBGs have not excelled much due to the lack of commercially available single-mode POFs and issues regarding their standardization, high losses, fabrication difficulties, etc. Application-oriented research on POF gratings is still in the initial phase where device level studies have not been carried out in detail, even though some of fundamental properties and characteristics are still unknown for polymer FBGs. Therefore, further research on the many fundamental aspects of polymer fibers and gratings are necessary to exploit their sensing capabilities. This chapter will summarize the recent research outcomes of solid core single-mode polymer FBGs and its sensing applications. Finally, comments are made on the outlook of POF sensors and suggestions for future research directions to enable new sensing capabilities.

Polymer Optical Fiber

Poly(methyl methacrylate) (PMMA) is the most commonly used polymer for POFs, and the development of POF began in the 1970s. At that time, two companies that worked on developing commercial POFs were Du Pont in the United States of America and Mitsubishi Rayon (M. R.) in Japan. Mitsubishi filed a patent for the melt spinning of a simple light-guiding core-cladding structure in 1974. In this structure, PMMA or polystyrene (PS) was employed as the core material and various fluorinated polymers used as cladding. The minimum attenuation measured was 3500 dB/km at an unspecified wavelength. Several months later, another patent for high-purity acrylics, which led to the development of Eska™ was reported. This was the first reported PMMA core with a poly(fluroalkylmethacrylate)-cladding POF, which led to a market for the fabrication of POFs, with Eska showing a minimum loss of 125 dB/km at a wavelength of 567 nm. Since the 1980s, the main developments in the field of POFs have been led by Kaino and his groups at NTT's Ibaraki Laboratories. Their most impressive work was a reduction in attenuation for PMMA-based POFs to 55 dB/km at a wavelength of 567 nm and also developed a technique for fabricating a low-loss POF with a PMMA-D5 core and fluorinated alkyl methacrylate copolymer cladding. The step-index (SI) POFs reported had some limitations of bandwidth due to their large multimode core, which resulted in increased modal dispersion. To solve this modal dispersion problem, the first graded index (GI) POF was reported by Ohtsuka and Hatanaka in 1976 (Ohtsuka and Hatanaka 1976). As these fibers had small nonlinearities in their responses to light, the fabrication of a highly nonlinear single-mode fiber would be beneficial for building fiber waveguide devices. Single-mode POFs would be advantageous for nonlinear optical effects because of their small effective core areas and long interaction lengths, together with higher nonlinearities achieved by doping their nonlinear organics.

When coming to single-mode fibers, silica fiber technology is abundant, while POF technology had focused exclusively on large core step and GI-MM fibers. Research on single-mode polymer was started in the early 1990s with the development of PMMA-based nonlinear POFs with dye-doped cores (Kuzyk et al. 1991). The core sizes of these fibers were approximately 8 μm and cladding diameters 125 μm. A single-mode fiber with copolymer for both core and cladding was first demonstrated by Bosc et al. (Bosc and Toinen 1992), where compositions of the copolymer in the core and cladding were adjusted to obtain a refractive index profile that yielded single-mode guidance. The poly(methyl methacrylate-co-ethyl methacrylate) copolymer was used as a core and poly(methyl methacrylate-co-trifluoroethyl methacrylate) copolymer as a cladding, and the reported core and cladding diameters of these fibers were 6 and 145 μm, respectively.

Later, research carried out by the University of New South Wales (UNSW) demonstrated various PMMA-based single-mode POFs (Peng et al. 1995; Xiong et al. 1999; Liu et al. 2002), where the fiber manufacture followed a multiple step process. Its basic monomer is methyl methacrylate (MMA) by which a POF is constructed through a polymerization process. In the POF fabrication process, while approximately the same amounts of two monomers, MMA and ethyl methacrylate (EMA), are used in both the core and cladding, a small amount (5–10%) of a third monomer, benzyl methacrylate (BzMA), is added in the core fabrication process for higher refractive index. This fiber has a core refractive index of 1.4793, and cladding index was 1.4707, and the difference in the refractive index of core and cladding is 0.0086. This index is capable of being tuned through adding a different methacrylate when necessary. Single-mode photosensitive POFs with different diameters ranging from 100 μm to 240 μm for the cladding and from 6 to 12 μm for the core have been reported (Xiong et al. 1999). These fibers are typically single mode in operation and suffer high transmission losses in the C-L band which are approximately 0.6 dB/cm, considerably higher compared to the silica fiber standards.

However, POFs have long been neglected compared with the far more popular and superior silica fibers. Only recently, research and development of POF have been greatly boosted by focusing on the sensing applications for which they are

Table 1 Comparison of relevant parameters of silica and polymer optical fibers

Property	Silica Fiber	Polymer Fiber
Attenuation (dB/km)	0.2∼0.3	10∼100
Young's modulus (GPa)	70	2∼3
Breakdown strain (%)	1∼2	5∼10

deemed to be able to compete with conventional silica optical fibers. Table 1 presents a summary of the relevant characteristics of typical silica and polymer fibers.

Due to the difficulties involved in fabricating single-mode POFs, commercially available POFs are typically multimode at their operating wavelengths with both step-index (SI) and graded-index (GI) configurations. The advantages of their large core sizes are the reason why single-mode POFs are given less attention, which has a core diameter of ∼6–12 μm. Also, they have experienced a surge in applications for short-haul telecommunication systems in which multimode POFs were previously used due to their ease of implementation. However, more interestingly, many researchers have realized their unique properties for sensing strain, temperature, humidity, etc. and looked into the sensing potential of single-mode POFs.

Polymer Optical Fiber Bragg Grating (POFBG)

When the core of anoptical fiber is irradiated with UV light, a permanent change occurs in its refractive index. This photorefractive effect, known as photosensitivity, has great practical significance due to the availability of increasing numbers of passive and active fiber devices. The photosensitivity of silica optical fibers and various photosensitization techniques for enhancing it, such as hydrogen loading or flame brushing, are well known. The photosensitivity of a POF is complex, where a variety of possible mechanisms dependent on the intensity, the inscription wavelength and the polymerization process used. Possible mechanisms for photosensitivity can be photopolymerization, chain scission, or cross-linking between polymer chains. However, the studies on the photosensitivity of a PMMA POF lead to the surge in the research on Bragg gratings based on POFs.

The first studies of the photosensitivity of bulk PMMA were conducted by Tomlinson et al. in the 1970s (Tomlinson et al. 1970). In this work, it was found that a properly prepared PMMA (through oxidation of the monomer) exhibited a substantial increase in refractive index irradiation with a UV light at 325 nm (He-Cd laser) or 365 nm (Hg arc). The polymer absorbed the UV radiation which produced free radicals that formed cross-links between adjacent polymer chains. It was assumed that, because a cross-link chain increased the density, the refractive index increased with index changes of up to 3×10^{-3}, accompanied by a density change of 0.8% reported. Then, three-dimensional holographic gratings with peak refractive index changes of 2.3×10^{-3} were introduced, but they required more than 100 h of illumination (Moran and Kaminow 1973). It was demonstrated that an observed refractive index change was due to polymerization of the residual monomer within the PMMA, not as a result of cross-linking, as had previously been reported (Bowden et al. 1974), with the role of monomers later confirmed by Marotz et al. (Marotz 1985). Subsequently, information related to the photosensitivity of PMMA was found in an article on a dye-doped polymer laser in which it was reported that a simple narrow-band fixed-frequency longitudinally pumped laser was made using a pair of gratings inside a large block of PMMA (Kaminow et al. 1971). These gratings had dimensions of $2 \times 2 \times 2$ mm^3 and a maximum reflectivity of 20%. However, it took a long time (>100 h) for any of the above mechanisms to modify the refractive index in a PMMA sample. In 1984, Kopietz et al. (1984) studied the evolution of inscribing gratings on bulk PMMA using a mercury UV lamp. In the first step of inscription, a decrease in the refractive index was observed due to creation of the initial monomer. But in the second step, it became positive as a result of photopolymerization of the residual monomers. The fabrication of integrated optical Bragg grating reflectors in organic waveguides using photolithography was reported (Aramaki et al. 1993).

Although the material preparation and fabrication processes for current POFs are considerably different from those in early works, they are useful as a reference. Research on the photosensitivity of a POF initially begun at the UNSW, Australia, where the photosensitivities of several PMMA-POFs made from undoped, dye-doped, or oxidated preforms were examined under varying irradiation wavelengths, intensities, and times. The large photosensitivity in a dye-doped POF with a small pump power (1 mW) at visible wavelength was demonstrated (Peng et al. 1999). It was confirmed that the refractive index in the core region could be changed under the irradiation of a pulse UV laser beam and shown that it was possible to write gratings in these multimode fibers.

Fabrication Techniques

The first report of writing a grating in a single-mode POF through photosensitivity was demonstrated by Xiong et al. (1999). The obtained reflectivity of the grating was 80% with an index change of 10^{-4} using a Nd:YAG pulse laser. Later, a grating with a 28 dB transmission rejection and bandwidth of less than 0.5 nm operating in near IR wavelengths was reported (Liu et al. 2002). Two distinct stages in the fabrication of POFBGs corresponding to low and high index modulation gratings were also reported (Liu et al. 2003a). It was referred as type-I and type-II gratings, where in the first stage, the index modulation grew slowly and linearly, leading to an index change of about 2×10^{-4} (when the exposure was below the threshold). But in the second stage, the index modulation increased rapidly and exponentially, leading to a maximum of around 2×10^{-3} (when the exposure was above that threshold). Microscope inspections revealed apparent damage at the core-cladding boundary. Then, a different growth behavior of a POFBG was observed, when it was exposed to low-power UV pulses (Liu et al. 2004). This behavior was quite different from those previous reported for type-I and type-II gratings, where its growth and erasure occurred when the UV irradiation was on and regrowth after the UV exposure was off. The reflectivity increased when the exposure was on for 28 min and remained roughly constant for another 20 min. Then, the reflectivity decreased by 88 min, and the UV exposure was blocked, increased again after approximately 8 h, and then remained constant. This was due to relaxation of the thermal stresses that might recover the grating's strength.

The grating inscription techniques for polymer optical fibers (similar to silica fibers) can be classified as:

(i) Phase mask technique (Hill et al. 1993; Anderson et al. 1993)
(ii) Point-by-point writing technique (Malo et al. 1993)
(iii) Interferometer fabrication technique (Meltz et al. 1989)

One of the most effective methods for writing gratings is phase mask approach which was first demonstrated in 1993 (Luo et al. 2011). The key advantages of this writing method are flexibility and repeatability. As its setup requires much simpler and less expensive equipment than other techniques, it incurs an overall lower fabrication cost. It reduces the complexity of the FBG inscription system through one optical component and, moreover, offers the capability to write different types of gratings using different phase masks. One of the main disadvantages of this technique is that a separate phase mask is required for each new specific Bragg wavelength, as the period of a Bragg grating is half of the phase mask. Also, the manufactured Bragg grating depends greatly on the quality of the phase mask.

High-Reflective POFBG Fabrication

Typical Bragg gratings in single-mode polymer fibers are fabricated using the standard phase mask technique. In a demonstrated example as shown in Fig. 1, a 50 mW Kimmon IK series He-Cd Laser emitting light at 325 nm was used to inscribe the gratings. The phase mask is 10 mm long and has a pitch of 1030 nm which is suitable for a 320 nm wavelength and can produce 1 cm long gratings with a peak-reflected wavelength circa 1530 nm in the single-mode POF used. The laser beam is irradiated onto the side of the fiber through the phase mask placed parallel to the fiber. An XYZ translation stage is used to adjust the position of the fiber, so that the UV light is uniformly focused onto the core of the fiber. Two plano-convex cylindrical lenses are also used in the setup, where one to expand the beam to cover approximately 10 mm of the fiber and the other to focus this expanded beam to the core region of the POF.

Due to the high transmission loss of polymer fiber in the wavelength of operation (1530 nm), the length of polymer fiber is typically restricted to approximately 10 cm and is coupled to a silica fiber pigtail. A high-power broadband source is used to observe the Bragg grating reflection spectrum, which operates in a wavelength range of 1520–1590 nm with a peak power at circa 1530 nm, and the reflected signal from the grating is directed via a fiber circulator to an optical spectrum analyzer (OSA)

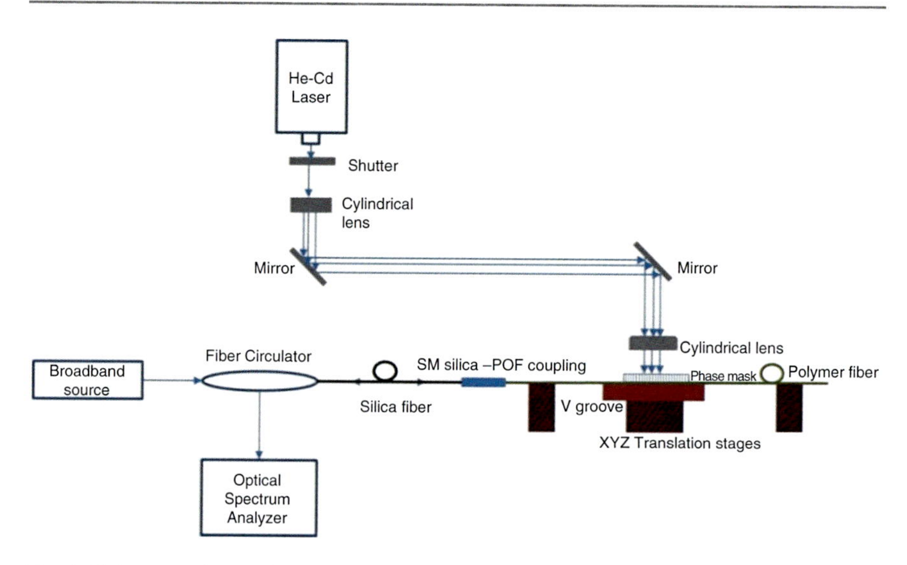

Fig. 1 Schematic of a standard polymer FBG fabrication setup

and is then monitored and recorded. A small drop of index-matching paraffin oil can be applied to join between the silica and POF to avoid back reflections. It's worth mentioning that the cleaving of a POF is not as easy as that of a silica fiber. A smooth and cross-sectional POF is important to avoid the joint loss between it and the silica fiber during grating fabrication and in pigtailed fabrications for applications. Since there are no commercialized POF cleavers available in the market, a hot plate and a sharp blade can be used to cleave the POF. To do this, the POF sample needs to be placed on the hot plate for 1–2 min which needs to be heated to approximately 60 °C. When the POF is properly heated, it can be easily cleaved by a new sharp blade. The quality of its cutting surface can be checked by a magnifying glass, and it depends mainly on the temperature of the fiber and the cutting speed.

Figure 2a demonstrates one of the examples of dynamic growth process of a POFBG under UV exposure where the transmission spectra at different UV exposure times are shown (Bhowmik et al. 2016a). In the transmission spectra, a change in transmission at around 1530 nm is observed after 30 s of exposure which is much quicker than previously reported other POF gratings (Liu et al. 2003a; Rajan et al. 2013a). The grating continues to grow stronger with further exposure and after 60 s of exposure, the highest dip (6.87 dB) in its transmission spectra corresponding to a reflectivity of 79.42%, and then decreases with further exposure. After 180 s of exposure, it appears stable, and after 300 s, this dip becomes 5.24 dB which corresponds to a 70.10% reflectivity. The blue shifts observed in the transmission spectra are due to the UV laser-induced temperature change during the grating inscription procedure.

The observed growth in the grating's reflection peaks at different UV exposure times for a POF is shown in Fig. 2b. A noticeable reflection peak is reported within 5 s of exposure, and then the grating continues to grow stronger with further

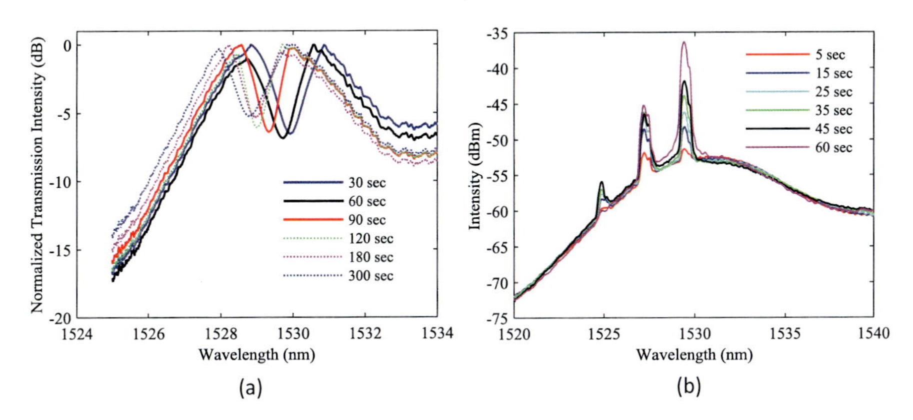

Fig. 2 (**a**) Transmission spectra of POFBG fabrication at different UV exposure times. (**b**) Reflection spectra of POFBG fabrication at different UV exposure times

exposure. From the figure, it is clear that the reflection spectrum increases quickly with UV exposure and the maximum reflection power is recorded for an exposure time of 60 s and the peak power decreases with further exposure. Therefore, in the growth pattern of the POFBG, the Bragg peak begins to appear just after the fiber is illuminated with UV exposure, grows continuously until reaching saturation, and then starts decreasing. The multiple peaks shown in the spectra are due to the effects of the coupling mismatch between the polymer and silica fibers and can be eliminated by adjusting the coupling.

To use this fabricated POFBGs for a particular application, POFBG pigtails need to be fabricated by gluing the polymer fiber to a silica fiber pigtail. A stable POFBG pigtail is highly important to reduce the error and the uncertainty due to fiber-coupling issues while characterizing the grating. A schematic diagram of the gluing procedure and a photograph of the glued region of the POF grating pigtail are shown in Fig. 3a and b, respectively. In this process, a 3 cm long transparent plastic capillary tube (approximately 2–3 mm diameter) with a small hole in the middle of the tube is first inserted to the silica fiber pigtails. Then the POF and single-mode silica fiber need to be aligned properly, and after that a small quantity of UV epoxy was applied at the junction to connect the fibers. The output spectrum needs to monitor continuously, and the fibers were slightly realigned to maintain the alignment, and the UV epoxy is cured for 10 min. After curing, more epoxy was added to enhance the strength of the connection and cured further. After this process the glued fiber was slightly lifted upwards, and the capillary tube is brought to the glued section. Now, more epoxy need to be added through the side hole in the middle of the tube until the tube is filled with epoxy. Then, the epoxy was cured for approximately 60 min. In that way a very stable and permanent POF grating pigtail can be made.

The interest in POF has increased recently, and since then, through the use of different fiber types and inscription processes, a considerable progress has

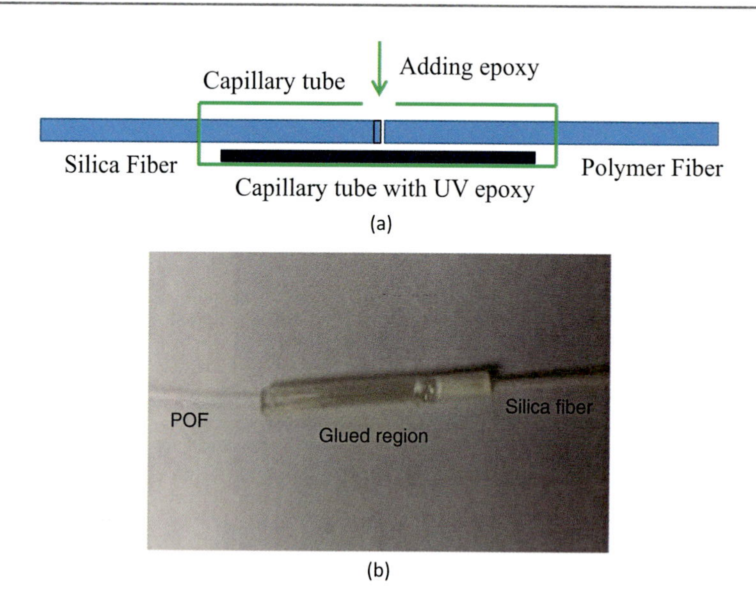

Fig. 3 (**a**) Schematic diagram showing the polymer-silica fiber connection procedure and (**b**) photograph of a POFBG pigtail

been realized toward the realization of high-quality gratings in polymer fibers. Previously, it has been reported that the inscription time for a grating fabrication in a POF was approximately 1 h (Liu et al. 2003a), and more recently together with the use of high photosensitivity POF and a 50 mW He-Cd laser with 325 nm emission, this inscription time has been reduced to approximately 5–10 min (Rajan et al. 2013a). However, this performance is still not sufficient for commercial bulk production for application research and also not competitive with gratings based on silica fibers. Therefore, it's important to optimize the grating fabrication procedure for a solid core single-mode POF, so that POFBGs can be used in its full potential in a range of application areas. To realize this, Bhowmik et al. (2016a) fabricated high-reflective Bragg gratings on single-mode POF using solvent etching technique.

Sensitivity of POFBG to Various Physical Parameters

Temperature

The first reported temperature sensitivity for a grating in a PMMA-based step-index fiber was −360 pm/°C which resulted from a 50 °C temperature rise (Liu et al. 2001), and the response was nonlinear with a greater thermal sensitivity at higher temperatures. After that, the authors reported a temperature sensitivity of −149 pm/°C (Liu et al. 2003b), this time a linear response and a working

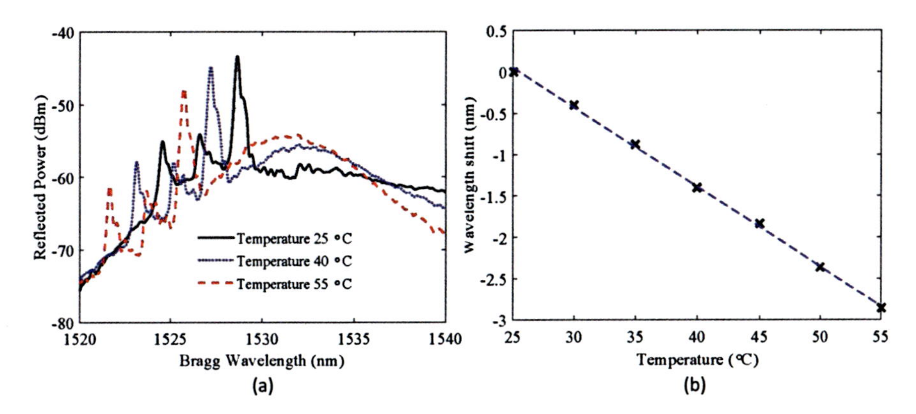

Fig. 4 (**a**) Reflection spectra of POFBG at different temperatures and (**b**) temperature response of single-mode POFBG

temperature up to 65 °C. Again this is significantly larger than that of silica and is in the negative direction due to the dominance of the negative thermo-optic coefficient of polymers.

To measure the temperature sensitivity of polymer FBGs, a Peltier cooler-based system can be used, which has an accuracy of ±0.10 °C, and the temperature of gratings are varied from 25 °C to 55 °C at 5 °C intervals (Bhowmik et al. 2015a). Five minutes time interval should be given between each temperature steps to ensure the stability of measurements. In each temperature step, the corresponding Bragg wavelength shift was recorded by an OSA, and the measured reflection spectra of the POFBG at room temperature (25 °C) and temperatures of 40 °C and 55 °C are shown in Fig. 4a. A negative Bragg wavelength shift is observed for an increase in temperature of the POFBG, and it is mainly due to the negative thermo-optic coefficient of the POF. The measured responses to temperature of the single-mode POFBG are presented in Fig. 4b. From the experimental results, it is proved that the temperature-induced wavelength shift is linear within the measurement range and the calculated temperature sensitivity of the single-mode POFBG is -95 ± 0.2 pm/°C with a diameter of 180 μm. The measured temperature sensitivity of the single-mode POFBG is nearly 9.5 times larger than that of a typical silica FBG, which is approximately 10 pm/°C.

Strain

Since the Young's modulus of a POF is much lower than that of a silica fiber, it is expected that the strain sensitivity of a POFBG will be higher than that of a silica FBG. A previous study by UNSW group's researchers demonstrated that a POFBG can be tuned to 20 nm using a simple tensile strain (Xiong et al. 1999). Moreover, a silica FBG can be tuned to only a few nanometers by a simple tensile stress and can

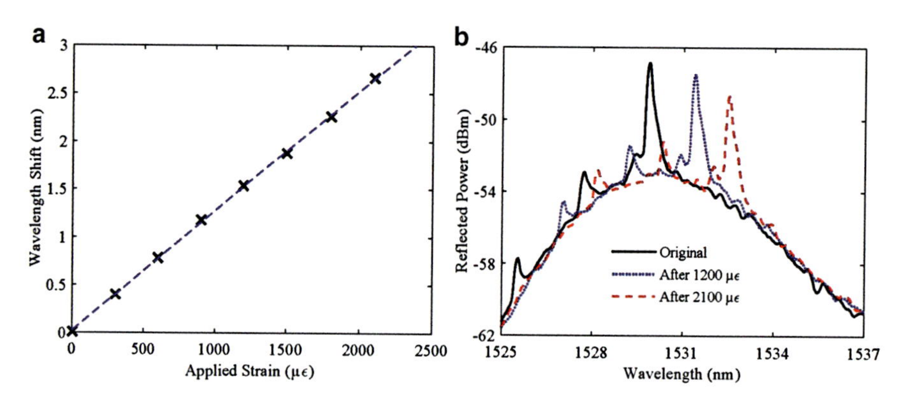

Fig. 5 (**a**) Strain response of single-mode POFBG and (**b**) reflection spectra of POFBG at two different tensile strains

be increased by compressive stress. A 32 nm single-frequency Bragg wavelength tuning under compression was first reported by Ball (Ball and Morey 1994). Then, a Bragg grating fast-tunable filter with a tuning range of 45 nm was reported by Iocco (Iocco et al. 1997, 1999). This tunable filter system is based on two piezoelectric actuators with the tuning speed increased up to 21 nm/ms. Although a broad tuning range with a high tuning speed have been achieved with silica FBGs through compression, the reversibility and reproducibility of this tuning are very low and require bulky components to perform this tuning action. On the other hand, due to its low Young's modulus, a larger breakdown strain sensitivity can be achieved by a POFBG. Liu (Liu et al. 2005) reported a very large Bragg wavelength tuning range (32 nm) with good reversibility, reproducibility, and repeatability using a POFBG. Therefore, a POF is far superior to a silica optical fiber due to its smaller Young's modulus, and, in this section, the tensile strain sensitivity of a single-mode POFBG is characterized.

To measure strain sensitivity, a standard procedure can be followed, where a known elongation is applied to the optical fiber using a translation stage separated by a distance l. The strain sensitivity of a single-mode POFBG is shown in Fig. 5a, and from the figure, it can be seen that the Bragg grating inscribed in the POF exhibits a strain sensitivity of 1.26 pm/$\mu\epsilon$. The reflection spectra of POFBG, under two different tensile strains (1200 and 2100 $\mu\epsilon$) also shown in Fig. 5b, indicate that a positive Bragg wavelength shift is observed with an increasing strain. Similar measurements were carried out for another diameter of a single-mode POFBG (243 μm), and here, the strain sensitivity was measured as 1.27 pm/$\mu\epsilon$ which is almost the same as that of the 180 μm diameter fiber. From these two results, it is clear that the strain response of the single-mode POFBG is linear and the calculated strain sensitivity is better than that of a silica FBG. Here it is noted that, to avoid hysteresis effects, the elongation of the POF was limited to less than 0.75% (Abang and Webb 2013); no significant hysteresis was found within the applied strain range.

Pressure

A pressure sensor based on a FBG is considered a very good alternative compared to traditional sensors due to its smaller size, lower weight, greater sensitivity, resistance to EMI, safety, etc. However, high sensitivity pressure measurements are challenging for silica FBGs because of the inherent limitations of the material properties of silica optical fiber (Yunqi et al. 2005). The hydrostatic pressure-induced wavelength shift reported for a silica FBG is –0.22 nm for a 0–70 MPa range which gives a pressure sensitivity of –0.003 pm/kPa (Johnson et al. 2012). To develop a pressure transducer, this relatively low intrinsic pressure sensitivity needs to be further enhanced and for which several techniques have been proposed (Hill and Cranch 1999; Song et al. 2009; Xu et al. 1993). However, as the requirement for suitable engineering and modified sensor designs make these techniques complicated in nature, the development of a fiber pressure transducer is an engineering research problem.

Bhowmik et al. (2014, 2015b) has recently reported the inherent pressure sensitivity of single-mode polymer FBG. When hydrostatic pressure is applied to an FBG, both the effective refractive index and grating period of the fiber will change and lead to the Bragg wavelength shift of:

$$\Delta\lambda_B = \lambda_B{}' - \lambda_B = \lambda_B\left(\frac{\Delta n_{eff}}{n_{eff}} + \frac{\Delta\wedge}{\wedge}\right)$$

where $\Delta n_{eff} = n_{eff}{}' - n_{eff}$ is change in effective index and $\Delta\wedge = \wedge' - \wedge$ is change in grating period.

Johnson et al. (2012) reported that for single-mode silica FBG, the pressure sensitivity was negative and is –0.00388 pm/kPa, and also for a multimode microstructured POF grating, the sensitivity was 0.10 pm/kPa. As the Young's modulus of the polymer fiber is much less than that of silica fiber, one would expect high-pressure sensitivity compared to silica fiber. To measure the hydrostatic pressure sensitivity, the compressed air from a cylinder can be used to increase the pressure within a perfectly sealed chamber. It should be noted that, inside the chamber, as the pressure increases, the temperature can decrease due to the pumping of cold dry air. Since a POFBG is sensitive to temperature changes, the temperature profile of the pressure chamber has to be measured in order to compensate the temperature-induced effect on the POFBG. In their reported result, –0.5 °C temperature change was observed for a pressure change of 0–200 kPa, and then the temperature remains approximately constant with only a – 0.08 °C change observed for a pressure change from 200 kPa to 1 MPa. The temperature profile of the chamber while applying pressure is shown in Fig. 6, and it can be seen that a – 0.58 °C nonlinear temperature change is observed for a pressure change from 0 kPa to 1 MPa within the duration of the experiment that needs to be eliminated.

The single-mode POFBG with a diameter of 180 μm shows sensitivity of 0.24 pm/kPa inclusive of the temperature effect for a pressure change from

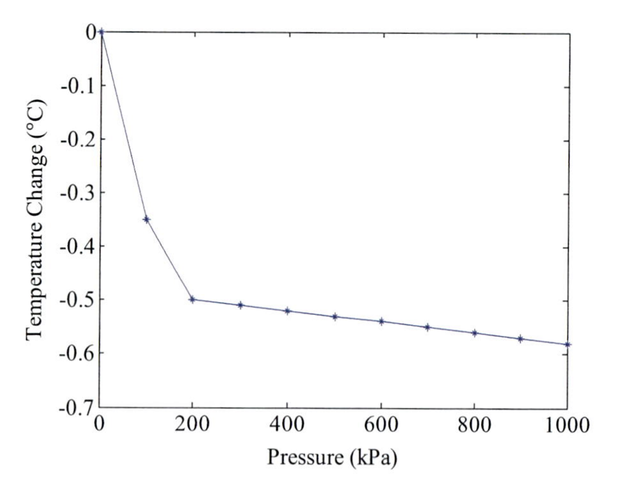

Fig. 6 Temperature profile of the chamber

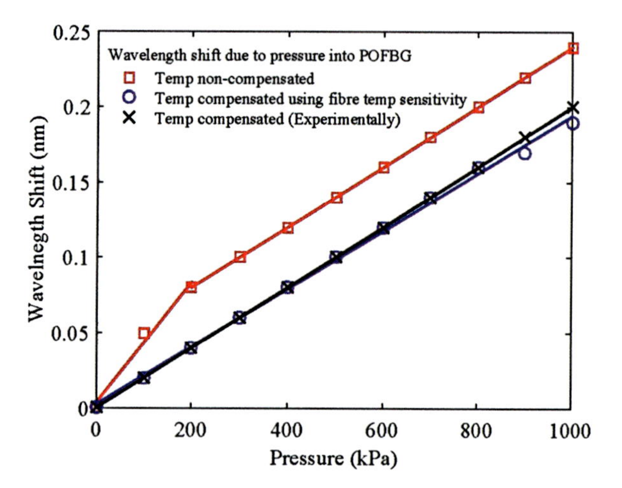

Fig. 7 Wavelength shifts of POFBG with applied hydrostatic pressure

0 kPa to 1 MPa. For POFBG, a positive sensitivity is observed when the hydrostatic pressure is increased, whereas it was negative for silica FBG. In the case of silica FBG, the hydrostatic pressure-induced grating period change (strain effect) dominates the change in refractive index, and as a result, a negative sensitivity is observed. While for a POFBG, the dominant effect is the refractive index change than the grating period change, and as a result, the sensitivity becomes positive. To experimentally eliminate the temperature and humidity effect, the chamber was filled with high viscous fluid, and one end of the fiber is immersed to the viscous fluid in the same way as in previous experiment, and the end was sealed properly after inserting the fiber. The result obtained using this approach is shown in Fig. 7, and the measured sensitivity of a POFBG with 180 μm diameter is 0.20 pm/kPa, and the change in wavelength exhibits a linear relation with change in hydrostatic pressure. Therefore it can be considered that the temperature effect is eliminated and

the sensitivity of 0.20 pm/kPa is the true pressure sensitivity of 180 μm diameter POFBG.

In order to verify this result, the temperature sensitivity of the POFBG can be used to extract the true inherent pressure sensitivity of the fiber. By using this sensitivity value and based on the temperature profile of the pressure chamber, the effect of temperature on the POFBG can be compensated to some extent. Using this approach, the estimated pressure sensitivity of the 180 μm diameter POFBG is 0.19 pm/kPa which is close to the value obtained using the first approach (viscous fluid method). Similar measurements were reported for another un-etched fiber of a different diameter of 243 μm, and the temperature compensated pressure sensitivity was measured the same as is 0.20 pm/kPa. Therefore from the results, it can be confirmed that the actual intrinsic pressure sensitivity of an un-etched POFBG is approximately 0.20 pm/kPa. This result also reveals that the intrinsic pressure sensitivity of POFBG is 52 times higher than that of silica FBG.

Etched POFBGs

Etching and Etching Effects on the Material Properties of POF

To reduce the cladding thickness, the POF can be etched using solvent etching technique as described in ref. (Merchant et al. 1999). A laboratory grade acetone with 99.5% concentration was mixed with methanol by 1:1 ratio, and then one end of the fiber was inserted into this mixture solution and then removed and cleaned. The etching time depends on the required cladding thickness, and to obtain the lowest diameter of 85 μm, the fiber needs to be immersed in the solution for approximately 7 min. Microscopic images of a fiber before (185 μm) and after etching (140, 110 and 85 μm) are shown in Fig. 8. From the results, it is confirmed that the etching rate of a POF in the acetone-methanol solution is approximately 14 μm/min as shown in Fig. 9.

Hu et al. (2014) reported that a cladding diameter decrease of 12% via etching is an ideal trade-off to produce highly reflective gratings by using trans-4-stilbenemethanol-doped photosensitive step-index poly(methyl methacrylate) (PMMA) POFs. However to investigate the effect of etching on the material properties of the polymer fiber, Young's modulus and absorption spectrum of the etched polymer fibers are experimentally obtained by Bhowmik et al. (2015a). In order to measure the Young's modulus, a standard procedure was adopted that is valid for fibers of dimension in the range of optical fibers (Standard test method for tensile strength and young's modulus of fibers 2003). In this case, each fiber sample was mounted in the testing machine and then stressed to failure at a constant cross-head displacement rate. The estimated Young's modulus from the stress-strain data obtained from the experiment for fibers with different diameters is shown in Table 2. For the 180 μm un-etched POF, the measured Young's modulus is 2.56, but when it is etched to 100 and 70 μm diameters, the Young's moduli are 2.07 and 1.71 GPa, respectively. To confirm these results, repeated tests were conducted using

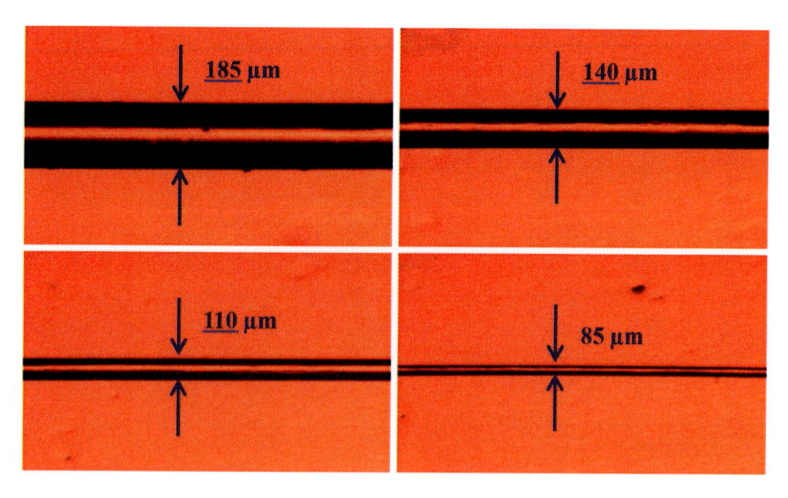

Fig. 8 Microscopic images of POFs before etching (185 µm) and after etching (140, 110, and 85 µm)

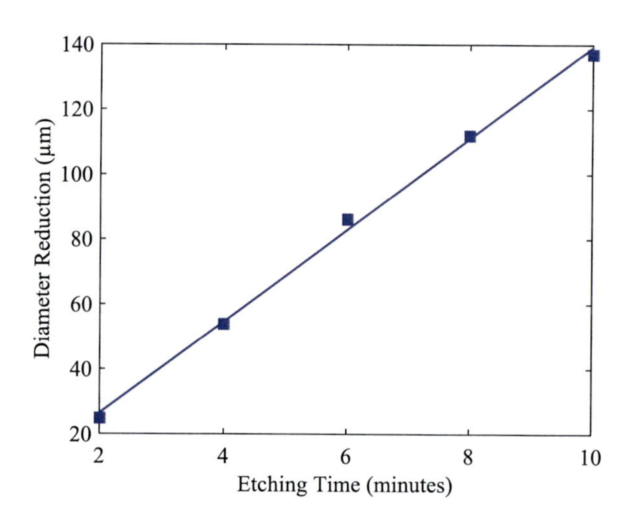

Fig. 9 Etching rate for polymer optical fiber using acetone-methanol solution

Table 2 Young's modulus of polymer fibers

Sample-1		Sample-2	
Fiber diameter (µm)	Young's modulus (GPa)	Fiber diameter (µm)	Young's modulus (GPa)
180 (un-etched)	2.56	450 (un-etched)	1.04
100 (etched)	2.07	324 (un-etched)	1.05
70 (etched)	1.71	340 (etched)	0.36

a different sample (sample-2) of PMMA fibers with different dopant concentrations and conducted the same experiment for that fibers. Two different un-etched fibers with diameters of 450 μm and 324 μm were tested which results the same Young's modulus of 1.05 GPa. But when it is etched down, the former 450 μm fiber to 340 μm, the Young's modulus reduces to 0.36 GPa.

From the table it is clear that Young's modulus of the polymer fiber has significantly reduced with reduction in the fiber diameter (Bhowmik et al. 2015a). The observed reduction in Young's modulus could be due to two reasons: the irreversible relaxation of the polymer chain of the material due to the solvent absorption and also due to the stress relaxation of fiber due to cladding diameter reduction. When the fiber is immersed in the etching solution, the internal stress distribution of the fiber varies with fiber diameter reduction (Lim et al. 2013) which changes the fiber material properties. So the combined effect of polymer chain relaxation and fiber stress relaxation can soften the fiber material that reduces the Young's modulus.

Etched Polymer Fiber Bragg Gratings

The grating can be fabricated under the same conditions as with the un-etched POFBG, and the same experimental setup can be used. In order to study the evolution of grating, the fiber was exposed to UV light for 5 min. Figure 10a shows a comparison of the reflectivities at different exposure times of three different etched POFBG and also for an un-etched POFBG (Bhowmik et al. 2016a). From the figure, it is clear that the growth patterns of all the gratings are similar, but the reflectivities and exposure times are different for the different etched polymer fibers. For 85 μm diameter etched fiber, the highest reflectivity was observed within 7 s of exposure with a grating transmission dip of 18.37 dB corresponding to a 98.54% reflectivity.

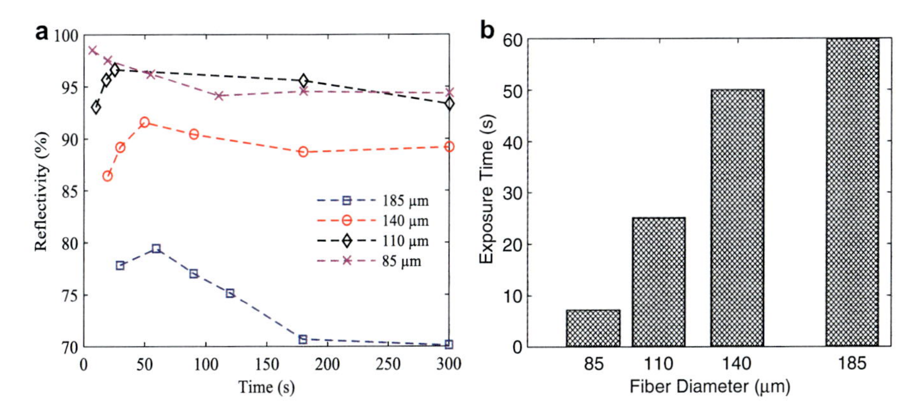

Fig. 10 (a) Reflectivity patterns of POFBGs with different diameters and (b) exposure times required to achieve highest reflectivities for different fiber diameters

This means that the reflectivity increases by 24.07% with a reduction in diameter of 54.05% through etching, and, at the same time, the exposure time reduces by 88.33%. On the other hand, the highest reflectivities achieved for the two other etched fibers are 91.59 and 96.63% for 140 and 110 μm diameters, respectively. Whereas the reflectivities after 5 min of exposure are 89.19, 93.33, and 94.36% for the etched fibers with 140, 110, and 85 μm diameters, respectively, but for the un-etched fiber, it was 70.10%.

Therefore, to ensure the highest reflectivity of a fabricated grating, it is important to know the exact exposure time required. Moreover, it is obvious that the exposure time required to get the highest reflectivity is less for an etched fiber than that of un-etched fiber which is shown in Fig. 10b. For etched fibers with 140, 110, and 85 μm diameters, the highest reflectivities were observed at 50, 25, and 7 s, respectively, compared with at 60 s for the un-etched fiber. Therefore, it is clear that the exposure time required to achieve the highest reflectivity of a grating is reduced through decreasing the diameter of a polymer fiber by etching.

From the above observations, it is clear that the reflectivity of a POFBG increases when the fiber diameter is reduced through etching. The exposure time is also reduced for an etched POF grating fabrication. This is due to the fact that, in the case of an etched fiber, the UV beam can be focused into the core more effectively, as its reduced cladding will considerably decrease its absorption of UV irradiation. Therefore, the UV interference pattern at its core will have a higher intensity than that of an un-etched fiber. Also due to solvent etching, penetration of the solvent can change the properties of the core that can alter its photosensitivity. Moreover, the fiber expansion that occurs during etching loosens the matrix of the fiber material, which can allow the UV light to react more in an etched fiber than that of an un-etched fiber. This results point to a direction that the etching procedure has a major role in enhancing reflectivity. Therefore, from the above results, it is proven that etching has a significant impact on a grating inscription, as it enhances reflectivity while reducing the required exposure time. This means that, to produce highly reflective gratings, there is an experimental trade-off between the fiber diameter and exposure time.

Stability of Etched POFBGs

In the previous section, it is demonstrated that highly reflective gratings can be fabricated into an etched POF. There is also a threshold point of exposure time for producing the highly reflective gratings for different diameter etched fibers. So it would be interesting to determine the stability of these highly reflective gratings (Bhowmik et al. 2016a). To study the stability of grating behavior, two scenarios can be considered: (i) the pattern of a grating after 300 s of exposure and (ii) the pattern of a grating when the exposure is turned OFF after obtaining the highest reflectivity.

In the first case, to observe the behavior of a grating after 5 min of inscription, its transmission spectra are continuously monitored. For example, the behavior of

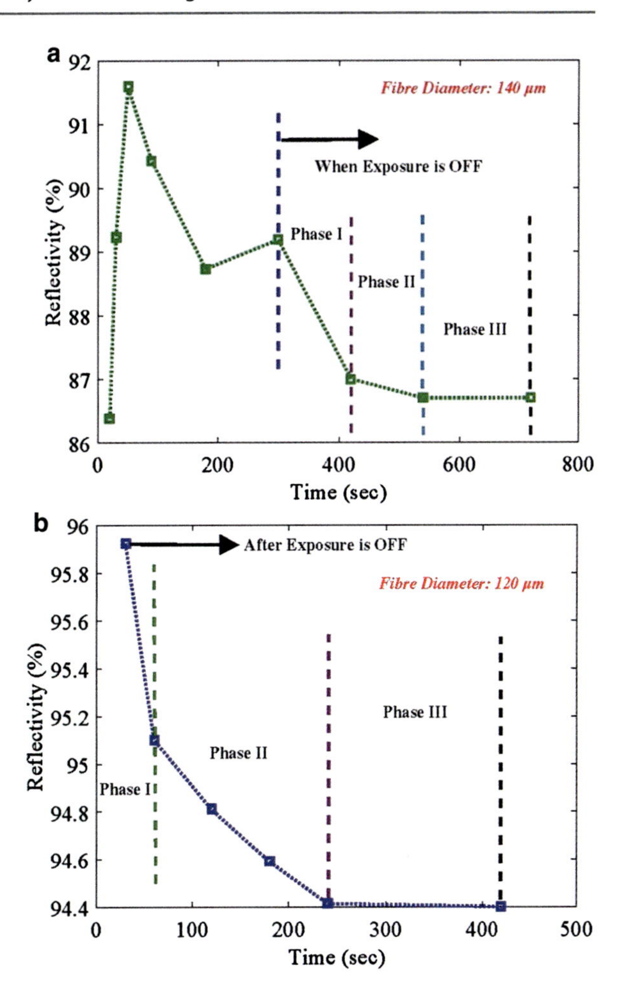

Fig. 11 (**a**) Grating pattern after 300 s of exposure and (**b**) grating pattern when exposure turned OFF after obtaining the highest reflectivity

the spectrum for the 140 μm diameter etched fiber is shown in Fig. 11a, in which it is clear that its reflectivity is reduced when the exposure is OFF after 5 min. This may be due to the chemical stabilization of the polymer materials occurring after the inscription on the polymer fiber. It is clearly observed that a sudden drop occurred when the UV irradiation is OFF which is shown in phase I and the reflectivity reduces by approximately 2.46% within 2-min time duration than the last observed reflectivity at 5-min exposure. Then the reduction rate is less than before as shown in phase II and finally reduces by approximately 2.80% within 7-min time duration and becomes stable as shown in phase III. This means that the total reflectivity reduces by approximately 5.34% from the highest reflectivity observed at 50 s of exposure.

In the second case, to observe the stability of a grating after obtaining the highest reflectivity, we fabricate a grating into a 120 μm diameter etched fiber. After obtaining the highest reflectivity at 30 s, further exposure is stopped and the behavior of the grating observed shown in Fig. 11b. From the figure, it can be clearly seen that the reflectivity of the grating reduces quickly after the UV exposure is turned off as the previous one and the reflectivity reduces by approximately 0.859% within 30 s time after stopping the exposure as shown in phase I. Then, the rate of reduction decreases (phase II), and within 5 min, it becomes constant (phase III). Finally, the reflectivity of the 120 μm diameter etched POFBG is 94.402%, which means that it reduces by approximately 1.59% during the 390 s observation time.

Long-Term Stability of Etched POFBGs

To test its long-term stability, the transmission spectra of the 120 μm etched POFBG were monitored for 30 days after its grating inscription is performed. Figure 12a displays the transmission spectra after 1, 10, 20, and 30 days of grating inscription. From the figure, it is clear that there are no noticeable changes in the transmission peak level over 30 days and the corresponding reflectivity behaviors are shown in Fig. 12b. After 1 day of inscription, the reflectivity was 94.40%, and after 30 days' time, the reflectivity remains 94.38%. This means that the reflectivity reduces only 0.02% in 30-day observation which is negligible. So this evidence indicates that the gratings fabricated on etched polymer fibers are stable over a long-time period.

Etched POFBG Sensing with Enhanced Intrinsic Sensitivity

It is well demonstrated that a sensor based on polymer fibers has a number of advantages compared to its silica counterparts. However for certain applications, such as in biomedical/biomechanical sensors, high sensitivities with high reflectivities are required. Due to superior material properties of polymer fiber compared to silica fiber, gratings based on polymer fiber could be a suitable alternative for these applications. For physical sensors, the most common parameters such as pressure, temperature, and strain are normally measured using fiber-optic sensors, and generally different extrinsic techniques are used to improve the sensitivity of them. Although there may be many other techniques available for extrinsically enhancing the sensitivities of different types of fiber sensors, they usually increase the size and cost and also involve additional physical complexities, which weaken the key advantage of a FBG sensor. Moreover, there is no any single technique available for improving the sensitivities of different measurands such as pressure, strain, and temperature in the same time. Therefore, any single technique to improve the sensitivity intrinsically for all measurands will be more highly valued than an extrinsic technique. An added advantage of high intrinsic sensitivity sensors is that devices based on such sensors can be less complex and miniature in size and can provide solutions to the much needed sectors such as in biomedical applications, a need that is still not addressed by existing sensors.

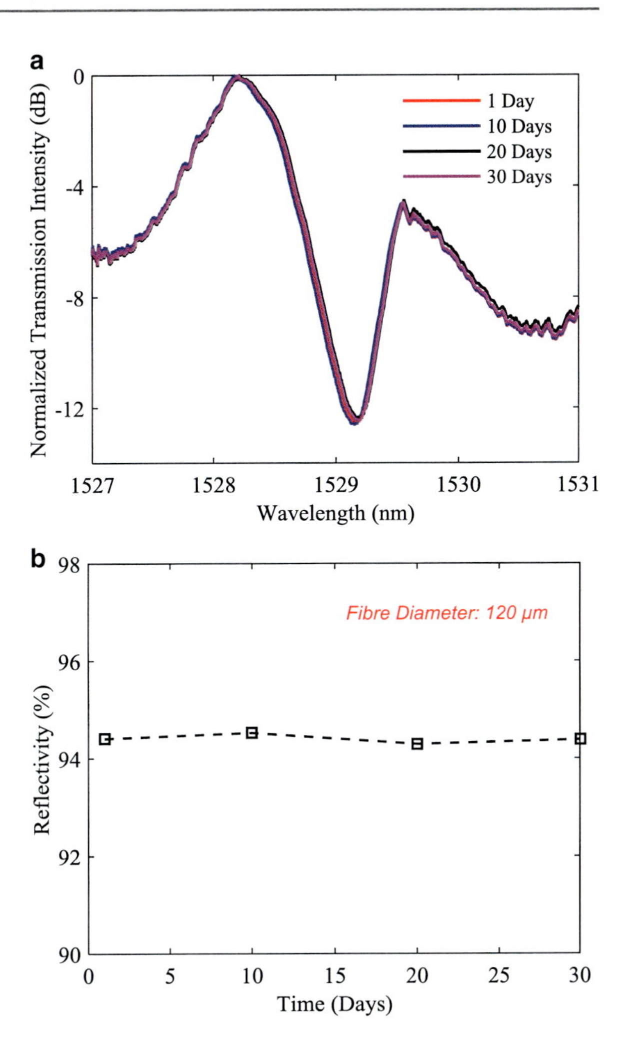

Fig. 12 (**a**) Transmission spectra of 120 μm-etched POFBG over 30 days observation and (**b**) grating pattern over 30 days after obtaining the highest reflectivity

Temperature Sensitivity of Etched POFBGs

It is known that the Young's modulus is inversely proportional to the thermal expansion coefficient; therefore, an enhanced intrinsic temperature sensitivity for etched POFBG can be expected. In order to demonstrate this concept, the fiber containing the FBG were etched to different diameters (130, 105, 80, and 55 μm), and the temperature sensitivity of each of the etched POFBG was measured under the same experimental conditions as an un-etched POFBG, and the same experimental setup is used. Figure 13a shows the measured wavelength shifts due to temperature for POFBGs with different diameters. From the experimental results, it is observed that there are Bragg wavelength shifts of 4.10, 4.60, and 5.10 nm for the 105, 80, and 55 μm etched POFBGs, respectively, for the same

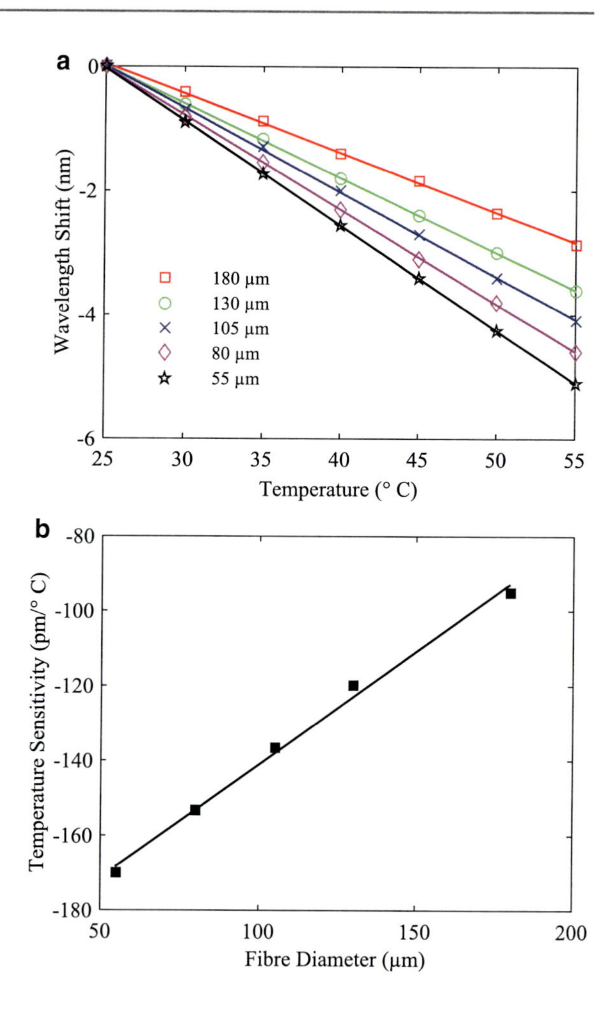

Fig. 13 (**a**) Wavelength shifts of POFBG with different fiber diameters. (**b**) Temperature sensitivities of POFBGs with different fiber diameters

30 °C temperature change. Therefore, the corresponding calculated temperature sensitivities are −136.7, −153.3, and −170 pm/°C, respectively, as shown in Fig. 13b. The highest measured temperature sensitivity is for 55 μm etched POFBG, which is −170 pm/°C, is 1.79 times larger than that of an un-etched POFBG. Moreover, this sensitivity value is 17 times larger than the value reported for a silica FBG, which was approximately 10 pm/°C. Therefore, it is evident that a Bragg grating in an etched fiber exhibits considerable increases in temperature sensitivity with decreases in the fiber's diameter through solvent etching.

Pressure Sensitivity of Etched POFBGs

It is known that the pressure sensitivity depends on the material properties of the fiber, especially Young's modulus of the fiber. If the Young's modulus of a fiber

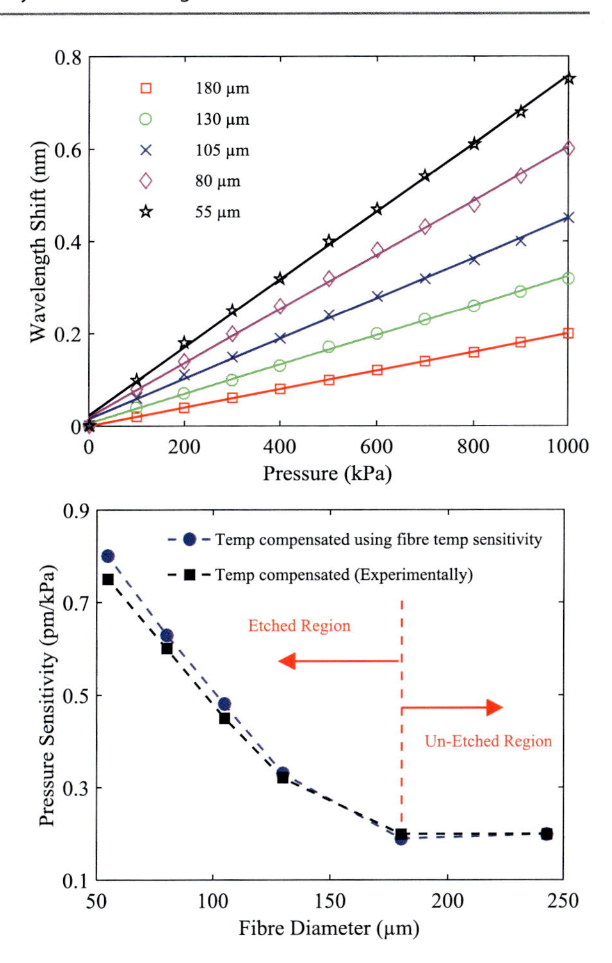

Fig. 14 Measured wavelength shifts of POFBGs with different fiber diameters (experimentally compensated) and (**b**) pressure sensitivities of POFBGs with different fiber diameters

decreases, the pressure sensitivity of the FBG increases. In order to demonstrate this concept, fibers containing FBGs were etched to different diameters (130, 105, 80, and 55 μm) to measure their intrinsic pressure sensitivities. The pressure sensitivity of each of the etched POFBGs was measured under the same hydrostatic pressure conditions as that of the un-etched POFBG using the same experimental setup. Even with the high temperature sensitivity of the etched POFBG, using the temperature compensation techniques, the effect of temperature was completely eliminated, and the measured temperature compensated wavelength shift due to hydrostatic pressure for POFBGs with different diameters is shown in Fig. 14a, and the corresponding calculated pressure sensitivity is shown in Fig. 14b. The calculated pressure sensitivity for 180, 130, 105, 80, and 55 μm diameter POFBGs were 0.20, 0.32, 0.45, 0.60, and 0.75 pm/kPa, respectively. It can be seen that the pressure sensitivity for un-etched POFBG (180 and 243 μm) is unchanged, while for etched POFBG, it increases with a decrease in fiber diameter through etching.

Strain Sensitivity of Etched POFBG

As the Young's modulus of an etched fiber is smaller than that of an un-etched one, it can be expected that the strain sensitivity of an etched POFBG would be better than that of an un-etched one. In order to demonstrate this, a fiber containing the FBG was etched to different diameters (150, 130, 80, 65, and 43 μm), and the strain sensitivity of each etched POFBG measured under the same experimental conditions as the un-etched POFBG using the same experimental setup which is shown in Fig. 15a. From the graph, it is clear that the strain response is linear for all POFBGs over a strain range of 2000 με. The observed strain sensitivities for the 130, 80, 65, and 43 μm-etched POFBGs are 1.53, 1.87, 1.95, and 2.07 pm/με, respectively. The corresponding strain sensitivities of different diameters are plotted in Fig. 15b which shows that the maximum one is 2.07 pm/με for the 43 μm-etched POFBG which is 1.64 times larger than that

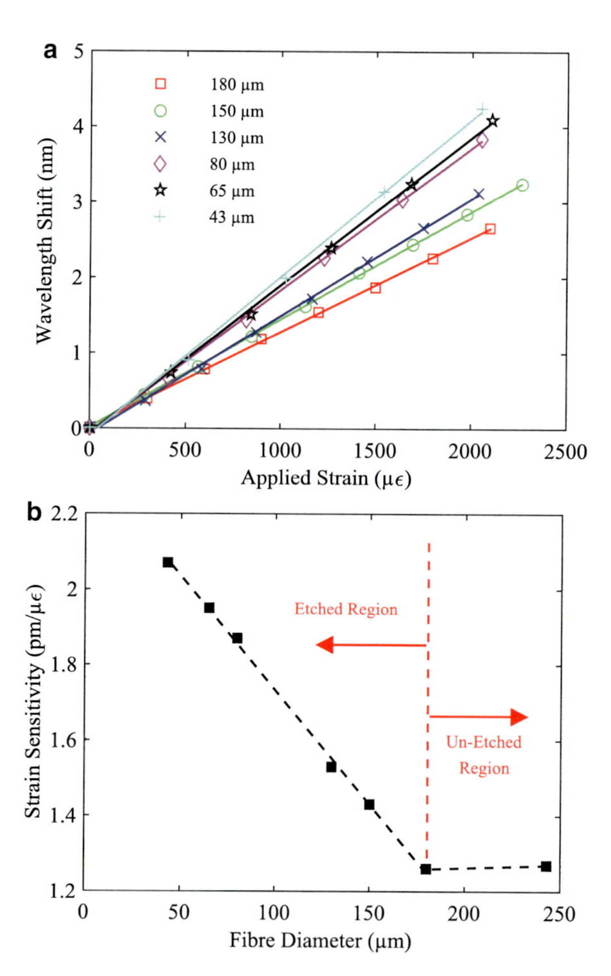

Fig. 15 (**a**) Measured wavelength shifts of POFBGs with different fiber diameters and (**b**) strain sensitivities of POFBGs with different fiber diameters

of an un-etched POFBG. Moreover this sensitivity value is 1.8 times higher than the value for the silica FBG, which is reported 1.15 pm/$\mu\epsilon$ at this wavelength. From the figure, it is evident that the measured strain sensitivities for the un-etched POFBGs with 180 and 243 μm diameters are similar, but a considerable increase in strain sensitivity is observed after a decrease in the fiber diameter through solvent etching.

POFBG Applications

The potential of fabricated POFBGs are demonstrated in a number of applications where the use of the polymer FBG can be of a distinct advantage. Given the low Young's modulus, high failure strain, and intrinsic humidity sensitivity, FBGs based on polymer fibers are generally used in application in these sensing areas. In this section we give an update of some of the recently reported research work on applications of polymer FBGs.

Force and Pressure Sensors

Fiber-optic methods for high sensitivity and low pressure measurements are considered as a challenging area due to the inherent limitations of the material properties of the silica optical fiber (Yunqi et al. 2005). Force and pressure sensors based on silica FBGs are reported by many research groups over the years. Due to low intrinsic sensitivity of silica FBGs, various methods are investigated to improve the sensitivity of FBGs to force and pressure but are complicated in nature. The unique material properties of the polymer optical fiber such as Young's modulus 30 times lower than that of silica fiber sensors based on polymer FBGs have shown higher sensitivity compared to silica fiber gratings (Luo et al. 2011; Wang et al. 2012).

Typically the force and pressure sensitivity of an FBG depends on the cross-sectional area of the fiber, which can be increased by reducing the diameter of the fiber. Rajan et al. (2013a) reported a sensitivity of 680 nm/N for a polymer FBG with a diameter of 30 μm. The focus of the measurement range was at the lower end of the spectrum ranging from 0.005 to 0.05 N. Force sensitivity of polymer FBG with different diameters is shown in Fig. 16. It is observed that sensitivity increases considerably with decrease in fiber diameter.

Rajan et al. (2013a) also demonstrated high sensitivity pressure using a POF grating-based sensor, where a thin vinyl diaphragm which has low Young's modulus is used to transfer the applied pressure to the POF grating. A pre-strained POF grating is attached to the diaphragm vertically as shown in the Fig. 17a. Due to very low Young's modulus (2–2.5 GPa) of the vinyl material, higher strain can be produced for very low pressure. The maximum deformation (δ) that can occur at the middle of the diaphragm for an applied pressure also depends on the diaphragm radius (r) and thickness (t) and can be expressed as:

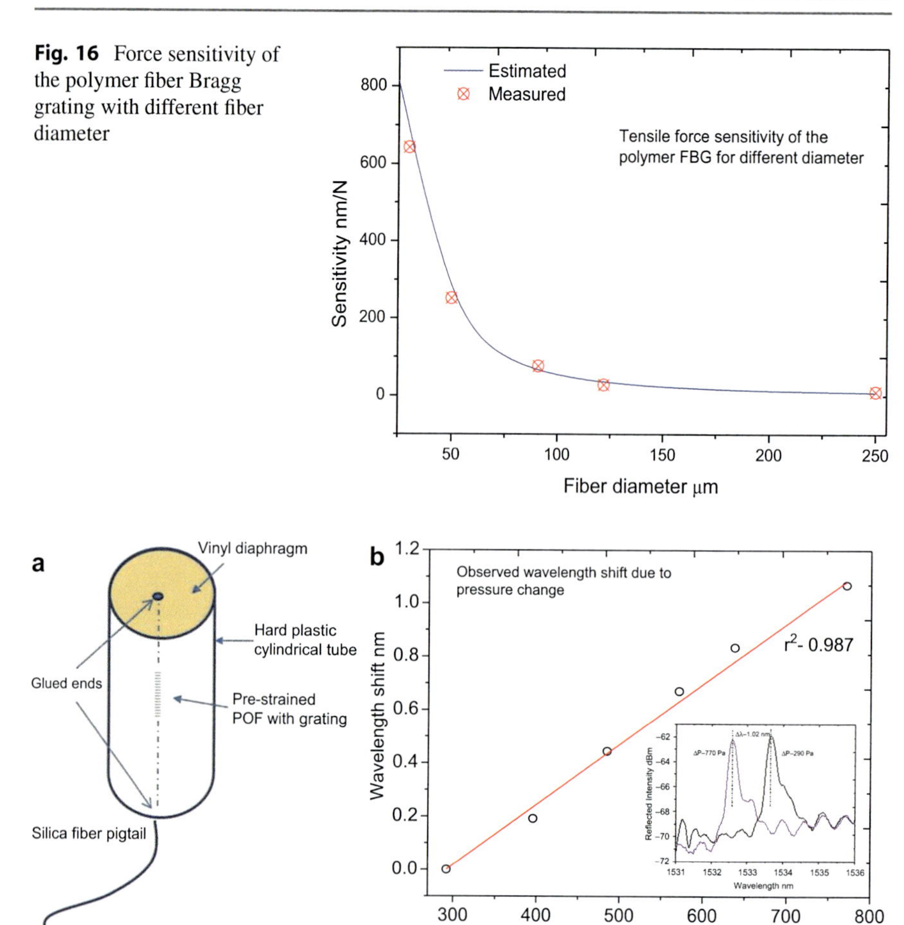

Fig. 16 Force sensitivity of the polymer fiber Bragg grating with different fiber diameter

Fig. 17 (**a**) Schematic of the pressure transducer and (**b**) pressure-induced wavelength shift (inset: pressure-induced spectral shift)

$$\delta = \frac{3P}{16t^3E}\left(1 - \mu^2\right) r^4$$

where P is the applied pressure, E is the Young's modulus, and μ is the Poisson ration of the vinyl diaphragm.

In this configuration a 91 μm diameter POF grating is attached perpendicular at the middle of an 8 mm diameter vinyl diaphragm. The total length of the fiber containing the grating was 7.5 cm, and the thickness of the diaphragm was approximately 60 μm. The diaphragm was attached to one end of a hard plastic cylindrical tube of length 8 cm with an inner diameter of 8 mm. Pre-strain is applied to the fiber, and the glued POF-silica junction is securely connected at the other end of the plastic cylindrical tube using an epoxy. The cylinder was sealed properly to have accurate pressure measurements. Low-value pressure changes ranging from 290 to 770 Pa were applied to the diaphragm, and the corresponding change in wavelength shift is shown in Fig. 17b and the estimated sensitivity approximately 1.32 pm/Pa. This sensitivity can be further improved by changing the diameter of the fiber. Such high sensitivity force/pressure sensors can find applications in biomedical field such as in artificial heart pumps, minimally invasive surgical devices, etc. That means for high sensitivity tensile force and pressure measurements, an etched polymer FBG will be a good alternative.

Humidity Sensors

A variety of optical fiber relative humidity sensors has been reported in the last decade, and most of them are based on externally coated moisture-sensitive reagents (Yeo et al. 2008). Some of the recent studies on photonic crystal fiber interferometer-based humidity sensors infiltrated with a hygroscopic material show a fast response time less than a second (Mathew et al. 2012). However in such sensors, the PCF holes are open to atmosphere; it can be contaminated and can affect the performance of the sensor for long-term monitoring of relative humidity. FBGs coated with a thermoplastic/polyimide material also show sensitivity to humidity and exhibits a response time in seconds (Huang et al. 2007; Yeo et al. 2005). However a very thin layer of coating is required to achieve this response time which reduces the sensitivity of the sensor. Due to water absorption properties of PMMA, the intrinsic humidity sensitivity of polymer FBG can make it superior compared to others. The humidity sensitivity of POFBGs was first demonstrated by Harbach (2008) and showed a linear response in the range 40–90% RH, though this breaks down at lower humidities. The sensitivity displayed by this device is 38 pm/% RH. The response times tend to be typically a few tens of minutes, being 30 min. Some applications would benefit from response times much faster than 30 min. To address this issue, Zhang et al. (2012a) reported that the diffusion time of water content into the polymer fiber can be improved from around 30 min down to 10 min by reducing the cladding thickness of the polymer fiber.

Rajan et al. (2013b) reported a simple, low-cost, and fast response time intrinsic relative humidity sensor system based on an etched polymer FBG. The time response of the polymer FBG humidity sensor with a thickness of 25 μm is shown in Fig. 18a; when the RH is changed from 90% to 60%, a voltage variation 9.8 mV is observed in 4.5 s, while for a RH change of 60% to 90%, a voltage variation of 8.2 mV is observed 5.5 s. The small kink observed in the response is due to the sudden temperature change in the chamber during the transition from wet air to dry air. The response time of POFBG with diameter varying from 25 μm to 125 μm is shown in Fig. 18b. From the figure it can be seen that the response time decreases exponentially as the diameter decreases in the measured range. The exponential nature of the change in response time is mainly attributed to the absorption mechanism of PMMA. The results demonstrate that polymer FBGs can respond to a change in humidity within seconds if etched to micron level diameter

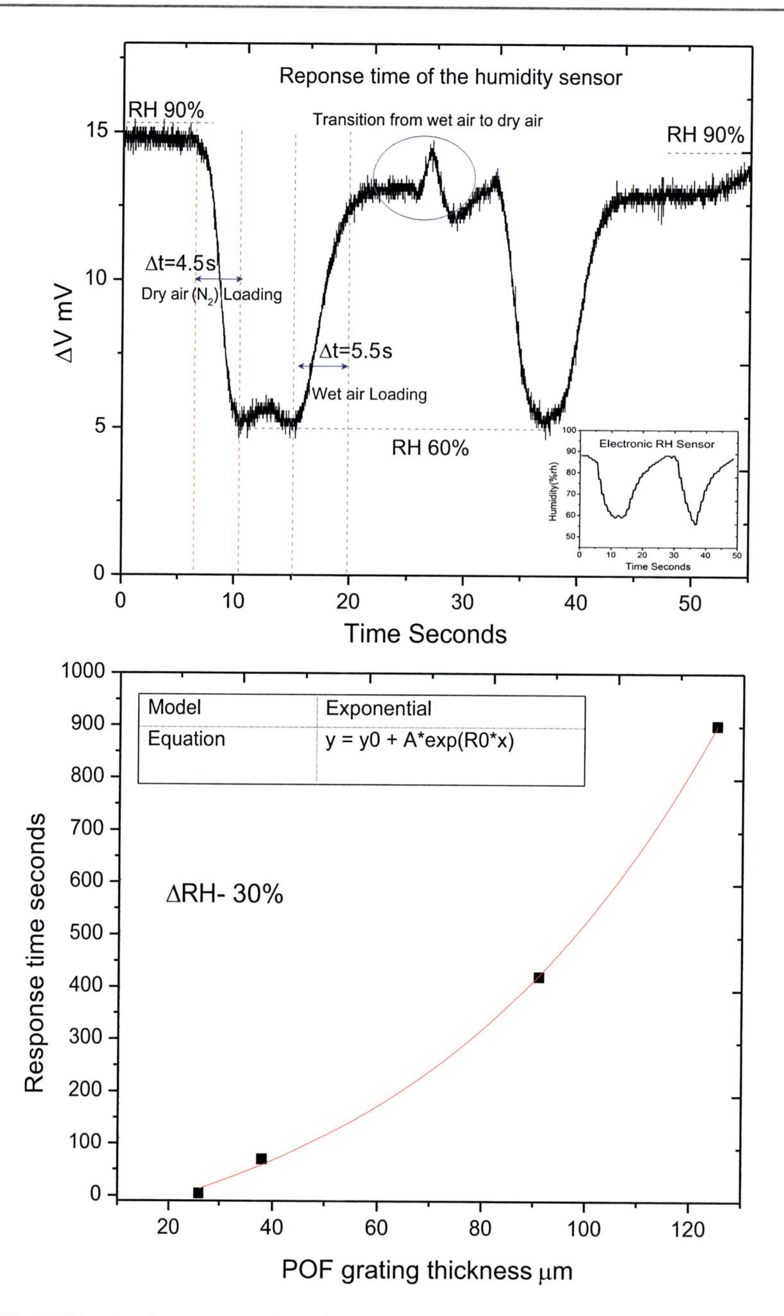

Fig. 18 (**a**) Plot showing response time of the humidity sensor and (**b**) response time of humidity sensor with different thickness

and could be ideal for many industrial applications where real-time humidity measurements are required.

Simultaneous Sensing

Simultaneous measurement is an indispensable aspect of fiber sensing, not only because it enables the cost and size of a global set of sensing systems to be minimized, with each addressing a particular measurand, but also removes the necessity to perform a parameter discrimination in situations in which cross-sensitivity is vital issue. The first sensing head reported for strain and temperature measurements was proposed and demonstrated by Xu et al. (1994) in 1994. There are many other concepts and techniques reported for obtaining such measurements or discriminating the cross-sensitivity effects, most of them rely on using two different types of fibers and FBGs or an extrinsic technique. Even some techniques involve either complex gratings or specially made fibers, while some even require other optical elements. As using more than one fiber element or other optical devices increases the system cost and complicates fabrication of the sensor head, it is highly desirable to use a single fiber without any extrinsic technique in order to distinguish between strain and temperature. Therefore, any technique using the same fiber without any extrinsic technique is more highly valued than others and is also cost-effective for simultaneous measurements.

Bhowmik et al. (2016b) reported a technique to simultaneously measure the strain and temperature utilizing the different Young's modulus in etched fiber than un-etched fiber, and the reported configuration of a POFBG sensor is shown in Fig. 19, where an FBG pair with one grating in the etched and another in the un-etched sections. Thus, by exploiting their different strain and temperature sensitivities and using this simple configuration, the strain and temperature can be simultaneously measured and also with very high sensitivity. Strain and temperature change can be measured simultaneously by utilizing the well-known characterization matrix as shown below:

$$\begin{bmatrix} \Delta\varepsilon \\ \Delta T \end{bmatrix} = \frac{1}{K_{1\varepsilon}K_{2T} - K_{2\varepsilon}K_{1T}} \begin{bmatrix} K_{2T} & -K_{1T} \\ -K_{2\varepsilon} & K_{1\varepsilon} \end{bmatrix} \begin{bmatrix} \Delta\lambda_1 \\ \Delta\lambda_2 \end{bmatrix} \tag{1}$$

where $\Delta\lambda_1$ and $\Delta\lambda_2$ are the net wavelength shift exhibited by the 185 μm and 105 μm FBGs.

To inscribe multiple FBGs on a single-mode polymer fiber using a single phase mask and by applying strain to the fiber, a method is described by Rajan et al. (2014). In this technique, initially a grating is inscribed at the far at zero strain condition, and then at each step, an elongation of 0.2% is applied to the fiber using the translation stage, and the position of the UV-exposed region is changed by moving the fiber using the motorized translation stage. In the case of gratings inscribed in a strained condition, after releasing the strain, the grating period of the fiber will change by

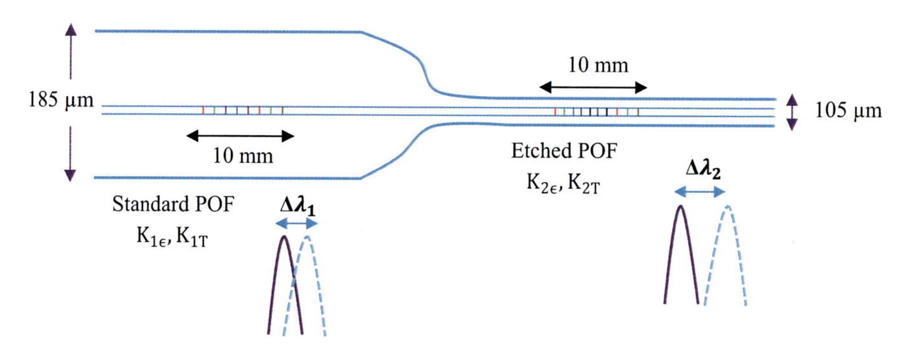

Fig. 19 Schematic diagram of fabricated FBG pair for simultaneous measurements

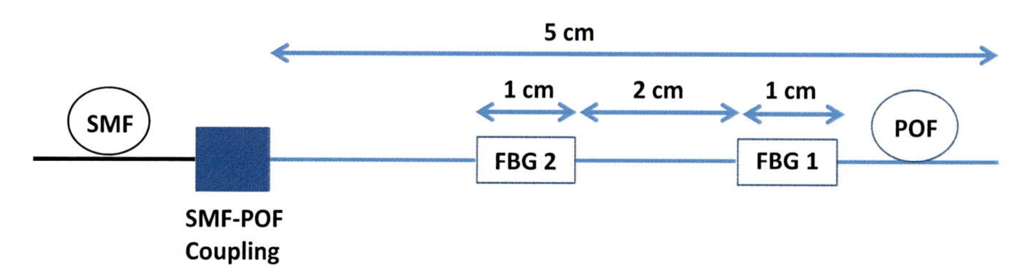

Fig. 20 Schematic of locations of POFBG pair on fiber

a factor of $\frac{\triangle E_l}{100}$, where E_l is the percentage of elongation applied to the fiber. As a result of this grating period change, the FBG exhibits a different peak wavelength. Upon releasing the tension, the spectrum will shift to the blue side, and given the capability of polymer fiber to apply larger elongation, grating with a different center wavelength can be inscribed. The schematic of the location of POFBG pair is shown in Fig. 20, in which it is seen that the total length of the POF used is approximately 5 cm, with the fabricated gratings 1 cm long and the gap between them 2 cm. The measured reflection spectra of the POFBG pair are shown in Fig. 21, in which that indicated by the dotted line is the one measured at the elongated condition after inscribing both gratings and that by the bold line is the reflected peaks of POFBG pair at the zero strain condition. The peak wavelengths of FBG_1 and FBG_2 are 1530 and 1526 nm, respectively, which gives a 4 nm separation between them. To use the fabricated POFBG pair for simultaneous sensing, one of the grating is etched to 105 µm to form a grating in the etched section and the other placed in the un-etched section and both having distinct peak-reflected wavelengths.

The Bragg grating inscribed on the un-etched fiber exhibits a strain sensitivity of 1.24 pm/µϵ, while a Bragg grating inscribed on the etched fiber with a diameter of 105 µm shows an enhanced sensitivity of 1.65 pm/µϵ. Similarly, the un-etched 185 µm POFBG exhibits a temperature sensitivity of −92 pm/°C, while a 105 µm diameter-etched POFBG exhibits a temperature sensitivity of −133 pm/°C. Since both POFBGs exhibited different sensitivity measurements, it is possible to use this sensing head for simultaneous sensing of strain and temperature variations.

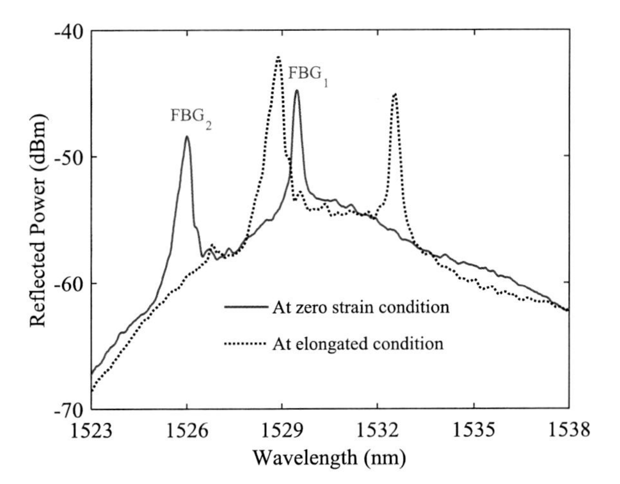

Fig. 21 Reflection spectra of fabricated POFBG pair at both zero strain and elongated conditions

To evaluate the performance of this scheme, the sensor head underwent strain variations up to 1500 $\mu\epsilon$ in 300 $\mu\epsilon$ step intervals at a constant temperature (T = 40 °C) and vice versa, i.e., temperature variations are in the range 0–40 °C at 10 °C intervals for a specific applied strain (ϵ = 500 $\mu\epsilon$). Inserting the values of the wavelength shifts exhibited by the 185 μm and 105 μm diameter POFBGs into Eq. 1, the average strain and temperature changes were calculated. Figure 22a shows the sensor outputs for temperature changes of 0–40 °C at a constant strain of 500 $\mu\epsilon$, and Fig. 22b shows the sensor results for strain changes of 0–1500 $\mu\epsilon$ at a constant temperature of 40 °C. When the strain is constant, there are ±0.50 °C and ± 35.05 $\mu\epsilon$ errors for the temperature and strain, respectively. Similarly when the temperature is constant, there are ±57.51 $\mu\epsilon$ and ± 0.77 °C errors for the strain and temperature, respectively.

To test the capability of the sensor to obtain simultaneous sensing, the temperature and strain were simultaneously applied experimentally, and the results are shown in Fig. 23. Firstly, a 10 °C temperature and 200 μm strain were applied to the sensor and measured their respective wavelength shift that corresponds to temperature and strain. Secondly, the temperature was increased to 15 °C at a 200 μm strain and then changed the strain to 400 μm and measured the respective wavelength shifts in each case. Then using Equation, these wavelength shifts were converted to their respective temperature and strain values. For example, initially, when 10 °C and 200 $\mu\epsilon$ were applied simultaneously, the measured temperature and strain are 10.61 °C and 213.66 $\mu\epsilon$, respectively. Then, when the temperature was changed to 15 °C with a constant 200 $\mu\epsilon$ strain, the measured temperature and strain are 15.08 °C and 193.62 $\mu\epsilon$, respectively. Following that, when the strain was changed to 400 $\mu\epsilon$ with the same temperature (15 °C), the measured temperature and strain are 15.46 °C and 402.72 $\mu\epsilon$, respectively. In that way, by changing the input temperature and strain, it can simultaneously measure both variables using the above matrix equation.

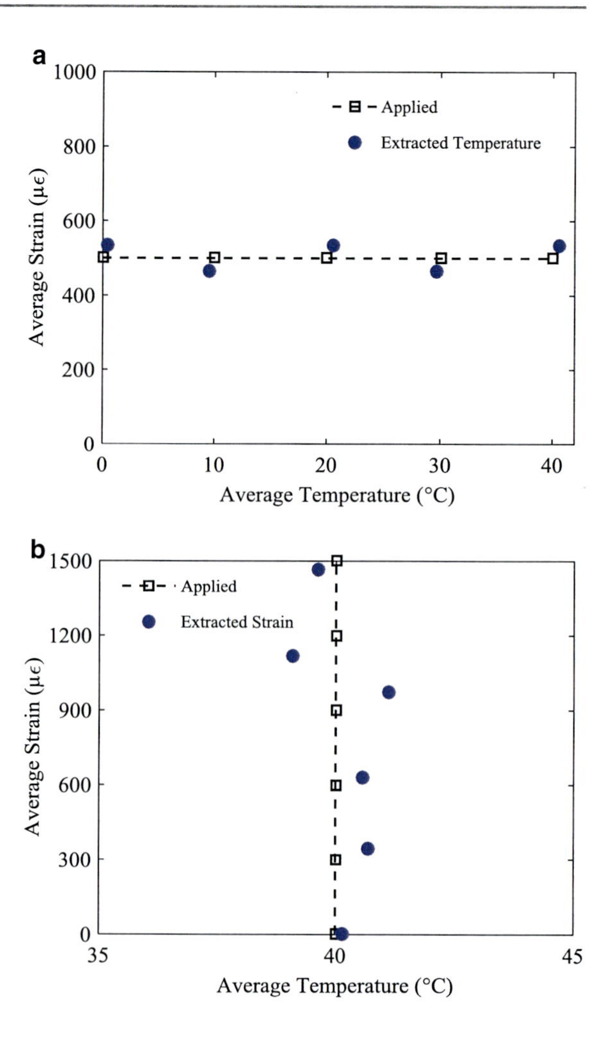

Fig. 22 (**a**) Sensor outputs for applied temperature at constant strain and (**b**) sensor outputs for applied strain at constant temperature

Dynamic Sensing

Due to low Young's modulus and high elastic limit, POFBG-based strain sensing has become more attractive and can be implemented for both static and dynamic measurements. Dynamic strain measurement applications include accelerometers, microphones, and vibration sensors (Stefani et al. 2012; Kuang and Cantwell 2003; Berkoff and Kersey 1996). However, as the POFs are made by viscoelastic materials, a varying response of strain/stress with frequency is observed due to the molecular rearrangement in the deformed polymer fiber. POFs always have a tendency to relax after a certain strain applied to it, and for bulk polymers, this behavior occurs at low frequencies, even less than 1 Hz (Read and Duncan 1981). A detailed investigation of the behaviors of mPOF and TOPAS polymer fibers under dynamic excitation, which showed elastic responses between a frequency range of 10 to

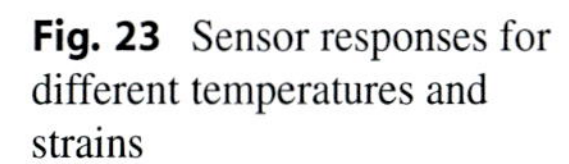

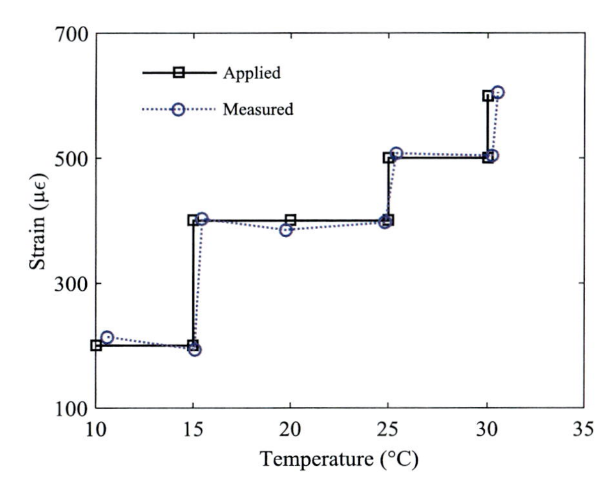

Fig. 23 Sensor responses for different temperatures and strains

100 Hz, is reported in ref. Stefani et al. (2012). It was also demonstrated that the Young's modulus is constant for strain up to 0.28%, and the viscoelastic effect increased with both applied strain and frequency. In recent years, considerable advancements in research on ultrasonic sensors based on silica FBGs have been made (Wild and Hinckley 2008). Ultrasound impinging on an FBG induces a subtle Bragg wavelength shift as the ultrasound signal incudes a small strain change in the FBG. However, the dynamic strain induced by the ultrasound signal is very small, and thus highly sophisticated interrogators are required to measure the ultrasound frequencies using silica FBG sensors. Therefore, more research has been focused on improving the interrogation system that allows the measurements of high-frequency sub-pico meter wavelength changes in silica FBGs. As a POF has superior mechanical properties to those of a silica fiber, it is expected that POFBGs could respond to low-amplitude ultrasonic signals more efficiently. It can be estimated that for the polymer fiber, the strain induced would be higher than that in a silica one for the same acoustic pressure and, together with the increased strain sensitivity achieved through etching the POFBGs, can exhibit a many fold increase in acoustic pressure wavelength sensitivity compared with those of silica FBGs.

Figure 24 shows the experimental setup for dynamic strain measurements of POFBGs. One end of the fiber containing the POFBG is attached to a piezoelectric transducer (PZT) by using a strong adhesive material, and the other end is glued to a fixed stage. It is assumed that the PZT is nearly in a zero load condition and controlled by a PZT controller driven by a function generator. The other side of the POFBG is connected to a light source and a commercial FBG interrogator (I-MON 512) through a circulator, and the corresponding changes are observed and recorded on a PC using the I-MON interrogator monitor software package. This I-MON interrogator can work up to a frequency of 3000 Hz, which operates in a wavelength range of 1510–1595 nm, where wavelength accuracy is 5 pm. Different magnitudes of dynamic strains are applied to the POFBG in a frequency range

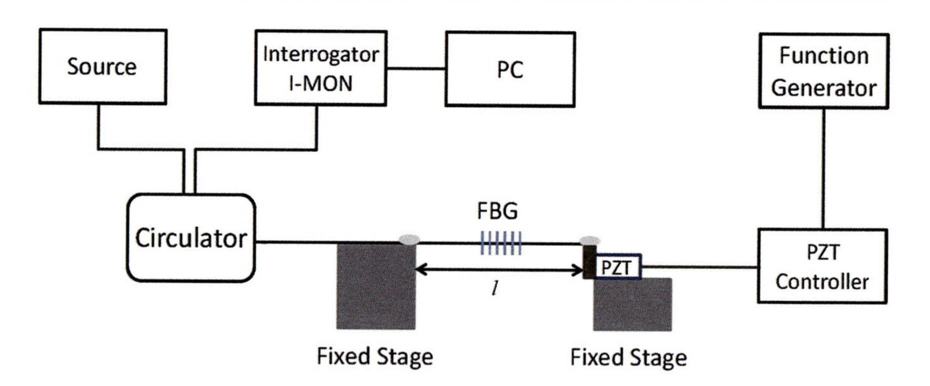

Fig. 24 Schematic of experimental setup for dynamic measurements

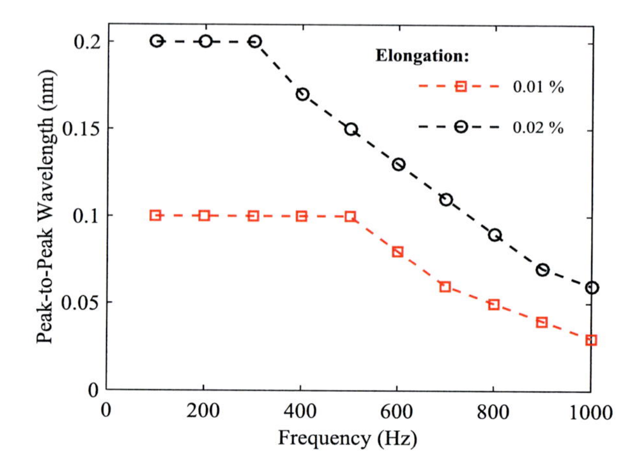

Fig. 25 Dynamic responses of un-etched POFBG for two different elongations

from 0 to 1000 Hz. All the experiments are conducted at room temperature in an air-conditioned environment. For dynamic measurements, a sinusoidal voltage is applied to the PZT, and the average voltage level of the detected signal is recorded.

Figure 25 shows the dynamic responses of an un-etched POFBG measured using the peak-to-peak wavelengths for frequencies ranging from 0 to 1000 Hz applied to the PZT at two different elongations of 0.01 and 0.02% of the fiber length. From the figure it can be seen that the frequency independent at responses are observed in the range of frequencies between 0 and 500 Hz and 0 and 300 Hz for the 0.01 and 0.02% elongations, respectively, indicating elastic behavior and low viscoelastic regime for these elongations. This indicates that at lower strain levels, the hysteresis is not impacted and the fiber relaxation time is not impacted till 500 Hz for 0.01% elongation and 300 Hz for 0.02% elongation. Since the sensitivity of POFBG is much higher, it can be measured at high frequencies as well. Even though peak-to-peak wavelength is dropping, it is still within the measurable range, because at 1000 Hz, the peak-to-peak wavelength is 0.03 and 0.06 nm for 0.01 and 0.02% elongation, respectively, which is still resolvable. Clearly, the fiber is capable of operating over this frequency range, with the calculated 3-dB frequencies 650 and 550 Hz for 0.01 and 0.02% elongations, respectively.

Biochemical Concentration Sensor

Another reported application of polymer FBG is a biochemical concentration sensor. Significant Bragg wavelength shifts can be obtained for different concentration of water-soluble substances. Zhang et al. (2012b) reported a sensor using a saline concentration as shown in Fig. 26 where it is seen that significant changes in Bragg wavelength occur with varying concentration and the process is reversible as well. The POFBG was first inserted into a test tube of distilled water for around 2 h, so that the POFBG was fully swelled by the water and then pulled out and inserted into a test tube of saline solution. Four samples of saline solution (8.5%, 13%, 17.5%, and 22%) were tested in order of ascending concentration and then in order of descending concentration. The POFBG device was left for 60 min in each sample to allow equilibrium to be reached. The POFBG wavelength was monitored with an interval of 60 s. If solute exits in the water, then the water content absorbed in the polymer fiber will change because of the osmotic effect. It is also reported that the sensor is reversible and thus can be used for in situ repeatable measurement and monitoring of concentration.

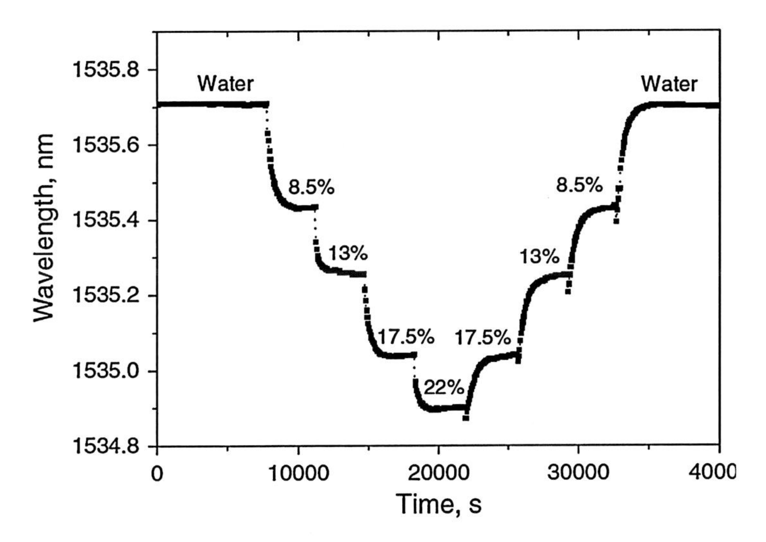

Fig. 26 Polymer FBG sensor response against step change of saline concentration (Zhang et al. 2012b)

Conclusions

In this chapter a review on the recent updates of research on solid core polymer fiber Bragg grating sensors and its applications were presented, and it is noted that it is an emerging area with potential engineering applications. The development of these sensors could be vital in a number of application areas, and some examples of the devices based on polymer FBGs are presented. Some valuable and important inventions on POF Bragg grating are presented in this chapter, but the research on POF Bragg grating is still in a primitive phase compared to silica FBGs.

The successful demonstration of intrinsic sensitivity enhancement technique will take the research on POFBG sensors and devices to a new level and lead to new research knowledge creation. One could explore further for the development of a range of novel sensors based on high sensitivity-etched POFBGs. Therefore, the demonstration of the potential of etched POFBG-based devices for use in biomedical applications is a worthwhile extension that could pave the way for new research and developments in this area. The investigation of the durability and ageing of POFBGs under various physical conditions could be a further potential, and the impact of thermal annealing on POFBGs could also be of interest. The needs for single-mode polymer fiber-compatible components, splicers, etc. are other areas of research that can provide acceleration to polymer FBG sensing. With the ongoing research activities on single-mode polymer FBGs, the research has a bright future outlook and potential applications in niche areas such as in biomedical.

References

A. Abang, D.J. Webb, Influence of mounting on the hysteresis of polymer fiber Bragg grating strain sensors. Opt. Lett. **38**, 1376–1378 (2013). 2013/05/01

D.Z. Anderson, V. Mizrahi, T. Erdogan, A.E. White, Production of in-fibre gratings using a diffractive optical element. Electron. Lett. **29**, 566–568 (1993)

S. Aramaki, G. Assanto, G.I. Stegeman, M. Marciniak, Realization of integrated Bragg reflectors in DANS-polymer waveguides. J. Lightwave Technol. **11**, 1189–1195 (1993)

G.A. Ball, W.W. Morey, Compression-tuned single-frequency Bragg grating fiber laser. Opt. Lett. **19**, 1979–1981 (1994). 1994/12/01

T.A. Berkoff, A.D. Kersey, Experimental demonstration of a fiber Bragg grating accelerometer. IEEE Photon. Technol. Lett. **8**, 1677–1679 (1996)

K. Bhowmik, G. Rajan, E. Ambikairajah, G.-D. Peng, Hydrostatic pressure sensitivity of standard polymer fibre Bragg gratings and etched polymer fibre Bragg gratings, in *OFS-23 Conference* (2014), pp. 91573G–91573G-4

K. Bhowmik, G. Peng, E. Ambikairajah, V. Lovric, W.R. Walsh, B.G. Prusty, et al., Intrinsic high-sensitivity sensors based on etched single-mode polymer optical fibers. Photon. Technol. Lett. IEEE **27**, 604–607 (2015a)

K. Bhowmik, G. Rajan, G. Peng, E. Ambikairajah, Y. Luo, B. Walsh, et al., Experimental study and analysis of hydrostatic pressure sensitivity of polymer fibre Bragg gratings. J. Lightwave Technol. **33**, 2456 (2015b). Accepted for publication

K. Bhowmik, G.D. Peng, Y. Luo, E. Ambikairajah, V. Lovric, W.R. Walsh, et al., Etching process related changes and effects on solid-core single-mode polymer optical fiber grating. IEEE Photon. J. **8**, 1–9 (2016a)

K. Bhowmik, G.D. Peng, Y. Luo, E. Ambikairajah, V. Lovric, W.R. Walsh, et al., High intrinsic sensitivity etched polymer fiber Bragg grating pair for simultaneous strain and temperature measurements. IEEE Sensors J. **16**, 2453–2459 (2016b)

R. Bogue, Fibre optic sensors: a review of today's applications. Sens. Rev. **31**, 304–309 (2011)

D. Bosc, C. Toinen, Full polymer single-mode optical fiber. IEEE Photon. Technol. Lett. **4**, 749–750 (1992)

M.J. Bowden, E.A. Chandross, I.P. Kaminow, Mechanism of the photoinduced refractive index increase in polymethyl methacrylate. Appl. Opt. **13**, 112–117 (1974). 1974/01/01

B. Culshaw, A. Kersey, Fiber-optic sensing: a historical perspective. J. Lightwave Technol. **26**, 1064–1078 (2008)

N.G. Harbach, *Fiber Bragg Gratings in Polymer Optical Fibers* (EPFL, Lausanne, 2008)

D.J. Hill, G.A. Cranch, Gain in hydrostatic pressure sensitivity of coated fibre Bragg grating. Electron. Lett. **35**, 1268–1269 (1999)

K.O. Hill, B. Malo, F. Bilodeau, D.C. Johnson, J. Albert, Bragg gratings fabricated in monomode photosensitive optical fiber by UV exposure through a phase mask. Appl. Phys. Lett. **62**, 1035–1037 (1993)

X. Hu, C.-F.J. Pun, H.-Y. Tam, P. Mégret, C. Caucheteur, Highly reflective Bragg gratings in slightly etched step-index polymer optical fiber. Opt. Express **22**, 18807–18817 (2014). 2014/07/28

X.F. Huang, D.R. Sheng, K.F. Cen, H. Zhou, Low-cost relative humidity sensor based on thermoplastic polyimide-coated fiber Bragg grating. Sensors Actuators B Chem. **127**, 518–524 (2007)

A. Iocco, H.G. Limberger, R.P. Salathe, Bragg grating fast tunable filter. Electron. Lett. **33**, 2147–2148 (1997)

A. Iocco, H.G. Limberger, R.P. Salathe, L.A. Everall, K.E. Chisholm, J.A.R. Williams, et al., Bragg grating fast tunable filter for wavelength division multiplexing. J. Lightwave Technol. **17**, 1217–1221 (1999)

I.P. Johnson, D.J. Webb, K. Kalli, Hydrostatic pressure sensing using a polymer optical fibre Bragg gratings (2012), pp. 835106–835106-7

I.P. Kaminow, H.P. Weber, E.A. Chandross, Poly(methyl methacrylate) dye laser with internal diffraction grating resonator. Appl. Phys. Lett. **18**, 497–499 (1971)

A.D. Kersey, M.A. Davis, H.J. Patrick, M. LeBlanc, K.P. Koo, C.G. Askins, et al., Fiber grating sensors. J. Lightwave Technol. **15**, 1442–1463 (1997)

M. Kopietz, M.D. Lechner, D.G. Steinmeier, J. Marotz, H. Franke, E. Krätzig, Light-induced refractive index changes in polymethylmethacrylate (PMMA) blocks. Polym. Photochem. **5**, 109–119 (1984). 1984/01/01

K.S.C. Kuang, W.J. Cantwell, The use of plastic optical fibre sensors for monitoring the dynamic response of fibre composite beams. Meas. Sci. Technol. **14**, 736 (2003)

C.R. Kurkjian, J.T. Krause, M.J. Matthewson, Strength and fatigue of silica optical fibers. J. Lightwave Technol. **7**, 1360–1370 (1989)

M.G. Kuzyk, U.C. Paek, C.W. Dirk, Guest-host polymer fibers for nonlinear optics. Appl. Phys. Lett. **59**, 902–904 (1991)

K.S. Lim, H.Z. Yang, W.Y. Chong, Y.K. Cheong, C.H. Lim, N.M. Ali, et al., Axial contraction in etched optical fiber due to internal stress reduction. Opt. Express **21**, 2551–2562 (2013)

H.Y. Liu, G.D. Peng, P.L. Chu, Thermal tuning of polymer optical fiber Bragg gratings. Photon Technol. Lett. IEEE **13**, 824–826 (2001)

H.Y. Liu, G.D. Peng, P.L. Chu, Polymer fiber Bragg gratings with 28-dB transmission rejection. IEEE Photon. Technol. Lett. **14**, 935–937 (2002)

H.Y. Liu, H.B. Liu, G.D. Peng, P.L. Chu, Observation of type I and type II gratings behavior in polymer optical fiber. Opt. Commun. **220**, 337–343 (2003a). 2003/05/15

H.B. Liu, H.Y. Liu, G.D. Peng, P.L. Chu, Strain and temperature sensor using a combination of polymer and silica fibre Bragg gratings. Opt. Commun. **219**, 139–142 (2003b)

H.B. Liu, H.Y. Liu, G.D. Peng, P.L. Chu, Novel growth behaviors of fiber Bragg gratings in polymer optical fiber under UV irradiation with low power. IEEE Photon. Technol. Lett. **16**, 159–161 (2004)

H.Y. Liu, H.B. Liu, G.D. Peng, Tensile strain characterization of polymer optical fibre Bragg gratings. Opt. Commun. **251**, 37–43 (2005). 2005/07/01

J.M. Lopez-Higuera, L.R. Cobo, A.Q. Incera, A. Cobo, Fiber optic sensors in structural health monitoring. J. Lightwave Technol. **29**, 587–608 (2011)

Y. Luo, B. Yan, M. Li, X. Zhang, W. Wu, Q. Zhang, et al., Analysis of multimode POF gratings in stress and strain sensing applications. Opt. Fiber Technol. **17**, 201–209 (2011)

B. Malo, K.O. Hill, F. Bilodeau, D.C. Johnson, J. Albert, Point-by-point fabrication of micro-Bragg gratings in photosensitive fibre using single excimer pulse refractive index modification techniques. Electron. Lett. **29**, 1668–1669 (1993)

J. Marotz, Holographic storage in sensitized polymethyl methacrylate blocks. Appl. Phys. B **37**, 181–187 (1985)

J. Mathew, Y. Semenova, G. Farrell, Relative humidity sensor based on an agarose-infiltrated photonic crystal fiber interferometer. IEEE J. Sel. Top. Quantum Electron. **18**, 1553–1559 (2012)

G. Meltz, W.W. Morey, W.H. Glenn, Formation of Bragg gratings in optical fibers by a transverse holographic method. Opt. Lett. **14**, 823–825 (1989). 1989/08/01

D.F. Merchant, P.J. Scully, N.F. Schmitt, Chemical tapering of polymer optical fibre. Sensors Actuators A Phys. **76**, 365–371 (1999). 1999/08/30

J. Mora, A. Diez, J.L. Cruz, M.V. Andres, A magnetostrictive sensor interrogated by fiber gratings for DC-current and temperature discrimination. IEEE Photon. Technol. Lett. **12**, 1680–1682 (2000)

J.M. Moran, I.P. Kaminow, Properties of holographic gratings photoinduced in polymethyl methacrylate. Appl. Opt. **12**, 1964–1970 (1973). 1973/08/01

P.M. Nellen, P. Mauron, A. Frank, U. Sennhauser, K. Bohnert, P. Pequignot, et al., Reliability of fiber Bragg grating based sensors for downhole applications. Sensors Actuators A Phys. **103**, 364–376 (2003). 2003/02/15

Y. Ohtsuka, Y. Hatanaka, Preparation of light-focusing plastic fiber by heat-drawing process. Appl. Phys. Lett. **29**, 735–737 (1976)

G.-D. Peng, P.L. Chu, X. Lou, R.A. Chaplin, Fabrication and characterization of polymer optical fibers. J. Electron. Electr. Eng. **15**, 289–296 (1995)

G.D. Peng, Z. Xiong, P.L. Chu, Photosensitivity and gratings in dye-doped polymer optical fibers. Opt. Fiber Technol. **5**, 242–251 (1999). 1999/04/01

G. Rajan, B. Liu, Y. Luo, E. Ambikairajah, G.D. Peng, High sensitivity force and pressure measurements using etched singlemode polymer fiber Bragg gratings. IEEE Sensors J. **13**, 1794–1800 (2013a)

G. Rajan, Y.M. Noor, B. Liu, E. Ambikairaja, D.J. Webb, G.-D. Peng, A fast response intrinsic humidity sensor based on an etched singlemode polymer fiber Bragg grating. Sensors Actuators A Phys. **203**, 107–111 (2013b)

G. Rajan, M.Y.M. Noor, E. Ambikairajah, G.D. Peng, Inscription of multiple Bragg gratings in a single-mode polymer optical fiber using a single phase mask and its analysis. IEEE Sensors J. **14**, 2384–2388 (2014)

B.E. Read, J.C. Duncan, Measurement of dynamic properties of polymeric glasses for different modes of deformation. Polym. Test. **2**, 135–150 (1981). 1981/04/01

D. Song, J. Zou, Z. Wei, S. Yang, H.-L. Cui, High-sensitivity fiber Bragg grating pressure sensor using metal bellows. Opt. Eng. **48**, 034403–034403-3 (2009)

Standard test method for tensile strength and young's modulus of fibers, ed Active Standard ASTM C1557, 2003

A. Stefani, S. Andresen, W. Yuan, O. Bang, Dynamic characterization of polymer optical fibers. IEEE Sensors J. **12**, 3047–3053 (2012)

W.J. Tomlinson, I.P. Kaminow, E.A. Chandross, R.L. Fork, W.T. Silfvast, Photoinduced refractive index increase in poly(methyl methacrylate) and its applications. Appl. Phys. Lett. **16**, 486–489 (1970)

T. Wang, Y. Luo, G.-D. Peng, Q. Zhang, High-sensitivity stress sensor based on Bragg grating in BDK-doped photosensitive polymer optical fiber (2012), pp. 83510M–83510M-8

G. Wild, S. Hinckley, Acousto-ultrasonic optical Fiber sensors: overview and state-of-the-art. IEEE Sensors J. **8**, 1184–1193 (2008)

Z. Xiong, G.D. Peng, B. Wu, P.L. Chu, Highly tunable Bragg gratings in single-mode polymer optical fibers. IEEE Photon. Technol. Lett. **11**, 352–354 (1999)

M.G. Xu, L. Reekie, Y.T. Chow, J.P. Dakin, Optical in-fibre grating high pressure sensor. Electron. Lett. **29**, 398–399 (1993)

M.G. Xu, J.L. Archambault, L. Reekie, J.P. Dakin, Discrimination between strain and temperature effects using dual-wavelength fibre grating sensors. Electron. Lett. **30**, 1085–1087 (1994)

T.L. Yeo, S. Tong, K.T.V. Grattan, D. Parry, R. Lade, B.D. Powell, Polymer-coated fiber Bragg grating for relative humidity sensing. Sensors J. IEEE **5**, 1082–1089 (2005)

T.L. Yeo, T. Sun, K.T.V. Grattan, Fibre-optic sensor technologies for humidity and moisture measurement. Sensors Actuators A Phys. **144**, 280–295 (2008)

L. Yunqi, C. Kin Seng, C. Pak Lim, Fiber-Bragg-grating force sensor based on a wavelength-switched self-seeded Fabry-Perot laser diode. IEEE Photon. Technol. Lett. **17**, 450–452 (2005)

W. Zhang, D.J. Webb, G.D. Peng, Investigation into time response of polymer fiber Bragg grating based humidity sensors. J. Lightwave Technol. **30**, 1090–1096 (2012a)

W. Zhang, D. Webb, G. Peng, Polymer optical fiber Bragg grating acting as an intrinsic biochemical concentration sensor. Opt. Lett. **37**, 1370–1372 (2012b)

Microstructured Polymer Optical Fiber Gratings and Sensors

52

Getinet Woyessa, Andrea Fasano, and Christos Markos

Contents

G. Woyessa (✉) · C. Markos
DTU Fotonik, Department of Photonics Engineering, Technical University of Denmark, Lyngby, Denmark
e-mail: gewoy@fotonik.dtu.dk; chmar@fotonik.dtu.dk

A. Fasano
DTU Mekanik, Department of Mechanical Engineering, Technical University of Denmark, Lyngby, Denmark
e-mail: andfas@kt.dtu.dk

G.-D. Peng (ed.), *Handbook of Optical Fibers*,
https://doi.org/10.1007/978-981-10-7087-7_2

Abstract

This chapter describes the realization of microstructured polymer optical fibers Bragg gratings (mPOFBGs) and their use in various sensing applications. Different grating inscription techniques based on different lasers used in recording grating in mPOF are presented. Grating inscription in mPOFs can be a challenging task compared to step index fibers because the microstructured cladding holes introduce scattering preventing the laser from reaching the core of the fiber easily. Inscription of gratings in mPOFs fabricated from different polymer materials such as Topas, Zeonex, Polycarbonate (PC) is discussed, and their optical and sensing performance is directly compared with the widely used poly(methylmethacrylate)(PMMA). The progress on fabrication of gratings in different types of mPOFs is presented in terms of grating inscription time, strength, and Bragg wavelengths. This chapter also describes the annealing process of mPOFs or mPOFBGs which is one of the curtail step in the development of stable mPOFBG sensors. The different annealing methods that have been applied by the research community are also presented. In addition, an overview on strain, humidity, temperature, pressure, and acceleration sensors developed from mPOFBGs is provided. A direct comparison in terms of their sensitivity, sensing range, and their performance in general is presented. Finally, the way to improve the development of stable mPOFBG sensors and widen their application areas is briefly discussed.

Microstructured Polymer Optical Fiber Bragg Gratings Inscription

Fiber Bragg gratings (FBGs) have been inscribed in microstructured polymer optical fibers (POFs) as in conventional step-index fibers based on different inscription techniques. The most common and used methods are the phase mask, point-by-point, as well as the interferometric method. Various types of lasers have been used for FBG inscription such as He-Cd laser, UV pulsed excimer laser, and ultrafast femtosecond lasers. The inscription of FBGs in microstructured polymer optical fibers (mPOFs) is not as straight forward as in step index POFs. The microstructured air hole pattern along the length of the fiber can significantly affect the writing process and the quality of the grating. Hence, the fiber should be aligned in such a way that the beam penetrates the highly scattering cladding and reaches the core easily, in order the inscription time to be reasonably short and the grating strength to be comparable with that of step index POFs. In this section, different types of FBG inscription methods in mPOFs and various types of laser used for inscription are reported as well as the effect of the microstructured cladding along the length of the fiber on the inscription process is discussed.

Interferometric Method

In this technique, UV laser beam from a single continuous wave (CW) laser source is split into two beams. This is commonly done by passing the beam through a 50:50 beam splitter. The two beams are then reflected by two mirrors and joined at angle θ after passing through two cylindrical lenses. The lenses are used to focus the UV laser beam and attain a high optical power density at the core of the mPOF. As the beams intersect, they produce periodic interference fringes, which, when they are focused onto the mPOF, effectively inscribe an FBG as shown in Fig. 1 (Meltz et al. 1989).

For a UV laser beam from a single source of wavelength λ and two beams incident on the fiber with an incidence angle θ with respect to the normal (Fig. 1), the period of the interference pattern formed as the two beams intersect, which corresponds to the period of the grating inscribed in the core of the mPOF fiber, is given by:

$$\Lambda = \frac{\lambda}{2\sin(\theta)} \tag{1}$$

From Eq. 1, it can be seen that, using this writing technique, it is possible to inscribe an FBG with the intended periodicity (pitch) by simply adjusting the angle of incident of the two intersecting UV beams. Although this method gives this flexibility to control the period of the grating, it also has some drawbacks. The major disadvantage is that it requires a highly coherent UV beam laser source. Moreover, this coherence must be maintained throughout the exposure, otherwise a shift in the interference fringes may occur and thus the grating may undergo erosion. In addition, the beam alignment is time consuming and the mechanical stability is

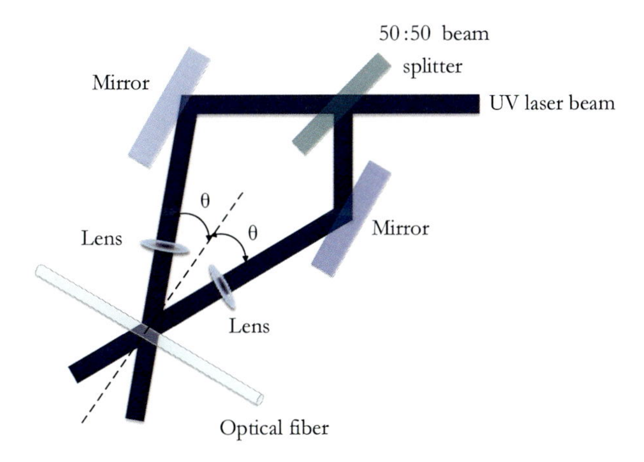

Fig. 1 FBG fabrication by transverse holographic technique

poor. For these reasons, the interferometric technique is not widely used in the FBG community even if it provides a great degree of flexibility.

Phase Mask Method

This inscription technique consists of a UV laser beam passing through the phase mask, which results the formation of a periodic interference pattern forming a grating in the mPOF. A phase mask is basically a plate of fused silica substrate which is transparent to the UV laser beam. One of the surfaces of the silica substrate is etched to create a surface relief structure which is periodic and has a square wave profile. In this technique, the phase mask is placed between the beam path and the mPOF as depicted in Fig. 2 (Hill et al. 1993). The UV light is directed to the phase mask, and after it passes through the periodic corrugations of the mask, the UV light splits into multiple diffracted beams in different directions, known as orders, defined as $m = 0, \pm 1, \pm 2, \pm 3$. These diffracted beams create a periodic interference pattern which, when exposed to an mPOF, photo-imprints a Bragg grating in the core of the fiber. The angle of diffraction, θ_m, can be calculated for each order based on the following diffraction Eq.:

$$m\lambda_{uv} = \Lambda_{pm}\left(\sin\theta_m + \sin\theta_i\right) \tag{2}$$

where, λ_{uv} is the wavelength of the inscription UV laser beam, Λ_{pm} is the period of the phase mask, and θ_i is the angle of incidence of the UV inscription beam to the top surface of the phase mask.

When the angle of incidence of the UV laser beam is normal to the top surface of the phase mask as shown in Fig. 2, which is the most commonly used configuration, Eq. 2 simplifies as follows:

$$m\lambda_{uv} = \Lambda_{pm}\sin\theta_m \tag{3}$$

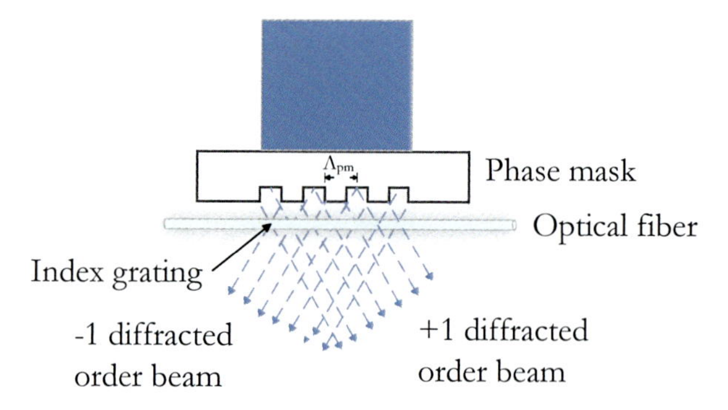

Fig. 2 UV laser beam passing through a phase mask and UV interference pattern generated from −1 and +1 diffraction orders of the phase mask

When the angle of incidence of the UV laser beam is perpendicular to the phase mask, the angles of diffraction are equal for both the positive and negative orders. For the majority of FBG fabrication processes, the diffracted light is limited to the orders -1, 0, and 1. Although most of the diffracted UV light beams are contained in -1, 0, and $+1$ diffracted orders, the most employed phase mask configuration normally suppresses the diffraction light beam contained in the zero order (Xiong et al. 1999). This can be achieved by controlling the depth of the corrugation in the phase mask. Thus, the period of the interference pattern formed by the -1 and $+1$ orders, hence the grating planes, Λ, is defined by both the UV laser beam wavelength, λ_{uv}, and the diffraction angle of the two beams.

$$\lambda_{uv} = \Lambda 2 \sin \theta_m \tag{4}$$

By using Eqs. 3 and 4, the period of the grating planes Λ is therefore one half of the period of the surface relief pattern of the mask Λ_{pm} as follows:

$$\begin{aligned} \lambda_{uv} &= \Lambda 2 \sin \theta_m = \Lambda_m \sin \theta_m \\ \Lambda &= \frac{\Lambda_{pm}}{2} \end{aligned} \tag{5}$$

From Eq. 5, it can be seen that the period of the grating is dependent only on the period of the phase mask and is independent of the wavelength of the UV laser beam irradiating the phase mask. Nevertheless, the corrugation depth required to suppress the diffracted beam of light contained in the zeroth order is a function of the wavelength of the UV laser beam (Williams 1997).

Figure 3 shows a complete schematic setup for microstructured polymer optical fiber Bragg grating (mPOFBG) inscription using the phase mask approach. As it can been seen, the main components of the setup are a UV laser, mirrors, cylindrical focusing lens, and phase mask. The UV laser light with a circular Gaussian beam profile is directed through a series of 4 mirrors and arrives at the focusing lens prior to the phase mask. The lens focus the beam through the phase mask down onto the fiber core. The phase mask is positioned parallel to the fiber surface as close as possible without actually touching it in order to protect the surface relief pattern of the mask from damage. The high intensity points of the pattern of the diffracted orders were projected into the mPOF core, resulting in a permanent periodic refractive index change within the mPOF core.

The most widely used UV laser for FBG inscription is He-Cd laser (IK5751I-G from Kimmon) operating at 325 nm. This laser has a circular Gaussian beam profile with a diameter of approximately 2 mm (Dobb et al. 2005; Johnson et al. 2010a; Stefani et al. 2011). Other types of lasers have been also used such as KrF excimer laser (Bragg Star *TM* Industrial-LN) operating at 248 nm (Oliveira et al. 2015). The KrF excimer laser has also Gaussian beam profile with a beam spot ~6 mm in diameter and pulse duration of ~15 ns. Femtosecond lasers have also been used for writing FBGs in mPOFs using phase mask technique. A 400 nm femtosecond laser pulse with a beam diameter of 6 mm has been used in the inscription of FBGs in doped poly (methylmethacrylate) (PMMA) mPOFs (Hu et al. 2017).

The phase mask approach has many distinct advantages over the other writing techniques. First, it allows the use of low-cost UV light source by lowering the requirement of the laser coherence, which is essentially required for transverse holographic technique. Second, it permits an easier arrangement of the fiber to be grated and also the stability requirements on the grating apparatus are reduced. This configuration set-up is mechanically stable and reliable, and FBGs are relatively easy to inscribe. Therefore, this method produces the most consistent and reproducible inscriptions, and hence it is the most common method of FBG writing in mPOFs. The main disadvantage of the phase mask technique is that to fabricate gratings with different pitches (i.e., different Bragg wavelengths), different phase masks are required, thereby increasing the production cost. It should be noted however that simple postprocessing approaches have been already reported which provide the ability to tune the grating to a certain degree. The most common one is by applying axial prestraining before inscription or postinscription annealing of the FBG. By using this method, it is possible to tune the grating by more than 250 nm (Yuan et al. 2012; Woyessa et al. 2016).

A typical interrogation system for the grating inscription in mPOF is also shown as a part of Fig. 3 which mainly consists of a broad band light source, a 3 dB coupler, and an optical spectrum analyzer (OSA). A broadband light source or a supercontinuum laser is used to couple the light into the core of the mPOF via the coupler and a silica fiber butt-coupled to the mPOF. Index matching oil is usually used in the gap between the silica fiber and the mPOF to minimize interference from the formed cavity and to reduce the reflection noise. The reflected signal from the

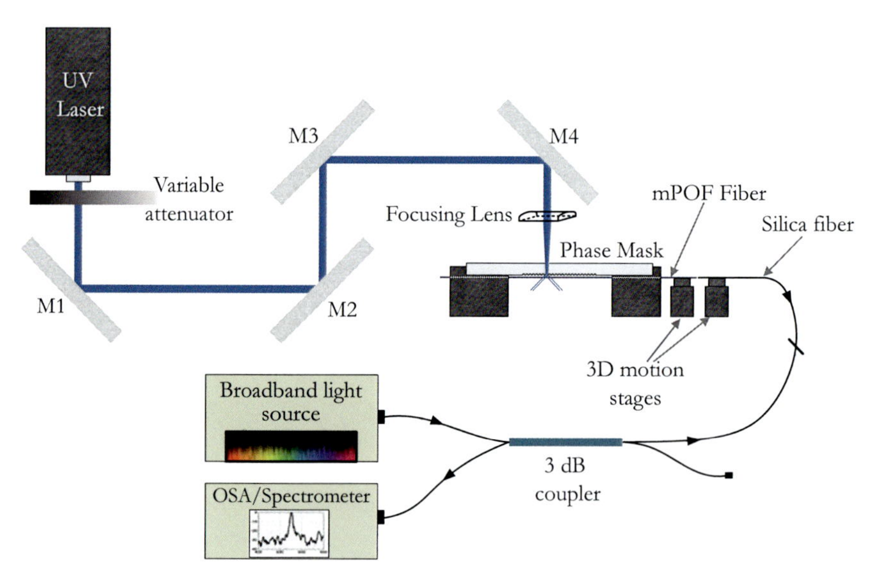

Fig. 3 One of the most commonly used FBG inscription and interrogation setups in mPOF using the phase mask technique

mPOFBG is coupled back through the 3 dB coupler, and finally it is collected from an optical spectrometer or the OSA.

Point-by-Point Method

In contrast to the inscriptions approaches described above, where the inscription process takes place in a single exposure to the laser beam, in the point-by-point approach, each grating plane in the mPOF core is photo-inscribed point by point by exposing the side of the optical fiber to a pulsed laser (Malo et al. 1993). Figure 4 shows the schematic representation of the FBG fabrication using the point-by-point method. In this method, the mPOF is first placed on a mechanical stage which enables the fiber to be translated along its length during the inscription process. Then a collimated UV or near infrared (near-IR) laser beam is focused on the core of the fiber from the side of the fiber to photo-imprint one grating plane at a time. The stage is then translated, hence the fiber, by a length which is the intended period of the grating, Λ, to write the next grating plane. The same procedure is repeated, until a periodic FBG is fabricated. The same technique has also been demonstrated with an ultrafast femtosecond pulsed laser (Martinez et al. 2004; Kalli et al. 2015).

A Ti:sapphire femtosecond laser system (Hurricane, Spectra-Physics) has been also used to inscribe gratings in PMMA mPOFs in the 1550 nm region (Stefani et al. 2012a). The output of the laser has 100 fs pulses with a central wavelength of 800 nm repetition rate of 1 kHz and average output power of 1 W. The significant advantage of the point-by-point grating technique is that it provides the choice of the length and pitch of the grating to be determined independently. This enables the location of the reflected wavelength to be controlled precisely and the use of expensive phase masks can also be avoided. Direct writing allows also the inscription of FBGs in nonphotosensitive materials. One of the main drawbacks of this fabrication approach is that it requires accurate alignment of the fiber core with the beam, while the continuous irradiation and translation process lead to extensive inscription durations compared, for example, to the previously described approaches.

The Effect of Microstructure in Grating Inscription

One of the most common challenges during FBG inscription in mPOFs is the scattering of the laser beam due to the presence of the cladding air-holes. This

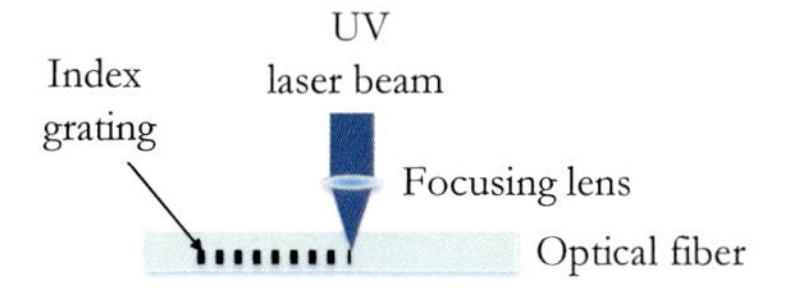

Fig. 4 FBG fabrication by point-by-point technique

usually prevents the laser beam from reaching the core of the mPOF. Thus, in order to achieve rapid inscription and high quality gratings in the mPOF, the orientation of the fiber has a crucial role in the process and it should be in such a way that the coupling of the laser beam to the core of the fiber is maximized. In a quest of finding the optimum fiber orientation, Marshall et al. preformed a thorough computational and experimental study to understand the effect of the microstructured air holes in the FBG inscription in microstructured optical fibers (Marshall et al. 2007). In their experiment, they focused a laser beam at 267 nm (3rd harmonic generated from 120 fs pulses at 800 nm, from a Spectra-Physics Hurricane) from the side on the core of the microstructured optical fiber. The laser beam was focused and coupled from the side to the core of the microstructured fiber, and then propagating to the tip of the fiber. From the tip of the fiber, the luminescence was measured and this process was repeated by rotating the fiber by one degree each time. It was found that the coupling of the laser beam to the core of the fiber was the strongest only for certain angles. Those are angles close to *X* but not exactly at *X* (see Fig. 5), the angular direction that offers direct access to the core for the laser light and hence the coupling to the core was the strongest. Another similar investigation on the effect of the microstructure on the grating inscription has been performed by T. Baghdasaryan et al. (2011). They used an ultrafast Ti:Sapphire laser at 800 nm by using different angular orientation and geometries of microstructured fibers. In their study, they also investigated the effect of the microstructured holes sizes and the hole to pitch ratio of the fiber. The results revealed that there is an orientation that gives the strongest coupling, which are angles close to *X*, and sufficiently small holes will result in a high quality grating, despite small air filling factor significantly reduces the strong confinement achievable in the microstructure optical fiber. Both of the above investigations were based on focusing the laser on the side of the fiber and measuring the photoluminescence at the tip of the fiber as the laser beam was guided through the core of the fiber.

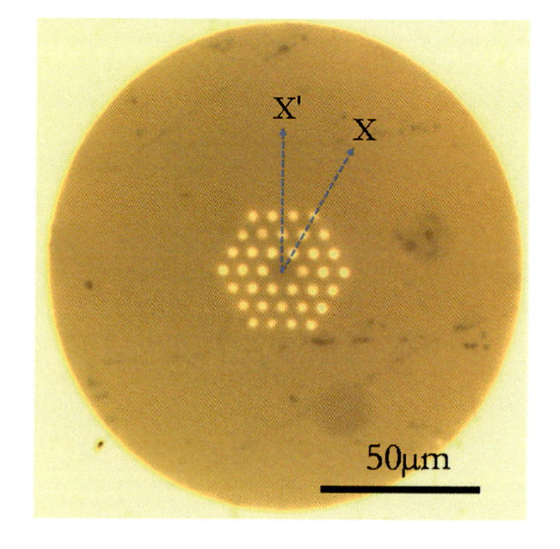

Fig. 5 Microscope image of mPOF with marked symmetry directions X' and X

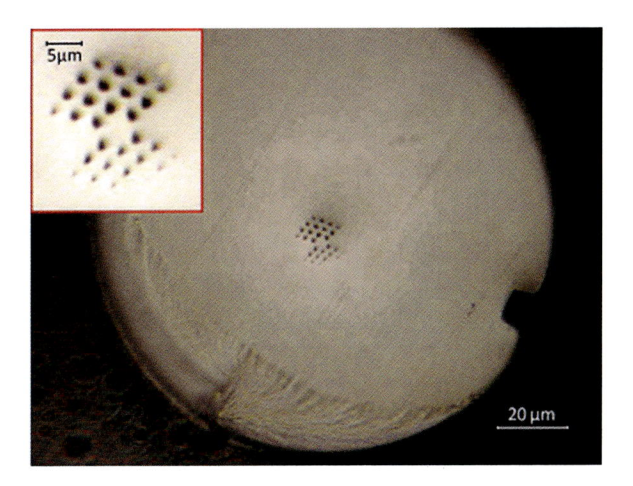

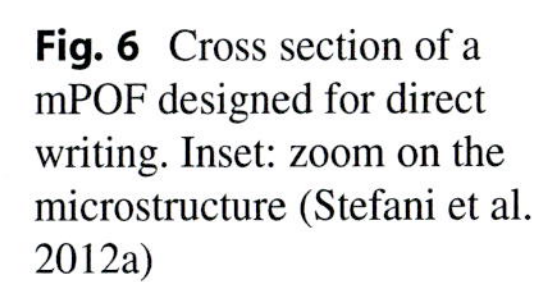

Fig. 6 Cross section of a mPOF designed for direct writing. Inset: zoom on the microstructure (Stefani et al. 2012a)

In 2015, Bundalo et al. experimentally demonstrated the above claims by inscribing a grating in mPOFs at various angles (Bundalo et al. 2015). They reached the same conclusion as Marshall et al. and T. Baghdasaryan et al. that certain angles can significantly assist in obtaining high quality gratings in mPOFs, which means fast inscription time and stronger gratings. They also showed that gratings could be obtained at almost any angle but the quality of the inscribed gratings was poor in terms of reflection strength and inscription time, if the fiber orientation was not in the proper range of angles, i.e., X'.

To minimize the effect of the microstructured air holes on the grating inscription and to access the core of the fiber with reduced difficulty, new fiber designs were proposed. For example, Stefani et al. designed and fabricated a PMMA mPOF for grating inscription using the point by point technique (Stefani et al. 2012a). The fiber was designed and fabricated to have the laser beam passing between the microstructure holes with minimum scattering before reaching the fiber core. To do so, six holes were removed from the three ring hexagonal microstructure, two from the outer ring and one from the second ring from both sides of the hexagonal configuration symmetrically as shown in Fig. 6. For the laser beam to pass the first ring of holes without encountering any interface, the pitch of the microstructure was designed to be bigger than the spot size of the laser beam. In addition, in order to facilitate the alignment of the microstructure, an alignment groove in the outer part of the fiber was also added.

Microstructured Polymer Optical Fiber Bragg Gratings

In this section, different mPOFs fabricated from different types of polymers are reported, and the progress on the fabrication of gratings in mPOFs in terms of inscription time, wavelength, and strength is presented. Currently existing mPOFs used for the grating fabrication are PMMA, Topas, Polycarbonate (PC), and Zeonex.

Fabrication of gratings in mPOF made from these polymer materials will be reported in the following sections.

PMMA mPOFBGs

The first FBG in mPOFs was inscribed by Dobb et al. in 2007 in the 1550 nm wavelength region (Dobb et al. 2005). The technique used for the inscription was the phase mask by using a 325 nm HeCd continuous wave UV laser with a power of 30 mW. The laser spot had a 1.8 mm diameter and it was expanded to 10 mm by cylindrical lens. The phase mask used for inscription had a uniform period of 1060.85 nm. Two types of PMMA mPOFs were used for the inscription: an endlessly single mode and a few modes fibers with hole to pitch ratio of 0.31 and 0.55, respectively. The FBGs inscribed in the single and few modes mPOFs have Bragg wavelength, FWHM, and reflection strength of 1569 nm, 0.5 nm and 5 dB, and 1570 nm, 1 nm and 7 dB, respectively. The length of the grating for both cases was 10 mm and the FBGs were saturated almost an hour after the exposure. Figure 7a and b shows the reflection spectrum and the growth dynamics of the fabricated gratings in a single and a few mode PMMA mPOF, respectively.

Some of the limitations of the FBG inscription process by Dobb et al. were the long writing time and the weak gratings. However, some years later (2014)

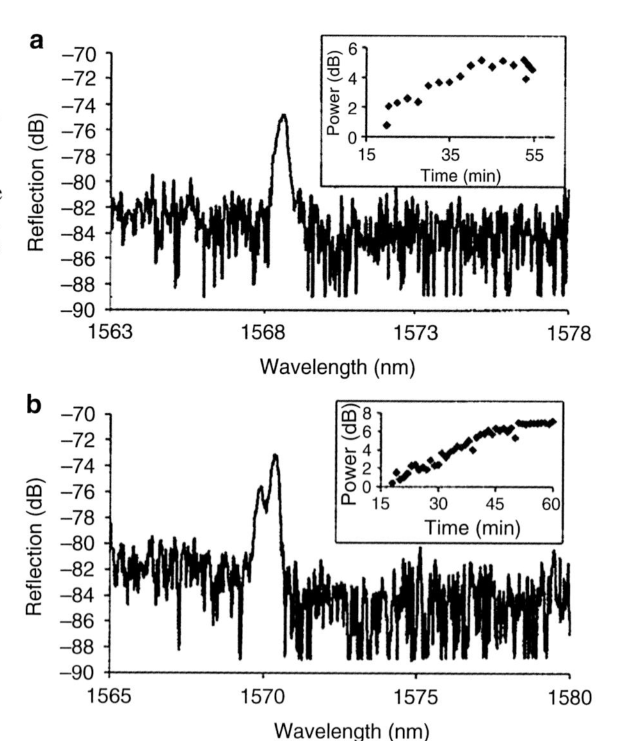

Fig. 7 (**a**) Reflection spectrum of an FBG fabricated in a single-mode PMMA mPOF. Inset: Growth of signal-to-noise ratio. (**b**) Reflection spectrum of an FBG fabricated in a few mode PMMA mPOF. Inset: Growth of signal-to-noise ratio (Dobb et al. 2005)

Bundalo et al. reported FBG inscription in PMMA mPOFs with a writing time 10 times shorter than it was previously reported (Bundalo et al. 2014). The setup used by Bundalo et al. was also a standard phase mask writing technique except that in this case the HeCd laser beam was not expanded. It was demonstrated that there was a strong correlation between laser power and the grating inscription time revealing that a high intensity of laser power in the core was required in order to obtain a fast writing time. The fiber used was an endlessly single mode 3 rings PMMA mPOF with a hole to pitch ratio of 0.26. The phase mask used in their experiments had a uniform pitch of 424.84 nm which is suitable for writing FBGs in the 650 nm wavelength region. The fastest inscription time achieved was less than 7 min with full laser power of 30 mW. The grating had the following characteristics: reflection strength, FWHM, and Bragg wavelength of 26 dB, 0.4 nm, and 632.6 nm, respectively. The length of the grating was 2 mm and it was shown that there was no dependence between the laser power and the grating strength. However, low laser inscription powers resulted in longer writing times. By using 72% of the full laser power, an FBG of the same strength but with longer inscription time, around 18 min, was obtained. Table 1 summarizes the comparison of the growth rates and the quality of the gratings for 6 inscription trials made in PMMA mPOFs. Bundalo et al. claimed that the most important factors leading to these results were the removal of the beam expansion lens and the careful alignment of the laser beam in the core of the mPOF. The laser power reaching the fiber without using the beam expanding lens was 7–8 times more intense than with the beam expansion lens.

A year later Oliveira et al. reported the inscription of FBGs in a PMMA mPOF in less than 30 s using the phase mask method (Oliveira et al. 2015). Despite they used the phase mask writing technique, the laser for inscription was not the same as the one used by Dobb et al. and Bundalo et al.; they rather used a KrF excimer laser (Bragg Star *TM* Industrial-LN) at 248 nm wavelength. They claimed that the key factors leading to such reduced writing time were the use of low number of pulses, low repetition rate, and low fluence. They revealed that by using energy density below the threshold for the PMMA ablation, it is possible to avoid the damage of the surface of the polymer fiber and increase the refractive index change in the core of the fiber. The fiber used in their experiments was a few mode 6-ring PMMA mPOF with a hole to pitch ratio of 0.52. The phase mask had a uniform

Table 1 Comparison of growth rates, grating saturation time, FWHM, and grating strength for six grating inscription attempts in PMMA mPOFs

FBG	Grating growth rate (dB/min)	Grating saturation time (mm:ss)	Grating FWHM (nm)	Grating strength (dB)
1	7.9	06:50	0.4	26
2	3.99	08:50	0.35	24
3	3.7	07:10	0.3	19
4	3.18	17:50	0.4	26
5	2.47	15:00	0.4	21
6	2.1	15:50	0.425	15

period of 1023 nm and the laser beam was not expanded. The laser beam passed through a slit of 4.5 mm width and was shaped before reaching the phase mask. The Bragg wavelength, the reflection strength, and 3 dB bandwidth of the grating were 1515 nm, 20 dB, and 0.16 nm, respectively. The length of the grating was 4.5 mm and it was inscribed with 1 Hz repetition rate, 33 mJ/cm^2 fluence, and 20 pulses. Figure 8 shows the growth dynamics of the grating during the grating inscription. This inscription time is the shortest writing time in POFs till now by using a UV laser. One of the advantages of having a fast inscription time is that the constraint related to the stability of the mechanical setup during the inscription process can be minimized.

The above-mentioned FBG inscriptions in PMMA mPOFs were done with a UV laser. However, PMMA mPOFs have also been inscribed with femtosecond laser by using the point by point inscription method. Stefani et al. reported the fabrication of FBGs in PMMA mPOF with 800 nm Ti:sapphire femtosecond laser system (Hurricane, Spectra-Physics) using direct-writing techniques (Stefani et al. 2012a). To avoid scattering before the laser reaching the core and facilitate the inscription process, they used a different fiber design which is shown in Fig. 6 in the previous section. The fiber had air filling factor in between 0.29 and 0.43. Despite this fiber design facilitated the inscription process, it is clear that it had higher confinement loss and high birefringence compared to the standard three ring fiber, thus reducing the quality of the grating as shown in Fig. 9. The grating was a fourth-order Bragg grating with a pitch of around 2 μm and Bragg wavelength at 1518.67 nm. The grating was 5 mm long and it took 2.5 s writing time for this length. This writing time is even much shorter than the one achieved with KrF laser for almost the same grating length.

Bragg gratings in PMMA mPOFs have also been fabricated at different wavelengths using different phase masks. Johnson et al. fabricated an FBG in a multimode (MM) PMMA mPOF with a core dimeter of 50 μm at 827 nm (Johnson et al. 2010a). The inscription system was a standard phase mask method and the

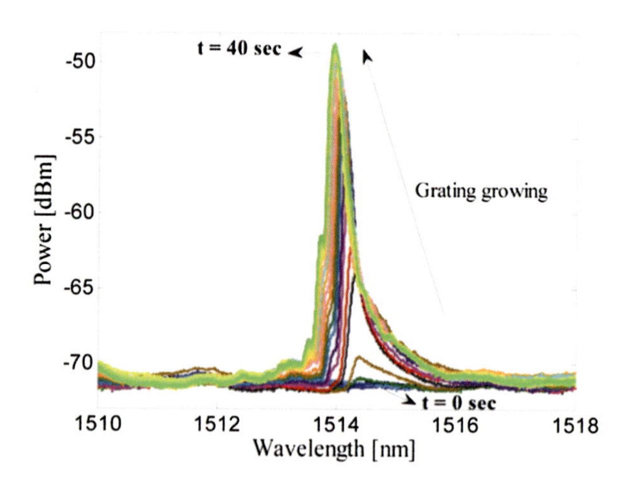

Fig. 8 Growth dynamic of the 4.5 mm grating in a few mode PMMA mPOF during writing using a KrF excimer laser (Oliveira et al. 2015)

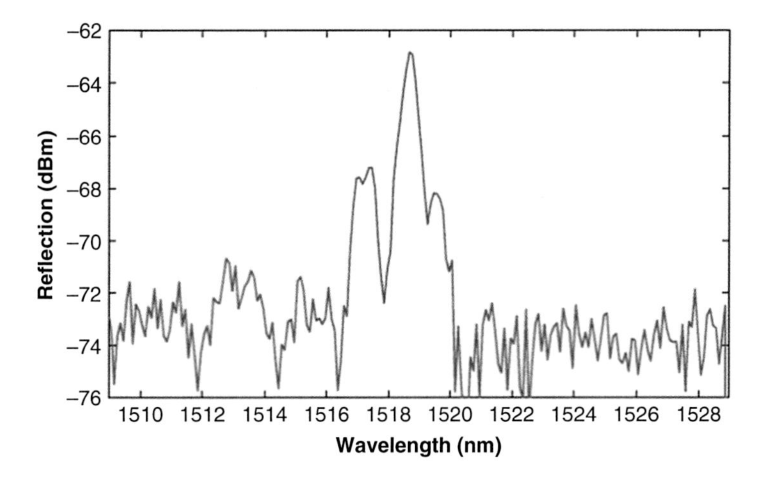

Fig. 9 Measured reflection spectrum of the fourth-order grating inscribed with the point by point method (Stefani et al. 2012a)

inscription laser was a 325 nm HeCd UV laser. The phase mask had a uniform period of 557.2 nm. The grating had 1.8 mm length and 7.5 dB strength in the reflection with FWHM of 2.45 nm. This grating was inscribed in 2 h. Stefani et al. inscribed FBGs in a few mode PMMA mPOFs at 850 nm (Stefani et al. 2011). The inscription system used was similar to the one used by Johnson et al. with the only difference that laser beam was expanded to 10 mm. The phase mask had a uniform period of 572.4 nm. The grating was 10 mm long and 10 dB strength in reflection with FWHM of 0.29 nm. The writing time was $\sim$185 min.

Most of the FBGs inscribed in mPOFs are single gratings which are the result of the ± 1 diffraction orders of the phase mask. Barabach et al. fabricated multiple FBGs in PMMA mPOFs using a phase mask with several diffraction orders (Statkiewicz-Barabach et al. 2013). They used a HeCd UV laser and a standard phase mask inscription technique. The fiber used was a standard three ring mPOF with hole to pitch ratio of 0.48, and it was fabricated from a commercial available PMMA rod. The phase mask used had a higher diffraction order ± 2 and ± 3 with large efficiency in addition to the ± 1 diffraction order. The phase mask had a period of 1.052 μm, and it gave a fundamental Bragg peak at $\lambda_B = 1555$ nm and higher order peaks at $\lambda_B/2$ and 2/3 λ_B. The detailed numerical simulation performed by Barabach et al. showed that the formation of the Bragg peaks at $\lambda_B/2$ and 2/3 λ_B could be attributed either to the first order reflection from the gratings with a periodicity of $\Lambda/4$ and $\Lambda/3$ or to the second and third order reflection from the grating of period Λ. The inscription of these multiple gratings took around 40 min. Figure 10 shows the reflection spectrum of the multiple gratings inscribed in PMMA mPOF using a single-phase mask with a single exposure.

All the PMMA mPOFBGs discussed above are inscribed in an undoped fiber. Sáez-Rodríguez et al. doped a core of PMMA mPOF with BDK to increase the

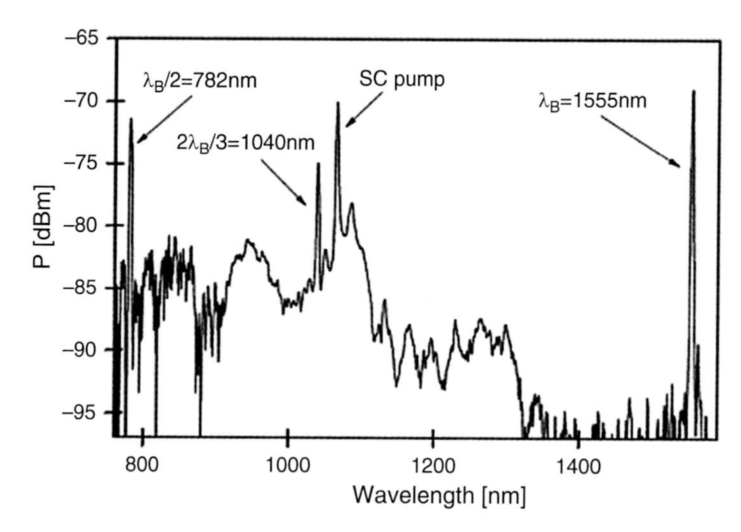

Fig. 10 Reflection spectrum recorded during the fabrication process (at the 28th min of exposure) of multiple Bragg gratings in PMMA mPOF (Statkiewicz-Barabach et al. 2013)

photosensitivity of the fiber and hence obtain a shorter writing time and a higher index change compared to the undoped one (Sáez-Rodríguez et al. 2013). The fiber fabricated had a hole to pitch ratio of 0.47. To inscribe the grating in the BDK doped fiber, they used a HeCd UV laser and a standard phase mask inscription technique. The laser beam had a diameter of 1.2 mm and the final grating length was 3.8 mm. However, to increase the length of the grating, the inscription was carried out using a mirror mounted on a motorized translation stage to scan a beam of 1.2 mm diameter focused with a cylindrical lens along the longitudinal axis of the fiber through a phase mask of pitch 557.5 nm, rather than expanding the beam. The grating was 23 dB strong in transmission with a 10 dB bandwidth of 0.3 nm. The corresponding inscription time was only 13 min. The microscope image of the end facet of the doped fiber used for the grating inscription and the grating growth dynamics is shown in Fig. 11a and b, respectively. Hu et al. also inscribed FBGs in BDK doped PMMA mPOF (Hu et al. 2017) using 400 nm femtosecond laser pulses with a beam diameter of 6 mm and a 1060 nm period uniform phase mask for the inscription. The fiber was fabricated purely from PMMA except that the core was BDK doped. The filling factor of the microstructure region was 0.4. Hu et al. achieved a grating with 40% reflectivity in 20 s. The microscope image of the end facet of the doped fiber used for the grating inscription and the grating growth dynamics is shown in Fig. 12a and b, respectively.

Topas mPOFBGs

In addition to PMMA mPOFs, FBGs have also been inscribed in mPOF made of different polymer materials such as Topas, PC, and Zeonex. The motivation behind

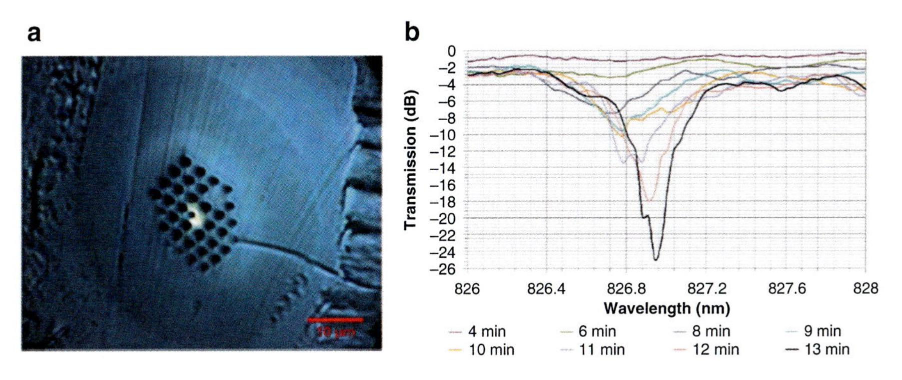

Fig. 11 (**a**) Microscope image of BDK doped PMMA mPOF. (**b**) The growth dynamics of the grating in transmission for the first 13 min (Sáez-Rodríguez et al. 2013)

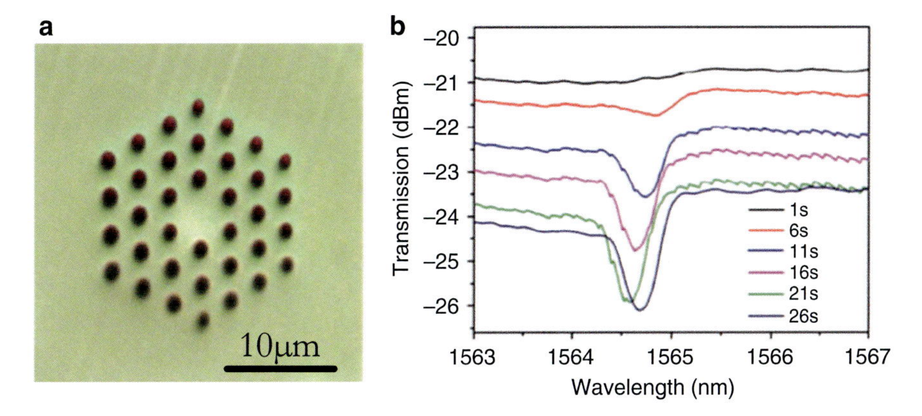

Fig. 12 (**a**) Microscope image of BDK doped PMMA mPOF. (**b**) Evolution of the transmitted amplitude spectrum during the photo-inscription process (Hu et al. 2017)

investigating new polymers for FBG sensors is due to the fact that different polymers have their own unique properties which are suitable for different FBG-based sensing applications. In this section, the fabrication of gratings in two different grades of Topas polymer is described. Topas belongs to the class of cyclic olefin copolymers (COCs), which is a class of optical thermoplastics that have a very low moisture uptake, chemical inertness to acids and bases, and many polar solvents (Topas COC 2014).

The first Topas mPOF was fabricated from Topas grade 8007F-04 (Johnson et al. 2011). This grade of Topas has a glass transition temperature of 78 °C and water absorption (saturation) at 23 °C less than 0.01% (Data Sheet Topas 8007F-04 2015). The fabricated fiber had 2 rings of holes with hole to pitch ratio of 0.44. A grating was also inscribed with a standard phase mask method and 30 mW 325 nm HeCd UV laser. The phase mask had a uniform pitch of 1034.2 nm. It required 45 min for the grating to reach a saturation level. The 1.8 mm grating had a Bragg wavelength of 1567.9 nm with a FWHM of 0.75 nm and 18 dB strength in the reflection.

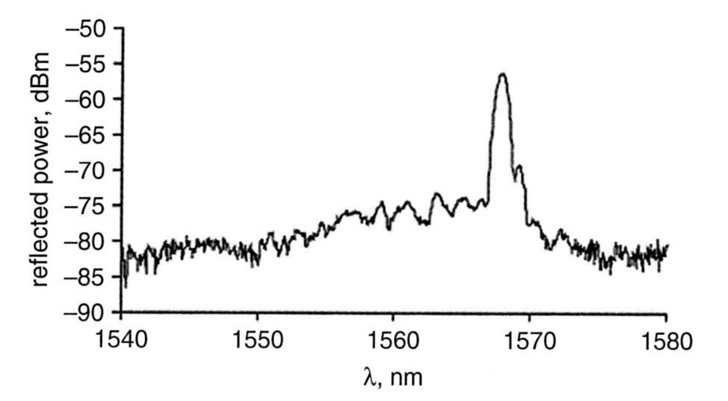

Fig. 13 Reflection spectrum of an FBG inscribed in Topas grade 8007F-04 mPOF (Johnson et al. 2011)

Figure 13 shows the reflection spectrum and the growth dynamics of the first grating inscribed in Topas-based mPOF.

Yuan et al. demonstrated also the inscription of an FBG in the same grade of Topas but in 850 nm spectral region, using a phase mask of a uniform period of 572.4 nm (Yuan et al. 2011). The main advantage of inscribing FBGs in the 850 nm region is that the loss of the fiber is very low compared to 1550 nm region which enables the use of longer fibers. The grating was inscribed with the same laser and technique as used by Johnson et al. (2011). The only difference is that here the UV laser beam was expanded from 1.2 mm to 12 mm. The resulted grating had a Bragg wavelength of 870 nm with FWHM less than 0.3 nm and reflection strength of 12 dB. The writing time was around 300 min.

Markos et al. reported the fabrication of FBGs in a different grade of Topas, commercially known as grade 5013S-04 (Markos et al. 2013). This grade of Topas had a glass transition of temperature (T_g) of 134 °C, 56 °C higher than Topas grade 8007F-04 and 25 °C higher than that of PMMA (Data Sheet Topas 5013S-04 2015). The water absorption capability is the same as Topas grade 8007F-04. But the main advantage of this grade of Topas is that it can be used to sense strain at high temperature as the T_g is relatively very high. The fabricated fiber had three rings of holes with hole to pitch ratio of 0.4, which is an endlessly single mode. The grating was fabricated with the same method and phase mask as used by Yuan et al. (Yuan et al. 2011). The limitation of this grade of Topas polymer is that it is difficult to draw fibers as it has a high flowability (Woyessa et al. 2017a).

Polycarbonate mPOFBGs

The first solid core polycarbonate (PC) mPOF was recently reported by Fasano et al. (2016). The motivation behind the fabrication of optical fibers from PC polymer lies in its T_g and the very good mechanical properties of the material. The T_g of PC is 145 °C and it is the highest among the currently existing mPOFs. Thus, FBGs fabricated in PC mPOFs can be used to sense temperature levels that PMMA and Topas mPOFBGs cannot reach. The other important advantage of PC is that it

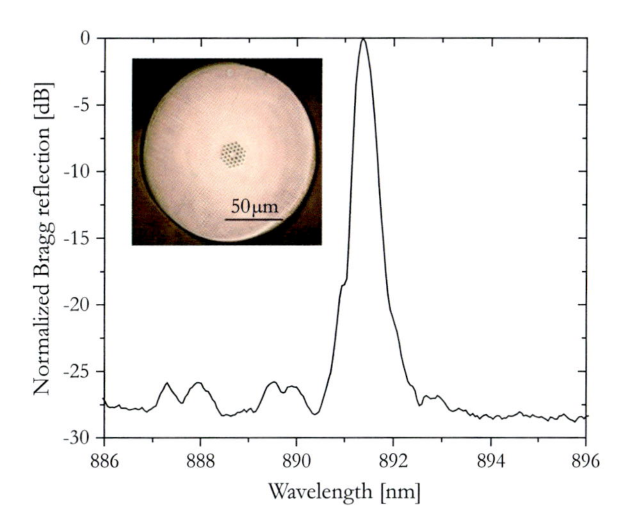

Fig. 14 PC mPOFBG reflection spectrum. Inset: Optical microscope image of the end facet of PC mPOF (Fasano et al. 2016)

usually yields and breaks at elevated values of strain (Optical properties of Makrolon and Apec 2014), and it is highly flexible in bending. The PC mPOF fabricated and characterized by Fasano et al. had a hole to pitch ratio of 0.4 ensuring thus endlessly single mode operation. The grating was inscribed with the phase mask technique and a 50 mW HeCd UV laser. For the grating inscription, the laser power was attenuated down to 4 mW which was found to be the suitable power for writing an FBG in the PC mPOF. The phase mask had a 572.4 nm uniform period. The grating had a Bragg wavelength of 892.4 nm with a FWHM of 0.46 nm and reflection peak of 25 dB. The required writing time was only 6 min by using only 4 mW power. Figure 14 shows the reflection of FBGs inscribed in the PC mPOF.

Zeonex 480R mPOFBGs

The polymer Zeonex 480R belongs to the class of cyclo-olefin polymers; in particular it is an amorphous homopolymer of norbornene. Zeonex 480R shares some of properties with Topas 5013S-04, such as low water absorption, high temperature resistance, and good chemical inertness to bases and acids (Khanarian and Celanese 2001). The T_g of Zeonex 480R is 138 °C (Technical data, Zeonex, Cyclo Olifen Polymer (COC) 2017). However, there are some fundamental differences regarding the chemical structure of Zeonex and Topas which make Zeonex better than Topas for the manufacturing of mPOFs. The main difference in their chemical structure is that Topas is not a homopolymer but an amorphous ethylene-norbornene copolymer with a high percentage of norbornene (Roy et al. 2012). Some of the unique advantages of Zeonex 480R are: its lower flowability compared to Topas 5013S-04, as well as its better transmittance, higher sensitivity to temperature, and better mechanical stability at high temperatures. Furthermore, Zeonex has very good compatibility with PMMA for co-drawing applications,

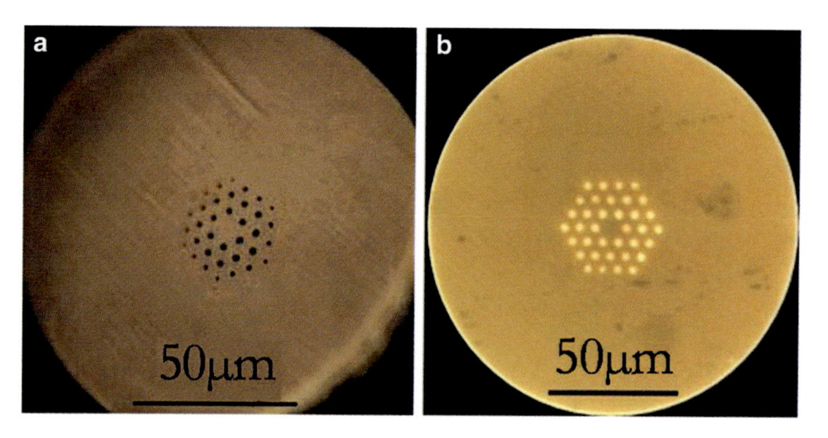

Fig. 15 Optical microscope image of the end facet of (**a**) Topas 5013S-04 (Markos et al. 2013). (**b**) Zeonex 480R mPOFs (Woyessa et al. 2017a)

low birefringence, and superior moldability (Technical data, Zeonex, Cyclo Olifen Polymer (COC) 2017; Leon-Saval et al. 2012). The first endlessly single mode Zeonex mPOF for Bragg grating sensing was fabricated and characterized by Woyessa et al. in 2017 (Woyessa et al. 2017a). It was fabricated from Zeonex grade 480R and the fiber had a hole to pitch ratio of 0.4. Woyessa et al. observed that during the drawing process of microstructured fibers, Zeonex 480R showed a superior drawability over Topas 5013S-04. This was due to the fact that the physical properties of Zeonex are well suitable for high quality fiber drawing. This fact is a direct consequence of Zeonex greater molecular weight. The weight-average molecular weight (M_w) of Zeonex 480R is six times larger than that of Topas 5013 (Roy et al. 2012; Torres et al. 2015), although their glass transition temperatures are similar, being 138 °C and 134 °C, respectively. For thermoplastic materials, the melt flow index defined as the flowability in general decreases with increasing M_w (Bremner et al. 1990). Therefore, Zeonex 480R preforms tend to flow slower than Topas 5013 under similar fiber drawing conditions, thereby ensuring highly controllable and stable fiber draw process. The other important advantage of having a lower melt flow rate is that it can allow getting a wide range of drawing temperature and stress. Therefore, the final mechanical properties of Zeonex 480R mPOFs can be tuned relatively more easily compared to that of Topas 5013S-04. As a result, Zeonex 480R allows a greater number of degrees of freedom in fiber design as the desired microstructures can be transferred to the final fiber more efficiently. For example, the cladding holes are uniform and symmetrical with minor distortions in their shape compared to Topas 5013 mPOFs (see Fig. 15). Due to the better stability of the drawing process, fluctuations in the fiber diameter are also reduced. Nevertheless, it should be noted that either too high or too low values of M_w can make the fiber drawing very challenging or even unfeasible. The M_w of Zeonex 480R is sufficiently low to avoid this potential problem.

A Bragg grating inscribed in the low loss spectral region of the Zeonex mPOF with the phase mask technique and a 50 mW HeCd UV laser has been also reported. For the grating inscription, the laser power was attenuated down to 5.5 mW which was suitable power for writing FBGs in Zeonex mPOFs. The phase mask had a uniform of 572.4 nm. The grating length was 2 mm and has a Bragg wavelength of 865.24 nm with a FWHM of 0.522 nm and reflection peak of 30 dB. The writing time was only 5 min by using a power of 5.5 mW.

The aforementioned polymers were shown to be photosensitive under 325 nm HeCd UV laser irradiation which enabled successful fabrication of gratings in the mPOFs. Materials alternative to PMMA, which is the standard material for polymeric Bragg gratings, have been developed to meet different needs, e.g., for insensitivity to humidity (Topas 8007 and 5013, Zeonex 480R), high thermal resistance (PC, but also Zeonex 480R and Topas 5013), and ease of fabrication (Zeonex 480R) (Johnson et al. 2011; Yuan et al. 2011; Markos et al. 2013; Woyessa et al. 2017a; Fasano et al. 2016).

Annealing of mPOFBGs

During the fabrication of an mPOF, the fiber is drawn under certain levels of drawing tension. This drawing stress aligns the molecular chain of the polymer along the fiber axis, thus leaving some residual stress. The alignment level depends on the amount of drawing tension applied during the fabrication process. When mPOFs are heated at a temperature close to their T_g, the polymer chains starts to relax from their original orientation formed during drawing. The amount of temperature that causes this process highly depends on the amount of stress applied when the fiber is fabricated, the thermal history of the preform from which the fiber is made, the amount of relative humidity (RH) in the vicinity of the fiber where the fiber is heating up, and the extent of UV exposure experienced by the fiber (Carroll et al. 2007; Ishigure et al. 2004; Shafee 1996). As the fiber is heated up, it shrinks and there will be a decrease in the length of the fiber and an increase in its diameter. This process is called annealing and leads to an irreversible change in the fiber dimensions. As mPOFBGs are annealed, the period of the Bragg grating decreases as the fiber shrinks and thus the grating experiences a permanent blue shift. The annealing of mPOFs either before or after inscription of a grating is a vital step in the development of stable mPOFBG sensor system. In this section, different methods of annealing of mPOFs and what advantages they can offer in the realm of mPOFBG sensors are discussed.

Temperature Assisted Annealing

In 2010 Johnson et al. showed for the first time the effect of thermal annealing on PMMA mPOFBGs (Johnson et al. 2010b). They first inscribed an FBG in MM PMMA mPOF using a phase mask with a period of 1057.2 nm producing a

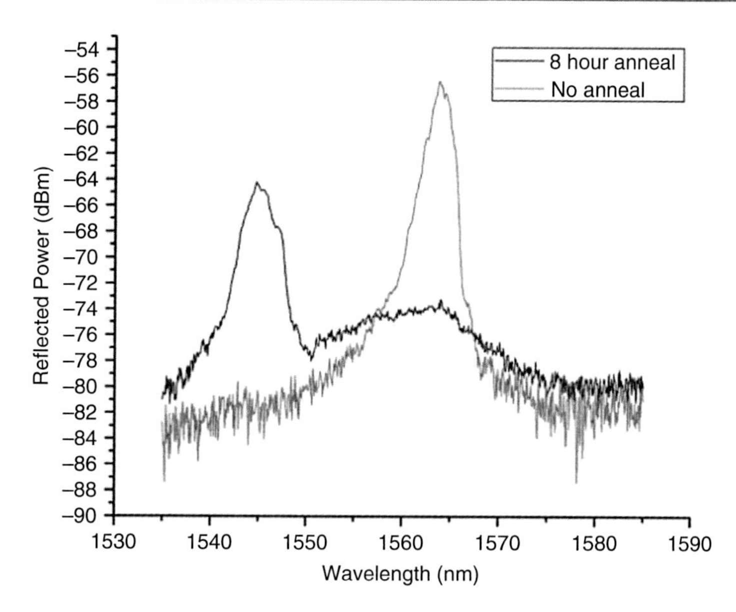

Fig. 16 Reflection spectrum of a MM PMMA mPOF before and after thermal annealing in a conventional oven at 80 °C for 8 h (Johnson et al. 2010b)

Bragg wavelength of 1562 nm. The grating was then annealed in a conventional oven at a constant temperature of 80 °C for 8 h. The thermal annealing induced a permanent blue shift of the reflected wavelength of ~18 nm. Figure 16 shows the reflection spectrum of the grating before and after the annealing. The permanent wavelength shift was probably induced by a permanent shrinkage of the MM mPOF, which resulted in a decrease of the grating pitch. Thus, thermal annealing offers the possibility of tuning of FBGs within a certain range. This feature of thermal annealing of mPOFBG was further used to develop the first wavelength division multiplexed (WDM) FBG sensor by using a single-phase mask.

Carroll et al. demonstrated how thermal annealing of PMMA mPOFBGs or annealing the fiber before the grating inscription affects the thermal response of the grating (Carroll et al. 2007). They showed that PMMA mPOFBGs that were not annealed had a linear temperature operational response only up to 50–55 °C. They demonstrated the maximum operating temperature range of the grating, by heating up the grating in cycles up to certain temperature levels. The grating was heated by heating element in three different cycles, up to 77 °C, 86 °C, and 92 °C for the first, second, and third cycle, respectively. In the first cycle, the reflection spectrum linearly shifted to the blue from 23 °C to 55 °C. However, after this temperature range, the grating was shift to the blue very rapidly up to 77 °C and the shift was nonlinear. They found out that the thermal-induced blue-shift was permanent when they measured the spectrum again at room temperature and the shift was −8.4 nm. In the second cycle, the linear region was up to 76 °C, which was very close to the maximum temperature applied to the grating during the first cycle. After this

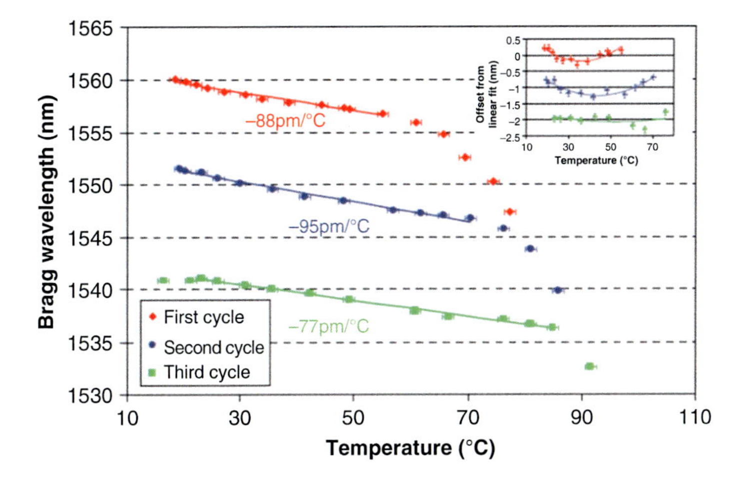

Fig. 17 Bragg wavelength shift with temperature for three consecutive heating cycles. Inset: Deviation from linear response for each cycle (data from successive cycles are offset by −1 nm for clarity) (Carroll et al. 2007)

temperature, the grating shifted to the blue again very rapidly up to 86 °C. When the temperature was decreased back to 23 °C from 86 °C, the grating shifted again permanently by −10.1 nm. In the third cycle, the grating operated linearly up to 86 °C. After this temperature, the observed rate of Bragg wavelength shift was fast up to 92 °C, as seen for the previous cycles. This process can be seen in Fig. 17. This investigation showed that thermal annealing is vital and essential step in order to achieve a linear and wide range of thermal response of a grating.

Temperature- and Humidity-Assisted Annealing

Annealing of mPOFBGs performed by both Johnson et al. and Carroll et al. was using conventional oven or simple heating elements. In this type of heating devices, humidity is not controlled. However, early investigations showed that humidity has a huge influence in the T_g of PMMA such that PMMA materials exposed to wet environment have a lower T_g than the one exposed to dry environment (Smith and Schmitz 1988). Woyessa et al. implemented a detailed experimental investigation on the combined effect of humidity and temperature on mPOFBGs (Woyessa et al. 2016). They reported that the RH has a great impact on the annealing of mPOFBGs. In particular they revealed that annealing mPOFBGs at high humidity and temperature resulted in an improved performance of mPOFBGs in terms of stability and sensitivity to humidity. In their investigation, an FBG was fabricated in an endlessly single mode PMMA mPOFs by using a phase mask with period 572.4 nm producing a Bragg wavelength at 850 nm. The annealing was done in a climate chamber which provides the ability to control

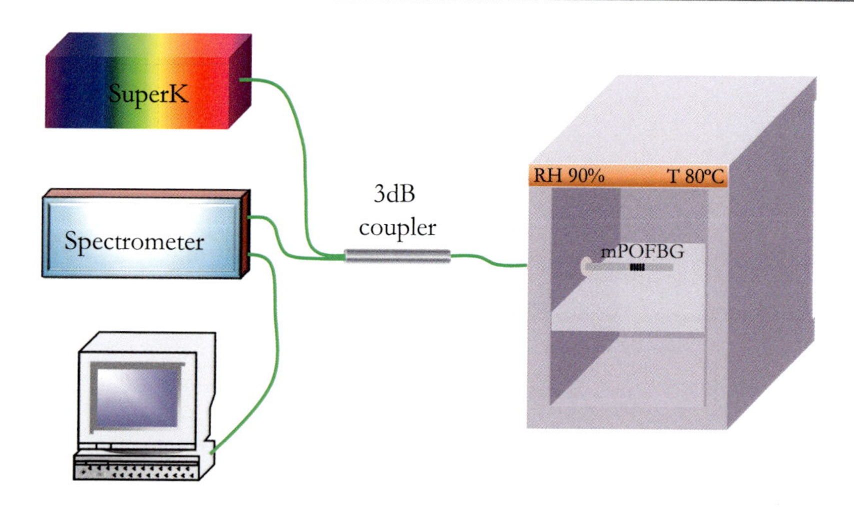

Fig. 18 Experimental setup used for temperature and humidity controlled annealing of PMMA mPOFBGs (Woyessa et al. 2016)

both temperature and humidity simultaneously. The experimental setup used for the annealing investigation is shown in the following Fig. 18.

In their experiments, firstly the climate chamber was set to a temperature and RH of 25 °C and 50%, respectively, which defined as the ambient conditions. Then after 2 h of ambient conditions, the RH of the chamber was decreased to 10% without changing the temperature. Three hours after 10% RH and 25 °C environmental conditions in the climate chamber, the temperature was ramped up to 80 °C at a fixed 10% RH. As the temperature was increasing to 80 °C, the mPOFBG was blue shifting rapidly. After 20 h the grating started to get stabilized. When the rate of the blue shift becomes 0.3 nm per hour (defined as the mPOFBG equilibrium condition), the RH of the chamber was increased by 20%, thus to 30%, yet at a fixed temperature of 80 °C. The grating was blue shifted by ~66.33 nm before the RH was increased by 20%. As the RH increased to 30%, the grating again started to blue shift rapidly. After 23 h the rate of the shift reached to 0.3 nm per hour and the chamber RH was increased to 50%, then to 70% and finally to 90%, for each case when the equilibrium condition was reached. This process is depicted in Fig. 19, whereas the amount of blue shift and the time taken for each RH level are summarized in Table 2.

As it can be seen from Fig. 19 and Table 2, as the temperature was ramped up from 25 °C to 80 °C there was a fast blue-shift of the reflected wavelength as initially the fiber was releasing the frozen-in stress induced during the fabrication process very fast. However, 20 h later, the rate of shift became low and again increased as the RH was increased. At higher levels of RH such as 70% and 90%, not only the rate of the blue-shift was faster but also the amount of the shift was larger. This phenomenon is due the fact that PMMA has a high water absorption

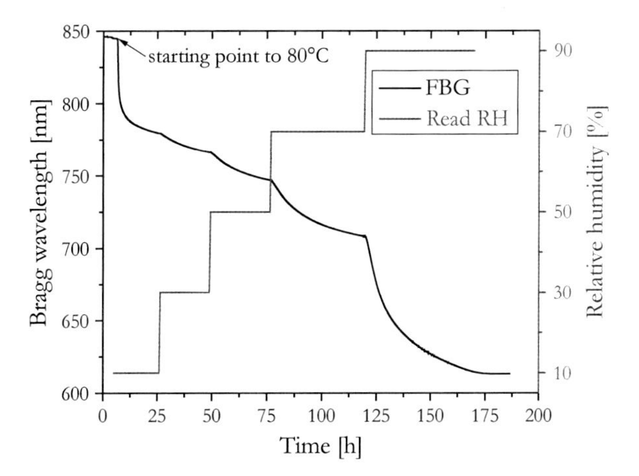

Fig. 19 Resonance wavelength of PMMA mPOFBGs during annealing at 80 °C and up to a RH of 90% starting from 10% and by 20% increment

Table 2 Summary of the amount of blue shift of the Bragg wavelength and the time taken during annealing of PMMA mPOFBGs at 80 °C and up to a RH of 90% starting from 10% and by 20% increment

RH (%)	Amount of blue shift (nm)	Time taken (hh:mm)
10	66.33	20:00
30	13.20	23:00
50	19.29	27:30
70	38.71	43:00
90	93.83	50:36

capability and water acts as a plasticizer for PMMA and hence decreases its T_g (Smith and Schmitz 1988). Thus, this investigation revealed for the first time that it is possible to tune the location of the Bragg wavelength in POFBGs by a relatively big amount by assisting the thermal annealing with humidity. The specific response of these PMMA-based mPOFBGs to humidity will be described in detail in section "Humidity Sensors." One of the potential advantages of humidity assisted thermal annealing is that it allows production of gratings at short wavelengths (large blue-shifts) in the visible range where POFs have lower loss but where the phase mask technique poses limitations in terms of efficiency and cost.

Chemical-Assisted Annealing at Room Temperature

The mPOF annealing methods discussed so far are activated thermally. It is the temperature that allows for the polymer chains to relax from their frozen in stresses. Furthermore, it has been shown that humidity facilitates the process as it lower the T_g of PMMA mPOFs. As the goal of annealing mPOFs is relaxing the frozen polymer chains, the question that may rise is, are there any other possible techniques that help the polymer chains to relax without increasing the temperature from the room condition? Fasano et al. interestingly demonstrated for the first time the possibility of annealing of mPOFBGs in methanol-water solutions at room temperature. This was based on early studies by Williams et al., which showed the effect of the presence of methanol on the T_g of PMMA (Williams et al. 1986). This investigation revealed that for methanol-equilibrated PMMA systems the T_g is ranging from 20 °C to 30 °C depending on the weight-average molecular weight (M_w). Fasano et al. investigated this effect systematically using gratings inscribed in PMMA mPOF (Fasano et al. 2017) from the same fiber draw and CHROMASOLV methanol (for HPLC, $\geq$ 99.9% by weight, Sigma-Aldrich)/Milli-Q water concentrations, 50:50%, 60:40%, and 70:30% v/v, where water was used as a diluent for methanol. For each solution, two different grating were prepared. The gratings were immersed in the solution and their Bragg wavelength was monitored throughout the experiments. In the methanol solution-assisted annealing experiments, two main phases were observed: (I) absorption/swelling when the grating was in the solution and (II) evaporation/shrinkage while the grating was removed from the solution. Figure 20 shows an example of the grating response over time for methanol/water concentrations of 50:50%, 60:40%, and 70:30% v/v. Once the mPOFBGs were immersed in the respective solutions, they immediately started absorbing the solution and getting swollen. This led to some initial red shift for all three cases. However, as soon as the solution started to swell the fibers (which corresponded to their T_g getting lower), the relaxation started to occur. The initial red shift can be seen as the result of a temporary positive balance between the red-shift caused by the solution-mediated swelling due to the solution absorption and the blue-shift induced by the chain alignment relaxation. After this initial lag-phase, the total shift referred to the initial Bragg wavelength became constantly negative and the recorded amount of blue-shift increased very rapidly. The higher the concentration of the methanol, the higher and the faster the relaxation of the fiber was. After the fast Bragg wavelength blue shift, an inflection point in all Bragg wavelength curves was observed and the rate of blue shift gradually became smaller and smaller. When the Bragg wavelength stabilized, the fibers were removed from the solutions. Upon removal, the solution that the fiber absorbed started to evaporate out rapidly. This caused the fiber to shrink further and hence the pitch of the grating was further reduced, leading to an abrupt blue shift. The total amount of blue shift that occurred in three phases for the PMMA mPOFBGs (duplicates) in the experiments at three different solution concentrations is summarized in Table 3. As can be seen from Table 3, the total amount of blue-shift at the end of the experiments increased with the methanol concentration. This proves that the real T_g of PMMA was decreased during the experiments, to an extent that depends on the methanol concentration, in particular increasing with the methanol concentration. The main advantage of this annealing method is that it is cheap compared to the others as it does not require any expensive climate chamber.

Annealing of mPOFBGs has several advantages and is a crucial phase of POFBGs sensor development. By annealing it is possible to fabricate a number of FBGs in a single fiber with a single-phase mask for WDM FBG sensors. In particular, it is possible to obtain an FBG at almost any wavelength with a

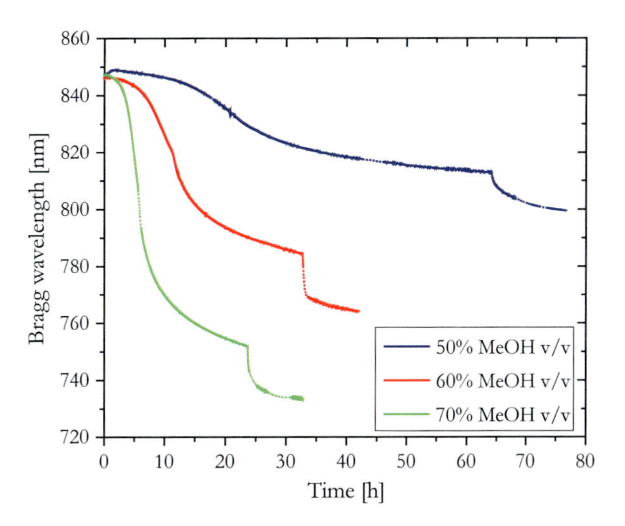

Fig. 20 Bragg wavelengths versus volumetric concentration of methanol (MeOH v/v). Note that the dotted lines indicate the experimental data missing due to high reflection noise and recovered by fitting. The sharp downward jump in the three curves corresponds to the time when the gratings were removed from the respective solutions

Table 3 The total amount of blue shift recorded for the three PMMA mPOFBGs in the three different solution

Methanol:water v/v(%)	Total Bragg wavelength shift (nm)
50:50	−50.0±3.0
60:40	−80.3±2.4
70:30	−111.6±3.2

single-phase mask. Also, annealing results in a linear response of gratings to temperature and humidity and increases the linear operating temperature range of FBGs (Woyessa et al. 2016; Carroll et al. 2007; Johnson et al. 2010b).

Microstructured Polymer Optical Fiber Grating Sensors

Microstructured polymer optical FBGs have been used for the development of various types of sensors such as strain, temperature, humidity, pressure, refractive index, accelerometer. In this section, an overview of the most important sensing applications that mPOFBGs have been employed for is provided.

Strain Sensors

FBGs fabricated in different types of mPOFs have been used for axial strain measurement. The most common way to determine the sensitivity of mPOFBGs to strain in a laboratory environment is by using two 3-D translation stages. First, the fiber is glued on v-grooves mounted on the 3D translation stages with a few centimeters separation from both sides of the grating. One of the translation stages is used to apply axial strain to the fiber (i.e., to the grating) and the other translation stage is used to butt couple light into the mPOF. The strain applied on the grating

Table 4 Strain sensitivity of gratings inscribed in different types of mPOFs and at different wavelengths

Base polymers	Fiber structure/ diameter/core/ hole-to-pitch ratio	Is the grating annealed or not?	Bragg wave-length (nm)	Stain range (%)	Sensitivity (pm/μm)	Temperature at which strain mea-surement is done
PMMA	3 ring hexagonal, 150 μm, 50 μm, MM (Johnson et al. 2010a)	No	827	0–1	0.71	Ambient
	3 ring hexagonal, 130 μm, 9 μm, 0.5 (Stefani et al. 2011)	No	847.6	0–2	0.71	Ambient
			1550		1.3	Ambient
	6 ring hexagonal, 250 μm, 18 μm, 0.52 (Oliveira et al. 2015)	No	1514	0–2	1.65	Ambient
Topas 8007F-04	2 ring hexagonal, 240 μm, 10.2 μm, 0.3 (Yuan et al. 2011)	No	870	0–2.17	0.64	Ambient
Topas 5013S-04	3 ring hexagonal, 130 μm, 9.2 μm, 0.4 (Markos et al. 2013)	Yes, at 80 °C for 3 h	853.4	0–0.3	0.76	50 °C
				0–0.3	0.8	100 °C
				0–0.16	1	110 °C
	3 ring hexagonal, 180 μm, 13 μm, 0.375 (Markos et al 2013)	No	869.6	0–3	0.75	Ambient
				0–2.2		50 °C
				0–2.5		98 °C
PC	3 ring hexagonal, 150 μm, 7 μm, 0.4 (Fasano et al. 2016)	Yes, at 120 °C for 24 h and at 130 °C for 12 h	875.7	0–3	0.70	Ambient
Zeonex 480R	3 ring hexagonal, 150 μm, 8.8 μm, 0.4 (Woyessa et al. 2017a)	Yes, at 120 °C for 36 h	831.57	0–3	0.77	Ambient

is then determined as the ratio of the elongated length of the fiber at the each strain level to the original length of the fiber defined as the length between the two glue points. By means of this method, strain measurements have been done and sensitivities to strain have been determined to PMMA, Topas 8007S-04, Topas 5013S-04, PC and Zeonex 480R mPOFBGs. Table 4 summarizes the main fiber materials used, whether the fiber is annealed or not before strain measurement, the Bragg grating wavelength region, the strain range, sensitivity to strain, and temperature in which strain measurement was performed.

As it can be seen from Table 4, the strain sensitivity is almost the same for all polymer types as long as the Bragg wavelengths are in the same wavelength region. The PC and the Zeonex mPOFs mentioned in Table 4 were annealed 20 °C below their T_g, and the annealing process was stopped when the blue shift saturated and shows a wider range of strain sensing up to 3%. Moreover, it can be seen that the sensitivity to strain is also dependent on the Bragg wavelength. Grating inscribed at 1550 nm region have about two times better sensitivity than those at 850 nm; nonetheless the propagation loss of POFs is higher in 1550 nm region.

Temperature Sensors

As mPOFBGs have been used to measure axial strain they have been also used to gauge temperature. The temperature operating range of mPOFBGs varies depending on the T_g of the base polymer material used to fabricate the fibers. The two most common setups used to characterize mPOFBGs are heating elements and climate chambers. In the former case, an mPOFBG is placed near the heating element and the temperature close to the grating is monitored by a thermocouple. Despite this characterization setup is cost effective; its major limitation is that the humidity in the vicinity of grating is uncontrolled. This causes a significant error particularly when the characterization involves FBGs fabricated from polymers that have affinity to water such as PMMA. This is due to the fact that humidity decreases as temperature increases and leading an additional blue shift of the Bragg wavelength. Characterizing the mPOFBGs in terms of temperature in a climate chamber is probably the best way to perform accurate thermal characterization of the sensor as the humidity is more accurately controlled and the resulting error is minimized. Table 5 summarizes the fiber materials used, whether the fiber is annealed or not before the temperature characterization, the Bragg wavelength region, the temperature range, sensitivity to temperature, setup used, and RH at which the temperature measurement was performed.

As can be seen from Table 5, unlike the strain sensitivity the temperature sensitivities of the gratings showed significantly variation, even for the same wavelength region. This inconsistent response can be attributed to several reasons such as different characterization setups employed, whether the grating is annealed or not and so on. Among the currently existing POFBGs, PC-based mPOFBGs are able to measure the highest temperature, i.e., up to 125 °C.

Humidity Sensors

Some of the polymers that they have been used to fabricate POFs have moisture absorbing capabilities. PMMA and PC are, for instance, humidity-sensitive polymers. When moisture is absorbed in mPOFs based on these polymers, two important phenomena occur in the fiber: a change in its size (length and diameter) and refractive index. Because the Bragg wavelength depends on both the

Table 5 Temperature sensitivity of gratings inscribed in different types of mPOFs and at different wavelengths

Base polymers	Fiber structure/diameter/core/hole-to-pitch ratio	Is the grating or fiber annealed or not?	Bragg wavelength (nm)	Temperature range (°C)	Sensitivity (pm/°C)	Set up	Humidity at which temperature measurement is done
PMMA	4 ring hexagonal, 150 μm, 14.6 μm, 0.3 (Carroll et al. 2007)	Yes, at 80 °C for 7 h	1560	20–89	52	Heating element	Ambient
Topas 8007F-04	2 ring hexagonal, 240 μm, 10.2 μm, 0.33 (Yuan et al. 2011)	No	870	23.5–32.6	−78	Heating element	Ambient
	2 ring hexagonal, 270 μm, 13.2 μm, 0.45 (Yuan et al. 2011)	No	1567.9	20–35	−36.5	Climate chamber	55% RH
Topas 5013S-04	3 ring hexagonal, 130 μm, 9.2 μm, 0.4 (Markos et al. 2013)	Yes, at 80 °C for 3 h	853.4	24–107	−14.46	Heating element	Ambient
Polycarbonate	3 ring hexagonal, 150 μm, 7 μm, 0.4 (Fasano et al. 2016)	Yes, at 120 °C for 24 h and at 130 °C for 12 h	875.7	23–125	−29.99 ± 0.17	Heating element	Ambient
	3 ring hexagonal, 125 μm, 10 μm, 0.4 (Woyessa et al. 2017b)	Yes 125 °C for 36 hr	880.19	20–100	−25.67 ± 0.6	Climate chamber	50%, 90% RH
Zeonex 480R	3 ring hexagonal, 150 μm, 8.8 μm, 0.4 (Woyessa et al. 2017a)	Yes, at 120 °C for 36 h	831.57	20–100	−24.01 ± 0.1	Climate chamber	50% RH

period of the grating and refractive index, when moisture is absorbed, the Bragg wavelength is consequently directly affected. By exploiting this crucial property of such polymers, mPOFBG humidity sensors have been developed. It is very important here to emphasize that PMMA and PC mPOFBGs are also sensitive to temperature. Therefore, when such mPOFBGs are used as humidity sensors, cross sensitivity to temperature affects the humidity response of the grating and vice versa. Woyessa et al. showed that in order to achieve temperature independent humidity responses, mPOFBGs have to be annealed not only at temperature close to the T_g of the polymer but also at high RH (Woyessa et al. 2016). For instance, at 80 °C and 90% RH for PMMA-based fibers to operate up to 75 °C and 90% RH with no hysteresis. Annealing the fiber at these conditions not only helps to develop temperature insensitive humidity sensor but also makes the sensor more stable, hysteresis free, and highly sensitive to humidity. PMMA mPOFBGs that have not been annealed or annealed at a very low RH, for instance, in a conventional oven, showed low sensitivity, highly temperature-dependent sensitivity, and highly nonlinear response to humidity. Figure 21a and b depicts how the response to humidity of PMMA mPOFBGs is highly affected by the RH level.

As it can be seen from Fig. 21a, only the mPOFBG that was annealed up to 90% RH at 80 °C exhibited the most stable response to humidity in the range 10–90%RH up to 75 °C. In contrast, the PMMA mPOFBGs annealed at a lower humidity showed a smaller sensitivity with hysteresis at 25 °C and 50 °C. Moreover, at 75 °C they showed a fast and high nonlinear decrease in the Bragg wavelength when the RH was raised just above the annealing RH level, as can be seen in Fig. 21b. The sensitivity to humidity for mPOFBG annealed up to 90% RH was around 35 pm/% RH for the whole range of operating temperature at a Bragg wavelength of 600 nm. It is possible to get a higher sensitivity if the Bragg wavelength falls within a longer wavelength yet still low loss region, such as the 850 nm region. To achieve this, rather than annealing the fiber after inscription and shifting the grating to shorter wavelength, the fiber can be annealed before inscription. Figure 22a and b shows the humidity response of a PMMA mPOFBG inscribed in a fiber that was annealed at 80 °C and 90% RH for 36 h before inscription. The humidity sensitivity of this grating was 45 pm/ %RH which is 10 pm/ % RH higher than the response measured at 600 nm Bragg grating wavelength.

However, considering that there are applications that require humidity measurements at temperatures beyond the one PMMA mPOFBGs can operate at, PC mPOFBGs humidity sensors have been developed instead. Polycarbonate has also moisture absorbing capability although it is not as strong as PMMA. The water absorption (saturation value) at 23 °C of PC is 0.3% (Optical properties of Makrolon and Apec 2014), whereas that of PMMA is 2.1% (GEHR PMMA (Acrylic), http://www.gehrplastics.com/pmma-acrylic.html). PC mPOFBGs were demonstrated to be able to measure RH in the range from 10% to 90% up to 100 °C. It is possible to operate PC mPOFBG beyond this temperature level as the T_g of PC is 145 °C, but the climate chamber used to measure humidity operates only up to 100 °C. The measured humidity sensitivity of PC mPOFBGs was 7.25 ± 0.08 pm/% RH

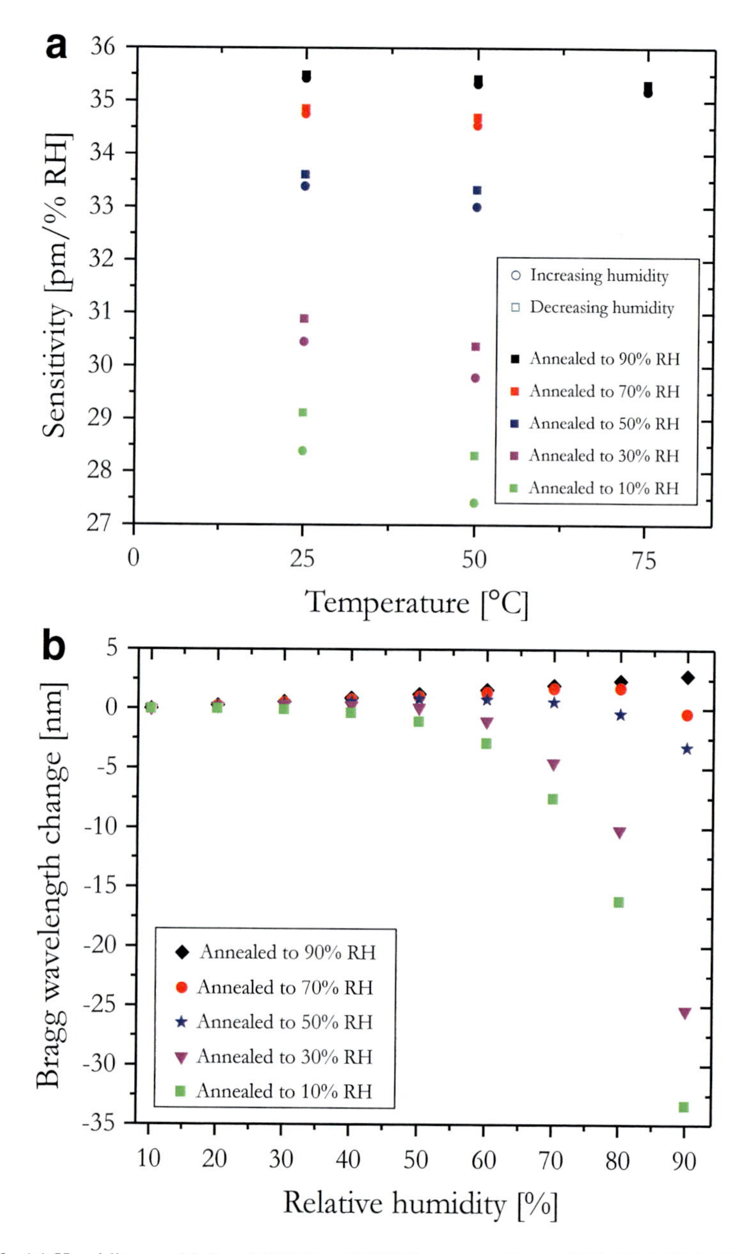

Fig. 21 (**a**) Humidity sensitivity of PMMA mPOFBGs annealed up to 90%, 70%,50%, 30%, and 10% RH at 25 °C and 50 °C and PMMA mPOFBG annealed to 90% at 75 °C. (**b**) Humidity responsivity of PMMA POFBGs annealed up to 90%, 70%, 50%, 30%, and 10% RH at 75 °C (Woyessa et al. 2016)

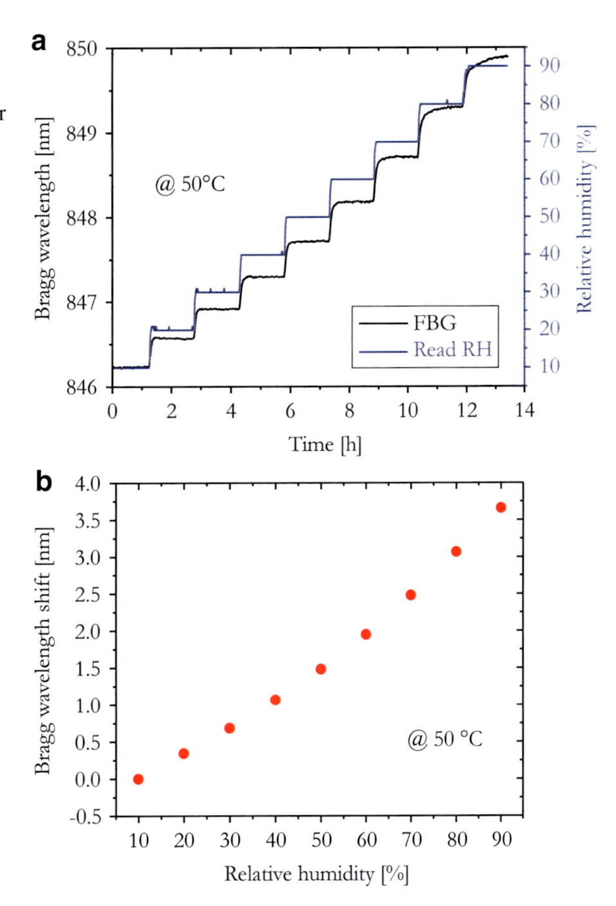

Fig. 22 (**a**) Humidity response of PMMA mPOFBG at 50 °C in the 850 nm region where the fiber annealed at 80 °C and 90% RH for 36 h before the grating inscription. (**b**) The corresponding stabilized response of the grating

in the range 10–90% RH at 100 °C. Polycarbonate mPOFBGs can be operated well beyond the PMMA mPOFBG operating temperature limit, but at the cost of having lower sensitivity than PMMA one. The RH sensitivity of PC mPOFBGs is indeed 7 times smaller than PMMA mPOFBGs. Despite the fact that the low moisture absorption property of PC can be considered as a limitation in humidity sensitivity, this turns out to be an advantage when a temperature measurement is performed. For instance, for a climate chamber that has RH precision of an order of 1% when PMMA mPOFBGs are characterized with regard to their temperature response, a single % of RH fluctuation in the chamber makes the temperature measurement very difficult as PMMA mPOFBGs are highly sensitive to humidity. Figure 23 shows the temperature response of PMMA mPOFBGs at 50% RH (set value). As it can be seen from the figure, it is hardly possible to get accurate temperature measurement due to 1% fluctuation in RH. In addition, PMMA mPOFBGs have a lower temperature sensitivity compared to PC mPOFBGs (Woyessa et al. 2017b).

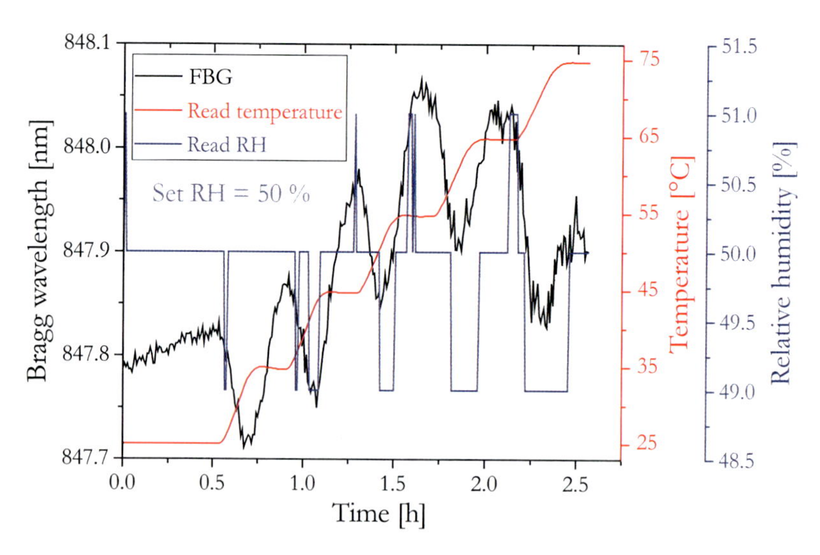

Fig. 23 Temperature responses of PMMA mPOFBGs at 50% RH. The fibers were annealed at 80 °C and 90% RH for 36 h before the grating inscription

This problem was not seen when temperature characterization was done for PC mPOFBG. This is due to the fact that PC has smaller moisture absorption capability, so that 1% RH fluctuation in the chamber does not affect the temperature response of the PC grating significantly. The temperature response of the PC mPOFG can be seen from Fig. 24a and b. Therefore, a more precise climate chamber is required to accurately calibrate the temperature response of PMMA mPOFBGs.

Pressure Sensors

Hydrostatic pressure is one of the most commonly monitored measurands in many applications. Microstructured polymer optical FBGs have been used to develop hydrostatic pressure sensors. Johnson et al. demonstrated the first hydrostatic pressure sensors using MM PMMA mPOFBGs (Johnson et al. 2012). The principle behind measuring hydrostatic pressure of the environment using mPOFBGs relies upon two factors. When the hydrostatic pressure of the environment in which the grating is exposed increases, the mPOF shrinks. Thus, the pitch of the grating decreases and this leads to a negative Bragg wavelength shift. The other effect relies upon the strain optic effect of the fiber which increases as the fiber is compressed resulting in a positive Bragg wavelength shift. The setup used by Johnson et al. to determine the hydrostatic pressure response of mPOFBG is shown in Fig. 25. The fiber used in their experiments was purely made from PMMA and had a core and an outer diameter of 50 μm and 150 μm, respectively. They measured pressure up 10 MPa (100 bar). The grating had a Bragg wavelength of 1561.5 nm

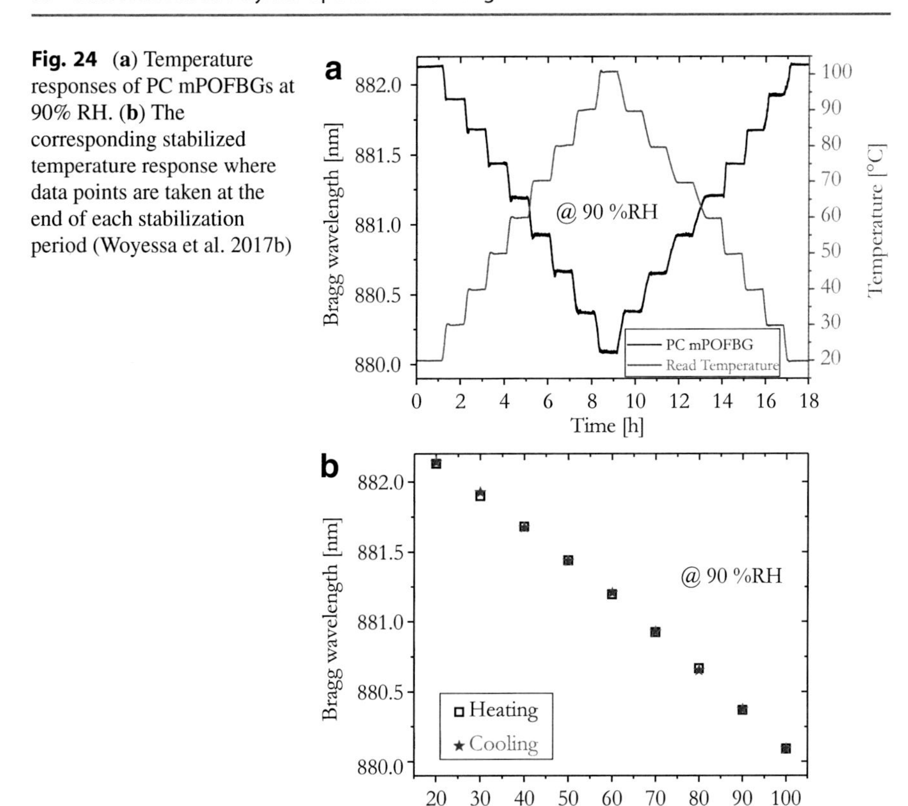

Fig. 24 (**a**) Temperature responses of PC mPOFBGs at 90% RH. (**b**) The corresponding stabilized temperature response where data points are taken at the end of each stabilization period (Woyessa et al. 2017b)

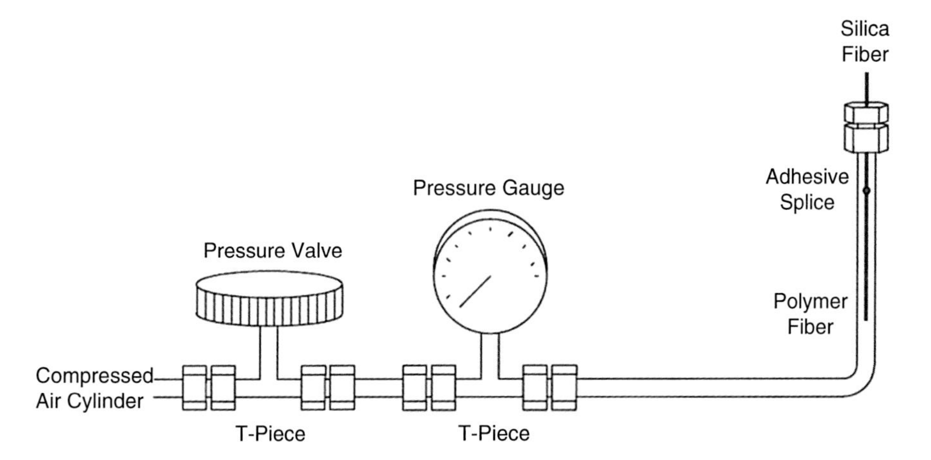

Fig. 25 Experimental setup to characterize the hydrostatic pressure responsivity of MM PMMA mPOFBG (Johnson et al. 2012)

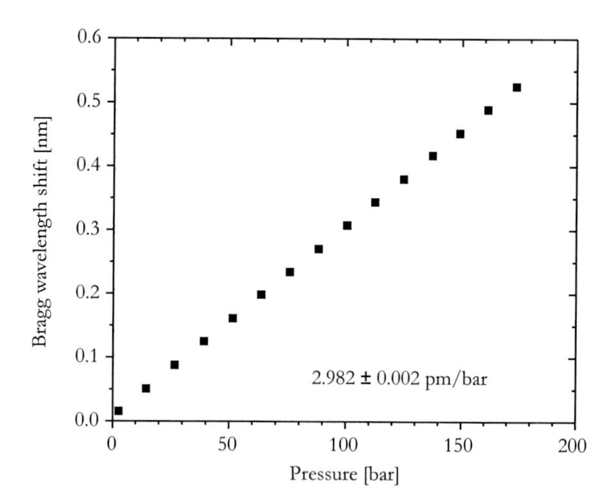

Fig. 26 Hydrostatic pressure response of a humidity insensitive single mode mPOFBG

with FWHM of 8.78 nm. In the experiment, it was seen that a ± 2 °C reversible temperature fluctuation when the pressure was ramped up from atmospheric to 10 MPa. By increasing the pressure from 2 MPa to 10 MPa with a step of by 1 MPa, a pressure sensitivity of 100 ± 9.38 pm/MPa was obtained. A similar experiment done on silica FBG with a Bragg wavelength of 1551.04 nm with FWHM of 0.22 nm gave a sensitivity of −3.88 ± 0.04 pm/MPa. This result showed that the MM mPOFBG was 25 times more sensitive than the silica FBG.

In 2016 Pedersen et al. demonstrated hydrostatic pressure sensor with humidity insensitive single-mode mPOFBGs up to 20 MPa (200 bar) (Pedersen et al. 2016). In their experiment in order to avoid the fluctuation of temperature and its effect on the pressure response of the FBG while the pressure was ramped up and down, the measurement was done in a gas free environment with the FBG saturated with water. As a result, the fluctuation in temperature was limited to only ±0.2 °C, which was around ±2 °C in Johnson et al.'s experiments. The grating used in their experiment had a Bragg wavelength of 865 nm with FWHM of 1 nm. The pressure sensitivity of the grating was 29.82 ± 0.02 pm/MPa with a resolution of 0.2 MPa. The response of the grating with pressure is shown in Fig. 26.

Acceleration Sensors

Microstructured polymer optical FBG sensors have also found application in the development of accelerometers. Stefani et al. demonstrated the first mPOFBG accelerometer both in the 850 nm and 1550 nm wavelength regions (Stefani et al. 2012b). The schematic diagram of the experimental setup for the characterization of the mPOFBG-based accelerometer is shown in Fig. 27. In this system, the acceleration was converted into strain by a mechanical transducer. The transducer was made in such way that the strain on the fiber was linearly

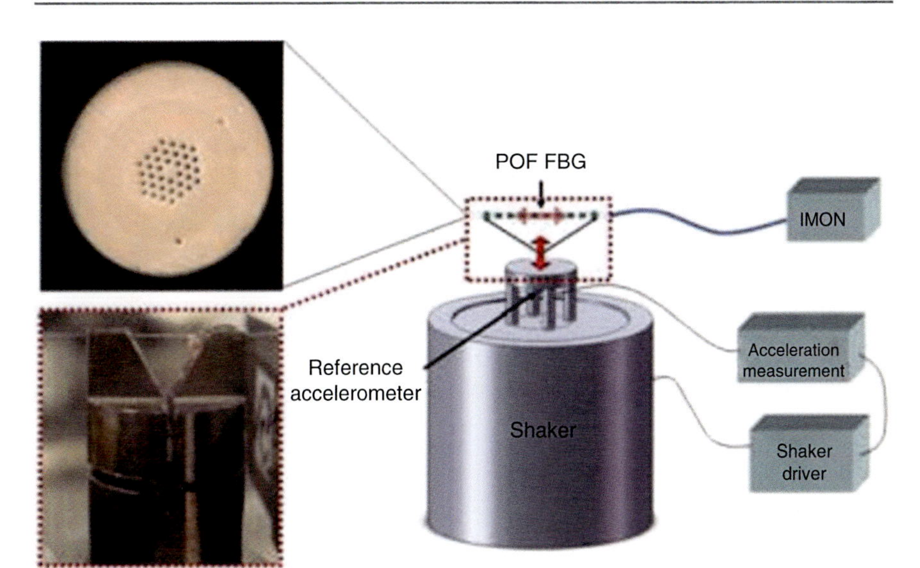

Fig. 27 Schematic of the accelerometer characterization setup. Top inset: mPOF cross section. Bottom inset: mPOF FBG-based accelerometer (Stefani et al. 2012b)

Table 6 The acceleration sensitivity of PMMA mPOFBGs and silica FBGs for two different fiber diameters and wavelength regions

Fiber type	Fiber diameter (μm)	Bragg wavelength (μm)	Sensitivity (pm/g)
PMMA mPOF	130	1550	15.2
Step index silica	80	1550	5.1
PMMA mPOF	130	850	5.9
Step index silica	125	850	1.8

dependent on the acceleration. Therefore, the fiber sensor was calibrated in way that the amount of the Bragg wavelength shift corresponded to certain levels of acceleration.

The gratings were fabricated from PMMA mPOFs with a hole to pitch ratio of 0.5 by using the phase mask technique. The phase masks had a pitch of 1024.7 nm for the 1550 nm FBGs and a pitch of 572.4 nm for the 850 nm FBGs. In their experiments, the Bragg wavelength change versus acceleration at a fixed frequency of 159.2 Hz was measured in the range from 0.1 g to 15 g acceleration. An acceleration of 15 g corresponded to a strain of 0.02%. POFs exhibit a larger linear strain operation range and thus can measure very high accelerations. The demonstration here was limited to only 15 g because of the shaker limit. For comparison purpose, the same experiment was repeated with silica FBGs in the same wavelength regions. The resulting sensitivities are summarized in Table 6. The response was found to be linear for both mPOF and silica. However, the mPOF displayed sensitivity three times higher than that of silica.

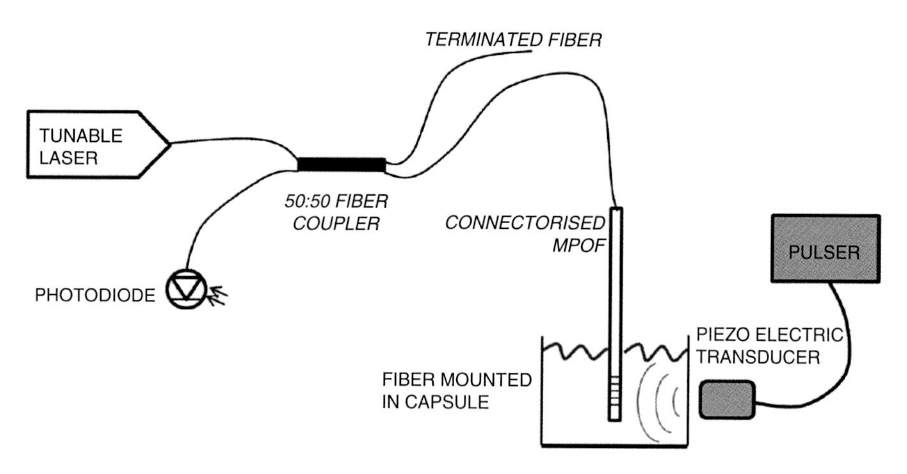

Fig. 28 Experimental setup for the ultrasonic detection experiments using PMMA mPOFBGs (Broadway et al. 2015)

Biomedical Sensor

One of the unique properties of polymers is their biocompatibility and mPOFBGs have been considered as potential candidate in biomedical areas such as for the detection of ultrasonic waves (Markos et al. 2011; Jensen et al. 2005; Emiliyanov et al. 2007; Broadway et al. 2015). Biocompatibility constitutes an important property for optoacoustic endoscopy. For endoscopic applications, one of the most important criteria is the size of the sensor, as the sensor has to be inserted in a cavity. In addition, high sensitivity is also required. The basic principle behind using mPOFBGs for ultrasonic wave detection again relies on the strain effect. The mPOFBGs have already proved to be highly sensitive to strain, and this makes them particularly suitable for optoacoustic endoscopy. The characterization setup used for the detection of ultrasonic waves in the laboratory is depicted as in Fig. 28 (Broadway et al. 2015).

First, the grating profile was examined by using a broadband light source and an OSA to determine the 3 dB point of the grating. Then the broadband source and the spectrum analyzer were replaced by a tunable laser and photodetector connected with an oscilloscope, respectively. The laser was then tuned to the 3 dB point of the mPOFBG. Therefore, when the ultrasonic wave perturbed the grating, the Bragg wavelength shifted and led to an amplitude variation at the detector output, which was then observed on the oscilloscope. In order to generate an ultrasonic wave, piezo electric transducer was used. A transducer was added to the setup and acoustically coupled using water as the acoustic medium. Figure 29 shows a typical ultrasonic response of PMMA mPOFBG immersed in water (acoustic medium) using transducers at 1, 5, and 10 MHz. Therefore, by this method a compact in-fiber ultrasonic detector based on mPOFBGs was developed.

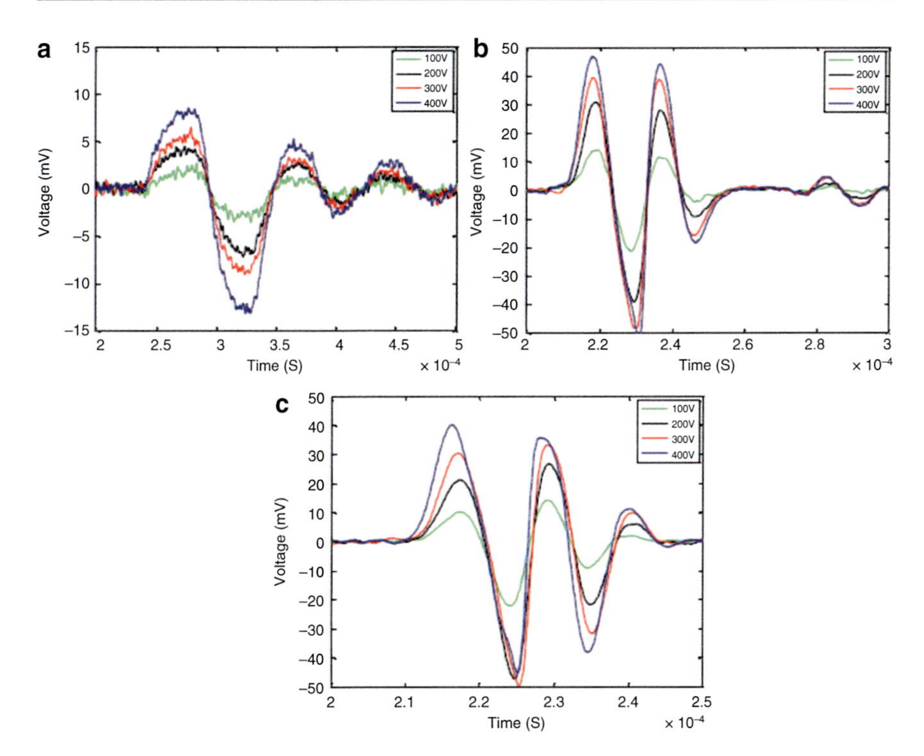

Fig. 29 Ultrasonic response PMMA mPOFBG for (**a**) 1 MHz, (**b**) 5 MHz, and (**c**) 10 MHz at excitation voltages between 100 and 400 V (Broadway et al. 2015)

Thermo-Hygrometer

Grating fabricated in a specially designed mPOF has been proposed by Woyessa et al. towards the development of a novel thermo-hygrometer (Woyessa et al. 2017c). The fiber was fabricated from two polymer materials and involved an over cladding as well. The core and the cladding of the fiber consisted of Zeonex 480R and PMMA over cladding (see Fig. 30). The sensing probe was based on two separate in-line FBGs inscribed in the fabricated mPOF. To inscribe the two gratings, only a single-phase mask was used in combination with thermal annealing technique.

In a piece of the fabricated fiber, at one end, the PMMA over cladding was etched out for few centimeters and the first grating was inscribed in the etched section of the fiber. Then the grating was annealed to shift the Bragg wavelength by few nanometers and followed by inscribing the second grating in the unetched section of the fiber few centimeters away from the first grating. The gratings configuration and spectrum is depicted in Fig. 31a and b, respectively.

The materials used for the sensor fabrication have different properties. PMMA has very high affinity to water, whereas Zeonex has a very small moisture absorbing capability (Woyessa et al. 2016, 2017a). Thus, the basic principle behind humidity

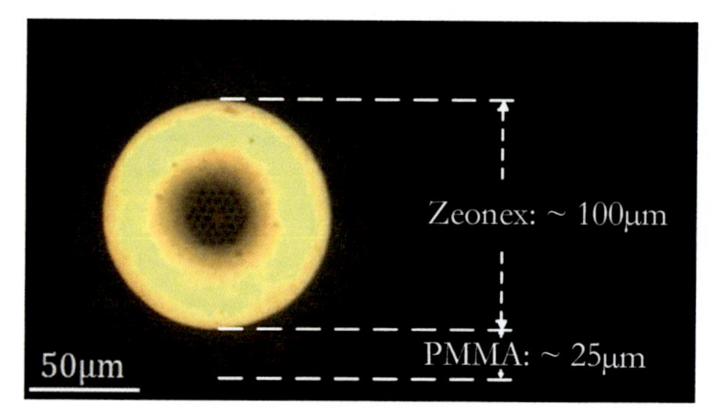

Fig. 30 Zeonex core and cladding and PMMA over-cladding mPOF

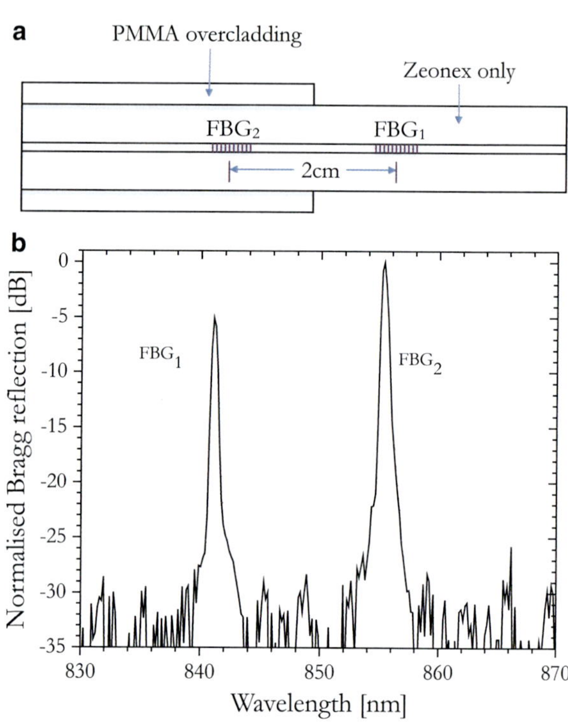

Fig. 31 Dual gratings (**a**) Configuration (**b**) reflection spectrum

measurement with the PMMA over cladding Zeonex mPOF relies on the swelling effect caused by PMMA. When PMMA absorbs moisture, it induces strain on the grating inscribed in the Zeonex core thus effectively leading to a shift in the Bragg wavelength. The temperature and the relative humidity response of the dual gratings are shown in Fig. 32a and b, respectively.

The responses of the two gratings are modeled as:

$$\Delta\lambda_1 = \alpha_1 \Delta T + \beta_1 \Delta H \tag{6}$$

$$\Delta\lambda_2 = \alpha_2 \Delta T + \beta_2 \Delta H + \gamma_2 \Delta H_2 \tag{7}$$

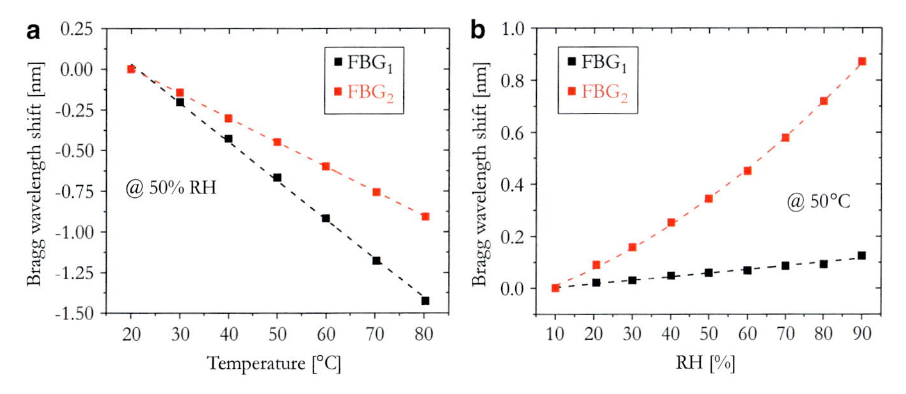

Fig. 32 (**a**) The temperature response of FBG_1 and FBG_2 at 50% RH. The dashed lines correspond to the linear fitting of the data. (**b**) The humidity response of FBG_1 and FBG_2 at 50 °C. The dashed lines correspond to a linear fit of FBG_1 data and a second-order polynomial fit of FBG_2 data

Table 7 Standard Errors of the Fitting Parameters

i	α_i (pm/°C)	β_i(Pm/% RH)	γ_i(Pm/% (RH)^2)
1	-23.9 ± 0.4	1.4 ± 0.1	–
2	-15.1 ± 0.1	6.4 ± 0.5	0.057 ± 0.004

where $\Delta\lambda_1$ and $\Delta\lambda_2$ are the Bragg wavelength shifts of FBG_1 and FBG_2, respectively, as a result of changes in RH ΔH and temperature ΔT and the coefficients α_i, β_i, and γ_i are listed in Table 7, which are determined by fitting the polynomial equations, Eqs. 6 and 7, to the calibration data.

These coefficients are used to reconstruct the values of temperature and relative humidity from the wavelength and in the range 10–90% RH and 20–80 °C, a root mean square deviation of 0.8% RH and 0.6 °C was found. These investigations showed that the proposed sensor system is a viable way of effectively separating the responses to temperature and humidity. The potential advantages of the developed device are being easy to fabricate, low cost, compact, all polymer based, and mechanically stable. In addition, it has also multiplexing capability as the two gratings are very close in wavelength.

Conclusion

In conclusion, this chapter overviewed the most important efforts from several research groups around the globe to develop FBG-based optical sensors using mPOFs. Initially the main three different FBG inscription techniques (interferometric, phase-mask, and point-by-point) based on different inscription lasers (CW UV, pulsed UV, near-IR lasers) are presented and directly compared. Then the crucial impact of microstructured cladding of a fiber on the inscription process for the development of high-quality FBGs was discussed. The main results reported so far regarding the optimal conditions in terms of angle and orientation of the mPOF during the inscription are also summarized. Section "Microstructured Polymer Optical Fiber Bragg Gratings" focused on how different polymer materials, such as PMMA, Topas, Zeonex and PC, that have been so far used for the development of novel mPOFBGs suitable for many environmental measurands monitoring applications, significantly extending their functionality in several sensing applications. In the last section, different case studies utilizing mPOFBGs towards the development of strain, temperature, humidity, pressure, acceleration, biomedical sensors as well thermo-hygrometer are discussed.

Acknowledgements The authors would like to acknowledge the People Programme (Marie Curie Actions) of the European Union's Seventh Framework Programme FP7/2007-2013/ under REA grant agreement n° 608382 and Danish Council for Independent Research (FTP Case No. 4184-00359B).

References

T. Baghdasaryan, T. Geernaert, F. Berghmans, H. Thienpont, Geometrical study of a hexagonal lattice photonic crystal fiber for efficient femtosecond laser grating inscription. Opt. Express **19**(8), 7705–7716 (2011)

T. Bremner, A. Rudin, D.G. Cook, Melt flow index values and molecular weight distributions of commercial thermoplastics. J. Appl. Polym. Sci. **41**(78), 1617–1627 (1990)

C. Broadway, D. Gallego, G. Woyessa, A. Pospori, O. Bang, D.J. Webb, G. Carpintero, H. Lamela, in Polymer optical fibre sensors for endoscopic opto-acoustic imaging, *Proceedings of SPIE 9539, Opto-Acoustic Methods and Applications in Biophotonics II*, 953907 (2015)

I.-L. Bundalo, K. Nielsen, C. Markos, O. Bang, Bragg grating writing in PMMA microstructured polymer optical fibers in less than 7 minutes. Opt. Express **22**(5), 5270–5276 (2014)

I.-L. Bundalo, K. Nielsen, O. Bang, Angle dependent fiber Bragg grating inscription in microstructured polymer optical fibers. Opt. Express **23**(3), 3699–3707 (2015)

K.E. Carroll, C. Zhang, D.J. Webb, K. Kalli, A. Argyros, M.C.J. Large, Thermal response of Bragg gratings in PMMA microstructured optical fibers. Opt. Express **15**(14), 8844–8850 (2007)

Data Sheet Topas 5013S-04, Topas Advanced Polymers Inc., 2015, http://www.topas.com/sites/default/files/TDS_5013S_04_e_1.pdf

Data Sheet Topas 8007F-04, Topas Advanced Polymers Inc., 2015, http://www.topas.com/sites/default/files/TDS_8007F-04_english%20units_0.pdf

H. Dobb, D.J. Webb, K. Kalli, A. Argyros, M.C.J. Large, M.A. van Eijkelenborg, Continuous wave ultraviolet light-induced fiber Bragg gratings in few- and single-mode microstructured polymer optical fibers. Opt. Lett. **30**(24), 3296–3298 (2005)

G. Emiliyanov, J.B. Jensen, O. Bang, P.E. Hoiby, L.H. Pedersen, E.M. Kjaer, L. Lindvold, Localized biosensing with Topas microstructured polymer optical fiber. Opt. Lett. **32**(5), 460–462 (2007)

A. Fasano, G. Woyessa, P. Stajanca, C. Markos, A. Stefani, K. Nielsen, H.K. Rasmussen, K. Krebber, O. Bang, Fabrication and characterization of polycarbonate microstructured polymer optical fibers for high temperature resistant fiber Bragg grating strain sensors. Opt. Mater. Express **6**(2), 649–659 (2016)

A. Fasano, G. Woyessa, J. Janting, H.K. Rasmussen, O. Bang, Solution-mediated annealing of polymer optical fiber Bragg gratings at room temperature. IEEE Photon. Technol. Lett. **29**(8), 687–690 (2017)

GEHR PMMA (Acrylic), http://www.gehrplastics.com/pmma-acrylic.html

K.O. Hill, B. Malo, F. Bilodeau, D.C. Johnson, J. Albert, Bragg gratings fabricated in monomode photosensitive optical fibre by UV exposure through a phase mask. Appl. Phys. Lett. **62**(10), 1035–1037 (1993)

X. Hu, G. Woyessa, D. Kinet, J. Janting, K. Nielsen, O. Bang, C. Caucheteur, BDK-doped core microstructured PMMA optical fiber for effective Bragg grating photo-inscription. Opt. Lett. **42**(11), 2206–2212 (2017)

T. Ishigure, M. Hirai, M. Sato, Y. Koike, Graded-index plastic optical fiber with high mechanical properties enabling easy network installations. II. J. Appl. Polym. Sci. **91**(1), 404–409 (2004)

J. Jensen, P. Hoiby, G. Emiliyanov, O. Bang, L. Pedersen, A. Bjarklev, Selective detection of antibodies in microstructured polymer optical fibers. Opt. Express **13**(15), 5883–5889 (2005)

I.P. Johnson, K. Kalli, D.J. Webb, 827 nm Bragg grating sensor in multimode microstructured polymer optical fibre. Electron. Lett. **46**(17), 1217–1218 (2010a)

I.P. Johnson, D.J. Webb, K. Kalli, M.C.J. Large, A. Argyros, Multiplexed FBG sensor recorded in multimode microstructured polymer optical fiber. Proc. SPIE **7714**, 77140D (2010b)

I.P. Johnson, W. Yuan, A. Stefani, K. Nielsen, H.K. Rasmussen, L. Khan, D.J. Webb, K. Kalli, O. Bang, Optical fibre Bragg grating recorded in TOPAS cyclic olefin copolymer. Electron. Lett. **47**(4), 271–272 (2011)

I.P. Johnson, D.J. Webb, K. Kalli, Hydrostatic pressure sensing using a polymer optical fiber Bragg gratings. Proc. SPIE **8351**, 835106 (2012)

K. Kalli, C. Riziotis, A. Posporis, C. Markos, C. Koutsides, S. Ambran, A.S. Webb, C. Holmes, J.C. Gates, J.K. Sahu, P.G.R. Smith, Flat fibre and femtosecond laser technology as a novel photonic integration platform for optofluidic based biosensing devices and lab-on-chip applications: Current results and future perspectives. Sens. Actuators B. Chem. **209**, 1030–1040 (2015)

G. Khanarian, H. Celanese, Optical properties of cyclic olefin copolymers. Opt. Eng. **40**(6), 1024–1029 (2001)

S.G. Leon-Saval, R. Lwin, A. Argyros, Multicore composite single-mode polymer fiber. Opt. Express **20**(1), 141–148 (2012)

B. Malo, K.O. Hill, F. Bilodeau, D.C. Johnson, J. Albert, Point-by-point fabrication of micro-Bragg grating in photosensitive fibre using single excimer pulse refractive index modification techniques. Electron. Lett. **29**(18), 1668–1669 (1993)

C. Markos, W. Yuan, K. Vlachos, G.E. Town, O. Bang, Label-free biosensing with high sensitivity in dual-core microstructured polymer optical fibers. Opt. Express **19**(8), 7790–7798 (2011)

C. Markos, A. Stefani, K. Nielsen, H.K. Rasmussen, W. Yuan, O. Bang, High-Tg TOPAS microstructured polymer optical fiber for fiber Bragg grating strain sensing at 110 degrees. Opt. Express **21**(4), 4758–4765 (2013)

G.D. Marshall, D.J. Kan, A.A. Asatryan, L.C. Botten, M.J. Withford, Transverse coupling to the core of a photonic crystal fiber: The photo-inscription of gratings. Opt. Express **15**(12), 7876–7887 (2007)

A. Martinez, M. Dubov, I. Khrushchev, I. Bennion, Direct writing of fibre Bragg gratings by femtosecond laser. Electron. Lett. **40**(19), 1170–1172 (2004)

G. Meltz, W.W. Morey, W.H. Glenn, Formation of Bragg gratings in optical fibres by a transverse holographic method. Opt. Lett. **14**(15), 823–825 (1989)

R. Oliveira, L. Bilro, R. Nogueira, Bragg gratings in a few mode microstructured polymer optical fiber in less than 30 seconds. Opt. Express **23**(8), 10181–10187 (2015)

Optical properties of Makrolon and Apec for non-imaging optics, Bayer Material ScienceAG, 2014, http://www.plastics.covestro.com/Products/~/media/B6555362438341FF9804F21A253E5B23.ashx?la=en

J.K.M. Pedersen, G. Woyessa, K. Nielsen, O. Bang, Intrinsic pressure response of a single-mode cyclo olefin polymer microstructured optical fibre Bragg grating, in *Proceedings of the International Plastic Optical Fibres Conference, Birmingham* (2016). ISBN: 978 1 85449 408 5

S. Roy, C.Y. Yue, Z.Y. Wang, L. Anand, Thermal bonding of microfluidic devices: Factors that affect interfacial strength of similar and dissimilar cyclic olefin copolymers. Sens. Actuators B Chem. **161**(1), 1067–1073 (2012)

D. Sáez-Rodríguez, K. Nielsen, H.K. Rasmussen, O. Bang, D.J. Webb, Highly photosensitive polymethyl methacrylate microstructured polymer optical fiber with doped core. Opt. Lett. **38**(19), 3769–3772 (2013)

E.E. Shafee, Effect of photodegradation on the [beta]-relaxation in poly(methylmethacrylate). Polym. Degrad. Stab. **53**, 57–61 (1996)

L.S.A. Smith, V. Schmitz, The effect of water on the glass transition temperature of poly (methyl methacrylate). Polymer **29**(10), 1871–1878 (1988)

G. Statkiewicz-Barabach, K. Tarnowski, D. Kowal, P. Mergo, W. Urbanczyk, Fabrication of multiple Bragg gratings in microstructured polymer fibers using a phase mask with several diffraction orders. Opt. Express **21**(7), 8521–8534 (2013)

A. Stefani, W. Yuan, C. Markos, O. Bang, Narrow bandwidth 850 nm fiber Bragg gratings in few-mode polymer optical fibers. IEEE Photon. Technol. Lett. **23**(10), 660–662 (2011)

A. Stefani, M. Stecher, G.E. Town, O. Bang, Direct writing of fiber Bragg grating in microstructured polymer optical fiber. IEEE Photon. Technol. Lett. **24**(13), 1148–1150 (2012a)

A. Stefani, S. Andresen, W. Yuan, N. Herholdt-Rasmussen, O. Bang, High sensitivity polymer optical fiber-Bragg-grating-based accelerometer. IEEE Photon. Technol. Lett. **24**(9), 763–765 (2012b)

Technical data, Zeonex, Cyclo Olifen Polymer (COC), 2017, http://www.zeonex.com/optics.aspx

Topas COC product overview, Topas Advanced Polymers Inc., 2014, http://www.topas.com/sites/default/files/files/TOPAS_Brochure_E_2014_06(1).pdf

É. Torres, M.N. Berberan-Santos, M.J. Brites, Synthesis, photophysical and electrochemical properties of perylene dyes. Dyes Pigments **112**, 298–304 (2015)

D.L. Williams, Photosensitivity: The phenomenon and its applications, *Advanced Photonic Topics*, (Universidad de Cantabria, Santander, 1997). books.google.com

D.R.G. Williams, P.E.M. Allen, V.T. Truong, Glass transition temperature and stress relaxation of methanol equilibrated poly (methyl methacrylate). Eur. Polym. J. **22**(11), 911–919 (1986)

G. Woyessa, K. Nielsen, A. Stefani, C. Markos, O. Bang, Temperature insensitive hysteresis free highly sensitive polymer optical fiber Bragg grating humidity sensor. Opt. Express **24**(2), 1206–1213 (2016)

G. Woyessa, A. Fasano, C. Markos, A. Stefani, H.K. Rasmussen, O. Bang, Zeonex microstructured polymer optical fiber: Fabrication friendly fibers for high temperature and humidity insensitive Bragg grating sensing. Opt. Mater. Express **7**(1), 286–295 (2017a)

G. Woyessa, A. Fasano, C. Markos, H.K. Rasmussen, O. Bang, Low loss polycarbonate polymer optical fiber for high temperature FBG humidity sensing. IEEE Photon. Technol. Lett. **29**(7), 575–578 (2017b)

G. Woyessa, J.M. Pedersen, A. Fasano, K. Nielsen, C. Markos, H.K. Rasmussen, O. Bang, Zeonex-PMMA microstructured polymer optical FBGs for simultaneous humidity and temperature sensing. Opt. Lett. **42**(6), 1161–1164 (2017c)

Z. Xiong, G.D. Peng, B. Wu, P.L. Chu, Effects of the zeroth-order diffraction of a phase mask on Bragg gratings. J. Lightwave Technol. **17**(11), 2361–2365 (1999)

W. Yuan, L. Khan, D.J. Webb, K. Kalli, H.K. Rasmussen, A. Stefani, O. Bang, Humidity insensitive TOPAS polymer fiber Bragg grating sensor. Opt. Express **19**(20), 19731–19739 (2011)

W. Yuan, A. Stefani, O. Bang, Tunable polymer fiber Bragg grating (FBG) inscription: Fabrication of dual-FBG temperature compensated polymer optical fiber strain sensors. IEEE Photon. Technol. Lett. **24**(5), 401–403 (2012)

Polymer Fiber Sensors for Structural and Civil Engineering Applications

53

Sascha Liehr

Contents

Abstract

This chapter gives an overview about polymer optical fiber (POF) sensors with the focus on structural and civil engineering applications. POF properties such as the high-strain range, the low Young's modulus, and specific scattering effects open new fields for fiber-optic sensing applications. POF properties, sensitivities, and cross-sensitivities that are relevant for sensing are introduced.

S. Liehr (✉)
Division 8.6 "Fibre Optic Sensors", Bundesanstalt für Materialforschung und –prüfung (BAM), Berlin, Germany
e-mail: sascha.liehr@bam.de

G.-D. Peng (ed.), *Handbook of Optical Fibers*,
https://doi.org/10.1007/978-981-10-7087-7_3

Advantages and limitations are discussed. State-of-the-art POF sensors and application examples are presented in subsections with regard to their underlying measurement principles.

Introduction

This chapter summarizes the recent developments, technologies, and practical implementations of polymer optical fiber (POF) sensors for structural and civil engineering applications. The general advantages of optical fiber sensors over traditional techniques, such as electrical transducers, are their small size and low weight, immunity to electromagnetic interference, galvanic isolation, chemical robustness, and the ability to be integrated into various structures and materials. This has led to an increasing scope of applications in the structural health monitoring (SHM) and civil engineering sector. Silica fiber sensors dominate the optical fiber-based share of these markets and are being used for the measurement of strain, temperature, and vibrations. Although POFs have a considerably higher attenuation and are not widely available for single-mode (SM) operation and thus for interferometric and high-resolution sensing, they are in the process to occupy market niches for specific sensing applications.

POFs already appeared as soon as in the 1960s, at the same time when silica fibers emerged. However, since low attenuation and high bandwidth were major driving forces for the development in data communication, POFs have not been a subject of major research until the 1990s. Around that time, they became more widely used for short-distance data links, and various POF sensing principles have been proposed and implemented. Although silica fibers had a head start, certain beneficial features of POFs helped to open new fields in fiber-optic sensing for specialty applications. POF-specific properties and advantages for sensing applications are their mechanical properties in terms of low Young's modulus, high strainability and ruggedness. These features provide solutions for sensing tasks that could not, or only insufficiently, be accomplished with electrical transducers or silica fiber-based techniques. Compared to a practical long-term strain range of about 1–2% for silica fibers, POFs can endure and measure strain, depending on the fiber type, up to 45% (Liehr et al. 2008) and even 100% (Liehr et al. 2009b). The typically large POF core diameters and the high numerical aperture (NA) allow for easy interconnection, handling and simple low-cost implementations of various sensor principles. POFs also exhibit specific scattering effects that can be used for distributed strain and humidity sensing. Inexpensive standard polymer fibers and cables for optical communication are available and low-cost transceivers in the visible wavelength range based on LEDs can readily be used for a multitude of sensor applications.

In recent years, a great number of POF-based sensing principles has been developed and applied in various fields such as medical, biological, environmental, chemical, and health monitoring applications (Bilro et al. 2012; Peters 2011).

The emphasis of this chapter is on principles and techniques that are relevant for structural and engineering applications. So far, relatively few systems have been commercialized in this field and also have a long record of field application. However, the performance improvement is dynamic and the scope of applications is constantly growing. The focus of this chapter is on sensor techniques that are already an option for engineering applications as well as those that have the potential to be applied in the near future. After a brief introduction of important POF properties and their relevance for sensing, promising technologies and representative application examples are presented. They are discussed in separate sections and are categorized into the underlying measurement principles rather than measurands or specific applications.

Properties of POFs

The main reasons for using POFs as a sensor medium are their mechanical and optical properties as well as their specific scattering effects. Relevant parameters and limitations regarding sensor applicability are in the following introduced for the most widely used fiber types.

Optical Properties

Various standard fiber types can be used for sensing applications. Typical POF core materials are poly(methyl methacrylate) (PMMA), polystyrene (PS), and polycarbonate (PC), all of which exhibit attenuation values significantly higher than silica fibers. Useful spectral transmission windows are around 500 nm, 650 nm, and 850 nm. Step-index (SI) and gradient-index (GI) profiles are available with core diameters typically ranging from 50 μm to 1 mm. Multimode (MM) fibers are standard; SM POFs are rather used on the level of research. For more detailed information on POF materials and properties, see Ziemann et al. (2008) and Zubia and Arrue (2001).

The most widely used POF type for sensor applications is based on a SI PMMA core. Typical attenuation values are about 90 dB/km at 500 nm and 150 dB/km at 650 nm wavelength. Propagation loss is mainly caused by the absorption of the polymer's CH vibrational bonds. A more recent development is a low-loss MM perfluorinated (PF) POF. The replacement of hydrogen atoms by fluorine atoms shifts the vibrational absorptions to lower frequencies. This fiber type, based on the amorphous fluoropolymer CYTOP™, exhibits attenuation values down to 10 dB/km at 1000 nm (Koike and Asai 2009), 15 dB/km at 1310 nm, and about 250 dB/km at 1550 nm and can therefore be used for extended sensor lengths. Typical current attenuation values of commercial PF POFs are around 30–45 dB/km. The theoretical lower attenuation limit of CYTOP-based POFs is comparable to that of silica fibers (Koike and Asai 2009). Since the attenuation is mainly caused by

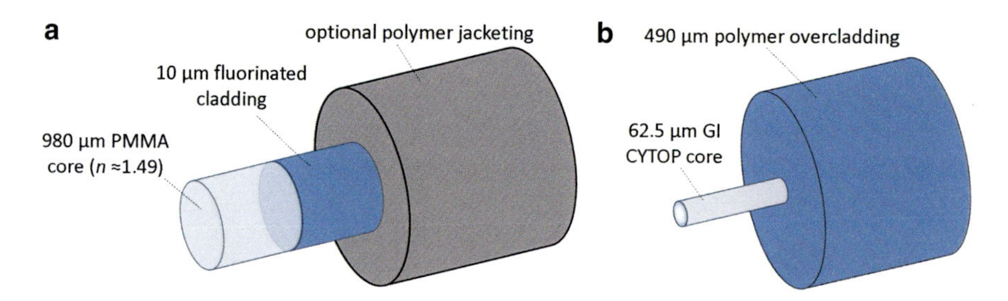

Fig. 1 Cross sections of the two most common POF types for sensing applications: 1 mm SI PMMA POF (**a**) and PF GI POF (**b**)

impurities and inhomogeneities that are induced during the fiber drawing process, optimized production processes may further reduce the transmission loss of PF POFs. CYTOP-based fibers are commercially available with GI structure and negligible modal dispersion.

The two most widely used polymer sensor fiber types, SI PMMA POFs and low-loss PF GI POFs, are exemplarily depicted in Fig. 1.

The typical core diameter of SI PMMA POFs for sensing applications is 980 μm, but also 250 μm and 500 μm core fibers are being used for sensing. The SI POF's high numerical aperture of up to 0.5 ensures easy interconnection but also causes significant modal dispersion which reduces the precision of time-of-flight measurement techniques and the resolution of distributed sensing techniques over longer sensor lengths. The GI structure of PF POFs prevents these dispersion-related limitations. They are, with typical core diameters of 50 μm or 62.5 μm and transmission window around 1310 nm, compatible with standard MM silica fiber components and devices. Specialty fibers such as microstructured POFs (mPOFs) and solid-core SM fibers are briefly introduced in section "Spectrometric and Interferometric POF Sensors" but are covered in more detail in separate chapters.

One aspect that may have important implications, especially for distributed sensing, is that POFs generally exhibit strong intragroup and intergroup mode coupling. This means that excited propagation modes are not as stable as in MM silica fibers. Optical power is constantly redistributed between different mode groups at short coupling lengths, typically in the meter range. Although this effect practically improves the bandwidth and decreases the modal dispersion, it largely prevents the practical use of interferometric sensing principles, such as swept wavelength interferometry (SWI). Most POF sensor principles are therefore based on incoherent measurement principles.

Besides the differing fiber core diameters, index structures, attenuation values, and usable wavelength ranges, PMMA POFs and PF POFs differ with respect to their mechanical properties, thermal properties, humidity dependence, and scattering effects. These properties and effects are described in more detail in the following subsections.

Mechanical Properties

The mechanical properties of POFs are highly suitable for strain sensing applications. The low Young's modulus (PMMA in the order of 3.2 GPa compared to 72 GPa for silica) is beneficial when strain is to be measured in highly ductile materials or structures. The integration of a continuous and compliant sensor fiber into a ductile structure does not alter the structure's response to strain and prevents measurement inaccuracies due to mechanical interaction of the fiber with the structure itself. Moreover, depending on the fiber type, POFs can endure and measure extreme strain exceeding 45% for PMMA POFs (Liehr et al. 2009b) or even 100% for PF GI POFs (Liehr et al. 2010b) with tolerable optical loss. Unlike large-core silica fibers, they can be bent and installed with small curvature radii. The mechanical properties, however, deviate toward the limits of the usable temperature range: They become more brittle at very low temperatures and exhibit differing stress-strain behavior at elevated temperatures. Elastic and viscoelastic strain limits are determined by the fiber's core, cladding and jacketing polymers, and typically range somewhere between 2% strain and 5% strain. These limits should be considered when cyclic high-strain loads are to be measured or dynamic loading-unloading applications are implemented. However, the hysteresis effect in the viscoelastic or plastic strain range is generally not problematic when the POF is integrated into a structure and the low-stiffness fiber is forced to comply to the strain of the surrounding material.

Thermal Properties

POFs are restricted to moderate-temperature environments since they lose their transparency and rigidity at elevated temperatures. Standard PMMA fibers have a long-term stability up to 85 °C (high-temperature types up to 105 °C), and PF POFs can be used up to 75 °C. Fibers with moderately increased temperature stability, but higher attenuation, are under development. Experiments with a fiber based on the polymer TOPAS® have demonstrated temperature stability up to 110 °C (Markos et al. 2013).

In comparison with fused silica, polymer materials have generally a significantly higher thermal expansion coefficient, for instance, about $75 \cdot 10^{-6}\ K^{-1}$ for PMMA (Goods et al. 2003). In contrast to silica fibers, the thermo-optic coefficient of most POF core materials is negative (about $-137 \cdot 10^{-6}\ K^{-1}$ for bulk PMMA (Beadie et al. 2015)) but can vary significantly for different polymers and dopants and also depends on the water content in the fiber core. The opposite signs of thermal expansion coefficient and thermo-optic coefficient must be taken into account for temperature and strain sensing applications. On the other hand, they also offer the potential for a temperature-compensated strain sensor design.

Care has also to be taken if long-term strain and temperature sensing is intended since the fiber drawing process usually induces a dominant orientation of the polymer chains along the fiber axis. The fiber material can therefore be assumed to be anisotropic with respect to its expansion coefficient and thermo-optic coefficient. Occurring long-term signal drifts due to temporal, mechanical, or temperature-induced molecular realignment can be largely prevented by annealing the fiber at elevated temperatures. This leads to a relaxation of the molecular alignment, and the fiber shrinks in axial direction. Optimized annealing processes are also crucial for the stabilization of fiber gratings in POFs and their response to humidity impact.

Humidity Dependence

As opposed to silica fibers, POFs are far more susceptible to relative humidity (RH) changes or moisture impact. Especially PMMA has an affinity to water and can absorb up to 2 wt%. This leads to increased attenuation at specific wavelength ranges due to OH absorption (Liehr et al. 2017). Figure 2 shows spectral transmission changes for saturated water content at 60% RH and 90% RH relative to 30% RH.

The wavelength range between 500 and 570 nm does not experience significant absorption changes and is clearly more suitable for humidity-independent power change sensing applications. The absorption of water into the fiber has also an effect on its volume and causes swelling and length change. Moreover, it affects the refractive index which has a combined impact on the optical signal runtime (Liehr et al. 2017). The refractive index change with the water content in PMMA has an additional dependency on the temperature.

The polymer's mechanical properties are also affected by the water content in the fiber: Increasing water content leads to a decrease of tensile strength. All these effects should be considered for sensor design, calibration, and operation of optical power change sensors, grating-based sensors, and time-of-flight measurement techniques. PF GI POFs exhibit measurable but lower water uptake which mainly affects the attenuation levels above 1300 nm (Liehr 2015). The humidity impact on POFs in general can be a significant cross-sensitivity, but it has also the potential for RH and moisture sensing applications, some of which are introduced

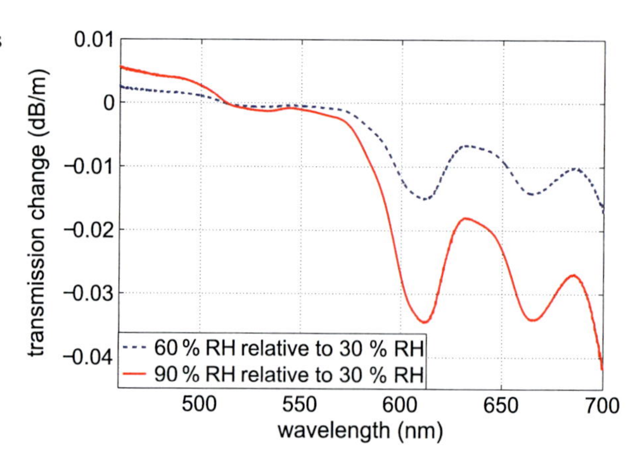

Fig. 2 Transmission changes of 1 mm diameter SI PMMA POF for RH values of 60% and 90% relative to a 30% RH reference measurement

in the following sections. Humidity-insensitive polymers, such as the cyclic olefin copolymer TOPAS®, have been developed and used for sensing with considerably reduced humidity impact (Yuan et al. 2011).

Organic polymers for POFs can be specifically produced or processed to be sensitive to chemical substances or parameters. This rather wide field of chemical sensing is, however, not covered in this chapter.

Optical Power Change Sensors

The first fiber-optic sensor, already proposed in the 1960s, was based on simple optical power change measurement. This most basic and low-cost sensing principle, typically comprising a light source, one or more optical fibers, commonly some kind of transducer, and a photodetector, has been implemented in a multitude of modifications. Especially POF-based sensors are advantageous since low-cost transmitters and receivers are available for POF-compatible wavelength ranges and the coupling of large-core POFs with high NA is straightforward.

POF optical power change sensors may be classified into two categories: *Intrinsic power change sensors* are sensors where the power of the transmitted optical signal along an intact optical fiber is modulated as a function of an external measurand or impact. *Extrinsic power change sensors* are more widely used and often involve some kind of external transducer which induces optical power changes, for example, due to the relative offset of the light-guiding fibers or power decoupling by means of physical or chemical impact. Some extrinsic principles involve structuring of a fiber section by side-polishing, grooving the fiber, or removing the cladding material. This way, the fiber core is exposed and can interact with the environment or decouple light depending on strain, fiber bending, or refractive index change of the surrounding medium. POFs are most suitable for structuring since they are not as brittle as silica fibers and their large core diameters can easily be mechanically and chemically processed. A great number of principles, measuring power changes either in transmission or in reflection, have been realized. Figure 3 shows exemplarily some of the most common implementations.

A simple voltage reading after transmitted power detection at a photoreceiver can be converted into a change of the measurement value. More advanced implementations may feature an additional reference arm or other mechanisms for source power change compensation or connector degradation compensation. These sensors can, after calibration, be relatively precise and can generally be designed for high measurement repetition rates. A variety of principles has been employed for structural monitoring applications, most of which are based on SI PMMA POF.

Several intrinsic crack detection and crack width measurement implementations make use of the extremely high-strain range of POFs and can, for example, be applied to measure the integrity of composite laminates. Small-diameter (250 μm) SI PMMA POFs have been integrated into glass fiber/epoxy composites to detect power loss (Fig. 3, implementation (a)) due to crack evolution in transverse direction (Takeda 2002). Figure 4a shows the transmitted power loss dependence during strain

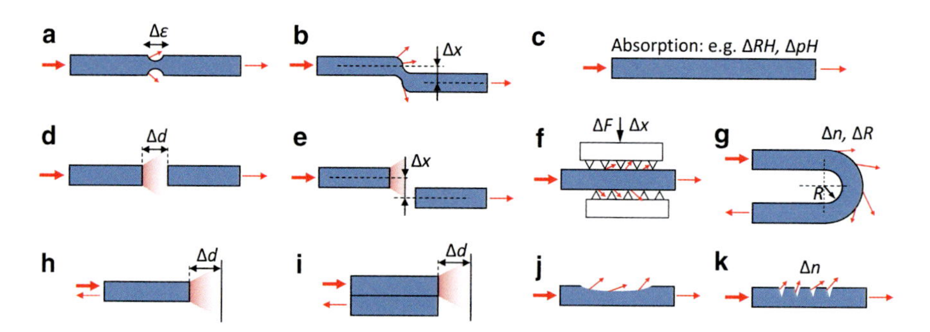

Fig. 3 Schematic of common intrinsic (**a–c**) and extrinsic (**d–k**) POF power change sensor implementations

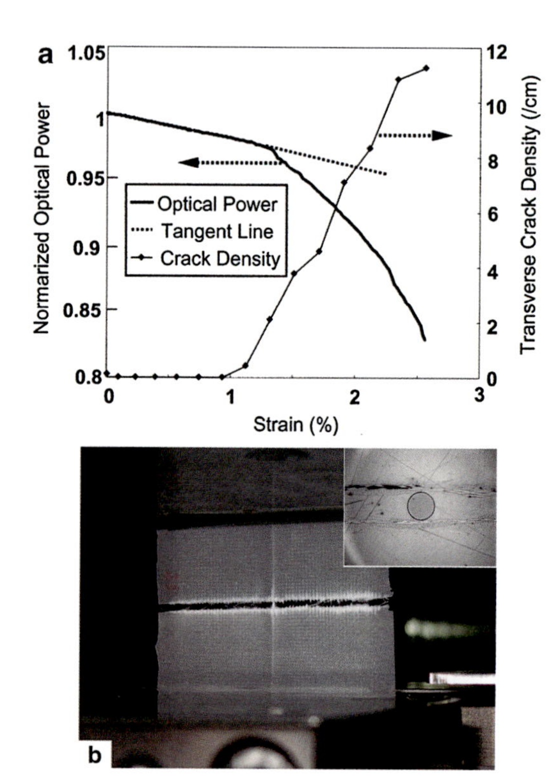

Fig. 4 Optical power change and crack density during increasing strain of the composite laminate (**a**) (Takeda 2002) (This figure has been reproduced with permission from Elsevier) and (**b**) photograph of an intact 500 μm PMMA POF after rupture of a glass fiber epoxy laminate and cross section of a sample with an integrated 250 μm PMMA POF

increase, exhibiting a deviation from the linear decay after crack initiation when exceeding about 1% strain. A similar tensile test with a 500 μm SI PMMA POF in a glass fiber epoxy laminate proved the integrity of the fiber, even exceeding the failure strain of the composite of 1.6% (Schukar et al. 2009); see Fig. 4b.

POFs can withstand extreme local strain values. Their potential for composite-integrated applications is, however, restricted by their typically large core diameter and limited temperature range.

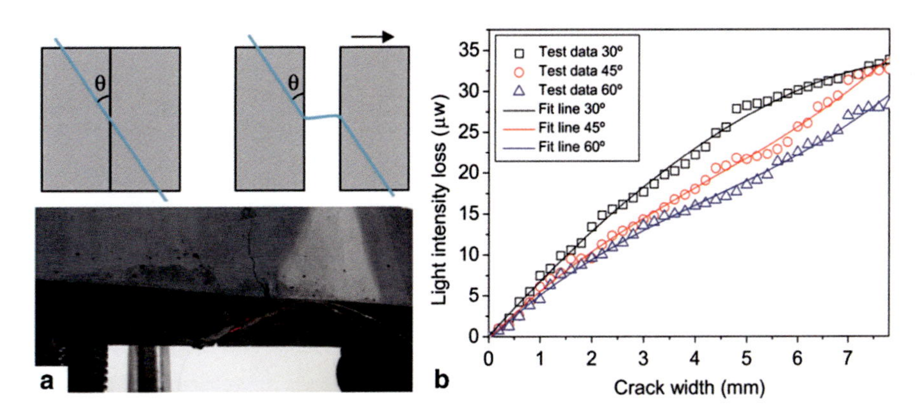

Fig. 5 Crack detection due to fiber bend power decoupling in a three-point bend setup (**a**) and loss characteristic for different angular orientations to the crack (**b**) (Zhao et al. 2015) (This figure has been reproduced with permission from Elsevier)

A more predictable crack response and loss characteristic can be obtained by installing the sensor fiber at a lower angle to the expected crack orientation (implementation (b) in Fig. 3). Figure 5 shows such a crack detection principle based on a single-crack fiber bend loss of a 500 μm SI PMMA POF glued onto the surface of a concrete specimen in a three-point bend setup (Zhao et al. 2015).

Crack widths well below 1 mm can be detected with these intrinsic power change sensor approaches. Their advantage is that they can respond to any crack appearing along the installed fiber and the structure. However, the crack location cannot be determined, and the precise measurement of the crack width, the number of cracks, or the displacement is often not possible since the power change characteristic is not linear and the light decoupling also depends on local conditions. The use of this sensor type is also restricted if high-strain values beyond the elastic deformation limit of the POF are to be measured under loading-unloading conditions. The closing of wider cracks after plastic deformation of the fiber may result in uncorrelated loss signals.

More precise and controlled crack width or displacement measurement can be obtained with extrinsic power change sensors based on light coupling changes between fibers (implementations (d), (e), (h), and (i) in Fig. 3). For this sensor type, rather the structural features such as the large core and high NA than the mechanical properties are the decisive advantages of POFs. These sensors have to be installed at specific locations of interest and can provide displacement change resolutions in the μm range. Strain response results of a typical implementation (Fig. 3d) are shown in Fig. 6. Here, the displacement measurement is based on the longitudinal separation between SI PMMA fibers in a high-opacity liquid-filled housing enabling resolutions in the μm range (Kuang et al. 2004).

A long-term application test of an extrinsic crack width sensor with a similar architecture, but an air gap, has been demonstrated over a period of 400 days (Perrone et al. 2008). Experience shows that power referencing and temperature

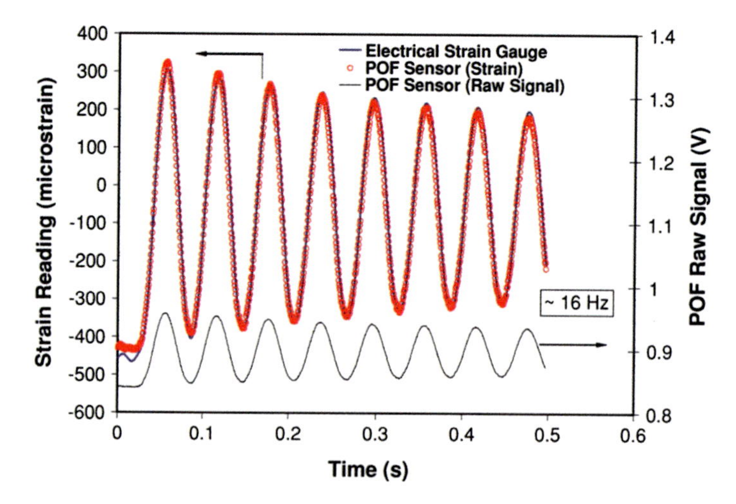

Fig. 6 Strain response from displacement measurement compared to electrical strain gauge (Kuang et al. 2004) (This figure has been reproduced with permission from IOP)

compensation are crucial for precise long-term measurement applications. More elaborated detection principles based on power modulation of the optical source have been implemented. Such a sensor in reflection configuration, Fig. 3h has been demonstrated to measure sub-μm vibrating surfaces into the ultrasonic frequency range (Perrone and Vallan 2009). Simple power-coupling dependency between misaligned POFs (configuration (e) in Fig. 3) can also be used for acceleration sensing and structural vibration evaluation (Costa Antunes et al. 2014). Power-coupling sensors (Fig. 3d, e, h, i) can be generally realized in various designs and at low costs. The displacement measurement range of these sensors is, depending on the implementation, typically limited to a few millimeters.

Structuring sections of the fiber core or the cladding material is another way to sensitize the fiber for transmission change measurement applications. Various sensor designs have been implemented for strain or displacement measurement. This can be done by grooving the fiber or removing segments of the cladding and core material (implementations (j) and (k) in Fig. 3). One way is to use the light decoupling dependence of the fiber at a structured section on the curvature and orientation. The cause is the change of the ratio of modes that experience total internal reflection. Flexural strain measurement of such a loss-based implementation (1 mm SI PMMA POF), where the cladding of a fiber segment has been removed, has been demonstrated by installing the fiber at the bottom of a three-point bend test setup (Kuang et al. 2002). Reproducible tensile strain sensing has also been conducted with the same sensor type by installing the fiber in a slightly curved shape which straightens out when tensile load is applied to the specimen; see Fig. 7.

The same approach has been used to monitor crack initiation perpendicular to the segmented fiber at the bottom of a three-point bend setup. Accumulated power loss

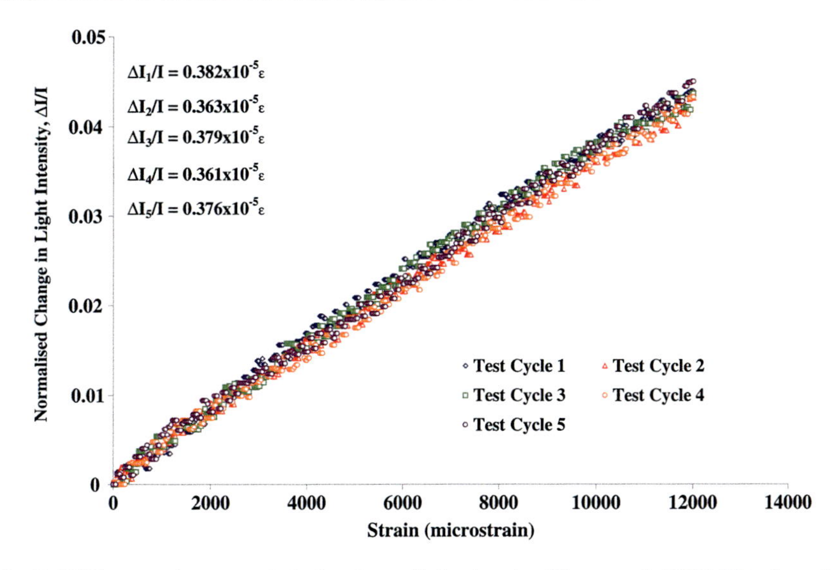

Fig. 7 POF power change result during six tensile load cycles (Kuang et al. 2002) (This figure has been reproduced with permission from IOP)

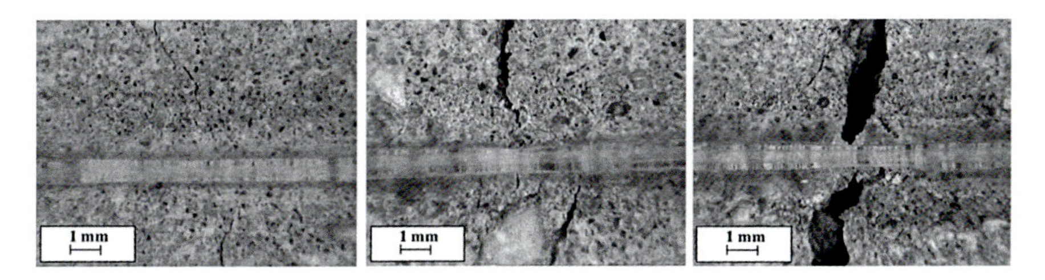

Fig. 8 Photograph of crack propagation of a concrete specimen with bonded POF sensor (Kuang et al. 2003) (This figure has been reproduced with permission from IOP)

along the sensitized region has been measured and hairline crack initiation could be detected, Fig. 8.

This approach can be used for the detection of crack initiation and post-crack vertical deflection for high local strain values. Intact, not structured POFs, however, exhibit negligible propagation loss up to high-strain values.

Similar concepts, based on local light decoupling, have been realized for liquid-level sensing by detecting the transmission loss of U-bent fibers (implementation (g) in Fig. 3) (Kuang et al. 2008), of side-polished curved fibers (Lomer et al. 2007), or along grooved fibers (Fig. 3, configuration (k)). The transmission loss is mainly caused by the refractive index change due to the presence of water at the sensitized region. This approach may also have potential for groundwater-level monitoring in civil engineering applications (Mesquita et al. 2016). An experimental implementation with a 1 mm SI PMMA POF with multiple grooves along the sensor section is shown in Fig. 9.

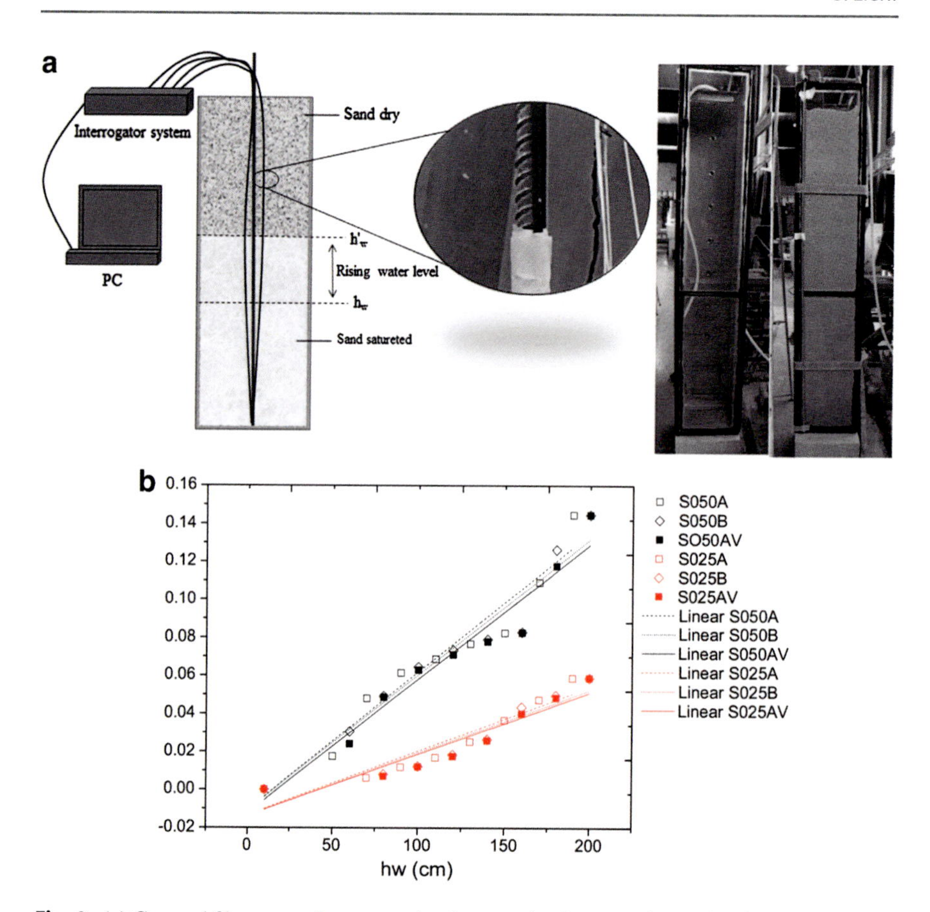

Fig. 9 (**a**) Grooved fiber power loss sensor implementation for groundwater-level monitoring and (**b**) optical power change results during increasing water level for 250 μm deep grooves (red) and 500 μm deep grooves (black) (Mesquita et al. 2016) (These figures have been reproduced with permission from Elsevier)

Raising water levels can be measured by evaluating power increase due to an increasing number of "water-filled" grooves. A similar design with grooved SI PMMA POFs has been used to monitor the curing of concrete by indirectly measuring the water content over a curing time of 28 days (André et al. 2012). The optimization of the concrete curing process and the information on the setting state are important issues in civil engineering, especially for the construction of large infrastructures. POF-based sensors can be used to retrieve information on the concrete setting state to optimize construction planning as well as the curing conditions for optimum mechanical properties of the concrete structure. Moreover, intrinsic implementations have been proposed to analyze the setting behavior of concrete, for example, by measuring the transmission loss along a SI PMMA POF caused by the formation of micro- and macrobends during concrete setting (Rajesh et al. 2007). Transmission change measurement due to OH absorption in SI

PMMA POF (Fig. 3c) has also potential for curing monitoring and moisture content measurement in concrete, creed, and soil (Liehr et al. 2017). Other transmission loss configurations target, for instance, the measurement of chemical quantities: Specially sensitized fiber sections or cladding sections are used for chemical sensing such as pH and ammonia. This is, however, beyond the scope of this more structural application-oriented chapter.

In summary, POF power change sensors are a low-cost solution for various static and dynamic monitoring applications. Two major groups of sensor implementations can be identified: One group is based on integrated optical power loss measurement along the whole fiber. These sensors are typically based on intrinsic effects and signal loss due to light decoupling or absorption and cannot provide specific information on the measurand distribution along the fiber or exact values of the measurand. This sensor type is most useful for early damage detection at critical locations of a structure or the measurement of general properties such as concrete curing progression or moisture content. The second major sensor implementation group is predominantly based on extrinsic configurations at predetermined locations and can, after calibration, be relatively precise. Power change referencing and temperature referencing as well as drift compensation may have to be implemented if precise long-term monitoring is to be conducted. Commonly, each power change sensor occupies at least one detection channel. Some principles have the potential to be multiplexed along the fiber. This can, for example, be accomplished by using distributed backscatter measurement techniques and detecting the Rayleigh scattering loss between sensor sections along the fiber. This approach, however, is only suitable if the accumulated optical loss does not exceed the dynamic range of the detection technique. The utilization of an optical time-domain reflectometer (OTDR) increases the instrumentation cost considerably, which is generally not tolerable for quasi-distributed power change sensor applications. Various intrinsic backscattering-based principles for the continuously distributed measurement in POFs with negligible loss have commercial potential and are presented in section "Distributed POF Sensors."

Spectrometric and Interferometric POF Sensors

This section summarizes principles and applications based on spectrometric measurement and/or interferometric measurement. Many of the spectrometric sensing principles in POF are actually based on interferometric principles, such as fiber Bragg gratings (FBGs) and differential mode interference methods. Other techniques use incoherent principles such as spectral decoupling of light in long-period gratings (LPGs). Most of the techniques presented here require specialty fibers or specifically processed standard fibers, which are still rather a subject of research. Also, absorption-based and fluorescence-based spectrometric sensing using POFs is an emerging topic for the detection of chemicals and biosensing. These applications and techniques, however, will not be discussed in this chapter.

Spectrometric sensor systems commonly consist of a broadband light source or a tunable laser, a sensor fiber that modulates the transmitted or reflected power spectrum, and some kind of spectrometer. Dynamic measurement is generally possible, and sensors can often be multiplexed by spectral separation. The steady cost decrease of the key components makes this technique potentially more widely applicable. However, since most sensor implementations are based on fibers or manufacturing techniques at research level, only few actual applications or demonstration examples have been published so far. Techniques and trends are therefore only briefly summarized in this section with a focus on POF interferometers, POF Bragg gratings (POFBGs), and POF LPGs.

POF Interferometers

As mentioned previously, the unavailability of low-loss SM POFs has been an obstacle and made it impossible to benefit from the mechanical and optical advantages of polymer fibers and to make full use of existing high-resolution interferometric measurement techniques. Still, some research has been done in this field: SM PMMA POFs have, for example, been characterized in a Mach-Zehnder interferometer setup for strains up to 15.8% (Kiesel et al. 2008, 2009). It should be noted that photoelastic nonlinearities cannot be neglected and the sensor must be calibrated for high-strain measurement. The low Young's modulus and the low acoustic impedance of POFs have also been utilized to significantly improve the sensitivity for ultrasound and acoustic sensing applications in a Mach-Zehnder interferometer by more than one order of magnitude in comparison to silica fibers (Gallego and Lamela 2009).

More recently, differential mode interference has been investigated in single-mode-multimode-single-mode (SMS) configuration sensors. Here, the interference of different guided modes in a PF GI POF section between two silica SM fibers has been used for high-sensitivity strain and temperature characterization (Numata et al. 2014). Strain sensitivities are about -112 pm/$\mu\varepsilon$/m and temperature sensitivities about 50 nm/K/m in a 62.5 μm diameter PF POF. Compared to silica GI SMS structures, this is about one order of magnitude higher for strain and more than three orders of magnitude higher for temperature. Sensing applications using this approach remain to be demonstrated.

A wider use of coherent (not grating-based) techniques in single-mode POF for structural applications can currently not be observed.

POF Bragg Gratings

Gratings in POFs are a relatively recent development but have experienced a more dynamic evolution with significant improvement of stability during the last few years. Again, the motivations for using POF-based Bragg gratings (POFBGs) are the extended strain range, the low Young's modulus as well as the susceptibility of most polymers to water for humidity sensing applications. POFBGs are predominantly based on single-mode operation in solid-core POF or mPOF with a relatively high attenuation in the scale of 1 dB/m or more. Details on these fiber types, technologies, and effects are presented in chapter ▶ "Microstructured Polymer Optical Fiber Gratings and Sensors" by Woyessa and Marcos and chapter ▶ "Solid Core Single-Mode Polymer Fiber Gratings and Sensors" by Bhowmik. It has to be mentioned that the interconnection of SM POF is considerably more challenging than connecting standard MM POF (Peters 2011).

The inscription of refractive index perturbations into the POF core is either achieved by means of photosensitivity using phase masks or, a more recent development, triggered by nonlinear absorption by focused femtosecond laser pulses. The flexibility and excellent focusing capabilities of the femtosecond inscription approach also allowed for inscription of gratings into low-loss MM PF GI POFs using a point-by-point and line-by-line inscription technique. Such gratings have, for example, been used to demonstrate mode-shape measurement of a vibrating beam (Theodosiou et al. 2016). Multimode PF POFBGs have the advantage of lower attenuation and more straightforward fiber coupling.

Most POFBGs, however, are fabricated by means of photosensitivity. Grating stability has been an issue from the beginning of research but significant progress in recent years on inscription parameters and annealing processes enabled the production of relatively long-term stable gratings. Strain measurement is an obvious application for POFBGs due to the extended strain measurement range, but polymer materials also exhibit cross-sensitivities to temperature and humidity. Humidity-insensitive polymers such as TOPAS® have been developed and demonstrated for humidity-insensitive sensing in microstructured POFBGs (mPOFBGs) (Yuan et al. 2011). The humidity sensitivity remains for most polymers and has recently been used in PMMA mPOFBGs for sensitive and stable humidity sensing (Woyessa et al. 2016); see Fig. 10. High stability has been achieved by annealing the fiber at high humidity and high temperature. This annealing process also improves the RH sensitivity.

POFBGs may find use in humidity or moisture measurement for civil engineering applications. The response time of POF RH sensors is determined by the sorption and desorption kinetics of the moisture into the fiber core and very much depends on the fiber diameter. Response times can range from hours to a few minutes (Zhang et al. 2012b) or even seconds (Rajan et al. 2013b) for very small-diameter fibers. POFBGs may even be used to measure the saline concentration in water using the osmotic effect (Zhang et al. 2012a).

Additionally down-etching the diameter of POFBGs of the already flexible polymers has the advantage of further increasing the sensitivity to external forces. A force sensitivity of 680 nm/N has been realized with a 30 μm diameter POFBG (Rajan et al. 2013a), and a pressure sensor with a sensitivity of 1.32 pm/Pa based on a vinyl diaphragm transducer has been demonstrated (Rajan et al. 2013a).

In summary, POFBGs are still at research level but generally have the potential for high-strain sensing, high-sensitivity force sensing, and humidity sensing for engineering applications once the stability issues are solved. The strain sensitivity

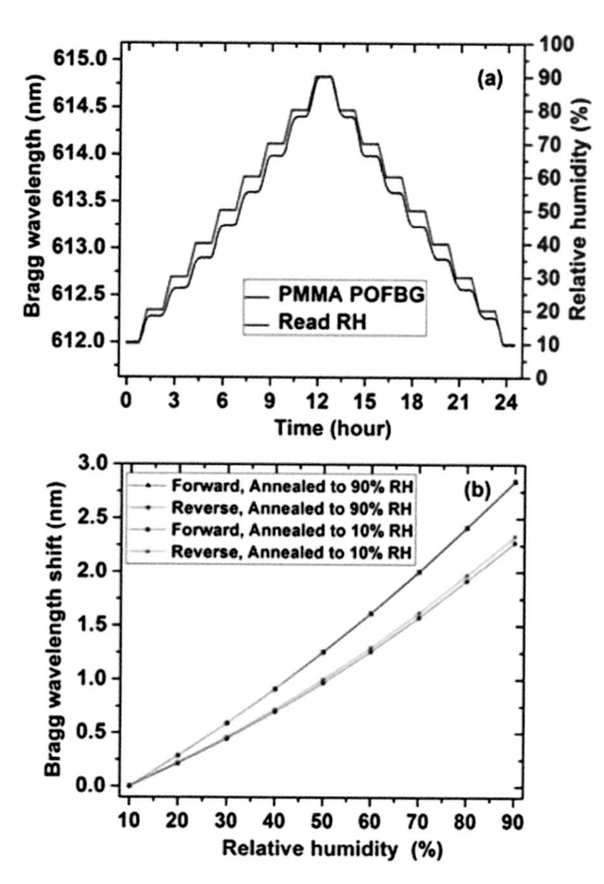

Fig. 10 Wavelength shift during humidity cycle (**a**) and wavelength shift versus humidity (**b**) for PMMA mPOFBGs with differing annealing history (Woyessa et al. 2016) (Reproduced with permission from OSA)

of POFBGs typically ranges from 1.1 to 1.5 pm/$\mu\varepsilon$. FBGs in low-loss MM PF POFs considerably extend the measurement distance in a single fiber, but the need for a low-loss SM POFs remains.

POF Long-Period Gratings

POF LPGs exhibit minima in their transmission spectra at wavelengths that are characterized by periodic variations along a fiber section. These variations, for example, due to mechanical perturbations, cause coupling between core and cladding modes of the fiber. The grating period is generally much larger than that of FBGs, which makes it possible to use various simple fabrication techniques other than photo inscription. POF LPGs can, for example, be realized by mechanical imprinting and applying mechanical stress onto a heated fiber. The core material does not have to be specifically doped to achieve photosensitivity, as it may be the case for FBGs. Most POF LPGs are based on microstructured fibers. Figure 11 shows a typical transmission spectrum of a mPOF LPG inscribed by heat imprinting.

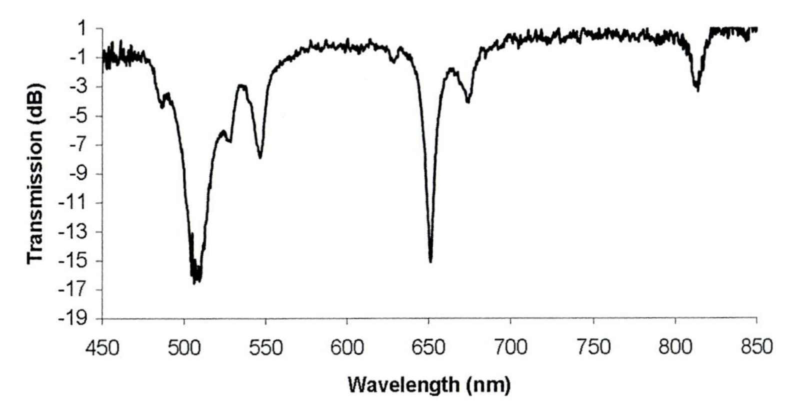

Fig. 11 Transmission spectrum of a mPOF LPG inscribed by heat imprinting of a 1 mm period grating (Hiscocks et al. 2006) (Reproduced with permission from OSA)

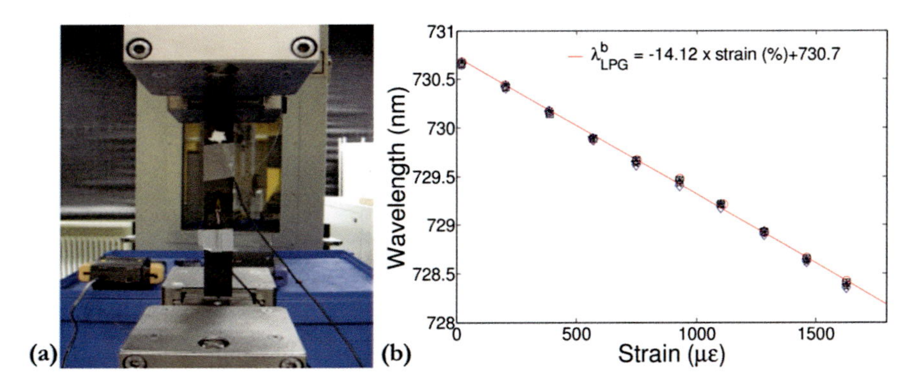

Fig. 12 Carbon fiber reinforced polymer laminated with integrated mPOF LPG in strain testing machine (**a**) and measured wavelength shift with strain during four quasi-static strain cycles (**b**) (Reproduced from de Oliveira et al. (2010))

Few structural sensing applications have been demonstrated. One example of a PMMA mPOF LPG that has been integrated into a unidirectional carbon fiber reinforced polymer (CFRP) is shown in Fig. 12a. Quasi-static tensile tests of the specimen showed good reproducibility and a linear strain response, Fig. 12b.

The strain sensitivity of the tested mPOF LPGs is in the order of 1–1.5 pm/$\mu\varepsilon$. POF LPGs are especially sensitive to humidity changes since the refractive index of the cladding has a large impact on the signal drift. As it is the case for POFBGs, the choice of polymer materials and dopants as well as annealing of the fiber can reduce this impact. In summary, POF LPGs exhibit strain sensitivities comparable to POFBGs, are relatively easy to produce, and have the potential for high-strain measurement. They are, however, currently not widely used for structural and civil engineering applications.

Time-of-Flight Measurement Techniques

Time-of-flight measurement techniques utilize the change of the light's runtime along an optical fiber to measure length changes. Interferometric length change measurement can currently not be used over longer distances due to the unavailability of low-loss SM POFs. However, interferometric length change resolution is not necessarily required for a wide range of applications, especially for measurement tasks where POFs are chosen because of their high-strain sensing range. The achievable strain resolution of incoherent detection in the μm-range is generally sufficient. Incoherent techniques also have the advantage that MM fibers can be deployed and polarization is not an issue.

Various incoherent sensing principles have been demonstrated for integral length change measurement of a fiber or sections of a fiber in transmission or reflection. The highest resolution is achieved by measuring the phase delay of a harmonic power-modulated optical signal. The most common basic principle of this power-modulated phase measurement technique is shown in Fig. 13. A voltage-controlled oscillator (VCO) provides a harmonic electrical signal with the frequency f_m for the output power modulation of a transmitter. The sinusoidal modulated optical signal is then split into the measurement path and a reference path (both POFs). The signals in both paths experience a phase delay and are detected by the photo receivers. A phase comparator transfers the phase difference $\Delta\varphi$ of the two photo receiver output signals into a voltage signal which is, after analogue-to-digital conversion (ADC), converted into the length change of the measurement path.

The phase resolution is scaled proportionally to f_m, whereas the unambiguous length change measurement range is inversely proportional to f_m. This technique is typically used in transmission and allows for dynamic long-gauge measurement with μm resolution. The use of low-cost transmitters and detectors makes this technique more cost-efficient than a fully distributed sensor system. Multiplexing of various channels is, however, not possible. Single sensor sections are typically installed in loop configuration and are bonded onto the section of interest.

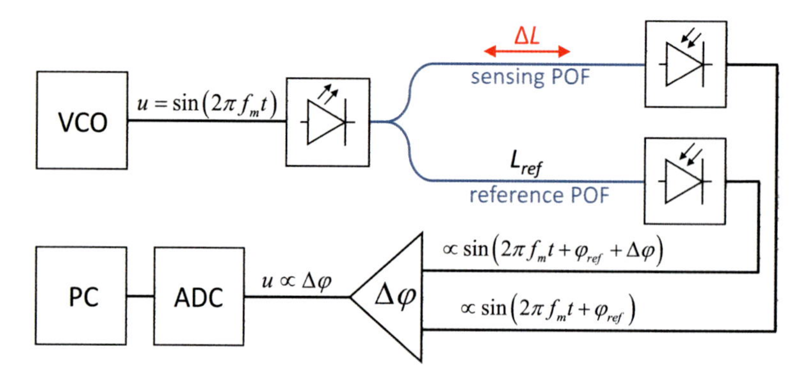

Fig. 13 Basic principle of a harmonic power-modulated time-of-flight measurement technique

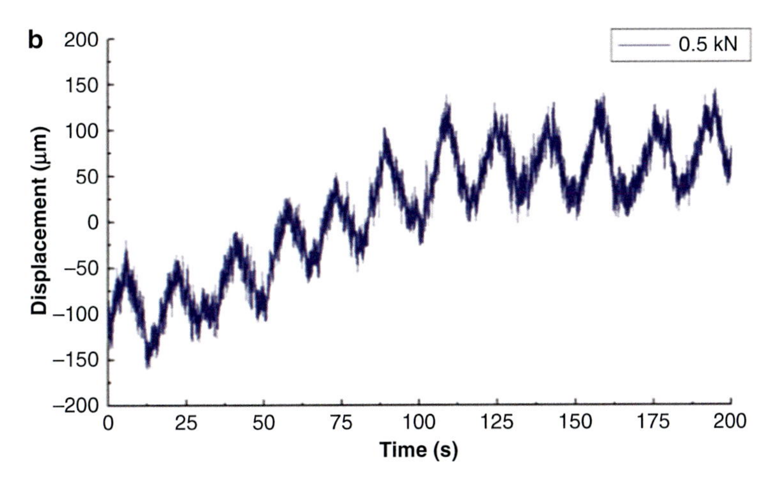

Fig. 14 Application of helicopter tail boom deflection measurement with a POF bonded onto the structure (**a**) and measured POF elongation during load cycles (**b**), CC BY Iker García (García et al. 2015)

The sensor measures the length change of the entire measurement path and the reference path is often used for temperature compensation.

This technique has been demonstrated and used for various structural applications. One example is the measurement of the deflection of an helicopter tail boom, as depicted in Fig. 14a (García et al. 2015).

A SI PMMA POF has been bonded onto the structure, and the integrated length change has been measured during flexural-tape loading conditions as shown in Fig. 14b. Other applications involve, for example, the dynamic deflection measurement of rotor blades, wind turbine blades, and the cumulative length change monitoring of cracks in a tunnel. Figure 15a, b shows a sensor fiber that has been

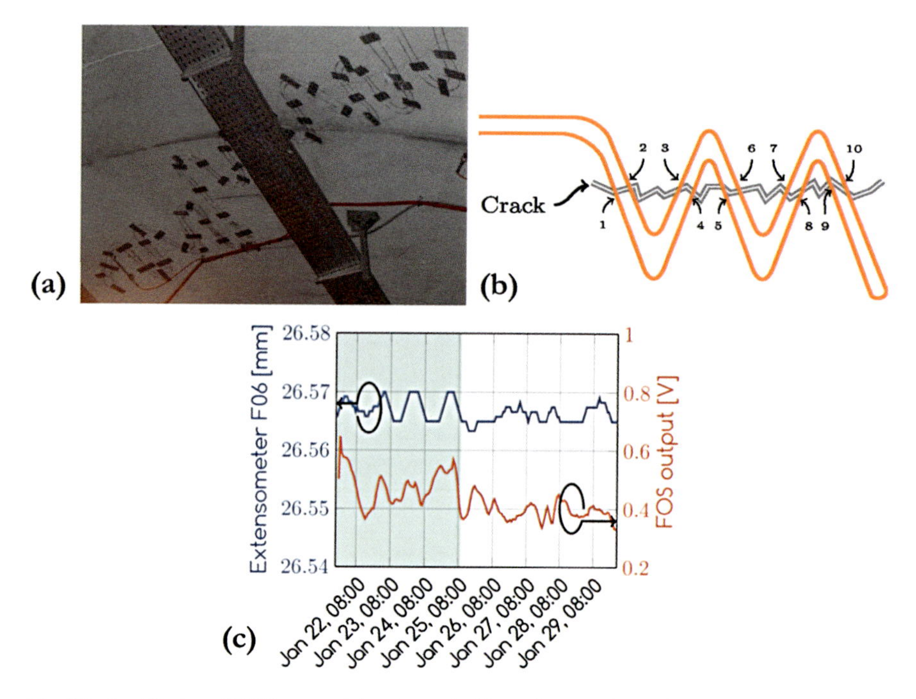

Fig. 15 Installation of POF sensor (**a**); schematic (**b**); and comparison of POF sensor results with extensometer measurement (**c**) (Reproduced from Schenato et al. (2016))

installed in meandering shape crossing a crack in a tunnel ceiling several times. Figure 15c shows the sensor response over a period of several days in comparison to an extensometer measurement.

Alternative techniques for the measurement of length changes in POF are based on distributed sensing using an optical time-domain reflectometer (OTDR) or incoherent optical frequency domain reflectometry (I-OFDR). The distance-resolved detection allows for relative position change sensing of reflective events along a fiber. The necessity of distributed backscatter devices considerably increases the instrumentation costs, but multiple sensor sections can be multiplexed (segmented along a single fiber or multiplexed in several fibers). The measurement repetition rate commonly allows only for static or low dynamic sensing with measurement resolutions into the sub-mm range.

Dynamic and quasi-distributed length change measurement between reflection points in an optical fiber has been demonstrated using the I-OFDR technique (Liehr and Krebber 2012). This approach allows for measurement repetition rates into the kHz range by calculating reflection position shifts and reflected power changes from incomplete frequency response measurements of the sensor fiber (Liehr and Krebber 2012). Simultaneous length change measurement between reflection points in the fiber and reflected power change measurement between the reflections can be conducted. Any strongly reflecting event in the fiber, such as physical contact

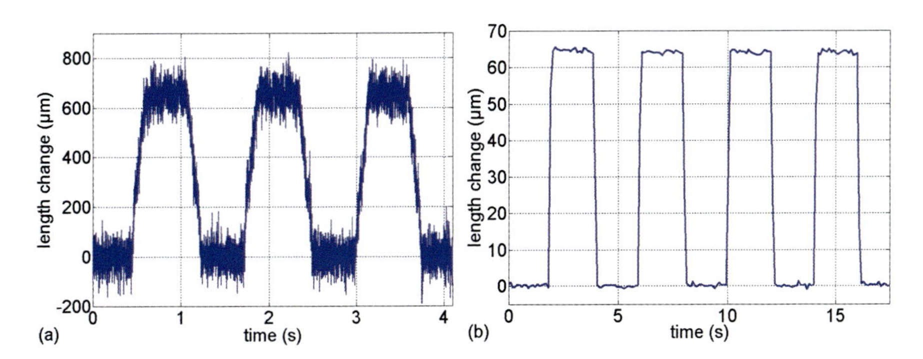

Fig. 16 High repetition rate length change measurement at 2 kHz (**a**) and high-resolution measurement at 10 Hz repetition rate in MM PF POF (**b**)

connectors or Fresnel reflections at fiber ends, can be used as reference points. SM fibers or MM fibers, including POFs, can be used and the gauge lengths can be freely chosen from centimeters to kilometers length. Depending on the chosen measurement parameters, measurement repetition rates up to 2 kHz can be achieved, or precise length change sensing with standard deviations below 1 μm can be realized in PF POFs (Liehr 2015); see Fig. 16.

This technique has been utilized to measure the dynamic deformation of a textile-reinforced two-story masonry building that has been constructed on a seismic shaking table (Liehr and Krebber 2012). Three sensor sections have been installed on the building, as indicated in Fig. 17. Two different fiber types, PF GI POF and MM silica fibers, have been used in the same setup and have been glued and screwed at different positions in pre-strained condition. Figure 17 shows the installed sensor network comprising POF and silica fiber section separated by strongly reflecting connectors.

Seismic loads of 45 s duration have been applied to the structure in direction of sensor 1 and perpendicular to the wall where sensors 2 and 3 are installed. Figure 18 shows the measured length changes and reflected power changes of all three sensor sections during a seismic load test of the already pre-damaged building.

The POF (sensor 1) measured the strongest deformation. Sensor 1 and sensor 3 show a permanent length change due to the opening of cracks in the walls. The measured optical power changes indicate oscillations of the cracked walls during the test and the associated power loss due to macrobend decoupling. The additional evaluation of power changes can provide important information for deformation and damage analysis of the structure. Any optical power change or length change sensor principle can be interrogated and multiplexed with this I-OFDR approach in transmission or reflection.

In summary, time-of-flight measurement techniques can be readily used for static and dynamic long-gauge sensing applications with length change resolutions in the μm-range. Fields of applications are primarily the dynamic deformation analysis, for example, of wind turbine blades, structural components, and civil

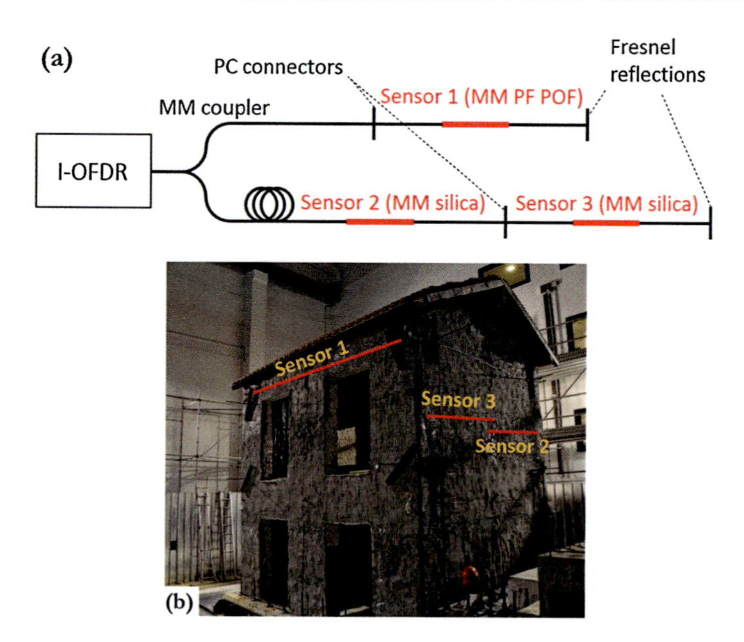

Fig. 17 Schematic of the installed fiber sections with four reflections in the fiber (**a**) and masonry building on seismic shaking table with indications of the installed sensor sections (**b**)

engineering structures such as bridges. The long gauge length of the sensor is advantageous for applications where accumulated length change is to be measured. Due to the dynamic measurement capabilities, mode-shape analysis can be conducted to detect possible eigenfrequency changes for early damage detection of a structure.

Distributed POF Sensors

Optical backscatter measurement techniques provide a spatially resolved measurement of Rayleigh backscattered power along the fiber. Incoherent techniques such as OTDR and I-OFDR are preferable for stable backscatter analysis and sensing applications in MM POFs. Integration times for OTDR and I-OFDR measurements are typically several seconds to minutes. The Rayleigh backscatter coefficient in silica fibers is practically independent of fundamental external quantities such as strain, temperature, and humidity. POFs, however, exhibit a backscattered power dependence on more or less all these quantities. The reason for this dependence is the strong inhomogeneity of the refractive index along the POF core material and its susceptibility to mechanical, thermal, or humidity impact. Effects, techniques, and sensing applications for distributed Rayleigh-based measurement of strain and humidity are presented in the following subsections, and a short outlook on distributed sensing in POFs using Brillouin scattering is given.

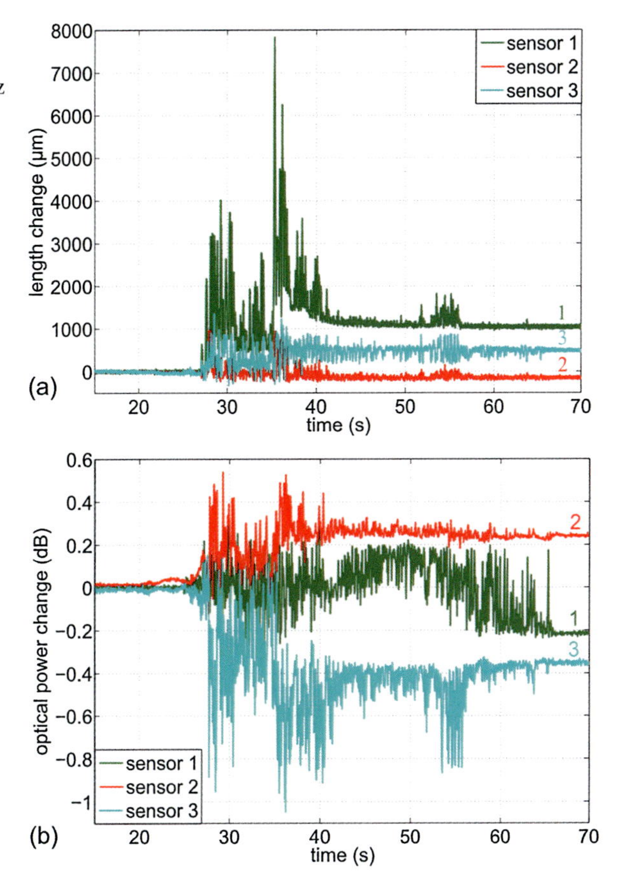

Fig. 18 Length change results of the three sensors during seismic load at 160 Hz measurement repetition rate (**a**) and optical power change results of the same measurement at 27 Hz evaluation (**b**)

Distributed Strain Sensing Using Rayleigh Scattering

The Rayleigh backscattered power level in PMMA POFs exhibits a significant dependency on strain. This effect has first been demonstrated in 2004 (Husdi et al. 2004), has later been characterized in detail (Lenke et al. 2007; Liehr et al. 2008), and has been utilized for distributed strain sensing applications (Liehr et al. 2009b). SI PMMA POFs can, depending on the rate of the strain increase and the temperature, be stretched up to more than 45%. The relatively strong modal dispersion of this fiber type decreases the spatial resolution with increasing sensor length. Large-core GI PMMA fibers would prevent this loss of spatial information but the high brittleness of this fiber type imposes limitations on high-strain sensing applications.

Photon counting OTDRs are commonly used for distributed detection of strained fiber sections. Figure 19a shows OTDR traces of a fiber with a local backscatter increase at a strained section and the backscatter-strain characteristic (integrated backscatter increase relative to a reference measurement) up to 45% strain, Fig. 19b.

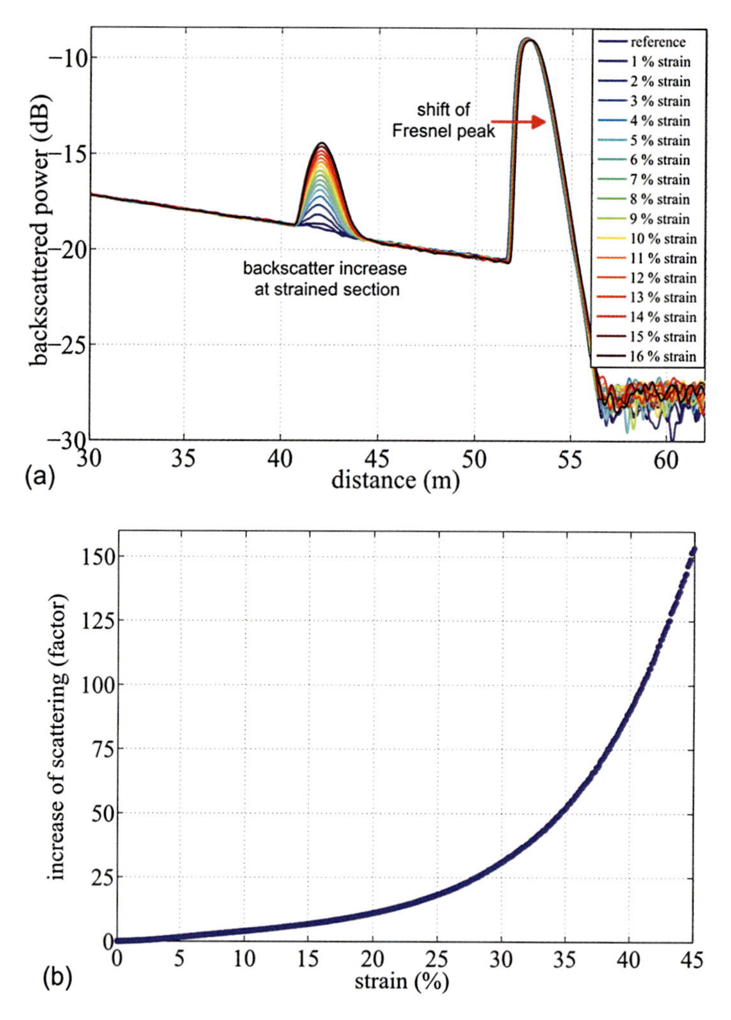

Fig. 19 OTDR traces of a 1.4-m-long fiber section at 42 m distance strained from 0% to 16% (**a**) and sensor characteristic (change of backscattered power with strain) up to 45% strain (**b**)

The shift of the Fresnel reflection peak at the fiber end can also be used for integral length change measurement of the entire fiber.

Also, more rugged versions of the 1 mm PMMA POF with a protective 2.2 mm PE jacketing can be used for applications exceeding 40% strain. The sensor characteristic is reproducible for the same strain rate, and the loss due to strain is negligible for strain up to 15% and less than 0.4 dB at 40% strain (Liehr et al. 2008). Under laboratory conditions, strain events can be measured with an accuracy of 0.1% (Liehr et al. 2008). However, relaxation in the fiber core leads to a decrease of the backscatter level over time. This behavior considerably decreases the accuracy of a strain measurement from backscatter change evaluation when the temporal history of the strain exposure is unknown. The actual strain distribution along the fiber with an unknown temporal strain history can then only be estimated. For a wide range of high-strain measurement applications, however, the knowledge of the exact strain value is not critical.

Other cross-sensitivities in PMMA POFs are humidity and temperature. Both quantities have a considerably lower impact on the backscatter level than strain: temperature-induced Rayleigh scattering change is relatively linear and corresponds to about 0.2% strain signal per 10 K temperature change. The relative humidity (RH) impact and its applications for distributed humidity sensing are summarized in section "Distributed Humidity Sensing" and correspond to about 0.2% strain signal per 12% RH change.

The greatest potential for distributed POF sensors lies in the civil engineering sector. Most relevant are applications where very high-strain values can be expected, for example, in the geotechnical field for the deformation measurement of earthwork structures or crack detection in concrete and masonry structures. These applications have been targeted by means of integrating POF sensors into geotextiles and architectural fabrics for health monitoring and damage detection purposes in various fields. The integration of optical fibers into technical textiles as an additional fiber thread has various advantages: the sensor-equipped textiles do not only fulfill their designated tasks, for example, as reinforcement element, but also protect the sensor fiber, facilitate easy installation in the area of interest, and guarantee optimal strain transfer to the sensor fiber. The integration of POF sensors into geotextiles and architectural fabrics has been achieved in industrial process scale. Such a sensor-integrated "smart textile" with POF and silica fibers is shown in Fig. 20a. Several laboratory and field tests with grid-like textiles, ropes, and fleece textiles have been demonstrated, for example, to detect the local settlement of soil (Liehr et al. 2008, 2009b). A long-term field test with a textile-integrated POF strain sensor has been conducted in an open brown coal pit: A 10-m-long geogrid with an integrated PE-jacketed PMMA POF has been installed at the top end of a creeping slope bridging the tear-off edge of the slope perpendicular to the sensor fiber (Liehr et al. 2009a); see Fig. 20b.

The sensor textile has been covered with a 10 cm layer of sand, and OTDR measurements have been conducted at irregular intervals. Scattering change results of the first three measurements in Fig. 21a indicate a steady increase of strain over a length of 7 m when the textile is stretched. The last two measurements exhibit a very strong local backscatter increase due to the rupture of the textile. The backscatter change indicates local strain of more than 20%, far more than silica fibers could endure.

The total length change of the sensor has been obtained by evaluating the shift of the fiber end reflection peak, as depicted in Fig. 19a. A relatively constant creep velocity with an average rate of about 2.5 mm per day has been determined. This field test demonstrates that extreme deformation can be localized and monitored with distributed POF sensors over an extended period of time. The distance range of PMMA-based POFs is attenuation-limited to about 100–200 m, but multiplexing

Fig. 20 Technical textiles with integrated POF and silica fibers (**a**) and installation of geogrid with integrated POF sensor bridging a cleft of a creeping slope at an open coal pit (**b**)

of multiple POF sensors can be implemented to cover a wider area with a single OTDR unit.

Crack detection and localization is another application that can be accomplished with distributed Rayleigh scattering analysis in POFs. The necessity for crack detection in concrete and masonry structures is motivated by the need for structural integrity assessment, for example, in structures or locations that cannot be visibly inspected. For the purpose of earthquake protection and the monitoring of historical buildings, technical textiles with integrated POFs have been developed. The sensor textiles have been applied to structures for reinforcement purposes to increase the ductility during seismic loads and enable distributed strain sensing. The extremely high-strain values at the locations of occurring cracks can be reliably detected by distributed backscatter analysis. A series of model tests with various grid-like textiles with integrated SI PMMA POF sensors bonded with mortar to the bottom side of two-stone samples have been conducted (Liehr et al. 2009b). The specimens were placed in a three-point bend setup, and a stepwise crack opening perpendicular to the sensor fiber has been provoked; see Fig. 22.

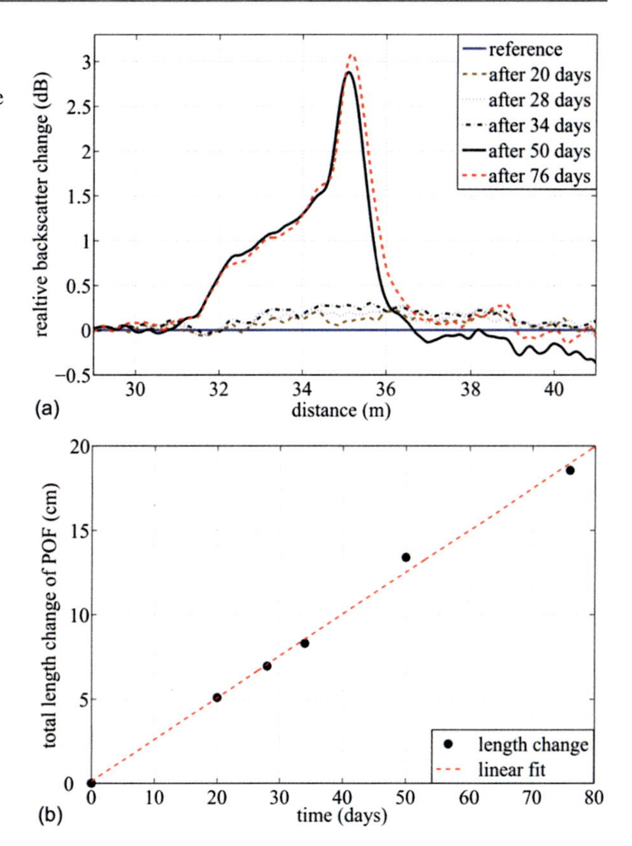

Fig. 21 Change of backscattering relative to a reference measurement at the location of the cleft (**a**) and total elongation of the POF derived from the fiber end reflection shift (**b**)

Crack openings below 1 mm could be detected from relative backscatter change analysis and similar sensor characteristics for different textile types have been demonstrated. The backscatter response to crack growth very much depends on how the fiber is bonded to the structure. Direct bonding without jacketing materials causes short gauge lengths which will result in a higher backscatter signal due to the nonlinear strain-backscatter dependence (Fig. 19b) but will rupture and induce loss at smaller crack widths. A trade-off between high-sensitivity crack detection and high crack width endurance is to be found, and optimized fiber integration techniques are to be chosen depending on the application.

Destructive large-scale test with SI PMMA POFs with PE jacketing, directly bonded onto masonry buildings on seismic shaking tables have been conducted (Liehr et al. 2009c). Figure 23 shows a one-story masonry building with indications where the POF sensors have been installed.

Strong seismic shocks resulted in structural damage and the opening of cracks that have been clearly detected and localized at the sensor positions at 27 m and 28.5 m.

Another example for local backscatter change detection, but at predetermined locations along a SI PMMA POF, has been demonstrated for deformation

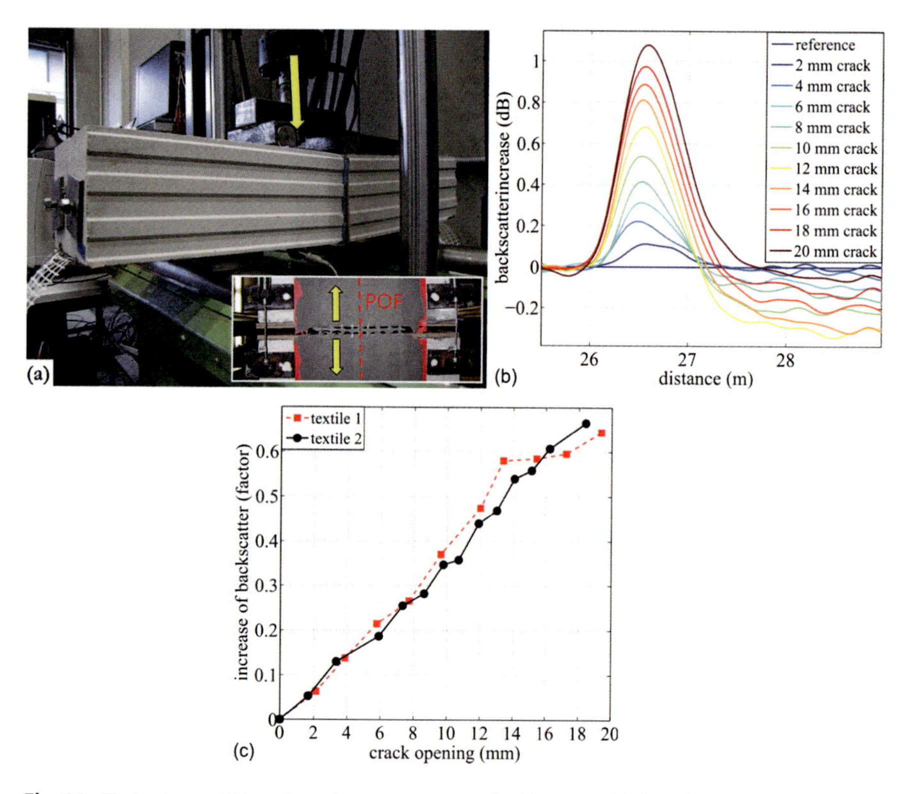

Fig. 22 Test setup and forced crack opening perpendicular to the POF at the bottom of the two-stone sample (**a**); relative change of backscatter signal for increasing crack width (**b**); and sensor characteristic (integrated backscatter increase vs. crack width) for two different textile types (**c**)

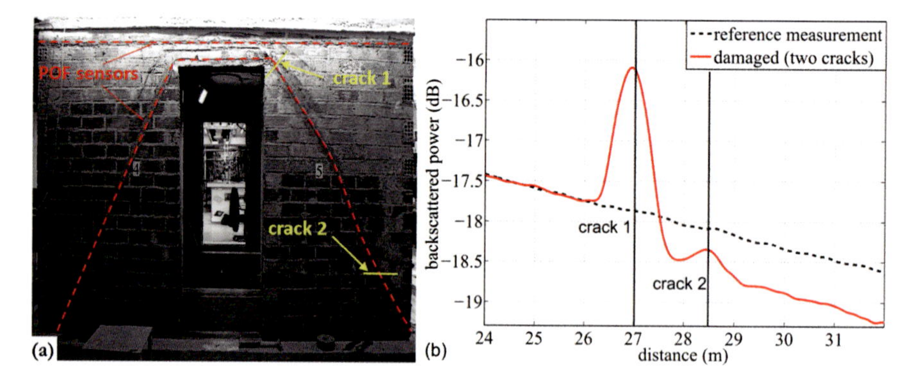

Fig. 23 Masonry building with SI PMMA POFs directly bonded as indicated (**a**) and OTDR backscatter traces showing the positions of the cracks at 27.0 m and 28.5 m (**b**)

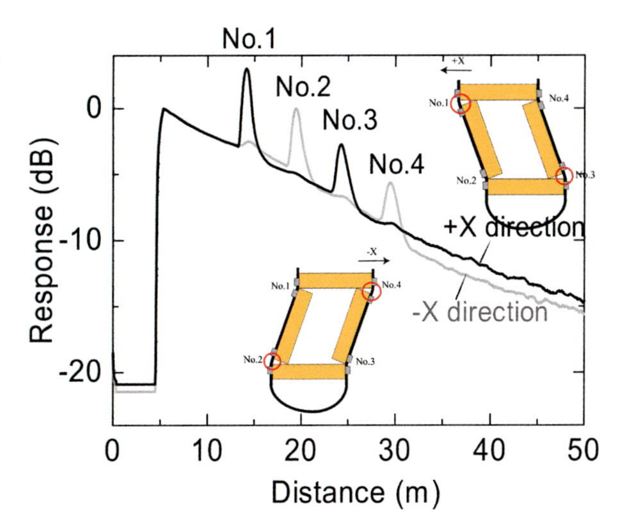

Fig. 24 OTDR traces during deformations of a rectangular frame in different directions (Fukumoto et al. 2008) (Reprinted with permission from SPIE "A POF-based distributed strain sensor for detecting deformation of wooden structures" by Fukumoto et al. (2008))

measurement of a wooden structure (Fukumoto et al. 2008). Short gauge lengths at the joints of the corners of a rectangular frame have been chosen for strain evaluation by locally bonding sections of the fiber. The direction and magnitude of the frame deformation could be determined from the backscatter increases of the single sensor sections; see Fig. 24. This quasi-distributed sensing approach requires a distributed backscatter measurement but has the potential to multiplex a large number of single sensors.

Also, various of the optical power change principles presented in section "Optical Power Change Sensors" may be analogously multiplexed to measure Rayleigh scattering loss between the single sensor points or sensor sections. These loss-based techniques, however, cause a rapid deterioration of the dynamic range budget of the OTDR. The number of power loss sensors in a single sensor is therefore considerably reduced, which limits the commercial applicability of OTDR-based power loss sensors. Other distributed techniques that are based on backscatter change evaluation rather than on power loss measurement can be efficiently multiplexed over longer distance. Quasi-distributed and intrinsic fiber curvature sensing has, for example, been demonstrated by analyzing backscatter changes caused by mode power distribution analysis and the interaction at off-center femtosecond laser pulse-inscribed scattering damage in SI PMMA POFs (Liehr et al. 2013).

The main advantages of PF GI POFs over SI PMMA POFs are their lower attenuation and negligible modal dispersion. They have the potential for high spatial resolution sensing over distances exceeding 500 m. However, the strong backscatter increase effect that has been observed in PMMA POFs is considerably weaker in PF GI POFs, has a nonlinear characteristic (Liehr et al. 2010b), and also exhibits relaxation (Liehr 2015). Nevertheless, this fiber type has the potential for distributed sensing applications. The incoherent optical frequency domain reflectometry (I-OFDR) has been advanced and adapted for distributed sensing applications in PF GI POFs (Liehr et al. 2010a, b). The spatial resolution, measurement speed, and backscatter signal stability have been considerably improved using the I-OFDR approach compared to photon-counting OTDRs. The high resolution and signal stability make it possible to measure the displacement of the typical randomly distributed scattering centers in PF GI POFs. These strong scattering events are mainly caused by particle contamination and the introduction of inhomogeneities during the fiber production process. Figure 25a shows several I-OFDR traces of a fiber section between 27 and 28 m that has been stepwise strained. The strained section with increased scattering can be located with cm-spatial resolution. The randomly distributed scattering centers along the fiber can be resolved with high resolution and experience a spatial shift at distances exceeding 27 m due to the strain of the fiber. Since the characteristic backscatter profile along the POF is permanent, this shift, or length change, along the fiber can be calculated by applying a cross-correlation algorithm to overlapping sections of a reference measurement and a new measurement section (Liehr et al. 2010b). By shifting this "correlation window" of a certain correlation length along the fiber, local displacements of the scattering centers due to strain can be calculated and evaluated as length change along the fiber. The length change resolution depends on the spatial resolution and the chosen correlation length. Figure 25b shows the distributed length change results of the same fiber for a 1-m-long section that has been strained in steps of 5 mm up to 35 mm. The correlation has been conducted with a correlation length of 2 m.

The measurement resolution can be improved by choosing longer correlation lengths or increasing the measurement time. Since the characteristic backscatter profile of the PF POF is permanent, the correlation technique is insensitive to optical loss occurring along the fiber. The combined evaluation of backscatter change evaluation and cross-correlation analysis makes it possible to localize strained fiber sections with cm-resolution and to measure length changes along the fiber with resolutions better than 1 mm. The measurement of length change distributions along the sensor is actually the more interesting parameter for a wide range of applications. Promising fields are high-strain and high-displacement measurement tasks where absolute displacement information is required, for example, in the geotechnical sector as well as for crack detection in the civil engineering sector. Sensor lengths up to a few hundred meters can be covered with PF GI POFs without degradation of the spatial resolution due to the absence of modal dispersion.

In summary, distributed strain sensing can be conducted in PMMA POFs as well as GI PF POFs. SI PMMA fibers exhibit a strong backscatter increase response to strain and can be used for high-strain sensing applications, but with reduced spatial resolution for longer distances due to the modal dispersion. The strain accuracy is limited due to the temporal relaxation of the backscatter signal. PF GI POFs can be deployed for extended distances up to several hundred meters with constantly high spatial resolution. The use of high-resolution I-OFDR in combination with the backscatter correlation analysis approach makes it possible to measure length changes along PF GI POFs and to localize strained fiber sections.

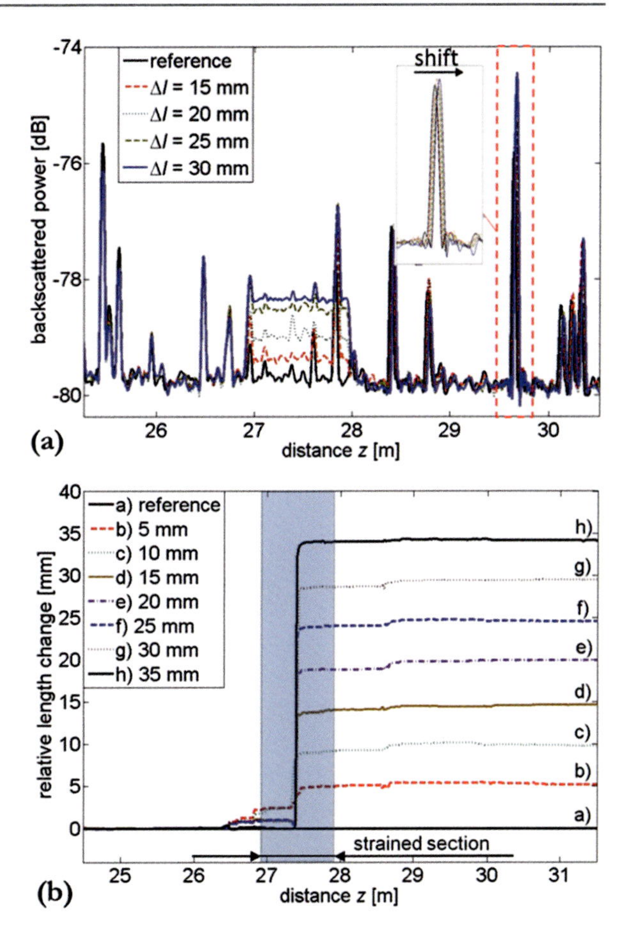

Fig. 25 I-OFDR measurements of stepwise strained 1-m-long PF POF section (**a**) and distributed length change results of the same fiber, strained in steps of 5 mm up to 35 mm, and a correlation length of 2 m (**b**)

Distributed Humidity Sensing

Strain has by far the highest impact on the backscatter level of PMMA POFs. Humidity changes exhibit, in comparison, a considerably lower impact on the Rayleigh backscatter level. However, this sensitivity is significant enough to conduct distributed humidity sensing along PMMA POFs. Depending on the chosen measurement wavelength, is the backscatter trace superimposed by additional attenuation along the fiber due to the OH vibrational absorption spectra (Liehr et al. 2017). The spectral absorption dependency is shown in Fig. 2. Figure 26a depicts the measurement setup that has been used for relative humidity (RH) characterization of SI PMMA POFs in a climate chamber. Two fiber sections (21–37 m and 58–69 m) are held at saturated RH values in water, and the rest of the fiber is subjected to RH changes. OTDR measurements have been conducted at 500 nm and 650 nm wavelength. Relative backscatter level evaluation in Fig. 26b shows a RH dependency at both wavelengths. The attenuation dependency, however,

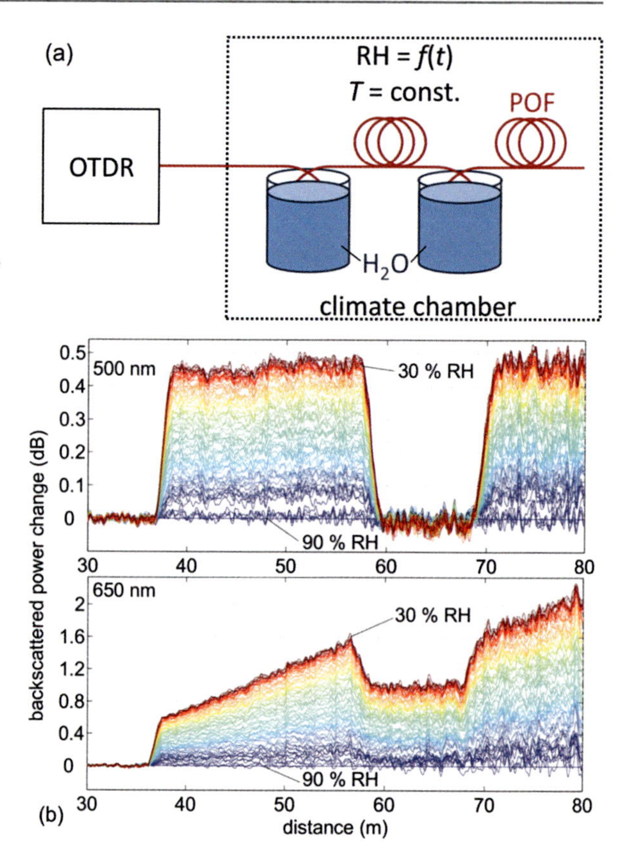

Fig. 26 Measurement setup with the fiber sections from 21–37 to 58–69 m in water (**a**) and relative backscatter changes during RH changes from 90% to 30% at 500 nm and 650 nm wavelengths relative to 90% RH (**b**), CC BY Sascha Liehr (Liehr et al. 2017)

exhibits negligible OH absorption at 500 nm, whereas 650 nm results exhibit an attenuation increase with increasing RH.

These dependencies can be used to conduct distributed humidity sensing along the fiber. Figure 27 shows the measured backscatter level changes as well as transmission changes due to absorption relative to the saturated fiber section for 500 nm and 650 nm OTDR wavelengths.

The humidity response characteristics are reproducible and exhibit a response time of about 31 h to reach 90% of the saturated humidity value in a 1 mm SI PMMA POF (Liehr et al. 2017). As discussed in section "POF Bragg Gratings," the response time can be considerably decreased by using fibers with smaller diameters. The combination of the differing RH impacts on absorption and scattering at 500 nm and 650 nm wavelengths, respectively, makes it possible to unambiguously distinguish RH impact from other backscatter sensitivities such as to strain. The scattering and attenuation dependencies along the fiber can be used for distributed RH measurement but also for moisture measurement in materials that are relevant in civil engineering such as soil, concrete, or screed. Distributed measurement of moisture can, for example, be used to localize and measure water ingress, leakage,

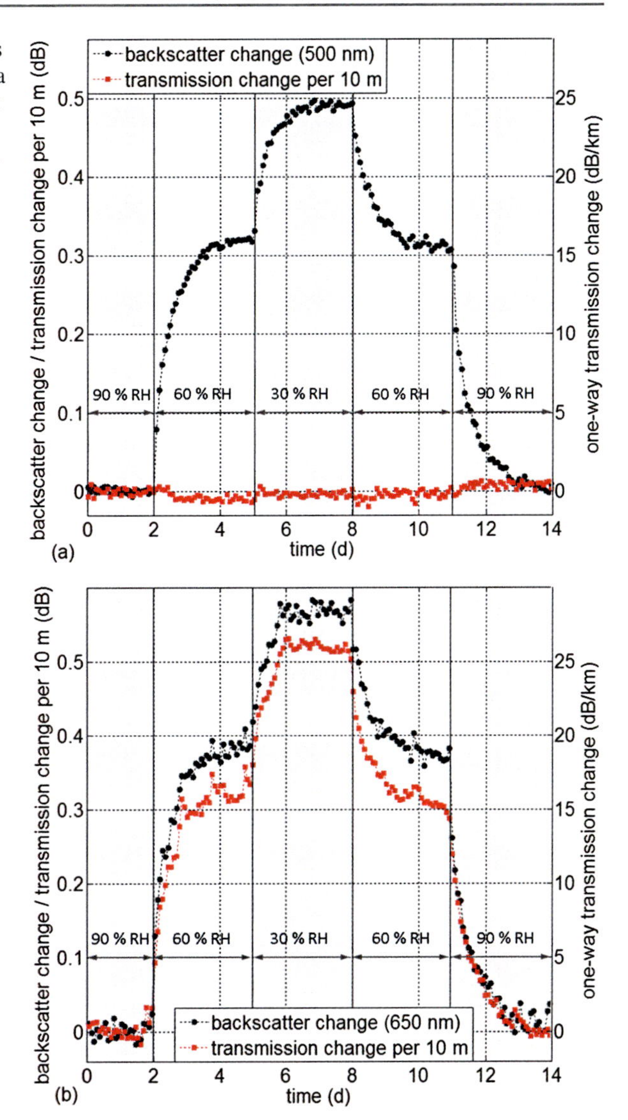

Fig. 27 Backscatter changes and transmission changes in a 1 mm SI PMMA POF during stepwise RH changes for OTDR emission wavelengths of 500 nm (**a**) and 650 nm (**b**), CC BY Sascha Liehr (Liehr et al. 2017)

or moisture penetration in building foundations, dykes, embankments, or landfills. Moreover, it has the potential to be used for the monitoring of moisture content in concrete or creed to optimize the setting process and construction planning.

Distributed Brillouin Sensing

Brillouin scattering is an inelastic scattering effect that causes a frequency shift of the scattered light. This Brillouin frequency shift depends on the fiber's strain and temperature and is one of the most widely used effects for strain and temperature measurement along silica optical fibers. The effect becomes only significant for high electric field strengths when stimulated Brillouin scattering occurs. The verification of Brillouin scattering in POFs has therefore been more challenging due to the higher attenuation of POFs, the multimode fiber core and, related with that, the increased effective area as well as the strong mode coupling. Stimulated Brillouin scattering and temperature- and strain-dependent Brillouin frequency shifts have eventually been characterized for MM PF GI POFs. The temperature dependence is about three times higher compared to silica fibers, and the strain coefficient is about one fifth of standard silica SM fibers (Hayashi et al. 2014). Distributed Brillouin sensing in PF GI POF has been demonstrated for short fiber sections using Brillouin optical frequency domain analysis (Minardo et al. 2014), Brillouin optical correlation domain reflectometry (Hayashi et al. 2014) and Brillouin optical time-domain analysis (Dong et al. 2014). However, the applicability and true potential of distributed Brillouin-based sensing in POFs for structural monitoring applications remains to be demonstrated.

Conclusion

Although POF sensors do not feature as long sensor lengths as their silica counterparts and they cannot yet be widely used for elaborated interferometric and nonlinear scattering measurement principles, they have qualified for niche applications due to their extraordinary properties. The greatest advantages of POF-based sensors are their extreme strain measurement range; their large core diameters and low Young's modulus; their ruggedness, ease of handling, and interconnection; their low cost; as well as their specific Rayleigh scattering dependencies. Limitations are their restricted temperature range and the relatively strong cross-sensitivities to other extrinsic parameters. As it is the case for most sensor principles, the long-term stability can be an issue and must be taken into account for specific sensing applications.

Extrinsic optical loss-based sensors can, for instance, be used for low-cost, precise, and dynamic measurement of strain, acceleration, and deformation in transmission or reflection. Accumulated loss measurement is a cost-effective option for early damage detection along the installed sensor fiber. Time-of-flight measurement techniques have proven their potential for precise and dynamic monitoring of extended structures and structural components but also for static measurement applications. Distributed sensing approaches can be used for high-strain and length change measurement up to hundreds of meters. Also, spatially resolved humidity sensing can be conducted with PMMA POFs. The recent advances in performance and stability improvement of grating-based POF sensors improved the prospects for precise high-strain sensing applications and humidity measurement.

The dynamic improvement of POF properties, polymer materials as well as the performance enhancement of interrogation techniques can be expected to further extend the scope of applications toward higher resolution and longer distance range. Various POF sensor systems, mostly loss-based techniques and time-of-flight sensors, became rather recently commercially available and are being employed for structural monitoring applications. The field of commercial monitoring applications can be expected to widen as distributed techniques are maturing and if low-loss POFs for interferometric and spectrometric sensing can be produced.

References

P.S. André, H. Varum, P. Antunes, L. Ferreira, M.G. Sousa, Measurement **45**, 556 (2012)
G. Beadie, M. Brindza, R.A. Flynn, A. Rosenberg, J.S. Shirk, Appl. Opt. **54**, F139 (2015)
L. Bilro, N. Alberto, J.L. Pinto, R. Nogueira, Sensors **12**, 12184 (2012)
P. Costa Antunes, J. Miguel Dias, H. Varum, P. André, Sens. Rev. **34**, 36 (2014)
Y. Dong, P. Xu, H. Zhang, Z. Lu, L. Chen, X. Bao, Opt. Express **22**, 26510 (2014)
T. Fukumoto, K. Nakamura, S. Ueha, in *Proceedings of the SPIE, 19st International Conference on Optical Fiber Sensors* (2008), pp. 700469–700469-4
D. Gallego, H. Lamela, Opt. Lett. **34**, 1807 (2009)
I. García, J. Zubia, G. Durana, G. Aldabaldetreku, M.A. Illarramendi, J. Villatoro, Sensors **15**, 15494 (2015)
S. H. Goods, R. M. Watson, M. Yi, *Thermal Expansion and Hydration Behavior of PMMA Molding Materials for LIGA Applications* (Prepared by Sandia National Laboratories, Albuquerque, 2003), pp. 1–57
N. Hayashi, Y. Mizuno, K. Nakamura, J. Lightwave Technol. **32**, 3397 (2014)
M.P. Hiscocks, M.A. van Eijkelenborg, A. Argyros, M.C.J. Large, Opt. Express **14**, 4644 (2006)
I.R. Husdi, K. Nakamura, S. Ueha, Meas. Sci. Technol. **15**, 1553 (2004)
S. Kiesel, K. Peters, T. Hassan, M. Kowalsky, IEEE Photon. Technol. Lett. **20**, 416 (2008)
S. Kiesel, K. Peters, T. Hassan, M. Kowalsky, Meas. Sci. Technol. **20**, 034016 (2009)
Y. Koike, M. Asai, APG Asia Mater. **1**, 22 (2009)
K.S.C. Kuang, W.J. Cantwell, P.J. Scully, Meas. Sci. Technol. **13**, 1523 (2002)
K.S.C. Kuang, W.J.C. Akmaluddin, C. Thomas, Meas. Sci. Technol. **14**, 205 (2003)
K.S.C. Kuang, S.T. Quek, M. Maalej, Meas. Sci. Technol. **15**, 2133 (2004)
K.S.C. Kuang, S.T. Quek, M. Maalej, Sens. Actuators, A **147**, 449 (2008)
P. Lenke, K. Krebber, M. Muthig, F. Weigand, E. Thiele, in *Proceedings of the III ECCOMAS Thematic Conference on Smart Structures and Materials* (Gdansk, Poland, 2007)
S. Liehr, *Fibre Optic Sensing Techniques Based on Incoherent Optical Frequency Domain Reflectometry* (BAM Dissertationsreihe, 2015)
S. Liehr, K. Krebber, IEEE Sensors J. **12**, 237 (2012)
S. Liehr, P. Lenke, K. Krebber, M. Seeger, E. Thiele, H. Metschies, B. Gebreselassie, J.C. Münich, L. Stempniewski, in *Proceedings of the SPIE, Photonics Europe* (Strasbourg, 2008), pp. 700302.1–700302.15
S. Liehr, P. Lenke, M. Wendt, K. Krebber, R. Gloetzl, J. Schneider-Gloetzl, L. Gabino, L. Krywult, in *Proceedings of the 4th International Conference on Struct. Health Monitoring of Intelligent Infrastructure (SHMII-4)* (Empa-Akademie, Zurich, 2009a)
S. Liehr, P. Lenke, M. Wendt, K. Krebber, M. Seeger, E. Thiele, H. Metschies, B. Gebreselassie, J.C. Munich, IEEE Sensors J. **9**, 1330 (2009b)
S. Liehr, M. Wendt, J. C. Münich, L. Stempniewski, H. Metschies, in *Proceedings of the 4th International Conference on Structural Health Monitoring of Intelligent Infrastructure (SHMII-4)* (Empa-Akademie, Zurich, 2009c)
S. Liehr, N. Nöther, K. Krebber, Meas. Sci. Technol. **21**, 017001.1 (2010a)
S. Liehr, M. Wendt, K. Krebber, Meas. Sci. Technol. **21**, 094023.1 (2010b)
S. Liehr, J. Burgmeier, K. Krebber, W. Schade, J. Lightwave Technol. **31**, 1418 (2013)
S. Liehr, M. Breithaupt, K. Krebber, Sensors **17**, 735 (2017)

M. Lomer, A. Quintela, M. López-Amo, J. Zubia, J.M. López-Higuera, Meas. Sci. Technol. **18**, 2261 (2007)
C. Markos, A. Stefani, K. Nielsen, H.K. Rasmussen, W. Yuan, O. Bang, Opt. Express **21**, 4758 (2013)
E. Mesquita, T. Paixão, P. Antunes, F. Coelho, P. Ferreira, P. André, H. Varum, Sens. Actuators, A **240**, 138 (2016)
A. Minardo, R. Bernini, L. Zeni, IEEE Photon. Technol. Lett. **26**, 387 (2014)
G. Numata, N. Hayashi, M. Tabaru, Y. Mizuno, K. Nakamura, IEEE Photon. J. **6**, 1 (2014)
R. de Oliveira, M. Schukar, K. Krebber, V. Michaud, in *Proceedings of the 19th International Conference on Plastic Optical Fibers* (Yokohama, 2010)
G. Perrone, A. Vallan, IEEE Trans. Instrum. Meas. **58**, 1650 (2009)
G. Perrone, M. Olivero, A. Vallan, A. Carullo, A. Neri, in *Proceedings of the 2008 IEEE Sensors* (2008), pp. 325–328
K. Peters, Smart Mater. Struct. **20**, 013002 (2011)
G. Rajan, B. Liu, Y. Luo, E. Ambikairajah, G.D. Peng, IEEE Sensors J. **13**, 1794 (2013a)
G. Rajan, Y.M. Noor, B. Liu, E. Ambikairaja, D.J. Webb, G.-D. Peng, Sens. Actuators, A **203**, 107 (2013b)
M. Rajesh, M. Sheeba, V.P.N. Nampoori, J. Phys. Conf. Ser. **85**, 012016 (2007)
L. Schenato, G. Bossi, G. Marcato, A. Pasuto, Rend. Online Soc. Geol. Ital. **39**, 19 (2016)
M. Schukar, B. Grzemba, K. Krebber, M. Luber, in *Proceedings of the 18th International Conference on Plastic Optical Fibers* (Sydney, 2009)
N. Takeda, Int. J. Fatigue **24**, 281 (2002)
A. Theodosiou, A. Lacraz, M. Polis, K. Kalli, M. Tsangari, A. Stassis, M. Komodromos, IEEE Photon. Technol. Lett. **28**, 1509 (2016)
G. Woyessa, K. Nielsen, A. Stefani, C. Markos, O. Bang, Opt. Express **24**, 1206 (2016)
W. Yuan, L. Khan, D.J. Webb, K. Kalli, H.K. Rasmussen, A. Stefani, O. Bang, Opt. Express **19**, 19731 (2011)
W. Zhang, D. Webb, G. Peng, Opt. Lett. **37**, 1370 (2012a)
W. Zhang, D.J. Webb, G.-D. Peng, J. Lightwave Technol. **30**, 1090 (2012b)
J. Zhao, T. Bao, R. Chen, Opt. Fiber Technol. **24**, 70 (2015)
O. Ziemann, J. Krauser, P. E. Zamzow, W. Daum, *POF Handbook: Optical Short Range Transmission Systems* (Springer, 2008)
J. Zubia, J. Arrue, Opt. Fiber Technol. **7**, 101 (2001)

Part XII
Photonic Crystal Fiber Sensing

Photonic Microcells for Sensing Applications

54

Chao Wang, Wei Jin, Hoi Lut Ho, and Fan Yang

Contents

Abstract

This chapter presents hollow-core and suspended-core photonic microcells (PMCs) made from commercial photonic crystal fibers and single mode fibers (SMFs). These PMCs are in-fiber platforms for strong light-matter interaction and can be connected into standard SMF systems with low loss. The fabrication process and basic properties of the PMCs are introduced. The use of the PMC as gas sensors, liquid-filled temperature sensors, in-fiber micro-cantilever accelerometers, and grating-based sensors is presented. In combination with

C. Wang
School of Electrical Engineering, Wuhan University, Wuhan, Hubei, China
e-mail: eecwang@whu.edu.cn

W. Jin (✉) · H. L. Ho · F. Yang
Department of Electrical Engineering, The Hong Kong Polytechnic University, Hong Kong, China
e-mail: eewjin@polyu.edu.hk

G.-D. Peng (ed.), *Handbook of Optical Fibers*,
https://doi.org/10.1007/978-981-10-7087-7_23

novel functional materials, the PMCs exhibit great potentials for lab-in/on-fiber and tunable photonic devices.

Introduction

Among the many unusual properties of photonic crystal fibers (PCFs), the air columns within the PCF offer a range of new possibilities. Via these air columns, different materials such as gas (Russell 2006), liquid (De Matos et al. 2007), metal (Hou et al. 2008), semiconductor (Sparks et al. 2011), polymer (Westbrook et al. 2000), and particles (Benabid et al. 2002) could be compactly integrated into optical fibers, enabling novel devices for sensing and other applications based on light-matter interaction within the hollow columns via a constrained propagating mode or the evanescent field of a mode. The air columns can also be enlarged, narrowed, or collapsed by post thermal treatment to create a region optimized for certain special functions such as enhanced light-matter interaction and in-fiber opto-mechanical light modulation.

The term photonic microcell (PMC) was firstly used for an all-fiber structure made of a gas-filled hollow-core photonic bandgap fiber (HC-PBF) with both ends spliced to standard single-mode fibers (SMFs) (Benabid et al. 2005, 2009). This PMC inherits the advantages of HC-PBF but can be conveniently connected to standard fiber-optic systems without bulky optical components. Recently, PMCs with different structures were reported, including suspended-core PMCs (SC-PMCs) (Wang et al. 2013a) and the micro-/nanofiber (MNF) PMCs (MNF-PMCs) (Jin et al. 2014). The SC- and MNF-PMCs are evanescent field-based devices comprising of a micro-/nano-waveguiding core embedded/encapsulated within an outer capillary jacket. These PMCs exhibit strong evanescent field comparable to that of the freestanding optical MNFs, but have additional advantages of robustness, free from environmental contamination, and low-loss connection into SMF systems in which standard fiber-optic components such as couplers, filters, and modulators can be used. The all-fiber PMCs would be useful for a range of applications in sensors, lasers, amplifiers, nonlinear optics, optical storage and quantum optics, and lab-on-fiber systems (Jin et al. 2015; Jones et al. 2011; Murari et al. 2016; Finger et al. 2015; Sprague et al. 2014; Wang et al. 2016). In this section, we review different types of PMCs and their applications in sensing.

Photonic Microcells

Hollow-Core PMCs

As shown in Fig. 1a, a hollow-core PMC (HC-PMC) is formed by splicing the two ends of a hollow-core fiber (HCF) to standard optical fibers (e.g., SMFs). A range of HCFs may be used to build a HC-PMC, and Fig. 1b–d shows three examples of HCFs, namely, the HC-PBF, the Kagome, and antiresonant fibers. The HC-PBF

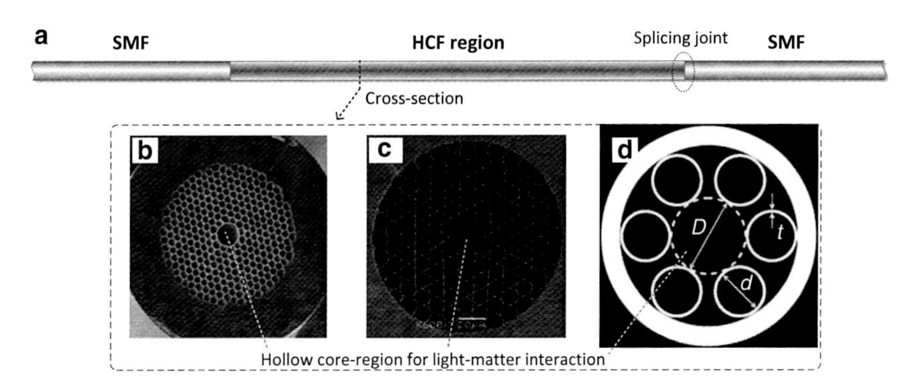

Fig. 1 (**a**) Basic structure of a HC-PMC, in which the HCF can be (**b**) a HC-PBF, (**c**) a Kagome lattice fiber (Couny et al. 2006), or (**d**) an antiresonant HCF (Uebel et al. 2016)

has a full two-dimensional photonic bandgap in its periodic cladding; light with wavelength within photonic bandgap is prohibited to travel transversely but confined to propagate in the hollow core (Cregan et al. 1999).

The HCF shown in Fig. 1b is a commercial 7-cell HC-PBF from NKT Photonics (model HC-1550-02). It supports a Gaussian-like fundamental mode with diameter comparable to that of a standard SMF (i.e., ~9 μm), which enables low-loss splicing between them. As shown in Fig. 2a, the fiber has a transmission window of ~200 nm centered around 1550 nm. The fiber attenuation is less than 30 dB/km, allowing strong light-matter interaction in the hollow-core over a long distance (e.g., tens of meters or longer (Chen et al. 2015)). There are commercial HC-PBFs with transmission bands ranging from 440 to 2000 nm, which can be used for HC-PMCs with specific operating wavelength bands as required by different applications (Benabid et al. 2005).

The Kagome lattice and antiresonant HCFs are based on antiresonant or inhibited coupling between the core and the cladding modes and typically exhibit a wider transmission window. Figure 2b shows the transmission spectrum of the Kagome fiber in Fig. 1c. The antiresonant fiber in Fig. 1d has a simpler cladding structure and could be optimized to achieve broadband single-mode operation via resonant filtering of higher-order modes (Uebel et al. 2016). These HCFs are however quite different in core sizes and modal properties and would be difficult to be spliced to standard SMFs to form all-fiber devices.

To build a high-performance HC-PMC, several issues should be considered. The first one is the splicing loss between the HCF and the SMF pigtails. Several factors such as the mode field mismatch, refractive index mismatch, and deformation/collapsing of air holes would affect the splicing loss. By using a commercial fusion splicer with offset heating and small fusion current with short duration, splicing loss of <1.5 dB from SMF to HC-PBF has been obtained (Xiao et al. 2007).

The second issue is the multipath interference that causes fluctuation (noise) in the transmitted light intensity. Current HCFs are typical multimode fibers, in

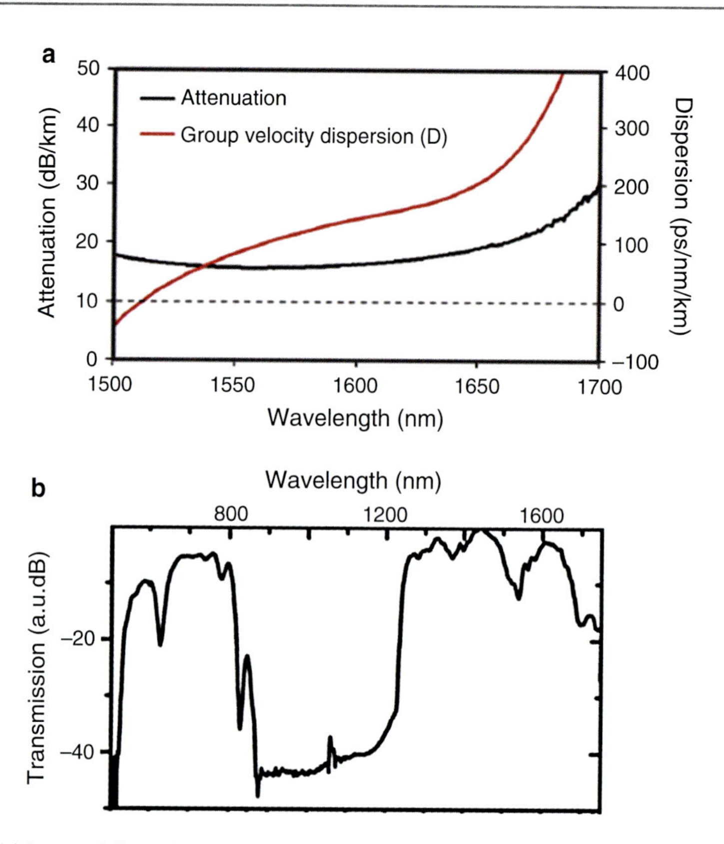

Fig. 2 (**a**) Loss and dispersion of the HC-PBF (HC-1550-02) from NKT Photonics. (**b**) Transmission spectrum of the Kagome fiber shown in Fig. 1c (Couny et al. 2006)

which the propagation of several core modes and surface modes are supported. The interference between these modes as well as the reflected waves from the HCF/SMF splicing joints results in noises. Figure 3a shows the spectral ripples in the transmitted spectra for a HC-PMC made with a HC-PBF (HC-1550-02) with different lengths. The Fourier transform of the transmission spectrum (the red curve in Fig. 3a) for a PMC made of 0.3-m-long HC-PBF with SMF pigtails is shown Fig. 3b, in which the existence of high-order core modes (LP_{11}, LP_{21}, LP_{02}), cladding modes (surface modes), and reflections at the fiber joints is clearly observable. The use of a longer HC-PBF reduces the mode interference (MI) because the higher-order modes and cladding modes experience higher transmission loss than the fundamental mode (Yang et al. 2014). The green line in Fig. 3a is the transmission spectrum of a PMC made with 13-m-long HC-PBF; the spectral ripples are clearly much weaker.

The effective refractive indexes of the modes in a HC-PBF are quite different ($\sim 10^{-2}$ or larger); this means that mode coupling would not occur under weak disturbance (Jin et al. 2010). Hence, MI would be reduced by avoiding the excitation

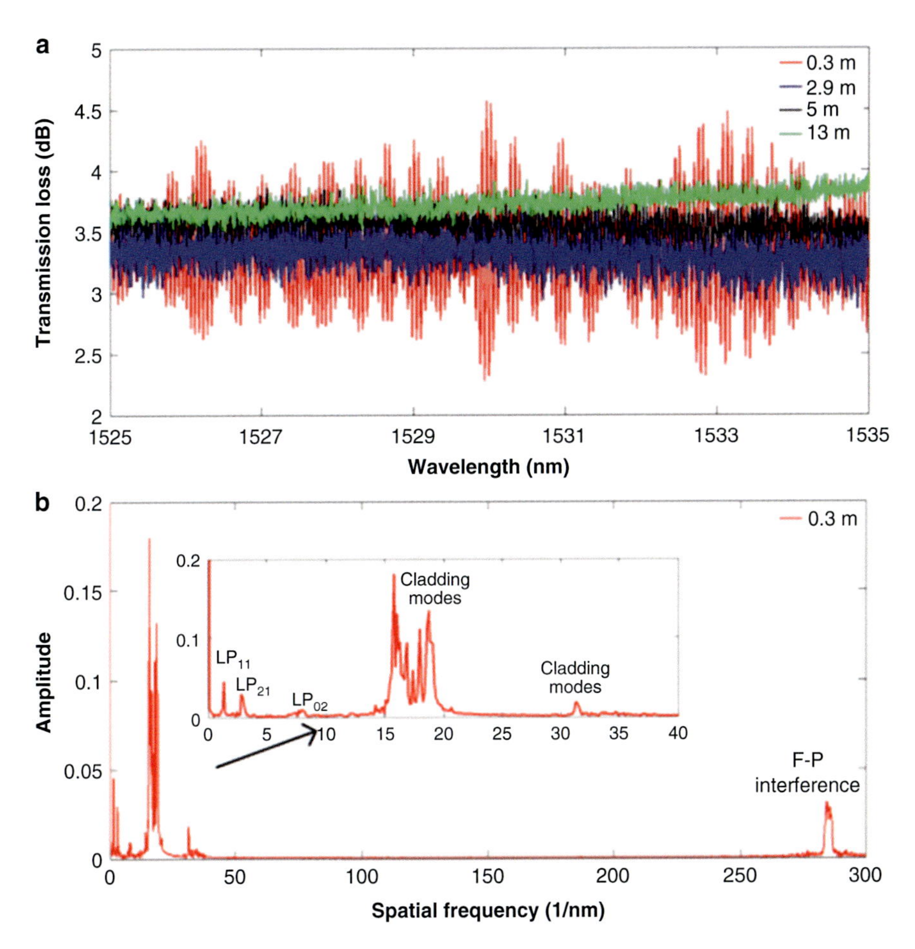

Fig. 3 (**a**) Transmission spectra of HC-PMCs made with HC-1550-02 fiber with different lengths. (**b**) The Fourier transforms of the transmission spectrums in (**a**) (Yang 2015)

of higher-order modes at the input end of the HC-PBF, which may be achieved by keeping a proper gap between the input SMF and HC-PBF (Yang et al. 2014), using angle-cleaved splicing (Miller and Cranch 2016) and controlled collapsing of the cladding holes or using a transition fiber as a modal filter (Yang 2015).

The third issue is the response time of the HC-PMC. The use of a long HC-PBF increases signal level (and hence detection sensitivity for sensors) as well as reduces MI; it results in a slow response due to the time taken to fill material (gas or liquid) into the hollow core (Hoo et al. 2005). Applying pressure differential would speed up the filling process (Ritari et al. 2004; Wynne et al. 2009), but it could be impractical for real-time sensing applications. A more practical approach is to introduce side holes or micro-channels along the sensing fiber (van Brakel et al. 2007). Then the response time would be determined by the separation between two adjacent micro-channels. Several micromachining techniques such as focused ion beam milling (Li et al. 2009) and femtosecond (fs) laser drilling (Hensley et al. 2007; Hoo et al. 2010) may be used to fabricate such openings. The fs laser drilling has been demonstrated to be an effective technique for fabricating a large number of low-loss micro-channels in a short time. With an 800-nm focused fs laser (120 fs, 1 kHz) and high-resolution translation stages, we fabricated 144 groups of micro-channels along a 3.2-m-long HC-PBF sample. The spacing between adjacent groups is ~2 cm, with each group further contains three micro-channels with spacing of 50 μm between them. The inset in Fig. 4a shows the cross-sectional image of a micro-channel, which has a diameter of about 5–7 μm and a depth of ~60 μm.

Figure 4a shows the normalized transmission spectra for the different numbers of drilled hole groups. The average loss per micro-channel is ~0.01 dB. The 3.2-m-long HC-PMC sample was coiled and fixed to a 5 × 10 cm PMMA board as shown in Fig. 4b and then tested with 0.5% acetylene in nitrogen. A response time of ~40 s was obtained. In comparison, it would take many hours to fill the same length of HC-PBF without micro-channels (Hoo et al. 2003). The long HC-PMC could be coiled to small diameters (<1 cm) with minimum additional loss (Hansen et al. 2004), enabling highly sensitive, compact all-fiber sensors with fast response.

Suspended-Core PMCs

Figure 5a sketches an SC-PMC, which consists of a SCF region and two transition regions. Figure 5b, c shows, respectively, the cross-sectional images of the SCF and transition regions of an SC-PMC made by inflating three selected air holes of a commercial PCF. In the SCF region, a wavelength or sub-wavelength scale core is supported by three thin struts, resembling a micro-/nanofiber (MNF) suspended within a silica capillary jacket, allowing strong interaction between the evanescent field around the MNF and the matter filling the SCF region. The transition from the PCF to the SCF region is adiabatic, making it possible for a light mode to propagate in/out the SCF region with low loss (<0.2 dB) and minimal mode perturbation.

Similar to the HC-PMC, the two ends of an SC-PMC can be spliced to SMFs with low-loss by using a PCF with its mode field approximately matched to that of the SMFs. For example, the splicing loss between NKT ESM-12 PCF and SMF-28 can be made down to about 0.1 dB (Xiao et al. 2007). Compared with a HC-PBF PMC, the SC-PMC supports wider-band spectral transmission with better mode purity.

The formation of SCF and transition regions in an SC-PMC is the result of a selectively inflation process, which includes the three major steps as shown in Fig. 6: (1) selective opening of air columns, (2) pressurizing of the air columns, and (3) heating and tapering the pressurized PCF.

In step (1), the air columns in a PCF are firstly sealed at both ends by splicing the PCF to SMF pigtails. Then, an fs laser-assisted selective opening technique is used to open selected air columns of the PCF (Ju et al. 2010; Wang et al. 2010). In this technique, a focused fs laser beam is firstly used to cut the SMF near one of the PCF/SCF splicing joint and then drill holes in the SMF remnant along the axial

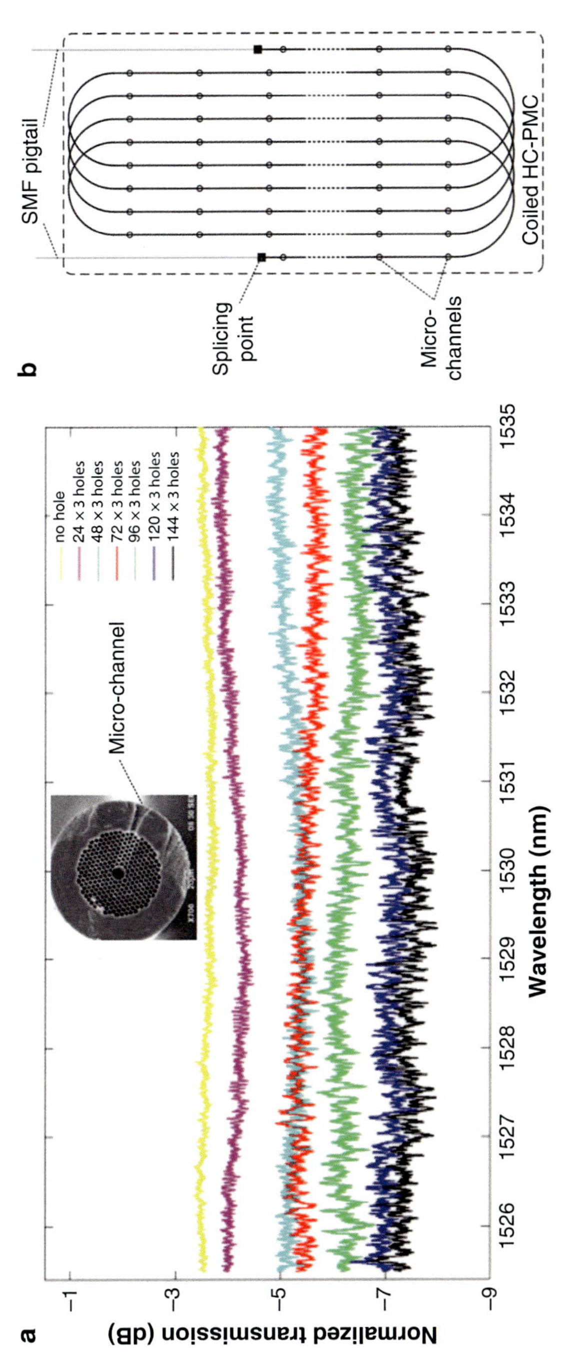

Fig. 4 (**a**) Normalized transmission spectrums of a 3.2-m-long coiled HC-PMC with 144 × 3 micro-channels or holes drilled on the HC-PBF (Yang et al. 2016a). Inset: image of a micro-channel drill on HC-PBF (Jin et al. 2013). (**b**) Sketch of the coiled HC-PMC

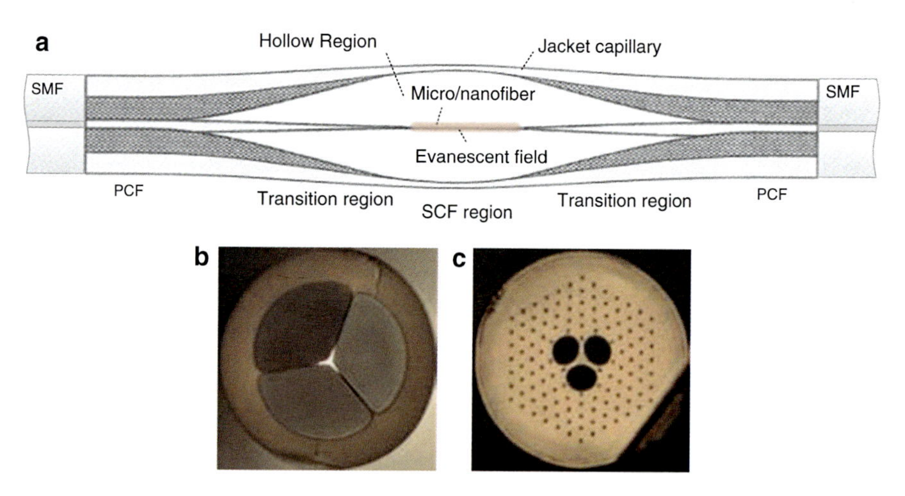

Fig. 5 (**a**) Sketch of an SC-PMC. Cross-sectional images of (**b**) the SCF region and (**c**) the transition region of a three-hole SC-PMC

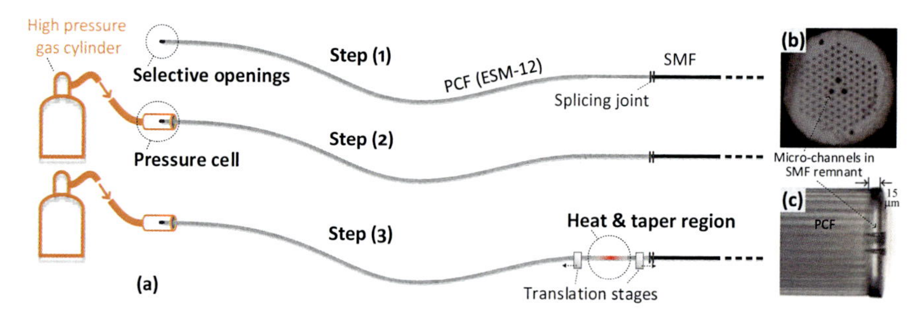

Fig. 6 (**a**) The fabrication procedure of SC-PMC. The (**b**) transverse and (**c**) axial images of the selective openings on a PCF tip for three-hole SC-PMC fabrication

direction of the fiber. An image of the fiber end with three holes drilled is shown in Fig. 6b. The depth of the holes can be viewed from the side of the fiber, as shown in Fig. 6c.

In step (2), the PCF end with selectively opened air columns is sealed into a high-pressure cell, which is connected to a pressure source (e.g., a pressurized gas cylinder or a pump). The pressure in the opened air columns increases, while the other columns remain at atmospheric pressure. Since the bore size of the PCF air columns is small (e.g., ~3.5 µm for NKT ESM-12), it would take a long time for the internal pressure of the selected columns to reach equilibrium. For example, it would take about 1 h for the internal pressure of a 1-meter-long PCF to reach eight bars.

In step (3), the pressurized PCF is heated locally to the soften point of silica and tapered slowly at the same time. During the process, the pressurized air columns would expand, while other columns collapse and eventually generate a

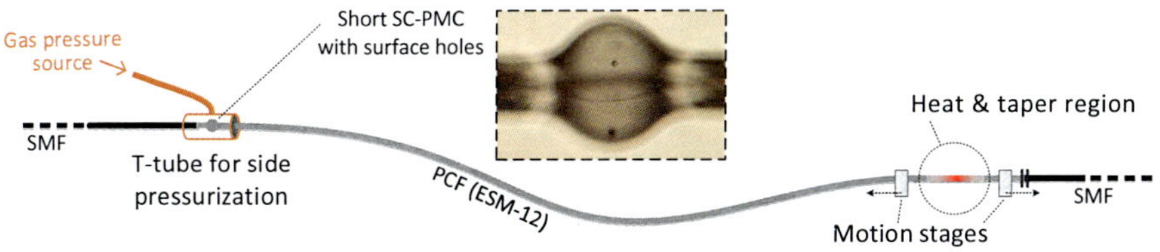

Gas pressure source
SMF
T-tube for side pressurization
Short SC-PMC with surface holes
PCF (ESM-12)
Heat & taper region
Motion stages
SMF

Fig. 7 An updated setup allowing monitoring the performance of the SC-PMC during fabrication process

suspended-core region. In this step, the heat source could be a brushing flame torch for fabricating long PMCs (e.g., centimeters or longer, depending on the traveling length of the torch) or electrical discharges for shorter PMC (few hundred microns). Finally, the fabricated PMC is cut from the PCF and fusion spliced to a SMF pigtail to produce the SC-PMC shown in Fig. 5a. Multiple SC-PMCs can be fabricated in serial in a single PCF by repeating step (3) at different locations along a PCF, which could be useful for multipoint sensing applications.

An update to the procedure can be made by applying gas pressure via an already made SC-PMC. As shown in Fig. 7, a short SC-PMC is firstly made by following the procedure shown in Fig. 6, and side holes are drilled on the PMC by a focused fs laser beam, as shown in the inset of Fig. 7. This PMC is then placed inside a three-port pressure cell, which serves as a chamber to apply gas pressure from the side of the microcell. This update would enable in situ monitoring the performance of new SC-PMCs made along the PCF by connecting the input SMF pigtail to a light source and the output SMF pigtail to an optical spectral analyzer (Wang et al. 2016).

By pressurizing different air columns, SC-PMCs with different internal structures can be made. Figure 8 shows some of the structures, which were made by selectively inflating those air columns marked in white in cross-sectional images of the PCF. These PMCs have different evanescent field, birefringent, core number, and other properties and would be useful for different applications.

For applications based on evanescent field interaction, the size of the waveguiding core is an important parameter. In general, a smaller core diameter would enable large fractional evanescent power propagating in the hollow region of an SC-PMC, hence enhanced light-matter interaction. Figure 9a shows the calculated fractional evanescent power in air for a three-hole SC-PMC at the wavelength of 1550 nm. Here the core radius is defined as radius of the inscribed circle as shown in the inset image of Fig. 9a. If this radius is smaller than 0.75 μm, the factional evanescent power would be larger than 10% and increase rapidly with further reduction in core size.

The birefringence of a high-aspect-ratio core would change significantly with its core size. Figure 9b shows the calculated phase birefringence as a function of the normalized radius of an SC-PMC with rhombus-like core as shown in Fig. 8a at the wavelength of 1550 nm. The normalized radius $\mathbf{r_N}$ is defined by the square

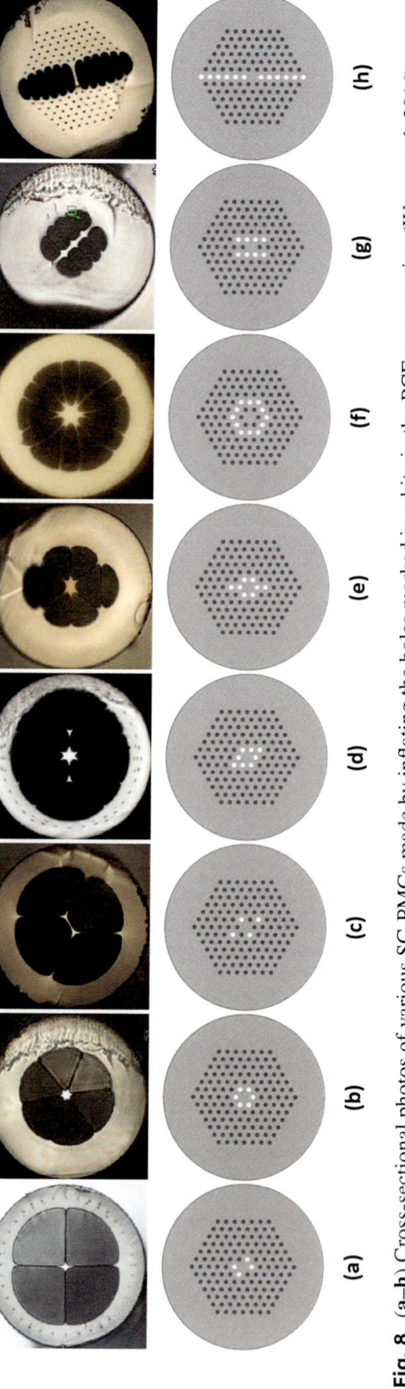

Fig. 8 (a–h) Cross-sectional photos of various SC-PMCs made by inflating the holes marked in white in the PCF cross section (Wang et al. 2016)

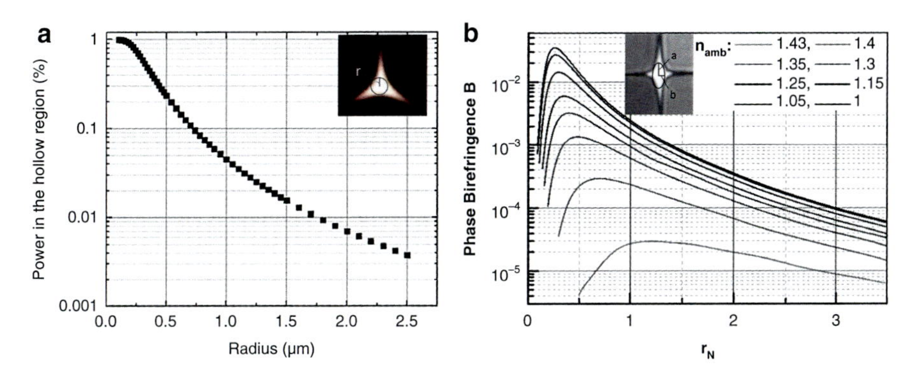

Fig. 9 Calculated relationship between (**a**) fractional evanescent power in the hollow region and core radius of the three-hole SC-PMC, inset: definition of the core radius; and (**b**) birefringence and core radius of the SC-PMC with rhombus-like core, inset – definition of parameters a and b, a/b = 2 (Wang et al. 2014)

root of *ab*, where *a* and *b* are, respectively, the long and short half axis of the ellipse in the core region as shown in the inset image of Fig. 9b. The different curves in the chart represent the relationships when the refractive index in the hollow region n_{amb} is varied from 1 to 1.43. In air environment ($n_{amb} = 1$), the phase birefringence reaches a theoretical maximum of $\sim 3 \times 10^{-2}$ when the r_N is around 0.257.

Applying axial stretching during the fabrication process of the PMC (i.e., step (3) in Fig. 6) would reduce the core size of the SC-PMC. However, with the current commercial PCFs, it is actually difficult to reduce the diameter of the suspended core to submicron dimensions. In building a three-hole SC-PMC, when the suspended core is smaller than 1.5 μm, the jacket capillary would become very flimsy and would easily break under internal pressurization. To overcome the problem, the PCF can be placed inside an additional silica tube to form a thicker jacket capillary (Wang et al. 2016); hence, the SC-PMC could support a much longer axial stretching during heating and pressurization and would generate a suspended core with small diameter. SC-PMCs with core diameter below 900 nm were fabricated for three-hole SC-PMCs, in which the fractional evanescent field is calculated to be about 30%.

Other Microcells

There are reports of other types of microcells. W. Jin et al. reported a practical micro-/nanofiber photonic microcell (MNF-PMC) as shown in Fig. 10a. It is made by encapsulating a tapered MNF inside a glass tube, in which micro-channels for fluid inlet and outlet are made by micromachining techniques. The device is cost-effective and could be used in many evanescent field-based sensor and device applications (Jin et al. 2014). G. Huyang et al. reported a microcell made by splicing a segment

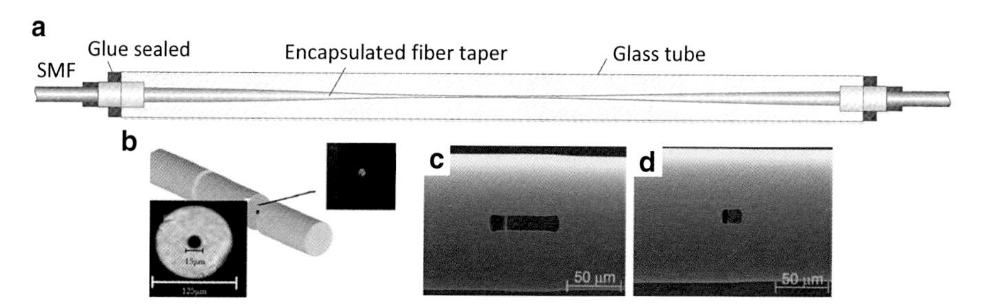

Fig. 10 Microcells made with (**a**) capsulated tapered SMF (Wang et al. 2016), (**b**) hollow capillary (Huyang et al. 2010), (**c**, **d**) hydrofluoric (*HF*) acid-etched SMF with different lengths (Donlagic 2011)

of hollow capillary between two SMFs as in Fig. 10b (Huyang et al. 2010). When the capillary was filled with high index liquid, light could propagate in and interact with the liquid core. The capillary-based PMC structure could be used in solvent analysis. D. Donlagic proposed a method to make microcell directly on solid fiber by anisotropic hydrofluoric (HF) acid etching (Donlagic 2011). Figure 10c, d shows two samples of such microcells with different lengths. These microcells are typically short and can be used as a transmission cell or as a miniature Fabry-Perot resonator (Donlagic 2011).

Sensors Based on PMCs

PMCs provide low-loss, compact, contamination-free platforms for light-matter interaction, which would enable practical all-fiber sensors and lab-on-fiber devices. HC-PMCs have been applied in many gas-phase photonic devices such as gas lasers (Benabid 2012), coherent Raman combs (Benabid 2011), frequency references (Wang et al. 2013b), and trace gas and liquid sensors (Jin et al. 2015; Han et al. 2008). The SC-PMCs possess many useful properties of SCFs and MNFs and would have potential applications in chemical and biological sensing (Monro et al. 2010), single-photon sources (Fujiwara et al. 2011), surface plasmon resonance devices (Hautakorpi et al. 2008), and surface absorption spectroscopy (Warken et al. 2007). In this section, some applications of PMCs in sensing are discussed.

Gas Sensing with HC-PMCs

The potential of HC-PBFs for gas detection was proposed as early as the first PBF is reported by Cregan et al. (Cregan et al. 1999) and experimentally demonstrated for spectroscopic gas detection by Hoo et al. (Hoo et al. 2005). Since then, HC-PMCs have been applied to the detection of different gases including C_2H_2 (Brakel et al.

2008; Lee et al. 2011), C_2H_6 (Li et al. 2012), CH_4 (Cubillas et al. 2007; Carvalho et al. 2009), CO_2 (Nwaboh et al. 2013), and H_2 (Yang and Jin 2017), by the use of different techniques such as laser absorption spectroscopy (LAS), photothermal spectroscopy, and Raman spectroscopy.

LAS, which relies on direct absorption of gas molecules to a laser beam with a specific wavelength (or frequency v), has been widely used in conventional free-space and fiber-based gas sensors. Light absorption in a medium with absorption coefficient α and concentration C of absorptive molecules is governed by the Beer-Lambert law $I(v) = I_0(v) \cdot \exp.[-\alpha(v) \cdot L \cdot C]$, where I and I_0 are, respectively, the intensities of the transmitted and incident light, and L is the absorption distance. HC-PBF-based gas cells provide a platform for strong light-gas interaction over a long length and also avoid the size and alignment problems associated with bulk optic cells. The lower limit of detection (LOD) of the HC-PBF-based sensors is however limited to the level of ppm (parts per million) in terms of noise equivalent concentration (NEC) (Carvalho et al. 2009). Further improvement of the LOD is mainly constrained by the weak overtone absorption within HC-PBF's transmission band and the mode interference (MI) noise as discussed in section "Hollow-Core PMCs" (Parry et al. 2009).

Recently, an all-fiber HC-PMC gas sensor based on photothermal (PT) spectroscopy in combination with interferometric detection was reported (Jin et al. 2015), which achieved a record-low LOD (ppb level of acetylene in nitrogen) and a unprecedented dynamic range of six orders of magnitude. Figure 11 sketches the setup of the detection system. A modulated pump laser with its wavelength tuned to the P(9) line of acetylene is coupled into the HC-PMC to trigger periodic spectral absorption in the microcell. This generates local heating; changes the temperature, density, and pressure distribution of gases in the HC-PMC; and modulated the refractive index (RI) of gas and hence the phase of a probe laser propagating along the same HC-PMC. An optical fiber Mach-Zehnder interferometer (MZI) transfers the variation of the probe phase modulation into intensity modulation, and the harmonics of the modulation signal is detected by a lock-in amplifier. The PT phase modulation can be maximized by optimizing the pulse duration of pump laser (Lin et al. 2016).

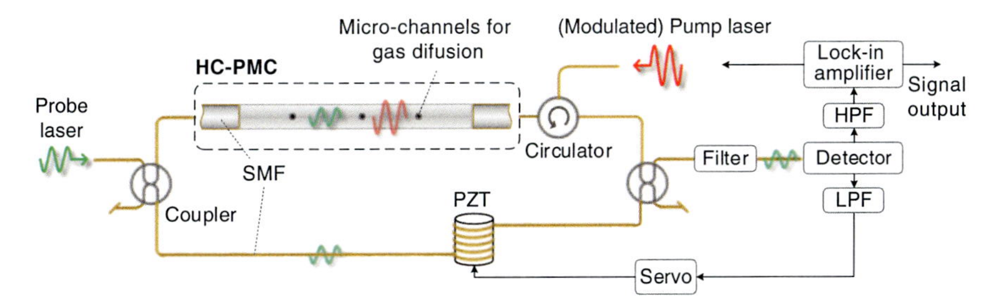

Fig. 11 MZI setup for photothermal gas detection with a HC-PMC. *HPF* high-pass filter, *LPF* low-pass filter, *PZT* piezoelectronic transducer

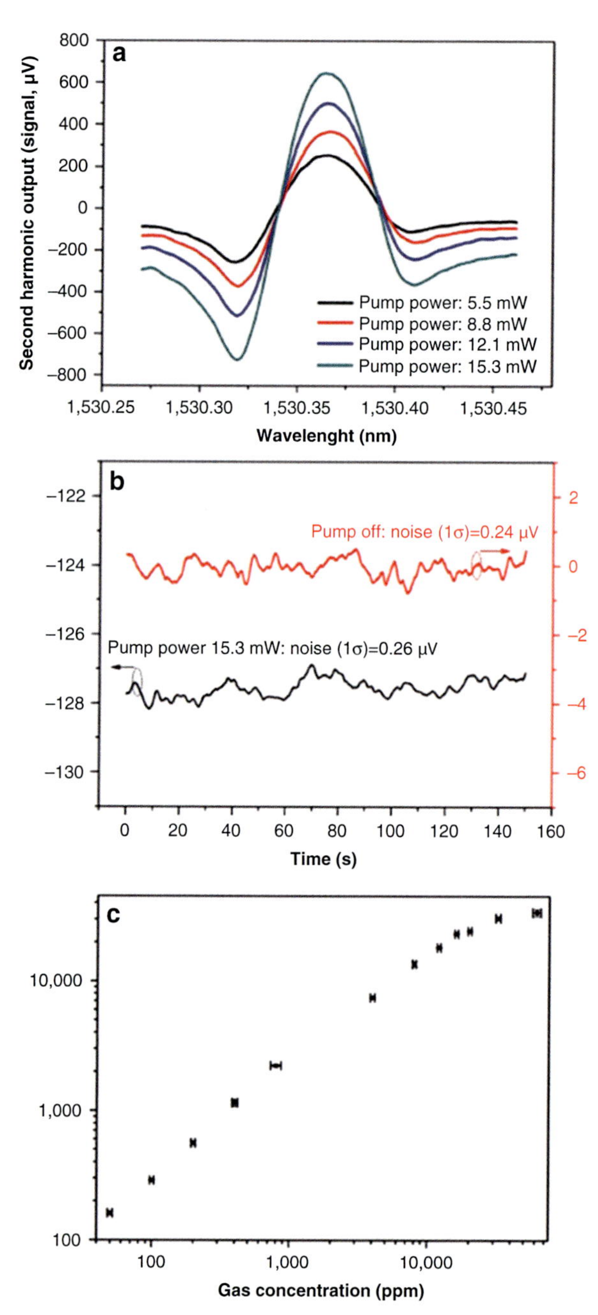

Fig. 12 (**a**) Second harmonic lock-in output when pump laser is tuned across the absorption line of acetylene. (**b**) Second harmonic lock-in output when the pump is on but tuned away from the absorption (*black*) and off (*red*). (**c**) Second harmonic lock-in output (μV) as a function of acetylene gas concentration (Jin et al. 2015)

Figure 12a shows the second harmonic lock-in signals for different pump power levels delivered to a 10-m-long HC-PBF filled with 10-ppm acetylene balanced by nitrogen. The peak-to-peak amplitude of the second harmonic signal is ~1.37 mV when the pump power is 15.3 mW. The lower LOD can be determined by comparing this signal with the second harmonic output (noise) when the pump is tuned away from the absorption peak. As shown in Fig. 12b, the standard deviation of the noise over 2-min duration is 0.26 mV, which is slightly larger than the noise level when the pump is off. Hence, the NEC of the sensor is estimated to be ~2 ppb (parts per billion) for a detection bandwidth of 0.094 Hz. With a similar setup but a 0.62-m-long HC-PBF gas cell, the dynamic range of the system was test and shown in Fig. 12c. An approximately linear relationship was obtained for acetylene concentration from ~30 ppm to ~1.6%, which is about six orders of magnitude (5.3×10^5).

The optical fiber MZI configuration has very high phase detection sensitivity, but it is susceptible to environmental disturbance, such as temperature or strain fluctuation. In Fig. 11, the MZI is stabilized at quadrature by using a piezoelectric transducer (PZT) phase modulator-based electronic servo loop, around which the PT phase modulation can be linearly and efficiently converted to intensity modulation at the interferometer output. The stable linear phase to intensity conversation can be achieved passively with an all-fiber Sagnac loop interferometer (SLI) with a 3 × 3 coupler and a balanced detector. The SLI is a single-fiber interferometer, which has zero optical path difference between interference beams and hence immune environmental disturbance. With a pulsed pump laser of 100 nJ energy and 10 ns duration, it is theoretically possible to achieve ppb level acetylene detection with a 1-m-long HC-PBF (Lin et al. 2016).

A simple configuration is to use a single-fiber Fabry-Perot interferometer (FPI) for phase detection. The reflections of the probe occurring at the two HC-PBF/SMF splicing points of a HC-PMC form a low-finesse FPI. One of the reflections would pass the HC-PBF gas cell twice and carries the information of gas absorption-induced PT phase modulation. With a 2-cm-long HC-PBF gas cell, we have achieved a lower LOD of ~440 ppb with a lock-in time of 1 s and a response time smaller than 19 s. The LOD goes down to 117 ppb for 77 s averaging time (Yang et al. 2016b).

Sensing with Liquid-Filled SC-PMCs

Liquid materials can be filled into an SC-PMC through micro-channels on the jacket and interact with the propagating light via evanescent field. This would enable a range of applications such as trace material detection in liquid (Heng et al. 2013) and liquid-based refractive index sensors (Wang et al. 2016) and functional devices (Moura et al. 2014). The region for light-material interaction is protected by a jacket capillary, which enables robust and compact devices free from external environmental contamination.

Figure 13 is a temperature sensor based on a high-birefringence (Hi-Bi) SC-PMC filled with a liquid with large thermo-optic coefficient. The SC-PMC has a rhombus-like core as shown in Fig. 8a; temperature change results in change in the liquid RI and hence the birefringence of the SC-PMC, which is detected by a SLI. The temperature sensitivity is affected by a number of factors such as the size and shape of the core and the optical properties of the liquid (Wang et al. 2014).

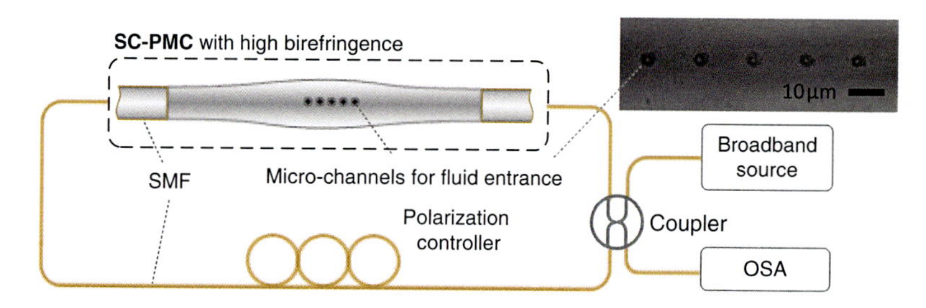

Fig. 13 A temperature sensor setup based on SLI and liquid-filled Hi-Bi SC-PMC. Inset: micro-channels drilled on the jacket capillary of the SC-PMC (Wang et al. 2014)

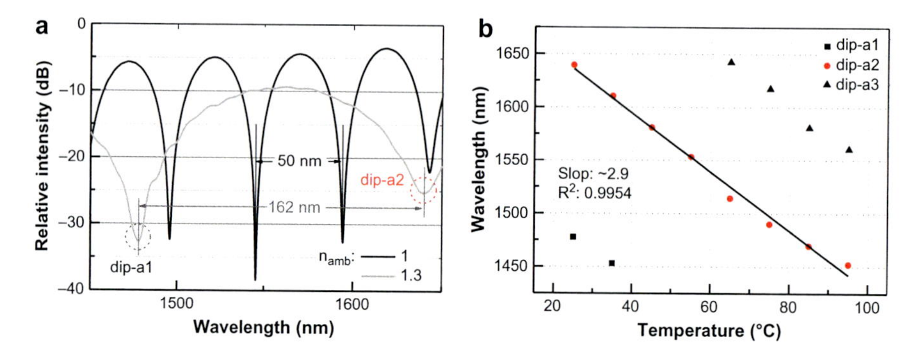

Fig. 14 (**a**) Measured transmission spectra of a highly birefringent (Hi-Bi) SC-PMC filled with air (*black*) and liquid with RI = 1.3 (*gray*). (**b**) Dip wavelengths of the liquid-filled Hi-Bi SC-PMC as functions of temperature (Wang et al. 2014)

Figure 14a shows the measured output spectra of the SLI with a Hi-Bi SC-PMC of waist length of ~1 cm and normalized core radius of ~1.6 μm. The black and gray lines are the results for the PMC filled with air and a liquid with RI = 1.3. The fringe spacing for the air-filled PMC is ~50 nm, corresponding to a birefringence of $\sim 10^{-3}$. The sensitivity of the dip wavelength to temperature for the air-filled PMC is ~3.5 pm/°C, which is quite small due to the low thermo-optic coefficient of silica.

The fringe span becomes much wider (~162 nm) for the liquid-filled PMC, indicating a reduction of birefringence. The fringe is much more sensitive to temperature because of the large thermo-optic coefficient of the liquid (3.34×10^{-4} RIU/°C). Figure 14b shows the results of temperature tests for the liquid-filled PMC from 25 °C to 95 °C. The wavelengths of the dips change linearly with temperature, with a temperature sensitivity of ~3 nm/°C, corresponding to a RI sensitivity of $\sim 9.1 \times 10^3$ nm/RIU. The sensitivity of the Hi-Bi SC-PMC can be further enhanced by optimizing the core diameter of Hi-Bi SC-PMC (Wang et al. 2014).

In-Fiber Accelerometers with SC-PMCs

The flexible structure of SC-PMC could also be used for other applications. Figure 15a shows a micro-cantilever structure built within an SC-PMC, which can be used as an accelerometer. The cantilever structure is made from the six-hole SC-PMC as in Fig. 8b by fs laser micromachining of the suspended core and the supporting thin struts. The suspended core is firstly cut, and then the supporting struts over a short section (~1 mm) were removed by scanning the focused fs laser beam across them to form a cantilever with a free end, as shown in Fig. 15a.

When a transverse acceleration is applied, the intensity of transmission light will be modulated due to the acceleration-induced bending of cantilever, which results in misalignment of the two sections of the core at the cutting location, as shown Fig. 15c. To obtain a linear response to acceleration, an offset at the free end is purposely introduced by removing large portion of struts at one side of the cantilever, as shown in Fig. 15b.

Figure 16a shows the frequency response of a micro-cantilever-based SC-PMC accelerometer with beam length of ~1.3 mm and diameter of ~8 μm. The response is flat until ~2.5 kHz with a sensitivity of ~2.6 mV/g. The sensitivity increase ~13 times at ~6.6 kHz, which is one of the resonances of the cantilever beam. From the inset in Fig. 16a, the minimum detectable acceleration corresponding to SNR = 1 is determined to be 119 $\mu g/(Hz)^{1/2}$ (Wang et al. 2013a). Linear relationships between the output and acceleration are obtained for the test range from 0.01 to 5 g, as shown in Fig. 16b. The performance of the accelerometer can be engineered by fabricating micro-cantilevers with different lengths and shapes.

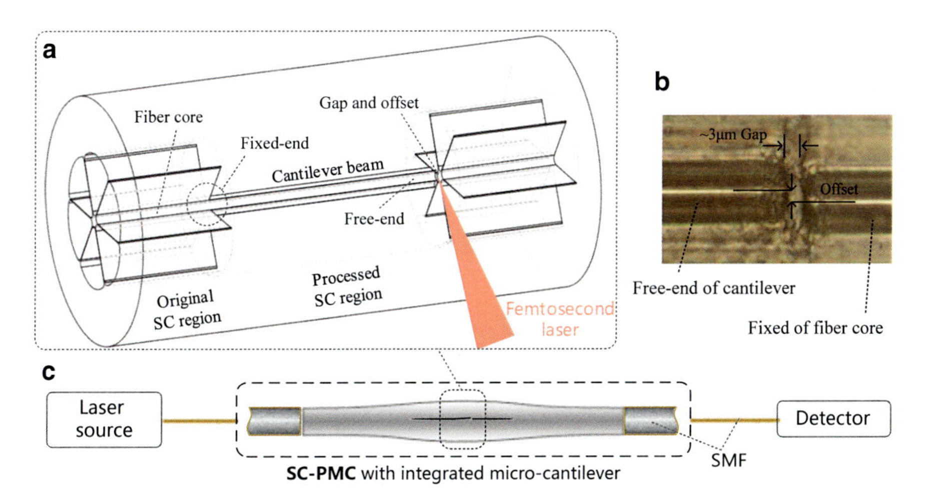

Fig. 15 (**a**) Structure of the micro-cantilever build in microcell. (**b**) Microimage of the gap and transverse offset between cantilever's free end and fixed fiber core (Wang et al. 2013a). (**c**) A simplified setup of cantilever-integrated SC-PMC accelerometer

Grating-Based Sensors and Devices

Gratings are important devices for optical sensing, communication, and laser applications. The integration of grating functions into the PMCs would extend the applications of the PMCs.

Long-period gratings (LPGs) were made on HC-PBF by using a high-frequency, short-duration CO_2 laser (Wang et al. 2008). As illustrated in Fig. 17a, periodic notches along the surface of a HC-PBF can be created by scanning a CO_2 laser beam over the fiber surface repeatedly. At the notch positions, local deformation of cladding microstructure perturbs the field distribution and the propagation constants of the guided modes; periodic perturbation couples the fundamental mode to higher-order modes at phase-matching wavelength, resulting resonant dips in the transmission spectrum (Jin et al. 2011).

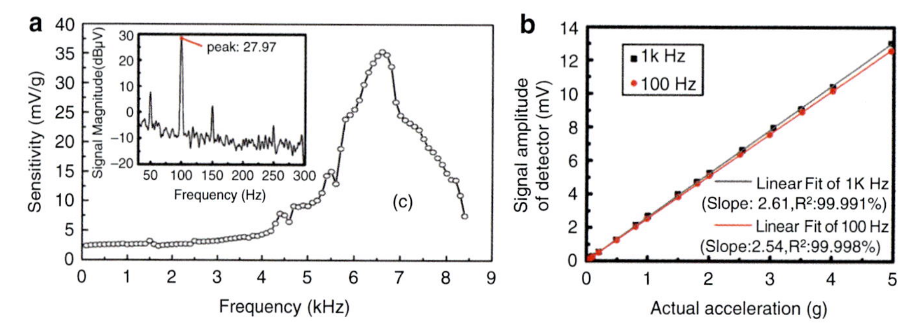

Fig. 16 (**a**) Frequency response of an SC-PMC accelerometer. Inset: output spectrum when acceleration of 10 mg (amplitude) and 100 Hz is applied. (**b**) Accelerometer output as functions of applied acceleration (Wang et al. 2013a)

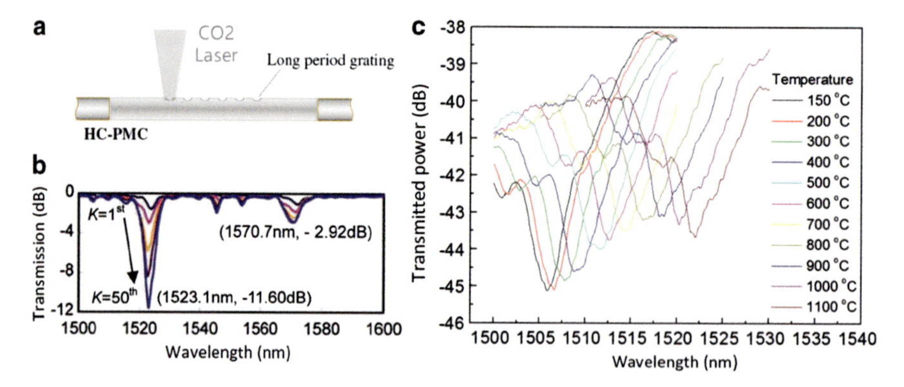

Fig. 17 (**a**) Grating inscription on a HC-PMC with a CO_2 laser beam. (**b**) Evolution of the transmitted spectrum of a LPG with 40 periods and a pitch of 430 μm with increasing scan cycles K (Wang et al. 2008). (**c**) Temperature response of a different LPG made on a HC-PBF

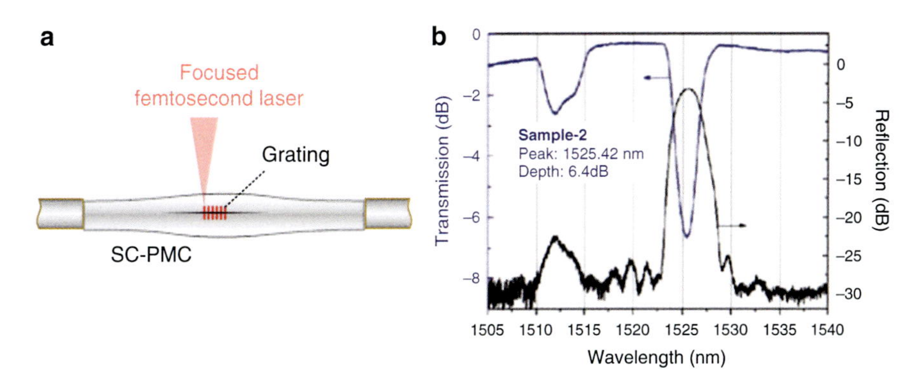

Fig. 18 (**a**) Bragg grating inscription in an SC-PMC by the use of a focused fs laser beam. (**b**) The transmission and reflection spectra of a FBG inscribed in an SC-PMC (Wang 2016)

Figure 17b shows the growth of a 40-period LPG made on the HC-1550-02 HC-PBF with a CO_2 laser. The LPG has a narrow 3 dB bandwidth of ~5.6 nm, which can be narrowed down by increasing the grating periods. The resonant peak of the LPG exhibits a low-temperature sensitivity of ~2.9 pm/°C and a strain sensitivity of −0.83 nm/mε and is insensitive to bending and external RI. LPGs made in HC-PBFs are structural gratings that would not be erased at elevated temperatures. Figure 17c shows the transmission spectrums of another LPG in a HC-PBF; the resonant dip remains at high temperatures up to 1100 °C.

LPGs and fiber Bragg gratings (FBGs) have been made in SC-PMCs, by the use of a point-by-point (Wang et al. 2013c) and phase-mask (Wang 2017) inscription techniques, respectively. The relatively simple structure of the PMC makes it possible to focus the inscription beam directly to the core region from the side of the fiber with little light scattering, as shown in Fig. 18a. Figure 18b shows the transmission and reflection spectra of a FBG fabricated in the three-hole SC-PMC by the use of a focused 800-nm fs laser beam. The core diameter of the PMC is ~4 μm. The 3 dB bandwidth of the FBG is ~2.5 nm, which may be due to the small sport size of the inscription beam. The FBG exhibits temperature and strain sensitivity similar to conventional FBGs in SMFs and operates well at temperature as high as 1100 °C. The FBG is also sensitive to refractive index variation surrounding the suspended core.

Conclusion

In this chapter, we have introduced two types of in-fiber photonic devices, the HC-PMCs and SC-PMCs. They are made from commercial HC-PBFs and solid-core PCFs, respectively, and have the advantages of compact in size, efficient in light-matter interaction, free from external contamination, and easy to be connected to standard SMF systems. Applications of these microcells in gas sensors, liquid-based temperature sensors, accelerometers, and grating-based sensors are discussed.

Integration of various gas-, liquid-, and solid-phase materials into the PMCs would enable novel optical tunable devices, sensors, and lab-on-fiber applications.

Acknowledgments The Natural Science Fundament of China (Grant Nos. 61405125, 61290313, and 61535004) supported this work.

References

F. Benabid, Photonic microcells based on hollow-core PCF, in *(C) Optical Fiber Communication Conference: Optical Society of America*, 2011

F. Benabid, Photonic microcell: a revival tool for gas lasers, in *(C) CLEO-12: Science and Innovations: Optical Society of America*, 2012

F. Benabid, F. Couny, J. Knight, T. Birks, P.S.J. Russell, Compact, stable and efficient all-fibre gas cells using hollow-core photonic crystal fibres. Nature **434**(7032), 488–491 (2005)

F. Benabid, P. Roberts, F. Couny, P.S. Light, Light and gas confinement in hollow-core photonic crystal fibre based photonic microcells. J. Eur. Opt. Soc. Rapid Publ. **4**, 09904-1 (2009)

F. Benabid, J.C. Knight, P.s.J. Russell, Particle levitation and guidance in hollow-core photonic crystal fiber. Opt. Express **10**(21), 1195–1203 (2002)

A.V. Brakel et al., Cavity ring-down in a photonic bandgap fiber gas cell, in *Lasers and Electro-Optics, 2008 and 2008 Conference on Quantum Electronics and Laser Science. CLEO/QELS 2008*. Conference on, 2008, pp. 1–2

J.P. Carvalho et al., Remote system for detection of low-levels of methane based on photonic crystal fibres and wavelength modulation spectroscopy. J. Sens. **2009**(2), 10 (2009)

Y. Chen et al., Demonstration of an 11km hollow core photonic bandgap fiber for broadband low-latency data transmission, in *(C) Optical Fiber Communication Conference Post Deadline Papers*, Los Angeles, California: Optical Society of America, 2015, p. Th5A.1

F. Couny, F. Benabid, P. Light, Large-pitch kagome-structured hollow-core photonic crystal fiber. Opt. Lett. **31**(24), 3574–3576 (2006)

R. Cregan et al., Single-mode photonic band gap guidance of light in air. Science **285**(5433), 1537–1539 (1999)

A.M. Cubillas, M. Silva-Lopez, J.M. Lazaro, O.M. Conde, M.N. Petrovich, J.M. Lopez-Higuera, Methane detection at 1670-nm band using a hollow-core photonic bandgap fiber and a multiline algorithm. Opt. Express **15**(26), 17570–17576 (2007)

C.J. De Matos, C.M.B. Cordeiro, E.M. Dos Santos, J.S. Ong, A. Bozolan, C.H. Brito Cruz, Liquid-core, liquid-cladding photonic crystal fibers. Opt. Express **15**(18), 11207–11212 (2007)

D. Donlagic, All-fiber micromachined microcell. Opt. Lett. **36**(16), 3148–3150 (2011)

M.A. Finger, T.S. Iskhakov, N.Y. Joly, M.V. Chekhova, P.S.J. Russell, Raman-free, noble-gas-filled photonic-crystal fiber source for ultrafast, very bright twin-beam squeezed vacuum. Phys. Rev. Lett. **115**(14), 143602 (2015)

M. Fujiwara, K. Toubaru, T. Noda, H.-Q. Zhao, S. Takeuchi, Highly efficient coupling of photons from nanoemitters into single-mode optical fibers. Nano Lett. **11**(10), 4362–4365 (2011)

Y. Han et al., Index-guiding liquid-core photonic crystal fiber for solution measurement using normal and surface-enhanced Raman scattering. Opt. Eng. **47**(4), 040502–040502-3 (2008)

T.P. Hansen et al., Air-guiding photonic bandgap fibers: spectral properties, macrobending loss, and practical handling. J. Lightwave Technol. **22**(1), 11 (2004)

M. Hautakorpi, M. Mattinen, H. Ludvigsen, Surface-plasmon-resonance sensor based on three-hole microstructured optical fiber. Opt. Express **16**(12), 8427–8432 (2008)

S. Heng et al., Microstructured optical fibers and live cells: a water-soluble, photochromic zinc sensor. Biomacromolecules **14**(10), 3376–3379 (2013)

C.J. Hensley, D.H. Broaddus, C.B. Schaffer, A.L. Gaeta, Photonic band-gap fiber gas cell fabricated using femtosecond micromachining. Opt. Express **15**(11), 6690–6695 (2007)

Y.L. Hoo, W. Jin, C. Shi, H.L. Ho, D.N. Wang, S.C. Ruan, Design and modeling of a photonic crystal fiber gas sensor. Appl. Opt. **42**(18), 3509–3515 (2003)
Y. Hoo, W. Jin, H. Ho, J. Ju, D. Wang, Gas diffusion measurement using hollow-core photonic bandgap fiber. Sensors Actuators B Chem. **105**(2), 183–186 (2005)
Y. Hoo, S. Liu, H.L. Ho, W. Jin, Fast response microstructured optical fiber methane sensor with multiple side-openings. Photonics Technol. Lett. IEEE **22**(5), 296–298 (2010)
J. Hou, D. Bird, A. George, S. Maier, B. Kuhlmey, J.C. Knight, Metallic mode confinement in microstructured fibres. Opt. Express **16**(9), 5983–5990 (2008)
G. Huyang, J. Canning, M.L. Åslund, D. Stocks, T. Khoury, M.J. Crossley, Evaluation of optical fiber microcell reactor for use in remote acid sensing. Opt. Lett. **35**(6), 817–819 (2010)
W. Jin, H. Xuan, H.L. Ho, Sensing with hollow-core photonic bandgap fibers. Meas. Sci. Technol. **21**(9), 094014 (2010)
L. Jin, W. Jin, J. Ju, Y. Wang, Investigation of long-period grating resonances in hollow-core photonic bandgap fibers. J. Lightwave Technol. **29**(11), 1707–1713 (2011)
W. Jin, H.L. Ho, Y. Cao, J. Ju, L. Qi, Gas detection with micro- and nano-engineered optical fibers. Opt. Fiber Technol. **19**(6), 741–759 (2013)
W. Jin, H. Xuan, C. Wang, W. Jin, Y. Wang, Robust microfiber photonic microcells for sensor and device applications. Opt. Express **22**(23), 28132–28141 (2014)
W. Jin, Y. Cao, F. Yang, H.L. Ho, Ultra-sensitive all-fibre photothermal spectroscopy with large dynamic range. Nat. Commun. **6**, 6767 (2015)
A.M. Jones et al., Mid-infrared gas filled photonic crystal fiber laser based on population inversion. Opt. Express **19**(3), 2309–2316 (2011)
J. Ju, H.F. Xuan, W. Jin, S. Liu, H.L. Ho, Selective opening of airholes in photonic crystal fiber. Opt. Lett. **35**(23), 3886–3888 (2010)
K.S. Lee, Y.K. Lee, H.J. Si, A novel grating modulation technique for photonic bandgap fiber gas sensors. IEEE Photon. Technol. Lett. **23**(10), 624–626 (2011)
X. Li, J. At, J. Liang, G. Xu, T. Ueda, Fabrication of photonic bandgap fiber gas cell using focused ion beam cutting. Jpn. J. Appl. Phys. **48**(6), 06FK05–06FK05 (2009)
X. Li, J. Liang, S. Lin, Y. Zimin, Y. Zhang, T. Ueda, NIR Spectrum analysis of natural gas based on hollow-core photonic bandgap fiber. Sens. J. IEEE **12**(7), 2362–2367 (2012)
Y. Lin et al., Pulsed photothermal interferometry for spectroscopic gas detection with hollow-core optical fibre. Sci. Rep. **6**, 39410 (2016)
G.A. Miller, G.A. Cranch, Reduction of intensity noise in hollow core optical fiber using angle-cleaved splices. IEEE Photon. Technol. Lett. **28**(4), 414–417 (2016)
T.M. Monro et al., Sensing with suspended-core optical fibers. Opt. Fiber Technol. **16**(6), 343–356 (2010)
J. Moura et al., Evaporation of volatile compounds in suspended-core fibers. Opt. Lett. **39**(13), 3868–3871 (2014)
K. Murari et al., Kagome-fiber-based pulse compression of mid-infrared picosecond pulses from a Ho:YLF amplifier. Optica **3**(8), 816–822 (2016)
J.A. Nwaboh, J. Hald, J.K. Lyngsø, J.C. Petersen, O. Werhahn, Measurements of CO2 in a multipass cell and in a hollow-core photonic bandgap fiber at 2 μm. Appl. Phys. B **110**(2), 187–194 (2013)
J.P. Parry et al., Towards practical gas sensing with micro-structured fibres. Meas. Sci. Technol. **20**(7), 075301 (2009)
T. Ritari et al., Gas sensing using air-guiding photonic bandgap fibers. Opt. Express **12**(17), 4080–4087 (2004)
P.S.J. Russell, Photonic-crystal fibers. J. Lightwave Technol. **24**(12), 4729–4749 (2006)
J.R. Sparks et al., Selective semiconductor filling of microstructured optical fibers. J. Lightwave Technol. **29**(13), 2005–2008 (2011)
M. Sprague et al., Broadband single-photon-level memory in a hollow-core photonic crystal fibre. Nat. Photonics **8**, 287 (2014)
P. Uebel et al., Broadband robustly single-mode hollow-core PCF by resonant filtering of higher-order modes. Opt. Lett. **41**(9), 1961–1964 (2016)

A. van Brakel, C. Grivas, M.N. Petrovich, D.J. Richardson, Micro-channels machined in microstructured optical fibers by femtosecond laser. Opt. Express **15**(14), 8731–8736 (2007)

C. Wang, Fiber Bragg gratings inscribed in all-silica suspended-core photonic microcells, in *Photonics and Fiber Technology 2016 (ACOFT, BGPP, NP)*, Sydney, p. BTh4B.5: Optical Society of America, 2016

C. Wang, J. He, J. Zhang, C. Liao, Y. Wang, W. Jin, Y. Wang, and J. Wang, Bragg gratings inscribed in selectively inflated photonic crystal fibers, Opt. Express **25**(23), 28442–28450 (2017)

Y. Wang et al., Long period gratings in air-core photonic bandgap fibers. Opt. Express **16**(4), 2784–2790 (2008)

Y. Wang, C. Liao, D. Wang, Femtosecond laser-assisted selective infiltration of microstructured optical fibers. Opt. Express **18**(17), 18056–18060 (2010)

C. Wang, W. Jin, J. Ma, Y. Wang, H.L. Ho, X. Shi, Suspended core photonic microcells for sensing and device applications. Opt. Lett. **38**(11), 1881–1883 (2013a)

C. Wang et al., Acetylene frequency references in gas-filled hollow optical fiber and photonic microcells. Appl. Opt. **52**(22), 5430–5439 (2013b)

C. Wang, W. Jin, J. Ma, W. Jin, H.L. Ho, Photonic microcells for novel devices and sensor applications, in *(C) APOS 2013*, Wuhan, China, 2013c, vol. 8924, p. 27

C. Wang et al., Highly birefringent suspended-core photonic microcells for refractive-index sensing. Appl. Phys. Lett. **105**(6), 061105 (2014)

C. Wang, W. Jin, W. Jin, J. Ju, J. Ma, H.L. Ho, Evanescent-field photonic microcells and their applications in sensing. Measurement **79**, 172–181 (2016)

F. Warken, E. Vetsch, D. Meschede, M. Sokolowski, A. Rauschenbeutel, Ultra-sensitive surface absorption spectroscopy using sub-wavelength diameter optical fibers. Opt. Express **15**(19), 11952–11958 (2007)

P. Westbrook, B. Eggleton, R. Windeler, A. Hale, T. Strasser, G. Burdge, Cladding-mode resonances in hybrid polymer-silica microstructured optical fiber gratings. Photonics Technol. Lett. IEEE **12**(5), 495–497 (2000)

R.M. Wynne, B. Barabadi, K.J. Creedon, A. Ortega, Sub-minute response time of a hollow-core photonic bandgap fiber gas sensor. J. Lightwave Technol. **27**(11), 1590–1596 (2009)

L. Xiao, M. Demokan, W. Jin, Y. Wang, C.-L. Zhao, Fusion splicing photonic crystal fibers and conventional single-mode fibers: microhole collapse effect. J. Lightwave Technol. **25**(11), 3563–3574 (2007)

F. Yang, Novel hollow-core optical fiber gas and acoustic sensors, Ph.D, (PolyU) Department of Electrical Engineering, The HK Polytechnic University, 2015

F. Yang, W. Jin, All-fiber hydrogen sensor based on stimulated Raman gain spectroscopy with a 1550-nm hollow-core fiber, in 25th International Conference on Optical Fiber Sensors, 2017, vol. 10323, p. 4: SPIE.

F. Yang, W. Jin, Y. Lin, C. Wang, H.L. Ho, Y. Tan, Hollow-core microstructured optical fiber gas sensors. J. Lightwave Technol. **PP**(99), 1–1 (2016a)

F. Yang, Y. Tan, W. Jin, Y. Lin, Y. Qi, H.L. Ho, Hollow-core fiber Fabry–Perot photothermal gas sensor. Opt. Lett. **41**(13), 3025–3028 (2016b)

F. Yang, W. Jin, Y. Cao, H.L. Ho, Y. Wang, Towards high sensitivity gas detection with hollow-core photonic bandgap fibers. Opt. Express **22**(20), 24894–24907 (2014)

Filling Technologies of Photonic Crystal Fibers and Their Applications

55

Chun-Liu Zhao, D. N. Wang, and Limin Xiao

Contents

C.-L. Zhao (✉) · D. N. Wang
College of Optical and Electrical Technology, China Jiliang University, Hangzhou, China
e-mail: clzhao@cjlu.edu.cn; dnwang@cjlu.edu.cn; eednwang@163.com

L. Xiao
Advanced Fiber Devices and Systems Group, Key Laboratory of Micro and Nano Photonic Structures (MoE), Department of Optical Science and Engineering Fudan University, Shanghai, China

Key Laboratory for Information Science of Electromagnetic Waves (MoE), Fudan University, Shanghai, China

Shanghai Engineering Research Center of Ultra-Precision Optical Manufacturing, Fudan University, Shanghai, China
e-mail: Liminxiao@fudan.edu.cn

G.-D. Peng (ed.), *Handbook of Optical Fibers*,
https://doi.org/10.1007/978-981-10-7087-7_13

Abstract

In this chapter, we shortly review the filling technologies of photonic crystal fibers (PCFs), including the selectively filling method by collapsing air holes, by splicing to a SMF with a lateral offset, and by femtosecond (fs) laser micromachining. Moreover, we demonstrate their applications in optical fiber devices, including a partially liquid-filled hollow-core PCF polarizer, a fiber in-line Mach-Zehnder interferometer constructed by selective infiltration of two air holes in PCF, and an embedded coupler based on selectively infiltrated PCF. For the applications in optical fiber temperature sensors, we demonstrate temperature sensors based on an alcohol-filled PCF Sagnac loop interferometers, temperature sensors with excellent temporal stability based on PCF with two infiltrated air holes, and a selectively infiltrated PCF with ultrahigh temperature sensitivity. Further, for the applications in gas sensors, we demonstrate a PCF loop mirror-based chemical vapor sensor, a chemical vapor sensor based on a simplified hollow-core PCF, and a hydrogen sensor based on selectively infiltrated PCF with Pt-loaded WO3 coating and discuss their sensing mechanisms. Finally, we will demonstrate a passively mode-locked fiber laser based on a hollow-core photonic crystal fiber filled with few-layered graphene oxide solution.

Keywords

PCF filling · Selective filling · Partial filling and selective infiltration

Introduction

Photonic crystal fibers (PCFs) have attracted significant attention recently. Various types of optical fiber devices and sensors have been fabricated by use of PCFs (Kerbage and Eggleton 2002; Domachuk et al. 2004; Huang et al. 2004; Benabid et al. 2005; Xiao et al. 2007; De Matos et al. 2007; Qian et al. 2011a; Canning et al. 2008; Wu et al. 2009). One of the attractive features of PCF is the selective infiltration, which allows different types of liquids or gases fill into the selected air holes, thus changing the guidance properties of the fiber. A number of functional PCFs and applications, such as birefringent-tunable PCF (Huang et al. 2004), bending-sensitive PCF (Xiao et al. 2007), polarized PCF (Qian et al. 2011a), and high sensitivity refractive index sensor (Wu et al. 2009), are realized by selective filling of the air holes.

One of the commonly used methods of selective infiltration is to block the air holes that are not to be infiltrated and then fill those remain opened (Kuhlmey et al.

2009), such as collapsing air holes (Nielsen et al. 2005; Xiao et al. 2005), injection-cure-cleaving (Huang et al. 2004), splicing PCF with a single-hole hollow-core fiber (Canning et al. 2008; Martelli et al. 2005), lateral access to the air holes (Cordeiro et al. 2006, 2007), direct manual gluing (Qian et al. 2011a; Wu et al. 2009; Kuhlmey et al. 2009), and femtosecond (fs) laser micromachining (Marcinkevi et al. 2001). The central air hole can be selectively infiltrated by collapsing other air holes of the PCF with arc fusion (Xiao et al. 2005), which only needs using a conventional fusion splicer. Injection-cure-cleaving can infiltrate the central hollow core with differential filling speed that depends on the size of the air holes (Huang et al. 2004), and the usage of curable glue in combination with differential capillary forces makes it a general method of filling any group of air holes with similar size. Selective infiltration also can be realized by splicing a single-hole hollow-core fiber to the PCF (Canning et al. 2008; Martelli et al. 2005). The single-hole is aligned to the selected air hole of the PCF, and the other holes are blocked by the solid cladding of the single-hole fiber. The technique of lateral access to the air holes of the PCFs can be performed by blowing a hole through the fiber wall using a fusion splicer and air pressure (Cordeiro et al. 2006) or by focused ion-beam milling (Cordeiro et al. 2007), and the materials are subsequently infiltrated into the PCFs through the side-opened hole. A more versatile approach is the direct manual gluing, typically under a microscope. In this method, a glass tip of submicrometer scale is used to drop UV curable polymers into the air hole before blocking it, and hence a flexible and well-controlled air hole infiltration can be realized (Wu et al. 2009; Kuhlmey et al. 2009). Recently, selectively infiltrating PCFs with the assistance of femtosecond laser micromachining have been proposed (Marcinkevi et al. 2001). With this technique, any type of air holes in the cross section of the PCFs can be selectively infiltrated, which opens up a highly efficient, precise, flexible, and reliable way of selective infiltrating.

Besides on the selected filling technologies, selectively coating the holes of a PCF also is attracting growing interest. Recently, Sazio et al. (2006) showed the experiments that various materials including semiconductors could be incorporated into PCFs using high-pressure microfluidic chemical deposition. Chemical vapor deposition (CVD) is in principle capable of achieving very high-quality films. Zhang et al. (2007) demonstrated selective coating of PCFs with silver and use it to fabricate an in-fiber absorptive polarizer. Technique improvements in coating the holes of a PCF with metal will be a benefit to the combination of plasmonics and photonics.

In this chapter, we will first introduce the filling technologies of PCFs, which mainly include the selectively filling method by collapsing air holes, the selectively filling method by splicing to a single-mode fiber (SMF) with a lateral offset, and the selectively filling method by femtosecond (fs) laser micromachining. Following, we will move on to their applications. We will demonstrate their applications in optical fiber devices, optical fiber sensors, and optical fiber lasers. In the section "Optical Devices Based on Filled PCFs," we will demonstrate their applications in optical fiber devices, including a partially liquid-filled hollow-core PCF polarizer, a fiber in-line Mach-Zehnder interferometer constructed by selective infiltration of two air holes in PCF, and an embedded coupler based on selectively infiltrated PCF. In the section "Temperature Sensors Based on Filled PCFs," we will focus on the applications in optical fiber temperature sensors and will demonstrate temperature sensors based on an alcohol-filled PCF Sagnac loop interferometers, temperature sensors with excellent temporal stability based on PCF with two infiltrated air holes, and a selectively infiltrated PCF with ultrahigh temperature sensitivity. In the section "Chemical Vapor and Gas Sensors Based on Infiltrated PCFs," we will demonstrate a PCF loop mirror-based chemical vapor sensor, a chemical vapor sensor based on a simplified hollow-core PCF, and a hydrogen sensor based on selectively infiltrated PCF with Pt-loaded WO3 coating and discuss their sensing mechanisms. Finally, in the section "Passively Mode-Locked Fiber Laser Based on a Hollow-Core PCF Filled with Few-Layered Graphene Oxide Solution," we will demonstrate passively mode-locked fiber laser based on a hollow-core photonic crystal fiber filled with few-layered graphene oxide solution.

PCF Filling-Related Technologies

Selective PCF Infiltration by Collapsing Air Holes

Xiao et al. (2005) first proposed the method by use of a conventional fusion splicer to selective collapse some or all of the holes in the cladding but leave the central hole remain open.

Figure 1a shows the end face of a hollow-core fiber. The arc discharge of the conventional fusion splicer is hottest at the electrode tips, and the temperature at the midpoint between the electrode tips falls to a minimum along the electrode's axis; hence the temperature distribution of the end face of PCF is that the inner cladding temperature is lower than the outer cladding temperature when the tip of the PCF is not too far from the electrode axis, i.e., when the offset is small. The heat transfer from the solid silica ring to the center of the holey region is much slower because of the presence of large air holes and the relative slow heat transfer in air than in solid silica. Hence it is believed that temperature of the inner cladding holes will keep lower than that of the outer cladding holes during the discharge duration which is typically a few hundreds milliseconds. When the temperature of heated fiber exceeds the softening point which is around 1670 °C, the surface tension will overcome the viscosity and cause the PCF's cylindrical air holes to begin collapsing.

Figure 1b, c shows the end face of the PCF for two typical discharge conditions when we varied the fusion current in steps of 0.5 mA, kept the fusion duration constant at 0.3 s, and offset distance constant at 50 μm. The surface tension makes the PCF's periphery less sharp and causes the central region to cave in. The detailed hole collapse pictures can be found in Fig. 2. When arc current is increased, the outer cladding holes collapse and close first when compared to the inner cladding holes.

Figure 3 shows the images of one of the ends of the hollow-core fiber after the central hole filling with polymer NOA74. The central hole is perfectly filled and

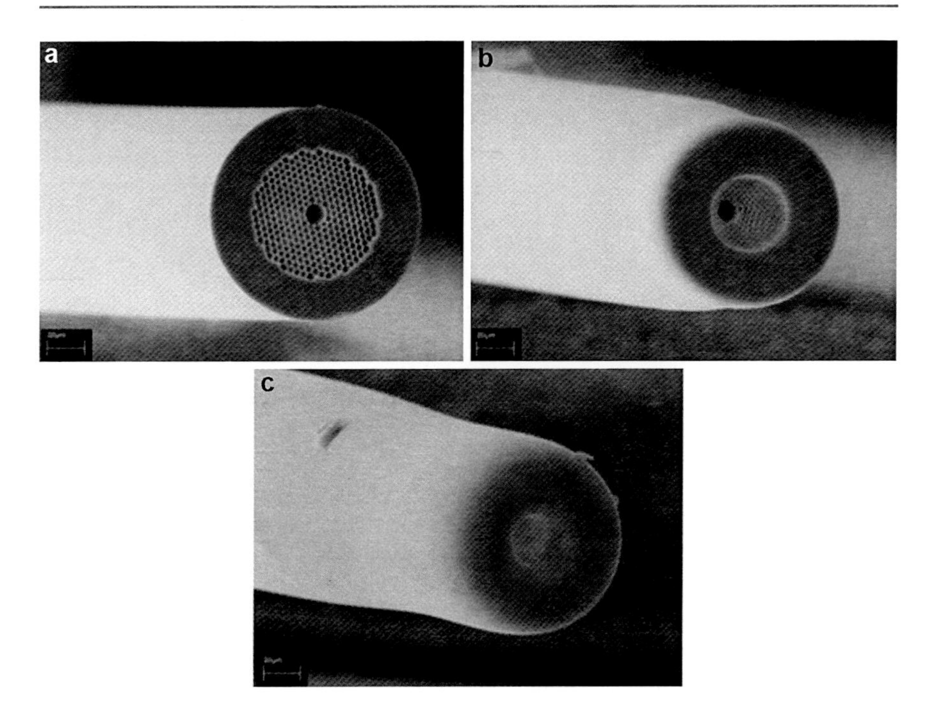

Fig. 1 End face of the PCF. (**a**) Without arc discharge; (**b**) arc current, 12.5 mA; (**c**) arc current, 14.5 mA. The discharge duration and offset distance are kept constant at 0.3 s and 50 μm, respectively

the holes in the cladding are not filled. The process includes, firstly, treat the two ends of the PCF with an Ericsson FSU-975 fusion splicer with arc duration, fusion current, and offset distance set, respectively, to 0.3 s, 14 mA and 50 μm. After thermal treatment, both ends of the PCF are like Fig. 2d, that is, the central hole is still partially open, and all the cladding holes are totally closed. The central hole is then filled with a type of polymer NOA74 by capillary action, the polymer is cured by a UV lamp, and the ends of the PCF are cleaved.

Selective PCF Infiltration by Splicing Single-Mode Fiber with Lateral Offset

Figure 4 illustrates the partial liquid filling technique by splicing a SMF to a PCF with a lateral offset (Qian et al. 2011a). In this illustration, a solid-core PCF is shown, but the technique can be used for both solid-core and hollow- core PCFs. The PCF is firstly spliced to a SMF with a lateral offset. Most of the cladding air holes are sealed by the splice, while some of the air holes are left open and used for subsequent liquid filling. The size of the unsealed region (indicated as S in Fig. 4b) is controlled by the lateral offset at the splice joint. The splice joint is then immersed into a

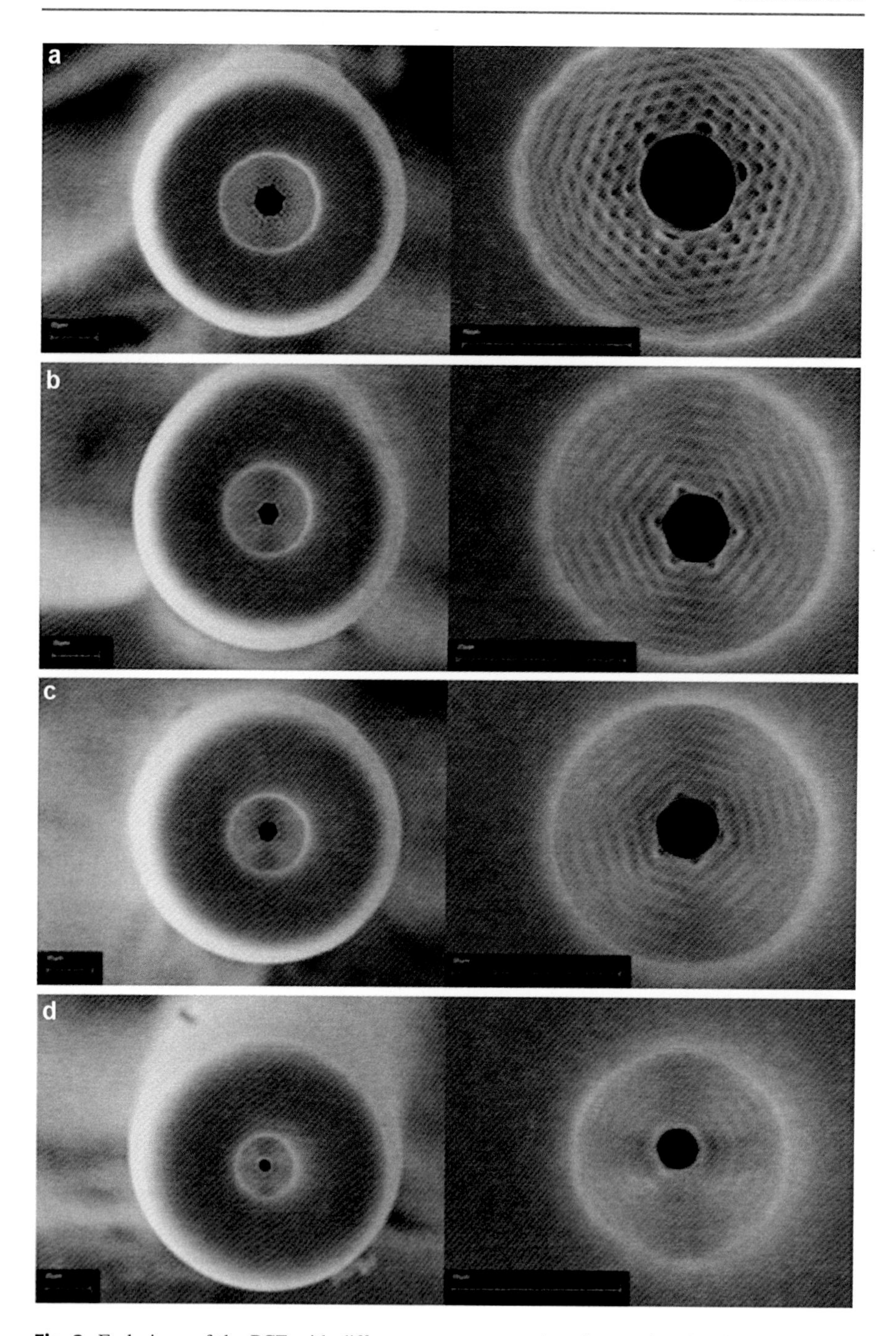

Fig. 2 End views of the PCF with different arc currents when the arc duration is 0.3 s and the offset distance is 50 μm. The right picture is the close-up of the center part of the left picture. (**a**) 12.5 mA, (**b**) 13 mA, (**c**) 13.5 mA, (**d**) 14 mA

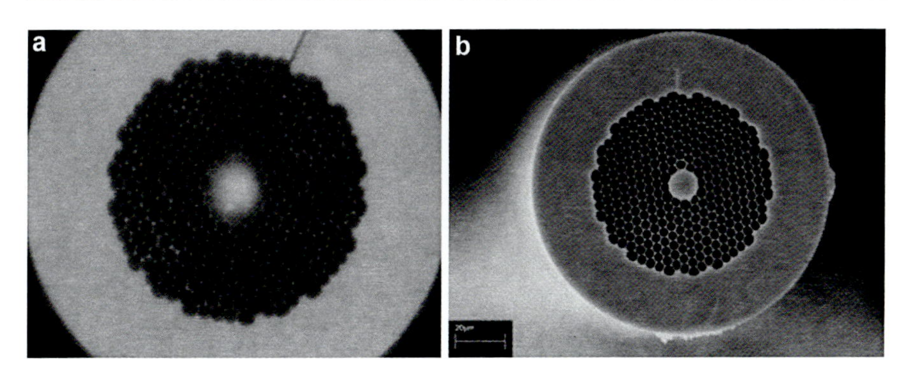

Fig. 3 (**a**) Optical microscope image and (**b**) SEM image of the PCF with the central hole filled with NOA74

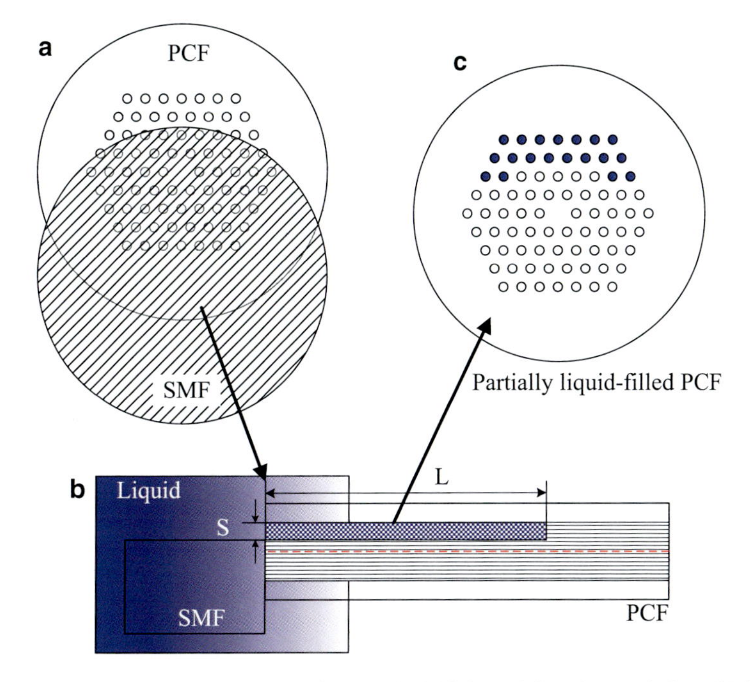

Fig. 4 Schematics showing the process of partial liquid filling of air holes. (**a**) Splice of PCF with SMF with a lateral offset. (**b**) Side view showing the liquid filling process. (**c**) Cross-sectional of the partially liquid-filled PCF

liquid, and the unsealed holes are filled with the liquid by capillarity action. The filling length (L) is controlled by filling time (Benabid et al. 2005). This technique is more easily operated and has a lower cost.

Figure 5a shows a schematic of the fusion splicing process. The lateral offset was controlled by adjusting the SMF position in the X-field view by manually operating the fusion splicer. To avoid or minimize the collapse of the air holes in the filling

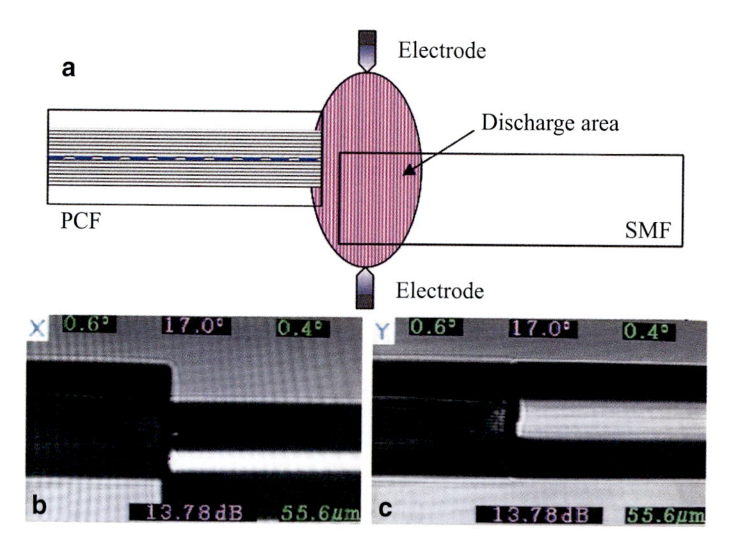

Fig. 5 (**a**) Schematic showing the splicing process. (**b**) X-field and (**c**) Y-field views of a PCF/SMF splice joint

region, the gap set position was fixed to the right side of the PCF/SMF joint, as shown in Fig. 5a. In this position, the PCF was far away from the electrodes, and it was affected only weakly by the arc discharge during the splicing process. At the same time, a lower discharge current and a shorter discharge duration were used, and discharge parameters were adjusted repeatedly to obtain a better performance splice. A good mismatched fusion splicing can be realized to keep air holes of the PCF's partial filling region open, by properly selecting gapset position, discharge current, and duration, combined with a suitable gap and overlap between the PCF and the SMF. Figure 5b, c shows fusion splicer images of a good quality splicing in the X-field and Y-field views.

Selective PCF Infiltration by Femtosecond Laser Micromachining

The selective infiltration technique (Wang et al. 2010a) includes the following steps, as shown in Fig. 6: (1) to block all the air holes by fusion splicing a section of PCF with a normal SMF; (2) to cut the SMF at the position close to the splicing point by use of femtosecond (fs) laser scanning; and (3) to drill through the selected air holes from the cutting end of the SMF to the splicing point by fs laser and allow liquid to be infiltrated into the PCF. For the laser micromachining in the first two steps, fs laser pulses ($\lambda = 800$ nm) of 120 fs duration with repetition rate of 1 kHz were used. As depicted in Fig. 7, the pulses were focused onto the fiber samples by microscopic objectives. The pulse energy was continuously adjustable in the range between 0 and 1 mJ by rotating the half-wave plate incorporated with a polarizer,

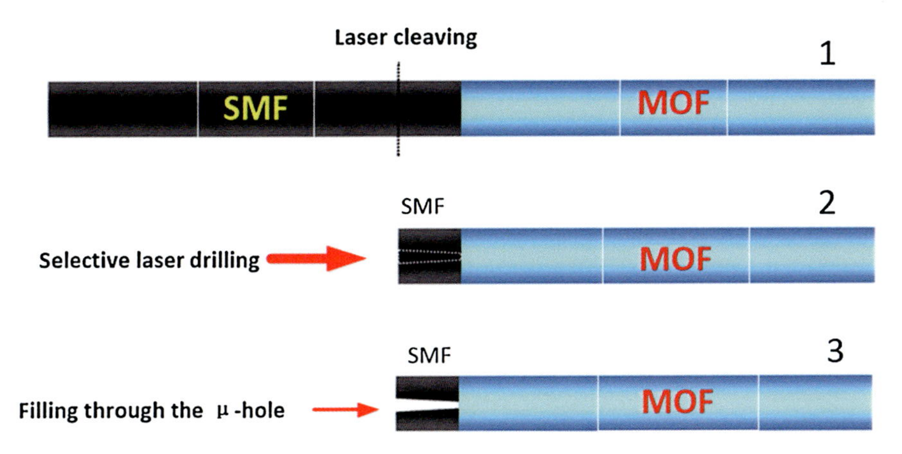

Fig. 6 Flowchart of selective infiltration with the assistance of femtosecond laser micromachining. (1) Fusion splicing and laser cleaving; (2) laser drilling; (3) infiltration

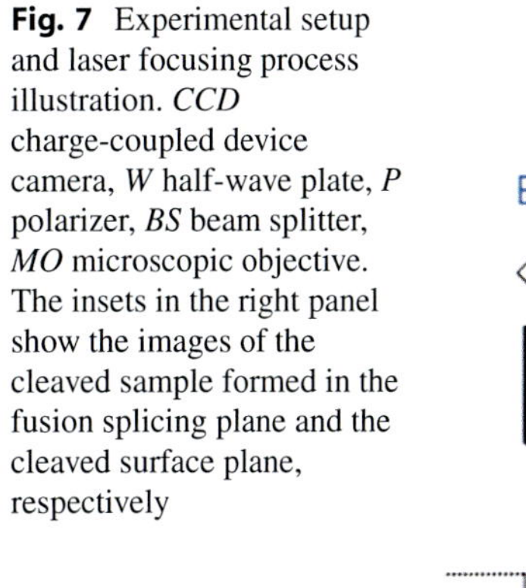

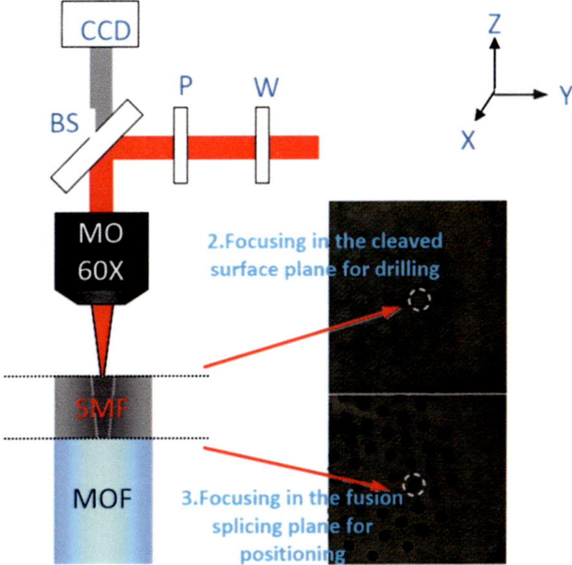

Fig. 7 Experimental setup and laser focusing process illustration. *CCD* charge-coupled device camera, *W* half-wave plate, *P* polarizer, *BS* beam splitter, *MO* microscopic objective. The insets in the right panel show the images of the cleaved sample formed in the fusion splicing plane and the cleaved surface plane, respectively

and hence the on-target pulse energy can be precisely controlled. A CCD camera was used to monitor the micromachining processes in real time. Fiber samples were mounted onto a computer-controlled three-dimensional (3-D) translation stage, of which the positioning accuracy was 40 nm.

Figure 8 shows the microscopic images of two selectively infiltrated LMA-10 fiber samples, which were filled with standard refractive index (RI) liquid with the RI value of 1.50 (from Cargille Laboratories). The samples with only one air hole infiltrated are referred to as A-hole and B-hole samples for short, respectively.

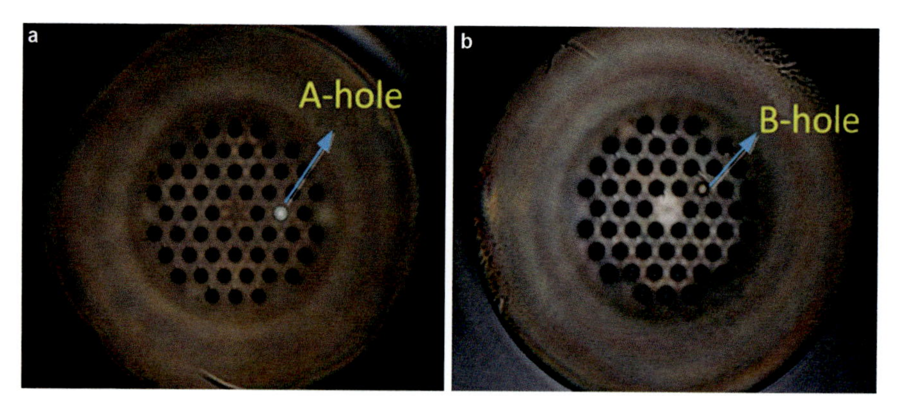

Fig. 8 Microscopic images of selectively infiltrated samples. (**a**) A-hole infiltrated; (**b**) B-hole infiltrated

One can see clearly from the images that the RI liquid has been injected into the selected air hole, and the neighbor holes are not affected, which implies that the fs laser-assisted selective infiltration method can be readily employed to create selectively infiltrated PCFs. With this technique, any type of air holes in the cross section of the microstructured optical fibers can be selectively infiltrated, which opens up a highly efficient, precise, flexible, and reliable way of selective infiltrating.

Optical Devices Based on Filled PCFs

Partially Liquid-Filled PCF Polarizer

In-line fiber polarizers with low insertion loss, high polarization extinction ratio, and wide operational wavelength range are often needed in optical fiber sensors and communication systems. Fiber-optic polarizers have been made by coating the flat side of a side-polished conventional SMF or a D-shaped optical fiber with a thin metal layer (Bergh et al. 1980; Dyott et al. 1987). However, the complex manufacturing process means higher component cost. Combination with the PCFs, a novel polarizer based on a long period grating written on a solid-core PCF is realized (Wang et al. 2007a). However, the device suffers from a narrow operational wavelength range ($\sim$10 nm). A novel fiber polarizer was made by using a pulsed CO_2 laser to deform the air holes of a hollow-core PCF (Xuan et al. 2008). The polarizer exhibits a polarization extinction ratio of more than 20 dB over an operational wavelength range of wider than 100 nm. However, the CO_2 laser irradiated region is fragile, which may be disadvantageous for practical applications.

In this part, we demonstrate a novel polarizer based on a partially ethanol-filled hollow-core PCF (Qian et al. 2011a). The partial filling of air holes by a liquid is

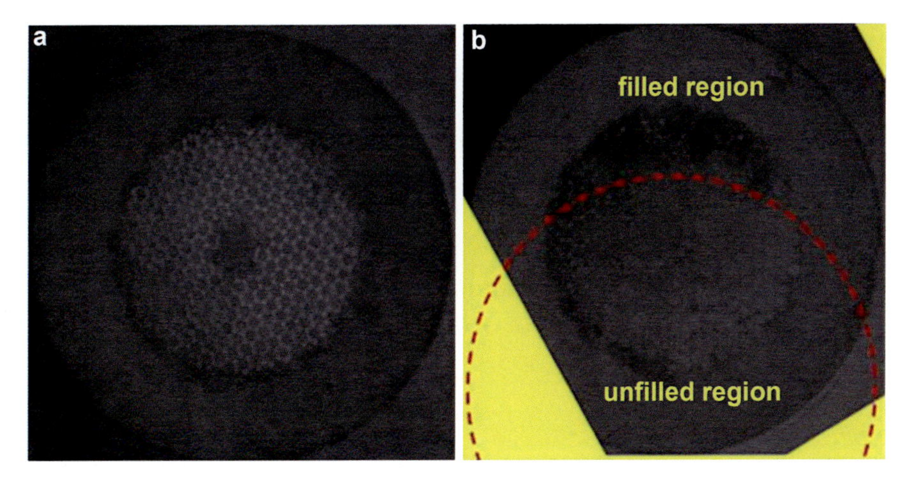

Fig. 9 (Color online) optical microscopic images of cross section of (**a**) the unfilled PCF and (**b**) the partially filled PCF

realized by a simple and practical technique as mentioned in section "Selective PCF Infiltration by Splicing Single-Mode Fiber with Lateral Offset." The partial filling of liquid results in leaking out of one polarization eigen mode while keeping the orthogonal polarization mode propagating along the fiber with relatively low loss.

In our experiment, the hollow-core PCF used is the Crystal Fiber's HC-1550-02 photonic bandgap fiber, and the diameters of the fiber cladding and the microstructured holey region are 125 and 80 μm, respectively. Figure 9a, b shows the cross-sectional images of the unfilled and the partially ethanol-filled PCF sections. The filled and unfilled regions are clearly distinguished in Fig. 9b, and ethanol filled most of the unsealed air holes.

A full-vector finite element method was used to analyze mode fields of the two eigen polarization states of the partially ethanol-filled PCF. The model used is shown in Fig. 10a, which closely resemble the partially filled PCF as shown in Fig. 9b. The pitch or spacing between air holes is $\Lambda = 3.9$ μm, and the air hole diameter to pitch ratio is $d/\Lambda \approx 97\%$. The red region indicates the liquid-filled region and the index of the liquid is taken as 1.36048 at room temperature (25 °C). Figure 10b, c shows the mode fields of X-polarization and Y-polarization modes at 1550 nm. Obviously the X-polarization mode is well confined to the core, while the Y-polarization mode has a significant leakage into the cladding region. The partially liquid-filled PCF has a large polarization dependence loss due to the asymmetric refractive index distribution in the cladding. A detail analysis of polarization-dependent loss of partially liquid-filled PCF has been reported previously (Martelli et al. 2005).

With a fiber cleaver, a 6 mm long PCF was then cut from a longer length of partially ethanol-filled PCF and then fusion spliced at both ends to standard SMF pigtails with a conventional splice condition (core to core). The total loss of the two splice joints is about 8 dB. A broadband light source with a polarizer was used as a

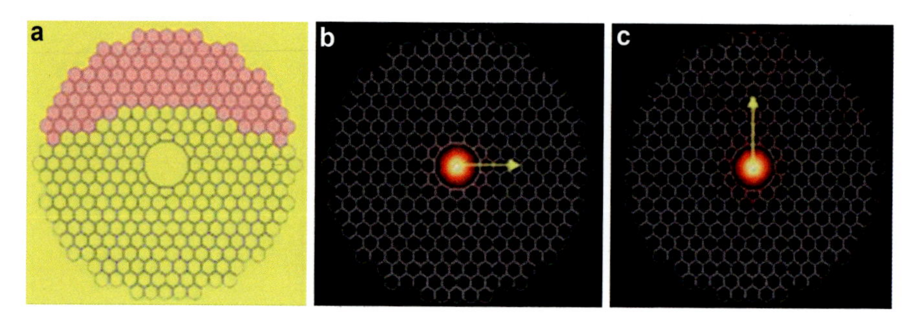

Fig. 10 (**a**) Model structure used in simulation. Red region: the ethanol-filled region. Mode field distribution of (**b**) X-polarization and (**c**) Y-polarization mode

linearly polarized source to test the performance of the fiber polarizer. The measured loss spectra for various angles of input polarization from 0 to 90° are shown in Fig. 11. The maximum and minimum losses are approximately 27.5 dB and 9.5 dB, at 90° and 0°, respectively. The polarization extinction ratio, which is the difference between the maximum and the minimum losses, is estimated to be approximately 18 dB in the wavelength range of 1480–1600 nm. The loss around 1470 nm has large variations, corresponding to the band edge of the hollow-core PCF transmission window. The insertion loss of the partially ethanol-filled section is about 1.5 dB, which is equal to the difference between the minimum loss (~9.5 dB) and the splicing loss (~8 dB). We also tested the temperature-dependent extinction ratio of the partially liquid-filled polarizer, and the results are shown in Fig. 12. The polarization extinction ratio has only weak dependence on temperature, and the variation is less than 2 dB for three different temperatures of 10 °C, 25 °C, and 40 °C.

In conclusion, a simple technique for partial filling liquid into cladding air holes of a PCF was demonstrated, in which only a conventional fusion splicer and capillary action are needed. With the technique, a compact fiber polarizer was fabricated. A fiber polarizer made of a 6 mm long partially ethanol-filled PCF exhibits a polarization extinction ratio of ~18 dB and an insertion loss ~1.5 dB over a wide operational wavelength range from 1480 to 1600 nm. By optimizing the filling length and the size of the liquid-filled region, such polarizers with better performances can be obtained. The proposed partial filling technique is simple and flexible to filling liquid into different air holes of PCFs and could be applied to fabricate other functional photonic devices and sensors.

Fiber In-Line Mach-Zehnder Interferometer

A novel type of fiber in-line Mach-Zehnder interferometer based on selective infiltration of two adjacent air holes in the innermost layer of the PCF was proposed (Yang et al. 2011). As the liquid infiltrated in the air holes has higher RI value than that of the background silica, the two-mode interference is generated by the PCF

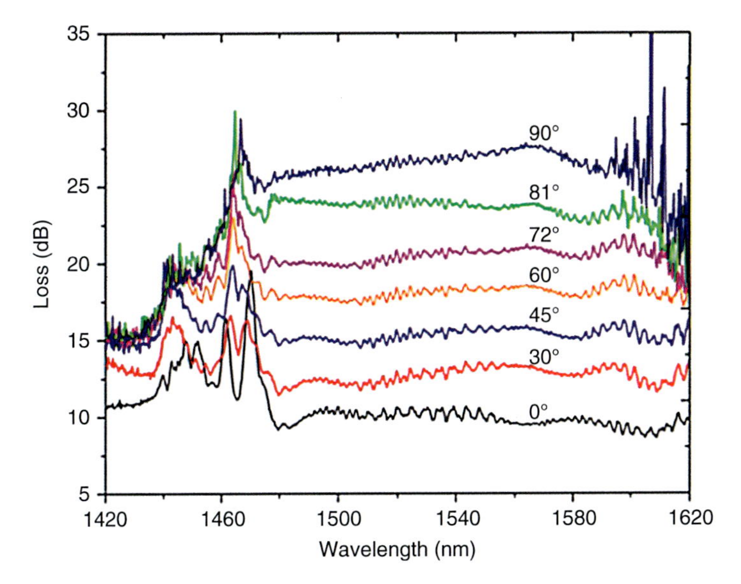

Fig. 11 Loss spectra for various linearly polarized light with different input polarization angles

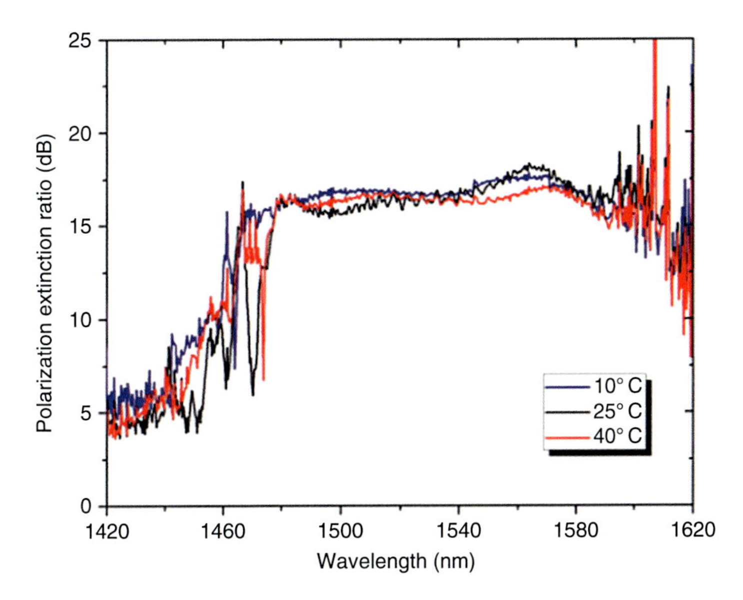

Fig. 12 Polarization extinction ratios of the partial ethanol-filled polarizer at different temperatures

core mode and the guide mode supported by the two high RI rods. Such a fiber in-line interferometer is robust and exhibits extremely high temperature sensitivity, and, in addition, a flexible sensor system operation can be achieved by tracking arbitrary fringe dip (peak) over a wide wavelength range.

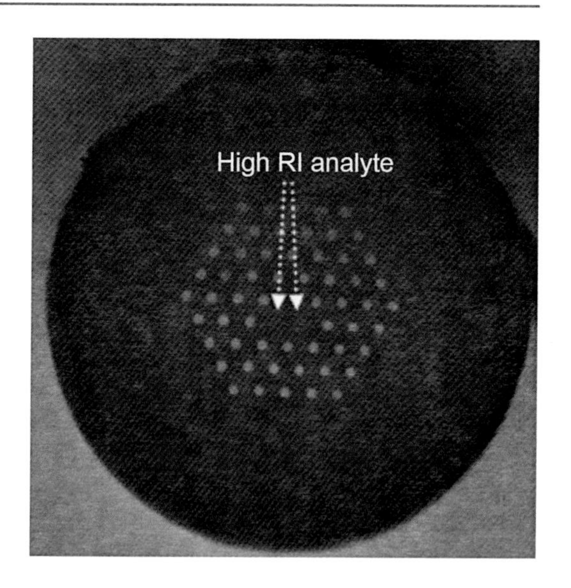

Fig. 13 Cross section of PCF with two holes infiltrated

In the experiment, by fs-assisted micromachining (Wang et al. 2010a), a section of PCF (NKT Photonics LMA-10, with hole diameter of 3.04 μm and hole pitch of 6.26 μm) is selectively infiltrated with liquids (Cargille Labs, $n = 1.47$, at 25 °C) with RI values higher than that of the background silica. Figure 13 shows the cross section of PCF with two adjacent air holes of the innermost layer infiltrated.

The motivation for infiltrating two air holes here is to generate a stable guide mode with smaller attenuation than the leaky mode generated by infiltrating only one air hole. If more than two holes are infiltrated, the interference fringe pattern may be complicated and not convenient for sensing applications. After the infiltration is completed, the PCF is cut back from the non-immersed fiber end with the help of a microscope, until all the PCF to be used is filled with RI liquid. The infiltrated PCF length is approximately 8.2 cm. Such a sample is then fused with SMF on both ends, and its transmission spectrum is measured. After each measurement, the sample is cut back again to allow measuring the transmission spectrum of the device with a different length, as shown in Fig. 14. It can be clearly seen from this figure that interference fringe pattern exists in each transmission spectrum and the FSR of the fringe pattern increases with the decrease of PCF length. For example, the FSR around 1550 nm is 8.7, 16.7, 24.2, and 40.2 nm for the PCF length of 8.2, 5.2, 3.4, and 1:8 cm, respectively. The relatively large insertion loss corresponding to the device length of 8.2 cm may be created in the splicing process.According to the theory of two-mode interference (Wang et al. 2010b),

$$\mathrm{FSR} = \frac{\lambda^2}{\Delta n_{\mathrm{eff}} L} \tag{1}$$

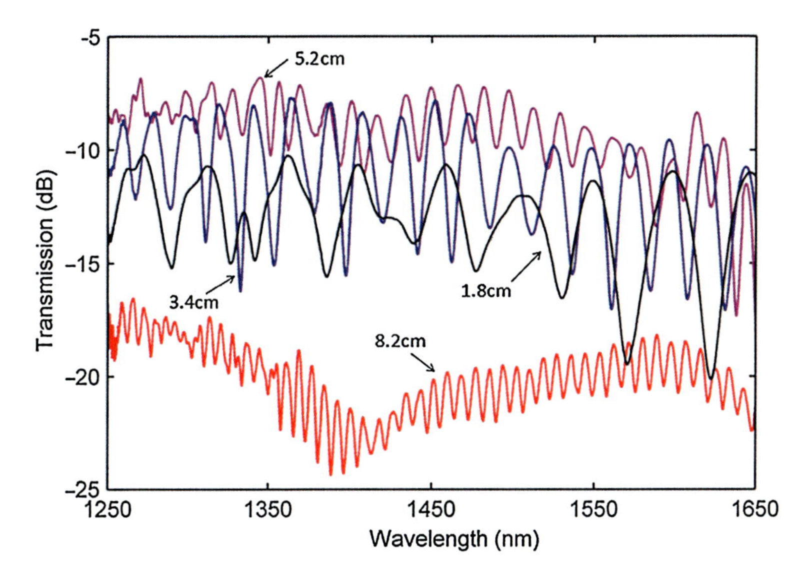

Fig. 14 Transmission spectra for different PCF lengths

where λ is the wavelength of fringe dip (or peak), L is the device length, and Δn_{eff} is the difference between the effective RI of the two modes, calculated as 3.4×10^{-3}, 2.8×10^{-3}, 3.0×10^{-3}, and 3.4×10^{-3}, respectively, for the PCF samples of different lengths. Such a Δn_{eff} difference may result from the measurement error of the PCF length. In Fig. 14, apart from the main fringe structure, a slowly varying envelope can also be found. This is likely due to the interference between the polarization modes, which has a much smaller Δn_{eff} and, hence, a larger FSR than that of the main fringe pattern. However, when the PCF length is reduced, the visibility of the main fringe pattern increases, and the slowly varying envelope has little significant effect.

In order to identify the modes involved in the main interference fringe pattern, the first four modes (each is two-folder degenerated) with the largest effective RI values are simulated by the finite element method (FEM), as shown in Fig. 15. For the sake of simplicity, only one polarization state is plotted. The arrows in Fig. 15 stand for the electrical fields. The modes in Fig. 15a, b are the bound modes of the high RI rods, but with different electric field direction, of the two LP01 modes: the modes in Fig. 15a have the same electric field direction, while in Fig. 15b, the directions of the two modes are opposite. Such a difference in electric field direction results in a difference in effective RI, as well as in mode coupling. We denote mode (c) as the PCF fundamental mode and mode (d) as the "virtual core" mode confined by the two high RI analyte rods. Being different from the real core mode supported by the second fiber core (Town et al. 2010; Yuan et al. 2010; Mangan et al. 2000; Wang et al. 2007b; Paddena et al. 2004; Du et al. 2008), the "virtual core" mode here is actually the overlap of the LP11 higher-order modes (which is the first higher-order

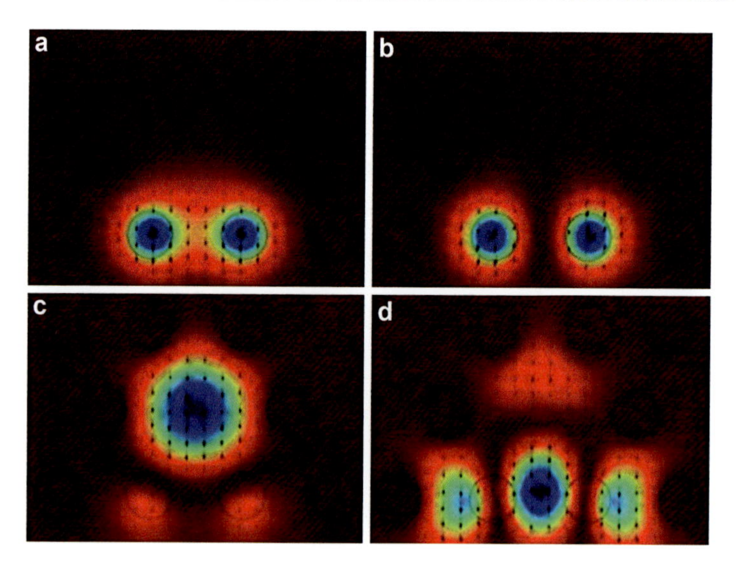

Fig. 15 Simulation of the first four mode fields of PCF with two adjacent holes infiltrate, only Y-polarization state is plotted

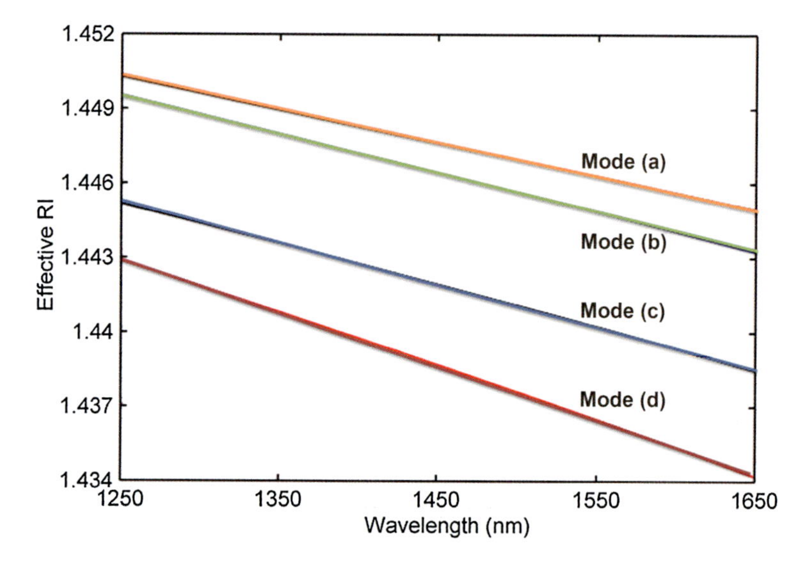

Fig. 16 Dispersion curves of fiber modes shown in Fig. 15

mode of the infiltrated holes) of the two high RI rods when they have the same electric field direction. This overlap makes the mode field intensity in the region between the two high RI rods stronger.

The dispersion curves of the modes appear in Fig. 15 are also simulated by use of FEM, with material dispersion being taken into account, as shown in Fig. 16. According to the simulated effective RI value of each mode and the previously

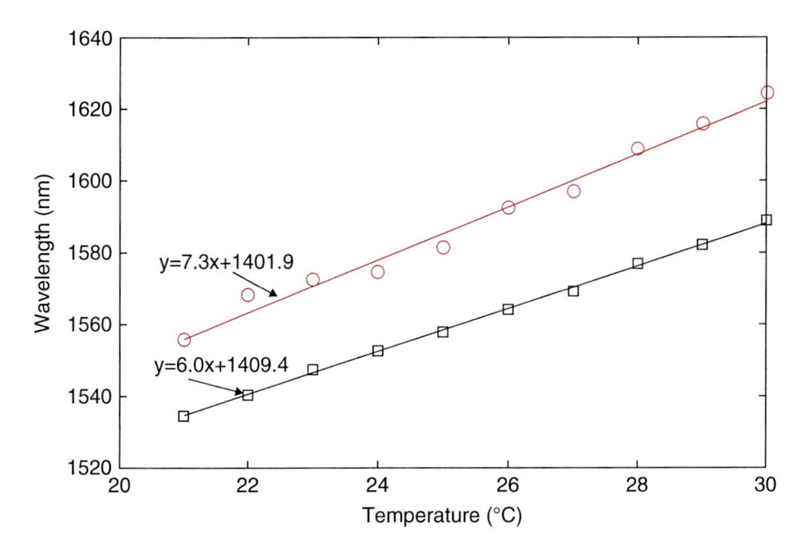

Fig. 17 Fringe wavelength shift with different PCF lengths: circles, 3:4 cm; squares, 1:8 cm; lines, linear fitted

calculated Δn_{eff}, the two modes involved in the main interference pattern should be the PCF fundamental mode and the "virtual core" mode. It is worth noting that the "virtual core" mode always has a smaller effective RI than that of the background silica, and, thus, is confined by the hole-silica structure (Wu et al. 2009). This is due to the fact that in the wavelength range between 1250 and 1650 nm, the dispersion curves of the PCF fundamental mode and the "virtual core" mode do not intersect, which means that no mode coupling will occur since the phase-matching condition cannot be satisfied, being different from selective infiltration of air holes in the outer layer.

The wavelength shifts of the fringe dip corresponding to different PCF lengths are shown in Fig. 17, where the temperature sensitivity obtained for the PCF lengths of 1.8 and 3.4 cm is 6.0 and 7.3 nm/°C, respectively. The sensitivity variation may come from the temperature fluctuation during the measurement: because of the high temperature sensitivity, a slight temperature fluctuation results in a large measurement error. Although any wavelength dip (peak) can be traced in the measurement, we choose the fringe dip around 1550 nm (the initial fringe dips are 1555.6 and 1534.5 nm for the PCF lengths of 3.4 and 1:8 cm, respectively), which simplifies the subsequent simulations.

For a Mach-Zehnder interferometer, the temperature sensitivity of a certain fringe dip (or peak) is (Wang et al. 2010b) where neff1 and neff2 are the effective RI of the two modes involved in interference. Since the high RI analyte usually has a much larger thermo-coefficient than that of silica, the temperature sensitivity of the device is much larger than that of the system based on changing RI of silica only. For example, the thermo-coefficient is 1.2×10^{-5} RIU/°C, where RIU is the refractive index unit, for silica (Toyoda and Yabe 1983), and -3.92×10^{-4} RIU/°C for the high RI liquid infiltrated in the air holes. Thus, the thermo-coefficient of the

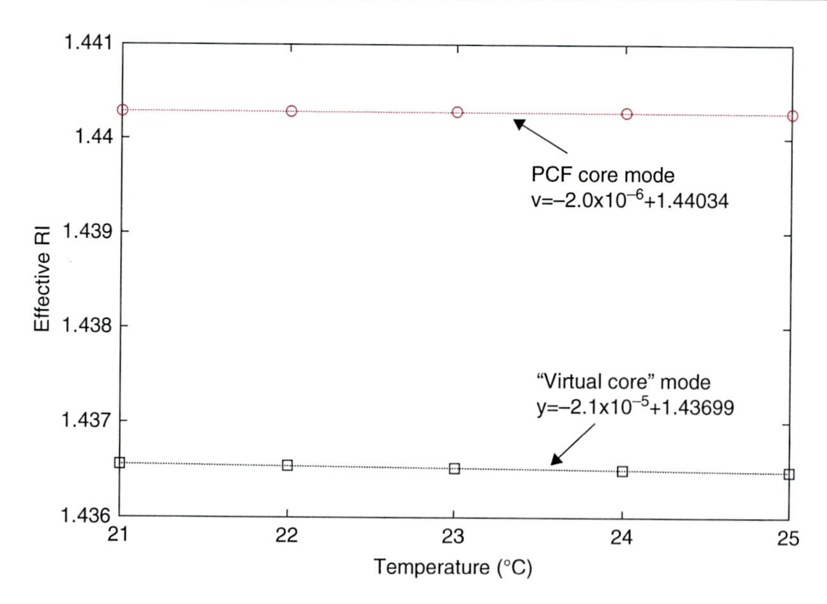

Fig. 18 Simulated thermo-coefficients of the PCF fundamental mode (circles) and the "virtual core" mode (squares); dotted lines, linear fit of the simulation results

PCF fundamental mode and the "virtual core" mode can be simulated accordingly (through FEM), and the results obtained are -2.0×10^{-6}/°C and -2.1×10^{-5}/°C, respectively, as shown in Fig. 18. The temperature sensitivity is approximately 7.7 nm/°C, reasonably close to the experimental results obtained in Fig. 17. The temperature sensitivity obtained from the experiment is much larger than that of the air cavity-based Mach-Zehnder interferometer (Wang et al. 2010b) or dual-core structured PCF (Du et al. 2008). Moreover, the close values obtained in the experiment and in the simulations also imply that the main interference pattern is most likely produced by the PCF fundamental mode and the "virtual core" mode.

In conclusion, by selective infiltrating of the two adjacent air holes of the innermost layer of the PCF, a "virtual core" mode can be supported by the electric field coupling between the higher-order modes of the high RI rods. This virtual mode can interfere with the PCF fundamental mode, thus producing interference fringe pattern in the transmission spectrum. Such a fiber in-line Mach-Zehnder interferometer device is robust, exhibits extremely high temperature sensitivity, and supports a flexible sensor system operation.

Embedded PCF Coupler for Strain Measurement

An ultrasensitive strain sensor based on PCF with embedded coupler structure was proposed (Wang et al. 2012), which is formed by selectively filling one of the PCF air holes with refractive index (RI) liquid by use of a femtosecond (fs) laser-assisted

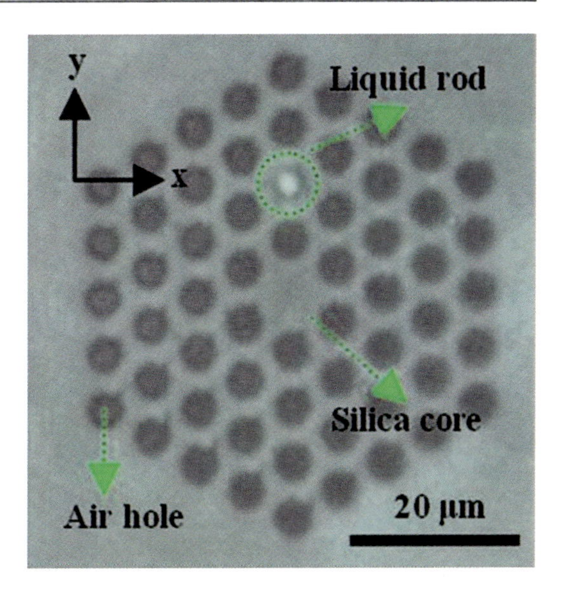

Fig. 19 Cross-sectional view of the selectively infiltrated PCF with embedded couplers. The red dotted circles highlight the hole filled with standard RI liquid

selective infiltration method (Wang et al. 2010a). The liquid rod created acts as a waveguide, which, together with PCF fiber core, essentially forms an embedded coupler. By simply changing the RI of the infiltrated liquid, the coupled modes of the device can be selected between the PCF core mode and the fundamental mode or higher-order mode of the liquid rod. The static strain sensitivity obtained in the experiment is ~−22 pm/με for the coupling between the PCF core mode and the fundamental mode of the liquid rod.

The cross-sectional view of the PCF with embedded coupler is shown in Fig. 19. One of the air holes in the cladding of the PCF (LMA-10, NKT Photonics) is filled with standard RI liquid (Cargille Laboratories, Inc.).

The operating principle of the proposed sensor can be explained by coupled-mode theory for the directional coupler. Figure 20a shows the phase-matching curves of the linearly polarized (LP) PCF core mode (core LP01) and the fundamental mode of the liquid rod (rod LP01), both in Y-polarization, at the temperature of 30 °C, with the hole filled with 1.46 RI liquid as shown in Fig. 19. A resonant wavelength of ~1485 nm can be determined in Fig. 20a with the help of the auxiliary gray-dashed lines in the avoided crossing region (marked with green lines in Fig. 20) where the modes existed cannot be identified as the core or the rod modes. This resonance originates from the coupling between the even and odd hybrid modes, and the even mode has a higher effective index than the odd mode over the avoided crossing. By applying an axial strain of 500 με, the resonant wavelength shifts to 1471.6 nm, and a strain sensitivity of −23.8 pm/με can be obtained.

The PCF core mode can also couple into a higher-order mode of the liquid rod provided that the air hole is filled with appropriate RI liquid. For instance, the resonance between the core mode and the LP11 mode of the liquid rod can be observed in the near-IR wavelength region by filling the air hole with liquid of RI of

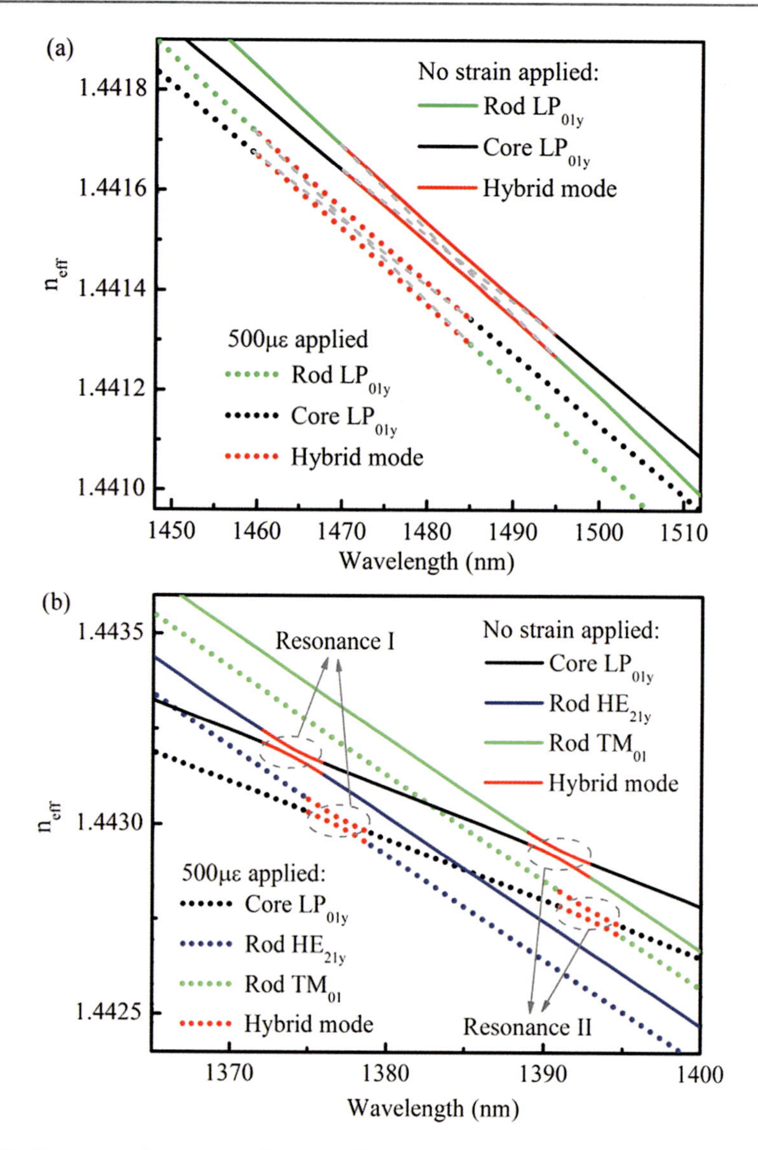

Fig. 20 Phase-matching curves of the coupling between the Y-polarized PCF core mode (**a**) fundamental mode and (**b**) LP11 mode of the infiltrated liquid rod waveguide at 30 °C before and after the axial strain is applied. Note that only the HE11, HE21, and TM01 modes are plotted in (**b**) for clarity, and the avoided crossings between the core LP01 and rod HE21 and between the core LP01 and TM01 modes are denoted as Resonance I and II, respectively

~1.50. Figure 20b displays the phase-matching curves of core LP01 and rod LP11 modes, which include TE, TM, and hybrid electromagnetic (HE) modes. Resonance I denotes the coupling between the core LP01y and rod HE21y modes, while the coupling between the core LP01y and rod TM01 modes is denoted as Resonance II.

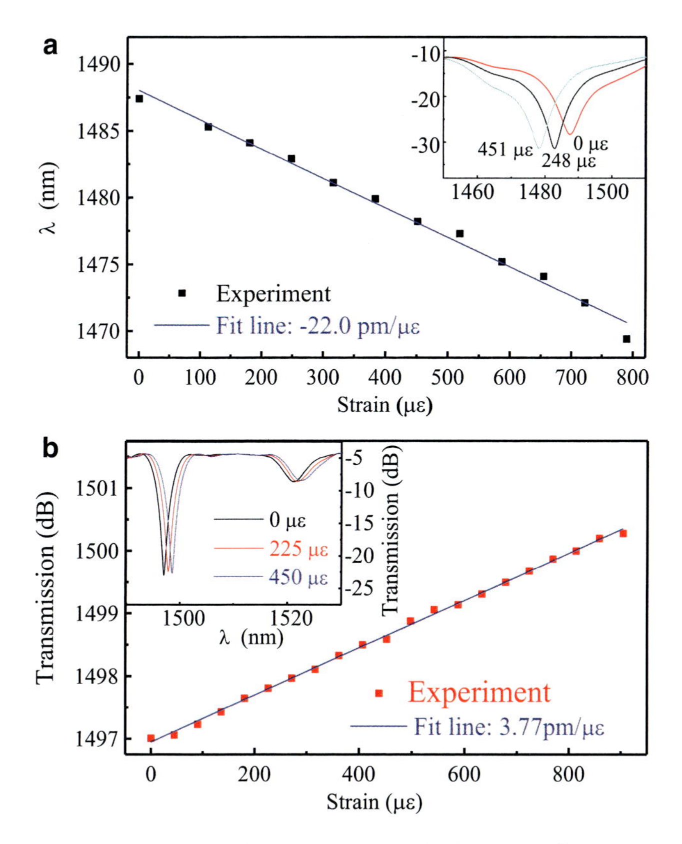

Fig. 21 Resonant wavelength shift with applied axial strain for (**a**) 31.3 °C, 1.46 RI liquid filled and (**b**) 25 °C, 1.506 RI liquid-filled PCF with lengths of 26 and 37 mm, respectively. The infiltrated holes of the samples are the same as that shown in Fig. 34. Insets show their transmission spectra at different strains

The strain sensitivities for all the mode combination cases mentioned above are calculated as ~5.9 pm/$\mu\varepsilon$.

In the experiments, PCF (LMA-10) was infiltrated following the procedures reported in Wang et al. (2010a). After infiltration, both ends of the selectively infiltrated PCF were cleaved and fusion spliced with single-mode fibers (SMFs). Standard 1.46 RI and 1.506 RI liquid were used in the experiment.

Two PCF samples with the infiltrated hole as shown in Fig. 19 were fabricated: sample I was filled with 1.46 RI liquid, had a length of ~26 mm, and was kept in a column oven at 31.3 °C; and sample II was filled with 1.506 RI liquid, had a length of ~37 mm, and was maintained at 25 °C. For sample I, it can be found from Fig. 21a and its inset that the resonant wavelength experiences a blueshift with the increase of strain and there is a linear relationship between the resonant wavelength and the axial strain applied. The average strain sensitivity obtained is ~−22 pm/$\mu\varepsilon$, while the largest resonant dip depth is >20 dB, with an insertion loss of ~11 dB, mainly due to the fusion splicing of PCF and SMF. Figure 21b demonstrates the strain sensing characteristics of sample II. The average strain sensitivity obtained is ~3.8 pm/$\mu\varepsilon$. The inset of Fig. 21b shows the transmission spectra of sample II at different strain values. The resonant dip around 1500 nm is due to the coupling between the PCF core LP01 and rod HE21 modes, while the other one around 1520 nm is due to the coupling between the core LP01 and rod TE01 and TM01 modes.

The strain sensitivity obtained in our device is one order higher than that of fiber interferometer and FBG and two or three times larger than that of an LPFG, which can be further enhanced if appropriate RI material can be utilized, as long as the slope of the rod mode dispersion is close to that of the silica core dispersion.

Temperature Sensors Based on Filled PCFs

Temperature Sensor with Alcohol Fully Filled Whole Length PCF FLM

By arranging the geometry or distribution of the core and the air hole cladding, PCFs can have ultrahigh birefringence, and therefore HiBi-PCFs have been the best choice to make fiber loop mirrors (FLMs) (Zhao et al. 2004; Dong et al. 2007; Fu et al. 2008; Qian et al. 2010). In strain sensors based on HiBi-PCF FLMs, the length of HiBi-PCF is shortened to ~8 cm, and thus the stability of the total system increases greatly (Dong et al. 2007; Qian et al. 2010). However, HiBi-PCF FLMs cannot be used to measure temperature, because HiBi-PCFs have a low thermo-optic and thermo-expansion coefficient. However, by inserting a short alcohol-filled HiBi-PCF with the whole length into a FLM, a temperature sensor with an extremely high sensitivity can be realized by measuring the wavelength shift of the resonant dips of the alcohol-filled HiBi-PCF FLM (Qian et al. 2011b).

The temperature sensor, as shown in Fig. 22, consists of a 3 dB coupler and a short alcohol-filled PCF with the whole length. Alcohol is chosen to fill into HiBi-PCF since it is an easy-filled liquid with a high temperature sensitivity. Here, an alcohol-filled HiBi-PCF is inserted into a FLM as a temperature sensing head. Birefringence change ΔB and length change ΔL of the alcohol-filled HiBi-PCF caused by temperature lead a wavelength shifting of the resonant dips. The relationship between the dip wavelength change $\Delta\lambda_{dip}$, ΔB, and ΔL is simply expressed as $\Delta\lambda_{dip} = (\Delta \mathrm{BL} + B\Delta L)/k$, where ΔB is the birefringence change caused by the thermo-optic effect, including that of the original HiBi-PCF and that of the filled alcohol, and ΔL is the length change caused by the thermo-expansion effect, which also includes the elongation of the original HiBi-PCF and the expansion of the filled alcohol.

We neglect ΔB and ΔL caused by the HiBi-PCF itself because of a good thermal independence of the HiBi-PCF. Further, ΔL caused by the thermo-expansion of the filled alcohol is also ignored since the volume of alcohol filled into the air holes of the HiBi-PCF is small. Thus, $\Delta\lambda_{dip}$ mainly depends on ΔB of the alcohol-filled HiBi-PCF. The birefringence-temperature dependence of the alcohol-filled

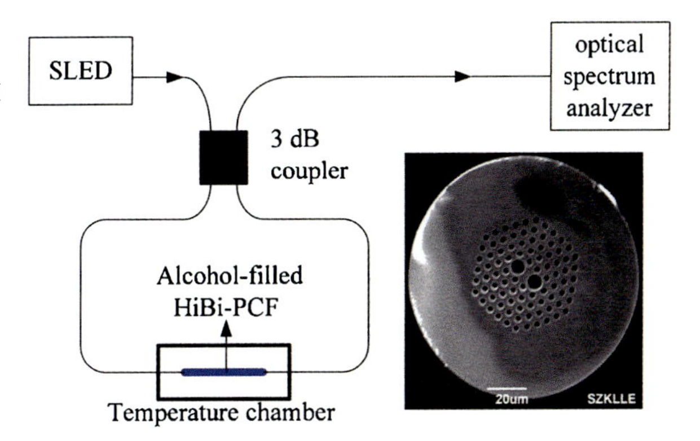

Fig. 22 Experimental setup for a temperature sensor based on an FLM. Inset: SEM of the used HiBi-PCF

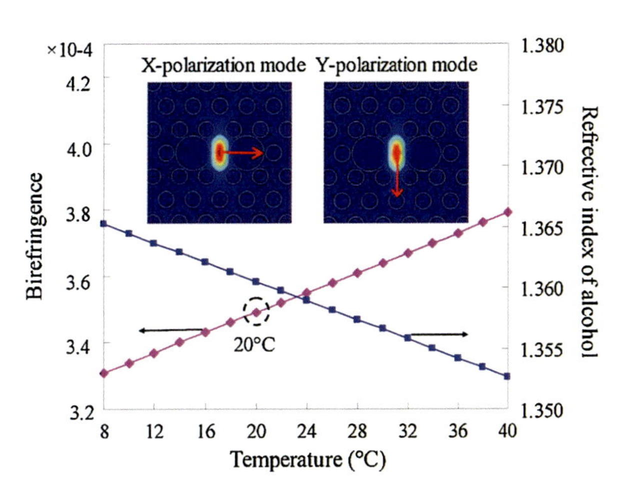

Fig. 23 Temperature dependence of the refractive index of alcohol and the birefringence of the alcohol-filled HiBi-PCF in theory. Insets: X and Y-polarization mode fields of the alcohol-filled HiBi-PCF at 20 °C

HiBi-PCF is analyzed by using a full-vector finite element method (FEM). The diameters of the bigger and smaller holes are 7 and 3.2 μm, respectively, and the pitch length between centers of two adjacent holes is 5.46 μm, according to the HiBi-PCF used in experiment. The refractive index of pure silica and the filled alcohol is taken as 1.4457.

Figure 23 shows the empirical temperature dependence of the refractive index of alcohol and the theoretical temperature dependence of the birefringence of the alcohol-filled HiBi-PCF. With the temperature rising, the refractive index of alcohol decreases linearly, while the birefringence of the alcohol-filled HiBi-PCF increases linearly. The mode fields of the two orthogonal polarizations at 20 °C are shown in the insertion of Fig. 23. The birefringence of the alcohol-filled HiBi-PCF is calculated at 3.5×10^{-4} at 20 °C. P_t is defined as a thermo-optic constant on the birefringence of the alcohol-filled HiBi-PCF, which equals to the slope of the temperature dependence curve of birefringence and is calculated at 1.5×10^{-6}/°C. The relationship between the resonant dip wavelengths shift $\Delta\lambda_{\text{dip}}$ and the temperature change ΔT can be gotten as

$$\Delta\lambda_{\mathrm{dip}} = \Delta \mathrm{BL}/k = \frac{\mathrm{LP}_t}{k}\Delta T = \frac{\lambda_{\mathrm{dip}}}{B}P_t\Delta T \quad (2)$$

Based on the above equation, the temperature sensitivity of the alcohol-filled HiBi-PCF FLM is related to λ_{dip}, P_t, and B. A high temperature sensitivity depends on a long wavelength λ_{dip} of the measured resonant dip, a high thermo-optic constant P_t, and a small birefringence B of the filled HiBi-PCF.

The HiBi-PCF used in the experiment is provided by Yangtze Optical Fibre and Cable Company, and the cross-sectional scanning electron micrograph (SEM) is shown in the insertion of Fig. 22. The HiBi-PCF has a birefringence of 10.2×10^{-4} at 1550 nm, and the length is 6.1 cm. After the HiBi-PCF filling with alcohol by air holes capillary force, the birefringence of the PCF reduces significantly, which bring advantages on a larger wavelength space between two resonant dips and on a wider measurement range. Both ends of the alcohol-filled HiBi-PCF are spliced to conventional single-mode fiber (SMF) by using a regular arc splicing machine (Fujikura FSM 60). The PCF-SMF splicing loss is ~3 dB, which is relatively large and caused by mismatching of mode field and numerical apertures between the HiBi-PCF and SMF. The total insertion loss of the FLM is ~8.5 dB. However, it wouldn't affect experimental results since we directly measure the resonant dip wavelength shift. A broadband SLED source with 200 nm wavelength range is used as an input light source. The transmission spectra of the FLM are measured by an optical spectrum analyzer with a wavelength resolution of 0.1 nm.

Figure 24 shows the transmission spectrum of the alcohol-filled HiBi-PCF FLM at room temperature (20 °C). Two resonant dips of the FLM display in the wavelength range from 1400 to 1600 nm. One is at the wavelength of 1455.8 nm (dip A) with 15.5 dB extinction ratio; the other is at about 1549.8 nm (dip B) with 10.5 dB extinction ratio. The wavelength spacing between these two dips is ~94 nm, and the corresponding birefringence of the alcohol-filled HiBi-PCF is $\sim 3.9 \times 10^{-4}$ at 20 °C, which is close to the theoretical value ($\sim 3.5 \times 10^{-4}$). The little difference between the experimental and theoretical values may be caused by the error of air holes geometry size of HiBi-PCF according the SEM.

In the experiment, the temperature characteristic of the alcohol-filled HiBi-PCF FLM is tested by placing the alcohol-filled HiBi-PCF of the FLM at a temperature-controlled container. Figure 25a, b shows the transmission spectra of the alcohol-filled HiBi-PCF FLM at temperature range of 20–34 °C and 8–20 °C, respectively. Dip A redshifts from 1455.8 to 1543.7 nm with temperature increasing gradually from 20 to 34 °C; at the same time, the extinction ratio of dip A decreases, while dip B blueshifts from 1549.8 to 1470.4 nm with the temperature decreasing gradually from 20 to 8 °C.

Figure 26 shows the experimental relationship between temperature and the resonant wavelength of dip A and dip B. The fitting curves can be expressed as $y = 6.2176x+1331.7$ for dip A and $y = 6.6335x+1416.7$ for dip B, and the high fitting degrees 0.9997 and 0.9995 mean the linearity of the resonant wavelength to temperature is excellent. The experimental temperature sensitivities of dip A and dip

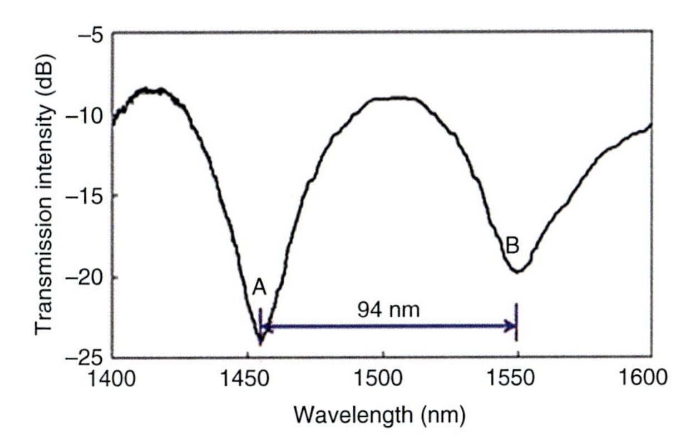

Fig. 24 Transmission spectrum of the alcohol-filled HiBi-PCF FLM at 20 °C

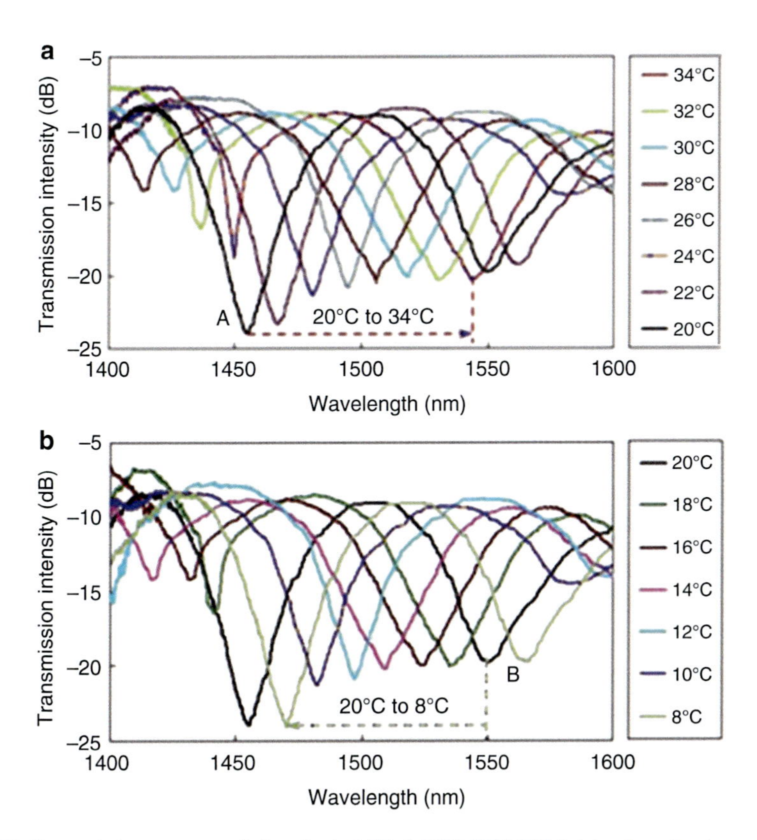

Fig. 25 Transmission spectra of the alcohol-filled HiBi-PCF FLM (**a**) when the temperature increases from 20 to 34 °C and (**b**) when the temperature decreases from 20 to 8 °C

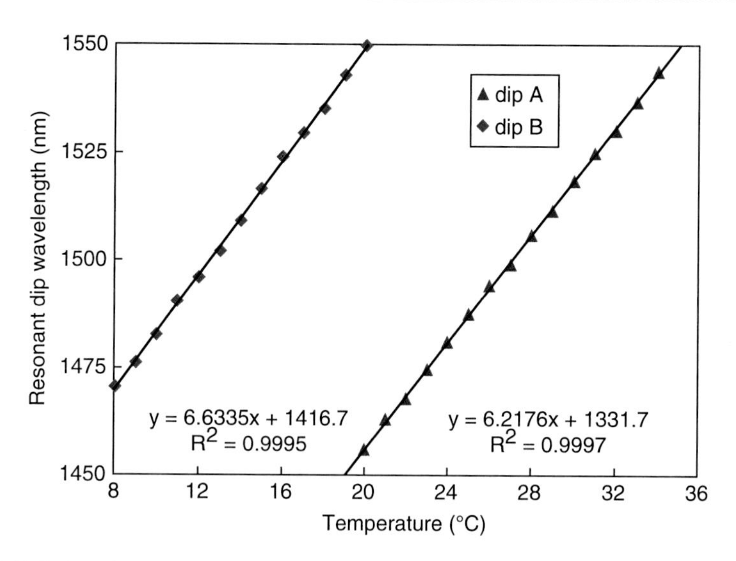

Fig. 26 Relationship between temperature and the resonant wavelength of dip A and dip B

B are ~6.2 nm/°C and ~6.6 nm/°C, respectively. And the theoretical sensitivities are ~6.1 nm/°C and ~6.5 nm/°C from Eq. (14). It is clear that the theoretical and the experimental results are in accordance. The temperature sensitivity of the alcohol-filled HiBi-PCF FLM is very high and reaches up to about 660 and seven times higher than that of a FBG (~0.01 nm/°C) and that of the FLM made of a conventional HiBi fiber with a 72 cm length (~0.94 nm/°C) (Liu et al. 2005).

In practical uses, for a wider measurement range of temperature, the length L of the HiBi-PCF can be shortened in order to widen the spacing between two resonant dips based on $S = \lambda^2/BL$. For example, when the alcohol-filled HiBi-PCF is 1 cm, the spacing of the proposed FLM sensor is ~564 nm. The measurement range of ~84 °C is provided with the same temperature sensitivity ~6.6 nm/°C according to Eq. (2), in which the length of the sensing fiber is the same as the length of FBG sensing head and is shortened 72 times than that of the conventional HiBi-FLM temperature sensor.

Temperature Sensor with Alcohol Fully Filled Partial Length PCF FLM

In the above temperature sensor, the birefringence change of the filled HiBi-PCF caused by the thermo-optic effect of filled alcohol was considered as the main reason for the high temperature sensitivity, and the length change of the filled alcohol in the HiBi-PCF, caused by the thermo-expansion of alcohol, was ignored since the alcohol was filled fully into the air holes of the HiBi-PCF with the whole length. However, what will happen if a fully filled HiBi-PCF with a partial filled length is used?

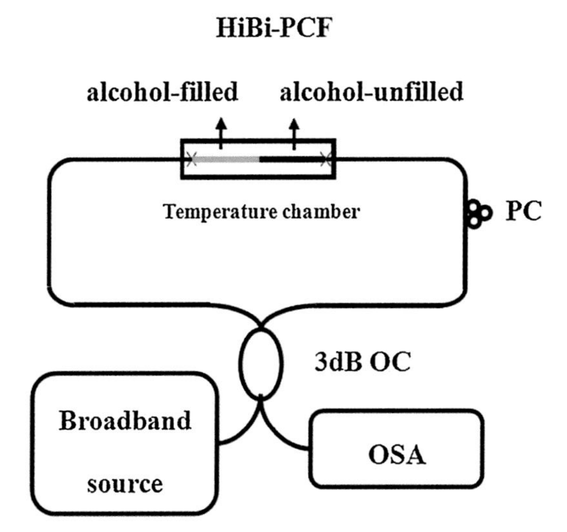

Fig. 27 Schematic setup of the alcohol fully filled partial length HiBi-PCF FLM. The blue line, the alcohol-filled HiBi-PCF, the black line, the alcohol-unfilled HiBi-PCF

In this part, we demonstrate an alcohol fully filled partial length HiBi-PCF temperature sensor based on an FLM and analyze its characteristics deeply in theory and experiment firstly (Zhao et al. 2012). A new phenomenon that the resonant dip wavelengths of the alcohol fully filled partial length HiBi-PCF FLM blueshift with temperature increasing is observed, which is contrary to what had been reported (Zhao et al. 2004; Qian et al. 2011b; Liu et al. 2005). By considering the influences of the group birefringence and the thermo-expansion of alcohol, we can explain this phenomenon very well.

Figure 27 shows the proposed alcohol fully filled partial length HiBi-PCF FLM temperature sensor, which contains an optical broadband source (BBS), an optical spectrum analyzer (OSA), and a FLM. The FLM consists of a 3 dB optical coupler (OC), a polarization controller (PC), and a section of an alcohol full-filled HiBi-PCF with a partial length. The alcohol-filled length is L_1 with a birefringence of B_1, and the alcohol-unfilled length is L_2 with a birefringence of B_2. The main principle of the FLM is summarized as follows. The input light is split into two counter-propagating beams equally by the 3 dB OC. And then, these two beams recombine at the 3 dB OC again after propagating around the loop. It exhibits interference due to an optical path difference after transmitting through the alcohol fully filled HiBi-PCF with a partial length because of its birefringence. Therefore, the output transmission intensity can be written as (Liu et al. 2005):

$$T(\lambda) = \left[1 - \cos\left(\varphi_1 + \varphi_2\right)\right]/2 \tag{3}$$

where $\varphi_i = 2\,\pi\,B_iL_i/\lambda\ (i = 1, 2)$ is the phase difference of the alcohol-filled (or unfilled) HiBi-PCF and λ is the wavelength. The transmission dip wavelengths are the resonant wavelengths satisfying $2\pi\,(B_1L_1 + B_2L_2)/\lambda_{\text{dip}} = 2k\pi$, where k is any integer. Thus, the resonant dip wavelengths can be described as follows:

$$\lambda_{\text{dip}} = \frac{B_1L_1 + B_2L_2}{k} \tag{4}$$

The changes of the HiBi-PCF birefringence and that of the length of the alcohol-filled/unfilled HiBi-PCF caused by temperature lead the wavelength of resonant dips shifting according to Eq. (4). The relationship between the change of the dip wavelength $\Delta\lambda_{\text{dip}}$, the birefringence change ΔB, and the length change ΔL can be expressed as follows:

$$\Delta\lambda_{\text{dip}} = \frac{\Delta B_1L_1 + \Delta L_1B_1 + \Delta B_2L_2 + \Delta L_2B_2}{k} \tag{5}$$

where ΔB_1 and ΔB_2 are birefringence changes of the filled-HiBi-PCF and the unfilled HiBi-PCF, caused by the thermo-optic effect, respectively. ΔB_1 includes the thermo-optic effect of the filled alcohol and that of the original HiBi-PCF, but ΔB_2 is only affected by the thermo-optic effect of the original HiBi-PCF. ΔL_1 and ΔL_2 are the alcohol-filled/unfilled length changes caused by the thermo-expansion effect, which includes the elongation of the HiBi-PCF and the expansion of the filled alcohol.

Because of the good thermal independence of the HiBi-PCF (Liu et al. 2005), the birefringence and the total length changes caused by the HiBi-PCF itself are neglected. Therefore, ΔB_1 is only affected by thermo-optic effect of the filled alcohol, and ΔB_2 is nearly zero. ΔL_1 equals to – ΔL_2 because the total length is a constant and $L_1 = L - L_2$. So, the dip wavelength shift with temperature can be given as:

$$\Delta\lambda_{\text{dip}} \approx \frac{\Delta B_1L_1 + \Delta L_1\left(B_1 - B_2\right)}{k} \tag{6}$$

with temperature T changing, the birefringence B_1 will change with both temperature, and dip wavelength changing. Therefore, except for considering the birefringence change $\mathrm{dB_1/dT}$ caused by temperature, the $\frac{\mathrm{dB_1}}{d\lambda}\frac{d\lambda}{\mathrm{dT}}$ also should be considered, and the same explanation should be for B_2. By substituting P_t, defined as a thermo-optic constant on the birefringence of the alcohol-filled HiBi-PCF which nearly equals to $\Delta B_1/\Delta T$, and α_t, a thermal expansion coefficient of alcohol which is $\Delta\mathrm{V}/(\mathrm{V}\Delta\mathrm{T})$, the final results about $\Delta\lambda_{\text{dip}}/\Delta T$ are shown below:

$$\frac{\Delta\lambda_{\text{dip}}}{\Delta T} = \frac{\lambda_{\text{dip}}}{B_{g1} + B_{g2}\frac{L_2}{L_1}}\left[P_t + \alpha_t\left(B_{g1} - B_{g2}\right)\right] \tag{7}$$

where ΔT is the temperature change. Therefore, the resonant dips may blueshift with temperature increasing in an alcohol fully filled partial length HiBi-PCF FLM when B_{g1} is less than B_{g2} in Eq. (7).

In Zhao et al. (2012), the authors used a 52.5 mm long HiBi-PCF provided by Yangtze Optical Fibre and Cable Company, and the cross-sectional scanning electron micrograph (SEM) is shown in the insertion of Fig. 27. The length of the alcohol-filled section is 35 mm, and that of the alcohol-unfilled section is 17.5 mm at room temperature 24 °C, respectively. The two ends of the alcohol un-full-filled HiBi-PCF are spliced to conventional SMFs by using the Fujikura FSM 60 arc splicing machine. The splice loss between the alcohol-filled HiBi-PCF and SMF is 3.5 dB, and that between the unfilled HiBi-PCF and SMF is 4 dB, which are caused by the mode field mismatching between the HiBi-PCFs and SMFs. The total loss of the FLM is 10.5 dB. The BBS with the wavelength range of 200 nm is used as the input light source. The transmission spectrum is measured by an OSA with a wavelength resolution of 0.1 nm.

The birefringence-temperature dependence of the alcohol-filled HiBi-PCF is analyzed using a full-vector finite element method. Figure 28 shows the birefringences B_1, B_2 and the group birefringences B_{g1}, B_{g2}, in response of the wavelength, respectively. The group birefringence B_{g2} of the unfilled HiBi-PCF increases largely compared with that birefringence B_2, but B_{g1} of the filled HiBi-PCF is less than that birefringence B_1. This causes B_{g1} less than B_{g2}. The HiBi-PCF has a birefringence of $\sim 10.8 \times 10^{-4}$, and the calculated birefringence of the alcohol-filled HiBi-PCF is $\sim 2.6 \times 10^{-4}$ at 32 °C for the wavelength of 1550 nm. P_t is $\sim 1 \times 10^{-6}$ which is just the slope of the temperature dependence curve of birefringence.

Figure 29a shows the transmission spectra of the alcohol fully filled partial length HiBi-PCF FLM when temperature increases from 32 to 64 °C. The temperature response of the resonant dips is measured by placing the alcohol fully filled HiBi-PCF with the partial length in a temperature-controlled container. We can clearly see that the wavelength spacing between two transmission minima is ~85.3 nm. The first transmission minimum (dip A) is at 1499.9 nm with an extinction ratio of 16.7 dB, and the other resonant dip B is at 1585.2 nm with a 15.4 dB extinction ratio when temperature is 32 °C. The corresponding length of

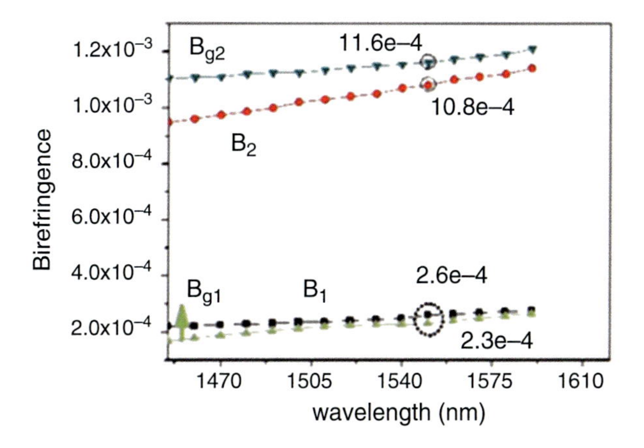

Fig. 28 The birefringence B and the group birefringence B_g of the HiBi-PCF in response of the wavelength, respectively

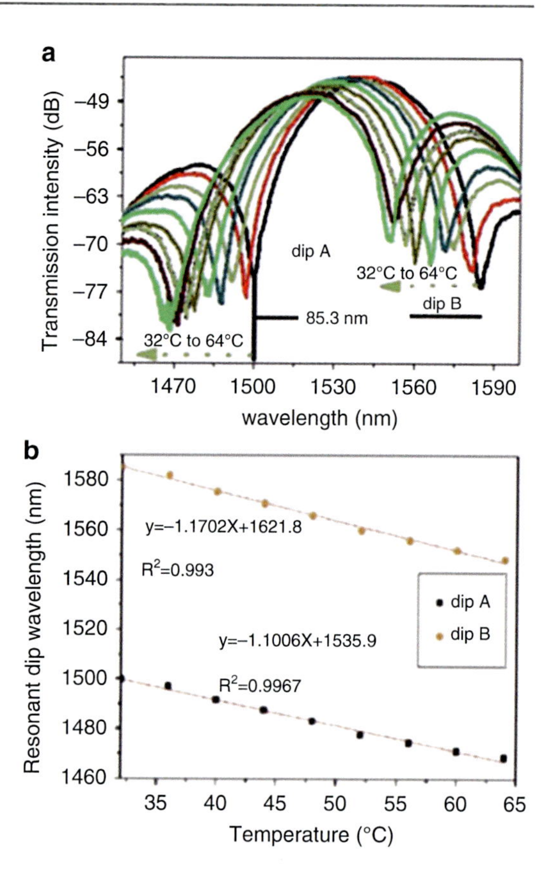

Fig. 29 (**a**) Transmission spectra of the alcohol fully filled partial length HiBi-PCF FLM when temperature is increasing. (**b**) Relationships between temperature and the wavelengths of the resonant dips

the alcohol-filled HiBi-PCF is calculated as ~35.54 mm, and that of the alcohol unfilled HiBi-PCF is ~16.96 mm, according to Eq. (7) (B_1 is 2.34×10^{-4} and 2.74×10^{-4} for the wavelength of 1499.9 and 1585.2 nm, and B_2 is 10.23×10^{-4} and 11.2×10^{-4} for the wavelength of 1499.9 and 1585.2 nm at 32 °C, respectively). Therefore, we can estimate α_t is 0.00193/°C, which is close to the empirical value of α_t (0.0011 at temperature of 20 °C). The difference may be caused by the inaccurate calculated values of the birefringence of the alcohol-filled/unfilled HiBi-PCF.

In addition, both of dip A and dip B blueshift with temperature increasing. The wavelength shift direction of resonant dips in the alcohol fully filled partial length HiBi-PCF FLM is totally opposite with that of the alcohol full-filled whole length HiBi-PCF FLM (Qian et al. 2011b), whose temperature sensitivity is positive since the thermal expansion of the filled alcohol can be ignored and the Eq. (7) becomes to $\Delta\lambda_{\text{dip}}/\Delta T = \lambda_{\text{dip}}\ P_t/B_{g1}$. However, due to the strong effect of the filled alcohol expansion on the alcohol fully filled HiBi-PCF with the partial length, the temperature sensitivity of the alcohol fully filled partial length HiBi-PCF

FLM is negative since $[P_t + \alpha_t (B_{g1} - B_{g2})] < 0$ based on the parameters obtained above. This phenomenon is consistent very well with our theoretical expectation.

Figure 29b shows the relationships between temperature and the resonant wavelengths of dip A and dip B when temperature increases. Both wavelengths of dip A and dip B vary linearly with temperature. The experimental wavelength temperature sensitivity of dip A is ~−1.1 nm/°C with a high fitting degree of 0.9967. For dip B, the wavelength temperature sensitivity of dip B is ~−1.17 nm/°C with a fitting degree of 0.993. Both the temperature sensitivity of dip A and that of dip B are close to the theoretical values of ~−1.51 nm/°C and ~−1.57 nm/°C according to Eq. (7). The difference between the experimental and the theoretical values is caused by the inaccurate values of the calculated birefringence of the alcohol-filled/unfilled HiBi-PCF and α_t.

In order to further understand the change of the filled alcohol in the alcohol fully filled HiBi-PCF with a partial length, we also measured the shifting of the resonant dips of the FLM with temperature decreasing. Figure 30a shows the transmission spectra of the alcohol fully filled partial length HiBi-PCF FLM from 64 to 32 °C. The resonant dips redshift with temperature decreasing, which is just contrary to the trend of the resonant dip shift in temperature rising process. The relationship between the resonant wavelengths and temperature keeps good linearity in the measured temperature range as shown in Fig. 30b. The temperature sensitivity of dip A and that of dip B are ~−1.05 nm/°C and ~−1.19 nm/°C, respectively, which are consistent with the temperature sensitivities in temperature rising process. That means the changing process of the filled alcohol in the full-filled HiBi-PCF with the partial length is reversible.

In the experiment, the resonant dips of the alcohol fully filled partial length HiBi-PCF FLM blueshift with temperature increasing, which are completely opposite to that of the alcohol full-filled whole length HiBi-PCF FLM (Qian et al. 2011b). The temperature sensitivities of the proposed sensor are negative and agree well with the theoretical expectations. The absolute values are one-sixth than that of the alcohol full-filled whole length HiBi-PCF FLM. Therefore, the existence of an alcohol fully filled HiBi-PCF with a partial length can have greatly effect on the temperature sensitivity of alcohol-filled HiBi-PCF FLMs.

In conclusion, a new phenomenon in the alcohol fully filled partial length HiBi-PCF FLM temperature sensor has been demonstrated and studied. Experimental results show that the temperature characteristic of the alcohol fully filled partial length HiBi-PCF FLM is unique, and the resonant dip wavelengths blueshift with temperature increasing. This new phenomenon is contrary to that in alcohol-filled whole length HiBi-PCF FLM and other reported FLMs and helpful to make researches pay much attention to the filling process of alcohol-filled HiBi-PCF FLM temperature sensors, and it has a directive significance to further investigate the properties of other liquid fully filled PCFs with partial lengths in some special occasions.

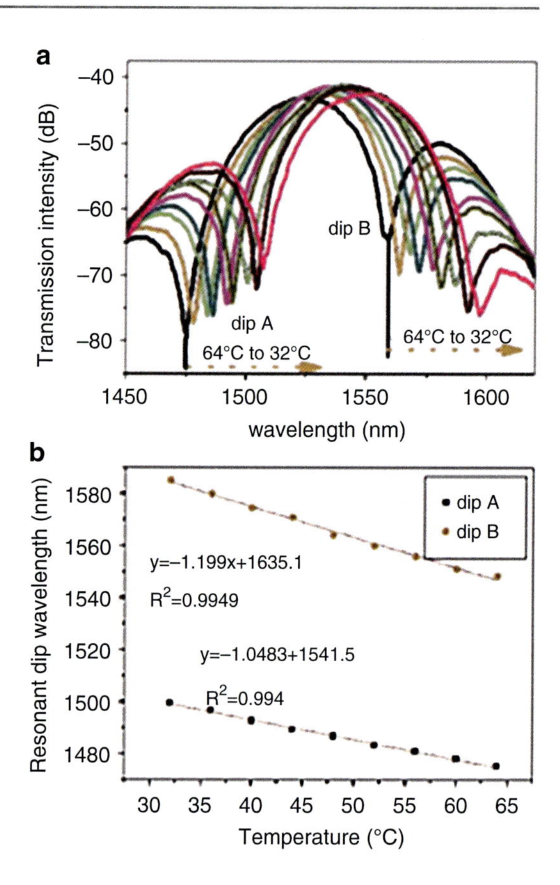

Fig. 30 (**a**) Transmission spectra of the alcohol fully filled partial length HiBi-PCF FLM when temperature is decreasing. (**b**) Relationships between temperature and the resonant wavelengths

Temperature Sensor Based on PCF with Two Infiltrated Air Holes

A liquid infiltrated PCF with ultrahigh temperature sensitivity and excellent temporal stability is proposed and demonstrated (Yang and Wang 2012). In such a device, the liquid with higher RI than background silica is infiltrated into two adjacent air holes at different layers (one of which is in the innermost layer) of PCF. The two liquid rods, together with the solid core, essentially form a three-parallel-waveguide structure with small separation, which leads to a strong and efficient mode coupling between the solid core and the liquid rods, thus enabling periodical mode energy beating and resulting in the interference fringe pattern in the transmission spectrum.

Figure 31 is the cross-sectional view of the device proposed. Two air holes of solid-core PCF (LMA-10, NKT Photonics, which has 60 air holes in the cladding, the hole diameter is 3.04 m, and the hole-to-hole pitch is 6.04 m) are infiltrated by the liquid with RI higher than that of silica. One of the air holes (H1) is in the innermost layer, next to the core, while the other (H2) is adjacent to H1, in the neighboring layer. The sample is spliced with SMFs at both ends.

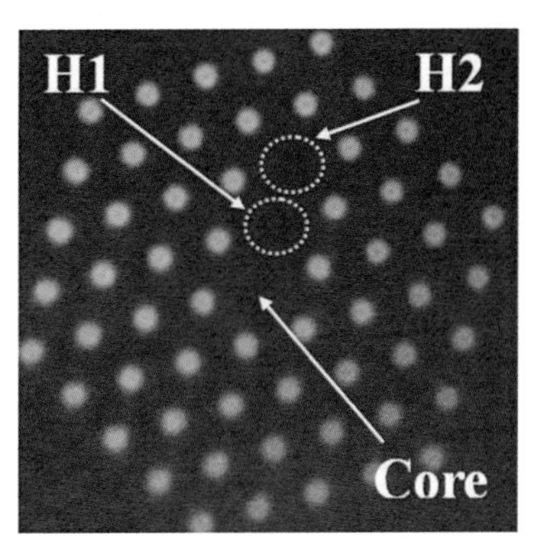

Fig. 31 Cross-sectional view of the proposed device structure around the core center of the PCF: two adjacent air holes in different layers are infiltrated with RI liquid, one of which (H1) is close to the solid core

When the device is heated, the liquid RI drops quickly to match that of silica, and a three-parallel-waveguide structure is formed, in which waveguide and cladding region between adjacent waveguides have the same RI value. Because of the small separation between adjacent waveguides, a strong mode-field overlap occurs, which leads to a significant enhancement of the coupling coefficient between adjacent waveguides and the reduction of the coupling length. Since the liquid usually possesses a relative large absorption of light energy, a small coupling length would enhance the spectral performance, such as the fringe visibility and the uniformity in free spectral range (FSR), when the device is used as a temperature sensor.

For different infiltration lengths, the transmission spectra of the sample at room temperature, 35 and 45 °C, are shown in Fig. 32a–c, respectively. The performance of the sample at different temperatures can be understood as follows. Since the liquid RI is higher than that of silica at room temperature, the phase-matching condition is not met, which results in a relatively small coupling coefficient between the core and other two liquid rods. The RI mismatch also prevents efficient energy oscillation between the eigen modes. In other words, the two liquid rods (which form a composite waveguide) and the core can be treated as the independent waveguides. Thus, the transmission spectra at low temperatures only exhibit an interference fringe pattern with small visibility. When the temperature is increased, the RI of the liquid drops quickly until it is close to that of silica. Light energy can then be easily transferred into liquid rods from the core and vice versa. Moreover, a small deviation from the strict phase-matching condition would not significantly reduce the coupling efficiency as the close distance effectively enhances the mode overlapping between H1 rod and the other two waveguides. As a result, the interference fringe pattern possesses a large and uniform visibility in a wide wavelength range compared with that at lower temperature.

In order to enhance the temporal stability of the proposed device, two small sections of UV curable glue (NOA84, Norland) are filled in the air hole before and after the RI liquid infiltration. The glue has a low viscosity in liquid state and does not dissolve within the RI liquid. The two types of liquids are put in the same tube with a small hole at the bottom but separated into two layers. The opened

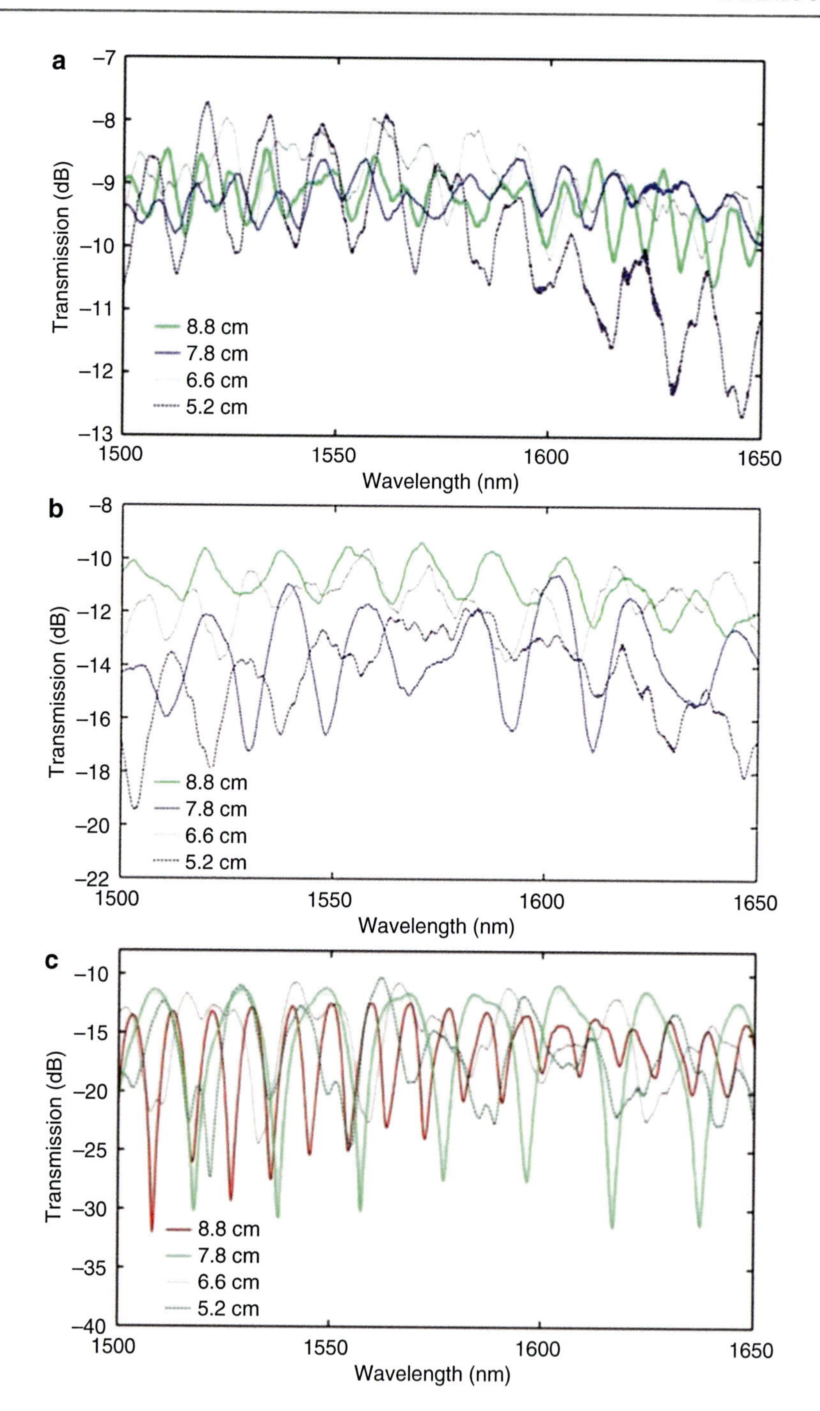

Fig. 32 Transmission spectra of different infiltrated PCF lengths at different temperatures. (**a**) Room temperature. (**b**) 35 °C. (**c**) 45 °C

end of the air hole is infiltrated by gravity first with glue for 30 s, then with RI liquid for 6 h, and finally with glue again for 30 min. A hypodermic needle helps to move and fix the position of the sample's open end. The second glue end of the PCF should not be too short as it would be cleaved to splice with SMF. During the infiltration process, the open end is always immersed in either one of the two liquids, which prevents generation of undesired bubble in the air holes. After infiltration, the sample is irradiated under a UV lamp for proper time to solidify the glue.

The temporal stability of the sample is tested in the temperature range between 45 and 50 °C, as shown in Fig. 33. When compared with the spectra demonstrated in Fig. 32, the shift of transmission wavelength dip can be clearly traced and a good temporal stability be maintained. Finally, by tracing one of the transmission dips (~1564 nm at 45 °C), the temperature sensitivity obtained is ~−8.8 nm/°C.

Selectively Infiltrated PCF with Ultrahigh Temperature Sensitivity

In this part, a selectively infiltrated PCF of 2.4 cm length with highly improved temperature sensitivity of ~54.3 nm/°C is demonstrated, which is based on the light coupling between the fiber core mode and the fundamental satellite rod mode (Wang et al. 2011). By selective filling of one of the air holes in the photonic crystal fiber, the fundamental core mode can be effectively coupled to the fundamental mode of the adjacent liquid rod waveguide at the resonant wavelength with extremely high temperature sensitivity. The spectral power of the rod mode can be filtered out by fusion splicing the selectively infiltrated photonic crystal fiber with conventional single-mode fiber, resulting in a sharp dip in the transmission spectrum.

The PCF used in the experiment is commercially available LMA-10 fiber (NKT Photonics) with the hole diameter of 3.04 m and the hole-to-hole spacing of 6.26 m. One of the air holes located in the cladding of the PCF is selectively infiltrated with standard 1.46 RI liquid (Cargille Labs), which is slightly higher than that of the background silica and possesses a temperature-RI coefficient of refractive index unit (RIU)/C. The selective infiltration of PCF is achieved through the fs laser-assisted infiltration procedure. The fiber geometry after selective infiltration process is shown in Fig. 34.

At room temperature, no apparent lossy dips can be observed in the wavelength region of 1250–1650 nm (of the broad band source we used), except weak interference fringes in the transmission spectrum of the sample, which is probably caused by the two splicing joints between the PCF and the SMF. The transmission spectra corresponding to the temperature range between 34.0 and 35.4 °C are demonstrated in Fig. 35a, where a dip appears around 1650 nm, which means that the 2.4 cm fiber length is still larger than the coupling length, although the resonant dip is unique enough to be traced. Meanwhile, the dip at 34.4 °C in Fig. 35a is much stronger than the dips at other temperatures. This is likely due to the measurement error since the spectra resolution was set to be 1 nm to reduce the measuring time.

To demonstrate the linear operation of the device, the dip wavelength variation with the temperature is plotted in Fig. 34b from which it can be observed that

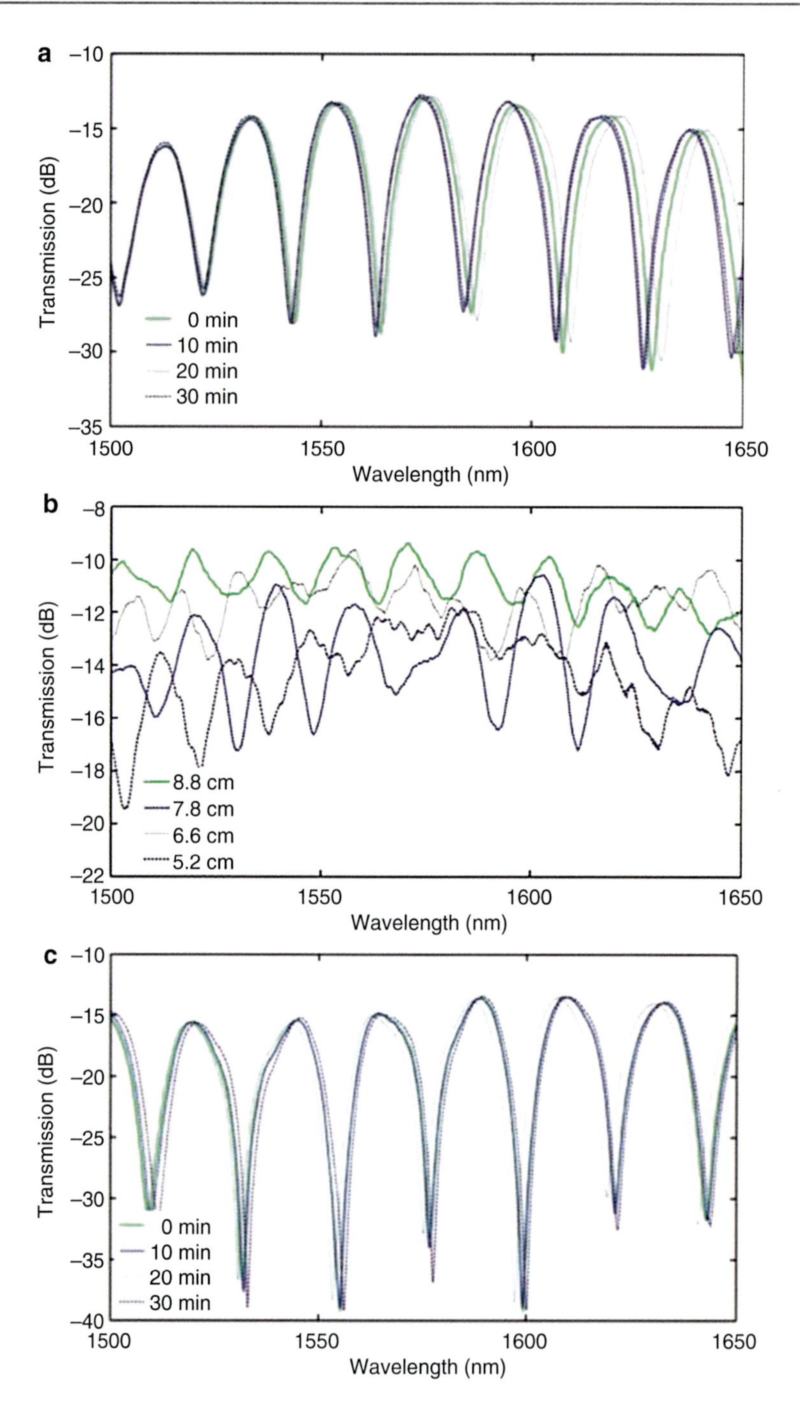

Fig. 33 Selected transmission spectra of temporal stability test of the UV glue encapsulated sample at (**a**) 45 (**b**) 47, and (**c**) 49 °C

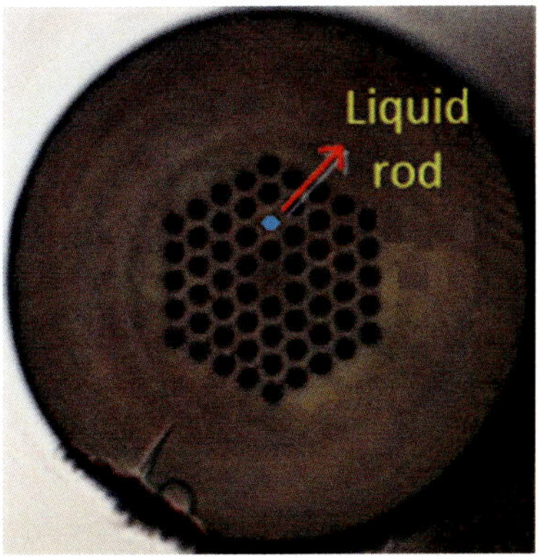

Fig. 34 Optical image (cross-sectional view) of the selectively infiltrated photonic crystal fiber. The white spot indicates the hole filled with standard 1.46 RI liquid, and the other dark spots in the cladding region represent the remained air holes

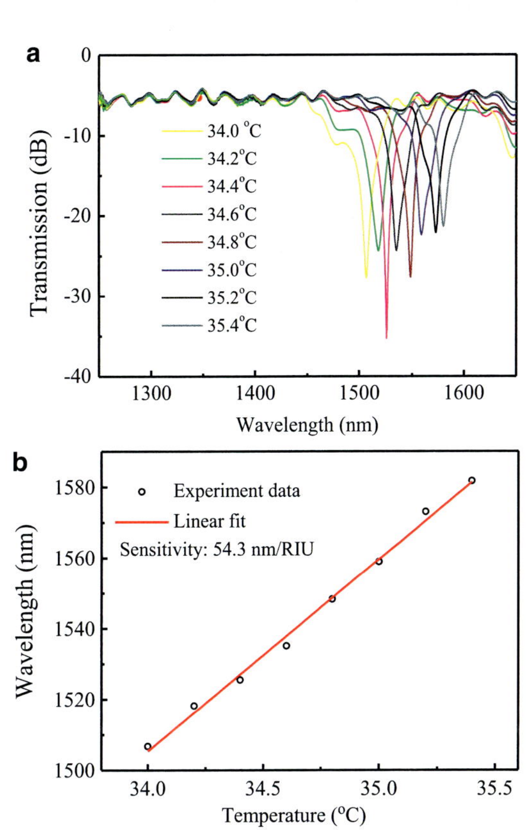

Fig. 35 (**a**) Transmission spectra and (**b**) dip wavelengths of the 2.4 cm length liquid-filled PCF device at temperatures from 34.0 to 35.4 °C

the dip wavelength is redshifted with the increase of temperature, and an average temperature sensitivity of ~54.3 nm/°C is obtained in the 1500–1600 nm region.

The appearance of the lossy dip has been explained with the standard coupled-mode theory for directional couplers (Wu et al. 2009). In our experiment, the lossy dip is believed to be the results of coupling between the fundamental core

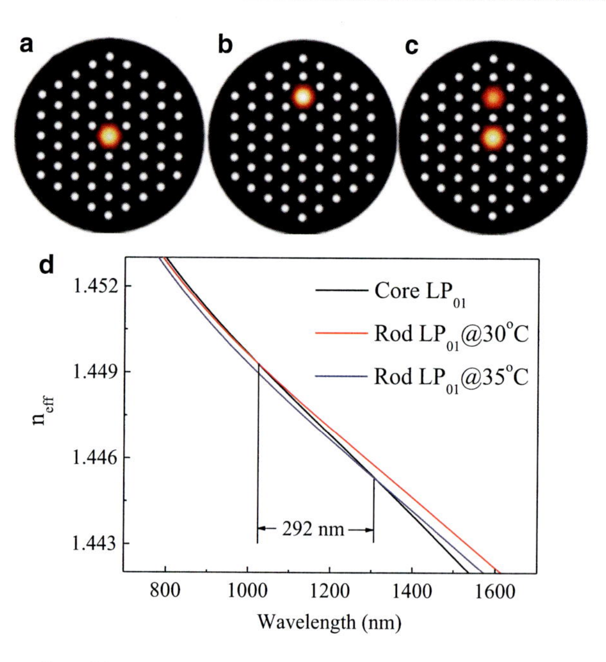

Fig. 36 Mode profiles of the core LP and rod LP modes at (**a**, **b**) nonresonant wavelengths and (**c**) resonant wavelengths; (**d**) phase-matching relationship of the coupling between the core LP and the rod LP modes at 30 °C and 35 °C, respectively

mode and the fundamental satellite rod mode. To confirm the coupling mechanism and the temperature sensitivity theoretically, the liquid selectively infiltrated PCF is modeled with the geometrical parameters aforementioned, and its modal and dispersion properties are calculated numerically by use of commercially available finite element analysis software. The material dispersion of fused silica and 1.46 RI liquid are both taken into account during the simulation. The calculated mode profile of the fundamental core mode and the fundamental rod mode for the off-resonant wavelengths are shown in Fig. 36a, b, and denoted as core LP and rod LP, respectively, where LP represents linearly polarized (Yariv 1997). Figure 36c reveals the mode profile at the resonant wavelength. The dispersion of the core LP and rod LP modes at 30 °C and 35 °C are calculated and plotted in Fig. 36d.

The rod LP mode is more temperature sensitive than that of the core LP mode, because the temperature-RI coefficient of the liquid is almost one order higher than that of silica. The dispersion curve of the core LP mode at 35 °C overlaps with that at 30 °C. According to Wu et al. (2009), mode coupling can only occur around the phase-matching point, where the effective RI of the rod mode is equal to that of the core LP mode. Since the RI of the liquid infiltrated is slightly larger than that of the background silica, when the temperature is increased, only the rod LP mode can meet the phase-matching condition, while the higher-order rod modes cannot do because they have smaller effective RI values than that of the core LP mode.

We have also calculated the dispersion of the LP mode of the liquid rod and found that it locates far below that of the core LP mode and hence does not intersect with the core LP curve in the wavelength region from 600 to 1700 nm. Thus the core LP mode cannot be coupled to the rod LP mode in our PCF device. Note that the dispersion of the rod LP mode is not plotted in this letter as it does not provide any extra information.

The coupling wavelength is determined by the avoided crossing between the core LP and the rod LP dispersion curves (Jansen et al. 2011). With the increase of temperature, the RI of the satellite liquid rod waveguide decreases, and hence the effective index decreases, resulting in a dramatic redshift of coupling wavelength. The coupling wavelength shifts for about 292 nm, from 1012 nm at 30 °C to 1304 nm at 35 °C, representing an average temperature sensitivity of nm ~54.3 nm/°C, which is in good agreement with the experimental results.

In conclusion, by selective infiltration of 1.46 RI liquid in one of the air holes in PCF, an ultrasensitive temperature sensing device has been demonstrated. The device operation is based on the light coupling between the fundamental core mode and the fundamental rod mode in the satellite waveguide. The two modes have similar behavior, and thus the device sensitivity can be significantly increased. The average temperature sensitivity achieved is nm/°C and shows a good agreement with that obtained in the numerical simulations, 5.43 nm/°C. The device is expected to have potential applications in miniaturized and highly accurate temperature and/or RI sensors.

Chemical Vapor and Gas Sensors Based on Infiltrated PCFs

Photonic Crystal Fiber Loop Mirror-Based Chemical Vapor Sensor

In this chapter, we demonstrate a novel chemical vapor sensor by inserting a short HiBi-PCF into a FLM without splicing (Niu et al. 2014a). As is known, the resonant dips of the HiBi-PCF FLM are sensitive to the phase difference between two counter-propagating waves in the FLM. So an extremely high sensitivity to chemical vapor can be realized by measuring the wavelength shift of the resonant dips when chemical vapor diffusing into the air holes of the HiBi-PCF. The sensor can work with a common broadband light source, without any preprocess to the fiber. Experimental results show that the sensitivity reaches up to 15.5 pm/ppm for ethanol when a 5.1 cm long PCF is used in the FLM. Three different possible mechanisms of the proposed sensor are also discussed in this chapter, in which a liquefaction model with a full-vector finite element method (FEM) is proposed. The simulation results indicate that the ethanol liquefaction contributes largely to the high sensitivity of the proposed sensor.

The proposed chemical vapor sensor, as shown in Fig. 37, consists of a 3 dB coupler and a short HiBi-PCF. The HiBi-PCF is fixed on a stable stage with a V-groove, and its both ends are aligned with two single-mode fibers (SMFs), leaving a very short air gap (AG) to allow the molecules of VOC to enter. The SMFs are

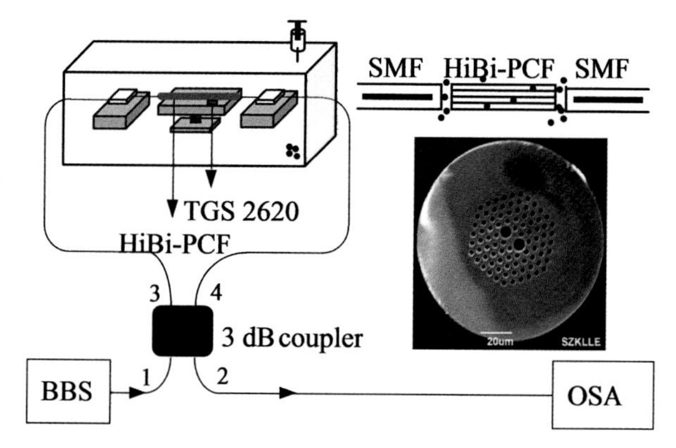

Fig. 37 Experimental setup for chemical vapor sensor based on a FLM. Insertion: the details of the sensor head and the SEM of the used HiBi-PCF. The dots represent VOC molecules

fixed on two 3-D translation stages, respectively, and both ends of the HiBi-PCF and SMFs are cleaved with a standard cleaving machine. All the stages are plated in a sealed acrylic glass chamber, and the concentration of VOCs in the chamber is controlled by pouring ethanol into the chamber with a pipette, whose resolution is 0.05 ml. A TGS 2620 (FIGARO Company) sensor head, based on a metal oxide semiconductor, is sensitive to VOC and can be applied as an organic vapor detector. We also placed this TGS 2620 sensor head in the chamber for the calibration of the concentration in our experiments. A broadband SLED source with 200 nm wavelength range is used as an input light source. The transmission spectra of the FLM are measured by an optical spectrum analyzer (OSA).

Here, the HiBi-PCF, which is inserted into a FLM, works as a chemical vapor sensing head. When the chemical vapors diffuse into the air holes of the HiBi-PCF through the AG, the birefringence of the HiBi-PCF B changes, while the length of the HiBi-PCF L remains the same due to its insensitivity to temperature. Therefore, the relationship between the dip wavelength change $\Delta\lambda_{dip}$ and the birefringence change ΔB can be simply expressed as:

$$\Delta\lambda_{\mathrm{dip}} = \Delta \mathrm{BL}/k \tag{8}$$

where λ is the wavelength, $B = n_x - n_y$ is the birefringence of the HiBi-PCF, n_x and n_y are the effective refractive index at the slow and fast axes, and L is the length of the HiBi-PCF. The resonant dip wavelengths satisfy $2\pi \mathrm{BL}/\lambda_{\mathrm{dip}} = 2k\pi$, where k is any integer. The HiBi-PCF used in our experiment is provided by Yangtze Optical Fibre and Cable Company, and the cross-sectional scanning electron micrograph (SEM) is shown in the insertion of Fig. 37. The diameters of the bigger and smaller holes are 7 and 3.2 μm, respectively, and the pitch length between centers of two adjacent holes is 5.46 μm. Here, a ~5.1 cm length of HiBi-PCF is alighted with two SMFs without splicing as a chemical vapor sensing head.

In our experiments, ethanol is used as the chemical vapor sample because it's easy to evaporate and the vapor is harmless for experimenters. The ethanol is injected into the chamber by 0.05 ml every time, and the corresponding ethanol

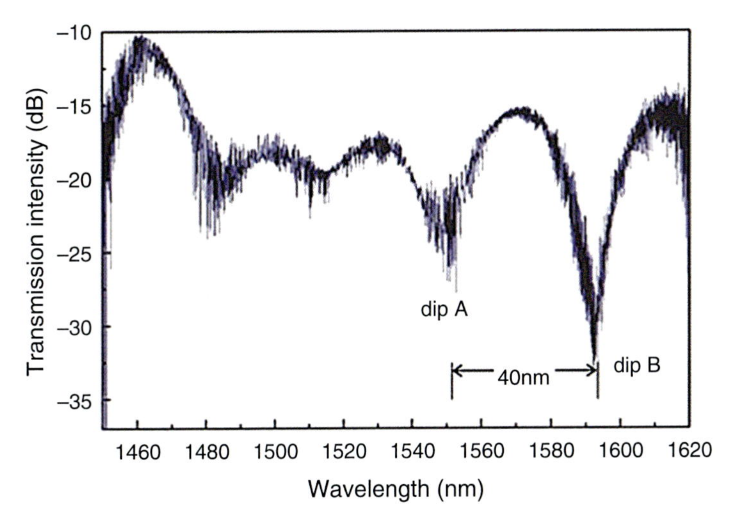

Fig. 38 Transmission spectrum of the HiBi-PCF FLM when the EVC in the chamber is 0 ppm

vapor concentration (EVC) in the chamber is changed by 250 ppm each time. When the response of the sensor reaches stable, the output spectrum will be recorded and more ethanol will be injected. Figure 38 shows the transmission spectrum of the HiBi-PCF FLM when the EVC is 0 ppm. Several resonant dips of the FLM can be observed in the wavelength ranging from 1450 to 1600 nm. In order to observe the wavelength shift, two main dips are chosen during our experiments, one of which is at the wavelength about 1548 nm (dip A), while the other is at about 1588 nm (dip B). The wavelength spacing between the two dips is ~40 nm. Meanwhile, there are also many interference dips with high frequency on the transmission spectrum due to the light interference in the two AGs.

Figure 39 shows the transmission spectra of the HiBi-PCF FLM when the response of the sensor is stable at 0, 250, 500, 750, 1000, and 1250 ppm, respectively. It is obvious that both dip A and dip B have a blueshift when the EVC increases. This is because the birefringence is inversely proportional with the increasing EVC in the air holes of the HiBi-PCF. Figure 40 shows the transmission spectra at each stable situation after the signal denoising, filtering the resonant dips with the high frequency. Apart from the evidently blueshift of dip A and dip B, the intensities of both dips also change as the EVC increases. The experimental relationships between EVC and the resonant wavelength of dip A and dip B are also shown in Fig. 41. The experimental EVC sensitivity of dip A and dip B are ~15.5 pm/ppm and ~13.2 pm/ppm, respectively. When an optical spectrum analyzer with 20 pm resolution is used, the EVC's resolutions are separately ~1.3 ppm and ~1.5 ppm in the range of 0 ppm to 1250 ppm.

The response characteristic of the proposed sensor has also been investigated. Figure 42 shows the measured dip A as a function of time when the EVC in the chamber changes from 0 to 1250 ppm, increased by 250 ppm each time. As shown

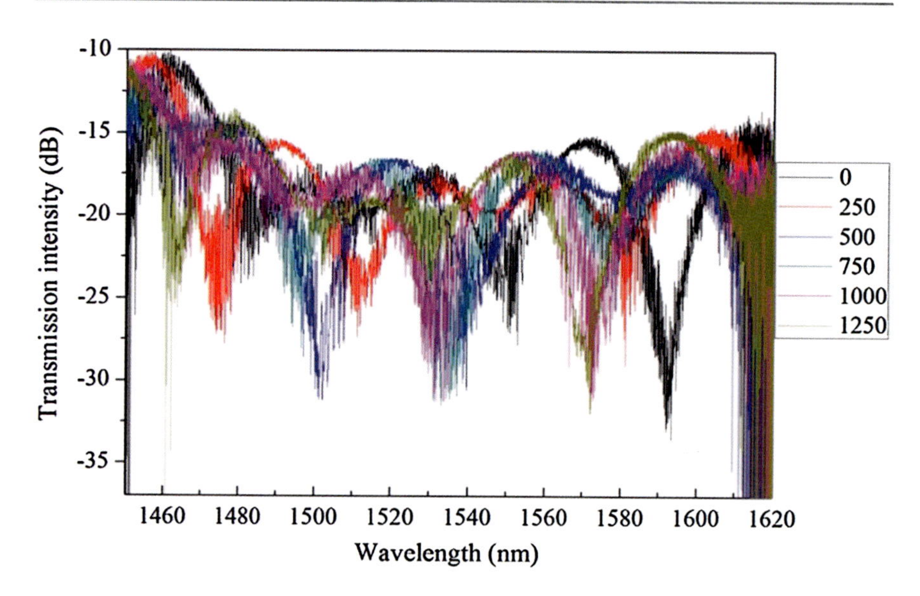

Fig. 39 Transmission spectrum of the HiBi-PCF FLM without filtering at each stable situation from 0 to 1250 ppm

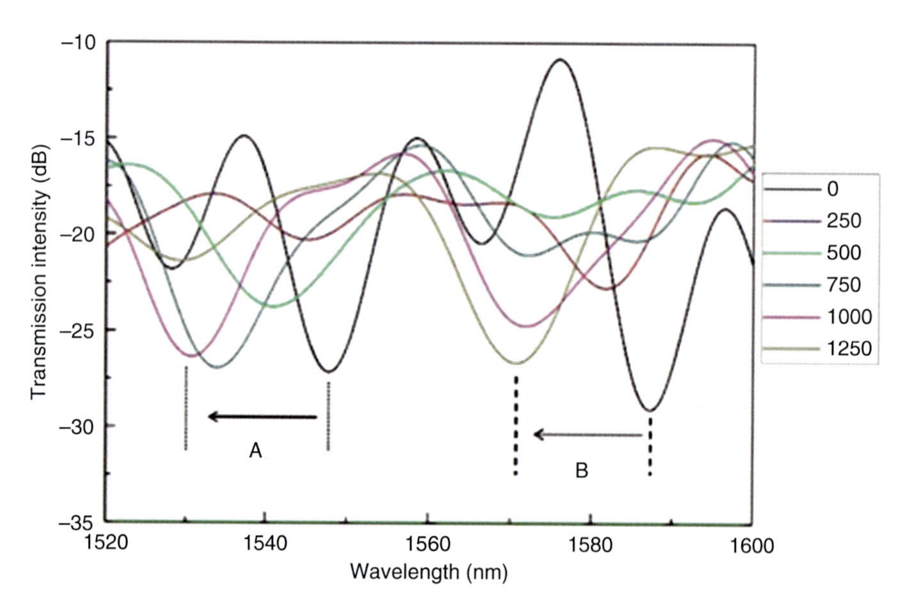

Fig. 40 Transmission spectra of the HiBi-PCF FLM at each stable situation after filtering

in Fig. 42, the dip shift becomes weaker, while the stability time becomes longer as the EVC increases. What is interesting is that the dip wavelength has a little redshift at the first 1–2 min after injecting ethanol. This may be caused by the evaporation process of the ethanol liquid.

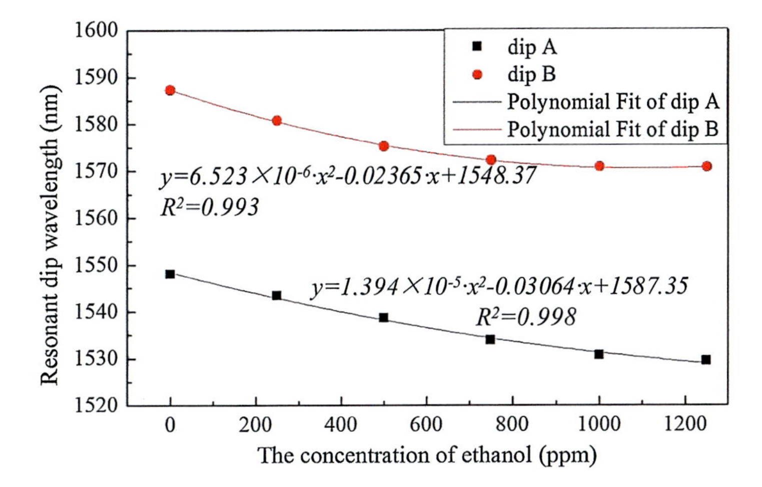

Fig. 41 The relationships between EVC and the resonant wavelength of dip A and dip B

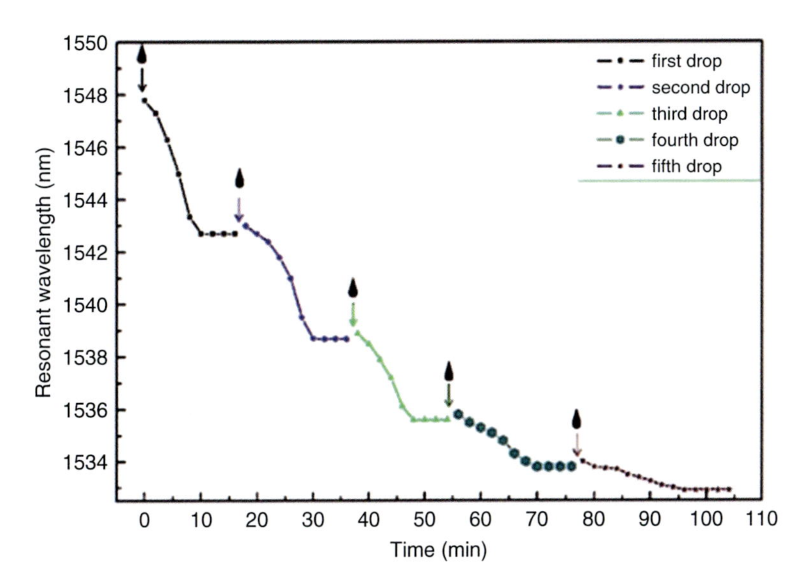

Fig. 42 The relationship between the measured dip A and time when the EVC in the chamber changes from 0 to 1250 ppm

In order to shorten the response time of the proposed sensor, a shorter HiBi-PCF with larger air holes is better, which needs less time to reach stable when the VOC cells diffuse. Moreover, a bigger birefringence change ΔB can be considered according to Eq. (3) for a higher sensitivity of the proposed sensor.

Why is the sensitivity so high? Here, we should discuss that the ethanol liquefaction model we propose contributes largely to ΔB, which may be the main reason causing the wavelength shift largely. We assume that the gaseous ethanol

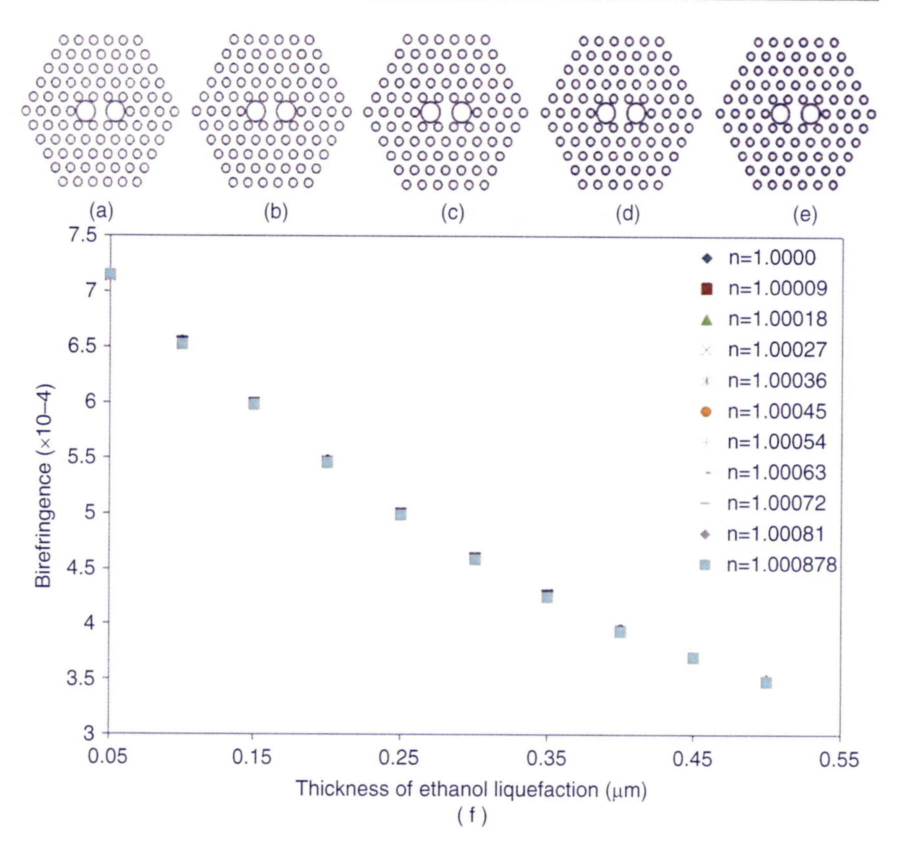

Fig. 43 (**a–f**) HiBi-PCF sections with different liquefaction thickness. (**f**) Relationships between the birefringence and different liquefaction thicknesses

vapor in the fiber holes will liquefy as the EVC increase in the chamber, and a liquefaction film will attach to the inside the fiber holes. As shown in Fig. 43a–e, the birefringence simulations of the HiBi-PCF with different thickness of ethanol liquefaction film are achieved by using FEM. The birefringence has an obvious change from 7.05×10^{-4} to 3.45×10^{-4} with a thickness of ethanol liquefaction film from 0.05 to 0.55 μm. Moreover, the change trend becomes flat gradually with an increase of the thickness, shown in Fig. 43f, which is similar to the trend of the experimental results, shown in Fig. 41. Figure 43f also shows that the RI variation has no effect on the birefringence.

The simulation results indicate that the ethanol liquefaction contributes largely to the birefringence changes ΔB, which will cause an evident wavelength shift in the FLM transmission. According to the experimental results and Eq. (8), the extreme thin ethanol liquefaction of less than 0.05 μm can induce a birefringence change which achieves a relatively large blueshift referring to Niu et al. (2014b) and Sun et al. (2008).

In conclusion, we have proposed a novel compact chemical vapor sensor based on HiBi-PCF FLM and analyzed the influence of three different mechanisms. Experimental results of the proposed HiBi-PCF FLM VOCs sensor show that the sensitivity for ethanol is as high as 15.5 pm/ppm. The liquefaction model is proposed to explain the reason of the proposed sensor with the extremely high sensitivity.

Chemical Vapor Sensor Based on Rayleigh Scattering Effect in Simplified Hollow-Core PCF

In this chapter, we propose a novel optical fiber sensor for chemical vapor content detection (Niu et al. 2014b), based on Rayleigh scattering effect in a simplified HC-PCF, which has a wider transmission bandwidth than those of conventional photonic bandgap HC-PCFs (Amezcua-Correa et al. 2008; Gérôme et al. 2010; Wu et al. 2011). Rayleigh scattering will be excited when VOCs molecules diffuse into the air holes, resulting in an intensity loss of transmission light. Therefore, we can measure the concentration of the chemical vapor by monitoring the transmission intensity loss. Ethanol is chosen as a VOC example, and a 5.1 cm long simplified HC-PCF is used. The experimental results prove that our proposal is feasible, and the sensitivity is as high as 0.022 dB/ppm. The sensitivity will be further improved when longer simplified HC-PCF is used.

The proposed chemical vapor sensor, as shown in Fig. 44, consists of a short simplified HC-PCF and two single-mode fibers (SMFs), and the insert shows the details of the sensor head. The cross-sectional scanning electron micrograph (SEM) of the simplified HC-PCF we used is also shown in the insertion of Fig. 44. The simplified HC-PCF is fixed on a stable stage with a V-groove, of which both the

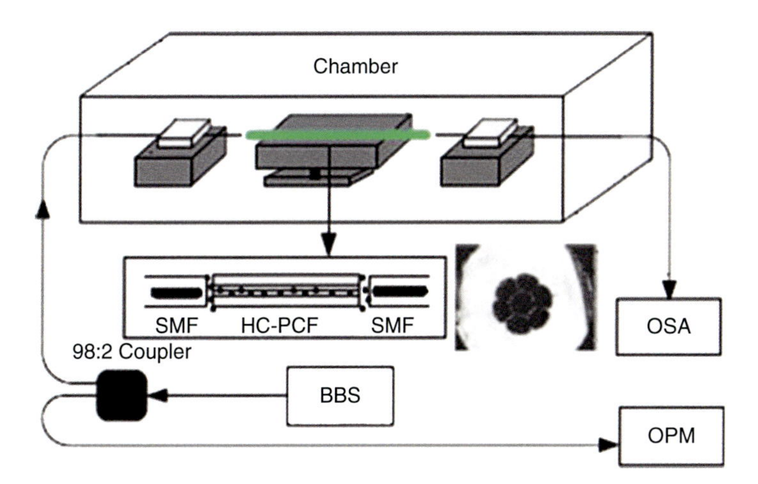

Fig. 44 The experimental setup for chemical vapor sensor. Insertion: the details of the sensor head and the SEM of the used simplified HC-PCF. The dots represent VOC molecules

ends are aligned with SMFs without splicing, leaving a very short air gap to allow the molecules of VOCs entering. The SMFs are fixed on two 3-D translation stages, and both the ends of PCF and SMFs are cleaved with a standard cleaving machine. The sensing head is plated in a sealed acrylic glass chamber, and the concentration of VOCs in the chamber is controlled by pouring ethanol into the chamber with a pipette, whose resolution is 0.05 ml. A broadband SLED source range from 1400 to 1640 nm is used as an input light source, which is split into two by a 98:2 fiber coupler. Ninety-eight percent of the light propagates into the simplified HC-PCF, while the remaining 2% is used to monitor the fluctuations of the light source. The transmission spectra are measured by an optical spectrum analyzer (OSA).

The main principle is summarized as follows. Light scattering is a complex phenomenon and can be used for investigation of air pollution, characterization of polymers and colloids, and study of proteins and bio-macromolecules (Fanguy et al. 2004). In our experiment, Rayleigh scattering will be excited when VOCs' molecules diffuse into the air holes of the simplified HC-PCF, resulting in a loss of transmission light. The beam propagation direction will change after the light beam meets some obstacles. In our cases, the VOCs' molecules can be viewed as the obstacles, and the transmission light will be scattered, leading to the transmission intensity decreasing to a large extent. Besides, the scattering will be enhanced when more molecules enter. Therefore, we can detect the concentration of the chemical vapor by monitoring the transmission intensity loss.The output transmission intensity I_{out} can be expressed as

$$I_{\text{out}} = I_{\text{in}} e^{-\frac{L}{\lambda^4}\text{kC}} \tag{9}$$

where I_{in} is the input transmission intensity, L is the length of the simplified HC-PCF, and k represents the coefficient between the Rayleigh scattering coefficient A and the sample concentration C.

The simplified HC-PCF used is provided by Yangtze Optical Fibre and Cable Company and composed of a hollow hexagonal core surrounded by six crown-like air holes as shown in the insertion of Fig. 44. The diameters of the core, air holes in the cladding and outer cladding are about ~22 μm, 70 μm, and 140 μm, respectively. And the thickness of the struts is about 370 nm. Here, a ~5.1 cm length of simplified HC-PCF is alighted with two SMFs without splicing as a chemical vapor sensing head, and the air gaps between SMF and the simplified HC-PCF are ~0.05 cm. This simplified HC-PCF has two main merits, one of which is that it can guide light in an extraordinary wide wavelength range, providing a broad transmission band. Therefore, the bandgap shifting effect can be excluded in our experiment. Another advantage is that the sensitivity can be improved when longer simplified HC-PCF is used because the output light power is also related to the length of the simplified HC-PCF from Eq. (9). Meanwhile, the light-gathering ability of the simplified HC-PCF is perfect. If we move the simplified HC-PCF away, light cannot be transmitted unless the two SMFs are very close or using a

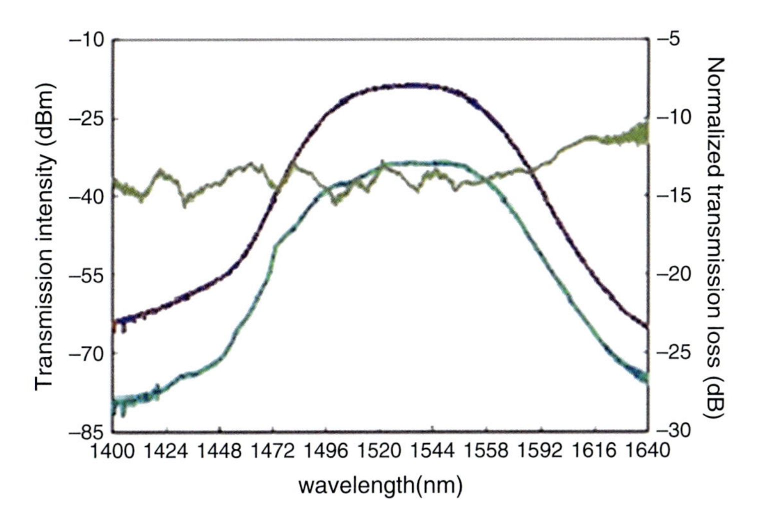

Fig. 45 Transmission spectra of the proposed sensor with a ~5.1 cm simplified HC-PCF when the ethanol concentration in the chamber is 0 ppm

GRIN lens (Gilsdorf and Palais 1994), which will limit the length of the gas cell or increase the complexity of fabrication.

During our experiment, ethanol is chosen as the chemical vapor sample because it's easy to evaporate and the vapor is harmless for experimenters. The ethanol is injected into the chamber by 0.05 ml every time, and the corresponding ethanol concentration in the chamber is changed by 250 ppm each time. When the response of the sensor reaches stable, the output spectrum will be recorded and more ethanol will be injected.

Figure 45 shows the transmission spectra of the proposed sensor with a ~5.1 cm simplified HC-PCF when the ethanol concentration in the chamber is 0 ppm. The red line and blue line represent the transmission spectrum of the input light source and the output of the sensor, respectively. The normalized transmission spectrum in green line shows a broad bandwidth in the range from 1400 to 1640 nm. The insertion loss, mainly caused by the alignment loss, is less than −15 dB.

Figure 46 shows the output transmission spectra of the proposed sensor when the response is stable at 0, 250, 500, 750, 1000, and 1250 ppm, respectively. Obviously, the transmission intensity decreases when the ethanol concentration increases from 0 to 1250 ppm. Meanwhile, the transparent ranges become smaller as the concentration becomes larger because of the increased transmission intensity loss. The decline of the transmission intensity is mainly caused by Rayleigh scattering. Rayleigh scattering will be excited and result in an intensity loss of transmission light as ethanol molecules diffuse into the air holes of the simplified HC-PCF. More ethanol molecules diffuse into the holes, more transmission intensity decline because of larger light ethanol molecules interactions. Therefore, the concentration of ethanol can be detected effectively by monitoring the transmission intensity loss.

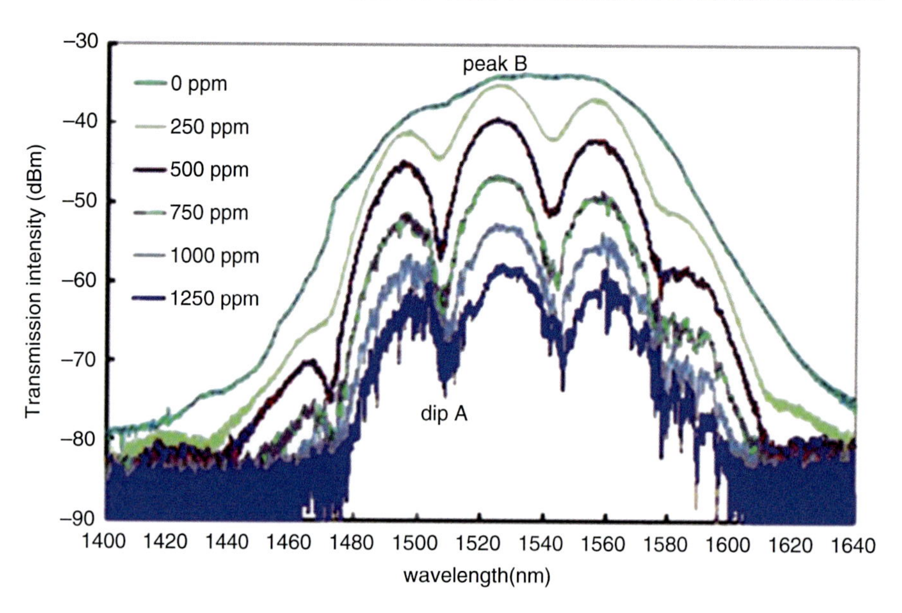

Fig. 46 Measured transmission spectra of the simplified HC-PCF when the ethanol concentration increases from 0 to 1250 ppm at 20 °C

In addition, we can see the interference fringe caused by the modal interference, as shown in Fig. 46, which exhibits fringe spacing of ~30 nm between the two adjacent transmission peaks after pouring 250 ppm ethanol into the chamber. The interference pattern is found no obvious shift when different amount of ethanol molecules diffuse into the simplified HC-PCF. On the contrary, the transmission intensity decreases while the fringe visibility increases. The visibility of interference fringe is about 16.2 dB at 500 ppm, which is nearly twice as much as at 250 ppm. Furthermore, many interference dips with high frequency on the transmission spectrum appear when the concentration of the ethanol increases, especially during 750–1250 ppm.

Light will be confined in the core of the simplified HC-PCF, which is mainly fundamental mode, when it's filled with the air. However, after pouring ethanol into the chamber, the ethanol molecules diffuse into the air holes of the simplified HC-PCF and the interference fringe appears. In order to investigate the number and energy distributions of the involved interfering modes, the spatial frequency spectra of the interferometers are analyzed through the fast Fourier transform (FFT) method. The spatial frequency spectrum is obtained and used to analyze the modal interferometer, as showed in Fig. 47. It shows there is only one dominant peak when the concentration is 0 ppm. After pouring the ethanol sample into the chamber, it has several peaks in moderate strengths induced by higher-order modes, and the strengths of the peaks increase as more ethanol molecules diffuse into the simplified HC-PCF. Due to the guidance mechanism of Kagomé-like fibers, which is known as "inhibited interaction," couplings between the core modes and the cladding modes

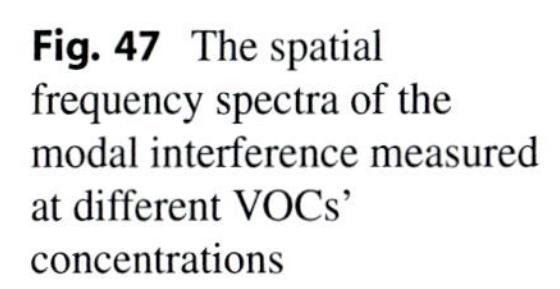

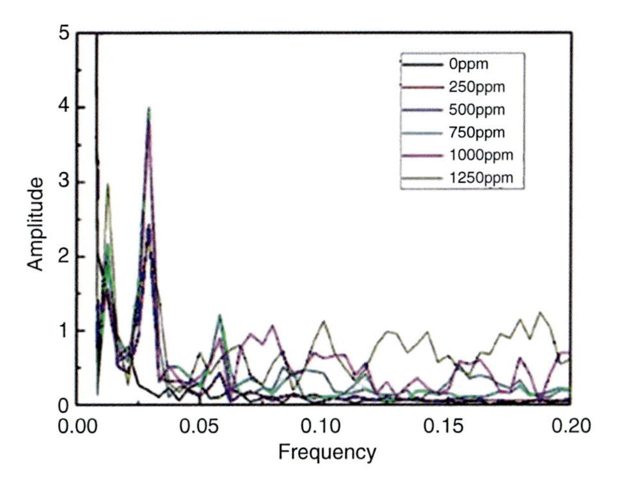

Fig. 47 The spatial frequency spectra of the modal interference measured at different VOCs' concentrations

are effectively inhibited. Therefore, the cladding modes can be excluded in our experiment, and the involved interfering modes are the fundamental core mode and high-order core modes. We can observe the emergence of the interference fringe caused by the coupling between the fundamental core mode and high-order core modes, and it became obviously as the ethanol concentration increased because of the increasing proportion of high-order modes.

Here, we choose two single wavelengths to measure the transmission intensity loss. One is at dip A, the wavelength of 1506 nm, and the other is at peak B, the wavelength of 1525 nm, as shown in Fig. 46. Figure 48 shows the experimental relationship between the ethanol concentration and the transmission intensity at dip A and peak B in the range of 0–1250 ppm. By taking logarithm on both sides of Eq. (4), the output transmission intensity (in dBm) should have a linear relationship with Rayleigh scattering coefficient *A(C)*. The figures in Fig. 48 illustrate that the transmission intensity decreases from −37.62 to −68.10 dBm at dip A, while it decreases from −33.97 to −58.06 dBm at peak B. The transmission intensity decreases approximately in a linear manner as the concentration of ethanol augments, which is in good agreement with theory.

The fitting curves can be expressed as $y = -6.0173x - 34.63$ and $y = -5.1892x - 26.232$ for the relationships at the dip A and the peak B, respectively. We can get the coefficient k by calculating the experimental data and considering the difference of the wavelengths, which is 5.79×10^{-10} at dip A while 6.38×10^{-10} at peak B. The difference k is small in our experiment. Therefore, the output transmission intensity is only related to the ethanol concentration C. Furthermore, the fitting degrees are relatively low, only 0.9569 for dip A and 0.9682 for peak B. This is due to the randomness and instability of modal interference, especially for dip A whose transmission intensity is lower. The experimental ethanol concentration sensitivity of dip A and peak B are ~0.024 dB/ppm and ~0.020 dB/ppm, respectively. The sensitivity at different wavelength differs because the output transmission intensity is also related to the wavelength λ, according to Eq. (4).

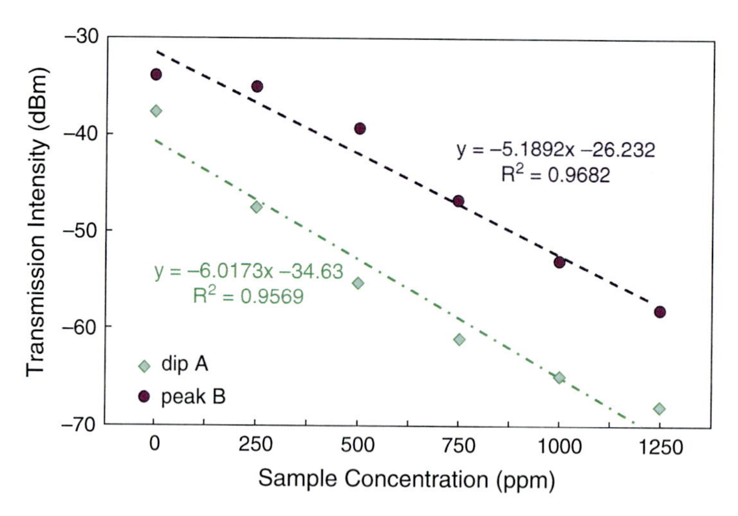

Fig. 48 The experimental relationship between the ethanol concentration and the transmission intensity at dip A (1506 nm) and peak B (1525 nm)

Considering the imprecision of the transmission intensity loss at single wavelengths mentioned above, we also measured the change of the total transmission intensity loss by using an optical power meter. It shows that the total transmission intensity is −1.13, −5.40, −10.51, −17.68, −23.71, and −28.95 dBm when the ethanol concentration is 0, 250, 500, 750, 1000, and 1250 ppm, respectively. The transmission intensity decreases ~27.82 dB when the ethanol concentration increases from 0 to 1250 ppm. Figure 49 shows the experimental relationship between the ethanol concentration and the total transmission intensity. Clearly, the total transmission intensity decreases in a good linear manner as the concentration increases. The fitting curve can be expressed as $y = -5.7495x+5.5595$, and the fitting degree is up to 0.995, much higher than the fitting degrees of the monitoring at single wavelengths. The experimental ethanol concentration sensitivity of the total transmission intensity reaches to ~0.022 dB/ppm. When an optical power meter with 0.01 dBm resolution is used, the ethanol concentration's resolution can reach to ~0.45 ppm by detecting the total transmission intensity loss. The sensitivity can be further improved when longer simplified HC-PCF is used according to Eq. (9). Thus we can get a compact sensing device which has higher sensitivity. In this way, minor VOC content can be detected.

Besides, we also repeated the experiment at the same environment by using the same simplified HC-PCF, after the ethanol molecules diffused out of the air holes, to ensure the sensor repeatability and reversibility. Figure 49 shows the relationships between the ethanol concentration and the total transmission intensity in the first and the second experiment. As shown in Fig. 49, we can see that the total transmission intensity also decreases in a linear manner as the concentration increases. However, the experimental data are different from the first one because of the fluctuation of the broadband SLED source, which leads to a different initial input power.

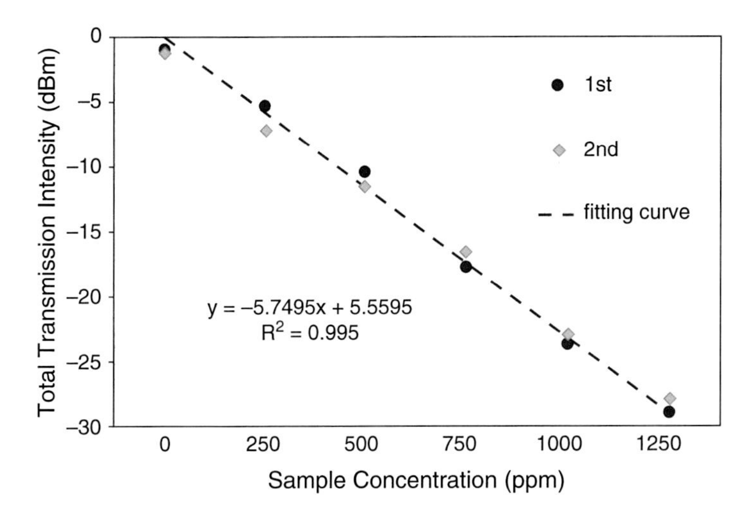

Fig. 49 The experimental relationships between the ethanol concentration and the total transmission intensity

The response characteristic of the proposed sensor for ethanol has also been investigated. Typical time-dependent response of the sensor is shown in Fig. 50. Obviously, the total transmission intensity changes as a function of time when the ethanol concentration in the chamber changes from 0 to 1250 ppm, increased by 250 ppm each time. For the first 70 min, the ethanol concentration in the chamber increases from 0 to 250 ppm. It changes little at first 5 min because the ethanol starts evaporating and diffusing in the chamber. After ~5 min later, the ethanol molecules diffuse into the air holes of the simplified HC-PCF, and the total transmission intensity begins to decrease. As time passes by, the transmission intensity declines to a large extent between 30 and 50 min. After that, the decline of the curve becomes gentle at 50–60 min. There is no obvious change after 60 min, and the response becomes stable. Similar plots are obtained when more ethanol is injected into the chamber, as shown in the insertion of Fig. 50. It is important to note that the response characteristic of the sensor would vary with different VOCs, because they have different evaporation rates. Meanwhile, the time for the molecules of different VOCs to diffuse into the air holes is also various for their own molecular sizes.

Such long response times are consequences of fact that the experiments are carried in the chamber with large volume at normal conditions. Obviously, the introduction of smaller chamber and shorter PCFs could accelerate the diffusion and speed up response times. Moreover, considering the response time will increase as the gas cell become longer, we can introduce a mass flow controller to control the diffusion rate of VOCs individually (Villatoro et al. 2009) or a mechanical roughing pump (Minkovich and Monzón-Hernández 2006). Besides, fabricating periodic microholes in the side of fiber by a femtosecond laser is also a good way to speed up the response time (Hensley et al. 2007; Hoo et al. 2010).

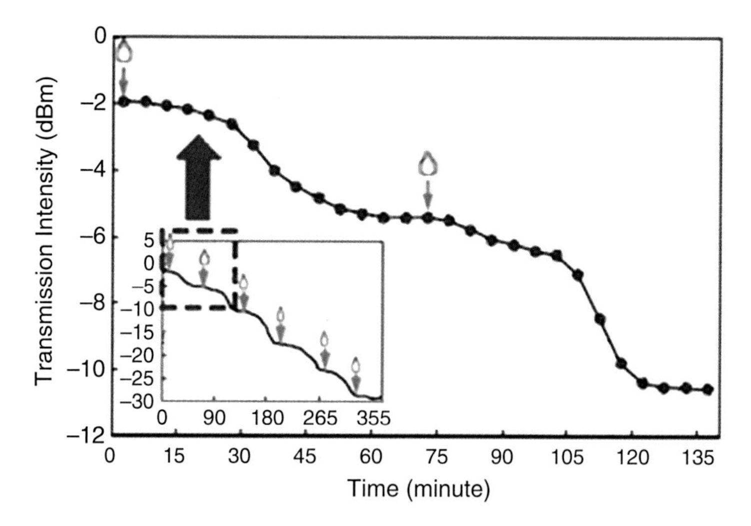

Fig. 50 The time-dependent response of the proposed sensor when the ethanol concentration in the chamber changes from 0 to 500 ppm. Insertion: the whole time-dependent response when the ethanol concentration in the chamber changes from 0 to 1250 ppm

In conclusion, we propose a novel and high-resolution chemical vapor sensor for minor chemical vapor content detecting, based on Rayleigh scattering effect in the simplified HC-PCF in this chapter. Both the ends of the simplified HC-PCF are aligned with SMFs without splicing, leaving a very short air gap to allow the molecules of VOCs to diffuse. The sensor can detect the concentration of VOCs by monitoring the transmission intensity loss. The experimental results show that the ethanol concentration sensitivity is as high as ~0.022 dB/ppm. When an optical power meter with 0.01 dBm resolution is used, the ethanol concentration's resolution can reach to ~0.45 ppm. The sensitivity will be further improved when longer simplified HC-PCF is used. This compact sensing device can be applied to detect and quantitative analysis the micro change of the chemical vapor content.

Hydrogen Sensor Based on Selectively Infiltrated PCF

A highly sensitive fiber device based on a selectively infiltrated PCF with Pt-loaded WO3 coating is demonstrated for hydrogen sensing (Wang et al. 2014). The selectively infiltrated PCF device exhibits extremely high sensitivity to temperature change and benefits from which a highly sensitive hydrogen sensor can be achieved via Pt-loaded WO3 coating. A 10 mm long sensor exhibits the maximum wavelength shift of ~98.5 nm, a response time of ~102 s for 4% (v/v) H2 concentration and a sensitivity of up to 32.3 nm/% (v/v) H2 within the range of 1%–4% (v/v) H2 in air at room temperature.

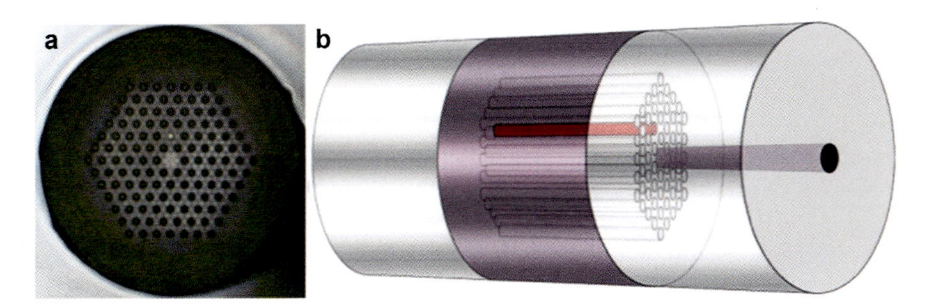

Fig. 51 (**a**) Cross-sectional view of the selectively infiltrated PCF (the white spot near the fiber core indicates the liquid-filled waveguide) and (**b**) schematic diagram of the proposed hydrogen sensor; the gray rendering represents Pt-loaded WO_3 coating, the red rod represents the liquid-filled waveguide, and the rightmost gray rod represents fiber core of single-mode fiber, respectively

In the experiment carried out, one air hole of an endlessly single-mode PCF (LMA-10, NKT Photonics) was filled with standard refractive index (RI) liquid (RI 1.508, Cargille Laboratory, Inc.) through the femtosecond laser-assisted selective infiltration technique. Figure 51a shows the cross-sectional view of the filled PCF. The liquid-filled waveguide forms an embedded coupler together with the fiber core based on the light coupling between the core LP01 and the liquid waveguide LP11-like modes at resonant wavelengths (Wang et al. 2012). The resonant wavelength is extremely sensitive to temperature due to the avoided crossing effect and the high thermal-optic coefficient of RI liquid (Wu et al. 2009; Han et al. 2010). The length of infiltration can be readily controlled by adjusting the liquid infiltration time. A 10 mm long selectively infiltrated PCF was fusion spliced to a standard single-mode fiber at both ends, and then the PCF section was coated with Pt-loaded WO3 for hydrogen sensing test, as depicted in Fig. 1b.

The chemical reaction that occurred in the Pt-loaded WO3 catalytic layer can be described as (Caucheteur et al. 2008; Yang et al. 2012

$$WO_3 + xH_2 \overset{Pt}{\rightarrow} WO_{3-x} + xH_2O \tag{10}$$

$$WO_{3-x} + \frac{x}{2}O_2 \overset{Pt}{\rightarrow} WO_3 \tag{11}$$

H2 can be oxidized by O_2 molecules existed in air and give rise to H_2O, which is an exothermic process. Under a constant hydrogen concentration, such a process continuously provides heat establishing a thermal equilibrium in the catalytic layer, finally resulting in a local temperature change of the fiber-optic sensor. Thus, hydrogen sensing can be achieved by monitoring the resonant wavelength shift induced by temperature change.

For hydrogen sensing, the PCF device was put into a tubular gas cell with the diameter of 1 cm and length of 12 cm. The transmission spectrum was monitored in real time by use of a BBS and an OSA. The sample was exposed to hydrogen concentrations ranging from 0% to 4% (v/v) H2 in air at the temperature of 18.6 °C. Three rounds of hydrogen concentration tests were conducted with a rising/falling step of 1% (v/v), and the corresponding resonant wavelengths are plotted in Fig. 52a. The resonant wavelength shifts to the shorter wavelength with the increase of hydrogen concentration, and the total wavelength shifts at 4% (v/v) H2 in air are 52.95, 50.22, and 51.88 nm, respectively. The hydrogen sensitivity can be estimated to be larger than 16.7 nm/% (v/v) H2 within the range of 1–4% (v/v) H2 in air at room temperature. In the process of hydrogen concentration decreasing from 4%

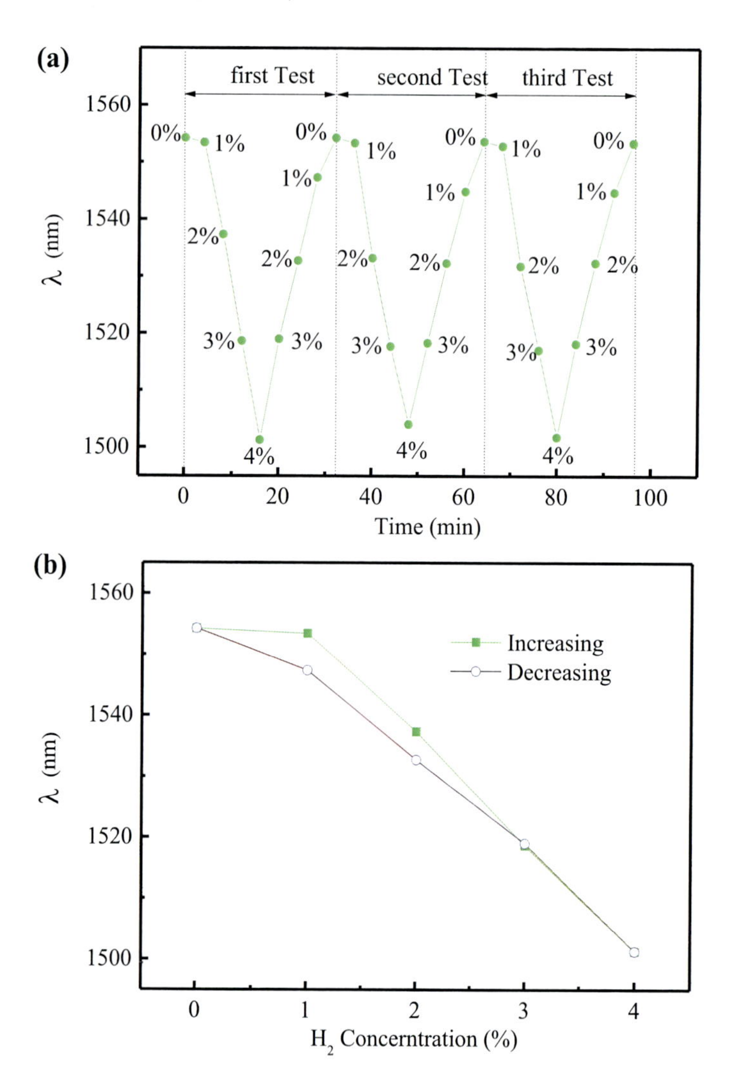

Fig. 52 (**a**) Wavelength shifts of the hydrogen sensor in successive hydrogenation cycles (between 0% and 4% H2 in air). (**b**) Hysteresis of the hydrogen sensor in the first round of test. The contrasts are 6.09 and 4.62 nm at hydrogen concentrations of 1% and 2% (v/v) H2 in air, respectively

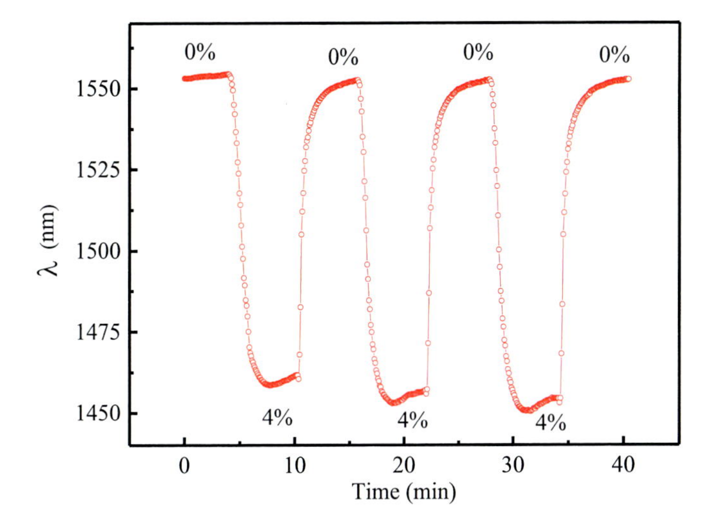

Fig. 53 Response of the Pt-loaded WO3-coated hydrogen sensor in successive hydrogenation cycles (at 0% and 4% H2 in air). The resonant wavelength shifts in the three cycles are 92.6, 96.1, and 98.5 nm, respectively, and the average rising and falling response times are 102 and 92 s, respectively

to 0% (v/v) H2 in air, the resonant wavelength shifts approximately linearly to the longer wavelength and returns to its original value with a fluctuation of less than 100 pm, which is dominated by the thermal response uncertainty of the PCF device.

Figure 53 displays the resonant wavelength response of the device to different hydrogen concentrations. The device responses are repeatable for the hydrogen concentration ranging from 0% to 4% (v/v) H2 in air. The resonant wavelength can restore to its initial value with a fluctuation of less than 100 pm. The average rising (from 0% to 4%) and falling (from 4% to 0%) response times are measured as 102 and 92 s, respectively (here we define them as the time taken from 0% to 90% of the final wavelength). The total wavelength shift (from 0% to 4% (v/v) H2 in air) of the device exceeds 90 nm in all the three cycles of hydrogen concentration test. The maximum resonant wavelength shift from 1% to 4% (v/v) H2 in air is measured to be 97.0 nm, which gives an average hydrogen sensitivity of 32.3 nm/% (v/v) H2 in air, which corresponds to a hydrogen resolution of 310 ppm within the range of 1–4% (v/v) H2 in air, by assuming a 1 nm optical spectral resolution. The resolution may slightly decay within the range of 1–2% (v/v) H2 in air due to the hysteresis of the hydrogen sensor. The measurement error is ~240 ppm.

Passively Mode-Locked Fiber Laser Based on a Hollow-Core PCF Filled with Few-Layered Graphene Oxide Solution

Passively mode-locked erbium-doped fiber lasers have many applications ranging from basic research to telecommunications, medicine, and materials processing because of their simple and compact design and high-quality pulse generation.

A saturable absorber (SA) is usually used as the mode-locking technique to turn the laser continuous wave into a train of short optical pulses. Semiconductor SA mirrors always dominate passive mode locking in spite of their complex fabrication and narrow tuning range (Steinmeyer et al. 1999; Okhotnikov et al. 2004). Recently, two simple and cost-effective alternatives, single-walled carbon nanotubes (SWCNTs) and graphene, have also attracted considerable attention. Graphene-polymer composites (Bao et al. 2010a), chemical vapor deposition-grown films (Zhang et al. 2010), functionalized graphene (Bao et al. 2009), and reduced graphene oxide (GO) flakes (Song et al. 2010) have been used for ultrafast lasers. The common method of integrating graphene SAs in laser cavities is to sandwich a graphene SA film between two fiber connectors with a fiber adaptor (Bao et al. 2010a, b, 2011; Zhang et al. 2010; Sun et al. 2010a, b; Popa et al. 2010). A solution-phase graphene with large SA is suitable as a broadband SA in laser cavities for ultrafast pulse generation, compared with GO. Here, by using a selective hole filling technique, an erbium-doped fiber laser was mode locked by a hollow-core photonic crystal fiber (HC-PCF) filled with few-layered graphene oxide (FGO) solution (Liu et al. 2011). This make SA easily be integrated into a range of photonic systems.

All-fiber mode-locked ring laser setup is shown in Fig. 54. A 1.5 m heavily erbium-doped fiber (OFS EDF-80) is used as the gain medium, pumped by a 1480 nm high-power laser diode through a wavelength division multiplexer coupler. A polarization controller (PC) is used to optimize the mode-locking operation, while a polarization independent isolator maintains the unidirectional laser pulse propagation. The mode-locked pulses can be directed out by use of a 90:10 coupler. The group velocity dispersion (GVD) is one of the key factors to maintain the fiber laser operation stability. The GVD of the dispersion compensation fiber used in the system is ~−30.5 ps/nm/km, and that of the EDF is ~−46.25 ps/nm/km, at the wavelength of 1560 nm. The whole cavity length is about 27.1 m. The rest of

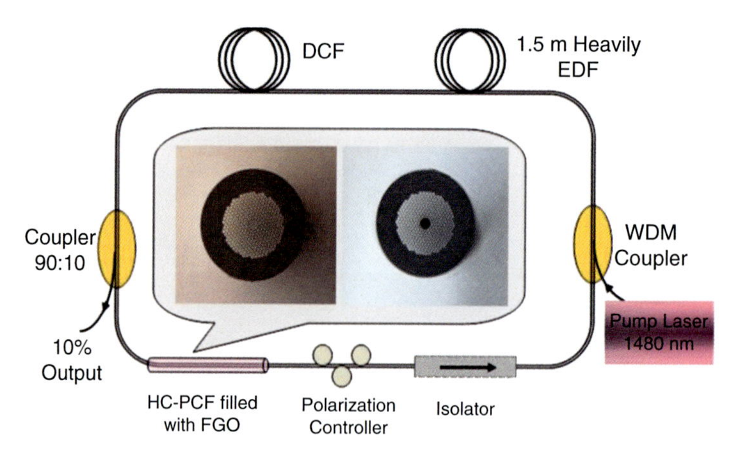

Fig. 54 Experimental setup for fiber pulsed laser with HC-PCF filled with FGO solution. Insets are the cross section images of the HC-PCF before (right) and after (left) infiltration of FGO solution

the cavity consists of single-mode fiber (SMF), which has anomalous dispersion at 1560 nm, with GVD of ~28 ps/nm/km. The HC-PCF filled with FGO solution is 12 cm in length and has a large normal dispersion of ~−186 ps/nm/km. Thus, the total intracavity dispersion obtained is ~0.53 ps2, typical of a dispersion-managed fiber laser cavity (Nelson et al. 1997). Two sections of SMF of 50 cm in length were spliced to the HC-PCF filled with FGO using a fusion splicer. The input/output splice loss is ~1.5 dB, and the total insertion loss of HC-PCF filled with FGO is ~4.1 dB.

FGO was dispersed in N,N-dimethyl methanamide (DMF) with concentration of 0.1 mg/ml and then was filled into the core of HC-PCF (HC1550-02, Crystal Fiber) through a selective hole filling process (Zhang et al. 2010). The insets of Fig. 54 are the cross section images of the HC-PCF before (right) and after (left) infiltration of FGO solution, observed by the use of a microscope (Nikon ECLIPSE 80i). The central hole is perfectly filled while the holes in the cladding are not filled. Figure 55b plots the measured transmission as a function of average pump power at the wavelength of 1550 nm (using a probe laser with the pulse width of 2 ps and

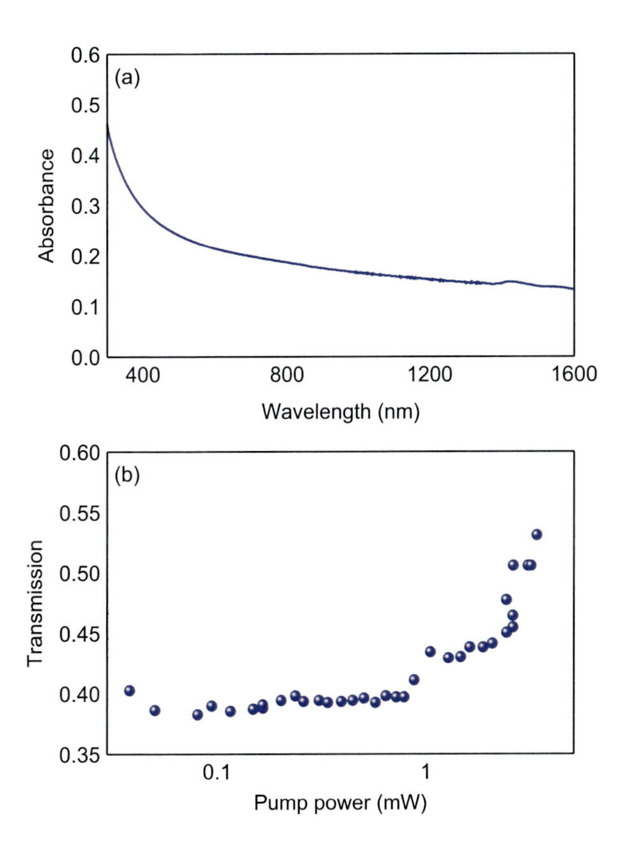

Fig. 55 (**a**) Absorption spectrum of FGO solution in DMF. (**b**) Typical transmission of the HC-PCF filled with FGO as a function of average pump power

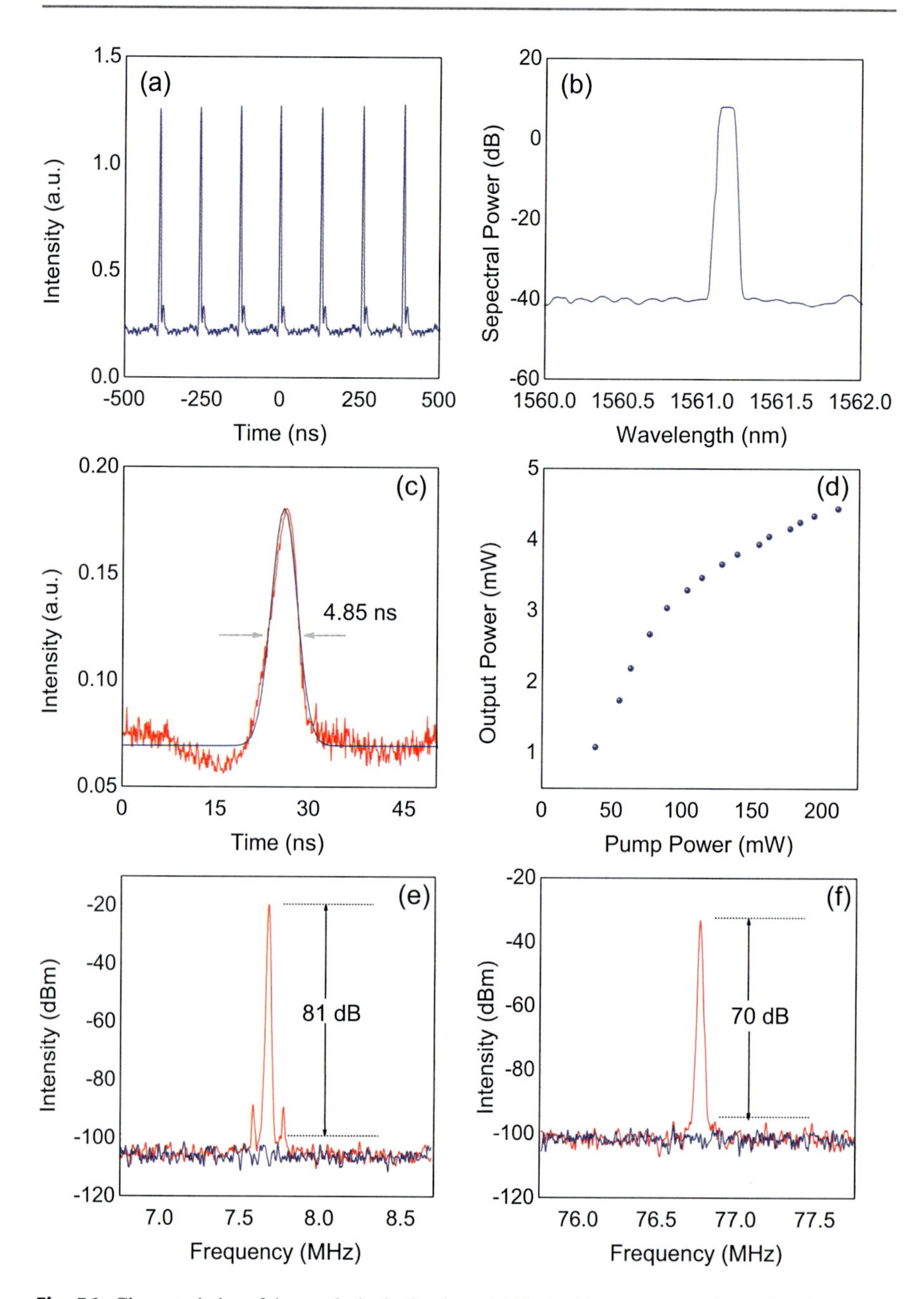

Fig. 56 Characteristics of the mode-locked pulses. (**a**) Typical laser output pulse train. (**b**) Output pulse spectrum, centered at 1561:2 nm. (**c**) Pulse shape (black line) with sech2 fit (red line). (**d**) Output power versus pump power. (**e**) RF spectrum measured around the fundamental repetition rate of 7:68 MHz. (**f**) RF spectrum measured around the tenth harmonic of the repletion rate at 76:8 MHz. The red traces in (**e**, **f**) depict the background when the laser is switched off

the repetition rate of 20 GHz). At low input power level, the transmission is almost independent of pump power.

Figure 56 summarizes the characteristics of the optical pulses emitted from the fiber laser mode locked by HC-PCF filled with FGO. The pulse train of the laser output shown in Fig. 56a has a period of 131 ns, which matches well with the cavity round-trip time and verifies that the laser is mode locked. Figure 56b shows the optical spectrum of the mode-locked pulses. The spectral width of the pulses has the full width at half maximum value of 0:11 nm and is centered at 1561.2 nm. Figure 56c demonstrates the pulse intensity profile, and the pulse width of 4:85 ns can be observed for a sech2 curve fit. When the pump power is increased to ~38 mW, stable mode locking can be initiated by adjusting the PC. The laser output power versus the pump power is shown in Fig. 56d. When the pump power is ~210 mW, the laser output power of ~4:3 mW with the pulse energy of ~0.56 nJ can be obtained. Such a long-pulse passive mode-locking regime has been recently identified by employing an SA (Wang et al. 2008; Renninger et al. 2008; Kobtsev et al. 2008; Tian et al. 2009). Figure 56e shows the radio frequency (RF) spectrum of the passively mode-locked fiber laser by real-time spectrum analyzer (Tektronix RSA 3303A, 3 GHz). Its fundamental peak locates at the cavity repetition rate (7:68 MHz) with the signal-to-noise ratio of 81 dB (108 contrast). Figure 56f reveals the RF spectrum measured around the tenth harmonic of the repetition rate at 76.8 MHz. The signal-to-noise ratio of >70 dB can also be observed, indicating good mode-locking stability (Linde 1986).

In conclusion, we have demonstrated a nanosecond pulse erbium-doped fiber laser that is passively mode locked by an HC-PCF filled with FGO solution. The flexibility offered by the graphene solution, along with the HC-PCF, can lead to a simple and efficient approach to integrate the graphene into the optical system.

Conclusions

In this chapter, we have shortly reviewed the filling technologies of PCFs, including the selectively filling method by collapsing air holes, by splicing to a SMF with a lateral offset, and by femtosecond (fs) laser micromachining. Then, we have demonstrated applications of the filled PCFs, including optical fiber devices, optical fiber sensors, and optical fiber lasers. Filling technologies of PCFs, especially selectively filling technologies, allow different types of liquids or gases to fill into the selected air holes of PCFs, thus changing the guidance properties of the fiber and providing wide and novel applications.

References

R. Amezcua-Correa, F. Gérôme, S.G. Leon-Saval, N.G.R. Broderick, T.A. Birks, J.C. Knight, Opt. Express **16**(2), 1142–1149 (2008)

Q.L. Bao, H. Zhang, Y. Wang, Z.H. Ni, Y.L. Yan, Z.X. Shen, K.P. Loh, D.Y. Tang, Adv. Funct. Mater. **19**, 3077 (2009)

Q.L. Bao, H. Zhang, J.X. Yang, S. Wang, D.Y. Tong, R. Jose, S. Ramakrishna, C.T. Lim, K.P. Loh, Adv. Funct. Mater. **20**, 782 (2010a)
Q.L. Bao, H. Zhang, J.X. Yang, S. Wang, D.Y. Tong, R. Jose, S. Ramakrishna, C.T. Lim, K.P. Loh, Adv. Funct. Mater. **20**, 782 (2010b)
Q. Bao, H. Zhang, Z. Ni, Y. Wang, L. Polavarapu, Z. Shen, Q.-H. Xu, D. Tang, K. Loh, Nano Res. **4**, 297 (2011)
F. Benabid, F. Couny, J.C. Knight, T.A. Birks, P. St, J. Russell, Nature **434**(7032), 488–491 (2005)
R.A. Bergh, H.C. Lefevre, H.J. Shaw, Opt. Lett. **5**, 479–481 (1980)
J. Canning, M. Stevenson, T.K. Yip, S.K. Lim, C. Martelli, Opt. Express **16**(20), 15700–15708 (2008)
C. Caucheteur, M. Debliquy, D. Lahem, P. Mégret, IEEE Photon. Technol. Lett. **20**(2), 96–98 (2008)
C.M.B. Cordeiro, E.M. Dos Santos, C.H. Brito Cruz, C.J.S. de Matos, D.S. Ferreiira, Opt. Express **14**(18), 8403–8412 (2006)
C.M.B. Cordeiro, C.J.S. de Matos, E.M. dos Santos, A. Bozolan, J.S.K. Ong, T. Facincani, G. Chesini, A.R. Vaz, C.H.B. Cruz, Meas. Sci. Technol. **18**(10), 3075–3081 (2007)
C.J.S. De Matos, C.M.B. Cordeiro, E.M. Dos Santos, J.S.K. Ong, A. Bozolan, C.H. Brito Cruz, Opt. Express **15**(18), 11207–11212 (2007)
P. Domachuk, H.C. Nguyen, B.J. Eggleton, M. Straub, M. Gu, Appl. Phys. Lett. **84**(11), 1838–1840 (2004)
X. Dong, H.Y. Tam, P. Shum, Appl. Phys. Lett. **90**, 151113 (2007)
J. Du, Y. Liu, Z. Wang, Z. Liu, B. Zou, L. Jin, B. Liu, G. Kai, X. Dong, Opt. Express **16**, 4263–4269 (2008)
R.B. Dyott, J. Bello, V.A. Handerek, Opt. Lett. **12**, 287–289 (1987)
J.C. Fanguy, L. Xu, K. Soni, S. Tao, Opt. Lett. **29**(11), 1191–1193 (2004)
H.Y. Fu, H.Y. Tam, L.-Y. Shao, X. Dong, P.K.A. Wai, C. Lu, S.K. Khijwania, Appl. Opt. **47**, 2835–1839 (2008)
F. Gérôme, R. Jamier, J.L. Auguste, G. Humbert, J.M. Blondy, Opt. Lett. **35**(8), 1157–1159 (2010)
R.W. Gilsdorf, J.C. Palais, Appl. Opt. **33**(16), 3440–3445 (1994)
T. Han, Y. Liu, Z. Wang, B. Zou, B. Tai, B. Liu, Opt. Lett. **35**(12), 2061–2063 (2010)
C.J. Hensley, D.H. Broaddus, et al., Opt. Express **15**, 6690–6695 (2007)
Y.L. Hoo, S. Liu, H.L. Ho, W. Jin, IEEE Photon. Technol. Lett. **22**, 296–298 (2010)
Y. Huang, Y. Xu, A. Yariv, Appl. Phys. Lett. **85**(22), 5182–5184 (2004)
F. Jansen, F. Stutzki, C. Jauregui, J. Limpert, A. Tünnermann, Opt. Express **19**(14), 13578–13589 (2011)
C. Kerbage, B.J. Eggleton, Opt. Express **10**(5), 246–255 (2002)
S. Kobtsev, S. Kukarin, Y. Fedotov, Opt. Express **16**, 21936 (2008)
B.T. Kuhlmey, B.J. Eggleton, D.K.C. Wu, J. Lightwave Technol. **27**(11), 1617–1630 (2009)
D. Linde, Appl. Phys. B **39**, 201 (1986)
Y. Liu, B. Liu, X. Feng, W. Zhang, G. Zhou, S. Yuan, G. Kai, X. Dong, Appl. Opt. **44**, 2382–2390 (2005)
Z.-B. Liu, X. He, D.N. Wang, Opt. Lett. **36**(16), 3024–3026 (2011)
B.J. Mangan, J.C. Knight, T.A. Birks, P. St, J. Russell, A.H. Greenaway, Electron. Lett. **36**, 1358–1359 (2000)
I.A. Marcinkevi, S. Juodkazis, M. Watanabe, M. Miwa, S. Matsuo, H. Misawa, J. Nishii, Opt. Lett. **26**(5), 277–279 (2001)
C. Martelli, J. Canning, K. Lyytikainen, N. Groothoff, Opt. Express **13**(10), 3890–3895 (2005)
V.P. Minkovich, D. Monzón-Hernández, Opt. Express **14**(18), 8413–8418 (2006)
L.E. Nelson, D.J. Jones, K. Tamura, H.A. Haus, E.P. Ippen, Appl. Phys. B **65**, 277 (1997)
K. Nielsen, D. Noordegraaf, T. Sørensen, A. Bjarklev, T.P. Hansen, J. Opt. A Pure Appl. Opt. **7**(8), L13–L20 (2005)
L. Niu, C.-L. Zhao, L. Qi, C.C. Chan, J. Kang, S. Jin, J. Guo, H. Wei, J. Lightwave Technol. **32**(22), 4416–4421 (2014a)
L. Niu, C.-L. Zhao, J. Kang, S. Jin, J. Guo, H. Wei, Opt. Commun. **313**, 243–247 (2014b)

O. Okhotnikov, A. Grudinin, M. Pessa, New J. Phys. **6**, 177 (2004)
W.E.P. Paddena, M.A. van Eijkelenborg, A. Argyros, N.A. Issa, Appl. Phys. Lett. **84**, 1689–1691 (2004)
D. Popa, Z. Sun, F. Torrisi, T. Hasan, F. Wang, A.C. Ferrari, Appl. Phys. Lett. **97**, 203106 (2010)
W. Qian, C.-L. Zhao, X. Dong, W. Jin, Opt. Commun. **283**, 5250–5254 (2010)
W. Qian, C.-L. Zhao, Y. Wang, C.C. Chan, S. Liu, W. Jin, Opt. Lett. **36**(16), 3296–3298 (2011a)
W. Qian, C.-L. Zhao, S. He, X. Dong, S. Zhang, Z. Zhang, S. Jin, J. Guo, H. Wei, Opt. Lett. **36**(9), 1548–1550 (2011b)
W.H. Renninger, A. Chong, F.W. Wise, Opt. Lett. **33**, 3025 (2008)
P.J.A. Sazio, A. Amezcua-Correa, C.E. Finlayson, J.R. Hayes, T.J. Scheidemantel, N.F. Baril, B.R. Jackson, D.-J. Won, F. Zhang, E.R. Margine, V. Gopalan, V.H. Crespi, J.V. Badding, Science **311**, 1583–1586 (2006)
Y.W. Song, S.Y. Jang, W.S. Han, M.K. Bae, Appl. Phys. Lett. **96**, 051122 (2010)
G. Steinmeyer, D.H. Sutter, L. Gallmann, N. Matuschek, U. Keller, Science **286**, 1507 (1999)
Y. Sun, S.I. Shopova, G. Frye-Mason, X. Fan, Opt. Lett. **33**, 788–790 (2008)
Z. Sun, T. Hasan, F. Torrisi, D. Popa, G. Privitera, F. Wang, F. Bonaccorso, D.M. Basko, A.C. Ferrari, ACS Nano **4**, 803 (2010a)
Z. Sun, D. Popa, T. Hasan, F. Torrisi, F. Wang, E.J.R. Kelleher, J.C. Travers, A.C. Ferrari, Nano Res. **3**, 653 (2010b)
X. Tian, M. Tang, P.P. Shum, Y. Gong, C. Lin, S. Fu, T. Zhang, Opt. Lett. **34**, 1432 (2009)
G.E. Town, W. Yuan, R. McCosker, O. Bang, Opt. Lett. **35**, 856–858 (2010)
T. Toyoda, M. Yabe, J. Phys. D Appl. Phys. **16**, L97–L100 (1983)
J. Villatoro, M.P. Kreuzer, R. Jha, V.P. Minkovich, V. Finazzi, G. Badenes, V. Pruneri, Opt. Express **17**(3), 1448–1453 (2009)
Y. Wang, L. Xiao, D. Wang, W. Jin, Opt. Lett. **32**, 1035 (2007a)
Z. Wang, T. Taru, T.A. Birks, J.C. Knight, Opt. Express **15**, 4795–4803 (2007b)
F. Wang, A.G. Rozhin, V. Scardaci, Z. Sun, F. Hennrich, I.H. White, W.I. Milne, A.C. Ferrari, Nat. Nanotechnol. **3**, 738 (2008)
Y. Wang, C.R. Liao, D.N. Wang, Opt. Express **18**(17), 18056–18060 (2010a)
Y. Wang, Y. Li, C. Liao, D.N. Wang, M. Yang, P. Lu, IEEE Photon. Lett. **22**, 39–41 (2010b)
Y. Wang, M. Yang, D.N. Wang, C.R. Liao, IEEE Photon. Technol. Lett. **23**(20), 1520–1522 (2011)
Y. Wang, C.R. Liao, D.N. Wang, Opt. Lett. **37**(22), 4747–4749 (2012)
Y. Wang, D.N. Wang, F. Yang, Z. Li, M. Yang, Opt. Lett. **39**(13), 3872–3875 (2014)
D.K.C. Wu, B.T. Kuhlmey, B.J. Eggleton, Opt. Lett. **34**, 322–324 (2009)
Z. Wu, Z. Wang, Y.-G. Liu, T. Han, S. Li, H. Wei, Opt. Express **19**(18), 17344–17349 (2011)
L. Xiao, W. Jin, M.S. Demokan, H.L. Ho, Y.L. Hoo, C. Zhao, Opt. Express **13**(22), 9014–9022 (2005)
L. Xiao, W. Jin, M.S. Demokan, Opt. Express **15**(24), 15637–15647 (2007)
H.F. Xuan, W. Jin, J. Ju, Y.P. Wang, M. Zhang, Y.B. Liao, M.H. Chen, Opt. Lett. **33**, 845–847 (2008)
M. Yang, D.N. Wang, J. Lightwave Technol. **30**(21), 3407–3412 (2012)
M. Yang, D.N. Wang, Y. Wang, C.R. Liao, Opt. Lett. **36**(5), 636–638 (2011)
M. Yang, Z. Yang, J. Dai, D. Zhang, Sensors Actuators B **166–167**, 632–636 (2012)
A. Yariv, Optical Electronics in Modern Communications, 5th ed. London, U.K.: Oxford Univ. Press, (1997)
W. Yuan, G.E. Town, O. Bang, IEEE Sensors J. **10**, 1192–1199 (2010)
X. Zhang, R. Wang, F.M. Cox, B.T. Kuhlmey, M.C.J. Large, Opt. Express **15**(24), 16270–16278 (2007)
H. Zhang, D.Y. Tang, R.J. Knize, L.M. Zhao, Q.L. Bao, K.P. Loh, Appl. Phys. Lett. **96**, 111112 (2010)
C.-L. Zhao, X. Yang, et al., IEEE Photon. Technol. Lett. **16**, 2535–2537 (2004)
C.-L. Zhao, Z. Wang, S. Zhang, L. Qi, C. Zhong, Z. Zhang, S. Jin, J. Guo, H. Wei, Opt. Lett. **37**(22), 4789–4791 (2012)

Photonic Crystal Fiber-Based Grating Sensors

56

Changrui Liao, Feng Zhu, and Chupao Lin

Contents

Abstract

Photonic crystal fibers support a powerful platform for the development of novel fiber devices. Combined with fiber grating, which is one of the most important fiber sensor configurations, the photonic crystal fibers exhibit extraordinary superiority in fiber sensing applications. In this chapter, we summarized the recent research works in the field of photonic crystal fiber - based fiber Bragg grating and long period grating and their sensing applications.

Introduction

Since the first publication by Knight et al. (1996) on photonic crystal fiber (PCF) (Knight et al. 1996), the optical fiber community has been continuously worked around these new fibers. A commonly classification of PCF divides into two main categories: index-guiding PCF (IG-PCF) and photonic-bandgap PCF (PBF). The

C. Liao (✉) · F. Zhu · C. Lin
College of Optoelectronic Engineering, Shenzhen University, Shenzhen, China
e-mail: cliao@szu.edu.cn; 2151190226@email.szu.edu.cn; linchupao2016@email.szu.edu.cn

G.-D. Peng (ed.), *Handbook of Optical Fibers*,
https://doi.org/10.1007/978-981-10-7087-7_12

IG-PCF usually contains a solid core, which is surrounded by a microstructured cladding. As a result of the presence of air holes, the effective refractive index of the cladding is smaller than that of the core, and light is guided along the core by the mechanism of total internal reflection. The PBF usually has a hollow core, and the light guidance mechanism is the result of the presence of photonic bandgap in the cladding region for a specific range of wavelengths. By adjusting the size and location of the cladding holes or the core diameter, the fiber transmission spectrum, mode shape, nonlinearity, dispersion, and birefringence can be tuned.

Photonic Crystal Fiber-Based Fiber Bragg Grating (FBG)

Index-Guiding Photonic Crystal Fiber-Based FBG Sensors

The First FBG Fabricated in IG-PCF

FBGs have been extensively investigated as sensing elements in single-mode fibers. The first demonstration of FBG in IG-PCF was made in Bell Labs in 1999 (Eggleton et al. 1999). The employed PCF consists of a Ge-doped core with radius of $\sim$1 μm and $\Delta = (n_{core} - n_{clad})/n_{core} \sim 0.5\%$, where n_{core} and n_{clad} are the refractive index of core and silica cladding. The fiber is loaded with deuterium to enhance the photosensitivity of the core and then exposed through phase mask from a dye laser. Transmission spectrum of IG-FBG is shown in Fig. 1, where some resonant dips from right to left are corresponding to the reflection of core mode and high-order modes.

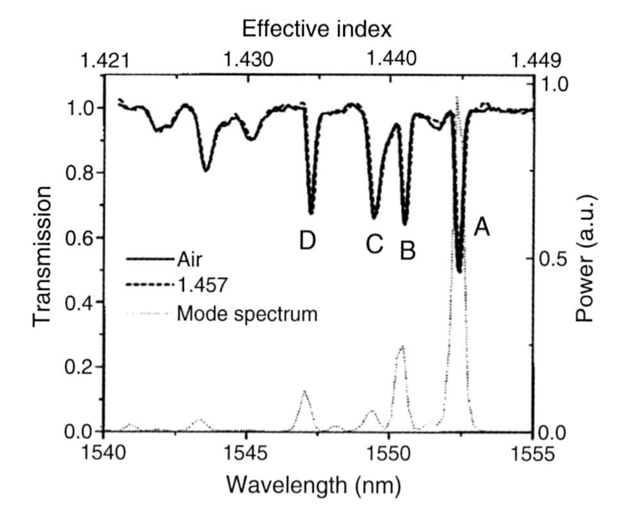

Fig. 1 Transmission spectrum of the FBG in IG-PCF (Eggleton et al. 1999)

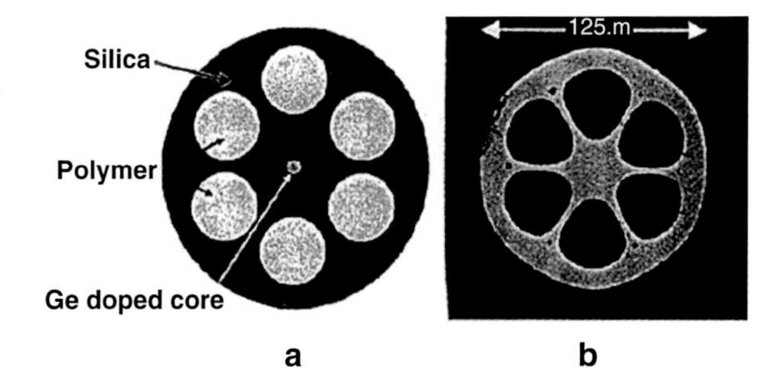

Fig. 2 (**a**) Cross section of the hybrid polymer-silica PCF. (**b**) Optical microscope image of real cross section before polymer infusion (Westbrook et al. 2000)

Temperature Sensing by Use of the FBG in Hybrid Polymer-Silica PCF

The holes in the cladding of IG-PCF allow the design of a wide range of guided-mode properties, which result in unique spectral characteristics of fiber gratings. In 2000, the FBG has been inscribed in a hybrid polymer-silica PCF, which is schematically shown in Fig. 2 (Westbrook et al. 2000). In this PCF, six holes with a diameter of ~40 μm form a hexagonal ring around the central silica region with a diameter of ~32 μm. Ge-doped core with a diameter of ~8 μm and Δ ~0.35% is in the center of the fiber. The fiber is deuterium loaded, and then the FBG is inscribed into the core by scanning exposure through a phase mask with 242-nm dye laser. Then, an acrylate-based polymer with similar refractive index of silica is filled into the air holes and UV cured to form a hybrid waveguide at the grating. The cladding spectrum will be changed with temperature due to the strong temperature dependence of the polymer refractive index.

Temperature Stability of the FBG in All-Silica IG-PCF

FBG inscription in SMFs usually relies on photosensitivity of core dopants (like germanium), and this photosensitivity can be enhanced by hydrogen loading treatment. FBGs in such doped fibers can be inscribed by use of single photon absorption at 242–248 nm using KrF excimer (248 nm) or frequency-doubled argon-ion (244 nm) lasers. However, pure silica is not photosensitive at 244 nm, so alternative grating inscriptions are required for pure silica IG-PCFs. In 2005, the FBG has been firstly inscribed in all-silica PCF with femtosecond laser at 267 nm (Fu et al. 2005). Gratings have been fabricated with a depth of 10 dB and an average of $\Delta n > 4 \times 10^{-4}$. An isochronal annealing has been done to evaluate the stability of the grating, and it is found that the refractive index monotonically decreased as a function of temperature. This decay performance indicates that the grating has an origin related to both defect formation and the compaction mechanism active in fused silica at 193 nm.

Strain Sensing by Use of the FBG in Holey IG-PCF

Han Y.-G et al. have investigated the effect of air-hole size on the strain response of the FBGs inscribed in four different holey fibers with a germanium-doped core and one-layered air holes in silica cladding, as shown in Fig. 3 (Han et al. 2006). The

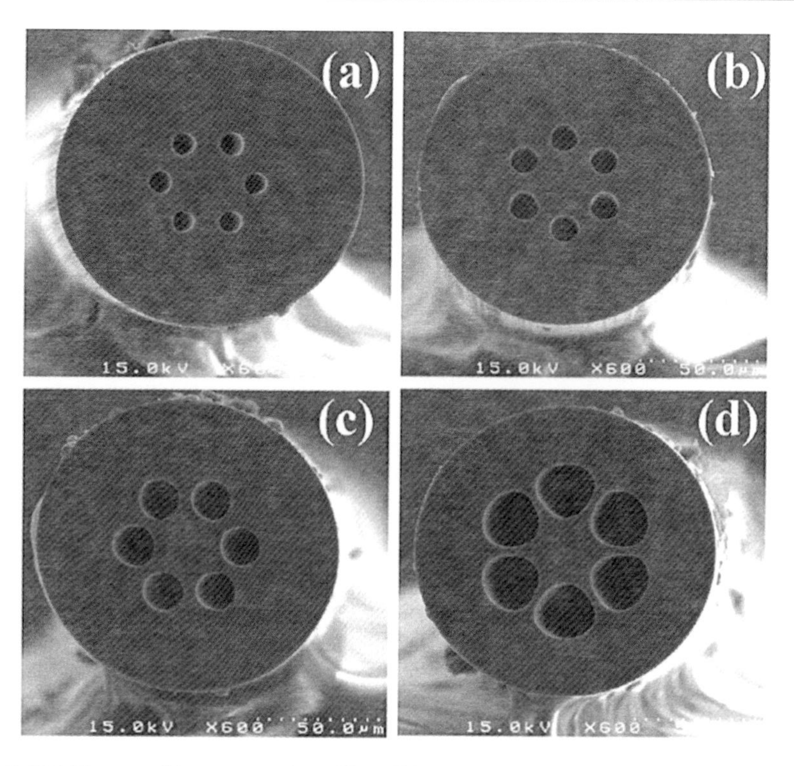

Fig. 3 SEM images of the cross section of four different holey IG-PCFs (Han et al. 2006)

FBGs have been fabricated by use of beam-scanning technique with 244 nm Ar^+ laser, and hydrogen loading treatment is taken to enhance the photosensitivity of the Ge-doped core. After fabricating FBGs, annealing process during 24 h is taken to remove unreacted hydrogen to stabilize the quality of FBGs.

The strain sensitivity of the FBGs in holey fibers corresponding to air-hole size has been experimentally investigated. The strain sensitivity of the FBGs can be written as

$$\frac{\partial \lambda_p}{\partial \varepsilon} = \lambda_p (1 - \rho)\, \varepsilon \tag{1}$$

where λp is the center wavelength of the FBG, ρ is the photoelastic coefficient, and ε is the applied strain. Since four types of holey fibers have the different cross-sectional area due to different sizes of air holes, their strain sensitivities become different corresponding to air-hole size. Figure 4 shows the center wavelength shift of FBGs with strain changes, and it can be found that the strain sensitivity is significantly enhanced as air-hole size increased because of small glass volume due to the air holes inside of silica cladding.

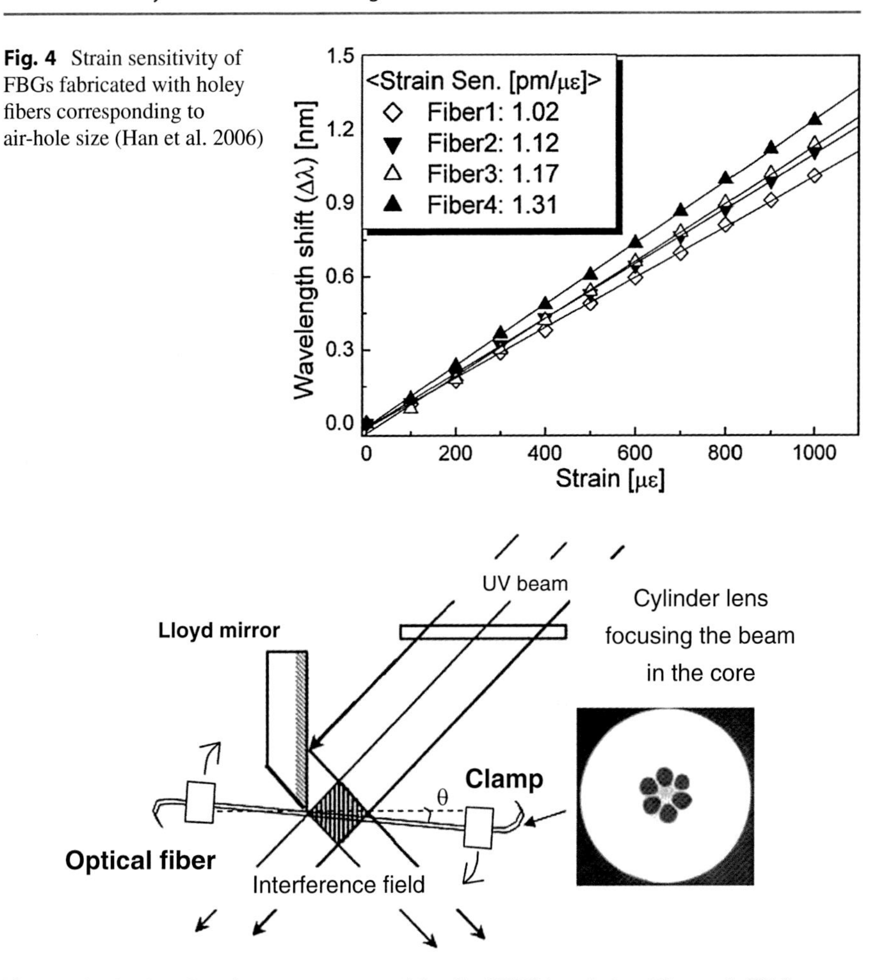

Fig. 4 Strain sensitivity of FBGs fabricated with holey fibers corresponding to air-hole size (Han et al. 2006)

Fig. 5 Lloyd mirror interferometer setup used for tilted FBG inscription (Huy et al. 2006)

Refractive Index Sensing by Use of the FBG in Holey IG-PCF

Huy et al. demonstrated a tilted FBG sensor inscribed in IG-PCF as an optical refractometer (Huy et al. 2006). The employed fiber can be described by a ring of six air holes (d $\sim$ 15 μm and $\Lambda \sim$ 15.8 μm), surrounding a noncircular and slightly decentered core of 11 μm in diameter, as shown in the inset of Fig. 5. Such a fiber is multimode in 1.5-μm spectral window. Fibers are required to be hydrogen-loaded to increase the photosensitivity of Ge-doped core. Tilted FBGs are inscribed by use of Lloyd mirror interferometer including a CW frequency-doubled argon laser emitting at 244 nm. The experimental setup is demonstrated in Fig. 5, where the fiber is held straight on between two clamps fixed on a rotary stage and the required angle θ between the fiber axis and the interference fringes pattern is precisely adjusted.

Spectral responses in transmission of tilted FBG inscribed in the six-hole PCF are shown in Fig. 6. For $\theta = 0°$, not only the Bragg resonance but also resonances to

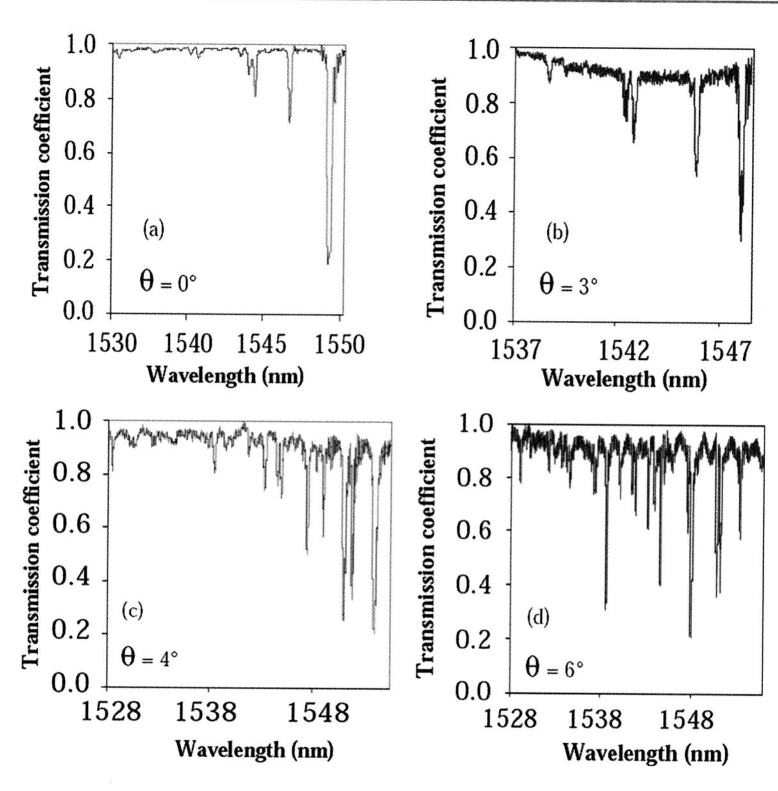

Fig. 6 Transmission spectra of (**a**) non-tilted, (**b**) 3°-tilted, (**c**) 4°-tilted, (**d**) 6°-tilted FBG written in the six-hole PCF (Huy et al. 2006)

higher-order modes can be observed. When θ increases, more and more transmission dips are distinguishable on the spectrum, indicating that more and more modes are involved in the coupling process.

Refractive index response experiment consists in introducing several calibrated refractive index liquids into the holes of the PCF and to determine the consequences on the spectral resonances of a tilted FBG. These liquids are successively inserted into the fiber holes of a unique fiber section, without removing the previous one. The spectral resonances experience a red shift, when a given liquid reaches the tilted FBG (see Fig. 7). The insertion of any liquid modifies the interaction with the evanescent field of the guided mode and hence the effective index. The effective index of the guided modes increases together with the refractive index of the inserted liquid.

Transversal Load Sensing by Use of the FBG in Hi-Bi PCF

T. Geernaert et al. presented an FBG written in Hi-Bi PCF to monitor stress inside a composite material (Geernaert et al. 2009). The microstructure of the PCF consists of three rows of air holes and a central Ge-doped core, as shown in Fig. 8. The

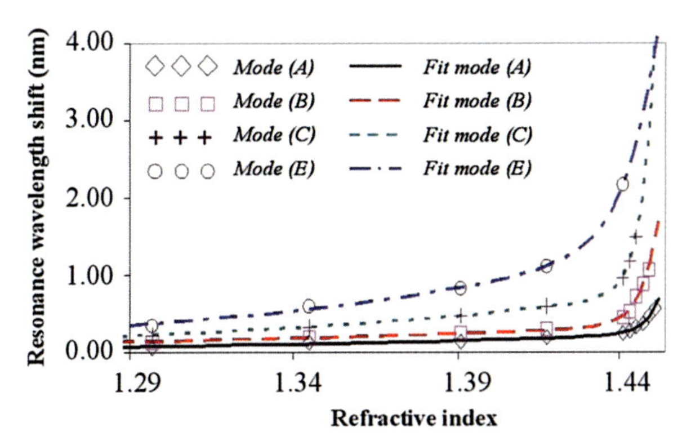

Fig. 7 Wavelength shift of the first four resonances versus the refractive index of the liquid (Huy et al. 2006)

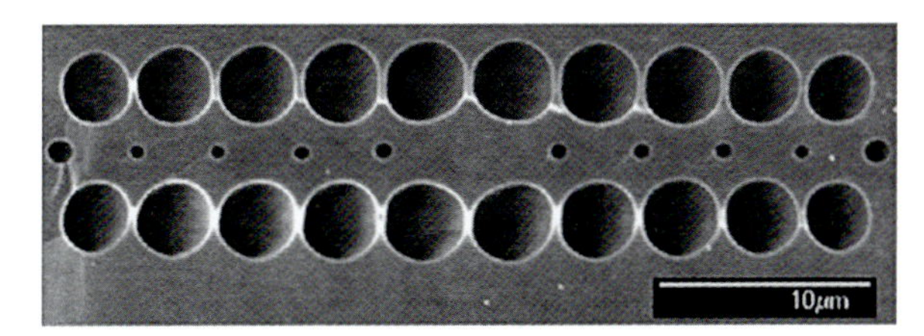

Fig. 8 SEM image of the cross section of the microstructure (Geernaert et al. 2009)

diameter of the PCF is 125 μm and the Ge-doped core is circular, and the lattice constant of the microstructure is 3.9 μm. The phase modal birefringence of this PCF is measured to be ~ 8×10^{-4} at 1550 nm. Since the PCF is birefringent, unpolarized light that is launched into the fiber yields two Bragg peaks, one for each orthogonally polarized mode propagating in the fiber. The peak separation is related to the phase modal birefringence by

$$\Delta\lambda = \lambda_{B2} - \lambda_{B1} = 2 \times B \times \Lambda \tag{2}$$

where Λ is the grating period in the core.

In the transversal load test, an 8-mm-long FBG is embedded between two carbon fiber reinforced epoxy laminates. The stress on the sample is continuously increased from 0 up to 4 MPa. The Bragg peak separation changes at a constant rate of 15.3 pm/MPa. The original Bragg peak separation of ~1 nm is large enough to guarantee that the two Bragg peaks remain well separated and can be detected, as shown in Fig. 9.

Gas Pressure Sensing by Use of the FBG in Grapefruit PCF

C. Wu et al. demonstrated FBGs written in grapefruit PCFs, which are used for pressure sensing (Wu et al. 2010). Figure 10 shows the cross sections of the two employed grapefruit PCFs. Both PCFs have outside diameter of 125 μm and core diameter of 8 μm. The large-hole PCF has hole diameter of 33.8 μm in radial direction. The small-hole PCF has hole diameter of 17 μm and hole pitch

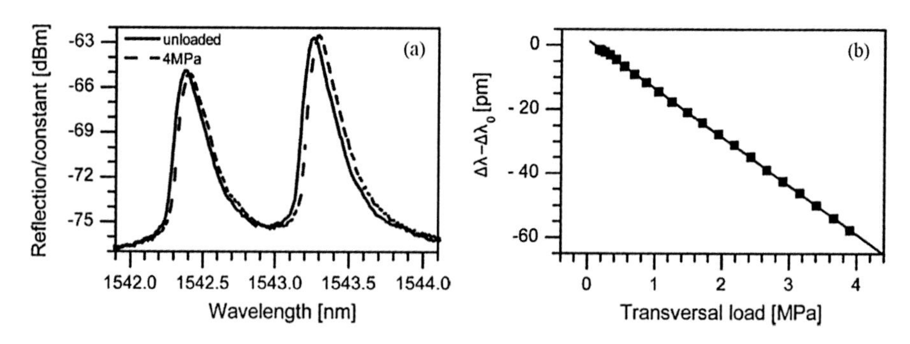

Fig. 9 (**a**) FBG reflection spectrum of the embedded FBG in the Hi-Bi PCF with transversally load and unloaded. (**b**) Variation of the Bragg peak separation versus transversal stress (Geernaert et al. 2009)

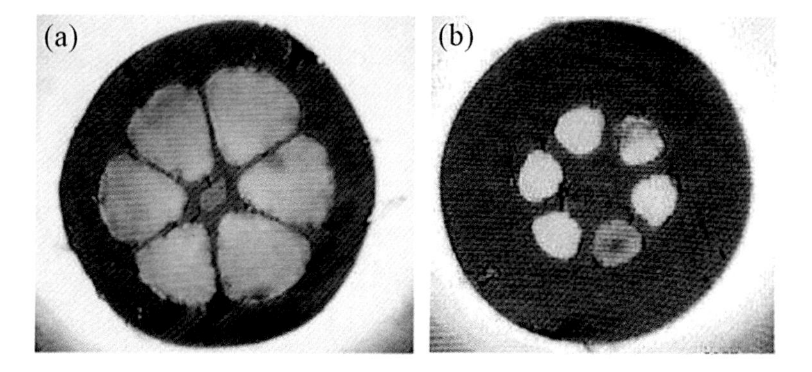

Fig. 10 Cross sections of (**a**) large-hole grapefruit PCF and (**b**) small-hole grapefruit PCF (Wu et al. 2010)

of 25.4 μm. The core of both PCFs is doped with germanium to enhance the photosensitivity.

FBGs have been inscribed in the grapefruit PCFs by use of 193-nm excimer laser and a phase mask. To characterize their response to pressure, the FBGs are located in a pressure chamber, and the pressure is changed from 0 to 25 MPa in a step of 2.5 MPa. Experimental results of wavelength shift as a function of pressure in two PCFs are shown in Fig. 11. It is found that the FBGs in grapefruit PCF show much higher pressure sensitivity than normal FBG.

Photonic-Bandgap Photonic Crystal Fiber-Based FBG Sensors

Bend Sensing Based on the FBG in All-Solid PBF

L. Jin et al. presented the FBGs inscribed into the all-solid PBF by forming a longitudinal periodic index modulation over the high-index rod lattice in the cladding (Jin et al. 2007). In the fiber, a Ge-doped high-index rod lattice of six layers

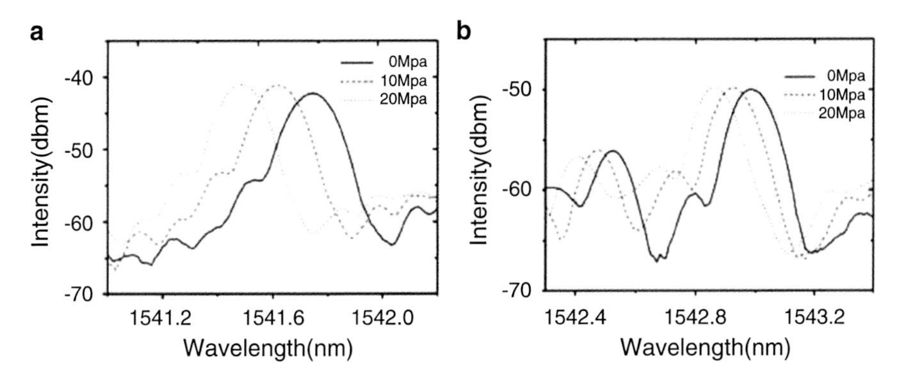

Fig. 11 Reflection spectra at different pressure of (**a**) large-hole PCF and (**b**) small-hole PCF (Wu et al. 2010)

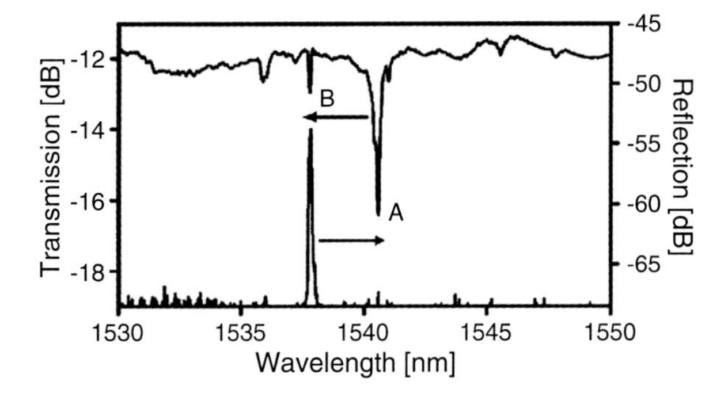

Fig. 12 Transmission and reflection spectra of the FBG in all-solid PBF (Jin et al. 2007)

is embedded in pure silica background (the index difference is 1%), and the core is formed by omitting a single rod. The pitch Λ of the high-index rods is 11 μm, and the nominal ratio of the diameter of the raised-index rods to the pitch is $d/\Lambda = 0.4$. The outer diameter is ~200 μm. The Ge-doped rod lattice in the cladding enables index modulation by UV light.

A length of PBF is firstly spliced to SMFs and then loaded in a hydrogen atmosphere at 100 atm, 100 °C for 48 h to enhance its photosensitivity. The PBF is then exposed by 248-nm KrF excimer laser at a repetition rate of 5 Hz for 15 min. The grating period is 532.5 nm, and the grating length is 1.3 cm. Figure 12 shows the transmission and reflection spectra of the FBG, from which peaks A (1540.58 nm, 4.4 dB) and B (1537.8 nm, 1.1 dB) are observed. However, only dip B finds its corresponding reflective peak in the reflection spectrum because the guided mode has the largest overlap with SMF among all the modes.

Figure 13a shows the transmission spectrum of the offset launched FBG. The fluctuation of the spectral background is caused by complicated couplings between the high-index rods and the defect core. No noticeable changes in the resonance depths of dips A and B are observed when the lateral offset is applied. Figure 13b

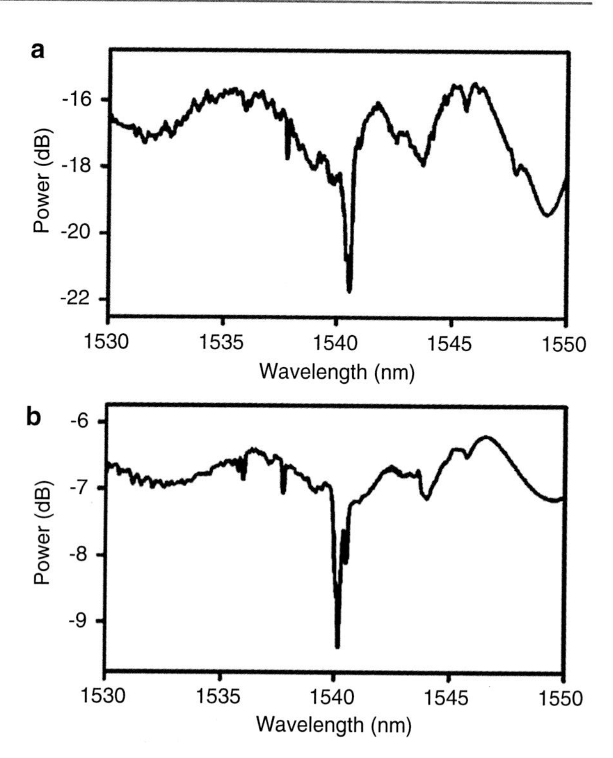

Fig. 13 (**a**) Transmission spectrum of the FBG with lateral offset between the launched SMF and the FBG. (**b**) Split of the super-mode resonance peak A when the FBG is bent (Jin et al. 2007)

shows the splits of the super-mode resonance peak A. The two separated peaks correspond to backward rod modes in stretched and compressed rods and the wavelength separation increases with curvature due to enhanced index gradient over the fiber. The FBG can be potentially applied for sensing directional bend.

High-Temperature Sensing by Use of the FBG in All-Solid PBF

Y. Li et al. reported the FBGs written in all-solid PBF by use of femtosecond laser through a phase mask (Li et al. 2009). The fiber is exposed to 800-nm 120-fs laser pulses from a Ti/sapphire amplifier with the repetition rate of 1 kHz and the 1/e Gaussian beam radius of 2 mm. The laser beam is focused using a cylindrical lens before passing through phase mask to illuminate the PBF. By carefully adjusting the focal line to be located in the fiber core, the grating is successfully formed due to refractive index modulation in the core, induced by multiphoton absorption. The transmission and reflection spectra are shown in Fig. 14.

Morphology of femtosecond laser-induced grating in all-solid PBF is shown in Fig. 15. The grating lines are located in the fiber core area as shown in Fig. 15a, where the laser incident direction is plotted. The influence area of the femtosecond pulses along the beam propagation direction was shown in Fig. 15b, where the cross section of the fiber is cleaved at the center of the grating. It can be observed that except for the all-silica core area, the Ge-doped rods are not affected by the laser pulses and remained unchanged.

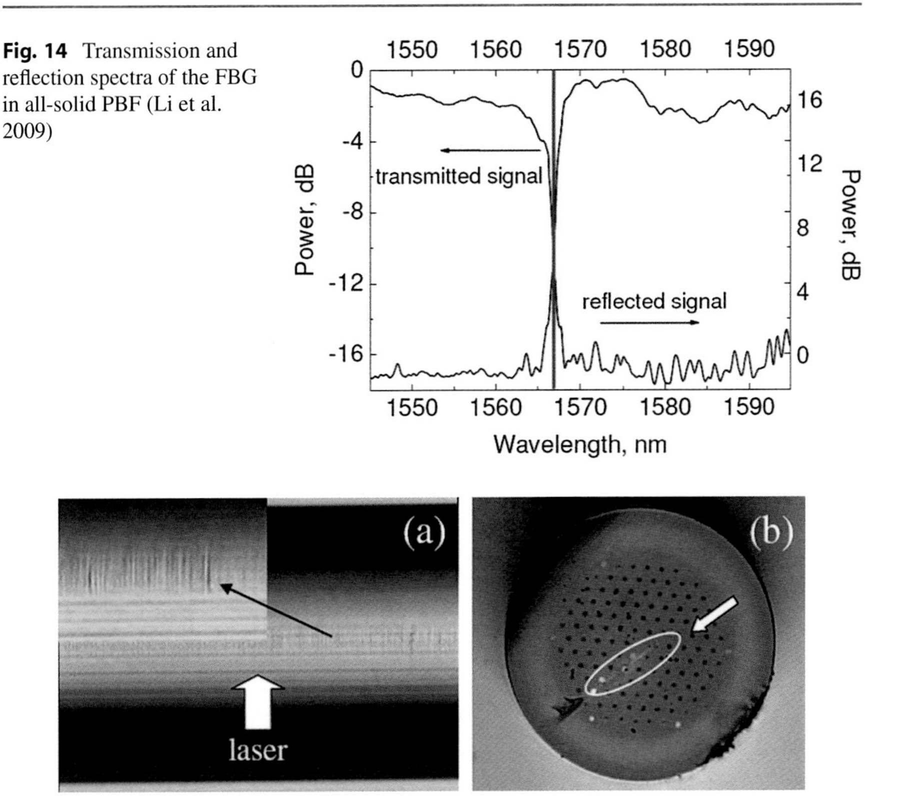

Fig. 14 Transmission and reflection spectra of the FBG in all-solid PBF (Li et al. 2009)

Fig. 15 Optical microscope images of the grating morphology in the (**a**) longitudinal and (**b**) the cross section of the fiber (Li et al. 2009)

The annealing characteristic of the FBG in all-solid PBF is shown in Fig. 16. The gratings are more thermally stable when compared with UV-laser-induced samples. The gratings are almost unaffected by the thermal exposure up to 700 °C, following which a slowly decay occurred at the elevated temperatures. The temperature corresponding to the 50% decrease in the normalized refractive index change is 900 °C, instead of 650 °C for the UV-laser-induced gratings in H2-free fibers, even higher than that of the type I femtosecond laser-induced gratings in Ge-doped SMF-28 fibers.

Photonic Crystal Fiber-Based Long Period Grating (LPG)

Index-Guiding Photonic Crystal Fiber-Based LPG Sensors

Long-period fiber grating (LPG) is another useful structure that can be written in PCF. Many different fabrication methods have been reported, and these fiber gratings have been used in diversified sensing applications.

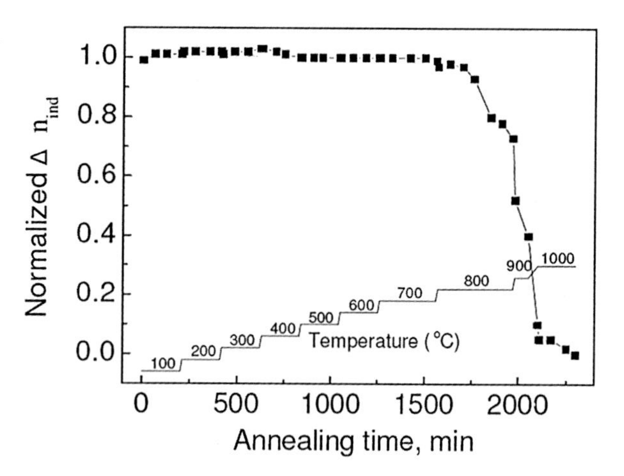

Fig. 16 Thermal degradation of the FBG written in all-solid PBF, with elapsed time over various temperature ranges (Li et al. 2009)

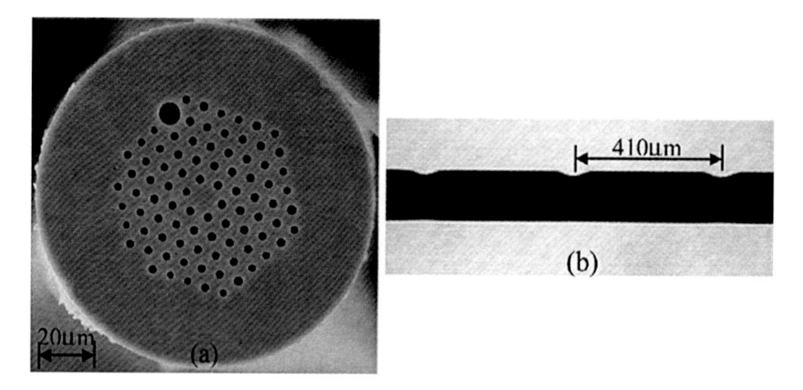

Fig. 17 (**a**) SEM image of the cross section of the PCF. (**b**) CCD photograph of the LPG with periodic grooves (Wang et al. 2006)

Strain Sensing by Use of the Carved LPG in IG-PCF

Y. P. Wang et al. demonstrated a PCF-based LPFG sensor with a high strain sensitivity of -7.6 pm/$\mu\varepsilon$, which is fabricated by use of focused CO_2 laser beam to carve periodic grooves on a large mode-area PCF (Wang et al. 2006). The SEM image of the cross section of the PCF employed is illustrated in Fig. 17b, where the PCF has a center-to-center distance between the air holes of 6.1 μm and an average air-hole diameter of $d = 1.8$ μm. The holes are arranged in a hexagonal pattern, which has a diameter of 65 μm.

The periodic grooves carved by CO_2 laser irradiation is shown in Fig. 17b. Such grooves can induce periodic refractive index modulations along the fiber axis due to the photoelastic effect, thus creating an LPFG. The groove's depth, which indicates the efficiency of CO_2 laser heating and the refractive index modulation, depends

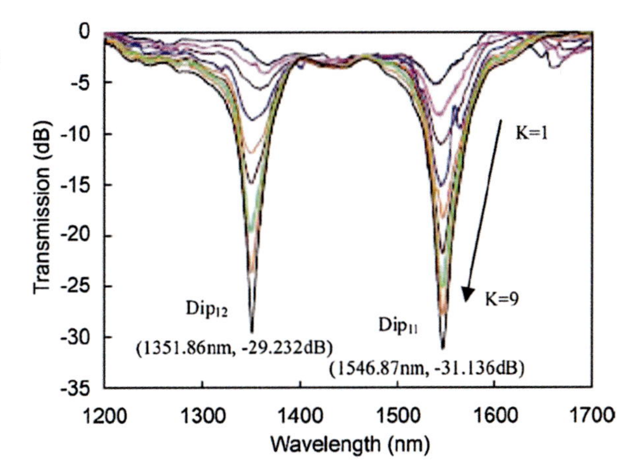

Fig. 18 Transmission spectral evolution of the LPG with the number, K, of scanning cycles being increased from 1 to 9, where $N = 40$ and $M = 5$ (Wang et al. 2006)

on the fabrication parameters. In the experiments, the diameter of the focused CO_2 laser beam spot is ~35 μm.

Figure 18 shows the transmission spectral evolution of the LPG with asymmetric periodic grooves, where a low insertion loss of ~2 dB and two attenuation dips (Dip11 and Dip12) are produced after nine scanning cycles, where the resonant wavelengths and the peak transmission attenuations of Dip11 and Dip12 are $\lambda 11 = 1546.87$ nm, $A11 = -31.136$ dB, $\lambda 12 = 1351.86$ nm, and $A12 = -29.232$ dB, respectively.

Figure 19 shows the schematic diagram of the carved LPG before and after a stretching force. The refractive index modulation in the carved LPFG that is stretched can be expressed as $\Delta n = \Delta n_{residual} + \Delta n_{groove} + \Delta n_{stretch}$, where $\Delta n_{residual}$ is the initial refractive index perturbation induced by the residual stress relaxation resulting from the high local temperature, which is similar to the case of the CO_2-laser-induced LPGs without periodic grooves; Δn_{groove} is the initial refractive index perturbation induced by the periodic grooves on the fiber; $\Delta n_{stretch}$ is the refractive index perturbation induced by the stretching force and can be expressed as $\Delta n_{stretch} = \Delta n_{strain} + \Delta n_{microbend}$, where Δn_{strain} is the refractive index perturbation induced by the difference between the stretch-induced tensile strains in the grooved and ungrooved regions via the photoelastic effect; $\Delta n_{microbend}$ is the refractive index perturbation induced by the stretch-induced microbends and this stretch-induced microbends effectively enhance refractive index modulation.

Tensile strain responses of two LPGs with and without periodic grooves have been compared. With the increase of the tensile strain, the resonant wavelengths of the LPG with periodic grooves shift rapidly toward the shorter wavelength and a good linearity is observed, whereas that of LPG without periodic grooves shifts slowly toward the shorter wavelength. The strain sensitivity of resonant wavelength of the LPGs written by CO_2 laser in the same fiber is increased by 25 times by means of carving periodic grooves on one side of the fiber.

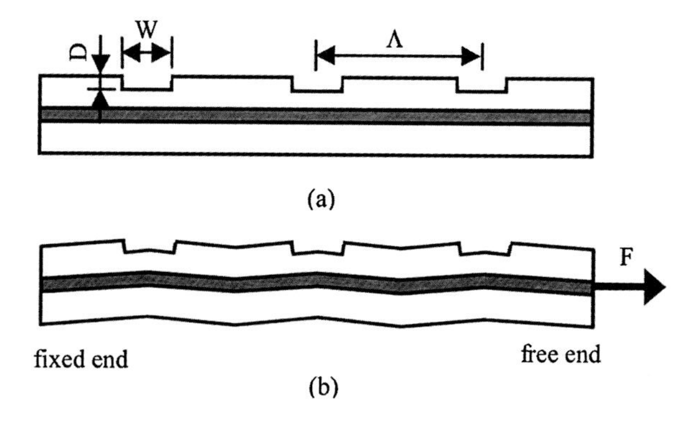

Fig. 19 Schematic of the CO_2-laser-carved LPFG (**a**) before and (**b**) after a stretching force (Wang et al. 2006)

Bend Sensing by Use of the LPG in IG-PCF

L. Jin et al. presented a directional bend sensor based on an LPG formed by introducing periodic grooves along one side of a PCF with a focused CO_2 laser beam (Jin et al. 2009). The fiber used for LPG fabrication is the crystal fiber's LMA-10 PCF, as shown in Fig. 20a. A focused CO_2 laser beam is controlled by a computer and scanned transversely across the fiber with a certain longitudinal period over a length of fiber, and the scanning may be repeated for a number of scanning cycles to produce LPGs with required strength of resonant coupling. The transmission spectrums and polarization-dependent loss (PDL) profiles of the LPG are shown in Fig. 20d, and the notches on one side of fiber surface and air-hole collapse and deformation are shown in Fig. 20b. The width and depth of the individual notches are estimated to be 70 and 12 μm, as shown in Fig. 20c.

Figure 21a shows the measured mode field of the LPG at the resonance wavelength under bend-free condition. The observed highly asymmetric mode intensity profile is a result of both structural change and asymmetric index modulation. The cladding mode intensity distribution around the core region is not easy to observe due to the existence of the uncoupled core mode; however, the shift of the cladding mode power to the unchanged side of the cladding is obvious. Figure 21b shows the measured bending response of LPG. The LPG has a directional bending response toward and against the notches and a relatively good linear response. The overall sensitivity is 2.26 nm/m^{-1} within the range of $-5 \sim +5\ m^{-1}$. The directional response mainly arises from the structural modifications in the PCF cladding, although the asymmetrical glass refractive-index modulation may also play a role. The good linear response and the simple fabrication process make the proposed sensor a suitable candidate for low-cost structural shape sensing in harsh environments.

Gas Pressure Sensing by Use of an Inflated LPG in IG-PCF

X. Y. Zhong et al. demonstrated the inscription of an inflated LPG (I-LPG) in a PCF by means of the pressure-assisted CO_2 laser beam-scanning technique to inflate periodically air holes along the fiber axis (Zhong et al. 2015). Figure 22 illustrates the detailed process for fabricating an I-LPG. First, as shown in Fig. 22a, an end of

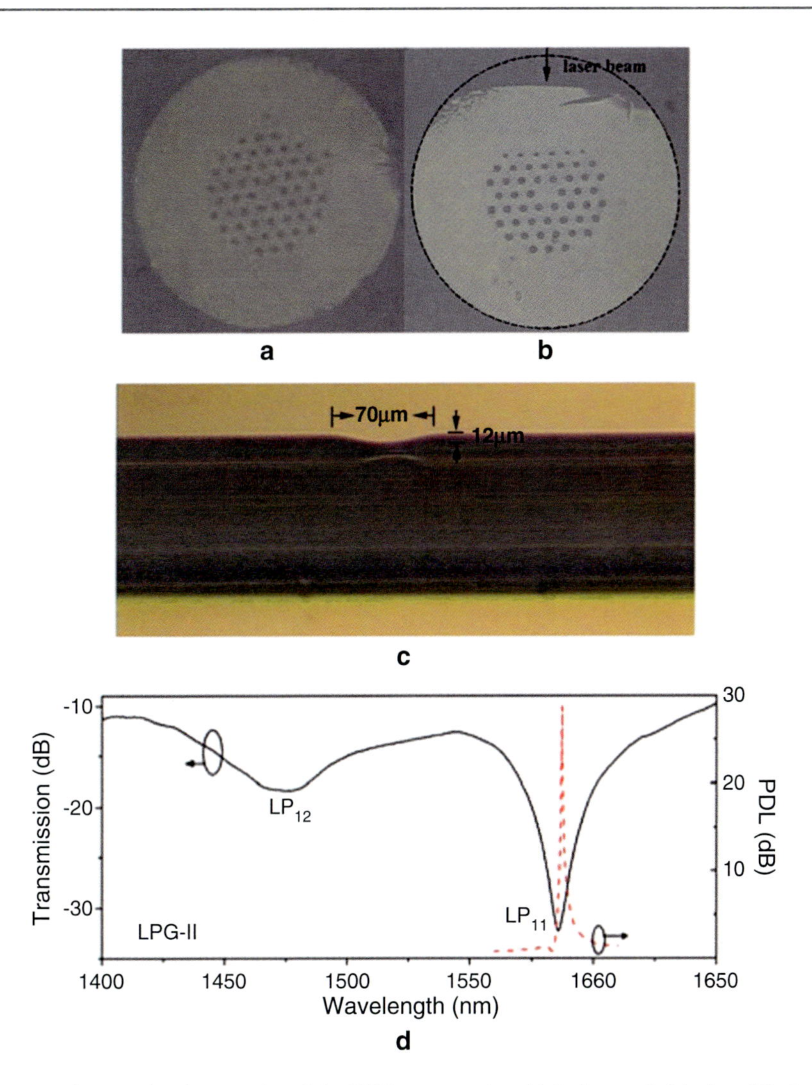

Fig. 20 Microscopic photographs of the PCF cross section (**a**) before and (**b**) after CO_2 laser irradiation. Air-hole collapse/deformation in the fiber cladding and the notch on the fiber surface is clearly observable in (**b**) and (**c**). The transmission spectrums and measured PDL profiles of the LPG in (**d**) (Jin et al. 2009)

a silica tube was spliced with a large-mode-area pure silica PCF (NKT ESM-12). Second, as shown in Fig. 22b, another end of the silica tube was cleaved to shorten its length to be about 80 μm to observe whether the air holes of the PCF end are open or not by use of a microscope. Third, as shown in Fig. 22c, the cleaved end of the silica tube was spliced with another SMF, thus achieving an air cavity with a length of 80 μm. And then a micro-channel (50*40 μm) was drilled through the sidewall of the silica tube by use of a femtosecond laser so that air pressure can access from

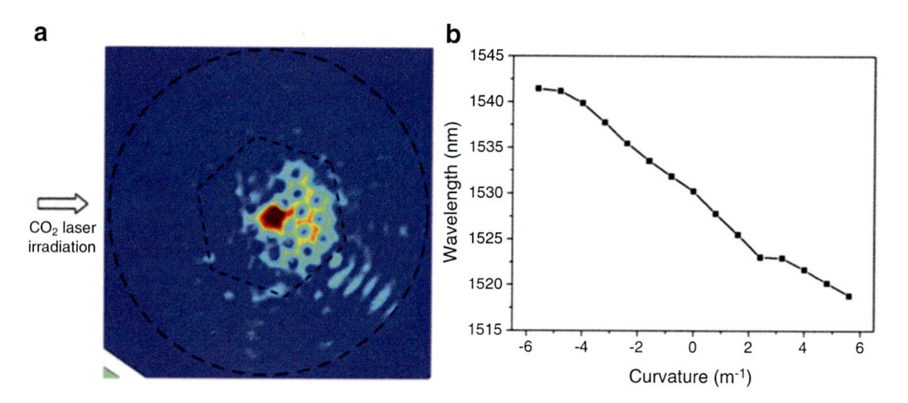

Fig. 21 (**a**) Measured near-field intensity profile at the output end of LPG at the resonance wavelength when the fiber is bend-free. (**b**) Measured bend response of the LPG. The measurement result shows a directional bending response with good linearity (Jin et al. 2009)

the silica tube into the air holes of the PCF. Fourth, as shown in Fig. 22d, the silica tube with a channel was placed into a gas chamber and sealed by use of strong glue. Then air with a pressure of ~1.5 MPa accessed the holes of the PCF via the gas chamber by use of a high-pressure air pump. Finally, as shown in Fig. 22e, f, the PCF was periodically heated along the fiber axis by use of a focused CO_2 laser beam. As a result, the holes of the PCF inflated periodically along the fiber axis due to high-pressure air and the CO_2-laser-induced high temperature.

In Fig. 23a, air holes of a PCF employed with a diameter of 3.3 μm are arranged in a hexagonal pattern with a pitch of 7.4 μm. The core and cladding diameters of the PCF are 10.4 and 125 μm, respectively. In Fig. 2b, asymmetric inflations of air holes were clearly created in the PCF along the fiber axis, resulting from the single-side irradiation of the focused CO_2 laser beam. Compared with the PCF diameter of 125 μm, the inflated region of the PCF has a diameter of 130 μm along the CO_2 laser irradiation direction. Uneven expansion of the air holes illustrated in Fig. 23b may be due to the inhomogeneity of the air holes in the PCF.

Figure 24 shows the transmission spectrum evolution of CO_2 laser-inscribed LPG with periodic inflations with scanning cycles increasing from 1 to 10. It can be seen from this figure that there is a low insertion loss of ~1 dB, and two attenuation dips (Dip11 and Dip12) are induced after ten scanning cycles.

Gas pressure response of the I-LPFG was measured with the gas pressure being increased from 0 to 10 MPa with a step of 1 MPa. As shown in Fig. 25, the resonant wavelength of I-LPG is shifted toward longer wavelength with a sensitivity of 1.68 nm/Mpa. For I-LPG, the silica wall within the inflated region is thinned due to the inflation of air holes in the PCF. As a result, in the case of a high gas pressure, physical deformations easily occur within the inflated region of the I-LPG, which changes the effective refractive index difference due to elasto-optical effect and induces a resonant wavelength shift. Thus periodic inflations in the I-LPG greatly enhance the pressure sensitivity of the grating.

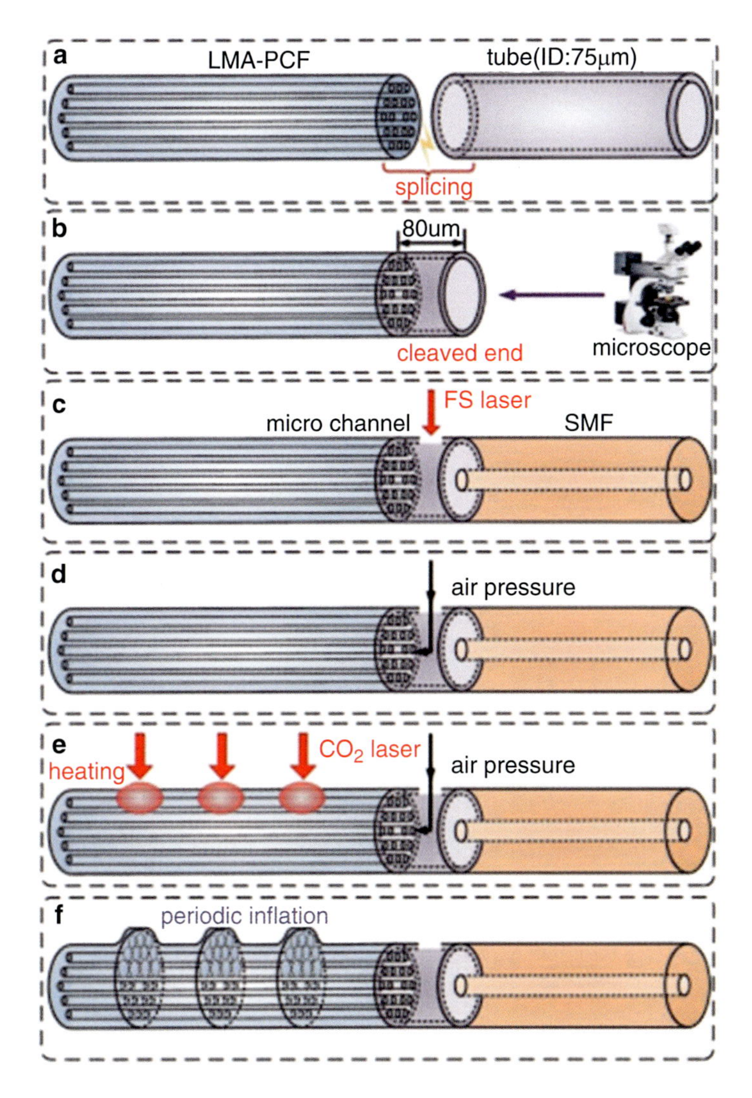

Fig. 22 Schematic diagram for fabricating an I-LPG (Zhong et al. 2015)

Refractive Index Sensing by Use of the Mechanical LPG in PCF

X Yu. et al. realized a refractive index sensor with the measurand infiltrated into the holes of single-mode PCF (Yu et al. 2008). The SEM image of an endlessly single-mode PCF Crystal-Fiber A/S) used for the experiment is shown in Fig. 26. It has a pitch size of 8 μm and air-hole diameter of 3.68 μm. The diameter of holey region is ~60 μm, and the outside diameter of fiber is 125 μm. A series of calibrated refractive index liquids are infiltrated into the fiber based on the capillary effect with a syringe pump.

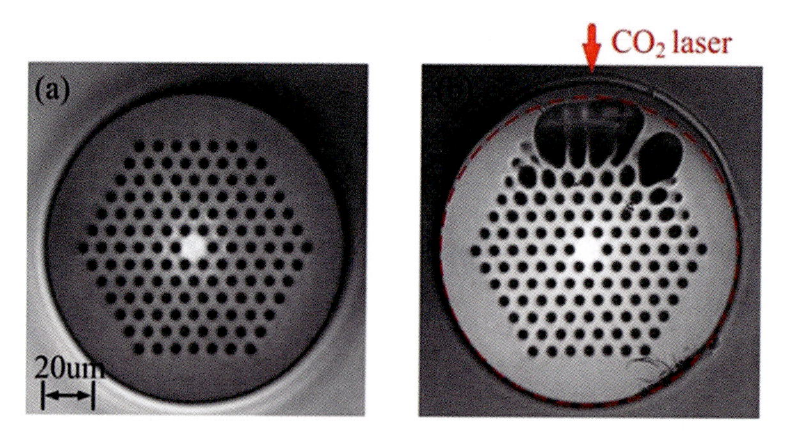

Fig. 23 Microscope image of the cross section of the PCF (**a**) before and (**b**) after CO_2 laser irradiation (Zhong et al. 2015)

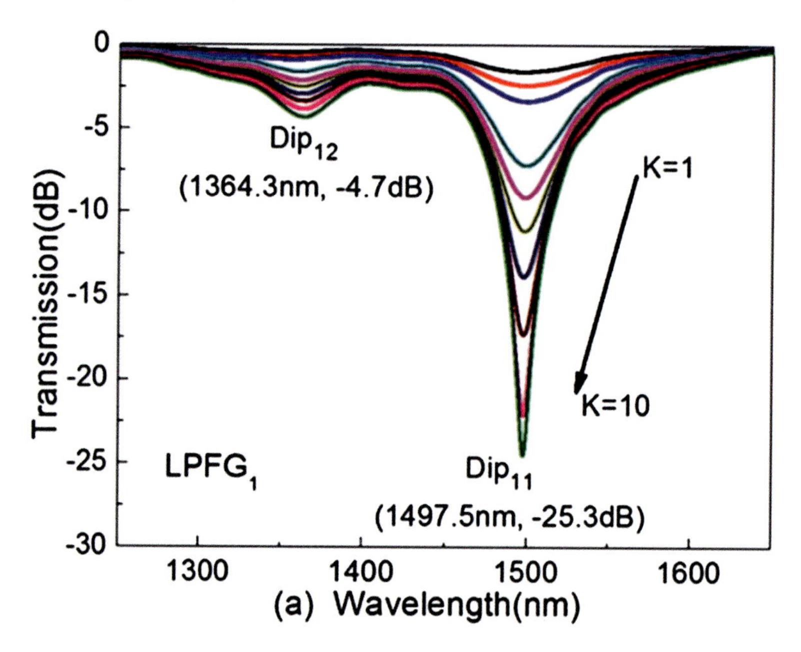

Fig. 24 Transmission spectral evolution of CO_2 laser-inscribed I-LPG with periodic inflations while scanning cycles increase (Zhong et al. 2015)

The LPG is fabricated by applying a transverse force on the section of PCF by a periodically grooved plate as shown in Fig. 26. The fixture is mechanically polished with rectangular grooves of period 600 μm. The length, width, and thickness of the grooved plate were 50, 20, and 5 mm, respectively. The transverse force is manually controlled by a stress gauge with 25 N to achieve a transmission dip ~15 dB. The grating period is not varied with the increase of the applied force since the grooved

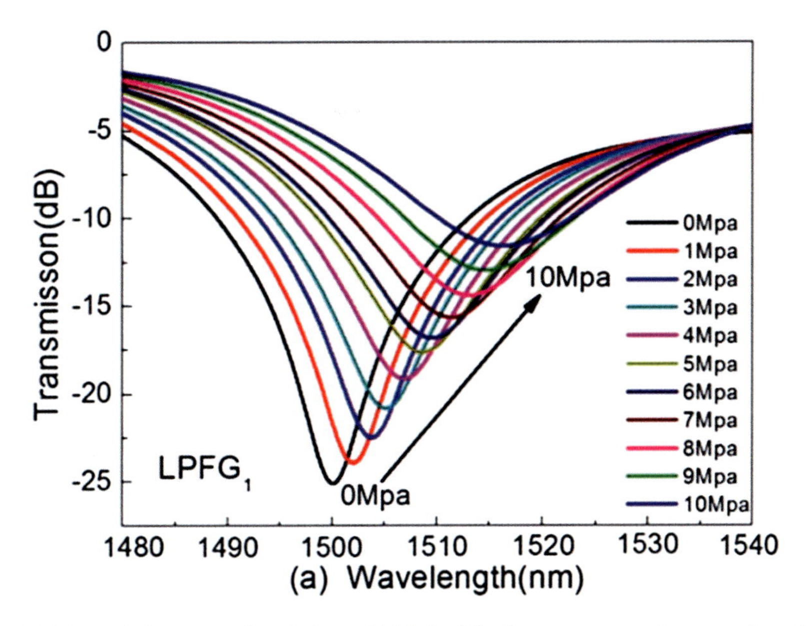

Fig. 25 Transmission spectral evolution of I-LPG while the gas pressure increases from 0 to 10 MPa (Zhong et al. 2015)

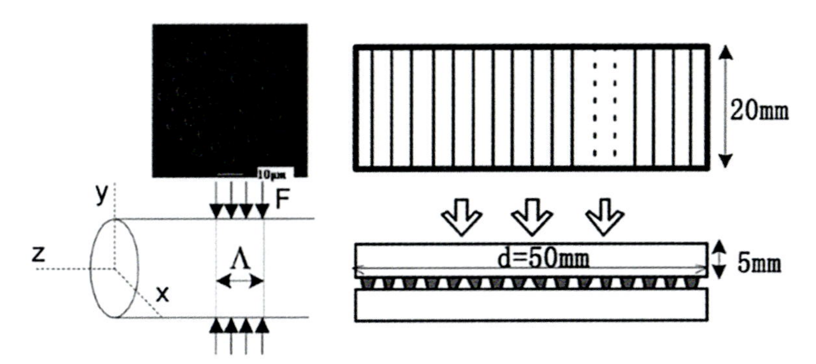

Fig. 26 SEM image of the PCF and grooved fixture and coordinate system for the fiber showing the applied external force (Yu et al. 2008)

plate is placed on top of the flat surface. Thus, the photoelastic effect induces index variations dominating the resonance position.

Figure 27a shows the transmission spectra of the PCF-based mechanical LPG with different infiltration index of 1.32, 1.36, and 1.38. Several measured transmission spectra ~1420 nm are achieved with different infiltration refractive index. The resonance wavelength is shifted from 1429.50 to 1414.80 nm when the infiltrated index oil changes from 1.32 to 1.38. The refractive index resolution can be obtained from the shift of transmission dip as shown in Fig. 27b. The resonance wavelength shift is 17.1 nm with a refractive index change of 0.07. A linear fitting to the

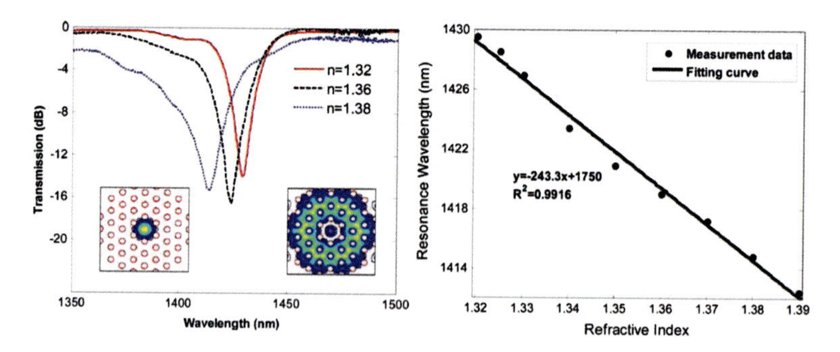

Fig. 27 (**a**) Transmission spectra of the PCF-based mechanical LPG with different infiltration index of 1.32, 1.36, and 1.38. (**b**) Measurement results and fitting curve for the resonance wavelength shift versus the infiltrated refractive index (Yu et al. 2008)

experimental data gives a wavelength-RI sensitivity of 2.4 pm for 1×10^{-5} change in the index proportionally.

Biochemical Sensing by Use of the LPG in PCF

L. Rindorf et al. reported the first label-free technique for detection of biomolecules using a PCF-based LPG (Rindorf et al. 2006). The LPG is inscribed by use of CO_2 laser scanning on the PCF (LMA10), which has a pitch of 7.2 μm and a relative (absolute) hole size of d/L = 0.47. The pitch of the PCF-LPG is 700 μm. The number of grating periods is chosen to be 26 making the length of grating 18.2 mm. The length of the fiber with the PCF-LPG in the middle is 30 cm in total. To introduce the solutions into the holes of the PCF, one end of the PCF is inserted into a pressure chamber.

As shown in Fig. 28, poly-L-lysine is used to immobilize negatively charged molecules such as DNA to a solid support. Poly-L-lysine has positively charged amino groups that can bind to the negatively charged silica surface through an ionic binding. DNA, on the other hand, has negatively charged phosphate groups in its backbone, and may thus be immobilized in a monolayer onto the poly-L-lysine, but cannot bind directly onto the silica.

Figure 29 shows the result of biochemical sensing. The resonant wavelength of the grating with air inside the holes is 753.6 nm. The introduction of the PBS into the holes caused the resonant wavelength to shift to 842.5 nm, giving a total shift of 88.9 nm. The immobilization of poly-L-lysine shifts the resonant wavelength to 849.2 nm, a shift of 6.7 nm. Finally, the DNA shifts the resonant wavelength to 851.4 nm, a shift of 2.3 nm.

Photonic-Bandgap Photonic Crystal Fiber-Based LPG Sensors

The First LPG in Air-Core PBF

In 2008, Y. P. Wang et al. reported the first LPG written in air-core PBF by use of CO_2 laser pulses to periodically modify the size, shape, and distribution of air holes

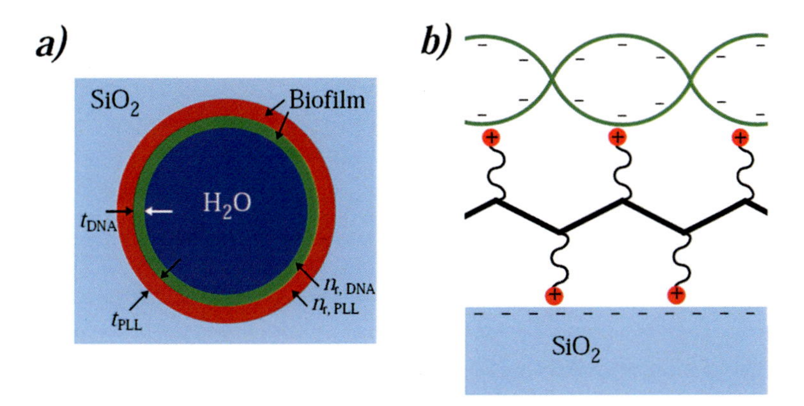

Fig. 28 (**a**) A hole of a photonic crystal fiber. The side is coated with poly-L-lysine and DNA in monolayers of various thickness (t_{DNA} and t_{PLL}) and refractive indices ($n_{r,DNA}$ and $n_{r,PLL}$). The thickness of the biofilms is vastly exaggerated compared to the hole diameter. (**b**) The molecular structure of poly-L-lysine (*red* and *black*) with positive charges immobilized onto the negatively charged silica surface (SiO_2). The negatively charged DNA (*green*) is immobilized on the poly-L-lysine (Rindorf et al. 2006)

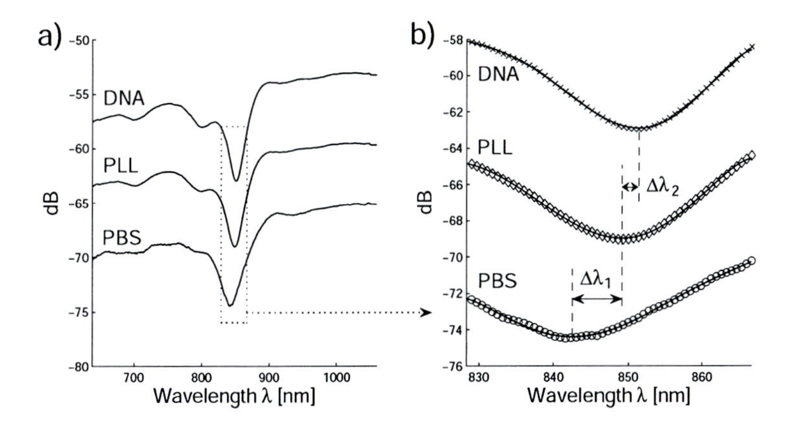

Fig. 29 Transmission spectra for PBS, poly-L-lysine (PLL), and DNA (Rindorf et al. 2006)

in the microstructured cladding (Wang et al. 2008). The fiber used is Crystal-Fiber's HC-1500-02 PBFs with a cross section shown in Fig. 30a. These high frequency, short duration, and CO_2 laser pulses hit repeatedly on one side of the PBF and induce a local high temperature, causing ablation of glass on the surface and change of shapes, sizes, locations, and even complete collapse of some of the air holes in the cladding, as shown in Fig. 30b. This results in a notch being created on the surface of the fiber, as shown in Fig. 30c. Figure 30d shows the measured transmitted spectrum of a 40-period LPG made by the above process. The 3 dB bandwidth is ~5.6 nm, which is much narrower than that of the LPGs in conventional SMFs. The insertion

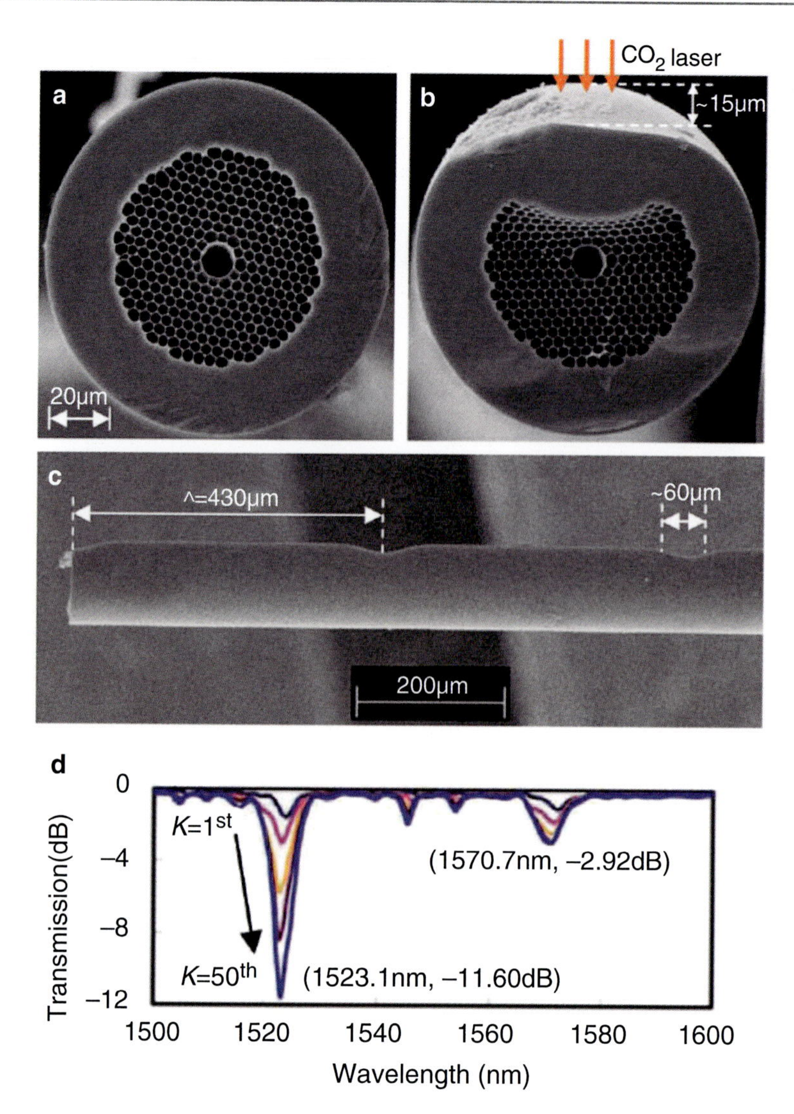

Fig. 30 SEM images of PBF cross sections (**a**) before and (**b**) after CO_2 laser irradiation. (**c**) Periodic notches on PBF after 50 scanning cycles. (**d**) Transmission spectrum of a 40-period LPG on PBF (Wang et al. 2008)

loss of the LPG is very low (<0.3 dB), because most light is guided in the hollow core where no deformation is observed.

Before the LPG is created, the light intensity is mainly in the fundamental mode, as shown in Fig. 31a. With an increase in the number of grating pitches, for the input light at 1523.1 nm, light energies near the core-cladding interface and in the holey cladding region are gradually enhanced whereas that in the fundamental mode is

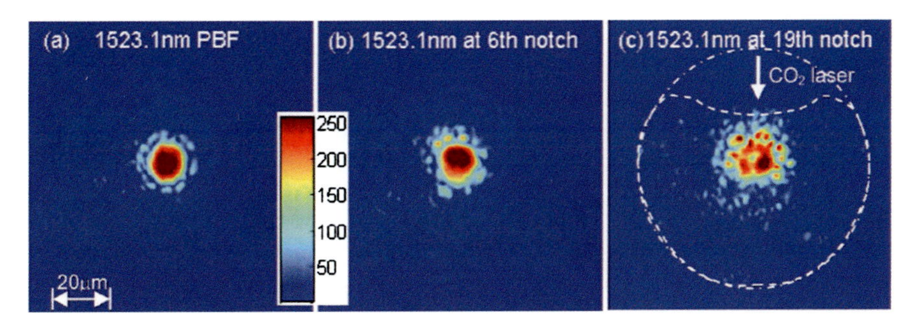

Fig. 31 (**a**) Near-field image in a PBF without LPG. (**b**) Near-field images of an LPG at the resonant wavelength of 1523.1 nm when the LPG is cut at 6th notch. (**c**) Near-field images of the same LPG observed at the 19th notch at 1523.1 nm (Wang et al. 2008)

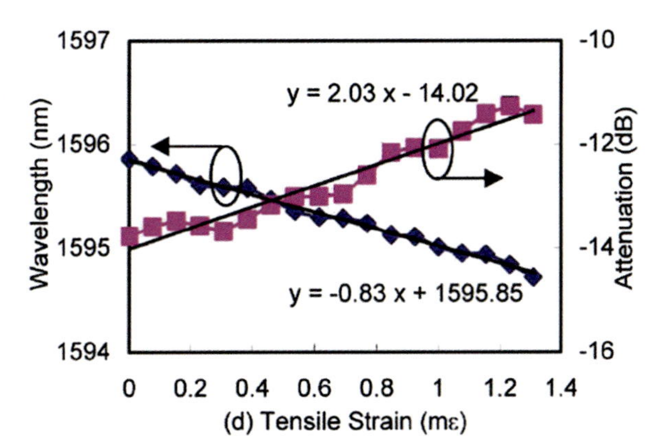

Fig. 32 Measured resonant wavelength and peak transmitted attenuation of the LPG versus tensile strain (Wang et al. 2008)

reduced, as shown in Fig. 31b, c. At the 19th notch, most energy in the fundamental mode is coupled out so that the light energies near the core-cladding interface and in the holey region are clearly observed and light intensity at the center of hollow core becomes very weak. Light coupled into the cladding region is mainly distributed within the holey cladding region as outlined by the dashed curve, and the energy in the side facing to CO_2 laser irradiation is stronger than that in the opposite side.

For the PBG-based LPG, the temperature sensitivity is ~2.9 pm/°C, which is one to two orders of magnitude less than those of the LPGs in SMFs. The curvature sensitivity is three to four orders of magnitude less than those of the LPGs in SMFs. With the increase of applied tensile strain, the resonant wavelength of the LPG is shifted linearly toward shorter wavelength with a strain sensitivity of −0.83 nm/mε, and the peak transmission attenuation is decreased with a sensitivity of 2.03 dB/mε, as shown in Fig. 32. The wavelength sensitivity to strain is two times higher than that of LPGs in SMFs, indicating that this LPG may be used as a strain sensor without cross-sensitivity to temperature and curvature.

Tensile Strain Sensing by Use of the LPG in All-Solid PBF

C. R. Liao et al. presented an LPG inscribed in all-solid PBF by use of femtosecond laser with line-scanning method (Liao et al. 2010). The employed PBF has an index-depressed layer (fluorine-doped) around the high-index rods (germanium-doped) in the unit cell of photonic crystal cladding, as shown in Fig. 33. The diameter of the fiber is 123 μm and the lattice spacing Λ is 9.21 μm. In the experiment, 5 cm of the PBF is spliced to SMF-28 fiber which are well aligned to restrain mode couplings between the high-index rods and the fiber core as well as to minimize the insertion loss.

LPGs are fabricated by use of an amplified 800-nm femtosecond laser system. The PBGF supports both LP_{01} and LP_{11} bandgap-guided core modes. At 1550 nm, n_{eff} of the fundamental mode (LP_{01}) and higher-order core mode (LP_{11}) are calculated to be 1.4419 and 1.4407 using finite element method. In order to satisfy phase-matching condition between the two core modes ~1550 nm, the period of the LPG is chosen as 640 μm, and the grating length is 24 mm. One piece of SMF-28 fiber was spliced onto the PBF to prevent the high-order core mode (LP_{11}) coupling from LP_{01}, and thus the LPG attenuation band can appear. Figure 34a shows the transmission spectrum evolution of the LPG within 38 periods. The resonance wavelength of the LPG created is located at 1542.3 nm, with a full-width at half-maximum (FWHM) value of ~24 nm, and the strongest resonance is measured as ~19 dB. Figure 34b, c illustrates the near-field images of the PBF at 1542.3 nm before and after the LPG inscription, respectively. Before the LPG is inscribed, the light intensity is mainly in the fundamental mode (LP_{01}) with a small part (supermodes) being distributed in the high-index rods, especially in the six rods being close to the core, as shown in Fig. 34b.

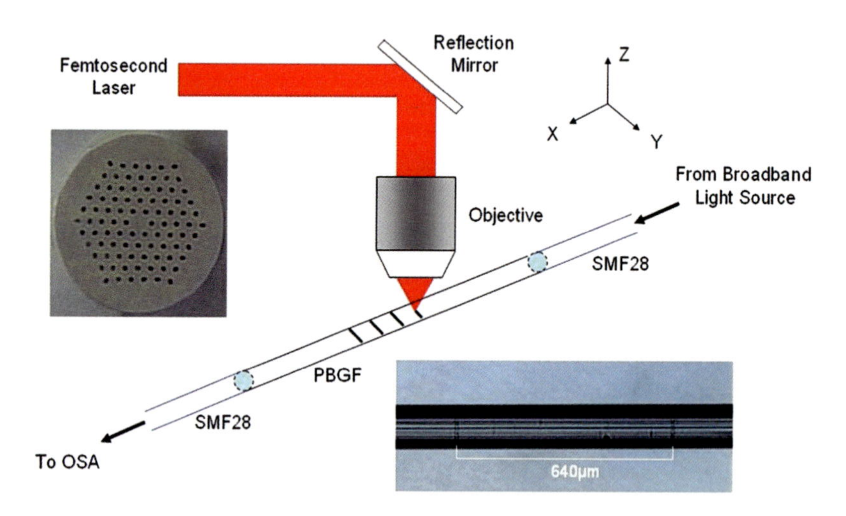

Fig. 33 Experimental setup for LPG fabrication by use of femtosecond laser with line-scanning method (Liao et al. 2010)

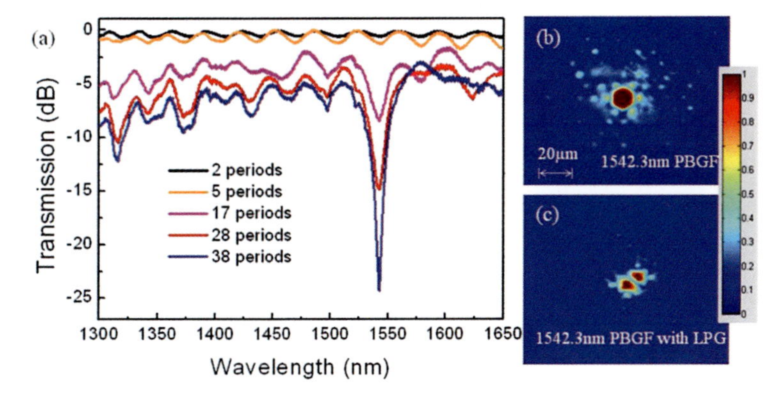

Fig. 34 (**a**) Evolution of the transmission spectrum of LPG with the increase of period number (2∼38), the grating pitch is 640 μm. (**b**) Near-field image in an all-solid PBF without LPG. (**c**) Near-field image at the resonant wavelength of 1542.3 nm with the period number of 38 (Liao et al. 2010)

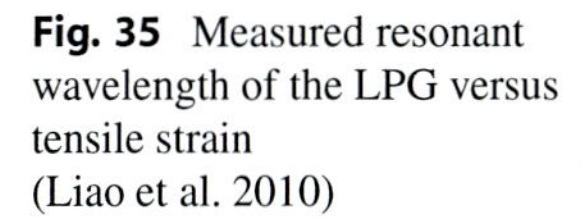

Fig. 35 Measured resonant wavelength of the LPG versus tensile strain (Liao et al. 2010)

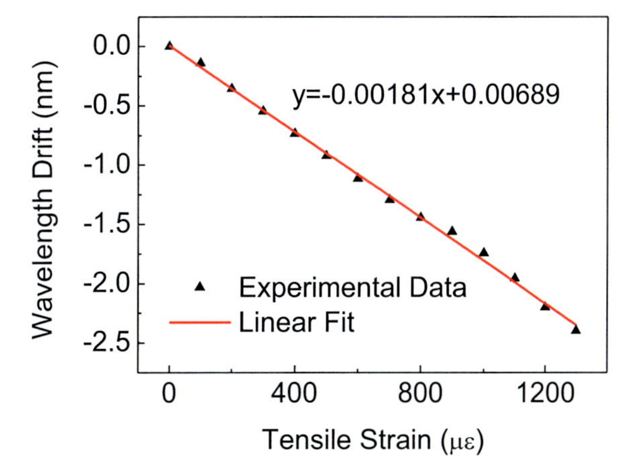

Figure 35 demonstrates the response of the LPG to tensile strain. There is a linear "blue" shifting for the resonant wavelength with the increase of applied tensile strain. The measured sensitivity is −1.8 nm/mε, this being four times higher than those of LPGs in conventional SMFs.

Gas Pressure Sensing by Use of the LPG in Air-Core PBF

J. Tang et al. demonstrated an LPG inscribed in an air-core PBF for gas pressure sensing (Tang et al. 2015). The LPG is inscribed by use of CO_2 laser beam-scanning technique, which greatly enhanced the writing efficiency and repeatability, and the fabrication setup is shown in Fig. 36a. An air-core PBF (HC-1500-02 PBF from Crystal-Fiber) is employed to inscribe a LPG, as shown inset of Fig. 36a. Figure 36b shows the microscope image of the LPG with periodic collapse face to the CO_2 laser

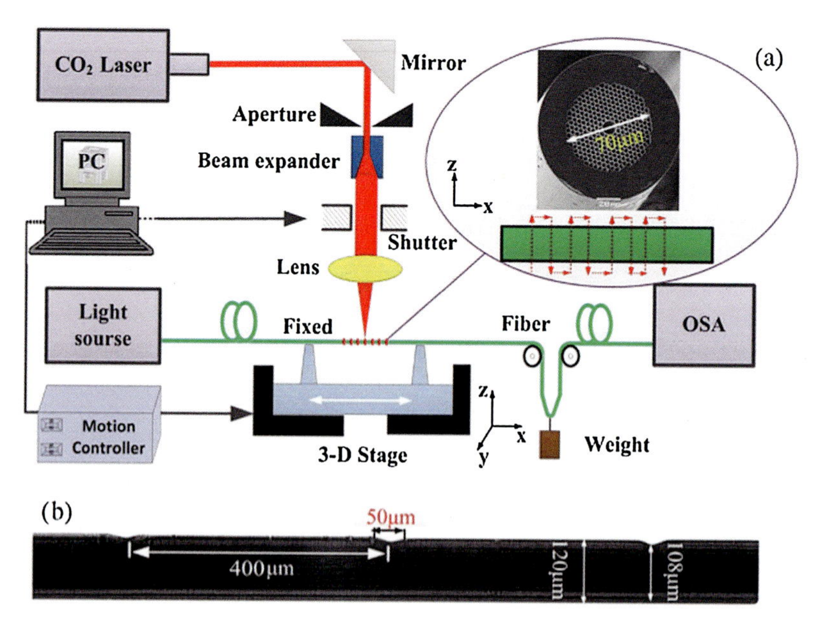

Fig. 36 (**a**) Experimental setup for LPG inscription with CO_2 laser beam-scanning technique. (Inset) SEM of the used HC-PBF. (**b**) Side view of the obtained LPG (Tang et al. 2015)

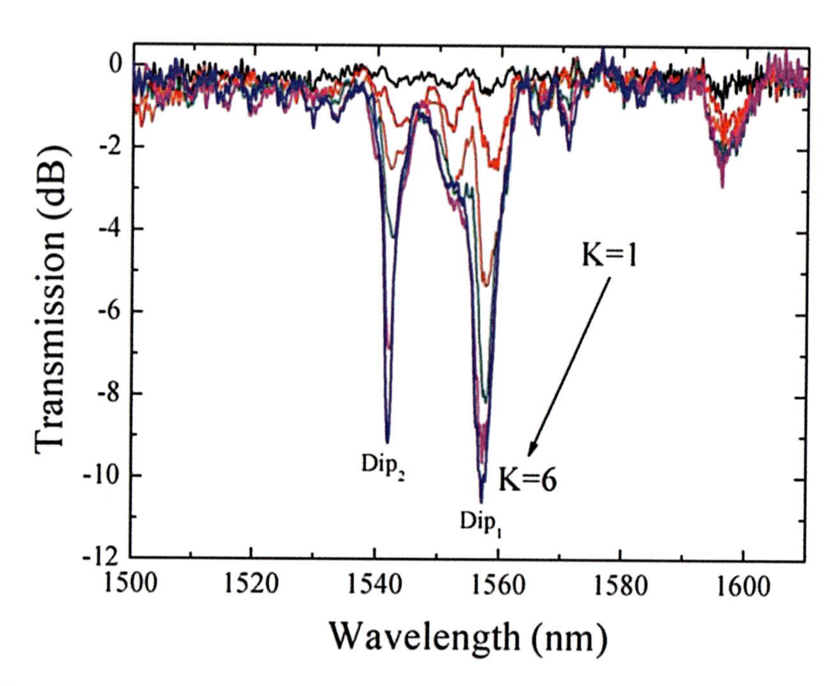

Fig. 37 Transmission spectral evolution of an LPG with grating pitch of 400 μm while the number of scanning cycles (K) increases from 1 to 6 (Tang et al. 2015)

exposure side. The width and depth of the collapsed region are 50 μm and 12 μm, respectively.

Figure 37 shows the transmission spectral evolution when the number of scanning cycles is increased from 1 to 6. Two resonant dips have been found at resonant wavelengths of 1555.9 nm (Dip1) and 1541.4 nm (Dip2), respectively. Two-step process might be involved: the fundamental core mode is coupled to discrete higher-order (core or surface-like) modes and then to lossy quasi-continuum of cladding and radiating modes. The depth of the two resonant dips gradually increases with an increase of scanning cycle and finally reached to −10.61 dB and −9.15 dB, respectively.

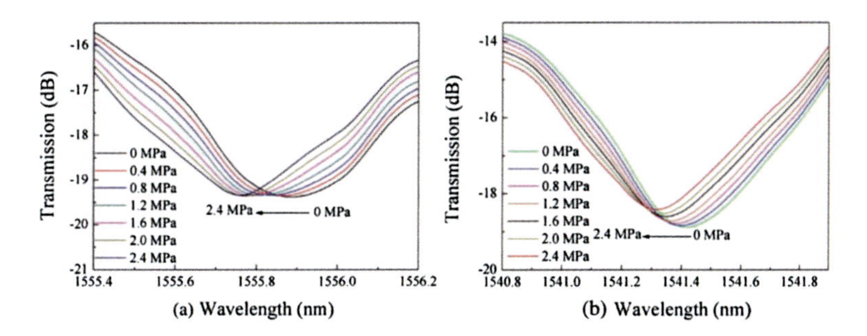

Fig. 38 Transmission spectrum evolution of (**a**) Dip1 and (**b**) Dip2 while the gas pressure increases from 0 to 2.4 MPa (Tang et al. 2015)

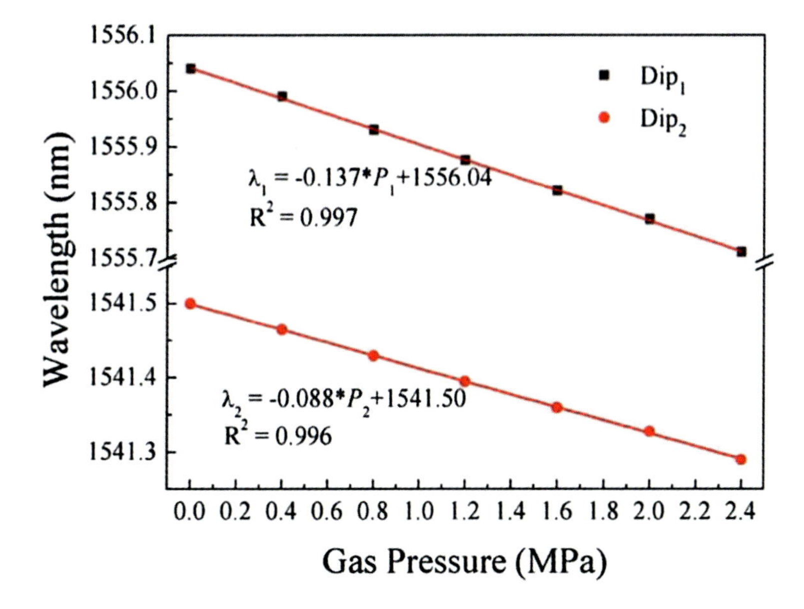

Fig. 39 Relationship between gas pressure and resonant wavelength of Dip1 and Dip2 while the gas pressure increases from 0 to 2.4 MPa (Tang et al. 2015)

The gas pressure responses of Dip1 and Dip2 are, respectively, monitored for better understanding pressure response of the LPGs written in the air-core PBF. As shown in Fig. 38a, b, both Dip1 and Dip2 shift toward shorter wavelengths with the pressure increasing gradually from 0 to 2.4 MPa with an interval of 0.4 MPa. The relationship between resonant wavelength and gas pressure is shown in Fig. 39. The experiment sensitivity of Dip1 and Dip2 are −137 pm/MPa and −88 pm/MPa, respectively. The PBF-based LPG exhibits a high-pressure sensitivity that is due to more significant physical deformation in air-core PBF.

References

B.J. Eggleton, P.S. Westbrook, R.S. Windeler, S. Spälter, T.A. Strasser, Grating resonances in air-silica microstructured optical fibers. Opt. Lett. **24**, 1460–1462 (1999)

L.B. Fu, G.D. Marshall, J.A. Bolger, P. Steinvurzel, E.C. Mägi, M.J. Mithford, B.J. Eggleton, Femtosecond laser writing Bragg gratings in pure silica photonic crystal fibres. Electron. Lett. **41** (2005)

T. Geernaert, G. Luyckx, E. Voet, T. Nasilowski, K. Chah, M. Becker, H. Bartelt, W. Urbanczyk, J. Wojcik, W.D. Waele, J. Degrieck, H. Terryn, F. Berghmans, H. Thienpont, Transversal load sensing with fiber Bragg gratings in microstructured optical fibers. IEEE Photon. Technol. Lett. **21**, 6–8 (2009)

Y.G. Han, Y.J. Lee, G.H. Kim, H.S. Cho, S.B. Lee, C.H. Jeong, C.H. Oh, H.J. Kang, Transmission characteristics of fiber Bragg gratings written in holey fibers corresponding to air-hole size and their application. IEEE Photon. Technol. Lett. **18**, 1783–1785 (2006)

M. Huy, G. Laffont, V. Dewynter, P. Ferfinand, L. Labonté, D. Pagnoux, P. Roy, W. Blanc, B. Dussardier, Tilted fiber Bragg grating photowritten in microstructured optical fiber for improved refractive index measurement. Opt. Express **14**, 10359–10370 (2006)

L. Jin, Z. Wang, Q. Fang, B. Liu, Y.G. Liu, G.Y. Kai, X.Y. Dong, B.O. Guan, Bragg grating resonances in all-solid bandgap fibers. Opt. Lett. **32**, 2717–2719 (2007)

L. Jin, W. Jin, J. Ju, Directional bend sensing with a CO_2 laser inscribed long period grating in a photonic crystal fiber. J. Lightwave Technol. **27**, 4884–4891 (2009)

J.C. Knight, T.A. Birks, P.S.J. Russell, D.M. Atkin, All-silica single-mode optical fiber with photonic crystal cladding. Opt. Lett. **21**, 1547–1549 (1996)

Y.H. Li, D.N. Wang, L. Jin, Single-mode grating reflection in all-solid photonic bandgap fibers inscribed by use of femtosecond laser pulse irradiation through a phase mask. Opt. Lett. **34**, 1264–1266 (2009)

C.R. Liao, Y. Wang, D.N. Wang, L. Jin, Femtosecond laser inscribed long-period gratings in all-solid photonic bandgap fibers. IEEE Photon. Technol. Lett. **22**, 425–427 (2010)

L. Rindorf, J.B. Jensen, M. Dufva, L.H. Pedersen, P.E. Hoiby, O. Bang, Photonic crystal fiber long-period gratings for biochemical sensing. Opt. Express **14**, 8224–8231 (2006)

J. Tang, G.L. Yin, S. Liu, X.Y. Zhong, C.R. Liao, Z.Y. Li, Q. Wang, J. Zhao, K.M. Yang, Y.P. Wang, Gas pressure sensor based on CO_2 laser induced long period fiber grating in air-core photonic bandgap fiber. IEEE Photon. J. **7**, 6803107 (2015)

Y.P. Wang, L.M. Xiao, D.N. Wang, W. Jin, Highly sensitive long-period fiber-grating strain sensor with low temperature sensitivity. Opt. Lett. **31**, 3414–3416 (2006)

Y.P. Wang, W. Jin, J. Ju, H.F. Xuan, H.L. Ho, L.M. Xiao, D.N. Wang, Long period gratings in air-core photonic bandgap fiber. Opt. Express **16**, 2784–2790 (2008)

P.S. Westbrook, B.J. Eggleton, R.S. Windeler, A. Hale, T.A. Strasser, G.L. Burdge, Cladding-mode resonances in hybrid polymer-silica microstructured optical fiber gratings. IEEE Photon. Technol. Lett. **12**, 495–497 (2000)

C. Wu, B.O. Guan, Z. Wang, X.H. Feng, Characterization of pressure response of Bragg gratings in grapefruit microstructured fibers. J. Lightwave Technol. **28**, 1392–1397 (2010)

X. Yu, P. Shum, G.B. Ren, Highly sensitive photonic crystal fiber-based refractive index sensing using mechanical long-period grating. IEEE Photon. Technol. Lett. **20**, 1688–1690 (2008)

X.Y. Zhong, Y.P. Wang, C.R. Liao, S. Liu, J. Tang, Q. Wang, Temperature-insensitive gas pressure sensor based on inflated long period fiber grating inscribed in photonic crystal fiber. Opt. Lett. **40**, 1791–1794 (2015)

Photonic Crystal Fiber-Based Interferometer Sensors

57

Min Wang, Jiankun Peng, Weijia Wang, and Minghong Yang

Contents

M. Wang
National Engineering Laboratory for Fiber Optic Sensing Technology (NEL-FOST), Wuhan University of Technology, Wuhan, China

School of Electronic and Electrical Engineering, Wuhan Textile University, Wuhan, China
e-mail: bluebluecherry@163.com

J. Peng · W. Wang · M. Yang (✉)
National Engineering Laboratory for Fiber Optic Sensing Technology (NEL-FOST), Wuhan University of Technology, Wuhan, China
e-mail: pengjiankun126@126.com; 18627049321@whut.edu.cn; weijiawang@whut.edu.cn; minghong.yang@whut.edu.cn

G.-D. Peng (ed.), *Handbook of Optical Fibers*,
https://doi.org/10.1007/978-981-10-7087-7_11

Abstract
A combination of interferometric architectures with photonic crystal fiber techniques is a new configuration in the field of sensing technologies. Since photonic crystal fibers have the extraordinary capability of light guiding and confinement, the photonic crystal fiber-based interferometers offer great potential for the realization of novel sensing applications. In this chapter, principal, fabrication, and typical applications of the photonic crystal fiber-based fiber interferometers are reviewed.

Introduction

The investigations of modal interferometers are booming with the development of photonic crystal fiber (PCF) techniques. The first photonic crystal fiber was fabricated by P. St. J. Russell et al. since early 1996 (Knight et al. 1996). Such fibers are manufactured by the same methods as conventional optical fibers. Before drawing to a small diameter, the photonic crystal fiber preform has to be constructed on the scale of centimeters in size. In the drawing process, the transverse structure is shrinking on the scale of micrometers in size but maintaining the original features. Due to the periodic transverse microstructure and the existence of linear defect, photonic crystal fibers have the capacity of guiding low-loss mode over a very broad spectral range. According to the different light-guiding mechanism (Hansen et al. 2001; Sun and Hu 2010), photonic crystal fibers have two fundamental types, i.e., index-guiding photonic crystal fiber (IG-PCF) and band gap photonic crystal fiber (BG-PCF), which is shown in Fig. 1. The IG-PCF has a solid silica core, which effective refractive index is higher than the micro-structured cladding. Its light-guiding mechanism is the same index-guiding principle as conventional optical fiber; therefore, it is also called multi-hole fibers or micro-structured fibers. However, its light confinement is stronger than conventional optical fiber on account of its periodic transverse micro-structured cladding. The BG-PCF has a core (generally a hollow core) which effective refractive index is lower than the micro-structured cladding. Its light-guiding mechanism is on account of the photonic band gap generated by the hollow core and the micro-structured cladding. Hollow-core photonic crystal fibers have the capability to confine and guide light in wavelengths for which transparent materials are not available since its core is air, not solid materials. An important advantage of photonic crystal fiber is that one can deliberately fill different materials into its micro-holes, which makes it more suitable for sensing applications than conventional optical fibers (Huang et al. 2004). Photonic crystal fibers overcome the limitations of conventional optical fiber optics, which is bringing new possibilities and opportunities to optical fiber sensing technology.

Traditional optical interferometry technologies include Fabry-Perot interferometer, Mach-Zehnder interferometer, Michelson interferometer, and Sagnac interferometer. Fabry-Perot interferometer is typically formed of a transparent medium with

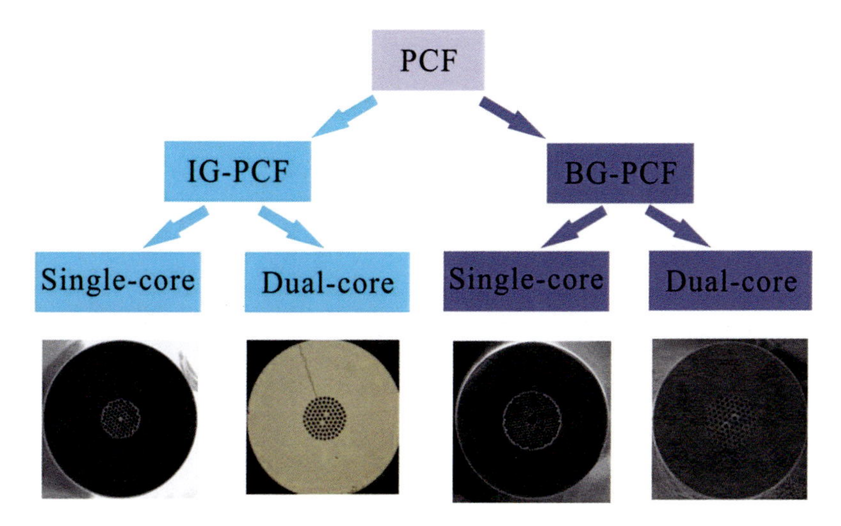

Fig. 1 Fundamental types of the photonic crystal fiber

two closely reflecting surfaces. For the fiber Fabry-Perot interferometer (Rao 2006), the reflecting surfaces are usually in the form of fusion splicing point between two different fibers. Each of the two surfaces reflects a beam of light, resulting in dual-beam interference. Mach-Zehnder interferometer is a configuration used to detect the relative phase shift variations between two split light beams derived from the same light source. For the fiber Mach-Zehnder interferometer (Tian et al. 2008a), a 3-dB fiber coupler is needed to split a single light into two light beams, these two light beams transmit through two fiber (i.e., the sensing arm and the reference arm), and then another 3-dB fiber coupler is used to couple the light beams to generate light interference. Michelson interferometer is a configuration used to detect the relative phase shift variations between two reflected light beams derived from splitting a single light source. For the fiber Michelson interferometer (Tian et al. 2008b), a 3-dB fiber coupler is needed to split a single light and couple lights into two different fiber arms. The split light will be reflected back to the coupler by the reflecting mirrors which are on the end face of the two fiber arms. Then the coupler collects the two reflected light beams to generate light interference. Sagnac interferometer is a configuration used to detect the rotation. For the fiber Sagnac interferometer (also called ring interferometer) (Kim and Kang 2004), a single light beam is split into two light beams and coupled into a fiber ring by a 3-dB coupler. The two light beams go through the same path (i.e., the optical fiber ring) but in opposite directions and then transmit back to the fiber coupler. Finally, the coupler collects the two light beams to generate light interference.

The combination of optical fiber with optical interferometers promotes the prosperity of sensing applications. Since the fabrication of photonic crystal fibers was realized, classical and novel interferometric architectures have been implemented with photonic crystal fibers. The fiber interferometers based on photonic crystal fiber exhibit excellent properties for sensing applications. In this chapter, the principal, fabrication, and typical applications of the photonic crystal fiber-based fiber interferometers are exhibited. The fiber interferometer configurations contain refractive index sensor, temperature sensor, strain sensor, pressure sensor, torsion sensor, curvature sensor, magnetic sensor, and their cross-sensing structure.

PCF-Based Fabry-Perot Interferometers

A Fabry-Perot interferometer (FPI) is generally composed of two parallel reflecting surfaces separated by a certain distance (Hercher 1968). The structure of FPI based on PCF can be classified into two categories: one is in-fiber and the other is tip.

Basic Structure of In-Fiber FPI

Principal

The basic schematic of in-fiber FPI on a PCF is given by Fig. 2. The air cavity is formed between the SMF and the PCF, including two reflective surfaces indicated by 1 and 2. Thus, such an air cavity could be used as a sensing element.

For FPI, the output intensity can be modeled using the two-beam optical interference equation (Yu et al. 2015):

$$I = I_1 + I_2 + 2\sqrt{I_1 I_2}\cos\left(\frac{4\pi n \cdot L}{\lambda}\right) \tag{1}$$

where I_1 and I_2 are the reflective intensities at the two cavity surfaces, L is the cavity length, λ is the operating wavelength, and n is the refractive index in the FPI cavity. For a certain dip, λ_m is in the spectrum under observation; its optical phase difference should be an odd number of π. That is, $4\pi nL/\lambda_m = (2\,m + 1)\pi$; here m is an integer. The wavelength of the dip is given by:

$$\lambda_m = \frac{4nL}{2m + 1} \tag{2}$$

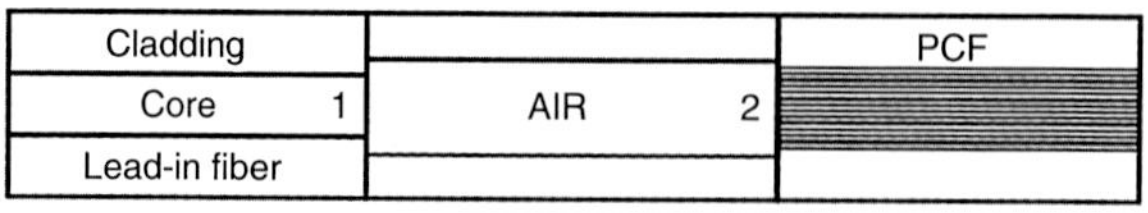

Fig. 2 Basic schematic of in-fiber FPI on a PCF

The air cavity can be used as a sensing element, while the wavelength shift of a given dip is induced by the variation of the cavity refractive index or the cavity length. The performance of the sensors based on in-fiber FPI can be detected by the wavelength shift of the reflective spectrum.

Fabrication

Several typical ways of fabricating the air cavity between the SMF and the PCF are illustrated as follows:

1. The self-enclosed air cavity can be created by splicing the SMF to a PCF with a precise hole micro-machined by a 157-nm laser at the end face of the PCF in Ran et al. (2015). First, a circular hole at the center of the cross section of the PCF is fabricated by using a 157-nm laser micromachining system; its diameter is larger than the core but smaller than the cladding. Then, the air cavity is formed by splicing the PCF with hole to the cleaved SMF. Next, the fiber is cleaved with a short distance from the air cavity, and another cleaved SMF is spliced to it for protecting the PCF.
2. The air cavity is an air bubble that can be formed by fusion splicing a PCF and an SMF with a commercialized fusion splicer in Deng et al. (2011). The SMF and the PCF fibers can be pressed together to form a permanent fusion splice joint, while the surface tension overcomes the viscosity of the glass (Chong and Rao 2003). The PCF with big air holes in the cladding is spliced to the SMF, and the surface tension will overcome the viscosity caused by the high temperature during the arc discharge which is high enough to exceed the MPCF softening point. The process not only forms a joint but also collapses the fine voids of MPCF as well (Yablon 2005; Xiao et al. 2005). As a result, a part of air originally inside the voids can be trapped, and the rapidly expanding gases can induce a microbubble.

Applications of In-Fiber FPI on a PCF

In order to perform physical sensing of the ambient medium, an in-fiber FPI on the PCF should change its reflective spectrum in response to the variation of the sample properties.

Refractive Index In-Fiber FPI Sensor

A miniaturized in-fiber FPI sensor for sensing liquid refractive index was demonstrated in Hu et al. (2011). The sensor was fabricated by a short length (less than 1 mm) of PCF spliced to a SMF. The SEM image of the PCF used in this work is shown in Fig. 3a. During the splicing process, an air bubble was formed in the fully collapsed zone as shown in Fig. 3b. The end of the PCF was fused to spherical shape which acted as a reflective mirror of the device. The air bubble diameter was 27.7 μm and the collapsed PCF length was 150.7 μm in Hu et al. (2011).

The response ofthe sensor was investigated with refractive indices of the liquids from 1.33 to 1.4. The corresponding spatial frequency spectrum was shown in

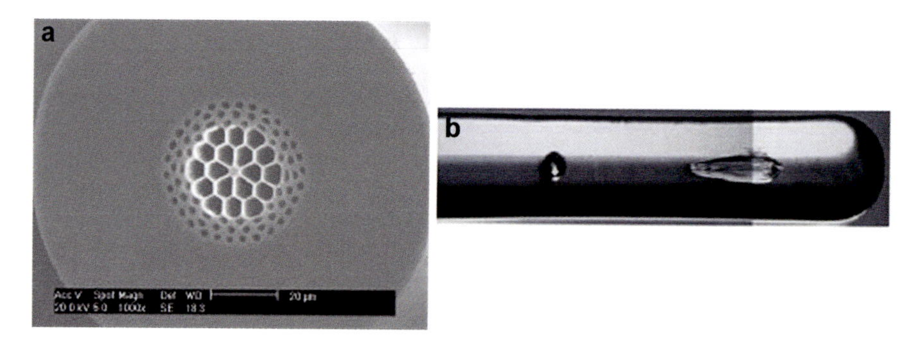

Fig. 3 (**a**) SEM image of the PCF cross section, (**b**) microscopic image of the sensor head

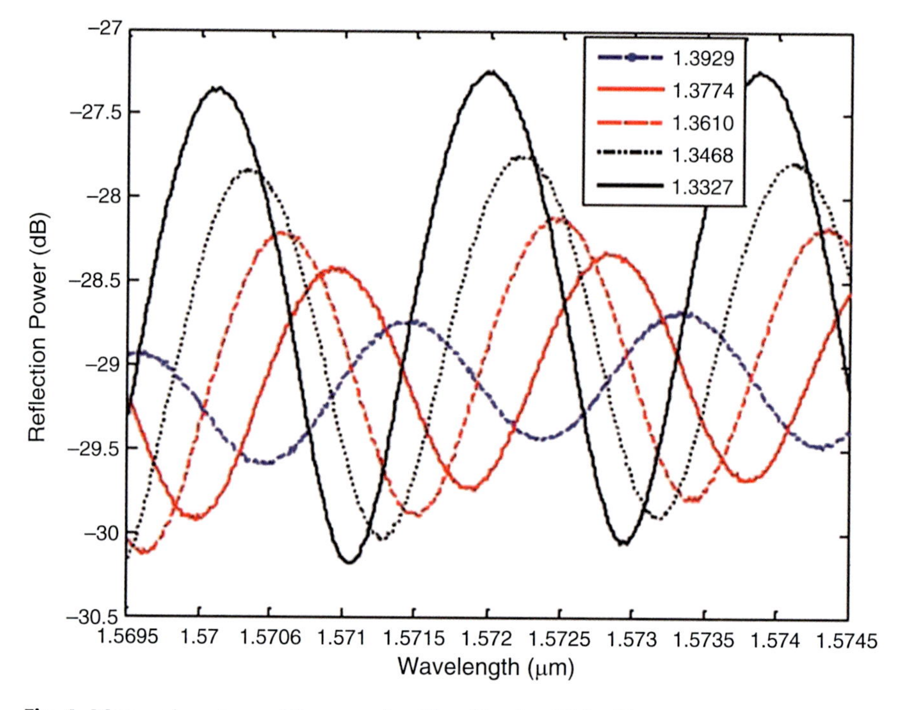

Fig. 4 Measured spectrum of the sensor head in calibrating RI liquid

Fig. 4. As RI increased, the fringes moved to longer wavelengths monotonically accompanied with decreasing extinction ratios. A strong higher frequency component in addition to the zero-frequency component was observed in the spectrum, which caused the modal interference sensitive to external fluid RI. In addition, although the FP cavity was not modified at different RI liquid, the reflection coefficient at the spherical fiber tip was varying corresponding to the bulk liquid RI values. As a result, the extinction ratio of the interfering spectrum was dependent on the external fluid RI. The linear curve fitting of the fringe dip wavelength shows that

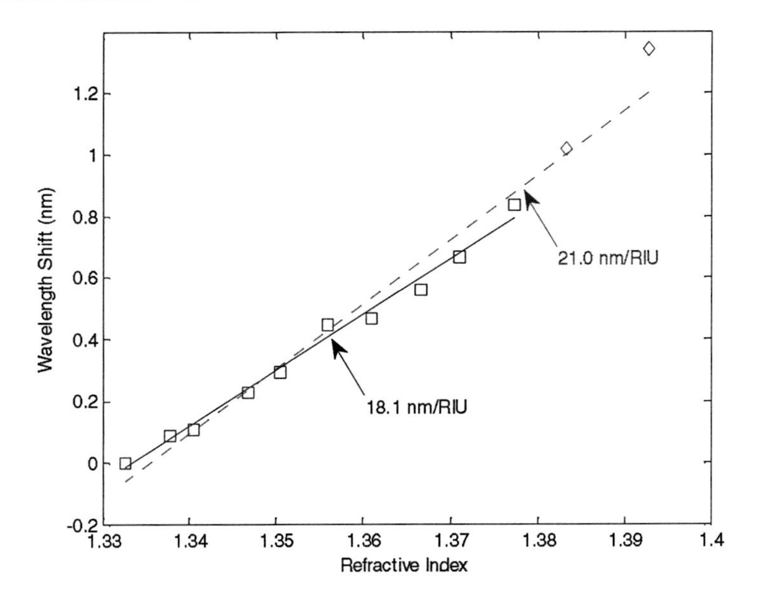

Fig. 5 Calibration curve of the sensor head based on fringe wavelength measurement

the sensitivity of the probe is around 21.4 nm/RIU (refractive index unit) ranging from 1.33 to 1.4 or 18.1 nm/RIU ranging from 1.33 to 1.37 shown in Fig. 5.

High-Temperature Pressure In-Fiber FPI Sensor

A high-temperature strain sensor based on an in-line FPI formed by splicing a multimode photonic crystal fiber (MPCF) to an SMF with a commercialized fusion splicer was demonstrated in Ran et al. (2015). The air microbubble inserted between the two fibers has two smooth glass-air interfaces separated by a distance as two reflective mirrors of the FPI. Three FPI structures including PCF FPI, SMF FPI, and etched PCF FPI are used in high-temperature and high-temperature pressure tests for comparison. The configuration of the FPI on a PCF was given by Fig. 6a. The microscopic images of the fabricated sensors were shown in Fig. 6b, c, respectively, for both etched and non-etched PCF FPIs. The self-enclosed air cavity could be created by splicing the SMF to a PCF with a precise hole micro-machined by a 157-nm laser at the end face of the PCF. A circular hole at the center of the cross section of the PCF by using a 157-nm laser micromachining system was larger than the core but smaller than the cladding. Then, the air cavity was formed by splicing the PCF with hole to the cleaved SMF. The fiber was cleaved with a short distance from the air cavity, and another cleaved SMF was spliced to it for protecting the PCF.

The temperature responses of the PCF FPI and etched PCF FPI were measured from 25 to 700 °C successively, while the temperature response of the SMF FPI was measured in the range of 25–400 °C. The temperature coefficients of the SMF FPI, PCF FPI, and etched PCF FPI were measured to be ~1.9 pm/°C, ~0.9 pm/°C,

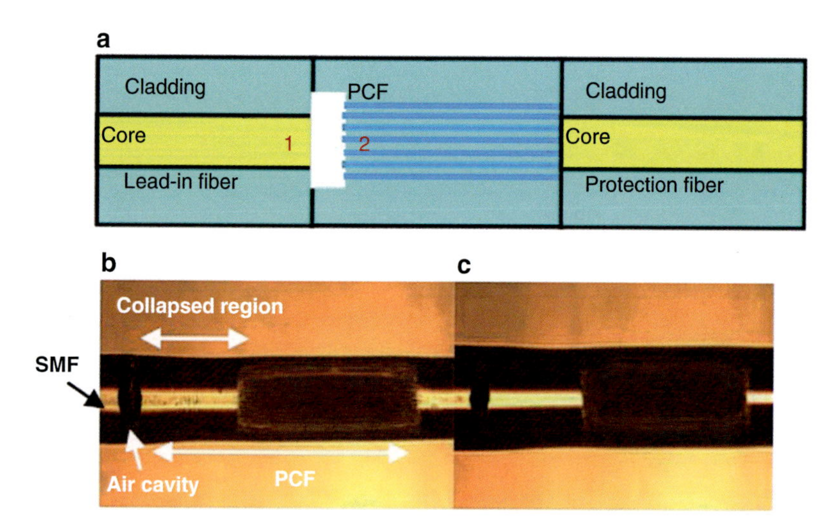

Fig. 6 (**a**) Schematic diagram of the PCF FPI sensor, (**b**) photo of the etched PCF FPI, (**c**) photo of non-etched PCF FPI

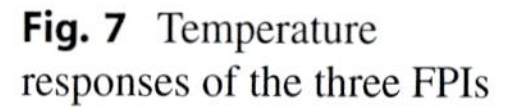

Fig. 7 Temperature responses of the three FPIs

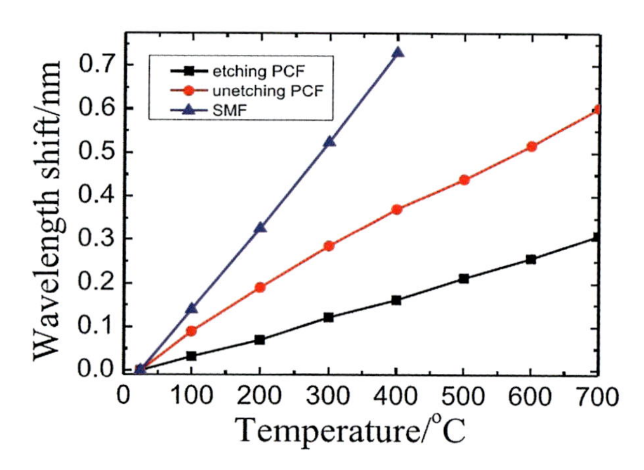

and ~0.45 pm/°C, respectively, as shown in Fig. 7. The pressure responses of the PCF FPI and etched PCF FPI were measured from 0 to 10 Mpa. The pressure and temperature coefficients of the SMF FPI, PCF FPI, and etched PCF FPI were measured as ~25.3 pm/MPa, ~39.3 pm/MPa, and ~54.7 pm/MPa, respectively, as shown in Fig. 8. The sensitivity of the etched PCF FPI in this paper was much higher than that of the SMF and PCF FPI. The pressure responses of the etched PCF FPI at different temperatures were measured in the range of 0–10 Mpa, as shown in Fig. 9. The pressure and temperature coefficients of the etched PCF FPI were 54.7 pm/MPa and 0.45 pm/°C, respectively. The response of the etched PCF FPI shows good linearity and constant sensitivity at different temperatures, even up to 700 °C.

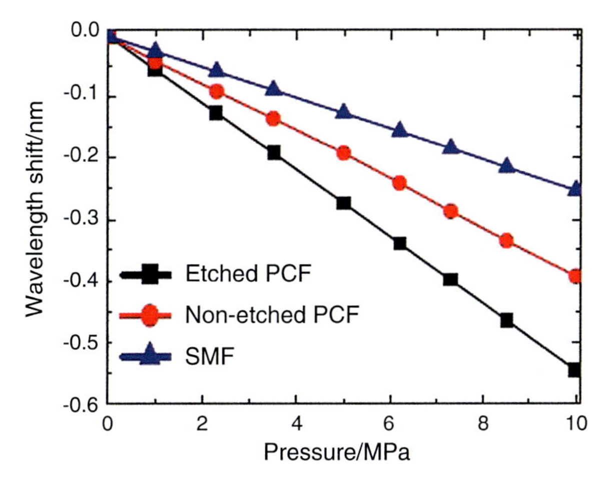

Fig. 8 Pressure responses of the three FPIs

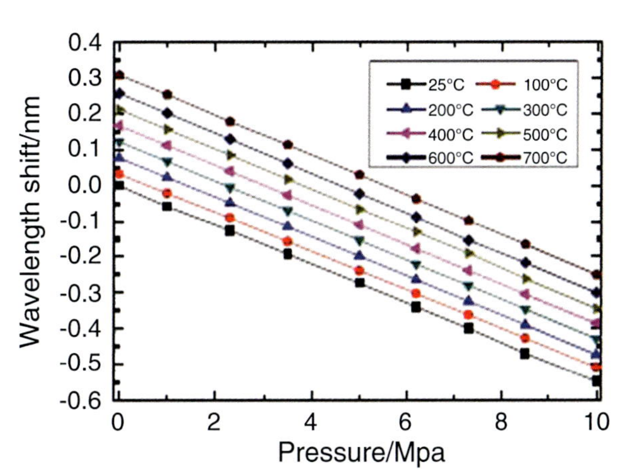

Fig. 9 Pressure responses of the etched PCF FPI at different temperatures

Strain In-Fiber FPI Sensor

A microscopic FPI whose cavity is an air bubble formed by fusion splicing an index-guiding PCF and a SMF was proposed in Villatoro et al. (2009). The cross section of the PCF and the diagram of the interrogation setup are shown in Fig. 10. The in-line FPI was fabricated with an air bubble with two smooth interfaces, and the reflectivity of each glass-air interface is less than 4%. Therefore, higher-order reflections are negligible, and the device can be considered as a two-beam interferometer. The phase of the FPI can change if it is perturbed by environmental variables; consequently, the interferometer can be used for optical sensing. This in-line FPI shows low-temperature sensitivity (less than 1 pm/°C), which is one order of magnitude lower than that of the popular fiber Bragg grating ~12 pm/°C.

The performance of the in-line FPI as a strain sensor had been investigated. The in-line FPI with 25-μm and 58-μm diameter bubble was subjected to strain ranging

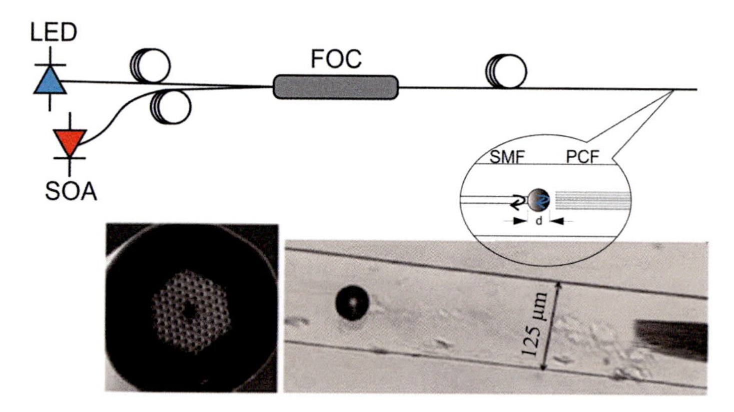

Fig. 10 Diagram of the interrogation setup highlighting the zone of the splice. *LED* light-emitting diode, *OSA* optical spectrum analyzer, *FOC* fiber-optic circulator, *SMF* single-mode fiber, *PCF* photonic crystal fiber. *d* is the diameter of the microcavity. The cross section of the PCF and a micrograph of the splice showing the microbubble are also shown

from 0 to 5000 $\mu\varepsilon$, respectively. The strain sensitivity of the interferometer with the smaller cavity was 0.62 pm/$\mu\varepsilon$, while the other one was 2.7 pm/$\mu\varepsilon$, as shown in Fig. 11. The sensitivity in the latter case was about 250% higher than that of FBG-based strain sensors (typically 1.2 pm/$\mu\varepsilon$) and that of more complex PCF-based strain sensors (Villatoro et al. 2007). As strain sensors based on the in-line FPI reported here exhibited wider dynamic ranges (between four to five times) than those based on hollow-core or PBG fibers (Rao et al. 2007a, b; Shi et al. 2008; Li et al. 2008), meanwhile, the cavities were shorter than those of FPIs built with such fibers (Rao et al. 2007a, b; Shi et al. 2008). These sensors can be important in space-constrained applications with the advantage that temperature compensation may not be required.

Basic Structure of Fiber-Tip FPI

Principal

The basic schematic of fiber-tip FPI on a PCF is given by Fig. 12. An intrinsic Fabry-Perot interferometer is formed by the end faces of the SMF and the PCF. As the modes excited in the PCF have different effective indices, the phase difference will depend on the length of the PCF and also on the wavelength of the light source (Ding et al. 2015).

The reflection can be expressed as:

$$I = I_1 + I_2 + 2\sqrt{I_1 I_2}\cos\left(\frac{2\pi\Delta n \cdot L}{\lambda}\right) \quad (3)$$

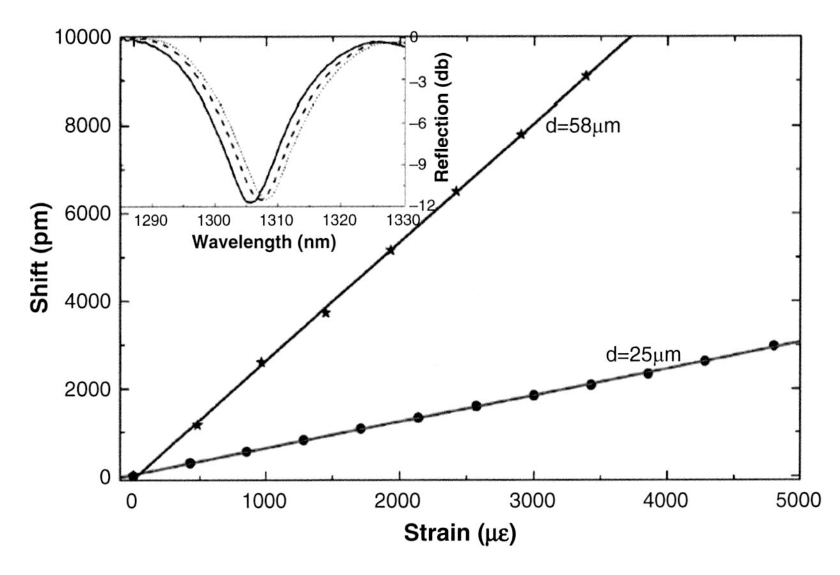

Fig. 11 Shift of the interference pattern as a function of strain observed in a 26-μm sample at 1290 ± 40 nm and in a 58-μm sample at 1550 ± 30 nm. The *continuous linear lines* are fitting to the data. The *inset* figure shows the shift of one of the interference dips at 0 (*solid curve*), 2570 (*dashed curve*), and 4288 με, (*dotted curve*) of the 26-μm sample

Fig. 12 Basic schematic of fiber-tip FPI on a PCF

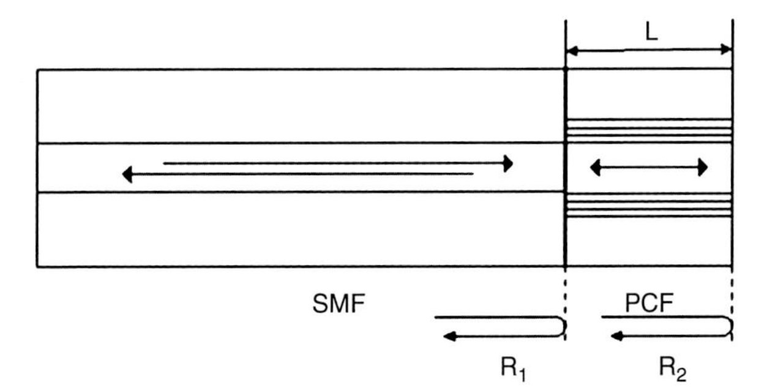

In Eq. 3, I_1 and I_2 are the intensities of the interfering modes, Δn is the difference between the effective indices of the interfering modes, L is the length of the PCF, and λ is the wavelength of the light source. Maxima of Eq. 3 appear when $2\pi \Delta nL/\lambda_m = 2m\pi$; here m is an integer. The wavelength of the dip is given by:

$$\lambda_m = \frac{\Delta n L}{m} \tag{4}$$

Fabrication

Fiber-tip FPI sensors are generally fabricated by fusion splicing a segment of PCF to an SMF. A section of endlessly single-mode PCF (SPCF) is used in Rao et al. (2008) and Wu et al. (2011), and a short length of a solid-core PCF is used in Du

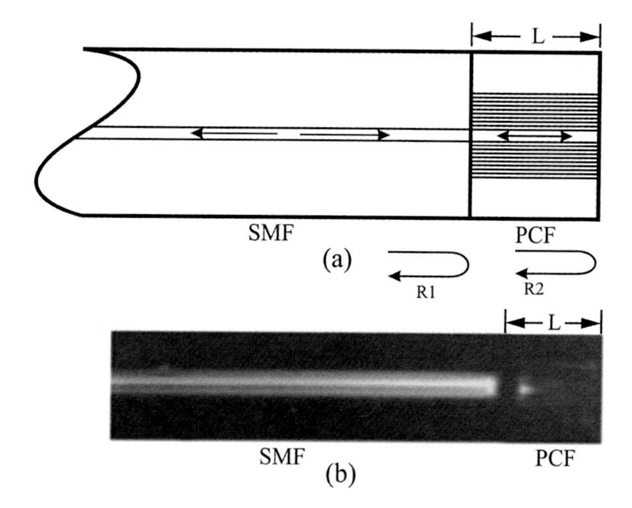

Fig. 13 (**a**) The operation principle, (**b**) the cross section of the ESM PCF

et al. (2014). The PCF was spliced to the SMF and then cleaved down to form the F-P structure. In some cases, another segment of SMF was spliced to the PCF to protect the F-P structure. The fiber-tip FPI can be used as high-temperature sensor, refractive index sensor, or high-pressure sensor.

Applications

High-Temperature Fiber-Tip FPI Sensor

A high-temperature fiber-tip FPI sensor on a PCF was proposed in Ding et al. (2015); the sensor was fabricated by fusion splicing a segment of the endless single-mode photonic crystal fiber (ESM PCF) to an SMF. The operation principle of the sensor is shown in Fig. 13a, and the microscope photograph of the sensor was illustrated in Fig. 13b.

The reflected spectrum of the FFPI was interrogated by using a micro-spectrometer. The temperature response of the sensor with a cavity length of 148 μm was experimentally investigated from the room temperature (17 °C) to 1200 °C. A peak wavelength shift with the change of the temperature is shown in Fig. 14. Experimental results show that the peak wavelength resolution of 10 pm and the temperature sensitivity of 10 pm/°C were achieved. The reflected spectrum kept stable even when the temperature exceeded 1200 °C, which is showing that the FFPI-based sensor exhibited a wide temperature measurement range.

Refractive Index and Temperature Fiber-Tip FPI Sensor

The authors of Rao et al. (2008) presented a fiber-tip FPI sensor formed by splicing a section of endlessly single-mode PCF (SPCF) to a SMF for simultaneous measurement of temperature and refractive index. The EPCF was spliced to the SMF and then cleaved down to splice to another SMF. The structure and the photo of the sensor are shown in Fig. 15.

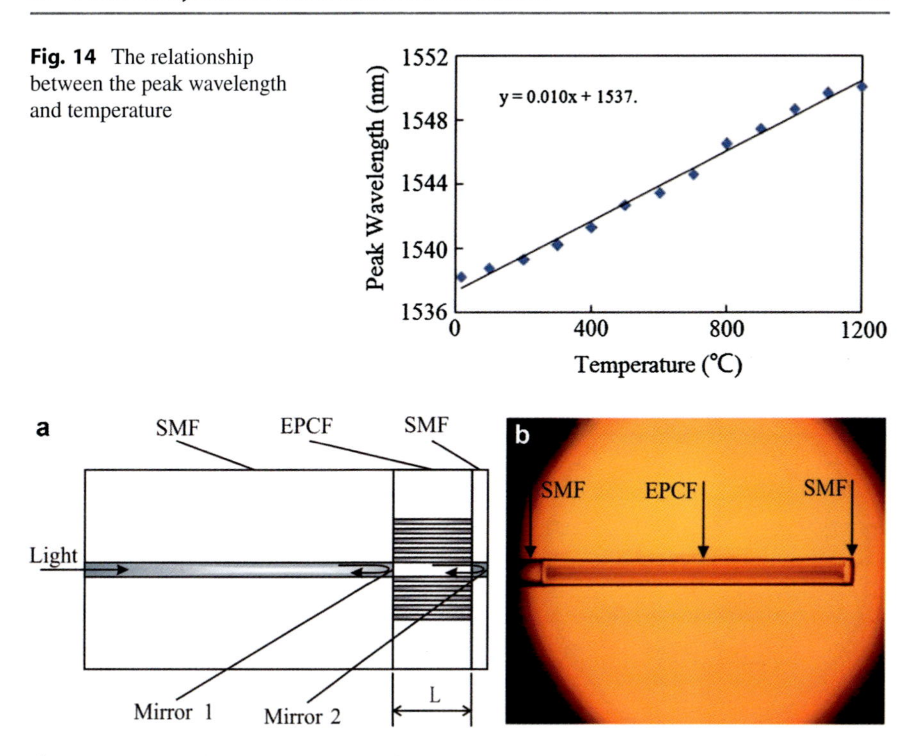

Fig. 14 The relationship between the peak wavelength and temperature

Fig. 15 (**a**, **b**) Configuration and photo of the sensor

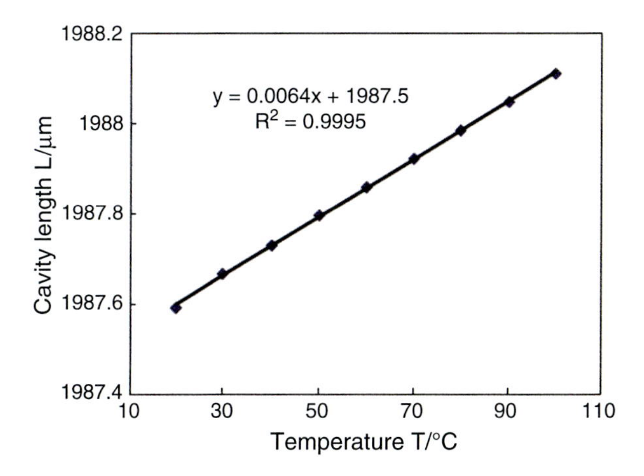

Fig. 16 Relationship between temperature and cavity length

In the experiment, the sensor was placed in a furnace, and the temperature was increased with a step of 100 °C from 200 to 1000 °C. As shown in Fig. 16, it shows that the cavity length linearly increased with the temperature increasing. The change of the cavity length is 517 nm and the temperature sensitivity is 6.4 nm/°C.

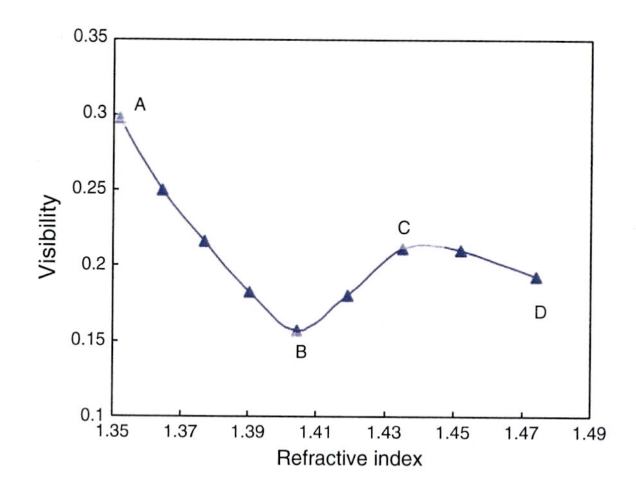

Fig. 17 Relationship between refractive index and visibility of the sensor

Fig. 18 (**a**) Schematic of the FPI sensor head, (**b**) photograph of the 2.1-mm-long FPI sensor

In addition, the sensor was immersed into glycerin in water whose refractive index modified the reflective coefficient at mirror 2. Hence the visibility of the interferometric signal changed with different indices ranging from 1.35 to 1.47. Figure 17 shows the relationship between the fringe visibility of the sensor and refractive index of the glycerin solution. It could be seen that the visibility reached its minimum and maximum at point B of ~1.4043 and point C of ~1.4345, respectively.

High-Pressure and High-Temperature Fiber-Tip FPI Sensor

The authors of Wu et al. (2011) presented a high-pressure and high-temperature fiber-tip FPI sensor fabricated by a short length of a solid-core PCF spliced to a SMF using the technique reported in Li et al. (2008). Figure 18 shows the schematic diagram and photograph of the sensor. It could be found that the wavelength shift can be changed with varying the refractive index and axial strain.

In Wu et al. (2011), two sensors with cavity lengths of 1 mm and 2.1 mm were investigated of the high-pressure and high-temperature characteristics. The temperature response of the two FPI sensors was investigated till the temperature reached 700 °C, as shown in Fig. 19a. The temperature sensitivities were 13.7 pm/°C and 13.1 pm/°C for the 2.1-mm sensor and 14.0 pm/°C and 13.5 pm/°C for the 1-mm sensor, respectively. The pressure was increased gradually with a step of

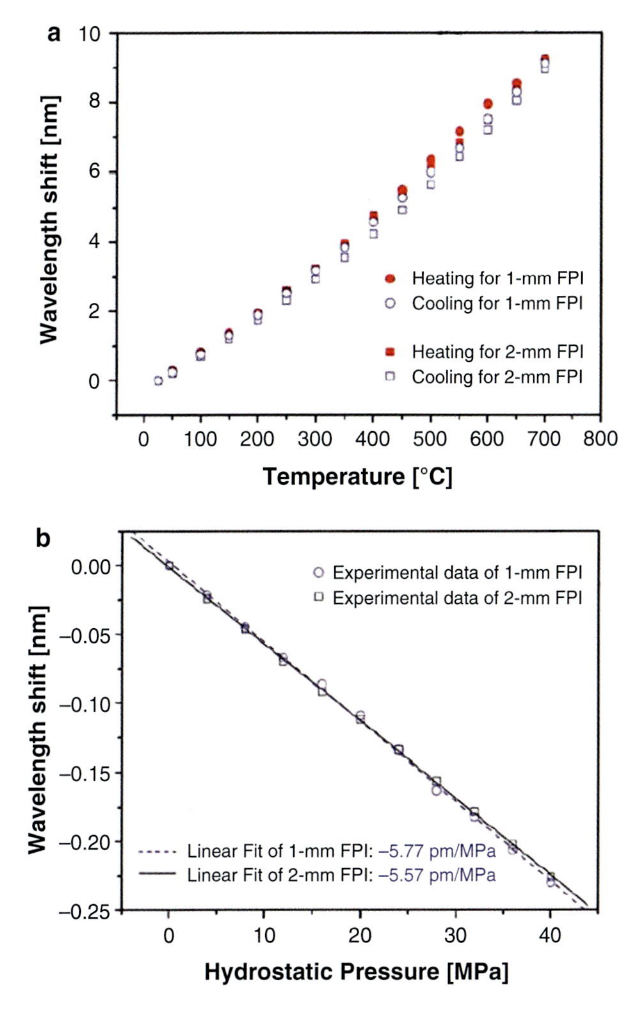

Fig. 19 (**a**) Wavelength shifts of the two sensors with temperatures increasing, (**b**) wavelength shift of the two sensors under different hydrostatic pressure

4 MPa in the range of 0–40 MPa. Figure 19b shows the optical spectrum experienced a blue shift with the increase of the pressure, and the pressure sensitivities of the two sensors were measured to be −5.57 pm/MPa and −5.77 pm/MPa, respectively.

Composite Structure

Dual-Core Photonic Crystal Fiber

The authors of Du et al. (2014) proposed a sensor fabricated by alignment splicing a small section of dual-core photonic crystal fiber (DC-PCF) to a SMF for

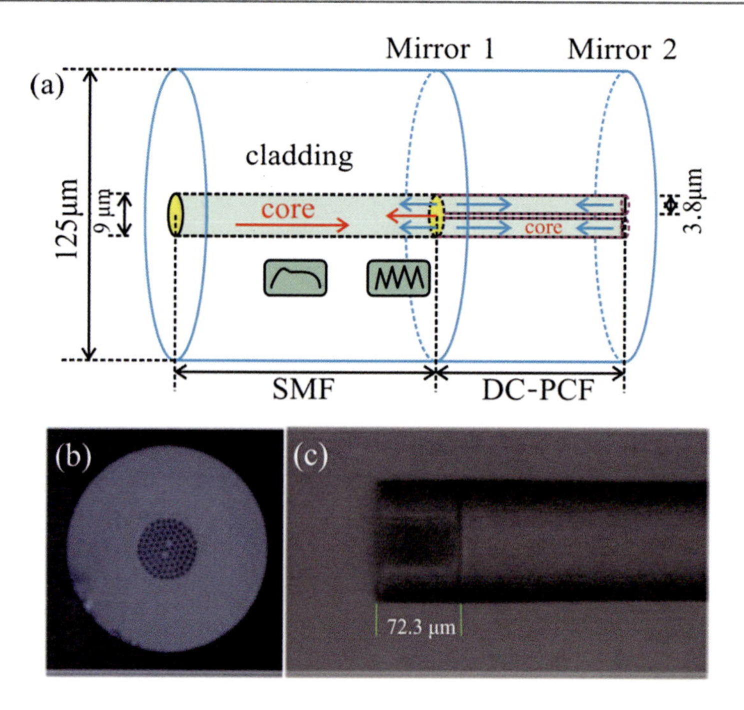

Fig. 20 (**a**) Schematic diagram of the DC-PCF-based interferometer, (**b**) cross section of the DC-PCF, (**c**) microscope image of the sensing head

high-temperature measurement up to 900 °C. Two reflective surfaces of glass-air at splicing region and DC-PCF end face formed the FP cavity with a 72.3-μm-long DC-PCF in Du et al. (2014). The schematic diagram, cross section, and microscope image of the sensor are shown in Fig. 20.

The sensor was inserted into a tube furnace to monitor the temperature, while the temperature was stepwise increased to 900 °C and then cooled down to the room temperature. Figure 21a, b shows the spectrum under different temperatures in heating and cooling processes. Figure 22 indicates the relationship between the wavelength and the temperature, and sensitivities of the heating/cooling processes are 13.88 pm/°C and 13.92 pm/°C, respectively, which is in well agreement with theoretical result.

Dual Hollow-Core Fibers (HCFs)

The sensor based on dual hollow-core fibers (HCFs) was demonstrated in Lee et al. (2015) for measuring the thermo-optic coefficients (TOCs) of liquids. A tiny segment of the HCF_1 (diameter $D = 30$ μm) filled with liquid was spliced with another section of HCF_2 (diameter $D = 5$ μm). The sensor was fabricated by a SMF spliced with two HCFs. Figure 23 shows the microphotograph of the sensor tip for measuring the TOCs of liquids.

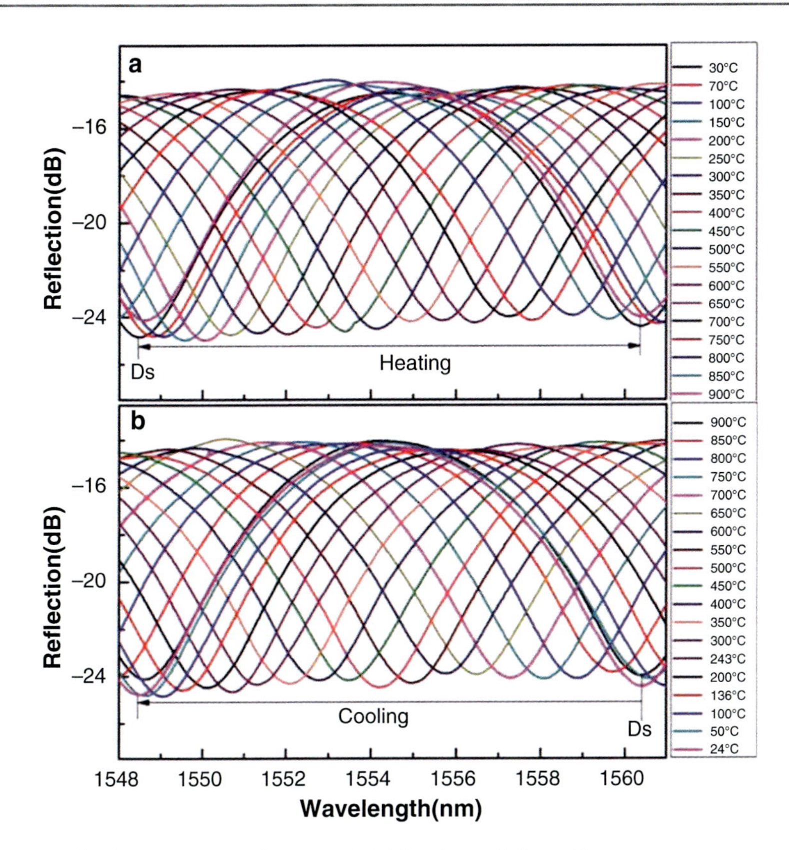

Fig. 21 Reflection spectrum of the sensor in (**a**) heating and (**b**) cooling processes

Figure 24 shows the wavelength shift of the liquid-filled sensor with the Cargille liquid, the DI water, and the ethanol in temperature from 20 to 60 °C. It shows that the wavelength is blue shift and the peak intensity is increased with the decreasing RI and the increasing T due to the thermal effect on the RI of the liquids.

PCF-Based Mach-Zehnder Interferometers

Single-Core Fiber MZI

Principal

Mach-Zehnder interferometer (MZI) has been commonly used in diverse sensing applications because of its flexible configuration. Early MZI had two independent arms, which are the reference arm and the sensing arm (Lee et al. 2012). An incident light is split into two arms by a fiber coupler and then recombined by another fiber coupler, while the recombined light has the interference component between the two

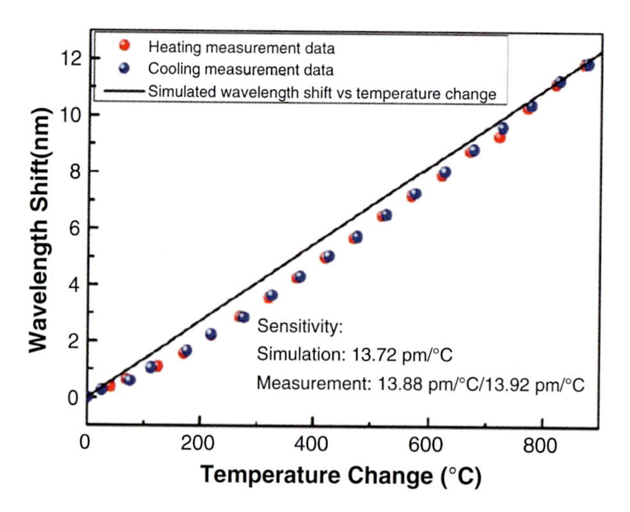

Fig. 22 Experimental and simulated wavelength shift as functions of temperature change

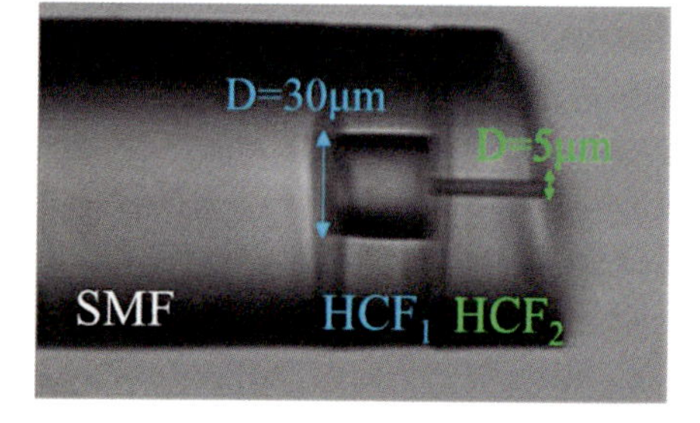

Fig. 23 Microphotograph of the MFFPI sensor tip. Here, HCF_1 has a diameter of 30 μm, and HCF_2 has a diameter of 5 μm. The cavity length of the HCF_1 herein is about 33.84 μm

arms. In Fig. 25, it shows the basic schematic of MZI on a PCF and the cross section of a single-core PCF (Sales et al. 2015).

The output light intensity can be expressed by:

$$I = I_{01} + I_{11} + 2\sqrt{I_{01}I_{11}}\cos(\varphi + \varphi_0) \tag{5}$$

where I_{01} and I_{11} are the intensity of LP_{01} and LP_{11} modes, respectively, and φ_0 is an initial phase. φ is the phase difference as a function of the LF-PCF physical length L and can be expressed as:

$$\varphi = \frac{2\pi \Delta n_{\mathrm{eff}} L}{\lambda} \tag{6}$$

When $\varphi = (2m + 1)\pi$, $m = 0, 1, 2\ldots$, the minimum wavelength value of output light intensity is located at:

$$\lambda_{\mathrm{dip}}^{m} = \frac{2n_{\mathrm{eff}}L}{2m + 1} \tag{7}$$

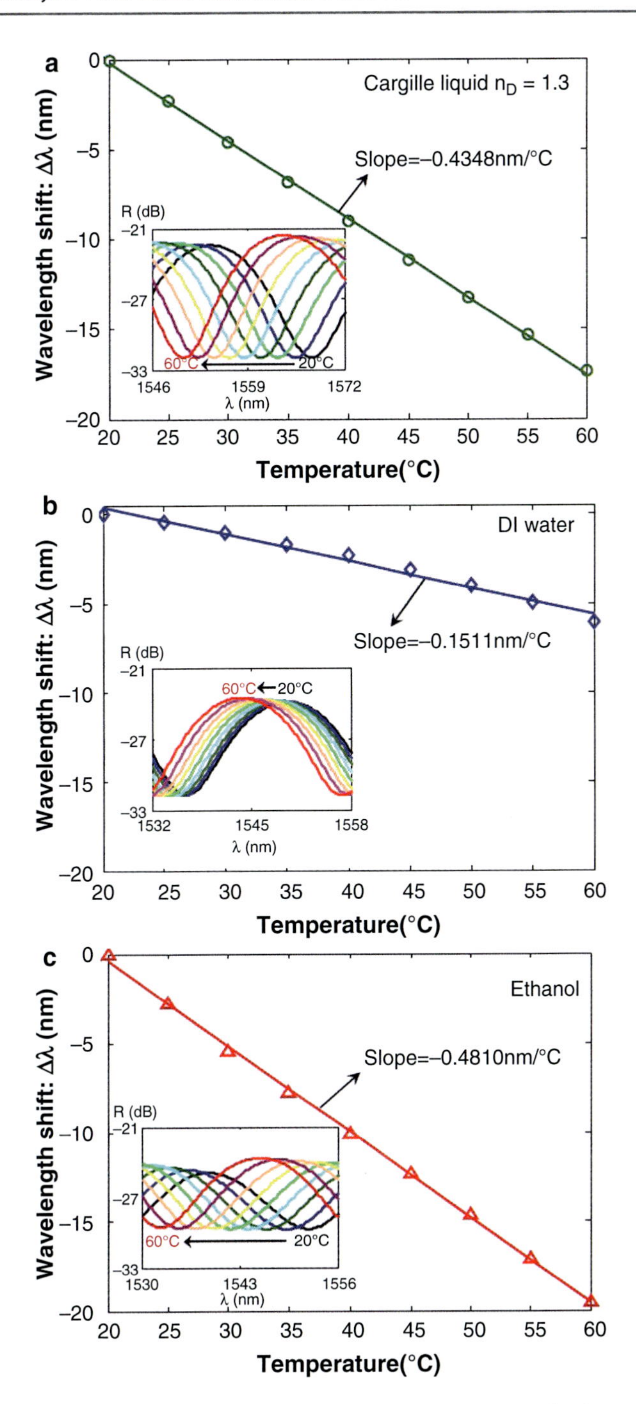

Fig. 24 Sensitivity of the wavelength shift in temperature from 20 to 60 °C for (**a**) the Cargille liquid where $n_D = 1.3$, (**b**) the DI water, and (**c**) the ethanol. *Inset* figures show corresponding spectra with an increasing T

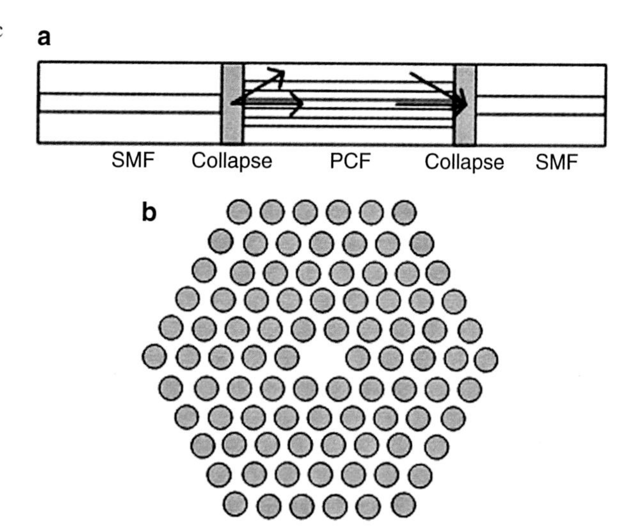

Fig. 25 (**a**) Basic schematic of MZI on a PCF, (**b**) cross section of a single-core PCF

where λ_{dip}^{m} is the wavelength with a minimum light intensity in the interference spectrum. When ambient temperature changes, due to the thermo-optic effect, the refractive index of the index-matching oil filled into the micro-holes of PCF significantly changes; therefore, it will inevitably cause the change of n_{eff} between the two modes.

Fabrication

Single-core photonic crystal fiber-based Mach-Zehnder interferometers are generally fabricated by fusion splicing a segment of PCF between two SMFs. The sensor is consisted of a short piece of single-core photonic crystal fiber and two single-mode fibers. The micro-holes in PCF can be filled by refractive index-matching oil or something else alike to enhance the sensor sensitivity. The PCF-based MZI fiber sensor can be used as temperature sensor or refractive index sensor.

Applications

Temperature MZI Sensor

A compact and ultrasensitive all-fiber temperature sensor was proposed in Geng et al. (2014); the proposed sensor is based on an in-line fully liquid-filled PCF MZI. The interferometer was consisted of a short piece of PCF and two SMFs. The PCF was filled with the refractive index-matching oil. Then both ends of PCF undergone heat treatment by the liquefied petroleum gas flame quickly, and the index-matching oil with around 1-mm length from the end of PCF was gasified. After that, the SMF and PCF were welded manually with accurate control over the micro-displacement in x and y directions. The structure of the sensor is shown in Fig. 26.

Figure 27 shows the sensor response to temperature. It could be seen from Fig. 27 that the sensitivities of the interference reach −1.83 nm/°C and −1.09 nm/°C

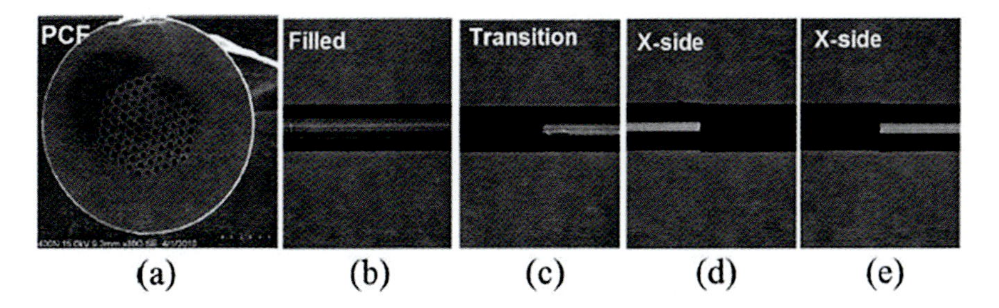

Fig. 26 (**a**) Cross section of PCF, (**b**) side view of completely filled PCF, (**c**) transition region of pretreated LF-PCF stub, (**d**, **e**) side views of splice point between SMF and LF-PCF

Fig. 27 Temperature response (**a**) and spectra shift (**b**) for the LF-PCF-based MZI with cavity length of $L = 7.5$ mm

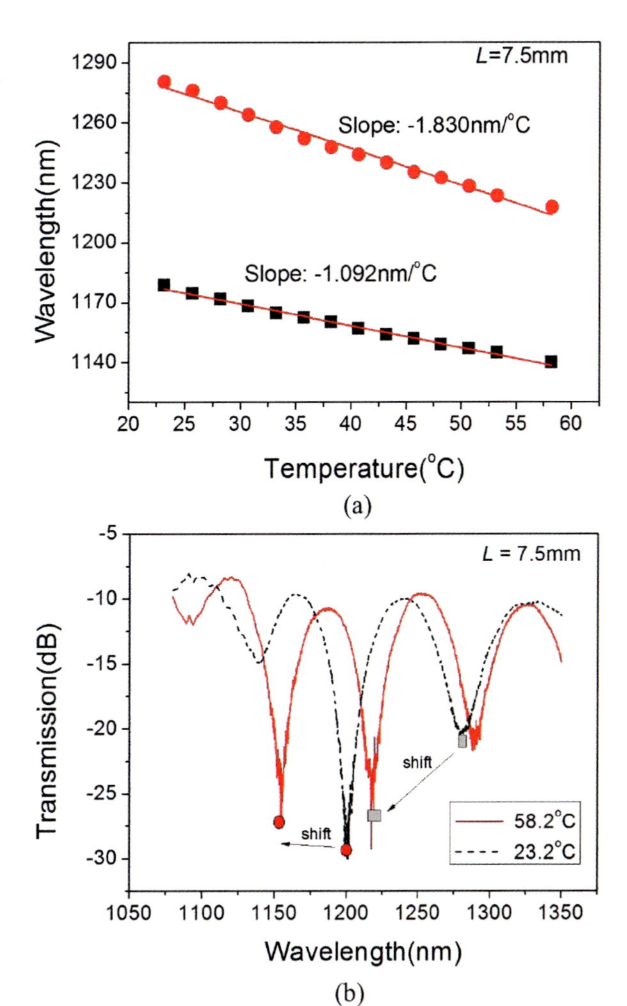

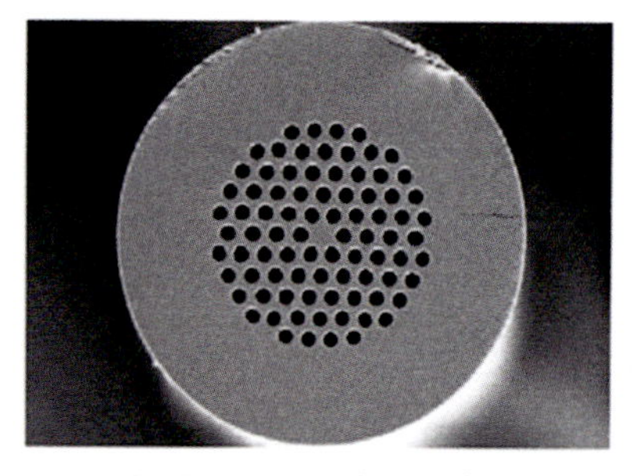

(a) Cross section of PCF

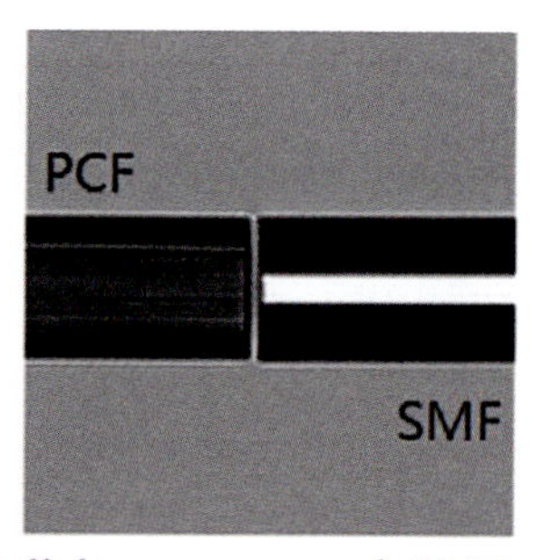

(b) Splicing process of SMF and PCF

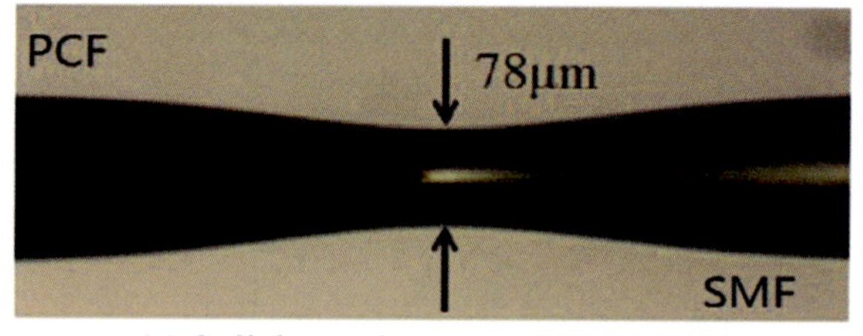

(c) Splicing point tapered SMF-PCF

Fig. 28 Photonic crystal fiber and SMF-PCF splicing and taper

for wavelength at 1281 nm and 1199 nm, respectively, and blue shift occurred with the temperature increased.

Refractive Index MZI Sensor

An MZI refractive index sensor was presented in Wang et al. (2016); the proposed sensor is composed of a segment of PCF and two segments of SMF. The cross section of PCF is shown in Fig. 28a. The single-core PCF was spliced with SMF at both of ends, and then the splicing points are tapered (Fig. 28b, c).

Figure 29 presents the relationship between the wavelength and the refractive index with the lengths of PCF are 2 cm, 3 cm, and 4 cm, respectively. It could be seen that while the length is 4 cm, the refractive index reached a maximum value of 224.2 nm/RIU.

Figure 30 shows the relationship between the wavelength and the surrounding refractive index of the sensor at different taper waist diameters. It could be concluded that the sensitivity increases with the decrease of the diameter.

Dual Core

Principal

Dual-core photonic crystal fiber-based Mach-Zehnder interferometer sensing principal is the same as single-core PCF-based MZI. The dual core is the two independent

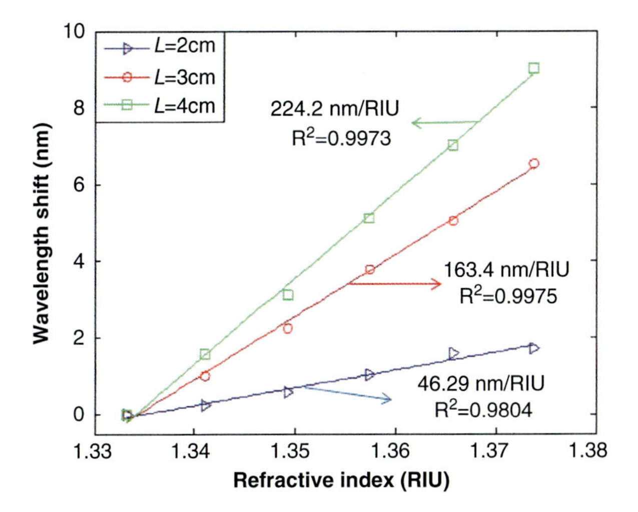

Fig. 29 Measuring sensitivity of the sensor at different PCF length

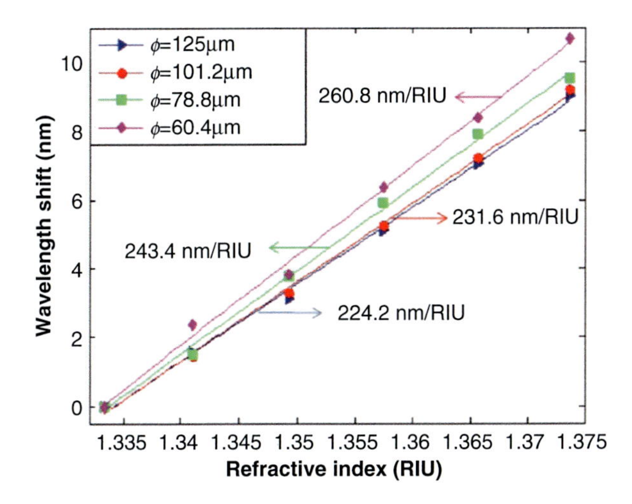

Fig. 30 Relations between wavelength shift and surrounding refractive index of the sensor

arms of the MZI. The output light intensity can also be calculated by Eqs. 5–7. Schematic structure of dual-core PCF is given in Fig. 31 (Sales et al. 2015).

Fabrication

A simple dual-core photonic crystal fiber-based Mach-Zehnder interferometer is generally fabricated by fusion splicing a segment of PCF between two SMFs. The sensor consisted of a short piece of dual-core photonic crystal fiber and two single-mode fibers. The two silicon cores of the PCF are used to form Mach-Zehnder interferometer. In order to create a sensitive structure, an air hole is induced by

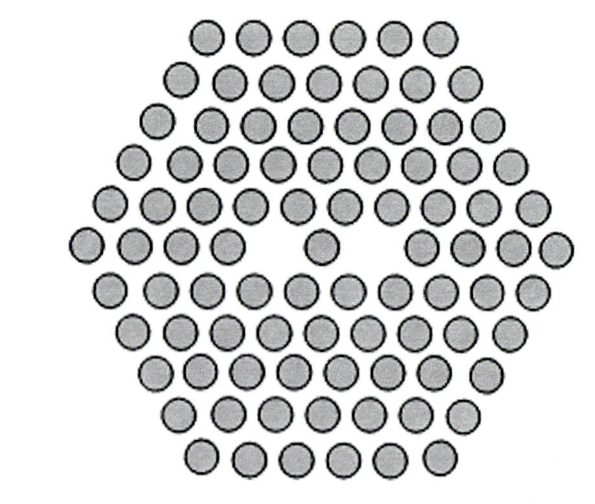

Fig. 31 Cross section of a double-core PCF

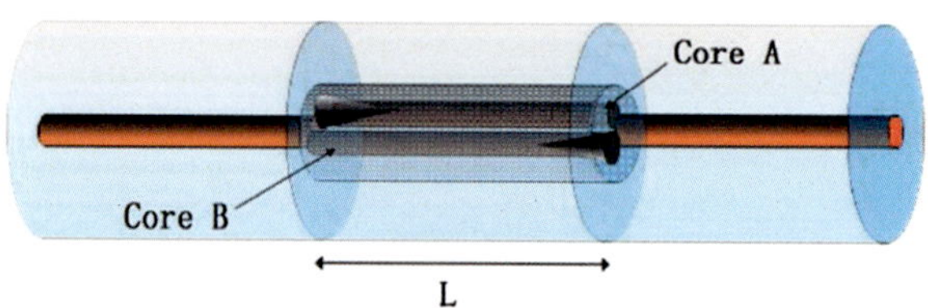

Fig. 32 Schematic structure of the sensor device

using the femtosecond laser technology (Liu et al. 2012), or refractive index is filled in some micro-holes of the PCF (Hou et al. 2016). The dual-core PCF-based MZI fiber sensor can be used as temperature and strain sensor.

Applications

Temperature and Stain MZI Sensor

A simple dual-core PCF-based MZI by use of femtosecond laser drilling and fiber splicing was demonstrated in Liu et al. (2012). Higher-order mode was excited in one fiber core by introducing a conical air hole at the splice point, which interfered with the fundamental mode in the other fiber core. The schematic structure of the sensor device is shown in Fig. 32. The two black cones were the femtosecond laser-induced air holes, and L was the length of the dual-core PCF. Figure 33 shows the cross-sectional view of the dual-core PCF before and after femtosecond laser drilling and the micrograph of the splice point between the single-mode fiber and PCF with a conical air hole.

The responds of the strain and temperature are shown in Fig. 34. Figure 34a demonstrates that the transmission peaks of the sensor shift to the shorter wavelength under a constant temperature of 22 °C with the increase of the strain from 0 to 4000 $\mu\varepsilon$. The strain sensitivities of the three samples with lengths of 7.5 cm,

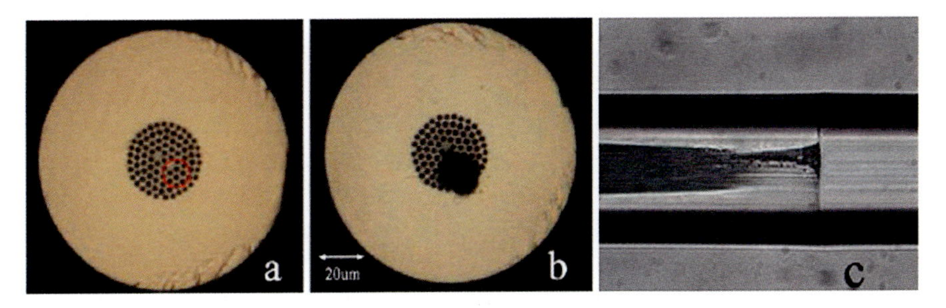

Fig. 33 Cross-sectional view of the dual-core PCF (**a**) before and (**b**) after femtosecond laser drilling. (*Red circle* indicates the location of the focusing point). (**c**) Micrograph of the splice point between the single-mode fiber and PCF with a conical air hole

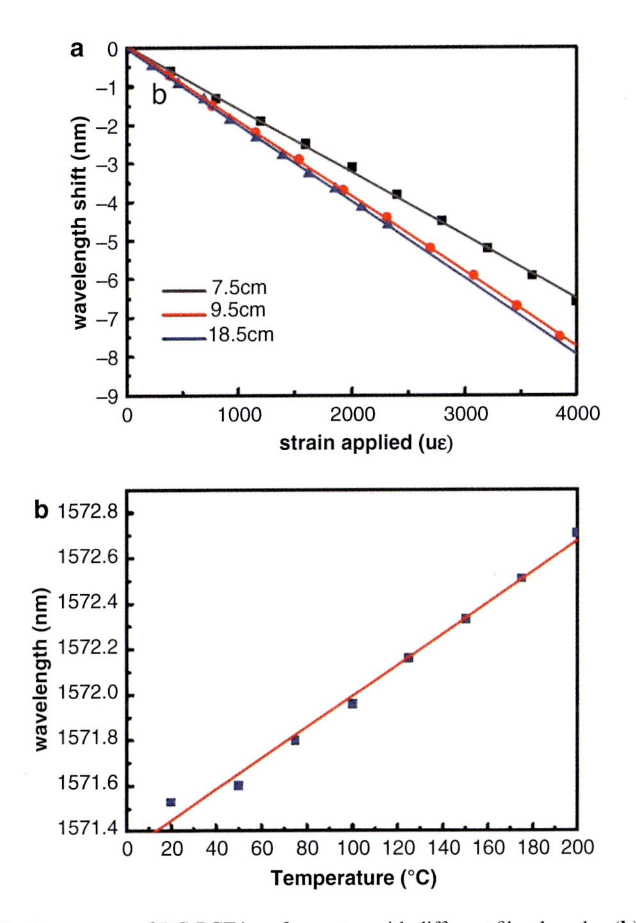

Fig. 34 (**a**) Strain response of DC-PCF interferometer with different fiber lengths, (**b**) wavelength variation against temperature

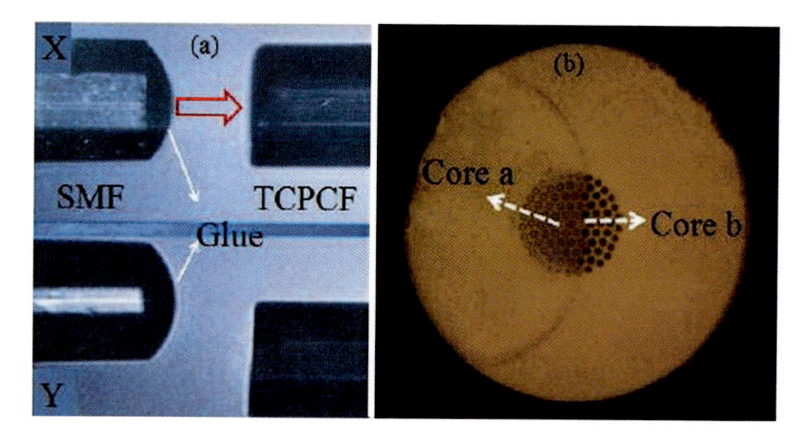

Fig. 35 (**a**) Microscopic image to explain the optical alignment and manual gluing procedure for the partially blocked process. The glue-dispensed fiber tip mounted on the V-groove of a fiber fusion splicer was moved and contacted the cleaved DC-PCF with a big offset. (**b**) The cross-sectional view of the partially blocked DC-PCF. The cladding air holes surrounding core *a* were blocked by glue, while air holes surrounding core *b* were remained

9.5 cm, and 18.5 cm are 1.64 pm/με, 1.95 pm/με, and 1.99 pm/με, respectively. It could be observed that the length of the DC-PCF has a significant impact on the strain sensitivity of the device, and the strain sensitivity of a 9.5-cm long DC-PCF around 1540-nm wavelength is higher than that around 1570 nm.

The temperature sensitivity of the sensor with the DC-PCF length of 9.5 cm was investigated from 20 to 200 °C by an increment of 25 °C. As shown in Fig. 34, the peak wavelength is found to shift toward longer wavelength due to the fiber expansion and thermo-optic effect, and the temperature sensitivity is estimated to be 6.8 pm/°C.

Temperature and Strain Sensing

A sensor based on partially filled dual-core photonic crystal fiber (DC-PCF) for measuring temperature and strain was demonstrated in Hou et al. (2016). The DC-PCF was prepared by manual gluing method, and the cladding air holes surrounding one core were selectively filled with RI liquid while, other air holes were unfilled. The DC-PCF had a diameter of 125 μm and five rings of circular air holes with 3-μm diameter and 3.7-μm pitch (Liu et al. 2012) hexagonally arranged in the cross section of the cladding and two solid cores located symmetrically on two sides of the fiber center.

Figure 35a shows the microscopic image of the optical alignment and manual gluing procedure for the partially blocked process. The glue-dispensed fiber tip mounted on the V-groove of a fiber fusion splicer was moved and contacted the cleaved DC-PCF with a big offset. Fig. 35b shows the cross-sectional view of the partially blocked DC-PCF. The cladding air holes surrounding core *a* were blocked by glue, while air holes surrounding core *b* were remained.

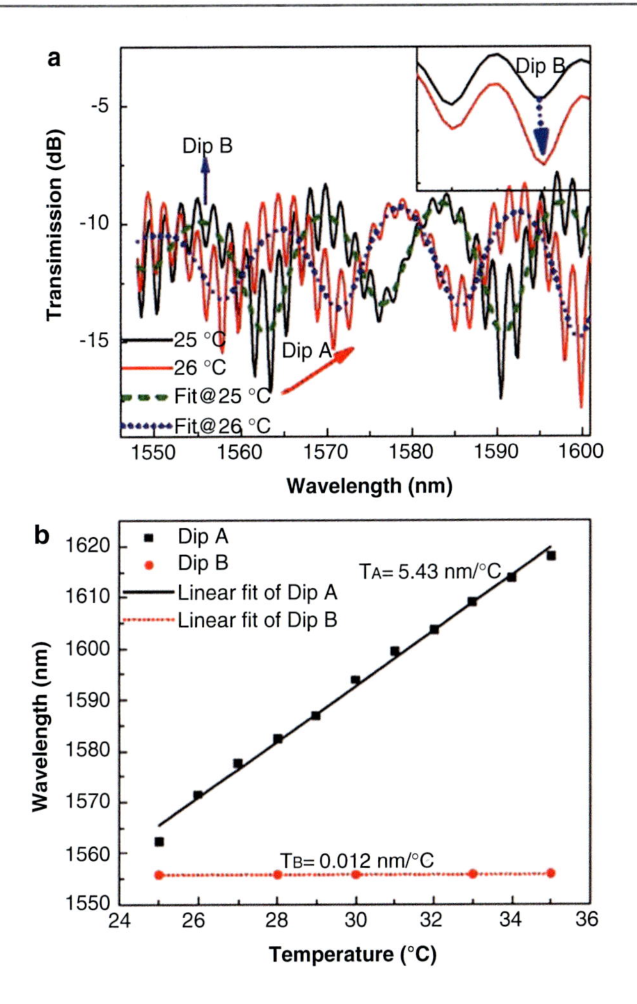

Fig. 36 (**a**) Transmission spectra of the proposed multicomponent interference sensor at 25 °C and 26 °C, with a PCF length of 6.8 cm, *inset* figure: magnified fringe of dip B in a small spectral range, (**b**) wavelength shifts of dip A and dip B as a function of temperature

The temperature response of the sensor tested in the oven increased gradually from 25 to 35 °C with a step of 1 °C. As shown in Fig. 36a, it shows the transmission spectra of the proposed multicomponent interference sensor at 25 °C and 26 °C. Figure 36b shows the wavelength shifts of dip A and dip B are functions of temperature. The temperature sensitivities are 5.43 nm/°C and 0.012 nm/°C for dip A and dip B, respectively.

The strain characteristics of the sensor were also investigated under a constant temperature of 25 °C with the applied strain ranging from 0 to 1400 $\mu\varepsilon$. Figure 37 shows the strain sensitivities of dips A and B are -1.95 pm/$\mu\varepsilon$ and -2.08 pm/$\mu\varepsilon$, respectively.

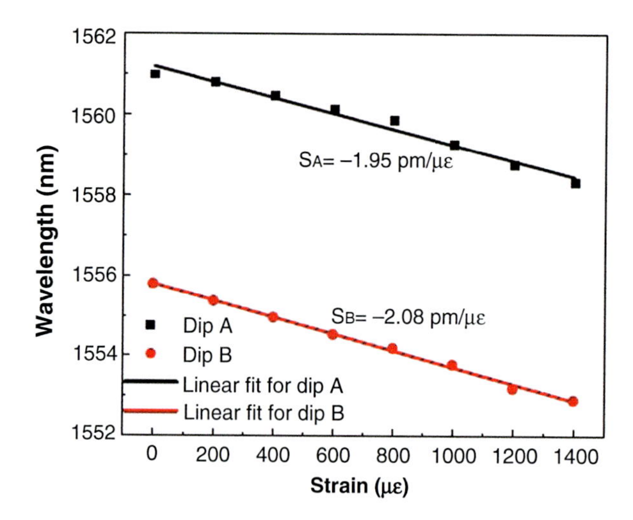

Fig. 37 Wavelength shifts of dip A and dip B as a function of strain

The simultaneous measurements of strain and temperature could be achieved by using the standard matrix demodulation method (Chen et al. 2013a); the cross sensitivity of temperature and strain could be calculated as below:

$$\begin{bmatrix} \Delta T \\ \Delta \varepsilon \end{bmatrix} = \begin{bmatrix} 5.43\text{nm}/^{\circ}C & -1.95\text{pm}/\mu\varepsilon \\ 0.012\text{nm}/^{\circ}C & -2.08\text{pm}/\mu\varepsilon \end{bmatrix}^{-1} \begin{bmatrix} \Delta_{\text{dip}A} \\ \Delta_{\text{dip}B} \end{bmatrix} \tag{8}$$

In the matrix, ΔT and $\Delta\varepsilon$ are the variations of the temperature and strain, and $\Delta\lambda_{\text{dipA}}$ and $\Delta\lambda_{\text{dipB}}$ are the wavelength shifts of dips A and B.

Composite Structure

Torsion Sensor with an Yb-Doped PCF Based on a MZI

A torsion sensor based on a MZI using a segment of ytterbium-doped double-cladding PCF between two SMFs by fusion splicing was proposed in Sierra-Hernandez et al. (2015). The layout of the MZI and the picture of the YbDPCF cross section are shown in Fig. 38. The YbDPCF core and claddings are the arms of the MZI (Fig. 38a).

The transmission spectrum is shown in Fig. 39, rotating in clockwise (CW) direction from 0 to 360° in steps of 60°. In order to determinate the responds of the sensor, four peaks at 1025, 1031, 1034, and 1039 nm were selected to observe the wavelength shifts.

The torsion sensitivities of the four peaks are 0.008, 0.006, 0.004, and 0.001 nm/°, respectively. Figure 40b shows the output power of the sensor at 1025, 1031, 1035, and 1039 nm, while they decrease as the applied torsion increase.

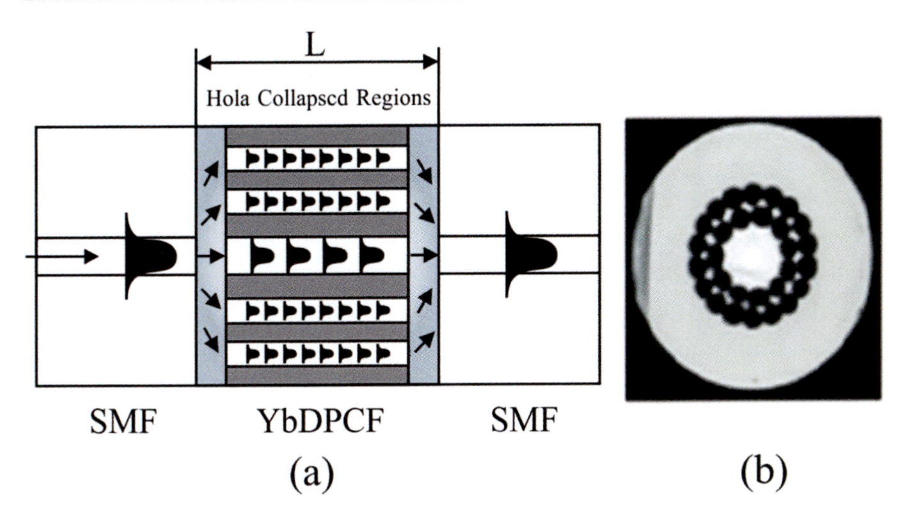

Fig. 38 (**a**) Layout of the MZI and (**b**) picture of the YbDPCF cross section

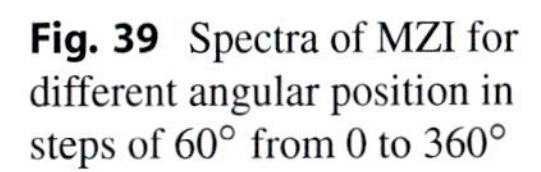
Fig. 39 Spectra of MZI for different angular position in steps of 60° from 0 to 360°

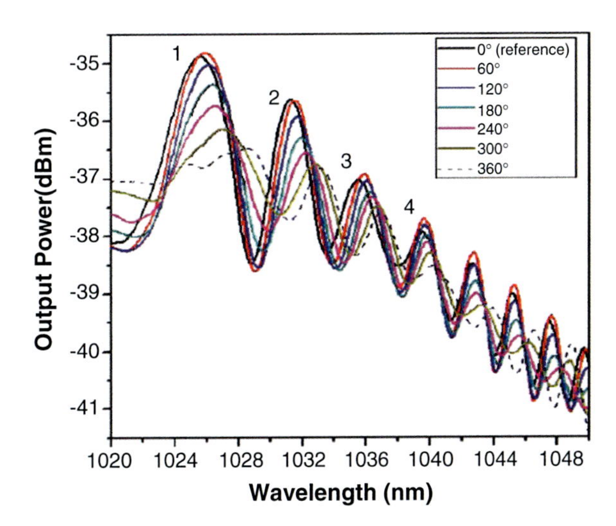

A Hybrid MZI for Refractive Index and Temperature Measurement

A hybrid structured in-line MZI composed of an embedded slender air cavity in a microfiber cascaded to a piece of PCF for simultaneous measurement of refractive index (RI) and temperature was demonstrated in Ni et al. (2016). The PCF was fused with the SMF with a little collapsing of air holes, as shown in Fig. 41a. The schematic diagram and microscopic image of the interferometer are shown in Fig. 41b, c.

The transmission spectra is examined by the fast Fourier transform (FFT) of the sensor with $L_1 = 10.84$ μm (46.08 μm after tapering) and $L_3 = 14$ mm in air at

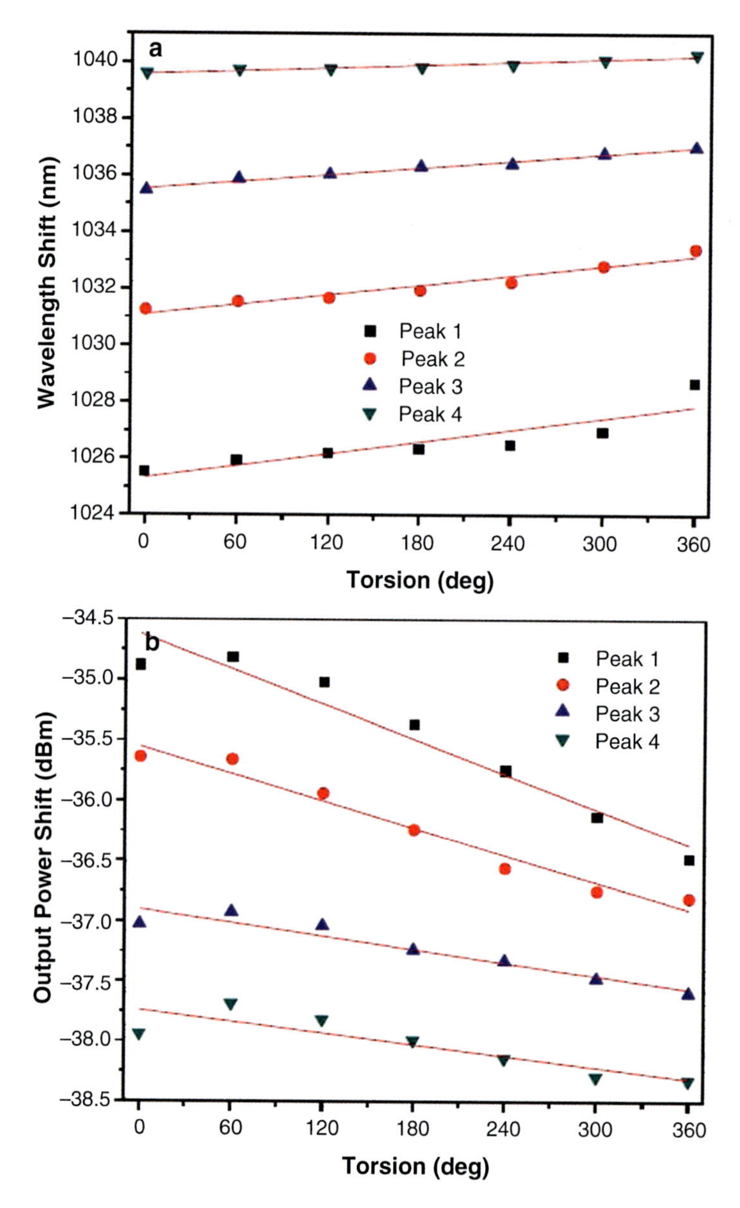

Fig. 40 (**a**) Spectra fringe wavelength shifting as a function of the applied torsion, (**b**) output power level as a function of the applied torsion

20 °C which are shown in Fig. 42. Three spatial frequency peaks are clearly found at 0.046, 0.089, and 0.135 nm^{-1} and designated by Peak1 to Peak3, respectively.

The RI response of the hybrid MZI was investigated with sugar solution of different mass fractions from 5% to 65% with increment of 10% at 20 °C. The

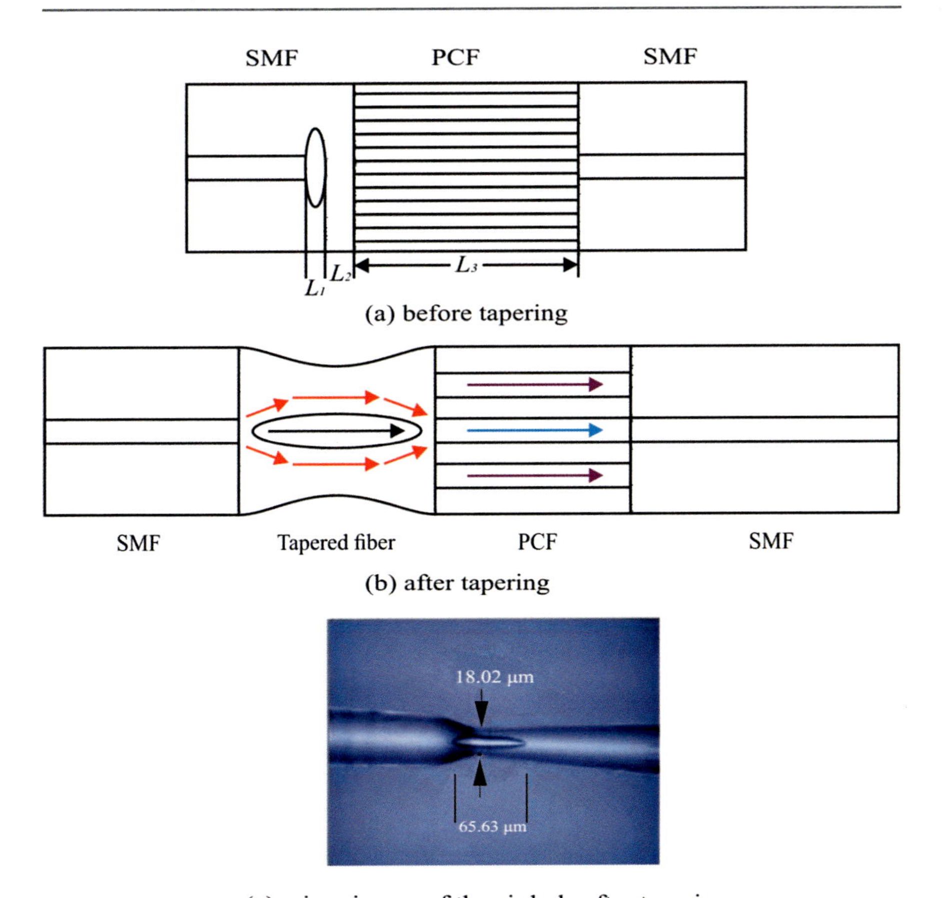

Fig. 41 Schematic diagram and microimage of the interferometer

RI responses of the PCF-based MZI and air-cavity-based MZI are shown in Fig. 43. It presents the RI sensitivities of the PCF-based and air-cavity-based MZI are 55.22 ± 3.32 nm/RIU and 55.84 ± 4.33 nm/RIU, respectively ($\lambda_1 =$ 1560.41 nm, $\lambda_2 =$ 1530.02 nm).

The temperature responses of the PCF-based and air-cavity-based MZI from 20 to 70 °C with a step of 10 °C are demonstrated in Fig. 44, and the temperature sensitivities are 0.045 ± 0.004 nm/°C and 0.143 ± 0.016 nm/°C, respectively.

Hence, the RI and temperature variations could be independently determined by using a sensitivity matrix as below:

$$\begin{bmatrix} \Delta \mathrm{RI} \\ \Delta T \end{bmatrix} = \begin{bmatrix} 55.220.045 \\ 55.840.143 \end{bmatrix}^{-1} \begin{bmatrix} \Delta\lambda_1 \\ \Delta\lambda_2 \end{bmatrix} \tag{9}$$

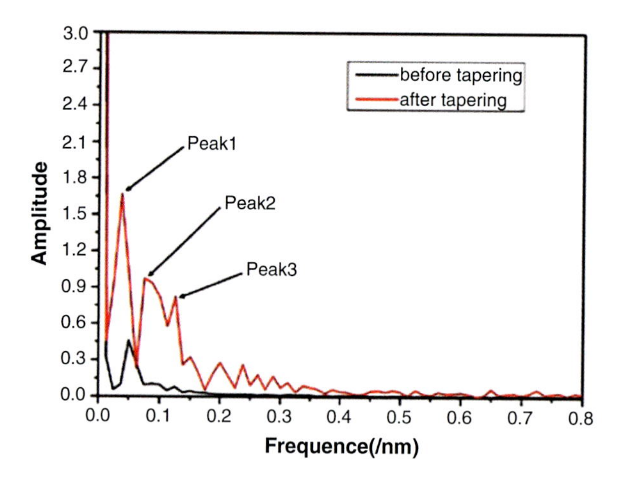

Fig. 42 Spatial frequency spectra of the interferometer before and after tapering

where ΔRI and ΔT are the RI and temperature variations, and $\Delta\lambda_1$ and $\Delta\lambda_2$ represent the wavelength shifts of the two fringe dips.

PCF-Based Michelson and Sagnac Interferometers

PCF-Based MI

Principal

Fiber-optic sensors based on Michelson interferometers (MIs) are quite similar to MZIs. The basic concept is the interference between the beams in two arms, but each beam is reflected at the end of each arm in an MI (Yuan et al. 2000; Kashyap and Nayar 1983; O'Mahoney et al. 2009; Zhao and Ansari 2001). In fact, an MI is like a half of an MZI in configuration, and the basic structure is shown in Fig. 45.

The interference mechanism of the proposed configuration is a two-mode (which are denoted as mode 1 and mode 2 with intensities of $I_{\mathrm{m}1}$ and $I_{\mathrm{m}2}$, respectively) interference due to the weak interference intensities of the modes. Thus, the intensity of the modal interference fringes can be easily expressed as:

$$I = I_{m1} + I_{m2} + 2\sqrt{I_{m1} I_{m2}} \cos(\varphi) \tag{10}$$

The optical phase difference between the two modes $\varphi = (2\pi/\lambda) \cdot \mathrm{OPD}$, where $\mathrm{OPD} = \Delta n_{\mathrm{eff}}^{m} 2L$ is the optical path difference. λ is the wavelength, L is the liquid-filled PCF section, and $\Delta n_{\mathrm{eff}}^{m}$ is the difference of effective index of the two modes as $\Delta n_{\mathrm{eff}}^{m} = n_{\mathrm{eff}}^{m1} - n_{\mathrm{eff}}^{m2}$. The n_{eff}^{m1} and n_{eff}^{m2} are effective indices of the mode 1 and mode 2 at a certain temperature (T), respectively.

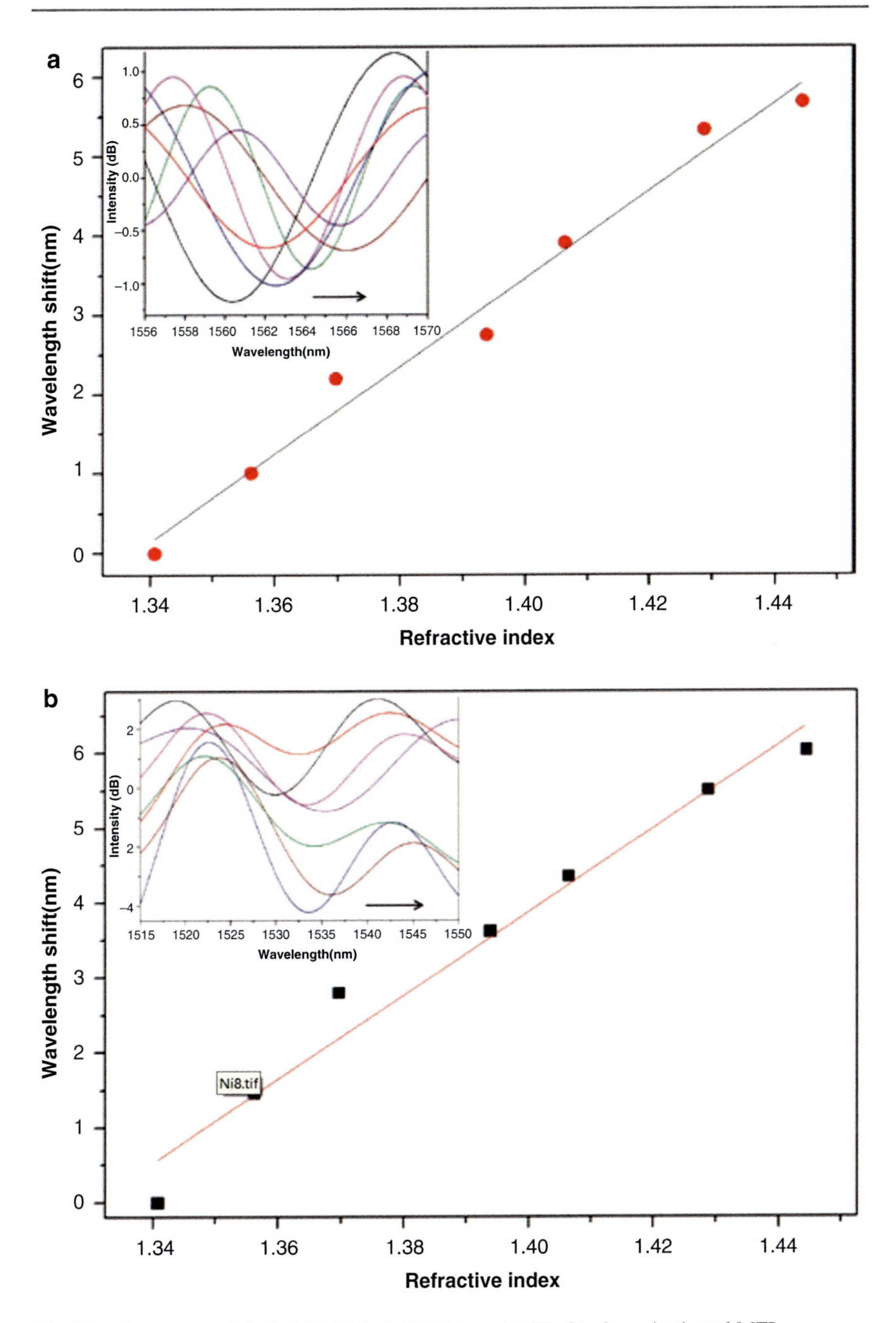

Fig. 43 RI response of the hybrid MZI: (**a**) PCF-based MZI, (**b**) air-cavity-based MZI

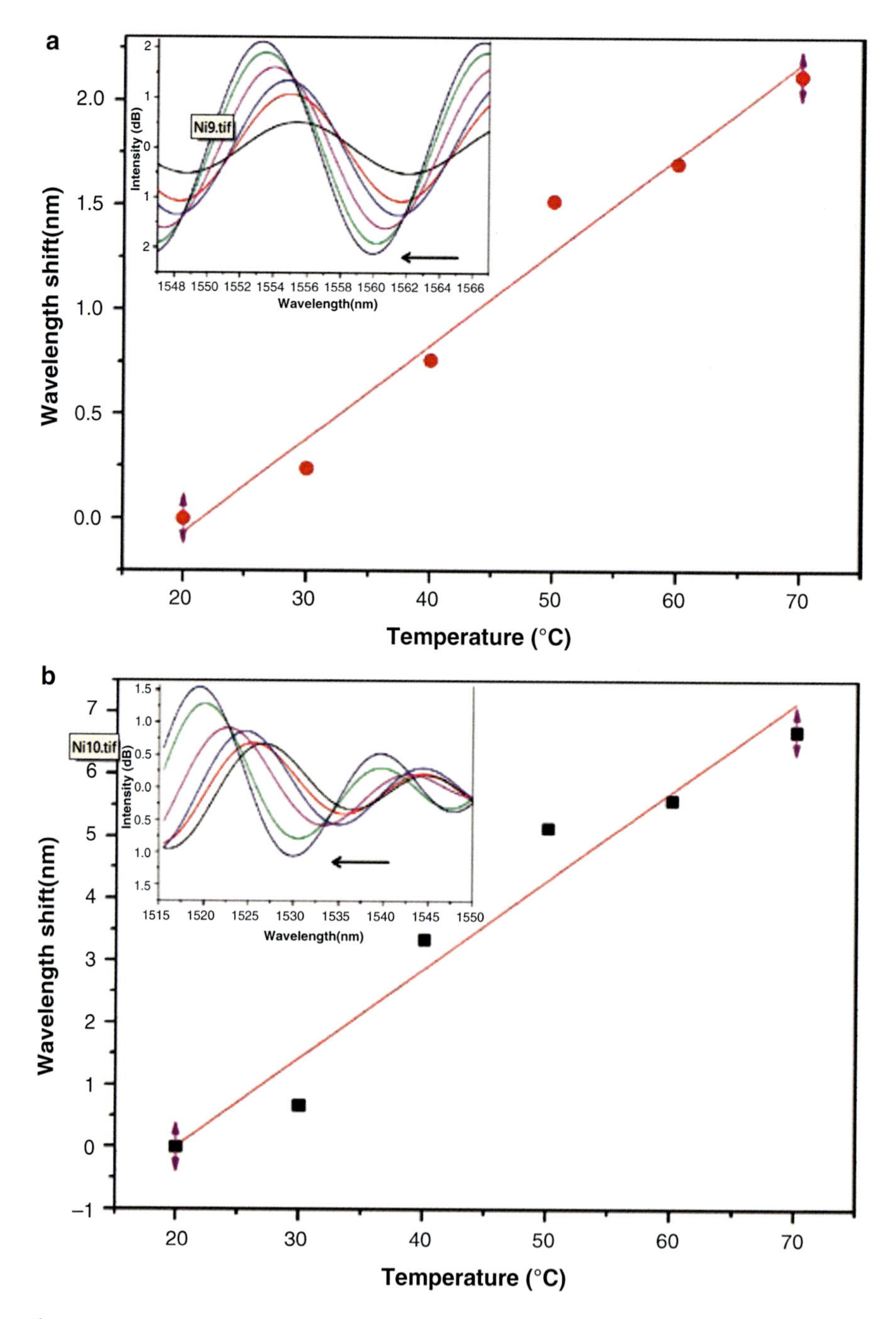

Fig. 44 Temperature response of the hybrid MZI: (**a**) PCF-based MZI, (**b**) air-cavity-based MZI

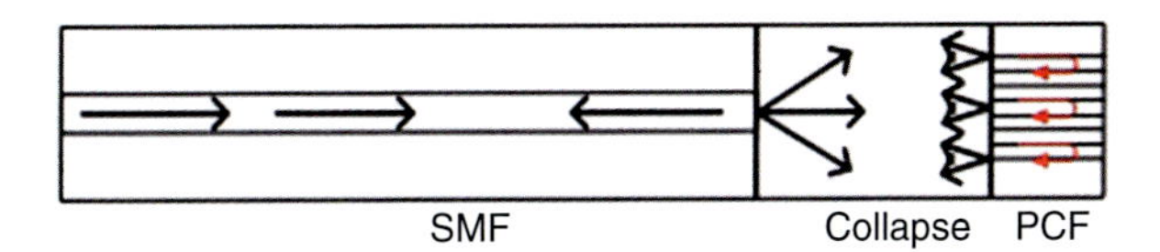

Fig. 45 Basic structure of PCF-based MI

Fabrication

Photonic crystal fiber-based Michelson interferometer is generally manufactured by fusion splicing a segment of PCF to an SMF. During the fusion splicing process, a collapsed region in the PCF is generated near the splicing point. Therefore the sensor is consisted of a single-mode fiber and a short piece of photonic crystal fiber which contains a collapsed region. In order to enhance sensor sensitivity, the micro-holes of photonic crystal can be filled with liquid. The PCF-based Michelson interferometer fiber sensor can be used as temperature sensor or strain sensor.

Applications

Temperature MI Sensor

An ultra-compact and highly sensitive liquid-filled photonic crystal fiber Michelson interferometer (LF-PCFMI) based on material dispersion engineering was proposed in Hsu et al. (2014). The sensor tip is composed of an SMF splicing with a small section of index-guiding PCF. Figure 46a presents the configuration of proposed LF-PCFMI. Figure 46b, c indicates the micrographs of the PCF tips without and with filled liquid, respectively.

Figure 47 shows the interference spectra of the liquid-filled ($n_D = 1.45$) LF-PCFMI filling length $L = 31.3$ μm of PCF in a temperature range of 25–30 °C. The experimental results show that wavelength shifts toward shorter wavelength side when ambient T increases. The agreement of the results between theoretical analysis and experimental measurements also demonstrates the effectiveness of the device. The sensitivity of the proposed sensor is ~5.4 nm/°C, which is much greater than that of the traditional long-period fiber grating (LPFG) (~0.05 nm/°C) in the air surrounding.

Strain MI Sensor

An in-line fiber quasi-Michelson interferometer (IFQMI) fabricated by splicing a section of polarization-maintaining photonic crystal fiber (PM-PCF) with a lead-in single-mode fiber (SMF) was proposed and experimentally demonstrated in Du et al. (2013). The schematic diagram of the experimental setup and the sensor is shown in Fig. 48. Some cladding modes are excited into the PM-PCF via the mismatch-core splicing interface between PM-PCF and SMF. Besides, two orthogonal polarized modes are formed due to the inherent multi-hole cladding structure of the PM-PCF.

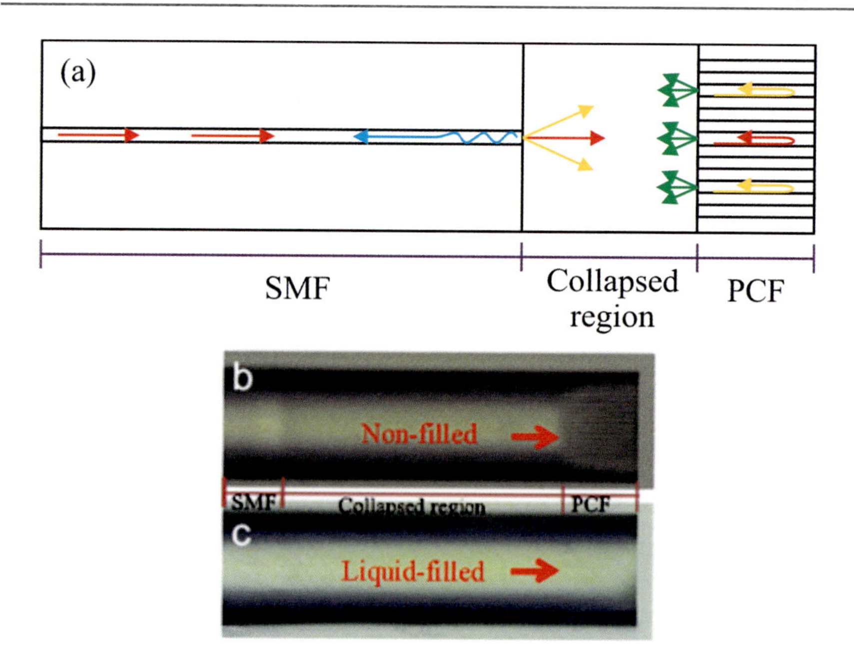

Fig. 46 (**a**) Configuration of the proposed LF-PCFMI. Photographs of the LF-PCFMI sensor tips with the (**b**) non-filled and (**c**) liquid-filled conditions, respectively

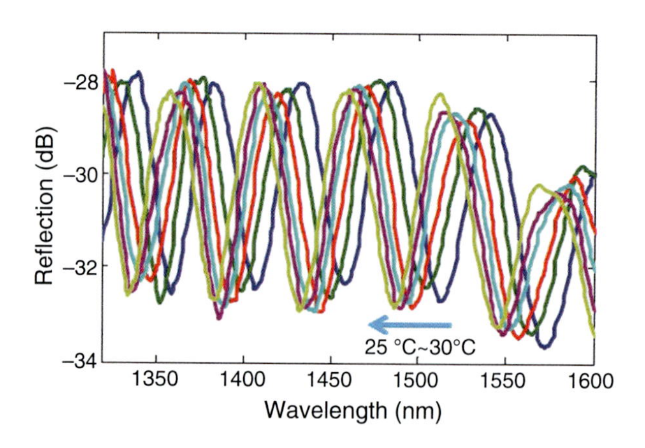

Fig. 47 Experimental interference spectra of the proposed LF-PCFMI when T varies

Figure 49 shows the strain and torsion responds of the sensor with 20-cm long PM-PCF. The strain and torsion sensitivities are -1.3 pm/$\mu\varepsilon$ and -19.17 pm/deg, respectively, as shown in Fig. 50. The proposed sensor with 10-cm-long PM-PCF exhibits a considered temperature sensitivity of 9.9 pm/°C. The IFQMI has a compact structure and small size, making it a good candidate for multiparameter measurements.

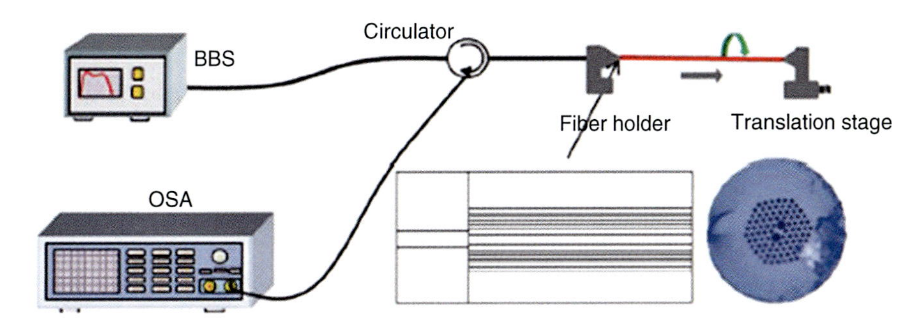

Fig. 48 Schematic diagram of the experimental setup

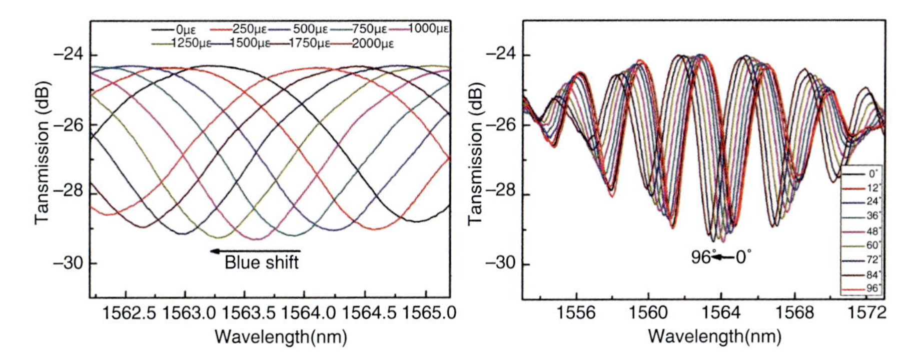

Fig. 49 Interference spectrum response to strain and torsion

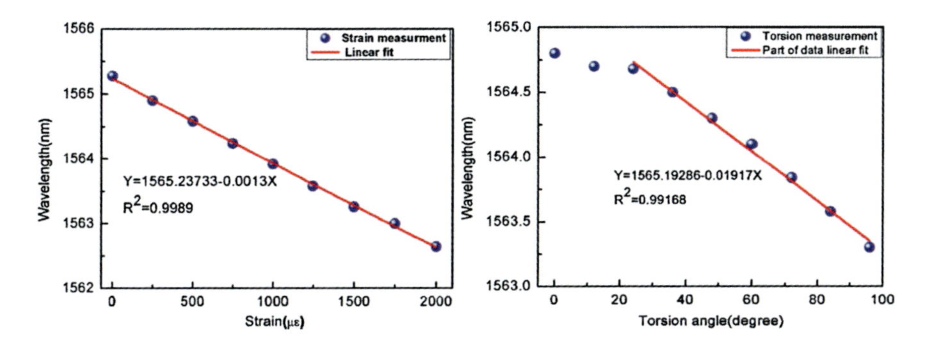

Fig. 50 Wavelength shift of the dip near 1565.28 nm versus the strain and torsion

PCF-Based SI

Principal

Figure 51 illustrates the schematic of the PCF sensor-based Sagnac interferometer (SI). The sensor consists of an optical fiber loop, along which two beams are propagating in counter directions with different polarization states. The input light

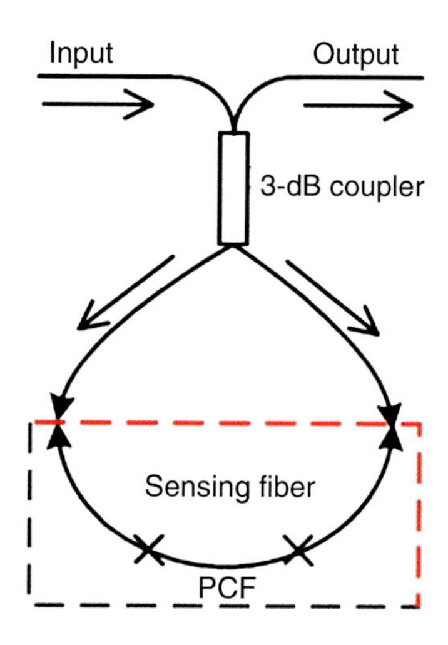

Fig. 51 Schematic of the PCF sensor based on SI

is split into two directions by a 3-dB fiber coupler, and the two counter-propagating beams are combined again at the same coupler.

Fabrication

Photonic crystal fiber-based Sagnac interferometer is generally manufactured by fusion splicing a segment of polarization-maintaining PCF into an SMF Sagnac ring. The sensor consists of a section of PM-PCF, two sections of SMF, and an SMF coupler. Filling liquid into the polarization-maintaining photonic crystal can enhance the sensor sensitivity (Cui et al. 2012). The PCF-based Sagnac interferometer fiber sensor can be used as pressure sensor or temperature sensor.

Applications

Pressure SI Sensor

A novel fiber SI pressure sensor realized by using a PM-PCF as the sensing element had been proposed and demonstrated in Fu et al. (2008). Figure 52 shows the experimental setup of the proposed pressure sensor with the PM-PCF-based SI. The sensor consisted of a section of PM-PCF, two sections of SMF, and an SMF coupler. The PM-PCF was slicing between the two sections of SMF, and the inset image in Fig. 52 shows the scanning electronic micrograph of the MF-PCF.

Experimental results and simplified theoretical analysis of the pressure sensor have been presented. The sensitivity of the pressure sensor is 3.42 nm/MPa. The proposed pressure sensor exhibits the advantages of high sensitivity, compact size, low-temperature sensitivity, and potentially low cost.

As shown in Fig. 53a, the wavelength shift changed 1.04 nm with the increase of the pressure by 0.3 MPa. Figure 53b shows the wavelength shift of the transmission minimum at 1551.86 nm against pressure with variation up to 0.3 MPa based on one atmospheric pressure, and the sensitivity of the pressure sensor is 3.42 nm/MPa.

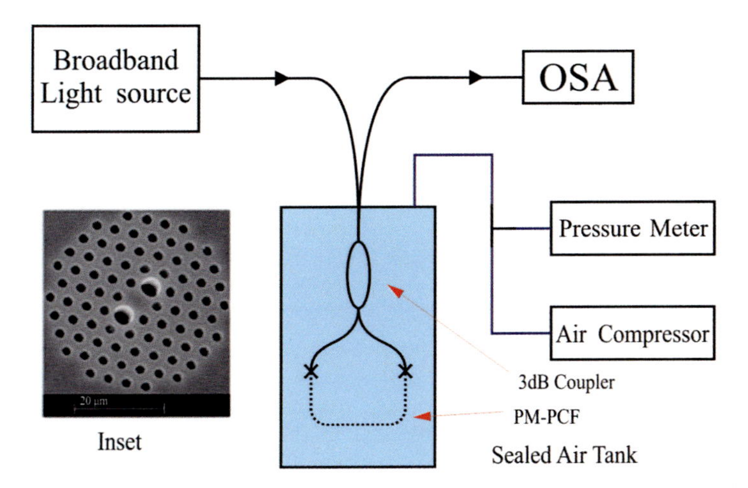

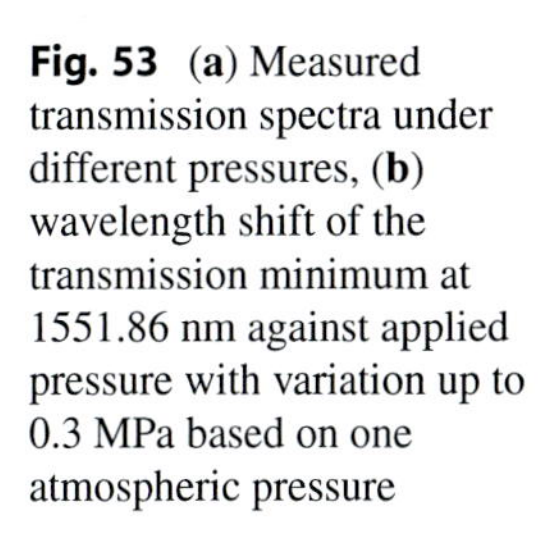

Fig. 52 Schematic diagram of the proposed pressure sensor constructed with PM-PCF-based SI

Fig. 53 (**a**) Measured transmission spectra under different pressures, (**b**) wavelength shift of the transmission minimum at 1551.86 nm against applied pressure with variation up to 0.3 MPa based on one atmospheric pressure

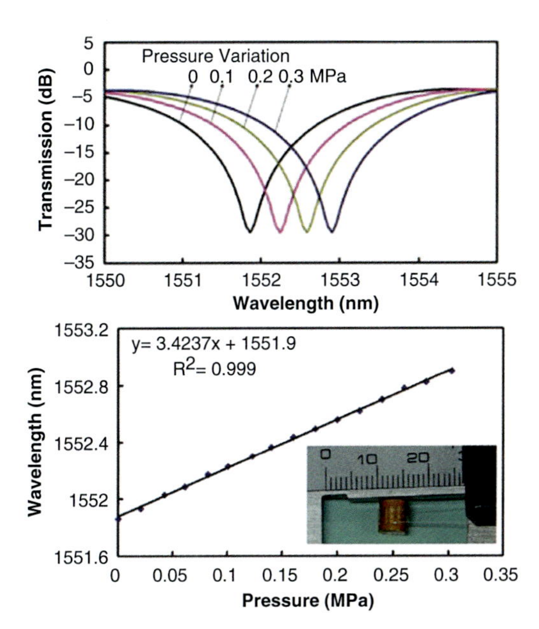

Temperature SI Sensor

An SI-based temperature sensor constructed by a selectively filled polarization-maintaining photonic crystal fiber (PM-PCF) was presented in Cui et al. (2012). Figure 54a shows the schematic diagram of the SI-based temperature sensor. L_1 is the infiltration length, and L is the total length of PM-PCF inside the fiber loop. Figure 54b shows the SEM image of the cross section of the PM-PCF used.

The proposed temperature sensor was fabricated by first blocking the small holes, then immersing the PM-PCF tip in water, and, after a while, splicing it with the

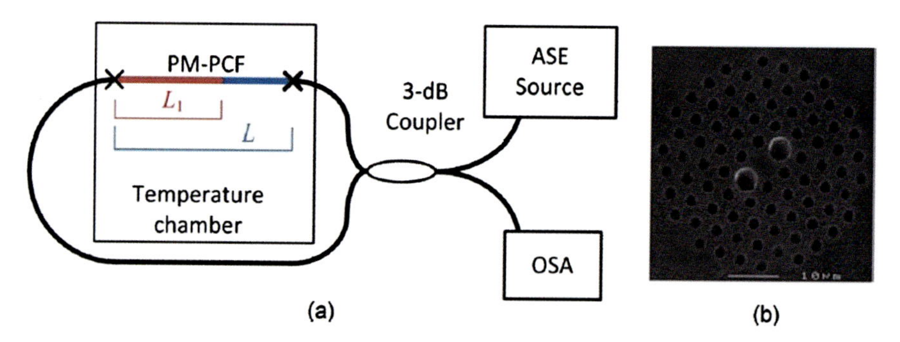

Fig. 54 (**a**) Schematic diagram of the SI-based temperature sensor, (**b**) SEM image of the cross section of the PM-PCF

3-dB coupler. The performance of the proposed sensor with a piece of selectively filled PM-PCF of 11.7 cm with the whole length infiltrated was investigated. The transmission spectra under different temperatures are shown in Fig. 55a with the increase in the temperature from 25 to 42 °C. The dip wavelength as a function of temperature is shown in Fig. 55b; the sensitivity of the proposed sensor is 2.58 nm/°C. The dip wavelengths of increasing and decreasing temperatures almost overlap with each other, indicating high stability and repeatability of the proposed sensor.

The sensitivity dependence on the infiltration length ratio was also investigated. Another sample with shorter infiltration length and longer total length was examined using the same configuration. Figure 56a shows the transmission spectra of the sensor of 44-cm total length infiltrated with water at 32 °C and 80 °C. The achieved sensitivity is ~0.15 nm/°C as shown in Fig. 56b, which agreed with the theoretical analysis that a higher infiltration length ratio provides a higher sensitivity.

Novel Interferometric Architectures in PCF and Their Applications

All-Photonic Crystal Fiber Interferometer

A highly sensitive tilt angle sensor based on an all-photonic crystal fiber interferometer (All-PCFI) was proposed and demonstrated in Zhao et al. (2016). Figure 57a shows the schematic diagram of the proposed reflected All-PCFI formed by a PCF with two collapse regions forming an MZI, as shown in Fig. 57b, c.

The core mode and cladding modes arrived to the second collapse region and silver film, and then the reflected cladding modes would recouple back to the core of the PCF and interfere with the reflected core mode light in the first collapse

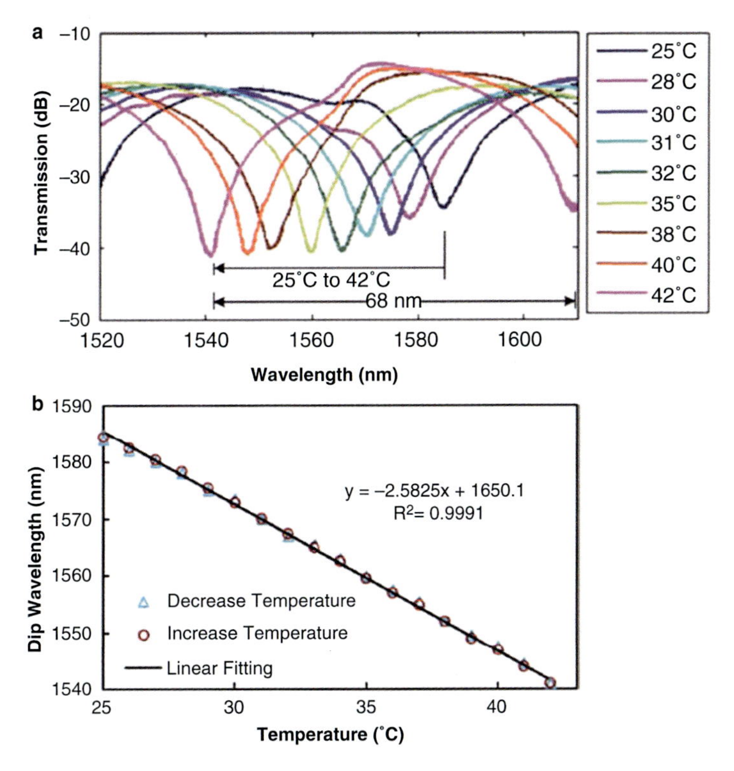

Fig. 55 (**a**) Transmission spectra of the sensor under temperatures, (**b**) dip wavelengths of the spectrum under different temperatures

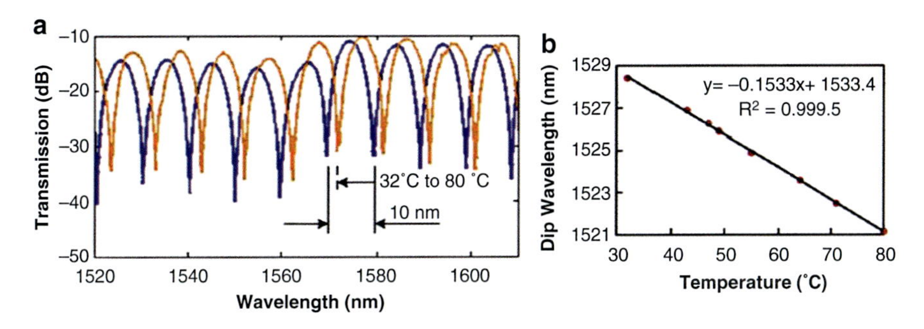

Fig. 56 (**a**) Transmission spectra of the sensor with 44-cm PM-PCF under different temperatures, (**b**) dip wavelength against temperature curve giving the sensitivity

region. The interference spectrum could be analyzed by using a simple two-mode interference model:

$$I = I_{\text{core}} + I_{\text{cladding}} + 2\sqrt{I_{\text{core}} I_{\text{cladding}}} \cos\left(\frac{4\pi \Delta n_{\text{eff}} \cdot L}{\lambda}\right) \tag{11}$$

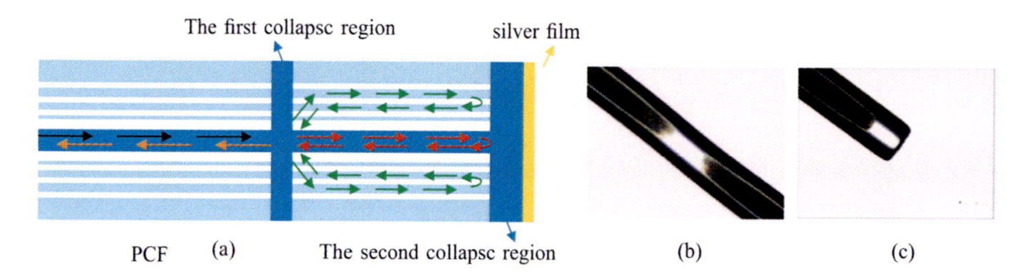

Fig. 57 (**a**) Schematic diagram of All-PCFI (**b**) the first collapse region of the PCF under the microscope (**c**) the second collapse region of the PCF under the microscope

where I_{core} and $I_{cladding}$ are light intensities of the core and cladding modes, and Δn_{neff} is the effective index difference, $\Delta n_{eff} = n_{core}^{neff} - n_{cladding}^{neff}$. Therefore, when the Δn_{neff} is changed, the dip wavelength of the interference spectrum would change.

Figure 58a is the device of the tilt angle measurement experiment formed by cantilever beams and iron ball designed to transform the tilt angle to the strain. The diagram of experiment system is shown in Fig. 58b.

The sensor was pasted in the cantilever beam to measure the tilt angle adjusted from 0 to 90°. Figure 59a shows that the spectrum is red shift with the increase of the tilt angle. The wavelength shift of the dip at 1557 nm is 3.27 nm, while the tilt angle is increased from 0 to 90°. The characteristic wavelength value is plotted and fitted as shown in Fig. 59b. In the measurement range of 0–45°, the linear measurement sensitivity is 55.67 pm/°.

SMF Cascaded Tapers with a Hollow-Core PCF-Based Microcavity for Curvature Sensing

A highly sensitive curvature sensor based on cascaded SMF tapers with a microcavity was proposed in Dass et al. (2016). The schematic showing the forward and reverse paths of light traveling in an MZI and microcavity is shown in Fig. 60.

The proposed sensor setup is a combination of two discrete structures, the in-line MZI and the microcavity. The in-line MZI consists of two dissimilar tapers fabricated using the flame and brush technique separated by a distance. The microcavity was created by splicing a small piece of hollow-core photonic crystal fiber (HCPCF) with one end of an SMF (Dash and Jha 2015).

Figure 61a shows the wavelength shifts of the sensors with a microcavity for different taper-2 diameters of 32 μm, 27 μm, and 18 μm. The dip wavelength shifts slightly toward shorter wavelength, and the wavelength sensitivities of the three MZIs are –0.79, –0.58, and –0.17 nm/m^{-1}, respectively, with the curvature range from 0 to 1 m^{-1}. The wavelength sensitivity is decreased with the decrease of the second taper diameter. It could be observed that the reflected intensity is decreased

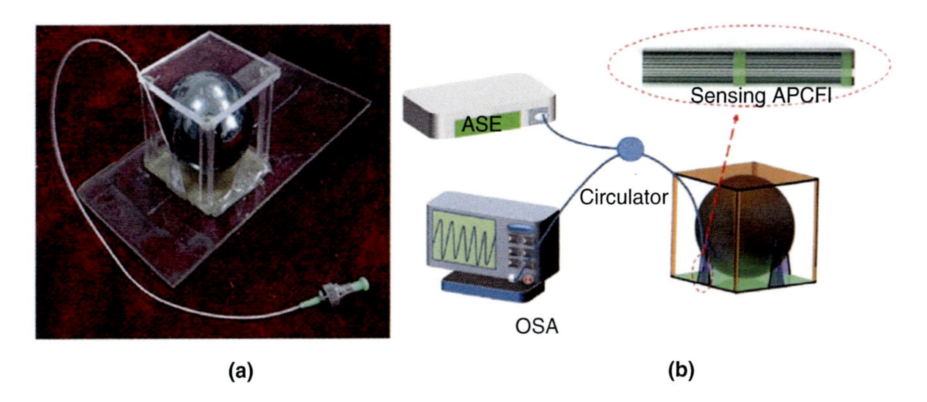

Fig. 58 (**a**) The tilt angle measuring device, (**b**) the diagram of experimental system

Fig. 59 (**a**) Spectra under different tilt angle, (**b**) fitting curve between wavelengths and tilt angle

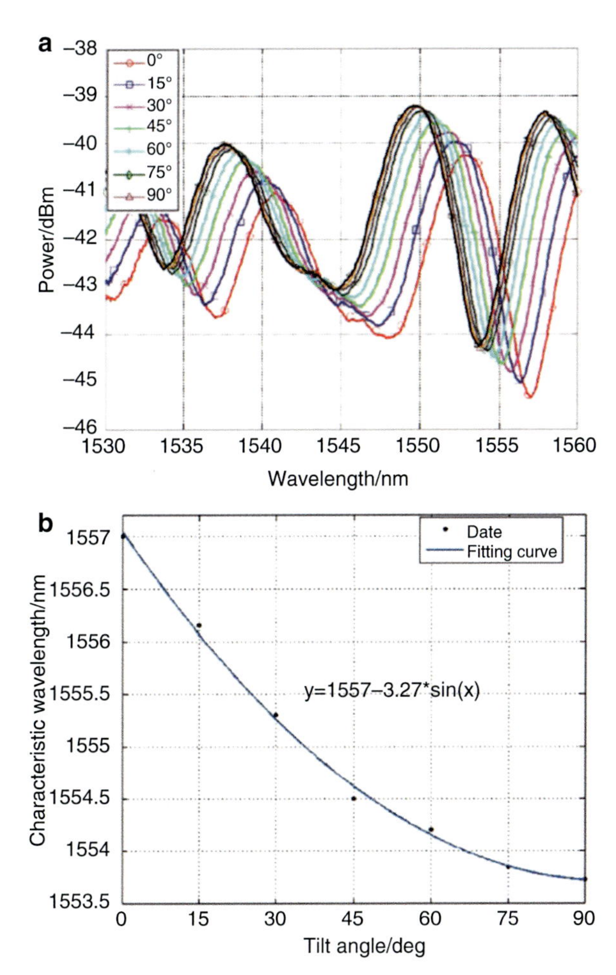

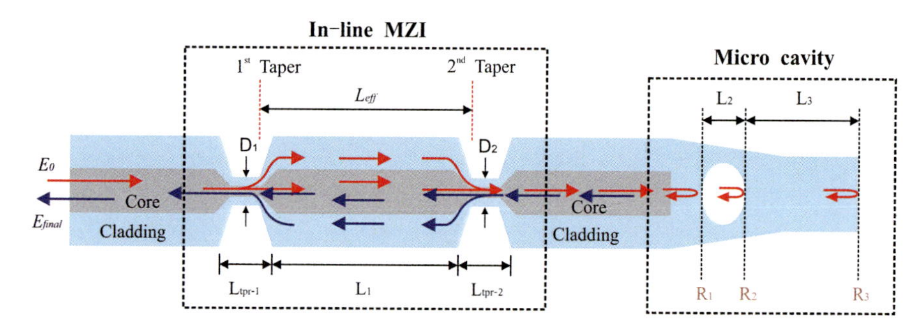

Fig. 60 Schematic showing the forward and reverse paths of light traveling in an MZI + microcavity structure

with the increase of the curvature, as shown in Fig. 61b. The amplitude sensitivities for an MZI with taper-2 diameters of 32 μm, 27 μm, and 18 μm are −9.3, −9.5, and -10.4 dB/m^{-1}, respectively, in the curvature range 0–1 m^{-1}.

Magnetic-Fluid-Coated Photonic Crystal Fiber and FBG for Magnetic Field and Temperature Sensing

A sensor composed of a cascaded PCF and fiber Bragg grating (FBG) for simultaneous measurement of magnetic field and temperature was proposed and demonstrated in Chen et al. (2016). Figure 62 shows the schematic diagram of the proposed sensor. An FBG close to the PCF spliced between two sections of SMF was inscribed in the lead-in SMF. Both the FBG and PCF were capsuled in a capillary tube full-filled with MF, and the two ends were sealed by UV glue. The sensor was fabricated with the PCF and the FBG lengths of 15 mm and 10 mm, respectively. The distance between the FBG and the PCF was ~5 mm. The MF consisted of aqueous Fe_3O_4 nanoparticles solution with a complex RI was filled into the glass capillary tube with an inner diameter of 0.3 mm.

Figure 63a, b shows the transmission spectra of the sensor and the magnified view of the spectra around Bragg wavelength under different H. It could see that the transmission spectrum of the sensor is changed with the increase of H caused by the H-related RI and absorption coefficient of the MF, while the Bragg wavelength is remained unchanged (Chen et al. 2013b).

The temperature response of the sensor exhibits an obvious change with the increase of temperature from 30 to 61.2 °C, which is because the MF is temperature-sensitive (Miao et al. 2011) and the cladding mode of PCF is significantly influenced by the MF. Figure 64 shows value shifts of the FBG and the dip at 1567 nm under different temperature. Fitting results indicate that both them have a good linear relationship with temperature. For the PCF interferometer and FBG, the temperature sensitivities are 0.0149 dB/°C and 8.8 pm/°C, respectively.

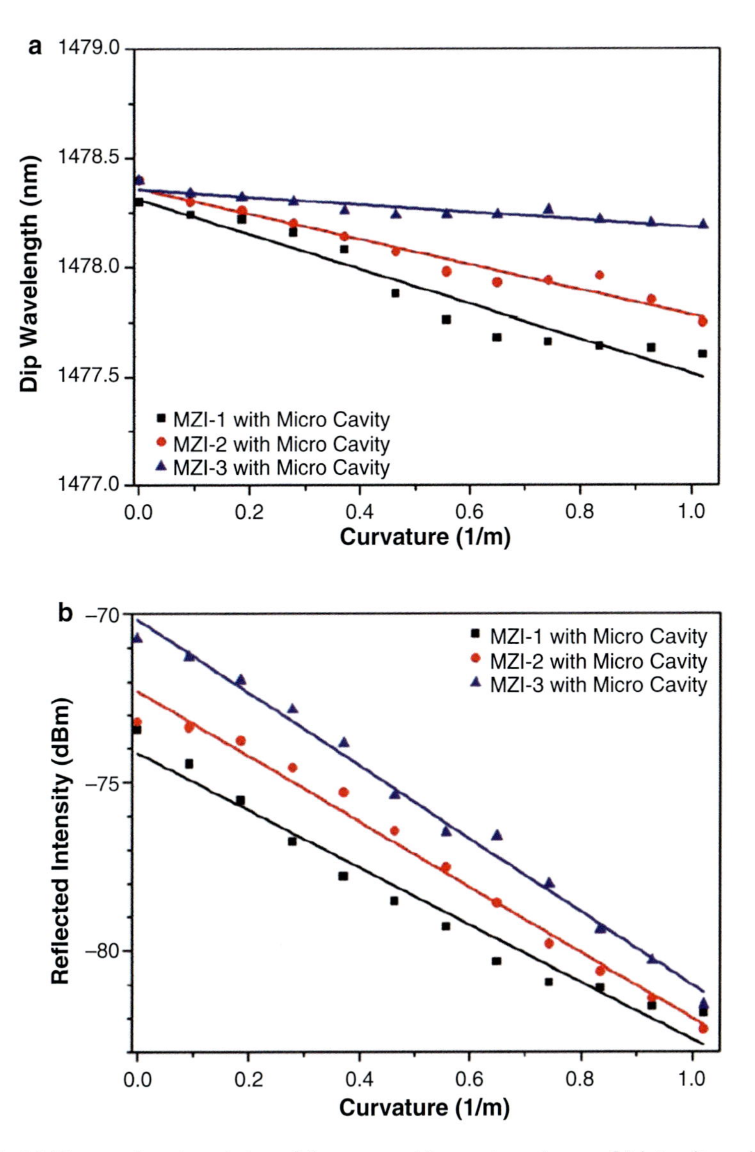

Fig. 61 (**a**) Dip wavelength variation of the sensor with curvature change, (**b**) intensity variation for all three sensor setups for different curvatures

After the above characterization, the demodulation equations can be displayed as follows:

$$\Delta T = 0.0149\Delta T_{\text{tem}} + 4.324 * [\coth(0.0098\Delta H) - 1/(0.0098\Delta H)] \tag{12}$$

$$\Delta\lambda_{\text{FBG}} = 8.8\Delta T_{\text{tem}} \tag{13}$$

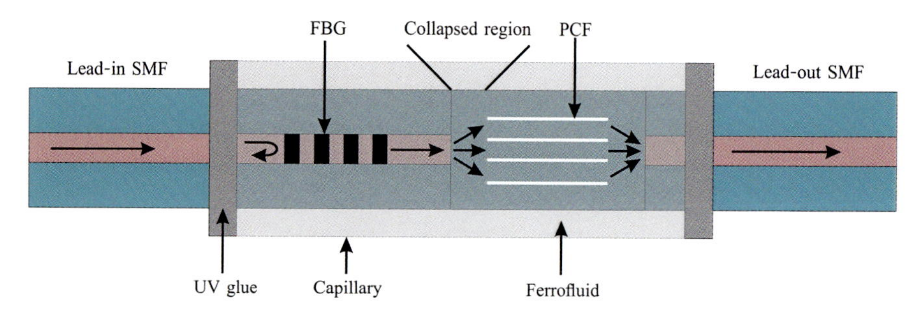

Fig. 62 Schematic diagram of the proposed sensor

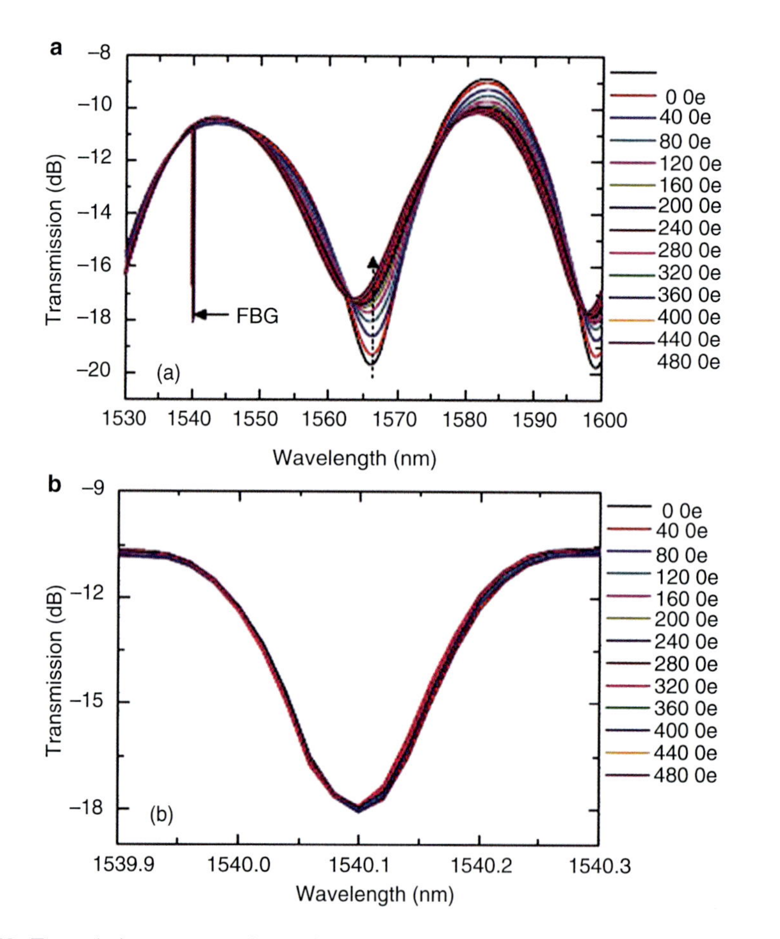

Fig. 63 Transmission spectra of (**a**) the sensor, (**b**) around the Bragg wavelength under different H

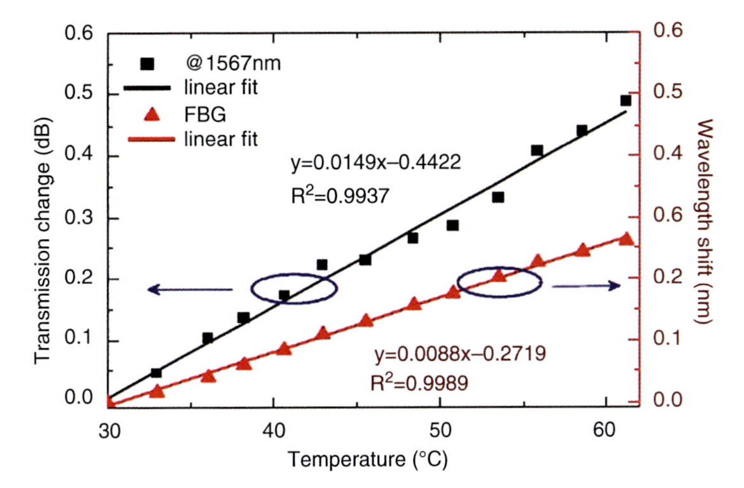

Fig. 64 Transmission changes at 1567 nm and the shifts of Bragg wavelength under different temperatures

The magnetic field change ΔH and the temperature change ΔT_{tem} could be simultaneously obtained by monitoring the value shifts, i.e., $\Delta\lambda_{FBG}$ (the FBG) and ΔT (the dip at 1567 nm).

Conclusion

Integration of interferometric architectures with photonic crystal fiber techniques is one of the new trends in the field of novel sensing technologies. The photonic crystal fiber-based interferometers exhibit excellent properties for novel sensors and offer great potential for the realization of sensing applications.

Fabry-Perot interferometer, Mach-Zehnder interferometer, Michelson interferometer, and Sagnac interferometer are traditional optical interferometry technologies. Optical fiber (especially the photonic crystal fiber) is a new technology. The combination of optical fiber with optical interferometers promotes the prosperity of sensing applications. The properties of the sensing applications are greatly enhanced by using photonic crystal fiber. The principal, fabrication, and typical applications of the photonic crystal fiber-based fiber interferometers are exhibited in this chapter. The fiber interferometer configurations contain refractive index sensor, temperature sensor, strain sensor, pressure sensor, torsion sensor, curvature sensor, magnetic sensor, and their composite structure. Due to their extraordinary performance, these sensors show great potential value in the field of sensing applications.

References

C. Chen, Y. Yu, X. Zhang, R. Yang, C. Zhu, C. Wang, Y. Xue, F. Zhu, Q. Chen, H. Sun, Opt. Lett. **38**, 17 (2013a)

Y. Chen, Q. Han, T. Liu, X. Lan, H. Xiao, Opt. Lett. **38**, 20 (2013b)

Y. Chen, Q. Han, W. Yan, Y. Yao, T. Liu, IEEE Photon. Technol. Lett. **28**, 23 (2016)

J.H. Chong, M.K. Rao, Opt. Express **11**, 12 (2003)

Y. Cui, P.P. Shum, D.J.J. Hu, G. Wang, G. Humbert, X. Dinh, IEEE Photonics J. **4**, 5 (2012)

J.N. Dash, R. Jha, IEEE Photon. Technol. Lett. **27**, 12 (2015)

S. Dass, J. N. Dash, R. Jha, J. Opt. **18**, 3 (2016)

M. Deng, C. Tang, T. Zhu, Y. Rao, IEEE Photon. Technol. Lett. **23**, 11 (2011)

W. Ding, Y. Jiang, R. Gao, Y. Liu, Rev. Sci. Instrum **86**, 5 (2015)

Y. Du, X. Qiao, Q. Rong, J. Zhang, D. Feng, R. Wang, H. Sun, M. Hu, Z. Feng, Appl. Opt. **52**, 15 (2013)

Y. Du, X. Qiao, Q. Rong, H. Yang, D. Feng, R. Wang, M. Hu, Z. Feng, IEEE Sensors J. **14**, 4 (2014)

H.Y. Fu, H.Y. Tam, L. Shao, X. Dong, P.K.A. Wai, C. Lu, S.K. Khijwania, Appl. Opt. **47**, 15 (2008)

Y. Geng, X. Li, X. Tan, Y. Deng, X. Hong, IEEE Sensors J. **14**, 1 (2014)

T.P. Hansen, J. Broeng, S.E.B. Libori, E. Knudsen, A. Bjarklev, J.R. Jensen, H. Simonsen, IEEE Photon. Technol. Lett. **13**, 6 (2001)

M. Hercher, Appl. Opt. **7**, 5 (1968)

M. Hou, Y. Wang, S. Liu, Z. Li, P. Lu, IEEE Sensors J. **16**, 16 (2016)

J. Hsu, J. Horng, C. Hsu, C. Lee, Opt. Commun. **331**, 331 (2014)

D.J.J. Hu, J.L. Lim, Y. Wang, Sensors **25**, 35 (2011)

Y. Huang, Y. Xu, A. Yariv, Appl. Phys. Lett. **85**, 22 (2004)

R. Kashyap, B. Nayar, J. Lightwave Technol. **LT-1**, 4 (1983)

D.H. Kim, J.U. Kang, Opt. Express **12**, 19 (2004)

J.C. Knight, T.A. Birks, P.S.J. Russell, D.M. Atkin, Opt. Lett. **21**, 19 (1996)

B. H. Lee, Y. H. Kim, K. S. Park, J. B. Eom, M. J. Kim, B. S. Rho, H. Y. Choi, Sensors, **12**, 3 (2012)

C.L. Lee, H.Y. Ho, J.H. Gu, T.Y. Yeh, C.H. Tseng, Opt. Lett. **40**, 4 (2015)

E. Li, G.D. Peng, X. Ding, Appl. Phys. Lett. **92**, 10 (2008)

S. Liu, N. Liu, Y. Wang, J. Guo, Z. Li, P. Lu, IEEE Photon. Technol. Lett. **24**, 19 (2012)

Y. Miao, B. Liu, K. Zhang, Y. Liu, H. Zhang, Appl. Phys. Lett. **98**, 2 (2011)

X. Ni, M. Wang, D. Guo, H. Hao, J. Zhu, IEEE Photon. Technol. Lett. **28**, 17 (2016)

K. T. O'Mahoney, R. P. O'Byrne, S. V. Sergeryev, L. Z. Zhang, I. Bennion, IEEE Sensors J. **9**, 10 (2009)

Z. Ran, S. Liu, Q. Liu, Y. Wang, H. Bao, Y. Rao, IEEE Sensors J. **15**, 7 (2015)

Y.J. Rao, Opt. Fiber Technol. **12**, 3 (2006)

Y. Rao, M. Deng, D. Duan, X. Yang, T. Zhu, G. Cheng, Opt. Express **15**, 21 (2007a)

Y. Rao, T. Zhu, X. Yang, D. Duan, Opt. Lett. **32**, 18 (2007b)

Y. Rao, M. Deng, D. Duan, T. Zhu, Sensors Actuators A: Phys. **148**, 1 (2008)

J.C. Sales, A.F.G.F. Filho, A.C. Ferreira, J.R.R. Sousa, K.M.V. Avila, J. Nonlinear Opt. Phys. Mater. **24**, 3 (2015)

Q. Shi, F.Y. Lv, Z. Wang, L. Jin, J.J. Hu, Z. Liu, G. Kai, X. Dong, IEEE Photon. Technol. Lett. **20**, 4 (2008)

J.M. Sierra-Hernandez, J.M. Estudillo-Ayala, D. Jauregui-Vazquez, R. Rojas-Laguna, Microw. Opt. Technol. Lett. **57**, 8 (2015)

X. Sun, D.J.J. Hu, IEEE Photon. Technol. Lett. **22**, 9 (2010)

Z. Tian, S.H. Yam, J. Barnes, W. Bock, P. Greig, J.M. Fraser, H. Loock, R.D. Oleschhuk, IEEE Photon. Technol. Lett. **20**, 8 (2008a)

Z. Tian, S.H. Yam, H. Loock, Opt. Lett. **33**, 10 (2008b)

J. Villatoro, V. Finazzi, V. P. Minkovich, V. Pruneri, G. Badenes, Appl. Phys. Lett. **91**, 9 (2007)

J. Villatoro, V. Finazzi, G. Coviello, V. Pruneri, Opt. Lett. **34**, 16 (2009)
Q. Wang, L. Kong, Y. Dang, F. Xia, Y. Zhang, Y. Zhao, H. Hu, J. Li, Sensors Actuators B: Chem. **225**, 3 (2016)
C. Wu, H.Y. Fu, K.K. Qureshi, B. Guan, H.Y. Tam, Opt. Lett. **36**, 3 (2011)
L.M. Xiao, W. Jin, M.S. Demokan, H.L. Ho, Y.L. Hoo, C.L. Zhao, Opt. Express **13**, 22 (2005)
A. D. Yablon (ed.), *Optical Fiber Fusion Splicing* (Springer-Verlag, Germany, 2005)
Y. Yu, X. Chen, Q. Huang, C. Du, S. Ruan, H. Wei, Appl. Phys. B **120**, 3 (2015)
L. B. Yuan, L. M. Zhou, J. S. Wu, Sensors Actuators A: Phys. **86**, 1 (2000)
Y. Zhao, F. Ansari, IEEE Photon Technol. Lett. **13**, 11 (2001)
Y. Zhao, X. Li, L. Cai, Sensors Actuators A: Phys. **244**, 6 (2016)

Part XIII
Optical Fiber Microfluidic Sensors

Optical Fiber Microfluidic Sensors Based on Opto-physical Effects

58

Chen-Lin Zhang, Chao-Yang Gong, Yuan Gong, Yun-Jiang Rao, and Gang-Ding Peng

Contents

Abstract

Microfluidics has been extensively investigated for biological and chemical applications such as biomolecule detection, drug screening, chemical synthesis, and analysis. The fusion of microfluidics and photonics has given birth to an exciting new area, optofluidics. Optofluidics could further broaden the application and extend the functionality of microfluidics. When a laser irradiates into a microfluid, opto-physical effects may happen due to the strong interaction

C.-L. Zhang · C.-Y. Gong · Y. Gong (✉) · Y.-J. Rao
Key Laboratory of Optical Fiber Sensing and Communications (Ministry of Education of China), University of Electronic Science and Technology of China, Chengdu, Sichuan, China
e-mail: c.zhang@std.uestc.edu.cn; gongcyde@std.uestc.edu.cn; ygong@uestc.edu.cn; gong.yuan@163.com; yjrao@uestc.edu.cn

G.-D. Peng
Photonics and Optical Communications, School of Electrical Engineering and Telecommunications, University of New South Wales, Sydney, NSW, Australia
e-mail: g.peng@unsw.edu.au

G.-D. Peng (ed.), *Handbook of Optical Fibers*,
https://doi.org/10.1007/978-981-10-7087-7_64

between light and liquid. Such opto-physical effects have great potential for optofluidic applications. In this chapter, the optical fiber optofluidic (**OF²**) sensors based on the opto-physical effects, including laser-induced force (optical force), and photothermal effects are introduced. One unique advantage of these sensors is the fabrication process that is very simple and cost-effective. Based on the opto-physical effects, a cleaved optical fiber is good enough to perform high-performance sensing, which is much simpler than microstructured optical fibers or micro-fabricated structures. The optical forces and photothermal effects in microfluidics are not only crucial for sensing applications but also promising for sorting cells or particles and for developing optofluidic devices.

Introduction

Optofluidics is a technology that integrates optical system with microfluidics. The advantages of optofluidics are threefold. First, liquid has unique advantages superior to the solid materials, making optofluidic system reconfigurable. Second, optical technology introduces new functionality, extends the flexibility, and enhances the performance of sensing applications (Psaltis et al. 2006; Fan and White 2011). Third, optofluidic technology can work at very small size or with an array of microfluidic channels. Overall, with the combination of optical elements and microfluidics, new portable and sensitive technologies became available for a wide range of applications such as environmental monitoring, medical diagnostics, and chemical detection.

The optical fiber-based optofluidics has several advantages (Domachuk and Eggleton 2007), which are fully reflected by the chapters in this section. The fiber-based optofluidic devices are cheap and simple thanks to the availability of mass production of optical fibers. The low loss transmission of optical fibers makes it easy for light to be coupled into the chip and interacted with the liquid sample. The diameter of optical fibers is around 125 μm that is perfect to be integrated in microfluidic chips with channels at submillimeter scale. Optical micro-/nanofibers can be as thin as sub-micrometer so that the evanescent wave can penetrate into and thus strongly interact with the microfluid. Microstructured optical fibers (MOFs), such as photonic crystal fiber (PCF), hollow-core fibers, suspended-core fibers, and air-clad fibers, can not only act as transmission medium for light but also serve as a microfluidic channels for liquid sampling (Cubillas et al. 2013). In addition to MOF, micro-holes or microchannels can be post-fabricated within common fibers by micromachining technologies and represent a new type of one-dimensional (1D) microfluidic channel which is even more compact than traditional microfluidic chips.

Previous research on optical fiber microfluidic sensors were based on integrating specific structures into microfluidics. For example, the fiber-optic interferometer is one of the frequently used technologies for optofluidic sensing (Tian et al. 2013,

2016). Refractive index (RI) sensors were developed based on fiber-optic Fabry-Perot interferometers (FFPIs) (Tian et al. 2013). The interferometer was fabricated by sandwiching a hollow capillary in between two cleaved microstructure fibers via fusion splicing. Micro-holes were fabricated by laser micromachining method and can be used to deliver the liquid samples into and out of the microfluidic channels with the assistance of pressure/vacuum pump. The RI sensing can be achieved by measuring the reflection spectrum. FFPI based on hollow-core PCF was also an excellent candidate for the effective sample delivery and optical transmission. The hollow-core PCF based RI sensors have more robust structures with short length and large sectional area of the channels for liquid sampling (Tian et al. 2016). The channels were naturally formed by the holes in the cladding of PCF, and the FFPI can be easily fabricated by splicing PCF to the single-mode fiber.

Besides utilizing the micro-interferometers, fiber grating is another option for optofluidic sensing. It can be used as a refractive index detection platform for label-free biosensing (He et al. 2011). A long-period fiber grating (LPFG) was inscribed in PCF, and further a unique microfluidic cell for biomolecular modification/binding was designed. The cell can not only couple light into and out of the PCF but also provide continuous liquid flow through the holes. The microfluidic cell provided an excellent platform to modify and monitor the binding interaction between biomolecules. Another example of fiber grating-based optofluidic sensing is for the detection of DNA (Bertucci et al. 2015). The components of sensing device are Bragg grating in MOF. The MOF contains microchannels to capture the genomic DNA. The detection of DNA can be achieved by measuring the wavelength shift of light. Streptavidin-coated gold nanoparticles were added into the microchannels for enhancing the sensing signal. This optofluidic system can achieve DNA detection without polymerase chain reaction (PCR) or other biochemical amplification. Besides, it is a sensitive label-free method and requires sample with very low volume.

Thanks to the special structure of PCF, it can achieve optofluidic sensing by itself without extra fabrication process (Liu et al. 2016; Wu et al. 2009). A solid-core PCF with holes in the cladding can be a direct coupler of core mode and an analyte mode for RI sensing. The analyte mode was formed in a hole that was close to the solid core and filled with high-index liquid. When the analyte mode became matched with the core mode, spectral notches can be observed in the transmission spectra of PCF, and the notch wavelength was dependent on the RI of analyte (Wu et al. 2009). Similarly, a low index sensor can be achieved based on the coupling between solid core and a composite core. The composite core can be formed by two parts. One was a hole in the second ring of the cladding from the core region. It was filled with matching liquid as a matching core. The other is a holy ring formed by six holes around the matching core as the microfluidic channel for analyte detection (Liu et al. 2016).

The progress in these optical fiber sensors has been reflected in detail by several excellent review articles. Note that optics can not only improve the performance of optofluidics by structural features but also open a new door for optofluidics by the laser-induced opto-physical effects. This chapter will focus on the recent advances in the fiber optofluidic technology based on opto-physical effects, especially the optical force and photothermal effect.

Laser can generate optical force by radiation pressure (Ashkin 1970). In 1970, *Ashkin* first observed that the particles could be accelerated by a single laser beam or optically trapped in counter-propagating dual beams. In 1986, *Ashkin, Steven Chu* et al. first achieved optical trapping of dielectric particles by a single-beam gradient force (Ashkin et al. 1986). This technology provided a new way to manipulating particles in the Rayleigh regime, like colloids, small aerosols, and biological molecules. Besides manipulating the micro-objects in microfluidics (Guck et al. 2000; Jess et al. 2006; Taguchi et al. 2000), laser-induced force can be used as a sensing mechanism (Gong et al. 2014; Zhang et al. 2014; Bykov et al. 2015). The optical manipulation is a non-contact and noninvasive method, so it has attracted much attention for biological applications (Bertucci et al. 2015; Guck et al. 2000).

On the other hand, photothermal effect in microfluid can be described as laser-induced temperature rise and further affects the refractive index and the volume of either liquid or the optical structure. It was first discovered by *Bell* in 1880 when he reported the photoacoustic effect (Tam and Patel 1979). In 2004, *Terazima* clarified the mechanism of photothermal effect (Terazima et al. 2004). The photothermal effect occurs when a laser beam irradiates on the medium. The irradiation induces variations of materials including molecular orientation, charge separation, charge carriers, electrostriction, radiation pressure, etc. These variations further produce thermal energy. The photothermal effect has several advantages for microfluidic applications. First, the photothermal device can be used for any kinds of materials including gas, liquid, and solid. Second, various parameters, including temperature, refractive index, flow rate, types of material, and concentration, can be detected as the photothermal performance of the sensor can change with these parameters. Third, the optical detection methods based on photothermal effect are often contactless.

In this chapter, optical fiber microfluidic sensors based on optical force and photothermal effect are introduced. Section "OF^2 Sensors Based on Laser-Induced Force" reviews the optical fiber optofluidic (OF^2) sensing by laser-induced force. First, the principle of laser-induced force is explained. The principle and applications of optical fiber tweezers, dual-beam optical manipulation, and single-beam optical manipulation are introduced. The single-beam optical manipulation can be performed with several kinds of fiber structures. Section "OF^2 Sensors Based on Photothermal Effect" introduces the OF^2 sensors based on photothermal effect. The sensing device can be performed with two structures, i.e., microring resonator and microbubble-on-tip. Based on the microring resonator, a flow rate sensor was developed. The OF^2 sensors based on microbubble-on-tip can achieve detection of temperature, flow rate, and concentration.

OF^2 Sensors Based on Laser-Induced Force

The fiber optofluidic technologies based on optical force can perform three-dimensional optical trapping and optical manipulation. They can also perform detection of flow rate in the microfluidic environment and achieve high sensitivity, thanks to the fine control of optical force at pico-Newton (**pN**) scale.

Laser-Induced Force

The optical force comes from the transfer of momentum of photons and can be divided into two components, i.e., the gradient force (F_g) and scattering force (F_s), for the trapping of a dielectric microsphere in the ray optics regime (Ashkin 1992). Optical forces are strongly dependent on the difference of refractive index on the interface between different materials. The direction of F_g is decided by the gradient of laser intensity and refractive index. That is, F_g attracts the objects with high refractive index to the higher intensity region and pushes away the objects with low refractive index. F_s is related with the laser power and often pushes the objects away from the light source.

In most cases, the refractive index of objects is higher than the surrounding medium, so the theoretical model will be discussed with an effective refractive index, defined by the ratio between refractive indices of the object and the environment, of $n > 1$. According to the size of the object to be trapped, the model will be discussed in the Rayleigh regime (Ashkin et al. 1986) and the Mie regime (Ashkin 1992), respectively.

In the Rayleigh regime with an object size of $2r < \lambda$, the scattering force can be determined by Ashkin et al. (1986)

$$F_s = \frac{n_b P_s}{c} = \frac{I_0}{c}\frac{128\pi^5 r^6}{3\lambda^4}\left(\frac{n^2-1}{n^2+1}\right)^2 n_b. \tag{1}$$

Here, P_s is the scattered power, I_0 is the incident intensity, and n and n_b are the effective index and the index of the surrounding medium, respectively. With effective refractive index of $n > 1$, F_g attracts objects to the location with higher laser intensity, often near the focus. The gradient force for a spherical Rayleigh object with a polarizability of α can be expressed as

$$F_g = -\frac{n_b}{2}\alpha\nabla E^2 = -\frac{n_b^3 r^3}{2}\left(\frac{n^2-1}{n^2+1}\right)\nabla E^2 \tag{2}$$

The optical force is generated due to the momentum transfer from photons to the microparticle. When the diameter of objects $2r \gg \lambda$, the momentum transformation

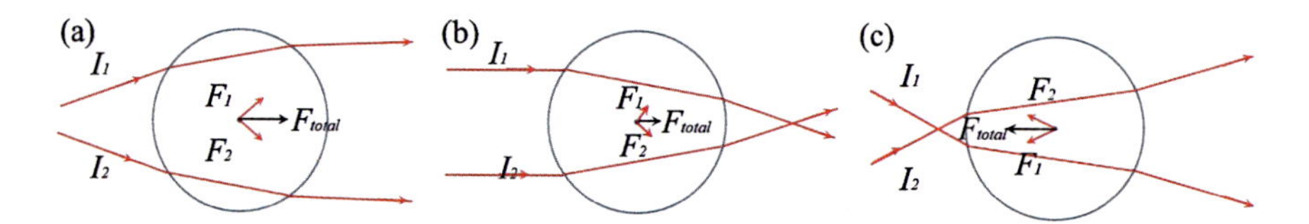

Fig. 1 Description of optical force on a dielectric particle

process can be simply described with ray optics in Mie regime (Ashkin 1992). Three types of beams, i.e., divergent beam, parallel beam, and convergent beam, are described, respectively, in Fig. 1a–c. I_1 and I_2 denote two symmetrical rays. The light-matter interaction occurs on the surface of microparticle, where the momentum transfers from photons to microparticle, inducing two relevant optical forces, F_1 and F_2. F_{total} denotes the total optical force induced by the laser beam and its direction is also given. In Fig. 1a, b, the direction of F_{total} is the same as the laser beam, resulting in a pushing force on the microparticle. F_{total} shows a reverse direction when the beam is convergent. In Fig. 1c, the microparticle will be pulled back and keep balanced near the focus of laser beam.

For microparticles with relatively larger diameter, optical force can be expressed in the Mie regime (Ashkin 1992). F_{total} can be broken into two components, F_Z and F_Y, which are determined by

$$F_s = \frac{nP}{c}\left\{1 + R\cos 2\theta - \frac{T^2\left[\cos\left(2\theta - 2\gamma\right) + R\cos 2\theta\right]}{1 + R^2 + 2R\cos 2\gamma}\right\} \tag{3}$$

$$F_g = \frac{nP}{c}\left\{R\sin 2\theta - \frac{T^2\left[\sin\left(2\theta - 2\gamma\right) + R\sin 2\theta\right]}{1 + R^2 + 2R\cos 2\gamma}\right\} \tag{4}$$

P is the power of the laser beam. θ is the angles of incidence and refraction. R and T are the Fresnel reflection and transmission coefficients. F_s is along with the optical axis and is regarded as the scattering force to push the microparticle away from the source. F_g is perpendicular to the beam and can be considered as the gradient force which can attract the microparticle close to the optical axis.

Optical Fiber Tweezers

Conventional optical tweezers employ the microscope objective with a high numerical aperture (NA) (Ashkin et al. 1986; Ashkin 1992). The bulky device make optical tweezers difficult to use and expensive. Optical fiber tweezers (OFT) offer advantages over conventional objective of low cost, flexibility, long transparent distance, and easy integration. However, it is a big challenge to enhance the trapping efficiency for OFT due to the low NA of the fiber. Three main categories, i.e., lensed

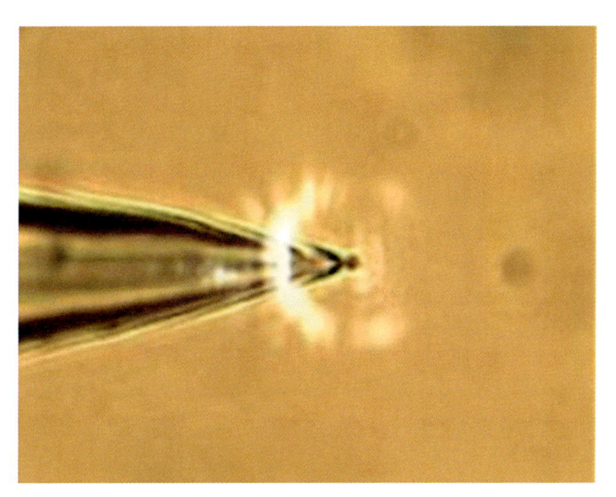

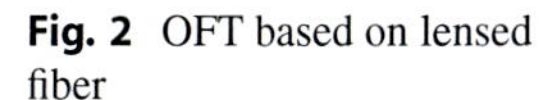

Fig. 2 OFT based on lensed fiber

fiber or fiber taper, dual-fiber, and special constructions, have been developed to resolve this problem.

Fiber taper is a straightforward way to enhance the efficiency of 3D trapping (Fig. 2). The tapered structure strongly focuses the laser beam so that the stable 3D trapping can be achieved (Liu et al. 2006). The fiber tapers can be fabricated by polishing (Kostovski et al. 2014), chemical etching (Gong et al. 2013, 2014; Chuang et al. 1998), or the heating-and-drawing methods (Liu et al. 2006). These methods are easy to fabricate and simple to operate. However, the sharp geometry corresponds to a short focus length and limits the operation range by 3D trapping the microparticle very close to the fiber end.

For resolving this problem, OFTs based on dual-fiber are proposed. The dual-fiber optical tweezers makes use of two aligned optical fibers with flat or lensed tips so that the laser beams emerging from the fibers are counter-propagating along a common optical axis (Guck et al. 2000). In this case, the trapping distance can be lengthened due to the balanced axial scattering forces (Jess et al. 2006). Since the fiber alignment is crucial for this arrangement, the two fibers are generally embedded in a substrate to facilitate the alignment, which limits the flexibility of OFTs. Moreover, due to the relatively large size of the optical fiber, it cannot trap particles at sub-wavelength scale or particles lying on the substrate. In 2000, an inclined dual-fiber optical tweezers were proposed by Taguchi et al. (2000), which can achieve levitation of a microscale object. It has better flexibility than the counter-propagating dual-fiber optical tweezers.

Accompanied with the development of fabrication technologies, OFTs with better performance based on special structures are proposed. Cristiani et al. proposed an OFTs based on a multi-core optical fiber (Liberale et al. 2007). The multi-cores were shaped with proper angles so that laser beams from these cores were reflected into a tight focus. Strong gradient force enables 3D trapping near the focus. It can trap and manipulate microparticle over a relative long distance with flexibility better than the dual-fiber optical tweezers. However, the complex manufacturing process hinders its broad applications in real world.

In 2009, Yu et al. developed an inclined dual-fiber optical tweezers, by which both manipulation and force sensing (Liu and Yu 2009) were achieved. Different from the traditional dual-fiber optical tweezers, there was an inclination angle (θ) between two fibers. With the help of the optical force applied by the two optical fibers, the object can be trapped below the intersection of the two beams. The inclined dual-fiber optical tweezers extended the distance of the equilibrium position, so it can trap larger particles of tens of micrometers without physical contact. The trapping efficiency changed linearly with the displacement in the range of -1 μm to $+1$ μm. Force sensing was demonstrated with the help of a position-sensitive photodiode, which was used for detecting the position of the trapped beads. By detecting the displacement, the external forces applied on the particle can be measured. The θ is an important parameter for this device. When $\theta \leq 45°$, the device can trap the bead in x and y axes, but cannot lift it. If $\theta \geq 50°$, it can achieve trapping in three dimensions including lifting. With larger θ, the OFT can achieve optical force strong enough in z axis for lifting. The inclined dual-fiber optical tweezers is more compact than the counter-propagation OFTs due to the less limitation on the fiber alignment.

Dual-Beam Optical Manipulation

The OFT can trap an object, but cannot control or adjust its position without moving the OFT. This weakness limits the functions and applications of OFTs. The dual-beam fiber trap (DFT) was first proposed to solve the problem. Comparing with the conventional structure based on two focused laser beams, the dual-beam optical manipulation has the advantages of easy fabrication and easy integration, which makes it suitable for interdisciplinary research. More importantly, the unfocused laser beams avoids thermal damage to the trapped object, making it suitable for biological use.

The DFT is composed by two counter-propagating laser beams with Gaussian intensity profile (Ashkin 1970). If the refractive index of the object is larger than the surrounding medium, it can be stabilized within the laser beams. One special advantage of DFT is that the trapped object can be stably and flexibly controlled between the two fiber ends. The schematic principle of DFT is shown in Fig. 3. Two counter-propagating beams from optical fibers formed a dual-beam trap. With the help of microfluid, objects can be loaded automatically and continuously, which could improve the throughput of the device. The object can be pushed and localized to different positions by controlling the laser power, P_1 and P_2.

The DFT can trap one or several objects and move them to any position on the axis between two fibers. Different numbers of yeast cells were trapped, as shown in Fig. 4, named as optical binding. For objects with ellipsoidal structure, they were often assembled along with their macro-axis. By controlling the power from each

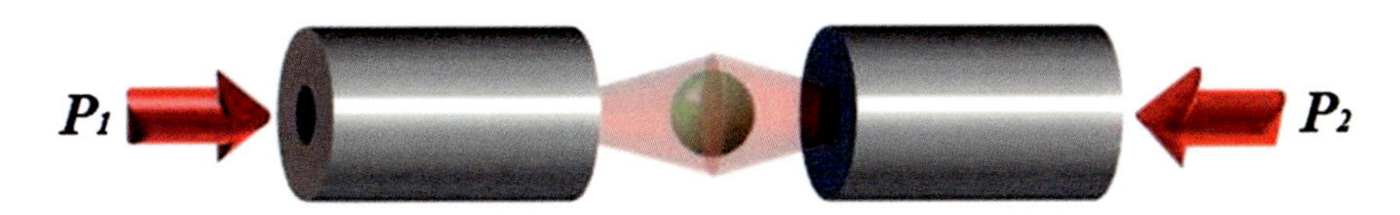

Fig. 3 Schematic principle for dual-beam optical fiber manipulation

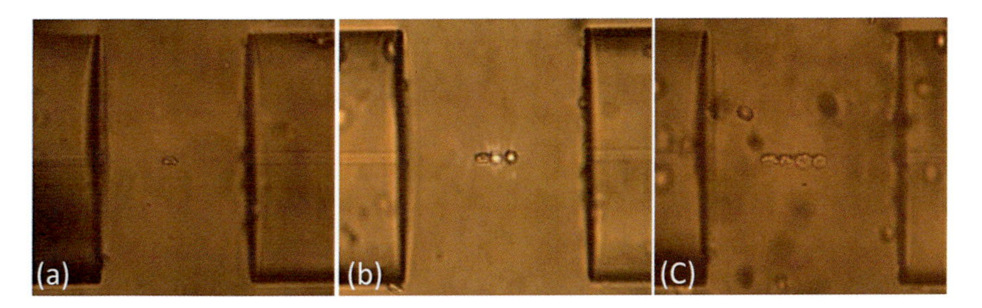

Fig. 4 Manipulation of different numbers of objects based on DFT

Fig. 5 Manipulation of a cell string to different positions based on DFT

fiber, the position of the trapped cell string can also be adjusted. The objects can be controlled to the right when the laser power from the left fiber is larger than the right (Fig. 5a). Increasing the power from the right fiber, the cell string can be gradually pushed to the left. When the power from the two fibers is equal, the objects were trapped in the middle (Fig. 5b). With increasing power from the right, the objects were pushed to the left (Fig. 5c).

In cytology, the Raman spectroscopy is very useful for cell identification by detecting the chemical composition. The function of Raman spectroscopy can be further enhanced by DFT. In 2011, scientists achieved tumor cell identification based on the combination of Raman spectroscopy and DFT (Guck et al. 2001). Two 1070 nm lasers were coupled into two single-mode fibers (SMFs) to construct a DFT and used for trapping the cell. With the help of microfluidics, the device can automatically distinguish the cell types including RBCs, leukocytes, acute myeloid leukemia cells, and breast tumor cells.

Besides optical trapping, the DFT can also deform the living cell, if the laser power is strong enough. Guck and co-workers reported a unique DFT, named optical stretcher (Dochow et al. 2011). They trapped single cell between two opposite, unfocused beams from SMF. Although the total force on the object is zero, the forces on the cytomembrane can be as high as several hundreds of pN. The trapped living cell can be stretched along the optical axis. Besides, due to the unfocused laser beams, the optical stretcher with a laser power as high as 1.7 W did not damage the living cell. It can be used to measure the cell elasticity, further to distinguish cancer cells from normal ones. Based on the strong stretching force, it can also be used for detecting the cytoskeleton quantitatively. As it is easy to be integrated with microfluidics, the optical stretcher has great potential for high throughput detection of the cell elasticity.

The DFT can also be used for temperature sensing (Zhang et al. 2014). They used etched fibers to form a more stable optical trap. The microparticle was trapped in a sealed cell that was fabricated by packaging a dual-fiber optical manipulation system in a quartz capillary. The fiber ends were chemically etched to a concave shape. For sensing application, they adjusted the position of the microparticle by tuning the power of 980 nm laser, and the shift of interference spectra near 1550 nm was used as sensing signals.

Hollow capillaries have the potential for guiding both light and liquid in it. However, high transmission losses limit the guiding distance. Hollow-core photonic crystal fiber (HC-PCF) is an excellent candidate for both optical waveguiding and liquid sampling. Russell and his co-workers introduced a tool for single-cell manipulation based on the liquid-filled HC-PCF (Unterkofler et al. 2013). First the cell was trapped in front of the fiber core with a focused laser beam vertically to the fiber. Then, a horizontal beam irradiated on the cell and pushed it into the fiber core. The cell was transferred along the HC-PCF in the core with a long distance of tens of centimeters. Besides, cell deformation was also observed during the optical manipulation due to shear force, making it promising for biomechanical detection of living cells. DFT based on HC-PCF is also an effective tool for sensing. Detection of multiple parameters with high spatial resolution was realized based on HC-PCF DFT (Bykov et al. 2015). A microparticle was trapped in the hollow core of HC-PCF, and the position was adjusted by the counter-propagating laser. Parameters of the environment around the fiber, including transverse mechanical vibration, temperature, and electric or magnetic field, can be detected by monitoring the reflected or back-scattered light.

Single-Beam Optical Manipulation

The single-beam optical manipulation (SOM) could improve the flexibility and integration, compared with the dual-beam optical manipulation. The SOM employs a single laser beam from an optical fiber, so it can be easier to move to the object to be manipulated, compact for integration with microfluidics, and has wide manipulation area. It might achieve internal detection of tiny object like cells by inserting a tapered fiber into it.

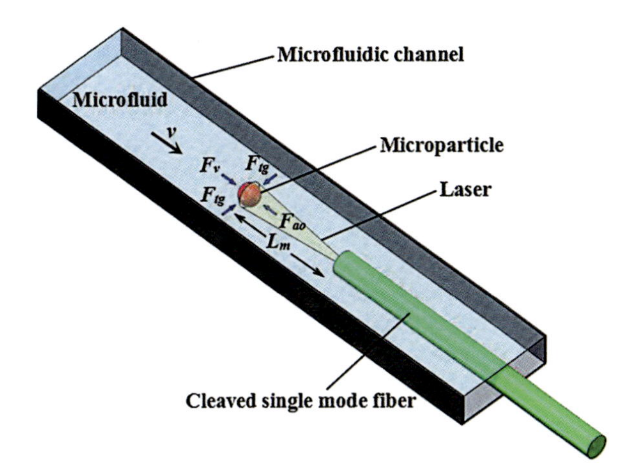

Fig. 6 The principle of the SOM based on a flat-facet SMF

Principles

The principle of the SOM is schematically shown in Fig. 6. The optical manipulation along with the optical axis is based on the force balance on the microparticle between the axial optical force, F_{ao}, and the microfluidic flow force, F_v. As the flow rate v increases, the flow force increases according to the Stokes law:

$$F_v = k_1 v \tag{5}$$

Here $k_1 = 6\pi\eta a$, with η the coefficient of viscosity of water and a the radius of the microparticle. Therefore, the F_v tends to push the microparticle toward the fiber facet. On the other hand, the F_{ao} increases as the manipulation length, L_m, decreases, which forms a counterforce to push the microparticle away from the fiber end. The manipulation length, L_m, is defined by the distance from the center of the microparticle to the fiber tip. The total force on the microparticle can be balanced at certain L_m, which uniquely corresponds to the flow rate. The force balance can be expressed as

$$F_v(v) = F_{ao}\,(L_m) \tag{6}$$

Thus the L_m, can be determined by both the optical laser power and the flow rate.

The gradient force points to the direction of the light intensity enhancement and can be divided into two components, the transverse gradient force, F_{tg}, and the axial gradient force, F_{ag}. F_{tg} compensates the difference between the gravitational force and the buoyancy force and confines the microparticle in the optical axis. In most cases, the density difference between the microparticle and the liquid is small so that the weak gradient force is more than enough to trap the microparticle on the axis. The F_{ao} consists of the scattering force, F_{as}, and the axial gradient force, F_{ag}. The F_{ag} is very small and is negligible compared with F_{as} and F_v and can be neglected in the axial direction. Therefore, the force microbalance can be rewritten approximately as

$$F_v(v) = F_{as}(L_m) \tag{7}$$

The scattering force is proportional to the laser intensity (Harada and Asakura 1996), i.e.,

$$F_{as} = k_2 I \tag{8}$$

Here k_2 is a constant related to the particle radius, the refractive index of the particle, and the environmental medium and the laser wavelength. I is the intensity distribution of the laser beam. The output from the fiber can be considered as a Gaussian beam, and the output intensity is determined by

$$I(r,z) = I_0\left(\frac{\omega_0}{\omega(z)}\right)^2 e^{-\frac{2r^2}{\omega^2(z)}} \tag{9}$$

Here I_0 is a constant intensity, and $\omega(z) = \omega_0\sqrt{1+(z/z_R)^2}$ is the beam radius at the axial position of z, with $z_R = \pi\omega_0^2/n_0\lambda$. ω_0 is the waist radius of the beam, and n_0 is the refractive index of the solution. Considering the microparticle on the z axis, i.e., $r = 0$, the intensity can be expressed as a function of the manipulation length, L_m:

$$I(0, L_m) \propto 1/\left[1+(L_m/z_R)^2\right] \tag{10}$$

Considering the Eqs. 5, 7, 8, and 10, the relationship between the flow rate and the manipulation length can be described as

$$\nu \propto 1/\left[1+(L_m/z_R)^2\right] \tag{11}$$

z_R is about 20 μm in the experiment. With a relatively large manipulation length, i.e., $(L_m/z_R)^2 \gg 1$, the relationship between the flow rate and the manipulation length can be simplified to be $\nu \propto (L_m/z_R)^{-2}$. In the log-log scale, the $L_m - \nu$ curve is approximately linear and the slope is

$$k = \log L_\mathrm{m}/\log v = -0.5 \tag{12}$$

Graded-Index Fiber Taper

Optical tweezers are mainly developed as a tool for 3D trapping the object. However, it suffers from the limited manipulation length (L_m), inherently owing to the small core diameter of the SMF and small radius of curvature of the fiber tip. The manipulation distance can be defined by the fiber end surface and the stable trapping point of the microparticle. It makes this kind of optical tweezers inflexible to be used, just like a short-working-distance objective in an optical microscope. It might even damage the biological samples due to a short L_m of the OFTs. Therefore, OFTs with a long L_m is meaningful for realizing a truly non-contact optical manipulation.

In 2014, Gong et al. proposed a single-fiber optical manipulation with a long adjustable distance of more than 40 μm by introducing the periodic focusing effect of graded-index fiber (GIF) into OFTs (Gong et al. 2014). As shown in Fig. 7, a fiber-coupled laser of 980 nm is employed as the light source for the optical manipulation. One percent of the light is used to monitor the laser power. The light transmitted from the 99% port of the coupler is coupled into the graded-index fiber (GIF), which is with a core diameter of 62.5 μm and an NA of 0.27. The single-mode fiber (SMF) and GIF are aligned by a silica capillary. An air cavity, introduced between the lead-in SMF and the GIF, can be precisely adjusted by a three-dimensional (3D) translation stage fixed with the lead-in SMF. The light distribution in the GIF can be optimized by adjusting the air cavity length. The light beam emerge from the GIF taper, which was fabricated by the process of the so-called two-step etching technique (Chuang et al. 1998). The microscopic image of the fabricated GIF taper is shown in the insert of Fig. 7, whose cone angle was measured to be about $\alpha = 58°$.

The performance of optical manipulation is improved mainly by two factors. One is by introducing the focusing effect of GIF taper. The other is by introducing an air gap with tunable cavity length (L_{air}) between the lead-in SMF and the GIF taper. The GIF has an index profile with quasi-parabolic function in the radial direction

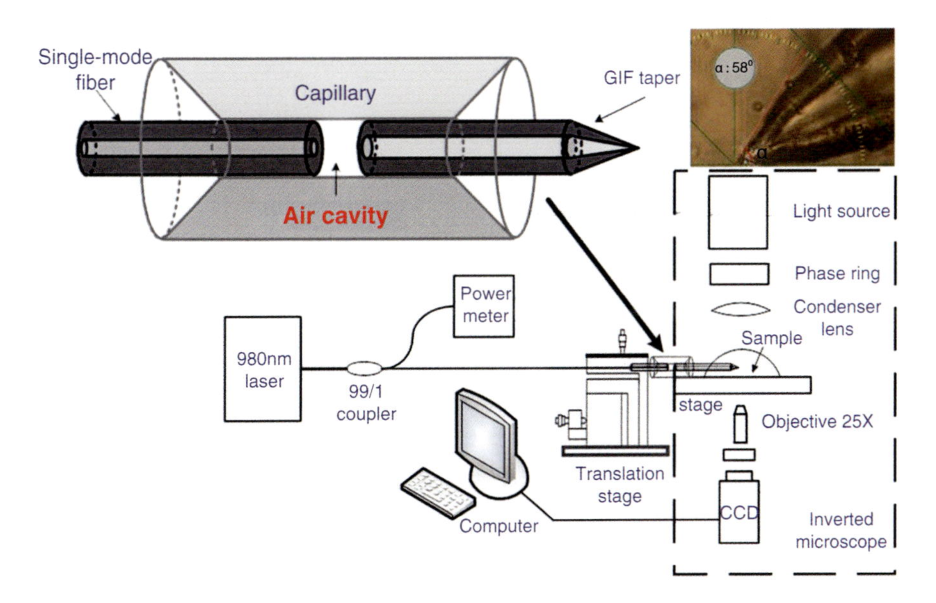

Fig. 7 Experimental setup for single-fiber optical manipulation by GIF taper

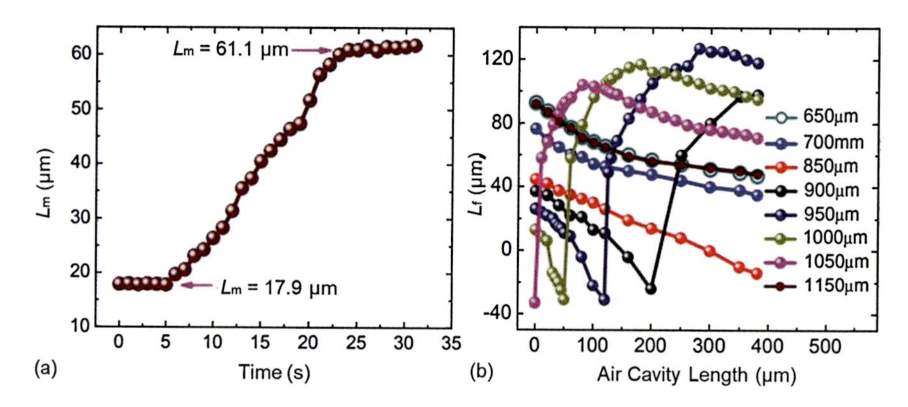

Fig. 8 (**a**) L_m versus time when increasing the laser power from 15.0 mW to 45.0 mW and (**b**) L_f versus the air cavity length at different GIF length

of the fiber core, which leads to periodic convergence and divergence of the light beam propagating in the GIF. So it has better focusing effect than SMF. Besides, it is fabricated with a large cone angle of 58° which is very helpful to obtain a large gradient force and to form a stable optical trap (Mohanty et al. 2008). With the help of the air cavity, stable optical trapping of yeast cells is observed in aqueous solution with a constant flow rate. It is also demonstrated that the manipulation distance can also be extended by changing the pump laser power.

Figure 8 shows the optical manipulation of a microparticle that was influenced by the laser power (Fig. 8a) and the air cavity length (Fig. 8b). The L_m versus time when increasing the power from 15.0 mW to 45.0 mW was shown in Fig. 8a. The cavity length is 100 μm, and the flow rate is 12 μm/s corresponding to a constant viscous force of 0.54 pN. At $t = 5$s, L_m was increased gradually from 17.9 μm once the laser power was increased. At $t = 20$s, while L_m was 47.4 μm and tended to approach a stable state, another yeast cell was trapped together with the first one and L_m further increased to 61.5 μm and then remained stable. The air cavity length can also manipulate the microparticle, and the results were reflected by L_f defined as the distance between the fiber tip and the focus point with the maximum intensity. With a GIF length (L_{GIF}) between 650 μm and 850 μm, one can see that the L_f decreases with L_{air}. With L_{GIF} between 900 μm and 1050 μm, the evolution of the L_f did not change monotonously with L_{air}, as L_f changes from a negative value to the positive. It is because in this range, $L_{eff} = L_{GIF} + L_{air}$ changes from smaller than to larger than two times of the GIF pitch. With the GIF length changing from 650 μm toward 1050 μm, the $L_f - L_{air}$ curve gradually shifts to the left, with a smaller L_f. For $L_{GIF} = 1150$ μm, the $L_f - L_{air}$ curve coincides with that of $L_{GIF} = 650$ μm. This indicates that increasing the air gap is equivalent to increase the GIF length. However, adjusting the L_{air} is much easier than precisely controlling the GIF length.

The manipulation length, L_m can be controlled by changing the balance between the optical force and the microfluidic flow force, as shown in Fig. 9 (Gong et al. 2015a). When fixing the laser power and the microcavity length during this

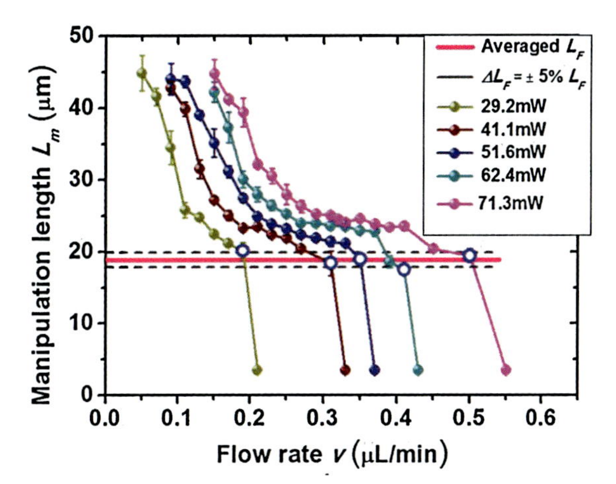

Fig. 9 Manipulation length versus flow rate at different laser powers

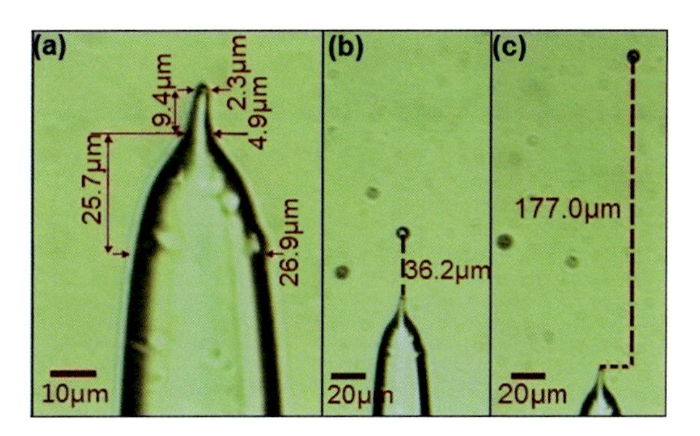

Fig. 10 Microscopic images of (**a**) a sharp GIF taper and optofluidic tunable manipulation with L_m of (**b**) 36.2 μm and (**c**) 177.0 μm

experiment, the light distribution, thus the optical force distribution, for trapping certain object were also fixed. L_m of the microparticle was tuned by the flow rate for breaking and reconstructing the force balance. This mechanism demonstrated the controllable optical manipulation and can also be considered as a prototype for flow rate sensing. Note that an abruptly fall of manipulation length from ~20 μm (hollow circles in Fig. 9) to 3.5 μm (in contact with the fiber tip) happened at all five laser powers. The last balanced position was independent with laser power and flow rate. Thus it is reasonable to consider it as the focus of light emitting from the GIF taper and is determined only by the light distribution.

By using a GIF taper with a special shape, the range of L_m can be tuned from 36.2 μm to 177 μm (Fig. 10). Different from previous GIF tweezers fabricated by self-terminated chemical etching method (Gong et al. 2013, 2014; Chuang et al. 1998), this GIF taper was fabricated by heating-and-drawing method. One advantage is that the loss near the GIF taper is lower so that the gradient force is strong enough to trap the object on the optical axis, even at long manipulation length.

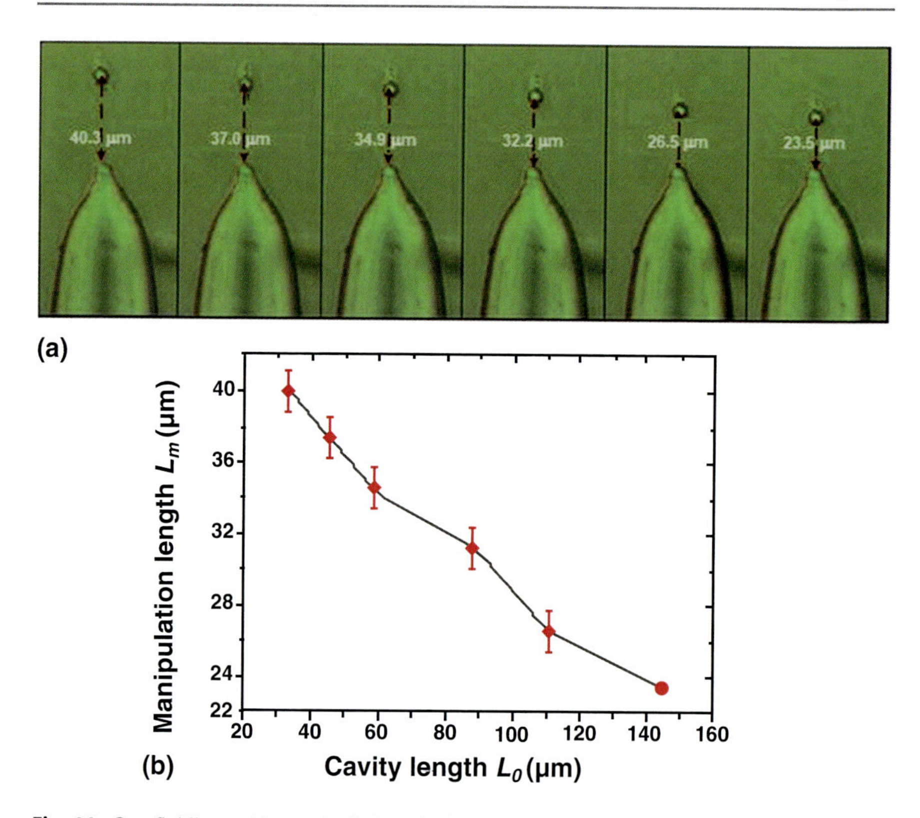

Fig. 11 Optofluidic tunable manipulation of microparticle by adjusting the air cavity

By introducing an air microcavity and adjusting its length, L_{air}, the periodic focusing effect of light in the GIF taper and therefore, L_m, can be adjusted. In this case, the microparticle can be balanced at the focus of the GIF taper by increasing the flow rate slowly. When the force is balanced at the focus, L_m can be controlled by tuning L_{air} (Fig. 11). During the tuning, the laser power and the flow rate were fixed at 43.5 mW and 450 nL/min, respectively.

Flat GIF

Previous optical manipulation were often based on fiber tapers, until Gong and his co-workers developed a new method based on GIF with a flat endface. The fabrication process is very simple by cleaving the GIF for optical manipulation, compared with tapering the fiber. Thanks to the quasi-parabolic index profile of GIF, a strain-controllable optofluidic manipulation technology based on GIF was developed. L_m of up to 1314.1 μm was demonstrated (Zhang et al. 2016).

The GIF-based controllable optical manipulation is achieved by directly stretching the GIF to change the light distribution, due to periodic convergence and divergence of the light beam propagating along axial direction of GIF. The periodic manipulation of microparticle is shown in Fig. 12. Figure 12a shows a sequence

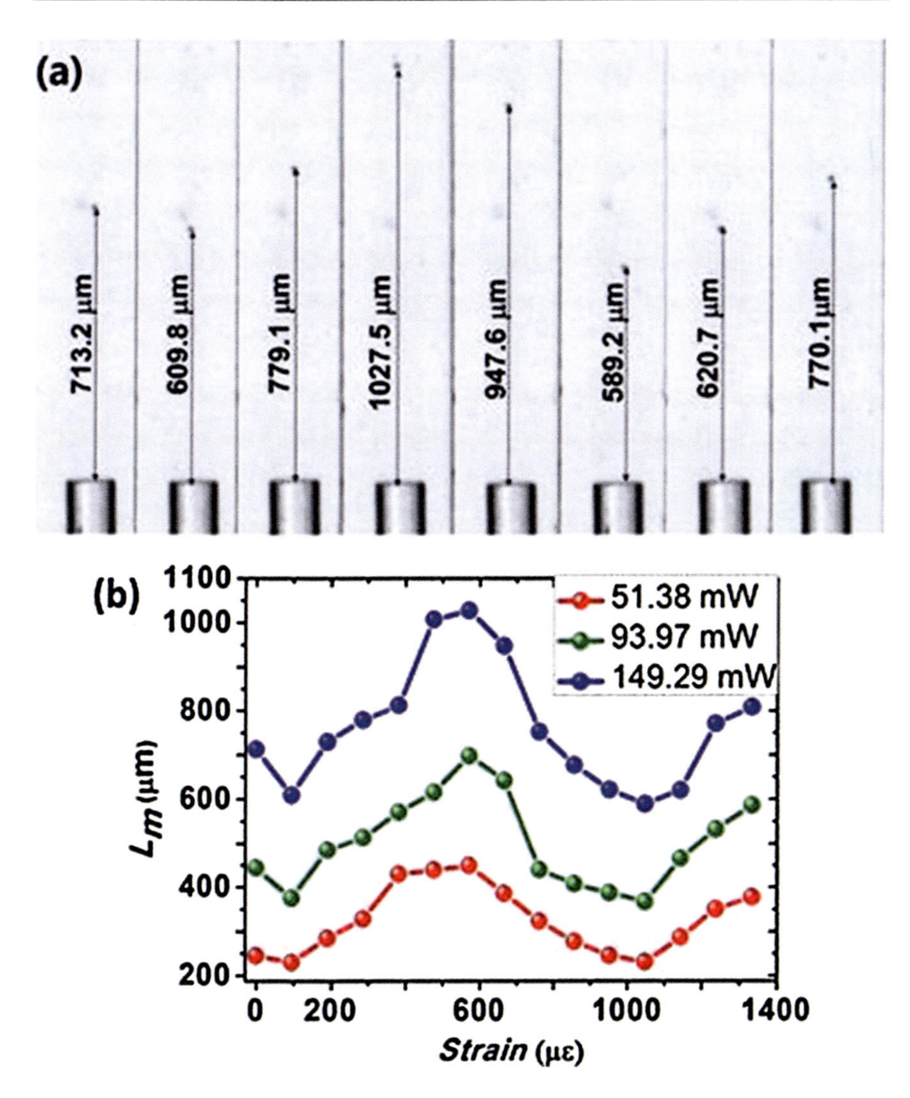

Fig. 12 (**a**) Microscopic images of optofluidic controllable manipulation of the single microsphere with strains of 0, 95 με, 286 με, 571 με, 667 με, 1048 με, 1143 με, and 1238 με, respectively, and (**b**) manipulation length, L_m, versus strain at flow rates of $v = 150$ nL/min

of the microscopic images with different strain on the GIF with a fixed power of 149.29 mW and a flow rate of 150 nL/min. L_m varied with the strain with a clear period around 953 με corresponding to a GIF length change of about 500 μm, which is in agreement with that of the GIF pitch.

The flow rate sensing was demonstrated by detecting the L_m with fixed strain and laser power. Figure 13a shows the L_m as a function of the flow rate in the log-log scale. The flow rate can be detected between 30 nL/min and 3 μL/min.

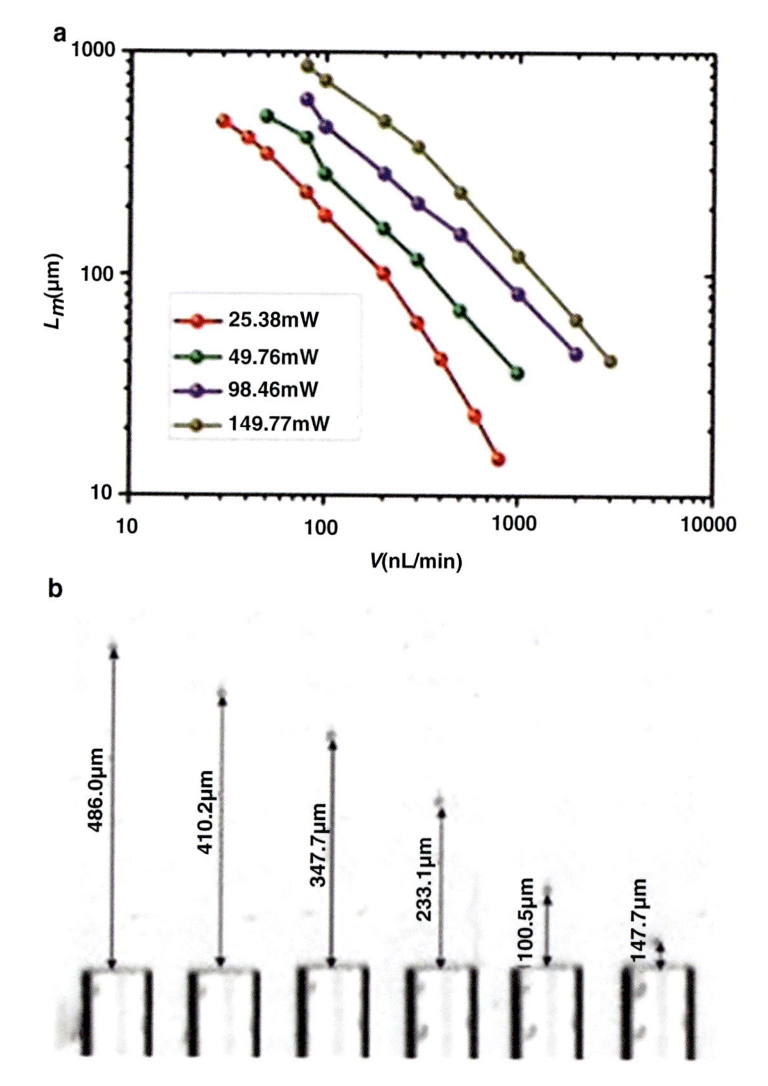

Fig. 13 (**a**) The result about flow rate sensor with 1048 $\mu\varepsilon$ and (**b**) selected microscopic images of green curve in (**a**) with flow rates of 30 nL/min, 40 nL/min, 50 nL/min, 80 nL/min, 200 nL/min, and 800 nL/min, respectively

Corresponding microscopic images of the optofluidic manipulation with different flow rate are shown in Fig. 13b. The laser power is fixed at 25.38 mW and the strain on the GIF is 1048 $\mu\varepsilon$.

Flat SMF

The fiber taper can focus the beam and form a 3D optical trap. However, it is difficult to fabricate fiber tapers with high repeatability. It is also difficult to precisely control the length of GIF in order to obtain the desired light distribution near the flat GIF tip. It is the simplest scheme to use a single flat SMF for optical manipulation. All that is needed is cleaving the SMF to form a flat fiber tip.

Combined with the flow force provided by the microfluid, the optical manipulation based on a flat SMF can be performed, which has the advantages of easy to fabricate and use, availability for mass production, and low cost. Thanks to these advantages, an optofluidic flow rate detection method was demonstrated (Gong et al. 2015b). A 980 nm fiber-coupled diode laser was used as the light source for the optofluidic manipulation. The absorption coefficient of both the water and PS microparticle is low at 980 nm. Therefore, the optical force dominates, while the photophoresis or negative photophoresis force can be neglected. The laser power was tunable for adjusting the optical force. One percent of the laser power was monitored for the calibration. Ninety-nine percent of the laser power was coupled directly into the cleaved SMF, which was inserted into the microfluidic channel for optical manipulation. The fiber was positioned against the flow. The microparticle with a diameter of 7 μm was flew through the channel and trapped by the gradient force and also the axial force microbalance. The flow rate was controlled by a syringe pump. The manipulation length was measured by an optical microscope.

According to Eq. 6 in section "Principles," the flow rate can be determined by detecting the manipulation length. Note that the manipulation length is inversely proportional to the flow rate, that is, L_m increases as v decreases. Therefore, the optofluidic manipulation technique is particularly useful to detect low flow rate.

The sensing performance of the device is shown in Fig. 14. The manipulation length as a function of the flow rate with different laser power is given in Fig. 14a. The maximum L_m was 715 μm with a laser power of 146.3 mW and flow rate of 75 nL/min. The minimum L_m was about 3 μm with a laser power of 11.4 mW and flow rate of 5000 nL/min.

At a laser power of 41 mW, a dynamic range of about three orders of magnitude, from 20 nL/min to 14,000 nL/min, was achieved. The dynamic ranges for the flow rate detection are similar with different laser powers, while a relative lower (higher) flow rate can be detected with a lower (higher) laser power. In conclusion, with the coordination of the laser power and flow rate, this device achieved flow rate sensing with a large dynamic range from 20 nL/min to 22 μL/min and manipulation range from 3 μm to 715 μm.

The flow rate was controlled between 40 nL/min and 80 nL/min with a step of 10 nL/min (red solid circles, Fig. 14b) and between 100 nL/min and 300 nL/min with a step of 50 nL/min (blue solid circles). The manipulation length was recorded at each flow rate for 15s. The temporal stability of the trapped microparticle is good during the flow rate detection and can be sustained for several hours. The step changes of the manipulation length can be clearly distinguished.

The sensing performance for flow rate was limited by L_m, which is one of the weaknesses of the flow rate sensor. The sensitivity became lower at shorter L_m. Although high sensitivity at long L_m, the optical trapping will be unstable due to the weak gradient force. To solve this problem, An idea of dual-mode detection was proposed (Gong et al. 2017).

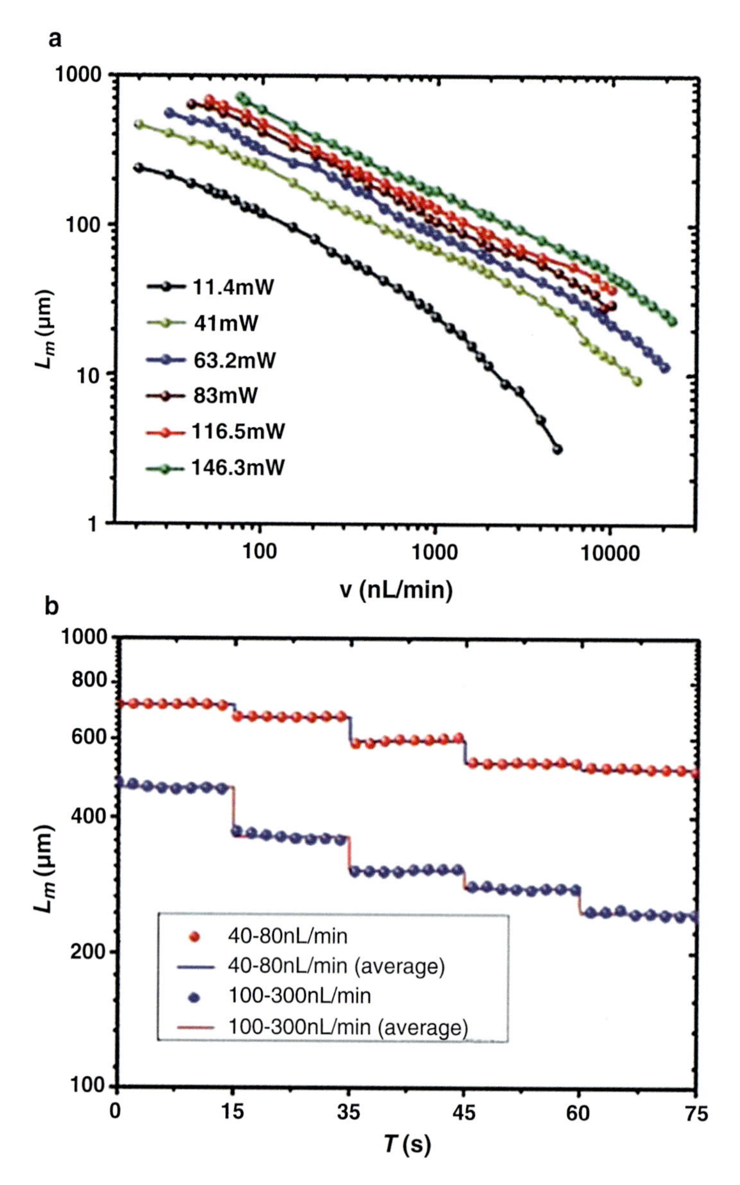

Fig. 14 The performance of the optofluidic flow rate detection based on the cleaved SMF optical manipulation

The dual-mode detection, i.e., open-loop mode and closed-loop mode, was developed as an optofluidic flowmeter with a large dynamic range of four orders of magnitude. The principle of open-loop mode is the same as that mentioned above. L_m is used as the sensing output, and the flowmeter has an inverse sensitivity so that it has higher sensitivity at lower flow rate. In the closed-loop mode, L_m is set to be

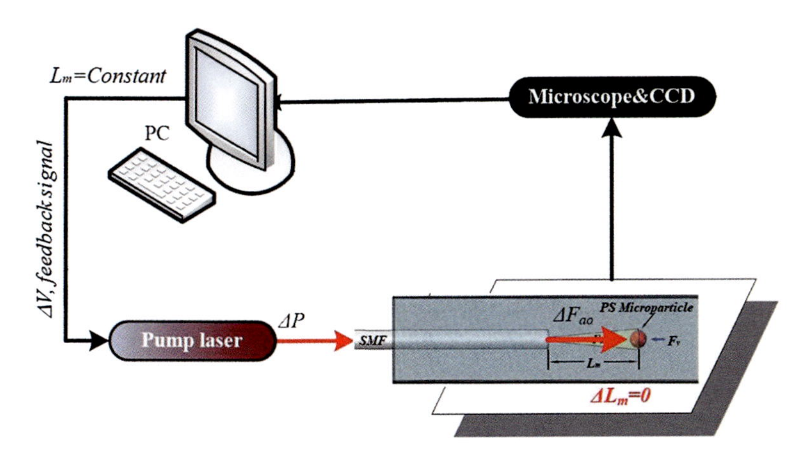

Fig. 15 Schematic experimental setup for dual-mode fiber optofluidic flowmeter

constant, and a feedback signal, tuning the laser power for the force balance, is used as the sensing output. The principle of the closed-loop mode is shown in Fig. 15. L_m is monitored by the CCD and calculated by a homemade software in real time. In the closed-loop mode, the L_m is set to be a constant value, L_0. Once there was a slight deviation, i.e., ΔL from L_0, a feedback signal was generated by tuning the laser power in order to keep the force balance on the microparticle and to keep the manipulation length constant at L_0. In this case, the laser power serves as the sensing output proportional to the flow rate. The closed-loop mode is helpful for extending the upper detection limit of flow rate and also enhancing the sensitivity at higher flow rate.

The calibration of the optofluidic flowmeter was performed in both open-loop and closed-loop detection modes. In the open-loop mode, the laser power was firstly fixed at 23.5 mW, and the manipulation length was measured as a function of flow rate. The manipulation length L_m decreased as a function of the flow rate (Fig. 16a). At 23.5 mW, the limit of detection (LOD) was pushed down to 10 nL/min, and an upper limit of 5000 nL/min was obtained.

The results indicated that lower LOD can be achieved with lower laser power, while the upper limit can be extended by using higher power. However, the gradient optical force was weak when using a small laser power, which may not be strong enough to trap the microparticle at a long manipulation length and the microparticle may flow away. And the flow rate may also have some fluctuations when the syringe pump was used to generate a very low flow rate. The stability of generating a low flow rate can be enhanced by using a low-volume syringe. However, a low-volume syringe was not helpful for detecting large flow rate as the solution in the syringe may run out in a short time.

Although the upper limit of the flow rate might be extended by using higher laser power, the sensitivity was rather low at high flow rate range in the open-loop mode. This can be seen from Fig. 16b, by plotting the L_m in the linear scale. One can see

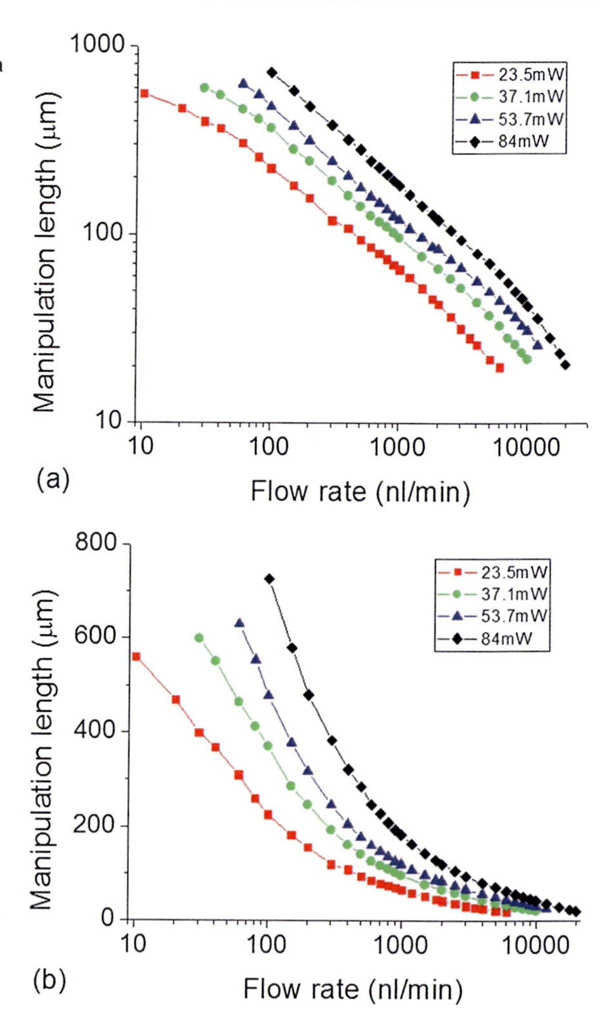

Fig. 16 Calibration of the optofluidic flow rate sensor in open-loop mode with y axis in (**a**) log-log and (**b**) semi-log scale

that the L_m decreases very quickly and nonlinearly with an increasing flow rate. This is an inverse sensitivity that is perfect for getting ultrahigh sensitivity at the low flow rate range. But the decrease in sensitivity limits the dynamic range in the high flow rate range. Therefore, the open-loop mode is good for measuring low flow rate with high sensitivity; thus we chose the laser power of 23.5 mW as an optimized power for the dual-mode flowmeter.

In the closed-loop mode, the compensated laser power was used as the sensing output and shown as a function of flow rate in Fig. 17. The L_m was first set at 60 μm and achieved a dynamic range of 1000–20,000 nL/min. By reducing the constant L_m to 30 μm and 15 μm, the dynamic range was extended from 2000 nL/min to 50,000 nL/min and 5000 nL/min to 100,000 nL/min, respectively.

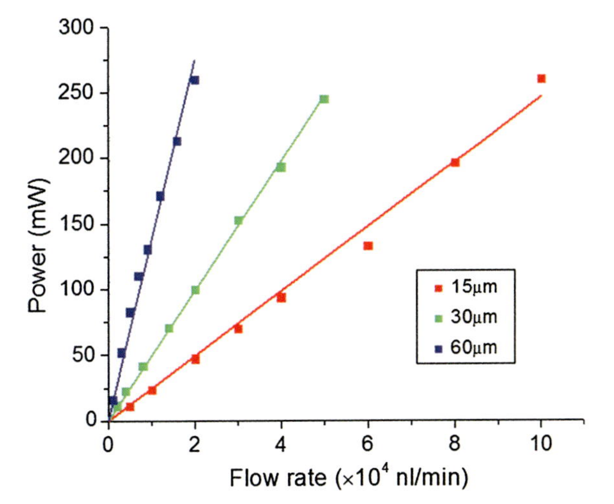

Fig. 17 Calibration of the optofluidic flow rate sensor in the closed-loop mode with manipulation length fixed at 15 μm, 30 μm, and 60 μm, respectively

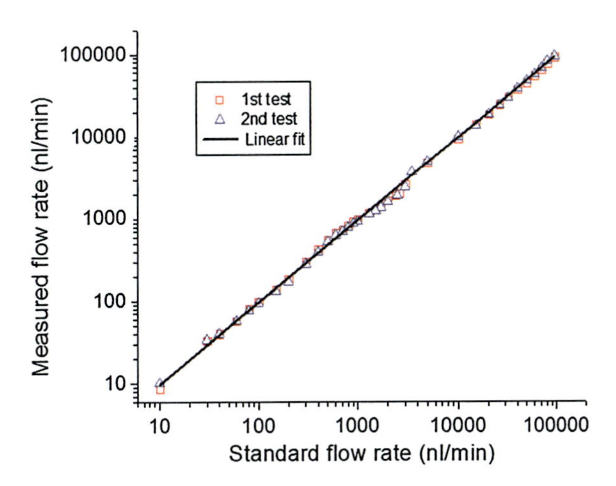

Fig. 18 Sensing performance of the optofluidic flow rate sensor

The dynamic ranges by using the closed-loop and the open-loop modes are complementary with each other. Therefore, by using the dual-mode flowmeter, the dynamic range can be greatly extended. A threshold was set as 5000 nL/min, corresponds to a threshold manipulation length, L_{m0}, and a threshold laser power, P_0, respectively. Initially, the laser power was set to be 23.5 mW for trapping a single microparticle, and the flowmeter was operated in the open-loop mode. If the initial flow rate was high so that the manipulation length was reduced down to L_{m0}, the flowmeter was switched to the closed-loop mode, and the laser power was adjusted by the feedback signal in order to maintain the manipulation length to be 15 μm. As shown in Fig. 18, the flowmeter achieved sensing with a large dynamic range of four orders of magnitude from 10 nL/min to 100,000 nL/min.

OF^2 Sensors Based on Photothermal Effect

In this section, two sensing schemes based on heat transfer in a microring resonator and the generation of microbubble-on-tip will be introduced. The fiber optofluidic technologies based on photothermal effects by heating the microfluidic channel via fiber-coupled laser can be applied for sensing applications. The heat transfer in liquid-core microring resonator can be employed for flow rate detection. Microbubbles can be photothermally generated on the fiber tip that is coated with carbon nanotubes or gold nanofilm. This type of gas microbubble generated in microfluid can be employed as a micro-interferometer in which its free spectral range (**FSR**) could be measured for sensitive temperature and flow rate detection. By using optical imaging to monitor the evolution of the microbubbles on the fiber tip coated with gold nanofilm, concentration detection with a large dynamic range can be realized.

Heat Transfer in Microring Resonator

Gong et al. reported an optofluidic flow rate sensor based on the heat transfer effect in a microfluidic channel for the lab-on-a-chip applications. Microscopic and SEM images of the flow rate sensor are shown in Fig. 19. The microfluid, whose geometrical structure is a hollow round capillary, acts as optofluidic ring resonator (OFRR). A fiber taper with a waist of about 3 μm is perpendicular to the capillary for optical coupling and light collection. A small fraction of the incident light is coupled from the taper into the OFRR and is kept circulating in the wall of the OFRR due to the total internal reflection. After each round trip, a small fraction of light is coupled out to the fiber taper. The multiple output beams interfere with each other, and the resonant dips are generated in the spectrum of the output beam. In case of the high Q factor resonator, the full width at half magnitude (FWHM) of the linewidth is narrow and is helpful for the high-performance sensing.

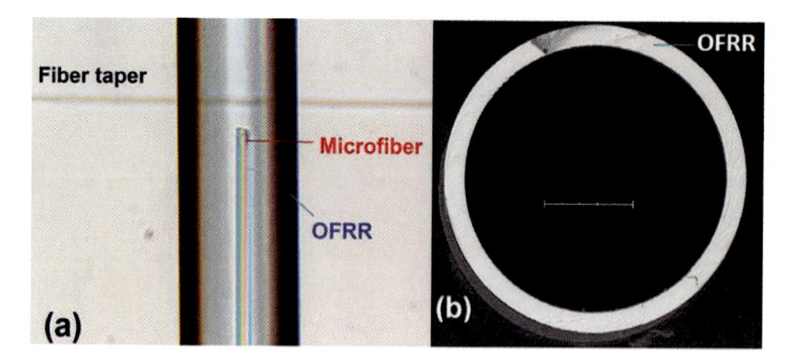

Fig. 19 (**a**) Microscopic image of the structure for the flow rate detection and (**b**) the SEM image of the cross section of the OFRR. Bar: 50 μm

A 1480 nm pump laser was used for heating the microfluid due to the high optical absorption coefficient of water in this spectral range. The laser was coupled into a fiber tip, which was fabricated by heating and drawing a SMF. The fundamental mode of the microring was often used for sensing which was located near the outer surface of the capillary with larger radius and the effective index (n_{eff}). The light intensity distribution, including the evanescent wave, of the fundamental radial mode will not penetrate into the inner liquid of the capillary. So the relation between the temperature and wavelength shift can be expressed as

$$\frac{\Delta\lambda}{\lambda} = \left(\alpha + \frac{\kappa_{wall}}{n_{eff}}\frac{\partial n_{eff}}{\partial n_{wall}}\right)\Delta T. \tag{13}$$

Here $\alpha = 1/r(\partial r/\partial T)$ is the thermal expansion coefficient of the resonator, and $\kappa_{wall} = \partial n/\partial T$ is the thermo-optic effect for the wall. The "wall" means the solid part of the capillary. Heat is at first transferred from the microfiber tip to the inner liquid core and then to the silica ring resonator. From Eq. 13, there are two factors, i.e., thermal expansion of the silica microring and thermo-optical effects, affecting the wavelength shift of the OFRR.

The transmission spectrum at a flow rate of 10 μL/min under different heating power is shown in Fig. 20. After tuning the laser power, the heat transfer got balanced in a few seconds, and then the spectrum was recorded. With the increase of heating power, the resonant dips show a right shift, indicating an increase in phase change in a round trip. The result agrees well with the theoretical prediction in Eq. 13.

As shown in Fig. 21 in linear scale, the wavelength changes nonlinearly with the flow rate under different heating power. The sensitivity of flow rate sensor increases as the heating power increases. The maximum sensitivity of 57.6 pm per μL/min was achieved at a heating power of 64.74 mW. Further increase in the heating power will led to unwanted microbubble generation. The data, replotted in log-log scale in Fig. 21b, show a good linearity.

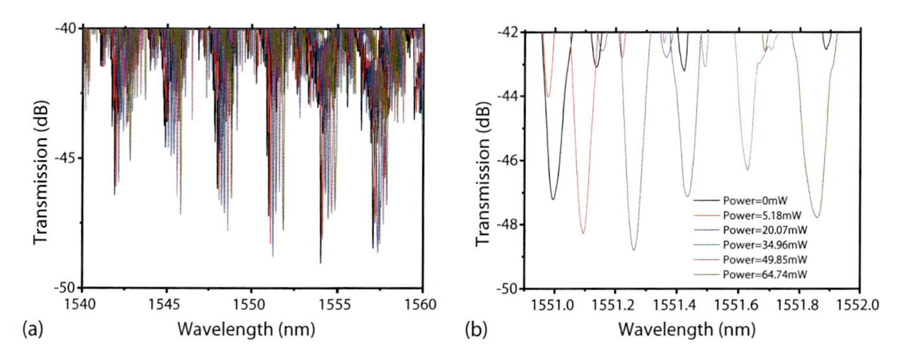

Fig. 20 (**a**) Transmission spectrum of OFRR under different heating power and (**b**) the enlargement of the spectrum

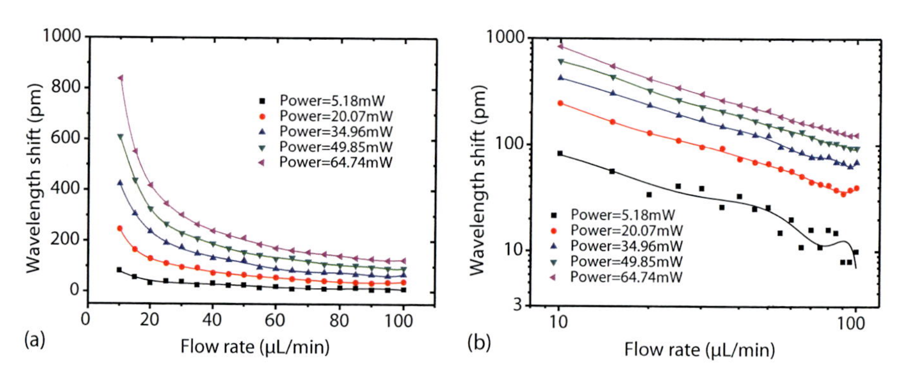

Fig. 21 Wavelength shift of the resonant dip for the OFRR as a function of the flow rate at different heating powers in (**a**) the linear scale and (**b**) the log-log scale

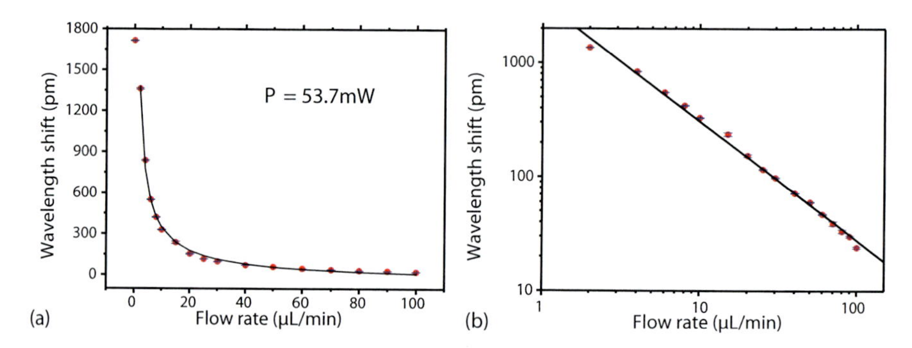

Fig. 22 Wavelength shift versus the flow rate in (**a**) linear scale and (**b**) log-log scale. Error bars are based on triplet measurements

The flow rate detection with a dynamic range of 2–100 μL/min was achieved (Fig. 22). With a heating power of 53.7 mW, the calibration curve of the wavelength shift versus the flow rate was measured and shown in the linear scale and the log-log scale in Fig. 22a, b, respectively. The error bar was calculated from the triplet measurements. The results indicate that the repeatability of the flow rate detection is good. The wavelength shift is set to be zero, $\Delta\lambda = 0$, at the initial condition with no laser heating for each flow rate and corresponds to the reference wavelength of λ_0. The data in Fig. 22a are fitted via $\Delta\lambda = av^b$. It is equivalent to the linear fitting of the data in Fig. 22b. The R square of linear fitting is about 0.993. The fitting curves are also shown in Fig. 22a, b as the black lines. The proposed sensor has the advantages of high flexibility and can perform local flow rate detection, especially for capillary-based microfluidics while not degrading other functionalities of the microring resonator including the biochemical sensing or optofluidic lasing.

Microbubble-on-Tip Structure

Laser-induced bubbles in microfluidics have been used as valves and pumps because they can be generated and cracked easily by a laser beam (Zhang et al. 2011). The generation of microbubbles is related to the properties of ambient environment, including temperature, viscosity, flow rate, and concentration, so that this kind of microbubble has the potential for sensing. In this section, two methods based on microbubble-on-tip (μBoT) structure will be described. The first one can achieve temperature or flow rate sensing based on microbubble interferometer generated on a flat fiber tip coated with carbon nanotube (CNT) film. The evolution of the microbubble was monitored by detecting the changes of the reflective interference spectra. The second method is based on microbubbles generated on a flat fiber tip coated with gold film, which is used as a concentration sensor. The diameter of the microbubble is monitored by imaging.

Temperature and Flow Rate Sensor

A fiber optofluidic interferometer for temperature and flow rate sensing in microfluid is developed based on the μBoT structure (Zhang et al. 2017). The generation process and sensing mechanism of this μBoT sensor are very different from traditional optical fiber sensors. Experimentally, two laser beams were delivered to the same fiber tip through a wavelength division multiplexer. A heating beam at 980 nm was employed to heat microfluid and generated the microbubble on the fiber tip that was coated with CNTs. The laser power was tunable from 0 to 300 mW. A tunable laser from an optical spectrum analyzer (OSA) (OPT162, 1505–1630 nm, Agilent) was used to monitor the changes of the reflective interference spectra that resulted from the evolution of the microbubble. The interference occurred between the Fresnel reflections from the solid/air surface and the other from the air/liquid surface.

CNT film was optically deposited on the fiber endface to increase the laser absorption, because it has a low conductivity of ~1.52 W/(m·K) in the radial direction and also a low specific heat capacity of 0.7 J/(g·K), which are helpful for increasing the absorption of pump laser and introducing a large temperature increment for the μBoT generation. Therefore, the μBoT structure can be generated with laser of 980 nm with a relative low power (Hepplestone et al. 2006). A microbubble was generated thanks to the increment of the gas content around the fiber tip, either from the separation of dissolved gas from the liquid or from the vaporization of the liquid by the laser heating. Then the microbubble kept on growing if the pump laser power was sustained.

The temporal evolution of the microbubble can be monitored by the charged-coupled device (CCD) and an example is given in Fig. 23. There was a sharp nonlinear increase in diameter at the beginning because a fast temperature rise was obtained when the laser was switched on. Then the diameter increased with time with a good linearity ($R^2 \sim 0.99$) between 50s and 150s. When the diameter was large, the temperature rise became slow and gradually got balanced. Since the FSR

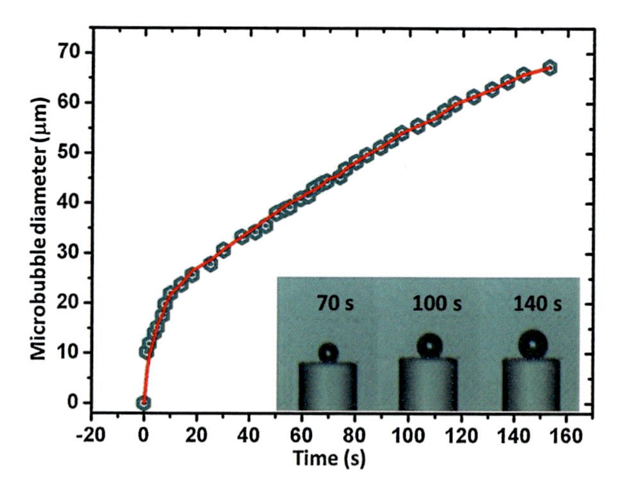

Fig. 23 The temporal change of the microbubble diameter

changed linearly over certain range of heating time, detecting FSR in this time range would not influence on the calibration curve for sensing.

The FSR can be expressed by $FSR = \lambda_1\lambda_2/(2nL_{eff})$, with n and L_{eff} as the refractive index and the effective length of the microbubble interferometer, respectively. The FSR is inversely proportional to the diameter of the microbubble so that it can be used to detect the generation of the microbubble.

The growth rate of the microbubble would change when the ambient parameters change. The FSR of the microbubble at a fixed heating time was used as the sensing signal. As the diameter of the μBoT interferometer increases with time, the free spectral range (FSR) decreases. By measuring the FSR, the temperature and flow rate sensing are demonstrated. For temperature sensing, the gas for microbubble mainly came from the vaporization of the liquid, similar to boiling the water by heating. For flow rate sensing, the gas mainly came from dissolved gas delivered together with the microfluid and can be refreshed as the fluid flows. High sensitivity was achieved thanks to the photothermal effect of CNT and the small volume of the microbubble.

The reflective spectra from the μBoT structure at heating time of 70s, 100s, and 140s are shown in Fig. 24a, while the ambient temperature was kept constant. The laser power for heating was 221 mW. The fringe contrast was not as high as solid fiber micro-interferometers but was sufficient to determine the free spectral range (FSR) of the interference spectra. A compromise between increasing the absorption by CNTs and keeping good interference fringes should be considered during the CNTs deposition. The FSR decreases with a good linearity ($R^2 \sim 0.98$) as the microbubble grows in time (Fig. 24b). This work might open a door to the development of novel reconfigurable fiber optofluidic sensors.

The temperature and flow rate were measured by detecting the FSR. The deviation of FSR was calculated by $D = \left(FSR - \overline{FSR}\right)/\overline{FSR} \times 100\%$, where $\overline{FSR}$ was the statistically averaged FSR over ten tests.

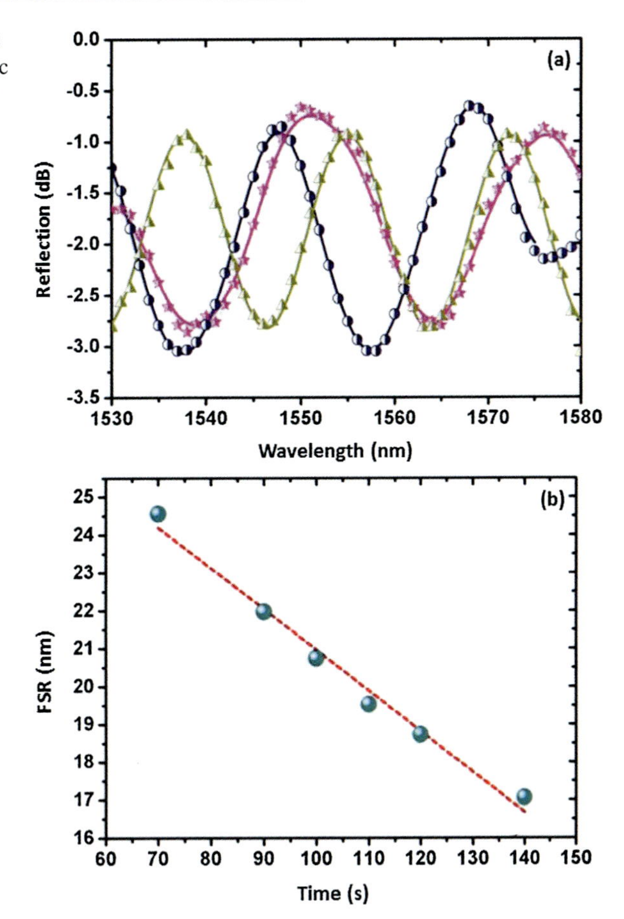

Fig. 24 (**a**) The interference spectra of the fiber optofluidic microbubble-on-tip structure and (**b**) FSR versus heating time

Temperature sensing was demonstrated and the results are shown in Fig. 25. At each temperature, the 980 nm laser was turned on with a power of 146 mW, and the reflective spectra of microbubble was recorded after heating for 60s. Figure 25a showed the FSR changes as a function of temperature. Temperature was measured between 25 °C and 45 °C with good linearity (R^2) of about 0.998 and sensitivity of −1146 pm/°C. This range covers the temperature range of the human body and is significant for the organ-on-a-chip applications. By using the same probe, good repeatability was confirmed by ten times of tests at 25 °C (Fig. 25b).

For the microfluidic flow rate sensing, a range of 0–150 nL/min was achieved with a laser power of 221 mW (Fig. 26a). A good linearity ($R^2 \sim 0.997$) is achieved, and the sensitivity is – 31 pm/(nL/min), which is determined by the slope of linear fitting. Ten times of tests were performed with a flow rate of 120 nL/min, and good repeatability was achieved, with a statistical deviation of ±0.94% (Fig. 26b).

The microfluidic flow takes away part of the laser-generated heat; therefore, the microbubble was supposed to grow slower at higher flow rate. However, the opposite

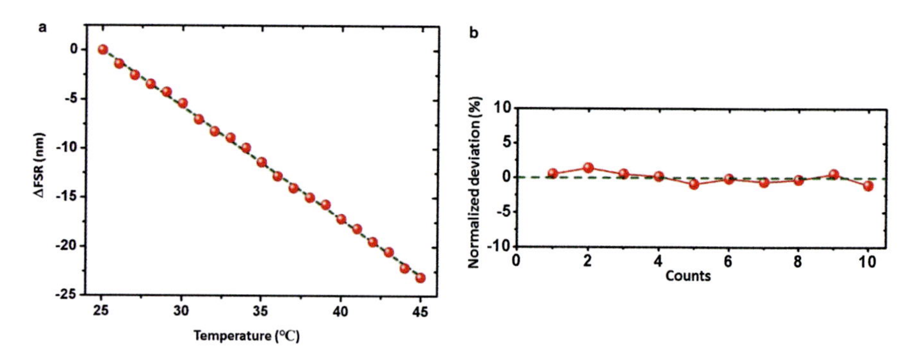

Fig. 25 (**a**) FSR changes versus temperature with a heating time of 60s at 146 mW and (**b**) the repeatability of the same probe for ten tests

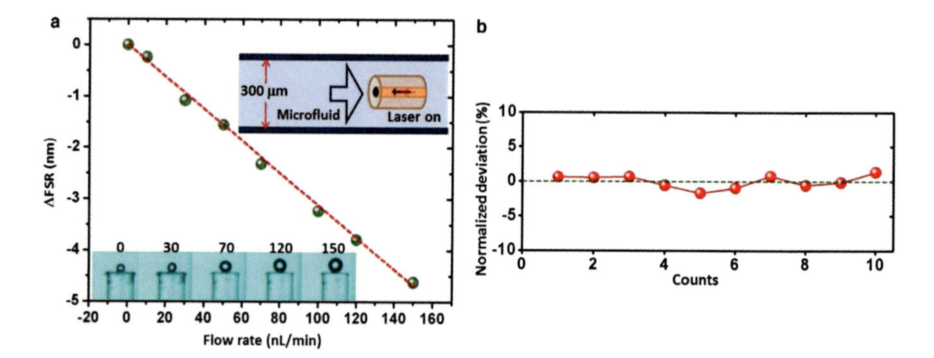

Fig. 26 (**a**) FSR changes as a function of flow rate with a heating time of 120s at 221 mW and (**b**) the repeatability of the same probe for ten tests

was surprisingly observed, that is, the microbubble grows faster at higher flow rate, leading to a smaller FSR at a fixed heating time (Fig. 26a). The potential explanation is that the dissolved air in the liquid was delivered more efficiently at higher flow rate, leading to a faster growth of the microbubble. Therefore, the air solubility was initialized by ultrasonic bathing in both temperature and flow rate sensing.

Concentration Sensor

In order to further enhance the lifetime of the μBoT sensor, gold nanofilm was coated on the fiber tip instead of CNT film. Beside the lifetime and stability enhancement, the gold nanofilm also enhanced the efficiency of microbubble generation. A fiber-coupled laser was used for heating the fiber tip, and single gas microbubbles can be photothermally generated. An easy-to-fabricate and easy-to-use μBoT concentration sensor was developed based on the μBoT structure.

Two methods were proposed to monitor the growth process as a function of the concentration. One was by the spectral detection, as described in the last section. After heating for 10s, the laser was switched off and recorded the reflective spectra from the microbubble using an optical spectrum analyzer (OSA). The microbubble acted as an interferometer based on the interference between the reflective beams from the two surfaces of the microbubble. The free spectral range (FSR) was measured so that the effective cavity length, ΔL, can be further calculated as a function of the concentration. The other was diameter detection by imaging. The microbubble diameter was monitored by imaging, and the diameter difference between heating time of 2s and 7s was calculated as a function of concentration.

The concentration of the solution can influence the growth rate of the gas microbubble in two different ways. One is the evaporation of liquid near the fiber tip, and the other is gas generated from the heat-induced chemical decomposition. Solutions of sucrose and hydrogen peroxide (H_2O_2) were used as samples for testing, and the experimental results confirmed the high sensing performance, including a large dynamic range and high sensitivity. This study is the first to report the development of the lab on tip (LoT) based on μBoT for concentration detection. This technique has many advantages such as high sensing performance, ease of fabrication and use, and low cost.

The schematic structure of LoT technology is shown in Fig. 27a. The coated fiber tip was inserted into a glass capillary with a square cross section of 1 mm × 1 mm, which was filled with deionized (DI) water or other kinds of liquid samples. Gas dissolved in the liquid may accelerate the generation speed of the microbubble. Therefore, the dissolved gas was removed by processing the liquid in an ultrasonic bath for 15 min. The liquid was withdrawn into the capillary by using a syringe pump or through the capillary force. During the experiment, the liquid was kept static. A 1550 nm laser was coupled into the fiber and heated the liquid near the core of the fiber tip. This wavelength was chosen because of its strong absorption in DI water. When the laser was switched on, a microbubble can be generated on the tip, and its diameter, d, increases with time, as shown in Fig. 27b. The growth process was monitored using an optical microscope with a low-cost CCD camera. The growth rate was determined to be about 11.61 μm/s. At 10s, the laser was switched off, and the microbubble remained stable for a relatively long time.

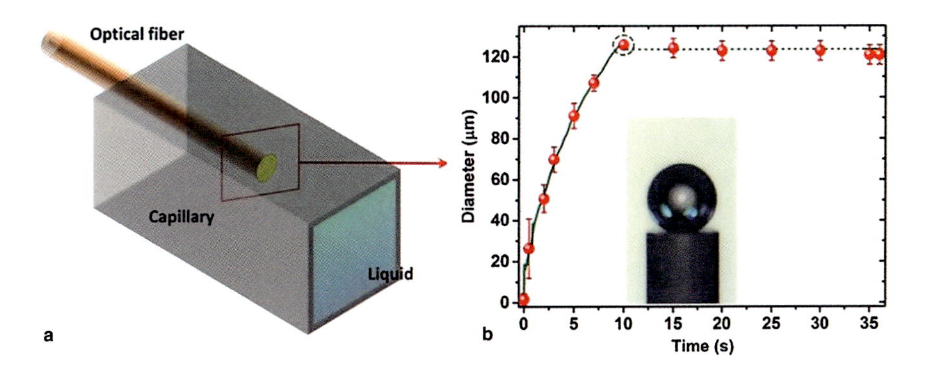

Fig. 27 (**a**) The schematic structure and (**b**) the temporal growth of LoT

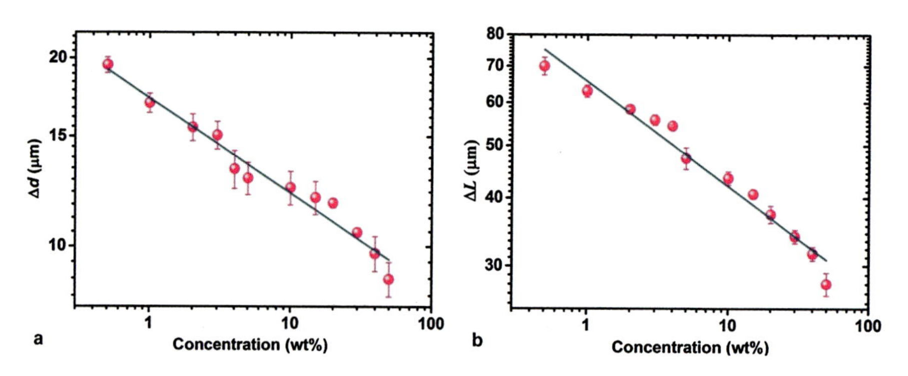

Fig. 28 Sucrose concentration detection based on (**a**) imaging and (**b**) spectral detection

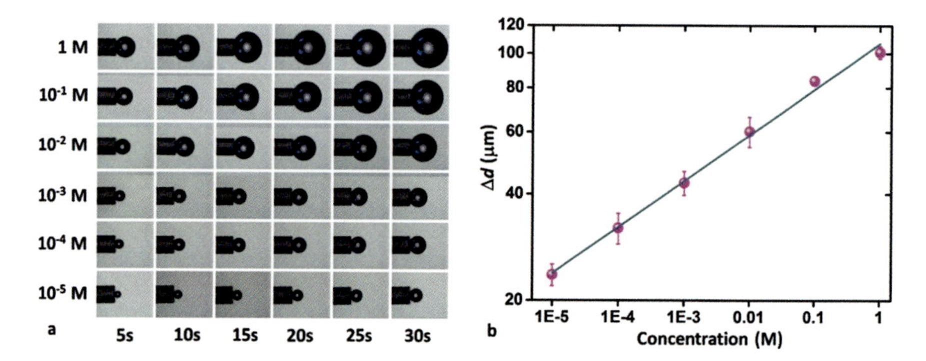

Fig. 29 (**a**) Microscopic images of the microbubbles and (**b**) the diameter difference as a function of H_2O_2 concentration

Figure 28 shows the performance of sucrose concentration sensing with a dynamic range of two orders of magnitude, from 0.5 wt% to 50.0 wt%. The diameter Δd and the FSR ΔL are shown as a function of the sucrose concentration, which was obtained by imaging and the spectral method, respectively. Results are shown in the log-log scale for clarifying the details in the lower concentration range. The diameter of the microbubbles was monitored after a fixed heating time. Obviously, Δd decreases with the increase of sucrose concentration. The results can be explained as below. As sucrose concentration increases, its boiling point increases, and it takes longer time to accumulate heat for the liquid near the tip to vaporize. Linear fits are also given, both with a linearity of 0.96. The maximum sensitivity was 5.1 μm/% at a concentration of 0.5 wt%. The averaged standard deviations were 0.6 μm (5.6%) and 1.3 μm (2.8%), respectively.

As H_2O_2 concentration increases, there are two contrary factors that affect the growth of the microbubble: (1) Similar to that for the sucrose solution, it takes a longer time for the liquid near the tip to vaporize at a higher concentration. Thus, the growth speed of the microbubble tends to decrease. (2) When heated by a laser, H_2O_2 decomposes and generates oxygen (O_2) gas. More gas can be generated with a higher concentration of H_2O_2, which tends to increase the growth speed of the microbubble. The results confirmed that the latter factor dominated in the experiment, which in principle was different from sucrose sensing. Therefore, the diameter difference increases as a function of H_2O_2 concentration.

For H_2O_2 detection, the imaging method was chosen due to low cost. The calibration results are shown in Fig. 29. The microscopic images of the microbubbles generated with different concentrations of H_2O_2 were recorded at different heating times (Fig. 29a). Δd as a function of the H_2O_2 concentration is shown in the log-log scale in Fig. 29b. A maximum sensitivity of 93.84 μm/μM was obtained. The concentration sensing based on μBoT structure achieved an unprecedented dynamic range of five orders of magnitude, from 10^{-5} M to 1 M.

Conclusion

In this chapter, optical fiber microfluidic sensors based on two types of opto-physical effects, i.e., laser-induced force and photothermal effect, are introduced. The laser-induced force is introduced for the application in optical tweezers and optical manipulation of microparticles. Controllable optical manipulation based on both graded-index fiber and single-mode fiber, either with a taper or flat endface, can be achieved by introducing a tunable microcavity or by adjusting the strain on GIF. Flow rate sensing based on the optical fiber manipulation methods can achieve a large dynamic range and high sensitivity. Photothermal effects, associated with the laser-induced heat and its transfer in a microring resonator, are introduced for microfluidic flow rate sensors. In addition the technique that gas microbubbles are generated on a fiber tip and used as a micro-interferometer is introduced for flow rate and temperature sensing, as well as concentration sensing with a large dynamic range.

The optical fiber microfluidic sensors have distinct advantages including easy to fabricate, miniaturization, low cost, high sensitivity, and large dynamic range. Besides the parameters such as flow rate, temperature, and concentration, as introduced in this chapter, we believe that many new optical fiber microfluidic sensors will be developed for other physical, chemical and biological parameters in the future.

References

A. Ashkin, Acceleration and trapping of particles by radiation pressure. Phys. Rev. Lett. **24**, 156 (1970)

A. Ashkin, Forces of a single-beam gradient laser trap on a dielectric sphere in the ray optics regime. Biophys. J. **61**, 569–582 (1992)

A. Ashkin, J.M. Dziedzic, J. Bjorkholm, S. Chu, Observation of a single-beam gradient force optical trap for dielectric particles. Opt. Lett. **11**, 288–290 (1986)

A. Bertucci, A. Manicardi, A. Candiani, S. Giannetti, A. Cucinotta, G. Spoto, M. Konstantaki, S. Pissadakis, S. Selleri, R. Corradini, Detection of unamplified genomic DNA by a PNA-based microstructured optical fiber (MOF) Bragg-grating optofluidic system. Biosens. Bioelectron. **63**, 248–254 (2015)

D.S. Bykov, O.A. Schmidt, T.G. Euser, P.S.J. Russell, Flying particle sensors in hollow-core photonic crystal fibre. Nat. Photonics **9**, 461 (2015)

Y.-H. Chuang, K.-G. Sun, C.-J. Wang, J. Huang, C.-L. Pan, A simple chemical etching technique for reproducible fabrication of robust scanning near-field fiber probes. Rev. Sci. Instrum. **69**, 437–439 (1998)

A.M. Cubillas, S. Unterkofler, T.G. Euser, B.J. Etzold, A.C. Jones, P.J. Sadler, P. Wasserscheid, P.S.J. Russell, Photonic crystal fibres for chemical sensing and photochemistry. Chem. Soc. Rev. **42**, 8629–8648 (2013)

S. Dochow, C. Krafft, U. Neugebauer, T. Bocklitz, T. Henkel, G. Mayer, J. Albert, J. Popp, Tumour cell identification by means of Raman spectroscopy in combination with optical traps and microfluidic environments. Lab Chip **11**, 1484 (2011)

P. Domachuk, B. Eggleton, Fiber-based optofluidics. Proc. SPIE **6588**, 65880C (2007)

X. Fan, I.M. White, Optofluidic microsystems for chemical and biological analysis. Nat. Photonics **5**, 591–597 (2011)

Y. Gong, A.-Y. Ye, Y. Wu, Y.-J. Rao, Y. Yao, S. Xiao, Graded-index fiber tip optical tweezers: numerical simulation and trapping experiment. Opt. Express **21**, 16181–16190 (2013)

Y. Gong, W. Huang, Q.-F. Liu, Y. Wu, Y. Rao, G.-D. Peng, J. Lang, K. Zhang, Graded-index optical fiber tweezers with long manipulation length. Opt. Express **22**, 25267–25276 (2014)

Y. Gong, C. Zhang, Q.-F. Liu, Y. Wu, H. Wu, Y. Rao, G.-D. Peng, Optofluidic tunable manipulation of microparticles by integrating graded-index fiber taper with a microcavity. Opt. Express **23**, 3762–3769 (2015a)

Y. Gong, Q.-F. Liu, C.-L. Zhang, Y. Wu, Y.-J. Rao, G.-D. Peng, Microfluidic flow rate detection with a large dynamic range by optical manipulation. IEEE Photon. Technol. Lett. **27**, 2508–2511 (2015b)

Y. Gong, L. Qiu, C. Zhang, Y. Wu, Y.-J. Rao, G.-D. Peng, Dual-mode fiber optofluidic flowmeter with a large dynamic range. J. Lightwave Technol. **35**, 2156–2160 (2017)

J. Guck, R. Ananthakrishnan, T. Moon, C. Cunningham, J. Käs, Optical deformability of soft biological dielectrics. Phys. Rev. Lett. **84**, 5451 (2000)

J. Guck, R. Ananthakrishnan, H. Mahmood, T.J. Moon, C.C. Cunningham, J. Käs, The optical stretcher: a novel laser tool to micromanipulate cells. Biophys. J. **81**, 767–784 (2001)

Y. Harada, T. Asakura, Radiation forces on a dielectric sphere in the Rayleigh scattering regime. Opt. Commun. **124**, 529–541 (1996)

Z. He, F. Tian, Y. Zhu, N. Lavlinskaia, H. Du, Long-period gratings in photonic crystal fiber as an optofluidic label-free biosensor. Biosens. Bioelectron. **26**, 4774–4778 (2011)

S. Hepplestone, A. Ciavarella, C. Janke, G. Srivastava, Size and temperature dependence of the specific heat capacity of carbon nanotubes. Surf. Sci. **600**, 3633–3636 (2006)

P. Jess, V. Garcés-Chávez, D. Smith, M. Mazilu, L. Paterson, A. Riches, C. Herrington, W. Sibbett, K. Dholakia, Dual beam fibre trap for Raman microspectroscopy of single cells. Opt. Express **14**, 5779–5791 (2006)

G. Kostovski, P.R. Stoddart, A. Mitchell, The optical fiber tip: an inherently light-coupled microscopic platform for micro-and nanotechnologies. Adv. Mater. **26**, 3798–3820 (2014)

C. Liberale, P. Minzioni, F. Bragheri, F. De Angelis, E. Di Fabrizio, I. Cristiani, Miniaturized all-fibre probe for three-dimensional optical trapping and manipulation. Nat. Photonics **1**, 723–727 (2007)

Y. Liu, M. Yu, Investigation of inclined dual-fiber optical tweezers for 3D manipulation and force sensing. Opt. Express **17**, 13624–13638 (2009)

Z. Liu, C. Guo, J. Yang, L. Yuan, Tapered fiber optical tweezers for microscopic particle trapping: fabrication and application. Opt. Express **14**, 12510–12516 (2006)

X. Liu, T. Gong, Y. Liu, Z. Wang, A novel refractometric sensor based on optofluidic integration of composite core photonic crystal fibers. J. Opt. **19**, 015301 (2016)

S.K. Mohanty, K.S. Mohanty, M.W. Berns, Manipulation of mammalian cells using a single-fiber optical microbeam. J. Biomed. Opt. **13**, 054049 (2008)

D. Psaltis, S.R. Quake, C. Yang, Developing optofluidic technology through the fusion of microfluidics and optics. Nature **442**, 381 (2006)

K. Taguchi, K. Atsuta, T. Nakata, R. Ikeda, Levitation of a microscopic object using plural optical fibers. Opt. Commun. **176**, 43–47 (2000)

A. Tam, C. Patel, Optical absorptions of light and heavy water by laser optoacoustic spectroscopy. Appl. Opt. **18**, 3348–3358 (1979)

M. Terazima, N. Hirota, S.E. Braslavsky, A. Mandelis, S.E. Bialkowski, G.J. Diebold, R. Miller, D. Fournier, R.A. Palmer, A. Tam, Quantities, terminology, and symbols in photothermal and related spectroscopies (IUPAC recommendations 2004). Pure Appl. Chem. **76**, 1083–1118 (2004)

J. Tian, Y. Lu, Q. Zhang, M. Han, Microfluidic refractive index sensor based on an all-silica in-line Fabry–Perot interferometer fabricated with microstructured fibers. Opt. Express **21**, 6633–6639 (2013)

J. Tian, Z. Lu, M. Quan, Y. Jiao, Y. Yao, Fast response Fabry–Perot interferometer microfluidic refractive index fiber sensor based on concave-core photonic crystal fiber. Opt. Express **24**, 20132–20142 (2016)

S. Unterkofler, M.K. Garbos, T.G. Euser, P.S.J. Russell, Long-distance laser propulsion and deformation-monitoring of cells in optofluidic photonic crystal fiber. J. Biophotonics **6**, 743–752 (2013)

D.K. Wu, B.T. Kuhlmey, B.J. Eggleton, Ultrasensitive photonic crystal fiber refractive index sensor. Opt. Lett. **34**, 322–324 (2009)

K. Zhang, A. Jian, X. Zhang, Y. Wang, Z. Li, H.-Y. Tam, Laser-induced thermal bubbles for microfluidic applications. Lab Chip **11**, 1389–1395 (2011)

Y. Zhang, P. Liang, Z. Liu, J. Lei, J. Yang, L. Yuan, A novel temperature sensor based on optical trapping technology. J. Lightwave Technol. **32**, 1394–1398 (2014)

C.-L. Zhang, Y. Gong, Q.-F. Liu, Y. Wu, Y.-J. Rao, G.-D. Peng, Graded-index fiber enabled strain-controllable optofluidic manipulation. IEEE Photon. Technol. Lett. **28**, 256–259 (2016)

C.-L. Zhang, Y. Gong, W.-L. Zou, Y. Wu, Y.-J. Rao, G.-D. Peng, X. Fan, Microbubble-based fiber optofluidic interferometer for sensing. J. Lightwave Technol. **35**, 2514–2519 (2017)

Micro-/Nano-optical Fiber Microfluidic Sensors

59

Lei Zhang

Contents

L. Zhang (✉)
College of Optical Science and Engineering, Zhejiang University, Hangzhou, China
e-mail: zhang_lei@zju.edu.cn

G.-D. Peng (ed.), *Handbook of Optical Fibers*,
https://doi.org/10.1007/978-981-10-7087-7_62

Abstract

Optical micro-/nanofiber (MNF) microfluidic sensors – the synergistic integration of MNF and microfluidics – provide a number of unique characteristics for enhancing the sensing performance and simplifying the design of microsystems. With diameter close to or below the wavelength of guided light and high index contrast between the MNF and the surrounding, an MNF shows a variety of interesting waveguiding properties, including widely tailorable optical confinement, evanescent fields, and waveguide dispersion. MNF sensor has been attracting increasing research interest due to its possibilities of realizing miniaturized fiber-optic sensors with small footprint, high sensitivity, fast response, high flexibility, and low optical power consumption. Note that most of the MNF sensors used MNFs suspended in air or mounted in a bulky volume flow chamber; thus, surface contamination and environmental factors are likely to affect the stability of these sensors. Microfluidics is the science and technology of systems that process or manipulate small amounts of fluids, using microchannels with dimensions of tens to hundreds of micrometers. The microfluidic chip can provide natural protection of the MNFs, small volume of samples, and new sensing mechanisms. The fusion of MNFs and microfluidics opens a door to the practical application of MNF sensors. This chapter describes the fundamentals of MNFs and microfluidics and reviews recent progress in MNF microfluidic sensors regarding their fabrication, waveguide properties, sensing structures, and sensing applications.

Introduction

In the past decades, optical fibers with diameters larger than the wavelength of transmitted light had quickly found extensive applications including chemical sensors, biosensors, and physical sensors (Lee 2003; Fan et al. 2008; Wolfbeis 2008). The optical fiber sensors have certain advantages that include immunity to electromagnetic interference, lightweight, small size, high sensitivity, large bandwidth, and ease in implementing multiplexed or distributed sensors (Lee 2003). Recently, with the rapid progresses in nanotechnology, there is an increasing demand for faster response, smaller footprint, higher sensitivity, and lower power consumption, which spurred great efforts for miniaturization of fiber-optic components and devices (Wu and Tong 2013). Although optical fibers with diameters close to the wavelength of propagating light had been fabricated, they had not been paid much attention until 2003. Tong and Mazur demonstrated low-loss optical waveguiding in micro-/nanofibers (MNFs) with diameters far below the wavelength of the guided light, which renewed research interests in MNFs (Tong et al. 2003). As potential building blocks for miniaturization of optical components and devices, MNFs have been attracting intensive research interests regarding their fabrication, properties, and applications. Among various MNF applications, optical sensing has been one of the most attractive research fields due to its possibilities of realizing miniaturized fiber-optic sensors with small footprint, high sensitivity, fast response, high flexibility, and low optical power consumption (Guo et al. 2014). Typical microfiber-based sensing structures, including biconical tapers, optical gratings, circular cavities, Mach–Zehnder interferometers, and functionally coated/doped microfibers, have been designed for refractive index, concentration, temperature, humidity, strain, and current measurement in gas or liquid environments (Lou et al. 2005).

The manipulation of fluids in channels with dimensions of tens of micrometers – microfluidics – has emerged as a distinct new field (Whitesides 2006). Microfluidics has demonstrated the capability to influence subject areas from analytical chemistry to optics and information technology (Psaltis et al. 2006; Fan and White 2011). Although quite a few commercialized biosensors based on microfluidics have been developed for clinical diagnosis and biological analysis, microfluidics is still at its early stage of development. For example, absorbance detection is the most universal detection mechanism in analytical chemistry. However, this popularity has not translated to the microfluidics mainly because of the poor sensitivity due to the shallow channel depth and the difficulties in coupling the light into and out of these channels.

The synergistic integration of MNF and microfluidics is a new sensing platform that provides a number of unique characteristics for enhancing the sensing performance and simplifying the design of microsystems. This chapter reviews recent progress in MNF microfluidic sensors regarding their fabrication, waveguide properties, and sensing applications. Finally, this chapter is concluded with an outlook for challenges and opportunities of optical MNF microfluidic sensors.

Fundamentals of MNFs

For optical waveguiding, excellent geometric uniformity and surface smoothness of the MNFs are critical for achieving low optical loss and high signal-to-noise ratio, and therefore the fabrication process of these tiny fibers is vitally important. Compared with many other techniques such as photo- or electron-beam lithography, chemical growth, and nano-imprint, high-temperature taper-drawing method yields MNFs with lowest surface roughness, largest length, and excellent diameter uniformity. Also, the amorphous structure of the glass material bestows the MNF with circular cross section, which is ideal for obtaining waveguiding modes by solving Maxwell's equations analytically. This section briefly introduces the fabrication of MNFs and guiding properties of MNFs.

Fabrication of MNFs

Flame-heated taper-drawing is widely used to draw MNFs from standard optical fibers. A typical illustration of flame-heated taper-drawing process is shown in Fig. 1a. A hydrogen–oxygen flame is used for heating the fiber. Under a certain pulling force, the fiber is stretched and elongated gradually with reduced diameter

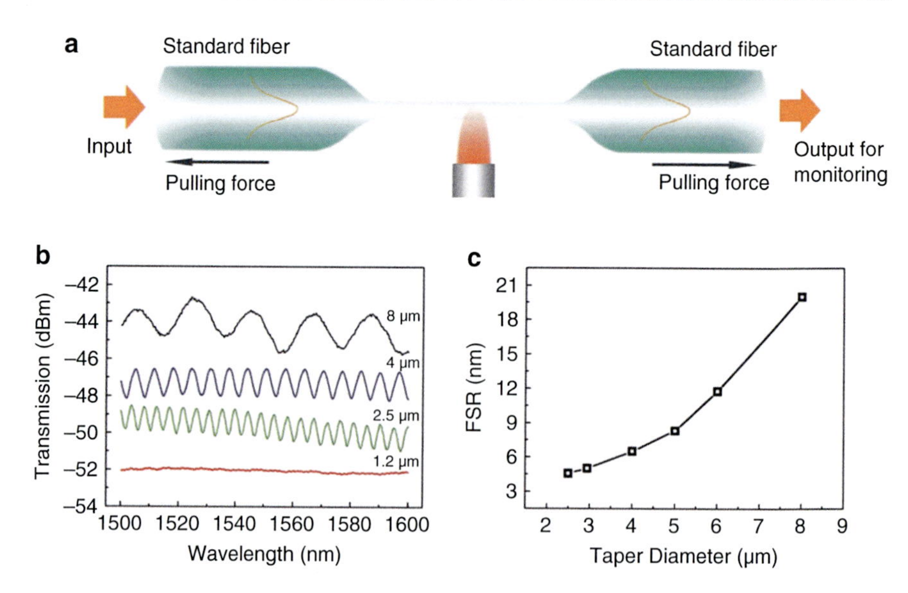

Fig. 1 (**a**) Schematic diagram of flame-heated taper-drawing of an MNF from a standard optical fiber (Adapted from Wu and Tong (2013), with permission from the Walter De Gruyter GmbH) (**b**) Transmission spectra (offset) of tapered microfibers with diameters of 8 μm, 4 μm, 2.5 μm, and 1.2 μm (corresponding to tapering lengths of 10 mm, 14 mm, 15 mm, and 16 mm) and (**c**) measured FSR monotonously decreases with the fiber diameter (Adapted from Li et al. (2014a), with permission from Elsevier)

until the desired length or diameter of the fiber taper is reached. Using this technique, the as-fabricated MNF is usually attached to the standard fiber through the tapering region at both ends and is usually named as a "biconical" fiber taper. Importantly, it is critical to in situ monitoring the waveguiding properties of the MNF during the pulling process in terms of propagation loss, multimode interference, and group velocity delay by measuring optical transmission via standard fiber. Typically, when the taper length is from 10 to 16 mm, the total transmittance of the multimode waveguiding light is higher than 80% around 1550-nm wavelength. By launching broadband light from a white light source (halogen lamp) into one side of the as-fabricated MNFs and collecting the output from the other side with an optical spectrum analyzer, the broadband transmission spectra of the microfibers with different diameters are measured within the wavelength range of 1500–1600 nm (Fig. 1b). For those microfibers with diameters larger than 1.2 μm (the single-mode cutoff diameter for an air-cladding silica microfiber operating at 1.55 μm wavelength), higher-order modes are supported. Experimentally, for microfibers with diameters larger than 2 μm, evident sinusoidal oscillatory transmission features are clearly seen. While for those with diameters close to or thinner than 1.2 μm, only the fundamental HE_{11} mode is supported, and thus the sinusoidal oscillation disappears. The relationship between the free spectrum range (FSR) and the diameter of the microfiber is shown in Fig. 1c. It shows that the FSR monotonously decreases with the fiber diameter (Li et al. 2014a).

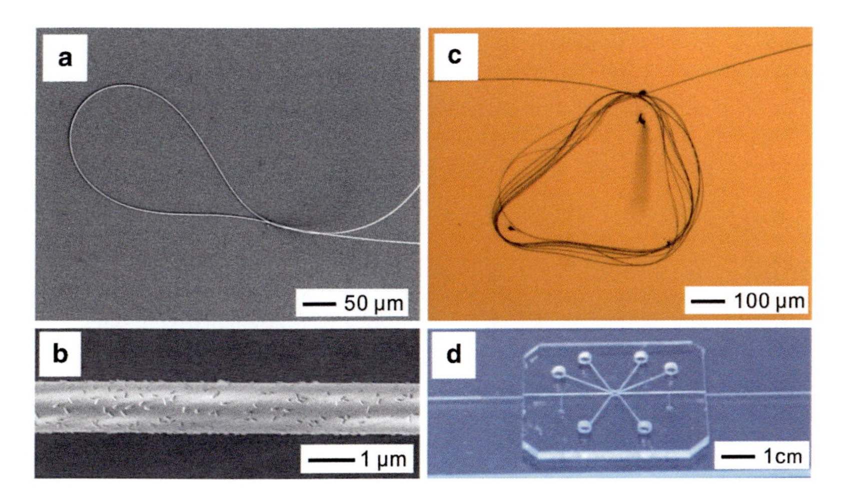

Fig. 2 Typical images of glass MNFs. (**a**) A 1.5-μm-diameter silica glass microfiber with a bending radius of 100 μm. (**b**) A 1-μmdiameter silica glass microfiber decorated with gold nanorods. (**c**) A coiled 1-μm-diameter silica glass microfiber with a total length of about 1 cm. (**d**) An 800-nm-diameter nanofiber integrated with a microfluidic chip with a 5-μm-wide detection channel

In some situations, flame-heated systems may present disadvantages such as the random turbulence of the flame and oxygen requirement in the burning process, leading to H_2O/OH contamination in MNFs. To avoid these issues, a CO_2 laser beam can be used as an alternative heating source. Usually, the direct laser heating tapering procedure shows a self-regulating effect, which automatically stops the stretching process when the fiber diameter goes down to a certain value (usually above 1 μm). By drawing MNFs in a microfurnace comprising a sapphire tube heated with a CO_2 laser, Sumetsky successfully fabricated sub-μm-diameter MNF with excellent surface smoothness and diameter uniformity (Sumetsky et al. 2004). Besides the abovementioned techniques, electrically heated taper-drawing approach is another simple and effective technique for fabricating high-quality MNFs (Shi et al. 2006). Usually, the electrical heater can be shaped into various geometries to precisely generate required temperature and temperature distribution, which makes it possible to draw MNFs with more flexibilities.

In order to bestow the MNFs with greater versatilities, a number of post-fabrication techniques including micromanipulation, plastic bend, coating, and embedding have been investigated in the past years. For reference, Fig. 2 shows typical images of twisted MNF, coiled MNF, gold nanorod-decorated MNF, and microfluidic chip embedded MNF, respectively.

Optical Properties and Opportunities of MNFs

Comparedwith a conventional optical fiber, a high index-contrast (Δn) MNF with wavelength (λ) or subwavelength (sub-λ) diameter offers a number of interesting optical properties and opportunities, including tight optical confinement, strong evanescent field, and small mass/weight (Guo et al. 2014). For basic investigation, a straight MNF is assumed to have a circular cross section, a smooth sidewall, a uniform diameter, and an infinite cladding with a step-index profile. The fiber diameter (D) is not very small (e.g., $D > 10$ nm) so that the statistic parameters permittivity (ε) and permeability (μ) can be used to describe the responses of a dielectric medium to an incident electromagnetic field. By solving Maxwell's equations, one can obtain the field distribution of a waveguiding MNF. Figure 3 shows power distribution (Z-direction Poynting vectors) of HE_{11} mode of silica MNFs with diameters of 800, 400, and 200 nm in 3D and 2D view, respectively. It is clear that, while an 800-nm-diameter MNF confines major energy inside the fiber, a 200-nm-diameter MNF leaves a large amount of light (>90%) guided outside as evanescent waves. Therefore, a freestanding MNF offers an opportunity to waveguide tightly confined optical fields with a high fraction of evanescent fields, which distinguishes the MNFs from many other waveguiding structures. For example, in Fig. 3b, the 400-nm-diameter nanowire guides a 633-nm-wavelength light with about 30% power outside as evanescent waves, while it maintains an effective mode area of about 560 nm in diameter.

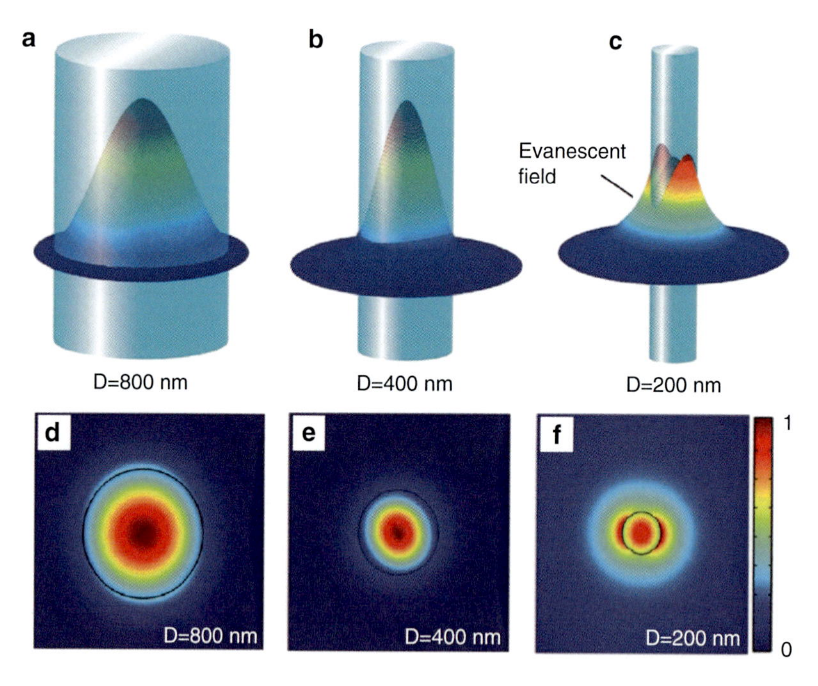

Fig. 3 Z-direction Poynting vectors of silica MNFs at 633-nm wavelength with diameters of (**a**) 800 nm, (**b**) 400 nm, and (**c**) 200 nm in 3D view and (**d**) 800 nm, (**e**) 400 nm, (**f**) 200 nm in 2D view (Adapted from Wu and Tong (2013), with permission from Walter De Gruyter GmbH)

Tight optical confinement bestows the MNF with small allowable bending radius (e.g., low loss when passing through sharp bends) and small mode area, which makes MNFs highly potential for compact circuits and devices with smaller footprint, faster response, and lower power consumption. As shown in Fig. 4a, b, Li et al. demonstrated Mach–Zehnder interferometers (MZIs) assembled with tellurite glass MNFs, achieving good interference fringes with extinction ratios of ~10 dB (Li and Tong 2008). Gu et al. reported highly versatile nanosensors using polymer single nanowires as shown in Fig. 4c. The reversible response of the polyacrylamide (PAM) nanowire is tested by alternately cycling 75% and 88% RH air inside the chamber, achieving an excellent reversibility as shown in Fig. 4d. The estimated response time (baseline to 90% signal saturation) of the sensor is about 30 ms, which is one or two orders of magnitude faster than those of existing RH sensors (Gu et al. 2008).

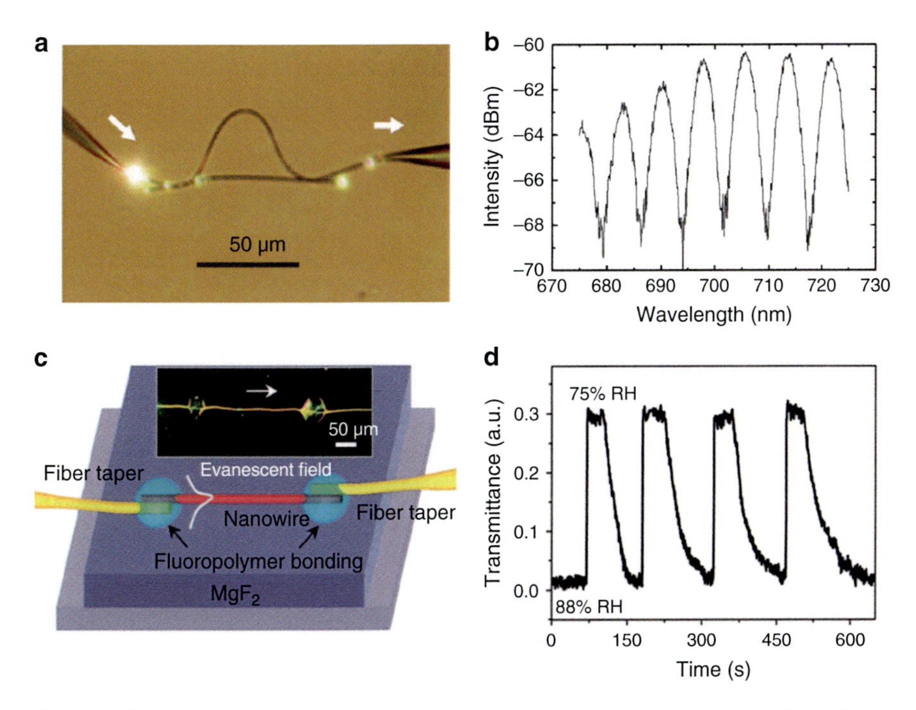

Fig. 4 MZI assembled with tellurite glass MNFs and humidity sensor assembled with PAM nanowire and fiber tapers. (**a**) Optical microscope image of an MZI assembled with two 480-nm-diameter tellurite MNFs. (**b**) Transmission spectrum of the MZI shown in Fig. 4a (Adapted from Li and Tong (2008), with permission from Optical Society of America). (**c**) Schematic illustration of the polymer single-nanowire humidity sensor. Inset, optical microscope image of an MgF_2-supported 410-nm-diameter PAM nanowire with a 532-nm-wavelength light launched from the left side. The white arrow indicates the direction of light propagation. (**d**) Reversible response of the nanowire tested by alternately cycling 75% and 88% RH air (Adapted with permission from Gu et al. (2008). Copyright (2008) American Chemical Society)

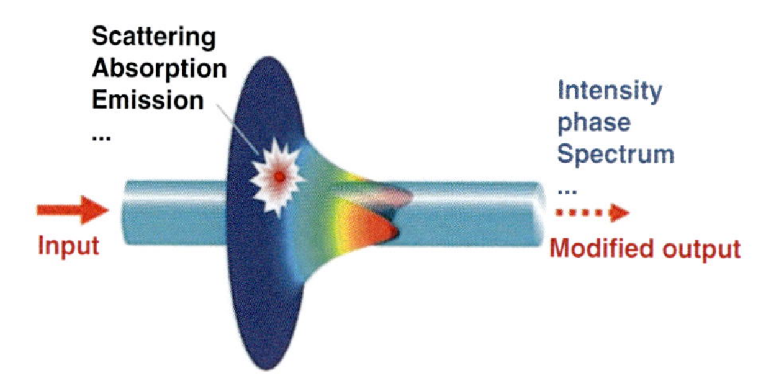

Fig. 5 Schematic illustration of a waveguiding MNF for optical sensing. The sample light interaction (e.g., scattering, absorption, and emission) is reflected by the output (e.g., intensity, phase, or spectrum) of the MNF (Adapted with permission from Guo et al. (2014). Copyright (2014) American Chemical Society)

Strong evanescent field offers strong near-field interaction between the MNF and its surroundings, making the MNF highly favorable for optical sensing. The mechanism of a typical MNF sensor is schematically illustrated in Fig. 5. Within the optical near field of an optical MNF, a slight fluctuation in dielectric constant might evidently modify the output of the MNF by scattering, absorption, and emission of the guided light. By measurement of the intensity, phase, polarization, or spectrum of the output, the cause behind the fluctuation can be identified (Guo et al. 2014).

In most situations, the measurand is either spatially distributed or highly localized. For sensing spatially distributed samples, which are usually in forms of or dissolved in liquids or gases, a typical approach is measuring the refractive index or absorption, which directly reflects the change of measurands (e.g., concentration, temperature, or pressure). The high fractional evanescent fields that are directly exposed to the surrounding sample may greatly enhance the sensitivity of the MNF sensors. Categorized by measurand, MNF sensors for refractive index, temperature, humidity, strain, and current measurement in gas or liquid environments have been reported (Lou et al. 2005).

Fundamentals of Microfluidics

Microfluidics, the manipulation of liquids in channels with cross-sectional dimensions on the order of 10–100 μm, has been a central technology in a number of miniaturized systems that are being developed for chemical, biological, medical, and optical applications (Arora et al. 2010). With the fusion of photonics and microfluidics, optofluidics has emerged as a distinct new field. Integration and reconfigurability are two major advantages associated with optofluidics. It provides new freedom to both photonics and microfluidics and permits the realization of optical and fluidic property manipulations at microscale. In the past decade, optofluidics has evolved into a multidisciplinary field, including chemical and biological sensing, lens-free imaging, particle manipulation, and tunable optical

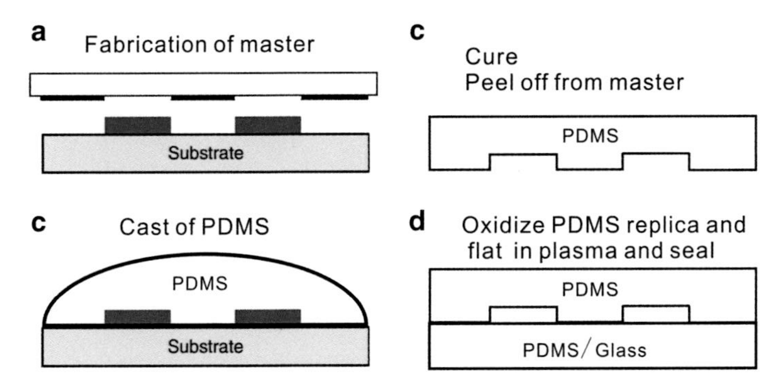

Fig. 6 Scheme describing replica molding of microfluidic chip. (**a**) A master with designed pattern is fabricated by lithography. (**b**) The prepolymer is cast on the master and cured. (**c**) The PDMS replica is removed from the master. (**d**) Exposing the replica and an appropriate material to an air plasma and placing the two surfaces in conformal contact make a tight, irreversible seal

devices, such as liquid core–liquid cladding waveguide, microlenses, gratings, and light source (Monat et al. 2007; Chen et al. 2012).

Fabrication of Microfluidic Chips

To date, a number of materials (e.g., glass, silicon, poly(dimethyl siloxane) (PDMS), and poly(methyl methacrylate) (PMMA)) have been used for fabricate microfluidic chips. PDMS is one of the most widely used materials to fabricate microfluidics, lab-on-a-chip, microelectromechanical systems, and flexible electronics/photonics device due to its high transparency, low refractive index, high flexibility, and high Poisson's ratio coefficient. Although any microfluidic fabrication method can, in principle, be adapted to fabricate a device for optical application, most of the implementations thus far have been with soft lithography. Soft lithography provides an approach to rapid prototyping of both microscale and nanoscale structures and devices on planar, curved, flexible, and soft substrates especially when low cost is required (Qin et al. 2010) Fig. 6 describes a typical procedure for fabricating PDMS microfluidic chip. Briefly, a master with designed pattern is fabricated by lithography. The prepolymer is then cast on the master and cured. When the PDMS replica is removed from the master, the replica and an appropriate substrate are exposed to an air plasma, and placing the two surfaces in conformal contact makes a tight, irreversible seal.

Manipulation of Fluids in Microchannels

In order to manipulate fluids in microchannels, one has to understand how fluids behave at microscale. In the microchannels, surface forces dominate over body and

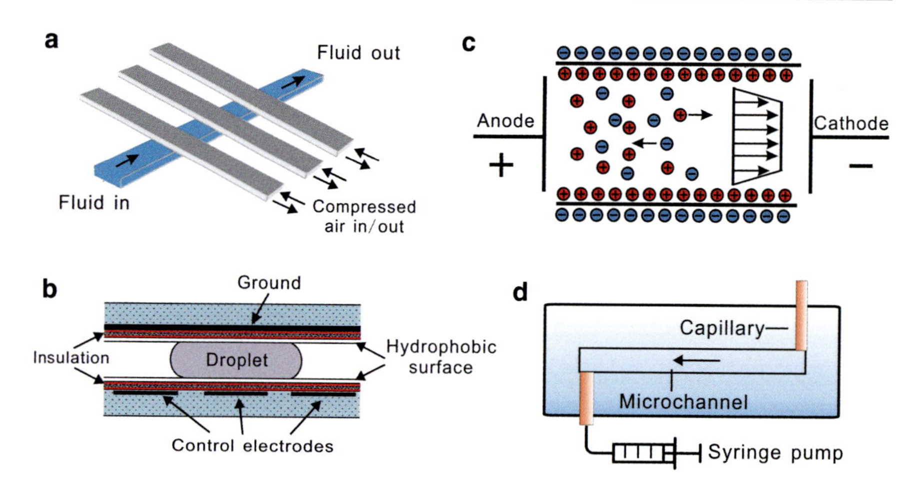

Fig. 7 Typical approaches to manipulate fluids in microchannels. (**a**) Pneumatic actuated peristaltic micropump. (**b**) Schematic illustration of electrowetting. (**c**) Schematic illustration of electroosmotic flow micropumps. (**d**) Schematic illustration of manipulation of fluid in microchannel via a syringe pump

inertial forces. As the channel dimensions decrease, the surface forces decrease proportional to the square of the length scale. The body forces, however, decrease as the cube of the length scale. Thus, a tenfold decrease in channel dimensions leads to a tenfold increase in the surface-to-volume ratio of the channel, 100-fold decrease in surface forces, and 1000-fold decrease in volume forces. Therefore, the surface forces become tenfold more important relative to the body forces. To date, a number of approaches have been used to manipulate fluid in microchannels, including peristaltic pumps, electrowetting, electroosmotic flow micropumps, negative pressure generated by syringe pump, and so on.

Peristaltic pumps are based upon using a series of actuation membranes to displace volume in the desired flow direction (Unger et al. 2000). By cycling membrane displacement, a peristaltic motion and volume displacement may be achieved. Peristaltic micropumps require at least three actuation membranes in series to obtain a nonreversible pump stroke. A typical pneumatic actuated peristaltic micropump is shown in Fig. 7a. The maximum pumping rate of 2.35 nL/s was attained at pump cycles of 75 Hz; above this rate, increasing numbers of pump cycles compete with incomplete valve opening and closing. The pumping rate was nearly constant until above 200 Hz and fell off slowly until 300 Hz.

Electrowetting is based upon a change in the liquid–solid surface tension by charging the electrical double layer at an electrode surface. Figure 7b shows a schematic illustration of manipulation of a droplet in microchannel via electrowetting technology. By electrically changing the wettability of each of the electrode on the surface, a droplet on these electrodes can be shaped and driven along the active electrodes. Currently, electrowetting mechanism has been used extensively for droplet manipulation and tunable microliquid prism (Lee et al. 2002, 2013).

Electroosmotic flow (EOF) pumping is another subset of electrokinetic phenomena related to the movement of electric charges in an applied electric field (Manz et al. 1994). Charge movement produces a shear on the surrounding fluid to produce flow. Electroosmotic pumps are used in small channels without a need for high pressures and may be combined with electrophoresis in bioanalytical separations. Note that electroosmotic flow is sensitive to the surface charge, Debye length, and applied electric field. As shown in Fig. 7c, in the presence of the electric field, positively charged ions will move toward the negatively charged cathode, and the negatively charge ions will move toward the positively charged anode. In the bulk region of the microchannel, charge neutrality is maintained, so there is no net charge movement. Although the abovementioned techniques have been widely used in chip-based chemical synthesis, electrophoresis, and manipulation of droplets, their application in microfluidic sensors is quite limited owing to the complicated fabrication process and control system.

Figure 7d shows a schematic illustration of manipulation of fluid in microchannel via a syringe pump; the negative pressure sampling method has been proven to be an efficient and convenient approach to manipulate small volume of fluid in the microchannel (Zhang et al. 2006). Two obvious advantages are their freedom from electrokinetic bias effects and their ease for fabrication and manipulation.

Planar Microfluidic Chip-Based Biconical MNF Sensors

One of the most straightforward approaches to assemble an MNF sensor is using a biconical MNF, which has been successfully applied in measuring a variety of samples. Due to its simple configuration, it can be integrated with planner microfluidic chip with ease. This kind of sensors usually measures the transmission intensity of the MNF and retrieves the sample information by concentration-dependent optical intensity.

Refractive Index Sensor

Measuring the refractive index of a sample is one of the most popular approaches in bio-/chemical analysis and label-free sensing. In particular, it is extremely useful for detecting samples that do not absorb in the UV-Vis range or without fluorescence. Because the refractive index signal scales with the analyte concentration or surface density, rather than with the total number of molecules, refractive index detection is attractive for microfluidic sensors that have extremely small detection volumes.

An MNF microfluidic sensor, which measures the refractive index of liquids propagating in a millimeter-scale channel, was demonstrated in Polynkin et al. (2005). The MNF was fabricated from a commercial single-mode fiber and embedded in PDMS. The sensing device, which is illustrated in Fig. 8a, was fabricated as follows. A rectangular cuvette serves as a mold for PDMS. A slit was machined through the bottom of the cuvette to accommodate a 3-mm-thick glass rod that

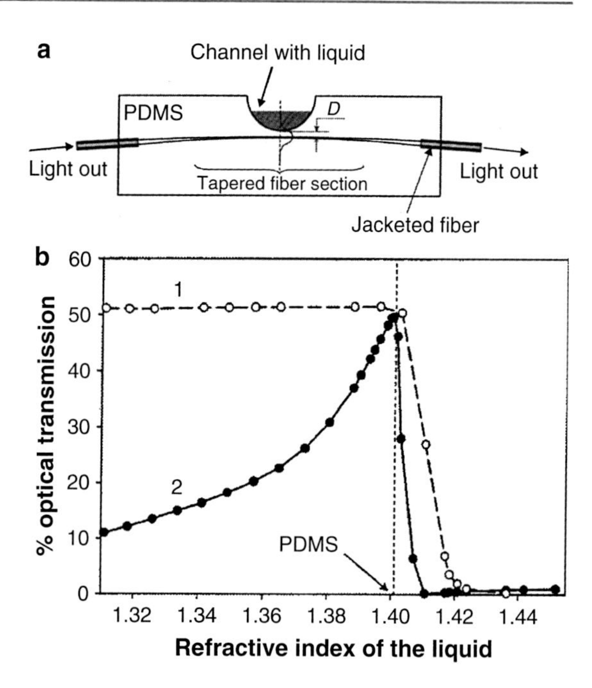

Fig. 8 (**a**) Illustration of the MNF sensor with liquid channel. (**b**) Measured optical transmission versus refractive index of the liquid in the channel for two sensors with different taper thicknesses at the waist: curve1–1.6 μm; curve 2–700 nm. Circles indicate the measurement of data points; the arrow shows the refractive index of the surrounding polymer (Adapted from Polynkin et al. (2005), with permission from the Optical Society of America)

defines the channel for the sample liquid in the sensor. When the rod was in place, it was sticking out of the bottom of the cuvette by approximately one half of its diameter. Thus the resulting 1-cm-long and 3-mm-wide channel has a semicircular cross section. After the mold was filled with liquid PDMS, the MNF oriented orthogonally to the glass rod was slowly lowered into the uncured PDMS from the top. After the PDMS is cured at room temperature, the device was flipped over and the glass rod was removed, exposing the channel for the sample liquid.

The sensing mechanism of this sensor is based on the fact that the shape of the fundamental optical mode, which travels through the sensing channel, was modified by the refractive index contrast between the sample liquid and the PDMS channel wall. When the refractive index of the sample liquid is higher than that of the PDMS (1.402 at 1.5-μm wavelength), the mode is attracted into the channel, where the light is effectively scattered because of the presence of the channel boundaries and is absorbed in the liquid. If the refractive index of the sample liquid is lower than the index of PDMS, the mode propagates mostly in the uniform, transparent polymer. Consequently, the change of the fundamental mode resulted in variation of the MNF transmission. The sensitivity of the sensor was investigated by measuring the optical transmission through the MNF with the channel filled with solutions of glycerol in water. The refractive index at 1.5-μm wavelength of the solutions can be varied from 1.311 for pure water to 1.459 for pure glycerol by using different concentrations of glycerol. The resulting calibration curves are shown in Fig. 8b for two different sensors, fabricated with (1.6 ± 0.2) μm-diameter and (700 ± 200) nm-diameter MNF. In both cases, the transmission is at maximum when the refractive index of the sample liquid matches that of the PDMS, and the thinner MNF-based sensing device is more sensitive, achieving a potential measurement accuracy of 5×10^{-4}.

Evanescent-Wave Absorption Sensor

Absorbance detection is one of the most universal detection mechanisms in analytical chemistry. However, this popularity has not been translated to the microfluidic sensors as the on-chip measurement of absorbing species has proven to be challenging, mainly because of the shallow channel depth (short optical path length) in microfluidic chip and the difficulties in coupling the light into and guiding the light out of these channels. When light is guided along an MNF, the evanescent fields outside the MNF interact with the analytes nearby. The theory of evanescent wave absorption (A) is given by Lambert–Beer's law:

$$A = \alpha L \eta$$

where α is the absorbance coefficient, L is the detection length, and η is the fractional power of evanescent fields outside the MNF. Most previously reported evanescent field absorption sensors adopted tapered fibers with a typical waist diameter larger than 10 μm, which results in low η and subsequently low sensitivity. Owing to the large achievable L and η, MNF-based evanescent field absorption has high potential for high sensitivity. Note that a great number of the reported evanescent field absorption sensors used MNFs suspended in air or mounted in a bulky volume flow chamber; thus, surface contamination and environmental factors are likely to affect the stability of these sensors.

Zhang et al. (2011) developed an MNF absorption sensor by using a 900-nm-diameter silica nanofiber embedded in a 125-μm-wide microchannel with a detection length of 2.5 cm. When an MNF was integrated with a microfluidic chip as shown in Fig. 9, the microfluidic chip can provide natural protection of the MNF and decrease the sample volume dramatically to 500 nL. Moreover, the surface of the MNF can be cleaned and renewed after a measurement by flushing water, ethanol, and air via the microchannels. The microfluidic chip was fabricated by bonding a PDMS replica to a piece of PDMS membrane to seal the microchannels. In order to embed an MNF into a microchannel, the width and depth of the microchannels were designed and fabricated to be 125 μm and 150 μm, respectively (Fig. 9b). The embedding process was manipulated by a 3D travel translation stage under an optical microscope. Two 2-cm-long capillaries (150 μm i.d., 375 μm o.d.) were inserted into the capillary channel; one served as sampling probe and the other connected to a syringe for sample delivery. As shown in Fig. 9c, a 1.5-μm-diameter MNF guiding a 473-nm-wavelength laser was embedded in a microchannel. When 0.01 mM fluorescein solution was injected into the microchannel, bright fluorescence excited by the 473-nm-wavelength laser can be seen as shown in Fig. 9d, indicating evanescent field outside the MNF.

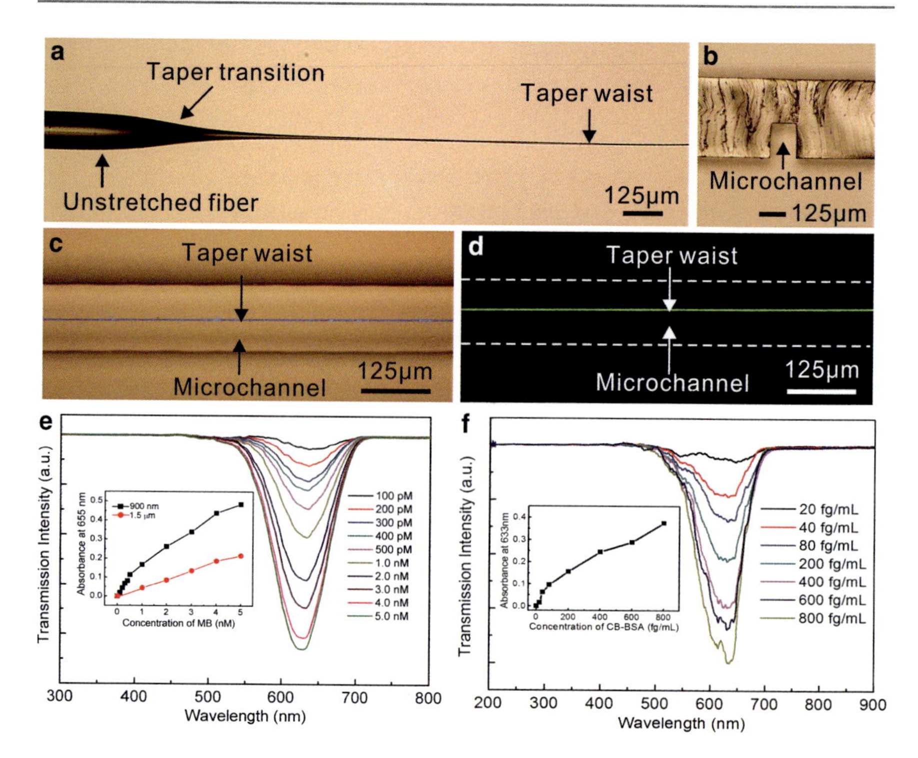

Fig. 9 Integrated MNF microfluidic optical sensors. (**a**) Optical microscope image of a tapered fiber with a waist of 900 nm. (**b**) Optical microscope image of the cross section of a PDMS microchannel. (**c**, **d**) Microfluidic structure before and after fluorescence excitation. (**e**, **f**) Transmission spectra of MB (**e**) and CB-BSA (**f**) obtained at different concentrations using a 900-nm-diameter nanofiber (Adapted from Zhang et al. (2011) with permission from Royal Society of Chemistry)

Investigated by measuring the absorbance of methylene blue (MB) around 630-nm wavelength, the sensor shows a detection limit down to 50 pM with excellent reversibility in a concentration range of 0–5 nM (Fig. 9e). The sensor has also been applied to bovine serum albumin (BSA) measurement, achieving a detection limit of 10 fg/mL (Fig. 9f). Note that the probing light power was about 150 nW, suggesting a promising route to low-power high-sensitivity biochemical sensors.

Evanescent-Wave Fluorescence Sensor

The evanescent wave fluorescence sensor is a well-developed tool for a wide range of applications, particularly for biological sensing (Leung et al. 2007). Fluorescence photons which can be efficiently coupled into guided modes of optical nanofibers offer a possibility for single-molecule or single-nanoparticle sensing (Yalla et al. 2012). Li et al. (2015) developed an MNF fluorescence sensor by using

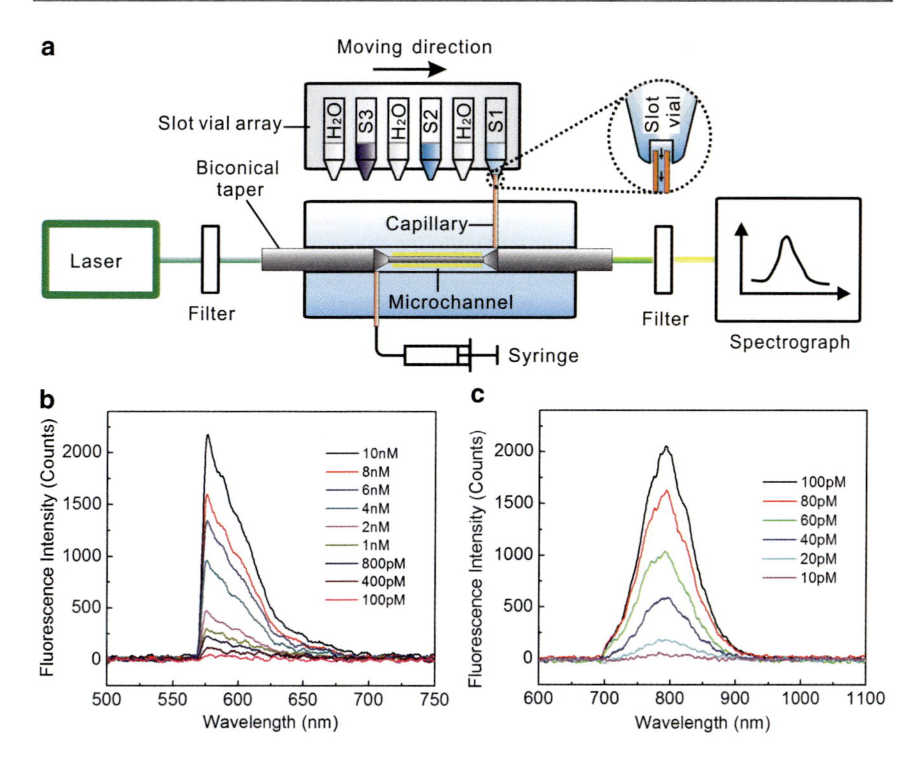

Fig. 10 (**a**) Schematic illustration of the experimental setup for the biconical MNF-based fluorescence sensing system. (**b**) Fluorescence spectra of R6G at different concentrations. (**c**) Fluorescence spectra of QD-labeled streptavidin at different QD concentrations (Adapted from Li et al. (2015) with permission from MDPI, Basel)

a biconical MNF with a diameter of 720 nm for both excitation and fluorescence collection.

Figure 10a shows the schematic experimental setup for fluorescence measurement. Excitation laser (532-nm wavelength) is coupled to the MNF from the left side of the unstretched fiber. A fiber spectrometer is used to record the fluorescence spectra. To enhance the signal-to-noise ratio, a 550-nm-wavelength short pass filter and a long pass filter (561 nm for rhodamine 6G (R6G) and 700 nm for quantum dot (QD)-labeled streptavidin) are used to remove any scattered or excitation light from the system. In the taper waist section (MNF), the mode of the unstretched fiber is adiabatically transformed into the strongly guided mode of the ultrathin section and back, resulting in a highly efficient coupling of light into and out of the taper waist.

It is worth mentioning that a slot sample vial array was adopted for sample introduction because of its high sampling throughput (Du et al. 2005). The sample vials were produced from 0.5 mL centrifuge tubes by cutting a 1.5-mm-wide, 2-mm-deep slot at the conical bottom of the tube for pass-through of the sampling capillary. The slotted sample vials were horizontally fixed on a glass slide in an array, with the slot of each vial positioned horizontally to allow free passage of the sampling capillary through all the vial slots sequentially by linearly moving the glass slide along the direction of the array. The vials were filled alternately with 100 μL samples and ultrapure water. Thus, the microchannel and the surface of the MNF can be cleaned before a new sample was introduced by linearly moving the glass slide. Samples in the slot vials were sucked into the microchannel by negative pressure generated by a syringe that connected with the sample outlet channel. In this case, the sampling throughput can be as high as 60 samples per hour.

The performance of the sensor was systematically investigated by measuring the fluorescence intensity of R6G and QD-labeled streptavidin, respectively. When a 532-nm-wavelength laser was guided into a 720-nm-diameter MNF to excite the fluorescence, the fractional power outside the core is about 15%, resulting in a strong evanescent field for exciting the fluorescence and a high efficiency for collecting the fluorescence. In this case, the estimated fluorescence collection efficiency for the sensing system was about 13.5%. Note that decreasing the diameter of an MNF is an effective method to increase the fractional power outside the core; however, when the fractional power outside the core increases to a critical value, the fluorescence collection efficiency will decrease. As shown in Fig. 10b, the fluorescence intensity increases obviously with the increases of R6G concentration. The detection limit for R6G based on three times the standard deviation of the blank values was about 100 pM. Compared to conventional organic fluorophores, QDs are virtually immune to the effects of photo-bleaching, and have a relatively higher absorption cross section, which may further enhance the sensitivity. Figure 10c shows the typical fluorescence spectra of QD-labeled streptavidin at different QD concentrations ranging from 10 to 100 pM. The detection limit for QDs was about 10 pM. Because there are three to five streptavidins covalently attached on one QD, the detection limit for streptavidin was about 30–50 pM.

With the advent of single-molecule detection and early disease diagnosis, there is a real need to reduce the detection volume with optical methods, because the concentrations of these samples are relatively high, but the amounts of these samples are quiet limit. To address this issue, total internal reflection with fluorescence correlation spectroscopy, confocal microscopy, zero-mode waveguides, and microstructured optical fibers have enabled the observation and/or detection of molecules and ions in reduced volumes ranging from atto- to nanoliter scale (Laurence and Weiss 2003). The abovementioned approaches have strengths in different areas and are suitable for different applications. However, they require expensive instruments and/or complicated fabrication processes, which prevent lots of researchers from studying and understanding biological processes on the molecular scale. MNF microfluidic sensors provide a cost-effective manner to achieve a small detection volume by manipulating the penetration depth of the evanescent field and the detection length of the MNF simultaneously (Zhang et al. 2015).

To achieve a detection volume of 1.0 fl, a typical detection volume for single-molecule analysis, it is necessary to reduce the channel width to less than 10 μm. Meanwhile, to embed an MNF and deliver liquid samples, the other channels'

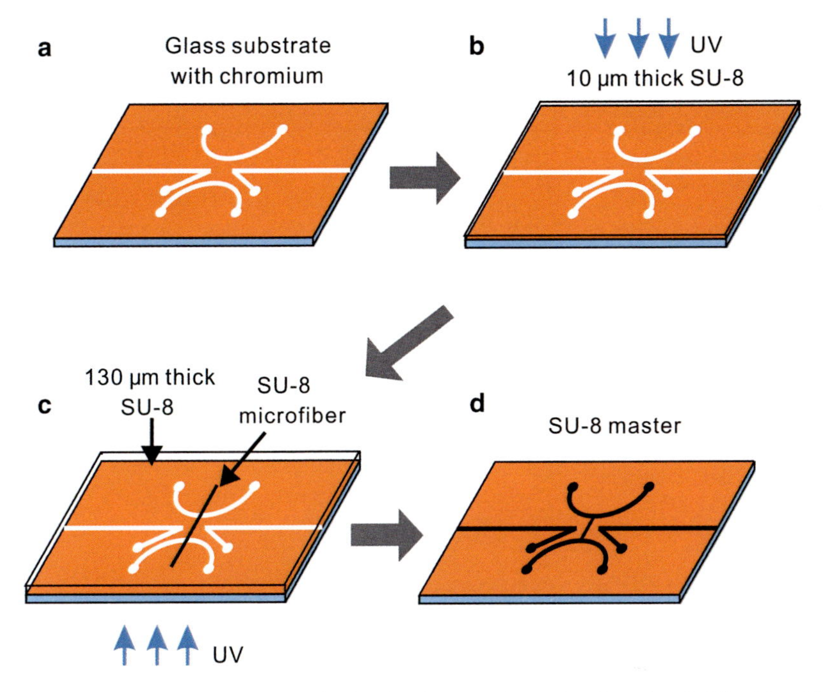

Fig. 11 Schematic illustration of fabricating SU-8 master for replicating microfluidic chip with a sub-10-μm-wide detection channel

dimension should be greater than that of the detection channel. It is a great challenge to fabricate a planar microfluidic chip with different channel depths, in particular, with a narrow and shallow detection channel by using standard lithography method.

Figure 11 shows a four-step procedure for fabricating a SU-8 master for replicating PDMS microfluidic chip with a sub-10-μm-wide detection channel. Firstly, a channel design without the detection channel on a mask is transferred onto a glass substrate (Fig. 11a). Secondly, a film of 10-μm-thick SU-8 photoresist is spin-coated on the substrate, followed by soft bake, exposure, post exposure bake, and hard bake (Fig. 11b). Thirdly, a SU-8 microfiber is located onto the substrate by a 3D travel translation stage under an optical microscope. After post exposure bake and hard bake, the SU-8 microfiber attaches tightly to the SU-8 film (Fig. 11c). Finally, a film of 130-μm-thick SU-8 photoresist is spin-coated on the substrate, followed by soft bake, expose from the bottom of the substrate, post exposure bake, developing, and hard bake (Fig. 11d).

Figure 12a shows a typical optical micrograph of an as-fabricated MNF microfluidic chip with a 500-μm-long, 8-μm-wide detection channel, which connects two 150-μm-wide channels for sample introduction. As shown in the inset of Fig. 12a, when 0.01 mM fluorescein solution was introduced into the 5-μm-wide detection channel, and a 473-nm-wavelength laser was launched to the nanofiber, a bright fluorescence spot excited by the evanescent field outside the nanofiber,

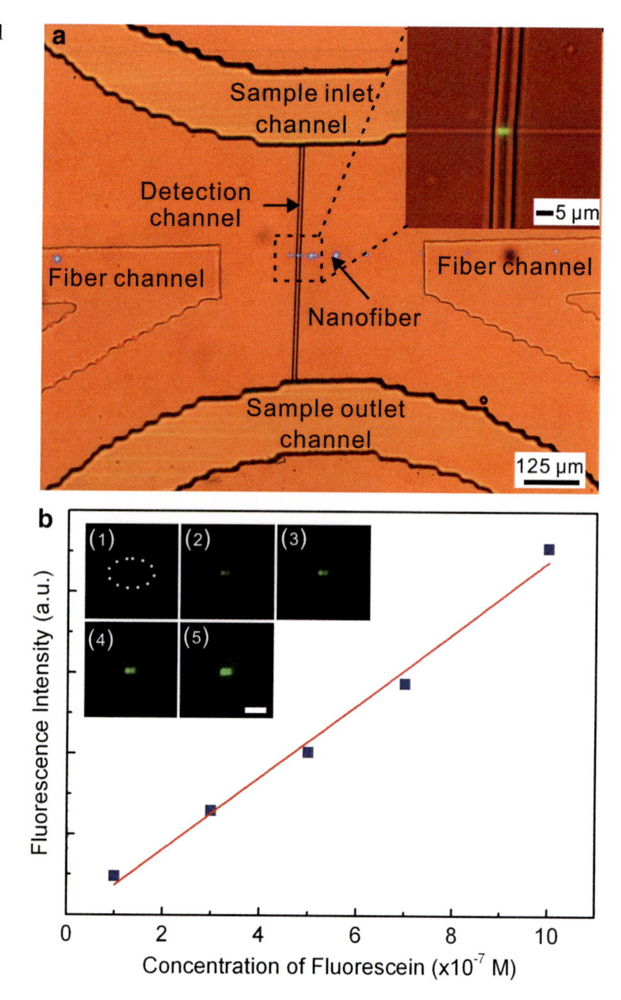

Fig. 12 (**a**) A typical optical micrograph of an MNF microfluidic chip with a nanofiber crossed a narrow detection channel. Inset: Optical micrograph of an 800-nm-diameter nanofiber crossed a 5-μm-wide channel. A bright fluorescence spot excited by evanescent field indicates a detection length of 2.5 μm. (**b**) Calculated fluorescence intensity versus the concentration of fluorescein. Insets: (1–5) Optical micrographs of the fluorescence spots excited by evanescent field outside an 800-nm-diameter nanofiber with fluorescein concentrations of 1×10^{-7}, 3×10^{-7}, 5×10^{-7}, 7×10^{-7}, and 1×10^{-6} M, respectively. Laser power, 5 mW; exposure time, 1.0 s. Scale bar: 5 μm (Adapted from Zhang et al. (2015) with permission from Optical Society of America)

indicating an effective detection length of 2.5 μm. When the penetration depth is defined as the length where the evanescent field intensity decays to 10% of the highest intensity outside the nanofiber, the penetration depth of an 800-nm-diameter silica nanofiber operated at 473-nm wavelength is about 150 nm, leading to an effective detection volume of ~ 1.0 fl. To investigate the sensitivity and linearity of the sensor, fluorescein solutions with concentrations were measured by calculating the fluorescence intensity. A linear concentration-dependent response was obtained as shown in Fig. 12b. When an electron multiplying charge-coupled device (EMCCD) or a fast-response, high-resolution spectrometer is used to record the signal, the femtoliter-scale MNF microfluidic sensor can provide a compact and versatile sensing platform for sensitive and fast detection of ultra-low volume samples, as well as studying the dynamics of single nanoparticles or single molecule.

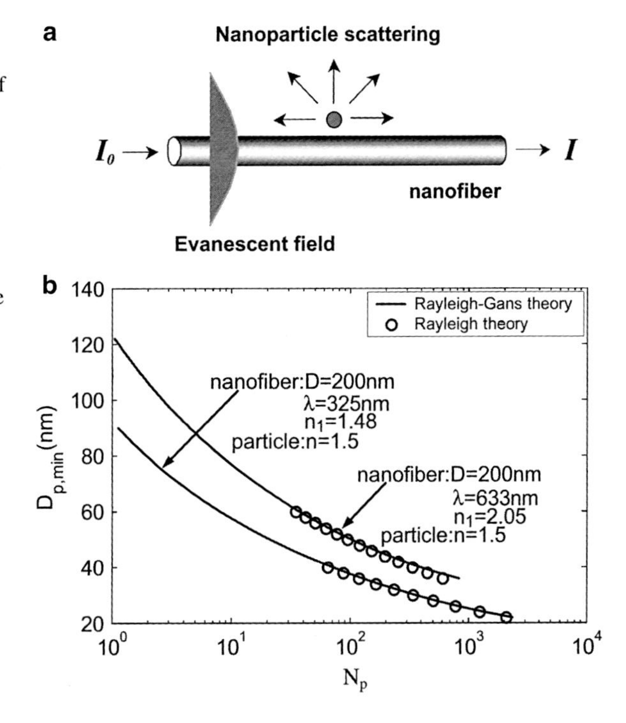

Fig. 13 Basic model of the light scattering of a nanoparticle in the vicinity of a nanofiber. The minimum required number (Np) of particles for being detectable with respect to the particle diameter (Dp, min). The refractive index (n) of the particle and the diameter (D) of the fiber are assumed to be 1.5 and 200 nm, respectively (Adapted from Wang et al. (2007) with permission from Elsevier)

Nanoparticle Sensors Based on the Scattering

Rapid detection and sizing of nanoparticles are becoming of increasing importance for applications in early-stage clinical diagnostics, process control of semiconductor manufacturing, and environmental monitoring. For MNF microfluidic sensor, the spectrum of scattered light provides an effective approach to detect nanoparticles with deep subwavelength cross section, which is imperceptible by ordinary light beams. It can reveal important information about the structure and dynamics of the material being examined. For example, the probing light guided along an MNF can be confined to a "thin light beam" with size comparable to the nanoparticles while maintaining strong evanescent fields for near-field interaction. In this case, a nanoparticle can cause detectable change to the transmission of an MNF by optimizing the wavelength of the probing light and the diameter of the MNF. Based on Rayleigh–Gans scattering theory, calculation showed that with a 325-nm-wavelength probing light guided in a 200-nm-diameter silica nanofiber, a single 90-nm-diameter particle (index of 1.5) can be detected (Fig. 13), indicating the possibility for single-molecule detection (Wang et al. 2007).

Experimentally, Yu et al. (2014) proposed a fast, compact, and inexpensive method to perform single-nanoparticle detection and sizing in an aqueous environment by using nanofiber pairs. By monitoring the step changes in the transmitted power of the fiber, detection and sizing of single polystyrene (PS) nanoparticles

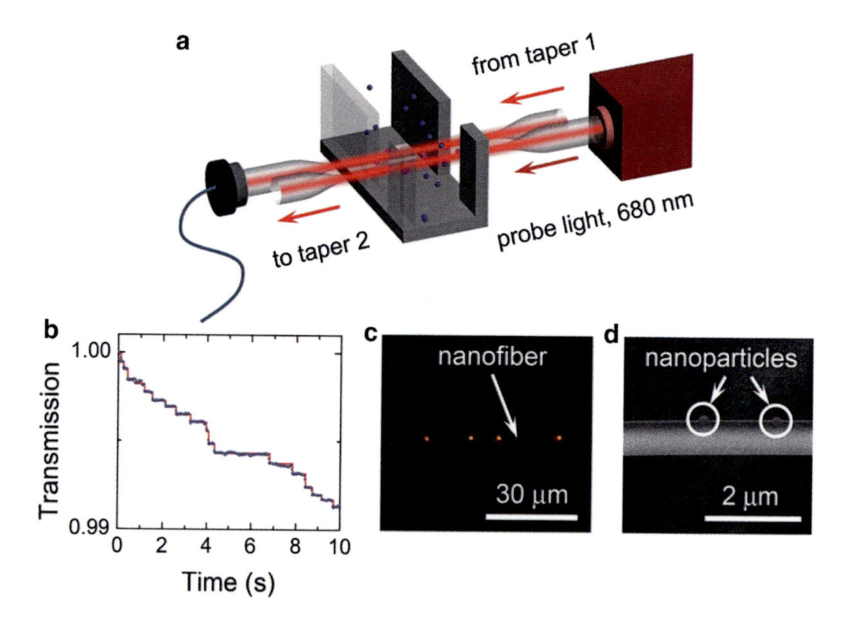

Fig. 14 (**a**) Schematic illustration of the nanofiber sensing system, where a pair of nanofibers is used. (**b**) A typical transmitted power of the fiber during a time interval of 10 s when the PS nanoparticles are binding to the surface of the nanofiber one by one. (**c**) The optical microscopy image of the nanofiber with single PS nanoparticles bound to its surface in an aqueous environment. (**d**) SEM image of a portion of the nanofiber, where PS nanoparticles ($R = 100$ nm) are bound on its surface (Adapted from Yu et al. (2014) with permission from Willey-VCH)

down to radius of 100 nm in an aqueous environment were demonstrated. Figure 14a shows a schematic illustration of the nanofiber sensing system. Since both nanofibers can be used to detect the nanoparticles, the sensing speed is doubled compared with a single nanofiber case. A pair of nanofibers with typical diameter down to 500 nm is fabricated simultaneously by heating and pulling a standard single-mode fiber; the two tapered parts possessing uniform diameter are then immersed in a sample channel filled with deionized water. A typical fiber transmission is shown in Fig. 13b, from which several obvious abrupt decreases in transmitted power can be seen. These step changes correspond to single PS nanoparticle binding events, also confirmed by the CCD image (Fig. 13c), in which each scattering point corresponds to one single nanoparticle bound to the surface of the nanofiber. The one-by-one attachment of the nanoparticles on the surface of the nanofiber was confirmed by the scanning electron microscope (SEM) image (Fig. 13d). In addition, sizing of nanoparticles with a single radius (100 nm) and of mixed nanoparticles with different radii (100 nm and 170 nm) has been realized, with the result agreeing well with theoretical predictions by Raleigh–Gans scattering model for certain-sized nanoparticles. Moreover, with plasmonic enhancement, the nanofiber holds the potential for the detection of single gold nanorods with much smaller sizes.

Low Refractive Index Polymer-Coated Coiled MNF Sensors

Compared to other optical waveguide sensors, a prominent advantage of MNF sensor is its larger area of interaction with the ambient medium and seamless connection to the light source and detector. Nevertheless, the fragileness of a freestanding MNF and surface containments restrict its applications. The design of low refractive index polymer-coated coiled MNFs can effectively increase the robustness of the device and provide more sensing mechanisms.

Evanescent-Wave Absorption Sensor

Lorenzi et al. (2011) developed an in-line absorption sensor, where the analyte flows in a fluidic channel whose walls were made of absorption-responsive coiled microfibers. Figure 15a describes the fabrication procedure: (1) coiling of a biconical MNF onto an expendable PMMA rod, which connects two Teflon tubes,

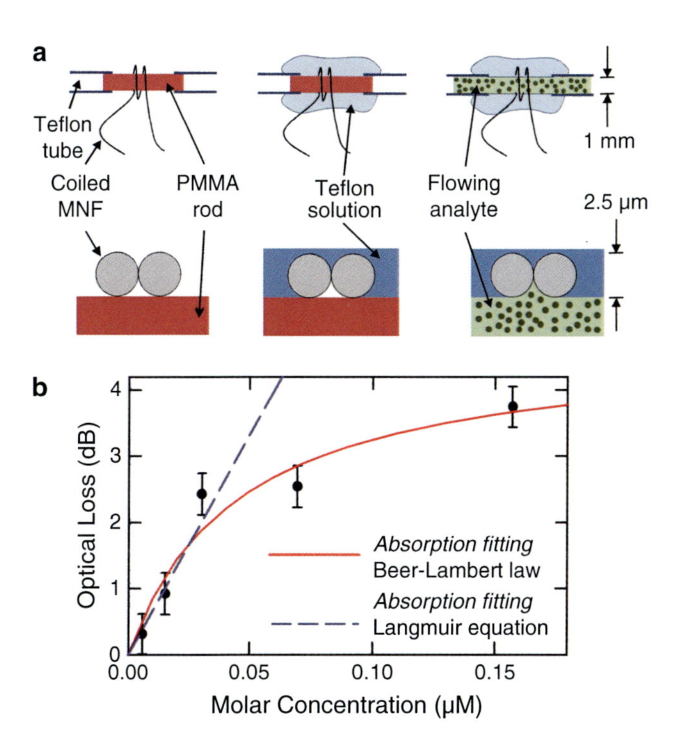

Fig. 15 (**a**) Schematic of the fabrication process. (**b**) Dependence of the measured optical losses at 630 nm on the analyte molar concentration. Dashed blue line represents the fitting curve calculated by the Gupta model of EW absorption in optical fibers. Solid red line represents the fitting result assuming adsorption mechanism and applying the Langmuir equation (Adapted from Lorenzi et al. (2011) with permission from AIP Publishing)

(2) embedding the PMMA rod and the coiled MNF with a Teflon solution and curing at 80 °C for 20 min, and (3) dissolving the PMMA rod by dipping the device in acetone for 1 day. In order to improve mechanical stability, the device is anchored on a microscope glass slide and covered with an UV-curable acrylate polymer as low-loss protective glue. Figure 15b shows the optical loss as a function of bright blue concentration. Dashed blue line represents the expected optical loss. Note that the linear relationship is a good approximation only for low concentrations. At low concentrations, the optical absorption will be proportional to the number of molecules effectively dispersed in the solution; since the number of molecules attached onto the fiber surface is too small to give any contribution. On the contrary, as the molarity increases, more and more molecules attach onto the fiber surface and the resulting optical absorption is dominated by these molecules. The proposed device may find applications as absorption-responsive tube when small volumes are needed.

Refractive Index Sensor Based on Coiled MNF Resonator

Resonating structures can provide a simple, cost-effective, and sensitive technology for refractive index measurement and label-free biosensing. MNFs are ideal elements for assembling high Q resonators because of their low loss and very large evanescent fields. Similar to the in-line absorption sensor mentioned above, the coiled MNF refractive index sensor was obtained by embedding MNF microcoil resonator with Teflon, followed by removing its supporting rod. Figure 16a shows a schematic of the cross section of the coiled MNF resonator. Because of the interface with the analyte, the mode properties are particularly affected by the microfiber radius (r) and the coating thickness (d) between the MNF and the fluidic channel. Based on calculations, the fundamental mode is the one with the largest propagation constant and the only mode that is well confined in the vicinity of the MNF. Generally, thinner MNFs and smaller d leads to a larger fraction of the mode propagating in the analyte. By measuring the wavelength shift, the sensitivity was about 40 nm/RIU.

Refractive Index Sensor Based on Coiled MNF Grating

Optical fiber gratings are widely used in fiber-optic sensors. The standard UV writing technique is inconvenient for an MNF due to the disappearance of the germanium core and the decrease of the field overlap with the core. To data, focused ion beam (FIB) milling and femtosecond infrared irradiation have been reported for the fabrication of MNF gratings; however, the expensive equipment and complicated procedure restricted their applications. Xu et al. (2009) suggested a novel fluidic MNF grating by wrapping an MNF on a microstructured rod, followed by coating them with a low refractive index polymer. Combining current enabling technologies on microstructured optical fibers and MNFs, it is possible to obtain periodic refractive index distribution, leading to the coupling between forward and

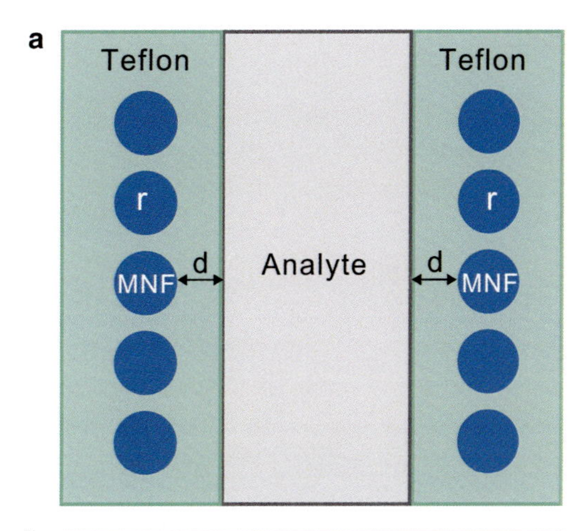

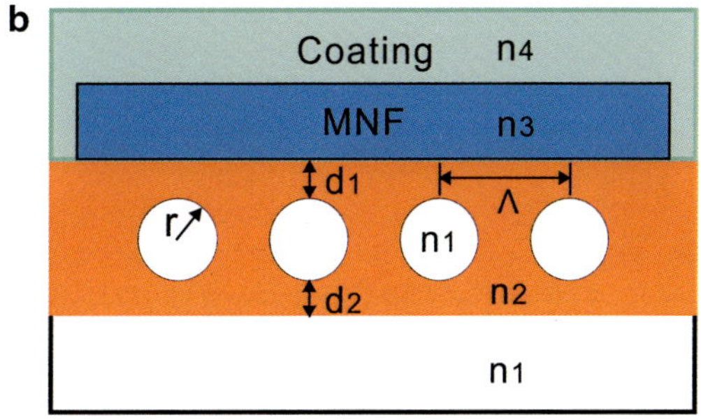

Fig. 16 (**a**) Schematic of the cross section of an OMCRS (Adapted from Xu and Brambilla (2008) with permission from AIP Publishing). (**b**) The equivalent planar structure; n_1, n_2, n_3, and n_4 are the refractive indexes of the hole, the rod, the OFM, and the coating, respectively; Λ is the distance between two adjacent holes; d_1 and d_2 are the thickness of the outer and inner walls of the one-ring microstructured rod; r is the hole radius (Adapted from Xu et al. (2009) with permission from Optical Society of America)

backward propagating waves. When a coordinate increasing along the microfiber is used, the surface corrugations experienced by the mode propagating along the curvilinear coordinate are similar to those experienced by a mode propagating straight in proximity of a conventional planar grating. Unfolding the coiled MNF and the microstructured rod, the equivalent MNF grating structure can be taken as a coated MNF on a planar substrate with air-hole corrugations (see Fig.16b). A refractometric sensor can be obtained by exploiting the holes of the microstructured rod as microfluidic channels. The sensitivity is highly dependent on thickness of the outer and inner walls and the diameter of the MNF. Generally, sensitivity increases with the decrease of the effective wall thickness or the diameter of the MNF, because the fraction of the mode field inside the fluidic channel increases. The calculated sensitivity can be as high as 1200 nm/RIU for a 400-nm-diameter nanofiber.

Flow Rate Sensor Based on Coiled MNF Coupler

Using an MNF coupler, Yan et al. (2016) demonstrated a "hot-wire" microfluidic flow rate sensor by wrapping an MNF coupler around a gold-coated capillary. The MNF coupler was fabricated by fusing and tapering two commercial single-mode

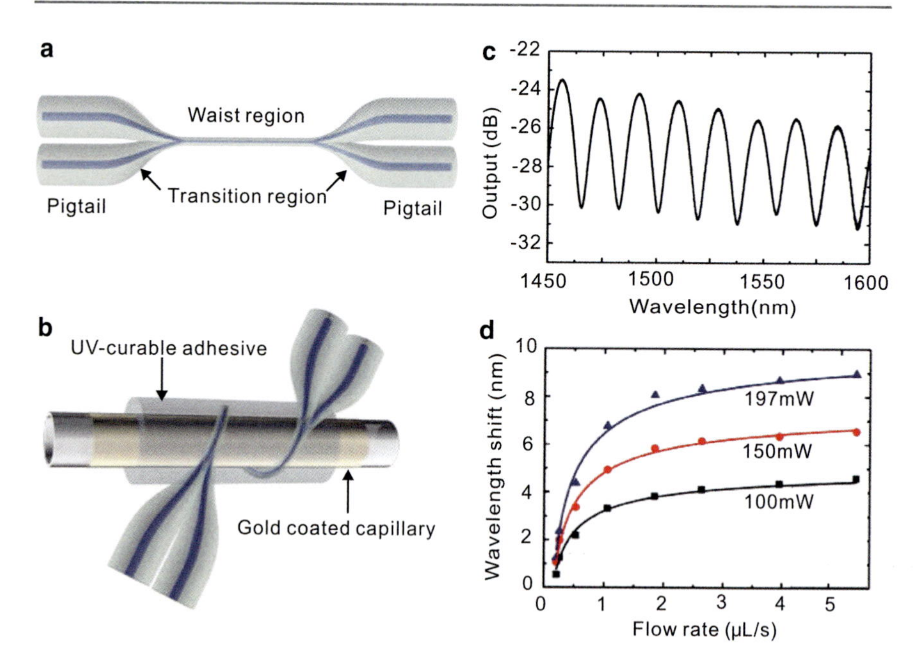

Fig. 17 (**a**) Schematic illustration of an MNF coupler. (**b**) Schematic illustration of an MNF coupler wrapping around the gold-coated glass capillary and embedding in the UV-curable adhesive. (**c**) Output spectra of the packaged MNF coupler sensor. (**d**) The wavelength shift with different microfluidic flow rates under different incident power values of 100, 150, and 197 mW (Adapted from Yan et al. (2016) with permission from Optical Society of America)

fibers (SMF-28, Corning, USA) with flame-brushing method. As shown in Fig. 17a, the MNF coupler includes one uniform waist region, two transition regions, and four pigtails. To fabricate a flow rate sensor as shown in Fig. 17b, a glass capillary (1 mm i.d.) with a wall thickness of ~120 μm was employed as a microfluidic channel. With vacuum-coating technology, a 100-nm-thick gold film was coated around the capillary. After that, an MNF coupler was wrapped around the capillary with one turn by a rotating stage. Finally, the sensing sections of the MNF coupler and the capillary were packaged with a low refractive index adhesive (EFiRON UVF PC-375, Luvantix). Figure 17c shows a typical output spectrum with multiple interference peaks of a packaged sensor. Each dip represents a characteristic wavelength of the MNF coupler, at which the power couples to another fiber at maximum efficiency. Due to the absorption of the evanescent field by gold film, heat is generated and temperature increases. When the microfluid flows through the channel, it takes away part of the heat and cools the MNF coupler, resulting in wavelength shift of the resonant peak. Generally, the larger the flow rate is, the more heat microfluid takes away, and the more the wavelength shifts. Based on calculation, when the flow rate is 3 μL/s, almost all the heat is taken away, and the wavelength shifts reach a saturation value. Figure 17d shows a typical output spectrum of the wavelength shift with different microfluidic flow rates under different incident power values of 100, 150, and 197 mW. Due to the long-distance interaction and high-temperature sensitivity, the proposed microfluidic flow rate sensor shows an ultrahigh flow rate sensitivity of 2.183 nm/(μL/s) at a flow rate of 1 μL/s.

Capillary-Based MNF Sensors

The capillary-based MNF sensors achieve dual use of the capillary as a sensor head and as a fluidic channel. As a result, it takes advantage of high sensitivity of ring resonators or interferometers and low sample consumption of microfluidic chips. In addition, they are highly compatible with well-developed capillary technologies for automated fluid delivery.

Refractive Index Sensor Based on Liquid Core Optical Ring Resonator

Liquid core optical ring resonator (LCORR) utilizes a fused silica capillary to carry the aqueous sample and to act as the ring resonator (White et al. 2006). The wall thickness of the LCORR is controlled to a few micrometers to expose the whispering gallery mode (WGM) to the aqueous core. Figure 18a shows a conceptual illustration of a LCORR sensor array. The WGM of each constituent ring resonator is launched through horizontally arranged MNFs, while the aqueous samples are conducted by the vertically positioned capillaries (see Fig. 18b). The LCORR uses the evanescent field of the WGM in the core to detect the refractive index change near the interior surface. Moreover, with the transverse arrangement, it is relatively easy for LCORRs to be integrated into a 2D array for simultaneous analysis of multiple samples.

To fabricate a LCORR, a fused silica capillary ($r1 = 0.45$ mm, $r2 = 0.6$ mm) was stretched under a flame until the outer radius reaches 35–50 μm, followed by further etching the capillary with low concentrations of HF to the desired wall thickness. And then, the LCORR was positioned in contact with an MNF of approximately 3–4 μm in diameter, as illustrated in Fig. 18b. Light from a tunable laser diode (980 nm) is coupled into the WGM through the evanescent coupling at the LCORR exterior surface. The tunable laser repeatedly scans across a wavelength range of approximately 100 pm. The WGM spectral positions are recorded at the output end of the taper. During the experiment, the liquid sample is delivered by a peristaltic pump and tubings attached to the LCORR. The Q factor for the 3 μm sensor is 4.1×10^5, which implies a WGM linewidth of 2.4 pm. Assuming one fiftieth of the linewidth can be resolved, the LCORR can theoretically detect a refractive index change of 1.8×10^{-5} RIU.

Besides its bulk refractive index sensing, LCORR is capable of detecting molecules on the interior capillary surface (White et al. 2007). When BSA prepared in the PBS buffer was pumped through the LCORR, the WGM spectral

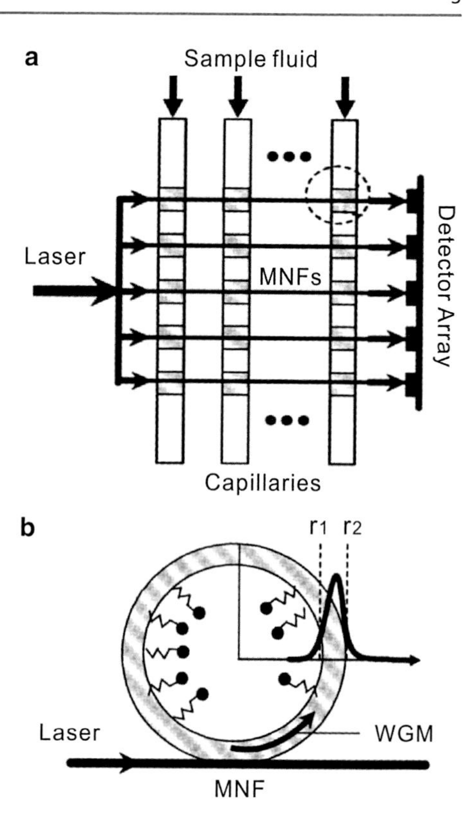

Fig. 18 (**a**) Conceptual illustration of a LCORR sensor array. (**b**) Cross section of a LCORR sensing element. The inner radius and the wall thickness are r_1 and $r_2 - r_1$, respectively (Adapted from White et al. (2006) with permission from Optical Society of America)

position shifts quickly to a longer wavelength and then levels off, indicating that the equilibrium is reached between the BSA molecules in solution and on the LCORR surface. With the increased BSA concentration, the equilibrium WGM shift increases and then becomes saturated when BSA concentration is higher than 200 nM. The experimental results were in good agreement with the theoretical prediction. Theoretically, the LCORR was capable of detecting BSA below 10 pM with sub-pg/mm^2 mass detection limit.

The development of surface-enhanced Raman scattering (SERS) detection has made Raman spectroscopy relevant for highly sensitive biological and chemical sensors. Despite the tremendous benefit in specificity that a Raman-based sensor can deliver, development of a lab-on-a-chip SERS tool has been limited thus far. Zhu et al. (2007) developed a SERS-based detection tool by utilizing an optofluidic ring resonator (OFRR) platform. The LCORR serves both as the microfluidic sample delivery and as a ring resonator, exciting the silver nanoclusters and target analytes as they pass through the channel. Using this OFRR approach, a measured detection limit is about 400 pM for R6G. The measured Raman signal in this case is likely generated by only a few hundred R6G molecules, which foreshadows the development of a SERS-based lab-on-a-chip bio-/chemical sensor capable of detecting a low number of target analyte molecules.

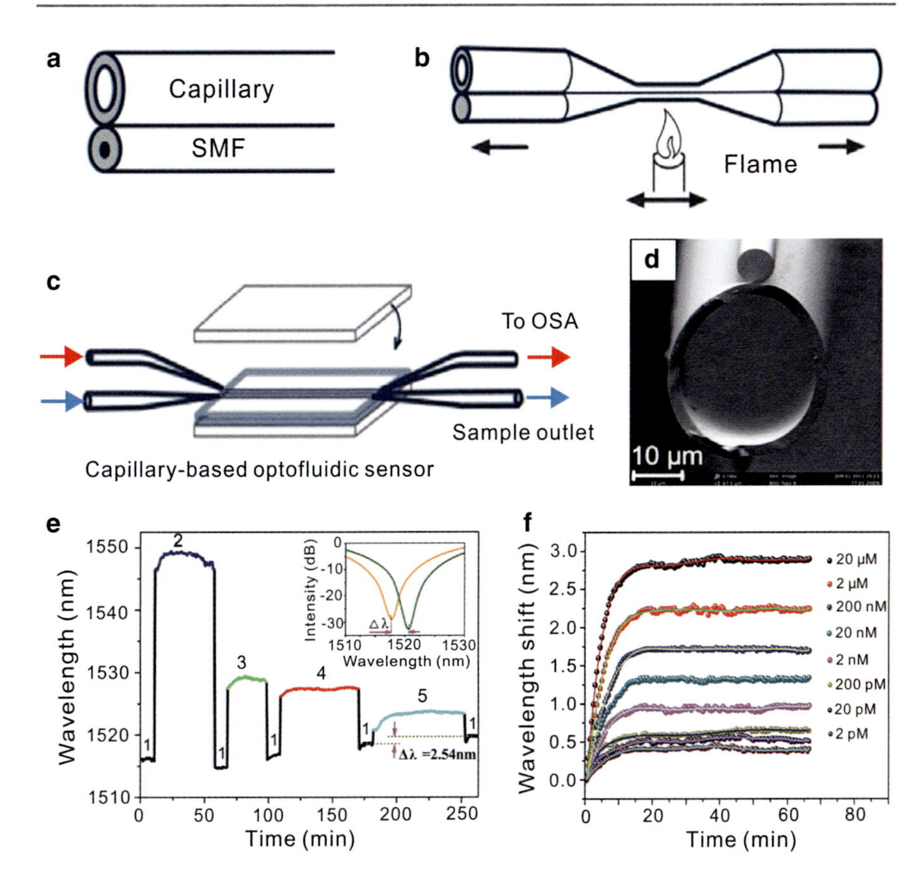

Fig. 19 (**a**–**c**) Schematic illustration of the fabrication procedure of the capillary-based optofluidic sensor. (**d**) Scanning electron microscopic (SEM) image of the capillary-based optofluidic sensor in cross-sectional view. (**e**) Sensor gram for surface activation and miRNA-let7a detection with a bulk concentration of 20 μM: (1) water-DEPC rinsing for 30 min, (2) APTES silanization, (3) glutaraldehyde cross-linking, (4) immobilization of amine-modified DNA probes, (5) hybridization with miRNA-let7a. Inset: Measured transmission spectra before and after miRNA hybridization. (**f**) The kinetic binding curves of the miRNA-let7a with different concentrations (Adapted from Liang et al. (2017) with permission from Elsevier)

Biomolecular Detection Based on Liquid Core Optical Interferometer

See Fig. 19.

Optical interferometer has attracted increasing research interest in biosensing owing to its high sensitivity, simple structure, and label-free sensing capability. By measuring the resonance peak shift, a number of biomaterials has been detected by various optical interferometers. However, most of the optical interferometers are limited to laboratory conditions due to the complexity in experiment setup or the difficulties in sample pretreatment. To design a robust and portable optical interferometer sensor for clinical applications, Liang et al. (2017) developed a liquid core optical interferometer by tapering silica capillary and optical fiber in parallel. Figure 19a–c schematically shows the fabrication procedure of the sensor. Briefly, a bare single-mode fiber is aligned with a silica capillary in lateral contact in the same direction. Both the single-mode fiber and the capillary are heated by a flame and stretched by two fiber holders. The tapered structure is then embedded in a low refractive index polymer and packaged between two cover-slides for better spatial stability. Figure 19d shows a typical as-fabricated attached MNF and capillary. Owing to the extremely thin wall, light guided in the MNF can penetrate the capillary and interact with the fluid in the capillary, leading to a spectral shift with a response to biomolecule binding event. The resonance wavelength shows an obvious shift during the hybridization with miRNA-let7a (see Fig. 19e). The kinetic binding curves of the miRNA-let7a with different concentrations are shown in Fig. 19f. With the pre-immobilization of DNA probes, the biosensor is capable of detecting single-stranded microRNA-let7a (molecular weight: 6.5 k). A log-linear response from 2 nM to 20 μM and a minimum detectable concentration of 212 pM (1.43 ng/mL) have been achieved. The sensor is promising for biomarker detection in preclinical applications owing to its advantages of high resistance to environmental perturbations, improved portability, easy fabrication and handling, and intrinsic connection to fiber-optic measurement.

Gold Nanoparticles Functionalized MNF Localized Surface Plasmon Resonance Sensors

The exploitation of localized surface plasmon resonance (LSPR) of noble metal nanoparticles in the development of plasmonic sensors has attracted considerable research interest for many years. LSPR is the collective oscillation of conduction electrons confined to metal nanoparticles, whose resonance frequency has been shown to be strongly dependent on the particle's size, shape, composition, and the dielectric properties of surrounding medium. Benefited from the favorable optical waveguiding properties of the MNFs, the LSPR of the noble metal nanoparticles could be effectively excited by a waveguiding approach. The highly efficient photon-to-plasmon conversion of noble metal nanoparticles on the surface of an MNF may greatly facilitate and enhance light-matter interactions within a highly localized area and open up opportunities for developing sensitive sensors with miniaturized sizes, high compactness, and low optical power consumption.

Refractive Index and Label-Free Biochemical Sensor

Lin et al. (2012) demonstrated an MNF LSPR sensor for refractive index sensing and label-free biochemical detection. The sensing strategy relies on the interrogation of the transmission intensity change due to the evanescent field absorption of immobilized gold nanoparticles (GNPs) on the MNF surface. Experimentally, the MNF was manufactured by tapering a standard single-mode fiber through a hydrogen–oxygen flame-brushing technique. The average waist diameter and waist length of the as-fabricated MNF was 48 μm and 1.25 mm, respectively. To immobilize GNPs on the MNF surface, the MNF was cleaned by a three-step process of ultrasonic bath in acetone for 20 min, soaking in Piranha solution for 30 min, followed by a thorough flushing with ultrapure water, and drying in an oven at the temperature of 70 °C. The clean MNF was then immersed in 1% solution of 3-mercaptopropyltrimethoxysilane (MPTMS) in ethanol overnight, leading to the formation of a thiol-terminated self-assembled monolayer (SAM) of MPTMS on the MNF surface. The thiol-functionalized MNF was subsequently rinsed by ethanol to remove unbound monomers from the surface and blow-dried with nitrogen gas. Afterward, the MNF was submerged in the solution of GNPs to form a monolayer of GNPs on the MNF surface. Finally, GNPs functionalized MNF which was mounted in a flow cell for refractive index sensing and label-free biochemical detection. When sucrose solutions with various refractive indexes ranging from 1.333 to 1.403 were successively pumped into the flow cell, the transmission at the peak wavelength exhibited obvious decreases and the peak red-shifted as the surrounding refractive index alters from low to high. The refractive index resolution based on the interrogation of transmission intensity change is calculated to be 3.2×10^{-5} RIU. To further investigate the feasibility of the MNF LSPR sensor for label-free biochemical detection, *N*-(2,4-dinitrophenyl)-6-aminohexanoic acid (DNP) was chemically functionalized on the MNF surface as the molecular recognition probe, and anti-DNP antibody was employed as analyte, achieving a LOD of 1.06×10^{-9} g/ml for anti-DNP antibody.

Cancer Biomarkers Sensor

Sensitive and selective detection for cancer biomarkers is critical in cancer clinical diagnostics. Li et al. (2014b) developed a novel MNF microfluidic biosensor using GNPs as amplification labels for the detection of alpha-fetoprotein (AFP) in serum samples. The main mechanism of this biosensor is the selective absorption of evanescent wave by GNPs when they approach the MNF surface. The sandwich assay shown in Fig. 20a was used for the detection of AFP molecules. First antibody was immobilized on the surface of the MNF as the capture antibody. Secondary antibody-functionalized GNPs were used as signal amplifiers. Owing to a prominent evanescent field with a depth of about several hundreds of nanometers around the MNF, GNPs binding to the MNF surface via specific interaction can lead to a great decrease in the output light intensity. Figure 20b, c shows a schematic illustration and photograph of an MNF sensor with an integrated PDMS chamber for sample delivery, respectively. The microfluidic chip can dramatically decrease sample consumption and improve the stability of the MNF sensor. To investigate the feasibility

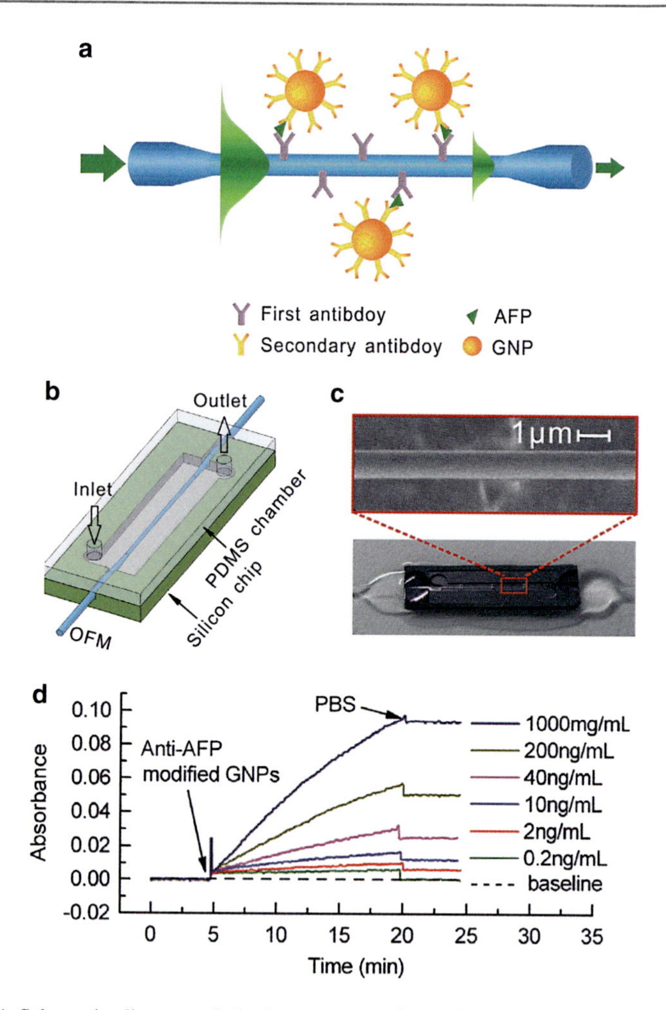

Fig. 20 (**a**) Schematic diagram of the immunoassay for α-fetoprotein (AFP) detection using GNPs as signal amplification labels. (**b**) Schematic diagram of an MNF sensor with an integrated PDMS chamber for sample delivery. (**c**) Image of a sensor cell and SEM image of a 1.0-mm-thick optical microfiber. (**d**) Secondary antibody-functionalized GNP enhanced sensor response to bovine serum samples spiked with different concentrations of AFP (Adapted from Li et al. (2014b) with permission from Elsevier)

of this sensing strategy for clinical applications, bovine serum samples spiked with different concentrations of AFP ranging from 0.2 to 1000 ng/mL was tested using the sandwich assay. Real-time response curves are shown in Fig. 20d. The limit of detection of this sensor for AFP is 0.2 ng/mL in PBS and 2 ng/mL in bovine serum, which is comparable to conventional assays. The advantages of this biosensor are simple detection scheme, fast response time, and ease of miniaturization, which might make this biosensor a promising platform for clinical cancer diagnosis and prognosis.

Conclusion

This chapter reviewed applications of MNF microfluidic sensors. It is shown that the MNF can serve as a basic sensing block for assembling compact and robust sensors. The fusion of microfluidic chip and MNFs provides a number of attractive advantages for enhancing the sensing performance and simplifying the design of microsystems. Compared with the sensors used, MNFs suspended in air or mounted in a bulky volume flow chamber, MNF microfluidic sensors are more stable, need less sample, and can be reused by cleaning the MNF surface via the microchannels. More importantly, the integration of MNF and microfluidics provides a new platform to develop novel sensing mechanisms and practical sensing structures for chemical, biological, medical, and optical applications. As a future outlook, there are a number of opportunities and challenges for MNF microfluidic sensing, including (1) higher sensitivity, detection of single molecule or single nanoparticle; (2) better selectivity, for example, enabling selective detection of target molecule by properly functionalizing the MNF structure with antibody; and (3) better robustness: as a tiny structure highly sensitive to environmental changes (e.g., temperature and displacement), long-term stability is highly desired for practical applications.

References

A. Arora, G. Simone, G.B. Salieb-Beugelaar, J.T. Kim, A. Manz, Anal. Chem. **82**, 4830 (2010)
Y.F. Chen, L. Jiang, M. Mancuso, A. Jain, V. Oncescu, D. Erickson, Nanoscale **4**, 4839 (2012)
W.B. Du, Q. Fang, Q.H. He, Z.L. Fang, Anal. Chem. **77**, 1330 (2005)
X. Fan, I.M. White, Nature Photon. **5**, 591 (2011)
X. Fan, I.M. White, S.I. Shopova, H. Zhu, J.D. Suter, Y. Sun, Anal. Chim. Acta **620**, 8 (2008)
F.X. Gu, L. Zhang, X.F. Yin, L.M. Tong, Nano Lett. **8**, 2757 (2008)
X. Guo, Y. Ying, L.M. Tong, Acc. Chem. Res. **47**, 656 (2014)
T.A. Laurence, S. Weiss, Science **299**, 667 (2003)
B. Lee, Opt. Fiber Technol. **9**, 57 (2003)
J. Lee, H. Moon, J. Fowler, T. Schoellhammer, C.J. Kim, Sens. Actuator A **95**, 259 (2002)
D.G. Lee, J. Park, J. Bae, H.Y. Kim, Lab Chip **13**, 274 (2013)
A. Leung, P.M. Shankar, R. Mutharasan, Sens. Actuator B **125**, 688 (2007)
Y.H. Li, L.M. Tong, Opt. Lett. **33**, 303 (2008)
W. Li, Z.F. Hu, X.Y. Li, W. Fang, X. Guo, L.M. Tong, J.Y. Lou, Opt. Commun. **314**, 28 (2014a)
K. Li, G. Liu, Y. Wu, P. Hao, W. Zhou, Z. Zhang, Talanta **120**, 419 (2014b)
Z. Li, Y. Xu, W. Fang, L.M. Tong, L. Zhang, Sensors **15**, 4890 (2015)
L.L. Liang, L. Jin, Y. Ran, L.P. Sun, B.O. Guan, Sens. Actuator B. **242**, 999 (2017)
H.-Y. Lin, C.-H. Huang, G.-L. Cheng, N.-K. Chen, H.-C. Chui, Opt. Express **20**, 21693 (2012)
R. Lorenzi, Y. Jung, G. Brambilla, Appl. Phys. Lett. **98**, 173504 (2011)
J.Y. Lou, L.M. Tong, Z.Z. Ye, Opt. Express **13**, 2135 (2005)
A. Manz, C.S. Effenhauser, N. Burggraf, D.J. Harrison, K. Seiler, K. Fluri, J. Micromech. Microeng. **4**, 257 (1994)
C. Monat, P. Domachuk, B.J. Eggleton, Nature Photon. **1**, 106 (2007)
P. Polynkin, A. Polynkin, N. Peyghambarian, M. Mansuripur, Opt. Lett. **30**, 1273 (2005)
D. Psaltis, S.R. Quake, C.H. Yang, Nature **442**, 381 (2006)
D. Qin, Y.N. Xia, G.M. Whitesides, Nature Protoc. **5**, 491 (2010)

L. Shi, X.F. Chen, H.J. Liu, Y.P. Chen, Z.Q. Ye, W.J. Liao, Y.X. Xia, Opt. Express **14**, 5055 (2006)
M. Sumetsky, Y. Dulashko, A. Hale, Opt. Express **12**, 3521 (2004)
L.M. Tong, R.R. Gattass, J.B. Ashcom, S.L. He, J.Y. Lou, M.Y. Shen, I. Maxwell, E. Mazur, Nature **426**, 816 (2003)
M.A. Unger, H.P. Chou, T. Thorsen, A. Scherer, S.R. Quake, Science **288**, 113 (2000)
S.S. Wang, X.Y. Pan, L.M. Tong, Opt. Commun. **276**, 293 (2007)
I.M. White, H. Oveys, X. Fan, Opt. Lett. **31**, 1319 (2006)
I.M. White, J. Gohring, X. Fan, Opt. Express **15**, 17433 (2007)
G.M. Whitesides, Nature **442**, 368 (2006)
O.S. Wolfbeis, Anal. Chem. **80**, 4269 (2008)
X.Q. Wu, L.M. Tong, Nanophotonics **2**, 407 (2013)
F. Xu, G. Brambilla, Appl. Phys. Lett. **92**, 101126 (2008)
F. Xu, G. Brambilla, Y.Q. Lu, Opt. Express **17**, 20866 (2009)
R. Yalla, F. Le Kien, M. Morinaga, K. Hakuta, Phys. Rev. Lett. **109**, 063602 (2012)
S. Yan, Z. Liu, C. Li, S. Ge, F. Xu, Y. Lu, Opt. Lett. **41**, 5680 (2016)
X.-C. Yu, B.-B. Li, P. Wang, L.M. Tong, X.-F. Jiang, Y. Li, Q. Gong, Y.-F. Xiao, Adv. Mater. **26**, 7462 (2014)
L. Zhang, X.F. Yin, Z.L. Fang, Lab Chip **6**, 258 (2006)
L. Zhang, P. Wang, Y. Xiao, H. Yu, L. Tong, Lab Chip **11**, 3720 (2011)
L. Zhang, Z. Li, J. Mu, W. Fang, L.M. Tong, Opt. Express **23**, 28408 (2015)
H.Y. Zhu, I.M. White, J.D. Suter, P.S. Dale, X. Fan, Opt. Express **15**, 9139 (2007)

All Optical Fiber Optofluidic or Ferrofluidic Microsensors Fabricated by Femtosecond Laser Micromachining

60

Hai Xiao, Lei Yuan, Baokai Cheng, and Yang Song

Contents

Abstract

Research and development in photonic micro-/nanostructures functioned as sensors have experienced significant growth in recent years, fueled by their broad applications in the fields of physical, chemical, and biological quantities. Compared with conventional sensors with bulky assemblies, recent progress

H. Xiao (✉) · L. Yuan · B. Cheng · Y. Song
Department of Electrical and Computer Engineering, Center for Optical Materials Science and Engineering Technologies (COMSET), Clemson University, Clemson, SC, USA
e-mail: haix@clemson.edu; yuan7@clemson.edu; baokaic@g.clemson.edu; song5@g.clemson.edu

G.-D. Peng (ed.), *Handbook of Optical Fibers*,
https://doi.org/10.1007/978-981-10-7087-7_63

in femtosecond (fs) laser three-dimensional (3D) micromachining technique has been proven an effective way for one-step fabrication of assembly-free microstructures in various transparent materials (i.e., fused silica). When used for fabrication, fs laser has many unique characteristics, such as negligible cracks, minimal heat-affected zone, low recast, high precision, and the capability of embedded 3D fabrication, compared with conventional long pulse lasers (i.e., ns laser). The merits of this advanced manufacturing technique enable the unique opportunity to fabricate integrated sensors with improved robustness, enriched functionality, enhanced intelligence, and unprecedented performance.

Recently, fiber-optic sensors have been widely used in many application areas, such as aeronautics and astronautics, petrochemical industry, chemical detection, biomedical science, homeland security, etc. In addition to the well-known advantages of miniaturized in size, high sensitivity, immunity to electromagnetic interference (EMI), and resistance to corrosion, fiber-optic sensors are becoming more and more desirable when designed with characteristics of assembly-free and operation in the reflection configuration. Additionally, such sensors are also needed in optofluidic/ferrofluidic systems for chemical/biomedical sensing applications.

In this chapter, liquid-assisted laser micromachining techniques were investigated for the fabrication of assembly-free, all-optical fiber sensor probes. All-in-fiber optofluidic sensor and fiber in-line ferrofluidic sensor were presented as examples with respect to these laser processing techniques.

Introduction

The invention of the ruby laser in 1960 (Maiman 1960) provided much higher laser intensities than previous candidates. Since then, laser has been developed rapidly and used for controllable material processing by high laser intensities. With the advances of mode-locking techniques (Spence et al. 1991) and chirped pulse amplification (Strickland and Mourou 1985), the intensities of commercially available femtosecond (fs) laser systems can be achieved of more than 10^{13} W/cm^2 (Perry and Mourou 1994). At such intensities, any materials, especially for transparent materials, will be ionized and exhibit nonlinear behavior, causing dielectric breakdown and structural change in transparent materials (Itoh et al. 2006).

Fs laser is also known as ultrafast laser, due to its unique advantages of ultrashort pulse width (<200fs, $1\text{fs} = 10^{-15}\text{s}$) and extremely high peak intensity (>10^{15} W/cm^2). Recently, fs laser micromachining has opened up a new avenue for material processing, especially for the transparent materials with large material bandgaps (i.e., fused silica (Eg = 9 eV)). When used for fabrication, fs laser can be used either to remove materials from the surface (ablation) or to modify the properties inside the materials (modification or irradiation). Compared with long pulse lasers (pulses longer than a few picoseconds) (Chichkov et al. 1996), fs laser has many unique characteristics, such as negligible cracks, minimal heat-affected zone, low recast, and high precision.

Initial studies of fs laser micromachining were first demonstrated in 1987, when ultrafast excimer UV lasers were used to ablate the surface of polymethyl methacrylate (PMMA) (Du et al. 1994). Later, micrometer-sized features on silica (Pronko et al. 1995) and silver surfaces (Chimmalgi et al. 2003) were performed using infrared fs laser systems. In less than 10 years, the resolution of surface ablation has improved to enable high precision with nanometer scale (Küper and Stuke 1987). For the material modification, Hirao's group firstly demonstrated fs laser processed in the bulk of transparent glass in 1996, and the material modification happened beneath the sample surface, forming waveguiding structures with a permanent refractive index change localized to the focal volume (Davis et al. 1996). This was followed by introducing two-photon polymerization into a resin (Kawata et al. 2001) and printing complex three-dimensional (3D) structures with nanometer-scale resolution. Over the past decade, fs laser micromachining has been used in a broad range of applications, from waveguide writing, cell ablation to biological sample modification (Gattass and Mazur 2008). As a result, fs laser was proven to be a unique and versatile contactless material modification tool.

The fiber-optic field has undergone tremendous growth and advancement over the past 50 years. Initially considered as a medium of transmitting light and imagery for medical endoscopic applications, optical fibers were later promoted as an information carrier for telecommunication applications in the mid-1960s (Bates 2001). C. Kao and G. Hockham of the British company Standard Telephones and Cables (STC) were the first to propose the idea that the attenuation in optical fibers could be controlled down to 20 dB/km, allowing fibers to be a good candidate for telecommunication applications (Hecht 2004). Since then, the research and development of optical fiber telecommunication applications have been widened immensely (Udd 1995). Today, optical telecommunication has proven to be the preferred method to transmit vast amounts of data and information with high speed and long-haul capability.

In addition to communication applications, optical fibers have been widely used for broad sensing applications in the fields of physical, chemical, and biological analyses, including structure health monitoring, harsh environment temperature sensing, biological and chemical refractive index/pH sensors, and medical imaging (Haque et al. 2014; Liu et al. 2013; Wei et al. 2008; Zhang et al. 2013; Zhou et al. 2010). Optical fiber, most of the time, is made of fused silica glass. It consists of fiber core, fiber cladding, and the outside buffer/jacket layer for protection, as shown in Fig. 1. Due to the small amount of doping elements (i.e., germanium-doped) in the fiber core area, the refractive index of fiber core (n_1) is slightly larger than the index of fiber cladding (n_2), so the light can propagate inside the fiber core due to the so-called total internal reflection (TIR). Compared with electrical sensors, fiber-optic sensors offer many intrinsic advantages, such as small size/lightweight, immunity to electromagnetic interference (EMI), resistance to chemical corrosion, high temperature capability, high sensitivity, and multiplexing and distributed sensing.

Fiber-optic sensors have been long envisioned as a cost-effective and reliable candidate for many applications. However, although the past half century has seen a

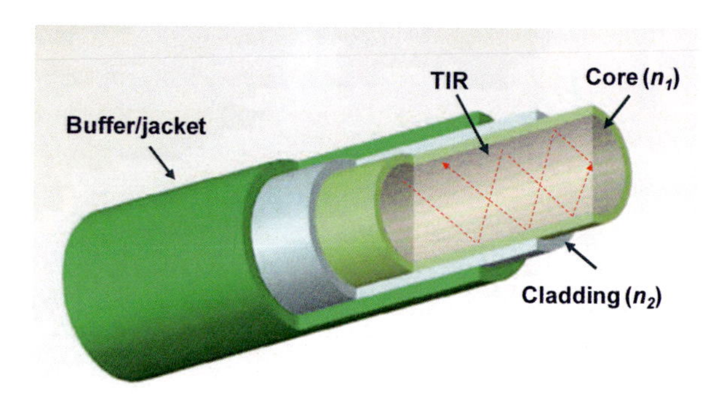

Fig. 1 Schematic representation of the optical fiber as a waveguide under TIR conditions

steady growth of appearance of optical fibers in various industry sectors, the market of fiber-optic sensors has not blew out as expected. On the contrary, many of the unique advantages and special requirements of advanced fiber-optic sensors and devices are yet to be fully harvested, including assembly-free, inexpensive, and miniaturized fiber-pigtailed sensor probes as well as long-haul, fully distributed fiber-optic sensor networks.

Special requirements for advanced fiber-optic sensors are highly needed in biomedical/chemical applications, i.e., the hot topic of optofluidic systems. Aiming to synergistically combine integrated optics and microfluidics, optofluidic-based systems have attracted much research interest because of their unique advantages toward biological/chemical sensing applications (Fan and White 2011). In an optofluidic system, the liquid of interest is constrained and manipulated in a small geometry to interact with the optics. As such, the physical, chemical, and biological properties of the liquid can be probed and analyzed effectively using optical means (Psaltis et al. 2006).

Most optofluidic systems have been constructed on a planar platform with microchannels in silica/polymetric materials and probed by a variety of optical means, such as absorbance, fluorescence, refractometry, Raman scattering, etc. (Monat et al. 2007). Objective lenses are commonly used to couple light into and out of the microfluidics (Woolley and Mathies 1994). However, the need of using a microscope to perform the optical alignment limits its field applicability. A number of efforts have been made to fabricate optical waveguides inside the substrate to confine and transport light in the substrate (Osellame et al. 2007) or directly integrate optical fibers with the fluidics for excitation and probing (Domachuk et al. 2006). However, the transmission efficiency of light coupling is still a challenge in most optofluidic configurations.

In addition to planar configurations, it has been suggested that the microfluidics can be directly fabricated on an optical fiber to form the so-called all-in-fiber optofluidics. The all-in-fiber configuration has the unique advantage of alignment-free optics and improved robustness. Examples include the photonic crystal fibers (PCFs) filled with functional fluids in their cladding air voids (Jensen et al.

2004), the capillary-based optofluidic ring resonator (OFRR) with a microchannel for sample delivery (Zhu et al. 2008), and a miniaturized microchannel directly fabricated on a conventional optical fiber for light-fluid interaction (Lai et al. 2006). While the PCF-based and OFRR-based optofluidic sensors utilize the evanescent fields to probe the fluid, the microchannel on a conventional fiber configuration allows a direct light passage through the liquid.

However, PCFs are still expensive, and huge transmission loss will be generated when fusion splicing PCFs with single-mode fibers (SMFs). OFRR structure needs to assemble with an ultrathin and fragile fiber-optic taper to read out the signal, limiting the application within the lab condition. Although the sensor with a microchannel in an SMF can probe the liquid, the sensing mechanism is still based on the intensity modulation, whose sensitivity and detection limit are much smaller and higher, respectively, than those of phase modulation-based sensors.

Similar to optofluidic systems, filling the embedded microchannels with functional liquids (i.e., ferrofluid) can also allow people to test surrounding magnetic field. Typically, varying magnetic field can change the properties of ferrofluid (i.e., the permittivity of the liquid) and then make the integrated device function as a ferrofluidic sensor for various applications. Additionally, it's highly desired that the ferrofluidic sensor can be directly implemented in an all-fiber form with minimum insertion loss and desired performance.

In this chapter, first of all, a brief overview of the background physics describing fs laser micromachining of transparent materials was presented. The discussion of laser/matter interactions was starting from atomic scale-free electron plasma formation to energy deposition and material modification. Secondly, the development of a typical home-integrated fs laser micromachining system and preliminary fabrication results were presented in details. After that, methods of liquid-assisted laser processing were illustrated. Buried microfluidic microchannels with sub-microresolution in an SMF can be fabricated either using laser-induced water breakdown technique directly or utilizing laser irradiation followed by chemical etching technique, resulting in fiber sensors for chemical/biomedical sensing application. Finally, an all-in-fiber optofluidic sensor and a fiber in-line ferrofluidic sensor are presented as examples of this technology.

Femtosecond Laser and Material Interactions

Free Electron Plasma Formation

The most popular femtosecond laser nowadays is Ti:sapphire laser which has its operating wavelength at around 800 nm. High-intensity focused fs laser pulses with the wavelength (λ) of 800 nm have insufficient photon energy ($E = h\nu = \hbar\omega = 1.55\text{eV}$, where $\omega = 2\pi\nu$) to be linearly absorbed in fused silica with the bandgap of material Eg, which is about 9 eV. As such, nonlinear photoionization is dominant to promote electrons from valence band to the conduction band and then generate free electrons. Typically, there are two classes of such ionization process: nonlinear photoionization and avalanche ionization (Beresna et al. 2014). Nonlinear photoionization refers to direct electron excitation by an electric laser field. Such process can generate seed electrons that will participate in the next process. Photoionization involves multiphoton ionization and/or tunneling ionization depending on the laser frequency and intensity (Stuart et al. 1995). Once the free electrons exist in the conduction band, free carrier absorption happens, leading to collisional ionization followed by avalanche ionization.

In the multiphoton absorption regime, photoionization rate dn_e/dt strongly depends on the laser intensity:

$$\frac{dn_e\left(t,r,z\right)}{dt}=\delta_m(I\left(t,r,z\right))^m \tag{1}$$

where t is the time; r, the distance to the Gaussian beam axis; z, the depth from the surface of the bulk material; n_e is free electron density; $I(t,\ r,\ z)$ is the laser intensity inside the bulk material; m is the order of the multiphoton process, i.e., the number of photons should be six in our case to satisfy the condition $E = m\hbar\omega \geq E_g$; and δ_m is the cross section of the m-photon absorption. The tunnelling rate, on the other hand, scales more weakly with the laser intensity than the multiphoton rate (Schaffer et al. 2001).

The strong dependence on the intensity also means that the photoionization process is more efficient for the laser with short pulse duration. For the long pulse laser (i.e., ns), the photoionization process cannot efficiently generate sufficient seed electrons, and the excitation process becomes strongly reliant on the low concentration of impurities with energy levels which are distributed randomly close to the conduction band. As such, the modification process becomes less deterministic, and precise machining is impossible for longer pulses (Beresna et al. 2014).

Avalanche ionization involves free carrier absorption followed by impact ionization. The free electrons in the conduction band oscillate in the electromagnetic field of the laser and gradually gain energy by collisions. After the conduction band electron's energy exceeds the minimum energy of the conduction band by more than the bandgap energy of the material, it can impact ionize another bound electron from the valence band via collision, resulting in two excited electrons near the bottom of the conduction band (Yablonovitch and Bloembergen 1972). Both electrons can undergo free carrier absorption, impact ionization, and repeat the described energy transfer cycle. Such process will not stop until the laser electric field is dissipated and not strong enough, leading to an electronic avalanche. The density of free electrons generated through the avalanche ionization is:

$$\frac{dn_e\left(t,r,z\right)}{dt}=a_i I\left(t,r,z\right)n_e\left(t,r,z\right) \tag{2}$$

where a_i is the avalanche ionization coefficient (Stuart et al. 1996). The original laser beam before it interacts with the material is assumed to be a Gaussian distribution in time and space. Obviously, at $z = 0$, it is assumed that the laser focus point is at the surface of the material.

The optical properties of the highly ionized dielectrics under an fs pulse can be well determined by plasma properties (Rethfeld et al. 2002). Generally, the plasma frequency ω_p is defined by (Mao et al. 2004):

$$\omega_p\left(n_e\right) = \sqrt{\frac{e^2 n_e(t)}{\varepsilon_0 m_e}} \tag{3}$$

where n_e is also denoted as free carrier density. When the free electron density excited by photoionization approaches a high density (i.e., $\omega_p \sim \omega \sim 10^{21}\ \text{cm}^{-3}$), a large fraction of the remaining fs laser pulse can be absorbed (Schaffer et al. 2001). Assumed 800 nm laser irradiation, the plasma frequency ω_p equals the laser frequency ω when the free carrier density n_e is approximately $1.7 \times 10^{21}\ \text{cm}^{-3}$, which is also known as the critical density of free electrons ($n_c(t) = \omega^2 \varepsilon_0 m_e / e^2$). Such critical value can be also used as a criterion in the model of laser-material interaction.

It should be noticed that avalanche ionization requires the presence of seed electrons, which can trigger the process and is more efficient with longer pulse durations. The following rate equation has been widely used to describe this nonlinear photoionization process (the combined action of multiphoton excitation and avalanche ionization) (Stuart et al. 1996):

$$\frac{dn_e\left(t,r,z\right)}{dt} = \delta_m (I\ (t,r,z))^m + a_i I\ (t,r,z)\, n_e\left(t,r,z\right) \tag{4}$$

However, free carrier losses (i.e., self-trapping and recombination) may occur on a time scale comparable to pulse duration (<40 fs) of fs laser beam in large bandgap materials, such as quartz and fused silica. As such, the rate equation shown above should be modified (Schaffer et al. 2001). For longer pulse duration that is more than 150 fs, such losses can be neglected.

For material with defects or some dopants, initial electrons can be excited from these low-lying levels via linear absorption or thermal excitation (Schaffer et al. 2001). For pure dielectrics, such electrons are generated by nonlinear photoionization (i.e., ultrashort pulses). Then, free electron plasma formation is realized.

To summarize, for ultrafast laser pulses (i.e., sub-picosecond or even smaller), photon-electron interaction occurs on a faster time scale than energy transfer from hot free electrons to the lattice, separating the absorption and lattice heating processes (Schaffer et al. 2001). The plasma will become strongly absorbing when the density of free electrons in the conduction band increases through avalanche ionization until $\omega_p \approx \omega$. In another word, it's generally assumed that material modification/optical breakdown occurs at this critical point. In fused silica, the laser intensity needs to be $\sim 10^{13}$ W/cm^2 (Schaffer et al. 2001) to achieve optical breakdown.

Energy Deposition and Material Modification

When the absorbed laser energy, via free electron plasma formation process, is high enough, it can be transferred to the lattice (phonon) and then deposited into the material via electron-phonon coupling, equalizing the temperature of the free electrons and the lattice, resulting in the permanent material modification subsequently. Since the lattice heating time is on the order of 10 ps, which is much longer than the pulse duration of the fs laser beam, less energy is needed to achieve the intensity for optical breakdown with respect to short pulses, and high-precision micromachining can be realized (Gattass and Mazur 2008).

Although the free electron plasma created by absorbing fs laser pulses for nonlinear photoionization in transparent materials is widely accepted, the physical mechanisms for material modification are not fully understood. Davis and coworkers (Davis et al. 1996) first observed morphology changes and then generally classified into three types of structural formations: an isotropic refractive index formation (i.e., waveguide) (Davis et al. 1996), a nanograting formation (Sudrie et al. 1999), and a confined microexplosion and void formation (Glezer and Mazur 1997). In fused silica, these three morphologies can be observed by adjusting the pulse energy of the incident fs laser beam (Itoh et al. 2006), as shown in Fig. 2.

It should be noted that the permanent formations in transparent materials with large bandgap energy also depend on many other exposure parameters, such as

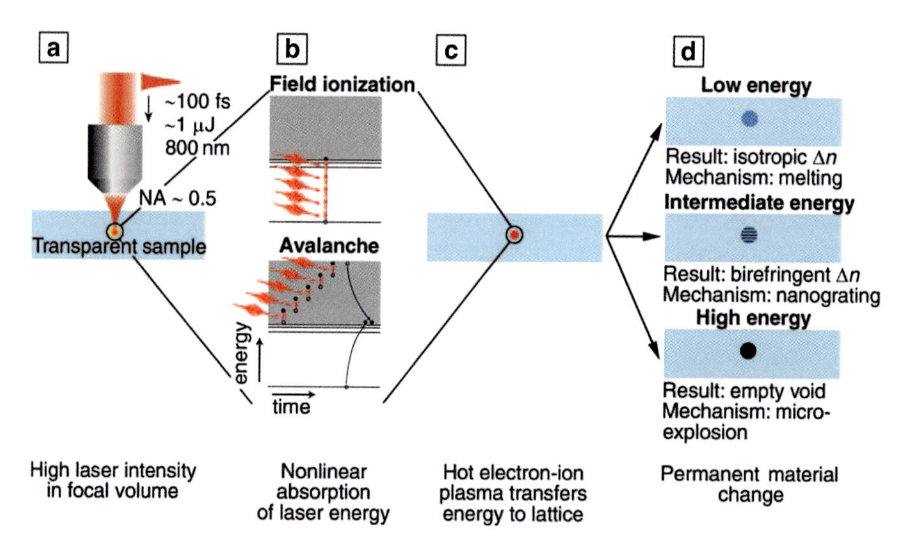

Fig. 2 Schematic illustration of physical mechanisms of intense fs laser pulses in transparent materials. (**a**) The laser is focused inside the sample leading to high laser intensity in focal volume. (**b**) The photon energy is nonlinearly absorbed, and the free electron plasma is generated by nonlinear photoionization. (**c**) The energy is transferred from hot electrons to lattice on a $\sim$10 ps time scale. (**d**) Three types of permanent material modification, including isotropic refractive index change at low pulse energy, birefringent nanograting formation at intermediate energy, and empty void formation at high pulse energy (Itoh et al. 2006)

pulse duration, repetition rate, laser wavelength, polarization state, tight/loose focusing condition, as well as the scanning speed of the translation stage. Further explanations of mechanisms will be found in the following discussions.

If the laser beam is focusing on the surface of the transparent materials, the basic processes during laser ablation such as excitation, melting, and material removal are temporally separated when fs laser pulses are applied, allowing a separate investigation in different time scales. Once free electron and free carrier excitations are finished, several mechanisms can be involved that may give rise to damage or optical breakdown of a non-defect transparent material under fs laser excitation. It's widely accepted that thermal phase change includes melting and vaporization of the solid, following strong phonon emission induced by free electron plasma formation (Mao et al. 2004). For the physical mechanisms of nonthermal phase change, Coulomb explosion was proposed to explain single-shot ablation by fs laser pulses (Stoian et al. 2000). All the phase changes are highly depending on the material properties (i.e., bandgap Eg) and applied laser parameters (i.e., pulse duration τ_p and intensity I).

For laser pulses shorter than a few picoseconds, a substantial amount of energy will be transferred to the lattice during pulse propagation (Schaffer et al. 2001). The excited lattice phonons are diffusing thermal energy to the vicinity of the laser focal volume and then melting and causing permanent damage (depending on the critical plasma concentration) to the transparent material. These regimes and the time scales of the corresponding processes (Rethfeld et al. 2004) are shown in Fig. 3. For pulse

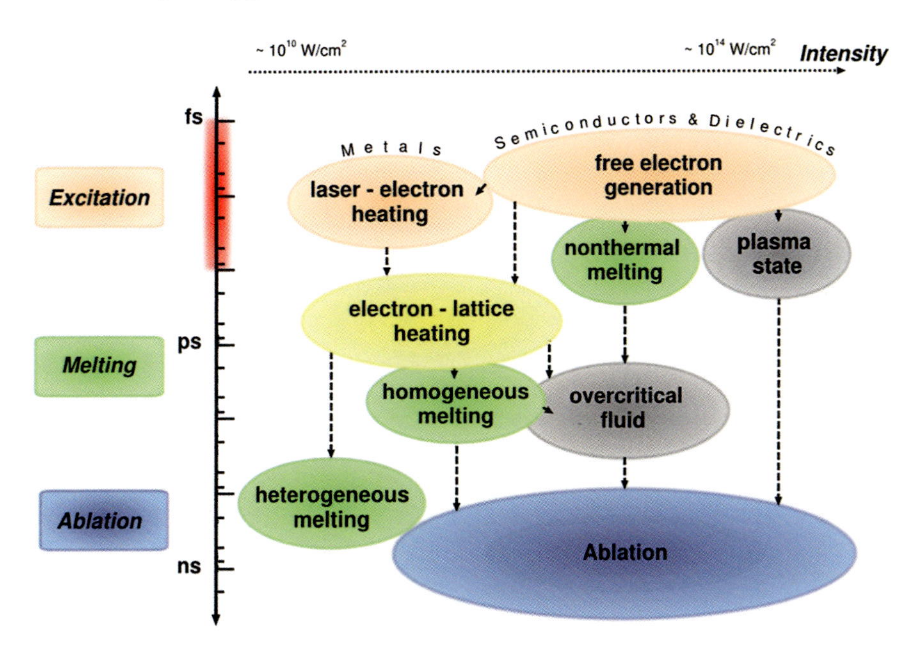

Fig. 3 Typical time scales versus intensity ranges of the phase changes occurring during and after irradiation of a solid with an fs laser pulse of about 100 fs duration. Excitation, melting, and ablation take place in the range of fs, ps, and ns regime, respectively (Rethfeld et al. 2004)

duration <10 ps, thermal diffusion and electron-ion interaction take place after the laser pulse; thus, the electrons can reach high temperatures, while keeping the lattice in the cold state during laser pulse irradiation, which is also called cold ablation. Such ablation mechanisms can make fs lasers an ideal tool to produce deterministic optical breakdown and damage near threshold and controllable material removal.

Our discussion of the structural changes induced by fs laser pulses on the surface or in the bulk of a transparent material was mainly based on single pulse interaction. However, when it comes to the multi-pulse condition within the same laser spot, the repetition rate may play a significant role during the structure formation. In general, two different modification regimes can be classified, including nonthermal and thermal regime (Gross and Withford 2015) (Fig. 4). If the repetition rate is low enough (i.e., 1 kHz), the generated heat in fused silica can be diffused out of the focal volume before the next pulse arrives. Typically, the thermal diffusion time in glass is $\sim$1 μs (Carslaw and Jaeger 1959), which is much shorter than the time between two consecutive pulses ($\sim$1 ms). In this situation, structure is modified pulse-by-pulse without heat accumulation effect. Furthermore, in order to create a smooth waveguide inside the transparent material, pulses need to be spatially overlapped, which may limit the translation speed and increase the fabrication time.

For high repetition rates from hundreds of kHz to several MHz regimes, the time between laser pulses is less than the heat diffusion time in transparent material, resulting in cumulative heating in the focal volume (Schaffer et al. 2003). Thus, the material within the focal volume will be locally melted, and the melted volume increases in size, until the laser is removed. After rapid cooling, the modified structure is formed with altered refractive index change. For a waveguide structure,

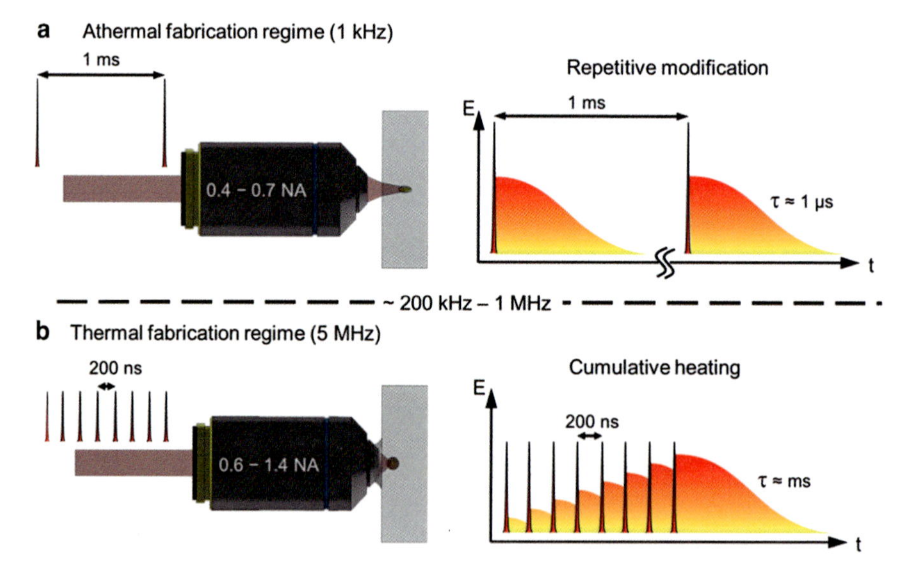

Fig. 4 (**a**) Nonthermal fabrication regime using low repetition rate and (**b**) thermal fabrication regime using high repetition regime (Gross and Withford 2015)

the size of melted volume can be controlled by the number of pulses N from laser exposure. Note that N can be expressed by $N = dR/v$, where d, R, and v are the spot size diameter (typically, $1/e^2$ of intensity), the repetition rate of the laser, and the scan speed of the translation stage, respectively. For cumulative heating, the morphology of the structural change is dominated by the heating, melting, and cooling dynamics of the material in and around the focal volume (Itoh et al. 2006).

Propagation Conditions

It is well known that linear effect includes dispersion, diffraction, and aberration during the light propagation in a medium. Typically, due to the large beam size (~3 mm), the intensity directly from the fs laser output port is not sufficient to achieve optical breakdown (~10^{13} W/cm^2) in transparent materials; as a fact, focused fs laser pulses are needed to obtain a small focal spot (micrometer-sized) via an external lens, resulting in a much higher laser intensity within the focal volume.

Before focusing inside the transparent materials, the spherical aberration and nonlinear propagation effects (i.e., self-focusing, self-phase modulation, etc.) can be ignored. The spatial intensity profile of an fs laser beam can be well represented by the paraxial wave equation and Gaussian optics. The intensity distribution of a Gaussian beam is:

$$I\,(r, z) = I_0\left(\frac{\omega_0}{\omega(z)}\right)^2 \exp\left(-\frac{2r^2}{\omega^2(z)}\right) \tag{5}$$

where the radial distance from the optical axis is $r = \sqrt{x^2 + y^2}$ and z is the axial distance from the beam waist and $I_0 = I(0{,}0)$ is the intensity at the center of the beam's tightest focus.

The waist radius is:

$$\omega(z) = \omega_0\sqrt{1 + \left(\frac{z}{z_0}\right)^2} \tag{6}$$

where ω_0 denotes the diffraction-limited minimum waist radius for a collimated Gaussian beam after focusing with lens characterized by NA and expressed by

$$\omega_0 = \frac{M^2\lambda}{\pi \mathrm{NA}} \tag{7}$$

where M^2 is the factor for non-perfect Gaussian beam (also indicating the beam quality), NA is the numerical aperture of the focusing microscope objective, and λ is the laser wavelength in free space. The Rayleigh range z_0 inside a transparent material with refractive index of n is given by:

$$z_0 = \frac{M^2\lambda n}{\pi \mathrm{NA}^2} \tag{8}$$

In our case, a Ti:sapphire fs laser with the bandwidth of $\sim$10.7 nm was applied for micromachining; such small value can reduce the effect of chromatic aberration caused by dispersion. Of course, one can employ a chromatic aberration-corrected microscope objective for the wavelength spectrum of interest, such as $\lambda = 800$ nm in our case. For the issue of spherical aberration coming from spherical shape of lens, it can be addressed by using multiple lenses (i.e., microscope objectives) or an aspherical focusing lens.

If the laser beam is focused inside the transparent materials, the light propagating through the air-glass interface (flat) may introduce additional spherical aberration. Even worse condition may happen with respect to optical fibers with curvature interface. Therefore, a common way of addressing this issue is employing an oil immersion microscope objective with the refractive index of matching oil equals to that of transparent materials.

In addition to the linear propagation, nonlinear effects can also happen when the laser intensity within the focal point inside the transparent materials is extremely high. Typically, the propagation of light in a medium is governed by Maxwell's equation:

$$\begin{aligned} \nabla \times E &= -\frac{\partial B}{\partial t} \\ \nabla \times H &= -\frac{\partial D}{\partial t} + J \\ \nabla \cdot D &= q \\ \nabla \cdot B &= 0 \end{aligned} \tag{9}$$

where q is the free charge density, J is the current density vector, **E** and **H** are the electric and magnetic field vectors, respectively, and **D** and **B** are the displacement vectors given by

$$\begin{aligned} D &= \varepsilon_0 E + P \\ B &= \mu_0 H + M \end{aligned} \tag{10}$$

where μ_0 is the permeability of free space. **P** and **M** are the laser-induced electric and magnetic polarizations, respectively. In our cases (i.e., fused silica and sapphire), $\mathbf{M} = 0$, indicating nonmagnetic dielectrics. J and q are also equal to 0 when the light is propagating in a dielectric. So the wave equation can be derived from Eqs. 9 and 10:

$$\nabla \times \nabla \times E + \frac{1}{c^2}\frac{\partial^2}{\partial t^2}E = -\mu_0\frac{\partial^2}{\partial t^2}P \tag{11}$$

where $c = 1/\sqrt{\mu_0 \varepsilon_0}$ is the speed of light in vacuum.

When light propagates through a dielectric material, it induces microscopic displacement of the bound charges, forming oscillating electric dipoles that add up to the macroscopic polarization which for glass with inversion symmetry is given by:

$$P = \varepsilon_0 \chi^{(1)} E + P^{\mathrm{NL}} = \varepsilon_0 \left[\chi^{(1)} + \frac{3}{4} \chi^{(3)} |E|^2 \right] E \tag{12}$$

where $\chi^{(i)}$ is the ith order susceptibility, with $\mathrm{P}^{(2)}$ and $\mathrm{P}^{(\mathrm{n})}$, where $n \geq 3$ left out of Eq. 12 due to even-order symmetry and negligible contribution when laser intensity is not extremely high. The refractive index can be identified from Eq. 12 as (Shen 1984):

$$n = \sqrt{1 + \chi^{(1)} + \frac{3}{4} \chi^{(3)} |E|^2} = n_0 + n_2 I \tag{13}$$

where $n_0 = \sqrt{1 + \chi^{(1)}}$, $n_2 = 3\chi^{(3)}/4\varepsilon_0 c n_0^2$, and $I = \frac{1}{2}\varepsilon_0 \mathrm{c} \mathrm{n}_0 |E|^2$ are the linear refractive index, nonlinear refractive index, and the laser intensity. A nonzero nonlinear refractive index n_2 (optical Kerr effect) gives rise to many nonlinear optical effects as an intense fs laser pulse propagates through transparent materials.

The Kerr nonlinearity $\chi^{(3)}$ results in an intensity-dependent refractive index change that follows the spatial intensity profile (i.e., Gaussian) of the laser. A Gaussian intensity profile results in the evolution of a positive or negative lens, depending on the sign of the nonlinear refractive index n_2. In the fs laser direct writing regime, n_2 is a positive value for most transparent materials. If the laser intensity is high enough, term $n_2 I$ in Eq. 13 plays a significant role to the refractive index n of the material, leading to focusing the light as passing through a positive lens. The phenomenon is also called self-focusing nonlinear effect. The strength of self-focusing depends only on the peak power of the laser, and the critical power for self-focusing can be expressed by (Shen 1984):

$$P_{\mathrm{cr}} = \frac{3.77\lambda^2}{8\pi n_0 n_2} \tag{14}$$

Obviously, if the peak power (P) of the fs laser pulse exceeds P_{cr}, the collapse of the pulse to a focal point is predicted. In addition, the initial point of self-focusing occurs at (Shen 1984):

$$Z_f = \frac{2n_0}{0.61} \frac{\omega_0}{\lambda} \frac{1}{(P/P_{\mathrm{cr}})^{1/2}} \tag{15}$$

Nevertheless, due to the free electron plasma formation at the focal point, such laser-induced plasma also has a defocusing effect for fs laser pulse propagation and can modify the real part of the refractive index as (Shen 1984):

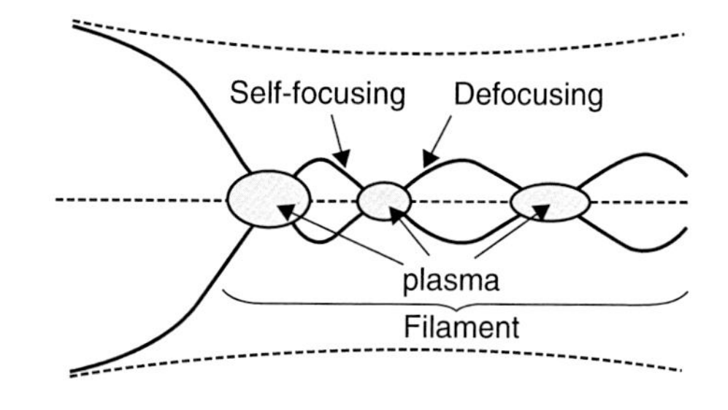

Fig. 5 Schematic representation of the focusing-defocusing cycles undergone by the intense center of a laser beam (Couairon and Mysyrowicz 2007)

$$n = n_0 - \frac{n(t)}{n_0 n_c(t)} \tag{16}$$

where $n_c(t)$ is the characteristic plasma density for which the plasma frequency equals to the laser frequency obtained from Eq. 3. The plasma-modified refractive index (Eq. 16) acts as a diverging lens to the laser beam and counters the Kerr lens self-focusing (Eq. 14). Such mechanism will always balance self-focusing and prevent pulse collapse inside transparent materials (Couairon and Mysyrowicz 2007), as shown in Fig. 5. In our cases, self-focusing is usually avoided in micromachining by tightly focusing the laser beam with an appropriate microscope objective to reach the intensity for optical breakdown ($\sim 10^{13}$ W/cm^2) without exceeding critical power for self-focusing. Typically, in fused silica, the critical power is $\sim$1.8 MW for the laser wavelength $\lambda = 800$ nm, $n_0 = 1.45$, and $n_2 = 3.5 \times 10^{-20}$ m^2/W (Sudrie et al. 2002).

Laser Processing System for Direct Writing in Optical Fiber

Femtosecond Laser Micromachining System

The optofluidic/ferrofluidic devices presented in this chapter are fabricated with our home-integrated fs laser micromachining system. The schematic of the system with direct writing capability is shown in Fig. 6. The full ultrafast laser system consists of a diode-pumped solid-state laser with high power CW output at 532 nm (Verdi V18, Coherent Inc.), a Ti:sapphire mode-locked laser (Mira 900, Coherent Inc.), and a Ti:sapphire regenerative amplifier laser (RegA 9000, Coherent Inc.). As mentioned above, such system was used for transparent material processing. The central wavelength, pulse width, and repetition rate of the fs laser are set at 800 nm, 200 fs, and 250 kHz, respectively. The maximum output power of the fs laser is 1 W.

A half-wave plate in combination with a Glan-Thompson calcite polarizer is used to precisely control the actual power used for fabrication, while an optional variable neutral density (ND) filter is used to tune the laser power more precisely. The laser exposure is switched on or off by an external mechanical shutter (SH05, Thorlabs) or electrically gating the internal clock via laser controller. In general case, the laser

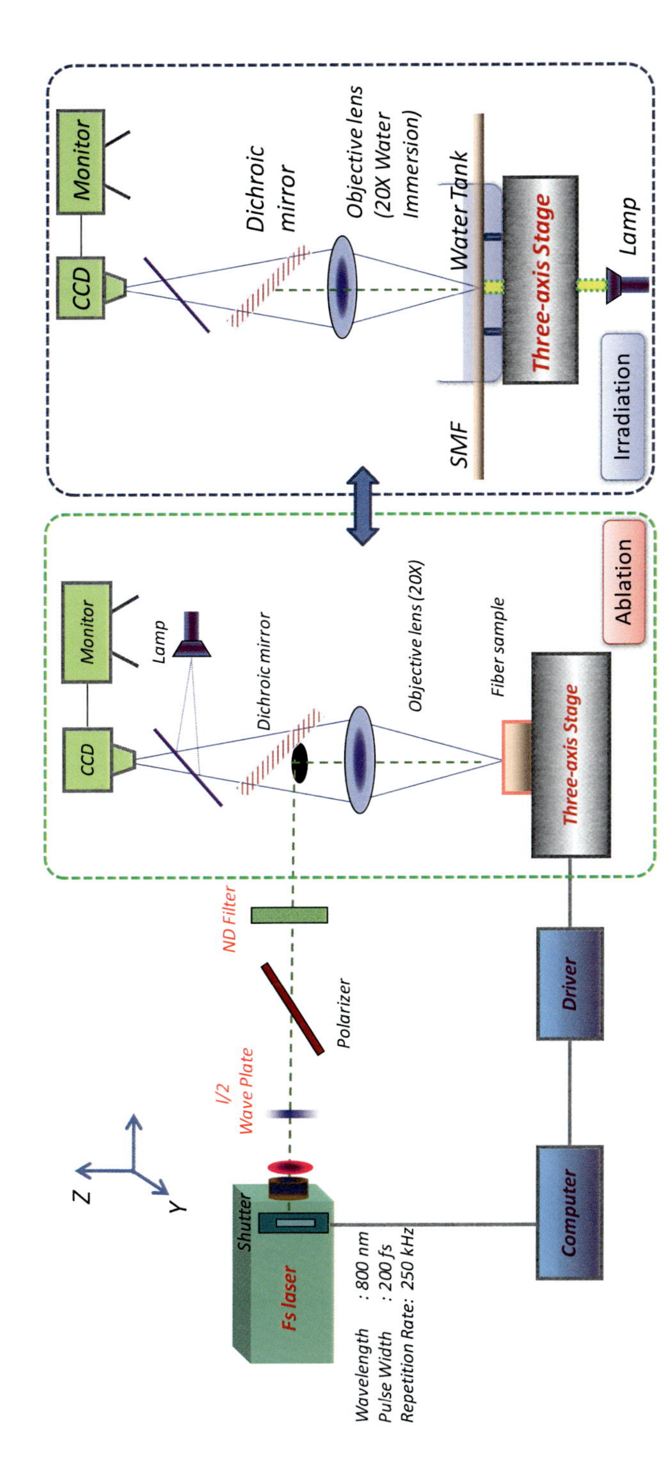

Fig. 6 Schematic diagram of the fs laser micromachining system

beam is passed through a commercial microscope objective to create a tightly focused spot. The actual laser energy used for processing silica glass and single crystal sapphire is approximately 0.4–0.5 μJ per pulse, respectively.

Silica glass substrates or optical fiber samples are mounted on a computer-controlled three-axis translation stage (PM500 series, Newport, Inc.) with a resolution of 0.1 μm. According to different fabrication cases, fiber mounting methods and illumination choices vary subsequently, so are the velocities of the translation stage. Details can be found in the following chapters. In our case, the fs laser beam is directly focused on the fiber surface (ablation) through an objective lens (20X) with a NA of 0.4 or focused inside the fiber sample (irradiation) using a water immersion lens (20X) with a NA of 0.4 or an oil immersion lens (100X) with a NA of 1.3. The diameter of the focused beam is varied from 0.5 to 1 μm with respect to different microscope objectives. A dichroic mirror and a charge-coupled device (CCD) camera connected to the computer are involved to in-situ monitor the fabrication process. Front-side and backward illumination methods can be used for different fabrication purposes. For example, both methods are suitable for transparent material ablation/irradiation, while for the nontransparent materials (i.e., silicon), only front-side method is acceptable.

During the laser processing of transparent materials, exposure conditions such as pulse energy, repetition rate, scanning speed, beam shape condition (spatio/temporal beam profile), and polarization state have significant impacts on the micromachined structures. For example, the polarization state of the laser beam can be well controlled using the combination of a half-wave plate and a polarizer, which is useful for nanostructure formation (Bhardwaj et al. 2006). Although such existing schematic is only used for low-throughput application, compared with parallel processing via spatial light modulator (SLM) (Wang et al. 2015), it is still suitable for most of research levels under laboratory conditions. Details of fs laser system and beam delivery path can be found in Fig. 7.

The movement of the three-axis translation stage can be well controlled through a home-developed graphical user interface (GUI) software based on MatLab.

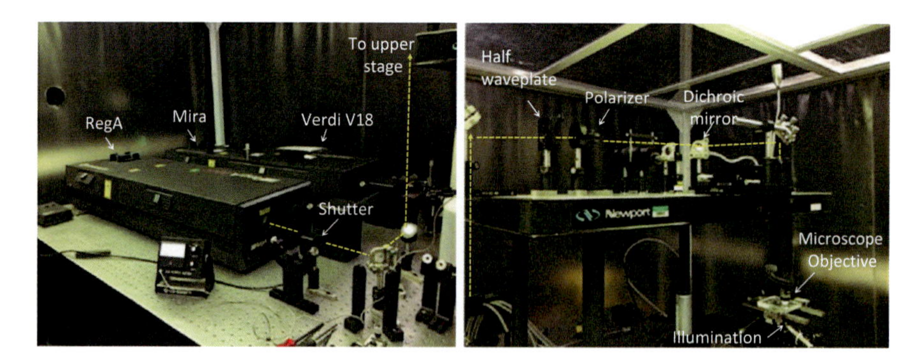

Fig. 7 Fs laser system and beam delivery path

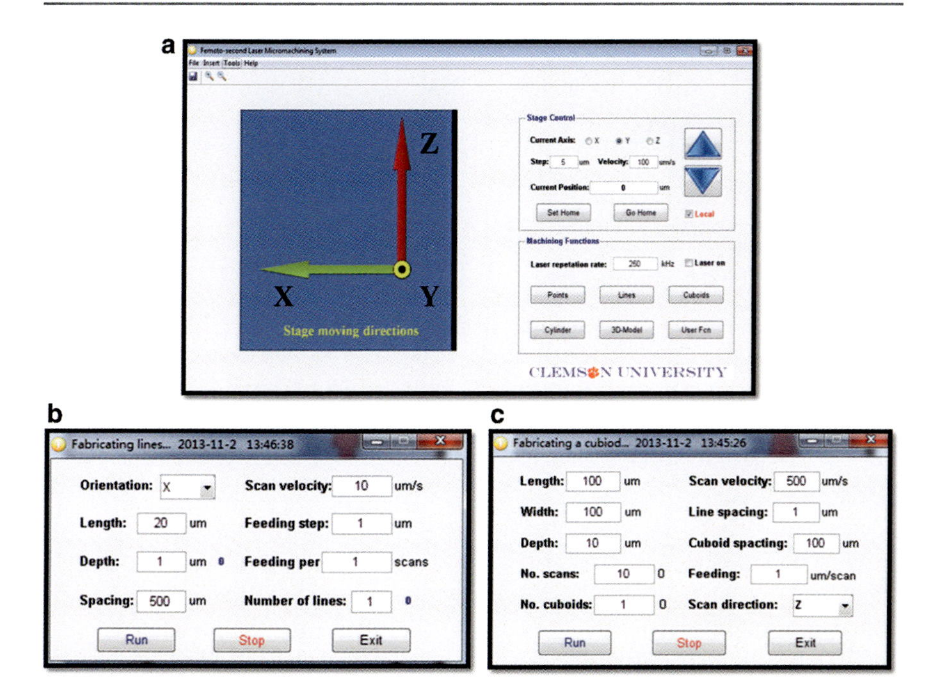

Fig. 8 Interfaces of the developed software. (**a**) Main GUI, (**b**) inscribing lines, and (**c**) fabricating cuboids

GPIB-USB connection method is adopted for the communication between the stage controller and a desktop. Multifunctional operations (i.e., stage control and machining functions) are involved within this software, showing great capabilities for complex structure micromachining. Details of the interfaces of this proposed software are shown in Fig. 8a–c.

Direct Femtosecond Laser Writing in Optical Fiber

Since we are mainly focusing on laser direct writing in optical fibers, two standard writing geometries can be applied for different purposes, including longitudinal mode and transverse mode, as shown in Fig. 9.

In longitudinal writing mode, the optical fiber is horizontally placed on a multi-axis translation stage, moving vertically either toward or away from the incident laser. Due to the transverse symmetry Gaussian intensity of the laser beam, the modifying structures have cylindrical symmetry. However, such configuration is highly depending on the working distance of the focusing lens, resulting in a short length of modifying structures.

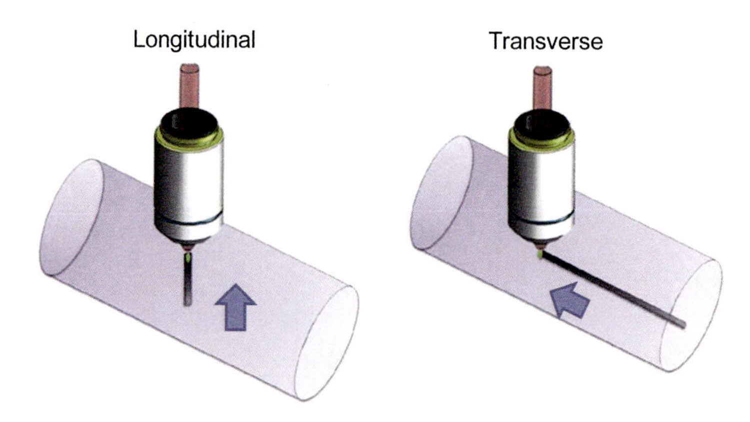

Fig. 9 Laser direct writing geometries in optical fibers, including longitudinal mode and transverse mode

In most cases, people employ transverse mode for waveguide direct writing (Davis et al. 1996), in which the writing direction is perpendicular to the laser propagation direction. The working distance is no longer an issue for this configuration, and 3D optical circuits can be realized using this method. However, the cross section of modifying structure is asymmetric (elliptical) due to the ratio between the focal depth ($2z_0$) and spot size ($2\omega_0$), which is related to the following relationship $2z_0/2\omega_0 = n/\mathrm{NA}$. To overcome this issue, using high NA objective lens is helpful, such as oil immersion lens.

The spot size of the focal point inside the transparent materials plays a significant role for high accuracy, high-precision 3D structure micromachining using fs laser. Typically, the radius r of the spot size can be roughly calculated by (Gamaly et al. 2006):

$$r = \frac{0.61}{\mathrm{NA}} \frac{\lambda}{n} \tag{20}$$

Here, λ is the wavelength of incident laser, and λ/n means the effective wavelength when the laser is focused inside media of index n. As such, the focal spot inside material is smaller by factor n than that in air ($n_{air}=1$). Obviously, once the wavelength of the laser and the refractive index of the material are fixed, r only depends on the NA of microscope objective. The larger the NA is, the smaller the spot size will be. In general, low NA (<0.1) microscope objective is used for large-area surface modification. Lenses with NAs from 0.1 to 0.5 can induce a confined permanent modification inside a transparent material or ablate the surface of materials with high precision. The immersion lenses (water immersion or oil immersion) usually have even larger NAs (NA > 0.9). The immersion optics not only tightly focused the laser beam but also compensates refractive index mismatch

Table 1 The pros and cons of each microscope objective

Microscope objective	Pros	Cons
20X (Zeiss EC Epiplan)	Small NA (0.4)	Not suitable for laser irradiation inside optical fibers
	Suitable for laser ablation with high precision	
20X water immersion (Olympus)	Less spherical aberration	Filament issue when used at high power; large spot size
	Suitable for laser-induced weak reflector/stresses	
60X water immersion (Olympus)	Small laser spot size	Limited working distance
	Laser-induced water breakdown	Spherical aberration
100X oil immersion (Olympus)	Small laser spot (~0.5 μm)	Limited working distance
	No spherical aberration high precision	Complicated operation

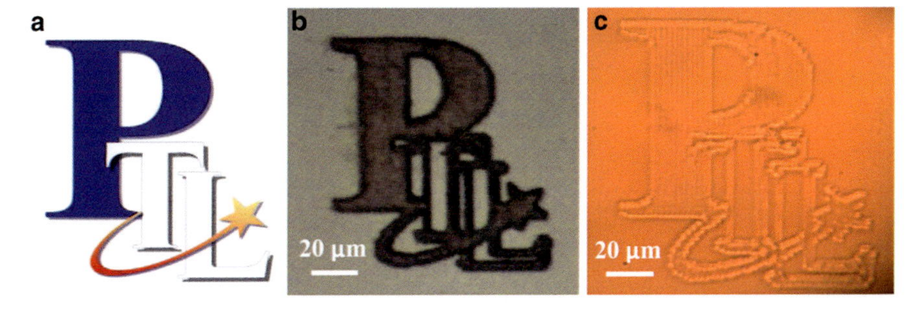

Fig. 10 The logos of Photonics Technology Lab. (**a**) Original file, (**b**) laser ablated on the surface of silica glass, and (**c**) laser irradiated inside the silica glass

at the surface of the sample. As a result, material can be modified with a tightly focused fs laser pulses even at the nJ energy level.

For different fabrication purposes, four types of microscope objectives are chosen in this work. The pros and cons of each microscope objective are discussed, as shown in Table 1.

With the proposed fabrication system, a lot of fancy and complex structures can be realized. Examples include Photonics Technology Lab (PTL) logos displayed on the surface and inside the silica glass (Fig. 10), helical shape formation inside a silica fiber (Fig. 11), line-by-line grating inscription inside a sapphire fiber (Fig. 12), and micro-cantilever beam ablated on the tip of a silica fiber (Fig. 13).

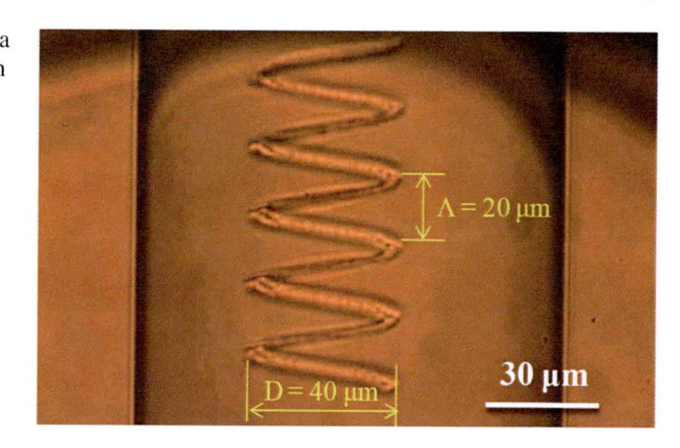

Fig. 11 Helical structure in a silica fiber. Pitch $\Lambda = 20\ \mu\text{m}$ and diameter $D = 40\ \mu\text{m}$

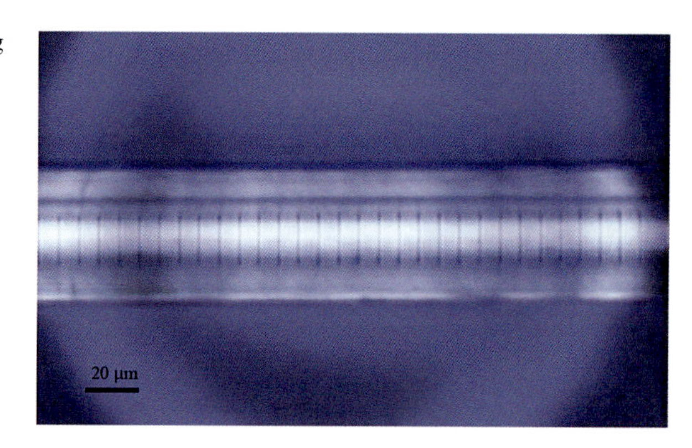

Fig. 12 Line-by-line grating inscription inside a sapphire fiber

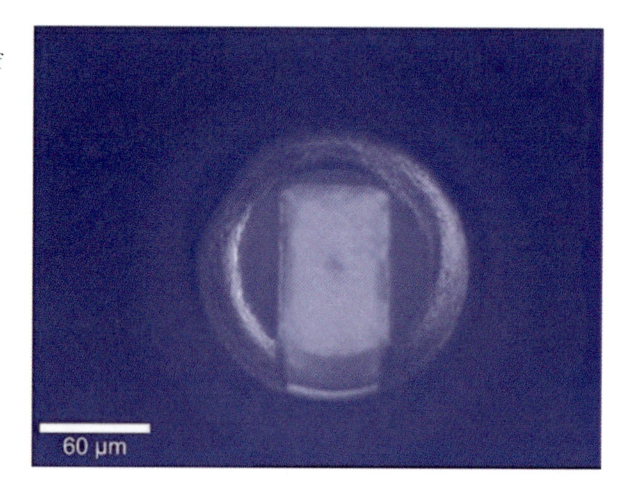

Fig. 13 Micro-cantilever structure ablated on the tip of silica fiber

Liquid-Assisted Laser Processing in Optical Fiber

Introduction to Liquid-Assisted Laser Processing

Recent technological advances have led to the development of optofluidic-based systems, in which waveguide and microfluidic architectures are synergistically integrated to provide enriched intelligence and enhanced functionalities for chemical and biological sensing applications (Schaffer et al. 2001). In an optofluidic system, buried microchannels can be formed in solid or soft transparent materials (i.e., silica or PMMA) through advanced manufacturing techniques. The liquid of interest (in a small volume) confined and manipulated inside the microchannels can be probed and analyzed using optical measurements (Yablonovitch and Bloembergen 1972).

In addition to the well-known planar configurations of optofluidic systems, the idea of all-in-fiber optofluidic devices (i.e., PCF-based and OFRR-based) has also been investigated (Glezer and Mazur 1997; Stoian et al. 2000). Additionally, taking advantage of fs laser micromachining technique in transparent materials, the alignment-free optics and improved robustness of all-in-fiber optofluidic devices have been proposed in the past years. Existing methods can be classified into two categories: (1) fs laser irradiation with chemical etching (i.e., HF or KOH), also known as FLICE technique, and (2) fs laser-induced water breakdown (FLIWD) technique.

In 2006, Y. Lai et al. from Bennion's group firstly demonstrated a microchannel formed in an SMF using FLICE for refractive index sensing application (Rethfeld et al. 2004). The intensity change as a function of refractive index variation was investigated. Later, their group created microfluidic networks in fiber cladding of SMFs using FLICE technique and made this device as a sensor for temperature and refractive index sensing applications (Zhou et al. 2010). However, due to the high roughness on the surfaces of microchannels, large optical insertion losses could exist. Recently, researchers from Hermann's group have improved the roughness on the FLICE-formed surfaces down to 10 nm (rms) in bulk fused silica (Ho et al. 2012) and presented an intricate and highly compact lab-in-fiber sensors fabricated by laser direct writing waveguides and FLICE techniques with a variety of functionalities, such as fluorescence, temperature, refractive index, and bending sensing applications (Haque et al. 2014). However, the uniformity (taper angle involved) of the long-length channel is an issue when using HF as etchant due to the etch selectivity between the laser-irradiated region and unexposed region. Although KOH solution can significantly enhance the etch selectivity, the etch rate is much lower than that of HF (Kiyama et al. 2009).

In addition to FLICE technique, a microchannel can be directly created in an SMF using FLIWD technique without using any toxic or hazard chemical solutions (Liu et al. 2013). Such device can be used for temperature and RI sensing applications. In this process, the interaction between the laser and the liquid (i.e., DI water) can cause the laser-induced water breakdown phenomenon with laser-induced bubbles, shockwaves, and a high-speed jet. Meanwhile, the water plays a significant role to efficiently remove debris from the ablated regions, resulting in the

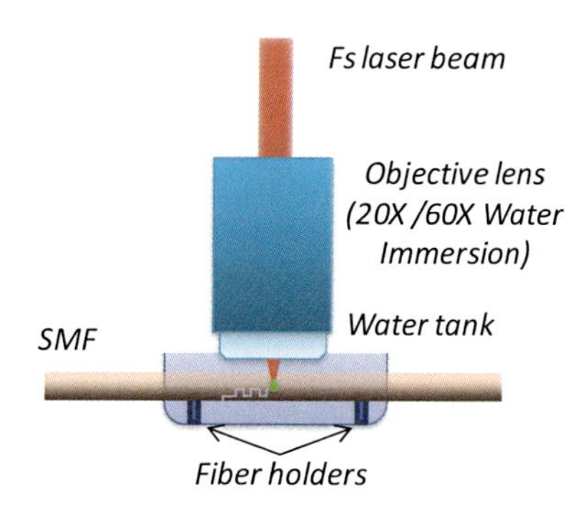

Fig. 14 Schematic of FLIWD for the fabrication of 3D arbitrary microchannels in an optic fiber

formation of microfluidic channels with arbitrary shapes. The schematic diagram of FLIWD for the fabrication of 3D microchannels in an SMF can be found in Fig. 14. Compared to FLICE, this technique can provide better uniformity of the channel diameter, which is mainly depending on the spot size of the laser beam (or the NA of objective lens). However, when the length of the microchannel reaches several hundreds of micrometers, the debris generated by fs laser ablation is hard to be washed out using water, which restricts the length of the fabricated microchannels (typically ~1 cm An et al. 2005). Additionally, the low-quality surface profiles can generate significant optical insertion loss, and relative low translation speed (~1 to 2 μm/s Liu et al. 2013) may cause time-consuming concern.

Fabrication of 3D Hollow Structure in Optical Fibers

Both FLIWD and FLICE can be used to fabricate 3D hollow structure in optical fibers. FLIWD technique is firstly investigated here. Figure 15a shows the schematic FLIWD micromachining system. The actual laser energy used for fabrication was approximately 0.8–1.2 μJ per pulse for low NA lens, which is much higher than the threshold of fused silica. Figure 15b shows the details of FLIWD fabrication process.

The fiber used in experiments was a single-mode optical fiber (Corning, SMF-28e) with the core and cladding diameters of 8.2 and 125 μm, respectively. After mechanically stripping off its buffer, the fiber was cleaned using acetone and clamped onto two bare fiber holders (Newport 561-FH). The optical fiber and fiber holders were immersed in distilled water during fabrication. The fiber assembly was mounted on a computer-controlled three-axis translation stage (Newport, Inc.) with a resolution of 0.1 μm. The fs laser beam was focused inside the optical fiber through a water immersion objective lens (Olympus UMPlanFL 20×/60×) with a numerical aperture (NA) of 0.4 or 0.9, respectively. The spot size of the focused beam was about 5 μm in fiber due to the much higher pulse energy for 20x lens, while for 60x lens, the spot size can be reduced down to 2 μm with the pulse energy of 0.4 μJ. The velocities of the stages were set at 1 μm/s during fabrication.

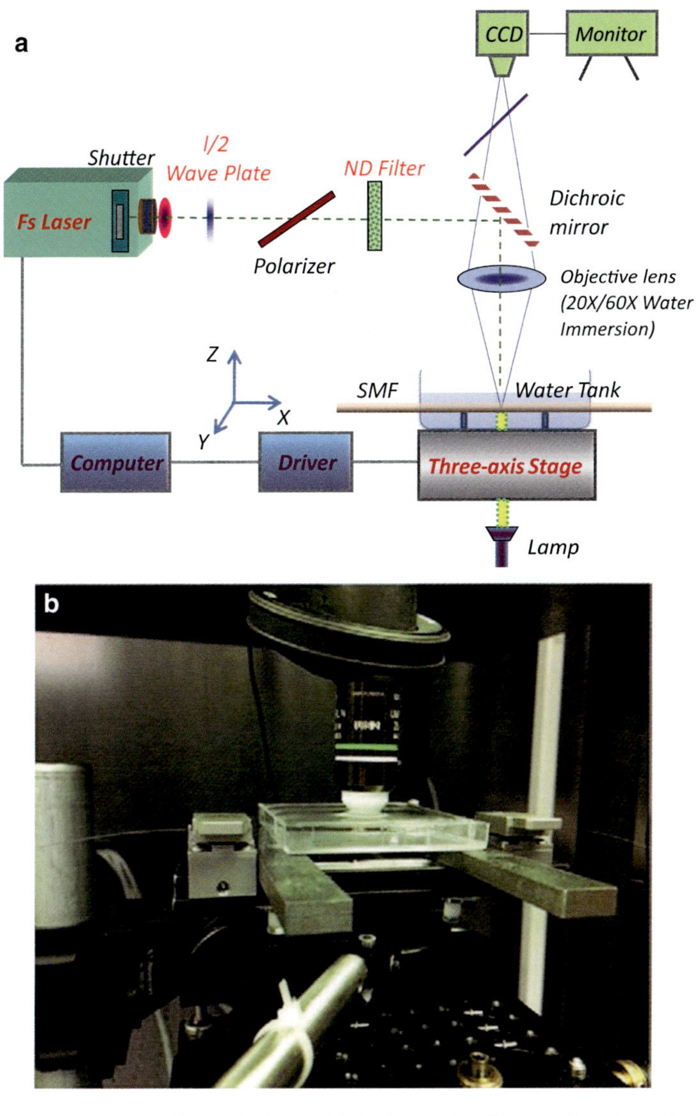

Fig. 15 (**a**) Block diagram of FLIWD for the fabrication of 3D arbitrary microchannels in an optic fiber. (**b**) Figure of details of FLIWD experiment setup

A straight microchannel with the diameter ~5 μm was fabricated in the x-y plane using FLIWD method, and the microscope image is shown in Fig. 16. Obviously, the roughness of the microchannel surface is not smooth enough, which means the optical insertion loss might be extremely high. If higher NA microscope objective was adopted, the result can be significantly improved. However, high NA lens is hard to operate and easy to be affected by spherical aberration issue.

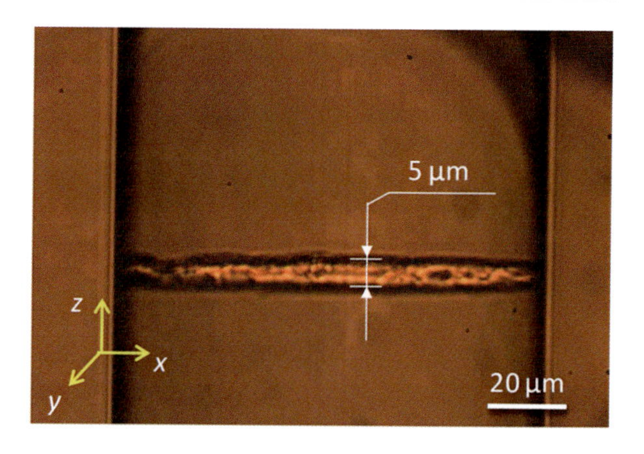

Fig. 16 Microscope image of a straight microchannel inside an SMF fabricated by FLIDW technique

Alternatively, FLICE technique was investigated to get a better result for the fabrication of 3D hollow structure inside an optical fiber. The block diagram of FLICE technique is much similar to laser irradiation micromachining system. In this case, the fs laser beam can be focused inside the optical fiber through either a water immersion objective lens (Olympus UMPlanFL 20×) with a numerical aperture (NA) of 0.4 or an oil immersion lens (Olympus UMPlanFLN 100X, NA = 1.3) with the index-matching oil of RI = 1.464 (Cargille Laboratories). The actual laser energy used for fabrication was approximately 0.4 μJ or 0.2 μJ per pulse for two different microscope objectives. So the spot size of the focused beam was about 2 μm or 0.5 μm in fiber. The velocities of the stages were set at 10–50 μm/s during fabrication. After laser irradiation, the fiber was well cleaned in acetone solution and DI water to fully remove the residue index-matching oil. Then the HF etching process was conducted. This technique offers great simplicity and flexibility to produce buried 3D structures with high aspect ratios in transparent materials (Bellouard et al. 2004). The etching rate of a laser-modified region was found to be two orders of magnitude higher than that of the unexposed region in silica materials. As a result, such hybrid technique has been adopted for fabrication of optofluidic devices in planar sample and optical fibers (Crespi et al. 2010).

It should be noticed that the etching rate in the germanium-doped core of the SMF is much faster than that in the fiber cladding. The respective reactions can be found as follows (Tuck et al. 2006):

$$SiO_2 + 4HF \rightarrow 2H_3O^+ + {SiF_6}^{2-} \tag{21}$$

$$GeO_2 + 4HF \rightarrow 2H_3O^+ + {GeF_6}^{2-} \tag{22}$$

The reactions are driven by different dissociation energies of the Si-O bond (799.6 kJ/mol) and Ge-O bond (660.3 kJ/mol) (Gong et al. 2009). Therefore, smaller pulse energy is needed in fiber core for exciting the electrons from valence band to conduction band, leading to a concave surface profile inside the fiber core during HF etching process. One possible solution is using HF solution (Acros Organics) with

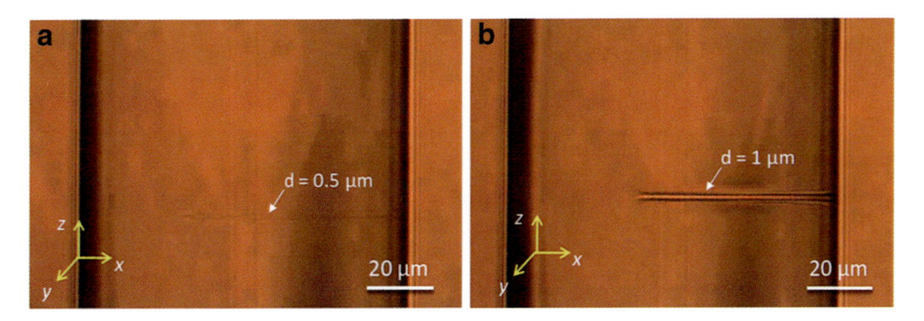

Fig. 17 Microscope images of a blind microhole inside an SMF fabricated by FLICE technique. (**a**) Before HF etching and (**b**) after HF etching

low concentration (i.e., 5%) for the etching of microchannels in SMFs. However, there is a trade-off between the quality of the channel and the consuming of the time. Typically, lower concentration of HF etchant will consume longer time, and then a better quality of microchannel will be formed and vice versa. Actually, by using HF solution as etchant solely, the etching rate difference between the fiber core and cladding is quite small due to the low doping concentration of GeO_2 in SMFs.

Here is an example to show how we can use FLICE to fabricate a blind microhole inside an SMF. High NA (1.3) oil immersion objective was used for laser irradiation, and low NA (0.4) water immersion lens was adopted for quality check. After laser irradiation, the fiber was cleaned with acetone and DI water and then etched with 5% aqueous HF solution for approximately 15 mins to fully open the blind microhole. Figure 17a shows the laser-irradiated structure before etching, and Fig. 17b shows the blind microhole formed inside the SMF after HF etching process. The diameter of the microhole is about 1 μm. In addition, we cannot see obvious difference between fiber core and cladding after HF etching. The proposed example indicates the potential of fabricating high-quality 3D hollow structure inside SMFs using our fs laser micromachining system.

All-in-Fiber Optofluidic Sensor

In this section, an assembly-free, all-in-fiber optofluidic sensor was proposed as an example for RI sensing in optofluidic systems. FLICE technique was adopted for the fabrication of such advanced fiber-optic sensor.

Operation Principle and Sensing Mechanism

The fs laser ablation technique has been used to micromachine FP cavities on optical fibers for various sensing applications such as RI (Wei et al. 2008) and pressure (Zhang et al. 2013) measurement. It has been proven that the formed

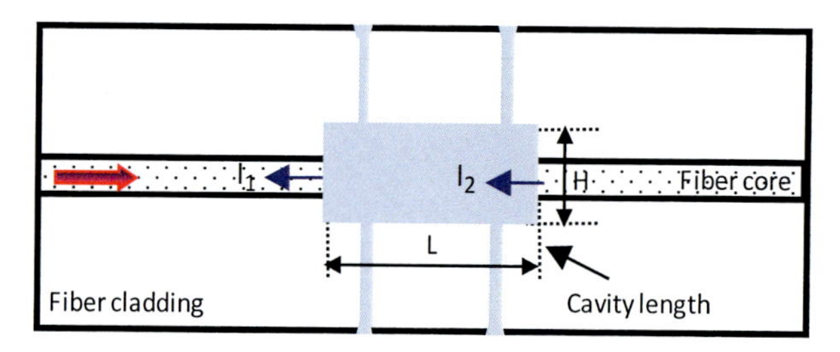

Fig. 18 Schematic of the all-in-fiber optofluidic device

FP cavity is insensitive to the ambient temperature. However, it was also found that the open cavity structure (Wei et al. 2008) was sensitive to fiber bending and easy to break. Here, we propose a prototype all-in-fiber 3D optofluidic microdevice fabricated by the FLICE technique. The FLICE technique has been proven for rapid fabrication of embedded 3D channels with flexible orientations inside an optical fiber (Fig. 17a, b). To demonstrate the feasibility, we fabricated and tested all-in-fiber optofluidic devices consisting of horizontal and vertical microchannels. The horizontal microchannel can be conceived as a FP cavity, while the vertical ones are the inlets/outlets to the cavity.

Figure 18 illustrates the schematic of the prototype all-in-fiber optofluidic device, where a horizontal fluid-holding cavity is embedded in the center of a single-mode fiber and four vertical microchannels are punched through the fiber cladding to allow liquid access to the cavity. The device is fabricated in two steps: (1) irradiation of the selected regions (the rectangular cavity and vertical channels) with focused fs laser pulses and (2) selective etching of the laser-modified zones using the HF solution.

Optically, the horizontal cavity also functions as a FP cavity. The light propagates inside the fiber and reflects at the two air/silica endfaces of the cavity. The light intensities of the reflected beams by the two surfaces of the FP cavity are denoted as I_1 and I_2I$_2$, respectively, superimpose to generate an interference pattern. The FP interferometer can be modeled using the following two-beam optical interference Eq. 20:

$$I = I_1 + I_2 + 2\sqrt{I_1 I_2}\cos\left(\frac{2\pi \cdot \mathrm{OPD}}{\lambda} + \phi_0\right) \tag{21}$$

where I is the intensity of the interference signal, ϕ_0 is the initial phase of the interference (normally equal to zero), and λ is the optical wavelength in vacuum. The round-trip OPD of the FP interferometer is given by:

$$\mathrm{OPD} = 2n_{\mathrm{cavity}}L \tag{22}$$

where n_{cavity} is the RI of the cavity medium and L is the length of the cavity length. At the valleys of the interferogram in spectrum domain, the phase difference of the two reflected light beams satisfies the condition of coherently destructive interference:

$$\frac{4\pi n_{\text{cavity}} L}{\lambda_m} + \phi_0 = (2m+1)\pi \tag{23}$$

where m is an integer and λ_m is the wavelength of the mth order interference valley. For two adjacent wavelength minima, the following condition is obtained:

$$\frac{4\pi n_{\text{core}} L}{\lambda_1} - \frac{4\pi n_{\text{core}} L}{\lambda_2} = (2m+1)\pi - [2(m+1)+1]\pi = 2\pi \tag{24}$$

The distance between two adjacent minima of the spectrum, defined as free spectrum range (FSR), can then be expressed as:

$$\text{FSR} = \frac{\lambda^2}{2n_{\text{cavity}} L} \tag{25}$$

According to Eq. 23, taking the derivative of n with respect to λ_m, one finds:

$$\frac{dn}{d\lambda_m} = \frac{[(2m+1)\pi - \phi_0]}{4\pi L} \tag{26}$$

Assuming the cavity length L is maintained constant during measurement, Eq. 26 indicates that the RI is a linear function of the wavelength at interference valley, and then the sensitivity of the FP interferometer is a constant. The amount of RI change (Δn) can thus be computed based on the wavelength shift of a specific interference valley using the following equation:

$$\frac{\Delta n}{n} = \frac{\Delta\lambda_m}{\lambda_m} \tag{27}$$

where the relative RI change is directly proportional to the spectral shift of the interferogram.

Sensor Fabrication

In order to fabricate a high-quality FP cavity, the following steps were adopted. First, a rectangle region, with the dimension of $L\times 10 \times H$ μm, was inscribed in the center of the fiber using fs irradiation from the bottom to the top (z direction, as shown in Fig. 15a), where L and H denote the length and height of the micro-cavity, respectively. The center of the inscribed region is aligned with the center of the fiber core. Second, the irradiation position of the vertical channels was chosen. The laser was initially focused at the center of the fiber core. Then the stage was moved along the z direction to make sure the fs laser beam can irradiate a straight line along the z direction. After one-side fabrication, the fiber was then rotated by 180° to

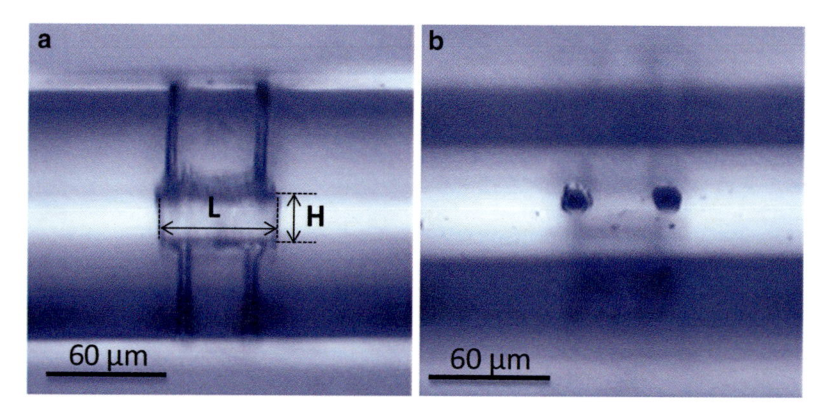

Fig. 19 (**a**) Microscope image of the fabricated all-in-fiber optofluidic device: the length L and the height H of the FP cavity are 55 and 20 μm, respectively. (**b**) Top view of the fiber device

fabricate the channels on the other side following the same procedure. After fs laser irradiations, the sample was dipped into a HF solution with a concentration of 20% for about 10 min. The actual etching time may depend on the HF concentration and cavity dimensions. As the final step, the etched sample was cleaned in an ultrasonic bath filled with distilled water and dried in air. The embedded cavity was close to the end of the fiber to minimize the possibility of device breakage due to fiber bending.

Figure 19a shows the microscope image of a fabricated all-in-fiber optofluidic device, where the cavity length L and the cavity height H are about 55 μm and 20 μm, respectively. Figure 19b shows the top view of the device. The diameter of the vertical microchannels is about 5 μm.

The fabricated all-in-fiber optofluidic device was interrogated using a broadband light source with wavelength range from 1520 to 1620 nm. A 3 dB fiber coupler was used to route the light into and out of the device. The interference spectrum was recorded by an OSA (AQ6319). Figure 20 shows the reflection spectra of the all-in-fiber optofluidic device with the FP cavity length of 35 μm and 55 μm, respectively. The FSRs of these two interferometers are 36 nm and 21 nm, respectively, which matched well with those calculated using Eq. 25. The clean interference pattern with a large fringe visibility of 20 dB was obtained with the cavity length of 55 μm. The device with 35 μm of cavity length had a fringe visibility of about 12 dB. The excess losses of both devices were about 18 dB, which was mainly caused by the roughness of the cavity surfaces. To a certain extent, the surface quality can be improved by tuning the laser irradiation power, adjusting the concentration of HF solution (i.e., 5%) and the etching time (i.e., ~45 mins).

Measurement of RI with the Optofluidic Sensor

The all-in-fiber optofluidic device with a cavity length of 55 μm was tested for its capability of RI measurement at room temperature. The vertically placed sensor

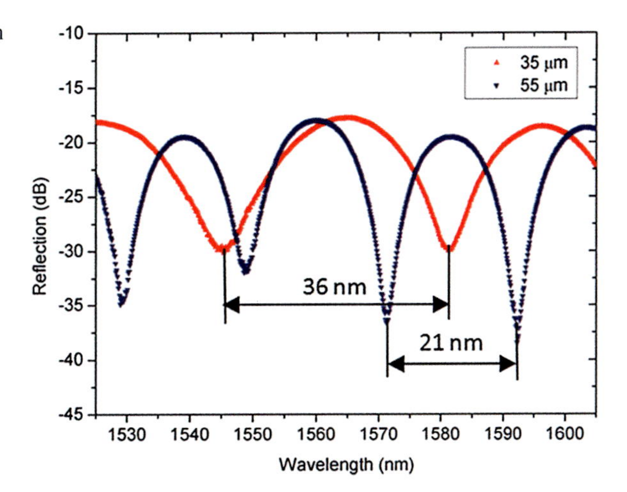

Fig. 20 Reflection spectra in air of the all-in-fiber optofluidic devices with cavity lengths of 35 and 55 μm, respectively

probe was directly immersed into the sucrose solutions with different concentrations of 0.00, 1.00, 2.00, 3.00, 4.00, 5.00, 6.00, 7.00, 8.00, and 9.00 (unit: percentage). The corresponding RIs are 1.3333, 1.3344, 1.3359, 1.3373, 1.3388, 1.3403, 1.3418, 1.3433, 1.3448, and 1.3463, respectively. The ultrasonic bath was used to assist the liquid to flow into the embedded FP cavity during tests and to ensure no air bubble left in the cavity. In each measurement cycle, the device was carefully cleaned using acetone and DI water and dried after each measurement to ensure there was no residual liquid left within the cavity, indicated by the interference spectrum restored to its original in air.

The time needed for ultrasonic bath-assisted sample loading was about 10 s. In comparison, it took about 16 min to load the cavity without the ultrasonic bath. It is worth noting that the purpose of this thesis is to demonstrate the potentials of using the FLICE technique to fabricate embedded 3D structures inside an optical fiber and the feasibility of integrating the fluidic channels with a fiber sensor toward optofluidic applications. It is envisioned that other microfluidic components can be fabricated on the surface or inside the optical fiber using the same technique to facilitate the fluidic transport and storage. When integrated with the extra fluidic components, the use of ultrasonic bath becomes unnecessary because the standard sample loading method, e.g., syringe pumping, shall be able to load the samples.

Figure 21a shows the interference spectra of all-in-fiber optofluidic device in the sucrose solutions at various concentrations of 0%, 2%, 4%, and 6%, respectively. As the RI increased, the interferogram shifted toward the long wavelength region (red shift). Figure 21b plots the center wavelength of an interference valley (1567.6 nm) as a function of the RI of the liquid. Linear regression was used to fit the response curve, and the slope of the fitted line was calculated as the RI sensitivity, which was estimated to be 1135.7 nm/RIU in the tested RI range. Similar responses were obtained using the device with the cavity length of 35 μm. The length of the cavity should be limited to several hundred microns to avoid the excess optical

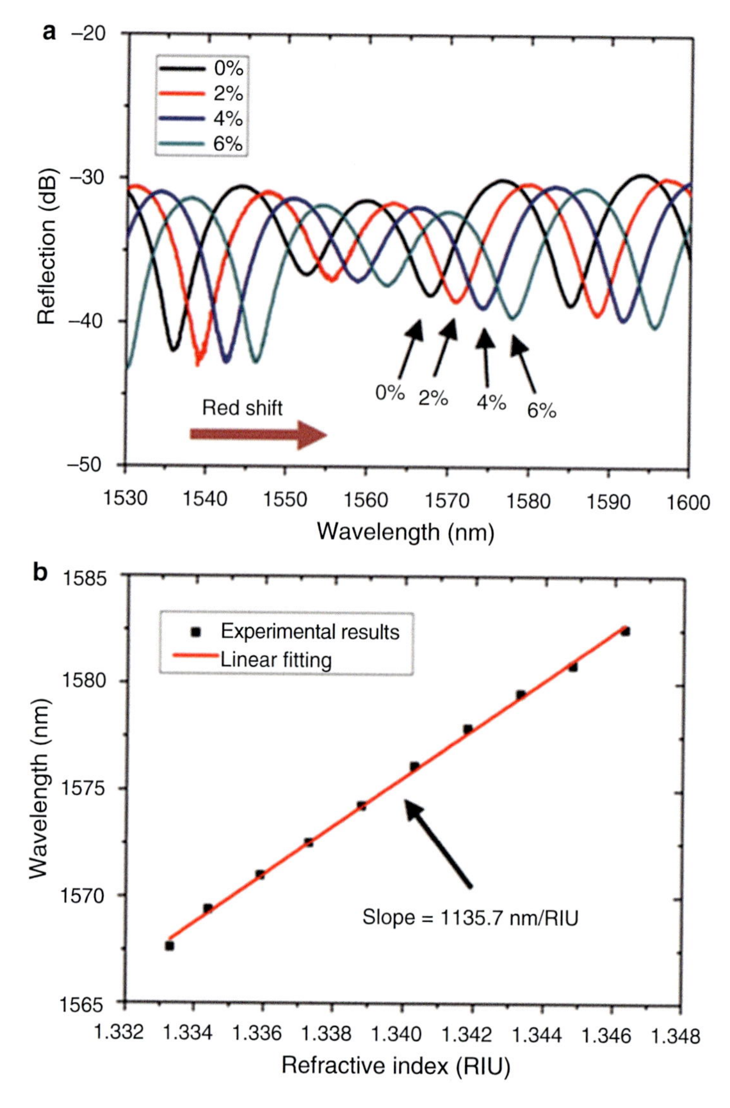

Fig. 21 (**a**) Interference spectra and (**b**) center wavelength of an interference valley (1567.6 nm) of the all-in-fiber device in sucrose solutions with different concentrations

loss caused by the divergence of the optical beam propagating inside the FP cavity (Zhang et al. 2010).

In summary, an all-in-fiber 3D optofluidic microdevice was fabricated by fs laser irradiation-assisted selective chemical etching. The proposed all-in-fiber optofluidic device is flexible in design, simple to fabricate, mechanically robust, and miniaturized in size, showing good potentials for chemical/biomedical sensing and integrated microfluidic applications.

Fiber In-Line Ferrofluidic Sensor

Ferrofluid is a type of stable colloidal suspension with single-domain magnetic nanoparticles that dispersed uniformly in a liquid carrier (Kruse et al. 2003). When exposed in external magnetic field, the nanoparticles inside the ferrofluid will be rearranged, resulting in change of properties of the solution such as refractive index and absorption coefficient of electromagnetic radiation (Yang et al. 2002). Based on its unique properties, a lot of magnetic field sensors have been proposed. Among them, optical fiber-based ferrofluidic sensors have attracted a lot of interests because of its compact size and high sensitivity. Various configurations of fiber-optic ferrofluid magnetic field sensors have been reported, such as singlemode-multimode-singlemode (SMS) structure (Chen et al. 2013), tapered microstructured optical fiber (Deng et al. 2015), fiber FP interferometer (Lv et al. 2014), etc.

In this section, we present a ferrofluid-based optical fiber magnetic field sensor (Fig. 22) fabricated by FLIWD technique. The glass tube (GT) was sandwiched by SMFs forming a typical FP cavity. The opening (μ-window), fabricated by FLIWD technique, offers sufficient access for the filling ferrofluid. The refractive index of the ferrofluid varies as the surrounding magnetic field strength changes, which can be optically probed by the FP interferometer. In our case, such FP interferometer is sensitive to RI change inside the cavity while insensitive to ambient temperature.

In this case, Ferrotec-Oil-based Ferrofluid EMG 909 is filled in the FP cavity. The n_{cavity} follows the equation:

$$\Delta n_{\text{cavity}} = \alpha_H \Delta H + \alpha_T \Delta T \tag{28}$$

where the Δn_{cavity} is the RI change when filling with ferrofluid liquid and α_H is the magnetic sensitivity of the ferrofluidic, while α_T is the temperature sensitivity. The thermal expansion can be ignored as the experiment environment is operated

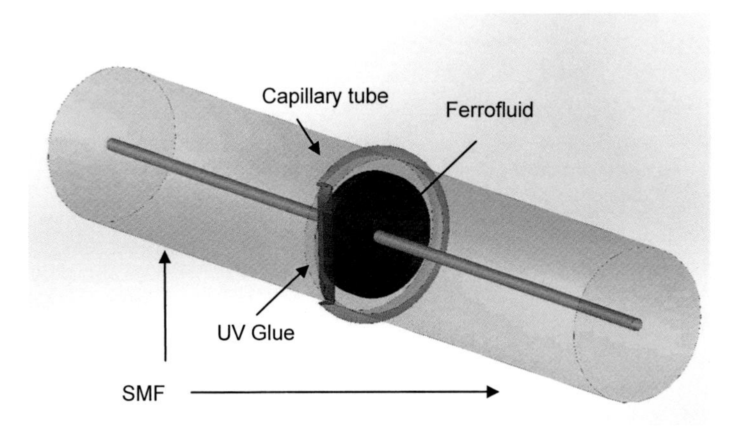

Fig. 22 The structure of sensor probe

at room temperature (21 °C). So the resonance wavelength shift of the interference spectrum is:

$$\Delta\lambda_m = \lambda_m \cdot \alpha_H \Delta H \quad (29)$$

Here we used SMF-28e (Corning) with core in 8.2 μm and cladding in 125 μm, respectively. The capillary tube used in experiment is glass tube with OD in 164 μm and ID in 100 μm. Both fiber and tube were mechanically stripped off its buffer and cleaned. First, an SMF was spliced with the capillary tube. Second, the assembly was mounted in fiber slicer and observed under measurement-controlled microscope (Nikon). The cutting point upon the capillary tube can be precisely controlled with the help of the microscope and a motorized translation stage. In order to get a highly performed FP cavity, the expected leaving length of tube is 30 μm.

To fabricate the compatible μ-window for the needle, the following steps were adopted. Figures 23 and 24 show the assembly before and after fs fabrication. The injection of ferrofluid was realized by a syringe pump and a fabricated microneedle. The approximate O.D. of the needle was around 20 μm. Then, the assembly was clamped horizontally onto two fiber holders (Newport 561-FH). The fabrication processed in water as the opening was fabricated from the surface to the deep. Then, a rectangle region with dimensions of 30 × 60 × 40 μm was inscribed on the capillary tube. The fs laser irradiation swept from the top to the bottom. The total layers of sweep were 40 and then an opening appeared on the chosen region.

Assisted by the microscope and manual three-dimension stage, the needle was inserted into the cavity though the μ-window. The syringe pump used in experiment was set in 1 μl per second. The ferrofluid was EMG 909 from Ferrotic with low viscosity of 3 cP (at 27 °C) and saturation magnetization (Ms) of 220 Gauss. After the cavity was fully filled up, the μ-window was sealed by small amount of UV glue. One side of the SMF was cut to the minimum length to reduce the device breakage from bending.

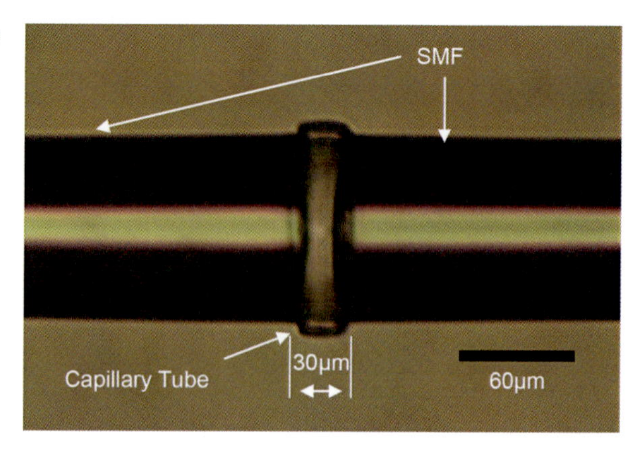

Fig. 23 The assembly before fs irradiation

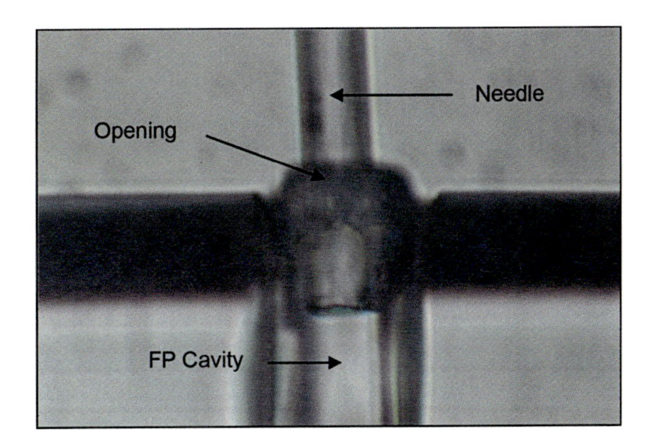

Fig. 24 The assembly after fs irradiation with needle

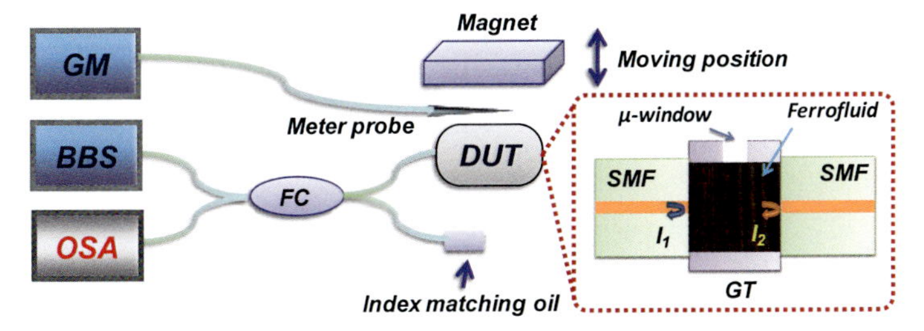

Fig. 25 The sensing system of the ferrofluidic sensor

Figure 25 shows the sensing system of the ferrofluidic sensor. The packaged device was interrogated by a broadband light source (BBS). The wavelength range is from 1520 to 1620 nm. The light was routed by a 3 dB fiber coupler (FC), and the interference spectrum was detected by an optical spectrum analyzer (OSA, AQ6319). The magnetic field is generated by a 20*20*20 mm rectangle magnet. Detector of the gauss meter was placed parallel and tightly to the sensor for precisely detecting the magnetic field intensity around. The range of magnetic field intensity applied is from 60 to 180 Gs.

The FP cavity dimension can be calculated from the FSR equation. The actual length is 31.6 μm. Figure 26 shows the interference spectrum of this ferrofluid-based magnetic field senor in magnetic field intensity at different magnetic field. As the magnetic intensity increases, the interferogram shifted toward the longer wavelength region. Figure 27 shows the spectrum valley shrift from 60 to 180 Gs with sensitivity at 0.121 nm/Gs. Linear relationship between the magnetic field intensity and the wavelength conformed the experiment principle. Similar result can be obtained with FP cavity at 50 μm as well.

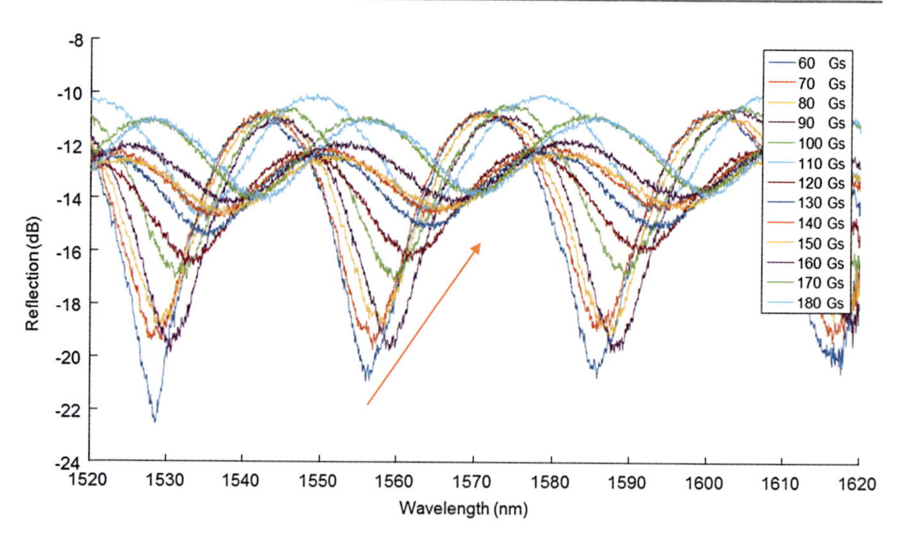

Fig. 26 The reflection spectrum of magnetic field sensor in magnetic field from 60 to 180 Gs

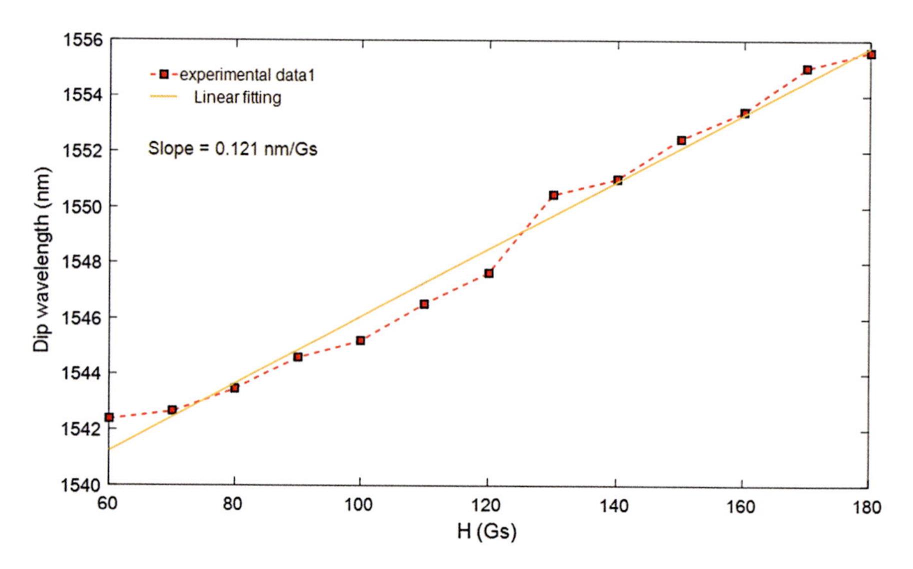

Fig. 27 Transmission spectrum at different magnetic field

In summary, a ferrofluid-based optical fiber magnetic field sensor was proposed, and the opening of the cavity was fabricated by FLIWD. The proposed sensor is novel at design and fabrication, showing impressive advantages in magnetic field sensing.

Summary

Fs laser micromachining has been proven a powerful tool for material removal and modification in transparent materials, especially in optical fibers. Due to its unique characteristic of ultrashort pulse width and extremely high peak intensity, fs laser can realize real 3D micromachining in silica or single crystal sapphire with large material bandgaps. Compared with other laser micromachining methods (i.e., UV laser, CO_2 laser, and long pulse lasers), fs laser can offer many unique advantages, such as negligible cracks, minimal heat-affected zone, low recast, high precision, and 3D structure's formation.

In this chapter, we first introduced the fundamental understanding of fs laser interaction with transparent materials for guiding the fabrication of high-performance fiber-optic sensors and devices. We then presented a typical fs laser micromachining system and the fabrication of 3D hollow structure in optical fibers with this system. As an example, an all-in-fiber 3D optofluidic microdevice was fabricated by fs laser irradiation-assisted selective chemical etching. Horizontal and vertical microchannels can be flexibly created into an optical fiber to form a fluidic cavity with inlets/outlets. The fluidic cavity also functions as an optical FP cavity in which the filled liquid can be probed. This assembly-free microdevice exhibited a high fringe visibility and demonstrated for measurement of the RI of the filling liquids. As another example, a ferrofluid-based optical fiber magnetic field sensor was fabricated by FLIWD. The structure was easy to make, and the FP cavity generated high fringe visibility. The RI changed linear theoretically with a high sensitivity of 0.121 Gs/nm.

References

R. An, Y. Li, Y. Dou, H. Yang, Q. Gong, Simultaneous multi-microhole drilling of soda-lime glass by water-assisted ablation with femtosecond laser pulses. Opt. Express **13**, 1855–1859 (2005)

R.J. Bates, *Optical Switching and Networking Handbook* (McGraw-Hill, New York, 2001)

Y. Bellouard, A. Said, M. Dugan, P. Bado, Fabrication of high-aspect ratio, micro-fluidic channels and tunnels using femtosecond laser pulses and chemical etching. Opt. Express **12**, 2120–2129 (2004)

M. Beresna, M. Gecevičius, P.G. Kazansky, Ultrafast laser direct writing and nanostructuring in transparent materials. Adv. Opt. Photon. **6**, 293–339 (2014)

V. Bhardwaj, E. Simova, P. Rajeev, C. Hnatovsky, R. Taylor, D. Rayner, P. Corkum, Optically produced arrays of planar nanostructures inside fused silica. Phys. Rev. Lett. **96**, 057404 (2006)

H.S. Carslaw, J.C. Jaeger, *Conduction of Heat in Solids*, 2nd edn. (Clarendon Press, Oxford, 1959)

Y. Chen, Q. Han, T. Liu, X. Lan, H. Xiao, Optical fiber magnetic field sensor based on single-mode-multimode-single-mode structure and magnetic fluid. Opt. Lett. **38**, 3999–4001 (2013)

B.N. Chichkov, C. Momma, S. Nolte, F. Von Alvensleben, A. Tünnermann, Femtosecond, picosecond and nanosecond laser ablation of solids. Appl. Phys. A **63**, 109–115 (1996)

A. Chimmalgi, T. Choi, C. Grigoropoulos, K. Komvopoulos, Femtosecond laser aperturless near-field nanomachining of metals assisted by scanning probe microscopy. Appl. Phys. Lett. **82**, 1146–1148 (2003)

A. Couairon, A. Mysyrowicz, Femtosecond filamentation in transparent media. Phys. Rep. **441**, 47–189 (2007)

A. Crespi, Y. Gu, B. Ngamsom, H.J. Hoekstra, C. Dongre, M. Pollnau, R. Ramponi, H.H. van den Vlekkert, P. Watts, G. Cerullo, Three-dimensional Mach-Zehnder interferometer in a microfluidic chip for spatially-resolved label-free detection. Lab. Chip **10**, 1167–1173 (2010)

K.M. Davis, K. Miura, N. Sugimoto, K. Hirao, Writing waveguides in glass with a femtosecond laser. Opt. Lett. **21**, 1729–1731 (1996)

M. Deng, C. Huang, D. Liu, W. Jin, T. Zhu, All fiber magnetic field sensor with Ferrofluid-filled tapered microstructured optical fiber interferometer. Opt. Express **23**, 20668–20674 (2015)

P. Domachuk, I. Littler, M. Cronin-Golomb, B. Eggleton, Compact resonant integrated microfluidic refractometer. Appl. Phys. Lett. **88**, 093513-1–093513-3 (2006)

D. Du, X. Liu, G. Korn, J. Squier, G. Mourou, Laser-induced breakdown by impact ionization in SiO2 with pulse widths from 7 ns to 150 fs. Appl. Phys. Lett. **64**, 3071–3073 (1994)

X. Fan, I.M. White, Optofluidic microsystems for chemical and biological analysis. Nat. Photonics **5**, 591–597 (2011)

E.G. Gamaly, S. Juodkazis, K. Nishimura, H. Misawa, B. Luther-Davies, L. Hallo, P. Nicolai, V.T. Tikhonchuk, Laser-matter interaction in the bulk of a transparent solid: confined microexplosion and void formation. Phys. Rev. B **73**, 214101 (2006)

R.R. Gattass, E. Mazur, Femtosecond laser micromachining in transparent materials. Nat. Photonics **2**, 219–225 (2008)

E.N. Glezer, E. Mazur, Ultrafast-laser driven micro-explosions in transparent materials. Appl. Phys. Lett. **71**, 882–884 (1997)

Y. Gong, Y.-J. Rao, Y. Guo, Z.-L. Ran, Y. Wu, Temperature-insensitive micro Fabry–Pérot strain sensor fabricated by chemically etching Er-doped fiber. IEEE Photon. Technol. Lett. **21**, 1725–1727 (2009)

S. Gross, M. Withford, Ultrafast-laser-inscribed 3D integrated photonics: challenges and emerging applications. Nanophotonics **4**, 332–352 (2015)

M. Haque, K.K. Lee, S. Ho, L.A. Fernandes, P.R. Herman, Chemical-assisted femtosecond laser writing of lab-in-fibers. Lab. Chip **14**, 3817–3829 (2014)

J. Hecht, *City of Light: The Story of Fiber Optics* (Oxford University Press on Demand, Oxford, 2004)

S. Ho, P.R. Herman, J.S. Aitchison, Single-and multi-scan femtosecond laser writing for selective chemical etching of cross section patternable glass micro-channels. Appl. Phys. A **106**, 5–13 (2012)

K. Itoh, W. Watanabe, S. Nolte, C.B. Schaffer, Ultrafast processes for bulk modification of transparent materials. MRS Bull. **31**, 620–625 (2006)

J.B. Jensen, L.H. Pedersen, P.E. Hoiby, L.B. Nielsen, T.P. Hansen, J.R. Folkenberg, J. Riishede, D. Noordegraaf, K. Nielsen, A. Carlsen, Photonic crystal fiber based evanescent-wave sensor for detection of biomolecules in aqueous solutions. Opt. Lett. **29**, 1974–1976 (2004)

S. Kawata, H.-B. Sun, T. Tanaka, K. Takada, Finer features for functional microdevices. Nature **412**, 697–698 (2001)

S. Kiyama, S. Matsuo, S. Hashimoto, Y. Morihira, Examination of etching agent and etching mechanism on femtosecond laser microfabrication of channels inside vitreous silica substrates. J. Phys. Chem. C **113**, 11560–11566 (2009)

T. Kruse, H.-G. Krauthäuser, A. Spanoudaki, R. Pelster, Agglomeration and chain formation in ferrofluids: two-dimensional x-ray scattering. Phys. Rev. B **67**, 094206 (2003)

S. Küper, M. Stuke, Femtosecond UV excimer laser ablation. Appl. Phys. B Lasers Opt. **44**, 199–204 (1987)

Y. Lai, K. Zhou, L. Zhang, I. Bennion, Microchannels in conventional single-mode fibers. Opt. Lett. **31**, 2559–2561 (2006)

Y. Liu, S. Qu, Y. Li, Single microchannel high-temperature fiber sensor by femtosecond laser-induced water breakdown. Opt. Lett. **38**, 335–337 (2013)

R.-Q. Lv, Y. Zhao, D. Wang, Q. Wang, Magnetic fluid-filled optical fiber Fabry–Pérot sensor for magnetic field measurement. IEEE Photon. Technol. Lett. **26**, 217–219 (2014)

T. H. Maiman, Stimulated optical radiation in ruby, 1960

S. Mao, F. Quéré, S. Guizard, X. Mao, R. Russo, G. Petite, P. Martin, Dynamics of femtosecond laser interactions with dielectrics. Appl. Phys. A Mater. Sci. Process. **79**, 1695–1709 (2004)

C. Monat, P. Domachuk, B. Eggleton, Integrated optofluidics: a new river of light. Nat. Photonics **1**, 106–114 (2007)

R. Osellame, V. Maselli, R.M. Vazquez, R. Ramponi, G. Cerullo, Integration of optical waveguides and microfluidic channels both fabricated by femtosecond laser irradiation. Appl. Phys. Lett. **90**, 231118-1–231118-3 (2007)

M.D. Perry, G. Mourou, Terawatt to petawatt subpicosecond lasers. Sci.-AAAS-Wkly. Pap. Ed.-Incl. Guide Sci. Inf. **264**, 917–923 (1994)

P. Pronko, S. Dutta, J. Squier, J. Rudd, D. Du, G. Mourou, Machining of sub-micron holes using a femtosecond laser at 800 nm. Opt. Commun. **114**, 106–110 (1995)

D. Psaltis, S.R. Quake, C. Yang, Developing optofluidic technology through the fusion of microfluidics and optics. Nature **442**, 381–386 (2006)

B. Rethfeld, A. Kaiser, M. Vicanek, G. Simon, Ultrafast dynamics of nonequilibrium electrons in metals under femtosecond laser irradiation. Phys. Rev. B **65**, 214303 (2002)

B. Rethfeld, K. Sokolowski-Tinten, D. Von Der Linde, S. Anisimov, Timescales in the response of materials to femtosecond laser excitation. Appl. Phys. A Mater. Sci. Process. **79**, 767–769 (2004)

C.B. Schaffer, A. Brodeur, E. Mazur, Laser-induced breakdown and damage in bulk transparent materials induced by tightly focused femtosecond laser pulses. Meas. Sci. Technol. **12**, 1784 (2001)

C.B. Schaffer, J.F. García, E. Mazur, Bulk heating of transparent materials using a high-repetition-rate femtosecond laser. Appl. Phys. A **76**, 351–354 (2003)

Y.-R. Shen, *The Principles of Nonlinear Optics*, vol 1 (Wiley-Interscience, New York, 1984), p. 575

D.E. Spence, P.N. Kean, W. Sibbett, 60-fsec pulse generation from a self-mode-locked Ti: sapphire laser. Opt. Lett. **16**, 42–44 (1991)

R. Stoian, D. Ashkenasi, A. Rosenfeld, E. Campbell, Coulomb explosion in ultrashort pulsed laser ablation of Al_2O_3. Phys. Rev. B **62**, 13167 (2000)

D. Strickland, G. Mourou, Compression of amplified chirped optical pulses. Opt. Commun. **55**, 447–449 (1985)

B. Stuart, M. Feit, A. Rubenchik, B. Shore, M. Perry, Laser-induced damage in dielectrics with nanosecond to subpicosecond pulses. Phys. Rev. Lett. **74**, 2248 (1995)

B.C. Stuart, M.D. Feit, S. Herman, A. Rubenchik, B. Shore, M. Perry, Nanosecond-to-femtosecond laser-induced breakdown in dielectrics. Phys. Rev. B **53**, 1749 (1996)

L. Sudrie, M. Franco, B. Prade, A. Mysyrowicz, Writing of permanent birefringent microlayers in bulk fused silica with femtosecond laser pulses. Opt. Commun. **171**, 279–284 (1999)

L. Sudrie, A. Couairon, M. Franco, B. Lamouroux, B. Prade, S. Tzortzakis, A. Mysyrowicz, Femtosecond laser-induced damage and filamentary propagation in fused silica. Phys. Rev. Lett. **89**, 186601 (2002)

C.J. Tuck, R. Hague, C. Doyle, Low cost optical fibre based Fabry–Perot strain sensor production. Meas. Sci. Technol. **17**, 2206 (2006)

E. Udd, An overview of fiber-optic sensors. Rev. Sci. Instrum. **66**, 4015–4030 (1995)

A. Wang, L. Jiang, X. Li, Y. Liu, X. Dong, L. Qu, X. Duan, Y. Lu, Mask-free patterning of high-conductivity metal nanowires in open air by spatially modulated femtosecond laser pulses. Adv. Mater. **27**, 6238–6243 (2015)

T. Wei, Y. Han, Y. Li, H.-L. Tsai, H. Xiao, Temperature-insensitive miniaturized fiber inline Fabry-Perot interferometer for highly sensitive refractive index measurement. Opt. Express **16**, 5764–5769 (2008)

A.T. Woolley, R.A. Mathies, Ultra-high-speed DNA fragment separations using microfabricated capillary array electrophoresis chips. Proc. Natl. Acad. Sci. **91**, 11348–11352 (1994)

E. Yablonovitch, N. Bloembergen, Avalanche ionization and the limiting diameter of filaments induced by light pulses in transparent media. Phys. Rev. Lett. **29**, 907 (1972)

S.Y. Yang, Y.F. Chen, H.E. Horng, C.-Y. Hong, W.S. Tse, H.C. Yang, Magnetically-modulated refractive index of magnetic fluid films. Appl. Phys. Lett. **81**, 4931 (2002)

Y. Zhang, Y. Li, T. Wei, X. Lan, Y. Huang, G. Chen, H. Xiao, Fringe visibility enhanced extrinsic Fabry–Perot interferometer using a graded index fiber collimator. IEEE Photonics J. **2**, 469–481 (2010)

Y. Zhang, L. Yuan, X. Lan, A. Kaur, J. Huang, H. Xiao, High-temperature fiber-optic Fabry–Perot interferometric pressure sensor fabricated by femtosecond laser. Opt. Lett. **38**, 4609–4612 (2013)

K. Zhou, L. Zhang, X. Chen, V. Mezentsev, I. Bennion, Microstructures made in optical fiber with femtosecond laser. Int. J. Smart Nano Mater. **1**, 237–248 (2010)

H. Zhu, I.M. White, J.D. Suter, M. Zourob, X. Fan, Opto-fluidic micro-ring resonator for sensitive label-free viral detection. Analyst **133**, 356–360 (2008)

Index

G.-D. Peng (ed.), *Handbook of Optical Fibers*,
https://doi.org/10.1007/978-981-10-7087-7

D

E

M

P

Q

R

S